W9-APR-216

A First Course in Differential Equations with Modeling Applications

Ninth Edition

Dennis G. Zill
Loyola Marymount University

Differential Equations with Boundary-Vary Problems

Seventh Edition

Dennis G. Zill
Loyola Marymount University

Michael R. Cullen
Late of Loyola Marymount University

By

Warren S. Wright
Loyola Marymount University

Carol D. Wright

BROOKS/COLE
CENGAGE Learning

Australia • Brazil • Japan • Korea • Mexico • Singapore • Spain • United Kingdom • United States

BROOKS/COLE
CENGAGE Learning

ISBN-13: 978-0-495-38609-4
ISBN-10: 0-495-38609-X

Brooks/Cole
10 Davis Drive
Belmont, CA 94002-3098
USA

Cengage Learning is a leading provider of customized learning solutions with office locations around the globe, including Singapore, the United Kingdom, Australia, Mexico, Brazil, and Japan. Locate your local office at: **international.cengage.com/region**

Cengage Learning products are represented in Canada by Nelson Education, Ltd.

For your course and learning solutions, visit **academic.cengage.com**

Purchase any of our products at your local college store or at our preferred online store **www.ichapters.com**

Printed in Canada
1 2 3 4 5 6 7 11 10 09 08

Table of Contents

1 Introduction to Differential Equations

1. Second order; linear

2. Third order; nonlinear because of $(dy/dx)^4$

3. Fourth order; linear

4. Second order; nonlinear because of $\cos(r + u)$

5. Second order; nonlinear because of $(dy/dx)^2$ or $\sqrt{1 + (dy/dx)^2}$

6. Second order; nonlinear because of R^2

7. Third order; linear

8. Second order; nonlinear because of $\dot{x}^2$

9. Writing the differential equation in the form $x(dy/dx) + y^2 = 1$, we see that it is nonlinear in y because of y^2. However, writing it in the form $(y^2 - 1)(dx/dy) + x = 0$, we see that it is linear in x.

10. Writing the differential equation in the form $u(dv/du) + (1 + u)v = ue^u$ we see that it is linear in v. However, writing it in the form $(v + uv - ue^u)(du/dv) + u = 0$, we see that it is nonlinear in u.

11. From $y = e^{-x/2}$ we obtain $y' = -\frac{1}{2}e^{-x/2}$. Then $2y' + y = -e^{-x/2} + e^{-x/2} = 0$.

12. From $y = \frac{6}{5} - \frac{6}{5}e^{-20t}$ we obtain $dy/dt = 24e^{-20t}$, so that

$$\frac{dy}{dt} + 20y = 24e^{-20t} + 20\left(\frac{6}{5} - \frac{6}{5}e^{-20t}\right) = 24.$$

13. From $y = e^{3x}\cos 2x$ we obtain $y' = 3e^{3x}\cos 2x - 2e^{3x}\sin 2x$ and $y'' = 5e^{3x}\cos 2x - 12e^{3x}\sin 2x$, so that $y'' - 6y' + 13y = 0$.

14. From $y = -\cos x \ln(\sec x + \tan x)$ we obtain $y' = -1 + \sin x \ln(\sec x + \tan x)$ and $y'' = \tan x + \cos x \ln(\sec x + \tan x)$. Then $y'' + y = \tan x$.

15. The domain of the function, found by solving $x + 2 \geq 0$, is $[-2, \infty)$. From $y' = 1 + 2(x+2)^{-1/2}$ we

have

$$(y - x)y' = (y - x)[1 + (2(x + 2)^{-1/2}]$$

$$= y - x + 2(y - x)(x + 2)^{-1/2}$$

$$= y - x + 2[x + 4(x + 2)^{1/2} - x](x + 2)^{-1/2}$$

$$= y - x + 8(x + 2)^{1/2}(x + 2)^{-1/2} = y - x + 8.$$

An interval of definition for the solution of the differential equation is $(-2, \infty)$ because y' is not defined at $x = -2$.

16. Since $\tan x$ is not defined for $x = \pi/2 + n\pi$, n an integer, the domain of $y = 5\tan 5x$ is $\{x \mid 5x \neq \pi/2 + n\pi\}$ or $\{x \mid x \neq \pi/10 + n\pi/5\}$. From $y' = 25\sec^2 5x$ we have

$$y' = 25(1 + \tan^2 5x) = 25 + 25\tan^2 5x = 25 + y^2.$$

An interval of definition for the solution of the differential equation is $(-\pi/10, \pi/10)$. Another interval is $(\pi/10, 3\pi/10)$, and so on.

17. The domain of the function is $\{x \mid 4 - x^2 \neq 0\}$ or $\{x \mid x \neq -2 \text{ or } x \neq 2\}$. From $y' = 2x/(4 - x^2)^2$ we have

$$y' = 2x\left(\frac{1}{4 - x^2}\right)^2 = 2xy.$$

An interval of definition for the solution of the differential equation is $(-2, 2)$. Other intervals are $(-\infty, -2)$ and $(2, \infty)$.

18. The function is $y = 1/\sqrt{1 - \sin x}$, whose domain is obtained from $1 - \sin x \neq 0$ or $\sin x \neq 1$. Thus, the domain is $\{x \mid x \neq \pi/2 + 2n\pi\}$. From $y' = -\frac{1}{2}(1 - \sin x)^{-3/2}(-\cos x)$ we have

$$2y' = (1 - \sin x)^{-3/2}\cos x = [(1 - \sin x)^{-1/2}]^3\cos x = y^3\cos x.$$

An interval of definition for the solution of the differential equation is $(\pi/2, 5\pi/2)$. Another interval is $(5\pi/2, 9\pi/2)$ and so on.

19. Writing $\ln(2X - 1) - \ln(X - 1) = t$ and differentiating implicitly we obtain

$$\frac{2}{2X - 1}\frac{dX}{dt} - \frac{1}{X - 1}\frac{dX}{dt} = 1$$

$$\left(\frac{2}{2X - 1} - \frac{1}{X - 1}\right)\frac{dX}{dt} = 1$$

$$\frac{2X - 2 - 2X + 1}{(2X - 1)(X - 1)}\frac{dX}{dt} = 1$$

$$\frac{dX}{dt} = -(2X - 1)(X - 1) = (X - 1)(1 - 2X).$$

Exponentiating both sides of the implicit solution we obtain

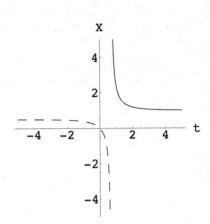

$$\frac{2X - 1}{X - 1} = e^t$$

$$2X - 1 = Xe^t - e^t$$

$$(e^t - 1) = (e^t - 2)X$$

$$X = \frac{e^t - 1}{e^t - 2}.$$

Solving $e^t - 2 = 0$ we get $t = \ln 2$. Thus, the solution is defined on $(-\infty, \ln 2)$ or on $(\ln 2, \infty)$. The graph of the solution defined on $(-\infty, \ln 2)$ is dashed, and the graph of the solution defined on $(\ln 2, \infty)$ is solid.

20. Implicitly differentiating the solution, we obtain

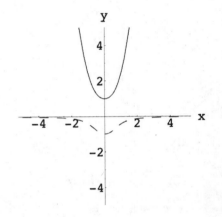

$$-2x^2 \frac{dy}{dx} - 4xy + 2y \frac{dy}{dx} = 0$$

$$-x^2 \, dy - 2xy \, dx + y \, dy = 0$$

$$2xy \, dx + (x^2 - y)dy = 0.$$

Using the quadratic formula to solve $y^2 - 2x^2 y - 1 = 0$ for y, we get $y = (2x^2 \pm \sqrt{4x^4 + 4})/2 = x^2 \pm \sqrt{x^4 + 1}$. Thus, two explicit solutions are $y_1 = x^2 + \sqrt{x^4 + 1}$ and $y_2 = x^2 - \sqrt{x^4 + 1}$. Both solutions are defined on $(-\infty, \infty)$. The graph of $y_1(x)$ is solid and the graph of y_2 is dashed.

21. Differentiating $P = c_1 e^t / \left(1 + c_1 e^t\right)$ we obtain

$$\frac{dP}{dt} = \frac{\left(1 + c_1 e^t\right) c_1 e^t - c_1 e^t \cdot c_1 e^t}{(1 + c_1 e^t)^2} = \frac{c_1 e^t}{1 + c_1 e^t} \frac{\left[\left(1 + c_1 e^t\right) - c_1 e^t\right]}{1 + c_1 e^t}$$

$$= \frac{c_1 e^t}{1 + c_1 e^t} \left[1 - \frac{c_1 e^t}{1 + c_1 e^t}\right] = P(1 - P).$$

22. Differentiating $y = e^{-x^2} \int_0^x e^{t^2} dt + c_1 e^{-x^2}$ we obtain

$$y' = e^{-x^2} e^{x^2} - 2xe^{-x^2} \int_0^x e^{t^2} dt - 2c_1 xe^{-x^2} = 1 - 2xe^{-x^2} \int_0^x e^{t^2} dt - 2c_1 xe^{-x^2}.$$

Substituting into the differential equation, we have

$$y' + 2xy = 1 - 2xe^{-x^2} \int_0^x e^{t^2} dt - 2c_1 xe^{-x^2} + 2xe^{-x^2} \int_0^x e^{t^2} dt + 2c_1 xe^{-x^2} = 1.$$

23. From $y = c_1 e^{2x} + c_2 x e^{2x}$ we obtain $\dfrac{dy}{dx} = (2c_1 + c_2)e^{2x} + 2c_2 x e^{2x}$ and $\dfrac{d^2y}{dx^2} = (4c_1 + 4c_2)e^{2x} + 4c_2 x e^{2x}$, so that

$$\frac{d^2y}{dx^2} - 4\frac{dy}{dx} + 4y = (4c_1 + 4c_2 - 8c_1 - 4c_2 + 4c_1)e^{2x} + (4c_2 - 8c_2 + 4c_2)xe^{2x} = 0.$$

24. From $y = c_1 x^{-1} + c_2 x + c_3 x \ln x + 4x^2$ we obtain

$$\frac{dy}{dx} = -c_1 x^{-2} + c_2 + c_3 + c_3 \ln x + 8x,$$

$$\frac{d^2y}{dx^2} = 2c_1 x^{-3} + c_3 x^{-1} + 8,$$

and

$$\frac{d^3y}{dx^3} = -6c_1 x^{-4} - c_3 x^{-2},$$

so that

$$x^3 \frac{d^3y}{dx^3} + 2x^2 \frac{d^2y}{dx^2} - x\frac{dy}{dx} + y = (-6c_1 + 4c_1 + c_1 + c_1)x^{-1} + (-c_3 + 2c_3 - c_2 - c_3 + c_2)x$$

$$+ (-c_3 + c_3)x \ln x + (16 - 8 + 4)x^2$$

$$= 12x^2.$$

25. From $y = \begin{cases} -x^2, & x < 0 \\ x^2, & x \geq 0 \end{cases}$ we obtain $y' = \begin{cases} -2x, & x < 0 \\ 2x, & x \geq 0 \end{cases}$ so that $xy' - 2y = 0.$

26. The function $y(x)$ is not continuous at $x = 0$ since $\lim\limits_{x \to 0^-} y(x) = 5$ and $\lim\limits_{x \to 0^+} y(x) = -5$. Thus, $y'(x)$ does not exist at $x = 0$.

27. From $y = e^{mx}$ we obtain $y' = me^{mx}$. Then $y' + 2y = 0$ implies

$$me^{mx} + 2e^{mx} = (m + 2)e^{mx} = 0.$$

Since $e^{mx} > 0$ for all x, $m = -2$. Thus $y = e^{-2x}$ is a solution.

28. From $y = e^{mx}$ we obtain $y' = me^{mx}$. Then $5y' = 2y$ implies

$$5me^{mx} = 2e^{mx} \quad \text{or} \quad m = \frac{2}{5}.$$

Thus $y = e^{2x/5}$ is a solution.

29. From $y = e^{mx}$ we obtain $y' = me^{mx}$ and $y'' = m^2 e^{mx}$. Then $y'' - 5y' + 6y = 0$ implies

$$m^2 e^{mx} - 5me^{mx} + 6e^{mx} = (m - 2)(m - 3)e^{mx} = 0.$$

Since $e^{mx} > 0$ for all x, $m = 2$ and $m = 3$. Thus $y = e^{2x}$ and $y = e^{3x}$ are solutions.

30. From $y = e^{mx}$ we obtain $y' = me^{mx}$ and $y'' = m^2 e^{mx}$. Then $2y'' + 7y' - 4y = 0$ implies

$$2m^2 e^{mx} + 7me^{mx} - 4e^{mx} = (2m - 1)(m + 4)e^{mx} = 0.$$

Since $e^{mx} > 0$ for all x, $m = \frac{1}{2}$ and $m = -4$. Thus $y = e^{x/2}$ and $y = e^{-4x}$ are solutions.

31. From $y = x^m$ we obtain $y' = mx^{m-1}$ and $y'' = m(m-1)x^{m-2}$. Then $xy'' + 2y' = 0$ implies

$$xm(m-1)x^{m-2} + 2mx^{m-1} = [m(m-1) + 2m]x^{m-1} = (m^2 + m)x^{m-1}$$

$$= m(m+1)x^{m-1} = 0.$$

Since $x^{m-1} > 0$ for $x > 0$, $m = 0$ and $m = -1$. Thus $y = 1$ and $y = x^{-1}$ are solutions.

32. From $y = x^m$ we obtain $y' = mx^{m-1}$ and $y'' = m(m-1)x^{m-2}$. Then $x^2y'' - 7xy' + 15y = 0$ implies

$$x^2 m(m-1)x^{m-2} - 7xmx^{m-1} + 15x^m = [m(m-1) - 7m + 15]x^m$$

$$= (m^2 - 8m + 15)x^m = (m-3)(m-5)x^m = 0.$$

Since $x^m > 0$ for $x > 0$, $m = 3$ and $m = 5$. Thus $y = x^3$ and $y = x^5$ are solutions.

In Problems 33–36 we substitute $y = c$ into the differential equations and use $y' = 0$ and $y'' = 0$

33. Solving $5c = 10$ we see that $y = 2$ is a constant solution.

34. Solving $c^2 + 2c - 3 = (c+3)(c-1) = 0$ we see that $y = -3$ and $y = 1$ are constant solutions.

35. Since $1/(c-1) = 0$ has no solutions, the differential equation has no constant solutions.

36. Solving $6c = 10$ we see that $y = 5/3$ is a constant solution.

37. From $x = e^{-2t} + 3e^{6t}$ and $y = -e^{-2t} + 5e^{6t}$ we obtain

$$\frac{dx}{dt} = -2e^{-2t} + 18e^{6t} \quad \text{and} \quad \frac{dy}{dt} = 2e^{-2t} + 30e^{6t}.$$

Then

$$x + 3y = (e^{-2t} + 3e^{6t}) + 3(-e^{-2t} + 5e^{6t}) = -2e^{-2t} + 18e^{6t} = \frac{dx}{dt}$$

and

$$5x + 3y = 5(e^{-2t} + 3e^{6t}) + 3(-e^{-2t} + 5e^{6t}) = 2e^{-2t} + 30e^{6t} = \frac{dy}{dt}.$$

38. From $x = \cos 2t + \sin 2t + \frac{1}{5}e^t$ and $y = -\cos 2t - \sin 2t - \frac{1}{5}e^t$ we obtain

$$\frac{dx}{dt} = -2\sin 2t + 2\cos 2t + \frac{1}{5}e^t \quad \text{and} \quad \frac{dy}{dt} = 2\sin 2t - 2\cos 2t - \frac{1}{5}e^t$$

and

$$\frac{d^2x}{dt^2} = -4\cos 2t - 4\sin 2t + \frac{1}{5}e^t \quad \text{and} \quad \frac{d^2y}{dt^2} = 4\cos 2t + 4\sin 2t - \frac{1}{5}e^t.$$

Then

$$4y + e^t = 4(-\cos 2t - \sin 2t - \frac{1}{5}e^t) + e^t = -4\cos 2t - 4\sin 2t + \frac{1}{5}e^t = \frac{d^2x}{dt^2}$$

and

5

$$4x - e^t = 4(\cos 2t + \sin 2t + \frac{1}{5}e^t) - e^t = 4\cos 2t + 4\sin 2t - \frac{1}{5}e^t = \frac{d^2 y}{dt^2}.$$

39. $(y')^2 + 1 = 0$ has no real solutions because $(y')^2 + 1$ is positive for all functions $y = \phi(x)$.

40. The only solution of $(y')^2 + y^2 = 0$ is $y = 0$, since if $y \neq 0$, $y^2 > 0$ and $(y')^2 + y^2 \geq y^2 > 0$.

41. The first derivative of $f(x) = e^x$ is e^x. The first derivative of $f(x) = e^{kx}$ is ke^{kx}. The differential equations are $y' = y$ and $y' = ky$, respectively.

42. Any function of the form $y = ce^x$ or $y = ce^{-x}$ is its own second derivative. The corresponding differential equation is $y'' - y = 0$. Functions of the form $y = c\sin x$ or $y = c\cos x$ have second derivatives that are the negatives of themselves. The differential equation is $y'' + y = 0$.

43. We first note that $\sqrt{1 - y^2} = \sqrt{1 - \sin^2 x} = \sqrt{\cos^2 x} = |\cos x|$. This prompts us to consider values of x for which $\cos x < 0$, such as $x = \pi$. In this case

$$\left.\frac{dy}{dx}\right|_{x=\pi} = \left.\frac{d}{dx}(\sin x)\right|_{x=\pi} = \cos x\big|_{x=\pi} = \cos \pi = -1,$$

but

$$\sqrt{1 - y^2}\big|_{x=\pi} = \sqrt{1 - \sin^2 \pi} = \sqrt{1} = 1.$$

Thus, $y = \sin x$ will only be a solution of $y' = \sqrt{1 - y^2}$ when $\cos x > 0$. An interval of definition is then $(-\pi/2, \pi/2)$. Other intervals are $(3\pi/2, 5\pi/2)$, $(7\pi/2, 9\pi/2)$, and so on.

44. Since the first and second derivatives of $\sin t$ and $\cos t$ involve $\sin t$ and $\cos t$, it is plausible that a linear combination of these functions, $A\sin t + B\cos t$, could be a solution of the differential equation. Using $y' = A\cos t - B\sin t$ and $y'' = -A\sin t - B\cos t$ and substituting into the differential equation we get

$$y'' + 2y' + 4y = -A\sin t - B\cos t + 2A\cos t - 2B\sin t + 4A\sin t + 4B\cos t$$

$$= (3A - 2B)\sin t + (2A + 3B)\cos t = 5\sin t.$$

Thus $3A - 2B = 5$ and $2A + 3B = 0$. Solving these simultaneous equations we find $A = \frac{15}{13}$ and $B = -\frac{10}{13}$. A particular solution is $y = \frac{15}{13}\sin t - \frac{10}{13}\cos t$.

45. One solution is given by the upper portion of the graph with domain approximately $(0, 2.6)$. The other solution is given by the lower portion of the graph, also with domain approximately $(0, 2.6)$.

46. One solution, with domain approximately $(-\infty, 1.6)$ is the portion of the graph in the second quadrant together with the lower part of the graph in the first quadrant. A second solution, with domain approximately $(0, 1.6)$ is the upper part of the graph in the first quadrant. The third solution, with domain $(0, \infty)$, is the part of the graph in the fourth quadrant.

47. Differentiating $(x^3 + y^3)/xy = 3c$ we obtain

$$\frac{xy(3x^2 + 3y^2 y') - (x^3 + y^3)(xy' + y)}{x^2 y^2} = 0$$

$$3x^3 y + 3xy^3 y' - x^4 y' - x^3 y - xy^3 y' - y^4 = 0$$

$$(3xy^3 - x^4 - xy^3)y' = -3x^3 y + x^3 y + y^4$$

$$y' = \frac{y^4 - 2x^3 y}{2xy^3 - x^4} = \frac{y(y^3 - 2x^3)}{x(2y^3 - x^3)}.$$

48. A tangent line will be vertical where y' is undefined, or in this case, where $x(2y^3 - x^3) = 0$. This gives $x = 0$ and $2y^3 = x^3$. Substituting $y^3 = x^3/2$ into $x^3 + y^3 = 3xy$ we get

$$x^3 + \frac{1}{2}x^3 = 3x\left(\frac{1}{2^{1/3}}x\right)$$

$$\frac{3}{2}x^3 = \frac{3}{2^{1/3}}x^2$$

$$x^3 = 2^{2/3}x^2$$

$$x^2(x - 2^{2/3}) = 0.$$

Thus, there are vertical tangent lines at $x = 0$ and $x = 2^{2/3}$, or at $(0, 0)$ and $(2^{2/3}, 2^{1/3})$. Since $2^{2/3} \approx 1.59$, the estimates of the domains in Problem 46 were close.

49. The derivatives of the functions are $\phi_1'(x) = -x/\sqrt{25 - x^2}$ and $\phi_2'(x) = x/\sqrt{25 - x^2}$, neither of which is defined at $x = \pm 5$.

50. To determine if a solution curve passes through $(0, 3)$ we let $t = 0$ and $P = 3$ in the equation $P = c_1 e^t / (1 + c_1 e^t)$. This gives $3 = c_1/(1 + c_1)$ or $c_1 = -\frac{3}{2}$. Thus, the solution curve

$$P = \frac{(-3/2)e^t}{1 - (3/2)e^t} = \frac{-3e^t}{2 - 3e^t}$$

passes through the point $(0, 3)$. Similarly, letting $t = 0$ and $P = 1$ in the equation for the one-parameter family of solutions gives $1 = c_1/(1 + c_1)$ or $c_1 = 1 + c_1$. Since this equation has no solution, no solution curve passes through $(0, 1)$.

51. For the first-order differential equation integrate $f(x)$. For the second-order differential equation integrate twice. In the latter case we get $y = \int(\int f(x)dx)dx + c_1 x + c_2$.

52. Solving for y' using the quadratic formula we obtain the two differential equations

$$y' = \frac{1}{x}\left(2 + 2\sqrt{1 + 3x^6}\right) \quad \text{and} \quad y' = \frac{1}{x}\left(2 - 2\sqrt{1 + 3x^6}\right),$$

so the differential equation cannot be put in the form $dy/dx = f(x, y)$.

53. The differential equation $yy' - xy = 0$ has normal form $dy/dx = x$. These are not equivalent because $y = 0$ is a solution of the first differential equation but not a solution of the second.

54. Differentiating we get $y' = c_1 + 3c_2 x^2$ and $y'' = 6c_2 x$. Then $c_2 = y''/6x$ and $c_1 = y' - xy''/2$, so

$$y = \left(y' - \frac{xy''}{2} \right) x + \left(\frac{y''}{6x} \right) x^3 = xy' - \frac{1}{3} x^2 y''$$

and the differential equation is $x^2 y'' - 3xy' + 3y = 0$.

55. (a) Since e^{-x^2} is positive for all values of x, $dy/dx > 0$ for all x, and a solution, $y(x)$, of the differential equation must be increasing on any interval.

(b) $\lim_{x \to -\infty} \dfrac{dy}{dx} = \lim_{x \to -\infty} e^{-x^2} = 0$ and $\lim_{x \to \infty} \dfrac{dy}{dx} = \lim_{x \to \infty} e^{-x^2} = 0$. Since dy/dx approaches 0 as x approaches $-\infty$ and ∞, the solution curve has horizontal asymptotes to the left and to the right.

(c) To test concavity we consider the second derivative

$$\frac{d^2 y}{dx^2} = \frac{d}{dx} \left(\frac{dy}{dx} \right) = \frac{d}{dx} \left(e^{-x^2} \right) = -2xe^{-x^2}.$$

Since the second derivative is positive for $x < 0$ and negative for $x > 0$, the solution curve is concave up on $(-\infty, 0)$ and concave down on $(0, \infty)$. ⓧ

(d)

56. (a) The derivative of a constant solution $y = c$ is 0, so solving $5 - c = 0$ we see that $c = 5$ and so $y = 5$ is a constant solution.

(b) A solution is increasing where $dy/dx = 5 - y > 0$ or $y < 5$. A solution is decreasing where $dy/dx = 5 - y < 0$ or $y > 5$.

57. (a) The derivative of a constant solution is 0, so solving $y(a - by) = 0$ we see that $y = 0$ and $y = a/b$ are constant solutions.

(b) A solution is increasing where $dy/dx = y(a - by) = by(a/b - y) > 0$ or $0 < y < a/b$. A solution is decreasing where $dy/dx = by(a/b - y) < 0$ or $y < 0$ or $y > a/b$.

(c) Using implicit differentiation we compute

$$\frac{d^2 y}{dx^2} = y(-by') + y'(a - by) = y'(a - 2by).$$

Solving $d^2y/dx^2 = 0$ we obtain $y = a/2b$. Since $d^2y/dx^2 > 0$ for $0 < y < a/2b$ and $d^2y/dx^2 < 0$ for $a/2b < y < a/b$, the graph of $y = \phi(x)$ has a point of inflection at $y = a/2b$.

(d)

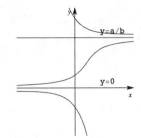

58. (a) If $y = c$ is a constant solution then $y' = 0$, but $c^2 + 4$ is never 0 for any real value of c.

(b) Since $y' = y^2 + 4 > 0$ for all x where a solution $y = \phi(x)$ is defined, any solution must be increasing on any interval on which it is defined. Thus it cannot have any relative extrema.

(c) Using implicit differentiation we compute $d^2y/dx^2 = 2yy' = 2y(y^2 + 4)$. Setting $d^2y/dx^2 = 0$ we see that $y = 0$ corresponds to the only possible point of inflection. Since $d^2y/dx^2 < 0$ for $y < 0$ and $d^2y/dx^2 > 0$ for $y > 0$, there is a point of inflection where $y = 0$.

(d)

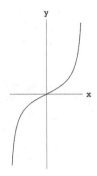

59. In *Mathematica* use

> **Clear[y]**
> **y[x_]:= x Exp[5x] Cos[2x]**
> **y[x]**
> **y''''[x] − 20y'''[x] + 158y''[x] − 580y'[x] +841y[x]//Simplify**

The output will show $y(x) = e^{5x} x \cos 2x$, which verifies that the correct function was entered, and 0, which verifies that this function is a solution of the differential equation.

60. In *Mathematica* use

> **Clear[y]**
> **y[x_]:= 20Cos[5Log[x]]/x − 3Sin[5Log[x]]/x**
> **y[x]**
> **x^3 y'''[x] + 2x^2 y''[x] + 20x y'[x] − 78y[x]//Simplify**

The output will show $y(x) = 20\cos(5\ln x)/x - 3\sin(5\ln x)/x$, which verifies that the correct function was entered, and 0, which verifies that this function is a solution of the differential equation.

Exercises 1.2   Initial-Value Problems

1. Solving $-1/3 = 1/(1 + c_1)$ we get $c_1 = -4$. The solution is $y = 1/(1 - 4e^{-x})$.

2. Solving $2 = 1/(1 + c_1 e)$ we get $c_1 = -(1/2)e^{-1}$. The solution is $y = 2/(2 - e^{-(x+1)})$.

3. Letting $x = 2$ and solving $1/3 = 1/(4 + c)$ we get $c = -1$. The solution is $y = 1/(x^2 - 1)$. This solution is defined on the interval $(1, \infty)$.

4. Letting $x = -2$ and solving $1/2 = 1/(4 + c)$ we get $c = -2$. The solution is $y = 1/(x^2 - 2)$. This solution is defined on the interval $(-\infty, -\sqrt{2}\,)$.

5. Letting $x = 0$ and solving $1 = 1/c$ we get $c = 1$. The solution is $y = 1/(x^2 + 1)$. This solution is defined on the interval $(-\infty, \infty)$.

6. Letting $x = 1/2$ and solving $-4 = 1/(1/4 + c)$ we get $c = -1/2$. The solution is $y = 1/(x^2 - 1/2) = 2/(2x^2 - 1)$. This solution is defined on the interval $(-1/\sqrt{2}\,, 1/\sqrt{2}\,)$.

In Problems 7–10 we use $x = c_1 \cos t + c_2 \sin t$ and $x' = -c_1 \sin t + c_2 \cos t$ to obtain a system of two equations in the two unknowns c_1 and c_2.

7. From the initial conditions we obtain the system

$$c_1 = -1$$

$$c_2 = 8.$$

The solution of the initial-value problem is $x = -\cos t + 8 \sin t$.

8. From the initial conditions we obtain the system

$$c_2 = 0$$

$$-c_1 = 1.$$

The solution of the initial-value problem is $x = -\cos t$.

9. From the initial conditions we obtain

$$\frac{\sqrt{3}}{2} c_1 + \frac{1}{2} c_2 = \frac{1}{2}$$

$$-\frac{1}{2} c_1 + \frac{\sqrt{3}}{2} c_2 = 0.$$

Solving, we find $c_1 = \sqrt{3}/4$ and $c_2 = 1/4$. The solution of the initial-value problem is $x = (\sqrt{3}/4) \cos t + (1/4) \sin t$.

10. From the initial conditions we obtain

$$\frac{\sqrt{2}}{2}c_1 + \frac{\sqrt{2}}{2}c_2 = \sqrt{2}$$

$$-\frac{\sqrt{2}}{2}c_1 + \frac{\sqrt{2}}{2}c_2 = 2\sqrt{2}.$$

Solving, we find $c_1 = -1$ and $c_2 = 3$. The solution of the initial-value problem is $x = -\cos t + 3\sin t$.

In Problems 11–14 we use $y = c_1 e^x + c_2 e^{-x}$ and $y' = c_1 e^x - c_2 e^{-x}$ to obtain a system of two equations in the two unknowns c_1 and c_2.

11. From the initial conditions we obtain

$$c_1 + c_2 = 1$$

$$c_1 - c_2 = 2.$$

Solving, we find $c_1 = \frac{3}{2}$ and $c_2 = -\frac{1}{2}$. The solution of the initial-value problem is $y = \frac{3}{2}e^x - \frac{1}{2}e^{-x}$.

12. From the initial conditions we obtain

$$e c_1 + e^{-1} c_2 = 0$$

$$e c_1 - e^{-1} c_2 = e.$$

Solving, we find $c_1 = \frac{1}{2}$ and $c_2 = -\frac{1}{2}e^2$. The solution of the initial-value problem is
$y = \frac{1}{2}e^x - \frac{1}{2}e^2 e^{-x} = \frac{1}{2}e^x - \frac{1}{2}e^{2-x}$.

13. From the initial conditions we obtain

$$e^{-1} c_1 + e c_2 = 5$$

$$e^{-1} c_1 - e c_2 = -5.$$

Solving, we find $c_1 = 0$ and $c_2 = 5e^{-1}$. The solution of the initial-value problem is $y = 5e^{-1}e^{-x} = 5e^{-1-x}$.

14. From the initial conditions we obtain

$$c_1 + c_2 = 0$$

$$c_1 - c_2 = 0.$$

Solving, we find $c_1 = c_2 = 0$. The solution of the initial-value problem is $y = 0$.

15. Two solutions are $y = 0$ and $y = x^3$.

16. Two solutions are $y = 0$ and $y = x^2$. (Also, any constant multiple of x^2 is a solution.)

17. For $f(x, y) = y^{2/3}$ we have $\dfrac{\partial f}{\partial y} = \dfrac{2}{3}y^{-1/3}$. Thus, the differential equation will have a unique solution in any rectangular region of the plane where $y \neq 0$.

18. For $f(x,y) = \sqrt{xy}$ we have $\partial f/\partial y = \frac{1}{2}\sqrt{x/y}$. Thus, the differential equation will have a unique solution in any region where $x > 0$ and $y > 0$ or where $x < 0$ and $y < 0$.

19. For $f(x,y) = \dfrac{y}{x}$ we have $\dfrac{\partial f}{\partial y} = \dfrac{1}{x}$. Thus, the differential equation will have a unique solution in any region where $x \neq 0$.

20. For $f(x,y) = x + y$ we have $\dfrac{\partial f}{\partial y} = 1$. Thus, the differential equation will have a unique solution in the entire plane.

21. For $f(x,y) = x^2/(4-y^2)$ we have $\partial f/\partial y = 2x^2 y/(4-y^2)^2$. Thus the differential equation will have a unique solution in any region where $y < -2$, $-2 < y < 2$, or $y > 2$.

22. For $f(x,y) = \dfrac{x^2}{1+y^3}$ we have $\dfrac{\partial f}{\partial y} = \dfrac{-3x^2 y^2}{(1+y^3)^2}$. Thus, the differential equation will have a unique solution in any region where $y \neq -1$.

23. For $f(x,y) = \dfrac{y^2}{x^2 + y^2}$ we have $\dfrac{\partial f}{\partial y} = \dfrac{2x^2 y}{(x^2 + y^2)^2}$. Thus, the differential equation will have a unique solution in any region not containing $(0,0)$.

24. For $f(x,y) = (y+x)/(y-x)$ we have $\partial f/\partial y = -2x/(y-x)^2$. Thus the differential equation will have a unique solution in any region where $y < x$ or where $y > x$.

In Problems 25–28 we identify $f(x,y) = \sqrt{y^2 - 9}$ *and* $\partial f/\partial y = y/\sqrt{y^2 - 9}$. *We see that* f *and* $\partial f/\partial y$ *are both continuous in the regions of the plane determined by* $y < -3$ *and* $y > 3$ *with no restrictions on* x.

25. Since $4 > 3$, $(1,4)$ is in the region defined by $y > 3$ and the differential equation has a unique solution through $(1,4)$.

26. Since $(5,3)$ is not in either of the regions defined by $y < -3$ or $y > 3$, there is no guarantee of a unique solution through $(5,3)$.

27. Since $(2,-3)$ is not in either of the regions defined by $y < -3$ or $y > 3$, there is no guarantee of a unique solution through $(2,-3)$.

28. Since $(-1,1)$ is not in either of the regions defined by $y < -3$ or $y > 3$, there is no guarantee of a unique solution through $(-1,1)$.

29. **(a)** A one-parameter family of solutions is $y = cx$. Since $y' = c$, $xy' = xc = y$ and $y(0) = c \cdot 0 = 0$.

    (b) Writing the equation in the form $y' = y/x$, we see that R cannot contain any point on the y-axis. Thus, any rectangular region disjoint from the y-axis and containing (x_0, y_0) will determine an interval around x_0 and a unique solution through (x_0, y_0). Since $x_0 = 0$ in part (a), we are not guaranteed a unique solution through $(0, 0)$.

(c) The piecewise-defined function which satisfies $y(0) = 0$ is not a solution since it is not differentiable at $x = 0$.

30. (a) Since $\dfrac{d}{dx} \tan(x + c) = \sec^2(x + c) = 1 + \tan^2(x + c)$, we see that $y = \tan(x + c)$ satisfies the differential equation.

(b) Solving $y(0) = \tan c = 0$ we obtain $c = 0$ and $y = \tan x$. Since $\tan x$ is discontinuous at $x = \pm\pi/2$, the solution is not defined on $(-2, 2)$ because it contains $\pm\pi/2$.

(c) The largest interval on which the solution can exist is $(-\pi/2, \pi/2)$.

31. (a) Since $\dfrac{d}{dx}\left(-\dfrac{1}{x + c}\right) = \dfrac{1}{(x + c)^2} = y^2$, we see that $y = -\dfrac{1}{x + c}$ is a solution of the differential equation.

(b) Solving $y(0) = -1/c = 1$ we obtain $c = -1$ and $y = 1/(1 - x)$. Solving $y(0) = -1/c = -1$ we obtain $c = 1$ and $y = -1/(1 + x)$. Being sure to include $x = 0$, we see that the interval of existence of $y = 1/(1 - x)$ is $(-\infty, 1)$, while the interval of existence of $y = -1/(1 + x)$ is $(-1, \infty)$.

(c) By inspection we see that $y = 0$ is a solution on $(-\infty, \infty)$.

32. (a) Applying $y(1) = 1$ to $y = -1/(x + c)$ gives

$$1 = -\frac{1}{1 + c} \quad \text{or} \quad 1 + c = -1.$$

Thus $c = -2$ and

$$y = -\frac{1}{x - 2} = \frac{1}{2 - x}.$$

(b) Applying $y(3) = -1$ to $y = -1/(x + c)$ gives

$$-1 = -\frac{1}{3 + c} \quad \text{or} \quad 3 + c = 1.$$

Thus $c = -2$ and

$$y = -\frac{1}{x - 2} = \frac{1}{2 - x}.$$

(c) No, they are not the same solution. The interval I of definition for the solution in part (a) is $(-\infty, 2)$; whereas the interval I of definition for the solution in part (b) is $(2, \infty)$. See the figure.

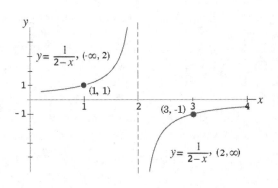

33. (a) Differentiating $3x^2 - y^2 = c$ we get $6x - 2yy' = 0$ or $yy' = 3x$.

(b) Solving $3x^2 - y^2 = 3$ for y we get

$$y = \phi_1(x) = \sqrt{3(x^2 - 1)}, \qquad 1 < x < \infty,$$

$$y = \phi_2(x) = -\sqrt{3(x^2 - 1)}, \qquad 1 < x < \infty,$$

$$y = \phi_3(x) = \sqrt{3(x^2 - 1)}, \qquad -\infty < x < -1,$$

$$y = \phi_4(x) = -\sqrt{3(x^2 - 1)}, \qquad -\infty < x < -1.$$

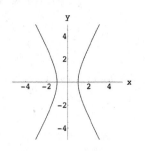

(c) Only $y = \phi_3(x)$ satisfies $y(-2) = 3$.

34. (a) Setting $x = 2$ and $y = -4$ in $3x^2 - y^2 = c$ we get $12 - 16 = -4 = c$, so the explicit solution is

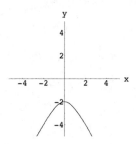

$$y = -\sqrt{3x^2 + 4}, \quad -\infty < x < \infty.$$

(b) Setting $c = 0$ we have $y = \sqrt{3}x$ and $y = -\sqrt{3}x$, both defined on $(-\infty, \infty)$.

In Problems 35–38 we consider the points on the graphs with x-coordinates $x_0 = -1$, $x_0 = 0$, and $x_0 = 1$. The slopes of the tangent lines at these points are compared with the slopes given by $y'(x_0)$ in (a) through (f).

35. The graph satisfies the conditions in (b) and (f).

36. The graph satisfies the conditions in (e).

37. The graph satisfies the conditions in (c) and (d).

38. The graph satisfies the conditions in (a).

39. Integrating $y' = 8e^{2x} + 6x$ we obtain

$$y = \int (8e^{2x} + 6x)dx = 4e^{2x} + 3x^2 + c.$$

Setting $x = 0$ and $y = 9$ we have $9 = 4 + c$ so $c = 5$ and $y = 4e^{2x} + 3x^2 + 5$.

40. Integrating $y'' = 12x - 2$ we obtain

$$y' = \int (12x - 2)dx = 6x^2 - 2x + c_1.$$

Then, integrating y' we obtain

$$y = \int (6x^2 - 2x + c_1)dx = 2x^3 - x^2 + c_1x + c_2.$$

At $x = 1$ the y-coordinate of the point of tangency is $y = -1+5 = 4$. This gives the initial condition $y(1) = 4$. The slope of the tangent line at $x = 1$ is $y'(1) = -1$. From the initial conditions we obtain

$$2 - 1 + c_1 + c_2 = 4 \quad \text{or} \quad c_1 + c_2 = 3$$

and

$$6 - 2 + c_1 = -1 \quad \text{or} \quad c_1 = -5.$$

Thus, $c_1 = -5$ and $c_2 = 8$, so $y = 2x^3 - x^2 - 5x + 8$.

41. When $x = 0$ and $y = \frac{1}{2}$, $y' = -1$, so the only plausible solution curve is the one with negative slope at $(0, \frac{1}{2})$, or the ~~black~~ curve. *red*

42. If the solution is tangent to the x-axis at $(x_0, 0)$, then $y' = 0$ when $x = x_0$ and $y = 0$. Substituting these values into $y' + 2y = 3x - 6$ we get $0 + 0 = 3x_0 - 6$ or $x_0 = 2$.

43. The theorem guarantees a unique (meaning single) solution through any point. Thus, there cannot be two distinct solutions through any point.

44. When $y = \frac{1}{16}x^4$, $y' = \frac{1}{4}x^3 = x(\frac{1}{4}x^2) = xy^{1/2}$, and $y(2) = \frac{1}{16}(16) = 1$. When

$$y = \begin{cases} 0, & x < 0 \\ \frac{1}{16}x^4, & x \geq 0 \end{cases}$$

we have

$$y' = \begin{cases} 0, & x < 0 \\ \frac{1}{4}x^3, & x \geq 0 \end{cases} = x \begin{cases} 0, & x < 0 \\ \frac{1}{4}x^2, & x \geq 0 \end{cases} = xy^{1/2},$$

and $y(2) = \frac{1}{16}(16) = 1$. The two different solutions are the same on the interval $(0, \infty)$, which is all that is required by Theorem 1.2.1.

45. At $t = 0$, $dP/dt = 0.15P(0) + 20 = 0.15(100) + 20 = 35$. Thus, the population is increasing at a rate of 3,500 individuals per year.

If the population is 500 at time $t = T$ then

$$\frac{dP}{dt}\bigg|_{t=T} = 0.15P(T) + 20 = 0.15(500) + 20 = 95.$$

Thus, at this time, the population is increasing at a rate of 9,500 individuals per year.

Exercises 1.3

Differential Equations as Mathematical Models

1. $\dfrac{dP}{dt} = kP + r;$ $\dfrac{dP}{dt} = kP - r$

2. Let b be the rate of births and d the rate of deaths. Then $b = k_1 P$ and $d = k_2 P$. Since $dP/dt = b - d$, the differential equation is $dP/dt = k_1 P - k_2 P$.

3. Let b be the rate of births and d the rate of deaths. Then $b = k_1 P$ and $d = k_2 P^2$. Since $dP/dt = b - d$, the differential equation is $dP/dt = k_1 P - k_2 P^2$.

4. $\dfrac{dP}{dt} = k_1 P - k_2 P^2 - h,\;\; h > 0$

5. From the graph in the text we estimate $T_0 = 180°$ and $T_m = 75°$. We observe that when $T = 85$, $dT/dt \approx -1$. From the differential equation we then have

$$k = \frac{dT/dt}{T - T_m} = \frac{-1}{85 - 75} = -0.1.$$

6. By inspecting the graph in the text we take T_m to be $T_m(t) = 80 - 30\cos \pi t/12$. Then the temperature of the body at time t is determined by the differential equation

$$\frac{dT}{dt} = k\left[T - \left(80 - 30\cos\frac{\pi}{12}t\right)\right], \quad t > 0.$$

7. The number of students with the flu is x and the number not infected is $1000 - x$, so $dx/dt = kx(1000 - x)$.

8. By analogy, with the differential equation modeling the spread of a disease, we assume that the rate at which the technological innovation is adopted is proportional to the number of people who have adopted the innovation and also to the number of people, $y(t)$, who have not yet adopted it. Then $x + y = n$, and assuming that initially one person has adopted the innovation, we have

$$\frac{dx}{dt} = kx(n - x), \quad x(0) = 1.$$

9. The rate at which salt is leaving the tank is

$$R_{out}\,(3\text{ gal/min}) \cdot \left(\frac{A}{300}\text{ lb/gal}\right) = \frac{A}{100}\text{ lb/min}.$$

Thus $dA/dt = -A/100$ (where the minus sign is used since the amount of salt is decreasing. The initial amount is $A(0) = 50$.

10. The rate at which salt is entering the tank is

$$R_{in} = (3\text{ gal/min}) \cdot (2\text{ lb/gal}) = 6\text{ lb/min}.$$

Since the solution is pumped out at a slower rate, it is accumulating at the rate of $(3 - 2)\text{gal/min} = 1 \text{ gal/min}$. After t minutes there are $300 + t$ gallons of brine in the tank. The rate at which salt is leaving is

$$R_{out} = (2 \text{ gal/min}) \cdot \left(\frac{A}{300 + t} \text{ lb/gal} \right) = \frac{2A}{300 + t} \text{ lb/min}.$$

The differential equation is

$$\frac{dA}{dt} = 6 - \frac{2A}{300 + t}.$$

11. The rate at which salt is entering the tank is

$$R_{in} = (3 \text{ gal/min}) \cdot (2 \text{ lb/gal}) = 6 \text{ lb/min}.$$

Since the tank loses liquid at the net rate of

$$3 \text{ gal/min} - 3.5 \text{ gal/min} = -0.5 \text{ gal/min},$$

after t minutes the number of gallons of brine in the tank is $300 - \frac{1}{2}t$ gallons. Thus the rate at which salt is leaving is

$$R_{out} = \left(\frac{A}{300 - t/2} \text{ lb/gal} \right) \cdot (3.5 \text{ gal/min}) = \frac{3.5A}{300 - t/2} \text{ lb/min} = \frac{7A}{600 - t} \text{ lb/min}.$$

The differential equation is

$$\frac{dA}{dt} = 6 - \frac{7A}{600 - t} \quad \text{or} \quad \frac{dA}{dt} + \frac{7}{600 - t} A = 6.$$

12. The rate at which salt is entering the tank is

$$R_{in} = (c_{in} \text{ lb/gal}) \cdot (r_{in} \text{ gal/min}) = c_{in} r_{in} \text{ lb/min}.$$

Now let $A(t)$ denote the number of pounds of salt and $N(t)$ the number of gallons of brine in the tank at time t. The concentration of salt in the tank as well as in the outflow is $c(t) = x(t)/N(t)$. But the number of gallons of brine in the tank remains steady, is increased, or is decreased depending on whether $r_{in} = r_{out}$, $r_{in} > r_{out}$, or $r_{in} < r_{out}$. In any case, the number of gallons of brine in the tank at time t is $N(t) = N_0 + (r_{in} - r_{out})t$. The output rate of salt is then

$$R_{out} = \left(\frac{A}{N_0 + (r_{in} - r_{out})t} \text{ lb/gal} \right) \cdot (r_{out} \text{ gal/min}) = r_{out} \frac{A}{N_0 + (r_{in} - r_{out})t} \text{ lb/min}.$$

The differential equation for the amount of salt, $dA/dt = R_{in} - R_{out}$, is

$$\frac{dA}{dt} = c_{in} r_{in} - r_{out} \frac{A}{N_0 + (r_{in} - r_{out})t} \quad \text{or} \quad \frac{dA}{dt} + \frac{r_{out}}{N_0 + (r_{in} - r_{out})t} A = c_{in} r_{in}.$$

13. The volume of water in the tank at time t is $V = A_w h$. The differential equation is then

$$\frac{dh}{dt} = \frac{1}{A_w} \frac{dV}{dt} = \frac{1}{A_w} \left(-cA_h \sqrt{2gh} \right) = -\frac{cA_h}{A_w} \sqrt{2gh}.$$

17

Using $A_h = \pi \left(\frac{2}{12}\right)^2 = \frac{\pi}{36}$, $A_w = 10^2 = 100$, and $g = 32$, this becomes

$$\frac{dh}{dt} = -\frac{c\pi/36}{100}\sqrt{64h} = -\frac{c\pi}{450}\sqrt{h}.$$

14. The volume of water in the tank at time t is $V = \frac{1}{3}\pi r^2 h$ where r is the radius of the tank at height h. From the figure in the text we see that $r/h = 8/20$ so that $r = \frac{2}{5}h$ and $V = \frac{1}{3}\pi\left(\frac{2}{5}h\right)^2 h = \frac{4}{75}\pi h^3$. Differentiating with respect to t we have $dV/dt = \frac{4}{25}\pi h^2\, dh/dt$ or

$$\frac{dh}{dt} = \frac{25}{4\pi h^2}\frac{dV}{dt}.$$

From Problem 13 we have $dV/dt = -cA_h\sqrt{2gh}$ where $c = 0.6$, $A_h = \pi\left(\frac{2}{12}\right)^2$, and $g = 32$. Thus $dV/dt = -2\pi\sqrt{h}/15$ and

$$\frac{dh}{dt} = \frac{25}{4\pi h^2}\left(-\frac{2\pi\sqrt{h}}{15}\right) = -\frac{5}{6h^{3/2}}.$$

15. Since $i = dq/dt$ and $L\,d^2q/dt^2 + R\,dq/dt = E(t)$, we obtain $L\,di/dt + Ri = E(t)$.

16. By Kirchhoff's second law we obtain $R\dfrac{dq}{dt} + \dfrac{1}{C}q = E(t)$.

17. From Newton's second law we obtain $m\dfrac{dv}{dt} = -kv^2 + mg$.

18. Since the barrel in Figure 1.3.16(b) in the text is submerged an additional y feet below its equilibrium position the number of cubic feet in the additional submerged portion is the volume of the circular cylinder: $\pi\times(\text{radius})^2\times\text{height}$ or $\pi(s/2)^2 y$. Then we have from Archimedes' principle

$$\text{upward force of water on barrel} = \text{weight of water displaced}$$

$$= (62.4) \times (\text{volume of water displaced})$$

$$= (62.4)\pi(s/2)^2 y = 15.6\pi s^2 y.$$

It then follows from Newton's second law that

$$\frac{w}{g}\frac{d^2y}{dt^2} = -15.6\pi s^2 y \qquad \text{or} \qquad \frac{d^2y}{dt^2} + \frac{15.6\pi s^2 g}{w}y = 0,$$

where $g = 32$ and w is the weight of the barrel in pounds.

19. The net force acting on the mass is

$$F = ma = m\frac{d^2x}{dt^2} = -k(s+x) + mg = -kx + mg - ks.$$

Since the condition of equilibrium is $mg = ks$, the differential equation is

$$m\frac{d^2x}{dt^2} = -kx.$$

20. From Problem 19, without a damping force, the differential equation is $m\,d^2x/dt^2 = -kx$. With a damping force proportional to velocity, the differential equation becomes

$$m\frac{d^2x}{dt^2} = -kx - \beta\frac{dx}{dt} \qquad \text{or} \qquad m\frac{d^2x}{dt^2} + \beta\frac{dx}{dt} + kx = 0.$$

21. From $g = k/R^2$ we find $k = gR^2$. Using $a = d^2r/dt^2$ and the fact that the positive direction is upward we get

$$\frac{d^2r}{dt^2} = -a = -\frac{k}{r^2} = -\frac{gR^2}{r^2} \qquad \text{or} \qquad \frac{d^2r}{dt^2} + \frac{gR^2}{r^2} = 0.$$

22. The gravitational force on m is $F = -kM_rm/r^2$. Since $M_r = 4\pi\delta r^3/3$ and $M = 4\pi\delta R^3/3$ we have $M_r = r^3M/R^3$ and

$$F = -k\frac{M_rm}{r^2} = -k\frac{r^3Mm/R^3}{r^2} = -k\frac{mM}{R^3}r.$$

Now from $F = ma = d^2r/dt^2$ we have

$$m\frac{d^2r}{dt^2} = -k\frac{mM}{R^3}r \quad \text{or} \quad \frac{d^2r}{dt^2} = -\frac{kM}{R^3}r.$$

23. The differential equation is $\dfrac{dA}{dt} = k(M - A)$.

24. The differential equation is $\dfrac{dA}{dt} = k_1(M - A) - k_2A$.

25. The differential equation is $x'(t) = r - kx(t)$ where $k > 0$.

26. By the Pythagorean Theorem the slope of the tangent line is $y' = \dfrac{-y}{\sqrt{s^2 - y^2}}$.

27. We see from the figure that $2\theta + \alpha = \pi$. Thus

$$\frac{y}{-x} = \tan\alpha = \tan(\pi - 2\theta) = -\tan 2\theta = -\frac{2\tan\theta}{1 - \tan^2\theta}.$$

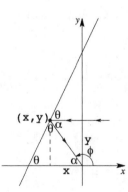

Since the slope of the tangent line is $y' = \tan\theta$ we have $y/x = 2y'/[1 - (y')^2]$ or $y - y(y')^2 = 2xy'$, which is the quadratic equation $y(y')^2 + 2xy' - y = 0$ in y'. Using the quadratic formula, we get

$$y' = \frac{-2x \pm \sqrt{4x^2 + 4y^2}}{2y} = \frac{-x \pm \sqrt{x^2 + y^2}}{y}.$$

Since $dy/dx > 0$, the differential equation is

$$\frac{dy}{dx} = \frac{-x + \sqrt{x^2 + y^2}}{y} \qquad \text{or} \qquad y\frac{dy}{dx} - \sqrt{x^2 + y^2} + x = 0.$$

28. The differential equation is $dP/dt = kP$, so from Problem 41 in Exercises 1.1, $P = e^{kt}$, and a one-parameter family of solutions is $P = ce^{kt}$.

29. The differential equation in (3) is $dT/dt = k(T - T_m)$. When the body is cooling, $T > T_m$, so $T - T_m > 0$. Since T is decreasing, $dT/dt < 0$ and $k < 0$. When the body is warming, $T < T_m$, so $T - T_m < 0$. Since T is increasing, $dT/dt > 0$ and $k < 0$.

30. The differential equation in (8) is $dA/dt = 6 - A/100$. If $A(t)$ attains a maximum, then $dA/dt = 0$ at this time and $A = 600$. If $A(t)$ continues to increase without reaching a maximum, then $A'(t) > 0$ for $t > 0$ and A cannot exceed 600. In this case, if $A'(t)$ approaches 0 as t increases to infinity, we see that $A(t)$ approaches 600 as t increases to infinity.

31. This differential equation could describe a population that undergoes periodic fluctuations.

32. (a) As shown in Figure 1.3.22(b) in the text, the resultant of the reaction force of magnitude F and the weight of magnitude mg of the particle is the centripetal force of magnitude $m\omega^2 x$. The centripetal force points to the center of the circle of radius x on which the particle rotates about the y-axis. Comparing parts of similar triangles gives

$$F \cos \theta = mg \quad \text{and} \quad F \sin \theta = m\omega^2 x.$$

(b) Using the equations in part (a) we find

$$\tan \theta = \frac{F \sin \theta}{F \cos \theta} = \frac{m\omega^2 x}{mg} = \frac{\omega^2 x}{g} \quad \text{or} \quad \frac{dy}{dx} = \frac{\omega^2 x}{g}.$$

33. From Problem 21, $d^2r/dt^2 = -gR^2/r^2$. Since R is a constant, if $r = R + s$, then $d^2r/dt^2 = d^2s/dt^2$ and, using a Taylor series, we get

$$\frac{d^2s}{dt^2} = -g\frac{R^2}{(R+s)^2} = -gR^2(R+s)^{-2} \approx -gR^2[R^{-2} - 2sR^{-3} + \cdots] = -g + \frac{2gs}{R^3} + \cdots.$$

Thus, for R much larger than s, the differential equation is approximated by $d^2s/dt^2 = -g$.

34. (a) If ρ is the mass density of the raindrop, then $m = \rho V$ and

$$\frac{dm}{dt} = \rho\frac{dV}{dt} = \rho\frac{d}{dt}\left[\frac{4}{3}\pi r^3\right] = \rho\left(4\pi r^2\frac{dr}{dt}\right) = \rho S\frac{dr}{dt}.$$

If dr/dt is a constant, then $dm/dt = kS$ where $\rho\, dr/dt = k$ or $dr/dt = k/\rho$. Since the radius is decreasing, $k < 0$. Solving $dr/dt = k/\rho$ we get $r = (k/\rho)t + c_0$. Since $r(0) = r_0$, $c_0 = r_0$ and $r = kt/\rho + r_0$.

(b) From Newton's second law, $\dfrac{d}{dt}[mv] = mg$, where v is the velocity of the raindrop. Then

$$m\frac{dv}{dt} + v\frac{dm}{dt} = mg \quad \text{or} \quad \rho\left(\frac{4}{3}\pi r^3\right)\frac{dv}{dt} + v(k4\pi r^2) = \rho\left(\frac{4}{3}\pi r^3\right)g.$$

Dividing by $4\rho\pi r^3/3$ we get

$$\frac{dv}{dt} + \frac{3k}{\rho r}v = g \quad \text{or} \quad \frac{dv}{dt} + \frac{3k/\rho}{kt/\rho + r_0}v = g, \quad k < 0.$$

20

35. We assume that the plow clears snow at a constant rate of k cubic miles per hour. Let t be the time in hours after noon, $x(t)$ the depth in miles of the snow at time t, and $y(t)$ the distance the plow has moved in t hours. Then dy/dt is the velocity of the plow and the assumption gives

$$wx\frac{dy}{dt} = k,$$

where w is the width of the plow. Each side of this equation simply represents the volume of snow plowed in one hour. Now let t_0 be the number of hours before noon when it started snowing and let s be the constant rate in miles per hour at which x increases. Then for $t > -t_0$, $x = s(t + t_0)$. The differential equation then becomes

$$\frac{dy}{dt} = \frac{k}{ws}\frac{1}{t + t_0}.$$

Integrating, we obtain

$$y = \frac{k}{ws}\left[\ln(t + t_0) + c\right]$$

where c is a constant. Now when $t = 0$, $y = 0$ so $c = -\ln t_0$ and

$$y = \frac{k}{ws}\ln\left(1 + \frac{t}{t_0}\right).$$

Finally, from the fact that when $t = 1$, $y = 2$ and when $t = 2$, $y = 3$, we obtain

$$\left(1 + \frac{2}{t_0}\right)^2 = \left(1 + \frac{1}{t_0}\right)^3.$$

Expanding and simplifying gives $t_0^2 + t_0 - 1 = 0$. Since $t_0 > 0$, we find $t_0 \approx 0.618$ hours $\approx$ 37 minutes. Thus it started snowing at about 11:23 in the morning.

36. (1): $\dfrac{dP}{dt} = kP$ is linear
(2): $\dfrac{dA}{dt} = kA$ is linear

(3): $\dfrac{dT}{dt} = k(T - T_m)$ is linear
(5): $\dfrac{dx}{dt} = kx(n + 1 - x)$ is nonlinear

(6): $\dfrac{dX}{dt} = k(\alpha - X)(\beta - X)$ is nonlinear
(8): $\dfrac{dA}{dt} = 6 - \dfrac{A}{100}$ is linear

(10): $\dfrac{dh}{dt} = -\dfrac{A_h}{A_w}\sqrt{2gh}$ is nonlinear
(11): $L\dfrac{d^2q}{dt^2} + R\dfrac{dq}{dt} + \dfrac{1}{C}q = E(t)$ is linear

(12): $\dfrac{d^2s}{dt^2} = -g$ is linear
(14): $m\dfrac{dv}{dt} = mg - kv$ is linear

(15): $m\dfrac{d^2s}{dt^2} + k\dfrac{ds}{dt} = mg$ is linear

(16): linearity or nonlinearity is determined by the manner in which W and T_1 involve x.

Chapter 1 in Review

1. $\dfrac{d}{dx} c_1 e^{10x} = 10 c_1 e^{10x}$; $\qquad \dfrac{dy}{dx} = 10y$

2. $\dfrac{d}{dx}(5 + c_1 e^{-2x}) = -2 c_1 e^{-2x} = -2(5 + c_1 e^{-2x} - 5)$; $\qquad \dfrac{dy}{dx} = -2(y - 5)$ or $\dfrac{dy}{dx} = -2y + 10$

3. $\dfrac{d}{dx}(c_1 \cos kx + c_2 \sin kx) = -k c_1 \sin kx + k c_2 \cos kx$;

$\dfrac{d^2}{dx^2}(c_1 \cos kx + c_2 \sin kx) = -k^2 c_1 \cos kx - k^2 c_2 \sin kx = -k^2(c_1 \cos kx + c_2 \sin kx)$;

$\dfrac{d^2 y}{dx^2} = -k^2 y$ or $\dfrac{d^2 y}{dx^2} + k^2 y = 0$

4. $\dfrac{d}{dx}(c_1 \cosh kx + c_2 \sinh kx) = k c_1 \sinh kx + k c_2 \cosh kx$;

$\dfrac{d^2}{dx^2}(c_1 \cosh kx + c_2 \sinh kx) = k^2 c_1 \cosh kx + k^2 c_2 \sinh kx = k^2(c_1 \cosh kx + c_2 \sinh kx)$;

$\dfrac{d^2 y}{dx^2} = k^2 y$ or $\dfrac{d^2 y}{dx^2} - k^2 y = 0$

5. $y = c_1 e^x + c_2 x e^x$; $\qquad y' = c_1 e^x + c_2 x e^x + c_2 e^x$; $\qquad y'' = c_1 e^x + c_2 x e^x + 2 c_2 e^x$;

$y'' + y = 2(c_1 e^x + c_2 x e^x) + 2 c_2 e^x = 2(c_1 e^x + c_2 x e^x + c_2 e^x) = 2y'$; $\qquad y'' - 2y' + y = 0$

6. $y' = -c_1 e^x \sin x + c_1 e^x \cos x + c_2 e^x \cos x + c_2 e^x \sin x$;

$y'' = -c_1 e^x \cos x - c_1 e^x \sin x - c_1 e^x \sin x + c_1 e^x \cos x - c_2 e^x \sin x + c_2 e^x \cos x + c_2 e^x \cos x + c_2 e^x \sin x$

$\qquad = -2 c_1 e^x \sin x + 2 c_2 e^x \cos x$;

$y'' - 2y' = -2 c_1 e^x \cos x - 2 c_2 e^x \sin x = -2y$; $\qquad y'' - 2y' + 2y = 0$

7. a,d $\qquad\qquad$ **8.** c $\qquad\qquad$ **9.** b $\qquad\qquad$ **10.** a,c $\qquad\qquad$ **11.** b $\qquad\qquad$ **12.** a,b,d

13. A few solutions are $y = 0$, $y = c$, and $y = e^x$.

14. Easy solutions to see are $y = 0$ and $y = 3$.

15. The slope of the tangent line at (x, y) is y', so the differential equation is $y' = x^2 + y^2$.

16. The rate at which the slope changes is $dy'/dx = y''$, so the differential equation is $y'' = -y'$ or $y'' + y' = 0$.

17. (a) The domain is all real numbers.

(b) Since $y' = 2/3 x^{1/3}$, the solution $y = x^{2/3}$ is undefined at $x = 0$. This function is a solution of the differential equation on $(-\infty, 0)$ and also on $(0, \infty)$.

18. **(a)** Differentiating $y^2 - 2y = x^2 - x + c$ we obtain $2yy' - 2y' = 2x - 1$ or $(2y - 2)y' = 2x - 1$.

    (b) Setting $x = 0$ and $y = 1$ in the solution we have $1 - 2 = 0 - 0 + c$ or $c = -1$. Thus, a solution of the initial-value problem is $y^2 - 2y = x^2 - x - 1$.

    (c) Solving $y^2 - 2y - (x^2 - x - 1) = 0$ by the quadratic formula we get $y = (2 \pm \sqrt{4 + 4(x^2 - x - 1)})/2$
    $= 1 \pm \sqrt{x^2 - x} = 1 \pm \sqrt{x(x - 1)}$. Since $x(x - 1) \geq 0$ for $x \leq 0$ or $x \geq 1$, we see that neither $y = 1 + \sqrt{x(x - 1)}$ nor $y = 1 - \sqrt{x(x - 1)}$ is differentiable at $x = 0$. Thus, both functions are solutions of the differential equation, but neither is a solution of the initial-value problem.

19. Setting $x = x_0$ and $y = 1$ in $y = -2/x + x$, we get

    $$1 = -\frac{2}{x_0} + x_0 \qquad \text{or} \qquad x_0^2 - x_0 - 2 = (x_0 - 2)(x_0 + 1) = 0.$$

    Thus, $x_0 = 2$ or $x_0 = -1$. Since $x = 0$ in $y = -2/x + x$, we see that $y = -2/x + x$ is a solution of the initial-value problem $xy' + y = 2x$, $y(-1) = 1$, on the interval $(-\infty, 0)$ and $y = -2/x + x$ is a solution of the initial-value problem $xy' + y = 2x$, $y(2) = 1$, on the interval $(0, \infty)$.

20. From the differential equation, $y'(1) = 1^2 + [y(1)]^2 = 1 + (-1)^2 = 2 > 0$, so $y(x)$ is increasing in some neighborhood of $x = 1$. From $y'' = 2x + 2yy'$ we have $y''(1) = 2(1) + 2(-1)(2) = -2 < 0$, so $y(x)$ is concave down in some neighborhood of $x = 1$.

21. **(a)**

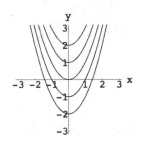 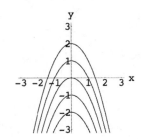

$$y = x^2 + c_1 \qquad\qquad y = -x^2 + c_2$$

    (b) When $y = x^2 + c_1$, $y' = 2x$ and $(y')^2 = 4x^2$. When $y = -x^2 + c_2$, $y' = -2x$ and $(y')^2 = 4x^2$.

    (c) Pasting together x^2, $x \geq 0$, and $-x^2$, $x \leq 0$, we get $y = \begin{cases} -x^2, & x \leq 0 \\ x^2, & x > 0. \end{cases}$

22. The slope of the tangent line is $y'\,|_{(-1,4)} = 6\sqrt{4} + 5(-1)^3 = 7$.

23. Differentiating $y = x \sin x + x \cos x$ we get

    $$y' = x \cos x + \sin x - x \sin x + \cos x$$

    and

    $$y'' = -x \sin x + \cos x + \cos x - x \cos x - \sin x - \sin x$$

    $$= -x \sin x - x \cos x + 2 \cos x - 2 \sin x.$$

Thus

$$y'' + y = -x \sin x - x \cos x + 2 \cos x - 2 \sin x + x \sin x + x \cos x = 2 \cos x - 2 \sin x.$$

An interval of definition for the solution is $(-\infty, \infty)$.

24. Differentiating $y = x \sin x + (\cos x) \ln(\cos x)$ we get

$$\begin{aligned} y' &= x \cos x + \sin x + \cos x \left(\frac{-\sin x}{\cos x} \right) - (\sin x) \ln(\cos x) \\ &= x \cos x + \sin x - \sin x - (\sin x) \ln(\cos x) \\ &= x \cos x - (\sin x) \ln(\cos x) \end{aligned}$$

and

$$\begin{aligned} y'' &= -x \sin x + \cos x - \sin x \left(\frac{-\sin x}{\cos x} \right) - (\cos x) \ln(\cos x) \\ &= -x \sin x + \cos x + \frac{\sin^2 x}{\cos x} - (\cos x) \ln(\cos x) \\ &= -x \sin x + \cos x + \frac{1 - \cos^2 x}{\cos x} - (\cos x) \ln(\cos x) \\ &= -x \sin x + \cos x + \sec x - \cos x - (\cos x) \ln(\cos x) \\ &= -x \sin x + \sec x - (\cos x) \ln(\cos x). \end{aligned}$$

Thus

$$y'' + y = -x \sin x + \sec x - (\cos x) \ln(\cos x) + x \sin x + (\cos x) \ln(\cos x) = \sec x.$$

To obtain an interval of definition we note that the domain of $\ln x$ is $(0, \infty)$, so we must have $\cos x > 0$. Thus, an interval of definition is $(-\pi/2, \pi/2)$.

25. Differentiating $y = \sin(\ln x)$ we obtain $y' = \cos(\ln x)/x$ and $y'' = -[\sin(\ln x) + \cos(\ln x)]/x^2$. Then

$$x^2 y'' + x y' + y = x^2 \left(-\frac{\sin(\ln x) + \cos(\ln x)}{x^2} \right) + x \frac{\cos(\ln x)}{x} + \sin(\ln x) = 0.$$

An interval of definition for the solution is $(0, \infty)$.

26. Differentiating $y = \cos(\ln x) \ln(\cos(\ln x)) + (\ln x) \sin(\ln x)$ we obtain

$$\begin{aligned} y' &= \cos(\ln x) \frac{1}{\cos(\ln x)} \left(-\frac{\sin(\ln x)}{x} \right) + \ln(\cos(\ln x)) \left(-\frac{\sin(\ln x)}{x} \right) + \ln x \frac{\cos(\ln x)}{x} + \frac{\sin(\ln x)}{x} \\ &= -\frac{\ln(\cos(\ln x)) \sin(\ln x)}{x} + \frac{(\ln x) \cos(\ln x)}{x} \end{aligned}$$

and

$$y'' = -x\left[\ln(\cos(\ln x))\frac{\cos(\ln x)}{x} + \sin(\ln x)\frac{1}{\cos(\ln x)}\left(-\frac{\sin(\ln x)}{x}\right)\right]\frac{1}{x^2}$$

$$+ \ln(\cos(\ln x))\sin(\ln x)\frac{1}{x^2} + x\left[(\ln x)\left(-\frac{\sin(\ln x)}{x}\right) + \frac{\cos(\ln x)}{x}\right]\frac{1}{x^2} - (\ln x)\cos(\ln x)\frac{1}{x^2}$$

$$= \frac{1}{x^2}\left[-\ln(\cos(\ln x))\cos(\ln x) + \frac{\sin^2(\ln x)}{\cos(\ln x)} + \ln(\cos(\ln x))\sin(\ln x)\right.$$

$$\left.- (\ln x)\sin(\ln x) + \cos(\ln x) - (\ln x)\cos(\ln x)\right].$$

Then

$$x^2 y'' + xy' + y = -\ln(\cos(\ln x))\cos(\ln x) + \frac{\sin^2(\ln x)}{\cos(\ln x)} + \ln(\cos(\ln x))\sin(\ln x) - (\ln x)\sin(\ln x)$$

$$+ \cos(\ln x) - (\ln x)\cos(\ln x) - \ln(\cos(\ln x))\sin(\ln x)$$

$$+ (\ln x)\cos(\ln x) + \cos(\ln x)\ln(\cos(\ln x)) + (\ln x)\sin(\ln x)$$

$$= \frac{\sin^2(\ln x)}{\cos(\ln x)} + \cos(\ln x) = \frac{\sin^2(\ln x) + \cos^2(\ln x)}{\cos(\ln x)} = \frac{1}{\cos(\ln x)} = \sec(\ln x).$$

To obtain an interval of definition, we note that the domain of $\ln x$ is $(0, \infty)$, so we must have $\cos(\ln x) > 0$. Since $\cos x > 0$ when $-\pi/2 < x < \pi/2$, we require $-\pi/2 < \ln x < \pi/2$. Since e^x is an increasing function, this is equivalent to $e^{-\pi/2} < x < e^{\pi/2}$. Thus, an interval of definition is $(e^{-\pi/2}, e^{\pi/2})$. (Much of this problem is more easily done using a computer algebra system such as *Mathematica* or *Maple*.)

In Problems 27 – 30 we have $y' = 3c_1 e^{3x} - c_2 e^{-x} - 2$.

27. The initial conditions imply

$$c_1 + c_2 = 0$$

$$3c_1 - c_2 - 2 = 0,$$

so $c_1 = \frac{1}{2}$ and $c_2 = -\frac{1}{2}$. Thus $y = \frac{1}{2}e^{3x} - \frac{1}{2}e^{-x} - 2x$.

28. The initial conditions imply

$$c_1 + c_2 = 1$$

$$3c_1 - c_2 - 2 = -3,$$

so $c_1 = 0$ and $c_2 = 1$. Thus $y = e^{-x} - 2x$.

29. The initial conditions imply

$$c_1 e^3 + c_2 e^{-1} - 2 = 4$$

$$3c_1 e^3 - c_2 e^{-1} - 2 = -2,$$

so $c_1 = \frac{3}{2}e^{-3}$ and $c_2 = \frac{9}{2}e$. Thus $y = \frac{3}{2}e^{3x-3} + \frac{9}{2}e^{-x+1} - 2x$.

30. The initial conditions imply

$$c_1 e^{-3} + c_2 e + 2 = 0$$

$$3c_1 e^{-3} - c_2 e - 2 = 1,$$

so $c_1 = \frac{1}{4}e^3$ and $c_2 = -\frac{9}{4}e^{-1}$. Thus $y = \frac{1}{4}e^{3x+3} - \frac{9}{4}e^{-x-1} - 2x$.

31. From the graph we see that estimates for y_0 and y_1 are $y_0 = -3$ and $y_1 = 0$.

32. The differential equation is

$$\frac{dh}{dt} = -\frac{cA_0}{A_w}\sqrt{2gh}.$$

Using $A_0 = \pi(1/24)^2 = \pi/576$, $A_w = \pi(2)^2 = 4\pi$, and $g = 32$, this becomes

$$\frac{dh}{dt} = -\frac{c\pi/576}{4\pi}\sqrt{64h} = \frac{c}{288}\sqrt{h}.$$

33. Let $P(t)$ be the number of owls present at time t. Then $dP/dt = k(P - 200 + 10t)$.

34. Setting $A'(t) = -0.002$ and solving $A'(t) = -0.0004332A(t)$ for $A(t)$, we obtain

$$A(t) = \frac{A'(t)}{-0.0004332} = \frac{-0.002}{-0.0004332} \approx 4.6 \text{ grams.}$$

Exercises 2.1

Solution Curves Without a Solution

1.

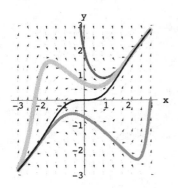

2.

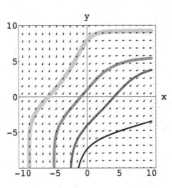

3.

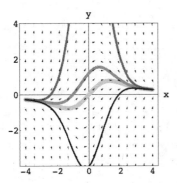

4.

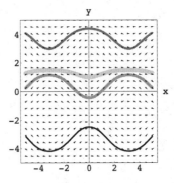

5.

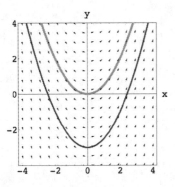

6.

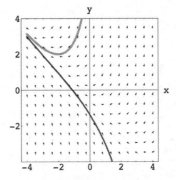

7.

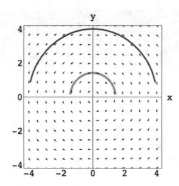

8.

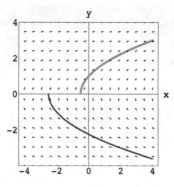

9.

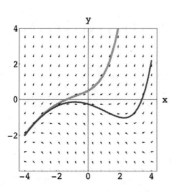

10.

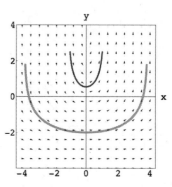

11.

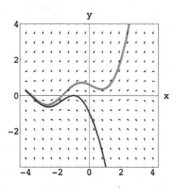

12.

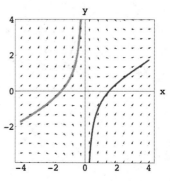

13.

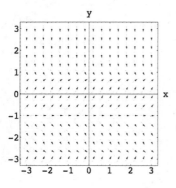

14.

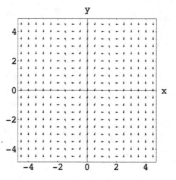

15. (a) The isoclines have the form $y = -x + c$, which are straight lines with slope -1.

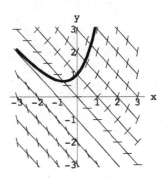

(b) The isoclines have the form $x^2 + y^2 = c$, which are circles centered at the origin.

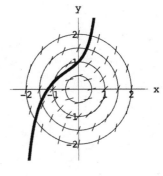

16. (a) When $x = 0$ or $y = 4$, $dy/dx = -2$ so the lineal elements have slope -2. When $y = 3$ or $y = 5$, $dy/dx = x - 2$, so the lineal elements at $(x, 3)$ and $(x, 5)$ have slopes $x - 2$.

(b) At $(0, y_0)$ the solution curve is headed down. If $y \to \infty$ as x increases, the graph must eventually turn around and head up, but while heading up it can never cross $y = 4$ where a tangent line to a solution curve must have slope -2. Thus, y cannot approach ∞ as x approaches ∞.

17. When $y < \frac{1}{2}x^2$, $y' = x^2 - 2y$ is positive and the portions of solution curves "outside" the nullcline parabola are increasing. When $y > \frac{1}{2}x^2$, $y' = x^2 - 2y$ is negative and the portions of the solution curves "inside" the nullcline parabola are decreasing.

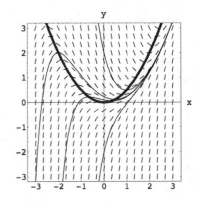

18. (a) Any horizontal lineal element should be at a point on a nullcline. In Problem 1 the nullclines are $x^2 - y^2 = 0$ or $y = \pm x$. In Problem 3 the nullclines are $1 - xy = 0$ or $y = 1/x$. In Problem 4 the nullclines are $(\sin x) \cos y = 0$ or $x = n\pi$ and $y = \pi/2 + n\pi$, where n is an integer. The graphs on the next page show the nullclines for the differential equations in Problems 1, 3, and 4 superimposed on the corresponding direction field.

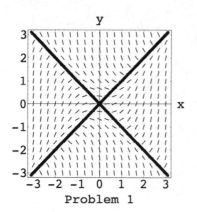

Problem 1

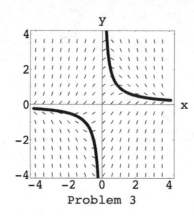

Problem 3

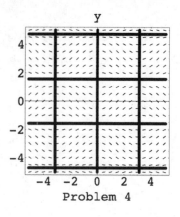

Problem 4

(b) An autonomous first-order differential equation has the form $y' = f(y)$. Nullclines have the form $y = c$ where $f(c) = 0$. These are the graphs of the equilibrium solutions of the differential equation.

19. Writing the differential equation in the form $dy/dx = y(1 - y)(1 + y)$ we see that critical points are located at $y = -1$, $y = 0$, and $y = 1$. The phase portrait is shown at the right.

(a)

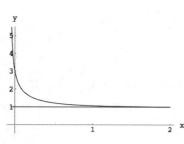

(b)

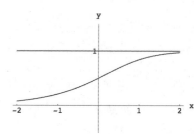

(c)

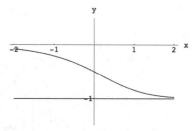

(d)

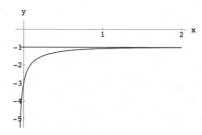

20. Writing the differential equation in the form $dy/dx = y^2(1 - y)(1 + y)$ we see that critical points are located at $y = -1$, $y = 0$, and $y = 1$. The phase portrait is shown at the right.

(a)

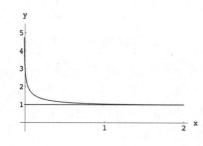

(b)

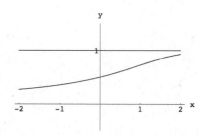

(c)

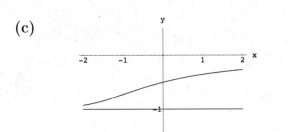

(d)

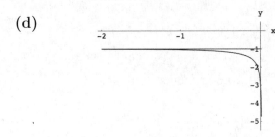

21. Solving $y^2 - 3y = y(y - 3) = 0$ we obtain the critical points 0 and 3. From the phase portrait we see that 0 is asymptotically stable (attractor) and 3 is unstable (repeller).

22. Solving $y^2 - y^3 = y^2(1 - y) = 0$ we obtain the critical points 0 and 1. From the phase portrait we see that 1 is asymptotically stable (attractor) and 0 is semi-stable.

23. Solving $(y - 2)^4 = 0$ we obtain the critical point 2. From the phase portrait we see that 2 is semi-stable.

24. Solving $10 + 3y - y^2 = (5 - y)(2 + y) = 0$ we obtain the critical points -2 and 5. From the phase portrait we see that 5 is asymptotically stable (attractor) and -2 is unstable (repeller).

25. Solving $y^2(4 - y^2) = y^2(2 - y)(2 + y) = 0$ we obtain the critical points -2, 0, and 2. From the phase portrait we see that 2 is asymptotically stable (attractor), 0 is semi-stable, and -2 is unstable (repeller).

26. Solving $y(2-y)(4-y) = 0$ we obtain the critical points 0, 2, and 4. From the phase portrait we see that 2 is asymptotically stable (attractor) and 0 and 4 are unstable (repellers).

27. Solving $y \ln(y + 2) = 0$ we obtain the critical points -1 and 0. From the phase portrait we see that -1 is asymptotically stable (attractor) and 0 is unstable (repeller).

28. Solving $ye^y - 9y = y(e^y - 9) = 0$ we obtain the critical points 0 and $\ln 9$. From the phase portrait we see that 0 is asymptotically stable (attractor) and $\ln 9$ is unstable (repeller).

29. The critical points are 0 and c because the graph of $f(y)$ is 0 at these points. Since $f(y) > 0$ for $y < 0$ and $y > c$, the graph of the solution is increasing on $(-\infty, 0)$ and (c, ∞). Since $f(y) < 0$ for $0 < y < c$, the graph of the solution is decreasing on $(0, c)$.

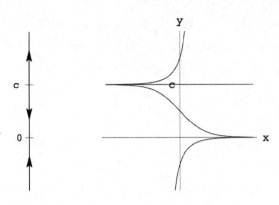

30. The critical points are approximately at $-2, 2, 0.5$, and 1.7. Since $f(y) > 0$ for $y < -2.2$ and $0.5 < y < 1.7$, the graph of the solution is increasing on $(-\infty, -2.2)$ and $(0.5, 1.7)$. Since $f(y) < 0$ for $-2.2 < y < 0.5$ and $y > 1.7$, the graph is decreasing on $(-2.2, 0.5)$ and $(1.7, \infty)$.

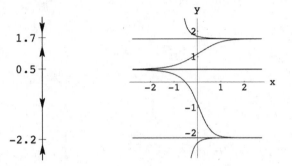

31. From the graphs of $z = \pi/2$ and $z = \sin y$ we see that $(\pi/2)y - \sin y = 0$ has only three solutions. By inspection we see that the critical points are $-\pi/2$, 0, and $\pi/2$.

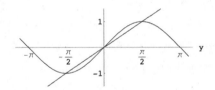

From the graph at the right we see that

$$\frac{2}{\pi}y - \sin y \begin{cases} < 0 & \text{for} \quad y < -\pi/2 \\ > 0 & \text{for} \quad y > \pi/2 \end{cases}$$

$$\frac{2}{\pi}y - \sin y \begin{cases} > 0 & \text{for} \quad -\pi/2 < y < 0 \\ < 0 & \text{for} \quad 0 < y < \pi/2. \end{cases}$$

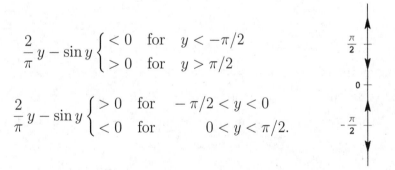

This enables us to construct the phase portrait shown at the right. From this portrait we see that $\pi/2$ and $-\pi/2$ are unstable (repellers), and 0 is asymptotically stable (attractor).

32. For $dy/dx = 0$ every real number is a critical point, and hence all critical points are nonisolated.

33. Recall that for $dy/dx = f(y)$ we are assuming that f and f' are continuous functions of y on some interval I. Now suppose that the graph of a nonconstant solution of the differential equation crosses the line $y = c$. If the point of intersection is taken as an initial condition we have two distinct solutions of the initial-value problem. This violates uniqueness, so the graph of any nonconstant solution must lie entirely on one side of any equilibrium solution. Since f is continuous it can only change signs at a point where it is 0. But this is a critical point. Thus, $f(y)$ is completely positive or completely negative in each region R_i. If $y(x)$ is oscillatory or has a relative extremum, then it must have a horizontal tangent line at some point (x_0, y_0). In this case y_0 would be a critical point of the differential equation, but we saw above that the graph of a nonconstant solution cannot intersect the graph of the equilibrium solution $y = y_0$.

34. By Problem 33, a solution $y(x)$ of $dy/dx = f(y)$ cannot have relative extrema and hence must be monotone. Since $y'(x) = f(y) > 0$, $y(x)$ is monotone increasing, and since $y(x)$ is bounded above by c_2, $\lim_{x \to \infty} y(x) = L$, where $L \le c_2$. We want to show that $L = c_2$. Since L is a horizontal asymptote of $y(x)$, $\lim_{x \to \infty} y'(x) = 0$. Using the fact that $f(y)$ is continuous we have

$$f(L) = f(\lim_{x \to \infty} y(x)) = \lim_{x \to \infty} f(y(x)) = \lim_{x \to \infty} y'(x) = 0.$$

But then L is a critical point of f. Since $c_1 < L \le c_2$, and f has no critical points between c_1 and c_2, $L = c_2$.

35. Assuming the existence of the second derivative, points of inflection of $y(x)$ occur where $y''(x) = 0$. From $dy/dx = f(y)$ we have $d^2y/dx^2 = f'(y)\, dy/dx$. Thus, the y-coordinate of a point of inflection can be located by solving $f'(y) = 0$. (Points where $dy/dx = 0$ correspond to constant solutions of the differential equation.)

36. Solving $y^2 - y - 6 = (y - 3)(y + 2) = 0$ we see that 3 and -2 are critical points. Now $d^2y/dx^2 = (2y - 1)\, dy/dx = (2y - 1)(y - 3)(y + 2)$, so the only possible point of inflection is at $y = \frac{1}{2}$, although the concavity of solutions can be different on either side of $y = -2$ and $y = 3$. Since $y''(x) < 0$ for $y < -2$ and $\frac{1}{2} < y < 3$, and $y''(x) > 0$ for $-2 < y < \frac{1}{2}$ and $y > 3$, we see that solution curves are concave down for $y < -2$ and $\frac{1}{2} < y < 3$ and concave up for $-2 < y < \frac{1}{2}$ and $y > 3$. Points of inflection of solutions of autonomous differential equations will have the same y-coordinates because between critical points they are horizontal translates of each other.

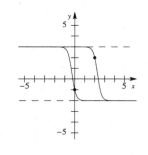

37. If (1) in the text has no critical points it has no constant solutions. The solutions have neither an upper nor lower bound. Since solutions are monotonic, every solution assumes all real values.

38. The critical points are 0 and b/a. From the phase portrait we see that 0 is an attractor and b/a is a repeller. Thus, if an initial population satisfies $P_0 > b/a$, the population becomes unbounded as t increases, most probably in finite time, i.e. $P(t) \to \infty$ as $t \to T$. If $0 < P_0 < b/a$, then the population eventually dies out, that is, $P(t) \to 0$ as $t \to \infty$. Since population $P > 0$ we do not consider the case $P_0 < 0$.

39. The only critical point of the autonomous differential equation is the positive number h/k. A phase portrait shows that this point is unstable, so h/k is a repeller. For any initial condition $P(0) = P_0 < h/k, dP/dt < 0$, which means $P(t)$ is monotonic decreasing and so the graph of $P(t)$ must cross the t-axis or the line $P = 0$ at some time $t_1 > 0$. But $P(t_1) = 0$ means the population is extinct at time t_1.

40. Writing the differential equation in the form

$$\frac{dv}{dt} = \frac{k}{m}\left(\frac{mg}{k} - v\right)$$

we see that a critical point is mg/k.

From the phase portrait we see that mg/k is an asymptotically stable critical point. Thus, $\lim_{t\to\infty} v = mg/k$.

41. Writing the differential equation in the form

$$\frac{dv}{dt} = \frac{k}{m}\left(\frac{mg}{k} - v^2\right) = \frac{k}{m}\left(\sqrt{\frac{mg}{k}} - v\right)\left(\sqrt{\frac{mg}{k}} + v\right)$$

we see that the only physically meaningful critical point is $\sqrt{mg/k}$.

From the phase portrait we see that $\sqrt{mg/k}$ is an asymptotically stable critical point. Thus, $\lim_{t\to\infty} v = \sqrt{mg/k}$.

42. (a) From the phase portrait we see that critical points are α and β. Let $X(0) = X_0$. If $X_0 < \alpha$, we see that $X \to \alpha$ as $t \to \infty$. If $\alpha < X_0 < \beta$, we see that $X \to \alpha$ as $t \to \infty$. If $X_0 > \beta$, we see that $X(t)$ increases in an unbounded manner, but more specific behavior of $X(t)$ as $t \to \infty$ is not known.

(b) When $\alpha = \beta$ the phase portrait is as shown. If $X_0 < \alpha$, then $X(t) \to \alpha$ as $t \to \infty$. If $X_0 > \alpha$, then $X(t)$ increases in an unbounded manner. This could happen in a finite amount of time. That is, the phase portrait does not indicate that X becomes unbounded as $t \to \infty$.

(c) When $k = 1$ and $\alpha = \beta$ the differential equation is $dX/dt = (\alpha - X)^2$. For $X(t) = \alpha - 1/(t+c)$ we have $dX/dt = 1/(t+c)^2$ and

$$(\alpha - X)^2 = \left[\alpha - \left(\alpha - \frac{1}{t+c}\right)\right]^2 = \frac{1}{(t+c)^2} = \frac{dX}{dt}.$$

For $X(0) = \alpha/2$ we obtain

$$X(t) = \alpha - \frac{1}{t + 2/\alpha}.$$

For $X(0) = 2\alpha$ we obtain

$$X(t) = \alpha - \frac{1}{t - 1/\alpha}.$$

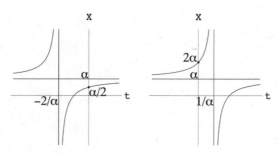

For $X_0 > \alpha$, $X(t)$ increases without bound up to $t = 1/\alpha$. For $t > 1/\alpha$, $X(t)$ increases but $X \to \alpha$ as $t \to \infty$.

Exercises 2.2

Separable Variables

In many of the following problems we will encounter an expression of the form $\ln |g(y)| = f(x) + c$. To solve for $g(y)$ we exponentiate both sides of the equation. This yields $|g(y)| = e^{f(x)+c} = e^{c}e^{f(x)}$ which implies $g(y) = \pm e^{c}e^{f(x)}$. Letting $c_1 = \pm e^{c}$ we obtain $g(y) = c_1 e^{f(x)}$.

1. From $dy = \sin 5x\, dx$ we obtain $y = -\frac{1}{5}\cos 5x + c$.

2. From $dy = (x+1)^2\, dx$ we obtain $y = \frac{1}{3}(x+1)^3 + c$.

3. From $dy = -e^{-3x}\, dx$ we obtain $y = \frac{1}{3}e^{-3x} + c$.

4. From $\dfrac{1}{(y-1)^2}\, dy = dx$ we obtain $-\dfrac{1}{y-1} = x + c$ or $y = 1 - \dfrac{1}{x+c}$.

5. From $\dfrac{1}{y}\, dy = \dfrac{4}{x}\, dx$ we obtain $\ln |y| = 4\ln |x| + c$ or $y = c_1 x^4$.

6. From $\dfrac{1}{y^2}\, dy = -2x\, dx$ we obtain $-\dfrac{1}{y} = -x^2 + c$ or $y = \dfrac{1}{x^2 + c_1}$.

7. From $e^{-2y}dy = e^{3x}dx$ we obtain $3e^{-2y} + 2e^{3x} = c$.

8. From $ye^{y}dy = \left(e^{-x} + e^{-3x}\right)dx$ we obtain $ye^{y} - e^{y} + e^{-x} + \dfrac{1}{3}e^{-3x} = c$.

9. From $\left(y + 2 + \dfrac{1}{y}\right)dy = x^2 \ln x\, dx$ we obtain $\dfrac{y^2}{2} + 2y + \ln |y| = \dfrac{x^3}{3}\ln |x| - \dfrac{1}{9}x^3 + c$.

10. From $\dfrac{1}{(2y+3)^2}\, dy = \dfrac{1}{(4x+5)^2}\, dx$ we obtain $\dfrac{2}{2y+3} = \dfrac{1}{4x+5} + c$.

11. From $\dfrac{1}{\csc y}\, dy = -\dfrac{1}{\sec^2 x}\, dx$ or $\sin y\, dy = -\cos^2 x\, dx = -\frac{1}{2}(1 + \cos 2x)\, dx$ we obtain
    $-\cos y = -\frac{1}{2}x - \frac{1}{4}\sin 2x + c$ or $4\cos y = 2x + \sin 2x + c_1$.

12. From $2y\, dy = -\dfrac{\sin 3x}{\cos^3 3x}\, dx$ or $2y\, dy = -\tan 3x \sec^2 3x\, dx$ we obtain $y^2 = -\frac{1}{6}\sec^2 3x + c$.

13. From $\dfrac{e^{y}}{(e^{y}+1)^2}\, dy = \dfrac{-e^{x}}{(e^{x}+1)^3}\, dx$ we obtain $-\left(e^{y}+1\right)^{-1} = \frac{1}{2}\left(e^{x}+1\right)^{-2} + c$.

14. From $\dfrac{y}{(1+y^2)^{1/2}}\, dy = \dfrac{x}{(1+x^2)^{1/2}}\, dx$ we obtain $\left(1+y^2\right)^{1/2} = \left(1+x^2\right)^{1/2} + c$.

15. From $\dfrac{1}{S}\, dS = k\, dr$ we obtain $S = ce^{kr}$.

16. From $\dfrac{1}{Q-70}\, dQ = k\, dt$ we obtain $\ln |Q - 70| = kt + c$ or $Q - 70 = c_1 e^{kt}$.

17. From $\dfrac{1}{P-P^2}dP = \left(\dfrac{1}{P}+\dfrac{1}{1-P}\right)dP = dt$ we obtain $\ln|P| - \ln|1-P| = t+c$ so that $\ln\left|\dfrac{P}{1-P}\right| = t+c$ or $\dfrac{P}{1-P} = c_1 e^t$. Solving for P we have $P = \dfrac{c_1 e^t}{1+c_1 e^t}$.

18. From $\dfrac{1}{N}dN = \left(te^{t+2}-1\right)dt$ we obtain $\ln|N| = te^{t+2} - e^{t+2} - t + c$ or $N = c_1 e^{te^{t+2}-e^{t+2}-t}$.

19. From $\dfrac{y-2}{y+3}dy = \dfrac{x-1}{x+4}dx$ or $\left(1-\dfrac{5}{y+3}\right)dy = \left(1-\dfrac{5}{x+4}\right)dx$ we obtain $y - 5\ln|y+3| =$

$x - 5\ln|x+4| + c$ or $\left(\dfrac{x+4}{y+3}\right)^5 = c_1 e^{x-y}$.

20. From $\dfrac{y+1}{y-1}dy = \dfrac{x+2}{x-3}dx$ or $\left(1+\dfrac{2}{y-1}\right)dy = \left(1+\dfrac{5}{x-3}\right)dx$ we obtain $y + 2\ln|y-1| =$

$x + 5\ln|x-3| + c$ or $\dfrac{(y-1)^2}{(x-3)^5} = c_1 e^{x-y}$.

21. From $x\,dx = \dfrac{1}{\sqrt{1-y^2}}dy$ we obtain $\tfrac{1}{2}x^2 = \sin^{-1}y + c$ or $y = \sin\left(\dfrac{x^2}{2}+c_1\right)$.

22. From $\dfrac{1}{y^2}dy = \dfrac{1}{e^x + e^{-x}}dx = \dfrac{e^x}{(e^x)^2 + 1}dx$ we obtain $-\dfrac{1}{y} = \tan^{-1}e^x + c$ or $y = -\dfrac{1}{\tan^{-1}e^x + c}$.

23. From $\dfrac{1}{x^2+1}dx = 4\,dt$ we obtain $\tan^{-1}x = 4t + c$. Using $x(\pi/4) = 1$ we find $c = -3\pi/4$. The

solution of the initial-value problem is $\tan^{-1}x = 4t - \dfrac{3\pi}{4}$ or $x = \tan\left(4t - \dfrac{3\pi}{4}\right)$.

24. From $\dfrac{1}{y^2-1}dy = \dfrac{1}{x^2-1}dx$ or $\dfrac{1}{2}\left(\dfrac{1}{y-1}-\dfrac{1}{y+1}\right)dy = \dfrac{1}{2}\left(\dfrac{1}{x-1}-\dfrac{1}{x+1}\right)dx$ we obtain

$\ln|y-1| - \ln|y+1| = \ln|x-1| - \ln|x+1| + \ln c$ or $\dfrac{y-1}{y+1} = \dfrac{c(x-1)}{x+1}$. Using $y(2) = 2$ we find

$c = 1$. A solution of the initial-value problem is $\dfrac{y-1}{y+1} = \dfrac{x-1}{x+1}$ or $y = x$.

25. From $\dfrac{1}{y}dy = \dfrac{1-x}{x^2}dx = \left(\dfrac{1}{x^2}-\dfrac{1}{x}\right)dx$ we obtain $\ln|y| = -\dfrac{1}{x} - \ln|x| = c$ or $xy = c_1 e^{-1/x}$. Using

$y(-1) = -1$ we find $c_1 = e^{-1}$. The solution of the initial-value problem is $xy = e^{-1-1/x}$ or

$y = e^{-(1+1/x)}/x$.

26. From $\dfrac{1}{1-2y}dy = dt$ we obtain $-\tfrac{1}{2}\ln|1-2y| = t + c$ or $1 - 2y = c_1 e^{-2t}$. Using $y(0) = 5/2$ we find

$c_1 = -4$. The solution of the initial-value problem is $1 - 2y = -4e^{-2t}$ or $y = 2e^{-2t} + \tfrac{1}{2}$.

27. Separating variables and integrating we obtain

$$\dfrac{dx}{\sqrt{1-x^2}} - \dfrac{dy}{\sqrt{1-y^2}} = 0 \quad \text{and} \quad \sin^{-1}x - \sin^{-1}y = c.$$

Setting $x = 0$ and $y = \sqrt{3}/2$ we obtain $c = -\pi/3$. Thus, an implicit solution of the initial-value problem is $\sin^{-1} x - \sin^{-1} y = -\pi/3$. Solving for y and using an addition formula from trigonometry, we get

$$y = \sin\left(\sin^{-1} x + \frac{\pi}{3}\right) = x \cos\frac{\pi}{3} + \sqrt{1 - x^2}\sin\frac{\pi}{3} = \frac{x}{2} + \frac{\sqrt{3}\sqrt{1 - x^2}}{2}.$$

28. From $\dfrac{1}{1 + (2y)^2}\, dy = \dfrac{-x}{1 + (x^2)^2}\, dx$ we obtain

$$\frac{1}{2}\tan^{-1} 2y = -\frac{1}{2}\tan^{-1} x^2 + c \quad \text{or} \quad \tan^{-1} 2y + \tan^{-1} x^2 = c_1.$$

Using $y(1) = 0$ we find $c_1 = \pi/4$. Thus, an implicit solution of the initial-value problem is $\tan^{-1} 2y + \tan^{-1} x^2 = \pi/4$. Solving for y and using a trigonometric identity we get

$$2y = \tan\left(\frac{\pi}{4} - \tan^{-1} x^2\right)$$

$$y = \frac{1}{2}\tan\left(\frac{\pi}{4} - \tan^{-1} x^2\right)$$

$$= \frac{1}{2}\frac{\tan\frac{\pi}{4} - \tan(\tan^{-1} x^2)}{1 + \tan\frac{\pi}{4}\tan(\tan^{-1} x^2)}$$

$$= \frac{1}{2}\frac{1 - x^2}{1 + x^2}.$$

29. Separating variables, integrating from 4 to x, and using t as a dummy variable of integration gives

$$\int_4^x \frac{1}{y}\frac{dy}{dt}\, dt = \int_4^x e^{-t^2}\, dt$$

$$\ln y(t)\Big|_4^x = \int_4^x e^{-t^2}\, dt$$

$$\ln y(x) - \ln y(4) = \int_4^x e^{-t^2}\, dt$$

Using the initial condition we have

$$\ln y(x) = \ln y(4) + \int_4^x e^{-t^2}\, dt = \ln 1 + \int_4^x e^{-t^2}\, dt = \int_4^x e^{-t^2}\, dt.$$

Thus,

$$y(x) = e^{\int_4^x e^{-t^2}\, dt}.$$

30. Separating variables, integrating from -2 to x, and using t as a dummy variable of integration gives

$$\int_{-2}^{x} \frac{1}{y^2} \frac{dy}{dt} \, dt = \int_{-2}^{x} \sin t^2 \, dt$$

$$-y(t)^{-1} \Big|_{-2}^{x} = \int_{-2}^{x} \sin t^2 \, dt$$

$$-y(x)^{-1} + y(-2)^{-1} = \int_{-2}^{x} \sin t^2 \, dt$$

$$-y(x)^{-1} = -y(-2)^{-1} + \int_{-2}^{x} \sin t^2 \, dt$$

$$y(x)^{-1} = 3 - \int_{-2}^{x} \sin t^2 \, dt.$$

Thus

$$y(x) = \frac{1}{3 - \int_{-2}^{x} \sin t^2 \, dt} \, .$$

31. (a) The equilibrium solutions $y(x) = 2$ and $y(x) = -2$ satisfy the initial conditions $y(0) = 2$ and $y(0) = -2$, respectively. Setting $x = \frac{1}{4}$ and $y = 1$ in $y = 2(1 + ce^{4x})/(1 - ce^{4x})$ we obtain

$$1 = 2\frac{1 + ce}{1 - ce}, \quad 1 - ce = 2 + 2ce, \quad -1 = 3ce, \quad \text{and} \quad c = -\frac{1}{3e} \, .$$

The solution of the corresponding initial-value problem is

$$y = 2\frac{1 - \frac{1}{3}e^{4x-1}}{1 + \frac{1}{3}e^{4x-1}} = 2\frac{3 - e^{4x-1}}{3 + e^{4x-1}} \, .$$

(b) Separating variables and integrating yields

$$\frac{1}{4} \ln|y - 2| - \frac{1}{4} \ln|y + 2| + \ln c_1 = x$$

$$\ln|y - 2| - \ln|y + 2| + \ln c = 4x$$

$$\ln \left| \frac{c(y - 2)}{y + 2} \right| = 4x$$

$$\pm c \frac{y - 2}{y + 2} = e^{4x} \, . \qquad \textit{let } \pm c \textit{ be } c_2$$

Solving for y we get $y = 2(c + e^{4x})/(c - e^{4x})$. The initial condition $y(0) = -2$ implies $2(c + 1)/(c - 1) = -2$ which yields $c = 0$ and $y(x) = -2$. The initial condition $y(0) = 2$ does not correspond to a value of c, and it must simply be recognized that $y(x) = 2$ is a solution of the initial-value problem. Setting $x = \frac{1}{4}$ and $y = 1$ in $y = 2(c + e^{4x})/(c - e^{4x})$ leads to $c = -3e$. Thus, a solution of the initial-value problem is

$$y = 2\frac{-3e + e^{4x}}{-3e - e^{4x}} = 2\frac{3 - e^{4x-1}}{3 + e^{4x-1}} \, .$$

32. Separating variables, we have

$$\frac{dy}{y^2 - y} = \frac{dx}{x} \qquad \text{or} \qquad \int \frac{dy}{y(y-1)} = \ln|x| + c.$$

Using partial fractions, we obtain

$$\int \left(\frac{1}{y-1} - \frac{1}{y} \right) dy = \ln|x| + c$$

$$\ln|y-1| - \ln|y| = \ln|x| + c$$

$$\ln\left| \frac{y-1}{xy} \right| = c$$

$$\frac{y-1}{xy} = \overset{\pm}{\underset{\wedge}{e^c}} = c_1.$$

Solving for y we get $y = 1/(1 - c_1 x)$. We note by inspection that $y = 0$ is a singular solution of the differential equation.

(a) Setting $x = 0$ and $y = 1$ we have $1 = 1/(1 - 0)$, which is true for all values of c_1. Thus, solutions passing through $(0, 1)$ are $y = 1/(1 - c_1 x)$.

(b) Setting $x = 0$ and $y = 0$ in $y = 1/(1 - c_1 x)$ we get $0 = 1$. Thus, the only solution passing through $(0, 0)$ is $y = 0$.

(c) Setting $x = \frac{1}{2}$ and $y = \frac{1}{2}$ we have $\frac{1}{2} = 1/(1 - \frac{1}{2} c_1)$, so $c_1 = -2$ and $y = 1/(1 + 2x)$.

(d) Setting $x = 2$ and $y = \frac{1}{4}$ we have $\frac{1}{4} = 1/(1 - 2c_1)$, so $c_1 = -\frac{3}{2}$ and $y = 1/(1 + \frac{3}{2}x) = 2/(2 + 3x)$.

33. Singular solutions of $dy/dx = x\sqrt{1 - y^2}$ are $y = -1$ and $y = 1$. A singular solution of $(e^x + e^{-x})dy/dx = y^2$ is $y = 0$.

34. Differentiating $\ln(x^2 + 10) + \csc y = c$ we get

$$\frac{2x}{x^2 + 10} - \csc y \, \cot y \, \frac{dy}{dx} = 0,$$

$$\frac{2x}{x^2 + 10} - \frac{1}{\sin y} \cdot \frac{\cos y}{\sin y} \frac{dy}{dx} = 0,$$

or

$$2x \sin^2 y \, dx - (x^2 + 10) \cos y \, dy = 0.$$

Writing the differential equation in the form

$$\frac{dy}{dx} = \frac{2x \sin^2 y}{(x^2 + 10) \cos y}$$

we see that singular solutions occur when $\sin^2 y = 0$, or $y = k\pi$, where k is an integer.

35. The singular solution $y = 1$ satisfies the initial-value problem.

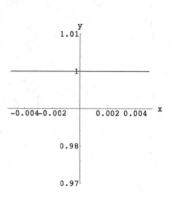

36. Separating variables we obtain $\dfrac{dy}{(y-1)^2} = dx$. Then

$$-\frac{1}{y-1} = x + c \quad \text{and} \quad y = \frac{x + c - 1}{x + c}.$$

Setting $x = 0$ and $y = 1.01$ we obtain $c = -100$. The solution is

$$y = \frac{x - 101}{x - 100}.$$

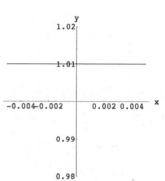

37. Separating variables we obtain $\dfrac{dy}{(y-1)^2 + 0.01} = dx$. Then

$$10 \tan^{-1} 10(y-1) = x + c \quad \text{and} \quad y = 1 + \frac{1}{10} \tan \frac{x + c}{10}.$$

Setting $x = 0$ and $y = 1$ we obtain $c = 0$. The solution is

$$y = 1 + \frac{1}{10} \tan \frac{x}{10}.$$

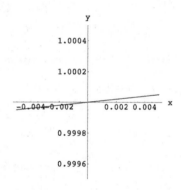

38. Separating variables we obtain $\dfrac{dy}{(y-1)^2 - 0.01} = dx$. Then, from (11) in this section of the manual with $u = y - 1$ and $a = \frac{1}{10}$, we get

$$5 \ln \left| \frac{10y - 11}{10y - 9} \right| = x + c.$$

Setting $x = 0$ and $y = 1$ we obtain $c = 5 \ln 1 = 0$. The solution is

$$5 \ln \left| \frac{10y - 11}{10y - 9} \right| = x.$$

Solving for y we obtain

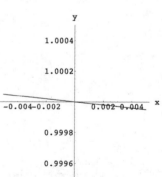

$$y = \frac{11 + 9e^{x/5}}{10 + 10e^{x/5}}.$$

Alternatively, we can use the fact that

$$\int \frac{dy}{(y-1)^2 - 0.01} = -\frac{1}{0.1} \tanh^{-1} \frac{y-1}{0.1} = -10 \tanh^{-1} 10(y-1).$$

(We use the inverse hyperbolic tangent because $|y - 1| < 0.1$ or $0.9 < y < 1.1$. This follows from the initial condition $y(0) = 1$.) Solving the above equation for y we get $y = 1 + 0.1 \tanh(x/10)$.

39. Separating variables, we have

$$\frac{dy}{y - y^3} = \frac{dy}{y(1-y)(1+y)} = \left(\frac{1}{y} + \frac{1/2}{1-y} - \frac{1/2}{1+y} \right) dy = dx.$$

Integrating, we get

$$\ln |y| - \frac{1}{2} \ln |1 - y| - \frac{1}{2} \ln |1 + y| = x + c.$$

When $y > 1$, this becomes

$$\ln y - \frac{1}{2} \ln(y - 1) - \frac{1}{2} \ln(y + 1) = \ln \frac{y}{\sqrt{y^2 - 1}} = x + c.$$

Letting $x = 0$ and $y = 2$ we find $c = \ln(2/\sqrt{3})$. Solving for y we get $y_1(x) = 2e^x/\sqrt{4e^{2x} - 3}$, where $x > \ln(\sqrt{3}/2)$.

When $0 < y < 1$ we have

$$\ln y - \frac{1}{2} \ln(1 - y) - \frac{1}{2} \ln(1 + y) = \ln \frac{y}{\sqrt{1 - y^2}} = x + c.$$

Letting $x = 0$ and $y = \frac{1}{2}$ we find $c = \ln(1/\sqrt{3})$. Solving for y we get $y_2(x) = e^x/\sqrt{e^{2x} + 3}$, where $-\infty < x < \infty$.

When $-1 < y < 0$ we have

$$\ln(-y) - \frac{1}{2} \ln(1 - y) - \frac{1}{2} \ln(1 + y) = \ln \frac{-y}{\sqrt{1 - y^2}} = x + c.$$

Letting $x = 0$ and $y = -\frac{1}{2}$ we find $c = \ln(1/\sqrt{3})$. Solving for y we get $y_3(x) = -e^x/\sqrt{e^{2x} + 3}$, where $-\infty < x < \infty$.

When $y < -1$ we have

$$\ln(-y) - \frac{1}{2} \ln(1 - y) - \frac{1}{2} \ln(-1 - y) = \ln \frac{-y}{\sqrt{y^2 - 1}} = x + c.$$

Letting $x = 0$ and $y = -2$ we find $c = \ln(2/\sqrt{3})$. Solving for y we get $y_4(x) = -2e^x/\sqrt{4e^{2x} - 3}$, where $x > \ln(\sqrt{3}/2)$.

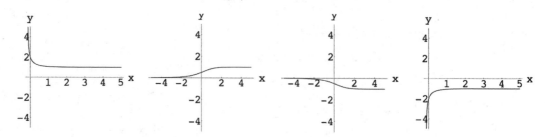

40. (a) The second derivative of y is

$$\frac{d^2y}{dx^2} = -\frac{dy/dx}{(y-1)^2} = -\frac{1/(y-3)}{(y-3)^2} = -\frac{1}{(y-3)^3}.$$

The solution curve is concave down when $d^2y/dx^2 < 0$ or $y > 3$, and concave up when $d^2y/dx^2 > 0$ or $y < 3$. From the phase portrait we see that the solution curve is decreasing when $y < 3$ and increasing when $y > 3$.

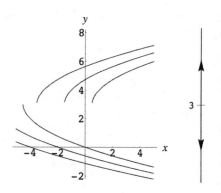

(b) Separating variables and integrating we obtain

$$(y-3)\,dy = dx$$

$$\frac{1}{2}y^2 - 3y = x + c$$

$$y^2 - 6y + 9 = 2x + c_1$$

$$(y-3)^2 = 2x + c_1$$

$$y = 3 \pm \sqrt{2x + c_1}.$$

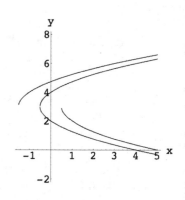

The initial condition dictates whether to use the plus or minus sign.

When $y_1(0) = 4$ we have $c_1 = 1$ and $y_1(x) = 3 + \sqrt{2x + 1}$.

When $y_2(0) = 2$ we have $c_1 = 1$ and $y_2(x) = 3 - \sqrt{2x + 1}$.

When $y_3(1) = 2$ we have $c_1 = -1$ and $y_3(x) = 3 - \sqrt{2x - 1}$.

When $y_4(-1) = 4$ we have $c_1 = 3$ and $y_4(x) = 3 + \sqrt{2x + 3}$.

41. (a) Separating variables we have $2y\,dy = (2x + 1)dx$. Integrating gives $y^2 = x^2 + x + c$. When $y(-2) = -1$ we find $c = -1$, so $y^2 = x^2 + x - 1$ and $y = -\sqrt{x^2 + x - 1}$. The negative square root is chosen because of the initial condition.

44

(b) From the figure, the largest interval of definition appears to be approximately $(-\infty, -1.65)$.

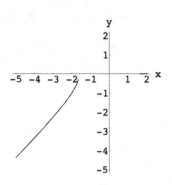

(c) Solving $x^2 + x - 1 = 0$ we get $x = -\frac{1}{2} \pm \frac{1}{2}\sqrt{5}$, so the largest interval of definition is $(-\infty, \ -\frac{1}{2} - \frac{1}{2}\sqrt{5})$. The right-hand endpoint of the interval is excluded because $y = -\sqrt{x^2 + x - 1}$ is not differentiable at this point.

42. **(a)** From Problem 7 the general solution is $3e^{-2y} + 2e^{3x} = c$. When $y(0) = 0$ we find $c = 5$, so $3e^{-2y} + 2e^{3x} = 5$. Solving for y we get $y = -\frac{1}{2} \ln \frac{1}{3}(5 - 2e^{3x})$.

(b) The interval of definition appears to be approximately $(-\infty, 0.3)$.

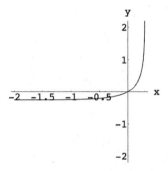

(c) Solving $\frac{1}{3}(5 - 2e^{3x}) = 0$ we get $x = \frac{1}{3}\ln(\frac{5}{2})$, so the exact interval of definition is $(-\infty, \frac{1}{3}\ln\frac{5}{2})$.

43. **(a)** While $y_2(x) = -\sqrt{25 - x^2}$ is defined at $x = -5$ and $x = 5$, $y_2'(x)$ is not defined at these values, and so the interval of definition is the open interval $(-5, 5)$.

(b) At any point on the x-axis the derivative of $y(x)$ is undefined, so no solution curve can cross the x-axis. Since $-x/y$ is not defined when $y = 0$, the initial-value problem has no solution.

44. **(a)** Separating variables and integrating we obtain $x^2 - y^2 = c$. For $c \neq 0$ the graph is a hyperbola centered at the origin. All four initial conditions imply $c = 0$ and $y = \pm x$. Since the differential equation is not defined for $y = 0$, solutions are $y = \pm x$, $x < 0$ and $y = \pm x$, $x > 0$. The solution for $y(a) = a$ is $y = x$, $x > 0$; for $y(a) = -a$ is $y = -x$; for $y(-a) = a$ is $y = -x$, $x < 0$; and for $y(-a) = -a$ is $y = x$, $x < 0$.

(b) Since x/y is not defined when $y = 0$, the initial-value problem has no solution.

(c) Setting $x = 1$ and $y = 2$ in $x^2 - y^2 = c$ we get $c = -3$, so $y^2 = x^2 + 3$ and $y(x) = \sqrt{x^2 + 3}$, where the positive square root is chosen because of the initial condition. The domain is all real numbers since $x^2 + 3 > 0$ for all x.

45. Separating variables we have $dy/(\sqrt{1+y^2}\sin^2 y) = dx$ which is not readily integrated (even by a CAS). We note that $dy/dx \geq 0$ for all values of x and y and that $dy/dx = 0$ when $y = 0$ and $y = \pi$, which are equilibrium solutions.

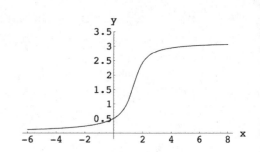

46. Separating variables we have $dy/(\sqrt{y}+y) = dx/(\sqrt{x}+x)$. To integrate $\int dx/(\sqrt{x}+x)$ we substitute $u^2 = x$ and get

$$\int \frac{2u}{u+u^2}\,du = \int \frac{2}{1+u}\,du = 2\ln|1+u| + c = 2\ln(1+\sqrt{x}\,) + c.$$

Integrating the separated differential equation we have

$$2\ln(1+\sqrt{y}\,) = 2\ln(1+\sqrt{x}\,) + c \quad \text{or} \quad \ln(1+\sqrt{y}\,) = \ln(1+\sqrt{x}\,) + \ln c_1.$$

Solving for y we get $y = [c_1(1+\sqrt{x}\,) - 1]^2$.

47. We are looking for a function $y(x)$ such that

$$y^2 + \left(\frac{dy}{dx}\right)^2 = 1.$$

Using the positive square root gives

$$\frac{dy}{dx} = \sqrt{1-y^2} \implies \frac{dy}{\sqrt{1-y^2}} = dx \implies \sin^{-1} y = x + c.$$

Thus a solution is $y = \sin(x+c)$. If we use the negative square root we obtain

$$y = \sin(c - x) = -\sin(x - c) = -\sin(x + c_1).$$

Note that when $c = c_1 = 0$ and when $c = c_1 = \pi/2$ we obtain the well known particular solutions $y = \sin x$, $y = -\sin x$, $y = \cos x$, and $y = -\cos x$. Note also that $y = 1$ and $y = -1$ are singular solutions.

48. (a)

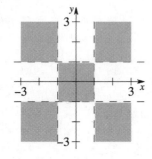

46

(b) For $|x| > 1$ and $|y| > 1$ the differential equation is $dy/dx = \sqrt{y^2 - 1}/\sqrt{x^2 - 1}$. Separating variables and integrating, we obtain

$$\frac{dy}{\sqrt{y^2 - 1}} = \frac{dx}{\sqrt{x^2 - 1}} \quad \text{and} \quad \cosh^{-1} y = \cosh^{-1} x + c.$$

Setting $x = 2$ and $y = 2$ we find $c = \cosh^{-1} 2 - \cosh^{-1} 2 = 0$ and $\cosh^{-1} y = \cosh^{-1} x$. An explicit solution is $y = x$.

49. Since the tension T_1 (or magnitude T_1) acts at the lowest point of the cable, we use symmetry to solve the problem on the interval $[0, L/2]$. The assumption that the roadbed is uniform (that is, weighs a constant ρ pounds per horizontal foot) implies $W = \rho x$, where x is measured in feet and $0 \leq x \leq L/2$. Therefore (10) in the text becomes $dy/dx = (\rho/T_1)x$. This last equation is a separable equation of the form given in (1) of Section 2.2 in the text. Integrating and using the initial condition $y(0) = a$ shows that the shape of the cable is a parabola: $y(x) = (\rho/2T_1)x^2 + a$. In terms of the sag h of the cable and the span L, we see from Figure 2.2.5 in the text that $y(L/2) = h + a$. By applying this last condition to $y(x) = (\rho/2T_1)x^2 + a$ enables us to express $\rho/2T_1$ in terms of h and L: $y(x) = (4h/L^2)x^2 + a$. Since $y(x)$ is an even function of x, the solution is valid on $-L/2 \leq x \leq L/2$.

50. (a) Separating variables and integrating, we have $(3y^2 + 1)dy = -(8x + 5)dx$ and $y^3 + y = -4x^2 - 5x + c$. Using a CAS we show various contours of $f(x, y) = y^3 + y + 4x^2 + 5x$. The plots shown on $[-5, 5] \times [-5, 5]$ correspond to c-values of 0, ± 5, ± 20, ± 40, ± 80, and ± 125.

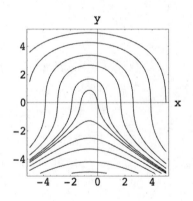

(b) The value of c corresponding to $y(0) = -1$ is $f(0, -1) = -2$; to $y(0) = 2$ is $f(0, 2) = 10$; to $y(-1) = 4$ is $f(-1, 4) = 67$; and to $y(-1) = -3$ is -31.

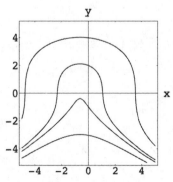

51. (a) An implicit solution of the differential equation $(2y + 2)dy - (4x^3 + 6x)dx = 0$ is

$$y^2 + 2y - x^4 - 3x^2 + c = 0.$$

The condition $y(0) = -3$ implies that $c = -3$. Therefore $y^2 + 2y - x^4 - 3x^2 - 3 = 0$.

(b) Using the quadratic formula we can solve for y in terms of x:

$$y = \frac{-2 \pm \sqrt{4 + 4(x^4 + 3x^2 + 3)}}{2}.$$

The explicit solution that satisfies the initial condition is then

$$y = -1 - \sqrt{x^4 + 3x^3 + 4}.$$

(c) From the graph of $f(x) = x^4 + 3x^3 + 4$ below we see that $f(x) \leq 0$ on the approximate interval $-2.8 \leq x \leq -1.3$. Thus the approximate domain of the function

$$y = -1 - \sqrt{x^4 + 3x^3 + 4} = -1 - \sqrt{f(x)}$$

is $x \leq -2.8$ or $x \geq -1.3$. The graph of this function is shown below.

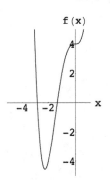

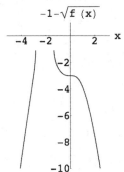

(d) Using the root finding capabilities of a CAS, the zeros of f are found to be -2.82202 and -1.3409. The domain of definition of the solution $y(x)$ is then $x > -1.3409$. The equality has been removed since the derivative dy/dx does not exist at the points where $f(x) = 0$. The graph of the solution $y = \phi(x)$ is given on the right.

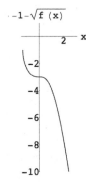

52. (a) Separating variables and integrating, we have

$$(-2y + y^2)\,dy = (x - x^2)\,dx$$

and

$$-y^2 + \frac{1}{3}y^3 = \frac{1}{2}x^2 - \frac{1}{3}x^3 + c.$$

Using a CAS we show some contours of $f(x, y) = 2y^3 - 6y^2 + 2x^3 - 3x^2$. The plots shown on $[-7, 7] \times [-5, 5]$ correspond to c-values of -450, -300, -200, -120, -60, -20, -10, -8.1, -5, -0.8, 20, 60, and 120.

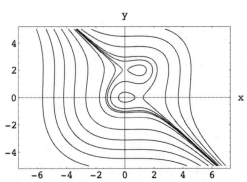

48

(b) The value of c corresponding to $y(0) = \frac{3}{2}$ is $f(0, \frac{3}{2}) = -\frac{27}{4}$. The portion of the graph between the dots corresponds to the solution curve satisfying the intial condition. To determine the interval of definition we find dy/dx for $2y^3 - 6y^2 + 2x^3 - 3x^2 = -\frac{27}{4}$. Using implicit differentiation we get $y' = (x - x^2)/(y^2 - 2y)$, which is infinite when $y = 0$ and $y = 2$. Letting $y = 0$ in $2y^3 - 6y^2 + 2x^3 - 3x^2 = -\frac{27}{4}$ and using a CAS to solve for x we get $x = -1.13232$. Similarly, letting $y = 2$, we find $x = 1.71299$. The largest interval of definition is approximately $(-1.13232, 1.71299)$.

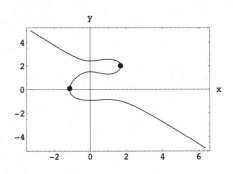

(c) The value of c corresponding to $y(0) = -2$ is $f(0, -2) = -40$. The portion of the graph to the right of the dot corresponds to the solution curve satisfying the initial condition. To determine the interval of definition we find dy/dx for $2y^3 - 6y^2 + 2x^3 - 3x^2 = -40$. Using implicit differentiation we get $y' = (x - x^2)/(y^2 - 2y)$, which is infinite when $y = 0$ and $y = 2$. Letting $y = 0$ in $2y^3 - 6y^2 + 2x^3 - 3x^2 = -40$ and using a CAS to solve for x we get $x = -2.29551$. The largest interval of definition is approximately $(-2.29551, \infty)$.

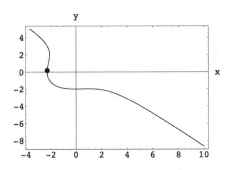

Exercises 2.3   Linear Equations

1. For $y' - 5y = 0$ an integrating factor is $e^{-\int 5\,dx} = e^{-5x}$ so that $\dfrac{d}{dx}\left[e^{-5x}y\right] = 0$ and $y = ce^{5x}$ for $-\infty < x < \infty$. There is no transient term.

2. For $y' + 2y = 0$ an integrating factor is $e^{\int 2\,dx} = e^{2x}$ so that $\dfrac{d}{dx}\left[e^{2x}y\right] = 0$ and $y = ce^{-2x}$ for $-\infty < x < \infty$. The transient term is ce^{-2x}.

3. For $y' + y = e^{3x}$ an integrating factor is $e^{\int dx} = e^x$ so that $\dfrac{d}{dx}[e^x y] = e^{4x}$ and $y = \frac{1}{4}e^{3x} + ce^{-x}$ for $-\infty < x < \infty$. The transient term is ce^{-x}.

4. For $y' + 4y = \frac{4}{3}$ an integrating factor is $e^{\int 4\,dx} = e^{4x}$ so that $\dfrac{d}{dx}\left[e^{4x}y\right] = \frac{4}{3}e^{4x}$ and $y = \frac{1}{3} + ce^{-4x}$ for $-\infty < x < \infty$. The transient term is ce^{-4x}.

5. For $y' + 3x^2 y = x^2$ an integrating factor is $e^{\int 3x^2\,dx} = e^{x^3}$ so that $\dfrac{d}{dx}\left[e^{x^3}y\right] = x^2 e^{x^3}$ and $y = \frac{1}{3} + ce^{-x^3}$ for $-\infty < x < \infty$. The transient term is ce^{-x^3}.

6. For $y' + 2xy = x^3$ an integrating factor is $e^{\int 2x\,dx} = e^{x^2}$ so that $\dfrac{d}{dx}\left[e^{x^2}y\right] = x^3 e^{x^2}$ and $y = \frac{1}{2}x^2 - \frac{1}{2} + ce^{-x^2}$ for $-\infty < x < \infty$. The transient term is ce^{-x^2}.

7. For $y' + \dfrac{1}{x}y = \dfrac{1}{x^2}$ an integrating factor is $e^{\int(1/x)dx} = x$ so that $\dfrac{d}{dx}\left[xy\right] = \dfrac{1}{x}$ and $y = \dfrac{1}{x}\ln x + \dfrac{c}{x}$ for $0 < x < \infty$. The entire solution is transient.

8. For $y' - 2y = x^2 + 5$ an integrating factor is $e^{-\int 2\,dx} = e^{-2x}$ so that $\dfrac{d}{dx}\left[e^{-2x}y\right] = x^2 e^{-2x} + 5e^{-2x}$ and $y = -\frac{1}{2}x^2 - \frac{1}{2}x - \frac{11}{4} + ce^{2x}$ for $-\infty < x < \infty$. There is no transient term.

9. For $y' - \dfrac{1}{x}y = x\sin x$ an integrating factor is $e^{-\int(1/x)dx} = \dfrac{1}{x}$ so that $\dfrac{d}{dx}\left[\dfrac{1}{x}y\right] = \sin x$ and $y = cx - x\cos x$ for $0 < x < \infty$. There is no transient term.

10. For $y' + \dfrac{2}{x}y = \dfrac{3}{x}$ an integrating factor is $e^{\int(2/x)dx} = x^2$ so that $\dfrac{d}{dx}\left[x^2 y\right] = 3x$ and $y = \dfrac{3}{2} + cx^{-2}$ for $0 < x < \infty$. The transient term is cx^{-2}.

11. For $y' + \dfrac{4}{x}y = x^2 - 1$ an integrating factor is $e^{\int(4/x)dx} = x^4$ so that $\dfrac{d}{dx}\left[x^4 y\right] = x^6 - x^4$ and $y = \frac{1}{7}x^3 - \frac{1}{5}x + cx^{-4}$ for $0 < x < \infty$. The transient term is cx^{-4}.

12. For $y' - \dfrac{x}{(1+x)}y = x$ an integrating factor is $e^{-\int[x/(1+x)]dx} = (x+1)e^{-x}$ so that $\dfrac{d}{dx}\left[(x+1)e^{-x}y\right] = x(x+1)e^{-x}$ and $y = -x - \dfrac{2x+3}{x+1} + \dfrac{ce^x}{x+1}$ for $-1 < x < \infty$. There is no transient term.

13. For $y' + \left(1 + \dfrac{2}{x}\right)y = \dfrac{e^x}{x^2}$ an integrating factor is $e^{\int[1+(2/x)]dx} = x^2 e^x$ so that $\dfrac{d}{dx}\left[x^2 e^x y\right] = e^{2x}$ and $y = \dfrac{1}{2}\dfrac{e^x}{x^2} + \dfrac{ce^{-x}}{x^2}$ for $0 < x < \infty$. The transient term is $\dfrac{ce^{-x}}{x^2}$.

14. For $y' + \left(1 + \dfrac{1}{x}\right)y = \dfrac{1}{x}e^{-x}\sin 2x$ an integrating factor is $e^{\int[1+(1/x)]dx} = xe^x$ so that $\dfrac{d}{dx}\left[xe^x y\right] = \sin 2x$ and $y = -\dfrac{1}{2x}e^{-x}\cos 2x + \dfrac{ce^{-x}}{x}$ for $0 < x < \infty$. The entire solution is transient.

15. For $\dfrac{dx}{dy} - \dfrac{4}{y}x = 4y^5$ an integrating factor is $e^{-\int(4/y)dy} = e^{\ln y^{-4}} = y^{-4}$ so that $\dfrac{d}{dy}\left[y^{-4}x\right] = 4y$ and $x = 2y^6 + cy^4$ for $0 < y < \infty$. There is no transient term.

16. For $\dfrac{dx}{dy} + \dfrac{2}{y}x = e^y$ an integrating factor is $e^{\int(2/y)dy} = y^2$ so that $\dfrac{d}{dy}\left[y^2x\right] = y^2e^y$ and

$x = e^y - \dfrac{2}{y}e^y + \dfrac{2}{y^2}e^y + \dfrac{c}{y^2}$ for $0 < y < \infty$. The transient term is $\dfrac{c}{y^2}$.

17. For $y' + (\tan x)y = \sec x$ an integrating factor is $e^{\int \tan x\, dx} = \sec x$ so that $\dfrac{d}{dx}\left[(\sec x)\,y\right] = \sec^2 x$ and

$y = \sin x + c\cos x$ for $-\pi/2 < x < \pi/2$. There is no transient term.

18. For $y' + (\cot x)y = \sec^2 x\csc x$ an integrating factor is $e^{\int \cot x\, dx} = e^{\ln|\sin x|} = \sin x$ so that

$\dfrac{d}{dx}\left[(\sin x)\,y\right] = \sec^2 x$ and $y = \sec x + c\csc x$ for $0 < x < \pi/2$. There is no transient term.

19. For $y' + \dfrac{x+2}{x+1}y = \dfrac{2xe^{-x}}{x+1}$ an integrating factor is $e^{\int [(x+2)/(x+1)]dx} = (x+1)e^x$, so $\dfrac{d}{dx}\left[(x+1)e^x y\right] =$

$2x$ and $y = \dfrac{x^2}{x+1}e^{-x} + \dfrac{c}{x+1}e^{-x}$ for $-1 < x < \infty$. The entire solution is transient.

20. For $y' + \dfrac{4}{x+2}y = \dfrac{5}{(x+2)^2}$ an integrating factor is $e^{\int [4/(x+2)]dx} = (x+2)^4$ so that $\dfrac{d}{dx}\left[(x+2)^4 y\right] =$

$5(x+2)^2$ and $y = \dfrac{5}{3}(x+2)^{-1} + c(x+2)^{-4}$ for $-2 < x < \infty$. The entire solution is transient.

21. For $\dfrac{dr}{d\theta} + r\sec\theta = \cos\theta$ an integrating factor is $e^{\int \sec\theta\, d\theta} = e^{\ln|\sec x + \tan x|} = \sec\theta + \tan\theta$ so that

$\dfrac{d}{d\theta}\left[(\sec\theta + \tan\theta)r\right] = 1 + \sin\theta$ and $(\sec\theta + \tan\theta)r = \theta - \cos\theta + c$ for $-\pi/2 < \theta < \pi/2$.

22. For $\dfrac{dP}{dt} + (2t-1)P = 4t - 2$ an integrating factor is $e^{\int(2t-1)\,dt} = e^{t^2-t}$ so that $\dfrac{d}{dt}\left[e^{t^2-t}P\right] =$

$(4t-2)e^{t^2-t}$ and $P = 2 + ce^{t-t^2}$ for $-\infty < t < \infty$. The transient term is ce^{t-t^2}.

23. For $y' + \left(3 + \dfrac{1}{x}\right)y = \dfrac{e^{-3x}}{x}$ an integrating factor is $e^{\int[3+(1/x)]dx} = xe^{3x}$ so that $\dfrac{d}{dx}\left[xe^{3x}y\right] = 1$ and

$y = e^{-3x} + \dfrac{ce^{-3x}}{x}$ for $0 < x < \infty$. The entire solution is transient.

24. For $y' + \dfrac{2}{x^2-1}y = \dfrac{x+1}{x-1}$ an integrating factor is $e^{\int[2/(x^2-1)]dx} = \dfrac{x-1}{x+1}$ so that $\dfrac{d}{dx}\left[\dfrac{x-1}{x+1}y\right] = 1$

and $(x-1)y = x(x+1) + c(x+1)$ for $-1 < x < 1$.

25. For $y' + \dfrac{1}{x}y = \dfrac{1}{x}e^x$ an integrating factor is $e^{\int(1/x)dx} = x$ so that $\dfrac{d}{dx}[xy] = e^x$ and $y = \dfrac{1}{x}e^x + \dfrac{c}{x}$

for $0 < x < \infty$. If $y(1) = 2$ then $c = 2 - e$ and $y = \dfrac{1}{x}e^x + \dfrac{2-e}{x}$.

26. For $\dfrac{dx}{dy} - \dfrac{1}{y}x = 2y$ an integrating factor is $e^{-\int(1/y)dy} = \dfrac{1}{y}$ so that $\dfrac{d}{dy}\left[\dfrac{1}{y}x\right] = 2$ and $x = 2y^2 + cy$

for $0 < y < \infty$. If $y(1) = 5$ then $c = -49/5$ and $x = 2y^2 - \dfrac{49}{5}y$.

27. For $\dfrac{di}{dt} + \dfrac{R}{L}i = \dfrac{E}{L}$ an integrating factor is $e^{\int(R/L)\,dt} = e^{Rt/L}$ so that $\dfrac{d}{dt}\left[e^{Rt/L}\,i\right] = \dfrac{E}{L}e^{Rt/L}$ and

$$i = \frac{E}{R} + ce^{-Rt/L} \text{ for } -\infty < t < \infty. \text{ If } i(0) = i_0 \text{ then } c = i_0 - E/R \text{ and } i = \frac{E}{R} + \left(i_0 - \frac{E}{R}\right)e^{-Rt/L}.$$

28. For $\dfrac{dT}{dt} - kT = -T_m k$ an integrating factor is $e^{\int(-k)dt} = e^{-kt}$ so that $\dfrac{d}{dt}\left[e^{-kt}T\right] = -T_m k e^{-kt}$ and $T = T_m + ce^{kt}$ for $-\infty < t < \infty$. If $T(0) = T_0$ then $c = T_0 - T_m$ and $T = T_m + (T_0 - T_m)e^{kt}$.

29. For $y' + \dfrac{1}{x+1}y = \dfrac{\ln x}{x+1}$ an integrating factor is $e^{\int[1/(x+1)]dx} = x+1$ so that $\dfrac{d}{dx}[(x+1)y] = $

$\ln x$ and $y = \dfrac{x}{x+1}\ln x - \dfrac{x}{x+1} + \dfrac{c}{x+1}$ for $0 < x < \infty$. If $y(1) = 10$ then $c = 21$ and

$$y = \frac{x}{x+1}\ln x - \frac{x}{x+1} + \frac{21}{x+1}.$$

30. For $y' + (\tan x)y = \cos^2 x$ an integrating factor is $e^{\int \tan x\, dx} = e^{\ln|\sec x|} = \sec x$ so that $\dfrac{d}{dx}[(\sec x)\,y] = $
$\cos x$ and $y = \sin x \cos x + c\cos x$ for $-\pi/2 < x < \pi/2$. If $y(0) = -1$ then $c = -1$ and $y = \sin x \cos x - \cos x$.

31. For $y' + 2y = f(x)$ an integrating factor is e^{2x} so that

$$ye^{2x} = \begin{cases} \frac{1}{2}e^{2x} + c_1, & 0 \le x \le 3 \\ c_2, & x > 3. \end{cases}$$

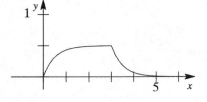

If $y(0) = 0$ then $c_1 = -1/2$ and for continuity we must have $c_2 = \frac{1}{2}e^6 - \frac{1}{2}$ so that

$$y = \begin{cases} \frac{1}{2}(1 - e^{-2x}), & 0 \le x \le 3 \\ \frac{1}{2}(e^6 - 1)e^{-2x}, & x > 3. \end{cases}$$

32. For $y' + y = f(x)$ an integrating factor is e^x so that

$$ye^x = \begin{cases} e^x + c_1, & 0 \le x \le 1 \\ -e^x + c_2, & x > 1. \end{cases}$$

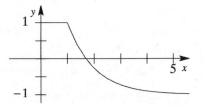

If $y(0) = 1$ then $c_1 = 0$ and for continuity we must have $c_2 = 2e$ so that

$$y = \begin{cases} 1, & 0 \le x \le 1 \\ 2e^{1-x} - 1, & x > 1. \end{cases}$$

33. For $y' + 2xy = f(x)$ an integrating factor is e^{x^2} so that

$$ye^{x^2} = \begin{cases} \frac{1}{2}e^{x^2} + c_1, & 0 \le x \le 1 \\ c_2, & x > 1. \end{cases}$$

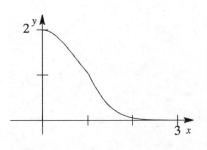

If $y(0) = 2$ then $c_1 = 3/2$ and for continuity we must have $c_2 = \frac{1}{2}e + \frac{3}{2}$ so that

$$y = \begin{cases} \frac{1}{2} + \frac{3}{2}e^{-x^2}, & 0 \le x \le 1 \\ \left(\frac{1}{2}e + \frac{3}{2}\right)e^{-x^2}, & x > 1. \end{cases}$$

34. For

$$y' + \frac{2x}{1+x^2}\,y = \begin{cases} \dfrac{x}{1+x^2}, & 0 \le x \le 1 \\ \dfrac{-x}{1+x^2}, & x > 1, \end{cases}$$

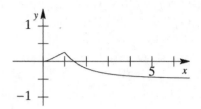

an integrating factor is $1 + x^2$ so that

$$\left(1+x^2\right)y = \begin{cases} \frac{1}{2}x^2 + c_1, & 0 \le x \le 1 \\ -\frac{1}{2}x^2 + c_2, & x > 1. \end{cases}$$

If $y(0) = 0$ then $c_1 = 0$ and for continuity we must have $c_2 = 1$ so that

$$y = \begin{cases} \dfrac{1}{2} - \dfrac{1}{2\left(1+x^2\right)}, & 0 \le x \le 1 \\ \dfrac{3}{2\left(1+x^2\right)} - \dfrac{1}{2}, & x > 1. \end{cases}$$

35. We first solve the initial-value problem $y' + 2y = 4x$, $y(0) = 3$ on the interval $[0, 1]$. The integrating factor is $e^{\int 2\,dx} = e^{2x}$, so

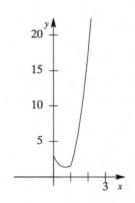

$$\frac{d}{dx}[e^{2x}y] = 4xe^{2x}$$

$$e^{2x}y = \int 4xe^{2x}\,dx = 2xe^{2x} - e^{2x} + c_1$$

$$y = 2x - 1 + c_1e^{-2x}.$$

Using the initial condition, we find $y(0) = -1 + c_1 = 3$, so $c_1 = 4$ and $y = 2x - 1 + 4e^{-2x}$, $0 \le x \le 1$. Now, since $y(1) = 2 - 1 + 4e^{-2} = 1 + 4e^{-2}$, we solve the initial-value problem $y' - (2/x)y = 4x$, $y(1) = 1 + 4e^{-2}$ on the interval $(1, \infty)$. The integrating factor is $e^{\int(-2/x)dx} = e^{-2\ln x} = x^{-2}$, so

$$\frac{d}{dx}[x^{-2}y] = 4xx^{-2} = \frac{4}{x}$$

$$x^{-2}y = \int \frac{4}{x}\,dx = 4\ln x + c_2$$

$$y = 4x^2\ln x + c_2x^2.$$

(We use $\ln x$ instead of $\ln|x|$ because $x > 1$.) Using the initial condition we find $y(1) = c_2 = 1 + 4e^{-2}$, so $y = 4x^2\ln x + (1 + 4e^{-2})x^2$, $x > 1$. Thus, the solution of the original initial-value problem is

$$y = \begin{cases} 2x - 1 + 4e^{-2x}, & 0 \leq x \leq 1 \\ 4x^2 \ln x + (1 + 4e^{-2})x^2, & x > 1. \end{cases}$$

See Problem 42 in this section.

36. For $y' + e^x y = 1$ an integrating factor is e^{e^x}. Thus

$$\frac{d}{dx}[e^{e^x} y] = e^{e^x} \quad \text{and} \quad e^{e^x} y = \int_0^x e^{e^t} dt + c.$$

From $y(0) = 1$ we get $c = e$, so $y = e^{-e^x} \int_0^x e^{e^t} dt + e^{1-e^x}$.

When $y' + e^x y = 0$ we can separate variables and integrate:

$$\frac{dy}{y} = -e^x \, dx \quad \text{and} \quad \ln|y| = -e^x + c.$$

Thus $y = c_1 e^{-e^x}$. From $y(0) = 1$ we get $c_1 = e$, so $y = e^{1-e^x}$.

When $y' + e^x y = e^x$ we can see by inspection that $y = 1$ is a solution.

37. An integrating factor for $y' - 2xy = 1$ is e^{-x^2}. Thus

$$\frac{d}{dx}[e^{-x^2} y] = e^{-x^2}$$

$$e^{-x^2} y = \int_0^x e^{-t^2} dt = \frac{\sqrt{\pi}}{2} \operatorname{erf}(x) + c$$

$$y = \frac{\sqrt{\pi}}{2} e^{x^2} \operatorname{erf}(x) + ce^{x^2}.$$

From $y(1) = (\sqrt{\pi}/2)e \operatorname{erf}(1) + ce = 1$ we get $c = e^{-1} - \frac{\sqrt{\pi}}{2} \operatorname{erf}(1)$. The solution of the initial-value problem is

$$y = \frac{\sqrt{\pi}}{2} e^{x^2} \operatorname{erf}(x) + \left(e^{-1} - \frac{\sqrt{\pi}}{2} \operatorname{erf}(1)\right) e^{x^2}$$

$$= e^{x^2 - 1} + \frac{\sqrt{\pi}}{2} e^{x^2} (\operatorname{erf}(x) - \operatorname{erf}(1)).$$

38. We want 4 to be a critical point, so we use $y' = 4 - y$.

39. (a) All solutions of the form $y = x^5 e^x - x^4 e^x + cx^4$ satisfy the initial condition. In this case, since $4/x$ is discontinuous at $x = 0$, the hypotheses of Theorem 1.2.1 are not satisfied and the initial-value problem does not have a unique solution.

(b) The differential equation has no solution satisfying $y(0) = y_0$, $y_0 > 0$.

(c) In this case, since $x_0 > 0$, Theorem 1.2.1 applies and the initial-value problem has a unique solution given by $y = x^5 e^x - x^4 e^x + cx^4$ where $c = y_0/x_0^4 - x_0 e^{x_0} + e^{x_0}$.

40. On the interval $(-3, 3)$ the integrating factor is

$$e^{\int x \, dx/(x^2 - 9)} = e^{-\int x \, dx/(9 - x^2)} = e^{\frac{1}{2} \ln(9 - x^2)} = \sqrt{9 - x^2}$$

[handwritten: in DE10e/DEBVP8e $ye^{dy}p$ are no longer]

and so

$$\frac{d}{dx}\left[\sqrt{9-x^2}\,y\right] = 0 \quad \text{and} \quad y = \frac{c}{\sqrt{9-x^2}}.$$

41. We want the general solution to be $y = 3x - 5 + ce^{-x}$. (Rather than e^{-x}, any function that approaches 0 as $x \to \infty$ could be used.) Differentiating we get *[handwritten: discussed in Section 2.3]*

$$y' = 3 - ce^{-x} = 3 - (y - 3x + 5) = -y + 3x - 2,$$

so the differential equation $y' + y = 3x - 2$ has solutions asymptotic to the line $y = 3x - 5$. *[handwritten: this is now]*

42. The left-hand derivative of the function at $x = 1$ is $1/e$ ~~and the right-hand~~ derivative at $x = 1$ is $1 - 1/e$. ~~Thus, y is not differentiable at $x = 1$.~~ *[handwritten: prob. 4]*

43. (a) Differentiating $y = c/x^3$ we get *[handwritten checkmarks]*

$$y' = -\frac{3c}{x^4} = -\frac{3}{x}\frac{c}{x^3} = -\frac{3}{x}y$$

so a differential equation with general solution $y = c/x^3$ is $xy' + 3y = 0$. Now

$$xy_p' + 3y_p = x(3x^2) + 3(x^3) = 6x^3$$

so a differential equation with general solution $y = c/x^3 + x^3$ is $xy' + 3y = 6x^3$. This will be a general solution on $(0, \infty)$ *[handwritten: or $(-\infty, 0)$.]*

(b) Since $y(1) = 1^3 - 1/1^3 = 0$, an initial condition is $\boxed{y(1) = 0.}$ Since $y(1) = 1^3 + 2/1^3 = 3$, an initial condition is $y(1) = 3$. In each case the interval of definition is $(0, \infty)$. The initial-value problem $xy' + 3y = 6x^3$, $y(0) = 0$ has solution $y = x^3$ for $-\infty < x < \infty$. In the figure the lower curve is the graph of $y(x) = x^3 - 1/x^3$, while the upper curve is the graph of $y = x^3 - 2/x^3$.

[handwritten: $y(-1) = -1+1 = 0$ so could also use $y(-1) = 0$ on $(-\infty, 0)$]

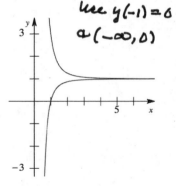

(c) The first two initial-value problems in part (b) are not unique. For example, setting $y(2) = 2^3 - 1/2^3 = 63/8$, we see that $y(2) = 63/8$ is also an initial condition leading to the solution $y = x^3 - 1/x^3$.

44. Since $e^{\int P(x)dx + c} = e^c e^{\int P(x)dx} = c_1 e^{\int P(x)dx}$, we would have

$$c_1 e^{\int P(x)dx} y = c_2 + \int c_1 e^{\int P(x)dx} f(x)\, dx \quad \text{and} \quad e^{\int P(x)dx} y = c_3 + \int e^{\int P(x)dx} f(x)\, dx,$$

which is the same as (6) in the text.

45. We see by inspection that $y = 0$ is a solution.

46. The solution of the first equation is $x = c_1 e^{-\lambda_1 t}$. From $x(0) = x_0$ we obtain $c_1 = x_0$ and so $x = x_0 e^{-\lambda_1 t}$. The second equation then becomes

$$\frac{dy}{dt} = x_0 \lambda_1 e^{-\lambda_1 t} - \lambda_2 y \quad \text{or} \quad \frac{dy}{dt} + \lambda_2 y = x_0 \lambda_1 e^{-\lambda_1 t}$$

which is linear. An integrating factor is $e^{\lambda_2 t}$. Thus

$$\frac{d}{dt}\left[e^{\lambda_2 t}y\right] = x_0\lambda_1 e^{-\lambda_1 t}e^{\lambda_2 t} = x_0\lambda_1 e^{(\lambda_2-\lambda_1)t}$$

$$e^{\lambda_2 t}y = \frac{x_0\lambda_1}{\lambda_2 - \lambda_1}\,e^{(\lambda_2-\lambda_1)t} + c_2$$

$$y = \frac{x_0\lambda_1}{\lambda_2 - \lambda_1}\,e^{-\lambda_1 t} + c_2 e^{-\lambda_2 t}.$$

From $y(0) = y_0$ we obtain $c_2 = (y_0\lambda_2 - y_0\lambda_1 - x_0\lambda_1)/(\lambda_2 - \lambda_1)$. The solution is

$$y = \frac{x_0\lambda_1}{\lambda_2 - \lambda_1}\,e^{-\lambda_1 t} + \frac{y_0\lambda_2 - y_0\lambda_1 - x_0\lambda_1}{\lambda_2 - \lambda_1}\,e^{-\lambda_2 t}.$$

47. Writing the differential equation as $\dfrac{dE}{dt} + \dfrac{1}{RC}E = 0$ we see that an integrating factor is $e^{t/RC}$. Then

$$\frac{d}{dt}[e^{t/RC}E] = 0$$

$$e^{t/RC}E = c$$

$$E = ce^{-t/RC}.$$

From $E(4) = ce^{-4/RC} = E_0$ we find $c = E_0 e^{4/RC}$. Thus, the solution of the initial-value problem is

$$E = E_0 e^{4/RC}e^{-t/RC} = E_0 e^{-(t-4)/RC}.$$

48. (a) An integrating factor for $y' - 2xy = -1$ is e^{-x^2}. Thus

$$\frac{d}{dx}[e^{-x^2}y] = -e^{-x^2}$$

$$e^{-x^2}y = -\int_0^x e^{-t^2}\,dt = -\frac{\sqrt{\pi}}{2}\,\text{erf}(x) + c.$$

From $y(0) = \sqrt{\pi}/2$, and noting that $\text{erf}(0) = 0$, we get $c = \sqrt{\pi}/2$. Thus

$$y = e^{x^2}\left(-\frac{\sqrt{\pi}}{2}\,\text{erf}(x) + \frac{\sqrt{\pi}}{2}\right) = \frac{\sqrt{\pi}}{2}\,e^{x^2}(1 - \text{erf}(x)) = \frac{\sqrt{\pi}}{2}\,e^{x^2}\,\text{erfc}(x).$$

(b) Using a CAS we find $y(2) \approx 0.226339$.

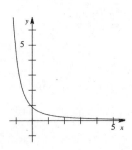

49. (a) An integrating factor for

$$y' + \frac{2}{x}y = \frac{10\sin x}{x^3}$$

56

is x^2. Thus

$$\frac{d}{dx}[x^2 y] = 10\frac{\sin x}{x}$$

$$x^2 y = 10 \int_0^x \frac{\sin t}{t}\, dt + c$$

$$y = 10x^{-2}\text{Si}(x) + cx^{-2}.$$

From $y(1) = 0$ we get $c = -10\text{Si}(1)$. Thus

$$y = 10x^{-2}\text{Si}(x) - 10x^{-2}\text{Si}(1) = 10x^{-2}(\text{Si}(x) - \text{Si}(1)).$$

(b)

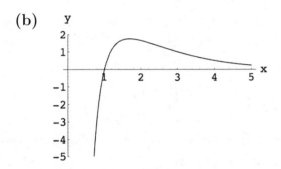

(c) From the graph in part (b) we see that the absolute maximum occurs around $x = 1.7$. Using the root-finding capability of a CAS and solving $y'(x) = 0$ for x we see that the absolute maximum is $(1.688, 1.742)$.

50. (a) The integrating factor for $y' - (\sin x^2)y = 0$ is $e^{-\int_0^x \sin t^2\, dt}$. Then

$$\frac{d}{dx}[e^{-\int_0^x \sin t^2 dt}y] = 0$$

$$e^{-\int_0^x \sin t^2\, dt}y = c_1$$

$$y = c_1 e^{\int_0^x \sin t^2 dt}.$$

Letting $t = \sqrt{\pi/2}\,u$ we have $dt = \sqrt{\pi/2}\,du$ and

$$\int_0^x \sin t^2\, dt = \sqrt{\frac{\pi}{2}} \int_0^{\sqrt{2/\pi}\,x} \sin\left(\frac{\pi}{2}u^2\right) du = \sqrt{\frac{\pi}{2}}\, S\left(\sqrt{\frac{2}{\pi}}\,x\right)$$

so $y = c_1 e^{\sqrt{\pi/2}\,S(\sqrt{2/\pi}\,x)}$. Using $S(0) = 0$ and $y(0) = c_1 = 5$ we have $y = 5e^{\sqrt{\pi/2}\,S(\sqrt{2/\pi}\,x)}$.

(b)

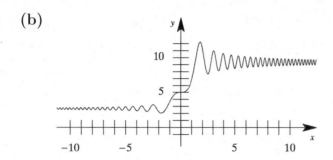

(c) From the graph we see that as $x \to \infty$, $y(x)$ oscillates with decreasing amplitudes approaching 9.35672. Since $\lim_{x \to \infty} 5S(x) = \frac{1}{2}$, $\lim_{x \to \infty} y(x) = 5e^{\sqrt{\pi/8}} \approx 9.357$, and since $\lim_{x \to -\infty} S(x) = -\frac{1}{2}$, $\lim_{x \to -\infty} y(x) = 5e^{-\sqrt{\pi/8}} \approx 2.672$.

(d) From the graph in part (b) we see that the absolute maximum occurs around $x = 1.7$ and the absolute minimum occurs around $x = -1.8$. Using the root-finding capability of a CAS and solving $y'(x) = 0$ for x, we see that the absolute maximum is $(1.772, 12.235)$ and the absolute minimum is $(-1.772, 2.044)$.

Exercises 2.4   Exact Equations

1. Let $M = 2x - 1$ and $N = 3y + 7$ so that $M_y = 0 = N_x$. From $f_x = 2x - 1$ we obtain $f = x^2 - x + h(y)$, $h'(y) = 3y + 7$, and $h(y) = \frac{3}{2}y^2 + 7y$. A solution is $x^2 - x + \frac{3}{2}y^2 + 7y = c$.

2. Let $M = 2x + y$ and $N = -x - 6y$. Then $M_y = 1$ and $N_x = -1$, so the equation is not exact.

3. Let $M = 5x + 4y$ and $N = 4x - 8y^3$ so that $M_y = 4 = N_x$. From $f_x = 5x + 4y$ we obtain $f = \frac{5}{2}x^2 + 4xy + h(y)$, $h'(y) = -8y^3$, and $h(y) = -2y^4$. A solution is $\frac{5}{2}x^2 + 4xy - 2y^4 = c$.

4. Let $M = \sin y - y \sin x$ and $N = \cos x + x \cos y - y$ so that $M_y = \cos y - \sin x = N_x$. From $f_x = \sin y - y \sin x$ we obtain $f = x \sin y + y \cos x + h(y)$, $h'(y) = -y$, and $h(y) = -\frac{1}{2}y^2$. A solution is $x \sin y + y \cos x - \frac{1}{2}y^2 = c$.

5. Let $M = 2y^2x - 3$ and $N = 2yx^2 + 4$ so that $M_y = 4xy = N_x$. From $f_x = 2y^2x - 3$ we obtain $f = x^2y^2 - 3x + h(y)$, $h'(y) = 4$, and $h(y) = 4y$. A solution is $x^2y^2 - 3x + 4y = c$.

6. Let $M = 4x^3 - 3y \sin 3x - y/x^2$ and $N = 2y - 1/x + \cos 3x$ so that $M_y = -3 \sin 3x - 1/x^2$ and $N_x = 1/x^2 - 3 \sin 3x$. The equation is not exact.

7. Let $M = x^2 - y^2$ and $N = x^2 - 2xy$ so that $M_y = -2y$ and $N_x = 2x - 2y$. The equation is not exact.

8. Let $M = 1 + \ln x + y/x$ and $N = -1 + \ln x$ so that $M_y = 1/x = N_x$. From $f_y = -1 + \ln x$ we obtain $f = -y + y\ln x + h(y)$, $h'(x) = 1 + \ln x$, and $h(y) = x\ln x$. A solution is $-y + y\ln x + x\ln x = c$.

9. Let $M = y^3 - y^2\sin x - x$ and $N = 3xy^2 + 2y\cos x$ so that $M_y = 3y^2 - 2y\sin x = N_x$. From $f_x = y^3 - y^2\sin x - x$ we obtain $f = xy^3 + y^2\cos x - \frac{1}{2}x^2 + h(y)$, $h'(y) = 0$, and $h(y) = 0$. A solution is $xy^3 + y^2\cos x - \frac{1}{2}x^2 = c$.

10. Let $M = x^3 + y^3$ and $N = 3xy^2$ so that $M_y = 3y^2 = N_x$. From $f_x = x^3 + y^3$ we obtain $f = \frac{1}{4}x^4 + xy^3 + h(y)$, $h'(y) = 0$, and $h(y) = 0$. A solution is $\frac{1}{4}x^4 + xy^3 = c$.

11. Let $M = y\ln y - e^{-xy}$ and $N = 1/y + x\ln y$ so that $M_y = 1 + \ln y + xe^{-xy}$ and $N_x = \ln y$. The equation is not exact.

12. Let $M = 3x^2y + e^y$ and $N = x^3 + xe^y - 2y$ so that $M_y = 3x^2 + e^y = N_x$. From $f_x = 3x^2y + e^y$ we obtain $f = x^3y + xe^y + h(y)$, $h'(y) = -2y$, and $h(y) = -y^2$. A solution is $x^3y + xe^y - y^2 = c$.

13. Let $M = y - 6x^2 - 2xe^x$ and $N = x$ so that $M_y = 1 = N_x$. From $f_x = y - 6x^2 - 2xe^x$ we obtain $f = xy - 2x^3 - 2xe^x + 2e^x + h(y)$, $h'(y) = 0$, and $h(y) = 0$. A solution is $xy - 2x^3 - 2xe^x + 2e^x = c$.

14. Let $M = 1 - 3/x + y$ and $N = 1 - 3/y + x$ so that $M_y = 1 = N_x$. From $f_x = 1 - 3/x + y$ we obtain $f = x - 3\ln|x| + xy + h(y)$, $h'(y) = 1 - \dfrac{3}{y}$, and $h(y) = y - 3\ln|y|$. A solution is $x + y + xy - 3\ln|xy| = c$.

15. Let $M = x^2y^3 - 1/\left(1 + 9x^2\right)$ and $N = x^3y^2$ so that $M_y = 3x^2y^2 = N_x$. From $f_x = x^2y^3 - 1/\left(1 + 9x^2\right)$ we obtain $f = \frac{1}{3}x^3y^3 - \frac{1}{3}\arctan(3x) + h(y)$, $h'(y) = 0$, and $h(y) = 0$. A solution is $x^3y^3 - \arctan(3x) = c$.

16. Let $M = -2y$ and $N = 5y - 2x$ so that $M_y = -2 = N_x$. From $f_x = -2y$ we obtain $f = -2xy + h(y)$, $h'(y) = 5y$, and $h(y) = \frac{5}{2}y^2$. A solution is $-2xy + \frac{5}{2}y^2 = c$.

17. Let $M = \tan x - \sin x\sin y$ and $N = \cos x\cos y$ so that $M_y = -\sin x\cos y = N_x$. From $f_x = \tan x - \sin x\sin y$ we obtain $f = \ln|\sec x| + \cos x\sin y + h(y)$, $h'(y) = 0$, and $h(y) = 0$. A solution is $\ln|\sec x| + \cos x\sin y = c$.

18. Let $M = 2y\sin x\cos x - y + 2y^2e^{xy^2}$ and $N = -x + \sin^2 x + 4xye^{xy^2}$ so that
$$M_y = 2\sin x\cos x - 1 + 4xy^3e^{xy^2} + 4ye^{xy^2} = N_x.$$

From $f_x = 2y\sin x\cos x - y + 2y^2e^{xy^2}$ we obtain $f = y\sin^2 x - xy + 2e^{xy^2} + h(y)$, $h'(y) = 0$, and $h(y) = 0$. A solution is $y\sin^2 x - xy + 2e^{xy^2} = c$.

19. Let $M = 4t^3y - 15t^2 - y$ and $N = t^4 + 3y^2 - t$ so that $M_y = 4t^3 - 1 = N_t$. From $f_t = 4t^3y - 15t^2 - y$ we obtain $f = t^4y - 5t^3 - ty + h(y)$, $h'(y) = 3y^2$, and $h(y) = y^3$. A solution is $t^4y - 5t^3 - ty + y^3 = c$.

20. Let $M = 1/t + 1/t^2 - y/\left(t^2 + y^2\right)$ and $N = ye^y + t/\left(t^2 + y^2\right)$ so that $M_y = \left(y^2 - t^2\right)/\left(t^2 + y^2\right)^2 = N_t$. From $f_t = 1/t + 1/t^2 - y/\left(t^2 + y^2\right)$ we obtain $f = \ln|t| - \dfrac{1}{t} - \arctan\left(\dfrac{t}{y}\right) + h(y)$, $h'(y) = ye^y$,

and $h(y) = ye^y - e^y$. A solution is

$$\ln|t| - \frac{1}{t} - \arctan\left(\frac{t}{y}\right) + ye^y - e^y = c.$$

21. Let $M = x^2 + 2xy + y^2$ and $N = 2xy + x^2 - 1$ so that $M_y = 2(x+y) = N_x$. From $f_x = x^2 + 2xy + y^2$ we obtain $f = \frac{1}{3}x^3 + x^2y + xy^2 + h(y)$, $h'(y) = -1$, and $h(y) = -y$. The solution is $\frac{1}{3}x^3 + x^2y + xy^2 - y = c$. If $y(1) = 1$ then $c = 4/3$ and a solution of the initial-value problem is $\frac{1}{3}x^3 + x^2y + xy^2 - y = \frac{4}{3}$.

22. Let $M = e^x + y$ and $N = 2 + x + ye^y$ so that $M_y = 1 = N_x$. From $f_x = e^x + y$ we obtain $f = e^x + xy + h(y)$, $h'(y) = 2 + ye^y$, and $h(y) = 2y + ye^y - y$. The solution is $e^x + xy + 2y + ye^y - e^y = c$. If $y(0) = 1$ then $c = 3$ and a solution of the initial-value problem is $e^x + xy + 2y + ye^y - e^y = 3$.

23. Let $M = 4y + 2t - 5$ and $N = 6y + 4t - 1$ so that $M_y = 4 = N_t$. From $f_t = 4y + 2t - 5$ we obtain $f = 4ty + t^2 - 5t + h(y)$, $h'(y) = 6y - 1$, and $h(y) = 3y^2 - y$. The solution is $4ty + t^2 - 5t + 3y^2 - y = c$. If $y(-1) = 2$ then $c = 8$ and a solution of the initial-value problem is $4ty + t^2 - 5t + 3y^2 - y = 8$.

24. Let $M = t/2y^4$ and $N = \left(3y^2 - t^2\right)/y^5$ so that $M_y = -2t/y^5 = N_t$. From $f_t = t/2y^4$ we obtain $f = \frac{t^2}{4y^4} + h(y)$, $h'(y) = \frac{3}{y^3}$, and $h(y) = -\frac{3}{2y^2}$. The solution is $\frac{t^2}{4y^4} - \frac{3}{2y^2} = c$. If $y(1) = 1$ then $c = -5/4$ and a solution of the initial-value problem is $\frac{t^2}{4y^4} - \frac{3}{2y^2} = -\frac{5}{4}$.

25. Let $M = y^2\cos x - 3x^2y - 2x$ and $N = 2y\sin x - x^3 + \ln y$ so that $M_y = 2y\cos x - 3x^2 = N_x$. From $f_x = y^2\cos x - 3x^2y - 2x$ we obtain $f = y^2\sin x - x^3y - x^2 + h(y)$, $h'(y) = \ln y$, and $h(y) = y\ln y - y$. The solution is $y^2\sin x - x^3y - x^2 + y\ln y - y = c$. If $y(0) = e$ then $c = 0$ and a solution of the initial-value problem is $y^2\sin x - x^3y - x^2 + y\ln y - y = 0$.

26. Let $M = y^2 + y\sin x$ and $N = 2xy - \cos x - 1/\left(1 + y^2\right)$ so that $M_y = 2y + \sin x = N_x$. From $f_x = y^2 + y\sin x$ we obtain $f = xy^2 - y\cos x + h(y)$, $h'(y) = \frac{-1}{1 + y^2}$, and $h(y) = -\tan^{-1}y$. The solution is $xy^2 - y\cos x - \tan^{-1}y = c$. If $y(0) = 1$ then $c = -1 - \pi/4$ and a solution of the initial-value problem is $xy^2 - y\cos x - \tan^{-1}y = -1 - \frac{\pi}{4}$.

27. Equating $M_y = 3y^2 + 4kxy^3$ and $N_x = 3y^2 + 40xy^3$ we obtain $k = 10$.

28. Equating $M_y = 18xy^2 - \sin y$ and $N_x = 4kxy^2 - \sin y$ we obtain $k = 9/2$.

29. Let $M = -x^2y^2\sin x + 2xy^2\cos x$ and $N = 2x^2y\cos x$ so that $M_y = -2x^2y\sin x + 4xy\cos x = N_x$. From $f_y = 2x^2y\cos x$ we obtain $f = x^2y^2\cos x + h(y)$, $h'(y) = 0$, and $h(y) = 0$. A solution of the differential equation is $x^2y^2\cos x = c$.

30. Let $M = \left(x^2 + 2xy - y^2\right)/\left(x^2 + 2xy + y^2\right)$ and $N = \left(y^2 + 2xy - x^2\right)/\left(y^2 + 2xy + x^2\right)$ so that $M_y = -4xy/(x + y)^3 = N_x$. From $f_x = \left(x^2 + 2xy + y^2 - 2y^2\right)/(x + y)^2$ we obtain

$f = x + \dfrac{2y^2}{x+y} + h(y)$, $h'(y) = -1$, and $h(y) = -y$. A solution of the differential equation is $x^2 + y^2 = c(x+y)$.

31. We note that $(M_y - N_x)/N = 1/x$, so an integrating factor is $e^{\int dx/x} = x$. Let $M = 2xy^2 + 3x^2$ and $N = 2x^2y$ so that $M_y = 4xy = N_x$. From $f_x = 2xy^2 + 3x^2$ we obtain $f = x^2y^2 + x^3 + h(y)$, $h'(y) = 0$, and $h(y) = 0$. A solution of the differential equation is $x^2y^2 + x^3 = c$.

32. We note that $(M_y - N_x)/N = 1$, so an integrating factor is $e^{\int dx} = e^x$. Let $M = xye^x + y^2e^x + ye^x$ and $N = xe^x + 2ye^x$ so that $M_y = xe^x + 2ye^x + e^x = N_x$. From $f_y = xe^x + 2ye^x$ we obtain $f = xye^x + y^2e^x + h(x)$, $h'(y) = 0$, and $h(y) = 0$. A solution of the differential equation is $xye^x + y^2e^x = c$.

33. We note that $(N_x - M_y)/M = 2/y$, so an integrating factor is $e^{\int 2dy/y} = y^2$. Let $M = 6xy^3$ and $N = 4y^3 + 9x^2y^2$ so that $M_y = 18xy^2 = N_x$. From $f_x = 6xy^3$ we obtain $f = 3x^2y^3 + h(y)$, $h'(y) = 4y^3$, and $h(y) = y^4$. A solution of the differential equation is $3x^2y^3 + y^4 = c$.

34. We note that $(M_y - N_x)/N = -\cot x$, so an integrating factor is $e^{-\int \cot x\, dx} = \csc x$. Let $M = \cos x \csc x = \cot x$ and $N = (1 + 2/y)\sin x \csc x = 1 + 2/y$, so that $M_y = 0 = N_x$. From $f_x = \cot x$ we obtain $f = \ln(\sin x) + h(y)$, $h'(y) = 1 + 2/y$, and $h(y) = y + \ln y^2$. A solution of the differential equation is $\ln(\sin x) + y + \ln y^2 = c$.

35. We note that $(M_y - N_x)/N = 3$, so an integrating factor is $e^{\int 3\,dx} = e^{3x}$. Let

$$M = (10 - 6y + e^{-3x})e^{3x} = 10e^{3x} - 6ye^{3x} + 1$$

and

$$N = -2e^{3x},$$

so that $M_y = -6e^{3x} = N_x$. From $f_x = 10e^{3x} - 6ye^{3x} + 1$ we obtain $f = \frac{10}{3}e^{3x} - 2ye^{3x} + x + h(y)$, $h'(y) = 0$, and $h(y) = 0$. A solution of the differential equation is $\frac{10}{3}e^{3x} - 2ye^{3x} + x = c$.

36. We note that $(N_x - M_y)/M = -3/y$, so an integrating factor is $e^{-3\int dy/y} = 1/y^3$. Let

$$M = (y^2 + xy^3)/y^3 = 1/y + x$$

and

$$N = (5y^2 - xy + y^3\sin y)/y^3 = 5/y - x/y^2 + \sin y,$$

so that $M_y = -1/y^2 = N_x$. From $f_x = 1/y + x$ we obtain $f = x/y + \frac{1}{2}x^2 + h(y)$, $h'(y) = 5/y + \sin y$, and $h(y) = 5\ln|y| - \cos y$. A solution of the differential equation is $x/y + \frac{1}{2}x^2 + 5\ln|y| - \cos y = c$.

37. We note that $(M_y - N_x)/N = 2x/(4 + x^2)$, so an integrating factor is $e^{-2\int x\,dx/(4+x^2)} = 1/(4 + x^2)$. Let $M = x/(4 + x^2)$ and $N = (x^2y + 4y)/(4 + x^2) = y$, so that $M_y = 0 = N_x$. From $f_x = x(4 + x^2)$ we obtain $f = \frac{1}{2}\ln(4 + x^2) + h(y)$, $h'(y) = y$, and $h(y) = \frac{1}{2}y^2$. A solution of the differential equation is $\frac{1}{2}\ln(4 + x^2) + \frac{1}{2}y^2 = c$.

38. We note that $(M_y - N_x)/N = -3/(1+x)$, so an integrating factor is $e^{-3\int dx/(1+x)} = 1/(1+x)^3$. Let $M = (x^2+y^2-5)/(1+x)^3$ and $N = -(y+xy)/(1+x)^3 = -y/(1+x)^2$, so that $M_y = 2y/(1+x)^3 = N_x$. From $f_y = -y/(1+x)^2$ we obtain $f = -\frac{1}{2}y^2/(1+x)^2 + h(x)$, $h'(x) = (x^2 - 5)/(1+x)^3$, and $h(x) = 2/(1+x)^2 + 2/(1+x) + \ln|1+x|$. A solution of the differential equation is

$$-\frac{y^2}{2(1+x)^2} + \frac{2}{(1+x)^2} + \frac{2}{(1+x)} + \ln|1+x| = c.$$

39. (a) Implicitly differentiating $x^3 + 2x^2y + y^2 = c$ and solving for dy/dx we obtain

$$3x^2 + 2x^2\frac{dy}{dx} + 4xy + 2y\frac{dy}{dx} = 0 \quad \text{and} \quad \frac{dy}{dx} = -\frac{3x^2 + 4xy}{2x^2 + 2y}.$$

By writing the last equation in differential form we get $(4xy + 3x^2)dx + (2y + 2x^2)dy = 0$.

(b) Setting $x = 0$ and $y = -2$ in $x^3 + 2x^2y + y^2 = c$ we find $c = 4$, and setting $x = y = 1$ we also find $c = 4$. Thus, both initial conditions determine the same implicit solution.

(c) Solving $x^3 + 2x^2y + y^2 = 4$ for y we get

$$y_1(x) = -x^2 - \sqrt{4 - x^3 + x^4}$$

and

$$y_2(x) = -x^2 + \sqrt{4 - x^3 + x^4}.$$

Observe in the figure that $y_1(0) = -2$ and $y_2(1) = 1$.

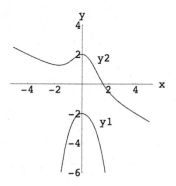

40. To see that the equations are not equivalent consider $dx = -(x/y)dy$. An integrating factor is $\mu(x, y) = y$ resulting in $y\, dx + x\, dy = 0$. A solution of the latter equation is $y = 0$, but this is not a solution of the original equation.

41. The explicit solution is $y = \sqrt{(3 + \cos^2 x)/(1 - x^2)}$. Since $3 + \cos^2 x > 0$ for all x we must have $1 - x^2 > 0$ or $-1 < x < 1$. Thus, the interval of definition is $(-1, 1)$.

42. (a) Since $f_y = N(x, y) = xe^{xy} + 2xy + 1/x$ we obtain $f = e^{xy} + xy^2 + \dfrac{y}{x} + h(x)$ so that $f_x = ye^{xy} + y^2 - \dfrac{y}{x^2} + h'(x)$. Let $M(x, y) = ye^{xy} + y^2 - \dfrac{y}{x^2}$.

(b) Since $f_x = M(x, y) = y^{1/2}x^{-1/2} + x\left(x^2 + y\right)^{-1}$ we obtain $f = 2y^{1/2}x^{1/2} + \dfrac{1}{2}\ln\left|x^2 + y\right| + g(y)$ so that $f_y = y^{-1/2}x^{1/2} + \dfrac{1}{2}\left(x^2 + y\right)^{-1} + g'(x)$. Let $N(x, y) = y^{-1/2}x^{1/2} + \dfrac{1}{2}\left(x^2 + y\right)^{-1}$.

43. First note that

$$d\left(\sqrt{x^2 + y^2}\right) = \frac{x}{\sqrt{x^2 + y^2}}\, dx + \frac{y}{\sqrt{x^2 + y^2}}\, dy.$$

Then $x\,dx + y\,dy = \sqrt{x^2 + y^2}\,dx$ becomes

$$\frac{x}{\sqrt{x^2 + y^2}}\,dx + \frac{y}{\sqrt{x^2 + y^2}}\,dy = d\left(\sqrt{x^2 + y^2}\right) = dx.$$

The left side is the total differential of $\sqrt{x^2 + y^2}$ and the right side is the total differential of $x + c$. Thus $\sqrt{x^2 + y^2} = x + c$ is a solution of the differential equation.

44. To see that the statement is true, write the separable equation as $-g(x)\,dx + dy/h(y) = 0$. Identifying $M = -g(x)$ and $N = 1/h(y)$, we see that $M_y = 0 = N_x$, so the differential equation is exact.

45. **(a)** In differential form we have $(v^2 - 32x)dx + xv\,dv = 0$. This is not an exact form, but $\mu(x) = x$ is an integrating factor. Multiplying by x we get $(xv^2 - 32x^2)dx + x^2v\,dv = 0$. This form is the total differential of $u = \frac{1}{2}x^2v^2 - \frac{32}{3}x^3$, so an implicit solution is $\frac{1}{2}x^2v^2 - \frac{32}{3}x^3 = c$. Letting $x = 3$ and $v = 0$ we find $c = -288$. Solving for v we get

$$v = 8\sqrt{\frac{x}{3} - \frac{9}{x^2}}\,.$$

(b) The chain leaves the platform when $x = 8$, so the velocity at this time is

$$v(8) = 8\sqrt{\frac{8}{3} - \frac{9}{64}} \approx 12.7\ \text{ft/s}.$$

46. **(a)** Letting

$$M(x, y) = \frac{2xy}{(x^2 + y^2)^2} \qquad \text{and} \qquad N(x, y) = 1 + \frac{y^2 - x^2}{(x^2 + y^2)^2}$$

we compute

$$M_y = \frac{2x^3 - 8xy^2}{(x^2 + y^2)^3} = N_x,$$

so the differential equation is exact. Then we have

$$\frac{\partial f}{\partial x} = M(x, y) = \frac{2xy}{(x^2 + y^2)^2} = 2xy(x^2 + y^2)^{-2}$$

$$f(x, y) = -y(x^2 + y^2)^{-1} + g(y) = -\frac{y}{x^2 + y^2} + g(y)$$

$$\frac{\partial f}{\partial y} = \frac{y^2 - x^2}{(x^2 + y^2)^2} + g'(y) = N(x, y) = 1 + \frac{y^2 - x^2}{(x^2 + y^2)^2}\,.$$

Thus, $g'(y) = 1$ and $g(y) = y$. The solution is $y - \dfrac{y}{x^2 + y^2} = c$. When $c = 0$ the solution is $x^2 + y^2 = 1$.

(b) The first graph below is obtained in *Mathematica* using $f(x, y) = y - y/(x^2 + y^2)$ and

ContourPlot[f[x, y], {x, -3, 3}, {y, -3, 3},

Axes−>True, AxesOrigin−>{0, 0}, AxesLabel−>{x, y},
Frame−>False, PlotPoints−>100, ContourShading−>False,
Contours−>{0, -0.2, 0.2, -0.4, 0.4, -0.6, 0.6, -0.8, 0.8}]

The second graph uses

$$x = -\sqrt{\frac{y^3 - cy^2 - y}{c - y}} \quad \text{and} \quad x = \sqrt{\frac{y^3 - cy^2 - y}{c - y}}.$$

In this case the x-axis is vertical and the y-axis is horizontal. To obtain the third graph, we solve $y - y/(x^2 + y^2) = c$ for y in a CAS. This appears to give one real and two complex solutions. When graphed in *Mathematica* however, all three solutions contribute to the graph. This is because the solutions involve the square root of expressions containing c. For some values of c the expression is negative, causing an apparent complex solution to actually be real.

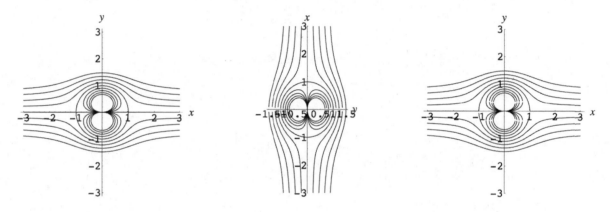

Exercises 2.5

Solutions by Substitutions

1. Letting $y = ux$ we have

$$(x - ux)\,dx + x(u\,dx + x\,du) = 0$$

$$dx + x\,du = 0$$

$$\frac{dx}{x} + du = 0$$

$$\ln|x| + u = c$$

$$x\ln|x| + y = cx.$$

2. Letting $y = ux$ we have

$$(x + ux)\,dx + x(u\,dx + x\,du) = 0$$

$$(1 + 2u)\,dx + x\,du = 0$$

$$\frac{dx}{x} + \frac{du}{1 + 2u} = 0$$

$$\ln|x| + \frac{1}{2}\ln|1 + 2u| = c$$

$$x^2\left(1 + 2\frac{y}{x}\right) = c_1$$

$$x^2 + 2xy = c_1.$$

3. Letting $x = vy$ we have

$$vy(v\,dy + y\,dv) + (y - 2vy)\,dy = 0$$

$$vy^2\,dv + y\left(v^2 - 2v + 1\right)dy = 0$$

$$\frac{v\,dv}{(v - 1)^2} + \frac{dy}{y} = 0$$

$$\ln|v - 1| - \frac{1}{v - 1} + \ln|y| = c$$

$$\ln\left|\frac{x}{y} - 1\right| - \frac{1}{x/y - 1} + \ln y = c$$

$$(x - y)\ln|x - y| - y = c(x - y).$$

4. Letting $x = vy$ we have

$$y(v\,dy + y\,dv) - 2(vy + y)\,dy = 0$$

$$y\,dv - (v + 2)\,dy = 0$$

$$\frac{dv}{v + 2} - \frac{dy}{y} = 0$$

$$\ln|v + 2| - \ln|y| = c$$

$$\ln\left|\frac{x}{y} + 2\right| - \ln|y| = c$$

$$x + 2y = c_1 y^2.$$

5. Letting $y = ux$ we have

$$\left(u^2 x^2 + ux^2\right) dx - x^2(u\,dx + x\,du) = 0$$

$$u^2\,dx - x\,du = 0$$

$$\frac{dx}{x} - \frac{du}{u^2} = 0$$

$$\ln|x| + \frac{1}{u} = c$$

$$\ln|x| + \frac{x}{y} = c$$

$$y\ln|x| + x = cy.$$

6. Letting $y = ux$ and using partial fractions, we have

$$\left(u^2 x^2 + ux^2\right) dx + x^2(u\,dx + x\,du) = 0$$

$$x^2\left(u^2 + 2u\right) dx + x^3\,du = 0$$

$$\frac{dx}{x} + \frac{du}{u(u+2)} = 0$$

$$\ln|x| + \frac{1}{2}\ln|u| - \frac{1}{2}\ln|u+2| = c$$

$$\frac{x^2 u}{u+2} = c_1$$

$$x^2 \frac{y}{x} = c_1\left(\frac{y}{x} + 2\right)$$

$$x^2 y = c_1(y + 2x).$$

7. Letting $y = ux$ we have

$$(ux - x)\,dx - (ux + x)(u\,dx + x\,du) = 0$$

$$\left(u^2 + 1\right) dx + x(u+1)\,du = 0$$

$$\frac{dx}{x} + \frac{u+1}{u^2+1}\,du = 0$$

$$\ln|x| + \frac{1}{2}\ln\left(u^2 + 1\right) + \tan^{-1} u = c$$

$$\ln x^2\left(\frac{y^2}{x^2} + 1\right) + 2\tan^{-1}\frac{y}{x} = c_1$$

$$\ln\left(x^2 + y^2\right) + 2\tan^{-1}\frac{y}{x} = c_1.$$

8. Letting $y = ux$ we have

$$(x + 3ux) \, dx - (3x + ux)(u \, dx + x \, du) = 0$$

$$\left(u^2 - 1\right) dx + x(u + 3) \, du = 0$$

$$\frac{dx}{x} + \frac{u + 3}{(u - 1)(u + 1)} \, du = 0$$

$$\ln |x| + 2 \ln |u - 1| - \ln |u + 1| = c$$

$$\frac{x(u - 1)^2}{u + 1} = c_1$$

$$x \left(\frac{y}{x} - 1\right)^2 = c_1 \left(\frac{y}{x} + 1\right)$$

$$(y - x)^2 = c_1(y + x).$$

9. Letting $y = ux$ we have

$$-ux \, dx + (x + \sqrt{u} \, x)(u \, dx + x \, du) = 0$$

$$(x^2 + x^2 \sqrt{u}) \, du + xu^{3/2} \, dx = 0$$

$$\left(u^{-3/2} + \frac{1}{u}\right) du + \frac{dx}{x} = 0$$

$$-2u^{-1/2} + \ln |u| + \ln |x| = c$$

$$\ln |y/x| + \ln |x| = 2\sqrt{x/y} + c$$

$$y(\ln |y| - c)^2 = 4x.$$

10. Letting $y = ux$ we have

$$\left(ux + \sqrt{x^2 - (ux)^2}\right) dx - x(udx + xdu) \, du = 0$$

$$\sqrt{x^2 - u^2 x^2} \, dx - x^2 \, du = 0$$

$$x\sqrt{1 - u^2} \, dx - x^2 \, du = 0, \quad (x > 0)$$

$$\frac{dx}{x} - \frac{du}{\sqrt{1 - u^2}} = 0$$

$$\ln x - \sin^{-1} u = c$$

$$\sin^{-1} u = \ln x + c_1$$

$$\sin^{-1}\frac{y}{x} = \ln x + c_2$$

$$\frac{y}{x} = \sin(\ln x + c_2)$$

$$y = x\sin(\ln x + c_2).$$

See Problem 33 in this section for an analysis of the solution.

11. Letting $y = ux$ we have

$$\left(x^3 - u^3 x^3\right)dx + u^2 x^3(u\,dx + x\,du) = 0$$

$$dx + u^2 x\,du = 0$$

$$\frac{dx}{x} + u^2\,du = 0$$

$$\ln|x| + \frac{1}{3}u^3 = c$$

$$3x^3\ln|x| + y^3 = c_1 x^3.$$

Using $y(1) = 2$ we find $c_1 = 8$. The solution of the initial-value problem is $3x^3\ln|x| + y^3 = 8x^3$.

12. Letting $y = ux$ we have

$$(x^2 + 2u^2 x^2)dx - ux^2(u\,dx + x\,du) = 0$$

$$x^2(1 + u^2)dx - ux^3\,du = 0$$

$$\frac{dx}{x} - \frac{u\,du}{1 + u^2} = 0$$

$$\ln|x| - \frac{1}{2}\ln(1 + u^2) = c$$

$$\frac{x^2}{1 + u^2} = c_1$$

$$x^4 = c_1(x^2 + y^2).$$

Using $y(-1) = 1$ we find $c_1 = 1/2$. The solution of the initial-value problem is $2x^4 = y^2 + x^2$.

13. Letting $y = ux$ we have

$$(x + uxe^u)\,dx - xe^u(u\,dx + x\,du) = 0$$

$$dx - xe^u\,du = 0$$

$$\frac{dx}{x} - e^u\,du = 0$$

$$\ln|x| - e^u = c$$

$$\ln|x| - e^{y/x} = c.$$

Using $y(1) = 0$ we find $c = -1$. The solution of the initial-value problem is $\ln|x| = e^{y/x} - 1$.

14. Letting $x = vy$ we have

$$y(v\,dy + y\,dv) + vy(\ln vy - \ln y - 1)\,dy = 0$$

$$y\,dv + v\ln v\,dy = 0$$

$$\frac{dv}{v\ln v} + \frac{dy}{y} = 0$$

$$\ln|\ln|v|| + \ln|y| = c$$

$$y\ln\left|\frac{x}{y}\right| = c_1.$$

Using $y(1) = e$ we find $c_1 = -e$. The solution of the initial-value problem is $y\ln\left|\frac{x}{y}\right| = -e$.

15. From $y' + \frac{1}{x}y = \frac{1}{x}y^{-2}$ and $w = y^3$ we obtain $\frac{dw}{dx} + \frac{3}{x}w = \frac{3}{x}$. An integrating factor is x^3 so that $x^3 w = x^3 + c$ or $y^3 = 1 + cx^{-3}$.

16. From $y' - y = e^x y^2$ and $w = y^{-1}$ we obtain $\frac{dw}{dx} + w = -e^x$. An integrating factor is e^x so that $e^x w = -\frac{1}{2}e^{2x} + c$ or $y^{-1} = -\frac{1}{2}e^x + ce^{-x}$.

17. From $y' + y = xy^4$ and $w = y^{-3}$ we obtain $\frac{dw}{dx} - 3w = -3x$. An integrating factor is e^{-3x} so that $e^{-3x}w = xe^{-3x} + \frac{1}{3}e^{-3x} + c$ or $y^{-3} = x + \frac{1}{3} + ce^{3x}$.

18. From $y' - \left(1 + \frac{1}{x}\right)y = y^2$ and $w = y^{-1}$ we obtain $\frac{dw}{dx} + \left(1 + \frac{1}{x}\right)w = -1$. An integrating factor is xe^x so that $xe^x w = -xe^x + e^x + c$ or $y^{-1} = -1 + \frac{1}{x} + \frac{c}{x}e^{-x}$.

19. From $y' - \frac{1}{t}y = -\frac{1}{t^2}y^2$ and $w = y^{-1}$ we obtain $\frac{dw}{dt} + \frac{1}{t}w = \frac{1}{t^2}$. An integrating factor is t so that $tw = \ln t + c$ or $y^{-1} = \frac{1}{t}\ln t + \frac{c}{t}$. Writing this in the form $\frac{t}{y} = \ln t + c$, we see that the solution can also be expressed in the form $e^{t/y} = c_1 t$.

20. From $y' + \frac{2}{3(1+t^2)}y = \frac{2t}{3(1+t^2)}y^4$ and $w = y^{-3}$ we obtain $\frac{dw}{dt} - \frac{2t}{1+t^2}w = \frac{-2t}{1+t^2}$. An integrating factor is $\frac{1}{1+t^2}$ so that $\frac{w}{1+t^2} = \frac{1}{1+t^2} + c$ or $y^{-3} = 1 + c\left(1+t^2\right)$.

69

21. From $y' - \dfrac{2}{x}y = \dfrac{3}{x^2}y^4$ and $w = y^{-3}$ we obtain $\dfrac{dw}{dx} + \dfrac{6}{x}w = -\dfrac{9}{x^2}$. An integrating factor is x^6 so that

$x^6 w = -\frac{9}{5}x^5 + c$ or $y^{-3} = -\frac{9}{5}x^{-1} + cx^{-6}$. If $y(1) = \frac{1}{2}$ then $c = \frac{49}{5}$ and $y^{-3} = -\frac{9}{5}x^{-1} + \frac{49}{5}x^{-6}$.

22. From $y' + y = y^{-1/2}$ and $w = y^{3/2}$ we obtain $\dfrac{dw}{dx} + \dfrac{3}{2}w = \dfrac{3}{2}$. An integrating factor is $e^{3x/2}$ so that

$e^{3x/2}w = e^{3x/2} + c$ or $y^{3/2} = 1 + ce^{-3x/2}$. If $y(0) = 4$ then $c = 7$ and $y^{3/2} = 1 + 7e^{-3x/2}$.

23. Let $u = x + y + 1$ so that $du/dx = 1 + dy/dx$. Then $\dfrac{du}{dx} - 1 = u^2$ or $\dfrac{1}{1+u^2}\,du = dx$. Thus $\tan^{-1}u = x + c$ or $u = \tan(x+c)$, and $x + y + 1 = \tan(x+c)$ or $y = \tan(x+c) - x - 1$.

24. Let $u = x + y$ so that $du/dx = 1 + dy/dx$. Then $\dfrac{du}{dx} - 1 = \dfrac{1-u}{u}$ or $u\,du = dx$. Thus $\frac{1}{2}u^2 = x + c$ or $u^2 = 2x + c_1$, and $(x+y)^2 = 2x + c_1$.

25. Let $u = x + y$ so that $du/dx = 1 + dy/dx$. Then $\dfrac{du}{dx} - 1 = \tan^2 u$ or $\cos^2 u\,du = dx$. Thus $\frac{1}{2}u + \frac{1}{4}\sin 2u = x + c$ or $2u + \sin 2u = 4x + c_1$, and $2(x+y) + \sin 2(x+y) = 4x + c_1$ or $2y + \sin 2(x+y) = 2x + c_1$.

26. Let $u = x + y$ so that $du/dx = 1 + dy/dx$. Then $\dfrac{du}{dx} - 1 = \sin u$ or $\dfrac{1}{1 + \sin u}\,du = dx$. Multiplying by $(1 - \sin u)/(1 - \sin u)$ we have $\dfrac{1 - \sin u}{\cos^2 u}\,du = dx$ or $(\sec^2 u - \sec u \tan u)du = dx$. Thus $\tan u - \sec u = x + c$ or $\tan(x+y) - \sec(x+y) = x + c$.

27. Let $u = y - 2x + 3$ so that $du/dx = dy/dx - 2$. Then $\dfrac{du}{dx} + 2 = 2 + \sqrt{u}$ or $\dfrac{1}{\sqrt{u}}\,du = dx$. Thus $2\sqrt{u} = x + c$ and $2\sqrt{y - 2x + 3} = x + c$.

28. Let $u = y - x + 5$ so that $du/dx = dy/dx - 1$. Then $\dfrac{du}{dx} + 1 = 1 + e^u$ or $e^{-u}du = dx$. Thus $-e^{-u} = x + c$ and $-e^{y-x+5} = x + c$.

29. Let $u = x + y$ so that $du/dx = 1 + dy/dx$. Then $\dfrac{du}{dx} - 1 = \cos u$ and $\dfrac{1}{1 + \cos u}\,du = dx$. Now

$$\frac{1}{1 + \cos u} = \frac{1 - \cos u}{1 - \cos^2 u} = \frac{1 - \cos u}{\sin^2 u} = \csc^2 u - \csc u \cot u$$

so we have $\int(\csc^2 u - \csc u \cot u)du = \int dx$ and $-\cot u + \csc u = x + c$. Thus $-\cot(x+y) + \csc(x+y) = x + c$. Setting $x = 0$ and $y = \pi/4$ we obtain $c = \sqrt{2} - 1$. The solution is

$$\csc(x+y) - \cot(x+y) = x + \sqrt{2} - 1.$$

30. Let $u = 3x + 2y$ so that $du/dx = 3 + 2\,dy/dx$. Then $\dfrac{du}{dx} = 3 + \dfrac{2u}{u+2} = \dfrac{5u+6}{u+2}$ and $\dfrac{u+2}{5u+6}\,du = dx$. Now by long division

$$\frac{u+2}{5u+6} = \frac{1}{5} + \frac{4}{25u + 30}$$

so we have

$$\int \left(\frac{1}{5} + \frac{4}{25u + 30} \right) du = dx$$

and $\frac{1}{5}u + \frac{4}{25} \ln |25u + 30| = x + c$. Thus

$$\frac{1}{5}(3x + 2y) + \frac{4}{25} \ln |75x + 50y + 30| = x + c.$$

Setting $x = -1$ and $y = -1$ we obtain $c = \frac{4}{25} \ln 95$. The solution is

$$\frac{1}{5}(3x + 2y) + \frac{4}{25} \ln |75x + 50y + 30| = x + \frac{4}{25} \ln 95$$

or

$$5y - 5x + 2 \ln |75x + 50y + 30| = 2 \ln 95.$$

31. We write the differential equation $M(x, y)dx + N(x, y)dy = 0$ as $dy/dx = f(x, y)$ where

$$f(x, y) = -\frac{M(x, y)}{N(x, y)}.$$

The function $f(x, y)$ must necessarily be homogeneous of degree 0 when M and N are homogeneous of degree α. Since M is homogeneous of degree α, $M(tx, ty) = t^\alpha M(x, y)$, and letting $t = 1/x$ we have

$$M(1, y/x) = \frac{1}{x^\alpha} M(x, y) \quad \text{or} \quad M(x, y) = x^\alpha M(1, y/x).$$

Thus

$$\frac{dy}{dx} = f(x, y) = -\frac{x^\alpha M(1, y/x)}{x^\alpha N(1, y/x)} = -\frac{M(1, y/x)}{N(1, y/x)} = F\left(\frac{y}{x}\right).$$

32. Rewrite $(5x^2 - 2y^2)dx - xy\, dy = 0$ as

$$xy \frac{dy}{dx} = 5x^2 - 2y^2$$

and divide by xy, so that

$$\frac{dy}{dx} = 5\frac{x}{y} - 2\frac{y}{x}.$$

We then identify

$$F\left(\frac{y}{x}\right) = 5\left(\frac{y}{x}\right)^{-1} - 2\left(\frac{y}{x}\right).$$

33. (a) By inspection $y = x$ and $y = -x$ are solutions of the differential equation and not members of the family $y = x \sin(\ln x + c_2)$.

(b) Letting $x = 5$ and $y = 0$ in $\sin^{-1}(y/x) = \ln x + c_2$ we get $\sin^{-1} 0 = \ln 5 + c$ or $c = -\ln 5$. Then $\sin^{-1}(y/x) = \ln x - \ln 5 = \ln(x/5)$. Because the range of the arcsine function is $[-\pi/2, \pi/2]$ we must have

$$-\frac{\pi}{2} \le \ln\frac{x}{5} \le \frac{\pi}{2}$$

$$e^{-\pi/2} \le \frac{x}{5} \le e^{\pi/2}$$

$$5e^{-\pi/2} \le x \le 5e^{\pi/2}.$$

The interval of definition of the solution is approximately [1.04, 24.05].

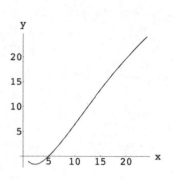

34. As $x \to -\infty$, $e^{6x} \to 0$ and $y \to 2x + 3$. Now write $(1 + ce^{6x})/(1 - ce^{6x})$ as $(e^{-6x} + c)/(e^{-6x} - c)$. Then, as $x \to \infty$, $e^{-6x} \to 0$ and $y \to 2x - 3$.

35. (a) The substitutions $y = y_1 + u$ and

$$\frac{dy}{dx} = \frac{dy_1}{dx} + \frac{du}{dx}$$

lead to

$$\frac{dy_1}{dx} + \frac{du}{dx} = P + Q(y_1 + u) + R(y_1 + u)^2$$

$$= P + Qy_1 + Ry_1^2 + Qu + 2y_1Ru + Ru^2$$

or

$$\frac{du}{dx} - (Q + 2y_1R)u = Ru^2.$$

This is a Bernoulli equation with $n = 2$ which can be reduced to the linear equation

$$\frac{dw}{dx} + (Q + 2y_1R)w = -R$$

by the substitution $w = u^{-1}$.

(b) Identify $P(x) = -4/x^2$, $Q(x) = -1/x$, and $R(x) = 1$. Then $\frac{dw}{dx} + \left(-\frac{1}{x} + \frac{4}{x}\right)w = -1$. An integrating factor is x^3 so that $x^3w = -\frac{1}{4}x^4 + c$ or $u = \left[-\frac{1}{4}x + cx^{-3}\right]^{-1}$. Thus, $y = \frac{2}{x} + u$.

36. Write the differential equation in the form $x(y'/y) = \ln x + \ln y$ and let $u = \ln y$. Then $du/dx = y'/y$ and the differential equation becomes $x(du/dx) = \ln x + u$ or $du/dx - u/x = (\ln x)/x$, which is first-order and linear. An integrating factor is $e^{-\int dx/x} = 1/x$, so that (using integration by parts)

$$\frac{d}{dx}\left[\frac{1}{x}u\right] = \frac{\ln x}{x^2} \quad \text{and} \quad \frac{u}{x} = -\frac{1}{x} - \frac{\ln x}{x} + c.$$

The solution is

$$\ln y = -1 - \ln x + cx \quad \text{or} \quad y = \frac{e^{cx-1}}{x}.$$

37. Write the differential equation as

$$\frac{dv}{dx} + \frac{1}{x}v = 32v^{-1},$$

72

and let $u = v^2$ or $v = u^{1/2}$. Then

$$\frac{dv}{dx} = \frac{1}{2}u^{-1/2}\frac{du}{dx},$$

and substituting into the differential equation, we have

$$\frac{1}{2}u^{-1/2}\frac{du}{dx} + \frac{1}{x}u^{1/2} = 32u^{-1/2} \qquad \text{or} \qquad \frac{du}{dx} + \frac{2}{x}u = 64.$$

The latter differential equation is linear with integrating factor $e^{\int (2/x)dx} = x^2$, so

$$\frac{d}{dx}\left[x^2 u\right] = 64x^2$$

and

$$x^2 u = \frac{64}{3}x^3 + c \qquad \text{or} \qquad v^2 = \frac{64}{3}x + \frac{c}{x^2}.$$

38. Write the differential equation as $dP/dt - aP = -bP^2$ and let $u = P^{-1}$ or $P = u^{-1}$. Then

$$\frac{dP}{dt} = -u^{-2}\frac{du}{dt},$$

slb cap. P

and substituting into the differential equation, we have

$$-u^{-2}\frac{du}{dt} - au^{-1} = -bu^{-2} \qquad \text{or} \qquad \frac{du}{dt} + au = b.$$

The latter differential equation is linear with integrating factor $e^{\int a\,dt} = e^{at}$, so

$$\frac{d}{dt}\left[e^{at}u\right] = be^{at}$$

and

$$e^{at}u = \frac{b}{a}e^{at} + c$$

$$e^{at}P^{-1} = \frac{b}{a}e^{at} + c$$

$$P^{-1} = \frac{b}{a} + ce^{-at}$$

$$P = \frac{1}{b/a + ce^{-at}} = \frac{a}{b + c_1 e^{-at}}.$$

Exercises 2.6 A Numerical Method

1. We identify $f(x, y) = 2x - 3y + 1$. Then, for $h = 0.1$,

$$y_{n+1} = y_n + 0.1(2x_n - 3y_n + 1) = 0.2x_n + 0.7y_n + 0.1,$$

and

$$y(1.1) \approx y_1 = 0.2(1) + 0.7(5) + 0.1 = 3.8$$

$$y(1.2) \approx y_2 = 0.2(1.1) + 0.7(3.8) + 0.1 = 2.98.$$

For $h = 0.05$,

$$y_{n+1} = y_n + 0.05(2x_n - 3y_n + 1) = 0.1x_n + 0.85y_n + 0.05,$$

and

$$y(1.05) \approx y_1 = 0.1(1) + 0.85(5) + 0.05 = 4.4$$

$$y(1.1) \approx y_2 = 0.1(1.05) + 0.85(4.4) + 0.05 = 3.895$$

$$y(1.15) \approx y_3 = 0.1(1.1) + 0.85(3.895) + 0.05 = 3.47075$$

$$y(1.2) \approx y_4 = 0.1(1.15) + 0.85(3.47075) + 0.05 = 3.11514.$$

2. We identify $f(x, y) = x + y^2$. Then, for $h = 0.1$,

$$y_{n+1} = y_n + 0.1(x_n + y_n^2) = 0.1x_n + y_n + 0.1y_n^2,$$

and

$$y(0.1) \approx y_1 = 0.1(0) + 0 + 0.1(0)^2 = 0$$

$$y(0.2) \approx y_2 = 0.1(0.1) + 0 + 0.1(0)^2 = 0.01.$$

For $h = 0.05$,

$$y_{n+1} = y_n + 0.05(x_n + y_n^2) = 0.05x_n + y_n + 0.05y_n^2,$$

and

$$y(0.05) \approx y_1 = 0.05(0) + 0 + 0.05(0)^2 = 0$$

$$y(0.1) \approx y_2 = 0.05(0.05) + 0 + 0.05(0)^2 = 0.0025$$

$$y(0.15) \approx y_3 = 0.05(0.1) + 0.0025 + 0.05(0.0025)^2 = 0.0075$$

$$y(0.2) \approx y_4 = 0.05(0.15) + 0.0075 + 0.05(0.0075)^2 = 0.0150.$$

3. Separating variables and integrating, we have

$$\frac{dy}{y} = dx \quad \text{and} \quad \ln|y| = x + c.$$

Thus $y = c_1 e^x$ and, using $y(0) = 1$, we find $c = 1$, so $y = e^x$ is the solution of the initial-value problem.

$h=0.1$

x_n	y_n	Actual Value	Abs. Error	% Rel. Error
0.00	1.0000	1.0000	0.0000	0.00
0.10	1.1000	1.1052	0.0052	0.47
0.20	1.2100	1.2214	0.0114	0.93
0.30	1.3310	1.3499	0.0189	1.40
0.40	1.4641	1.4918	0.0277	1.86
0.50	1.6105	1.6487	0.0382	2.32
0.60	1.7716	1.8221	0.0506	2.77
0.70	1.9487	2.0138	0.0650	3.23
0.80	2.1436	2.2255	0.0820	3.68
0.90	2.3579	2.4596	0.1017	4.13
1.00	2.5937	2.7183	0.1245	4.58

$h=0.05$

x_n	y_n	Actual Value	Abs. Error	% Rel. Error
0.00	1.0000	1.0000	0.0000	0.00
0.05	1.0500	1.0513	0.0013	0.12
0.10	1.1025	1.1052	0.0027	0.24
0.15	1.1576	1.1618	0.0042	0.36
0.20	1.2155	1.2214	0.0059	0.48
0.25	1.2763	1.2840	0.0077	0.60
0.30	1.3401	1.3499	0.0098	0.72
0.35	1.4071	1.4191	0.0120	0.84
0.40	1.4775	1.4918	0.0144	0.96
0.45	1.5513	1.5683	0.0170	1.08
0.50	1.6289	1.6487	0.0198	1.20
0.55	1.7103	1.7333	0.0229	1.32
0.60	1.7959	1.8221	0.0263	1.44
0.65	1.8856	1.9155	0.0299	1.56
0.70	1.9799	2.0138	0.0338	1.68
0.75	2.0789	2.1170	0.0381	1.80
0.80	2.1829	2.2255	0.0427	1.92
0.85	2.2920	2.3396	0.0476	2.04
0.90	2.4066	2.4596	0.0530	2.15
0.95	2.5270	2.5857	0.0588	2.27
1.00	2.6533	2.7183	0.0650	2.39

4. Separating variables and integrating, we have

$$\frac{dy}{y} = 2x\,dx \quad \text{and} \quad \ln|y| = x^2 + c.$$

Thus $y = c_1 e^{x^2}$ and, using $y(1) = 1$, we find $c = e^{-1}$, so $y = e^{x^2 - 1}$ is the solution of the initial-value problem.

$h=0.1$

x_n	y_n	Actual Value	Abs. Error	% Rel. Error
1.00	1.0000	1.0000	0.0000	0.00
1.10	1.2000	1.2337	0.0337	2.73
1.20	1.4640	1.5527	0.0887	5.71
1.30	1.8154	1.9937	0.1784	8.95
1.40	2.2874	2.6117	0.3243	12.42
1.50	2.9278	3.4903	0.5625	16.12

$h=0.05$

x_n	y_n	Actual Value	Abs. Error	% Rel. Error
1.00	1.0000	1.0000	0.0000	0.00
1.05	1.1000	1.1079	0.0079	0.72
1.10	1.2155	1.2337	0.0182	1.47
1.15	1.3492	1.3806	0.0314	2.27
1.20	1.5044	1.5527	0.0483	3.11
1.25	1.6849	1.7551	0.0702	4.00
1.30	1.8955	1.9937	0.0982	4.93
1.35	2.1419	2.2762	0.1343	5.90
1.40	2.4311	2.6117	0.1806	6.92
1.45	2.7714	3.0117	0.2403	7.98
1.50	3.1733	3.4903	0.3171	9.08

5.

$h=0.1$

x_n	y_n
0.00	0.0000
0.10	0.1000
0.20	0.1905
0.30	0.2731
0.40	0.3492
0.50	0.4198

$h=0.05$

x_n	y_n
0.00	0.0000
0.05	0.0500
0.10	0.0976
0.15	0.1429
0.20	0.1863
0.25	0.2278
0.30	0.2676
0.35	0.3058
0.40	0.3427
0.45	0.3782
0.50	0.4124

6.

$h=0.1$

x_n	y_n
0.00	1.0000
0.10	1.1000
0.20	1.2220
0.30	1.3753
0.40	1.5735
0.50	1.8371

$h=0.05$

x_n	y_n
0.00	1.0000
0.05	1.0500
0.10	1.1053
0.15	1.1668
0.20	1.2360
0.25	1.3144
0.30	1.4039
0.35	1.5070
0.40	1.6267
0.45	1.7670
0.50	1.9332

7.

$h=0.1$

x_n	y_n
0.00	0.5000
0.10	0.5250
0.20	0.5431
0.30	0.5548
0.40	0.5613
0.50	0.5639

$h=0.05$

x_n	y_n
0.00	0.5000
0.05	0.5125
0.10	0.5232
0.15	0.5322
0.20	0.5395
0.25	0.5452
0.30	0.5496
0.35	0.5527
0.40	0.5547
0.45	0.5559
0.50	0.5565

8.

$h=0.1$

x_n	y_n
0.00	1.0000
0.10	1.1000
0.20	1.2159
0.30	1.3505
0.40	1.5072
0.50	1.6902

$h=0.05$

x_n	y_n
0.00	1.0000
0.05	1.0500
0.10	1.1039
0.15	1.1619
0.20	1.2245
0.25	1.2921
0.30	1.3651
0.35	1.4440
0.40	1.5293
0.45	1.6217
0.50	1.7219

9.

$h=0.1$	
x_n	y_n
1.00	1.0000
1.10	1.0000
1.20	1.0191
1.30	1.0588
1.40	1.1231
1.50	1.2194

$h=0.05$	
x_n	y_n
1.00	1.0000
1.05	1.0000
1.10	1.0049
1.15	1.0147
1.20	1.0298
1.25	1.0506
1.30	1.0775
1.35	1.1115
1.40	1.1538
1.45	1.2057
1.50	1.2696

10.

$h=0.1$	
x_n	y_n
0.00	0.5000
0.10	0.5250
0.20	0.5499
0.30	0.5747
0.40	0.5991
0.50	0.6231

$h=0.05$	
x_n	y_n
0.00	0.5000
0.05	0.5125
0.10	0.5250
0.15	0.5375
0.20	0.5499
0.25	0.5623
0.30	0.5746
0.35	0.5868
0.40	0.5989
0.45	0.6109
0.50	0.6228

11. Tables of values were computed using the Euler and RK4 methods. The resulting points were plotted and joined using **ListPlot** in *Mathematica*. A somewhat simplified version of the code used to do this is given in the *Student Resource and Solutions Manual (SRSM)* under **Use of Computers** in Section 2.6.

$h = 0.25$

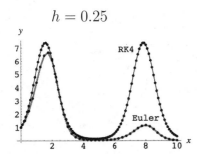

$h = 0.1$

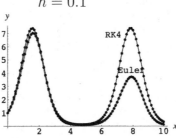

$h = 0.05$

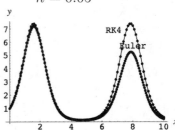

12. See the comments in Problem 11 above.

$h = 0.25$

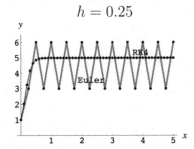

$h = 0.1$

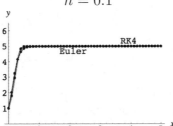

$h = 0.05$

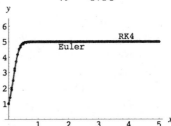

13. Tables of values, shown below, were first computed using Euler's method with $h = 0.1$ and $h = 0.05$ and then using the RK4 method with the same values of h. Using separation of variables we find that the solution of the differential equation is $y = 1/(1 - x^2)$, which is undefined at $x = 1$, where the graph has a vertical asymptote. Because the actual solution of the differential equation becomes unbounded at x approaches 1, very small changes in the inputs x will result in large changes in the corresponding outputs y. This can be expected to have a serious effect on numerical procedures.

$h=0.1$ (Euler)			$h=0.05$ (Euler)			$h=0.1$ (RK4)			$h=0.05$ (RK4)	
x_n	y_n		x_n	y_n		x_n	y_n		x_n	y_n
0.00	1.0000		0.00	1.0000		0.00	1.0000		0.00	1.0000
0.10	1.0000		0.05	1.0000		0.10	1.0101		0.05	1.0025
0.20	1.0200		0.10	1.0050		0.20	1.0417		0.10	1.0101
0.30	1.0616		0.15	1.0151		0.30	1.0989		0.15	1.0230
0.40	1.1292		0.20	1.0306		0.40	1.1905		0.20	1.0417
0.50	1.2313		0.25	1.0518		0.50	1.3333		0.25	1.0667
0.60	1.3829		0.30	1.0795		0.60	1.5625		0.30	1.0989
0.70	1.6123		0.35	1.1144		0.70	1.9607		0.35	1.1396
0.80	1.9763		0.40	1.1579		0.80	2.7771		0.40	1.1905
0.90	2.6012		0.45	1.2115		0.90	5.2388		0.45	1.2539
1.00	3.8191		0.50	1.2776		1.00	42.9931		0.50	1.3333
			0.55	1.3592					0.55	1.4337
			0.60	1.4608					0.60	1.5625
			0.65	1.5888					0.65	1.7316
			0.70	1.7529					0.70	1.9608
			0.75	1.9679					0.75	2.2857
			0.80	2.2584					0.80	2.7777
			0.85	2.6664					0.85	3.6034
			0.90	3.2708					0.90	5.2609
			0.95	4.2336					0.95	10.1973
			1.00	5.9363					1.00	84.0132

The graphs below were obtained as described above in Problem 11.

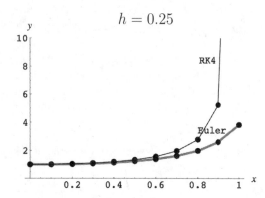

$h = 0.25$

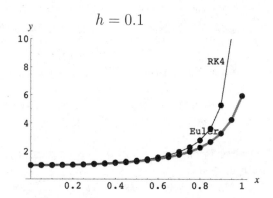

$h = 0.1$

14. (a) The graph to the right was obtained as described above
    in Problem 11 using $h = 0.1$.

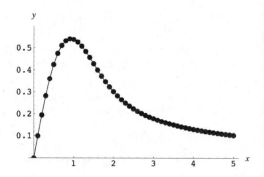

(b) Writing the differential equation in the form $y' + 2xy = 1$ we see that an integrating factor is $e^{\int 2x\,dx} = e^{x^2}$, so

$$\frac{d}{dx}[e^{x^2}y] = e^{x^2}$$

and

$$y = e^{-x^2}\int_0^x e^{t^2}\,dt + ce^{-x^2}.$$

This solution can also be expressed in terms of the inverse error function as

$$y = \frac{\sqrt{\pi}}{2}e^{-x^2}\operatorname{erfi}(x) + ce^{-x^2}.$$

Letting $x = 0$ and $y(0) = 0$ we find $c = 0$, so the solution of the initial-value problem is

$$y = e^{-x^2}\int_0^x e^{t^2}\,dt = \frac{\sqrt{\pi}}{2}e^{-x^2}\operatorname{erfi}(x).$$

(c) Using either **FindRoot** in *Mathematica* or `fsolve` in *Maple* we see that $y'(x) = 0$ when $x = 0.924139$. Since $y(0.924139) = 0.541044$, we see from the graph in part (a) that $(0.924139, 0.541044)$ is a relative maximum. Now, using the substitution $u = -t$ in the integral below, we have

$$y(-x) = e^{-(-x)^2}\int_0^{-x} e^{t^2}\,dt = e^{-x^2}\int_0^x e^{(-u)^2}(-du) = -e^{-x^2}\int_0^x e^{u^2}\,du = -y(x).$$

Thus, $y(x)$ is an odd function and $(-0.924139, -0.541044)$ is a relative minimum.

Chapter 2 in Review

1. Writing the differential equation in the form $y' = k(y + A/k)$ we see that the critical point $-A/k$ is a repeller for $k > 0$ and an attractor for $k < 0$.

2. Separating variables and integrating we have

$$\frac{dy}{y} = \frac{4}{x}\,dx$$

$$\ln y = 4\ln x + c = \ln x^4 + c$$

$$y = c_1 x^4.$$

We see that when $x = 0$, $y = 0$, so the initial-value problem has an infinite number of solutions for $k = 0$ and no solutions for $k \neq 0$.

3. True; $y = k_2/k_1$ is always a solution for $k_1 \neq 0$.

4. True; writing the differential equation as $a_1(x)\,dy + a_2(x)y\,dx = 0$ and separating variables yields

$$\frac{dy}{y} = -\frac{a_2(x)}{a_1(x)}\,dx.$$

5. $\dfrac{dy}{dx} = (y-1)^2(y-3)^2$

6. $\dfrac{dy}{dx} = y(y-2)^2(y-4)$

7. When n is odd, $x^n < 0$ for $x < 0$ and $x^n > 0$ for $x > 0$. In this case 0 is unstable. When n is even, $x^n > 0$ for $x < 0$ and for $x > 0$. In this case 0 is semi-stable.

   When n is odd, $-x^n > 0$ for $x < 0$ and $-x^n < 0$ for $x > 0$. In this case 0 is asymptotically stable. When n is even, $-x^n < 0$ for $x < 0$ and for $x > 0$. In this case 0 is semi-stable.

8. Using a CAS we find that the zero of f occurs at approximately $P = 1.3214$. From the graph we observe that $dP/dt > 0$ for $P < 1.3214$ and $dP/dt < 0$ for $P > 1.3214$, so $P = 1.3214$ is an asymptotically stable critical point. Thus, $\lim_{t\to\infty} P(t) = 1.3214$.

9.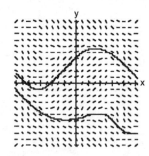

10. **(a)** linear in y, homogeneous, exact   **(b)** linear in x

    (c) separable, exact, linear in x and y **(d)** Bernoulli in x

    (e) separable   **(f)** separable, linear in x, Bernoulli

    (g) linear in x **(h)** homogeneous

    (i) Bernoulli   **(j)** homogeneous, exact, Bernoulli

    (k) linear in x and y, exact, separable, homogeneous

    (l) exact, linear in y **(m)** homogeneous

    (n) separable

11. Separating variables and using the identity $\cos^2 x = \frac{1}{2}(1 + \cos 2x)$, we have

$$\cos^2 x\,dx = \frac{y}{y^2 + 1}\,dy,$$

$$\frac{1}{2}x + \frac{1}{4}\sin 2x = \frac{1}{2}\ln\left(y^2 + 1\right) + c,$$

   and

$$2x + \sin 2x = 2\ln\left(y^2 + 1\right) + c.$$

12. Write the differential equation in the form

$$y\ln\frac{x}{y}\,dx = \left(x\ln\frac{x}{y} - y\right)dy.$$

This is a homogeneous equation, so let $x = uy$. Then $dx = u\,dy + y\,du$ and the differential equation becomes

$$y\ln u(u\,dy + y\,du) = (uy\ln u - y)\,dy \quad\text{or}\quad y\ln u\,du = -dy.$$

Separating variables, we obtain

$$\ln u\,du = -\frac{dy}{y}$$

$$u\ln|u| - u = -\ln|y| + c$$

$$\frac{x}{y}\ln\left|\frac{x}{y}\right| - \frac{x}{y} = -\ln|y| + c$$

$$x(\ln x - \ln y) - x = -y\ln|y| + cy.$$

13. The differential equation

$$\frac{dy}{dx} + \frac{2}{6x+1}y = -\frac{3x^2}{6x+1}y^{-2}$$

is Bernoulli. Using $w = y^3$, we obtain the linear equation

$$\frac{dw}{dx} + \frac{6}{6x+1}w = -\frac{9x^2}{6x+1}.$$

An integrating factor is $6x + 1$, so

$$\frac{d}{dx}[(6x+1)w] = -9x^2,$$

$$w = -\frac{3x^3}{6x+1} + \frac{c}{6x+1},$$

and

$$(6x+1)y^3 = -3x^3 + c.$$

(Note: The differential equation is also exact.)

14. Write the differential equation in the form $(3y^2 + 2x)dx + (4y^2 + 6xy)dy = 0$. Letting $M = 3y^2 + 2x$ and $N = 4y^2 + 6xy$ we see that $M_y = 6y = N_x$, so the differential equation is exact. From $f_x = 3y^2 + 2x$ we obtain $f = 3xy^2 + x^2 + h(y)$. Then $f_y = 6xy + h'(y) = 4y^2 + 6xy$ and $h'(y) = 4y^2$ so $h(y) = \frac{4}{3}y^3$. A one-parameter family of solutions is

$$3xy^2 + x^2 + \frac{4}{3}y^3 = c.$$

15. Write the equation in the form

$$\frac{dQ}{dt} + \frac{1}{t}Q = t^3 \ln t.$$

An integrating factor is $e^{\ln t} = t$, so

$$\frac{d}{dt}[tQ] = t^4 \ln t$$

$$tQ = -\frac{1}{25}t^5 + \frac{1}{5}t^5 \ln t + c$$

and

$$Q = -\frac{1}{25}t^4 + \frac{1}{5}t^4 \ln t + \frac{c}{t}.$$

16. Letting $u = 2x + y + 1$ we have

$$\frac{du}{dx} = 2 + \frac{dy}{dx},$$

and so the given differential equation is transformed into

$$u\left(\frac{du}{dx} - 2\right) = 1 \quad \text{or} \quad \frac{du}{dx} = \frac{2u+1}{u}.$$

Separating variables and integrating we get

$$\frac{u}{2u+1}\,du = dx$$

$$\left(\frac{1}{2} - \frac{1}{2}\frac{1}{2u+1}\right)du = dx$$

$$\frac{1}{2}u - \frac{1}{4}\ln|2u+1| = x + c$$

$$2u - \ln|2u+1| = 2x + c_1.$$

Resubstituting for u gives the solution

$$4x + 2y + 2 - \ln|4x + 2y + 3| = 2x + c_1$$

or

$$2x + 2y + 2 - \ln|4x + 2y + 3| = c_1.$$

17. Write the equation in the form

$$\frac{dy}{dx} + \frac{8x}{x^2+4}y = \frac{2x}{x^2+4}.$$

An integrating factor is $\left(x^2 + 4\right)^4$, so

$$\frac{d}{dx}\left[\left(x^2+4\right)^4 y\right] = 2x\left(x^2+4\right)^3$$

$$\left(x^2 + 4\right)^4 y = \frac{1}{4}\left(x^2 + 4\right)^4 + c$$

and

$$y = \frac{1}{4} + c\left(x^2 + 4\right)^{-4}.$$

18. Letting $M = 2r^2 \cos\theta \sin\theta + r\cos\theta$ and $N = 4r + \sin\theta - 2r\cos^2\theta$ we see that $M_r = 4r\cos\theta\sin\theta + \cos\theta = N_\theta$, so the differential equation is exact. From $f_\theta = 2r^2\cos\theta\sin\theta + r\cos\theta$ we obtain $f = -r^2\cos^2\theta + r\sin\theta + h(r)$. Then $f_r = -2r\cos^2\theta + \sin\theta + h'(r) = 4r + \sin\theta - 2r\cos^2\theta$ and $h'(r) = 4r$ so $h(r) = 2r^2$. The solution is

$$-r^2\cos^2\theta + r\sin\theta + 2r^2 = c.$$

19. The differential equation has the form $(d/dx)\left[(\sin x)y\right] = 0$. Integrating, we have $(\sin x)y = c$ or $y = c/\sin x$. The initial condition implies $c = -2\sin(7\pi/6) = 1$. Thus, $y = 1/\sin x$, where the interval $\pi < x < 2\pi$ is chosen to include $x = 7\pi/6$.

20. Separating variables and integrating we have

$$\frac{dy}{y^2} = -2(t+1)\,dt$$

$$-\frac{1}{y} = -(t+1)^2 + c$$

$$y = \frac{1}{(t+1)^2 + c_1}, \qquad \text{where } -c = c_1.$$

The initial condition $y(0) = -\frac{1}{8}$ implies $c_1 = -9$, so a solution of the initial-value problem is

$$y = \frac{1}{(t+1)^2 - 9} \qquad \text{or} \qquad y = \frac{1}{t^2 + 2t - 8},$$

where $-4 < t < 2$.

21. (a) For $y < 0$, $\sqrt{y}$ is not a real number.

(b) Separating variables and integrating we have

$$\frac{dy}{\sqrt{y}} = dx \quad \text{and} \quad 2\sqrt{y} = x + c.$$

Letting $y(x_0) = y_0$ we get $c = 2\sqrt{y_0} - x_0$, so that

$$2\sqrt{y} = x + 2\sqrt{y_0} - x_0 \quad \text{and} \quad y = \frac{1}{4}(x + 2\sqrt{y_0} - x_0)^2.$$

Since $\sqrt{y} > 0$ for $y \neq 0$, we see that $dy/dx = \frac{1}{2}(x + 2\sqrt{y_0} - x_0)$ must be positive. Thus, the interval on which the solution is defined is $(x_0 - 2\sqrt{y_0}, \infty)$.

22. (a) The differential equation is homogeneous and we let $y = ux$. Then

$$(x^2 - y^2)\, dx + xy\, dy = 0$$

$$(x^2 - u^2 x^2)\, dx + ux^2(u\, dx + x\, du) = 0$$

$$dx + ux\, du = 0$$

$$u\, du = -\frac{dx}{x}$$

$$\frac{1}{2}u^2 = -\ln|x| + c$$

$$\frac{y^2}{x^2} = -2\ln|x| + c_1.$$

The initial condition gives $c_1 = 2$, so an implicit solution is $y^2 = x^2(2 - 2\ln|x|)$.

(b) Solving for y in part (a) and being sure that the initial condition is still satisfied, we have $y = -\sqrt{2}\,|x|(1 - \ln|x|)^{1/2}$, where $-e \le x \le e$ so that $1 - \ln|x| \ge 0$. The graph of this function indicates that the derivative is not defined at $x = 0$ and $x = e$. Thus, the solution of the initial-value problem is $y = -\sqrt{2}\,x(1 - \ln x)^{1/2}$, for $0 < x < e$.

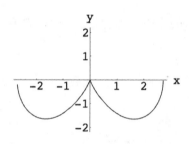

23. The graph of $y_1(x)$ is the portion of the closed black curve lying in the fourth quadrant. Its interval of definition is approximately $(0.7, 4.3)$. The graph of $y_2(x)$ is the portion of the left-hand black curve lying in the third quadrant. Its interval of definition is $(-\infty, 0)$.

24. The first step of Euler's method gives $y(1.1) \approx 9 + 0.1(1 + 3) = 9.4$. Applying Euler's method one more time gives $y(1.2) \approx 9.4 + 0.1(1 + 1.1\sqrt{9.4}) \approx 9.8373$.

25. Since the differential equation is autonomous, all lineal elements on a given horizontal line have the same slope. The direction field is then as shown in the figure at the right. It appears from the figure that the differential equation has critical points at -2 (an attractor) and at 2 (a repeller). Thus, -2 is an aymptotically stable critical point and 2 is an unstable critical point.

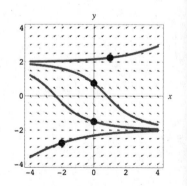

26. Since the differential equation is autonomous, all lineal elements on a given horizontal line have the same slope. The direction field is then as shown in the figure at the right. It appears from the figure that the differential equation has no critical points.

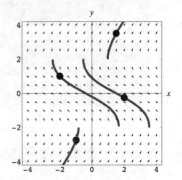

3 Modeling with First-Order Differential Equations

Exercises 3.1   Linear Models

1. Let $P = P(t)$ be the population at time t, and P_0 the initial population. From $dP/dt = kP$ we obtain $P = P_0 e^{kt}$. Using $P(5) = 2P_0$ we find $k = \frac{1}{5} \ln 2$ and $P = P_0 e^{(\ln 2)t/5}$. Setting $P(t) = 3P_0$ we have $3 = e^{(\ln 2)t/5}$, so

$$\ln 3 = \frac{(\ln 2)t}{5} \quad \text{and} \quad t = \frac{5 \ln 3}{\ln 2} \approx 7.9 \text{ years.}$$

   Setting $P(t) = 4P_0$ we have $4 = e^{(\ln 2)t/5}$, so

$$\ln 4 = \frac{(\ln 2)t}{5} \quad \text{and} \quad t \approx 10 \text{ years.}$$

2. From Problem 1 the growth constant is $k = \frac{1}{5} \ln 2$. Then $P = P_0 e^{(1/5)(\ln 2)t}$ and $10{,}000 = P_0 e^{(3/5) \ln 2}$. Solving for P_0 we get $P_0 = 10{,}000 e^{-(3/5) \ln 2} = 6{,}597.5$. Now

$$P(10) = P_0 e^{(1/5)(\ln 2)(10)} = 6{,}597.5 e^{2 \ln 2} = 4P_0 = 26{,}390.$$

   The rate at which the population is growing is

$$P'(10) = kP(10) = \frac{1}{5}(\ln 2)26{,}390 = 3658 \text{ persons/year.}$$

3. Let $P = P(t)$ be the population at time t. Then $dP/dt = kP$ and $P = ce^{kt}$. From $P(0) = c = 500$ we see that $P = 500 e^{kt}$. Since 15% of 500 is 75, we have $P(10) = 500 e^{10k} = 575$. Solving for k, we get $k = \frac{1}{10} \ln \frac{575}{500} = \frac{1}{10} \ln 1.15$. When $t = 30$,

$$P(30) = 500 e^{(1/10)(\ln 1.15)30} = 500 e^{3 \ln 1.15} = 760 \text{ years}$$

   and

$$P'(30) = kP(30) = \frac{1}{10}(\ln 1.15)760 = 10.62 \text{ persons/year.}$$

4. Let $P = P(t)$ be bacteria population at time t and P_0 the initial number. From $dP/dt = kP$ we obtain $P = P_0 e^{kt}$. Using $P(3) = 400$ and $P(10) = 2000$ we find $400 = P_0 e^{3k}$ or $e^k = (400/P_0)^{1/3}$. From $P(10) = 2000$ we then have $2000 = P_0 e^{10k} = P_0 (400/P_0)^{10/3}$, so

$$\frac{2000}{400^{10/3}} = P_0^{-7/3} \quad \text{and} \quad P_0 = \left(\frac{2000}{400^{10/3}} \right)^{-3/7} \approx 201.$$

5. Let $A = A(t)$ be the amount of lead present at time t. From $dA/dt = kA$ and $A(0) = 1$ we obtain $A = e^{kt}$. Using $A(3.3) = 1/2$ we find $k = \frac{1}{3.3} \ln(1/2)$. When 90% of the lead has decayed, 0.1 grams will remain. Setting $A(t) = 0.1$ we have $e^{t(1/3.3)\ln(1/2)} = 0.1$, so

$$\frac{t}{3.3} \ln \frac{1}{2} = \ln 0.1 \quad \text{and} \quad t = \frac{3.3 \ln 0.1}{\ln(1/2)} \approx 10.96 \text{ hours.}$$

6. Let $A = A(t)$ be the amount present at time t. From $dA/dt = kA$ and $A(0) = 100$ we obtain $A = 100e^{kt}$. Using $A(6) = 97$ we find $k = \frac{1}{6} \ln 0.97$. Then $A(24) = 100e^{(1/6)(\ln 0.97)24} = 100(0.97)^4 \approx 88.5$ mg.

7. Setting $A(t) = 50$ in Problem 6 we obtain $50 = 100e^{kt}$, so

$$kt = \ln \frac{1}{2} \quad \text{and} \quad t = \frac{\ln(1/2)}{(1/6)\ln 0.97} \approx 136.5 \text{ hours.}$$

8. **(a)** The solution of $dA/dt = kA$ is $A(t) = A_0 e^{kt}$. Letting $A = \frac{1}{2}A_0$ and solving for t we obtain the half-life $T = -(\ln 2)/k$.

   (b) Since $k = -(\ln 2)/T$ we have

   $$A(t) = A_0 e^{-(\ln 2)t/T} = A_0 2^{-t/T}.$$

   (c) Writing $\frac{1}{8}A_0 = A_0 2^{-t/T}$ as $2^{-3} = 2^{-t/T}$ and solving for t we get $t = 3T$. Thus, an initial amount A_0 will decay to $\frac{1}{8}A_0$ in three half-lives.

9. Let $I = I(t)$ be the intensity, t the thickness, and $I(0) = I_0$. If $dI/dt = kI$ and $I(3) = 0.25I_0$, then $I = I_0 e^{kt}$, $k = \frac{1}{3} \ln 0.25$, and $I(15) = 0.00098I_0$.

10. From $dS/dt = rS$ we obtain $S = S_0 e^{rt}$ where $S(0) = S_0$.

    (a) If $S_0 = \$5000$ and $r = 5.75\%$ then $S(5) = \$6665.45$.

    (b) If $S(t) = \$10,000$ then $t = 12$ years.

    (c) $S \approx \$6651.82$

11. Assume that $A = A_0 e^{kt}$ and $k = -0.00012378$. If $A(t) = 0.145A_0$ then $t \approx 15{,}600$ years.

12. From Example 3 in the text, the amount of carbon present at time t is $A(t) = A_0 e^{-0.00012378t}$. Letting $t = 660$ and solving for A_0 we have $A(660) = A_0 e^{-0.0001237(660)} = 0.921553A_0$. Thus, approximately 92% of the original amount of C-14 remained in the cloth as of 1988.

13. Assume that $dT/dt = k(T - 10)$ so that $T = 10 + ce^{kt}$. If $T(0) = 70°$ and $T(1/2) = 50°$ then $c = 60$ and $k = 2\ln(2/3)$ so that $T(1) = 36.67°$. If $T(t) = 15°$ then $t = 3.06$ minutes.

14. Assume that $dT/dt = k(T - 5)$ so that $T = 5 + ce^{kt}$. If $T(1) = 55°$ and $T(5) = 30°$ then $k = -\frac{1}{4}\ln 2$ and $c = 59.4611$ so that $T(0) = 64.4611°$.

15. We use the fact that the boiling temperature for water is $100°$ C. Now assume that $dT/dt = k(T - 100)$ so that $T = 100 + ce^{kt}$. If $T(0) = 20°$ and $T(1) = 22°$, then $c = -80$ and $k = \ln(39/40) \approx -0.0253$. Then $T(t) = 100 - 80e^{-0.0253t}$, and when $T = 90$, $t = 82.1$ seconds. If $T(t) = 98°$ then $t = 145.7$ seconds.

16. The differential equation for the first container is $dT_1/dt = k_1(T_1 - 0) = k_1 T_1$, whose solution is $T_1(t) = c_1 e^{k_1 t}$. Since $T_1(0) = 100$ (the initial temperature of the metal bar), we have $100 = c_1$ and $T_1(t) = 100 e^{k_1 t}$. After 1 minute, $T_1(1) = 100 e^{k_1} = 90°$C, so $k_1 = \ln 0.9$ and $T_1(t) = 100 e^{t \ln 0.9}$. After 2 minutes, $T_1(2) = 100 e^{2 \ln 0.9} = 100(0.9)^2 = 81°$C.

    The differential equation for the second container is $dT_2/dt = k_2(T_2 - 100)$, whose solution is $T_2(t) = 100 + c_2 e^{k_2 t}$. When the metal bar is immersed in the second container, its initial temperature is $T_2(0) = 81$, so

$$T_2(0) = 100 + c_2 e^{k_2(0)} = 100 + c_2 = 81$$

and $c_2 = -19$. Thus, $T_2(t) = 100 - 19 e^{k_2 t}$. After 1 minute in the second tank, the temperature of the metal bar is $91°$C, so

$$T_2(1) = 100 - 19 e^{k_2} = 91$$

$$e^{k_2} = \frac{9}{19}$$

$$k_2 = \ln \frac{9}{19}$$

and $T_2(t) = 100 - 19 e^{t \ln(9/19)}$. Setting $T_2(t) = 99.9$ we have

$$100 - 19 e^{t \ln(9/19)} = 99.9$$

$$e^{t \ln(9/19)} = \frac{0.1}{19}$$

$$t = \frac{\ln(0.1/19)}{\ln(9/19)} \approx 7.02.$$

Thus, from the start of the "double dipping" process, the total time until the bar reaches $99.9°$C in the second container is approximately 9.02 minutes.

17. Using separation of variables to solve $dT/dt = k(T - T_m)$ we get $T(t) = T_m + ce^{kt}$. Using $T(0) = 70$ we find $c = 70 - T_m$, so $T(t) = T_m + (70 - T_m)e^{kt}$. Using the given observations, we obtain

$$T\left(\frac{1}{2}\right) = T_m + (70 - T_m)e^{k/2} = 110$$

$$T(1) = T_m + (70 - T_m)e^k = 145.$$

Then, from the first equation, $e^{k/2} = (110 - T_m)/(70 - T_m)$ and

$$e^k = (e^{k/2})^2 = \left(\frac{110 - T_m}{70 - T_m}\right)^2 = \frac{145 - T_m}{70 - T_m}$$

$$\frac{(110 - T_m)^2}{70 - T_m} = 145 - T_m$$

$$12100 - 220T_m + T_m^2 = 10150 - 215T_m + T_m^2$$

$$T_m = 390.$$

The temperature in the oven is 390°.

18. **(a)** The initial temperature of the bath is $T_m(0) = 60°$, so in the short term the temperature of the chemical, which starts at 80°, should decrease or cool. Over time, the temperature of the bath will increase toward 100° since $e^{-0.1t}$ decreases from 1 toward 0 as t increases from 0. Thus, in the long term, the temperature of the chemical should increase or warm toward 100°.

(b) Adapting the model for Newton's law of cooling, we have

$$\frac{dT}{dt} = -0.1(T - 100 + 40e^{-0.1t}), \quad T(0) = 80.$$

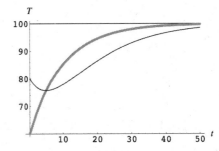

Writing the differential equation in the form

$$\frac{dT}{dt} + 0.1T = 10 - 4e^{-0.1t}$$

we see that it is linear with integrating factor $e^{\int 0.1\,dt} = e^{0.1t}$. Thus

$$\frac{d}{dt}[e^{0.1t}T] = 10e^{0.1t} - 4$$

$$e^{0.1t}T = 100e^{0.1t} - 4t + c$$

and

$$T(t) = 100 - 4te^{-0.1t} + ce^{-0.1t}.$$

Now $T(0) = 80$ so $100 + c = 80$, $c = -20$ and

$$T(t) = 100 - 4te^{-0.1t} - 20e^{-0.1t} = 100 - (4t + 20)e^{-0.1t}.$$

The thinner curve verifies the prediction of cooling followed by warming toward 100°. The wider curve shows the temperature T_m of the liquid bath.

19. Identifying $T_m = 70$, the differential equation is $dT/dt = k(T - 70)$. Assuming $T(0) = 98.6$ and separating variables we find $T(t) = 70 + 28.9e^{kt}$. If $t_1 > 0$ is the time of discovery of the body, then

$$T(t_1) = 70 + 28.6e^{kt_1} = 85 \quad \text{and} \quad T(t_1 + 1) = 70 + 28.6e^{k(t_1+1)} = 80.$$

Therefore $e^{kt_1} = 15/28.6$ and $e^{k(t_1+1)} = 10/28.6$. This implies

$$e^k = \frac{10}{28.6}e^{-kt_1} = \frac{10}{28.6} \cdot \frac{28.6}{15} = \frac{2}{3},$$

so $k = \ln\frac{2}{3} \approx -0.405465108$. Therefore

$$t_1 = \frac{1}{k} \ln\frac{15}{28.6} \approx 1.5916 \approx 1.6.$$

Death took place about 1.6 hours prior to the discovery of the body.

20. Solving the differential equation $dT/dt = kS(T - T_m)$ subject to $T(0) = T_0$ gives

$$T(t) = T_m + (T_0 - T_m)e^{kSt}.$$

The temperatures of the coffee in cups A and B are, respectively,

$$T_A(t) = 70 + 80e^{kSt} \quad \text{and} \quad T_B(t) = 70 + 80e^{2kSt}.$$

Then $T_A(30) = 70 + 80e^{30kS} = 100$, which implies $e^{30kS} = \frac{3}{8}$. Hence

$$T_B(30) = 70 + 80e^{60kS} = 70 + 80\left(e^{30kS}\right)^2$$

$$= 70 + 80\left(\frac{3}{8}\right)^2 = 70 + 80\left(\frac{9}{64}\right) = 81.25°\text{F}.$$

21. From $dA/dt = 4 - A/50$ we obtain $A = 200 + ce^{-t/50}$. If $A(0) = 30$ then $c = -170$ and $A = 200 - 170e^{-t/50}$.

22. From $dA/dt = 0 - A/50$ we obtain $A = ce^{-t/50}$. If $A(0) = 30$ then $c = 30$ and $A = 30e^{-t/50}$.

23. From $dA/dt = 10 - A/100$ we obtain $A = 1000 + ce^{-t/100}$. If $A(0) = 0$ then $c = -1000$ and $A(t) = 1000 - 1000e^{-t/100}$.

24. From Problem 23 the number of pounds of salt in the tank at time t is $A(t) = 1000 - 1000e^{-t/100}$. The concentration at time t is $c(t) = A(t)/500 = 2 - 2e^{-t/100}$. Therefore $c(5) = 2 - 2e^{-1/20} = 0.0975\,\text{lb/gal}$ and $\lim_{t\to\infty} c(t) = 2$. Solving $c(t) = 1 = 2 - 2e^{-t/100}$ for t we obtain $t = 100\ln 2 \approx 69.3\,\text{min}$.

25. From

$$\frac{dA}{dt} = 10 - \frac{10A}{500 - (10 - 5)t} = 10 - \frac{2A}{100 - t}$$

we obtain $A = 1000 - 10t + c(100 - t)^2$. If $A(0) = 0$ then $c = -\frac{1}{10}$. The tank is empty in 100 minutes.

26. With $c_{in}(t) = 2 + \sin(t/4)\,\text{lb/gal}$, the initial-value problem is

$$\frac{dA}{dt} + \frac{1}{100}A = 6 + 3\sin\frac{t}{4}, \quad A(0) = 50.$$

The differential equation is linear with integrating factor $e^{\int dt/100} = e^{t/100}$, so

$$\frac{d}{dt}[e^{t/100} A(t)] = \left(6 + 3\sin\frac{t}{4}\right)e^{t/100}$$

$$e^{t/100} A(t) = 600e^{t/100} + \frac{150}{313}e^{t/100}\sin\frac{t}{4} - \frac{3750}{313}e^{t/100}\cos\frac{t}{4} + c,$$

and

$$A(t) = 600 + \frac{150}{313}\sin\frac{t}{4} - \frac{3750}{313}\cos\frac{t}{4} + ce^{-t/100}.$$

Letting $t = 0$ and $A = 50$ we have $600 - 3750/313 + c = 50$ and $c = -168400/313$. Then

$$A(t) = 600 + \frac{150}{313}\sin\frac{t}{4} - \frac{3750}{313}\cos\frac{t}{4} - \frac{168400}{313}e^{-t/100}.$$

The graphs on $[0, 300]$ and $[0, 600]$ below show the effect of the sine function in the input when compared with the graph in Figure 3.1.4(a) in the text.

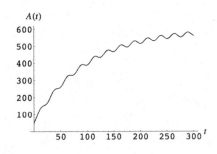
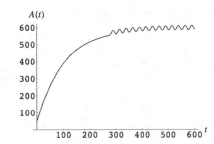

27. From

$$\frac{dA}{dt} = 3 - \frac{4A}{100 + (6 - 4)t} = 3 - \frac{2A}{50 + t}$$

we obtain $A = 50 + t + c(50 + t)^{-2}$. If $A(0) = 10$ then $c = -100{,}000$ and $A(30) = 64.38$ pounds.

28. (a) Initially the tank contains 300 gallons of solution. Since brine is pumped in at a rate of 3 gal/min and the mixture is pumped out at a rate of 2 gal/min, the net change is an increase of 1 gal/min. Thus, in 100 minutes the tank will contain its capacity of 400 gallons.

(b) The differential equation describing the amount of salt in the tank is $A'(t) = 6 - 2A/(300 + t)$ with solution

$$A(t) = 600 + 2t - (4.95 \times 10^7)(300 + t)^{-2}, \qquad 0 \le t \le 100,$$

as noted in the discussion following Example 5 in the text. Thus, the amount of salt in the tank when it overflows is

$$A(100) = 800 - (4.95 \times 10^7)(400)^{-2} = 490.625 \text{ lbs.}$$

(c) When the tank is overflowing the amount of salt in the tank is governed by the differential equation

$$\frac{dA}{dt} = (3 \text{ gal/min})(2 \text{ lb/gal}) - \left(\frac{A}{400} \text{ lb/gal}\right)(3 \text{ gal/min})$$

$$= 6 - \frac{3A}{400}, \qquad A(100) = 490.625.$$

Solving the equation, we obtain $A(t) = 800 + ce^{-3t/400}$. The initial condition yields $c = -654.947$, so that

$$A(t) = 800 - 654.947e^{-3t/400}.$$

When $t = 150$, $A(150) = 587.37$ lbs.

(d) As $t \to \infty$, the amount of salt is 800 lbs, which is to be expected since (400 gal)(2 lb/gal)= 800 lbs.

(e)

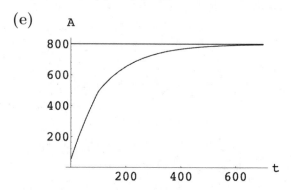

29. Assume $L\, di/dt + Ri = E(t)$, $L = 0.1$, $R = 50$, and $E(t) = 50$ so that $i = \frac{3}{5} + ce^{-500t}$. If $i(0) = 0$ then $c = -3/5$ and $\lim_{t\to\infty} i(t) = 3/5$.

30. Assume $L\, di/dt + Ri = E(t)$, $E(t) = E_0 \sin \omega t$, and $i(0) = i_0$ so that

$$i = \frac{E_0 R}{L^2\omega^2 + R^2} \sin \omega t - \frac{E_0 L \omega}{L^2\omega^2 + R^2} \cos \omega t + ce^{-Rt/L}.$$

Since $i(0) = i_0$ we obtain $c = i_0 + \dfrac{E_0 L \omega}{L^2\omega^2 + R^2}$.

31. Assume $R\, dq/dt + (1/C)q = E(t)$, $R = 200$, $C = 10^{-4}$, and $E(t) = 100$ so that $q = 1/100 + ce^{-50t}$. If $q(0) = 0$ then $c = -1/100$ and $i = \frac{1}{2}e^{-50t}$.

32. Assume $R\, dq/dt + (1/C)q = E(t)$, $R = 1000$, $C = 5 \times 10^{-6}$, and $E(t) = 200$. Then $q = \frac{1}{1000} + ce^{-200t}$ and $i = -200ce^{-200t}$. If $i(0) = 0.4$ then $c = -\frac{1}{500}$, $q(0.005) = 0.003$ coulombs, and $i(0.005) = 0.1472$ amps. We have $q \to \frac{1}{1000}$ as $t \to \infty$.

33. For $0 \le t \le 20$ the differential equation is $20\, di/dt + 2i = 120$. An integrating factor is $e^{t/10}$, so $(d/dt)[e^{t/10}i] = 6e^{t/10}$ and $i = 60 + c_1 e^{-t/10}$. If $i(0) = 0$ then $c_1 = -60$ and $i = 60 - 60e^{-t/10}$. For $t > 20$ the differential equation is $20\, di/dt + 2i = 0$ and $i = c_2 e^{-t/10}$. At $t = 20$ we want

$c_2 e^{-2} = 60 - 60e^{-2}$ so that $c_2 = 60\left(e^2 - 1\right)$. Thus

$$i(t) = \begin{cases} 60 - 60e^{-t/10}, & 0 \le t \le 20 \\ 60\left(e^2 - 1\right)e^{-t/10}, & t > 20. \end{cases}$$

34. Separating variables, we obtain

$$\frac{dq}{E_0 - q/C} = \frac{dt}{k_1 + k_2 t}$$

$$-C \ln\left|E_0 - \frac{q}{C}\right| = \frac{1}{k_2} \ln|k_1 + k_2 t| + c_1$$

$$\frac{(E_0 - q/C)^{-C}}{(k_1 + k_2 t)^{1/k_2}} = c_2.$$

Setting $q(0) = q_0$ we find $c_2 = (E_0 - q_0/C)^{-C}/k_1^{1/k_2}$, so

$$\frac{(E_0 - q/C)^{-C}}{(k_1 + k_2 t)^{1/k_2}} = \frac{(E_0 - q_0/C)^{-C}}{k_1^{1/k_2}}$$

$$\left(E_0 - \frac{q}{C}\right)^{-C} = \left(E_0 - \frac{q_0}{C}\right)^{-C}\left(\frac{k_1}{k + k_2 t}\right)^{-1/k_2}$$

$$E_0 - \frac{q}{C} = \left(E_0 - \frac{q_0}{C}\right)\left(\frac{k_1}{k + k_2 t}\right)^{1/Ck_2}$$

$$q = E_0 C + (q_0 - E_0 C)\left(\frac{k_1}{k + k_2 t}\right)^{1/Ck_2}.$$

35. (a) From $m\, dv/dt = mg - kv$ we obtain $v = mg/k + ce^{-kt/m}$. If $v(0) = v_0$ then $c = v_0 - mg/k$ and the solution of the initial-value problem is

$$v(t) = \frac{mg}{k} + \left(v_0 - \frac{mg}{k}\right)e^{-kt/m}.$$

(b) As $t \to \infty$ the limiting velocity is mg/k.

(c) From $ds/dt = v$ and $s(0) = 0$ we obtain

$$s(t) = \frac{mg}{k}t - \frac{m}{k}\left(v_0 - \frac{mg}{k}\right)e^{-kt/m} + \frac{m}{k}\left(v_0 - \frac{mg}{k}\right).$$

36. (a) Integrating $d^2 s/dt^2 = -g$ we get $v(t) = ds/dt = -gt + c$. From $v(0) = 300$ we find $c = 300$, and we are given $g = 32$, so the velocity is $v(t) = -32t + 300$.

(b) Integrating again and using $s(0) = 0$ we get $s(t) = -16t^2 + 300t$. The maximum height is attained when $v = 0$, that is, at $t_a = 9.375$. The maximum height will be $s(9.375) = 1406.25\,\text{ft}$.

37. When air resistance is proportional to velocity, the model for the velocity is $m\,dv/dt = -mg - kv$ (using the fact that the positive direction is upward.) Solving the differential equation using separation of variables we obtain $v(t) = -mg/k + ce^{-kt/m}$. From $v(0) = 300$ we get

$$v(t) = -\frac{mg}{k} + \left(300 + \frac{mg}{k}\right)e^{-kt/m}.$$

Integrating and using $s(0) = 0$ we find

$$s(t) = -\frac{mg}{k}t + \frac{m}{k}\left(300 + \frac{mg}{k}\right)(1 - e^{-kt/m}).$$

Setting $k = 0.0025$, $m = 16/32 = 0.5$, and $g = 32$ we have

$$s(t) = 1{,}340{,}000 - 6{,}400t - 1{,}340{,}000e^{-0.005t}$$

and

$$v(t) = -6{,}400 + 6{,}700e^{-0.005t}.$$

The maximum height is attained when $v = 0$, that is, at $t_a = 9.162$. The maximum height will be $s(9.162) = 1363.79\,\text{ft}$, which is less than the maximum height in Problem 36.

38. Assuming that the air resistance is proportional to velocity and the positive direction is downward with $s(0) = 0$, the model for the velocity is $m\,dv/dt = mg - kv$. Using separation of variables to solve this differential equation, we obtain $v(t) = mg/k + ce^{-kt/m}$. Then, using $v(0) = 0$, we get $v(t) = (mg/k)(1 - e^{-kt/m})$. Letting $k = 0.5$, $m = (125 + 35)/32 = 5$, and $g = 32$, we have $v(t) = 320(1 - e^{-0.1t})$. Integrating, we find $s(t) = 320t + 3200e^{-0.1t} + c_1$. Solving $s(0) = 0$ for c_1 we find $c_1 = -3200$, therefore $s(t) = 320t + 3200e^{-0.1t} - 3200$. At $t = 15$, when the parachute opens, $v(15) = 248.598$ and $s(15) = 2314.02$. At this time the value of k changes to $k = 10$ and the new initial velocity is $v_0 = 248.598$. With the parachute open, the skydiver's velocity is $v_p(t) = mg/k + c_2 e^{-kt/m}$, where t is reset to 0 when the parachute opens. Letting $m = 5$, $g = 32$, and $k = 10$, this gives $v_p(t) = 16 + c_2 e^{-2t}$. From $v(0) = 248.598$ we find $c_2 = 232.598$, so $v_p(t) = 16 + 232.598e^{-2t}$. Integrating, we get $s_p(t) = 16t - 116.299e^{-2t} + c_3$. Solving $s_p(0) = 0$ for c_3, we find $c_3 = 116.299$, so $s_p(t) = 16t - 116.299e^{-2t} + 116.299$. Twenty seconds after leaving the plane is five seconds after the parachute opens. The skydiver's velocity at this time is $v_p(5) = 16.0106\,\text{ft/s}$ and she has fallen a total of $s(15) + s_p(5) = 2314.02 + 196.294 = 2510.31\,\text{ft}$. Her terminal velocity is $\lim_{t\to\infty} v_p(t) = 16$, so she has very nearly reached her terminal velocity five seconds after the parachute opens. When the parachute opens, the distance to the ground is $15{,}000 - s(15) = 15{,}000 - 2{,}314 = 12{,}686\,\text{ft}$. Solving $s_p(t) = 12{,}686$ we get $t = 785.6\,\text{s} = 13.1$ min. Thus, it will take her approximately 13.1 minutes to reach the ground after her parachute has opened and a total of $(785.6 + 15)/60 = 13.34$ minutes after she exits the plane.

39. (a) The differential equation is first-order and linear. Letting $b = k/\rho$, the integrating factor is

$e^{\int 3b\,dt/(bt+r_0)} = (r_0 + bt)^3$. Then

$$\frac{d}{dt}[(r_0 + bt)^3 v] = g(r_0 + bt)^3 \quad \text{and} \quad (r_0 + bt)^3 v = \frac{g}{4b}(r_0 + bt)^4 + c.$$

The solution of the differential equation is $v(t) = (g/4b)(r_0 + bt) + c(r_0 + bt)^{-3}$. Using $v(0) = 0$ we find $c = -gr_0^4/4b$, so that

$$v(t) = \frac{g}{4b}(r_0 + bt) - \frac{gr_0^4}{4b(r_0 + bt)^3} = \frac{g\rho}{4k}\left(r_0 + \frac{k}{\rho}t\right) - \frac{g\rho r_0^4}{4k(r_0 + kt/\rho)^3}.$$

(b) Integrating $dr/dt = k/\rho$ we get $r = kt/\rho + c$. Using $r(0) = r_0$ we have $c = r_0$, so $r(t) = kt/\rho + r_0$.

(c) If $r = 0.007\,\text{ft}$ when $t = 10\,\text{s}$, then solving $r(10) = 0.007$ for k/ρ, we obtain $k/\rho = -0.0003$ and $r(t) = 0.01 - 0.0003t$. Solving $r(t) = 0$ we get $t = 33.3$, so the raindrop will have evaporated completely at 33.3 seconds.

40. Separating variables, we obtain $dP/P = k\cos t\,dt$, so

$$\ln|P| = k\sin t + c \quad \text{and} \quad P = c_1 e^{k\sin t}.$$

If $P(0) = P_0$, then $c_1 = P_0$ and $P = P_0 e^{k\sin t}$.

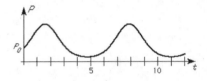

41. (a) From $dP/dt = (k_1 - k_2)P$ we obtain $P = P_0 e^{(k_1 - k_2)t}$ where $P_0 = P(0)$.

(b) If $k_1 > k_2$ then $P \to \infty$ as $t \to \infty$. If $k_1 = k_2$ then $P = P_0$ for every t. If $k_1 < k_2$ then $P \to 0$ as $t \to \infty$.

42. (a) The solution of the differential equation is $P(t) = c_1 e^{kt} + h/k$. If we let the initial population of fish be P_0 then $P(0) = P_0$ which implies that

$$c_1 = P_0 - \frac{h}{k} \quad \text{and} \quad P(t) = \left(P_0 - \frac{h}{k}\right)e^{kt} + \frac{h}{k}.$$

(b) For $P_0 > h/k$ all terms in the solution are positive. In this case $P(t)$ increases as time t increases. That is, $P(t) \to \infty$ as $t \to \infty$.

For $P_0 = h/k$ the population remains constant for all time t:

$$P(t) = \left(\frac{h}{k} - \frac{h}{k}\right)e^{kt} + \frac{h}{k} = \frac{h}{k}.$$

For $0 < P_0 < h/k$ the coefficient of the exponential function is negative and so the function decreases as time t increases.

(c) Since the function decreases and is concave down, the graph of $P(t)$ crosses the t-axis. That is, there exists a time $T > 0$ such that $P(T) = 0$. Solving

$$\left(P_0 - \frac{h}{k}\right)e^{kT} + \frac{h}{k} = 0$$

for T shows that the time of extinction is

$$T = \frac{1}{k} \ln \left(\frac{h}{h - kP_0} \right).$$

43. (a) Solving $r - kx = 0$ for x we find the equilibrium solution $x = r/k$. When $x < r/k$, $dx/dt > 0$ and when $x > r/k$, $dx/dt < 0$. From the phase portrait we see that $\lim_{t \to \infty} x(t) = r/k$.

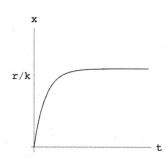

(b) From $dx/dt = r - kx$ and $x(0) = 0$ we obtain $x = r/k - (r/k)e^{-kt}$ so that $x \to r/k$ as $t \to \infty$. If $x(T) = r/2k$ then $T = (\ln 2)/k$.

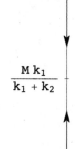

44. (a) Solving $k_1(M - A) - k_2 A = 0$ for A we find the equilibrium solution $A = k_1 M/(k_1 + k_2)$. From the phase portrait we see that $\lim_{t \to \infty} A(t) = k_1 M/(k_1 + k_2)$. Since $k_2 > 0$, the material will never be completely memorized and the larger k_2 is, the less the amount of material will be memorized over time.

(b) Write the differential equation in the form $dA/dt + (k_1 + k_2)A = k_1 M$. Then an integrating factor is $e^{(k_1 + k_2)t}$, and

$$\frac{d}{dt} \left[e^{(k_1+k_2)t} A \right] = k_1 M e^{(k_1+k_2)t}$$

$$e^{(k_1+k_2)t} A = \frac{k_1 M}{k_1 + k_2} e^{(k_1+k_2)t} + c$$

$$A = \frac{k_1 M}{k_1 + k_2} + c e^{-(k_1+k_2)t}.$$

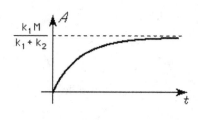

Using $A(0) = 0$ we find $c = -\dfrac{k_1 M}{k_1 + k_2}$ and $A = \dfrac{k_1 M}{k_1 + k_2}\left(1 - e^{-(k_1+k_2)t}\right)$. As $t \to \infty$,

$A \to \dfrac{k_1 M}{k_1 + k_2}$.

45. (a) For $0 \le t < 4$, $6 \le t < 10$ and $12 \le t < 16$, no voltage is applied to the heart and $E(t) = 0$. At the other times, the differential equation is $dE/dt = -E/RC$. Separating variables, integrating, and solving for e, we get $E = ke^{-t/RC}$, subject to $E(4) = E(10) = E(16) = 12$. These intitial conditions yield, respectively, $k = 12e^{4/RC}$, $k = 12e^{10/RC}$, $k = 12e^{16/RC}$, and $k = 12e^{22/RC}$. Thus

$$E(t) = \begin{cases} 0, & 0 \le t < 4, \;\; 6 \le t < 10, \;\; 12 \le t < 16 \\ 12e^{(4-t)/RC}, & 4 \le t < 6 \\ 12e^{(10-t)/RC}, & 10 \le t < 12 \\ 12e^{(16-t)/RC}, & 16 \le t < 18 \\ 12e^{(22-t)/RC}, & 22 \le t < 24. \end{cases}$$

(b)

46. (a) (i) Using Newton's second law of motion, $F = ma = m\,dv/dt$, the differential equation for the velocity v is

$$m\frac{dv}{dt} = mg\sin\theta \qquad \text{or} \qquad \frac{dv}{dt} = g\sin\theta,$$

where $mg\sin\theta$, $0 < \theta < \pi/2$, is the component of the weight along the plane in the direction of motion.

(ii) The model now becomes

$$m\frac{dv}{dt} = mg\sin\theta - \mu mg\cos\theta,$$

where $\mu mg\cos\theta$ is the component of the force of sliding friction (which acts perpendicular to the plane) along the plane. The negative sign indicates that this component of force is a retarding force which acts in the direction opposite to that of motion.

(iii) If air resistance is taken to be proportional to the instantaneous velocity of the body, the model becomes

$$m\frac{dv}{dt} = mg\sin\theta - \mu mg\cos\theta - kv,$$

where k is a constant of proportionality.

(b) (*i*) With $m = 3$ slugs, the differential equation is

$$3\frac{dv}{dt} = (96) \cdot \frac{1}{2} \qquad \text{or} \qquad \frac{dv}{dt} = 16.$$

Integrating the last equation gives $v(t) = 16t + c_1$. Since $v(0) = 0$, we have $c_1 = 0$ and so $v(t) = 16t$.

(*ii*) With $m = 3$ slugs, the differential equation is

$$3\frac{dv}{dt} = (96) \cdot \frac{1}{2} - \frac{\sqrt{3}}{4} \cdot (96) \cdot \frac{\sqrt{3}}{2} \qquad \text{or} \qquad \frac{dv}{dt} = 4.$$

In this case $v(t) = 4t$.

(*iii*) When the retarding force due to air resistance is taken into account, the differential equation for velocity v becomes

$$3\frac{dv}{dt} = (96) \cdot \frac{1}{2} - \frac{\sqrt{3}}{4} \cdot (96) \cdot \frac{\sqrt{3}}{2} - \frac{1}{4}v \qquad \text{or} \qquad 3\frac{dv}{dt} = 12 - \frac{1}{4}v.$$

The last differential equation is linear and has solution $v(t) = 48 + c_1 e^{-t/12}$. Since $v(0) = 0$, we find $c_1 = -48$, so $v(t) = 48 - 48e^{-t/12}$.

47. **(a)** (*i*) If $s(t)$ is distance measured down the plane from the highest point, then $ds/dt = v$. Integrating $ds/dt = 16t$ gives $s(t) = 8t^2 + c_2$. Using $s(0) = 0$ then gives $c_2 = 0$. Now the length L of the plane is $L = 50/\sin 30° = 100$ ft. The time it takes the box to slide completely down the plane is the solution of $s(t) = 100$ or $t^2 = 25/2$, so $t \approx 3.54$ s.

(*ii*) Integrating $ds/dt = 4t$ gives $s(t) = 2t^2 + c_2$. Using $s(0) = 0$ gives $c_2 = 0$, so $s(t) = 2t^2$ and the solution of $s(t) = 100$ is now $t \approx 7.07$ s.

(*iii*) Integrating $ds/dt = 48 - 48e^{-t/12}$ and using $s(0) = 0$ to determine the constant of integration, we obtain $s(t) = 48t + 576e^{-t/12} - 576$. With the aid of a CAS we find that the solution of $s(t) = 100$, or

$$100 = 48t + 576e^{-t/12} - 576 \qquad \text{or} \qquad 0 = 48t + 576e^{-t/12} - 676,$$

is now $t \approx 7.84$ s.

(b) The differential equation $m\,dv/dt = mg\sin\theta - \mu mg\cos\theta$ can be written

$$m\frac{dv}{dt} = mg\cos\theta(\tan\theta - \mu).$$

If $\tan\theta = \mu$, $dv/dt = 0$ and $v(0) = 0$ implies that $v(t) = 0$. If $\tan\theta < \mu$ and $v(0) = 0$, then integration implies $v(t) = g\cos\theta(\tan\theta - \mu)t < 0$ for all time t.

(c) Since $\tan 23° = 0.4245$ and $\mu = \sqrt{3}/4 = 0.4330$, we see that $\tan 23° < 0.4330$. The differential equation is $dv/dt = 32\cos 23°(\tan 23° - \sqrt{3}/4) = -0.251493$. Integration and the use of the initial condition gives $v(t) = -0.251493t + 1$. When the box stops, $v(t) = 0$ or $0 = -0.251493t + 1$ or $t = 3.976254$ s. From $s(t) = -0.125747t^2 + t$ we find $s(3.976254) = 1.988119$ ft.

(d) With $v_0 > 0$, $v(t) = -0.251493t + v_0$ and $s(t) = -0.125747t^2 + v_0 t$. Because two real positive solutions of the equation $s(t) = 100$, or $0 = -0.125747t^2 + v_0 t - 100$, would be physically meaningless, we use the quadratic formula and require that $b^2 - 4ac = 0$ or $v_0^2 - 50.2987 = 0$. From this last equality we find $v_0 \approx 7.092164$ ft/s. For the time it takes the box to traverse the entire inclined plane, we must have $0 = -0.125747t^2 + 7.092164t - 100$. *Mathematica* gives complex roots for the last equation: $t = 28.2001 \pm 0.0124458i$. But, for

$$0 = -0.125747t^2 + 7.092164691t - 100,$$

the roots are $t = 28.1999$ s and $t = 28.2004$ s. So if $v_0 > 7.092164$, we are guaranteed that the box will slide completely down the plane.

48. (a) We saw in part (b) of Problem 36 that the ascent time is $t_a = 9.375$. To find when the cannonball hits the ground we solve $s(t) = -16t^2 + 300t = 0$, getting a total time in flight of $t = 18.75$ s. Thus, the time of descent is $t_d = 18.75 - 9.375 = 9.375$. The impact velocity is $v_i = v(18.75) = -300$, which has the same magnitude as the initial velocity.

(b) We saw in Problem 37 that the ascent time in the case of air resistance is $t_a = 9.162$. Solving $s(t) = 1{,}340{,}000 - 6{,}400t - 1{,}340{,}000e^{-0.005t} = 0$ we see that the total time of flight is 18.466 s. Thus, the descent time is $t_d = 18.466 - 9.162 = 9.304$. The impact velocity is $v_i = v(18.466) = -290.91$, compared to an initial velocity of $v_0 = 300$.

Exercises 3.2   Nonlinear Models

1. (a) Solving $N(1 - 0.0005N) = 0$ for N we find the equilibrium solutions $N = 0$ and $N = 2000$. When $0 < N < 2000$, $dN/dt > 0$. From the phase portrait we see that $\lim_{t \to \infty} N(t) = 2000$. A graph of the solution is shown in part (b).

(b) Separating variables and integrating we have

$$\frac{dN}{N(1 - 0.0005N)} = \left(\frac{1}{N} - \frac{1}{N - 2000}\right)dN = dt$$

and

$$\ln N - \ln(N - 2000) = t + c.$$

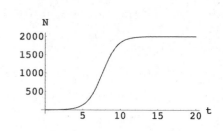

Solving for N we get $N(t) = 2000e^{c+t}/(1 + e^{c+t}) = 2000e^c e^t/(1 + e^c e^t)$. Using $N(0) = 1$ and solving for e^c we find $e^c = 1/1999$ and so $N(t) = 2000e^t/(1999 + e^t)$. Then $N(10) = 1833.59$, so 1834 companies are expected to adopt the new technology when $t = 10$.

2. From $dN/dt = N(a - bN)$ and $N(0) = 500$ we obtain

$$N = \frac{500a}{500b + (a - 500b)e^{-at}}.$$

Since $\lim_{t\to\infty} N = a/b = 50{,}000$ and $N(1) = 1000$ we have $a = 0.7033$, $b = 0.00014$, and $N = 50{,}000/(1 + 99e^{-0.7033t})$.

3. From $dP/dt = P\left(10^{-1} - 10^{-7}P\right)$ and $P(0) = 5000$ we obtain $P = 500/(0.0005 + 0.0995e^{-0.1t})$ so that $P \to 1{,}000{,}000$ as $t \to \infty$. If $P(t) = 500{,}000$ then $t = 52.9$ months.

4. (a) We have $dP/dt = P(a - bP)$ with $P(0) = 3.929$ million. Using separation of variables we obtain

$$P(t) = \frac{3.929a}{3.929b + (a - 3.929b)e^{-at}} = \frac{a/b}{1 + (a/3.929b - 1)e^{-at}}$$

$$= \frac{c}{1 + (c/3.929 - 1)e^{-at}},$$

where $c = a/b$. At $t = 60(1850)$ the population is 23.192 million, so

$$23.192 = \frac{c}{1 + (c/3.929 - 1)e^{-60a}}$$

or $c = 23.192 + 23.192(c/3.929 - 1)e^{-60a}$. At $t = 120(1910)$,

$$91.972 = \frac{c}{1 + (c/3.929 - 1)e^{-120a}}$$

or $c = 91.972 + 91.972(c/3.929 - 1)(e^{-60a})^2$. Combining the two equations for c we get

$$\left(\frac{(c - 23.192)/23.192}{c/3.929 - 1}\right)^2 \left(\frac{c}{3.929} - 1\right) = \frac{c - 91.972}{91.972}$$

or

$$91.972(3.929)(c - 23.192)^2 = (23.192)^2(c - 91.972)(c - 3.929).$$

The solution of this quadratic equation is $c = 197.274$. This in turn gives $a = 0.0313$. Therefore,

$$P(t) = \frac{197.274}{1 + 49.21e^{-0.0313t}}.$$

(b)

Year	Census Population	Predicted Population	Error	% Error
1790	3.929	3.929	0.000	0.00
1800	5.308	5.334	-0.026	-0.49
1810	7.240	7.222	0.018	0.24
1820	9.638	9.746	-0.108	-1.12
1830	12.866	13.090	-0.224	-1.74
1840	17.069	17.475	-0.406	-2.38
1850	23.192	23.143	0.049	0.21
1860	31.433	30.341	1.092	3.47
1870	38.558	39.272	-0.714	-1.85
1880	50.156	50.044	0.112	0.22
1890	62.948	62.600	0.348	0.55
1900	75.996	76.666	-0.670	-0.88
1910	91.972	91.739	0.233	0.25
1920	105.711	107.143	-1.432	-1.35
1930	122.775	122.140	0.635	0.52
1940	131.669	136.068	-4.399	-3.34
1950	150.697	148.445	2.252	1.49

The model predicts a population of 159.0 million for 1960 and 167.8 million for 1970. The census populations for these years were 179.3 and 203.3, respectively. The percentage errors are 12.8 and 21.2, respectively.

5. (a) The differential equation is $dP/dt = P(5 - P) - 4$. Solving $P(5 - P) - 4 = 0$ for P we obtain equilibrium solutions $P = 1$ and $P = 4$. The phase portrait is shown on the right and solution curves are shown in part (b). We see that for $P_0 > 4$ and $1 < P_0 < 4$ the population approaches 4 as t increases. For $0 < P < 1$ the population decreases to 0 in finite time.

(b) The differential equation is

$$\frac{dP}{dt} = P(5 - P) - 4 = -(P^2 - 5P + 4) = -(P - 4)(P - 1).$$

Separating variables and integrating, we obtain

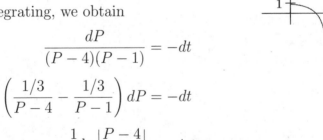

$$\frac{dP}{(P - 4)(P - 1)} = -dt$$

$$\left(\frac{1/3}{P - 4} - \frac{1/3}{P - 1}\right) dP = -dt$$

$$\frac{1}{3} \ln\left|\frac{P - 4}{P - 1}\right| = -t + c$$

$$\frac{P - 4}{P - 1} = c_1 e^{-3t}.$$

Setting $t = 0$ and $P = P_0$ we find $c_1 = (P_0 - 4)/(P_0 - 1)$. Solving for P we obtain

$$P(t) = \frac{4(P_0 - 1) - (P_0 - 4)e^{-3t}}{(P_0 - 1) - (P_0 - 4)e^{-3t}}.$$

(c) To find when the population becomes extinct in the case $0 < P_0 < 1$ we set $P = 0$ in

$$\frac{P-4}{P-1} = \frac{P_0-4}{P_0-1} e^{-3t}$$

from part (a) and solve for t. This gives the time of extinction

$$t = -\frac{1}{3} \ln \frac{4(P_0-1)}{P_0-4}.$$

6. Solving $P(5-P) - \frac{25}{4} = 0$ for P we obtain the equilibrium solution $P = \frac{5}{2}$. For $P \neq \frac{5}{2}$, $dP/dt < 0$. Thus, if $P_0 < \frac{5}{2}$, the population becomes extinct (otherwise there would be another equilibrium solution.) Using separation of variables to solve the initial-value problem, we get

$$P(t) = [4P_0 + (10P_0 - 25)t]/[4 + (4P_0 - 10)t].$$

To find when the population becomes extinct for $P_0 < \frac{5}{2}$ we solve $P(t) = 0$ for t. We see that the time of extinction is $t = 4P_0/5(5 - 2P_0)$.

7. Solving $P(5-P) - 7 = 0$ for P we obtain complex roots, so there are no equilibrium solutions. Since $dP/dt < 0$ for all values of P, the population becomes extinct for any initial condition. Using separation of variables to solve the initial-value problem, we get

$$P(t) = \frac{5}{2} + \frac{\sqrt{3}}{2} \tan\left[\tan^{-1}\left(\frac{2P_0 - 5}{\sqrt{3}}\right) - \frac{\sqrt{3}}{2}t\right].$$

Solving $P(t) = 0$ for t we see that the time of extinction is

$$t = \frac{2}{3}\left(\sqrt{3}\tan^{-1}(5/\sqrt{3}) + \sqrt{3}\tan^{-1}[(2P_0 - 5)/\sqrt{3}]\right).$$

8. (a) The differential equation is $dP/dt = P(1 - \ln P)$, which has the equilibrium solution $P = e$. When $P_0 > e$, $dP/dt < 0$, and when $P_0 < e$, $dP/dt > 0$.

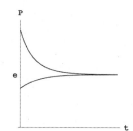

(b) The differential equation is $dP/dt = P(1 + \ln P)$, which has the equilibrium solution $P = 1/e$. When $P_0 > 1/e$, $dP/dt > 0$, and when $P_0 < 1/e$, $dP/dt < 0$.

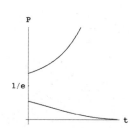

(c) From $dP/dt = P(a - b\ln P)$ we obtain $-(1/b)\ln|a - b\ln P| = t + c_1$ so that $P = e^{a/b}e^{-ce^{-bt}}$. If $P(0) = P_0$ then $c = (a/b) - \ln P_0$.

9. Let $X = X(t)$ be the amount of C at time t and $dX/dt = k(120 - 2X)(150 - X)$. If $X(0) = 0$ and $X(5) = 10$, then

$$X(t) = \frac{150 - 150e^{180kt}}{1 - 2.5e^{180kt}},$$

where $k = .0001259$ and $X(20) = 29.3$ grams. Now by L'Hôpital's rule, $X \to 60$ as $t \to \infty$, so that the amount of $A \to 0$ and the amount of $B \to 30$ as $t \to \infty$.

10. From $dX/dt = k(150 - X)^2$, $X(0) = 0$, and $X(5) = 10$ we obtain $X = 150 - 150/(150kt + 1)$ where $k = .000095238$. Then $X(20) = 33.3$ grams and $X \to 150$ as $t \to \infty$ so that the amount of $A \to 0$ and the amount of $B \to 0$ as $t \to \infty$. If $X(t) = 75$ then $t = 70$ minutes.

11. **(a)** The initial-value problem is $dh/dt = -8A_h\sqrt{h}/A_w$, $h(0) = H$. Separating variables and integrating we have

    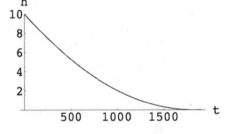

$$\frac{dh}{\sqrt{h}} = -\frac{8A_h}{A_w}\,dt \quad \text{and} \quad 2\sqrt{h} = -\frac{8A_h}{A_w}t + c.$$

Using $h(0) = H$ we find $c = 2\sqrt{H}$, so the solution of the initial-value problem is $\sqrt{h(t)} = (A_w\sqrt{H} - 4A_h t)/A_w$, where $A_w\sqrt{H} - 4A_h t \geq 0$. Thus,

$$h(t) = (A_w\sqrt{H} - 4A_h t)^2/A_w^2 \quad \text{for} \quad 0 \leq t \leq A_w\sqrt{H}/4A_h.$$

   (b) Identifying $H = 10$, $A_w = 4\pi$, and $A_h = \pi/576$ we have $h(t) = t^2/331{,}776 - (\sqrt{5/2}/144)t + 10$. Solving $h(t) = 0$ we see that the tank empties in $576\sqrt{10}$ seconds or 30.36 minutes.

12. To obtain the solution of this differential equation we use $h(t)$ from Problem 13 in Exercises 1.3. Then $h(t) = (A_w\sqrt{H} - 4cA_h t)^2/A_w^2$. Solving $h(t) = 0$ with $c = 0.6$ and the values from Problem 11 we see that the tank empties in 3035.79 seconds or 50.6 minutes.

13. **(a)** Separating variables and integrating gives

$$6h^{3/2}dh = -5dt \quad \text{and} \quad \frac{12}{5}h^{5/2} = -5t + c.$$

Using $h(0) = 20$ we find $c = 1920\sqrt{5}$, so the solution of the initial-value problem is $h(t) = (800\sqrt{5} - \frac{25}{12}t)^{2/5}$. Solving $h(t) = 0$ we see that the tank empties in $384\sqrt{5}$ seconds or 14.31 minutes.

   (b) When the height of the water is h, the radius of the top of the water is $r = h\tan 30° = h/\sqrt{3}$ and $A_w = \pi h^2/3$. The differential equation is

$$\frac{dh}{dt} = -c\frac{A_h}{A_w}\sqrt{2gh} = -0.6\frac{\pi(2/12)^2}{\pi h^2/3}\sqrt{64h} = -\frac{2}{5h^{3/2}}.$$

Separating variables and integrating gives

$$5h^{3/2}dh = -2\,dt \quad \text{and} \quad 2h^{5/2} = -2t + c.$$

Using $h(0) = 9$ we find $c = 486$, so the solution of the initial-value problem is $h(t) = (243 - t)^{2/5}$. Solving $h(t) = 0$ we see that the tank empties in 243 seconds or 4.05 minutes.

14. When the height of the water is h, the radius of the top of the water is $\frac{2}{5}(20 - h)$ and $A_w = 4\pi(20 - h)^2/25$. The differential equation is

$$\frac{dh}{dt} = -c\frac{A_h}{A_w}\sqrt{2gh} = -0.6\frac{\pi(2/12)^2}{4\pi(20-h)^2/25}\sqrt{64h} = -\frac{5}{6}\frac{\sqrt{h}}{(20-h)^2}.$$

Separating variables and integrating we have

$$\frac{(20-h)^2}{\sqrt{h}}\,dh = -\frac{5}{6}\,dt \quad \text{and} \quad 800\sqrt{h} - \frac{80}{3}h^{3/2} + \frac{2}{5}h^{5/2} = -\frac{5}{6}t + c.$$

Using $h(0) = 20$ we find $c = 2560\sqrt{5}/3$, so an implicit solution of the initial-value problem is

$$800\sqrt{h} - \frac{80}{3}h^{3/2} + \frac{2}{5}h^{5/2} = -\frac{5}{6}t + \frac{2560\sqrt{5}}{3}.$$

To find the time it takes the tank to empty we set $h = 0$ and solve for t. The tank empties in $1024\sqrt{5}$ seconds or 38.16 minutes. Thus, the tank empties more slowly when the base of the cone is on the bottom.

15. (a) After separating variables we obtain

$$\frac{m\,dv}{mg - kv^2} = dt$$

$$\frac{1}{g}\frac{dv}{1 - (\sqrt{k}\,v/\sqrt{mg}\,)^2} = dt$$

$$\frac{\sqrt{mg}}{\sqrt{k}\,g}\frac{\sqrt{k/mg}\,dv}{1 - (\sqrt{k}\,v/\sqrt{mg}\,)^2} = dt$$

$$\sqrt{\frac{m}{kg}}\tanh^{-1}\frac{\sqrt{k}\,v}{\sqrt{mg}} = t + c$$

$$\tanh^{-1}\frac{\sqrt{k}\,v}{\sqrt{mg}} = \sqrt{\frac{kg}{m}}\,t + c_1.$$

Thus the velocity at time t is

$$v(t) = \sqrt{\frac{mg}{k}}\tanh\left(\sqrt{\frac{kg}{m}}\,t + c_1\right).$$

Setting $t = 0$ and $v = v_0$ we find $c_1 = \tanh^{-1}(\sqrt{k}\,v_0/\sqrt{mg}\,)$.

Exercises 3.2 Nonlinear Models

(b) Since $\tanh t \to 1$ as $t \to \infty$, we have $v \to \sqrt{mg/k}$ as $t \to \infty$.

(c) Integrating the expression for $v(t)$ in part (a) we obtain an integral of the form $\int du/u$:

$$s(t) = \sqrt{\frac{mg}{k}} \int \tanh\left(\sqrt{\frac{kg}{m}}\, t + c_1\right) dt = \frac{m}{k} \ln\left[\cosh\left(\sqrt{\frac{kg}{m}}\, t + c_1\right)\right] + c_2.$$

Setting $t = 0$ and $s = 0$ we find $c_2 = -(m/k) \ln(\cosh c_1)$, where c_1 is given in part (a).

16. The differential equation is $m\, dv/dt = -mg - kv^2$. Separating variables and integrating, we have

$$\frac{dv}{mg + kv^2} = -\frac{dt}{m}$$

$$\frac{1}{\sqrt{mgk}} \tan^{-1}\left(\frac{\sqrt{k}\, v}{\sqrt{mg}}\right) = -\frac{1}{m} t + c$$

$$\tan^{-1}\left(\frac{\sqrt{k}\, v}{\sqrt{mg}}\right) = -\sqrt{\frac{gk}{m}}\, t + c_1$$

$$v(t) = \sqrt{\frac{mg}{k}} \tan\left(c_1 - \sqrt{\frac{gk}{m}}\, t\right).$$

Setting $v(0) = 300$, $m = \frac{16}{32} = \frac{1}{2}$, $g = 32$, and $k = 0.0003$, we find $v(t) = 230.94 \tan(c_1 - 0.138564t)$ and $c_1 = 0.914743$. Integrating

$$v(t) = 230.94 \tan(0.914743 - 0.138564t)$$

we get

$$s(t) = 1666.67 \ln|\cos(0.914743 - 0.138564t)| + c_2.$$

Using $s(0) = 0$ we find $c_2 = 823.843$. Solving $v(t) = 0$ we see that the maximum height is attained when $t = 6.60159$. The maximum height is $s(6.60159) = 823.843\,\text{ft}$.

17. (a) Let ρ be the weight density of the water and V the volume of the object. Archimedes' principle states that the upward buoyant force has magnitude equal to the weight of the water displaced. Taking the positive direction to be down, the differential equation is

$$m\frac{dv}{dt} = mg - kv^2 - \rho V.$$

(b) Using separation of variables we have

$$\frac{m\, dv}{(mg - \rho V) - kv^2} = dt$$

$$\frac{m}{\sqrt{k}} \frac{\sqrt{k}\, dv}{(\sqrt{mg - \rho V})^2 - (\sqrt{k}\, v)^2} = dt$$

$$\frac{m}{\sqrt{k}} \frac{1}{\sqrt{mg - \rho V}} \tanh^{-1} \frac{\sqrt{k}\, v}{\sqrt{mg - \rho V}} = t + c.$$

105

Thus

$$v(t) = \sqrt{\frac{mg - \rho V}{k}} \tanh\left(\frac{\sqrt{kmg - k\rho V}}{m} t + c_1\right).$$

(c) Since $\tanh t \to 1$ as $t \to \infty$, the terminal velocity is $\sqrt{(mg - \rho V)/k}$.

18. (a) Writing the equation in the form $(x - \sqrt{x^2 + y^2})dx + y\,dy = 0$ we identify $M = x - \sqrt{x^2 + y^2}$ and $N = y$. Since M and N are both homogeneous functions of degree 1 we use the substitution $y = ux$. It follows that

$$\left(x - \sqrt{x^2 + u^2 x^2}\right)dx + ux(u\,dx + x\,du) = 0$$

$$x\left[1 - \sqrt{1 + u^2} + u^2\right]dx + x^2 u\,du = 0$$

$$-\frac{u\,du}{1 + u^2 - \sqrt{1 + u^2}} = \frac{dx}{x}$$

$$\frac{u\,du}{\sqrt{1 + u^2}\,(1 - \sqrt{1 + u^2})} = \frac{dx}{x}.$$

Letting $w = 1 - \sqrt{1 + u^2}$ we have $dw = -u\,du/\sqrt{1 + u^2}$ so that

$$-\ln\left|1 - \sqrt{1 + u^2}\right| = \ln|x| + c$$

$$\frac{1}{1 - \sqrt{1 + u^2}} = c_1 x$$

$$1 - \sqrt{1 + u^2} = -\frac{c_2}{x} \qquad (-c_2 = 1/c_1)$$

$$1 + \frac{c_2}{x} = \sqrt{1 + \frac{y^2}{x^2}}$$

$$1 + \frac{2c_2}{x} + \frac{c_2^2}{x^2} = 1 + \frac{y^2}{x^2}.$$

Solving for y^2 we have

$$y^2 = 2c_2 x + c_2^2 = 4\left(\frac{c_2}{2}\right)\left(x + \frac{c_2}{2}\right)$$

which is a family of parabolas symmetric with respect to the x-axis with vertex at $(-c_2/2, 0)$ and focus at the origin.

(b) Let $u = x^2 + y^2$ so that

$$\frac{du}{dx} = 2x + 2y\frac{dy}{dx}.$$

Then
$$y\frac{dy}{dx} = \frac{1}{2}\frac{du}{dx} - x$$
and the differential equation can be written in the form
$$\frac{1}{2}\frac{du}{dx} - x = -x + \sqrt{u} \quad \text{or} \quad \frac{1}{2}\frac{du}{dx} = \sqrt{u}\,.$$
Separating variables and integrating gives
$$\frac{du}{2\sqrt{u}} = dx$$
$$\sqrt{u} = x + c$$
$$u = x^2 + 2cx + c^2$$
$$x^2 + y^2 = x^2 + 2cx + c^2$$
$$y^2 = 2cx + c^2.$$

19. **(a)** From $2W^2 - W^3 = W^2(2 - W) = 0$ we see that $W = 0$ and $W = 2$ are constant solutions.

    (b) Separating variables and using a CAS to integrate we get
    $$\frac{dW}{W\sqrt{4 - 2W}} = dx \quad \text{and} \quad -\tanh^{-1}\!\left(\frac{1}{2}\sqrt{4 - 2W}\right) = x + c.$$

    Using the facts that the hyperbolic tangent is an odd function and $1 - \tanh^2 x = \operatorname{sech}^2 x$ we have
    $$\frac{1}{2}\sqrt{4 - 2W} = \tanh(-x - c) = -\tanh(x + c)$$
    $$\frac{1}{4}(4 - 2W) = \tanh^2(x + c)$$
    $$1 - \frac{1}{2}W = \tanh^2(x + c)$$
    $$\frac{1}{2}W = 1 - \tanh^2(x + c) = \operatorname{sech}^2(x + c).$$

    Thus, $W(x) = 2\operatorname{sech}^2(x + c)$.

    (c) Letting $x = 0$ and $W = 2$ we find that $\operatorname{sech}^2(c) = 1$ and $c = 0$.

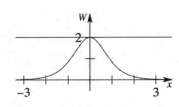

20. **(a)** Solving $r^2 + (10 - h)^2 = 10^2$ for r^2 we see that $r^2 = 20h - h^2$. Combining the rate of input of water, π, with the rate of output due to evaporation, $k\pi r^2 = k\pi(20h - h^2)$, we have $dV/dt =$

$\pi - k\pi(20h - h^2)$. Using $V = 10\pi h^2 - \frac{1}{3}\pi h^3$, we see also that $dV/dt = (20\pi h - \pi h^2)dh/dt$. Thus,

$$(20\pi h - \pi h^2)\frac{dh}{dt} = \pi - k\pi(20h - h^2) \quad \text{and} \quad \frac{dh}{dt} = \frac{1 - 20kh + kh^2}{20h - h^2}.$$

(b) Letting $k = 1/100$, separating variables and integrating (with the help of a CAS), we get

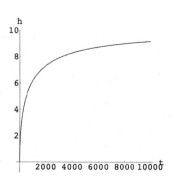

$$\frac{100h(h - 20)}{(h - 10)^2} dh = dt \quad \text{and} \quad \frac{100(h^2 - 10h + 100)}{10 - h} = t + c.$$

Using $h(0) = 0$ we find $c = 1000$, and solving for h we get $h(t) = 0.005(\sqrt{t^2 + 4000t} - t)$, where the positive square root is chosen because $h \geq 0$.

(c) The volume of the tank is $V = \frac{2}{3}\pi(10)^3$ feet, so at a rate of π cubic feet per minute, the tank will fill in $\frac{2}{3}(10)^3 \approx 666.67$ minutes ≈ 11.11 hours.

(d) At 666.67 minutes, the depth of the water is $h(666.67) = 5.486$ feet. From the graph in (b) we suspect that $\lim_{t\to\infty} h(t) = 10$, in which case the tank will never completely fill. To prove this we compute the limit of $h(t)$:

$$\lim_{t\to\infty} h(t) = 0.005 \lim_{t\to\infty}\left(\sqrt{t^2 + 4000t} - t\right) = 0.005 \lim_{t\to\infty} \frac{t^2 + 4000t - t^2}{\sqrt{t^2 + 4000t} + t}$$

$$= 0.005 \lim_{t\to\infty} \frac{4000t}{t\sqrt{1 + 4000/t} + t} = 0.005 \frac{4000}{1 + 1} = 0.005(2000) = 10.$$

21. (a)

t	P(t)	Q(t)
0	3.929	0.035
10	5.308	0.036
20	7.240	0.033
30	9.638	0.033
40	12.866	0.033
50	17.069	0.036
60	23.192	0.036
70	31.433	0.023
80	38.558	0.030
90	50.156	0.026
100	62.948	0.021
110	75.996	0.021
120	91.972	0.015
130	105.711	0.016
140	122.775	0.007
150	131.669	0.014
160	150.697	0.019
170	179.300	

(b) The regression line is $Q = 0.0348391 - 0.000168222P$.

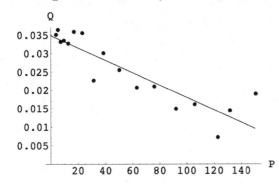

(c) The solution of the logistic equation is given in equation (5) in the text. Identifying $a = 0.0348391$ and $b = 0.000168222$ we have

$$P(t) = \frac{aP_0}{bP_0 + (a - bP_0)e^{-at}}.$$

(d) With $P_0 = 3.929$ the solution becomes

$$P(t) = \frac{0.136883}{0.000660944 + 0.0341781e^{-0.0348391t}}.$$

(e)

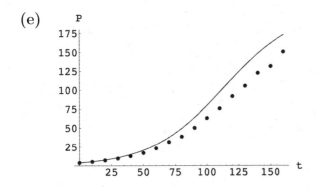

(f) We identify $t = 180$ with 1970, $t = 190$ with 1980, and $t = 200$ with 1990. The model predicts $P(180) = 188.661$, $P(190) = 193.735$, and $P(200) = 197.485$. The actual population figures for these years are 203.303, 226.542, and 248.765 millions. As $t \to \infty$, $P(t) \to a/b = 207.102$.

22. (a) Using a CAS to solve $P(1 - P) + 0.3e^{-P} = 0$ for P we see that $P = 1.09216$ is an equilibrium solution.

(b) Since $f(P) > 0$ for $0 < P < 1.09216$, the solution $P(t)$ of

$$dP/dt = P(1 - P) + 0.3e^{-P}, \quad P(0) = P_0,$$

is increasing for $P_0 < 1.09216$. Since $f(P) < 0$ for $P > 1.09216$, the solution $P(t)$ is decreasing for $P_0 > 1.09216$. Thus $P = 1.09216$ is an attractor.

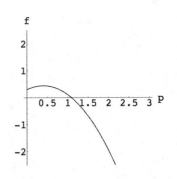

(c) The curves for the second initial-value problem are thicker. The equilibrium solution for the logic model is $P = 1$. Comparing 1.09216 and 1, we see that the percentage increase is 9.216%.

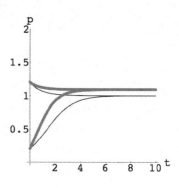

23. To find t_d we solve

$$m \frac{dv}{dt} = mg - kv^2, \qquad v(0) = 0$$

using separation of variables. This gives

$$v(t) = \sqrt{\frac{mg}{k}} \tanh \sqrt{\frac{kg}{m}} \, t.$$

Integrating and using $s(0) = 0$ gives

$$s(t) = \frac{m}{k} \ln \left(\cosh \sqrt{\frac{kg}{m}} \, t \right).$$

To find the time of descent we solve $s(t) = 823.84$ and find $t_d = 7.77882$. The impact velocity is $v(t_d) = 182.998$, which is positive because the positive direction is downward.

24. (a) Solving $v_t = \sqrt{mg/k}$ for k we obtain $k = mg/v_t^2$. The differential equation then becomes

$$m \frac{dv}{dt} = mg - \frac{mg}{v_t^2} v^2 \quad \text{or} \quad \frac{dv}{dt} = g \left(1 - \frac{1}{v_t^2} v^2 \right).$$

Separating variables and integrating gives

$$v_t \tanh^{-1} \frac{v}{v_t} = gt + c_1.$$

The initial condition $v(0) = 0$ implies $c_1 = 0$, so

$$v(t) = v_t \tanh \frac{gt}{v_t}.$$

We find the distance by integrating:

$$s(t) = \int v_t \tanh \frac{gt}{v_t} \, dt = \frac{v_t^2}{g} \ln \left(\cosh \frac{gt}{v_t} \right) + c_2.$$

The initial condition $s(0) = 0$ implies $c_2 = 0$, so

$$s(t) = \frac{v_t^2}{g} \ln \left(\cosh \frac{gt}{v_t} \right).$$

110

In 25 seconds she has fallen $20,000 - 14,800 = 5,200$ feet. Using a CAS to solve

$$5200 = (v_t^2/32) \ln\left(\cosh \frac{32(25)}{v_t}\right)$$

for v_t gives $v_t \approx 271.711$ ft/s. Then

$$s(t) = \frac{v_t^2}{g} \ln\left(\cosh \frac{gt}{v_t}\right) = 2307.08 \ln(\cosh 0.117772t).$$

(b) At $t = 15$, $s(15) = 2,542.94$ ft and $v(15) = s'(15) = 256.287$ ft/sec.

25. While the object is in the air its velocity is modeled by the linear differential equation $m\, dv/dt = mg - kv$. Using $m = 160$, $k = \frac{1}{4}$, and $g = 32$, the differential equation becomes $dv/dt + (1/640)v = 32$. The integrating factor is $e^{\int dt/640} = e^{t/640}$ and the solution of the differential equation is $e^{t/640}v = \int 32e^{t/640}dt = 20,480e^{t/640} + c$. Using $v(0) = 0$ we see that $c = -20,480$ and $v(t) = 20,480 - 20,480e^{-t/640}$. Integrating we get $s(t) = 20,480t + 13,107,200e^{-t/640} + c$. Since $s(0) = 0$, $c = -13,107,200$ and $s(t) = -13,107,200 + 20,480t + 13,107,200e^{-t/640}$. To find when the object hits the liquid we solve $s(t) = 500 - 75 = 425$, obtaining $t_a = 5.16018$. The velocity at the time of impact with the liquid is $v_a = v(t_a) = 164.482$. When the object is in the liquid its velocity is modeled by the nonlinear differential equation $m\, dv/dt = mg - kv^2$. Using $m = 160$, $g = 32$, and $k = 0.1$ this becomes $dv/dt = (51,200 - v^2)/1600$. Separating variables and integrating we have

$$\frac{dv}{51,200 - v^2} = \frac{dt}{1600} \quad \text{and} \quad \frac{\sqrt{2}}{640} \ln\left| \frac{v - 160\sqrt{2}}{v + 160\sqrt{2}} \right| = \frac{1}{1600}t + c.$$

Solving $v(0) = v_a = 164.482$ we obtain $c = -0.00407537$. Then, for $v < 160\sqrt{2} = 226.274$,

$$\left| \frac{v - 160\sqrt{2}}{v + 160\sqrt{2}} \right| = e^{\sqrt{2}t/5 - 1.8443} \quad \text{or} \quad -\frac{v - 160\sqrt{2}}{v + 160\sqrt{2}} = e^{\sqrt{2}t/5 - 1.8443}.$$

Solving for v we get

$$v(t) = \frac{13964.6 - 2208.29e^{\sqrt{2}t/5}}{61.7153 + 9.75937e^{\sqrt{2}t/5}}.$$

Integrating we find

$$s(t) = 226.275t - 1600 \ln(6.3237 + e^{\sqrt{2}t/5}) + c.$$

Solving $s(0) = 0$ we see that $c = 3185.78$, so

$$s(t) = 3185.78 + 226.275t - 1600 \ln(6.3237 + e^{\sqrt{2}t/5}).$$

To find when the object hits the bottom of the tank we solve $s(t) = 75$, obtaining $t_b = 0.466273$. The time from when the object is dropped from the helicopter to when it hits the bottom of the tank is $t_a + t_b = 5.62708$ seconds.

26. The velocity vector of the swimmer is

$$\mathbf{v} = \mathbf{v}_s + \mathbf{v}_r = (-v_s \cos\theta, -v_s \sin\theta) + (0, v_r) = (-v_s \cos\theta, -v_s \sin\theta + v_r) = \left(\frac{dx}{dt}, \frac{dy}{dt}\right).$$

Equating components gives

$$\frac{dx}{dt} = -v_s \cos\theta \quad \text{and} \quad \frac{dy}{dt} = -v_s \sin\theta + v_r$$

so

$$\frac{dx}{dt} = -v_s \frac{x}{\sqrt{x^2 + y^2}} \quad \text{and} \quad \frac{dy}{dt} = -v_s \frac{y}{\sqrt{x^2 + y^2}} + v_r.$$

Thus,

$$\frac{dy}{dx} = \frac{dy/dt}{dx/dt} = \frac{-v_s y + v_r \sqrt{x^2 + y^2}}{-v_s x} = \frac{v_s y - v_r \sqrt{x^2 + y^2}}{v_s x}.$$

27. (a) With $k = v_r/v_s$,

$$\frac{dy}{dx} = \frac{y - k\sqrt{x^2 + y^2}}{x}$$

is a first-order homogeneous differential equation (see Section 2.5). Substituting $y = ux$ into the differential equation gives

$$u + x\frac{du}{dx} = u - k\sqrt{1 + u^2} \quad \text{or} \quad \frac{du}{dx} = -k\sqrt{1 + u^2}.$$

Separating variables and integrating we obtain

$$\int \frac{du}{\sqrt{1 + u^2}} = -\int k\,dx \quad \text{or} \quad \ln\left(u + \sqrt{1 + u^2}\right) = -k\ln x + \ln c.$$

This implies

$$\ln x^k \left(u + \sqrt{1 + u^2}\right) = \ln c \quad \text{or} \quad x^k \left(\frac{y}{x} + \frac{\sqrt{x^2 + y^2}}{x}\right) = c.$$

The condition $y(1) = 0$ gives $c = 1$ and so $y + \sqrt{x^2 + y^2} = x^{1-k}$. Solving for y gives

$$y(x) = \frac{1}{2}\left(x^{1-k} - x^{1+k}\right).$$

(b) If $k = 1$, then $v_s = v_r$ and $y = \frac{1}{2}(1 - x^2)$. Since $y(0) = \frac{1}{2}$, the swimmer lands on the west beach at $(0, \frac{1}{2})$. That is, $\frac{1}{2}$ mile north of $(0,0)$.

If $k > 1$, then $v_r > v_s$ and $1 - k < 0$. This means $\lim_{x\to 0^+} y(x)$ becomes infinite, since $\lim_{x\to 0^+} x^{1-k}$ becomes infinite. The swimmer never makes it to the west beach and is swept northward with the current.

If $0 < k < 1$, then $v_s > v_r$ and $1 - k > 0$. The value of $y(x)$ at $x = 0$ is $y(0) = 0$. The swimmer has made it to the point $(0,0)$.

28. The velocity vector of the swimmer is

$$\mathbf{v} = \mathbf{v}_s + \mathbf{v}_r = (-v_s, 0) + (0, v_r) = \left(\frac{dx}{dt}, \frac{dy}{dt}\right).$$

Equating components gives

$$\frac{dx}{dt} = -v_s \quad \text{and} \quad \frac{dy}{dt} = v_r$$

so

$$\frac{dy}{dx} = \frac{dy/dt}{dx/dt} = \frac{v_r}{-v_s} = -\frac{v_r}{v_s}.$$

29. The differential equation

$$\frac{dy}{dx} = -\frac{30x(1-x)}{2}$$

separates into $dy = 15(-x + x^2)dx$. Integration gives $y(x) = -\frac{15}{2}x^2 + 5x^3 + c$. The condition $y(1) = 0$ gives $c = \frac{5}{2}$ and so $y(x) = \frac{1}{2}(-15x^2 + 10x^3 + 5)$. Since $y(0) = \frac{5}{2}$, the swimmer has to walk 2.5 miles back down the west beach to reach $(0, 0)$.

30. This problem has a great many components, so we will consider the case in which air resistance is assumed to be proportional to the velocity. By Problem 35 in Section 3.1 the differential equation is

$$m\frac{dv}{dt} = mg - kv,$$

and the solution is

$$v(t) = \frac{mg}{k} + \left(v_0 - \frac{mg}{k}\right)e^{-kt/m}.$$

If we take the initial velocity to be 0, then the velocity at time t is

$$v(t) = \frac{mg}{k} - \frac{mg}{k}e^{-kt/m}.$$

The mass of the raindrop is about $m = 62 \times 0.000000155/32 \approx 0.0000003$ and $g = 32$, so the volocity at time t is

$$v(t) = \frac{0.0000096}{k} - \frac{0.0000096}{k}e^{-3333333kt}.$$

If we let $k = 0.0000007$, then $v(100) \approx 13.7$ ft/s. In this case 100 is the time in seconds. Since 7 mph ≈ 10.3 ft/s, the assertion that the average velocity is 7 mph is not unreasonable. Of course, this assumes that the air resistance is proportional to the velocity, and, more importantly, that the constant of proportionality is 0.0000007. The assumption about the constant is particularly suspect.

31. (a) Letting $c = 0.6$, $A_h = \pi(\frac{1}{32} \cdot \frac{1}{12})^2$, $A_w = \pi \cdot 1^2 = \pi$, and $g = 32$, the differential equation in Proble 12 becomes $dh/dt = -0.00003255\sqrt{h}$. Separating variables and integrating, we get $2\sqrt{h} = -0.00003255t + c$, so $h = (c_1 - 0.00001628t)^2$. Setting $h(0) = 2$, we find $c = \sqrt{2}$, so $h(t) = (\sqrt{2} - 0.00001628t)^2$, where h is measured in feet and t in seconds.

113

(b) One hour is 3,600 seconds, so the hour mark should be placed at

$$h(3600) = [\sqrt{2} - 0.00001628(3600)]^2 \approx 1.838 \,\text{ft} \approx 22.0525 \,\text{in.}$$

up from the bottom of the tank. The remaining marks corresponding to the passage of 2, 3, 4, ..., 12 hours are placed at the values shown in the table. The marks are not evenly spaced because the water is not draining out at a uniform rate; that is, $h(t)$ is not a linear function of time.

time (seconds)	height (inches)
0	24.0000
1	22.0520
2	20.1864
3	18.4033
4	16.7026
5	15.0844
6	13.5485
7	12.0952
8	10.7242
9	9.4357
10	8.2297
11	7.1060
12	6.0648

32. (a) In this case $A_w = \pi h^2/4$ and the differential equation is

$$\frac{dh}{dt} = -\frac{1}{7680}\, h^{-3/2}.$$

Separating variables and integrating, we have

$$h^{3/2}\, dh = -\frac{1}{7680}\, dt$$

$$\frac{2}{5}\, h^{5/2} = -\frac{1}{7680}\, t + c_1.$$

Setting $h(0) = 2$ we find $c_1 = 8\sqrt{2}/5$, so that

$$\frac{2}{5}\, h^{5/2} = -\frac{1}{7680}\, t + \frac{8\sqrt{2}}{5},$$

$$h^{5/2} = 4\sqrt{2} - \frac{1}{3072}\, t,$$

and

$$h = \left(4\sqrt{2} - \frac{1}{3072}\, t\right)^{2/5}.$$

(b) In this case $h(4\,\text{hr}) = h(14{,}400\,\text{s}) = 11.8515$ inches and $h(5\,\text{hr}) = h(18{,}000\,\text{s})$ is not a real number. Using a CAS to solve $h(t) = 0$, we see that the tank runs dry at $t \approx 17{,}378\,\text{s} \approx 4.83$ hr. Thus, this particular conical water clock can only measure time intervals of less than 4.83 hours.

33. If we let r_h denote the radius of the hole and $A_w = \pi[f(h)]^2$, then the differential equation $dh/dt = -k\sqrt{h}$, where $k = cA_h\sqrt{2g}/A_w$, becomes

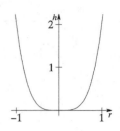

$$\frac{dh}{dt} = -\frac{c\pi r_h^2\sqrt{2g}}{\pi[f(h)]^2}\sqrt{h} = -\frac{8cr_h^2\sqrt{h}}{[f(h)]^2}.$$

For the time marks to be equally spaced, the rate of change of the height must be a constant; that is, $dh/dt = -a$. (The constant is negative because the height is decreasing.) Thus

$$-a = -\frac{8cr_h^2\sqrt{h}}{[f(h)]^2}, \qquad [f(h)]^2 = \frac{8cr_h^2\sqrt{h}}{a}, \qquad \text{and} \qquad r = f(h) = 2r_h\sqrt{\frac{2c}{a}}\,h^{1/4}.$$

Solving for h, we have

$$h = \frac{a^2}{64c^2r_h^4}\,r^4.$$

The shape of the tank with $c = 0.6$, $a = 2$ ft/12 hr $= 1$ ft/21,600 s, and $r_h = 1/32(12) = 1/384$ is shown in the above figure.

34. (*This is a Contributed Problem and the solution has been provided by the authors of the problem.*)

(a) Answers will vary

(b) Answers will vary. This sample data is from Data from "Growth of Sunflower Seeds" by H.S. Reed and R.H. Holland, Proc. Nat. Acad. Sci., Volume 5, 1919, page 140. as quoted in http://math.arizona.edu/~dsl/bflower.htm

day	height
7	17.93
14	36.36
21	67.76
28	98.10
35	131.00
42	169.50
49	205.50
56	228.30
63	247.10
70	250.50
77	253.80
84	254.50

(c)

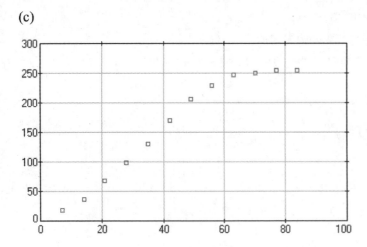

(d) In the case of the sample data, it looks more like logistic growth, with C = 255 cm. C is the height of the flower when it is fully grown.

(e) For our sample data:

day	height	dH/dt	k estimate
7	17.93	2.633	0.000619
14	36.36	3.559	0.000448
21	67.76	4.410	0.000348
28	98.10	4.517	0.000293
35	131.00	5.100	0.000314
42	169.50	5.321	0.000367
49	205.50	4.200	0.000413
56	228.30	2.971	0.000487
63	247.10	1.586	0.000812
70	250.50	0.479	0.000425
77	253.80	0.286	0.000938
84	254.50	0.100	0.000786

We average the k values to obtain $k \approx 0.000521$. An argument can be made for dropping the first two and last two estimates, to obtain $k \approx 0.000432$.

(f) The solution is $y = \dfrac{255}{1 + Ke^{-.133t}}$. We use the height of the sunflower at day 42 to

obtain $y = \dfrac{255}{1 + 133.697e^{-.133t}}$.

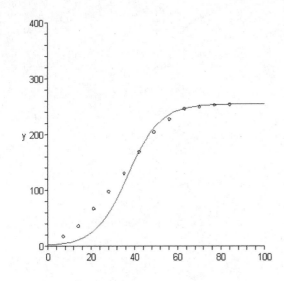

35. (*This is a Contributed Problem and the solution has been provided by the author of the problem.*)

(a) Direction field and the solution curve sketch together:

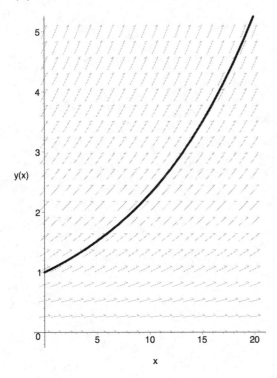

(b) The solution is $P(t) = e^{kt}, k = 1/12$, with graph:

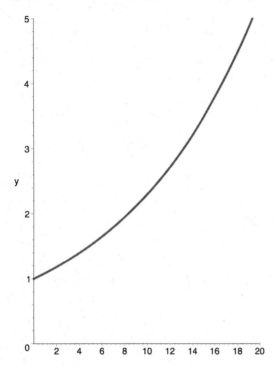

(c) the DE has the constant zero function as equilibrium.

(d) The population grows to infinity.

(e) If the initial population is P_0 then the resulting population would be $P(t) = P_0 e^{kt}, k = 1/12$,

(f) The solution would change from constant to exponential.

(g) Direction field with solution sketch.

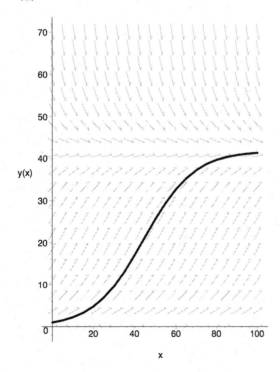

(**h**) The solution to the IVP is

$$P = \frac{125}{3 + 122e^{-t/12}}$$

and the graph is

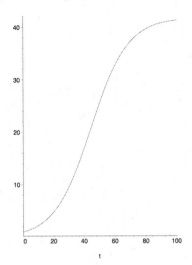

(**i**) the constant solutions to the DE are the zero function and the 125/3 function.

(**j**) solutions tend to 125/3.

(**k**) If the initial population is P_0 then the resulting population could be expressed by

$$P = \frac{125}{3 + 125Ce^{-t/12}}$$

where

$$C = \frac{1}{P_0} - \frac{3}{125}.$$

(**l**) the solution would no longer be constant but tend to 125/3.

(**m**) there would be little change...the new solution would still tend to 125/3.

(**n**) Direction field with solution sketch.

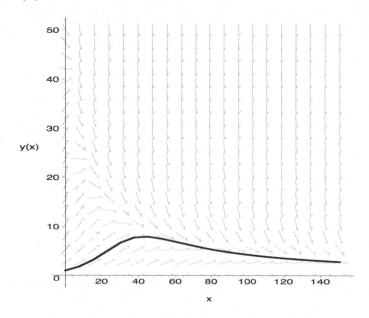

(**o**) the zero function is the only constant solution.
(**p**) The solution is slowly approaching 0; a change to $P(0)$ would still result in a solution curve which tends to 0.

Exercises 3.3

Modeling with Systems of First-Order DEs

1. The linear equation $dx/dt = -\lambda_1 x$ can be solved by either separation of variables or by an integrating factor. Integrating both sides of $dx/x = -\lambda_1 dt$ we obtain $\ln|x| = -\lambda_1 t + c$ from which we get $x = c_1 e^{-\lambda_1 t}$. Using $x(0) = x_0$ we find $c_1 = x_0$ so that $x = x_0 e^{-\lambda_1 t}$. Substituting this result into the second differential equation we have

$$\frac{dy}{dt} + \lambda_2 y = \lambda_1 x_0 e^{-\lambda_1 t}$$

which is linear. An integrating factor is $e^{\lambda_2 t}$ so that

$$\frac{d}{dt}\left[e^{\lambda_2 t} y\right] = \lambda_1 x_0 e^{(\lambda_2 - \lambda_1)t} + c_2$$

$$y = \frac{\lambda_1 x_0}{\lambda_2 - \lambda_1} e^{(\lambda_2 - \lambda_1)t} e^{-\lambda_2 t} + c_2 e^{-\lambda_2 t} = \frac{\lambda_1 x_0}{\lambda_2 - \lambda_1} e^{-\lambda_1 t} + c_2 e^{-\lambda_2 t}.$$

Using $y(0) = 0$ we find $c_2 = -\lambda_1 x_0/(\lambda_2 - \lambda_1)$. Thus

$$y = \frac{\lambda_1 x_0}{\lambda_2 - \lambda_1}\left(e^{-\lambda_1 t} - e^{-\lambda_2 t}\right).$$

Substituting this result into the third differential equation we have

$$\frac{dz}{dt} = \frac{\lambda_1 \lambda_2 x_0}{\lambda_2 - \lambda_1} \left(e^{-\lambda_1 t} - e^{-\lambda_2 t} \right).$$

Integrating we find

$$z = -\frac{\lambda_2 x_0}{\lambda_2 - \lambda_1} e^{-\lambda_1 t} + \frac{\lambda_1 x_0}{\lambda_2 - \lambda_1} e^{-\lambda_2 t} + c_3.$$

Using $z(0) = 0$ we find $c_3 = x_0$. Thus

$$z = x_0 \left(1 - \frac{\lambda_2}{\lambda_2 - \lambda_1} e^{-\lambda_1 t} + \frac{\lambda_1}{\lambda_2 - \lambda_1} e^{-\lambda_2 t} \right).$$

2. We see from the graph that the half-life of A is approximately 4.7 days. To determine the half-life of B we use $t = 50$ as a base, since at this time the amount of substance A is so small that it contributes very little to substance B. Now we see from the graph that $y(50) \approx 16.2$ and $y(191) \approx 8.1$. Thus, the half-life of B is approximately 141 days.

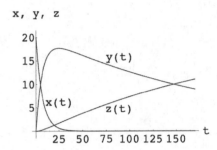

3. The amounts x and y are the same at about $t = 5$ days. The amounts x and z are the same at about $t = 20$ days. The amounts y and z are the same at about $t = 147$ days. The time when y and z are the same makes sense because most of A and half of B are gone, so half of C should have been formed.

4. Suppose that the series is described schematically by $W \Longrightarrow -\lambda_1 X \Longrightarrow -\lambda_2 Y \Longrightarrow -\lambda_3 Z$ where $-\lambda_1$, $-\lambda_2$, and $-\lambda_3$ are the decay constants for W, X and Y, respectively, and Z is a stable element. Let $w(t)$, $x(t)$, $y(t)$, and $z(t)$ denote the amounts of substances W, X, Y, and Z, respectively. A model for the radioactive series is

$$\frac{dw}{dt} = -\lambda_1 w$$

$$\frac{dx}{dt} = \lambda_1 w - \lambda_2 x$$

$$\frac{dy}{dt} = \lambda_2 x - \lambda_3 y$$

$$\frac{dz}{dt} = \lambda_3 y.$$

5. The system is

$$x_1' = 2 \cdot 3 + \frac{1}{50} x_2 - \frac{1}{50} x_1 \cdot 4 = -\frac{2}{25} x_1 + \frac{1}{50} x_2 + 6$$

$$x_2' = \frac{1}{50} x_1 \cdot 4 - \frac{1}{50} x_2 - \frac{1}{50} x_2 \cdot 3 = \frac{2}{25} x_1 - \frac{2}{25} x_2.$$

6. Let x_1, x_2, and x_3 be the amounts of salt in tanks A, B, and C, respectively, so that

$$x_1' = \frac{1}{100}x_2 \cdot 2 - \frac{1}{100}x_1 \cdot 6 = \frac{1}{50}x_2 - \frac{3}{50}x_1$$

$$x_2' = \frac{1}{100}x_1 \cdot 6 + \frac{1}{100}x_3 - \frac{1}{100}x_2 \cdot 2 - \frac{1}{100}x_2 \cdot 5 = \frac{3}{50}x_1 - \frac{7}{100}x_2 + \frac{1}{100}x_3$$

$$x_3' = \frac{1}{100}x_2 \cdot 5 - \frac{1}{100}x_3 - \frac{1}{100}x_3 \cdot 4 = \frac{1}{20}x_2 - \frac{1}{20}x_3.$$

7. (a) A model is

$$\frac{dx_1}{dt} = 3 \cdot \frac{x_2}{100 - t} - 2 \cdot \frac{x_1}{100 + t}, \qquad x_1(0) = 100$$

$$\frac{dx_2}{dt} = 2 \cdot \frac{x_1}{100 + t} - 3 \cdot \frac{x_2}{100 - t}, \qquad x_2(0) = 50.$$

(b) Since the system is closed, no salt enters or leaves the system and $x_1(t) + x_2(t) = 100 + 50 = 150$ for all time. Thus $x_1 = 150 - x_2$ and the second equation in part (a) becomes

$$\frac{dx_2}{dt} = \frac{2(150 - x_2)}{100 + t} - \frac{3x_2}{100 - t} = \frac{300}{100 + t} - \frac{2x_2}{100 + t} - \frac{3x_2}{100 - t}$$

or

$$\frac{dx_2}{dt} + \left(\frac{2}{100 + t} + \frac{3}{100 - t} \right) x_2 = \frac{300}{100 + t},$$

which is linear in x_2. An integrating factor is

$$e^{2 \ln(100+t) - 3 \ln(100-t)} = (100 + t)^2 (100 - t)^{-3}$$

so

$$\frac{d}{dt}[(100 + t)^2 (100 - t)^{-3} x_2] = 300(100 + t)(100 - t)^{-3}.$$

Using integration by parts, we obtain

$$(100 + t)^2 (100 - t)^{-3} x_2 = 300 \left[\frac{1}{2}(100 + t)(100 - t)^{-2} - \frac{1}{2}(100 - t)^{-1} + c \right].$$

Thus

$$x_2 = \frac{300}{(100 + t)^2} \left[c(100 - t)^3 - \frac{1}{2}(100 - t)^2 + \frac{1}{2}(100 + t)(100 - t) \right]$$

$$= \frac{300}{(100 + t)^2}[c(100 - t)^3 + t(100 - t)].$$

Using $x_2(0) = 50$ we find $c = 5/3000$. At $t = 30$, $x_2 = (300/130^2)(70^3 c + 30 \cdot 70) \approx 47.4$ lbs.

8. A model is

$$\frac{dx_1}{dt} = (4 \text{ gal/min})(0 \text{ lb/gal}) - (4 \text{ gal/min})\left(\frac{1}{200}x_1 \text{ lb/gal} \right)$$

$$\frac{dx_2}{dt} = (4 \text{ gal/min})\left(\frac{1}{200}x_1 \text{ lb/gal} \right) - (4 \text{ gal/min})\left(\frac{1}{150}x_2 \text{ lb/gal} \right)$$

$$\frac{dx_3}{dt} = (4 \text{ gal/min})\left(\frac{1}{150}x_2 \text{ lb/gal} \right) - (4 \text{ gal/min})\left(\frac{1}{100}x_3 \text{ lb/gal} \right)$$

or

$$\frac{dx_1}{dt} = -\frac{1}{50}x_1$$

$$\frac{dx_2}{dt} = \frac{1}{50}x_1 - \frac{2}{75}x_2$$

$$\frac{dx_3}{dt} = \frac{2}{75}x_2 - \frac{1}{25}x_3.$$

Over a long period of time we would expect x_1, x_2, and x_3 to approach 0 because the entering pure water should flush the salt out of all three tanks.

9. Zooming in on the graph it can be seen that the populations are first equal at about $t = 5.6$. The approximate periods of x and y are both 45.

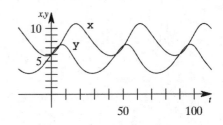

10. (a) The population $y(t)$ approaches 10,000, while the population $x(t)$ approaches extinction.

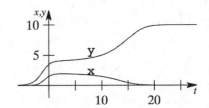

(b) The population $x(t)$ approaches 5,000, while the population $y(t)$ approaches extinction.

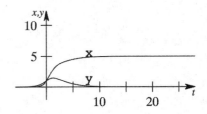

(c) The population $y(t)$ approaches 10,000, while the population $x(t)$ approaches extinction.

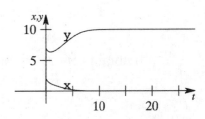

(d) The population $x(t)$ approaches 5,000, while the population $y(t)$ approaches extinction.

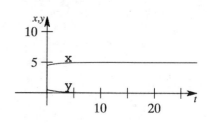

11. (a)

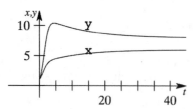

(b)

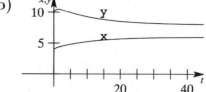

(c)

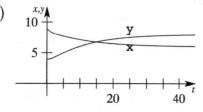

(d)

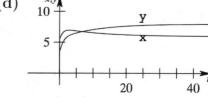

In each case the population $x(t)$ approaches 6,000, while the population $y(t)$ approaches 8,000.

12. By Kirchhoff's first law we have $i_1 = i_2 + i_3$. By Kirchhoff's second law, on each loop we have $E(t) = Li_1' + R_1 i_2$ and $E(t) = Li_1' + R_2 i_3 + q/C$ so that $q = CR_1 i_2 - CR_2 i_3$. Then $i_3 = q' = CR_1 i_2' - CR_2 i_3$ so that the system is

$$Li_2' + Li_3' + R_1 i_2 = E(t)$$

$$-R_1 i_2' + R_2 i_3' + \frac{1}{C} i_3 = 0.$$

13. By Kirchhoff's first law we have $i_1 = i_2 + i_3$. Applying Kirchhoff's second law to each loop we obtain

$$E(t) = i_1 R_1 + L_1 \frac{di_2}{dt} + i_2 R_2$$

and

$$E(t) = i_1 R_1 + L_2 \frac{di_3}{dt} + i_3 R_3.$$

Combining the three equations, we obtain the system

$$L_1 \frac{di_2}{dt} + (R_1 + R_2)i_2 + R_1 i_3 = E$$

$$L_2 \frac{di_3}{dt} + R_1 i_2 + (R_1 + R_3)i_3 = E.$$

124

14. By Kirchhoff's first law we have $i_1 = i_2 + i_3$. By Kirchhoff's second law, on each loop we have $E(t) = Li_1' + Ri_2$ and $E(t) = Li_1' + q/C$ so that $q = CRi_2$. Then $i_3 = q' = CRi_2'$ so that system is

$$Li' + Ri_2 = E(t)$$

$$CRi_2' + i_2 - i_1 = 0.$$

15. We first note that $s(t) + i(t) + r(t) = n$. Now the rate of change of the number of susceptible persons, $s(t)$, is proportional to the number of contacts between the number of people infected and the number who are susceptible; that is, $ds/dt = -k_1 si$. We use $-k_1 < 0$ because $s(t)$ is decreasing. Next, the rate of change of the number of persons who have recovered is proportional to the number infected; that is, $dr/dt = k_2 i$ where $k_2 > 0$ since r is increasing. Finally, to obtain di/dt we use

$$\frac{d}{dt}(s + i + r) = \frac{d}{dt}n = 0.$$

This gives

$$\frac{di}{dt} = -\frac{dr}{dt} - \frac{ds}{dt} = -k_2 i + k_1 si.$$

The system of differential equations is then

$$\frac{ds}{dt} = -k_1 si$$

$$\frac{di}{dt} = -k_2 i + k_1 si$$

$$\frac{dr}{dt} = k_2 i.$$

A reasonable set of initial conditions is $i(0) = i_0$, the number of infected people at time 0, $s(0) = n - i_0$, and $r(0) = 0$.

16. (a) If we know $s(t)$ and $i(t)$ then we can determine $r(t)$ from $s + i + r = n$.

(b) In this case the system is

$$\frac{ds}{dt} = -0.2si$$

$$\frac{di}{dt} = -0.7i + 0.2si.$$

We also note that when $i(0) = i_0$, $s(0) = 10 - i_0$ since $r(0) = 0$ and $i(t) + s(t) + r(t) = 0$ for all values of t. Now $k_2/k_1 = 0.7/0.2 = 3.5$, so we consider initial conditions $s(0) = 2$, $i(0) = 8$; $s(0) = 3.4$, $i(0) = 6.6$; $s(0) = 7$, $i(0) = 3$; and $s(0) = 9$, $i(0) = 1$.

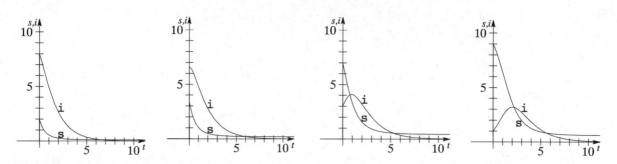

We see that an initial susceptible population greater than k_2/k_1 results in an epidemic in the sense that the number of infected persons increases to a maximum before decreasing to 0. On the other hand, when $s(0) < k_2/k_1$, the number of infected persons decreases from the start and there is no epidemic.

17. Since $x_0 > y_0 > 0$ we have $x(t) > y(t)$ and $y - x < 0$. Thus $dx/dt < 0$ and $dy/dt > 0$. We conclude that $x(t)$ is decreasing and $y(t)$ is increasing. As $t \to \infty$ we expect that $x(t) \to C$ and $y(t) \to C$, where C is a constant common equilibrium concentration.

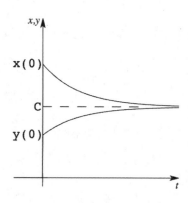

18. We write the system in the form

$$\frac{dx}{dt} = k_1(y - x)$$

$$\frac{dy}{dt} = k_2(x - y),$$

where $k_1 = \kappa/V_A$ and $k_2 = \kappa/V_B$. Letting $z(t) = x(t) - y(t)$ we have

$$\frac{dx}{dt} - \frac{dy}{dt} = k_1(y - x) - k_2(x - y)$$

$$\frac{dz}{dt} = k_1(-z) - k_2 z$$

$$\frac{dz}{dt} + (k_1 + k_2)z = 0.$$

This is a linear first-order differential equation with solution $z(t) = c_1 e^{-(k_1+k_2)t}$. Now

$$\frac{dx}{dt} = -k_1(y - x) = -k_1 z = -k_1 c_1 e^{-(k_1+k_2)t}$$

and

$$x(t) = c_1 \frac{k_1}{k_1 + k_2} e^{-(k_1+k_2)t} + c_2.$$

Since $y(t) = x(t) - z(t)$ we have

$$y(t) = -c_1 \frac{k_2}{k_1 + k_2} e^{-(k_1+k_2)t} + c_2.$$

The initial conditions $x(0) = x_0$ and $y(0) = y_0$ imply

$$c_1 = x_0 - y_0 \qquad \text{and} \qquad c_2 = \frac{x_0 k_2 + y_0 k_1}{k_1 + k_2}.$$

The solution of the system is

$$x(t) = \frac{(x_0 - y_0)k_1}{k_1 + k_2} e^{-(k_1+k_2)t} + \frac{x_0 k_2 + y_0 k_1}{k_1 + k_2}$$

$$y(t) = \frac{(y_0 - x_0)k_2}{k_1 + k_2} e^{-(k_1+k_2)t} + \frac{x_0 k_2 + y_0 k_1}{k_1 + k_2}.$$

As $t \to \infty$, $x(t)$ and $y(t)$ approach the common limit

$$\frac{x_0 k_2 + y_0 k_1}{k_1 + k_2} = \frac{x_0 \kappa/V_B + y_0 \kappa/V_A}{\kappa/V_A + \kappa/V_B} = \frac{x_0 V_A + y_0 V_B}{V_A + V_B}$$

$$= x_0 \frac{V_A}{V_A + V_B} + y_0 \frac{V_B}{V_A + V_B}.$$

This makes intuitive sense because the limiting concentration is seen to be a weighted average of the two initial concentrations.

19. Since there are initially 25 pounds of salt in tank A and none in tank B, and since furthermore only pure water is being pumped into tank A, we would expect that $x_1(t)$ would steadily decrease over time. On the other hand, since salt is being added to tank B from tank A, we would expect $x_2(t)$ to increase over time. However, since pure water is being added to the system at a constant rate and a mixed solution is being pumped out of the system, it makes sense that the amount of salt in both tanks would approach 0 over time.

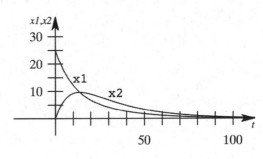

20. We assume here that the temperature, $T(t)$, of the metal bar does not affect the temperature, $T_A(t)$, of the medium in container A. By Newton's law of cooling, then, the differential equations for $T_A(t)$ and $T(t)$ are

$$\frac{dT_A}{dt} = k_A(T_A - T_B), \quad k_A < 0$$

$$\frac{dT}{dt} = k(T - T_A), \quad k < 0,$$

subject to the initial conditions $T(0) = T_0$ and $T_A(0) = T_1$. Separating variables in the first equation, we find $T_A(t) = T_B + c_1 e^{k_A t}$. Using $T_A(0) = T_1$ we find $c_1 = T_1 - T_B$, so

$$T_A(t) = T_B + (T_1 - T_B)e^{k_A t}.$$

Substituting into the second differential equation, we have

$$\frac{dT}{dt} = k(T - T_A) = kT - kT_A = kT - k[T_B + (T_1 - T_B)e^{k_A t}]$$

$$\frac{dT}{dt} - kT = -kT_B - k(T_1 - T_B)e^{k_A t}.$$

This is a linear differential equation with integrating factor $e^{\int -k \, dt} = e^{-kt}$. Then

$$\frac{d}{dt}[e^{-kt}T] = -kT_B e^{-kt} - k(T_1 - T_B)e^{(k_A - k)t}$$

$$e^{-kt}T = T_B e^{-kt} - \frac{k}{k_A - k}(T_1 - T_B)e^{(k_A - k)t} + c_2$$

$$T = T_B - \frac{k}{k_A - k}(T_1 - T_B)e^{k_A t} + c_2 e^{kt}.$$

Using $T(0) = T_0$ we find $c_2 = T_0 - T_B + \dfrac{k}{k_A - k}(T_1 - T_B)$, so

$$T(t) = T_B - \frac{k}{k_A - k}(T_1 - T_B)e^{k_A t} + \left[T_0 - T_B + \frac{k}{k_A - k}(T_1 - T_B)\right]e^{kt}.$$

21. (*This is a Contributed Problem and the solution has been provided by the authors of the problem.*)

(**a**) In the short term there is a mixing of an ethanol solution. In the long term, the system will contain a 20% solution of ethanol.

(**b**)

$$100P'' = \frac{1}{50}P - \frac{1}{10}Q - P'$$

(**c**) First write $Q = 50P' - 30 + P/2$ and then it's straightforward substitution into the equation in (**b**).

(**d**) From equation in (19) we find $P'(0) = 6/10 + 7/50 - 200/100 = -63/50$. The solution is

$$P(t) = \frac{-604}{19}e^{-t/400}\sin(\frac{\sqrt{95}t}{2000})\sqrt{95} - 100e^{-t/400}\cos(\frac{\sqrt{95}t}{2000}) + 100$$

(**e**) The solution is

$$Q(t) = \frac{-270}{19}e^{-t/400}\cos(\frac{\sqrt{95}t}{2000}) - \frac{130}{19}e^{-t/400}\sin(\frac{\sqrt{95}t}{2000})\sqrt{95} + 20 + \frac{23}{19}e^{-t/20}$$

(**f**) In both cases, the there is a concentration of 20% in each tank; $P(t) \to 100$ and $Q(t) \to 20$.

1. The differential equation is $dP/dt = 0.15P$.

2. True. From $dA/dt = kA$, $A(0) = A_0$, we have $A(t) = A_0 e^{kt}$ and $A'(t) = kA_0 e^{kt}$, so $A'(0) = kA_0$. At $T = -(\ln 2)k$,

$$A'(-(\ln 2)/k) = kA(-(\ln 2)/k) = kA_0 e^{k[-(\ln 2)/k]} = kA_0 e^{-\ln 2} = \frac{1}{2}kA_0.$$

3. From $\dfrac{dP}{dt} = 0.018P$ and $P(0) = 4$ billion we obtain $P = 4e^{0.018t}$ so that $P(45) = 8.99$ billion.

4. Let $A = A(t)$ be the volume of CO_2 at time t. From $dA/dt = 1.2 - A/4$ and $A(0) = 16 \text{ ft}^3$ we obtain $A = 4.8 + 11.2e^{-t/4}$. Since $A(10) = 5.7 \text{ ft}^3$, the concentration is 0.017%. As $t \to \infty$ we have $A \to 4.8 \text{ ft}^3$ or 0.06%.

5. Separating variables, we have

$$\frac{\sqrt{s^2 - y^2}}{y} \, dy = -dx.$$

Substituting $y = s \sin \theta$, this becomes

$$\frac{\sqrt{s^2 - s^2 \sin^2 \theta}}{s \sin \theta} (s \cos \theta) d\theta = -dx$$

$$s \int \frac{\cos^2 \theta}{\sin \theta} \, d\theta = -\int dx$$

$$s \int \frac{1 - \sin^2 \theta}{\sin \theta} \, d\theta = -x + c$$

$$s \int (\csc \theta - \sin \theta) d\theta = -x + c$$

$$-s \ln |\csc \theta + \cot \theta| + s \cos \theta = -x + c$$

$$-s \ln \left| \frac{s}{y} + \frac{\sqrt{s^2 - y^2}}{y} \right| + s \frac{\sqrt{s^2 - y^2}}{s} = -x + c.$$

Letting $s = 10$, this is

$$-10 \ln \left| \frac{10}{y} + \frac{\sqrt{100 - y^2}}{y} \right| + \sqrt{100 - y^2} = -x + c.$$

Letting $x = 0$ and $y = 10$ we determine that $c = 0$, so the solution is

$$-10 \ln \left| \frac{10 + \sqrt{100 - y^2}}{y} \right| + \sqrt{100 - y^2} = -x$$

or

$$x = 10 \ln \left| \frac{10 + \sqrt{100 - y^2}}{y} \right| - \sqrt{100 - y^2}.$$

6. From $V \, dC/dt = kA(C_s - C)$ and $C(0) = C_0$ we obtain $C = C_s + (C_0 - C_s)e^{-kAt/V}$.

7. (a) The differential equation

$$\frac{dT}{dt} = k(T - T_m) = k[T - T_2 - B(T_1 - T)]$$

$$= k[(1 + B)T - (BT_1 + T_2)] = k(1 + B)\left(T - \frac{BT_1 + T_2}{1 + B} \right)$$

is autonomous and has the single critical point $(BT_1 + T_2)/(1 + B)$. Since $k < 0$ and $B > 0$, by phase-line analysis it is found that the critical point is an attractor and

$$\lim_{t \to \infty} T(t) = \frac{BT_1 + T_2}{1 + B}.$$

Moreover,

$$\lim_{t \to \infty} T_m(t) = \lim_{t \to \infty} [T_2 + B(T_1 - T)] = T_2 + B\left(T_1 - \frac{BT_1 + T_2}{1 + B} \right) = \frac{BT_1 + T_2}{1 + B}.$$

(b) The differential equation is

$$\frac{dT}{dt} = k(T - T_m) = k(T - T_2 - BT_1 + BT)$$

or

$$\frac{dT}{dt} - k(1 + B)T = -k(BT_1 + T_2).$$

This is linear and has integrating factor $e^{-\int k(1+B)dt} = e^{-k(1+B)t}$. Thus,

$$\frac{d}{dt}[e^{-k(1+B)t}T] = -k(BT_1 + T_2)e^{-k(1+B)t}$$

$$e^{-k(1+B)t}T = \frac{BT_1 + T_2}{1 + B}e^{-k(1+B)t} + c$$

$$T(t) = \frac{BT_1 + T_2}{1 + B} + ce^{k(1+B)t}.$$

Since k is negative, $\lim_{t \to \infty} T(t) = (BT_1 + T_2)/(1 + B)$.

(c) The temperature $T(t)$ decreases to the value $(BT_1 + T_2)/(1 + B)$, whereas $T_m(t)$ increases to $(BT_1 + T_2)/(1 + B)$ as $t \to \infty$. Thus, the temperature $(BT_1 + T_2)/(1 + B)$, (which is a weighted average

$$\frac{B}{1 + B}T_1 + \frac{1}{1 + B}T_2$$

of the two initial temperatures), can be interpreted as an equilibrium temperature. The body cannot get cooler than this value whereas the medium cannot get hotter than this value.

8. By separation of variables and partial fractions,

$$\ln\left|\frac{T - T_m}{T + T_m}\right| - 2\tan^{-1}\left(\frac{T}{T_m}\right) = 4T_m^3 kt + c.$$

Then rewrite the right-hand side of the differential equation as

$$\frac{dT}{dt} = k(T^4 - T_m^4) = [(T_m + (T - T_m))^4 - T_m^4]$$

$$= kT_m^4\left[\left(1 + \frac{T - T_m}{T_m}\right)^4 - 1\right]$$

$$= kT_m^4\left[\left(1 + 4\frac{T - T_m}{T_m} + 6\left(\frac{T - T_m}{T_m}\right)^2 \cdots\right) - 1\right] \leftarrow \text{binomial expansion}$$

When $T - T_m$ is small compared to T_m, every term in the expansion after the first two can be ignored, giving

$$\frac{dT}{dt} \approx k_1(T - T_m), \quad \text{where} \quad k_1 = 4kT_m^3.$$

9. We first solve $(1 - t/10)di/dt + 0.2i = 4$. Separating variables we obtain $di/(40 - 2i) = dt/(10 - t)$. Then

$$-\frac{1}{2}\ln|40 - 2i| = -\ln|10 - t| + c \quad \text{or} \quad \sqrt{40 - 2i} = c_1(10 - t).$$

Since $i(0) = 0$ we must have $c_1 = 2/\sqrt{10}$. Solving for i we get $i(t) = 4t - \frac{1}{5}t^2$, $0 \le t < 10$. For $t \ge 10$ the equation for the current becomes $0.2i = 4$ or $i = 20$. Thus

$$i(t) = \begin{cases} 4t - \frac{1}{5}t^2, & 0 \le t < 10 \\ 20, & t \ge 10. \end{cases}$$

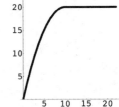

The graph of $i(t)$ is given in the figure.

10. From $y\left[1 + (y')^2\right] = k$ we obtain $dx = (\sqrt{y}/\sqrt{k - y})dy$. If $y = k\sin^2\theta$ then

$$dy = 2k\sin\theta\cos\theta\, d\theta, \quad dx = 2k\left(\frac{1}{2} - \frac{1}{2}\cos 2\theta\right)d\theta, \quad \text{and} \quad x = k\theta - \frac{k}{2}\sin 2\theta + c.$$

If $x = 0$ when $\theta = 0$ then $c = 0$.

11. From $dx/dt = k_1 x(\alpha - x)$ we obtain

$$\left(\frac{1/\alpha}{x} + \frac{1/\alpha}{\alpha - x}\right)dx = k_1\, dt$$

so that $x = \alpha c_1 e^{\alpha k_1 t}/(1 + c_1 e^{\alpha k_1 t})$. From $dy/dt = k_2 xy$ we obtain

$$\ln|y| = \frac{k_2}{k_1}\ln\left|1 + c_1 e^{\alpha k_1 t}\right| + c \quad \text{or} \quad y = c_2\left(1 + c_1 e^{\alpha k_1 t}\right)^{k_2/k_1}.$$

12. In tank A the salt input is

$$\left(7\,\frac{\text{gal}}{\text{min}}\right)\left(2\,\frac{\text{lb}}{\text{gal}}\right)+\left(1\,\frac{\text{gal}}{\text{min}}\right)\left(\frac{x_2}{100}\,\frac{\text{lb}}{\text{gal}}\right)=\left(14+\frac{1}{100}x_2\right)\frac{\text{lb}}{\text{min}}\,.$$

The salt output is

$$\left(3\,\frac{\text{gal}}{\text{min}}\right)\left(\frac{x_1}{100}\,\frac{\text{lb}}{\text{gal}}\right)+\left(5\,\frac{\text{gal}}{\text{min}}\right)\left(\frac{x_1}{100}\,\frac{\text{lb}}{\text{gal}}\right)=\frac{2}{25}x_1\,\frac{\text{lb}}{\text{min}}\,.$$

In tank B the salt input is

$$\left(5\,\frac{\text{gal}}{\text{min}}\right)\left(\frac{x_1}{100}\,\frac{\text{lb}}{\text{gal}}\right)=\frac{1}{20}x_1\,\frac{\text{lb}}{\text{min}}\,.$$

The salt output is

$$\left(1\,\frac{\text{gal}}{\text{min}}\right)\left(\frac{x_2}{100}\,\frac{\text{lb}}{\text{gal}}\right)+\left(4\,\frac{\text{gal}}{\text{min}}\right)\left(\frac{x_2}{100}\,\frac{\text{lb}}{\text{gal}}\right)=\frac{1}{20}x_2\,\frac{\text{lb}}{\text{min}}\,.$$

The system of differential equations is then

$$\frac{dx_1}{dt}=14+\frac{1}{100}x_2-\frac{2}{25}x_1$$

$$\frac{dx_2}{dt}=\frac{1}{20}x_1-\frac{1}{20}x_2.$$

13. From $y=-x-1+c_1e^x$ we obtain $y'=y+x$ so that the differential equation of the orthogonal family is

$$\frac{dy}{dx}=-\frac{1}{y+x}\qquad\text{or}\qquad\frac{dx}{dy}+x=-y.$$

This is a linear differential equation and has integrating factor $e^{\int dy}=e^y$, so

$$\frac{d}{dy}[e^yx]=-ye^y$$

$$e^yx=-ye^y+e^y+c_2$$

$$x=-y+1+c_2e^{-y}.$$

14. Differentiating the family of curves, we have

$$y' = -\frac{1}{(x+c_1)^2} = -\frac{1}{y^2}.$$

The differential equation for the family of orthogonal trajectories is then $y' = y^2$. Separating variables and integrating we get

$$\frac{dy}{y^2} = dx$$

$$-\frac{1}{y} = x + c_1$$

$$y = -\frac{1}{x+c_1}.$$

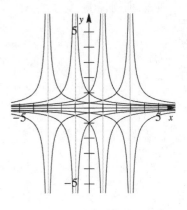

15. (*This is a Contributed Problem and the solution has been provided by the author of the problem.*)

(a) $p(x) = -\rho(x)g\left(y + \dfrac{1}{K}\displaystyle\int q(x)\,dx\right)$

(b) The ratio is increasing. The ratio is constant.

(c) $p(x) = ke^{-(\alpha g\rho/K)x}$

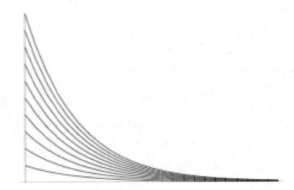

(d) When the pressure p is constant but the density ρ is a function of x then

$$\rho(x) = -\frac{Kp}{g\left(Ky + \int q(x)\,dx\right)}.$$

When the Darcy flux is proportional to the density then

$$\rho = \sqrt{\frac{Kp}{2(CKp - \beta gx)}},$$

where C is an arbitrary constant.

(e) As the density and Darcy velocity decreases, the pressure in the container initially increases but then decreases. The density change is less dramatic than the drop in the velocity and has a greater initial effect on the system. However, as the density of the fluid decreases, the effect is to decrease the pressure.

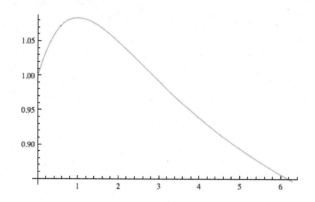

16. (*This is a Contributed Problem and the solution has been provided by the authors of the problem.*)

(**a**) Direction field and the solution curve sketch together:

(**b**) The solution is $P(t) = e^{kt}, k = 1/12$, with graph:

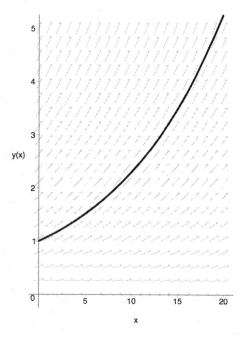

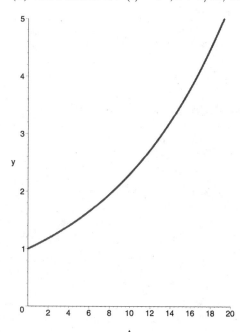

(**c**) the DE has the constant zero function as equilibrium.
(**d**) The population grows to infinity.
(**e**) If the initial population is P_0 then the resulting population would be
$P(t) = P_0 e^{kt}, k = 1/12$,
(**f**) The solution would change from constant to exponential.

(g) Direction field with solution sketch.

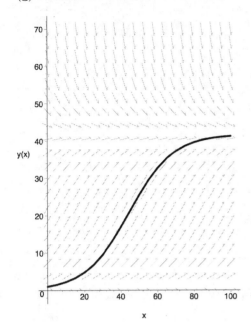

(h) The solution to the IVP is

$$P = \frac{125}{3 + 122e^{-t/12}}$$

and the graph is

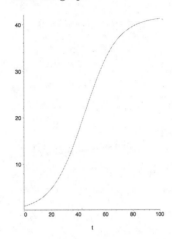

(i) the constant solutions to the DE are the zero function and the 125/3 function.

(j) solutions tend to 125/3.

(k) If the initial population is P_0 then the resulting population could be expressed by

$$P = \frac{125}{3 + 125Ce^{-t/12}}$$

where

$$C = \frac{1}{P_0} - \frac{3}{125}.$$

(l) the solution would no longer be constant but tend to 125/3.

(m) there would be little change...the new solution would still tend to 125/3.

(**n**) Direction field with solution sketch.

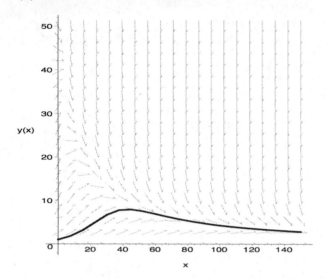

(**o**) the zero function is the only constant solution.

(**p**) The solution is slowly approaching 0; a change to $P(0)$ would still result in a solution curve which tends to 0.

4 Higher-Order Differential Equations

1. From $y = c_1 e^x + c_2 e^{-x}$ we find $y' = c_1 e^x - c_2 e^{-x}$. Then $y(0) = c_1 + c_2 = 0$, $y'(0) = c_1 - c_2 = 1$ so that $c_1 = \frac{1}{2}$ and $c_2 = -\frac{1}{2}$. The solution is $y = \frac{1}{2}e^x - \frac{1}{2}e^{-x}$.

2. From $y = c_1 e^{4x} + c_2 e^{-x}$ we find $y' = 4c_1 e^{4x} - c_2 e^{-x}$. Then $y(0) = c_1 + c_2 = 1$, $y'(0) = 4c_1 - c_2 = 2$ so that $c_1 = \frac{3}{5}$ and $c_2 = \frac{2}{5}$. The solution is $y = \frac{3}{5}e^{4x} + \frac{2}{5}e^{-x}$.

3. From $y = c_1 x + c_2 x \ln x$ we find $y' = c_1 + c_2(1 + \ln x)$. Then $y(1) = c_1 = 3$, $y'(1) = c_1 + c_2 = -1$ so that $c_1 = 3$ and $c_2 = -4$. The solution is $y = 3x - 4x \ln x$.

4. From $y = c_1 + c_2 \cos x + c_3 \sin x$ we find $y' = -c_2 \sin x + c_3 \cos x$ and $y'' = -c_2 \cos x - c_3 \sin x$. Then $y(\pi) = c_1 - c_2 = 0$, $y'(\pi) = -c_3 = 2$, $y''(\pi) = c_2 = -1$ so that $c_1 = -1$, $c_2 = -1$, and $c_3 = -2$. The solution is $y = -1 - \cos x - 2 \sin x$.

5. From $y = c_1 + c_2 x^2$ we find $y' = 2c_2 x$. Then $y(0) = c_1 = 0$, $y'(0) = 2c_2 \cdot 0 = 0$ and hence $y'(0) = 1$ is not possible. Since $a_2(x) = x$ is 0 at $x = 0$, Theorem 4.1 is not violated.

6. In this case we have $y(0) = c_1 = 0$, $y'(0) = 2c_2 \cdot 0 = 0$ so $c_1 = 0$ and c_2 is arbitrary. Two solutions are $y = x^2$ and $y = 2x^2$.

7. From $x(0) = x_0 = c_1$ we see that $x(t) = x_0 \cos \omega t + c_2 \sin \omega t$ and $x'(t) = -x_0 \sin \omega t + c_2 \omega \cos \omega t$. Then $x'(0) = x_1 = c_2 \omega$ implies $c_2 = x_1/\omega$. Thus

$$x(t) = x_0 \cos \omega t + \frac{x_1}{\omega} \sin \omega t.$$

8. Solving the system

$$x(t_0) = c_1 \cos \omega t_0 + c_2 \sin \omega t_0 = x_0$$

$$x'(t_0) = -c_1 \omega \sin \omega t_0 + c_2 \omega \cos \omega t_0 = x_1$$

for c_1 and c_2 gives

$$c_1 = \frac{\omega x_0 \cos \omega t_0 - x_1 \sin \omega t_0}{\omega} \quad \text{and} \quad c_2 = \frac{x_1 \cos \omega t_0 + \omega x_0 \sin \omega t_0}{\omega}.$$

Thus

$$x(t) = \frac{\omega x_0 \cos \omega t_0 - x_1 \sin \omega t_0}{\omega} \cos \omega t + \frac{x_1 \cos \omega t_0 + \omega x_0 \sin \omega t_0}{\omega} \sin \omega t$$

$$= x_0 (\cos \omega t \cos \omega t_0 + \sin \omega t \sin \omega t_0) + \frac{x_1}{\omega} (\sin \omega t \cos \omega t_0 - \cos \omega t \sin \omega t_0)$$

$$= x_0 \cos \omega (t - t_0) + \frac{x_1}{\omega} \sin \omega (t - t_0).$$

9. Since $a_2(x) = x - 2$ and $x_0 = 0$ the problem has a unique solution for $-\infty < x < 2$.

10. Since $a_0(x) = \tan x$ and $x_0 = 0$ the problem has a unique solution for $-\pi/2 < x < \pi/2$.

11. **(a)** We have $y(0) = c_1 + c_2 = 0$, $y(1) = c_1 e + c_2 e^{-1} = 1$ so that $c_1 = e/\left(e^2 - 1\right)$ and $c_2 = -e/\left(e^2 - 1\right)$. The solution is $y = e\left(e^x - e^{-x}\right)/\left(e^2 - 1\right)$.

   (b) We have $y(0) = c_3 \cosh 0 + c_4 \sinh 0 = c_3 = 0$ and $y(1) = c_3 \cosh 1 + c_4 \sinh 1 = c_4 \sinh 1 = 1$, so $c_3 = 0$ and $c_4 = 1/\sinh 1$. The solution is $y = (\sinh x)/(\sinh 1)$.

   (c) Starting with the solution in part (b) we have

$$y = \frac{1}{\sinh 1} \sinh x = \frac{2}{e^1 - e^{-1}} \frac{e^x - e^{-x}}{2} = \frac{e^x - e^{-x}}{e - 1/e} = \frac{e}{e^2 - 1}(e^x - e^{-x}).$$

12. In this case we have $y(0) = c_1 = 1$, $y'(1) = 2c_2 = 6$ so that $c_1 = 1$ and $c_2 = 3$. The solution is $y = 1 + 3x^2$.

13. From $y = c_1 e^x \cos x + c_2 e^x \sin x$ we find $y' = c_1 e^x(-\sin x + \cos x) + c_2 e^x(\cos x + \sin x)$.

   (a) We have $y(0) = c_1 = 1$, $y'(\pi) = -e^\pi(c_1 + c_2) = 0$ so that $c_1 = 1$ and $c_2 = -1$. The solution is $y = e^x \cos x - e^x \sin x$.

   (b) We have $y(0) = c_1 = 1$, $y(\pi) = -e^\pi = -1$, which is not possible.

   (c) We have $y(0) = c_1 = 1$, $y(\pi/2) = c_2 e^{\pi/2} = 1$ so that $c_1 = 1$ and $c_2 = e^{-\pi/2}$. The solution is $y = e^x \cos x + e^{-\pi/2} e^x \sin x$.

   (d) We have $y(0) = c_1 = 0$, $y(\pi) = c_2 e^\pi \sin \pi = 0$ so that $c_1 = 0$ and c_2 is arbitrary. Solutions are $y = c_2 e^x \sin x$, for any real numbers c_2.

14. **(a)** We have $y(-1) = c_1 + c_2 + 3 = 0$, $y(1) = c_1 + c_2 + 3 = 4$, which is not possible.

   (b) We have $y(0) = c_1 \cdot 0 + c_2 \cdot 0 + 3 = 1$, which is not possible.

   (c) We have $y(0) = c_1 \cdot 0 + c_2 \cdot 0 + 3 = 3$, $y(1) = c_1 + c_2 + 3 = 0$ so that c_1 is arbitrary and $c_2 = -3 - c_1$. Solutions are $y = c_1 x^2 - (c_1 + 3)x^4 + 3$.

   (d) We have $y(1) = c_1 + c_2 + 3 = 3$, $y(2) = 4c_1 + 16c_2 + 3 = 15$ so that $c_1 = -1$ and $c_2 = 1$. The solution is $y = -x^2 + x^4 + 3$.

15. Since $(-4)x + (3)x^2 + (1)(4x - 3x^2) = 0$ the set of functions is linearly dependent.

16. Since $(1)0 + (0)x + (0)e^x = 0$ the set of functions is linearly dependent. A similar argument shows that any set of functions containing $f(x) = 0$ will be linearly dependent.

17. Since $(-1/5)5 + (1)\cos^2 x + (1)\sin^2 x = 0$ the set of functions is linearly dependent.

18. Since $(1)\cos 2x + (1)1 + (-2)\cos^2 x = 0$ the set of functions is linearly dependent.

19. Since $(-4)x + (3)(x-1) + (1)(x+3) = 0$ the set of functions is linearly dependent.

20. From the graphs of $f_1(x) = 2 + x$ and $f_2(x) = 2 + |x|$ we see that the set of functions is linearly independent since they cannot be multiples of each other.

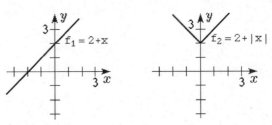

21. Suppose $c_1(1+x) + c_2 x + c_3 x^2 = 0$. Then $c_1 + (c_1 + c_2)x + c_3 x^2 = 0$ and so $c_1 = 0$, $c_1 + c_2 = 0$, and $c_3 = 0$. Since $c_1 = 0$ we also have $c_2 = 0$. Thus, the set of functions is linearly independent.

22. Since $(-1/2)e^x + (1/2)e^{-x} + (1)\sinh x = 0$ the set of functions is linearly dependent.

23. The functions satisfy the differential equation and are linearly independent since

$$W\left(e^{-3x}, e^{4x}\right) = 7e^x \neq 0$$

for $-\infty < x < \infty$. The general solution is

$$y = c_1 e^{-3x} + c_2 e^{4x}.$$

24. The functions satisfy the differential equation and are linearly independent since

$$W(\cosh 2x, \sinh 2x) = 2$$

for $-\infty < x < \infty$. The general solution is

$$y = c_1 \cosh 2x + c_2 \sinh 2x.$$

25. The functions satisfy the differential equation and are linearly independent since

$$W\left(e^x \cos 2x, e^x \sin 2x\right) = 2e^{2x} \neq 0$$

for $-\infty < x < \infty$. The general solution is $y = c_1 e^x \cos 2x + c_2 e^x \sin 2x$.

26. The functions satisfy the differential equation and are linearly independent since

$$W\left(e^{x/2}, xe^{x/2}\right) = e^x \neq 0$$

for $-\infty < x < \infty$. The general solution is

$$y = c_1 e^{x/2} + c_2 xe^{x/2}.$$

27. The functions satisfy the differential equation and are linearly independent since

$$W\left(x^3, x^4\right) = x^6 \neq 0$$

for $0 < x < \infty$. The general solution on this interval is

$$y = c_1 x^3 + c_2 x^4.$$

28. The functions satisfy the differential equation and are linearly independent since

$$W\left(\cos(\ln x), \sin(\ln x)\right) = 1/x \neq 0$$

for $0 < x < \infty$. The general solution on this interval is

$$y = c_1 \cos(\ln x) + c_2 \sin(\ln x).$$

29. The functions satisfy the differential equation and are linearly independent since

$$W\left(x, x^{-2}, x^{-2} \ln x\right) = 9x^{-6} \neq 0$$

for $0 < x < \infty$. The general solution on this interval is

$$y = c_1 x + c_2 x^{-2} + c_3 x^{-2} \ln x.$$

30. The functions satisfy the differential equation and are linearly independent since

$$W(1, x, \cos x, \sin x) = 1$$

for $-\infty < x < \infty$. The general solution on this interval is

$$y = c_1 + c_2 x + c_3 \cos x + c_4 \sin x.$$

31. The functions $y_1 = e^{2x}$ and $y_2 = e^{5x}$ form a fundamental set of solutions of the associated homogeneous equation, and $y_p = 6e^x$ is a particular solution of the nonhomogeneous equation.

32. The functions $y_1 = \cos x$ and $y_2 = \sin x$ form a fundamental set of solutions of the associated homogeneous equation, and $y_p = x \sin x + (\cos x) \ln(\cos x)$ is a particular solution of the nonhomogeneous equation.

33. The functions $y_1 = e^{2x}$ and $y_2 = xe^{2x}$ form a fundamental set of solutions of the associated homogeneous equation, and $y_p = x^2 e^{2x} + x - 2$ is a particular solution of the nonhomogeneous equation.

34. The functions $y_1 = x^{-1/2}$ and $y_2 = x^{-1}$ form a fundamental set of solutions of the associated homogeneous equation, and $y_p = \frac{1}{15}x^2 - \frac{1}{6}x$ is a particular solution of the nonhomogeneous equation.

35. (a) We have $y'_{p_1} = 6e^{2x}$ and $y''_{p_1} = 12e^{2x}$, so

$$y''_{p_1} - 6y'_{p_1} + 5y_{p_1} = 12e^{2x} - 36e^{2x} + 15e^{2x} = -9e^{2x}.$$

Also, $y'_{p2} = 2x + 3$ and $y''_{p2} = 2$, so

$$y''_{p2} - 6y'_{p2} + 5y_{p2} = 2 - 6(2x + 3) + 5(x^2 + 3x) = 5x^2 + 3x - 16.$$

(b) By the superposition principle for nonhomogeneous equations a particular solution of $y'' - 6y' + 5y = 5x^2 + 3x - 16 - 9e^{2x}$ is $y_p = x^2 + 3x + 3e^{2x}$. A particular solution of the second equation is

$$y_p = -2y_{p2} - \frac{1}{9}y_{p1} = -2x^2 - 6x - \frac{1}{3}e^{2x}.$$

36. (a) $y_{p1} = 5$

(b) $y_{p2} = -2x$

(c) $y_p = y_{p1} + y_{p2} = 5 - 2x$

(d) $y_p = \frac{1}{2}y_{p1} - 2y_{p2} = \frac{5}{2} + 4x$

37. (a) Since $D^2 x = 0$, x and 1 are solutions of $y'' = 0$. Since they are linearly independent, the general solution is $y = c_1 x + c_2$.

(b) Since $D^3 x^2 = 0$, x^2, x, and 1 are solutions of $y''' = 0$. Since they are linearly independent, the general solution is $y = c_1 x^2 + c_2 x + c_3$.

(c) Since $D^4 x^3 = 0$, x^3, x^2, x, and 1 are solutions of $y^{(4)} = 0$. Since they are linearly independent, the general solution is $y = c_1 x^3 + c_2 x^2 + c_3 x + c_4$.

(d) By part (a), the general solution of $y'' = 0$ is $y_c = c_1 x + c_2$. Since $D^2 x^2 = 2! = 2$, $y_p = x^2$ is a particular solution of $y'' = 2$. Thus, the general solution is $y = c_1 x + c_2 + x^2$.

(e) By part (b), the general solution of $y''' = 0$ is $y_c = c_1 x^2 + c_2 x + c_3$. Since $D^3 x^3 = 3! = 6$, $y_p = x^3$ is a particular solution of $y''' = 6$. Thus, the general solution is $y = c_1 x^2 + c_2 x + c_3 + x^3$.

(f) By part (c), the general solution of $y^{(4)} = 0$ is $y_c = c_1 x^3 + c_2 x^2 + c_3 x + c_4$. Since $D^4 x^4 = 4! = 24$, $y_p = x^4$ is a particular solution of $y^{(4)} = 24$. Thus, the general solution is $y = c_1 x^3 + c_2 x^2 + c_3 x + c_4 + x^4$.

38. By the superposition principle, if $y_1 = e^x$ and $y_2 = e^{-x}$ are both solutions of a homogeneous linear differential equation, then so are

$$\frac{1}{2}(y_1 + y_2) = \frac{e^x + e^{-x}}{2} = \cosh x \quad \text{and} \quad \frac{1}{2}(y_1 - y_2) = \frac{e^x - e^{-x}}{2} = \sinh x.$$

39. (a) From the graphs of $y_1 = x^3$ and $y_2 = |x|^3$ we see that the functions are linearly independent since they cannot be multiples of each other. It is easily shown that $y_1 = x^3$ is a solution of $x^2 y'' - 4xy' + 6y = 0$. To show that $y_2 = |x|^3$ is a solution let $y_2 = x^3$ for $x \geq 0$ and let $y_2 = -x^3$ for $x < 0$.

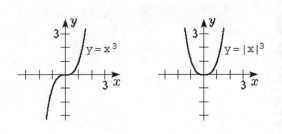

(b) If $x \geq 0$ then $y_2 = x^3$ and

$$W(y_1, y_2) = \begin{vmatrix} x^3 & x^3 \\ 3x^2 & 3x^2 \end{vmatrix} = 0.$$

If $x < 0$ then $y_2 = -x^3$ and

$$W(y_1, y_2) = \begin{vmatrix} x^3 & -x^3 \\ 3x^2 & -3x^2 \end{vmatrix} = 0.$$

This does not violate Theorem 4.1.3 since $a_2(x) = x^2$ is zero at $x = 0$.

(c) The functions $Y_1 = x^3$ and $Y_2 = x^2$ are solutions of $x^2 y'' - 4xy' + 6y = 0$. They are linearly independent since $W\left(x^3, x^2\right) = x^4 \neq 0$ for $-\infty < x < \infty$.

(d) The function $y = x^3$ satisfies $y(0) = 0$ and $y'(0) = 0$.

(e) Neither is the general solution on $(-\infty, \infty)$ since we form a general solution on an interval for which
$a_2(x) \neq 0$ for every x in the interval.

40. Since $e^{x-3} = e^{-3}e^x = (e^{-5}e^2)e^x = e^{-5}e^{x+2}$, we see that e^{x-3} is a constant multiple of e^{x+2} and the set of functions is linearly dependent.

41. Since $0y_1 + 0y_2 + \cdots + 0y_k + 1y_{k+1} = 0$, the set of solutions is linearly dependent.

42. The set of solutions is linearly dependent. Suppose n of the solutions are linearly independent (if not, then the set of $n+1$ solutions is linearly dependent). Without loss of generality, let this set be $y_1, y_2, \ldots, y_n$. Then $y = c_1 y_1 + c_2 y_2 + \cdots + c_n y_n$ is the general solution of the nth-order differential equation and for some choice, $c_1^*, c_2^*, \ldots, c_n^*$, of the coefficients $y_{n+1} = c_1^* y_1 + c_2^* y_2 + \cdots + c_n^* y_n$. But then the set $y_1, y_2, \ldots, y_n, y_{n+1}$ is linearly dependent.

Reduction of Order

In Problems 1-8 we use reduction of order to find a second solution. In Problems 9-16 we use formula (5) from the text.

1. Define $y = u(x)e^{2x}$ so

$$y' = 2ue^{2x} + u'e^{2x}, \quad y'' = e^{2x}u'' + 4e^{2x}u' + 4e^{2x}u, \quad \text{and} \quad y'' - 4y' + 4y = e^{2x}u'' = 0.$$

Therefore $u'' = 0$ and $u = c_1x + c_2$. Taking $c_1 = 1$ and $c_2 = 0$ we see that a second solution is $y_2 = xe^{2x}$.

2. Define $y = u(x)xe^{-x}$ so

$$y' = (1-x)e^{-x}u + xe^{-x}u', \quad y'' = xe^{-x}u'' + 2(1-x)e^{-x}u' - (2-x)e^{-x}u,$$

and

$$y'' + 2y' + y = e^{-x}(xu'' + 2u') = 0 \quad \text{or} \quad u'' + \frac{2}{x}u' = 0.$$

If $w = u'$ we obtain the linear first-order equation $w' + \frac{2}{x}w = 0$ which has the integrating factor $e^{2\int dx/x} = x^2$. Now

$$\frac{d}{dx}[x^2w] = 0 \quad \text{gives} \quad x^2w = c.$$

Therefore $w = u' = c/x^2$ and $u = c_1/x$. A second solution is $y_2 = \frac{1}{x}xe^{-x} = e^{-x}$.

3. Define $y = u(x)\cos 4x$ so

$$y' = -4u\sin 4x + u'\cos 4x, \quad y'' = u''\cos 4x - 8u'\sin 4x - 16u\cos 4x$$

and

$$y'' + 16y = (\cos 4x)u'' - 8(\sin 4x)u' = 0 \quad \text{or} \quad u'' - 8(\tan 4x)u' = 0.$$

If $w = u'$ we obtain the linear first-order equation $w' - 8(\tan 4x)w = 0$ which has the integrating factor $e^{-8\int \tan 4x\, dx} = \cos^2 4x$. Now

$$\frac{d}{dx}[(\cos^2 4x)w] = 0 \quad \text{gives} \quad (\cos^2 4x)w = c.$$

Therefore $w = u' = c\sec^2 4x$ and $u = c_1 \tan 4x$. A second solution is $y_2 = \tan 4x \cos 4x = \sin 4x$.

4. Define $y = u(x)\sin 3x$ so

$$y' = 3u\cos 3x + u'\sin 3x, \quad y'' = u''\sin 3x + 6u'\cos 3x - 9u\sin 3x,$$

143

and

$$y'' + 9y = (\sin 3x)u'' + 6(\cos 3x)u' = 0 \quad \text{or} \quad u'' + 6(\cot 3x)u' = 0.$$

If $w = u'$ we obtain the linear first-order equation $w' + 6(\cot 3x)w = 0$ which has the integrating factor $e^{6\int \cot 3x\, dx} = \sin^2 3x$. Now

$$\frac{d}{dx}[(\sin^2 3x)w] = 0 \quad \text{gives} \quad (\sin^2 3x)w = c.$$

Therefore $w = u' = c\csc^2 3x$ and $u = c_1 \cot 3x$. A second solution is $y_2 = \cot 3x \sin 3x = \cos 3x$.

5. Define $y = u(x)\cosh x$ so

$$y' = u \sinh x + u' \cosh x, \quad y'' = u'' \cosh x + 2u' \sinh x + u \cosh x$$

and

$$y'' - y = (\cosh x)u'' + 2(\sinh x)u' = 0 \quad \text{or} \quad u'' + 2(\tanh x)u' = 0.$$

If $w = u'$ we obtain the linear first-order equation $w' + 2(\tanh x)w = 0$ which has the integrating factor $e^{2\int \tanh x\, dx} = \cosh^2 x$. Now

$$\frac{d}{dx}[(\cosh^2 x)w] = 0 \quad \text{gives} \quad (\cosh^2 x)w = c.$$

Therefore $w = u' = c\,\text{sech}^2 x$ and $u = c \tanh x$. A second solution is $y_2 = \tanh x \cosh x = \sinh x$.

6. Define $y = u(x)e^{5x}$ so

$$y' = 5e^{5x}u + e^{5x}u', \quad y'' = e^{5x}u'' + 10e^{5x}u' + 25e^{5x}u$$

and

$$y'' - 25y = e^{5x}(u'' + 10u') = 0 \quad \text{or} \quad u'' + 10u' = 0.$$

If $w = u'$ we obtain the linear first-order equation $w' + 10w = 0$ which has the integrating factor $e^{10\int dx} = e^{10x}$. Now

$$\frac{d}{dx}[e^{10x}w] = 0 \quad \text{gives} \quad e^{10x}w = c.$$

Therefore $w = u' = ce^{-10x}$ and $u = c_1 e^{-10x}$. A second solution is $y_2 = e^{-10x}e^{5x} = e^{-5x}$.

7. Define $y = u(x)e^{2x/3}$ so

$$y' = \frac{2}{3}e^{2x/3}u + e^{2x/3}u', \quad y'' = e^{2x/3}u'' + \frac{4}{3}e^{2x/3}u' + \frac{4}{9}e^{2x/3}u.$$

and

$$9y'' - 12y' + 4y = 9e^{2x/3}u'' = 0.$$

Therefore $u'' = 0$ and $u = c_1 x + c_2$. Taking $c_1 = 1$ and $c_2 = 0$ we see that a second solution is $y_2 = xe^{2x/3}$.

8. Define $y = u(x)e^{x/3}$ so

$$y' = \frac{1}{3}e^{x/3}u + e^{x/3}u', \quad y'' = e^{x/3}u'' + \frac{2}{3}e^{x/3}u' + \frac{1}{9}e^{x/3}u$$

and

$$6y'' + y' - y = e^{x/3}(6u'' + 5u') = 0 \quad \text{or} \quad u'' + \frac{5}{6}u' = 0.$$

If $w = u'$ we obtain the linear first-order equation $w' + \frac{5}{6}w = 0$ which has the integrating factor $e^{(5/6)\int dx} = e^{5x/6}$. Now

$$\frac{d}{dx}\left[e^{5x/6}w\right] = 0 \quad \text{gives} \quad e^{5x/6}w = c.$$

Therefore $w = u' = ce^{-5x/6}$ and $u = c_1 e^{-5x/6}$. A second solution is $y_2 = e^{-5x/6}e^{x/3} = e^{-x/2}$.

9. Identifying $P(x) = -7/x$ we have

$$y_2 = x^4 \int \frac{e^{-\int(-7/x)\,dx}}{x^8}\,dx = x^4 \int \frac{1}{x}\,dx = x^4 \ln|x|.$$

A second solution is $y_2 = x^4 \ln|x|$.

10. Identifying $P(x) = 2/x$ we have

$$y_2 = x^2 \int \frac{e^{-\int(2/x)\,dx}}{x^4}\,dx = x^2 \int x^{-6}\,dx = -\frac{1}{5}x^{-3}.$$

A second solution is $y_2 = x^{-3}$.

11. Identifying $P(x) = 1/x$ we have

$$y_2 = \ln x \int \frac{e^{-\int dx/x}}{(\ln x)^2}\,dx = \ln x \int \frac{dx}{x(\ln x)^2} = \ln x \left(-\frac{1}{\ln x}\right) = -1.$$

A second solution is $y_2 = 1$.

12. Identifying $P(x) = 0$ we have

$$y_2 = x^{1/2}\ln x \int \frac{e^{-\int 0\,dx}}{x(\ln x)^2}\,dx = x^{1/2}\ln x \left(-\frac{1}{\ln x}\right) = -x^{1/2}.$$

A second solution is $y_2 = x^{1/2}$.

13. Identifying $P(x) = -1/x$ we have

$$y_2 = x\sin(\ln x) \int \frac{e^{-\int -dx/x}}{x^2 \sin^2(\ln x)}\,dx = x\sin(\ln x) \int \frac{x}{x^2 \sin^2(\ln x)}\,dx$$

$$= x\sin(\ln x) \int \frac{\csc^2(\ln x)}{x}\,dx = [x\sin(\ln x)]\,[-\cot(\ln x)] = -x\cos(\ln x).$$

A second solution is $y_2 = x\cos(\ln x)$.

14. Identifying $P(x) = -3/x$ we have

$$y_2 = x^2 \cos(\ln x) \int \frac{e^{-\int -3\,dx/x}}{x^4 \cos^2(\ln x)}\,dx = x^2 \cos(\ln x) \int \frac{x^3}{x^4 \cos^2(\ln x)}\,dx$$

$$= x^2 \cos(\ln x) \int \frac{\sec^2(\ln x)}{x}\,dx = x^2 \cos(\ln x)\tan(\ln x) = x^2 \sin(\ln x).$$

A second solution is $y_2 = x^2 \sin(\ln x)$.

15. Identifying $P(x) = 2(1+x)/\left(1 - 2x - x^2\right)$ we have

$$y_2 = (x+1) \int \frac{e^{-\int 2(1+x)dx/\left(1-2x-x^2\right)}}{(x+1)^2}\, dx = (x+1) \int \frac{e^{\ln\left(1-2x-x^2\right)}}{(x+1)^2}\, dx$$

$$= (x+1) \int \frac{1 - 2x - x^2}{(x+1)^2}\, dx = (x+1) \int \left[\frac{2}{(x+1)^2} - 1\right] dx$$

$$= (x+1)\left[-\frac{2}{x+1} - x\right] = -2 - x^2 - x.$$

A second solution is $y_2 = x^2 + x + 2$.

16. Identifying $P(x) = -2x/\left(1 - x^2\right)$ we have

$$y_2 = \int e^{-\int -2x\, dx/(1-x^2)}\, dx = \int e^{-\ln\left(1-x^2\right)}\, dx = \int \frac{1}{1-x^2}\, dx = \frac{1}{2}\ln\left|\frac{1+x}{1-x}\right|.$$

A second solution is $y_2 = \ln|(1+x)/(1-x)|$.

17. Define $y = u(x)e^{-2x}$ so

$$y' = -2ue^{-2x} + u'e^{-2x}, \quad y'' = u''e^{-2x} - 4u'e^{-2x} + 4ue^{-2x}$$

and

$$y'' - 4y = e^{-2x}u'' - 4e^{-2x}u' = 0 \quad \text{or} \quad u'' - 4u' = 0.$$

If $w = u'$ we obtain the linear first-order equation $w' - 4w = 0$ which has the integrating factor $e^{-4\int dx} = e^{-4x}$. Now

$$\frac{d}{dx}[e^{-4x}w] = 0 \quad \text{gives} \quad e^{-4x}w = c.$$

Therefore $w = u' = ce^{4x}$ and $u = c_1 e^{4x}$. A second solution is $y_2 = e^{-2x}e^{4x} = e^{2x}$. We see by observation that a particular solution is $y_p = -1/2$. The general solution is

$$y = c_1 e^{-2x} + c_2 e^{2x} - \frac{1}{2}.$$

18. Define $y = u(x) \cdot 1$ so

$$y' = u', \quad y'' = u'' \quad \text{and} \quad y'' + y' = u'' + u' = 1.$$

If $w = u'$ we obtain the linear first-order equation $w' + w = 1$ which has the integrating factor $e^{\int dx} = e^x$. Now

$$\frac{d}{dx}[e^x w] = e^x \quad \text{gives} \quad e^x w = e^x + c.$$

Therefore $w = u' = 1 + ce^{-x}$ and $u = x + c_1 e^{-x} + c_2$. The general solution is

$$y = u = x + c_1 e^{-x} + c_2.$$

19. Define $y = u(x)e^x$ so

$$y' = ue^x + u'e^x, \quad y'' = u''e^x + 2u'e^x + ue^x$$

and

$$y'' - 3y' + 2y = e^x u'' - e^x u' = 5e^{3x}.$$

If $w = u'$ we obtain the linear first-order equation $w' - w = 5e^{2x}$ which has the integrating factor $e^{-\int dx} = e^{-x}$. Now

$$\frac{d}{dx}[e^{-x}w] = 5e^x \quad \text{gives} \quad e^{-x}w = 5e^x + c_1.$$

Therefore $w = u' = 5e^{2x} + c_1 e^x$ and $u = \frac{5}{2}e^{2x} + c_1 e^x + c_2$. The general solution is

$$y = ue^x = \frac{5}{2}e^{3x} + c_1 e^{2x} + c_2 e^x.$$

20. Define $y = u(x)e^x$ so

$$y' = ue^x + u'e^x, \quad y'' = u''e^x + 2u'e^x + ue^x$$

and

$$y'' - 4y' + 3y = e^x u'' - e^x u' = x.$$

If $w = u'$ we obtain the linear first-order equation $w' - 2w = xe^{-x}$ which has the integrating factor $e^{-\int 2dx} = e^{-2x}$. Now

$$\frac{d}{dx}[e^{-2x}w] = xe^{-3x} \quad \text{gives} \quad e^{-2x}w = -\frac{1}{3}xe^{-3x} - \frac{1}{9}e^{-3x} + c_1.$$

Therefore $w = u' = -\frac{1}{3}xe^{-x} - \frac{1}{9}e^{-x} + c_1 e^{2x}$ and $u = \frac{1}{3}xe^{-x} + \frac{4}{9}e^{-x} + c_2 e^{2x} + c_3$. The general solution is

$$y = ue^x = \frac{1}{3}x + \frac{4}{9} + c_2 e^{3x} + c_3 e^x.$$

21. (a) For m_1 constant, let $y_1 = e^{m_1 x}$. Then $y_1' = m_1 e^{m_1 x}$ and $y_1'' = m_1^2 e^{m_1 x}$. Substituting into the differential equation we obtain

$$ay_1'' + by_1' + cy_1 = am_1^2 e^{m_1 x} + bm_1 e^{m_1 x} + ce^{m_1 x}$$

$$= e^{m_1 x}(am_1^2 + bm_1 + c) = 0.$$

Thus, $y_1 = e^{m_1 x}$ will be a solution of the differential equation whenever $am_1^2 + bm_1 + c = 0$. Since a quadratic equation always has ~~at least one real or complex roots~~ two roots (real or complex) the differential equation must have a solution of the form $y_1 = e^{m_1 x}$.

(b) Write the differential equation in the form

$$y'' + \frac{b}{a}y' + \frac{c}{a}y = 0,$$

147

and let $y_1 = e^{m_1 x}$ be a solution. Then a second solution is given by

$$y_2 = e^{m_1 x} \int \frac{e^{-bx/a}}{e^{2m_1 x}} \, dx$$

$$= e^{m_1 x} \int e^{-(b/a + 2m_1)x} dx$$

$$= -\frac{1}{b/a + 2m_1} e^{m_1 x} e^{-(b/a + 2m_1)x} \qquad (m_1 \neq -b/2a)$$

$$= -\frac{1}{b/a + 2m_1} e^{-(b/a + m_1)x}.$$

Thus, when $m_1 \neq -b/2a$, a second solution is given by $y_2 = e^{m_2 x}$ where $m_2 = -b/a - m_1$. When $m_1 = -b/2a$ a second solution is given by

$$y_2 = e^{m_1 x} \int dx = x e^{m_1 x}.$$

(c) The functions

$$\sin x = \frac{1}{2i}(e^{ix} - e^{-ix}) \qquad \cos x = \frac{1}{2}(e^{ix} + e^{-ix})$$

$$\sinh x = \frac{1}{2}(e^x - e^{-x}) \qquad \cosh x = \frac{1}{2}(e^x + e^{-x})$$

are all expressible in terms of exponential functions.

22. We have $y_1' = 1$ and $y_1'' = 0$, so $xy_1'' - xy_1' + y_1 = 0 - x + x = 0$ and $y_1(x) = x$ is a solution of the differential equation. Letting $y = u(x)y_1(x) = xu(x)$ we get

$$y' = xu'(x) + u(x) \quad \text{and} \quad y'' = xu''(x) + 2u'(x).$$

Then $xy'' - xy' + y = x^2 u'' + 2xu' - x^2 u' - xu + xu = x^2 u'' - (x^2 - 2x)u' = 0$. If we make the substitution $w = u'$, the linear first-order differential equation becomes $x^2 w' - (x^2 - x)w = 0$, which is separable:

$$\frac{dw}{dx} = \left(1 - \frac{1}{x}\right)w$$

$$\frac{dw}{w} = \left(1 - \frac{1}{x}\right)dx$$

$$\ln w = x - \ln x + c$$

$$w = c_1 \frac{e^x}{x}.$$

Then $u' = c_1 e^x / x$ and $u = c_1 \int e^x dx/x$. To integrate e^x/x we use the series representation for e^x.

Thus, a second solution is

$$y_2 = xu(x) = c_1 x \int \frac{e^x}{x} \, dx$$

$$= c_1 x \int \frac{1}{x} \left(1 + x + \frac{1}{2!}x^2 + \frac{1}{3!}x^3 + \cdots \right) dx$$

$$= c_1 x \int \left(\frac{1}{x} + 1 + \frac{1}{2!}x + \frac{1}{3!}x^2 + \cdots \right) dx$$

$$= c_1 x \left(\ln x + x + \frac{1}{2(2!)}x^2 + \frac{1}{3(3!)}x^3 + \cdots \right)$$

$$= c_1 \left(x \ln x + x^2 + \frac{1}{2(2!)}x^3 + \frac{1}{3(3!)}x^4 + \cdots \right).$$

An interval of definition is probably $(0, \infty)$ because of the $\ln x$ term.

23. (a) We have $y' = y'' = e^x$, so

$$xy'' - (x + 10)y' + 10y = xe^x - (x + 10)e^x + 10e^x = 0,$$

and $y = e^x$ is a solution of the differential equation.

(b) By (5) a second solution is

$$y_2 = y_1 \int \frac{e^{-\int P(x)\,dx}}{y_1^2} \, dx = e^x \int \frac{e^{\int \frac{x+10}{x}\,dx}}{e^{2x}} \, dx = e^x \int \frac{e^{\int (1+10/x)\,dx}}{e^{2x}} \, dx$$

$$= e^x \int \frac{e^{x+\ln x^{10}}}{e^{2x}} \, dx = e^x \int x^{10} e^{-x} \, dx$$

$$= e^x(-3{,}628{,}800 - 3{,}628{,}800x - 1{,}814{,}400x^2 - 604{,}800x^3 - 151{,}200x^4$$

$$- 30{,}240x^5 - 5{,}040x^6 - 720x^7 - 90x^8 - 10x^9 - x^{10})e^{-x}$$

$$= -3{,}628{,}800 - 3{,}628{,}800x - 1{,}814{,}400x^2 - 604{,}800x^3 - 151{,}200x^4$$

$$- 30{,}240x^5 - 5{,}040x^6 - 720x^7 - 90x^8 - 10x^9 - x^{10}.$$

(c) By Corollary **(A)** of Theorem 4.1.2, $-\dfrac{1}{10!} y_2 = \displaystyle\sum_{n=0}^{10} \frac{1}{n!} x^n$ is a solution.

Exercises 4.3

Homogeneous Linear Equations with Constant Coefficients

1. From $4m^2 + m = 0$ we obtain $m = 0$ and $m = -1/4$ so that $y = c_1 + c_2 e^{-x/4}$.

2. From $m^2 - 36 = 0$ we obtain $m = 6$ and $m = -6$ so that $y = c_1 e^{6x} + c_2 e^{-6x}$.

3. From $m^2 - m - 6 = 0$ we obtain $m = 3$ and $m = -2$ so that $y = c_1 e^{3x} + c_2 e^{-2x}$.

4. From $m^2 - 3m + 2 = 0$ we obtain $m = 1$ and $m = 2$ so that $y = c_1 e^x + c_2 e^{2x}$.

5. From $m^2 + 8m + 16 = 0$ we obtain $m = -4$ and $m = -4$ so that $y = c_1 e^{-4x} + c_2 x e^{-4x}$.

6. From $m^2 - 10m + 25 = 0$ we obtain $m = 5$ and $m = 5$ so that $y = c_1 e^{5x} + c_2 x e^{5x}$.

7. From $12m^2 - 5m - 2 = 0$ we obtain $m = -1/4$ and $m = 2/3$ so that $y = c_1 e^{-x/4} + c_2 e^{2x/3}$.

8. From $m^2 + 4m - 1 = 0$ we obtain $m = -2 \pm \sqrt{5}$ so that $y = c_1 e^{(-2+\sqrt{5})x} + c_2 e^{(-2-\sqrt{5})x}$.

9. From $m^2 + 9 = 0$ we obtain $m = 3i$ and $m = -3i$ so that $y = c_1 \cos 3x + c_2 \sin 3x$.

10. From $3m^2 + 1 = 0$ we obtain $m = i/\sqrt{3}$ and $m = -i/\sqrt{3}$ so that $y = c_1 \cos(x/\sqrt{3}) + c_2(\sin x/\sqrt{3})$.

11. From $m^2 - 4m + 5 = 0$ we obtain $m = 2 \pm i$ so that $y = e^{2x}(c_1 \cos x + c_2 \sin x)$.

12. From $2m^2 + 2m + 1 = 0$ we obtain $m = -1/2 \pm i/2$ so that

$$y = e^{-x/2}[c_1 \cos(x/2) + c_2 \sin(x/2)].$$

13. From $3m^2 + 2m + 1 = 0$ we obtain $m = -1/3 \pm \sqrt{2}\, i/3$ so that

$$y = e^{-x/3}[c_1 \cos(\sqrt{2}x/3) + c_2 \sin(\sqrt{2}x/3)].$$

14. From $2m^2 - 3m + 4 = 0$ we obtain $m = 3/4 \pm \sqrt{23}\, i/4$ so that

$$y = e^{3x/4}[c_1 \cos(\sqrt{23}x/4) + c_2 \sin(\sqrt{23}x/4)].$$

15. From $m^3 - 4m^2 - 5m = 0$ we obtain $m = 0$, $m = 5$, and $m = -1$ so that

$$y = c_1 + c_2 e^{5x} + c_3 e^{-x}.$$

16. From $m^3 - 1 = 0$ we obtain $m = 1$ and $m = -1/2 \pm \sqrt{3}\, i/2$ so that

$$y = c_1 e^x + e^{-x/2}[c_2 \cos(\sqrt{3}x/2) + c_3 \sin(\sqrt{3}x/2)].$$

17. From $m^3 - 5m^2 + 3m + 9 = 0$ we obtain $m = -1$, $m = 3$, and $m = 3$ so that

$$y = c_1 e^{-x} + c_2 e^{3x} + c_3 x e^{3x}.$$

18. From $m^3 + 3m^2 - 4m - 12 = 0$ we obtain $m = -2$, $m = 2$, and $m = -3$ so that

$$y = c_1 e^{-2x} + c_2 e^{2x} + c_3 e^{-3x}.$$

19. From $m^3 + m^2 - 2 = 0$ we obtain $m = 1$ and $m = -1 \pm i$ so that

$$u = c_1 e^t + e^{-t}(c_2 \cos t + c_3 \sin t).$$

20. From $m^3 - m^2 - 4 = 0$ we obtain $m = 2$ and $m = -1/2 \pm \sqrt{7}\,i/2$ so that

$$x = c_1 e^{2t} + e^{-t/2}[c_2 \cos(\sqrt{7}t/2) + c_3 \sin(\sqrt{7}t/2)].$$

21. From $m^3 + 3m^2 + 3m + 1 = 0$ we obtain $m = -1$, $m = -1$, and $m = -1$ so that

$$y = c_1 e^{-x} + c_2 x e^{-x} + c_3 x^2 e^{-x}.$$

22. From $m^3 - 6m^2 + 12m - 8 = 0$ we obtain $m = 2$, $m = 2$, and $m = 2$ so that

$$y = c_1 e^{2x} + c_2 x e^{2x} + c_3 x^2 e^{2x}.$$

23. From $m^4 + m^3 + m^2 = 0$ we obtain $m = 0$, $m = 0$, and $m = -1/2 \pm \sqrt{3}\,i/2$ so that

$$y = c_1 + c_2 x + e^{-x/2}[c_3 \cos(\sqrt{3}x/2) + c_4 \sin(\sqrt{3}x/2)].$$

24. From $m^4 - 2m^2 + 1 = 0$ we obtain $m = 1$, $m = 1$, $m = -1$, and $m = -1$ so that

$$y = c_1 e^x + c_2 x e^x + c_3 e^{-x} + c_4 x e^{-x}.$$

25. From $16m^4 + 24m^2 + 9 = 0$ we obtain $m = \pm\sqrt{3}\,i/2$ and $m = \pm\sqrt{3}\,i/2$ so that

$$y = c_1 \cos(\sqrt{3}x/2) + c_2 \sin(\sqrt{3}x/2) + c_3 x \cos(\sqrt{3}x/2) + c_4 x \sin(\sqrt{3}x/2).$$

26. From $m^4 - 7m^2 - 18 = 0$ we obtain $m = 3$, $m = -3$, and $m = \pm\sqrt{2}\,i$ so that

$$y = c_1 e^{3x} + c_2 e^{-3x} + c_3 \cos\sqrt{2}x + c_4 \sin\sqrt{2}x.$$

27. From $m^5 + 5m^4 - 2m^3 - 10m^2 + m + 5 = 0$ we obtain $m = -1$, $m = -1$, $m = 1$, and $m = 1$, and $m = -5$ so that

$$u = c_1 e^{-r} + c_2 r e^{-r} + c_3 e^r + c_4 r e^r + c_5 e^{-5r}.$$

28. From $2m^5 - 7m^4 + 12m^3 + 8m^2 = 0$ we obtain $m = 0$, $m = 0$, $m = -1/2$, and $m = 2 \pm 2i$ so that

$$x = c_1 + c_2 s + c_3 e^{-s/2} + e^{2s}(c_4 \cos 2s + c_5 \sin 2s).$$

29. From $m^2 + 16 = 0$ we obtain $m = \pm 4i$ so that $y = c_1 \cos 4x + c_2 \sin 4x$. If $y(0) = 2$ and $y'(0) = -2$ then $c_1 = 2$, $c_2 = -1/2$, and $y = 2\cos 4x - \frac{1}{2}\sin 4x$.

30. From $m^2 + 1 = 0$ we obtain $m = \pm i$ so that $y = c_1 \cos\theta + c_2 \sin\theta$. If $y(\pi/3) = 0$ and $y'(\pi/3) = 2$ then

$$\frac{1}{2}c_1 + \frac{\sqrt{3}}{2}c_2 = 0$$

$$-\frac{\sqrt{3}}{2}c_1 + \frac{1}{2}c_2 = 2,$$

so $c_1 = -\sqrt{3}$, $c_2 = 1$, and $y = -\sqrt{3}\cos\theta + \sin\theta$.

31. From $m^2 - 4m - 5 = 0$ we obtain $m = -1$ and $m = 5$, so that $y = c_1 e^{-t} + c_2 e^{5t}$. If $y(1) = 0$ and $y'(1) = 2$, then $c_1 e^{-1} + c_2 e^5 = 0$, $-c_1 e^{-1} + 5c_2 e^5 = 2$, so $c_1 = -e/3$, $c_2 = e^{-5}/3$, and $y = -\frac{1}{3}e^{1-t} + \frac{1}{3}e^{5t-5}$.

32. From $4m^2 - 4m - 3 = 0$ we obtain $m = -1/2$ and $m = 3/2$ so that $y = c_1 e^{-x/2} + c_2 e^{3x/2}$. If $y(0) = 1$ and $y'(0) = 5$ then $c_1 + c_2 = 1$, $-\frac{1}{2}c_1 + \frac{3}{2}c_2 = 5$, so $c_1 = -7/4$, $c_2 = 11/4$, and $y = -\frac{7}{4}e^{-x/2} + \frac{11}{4}e^{3x/2}$.

33. From $m^2 + m + 2 = 0$ we obtain $m = -1/2 \pm \sqrt{7}\,i/2$ so that $y = e^{-x/2}[c_1 \cos(\sqrt{7}\,x/2) + c_2 \sin(\sqrt{7}\,x/2)]$. If $y(0) = 0$ and $y'(0) = 0$ then $c_1 = 0$ and $c_2 = 0$ so that $y = 0$.

34. From $m^2 - 2m + 1 = 0$ we obtain $m = 1$ and $m = 1$ so that $y = c_1 e^x + c_2 x e^x$. If $y(0) = 5$ and $y'(0) = 10$ then $c_1 = 5$, $c_1 + c_2 = 10$ so $c_1 = 5$, $c_2 = 5$, and $y = 5e^x + 5xe^x$.

35. From $m^3 + 12m^2 + 36m = 0$ we obtain $m = 0$, $m = -6$, and $m = -6$ so that $y = c_1 + c_2 e^{-6x} + c_3 x e^{-6x}$. If $y(0) = 0$, $y'(0) = 1$, and $y''(0) = -7$ then

$$c_1 + c_2 = 0, \quad -6c_2 + c_3 = 1, \quad 36c_2 - 12c_3 = -7,$$

so $c_1 = 5/36$, $c_2 = -5/36$, $c_3 = 1/6$, and $y = \frac{5}{36} - \frac{5}{36}e^{-6x} + \frac{1}{6}xe^{-6x}$.

36. From $m^3 + 2m^2 - 5m - 6 = 0$ we obtain $m = -1$, $m = 2$, and $m = -3$ so that

$$y = c_1 e^{-x} + c_2 e^{2x} + c_3 e^{-3x}.$$

If $y(0) = 0$, $y'(0) = 0$, and $y''(0) = 1$ then

$$c_1 + c_2 + c_3 = 0, \quad -c_1 + 2c_2 - 3c_3 = 0, \quad c_1 + 4c_2 + 9c_3 = 1,$$

so $c_1 = -1/6$, $c_2 = 1/15$, $c_3 = 1/10$, and

$$y = -\frac{1}{6}e^{-x} + \frac{1}{15}e^{2x} + \frac{1}{10}e^{-3x}.$$

37. From $m^2 - 10m + 25 = 0$ we obtain $m = 5$ and $m = 5$ so that $y = c_1 e^{5x} + c_2 x e^{5x}$. If $y(0) = 1$ and $y(1) = 0$ then $c_1 = 1$, $c_1 e^5 + c_2 e^5 = 0$, so $c_1 = 1$, $c_2 = -1$, and $y = e^{5x} - xe^{5x}$.

38. From $m^2 + 4 = 0$ we obtain $m = \pm 2i$ so that $y = c_1 \cos 2x + c_2 \sin 2x$. If $y(0) = 0$ and $y(\pi) = 0$ then $c_1 = 0$ and $y = c_2 \sin 2x$.

39. From $m^2 + 1 = 0$ we obtain $m = \pm i$ so that $y = c_1 \cos x + c_2 \sin x$ and $y' = -c_1 \sin x + c_2 \cos x$. From $y'(0) = c_1(0) + c_2(1) = c_2 = 0$ and $y'(\pi/2) = -c_1(1) = 0$ we find $c_1 = c_2 = 0$. A solution of the boundary-value problem is $y = 0$.

40. From $m^2 - 2m + 2 = 0$ we obtain $m = 1 \pm i$ so that $y = e^x(c_1 \cos x + c_2 \sin x)$. If $y(0) = 1$ and $y(\pi) = 1$ then $c_1 = 1$ and $y(\pi) = e^\pi \cos\pi = -e^\pi$. Since $-e^\pi \neq 1$, the boundary-value problem has no solution.

41. The auxiliary equation is $m^2 - 3 = 0$ which has roots $-\sqrt{3}$ and $\sqrt{3}$. By (10) the general solution is $y = c_1 e^{\sqrt{3}x} + c_2 e^{-\sqrt{3}x}$. By (11) the general solution is $y = c_1 \cosh\sqrt{3}x + c_2 \sinh\sqrt{3}x$. For $y = c_1 e^{\sqrt{3}x} + c_2 e^{-\sqrt{3}x}$ the initial conditions imply $c_1 + c_2 = 1$, $\sqrt{3}c_1 - \sqrt{3}c_2 = 5$. Solving for c_1 and c_2 we find $c_1 = \frac{1}{2}(1 + 5\sqrt{3})$ and $c_2 = \frac{1}{2}(1 - 5\sqrt{3})$ so $y = \frac{1}{2}(1 + 5\sqrt{3})e^{\sqrt{3}x} + \frac{1}{2}(1 - 5\sqrt{3})e^{-\sqrt{3}x}$. For $y = c_1 \cosh\sqrt{3}x + c_2 \sinh\sqrt{3}x$ the initial conditions imply $c_1 = 1$, $\sqrt{3}c_2 = 5$. Solving for c_1 and c_2 we find $c_1 = 1$ and $c_2 = \frac{5}{3}\sqrt{3}$ so $y = \cosh\sqrt{3}x + \frac{5}{3}\sqrt{3}\sinh\sqrt{3}x$.

42. The auxiliary equation is $m^2 - 1 = 0$ which has roots -1 and 1. By (10) the general solution is $y = c_1 e^x + c_2 e^{-x}$. By (11) the general solution is $y = c_1 \cosh x + c_2 \sinh x$. For $y = c_1 e^x + c_2 e^{-x}$ the boundary conditions imply $c_1 + c_2 = 1$, $c_1 e - c_2 e^{-1} = 0$. Solving for c_1 and c_2 we find $c_1 = 1/(1 + e^2)$ and $c_2 = e^2/(1 + e^2)$ so $y = e^x/(1 + e^2) + e^2 e^{-x}/(1 + e^2)$. For $y = c_1 \cosh x + c_2 \sinh x$ the boundary conditions imply $c_1 = 1$, $c_2 = -\tanh 1$, so $y = \cosh x - (\tanh 1)\sinh x$.

43. The auxiliary equation should have two positive roots, so that the solution has the form $y = c_1 e^{k_1 x} + c_2 e^{k_2 x}$. Thus, the differential equation is (f).

44. The auxiliary equation should have one positive and one negative root, so that the solution has the form $y = c_1 e^{k_1 x} + c_2 e^{-k_2 x}$. Thus, the differential equation is (a).

45. The auxiliary equation should have a pair of complex roots $\alpha \pm \beta i$ where $\alpha < 0$, so that the solution has the form $e^{\alpha x}(c_1 \cos\beta x + c_2 \sin\beta x)$. Thus, the differential equation is (e).

46. The auxiliary equation should have a repeated negative root, so that the solution has the form $y = c_1 e^{-x} + c_2 x e^{-x}$. Thus, the differential equation is (c).

47. The differential equation should have the form $y'' + k^2 y = 0$ where $k = 1$ so that the period of the solution is 2π. Thus, the differential equation is (d).

48. The differential equation should have the form $y'' + k^2 y = 0$ where $k = 2$ so that the period of the solution is π. Thus, the differential equation is (b).

49. Since $(m - 4)(m + 5)^2 = m^3 + 6m^2 - 15m - 100$ the differential equation is $y''' + 6y'' - 15y' - 100y = 0$. The differential equation is not unique since any constant multiple of the left-hand side of the differential equation would lead to the auxiliary roots.

50. A third root must be $m_3 = 3 - i$ and the auxiliary equation is

$$\left(m + \frac{1}{2}\right)[m - (3 + i)][m - (3 - i)] = \left(m + \frac{1}{2}\right)(m^2 - 6x + 10) = m^3 - \frac{11}{2}m^2 + 7m + 5.$$

The differential equation is

$$y''' - \frac{11}{2}y'' + 7y' + 5y = 0.$$

51. From the solution $y_1 = e^{-4x}\cos x$ we conclude that $m_1 = -4 + i$ and $m_2 = -4 - i$ are roots of the auxiliary equation. Hence another solution must be $y_2 = e^{-4x}\sin x$. Now dividing the polynomial

$m^3 + 6m^2 + m - 34$ by $[m - (-4 + i)][m - (-4 - i)] = m^2 + 8m + 17$ gives $m - 2$. Therefore $m_3 = 2$ is the third root of the auxiliary equation, and the general solution of the differential equation is

$$y = c_1 e^{-4x} \cos x + c_2 e^{-4x} \sin x + c_3 e^{2x}.$$

52. Factoring the difference of two squares we obtain

$$m^4 + 1 = (m^2 + 1)^2 - 2m^2 = (m^2 + 1 - \sqrt{2}\,m)(m^2 + 1 + \sqrt{2}\,m) = 0.$$

Using the quadratic formula on each factor we get $m = \pm\sqrt{2}/2 \pm \sqrt{2}\,i/2$. The solution of the differential equation is

$$y(x) = e^{\sqrt{2}\,x/2}\left(c_1 \cos\frac{\sqrt{2}}{2}x + c_2 \sin\frac{\sqrt{2}}{2}x\right) + e^{-\sqrt{2}\,x/2}\left(c_3 \cos\frac{\sqrt{2}}{2}x + c_4 \sin\frac{\sqrt{2}}{2}x\right).$$

53. Using the definition of $\sinh x$ and the formula for the cosine of the sum of two angles, we have

$$y = \sinh x - 2\cos(x + \pi/6)$$

$$= \frac{1}{2}e^x - \frac{1}{2}e^{-x} - 2\left[(\cos x)\left(\cos\frac{\pi}{6}\right) - (\sin x)\left(\sin\frac{\pi}{6}\right)\right]$$

$$= \frac{1}{2}e^x - \frac{1}{2}e^{-x} - 2\left(\frac{\sqrt{3}}{2}\cos x - \frac{1}{2}\sin x\right)$$

$$= \frac{1}{2}e^x - \frac{1}{2}e^{-x} - \sqrt{3}\cos x + \sin x.$$

This form of the solution can be obtained from the general solution $y = c_1 e^x + c_2 e^{-x} + c_3 \cos x + c_4 \sin x$ by choosing $c_1 = \frac{1}{2}$, $c_2 = -\frac{1}{2}$, $c_3 = -\sqrt{3}$, and $c_4 = 1$.

54. The auxiliary equation is $m^2 + \alpha = 0$ and we consider three cases where $\lambda = 0$, $\lambda = \alpha^2 > 0$, and $\lambda = -\alpha^2 < 0$:

Case I When $\alpha = 0$ the general solution of the differential equation is $y = c_1 + c_2 x$. The boundary conditions imply $0 = y(0) = c_1$ and $0 = y(\pi/2) = c_2\pi/2$, so that $c_1 = c_2 = 0$ and the problem possesses only the trivial solution.

Case II When $\lambda = -\alpha^2 < 0$ the general solution of the differential equation is $y = c_1 e^{\alpha x} + c_2 e^{-\alpha x}$, or alternatively, $y = c_1 \cosh \alpha x + c_2 \sinh \alpha x$. Again, $y(0) = 0$ implies $c_1 = 0$ so $y = c_2 \sinh \alpha x$. The second boundary condition implies $0 = y(\pi/2) = c_2 \sinh \alpha \pi/2$ or $c_2 = 0$. In this case also, the problem possesses only the trivial solution.

Case III When $\lambda = \alpha^2 > 0$ the general solution of the differential equation is $y = c_1 \cos \alpha x + c_2 \sin \alpha x$. In this case also, $y(0) = 0$ yields $c_1 = 0$, so that $y = c_2 \sin \alpha x$. The second boundary condition implies $0 = c_2 \sin \alpha \pi/2$. When $\alpha \pi/2$ is an integer multiple of π, that is, when $\alpha = 2k$ for k a nonzero integer, the problem will have nontrivial solutions. Thus, for $\lambda = \alpha^2 = 4k^2$ the boundary-value problem will have nontrivial solutions $y = c_2 \sin 2kx$, where k is a nonzero integer.

On the other hand, when α is not an even integer, the boundary-value problem will have only the trivial solution.

55. Using a CAS to solve the auxiliary equation $m^3 - 6m^2 + 2m + 1$ we find $m_1 = -0.270534$, $m_2 = 0.658675$, and $m_3 = 5.61186$. The general solution is

$$y = c_1 e^{-0.270534x} + c_2 e^{0.658675x} + c_3 e^{5.61186x}.$$

56. Using a CAS to solve the auxiliary equation $6.11m^3 + 8.59m^2 + 7.93m + 0.778 = 0$ we find $m_1 = -0.110241$, $m_2 = -0.647826 + 0.857532i$, and $m_3 = -0.647826 - 0.857532i$. The general solution is

$$y = c_1 e^{-0.110241x} + e^{-0.647826x}(c_2 \cos 0.857532x + c_3 \sin 0.857532x).$$

57. Using a CAS to solve the auxiliary equation $3.15m^4 - 5.34m^2 + 6.33m - 2.03 = 0$ we find $m_1 = -1.74806$, $m_2 = 0.501219$, $m_3 = 0.62342 + 0.588965i$, and $m_4 = 0.62342 - 0.588965i$. The general solution is

$$y = c_1 e^{-1.74806x} + c_2 e^{0.501219x} + e^{0.62342x}(c_3 \cos 0.588965x + c_4 \sin 0.588965x).$$

58. Using a CAS to solve the auxiliary equation $m^4 + 2m^2 - m + 2 = 0$ we find $m_1 = 1/2 + \sqrt{3}\,i/2$, $m_2 = 1/2 - \sqrt{3}\,i/2$, $m_3 = -1/2 + \sqrt{7}\,i/2$, and $m_4 = -1/2 - \sqrt{7}\,i/2$. The general solution is

$$y = e^{x/2}\left(c_1 \cos \frac{\sqrt{3}}{2}x + c_2 \sin \frac{\sqrt{3}}{2}x\right) + e^{-x/2}\left(c_3 \cos \frac{\sqrt{7}}{2}x + c_4 \sin \frac{\sqrt{7}}{2}x\right).$$

59. From $2m^4 + 3m^3 - 16m^2 + 15m - 4 = 0$ we obtain $m = -4$, $m = \frac{1}{2}$, $m = 1$, and $m = 1$, so that $y = c_1 e^{-4x} + c_2 e^{x/2} + c_3 e^x + c_4 x e^x$. If $y(0) = -2$, $y'(0) = 6$, $y''(0) = 3$, and $y'''(0) = \frac{1}{2}$, then

$$c_1 + c_2 + c_3 = -2$$

$$-4c_1 + \frac{1}{2}c_2 + c_3 + c_4 = 6$$

$$16c_1 + \frac{1}{4}c_2 + c_3 + 2c_4 = 3$$

$$-64c_1 + \frac{1}{8}c_2 + c_3 + 3c_4 = \frac{1}{2},$$

so $c_1 = -\frac{4}{75}$, $c_2 = -\frac{116}{3}$, $c_3 = \frac{918}{25}$, $c_4 = -\frac{58}{5}$, and

$$y = -\frac{4}{75}e^{-4x} - \frac{116}{3}e^{x/2} + \frac{918}{25}e^x - \frac{58}{5}xe^x.$$

60. From $m^4 - 3m^3 + 3m^2 - m = 0$ we obtain $m = 0$, $m = 1$, $m = 1$, and $m = 1$ so that $y = c_1 + c_2 e^x + c_3 x e^x + c_4 x^2 e^x$. If $y(0) = 0$, $y'(0) = 0$, $y''(0) = 1$, and $y'''(0) = 1$ then

$$c_1 + c_2 = 0, \quad c_2 + c_3 = 0, \quad c_2 + 2c_3 + 2c_4 = 1, \quad c_2 + 3c_3 + 6c_4 = 1,$$

so $c_1 = 2$, $c_2 = -2$, $c_3 = 2$, $c_4 = -1/2$, and

$$y = 2 - 2e^x + 2xe^x - \frac{1}{2}x^2 e^x.$$

Exercises 4.4

Undetermined Coefficients – Superposition Approach

1. From $m^2 + 3m + 2 = 0$ we find $m_1 = -1$ and $m_2 = -2$. Then $y_c = c_1 e^{-x} + c_2 e^{-2x}$ and we assume $y_p = A$. Substituting into the differential equation we obtain $2A = 6$. Then $A = 3$, $y_p = 3$ and

$$y = c_1 e^{-x} + c_2 e^{-2x} + 3.$$

2. From $4m^2 + 9 = 0$ we find $m_1 = -\frac{3}{2}i$ and $m_2 = \frac{3}{2}i$. Then $y_c = c_1 \cos \frac{3}{2}x + c_2 \sin \frac{3}{2}x$ and we assume $y_p = A$. Substituting into the differential equation we obtain $9A = 15$. Then $A = \frac{5}{3}$, $y_p = \frac{5}{3}$ and

$$y = c_1 \cos \frac{3}{2}x + c_2 \sin \frac{3}{2}x + \frac{5}{3}.$$

3. From $m^2 - 10m + 25 = 0$ we find $m_1 = m_2 = 5$. Then $y_c = c_1 e^{5x} + c_2 x e^{5x}$ and we assume $y_p = Ax + B$. Substituting into the differential equation we obtain $25A = 30$ and $-10A + 25B = 3$. Then $A = \frac{6}{5}$, $B = \frac{3}{5}$, $y_p = \frac{6}{5}x + \frac{3}{5}$, and

$$y = c_1 e^{5x} + c_2 x e^{5x} + \frac{6}{5}x + \frac{3}{5}.$$

4. From $m^2 + m - 6 = 0$ we find $m_1 = -3$ and $m_2 = 2$. Then $y_c = c_1 e^{-3x} + c_2 e^{2x}$ and we assume $y_p = Ax + B$. Substituting into the differential equation we obtain $-6A = 2$ and $A - 6B = 0$. Then $A = -\frac{1}{3}$, $B = -\frac{1}{18}$, $y_p = -\frac{1}{3}x - \frac{1}{18}$, and

$$y = c_1 e^{-3x} + c_2 e^{2x} - \frac{1}{3}x - \frac{1}{18}.$$

5. From $\frac{1}{4}m^2 + m + 1 = 0$ we find $m_1 = m_2 = -2$. Then $y_c = c_1 e^{-2x} + c_2 x e^{-2x}$ and we assume $y_p = Ax^2 + Bx + C$. Substituting into the differential equation we obtain $A = 1$, $2A + B = -2$, and $\frac{1}{2}A + B + C = 0$. Then $A = 1$, $B = -4$, $C = \frac{7}{2}$, $y_p = x^2 - 4x + \frac{7}{2}$, and

$$y = c_1 e^{-2x} + c_2 x e^{-2x} + x^2 - 4x + \frac{7}{2}.$$

6. From $m^2 - 8m + 20 = 0$ we find $m_1 = 4 + 2i$ and $m_2 = 4 - 2i$. Then $y_c = e^{4x}(c_1 \cos 2x + c_2 \sin 2x)$ and we assume $y_p = Ax^2 + Bx + C + (Dx + E)e^x$. Substituting into the differential equation we

obtain

$$2A - 8B + 20C = 0$$

$$-6D + 13E = 0$$

$$-16A + 20B = 0$$

$$13D = -26$$

$$20A = 100.$$

Then $A = 5$, $B = 4$, $C = \frac{11}{10}$, $D = -2$, $E = -\frac{12}{13}$, $y_p = 5x^2 + 4x + \frac{11}{10} + \left(-2x - \frac{12}{13}\right)e^x$ and

$$y = e^{4x}(c_1 \cos 2x + c_2 \sin 2x) + 5x^2 + 4x + \frac{11}{10} + \left(-2x - \frac{12}{13}\right)e^x.$$

7. From $m^2 + 3 = 0$ we find $m_1 = \sqrt{3}\,i$ and $m_2 = -\sqrt{3}\,i$. Then $y_c = c_1 \cos \sqrt{3}\,x + c_2 \sin \sqrt{3}\,x$ and we assume $y_p = (Ax^2 + Bx + C)e^{3x}$. Substituting into the differential equation we obtain $2A + 6B + 12C = 0$, $12A + 12B = 0$, and $12A = -48$. Then $A = -4$, $B = 4$, $C = -\frac{4}{3}$, $y_p = \left(-4x^2 + 4x - \frac{4}{3}\right)e^{3x}$ and

$$y = c_1 \cos \sqrt{3}\,x + c_2 \sin \sqrt{3}\,x + \left(-4x^2 + 4x - \frac{4}{3}\right)e^{3x}.$$

8. From $4m^2 - 4m - 3 = 0$ we find $m_1 = \frac{3}{2}$ and $m_2 = -\frac{1}{2}$. Then $y_c = c_1 e^{3x/2} + c_2 e^{-x/2}$ and we assume $y_p = A \cos 2x + B \sin 2x$. Substituting into the differential equation we obtain $-19 - 8B = 1$ and $8A - 19B = 0$. Then $A = -\frac{19}{425}$, $B = -\frac{8}{425}$, $y_p = -\frac{19}{425} \cos 2x - \frac{8}{425} \sin 2x$, and

$$y = c_1 e^{3x/2} + c_2 e^{-x/2} - \frac{19}{425} \cos 2x - \frac{8}{425} \sin 2x.$$

9. From $m^2 - m = 0$ we find $m_1 = 1$ and $m_2 = 0$. Then $y_c = c_1 e^x + c_2$ and we assume $y_p = Ax$. Substituting into the differential equation we obtain $-A = -3$. Then $A = 3$, $y_p = 3x$ and $y = c_1 e^x + c_2 + 3x$.

10. From $m^2 + 2m = 0$ we find $m_1 = -2$ and $m_2 = 0$. Then $y_c = c_1 e^{-2x} + c_2$ and we assume $y_p = Ax^2 + Bx + Cxe^{-2x}$. Substituting into the differential equation we obtain $2A + 2B = 5$, $4A = 2$, and $-2C = -1$. Then $A = \frac{1}{2}$, $B = 2$, $C = \frac{1}{2}$, $y_p = \frac{1}{2}x^2 + 2x + \frac{1}{2}xe^{-2x}$, and

$$y = c_1 e^{-2x} + c_2 + \frac{1}{2}x^2 + 2x + \frac{1}{2}xe^{-2x}.$$

11. From $m^2 - m + \frac{1}{4} = 0$ we find $m_1 = m_2 = \frac{1}{2}$. Then $y_c = c_1 e^{x/2} + c_2 xe^{x/2}$ and we assume $y_p = A + Bx^2 e^{x/2}$. Substituting into the differential equation we obtain $\frac{1}{4}A = 3$ and $2B = 1$. Then $A = 12$, $B = \frac{1}{2}$, $y_p = 12 + \frac{1}{2}x^2 e^{x/2}$, and

$$y = c_1 e^{x/2} + c_2 xe^{x/2} + 12 + \frac{1}{2}x^2 e^{x/2}.$$

12. From $m^2 - 16 = 0$ we find $m_1 = 4$ and $m_2 = -4$. Then $y_c = c_1 e^{4x} + c_2 e^{-4x}$ and we assume $y_p = Axe^{4x}$. Substituting into the differential equation we obtain $8A = 2$. Then $A = \frac{1}{4}$, $y_p = \frac{1}{4} x e^{4x}$ and

$$y = c_1 e^{4x} + c_2 e^{-4x} + \frac{1}{4} x e^{4x}.$$

13. From $m^2 + 4 = 0$ we find $m_1 = 2i$ and $m_2 = -2i$. Then $y_c = c_1 \cos 2x + c_2 \sin 2x$ and we assume $y_p = Ax \cos 2x + Bx \sin 2x$. Substituting into the differential equation we obtain $4B = 0$ and $-4A = 3$. Then $A = -\frac{3}{4}$, $B = 0$, $y_p = -\frac{3}{4} x \cos 2x$, and

$$y = c_1 \cos 2x + c_2 \sin 2x - \frac{3}{4} x \cos 2x.$$

14. From $m^2 - 4 = 0$ we find $m_1 = 2$ and $m_2 = -2$. Then $y_c = c_1 e^{2x} + c_2 e^{-2x}$ and we assume that $y_p = (Ax^2 + Bx + C) \cos 2x + (Dx^2 + Ex + F) \sin 2x$. Substituting into the differential equation we obtain

$$-8A = 0$$

$$-8B + 8D = 0$$

$$2A - 8C + 4E = 0$$

$$-8D = 1$$

$$-8A - 8E = 0$$

$$-4B + 2D - 8F = -3.$$

Then $A = 0$, $B = -\frac{1}{8}$, $C = 0$, $D = -\frac{1}{8}$, $E = 0$, $F = \frac{13}{32}$, so $y_p = -\frac{1}{8} x \cos 2x + \left(-\frac{1}{8} x^2 + \frac{13}{32} \right) \sin 2x$, and

$$y = c_1 e^{2x} + c_2 e^{-2x} - \frac{1}{8} x \cos 2x + \left(-\frac{1}{8} x^2 + \frac{13}{32} \right) \sin 2x.$$

15. From $m^2 + 1 = 0$ we find $m_1 = i$ and $m_2 = -i$. Then $y_c = c_1 \cos x + c_2 \sin x$ and we assume $y_p = (Ax^2 + Bx) \cos x + (Cx^2 + Dx) \sin x$. Substituting into the differential equation we obtain $4C = 0$, $2A + 2D = 0$, $-4A = 2$, and $-2B + 2C = 0$. Then $A = -\frac{1}{2}$, $B = 0$, $C = 0$, $D = \frac{1}{2}$, $y_p = -\frac{1}{2} x^2 \cos x + \frac{1}{2} x \sin x$, and

$$y = c_1 \cos x + c_2 \sin x - \frac{1}{2} x^2 \cos x + \frac{1}{2} x \sin x.$$

16. From $m^2 - 5m = 0$ we find $m_1 = 5$ and $m_2 = 0$. Then $y_c = c_1 e^{5x} + c_2$ and we assume $y_p = Ax^4 + Bx^3 + Cx^2 + Dx$. Substituting into the differential equation we obtain $-20A = 2$, $12A - 15B = -4$, $6B - 10C = -1$, and $2C - 5D = 6$. Then $A = -\frac{1}{10}$, $B = \frac{14}{75}$, $C = \frac{53}{250}$, $D = -\frac{697}{625}$, $y_p = -\frac{1}{10} x^4 + \frac{14}{75} x^3 + \frac{53}{250} x^2 - \frac{697}{625} x$, and

$$y = c_1 e^{5x} + c_2 - \frac{1}{10} x^4 + \frac{14}{75} x^3 + \frac{53}{250} x^2 - \frac{697}{625} x.$$

17. From $m^2 - 2m + 5 = 0$ we find $m_1 = 1 + 2i$ and $m_2 = 1 - 2i$. Then $y_c = e^x(c_1 \cos 2x + c_2 \sin 2x)$ and we assume $y_p = Axe^x \cos 2x + Bxe^x \sin 2x$. Substituting into the differential equation we obtain $4B = 1$ and $-4A = 0$. Then $A = 0$, $B = \frac{1}{4}$, $y_p = \frac{1}{4}xe^x \sin 2x$, and

$$y = e^x(c_1 \cos 2x + c_2 \sin 2x) + \frac{1}{4}xe^x \sin 2x.$$

18. From $m^2 - 2m + 2 = 0$ we find $m_1 = 1 + i$ and $m_2 = 1 - i$. Then $y_c = e^x(c_1 \cos x + c_2 \sin x)$ and we assume $y_p = Ae^{2x} \cos x + Be^{2x} \sin x$. Substituting into the differential equation we obtain $A + 2B = 1$ and $-2A + B = -3$. Then $A = \frac{7}{5}$, $B = -\frac{1}{5}$, $y_p = \frac{7}{5}e^{2x} \cos x - \frac{1}{5}e^{2x} \sin x$ and

$$y = e^x(c_1 \cos x + c_2 \sin x) + \frac{7}{5}e^{2x} \cos x - \frac{1}{5}e^{2x} \sin x.$$

19. From $m^2 + 2m + 1 = 0$ we find $m_1 = m_2 = -1$. Then $y_c = c_1 e^{-x} + c_2 x e^{-x}$ and we assume $y_p = A \cos x + B \sin x + C \cos 2x + D \sin 2x$. Substituting into the differential equation we obtain $2B = 0$, $-2A = 1$, $-3C + 4D = 3$, and $-4C - 3D = 0$. Then $A = -\frac{1}{2}$, $B = 0$, $C = -\frac{9}{25}$, $D = \frac{12}{25}$, $y_p = -\frac{1}{2}\cos x - \frac{9}{25}\cos 2x + \frac{12}{25}\sin 2x$, and

$$y = c_1 e^{-x} + c_2 x e^{-x} - \frac{1}{2}\cos x - \frac{9}{25}\cos 2x + \frac{12}{25}\sin 2x.$$

20. From $m^2 + 2m - 24 = 0$ we find $m_1 = -6$ and $m_2 = 4$. Then $y_c = c_1 e^{-6x} + c_2 e^{4x}$ and we assume $y_p = A + (Bx^2 + Cx)e^{4x}$. Substituting into the differential equation we obtain $-24A = 16$, $2B + 10C = -2$, and $20B = -1$. Then $A = -\frac{2}{3}$, $B = -\frac{1}{20}$, $C = -\frac{19}{100}$, $y_p = -\frac{2}{3} - \left(\frac{1}{20}x^2 + \frac{19}{100}x\right)e^{4x}$, and

$$y = c_1 e^{-6x} + c_2 e^{4x} - \frac{2}{3} - \left(\frac{1}{20}x^2 + \frac{19}{100}x\right)e^{4x}.$$

21. From $m^3 - 6m^2 = 0$ we find $m_1 = m_2 = 0$ and $m_3 = 6$. Then $y_c = c_1 + c_2 x + c_3 e^{6x}$ and we assume $y_p = Ax^2 + B \cos x + C \sin x$. Substituting into the differential equation we obtain $-12A = 3$, $6B - C = -1$, and $B + 6C = 0$. Then $A = -\frac{1}{4}$, $B = -\frac{6}{37}$, $C = \frac{1}{37}$, $y_p = -\frac{1}{4}x^2 - \frac{6}{37}\cos x + \frac{1}{37}\sin x$, and

$$y = c_1 + c_2 x + c_3 e^{6x} - \frac{1}{4}x^2 - \frac{6}{37}\cos x + \frac{1}{37}\sin x.$$

22. From $m^3 - 2m^2 - 4m + 8 = 0$ we find $m_1 = m_2 = 2$ and $m_3 = -2$. Then $y_c = c_1 e^{2x} + c_2 x e^{2x} + c_3 e^{-2x}$ and we assume $y_p = (Ax^3 + Bx^2)e^{2x}$. Substituting into the differential equation we obtain $24A = 6$ and $6A + 8B = 0$. Then $A = \frac{1}{4}$, $B = -\frac{3}{16}$, $y_p = \left(\frac{1}{4}x^3 - \frac{3}{16}x^2\right)e^{2x}$, and

$$y = c_1 e^{2x} + c_2 x e^{2x} + c_3 e^{-2x} + \left(\frac{1}{4}x^3 - \frac{3}{16}x^2\right)e^{2x}.$$

23. From $m^3 - 3m^2 + 3m - 1 = 0$ we find $m_1 = m_2 = m_3 = 1$. Then $y_c = c_1 e^x + c_2 x e^x + c_3 x^2 e^x$ and we assume $y_p = Ax + B + Cx^3 e^x$. Substituting into the differential equation we obtain $-A = 1$,

$3A - B = 0$, and $6C = -4$. Then $A = -1$, $B = -3$, $C = -\frac{2}{3}$, $y_p = -x - 3 - \frac{2}{3}x^3 e^x$, and

$$y = c_1 e^x + c_2 x e^x + c_3 x^2 e^x - x - 3 - \frac{2}{3}x^3 e^x.$$

24. From $m^3 - m^2 - 4m + 4 = 0$ we find $m_1 = 1$, $m_2 = 2$, and $m_3 = -2$. Then $y_c = c_1 e^x + c_2 e^{2x} + c_3 e^{-2x}$ and we assume $y_p = A + Bxe^x + Cxe^{2x}$. Substituting into the differential equation we obtain $4A = 5$, $-3B = -1$, and $4C = 1$. Then $A = \frac{5}{4}$, $B = \frac{1}{3}$, $C = \frac{1}{4}$, $y_p = \frac{5}{4} + \frac{1}{3}xe^x + \frac{1}{4}xe^{2x}$, and

$$y = c_1 e^x + c_2 e^{2x} + c_3 e^{-2x} + \frac{5}{4} + \frac{1}{3}xe^x + \frac{1}{4}xe^{2x}.$$

25. From $m^4 + 2m^2 + 1 = 0$ we find $m_1 = m_3 = i$ and $m_2 = m_4 = -i$. Then $y_c = c_1 \cos x + c_2 \sin x + c_3 x \cos x + c_4 x \sin x$ and we assume $y_p = Ax^2 + Bx + C$. Substituting into the differential equation we obtain $A = 1$, $B = -2$, and $4A + C = 1$. Then $A = 1$, $B = -2$, $C = -3$, $y_p = x^2 - 2x - 3$, and

$$y = c_1 \cos x + c_2 \sin x + c_3 x \cos x + c_4 x \sin x + x^2 - 2x - 3.$$

26. From $m^4 - m^2 = 0$ we find $m_1 = m_2 = 0$, $m_3 = 1$, and $m_4 = -1$. Then $y_c = c_1 + c_2 x + c_3 e^x + c_4 e^{-x}$ and we assume $y_p = Ax^3 + Bx^2 + (Cx^2 + Dx)e^{-x}$. Substituting into the differential equation we obtain $-6A = 4$, $-2B = 0$, $10C - 2D = 0$, and $-4C = 2$. Then $A = -\frac{2}{3}$, $B = 0$, $C = -\frac{1}{2}$, $D = -\frac{5}{2}$, $y_p = -\frac{2}{3}x^3 - \left(\frac{1}{2}x^2 + \frac{5}{2}x\right)e^{-x}$, and

$$y = c_1 + c_2 x + c_3 e^x + c_4 e^{-x} - \frac{2}{3}x^3 - \left(\frac{1}{2}x^2 + \frac{5}{2}x\right)e^{-x}.$$

27. We have $y_c = c_1 \cos 2x + c_2 \sin 2x$ and we assume $y_p = A$. Substituting into the differential equation we find $A = -\frac{1}{2}$. Thus $y = c_1 \cos 2x + c_2 \sin 2x - \frac{1}{2}$. From the initial conditions we obtain $c_1 = 0$ and $c_2 = \sqrt{2}$, so $y = \sqrt{2} \sin 2x - \frac{1}{2}$.

28. We have $y_c = c_1 e^{-2x} + c_2 e^{x/2}$ and we assume $y_p = Ax^2 + Bx + C$. Substituting into the differential equation we find $A = -7$, $B = -19$, and $C = -37$. Thus $y = c_1 e^{-2x} + c_2 e^{x/2} - 7x^2 - 19x - 37$. From the initial conditions we obtain $c_1 = -\frac{1}{5}$ and $c_2 = \frac{186}{5}$, so

$$y = -\frac{1}{5}e^{-2x} + \frac{186}{5}e^{x/2} - 7x^2 - 19x - 37.$$

29. We have $y_c = c_1 e^{-x/5} + c_2$ and we assume $y_p = Ax^2 + Bx$. Substituting into the differential equation we find $A = -3$ and $B = 30$. Thus $y = c_1 e^{-x/5} + c_2 - 3x^2 + 30x$. From the initial conditions we obtain $c_1 = 200$ and $c_2 = -200$, so

$$y = 200e^{-x/5} - 200 - 3x^2 + 30x.$$

30. We have $y_c = c_1 e^{-2x} + c_2 x e^{-2x}$ and we assume $y_p = (Ax^3 + Bx^2)e^{-2x}$. Substituting into the differential equation we find $A = \frac{1}{6}$ and $B = \frac{3}{2}$. Thus $y = c_1 e^{-2x} + c_2 x e^{-2x} + \left(\frac{1}{6}x^3 + \frac{3}{2}x^2\right)e^{-2x}$. From the initial conditions we obtain $c_1 = 2$ and $c_2 = 9$, so

$$y = 2e^{-2x} + 9xe^{-2x} + \left(\frac{1}{6}x^3 + \frac{3}{2}x^2\right)e^{-2x}.$$

31. We have $y_c = e^{-2x}(c_1 \cos x + c_2 \sin x)$ and we assume $y_p = Ae^{-4x}$. Substituting into the differential equation we find $A = 7$. Thus $y = e^{-2x}(c_1 \cos x + c_2 \sin x) + 7e^{-4x}$. From the initial conditions we obtain $c_1 = -10$ and $c_2 = 9$, so

$$y = e^{-2x}(-10 \cos x + 9 \sin x) + 7e^{-4x}.$$

32. We have $y_c = c_1 \cosh x + c_2 \sinh x$ and we assume $y_p = Ax \cosh x + Bx \sinh x$. Substituting into the differential equation we find $A = 0$ and $B = \frac{1}{2}$. Thus

$$y = c_1 \cosh x + c_2 \sinh x + \frac{1}{2}x \sinh x.$$

From the initial conditions we obtain $c_1 = 2$ and $c_2 = 12$, so

$$y = 2 \cosh x + 12 \sinh x + \frac{1}{2}x \sinh x.$$

33. We have $x_c = c_1 \cos \omega t + c_2 \sin \omega t$ and we assume $x_p = At \cos \omega t + Bt \sin \omega t$. Substituting into the differential equation we find $A = -F_0/2\omega$ and $B = 0$. Thus $x = c_1 \cos \omega t + c_2 \sin \omega t - (F_0/2\omega)t \cos \omega t$. From the initial conditions we obtain $c_1 = 0$ and $c_2 = F_0/2\omega^2$, so

$$x = (F_0/2\omega^2) \sin \omega t - (F_0/2\omega)t \cos \omega t.$$

34. We have $x_c = c_1 \cos \omega t + c_2 \sin \omega t$ and we assume $x_p = A \cos \gamma t + B \sin \gamma t$, where $\gamma \neq \omega$. Substituting into the differential equation we find $A = F_0/(\omega^2 - \gamma^2)$ and $B = 0$. Thus

$$x = c_1 \cos \omega t + c_2 \sin \omega t + \frac{F_0}{\omega^2 - \gamma^2} \cos \gamma t.$$

From the initial conditions we obtain $c_1 = -F_0/(\omega^2 - \gamma^2)$ and $c_2 = 0$, so

$$x = -\frac{F_0}{\omega^2 - \gamma^2} \cos \omega t + \frac{F_0}{\omega^2 - \gamma^2} \cos \gamma t.$$

35. We have $y_c = c_1 + c_2 e^x + c_3 x e^x$ and we assume $y_p = Ax + Bx^2 e^x + Ce^{5x}$. Substituting into the differential equation we find $A = 2$, $B = -12$, and $C = \frac{1}{2}$. Thus

$$y = c_1 + c_2 e^x + c_3 x e^x + 2x - 12x^2 e^x + \frac{1}{2}e^{5x}.$$

From the initial conditions we obtain $c_1 = 11$, $c_2 = -11$, and $c_3 = 9$, so

$$y = 11 - 11e^x + 9xe^x + 2x - 12x^2 e^x + \frac{1}{2}e^{5x}.$$

36. We have $y_c = c_1 e^{-2x} + e^x(c_2 \cos \sqrt{3}\,x + c_3 \sin \sqrt{3}\,x)$ and we assume $y_p = Ax + B + Cxe^{-2x}$. Substituting into the differential equation we find $A = \frac{1}{4}$, $B = -\frac{5}{8}$, and $C = \frac{2}{3}$. Thus

$$y = c_1 e^{-2x} + e^x(c_2 \cos \sqrt{3}\,x + c_3 \sin \sqrt{3}\,x) + \frac{1}{4}x - \frac{5}{8} + \frac{2}{3}xe^{-2x}.$$

From the initial conditions we obtain $c_1 = -\frac{23}{12}$, $c_2 = -\frac{59}{24}$, and $c_3 = \frac{17}{72}\sqrt{3}$, so

$$y = -\frac{23}{12}e^{-2x} + e^x\left(-\frac{59}{24}\cos\sqrt{3}\,x + \frac{17}{72}\sqrt{3}\sin\sqrt{3}\,x\right) + \frac{1}{4}x - \frac{5}{8} + \frac{2}{3}xe^{-2x}.$$

37. We have $y_c = c_1\cos x + c_2\sin x$ and we assume $y_p = Ax^2 + Bx + C$. Substituting into the differential equation we find $A = 1$, $B = 0$, and $C = -1$. Thus $y = c_1\cos x + c_2\sin x + x^2 - 1$. From $y(0) = 5$ and $y(1) = 0$ we obtain

$$c_1 - 1 = 5$$

$$(\cos 1)c_1 + (\sin 1)c_2 = 0.$$

Solving this system we find $c_1 = 6$ and $c_2 = -6\cot 1$. The solution of the boundary-value problem is

$$y = 6\cos x - 6(\cot 1)\sin x + x^2 - 1.$$

38. We have $y_c = e^x(c_1\cos x + c_2\sin x)$ and we assume $y_p = Ax + B$. Substituting into the differential equation we find $A = 1$ and $B = 0$. Thus $y = e^x(c_1\cos x + c_2\sin x) + x$. From $y(0) = 0$ and $y(\pi) = \pi$ we obtain

$$c_1 = 0$$

$$\pi - e^\pi c_1 = \pi.$$

Solving this system we find $c_1 = 0$ and c_2 is any real number. The solution of the boundary-value problem is

$$y = c_2 e^x \sin x + x.$$

39. The general solution of the differential equation $y'' + 3y = 6x$ is $y = c_1\cos\sqrt{3}x + c_2\sin\sqrt{3}x + 2x$. The condition $y(0) = 0$ implies $c_1 = 0$ and so $y = c_2\sin\sqrt{3}x + 2x$. The condition $y(1) + y'(1) = 0$ implies $c_2\sin\sqrt{3} + 2 + c_2\sqrt{3}\cos\sqrt{3} + 2 = 0$ so $c_2 = -4/(\sin\sqrt{3} + \sqrt{3}\cos\sqrt{3})$. The solution is

$$y = \frac{-4\sin\sqrt{3}x}{\sin\sqrt{3} + \sqrt{3}\cos\sqrt{3}} + 2x.$$

40. Using the general solution $y = c_1\cos\sqrt{3}x + c_2\sin\sqrt{3}x + 2x$, the boundary conditions $y(0) + y'(0) = 0$, $y(1) = 0$ yield the system

$$c_1 + \sqrt{3}c_2 + 2 = 0$$

$$c_1\cos\sqrt{3} + c_2\sin\sqrt{3} + 2 = 0.$$

Solving gives

$$c_1 = \frac{2(-\sqrt{3} + \sin\sqrt{3})}{\sqrt{3}\cos\sqrt{3} - \sin\sqrt{3}} \quad\text{and}\quad c_2 = \frac{2(1 - \cos\sqrt{3})}{\sqrt{3}\cos\sqrt{3} - \sin\sqrt{3}}.$$

Thus,

$$y = \frac{2(-\sqrt{3}+\sin\sqrt{3})\cos\sqrt{3}x}{\sqrt{3}\cos\sqrt{3}-\sin\sqrt{3}} + \frac{2(1-\cos\sqrt{3})\sin\sqrt{3}x}{\sqrt{3}\cos\sqrt{3}-\sin\sqrt{3}} + 2x.$$

41. We have $y_c = c_1\cos 2x + c_2\sin 2x$ and we assume $y_p = A\cos x + B\sin x$ on $[0, \pi/2]$. Substituting into the differential equation we find $A = 0$ and $B = \frac{1}{3}$. Thus $y = c_1\cos 2x + c_2\sin 2x + \frac{1}{3}\sin x$ on $[0, \pi/2]$. On $(\pi/2, \infty)$ we have $y = c_3\cos 2x + c_4\sin 2x$. From $y(0) = 1$ and $y'(0) = 2$ we obtain

$$c_1 = 1$$

$$\frac{1}{3} + 2c_2 = 2.$$

Solving this system we find $c_1 = 1$ and $c_2 = \frac{5}{6}$. Thus $y = \cos 2x + \frac{5}{6}\sin 2x + \frac{1}{3}\sin x$ on $[0, \pi/2]$. Now continuity of y at $x = \pi/2$ implies

$$\cos\pi + \frac{5}{6}\sin\pi + \frac{1}{3}\sin\frac{\pi}{2} = c_3\cos\pi + c_4\sin\pi$$

or $-1 + \frac{1}{3} = -c_3$. Hence $c_3 = \frac{2}{3}$. Continuity of y' at $x = \pi/2$ implies

$$-2\sin\pi + \frac{5}{3}\cos\pi + \frac{1}{3}\cos\frac{\pi}{2} = -2c_3\sin\pi + 2c_4\cos\pi$$

or $-\frac{5}{3} = -2c_4$. Then $c_4 = \frac{5}{6}$ and the solution of the initial-value problem is

$$y(x) = \begin{cases} \cos 2x + \frac{5}{6}\sin 2x + \frac{1}{3}\sin x, & 0 \le x \le \pi/2 \\ \frac{2}{3}\cos 2x + \frac{5}{6}\sin 2x, & x > \pi/2. \end{cases}$$

42. We have $y_c = e^x(c_1\cos 3x + c_2\sin 3x)$ and we assume $y_p = A$ on $[0, \pi]$. Substituting into the differential equation we find $A = 2$. Thus, $y = e^x(c_1\cos 3x + c_2\sin 3x) + 2$ on $[0, \pi]$. On (π, ∞) we have $y = e^x(c_3\cos 3x + c_4\sin 3x)$. From $y(0) = 0$ and $y'(0) = 0$ we obtain

$$c_1 = -2, \qquad c_1 + 3c_2 = 0.$$

Solving this system, we find $c_1 = -2$ and $c_2 = \frac{2}{3}$. Thus $y = e^x(-2\cos 3x + \frac{2}{3}\sin 3x) + 2$ on $[0, \pi]$. Now, continuity of y at $x = \pi$ implies

$$e^\pi\left(-2\cos 3\pi + \frac{2}{3}\sin 3\pi\right) + 2 = e^\pi(c_3\cos 3\pi + c_4\sin 3\pi)$$

or $2 + 2e^\pi = -c_3 e^\pi$ or $c_3 = -2e^{-\pi}(1 + e^\pi)$. Continuity of y' at π implies

$$\frac{20}{3}e^\pi\sin 3\pi = e^\pi[(c_3 + 3c_4)\cos 3\pi + (-3c_3 + c_4)\sin 3\pi]$$

or $-c_3 e^\pi - 3c_4 e^\pi = 0$. Since $c_3 = -2e^{-\pi}(1 + e^\pi)$ we have $c_4 = \frac{2}{3}e^{-\pi}(1 + e^\pi)$. The solution of the initial-value problem is

$$y(x) = \begin{cases} e^x(-2\cos 3x + \frac{2}{3}\sin 3x) + 2, & 0 \le x \le \pi \\ (1 + e^\pi)e^{x-\pi}(-2\cos 3x + \frac{2}{3}\sin 3x), & x > \pi. \end{cases}$$

43. (a) From $y_p = Ae^{kx}$ we find $y_p' = Ake^{kx}$ and $y_p'' = Ak^2e^{kx}$. Substituting into the differential equation we get

$$aAk^2e^{kx} + bAke^{kx} + cAe^{kx} = (ak^2 + bk + c)Ae^{kx} = e^{kx},$$

so $(ak^2 + bk + c)A = 1$. Since k is not a root of $am^2 + bm + c = 0$, $A = 1/(ak^2 + bk + c)$.

(b) From $y_p = Axe^{kx}$ we find $y_p' = Akxe^{kx} + Ae^{kx}$ and $y_p'' = Ak^2xe^{kx} + 2Ake^{kx}$. Substituting into the differential equation we get

$$aAk^2xe^{kx} + 2aAke^{kx} + bAkxe^{kx} + bAe^{kx} + cAxe^{kx}$$

$$= (ak^2 + bk + c)Axe^{kx} + (2ak + b)Ae^{kx}$$

$$= (0)Axe^{kx} + (2ak + b)Ae^{kx} = (2ak + b)Ae^{kx} = e^{kx}$$

where $ak^2 + bk + c = 0$ because k is a root of the auxiliary equation. Now, the roots of the auxiliary equation are $-b/2a \pm \sqrt{b^2 - 4ac}/2a$, and since k is a root of multiplicity one, $k \neq -b/2a$ and $2ak + b \neq 0$. Thus $(2ak + b)A = 1$ and $A = 1/(2ak + b)$.

(c) If k is a root of multiplicity two, then, as we saw in part (b), $k = -b/2a$ and $2ak + b = 0$. From $y_p = Ax^2e^{kx}$ we find $y_p' = Akx^2e^{kx} + 2Axe^{kx}$ and $y_p'' = Ak^2x^2e^{kx} + 4Akxe^{kx} = 2Ae^{kx}$. Substituting into the differential equation, we get

$$aAk^2x^2e^{kx} + 4aAkxe^{kx} + 2aAe^{kx} + bAkx^2e^{kx} + 2bAxe^{kx} + cAx^2e^{kx}$$

$$= (ak^2 + bk + c)Ax^2e^{kx} + 2(2ak + b)Axe^{kx} + 2aAe^{kx}$$

$$= (0)Ax^2e^{kx} + 2(0)Axe^{kx} + 2aAe^{kx} = 2aAe^{kx} = e^{kx}.$$

Since the differential equation is second order, $a \neq 0$ and $A = 1/(2a)$.

44. Using the double-angle formula for the cosine, we have

$$\sin x \cos 2x = \sin x(\cos^2 x - \sin^2 x) = \sin x(1 - 2\sin^2 x) = \sin x - 2\sin^3 x.$$

Since $\sin x$ is a solution of the related homogeneous differential equation we look for a particular solution of the form $y_p = Ax\sin x + Bx\cos x + C\sin^3 x$. Substituting into the differential equation we obtain

$$2A\cos x + (6C - 2B)\sin x - 8C\sin^3 x = \sin x - 2\sin^3 x.$$

Equating coefficients we find $A = 0$, $C = \frac{1}{4}$, and $B = \frac{1}{4}$. Thus, a particular solution is

$$y_p = \frac{1}{4}x\cos x + \frac{1}{4}\sin^3 x.$$

45. (a) $f(x) = e^x \sin x$. We see that $y_p \to \infty$ as $x \to \infty$ and $y_p \to 0$ as $x \to -\infty$.

(b) $f(x) = e^{-x}$. We see that $y_p \to \infty$ as $x \to \infty$ and $y_p \to \infty$ as $x \to -\infty$.

(c) $f(x) = \sin 2x$. We see that y_p is sinusoidal.

(d) $f(x) = 1$. We see that y_p is constant and simply translates y_c vertically.

46. The complementary function is $y_c = e^{2x}(c_1 \cos 2x + c_2 \sin 2x)$. We assume a particular solution of the form $y_p = (Ax^3 + Bx^2 + Cx)e^{2x} \cos 2x + (Dx^3 + Ex^2 + F)e^{2x} \sin 2x$. Substituting into the differential equation and using a CAS to simplify yields

$$[12Dx^2 + (6A + 8E)x + (2B + 4F)]e^{2x} \cos 2x$$

$$+ [-12Ax^2 + (-8B + 6D)x + (-4C + 2E)]e^{2x} \sin 2x$$

$$= (2x^2 - 3x)e^{2x} \cos 2x + (10x^2 - x - 1)e^{2x} \sin 2x.$$

This gives the system of equations

$$12D = 2, \qquad 6A + 8E = -3, \qquad 2B + 4F = 0,$$

$$-12A = 10, \qquad -8B + 6D = -1, \qquad -4C + 2E = -1,$$

from which we find $A = -\frac{5}{6}$, $B = \frac{1}{4}$, $C = \frac{3}{8}$, $D = \frac{1}{6}$, $E = \frac{1}{4}$, and $F = -\frac{1}{8}$. Thus, a particular solution of the differential equation is

$$y_p = \left(-\frac{5}{6}x^3 + \frac{1}{4}x^2 + \frac{3}{8}x\right)e^{2x} \cos 2x + \left(\frac{1}{6}x^3 + \frac{1}{4}x^2 - \frac{1}{8}x\right)e^{2x} \sin 2x.$$

47. The complementary function is $y_c = c_1 \cos x + c_2 \sin x + c_3 x \cos x + c_4 x \sin x$. We assume a particular solution of the form $y_p = Ax^2 \cos x + Bx^3 \sin x$. Substituting into the differential equation and using a CAS to simplify yields

$$(-8A + 24B) \cos x + 3Bx \sin x = 2 \cos x - 3x \sin x.$$

This implies $-8A + 24B = 2$ and $-24B = -3$. Thus $B = \frac{1}{8}$, $A = \frac{1}{8}$, and $y_p = \frac{1}{8}x^2 \cos x + \frac{1}{8}x^3 \sin x$.

Exercises 4.5   Undetermined Coefficients - Annihilator Approach

1. $(9D^2 - 4)y = (3D - 2)(3D + 2)y = \sin x$

2. $(D^2 - 5)y = (D - \sqrt{5})(D + \sqrt{5})y = x^2 - 2x$

3. $(D^2 - 4D - 12)y = (D - 6)(D + 2)y = x - 6$

4. $(2D^2 - 3D - 2)y = (2D + 1)(D - 2)y = 1$

5. $(D^3 + 10D^2 + 25D)y = D(D + 5)^2 y = e^x$

6. $(D^3 + 4D)y = D(D^2 + 4)y = e^x \cos 2x$

7. $(D^3 + 2D^2 - 13D + 10)y = (D-1)(D-2)(D+5)y = xe^{-x}$

8. $(D^3 + 4D^2 + 3D)y = D(D+1)(D+3)y = x^2 \cos x - 3x$

9. $(D^4 + 8D)y = D(D+2)(D^2 - 2D + 4)y = 4$

10. $(D^4 - 8D^2 + 16)y = (D-2)^2(D+2)^2 y = (x^3 - 2x)e^{4x}$

11. $D^4 y = D^4(10x^3 - 2x) = D^3(30x^2 - 2) = D^2(60x) = D(60) = 0$

12. $(2D-1)y = (2D-1)4e^{x/2} = 8De^{x/2} - 4e^{x/2} = 4e^{x/2} - 4e^{x/2} = 0$

13. $(D-2)(D+5)(e^{2x} + 3e^{-5x}) = (D-2)(2e^{2x} - 15e^{-5x} + 5e^{2x} + 15e^{-5x}) = (D-2)7e^{2x} = 14e^{2x} - 14e^{2x} = 0$

14. $(D^2 + 64)(2\cos 8x - 5\sin 8x) = D(-16\sin 8x - 40\cos 8x) + 64(2\cos 8x - 5\sin 8x)$

$$= -128\cos 8x + 320\sin 8x + 128\cos 8x - 320\sin 8x = 0$$

15. D^4 because of x^3

16. D^5 because of x^4

17. $D(D-2)$ because of 1 and e^{2x}

18. $D^2(D-6)^2$ because of x and xe^{6x}

19. $D^2 + 4$ because of $\cos 2x$

20. $D(D^2 + 1)$ because of 1 and $\sin x$

21. $D^3(D^2 + 16)$ because of x^2 and $\sin 4x$

22. $D^2(D^2 + 1)(D^2 + 25)$ because of x, $\sin x$, and $\cos 5x$

23. $(D+1)(D-1)^3$ because of e^{-x} and $x^2 e^x$

24. $D(D-1)(D-2)$ because of 1, e^x, and e^{2x}

25. $D(D^2 - 2D + 5)$ because of 1 and $e^x \cos 2x$

26. $(D^2 + 2D + 2)(D^2 - 4D + 5)$ because of $e^{-x}\sin x$ and $e^{2x}\cos x$

27. 1, x, x^2, x^3, x^4

28. $D^2 + 4D = D(D+4)$; 1, e^{-4x}

29. e^{6x}, $e^{-3x/2}$

30. $D^2 - 9D - 36 = (D-12)(D+3)$; e^{12x}, e^{-3x}

31. $\cos\sqrt{5}\,x$, $\sin\sqrt{5}\,x$

32. $D^2 - 6D + 10 = D^2 - 2(3)D + (3^2 + 1^2)$; $e^{3x}\cos x$, $e^{3x}\sin x$

33. $D^3 - 10D^2 + 25D = D(D-5)^2$; 1, e^{5x}, xe^{5x}

34. 1, x, e^{5x}, e^{7x}

35. Applying D to the differential equation we obtain

$$D(D^2 - 9)y = 0.$$

166

Then

$$y = \underbrace{c_1 e^{3x} + c_2 e^{-3x}}_{y_c} + c_3$$

and $y_p = A$. Substituting y_p into the differential equation yields $-9A = 54$ or $A = -6$. The general solution is

$$y = c_1 e^{3x} + c_2 e^{-3x} - 6.$$

36. Applying D to the differential equation we obtain

$$D(2D^2 - 7D + 5)y = 0.$$

Then

$$y = \underbrace{c_1 e^{5x/2} + c_2 e^{x}}_{y_c} + c_3$$

and $y_p = A$. Substituting y_p into the differential equation yields $5A = -29$ or $A = -29/5$. The general solution is

$$y = c_1 e^{5x/2} + c_2 e^{x} - \frac{29}{5}.$$

37. Applying D to the differential equation we obtain

$$D(D^2 + D)y = D^2(D+1)y = 0.$$

Then

$$y = \underbrace{c_1 + c_2 e^{-x}}_{y_c} + c_3 x$$

and $y_p = Ax$. Substituting y_p into the differential equation yields $A = 3$. The general solution is

$$y = c_1 + c_2 e^{-3x} + 3x.$$

38. Applying D to the differential equation we obtain

$$D(D^3 + 2D^2 + D)y = D^2(D+1)^2 y = 0.$$

Then

$$y = \underbrace{c_1 + c_2 e^{-x} + c_3 x e^{-x}}_{y_c} + c_4 x$$

and $y_p = Ax$. Substituting y_p into the differential equation yields $A = 10$. The general solution is

$$y = c_1 + c_2 e^{-x} + c_3 x e^{-x} + 10x.$$

39. Applying D^2 to the differential equation we obtain

$$D^2(D^2 + 4D + 4)y = D^2(D+2)^2 y = 0.$$

Then
$$y = \underbrace{c_1 e^{-2x} + c_2 x e^{-2x}}_{y_c} + c_3 + c_4 x$$

and $y_p = Ax + B$. Substituting y_p into the differential equation yields $4Ax + (4A + 4B) = 2x + 6$. Equating coefficients gives

$$4A = 2$$

$$4A + 4B = 6.$$

Then $A = 1/2$, $B = 1$, and the general solution is

$$y = c_1 e^{-2x} + c_2 x e^{-2x} + \frac{1}{2}x + 1.$$

40. Applying D^2 to the differential equation we obtain

$$D^2(D^2 + 3D)y = D^3(D + 3)y = 0.$$

Then

$$y = \underbrace{c_1 + c_2 e^{-3x}}_{y_c} + c_3 x^2 + c_4 x$$

and $y_p = Ax^2 + Bx$. Substituting y_p into the differential equation yields $6Ax + (2A + 3B) = 4x - 5$. Equating coefficients gives

$$6A = 4$$

$$2A + 3B = -5.$$

Then $A = 2/3$, $B = -19/9$, and the general solution is

$$y = c_1 + c_2 e^{-3x} + \frac{2}{3}x^2 - \frac{19}{9}x.$$

41. Applying D^3 to the differential equation we obtain

$$D^3(D^3 + D^2)y = D^5(D + 1)y = 0.$$

Then

$$y = \underbrace{c_1 + c_2 x + c_3 e^{-x}}_{y_c} + c_4 x^4 + c_5 x^3 + c_6 x^2$$

and $y_p = Ax^4 + Bx^3 + Cx^2$. Substituting y_p into the differential equation yields

$$12Ax^2 + (24A + 6B)x + (6B + 2C) = 8x^2.$$

Equating coefficients gives

$$12A = 8$$

$$24A + 6B = 0$$

$$6B + 2C = 0.$$

Then $A = 2/3$, $B = -8/3$, $C = 8$, and the general solution is

$$y = c_1 + c_2 x + c_3 e^{-x} + \frac{2}{3}x^4 - \frac{8}{3}x^3 + 8x^2.$$

42. Applying D^4 to the differential equation we obtain

$$D^4(D^2 - 2D + 1)y = D^4(D - 1)^2 y = 0.$$

Then

$$y = \underbrace{c_1 e^x + c_2 x e^x}_{y_c} + c_3 x^3 + c_4 x^2 + c_5 x + c_6$$

and $y_p = Ax^3 + Bx^2 + Cx + E$. Substituting y_p into the differential equation yields

$$Ax^3 + (B - 6A)x^2 + (6A - 4B + C)x + (2B - 2C + E) = x^3 + 4x.$$

Equating coefficients gives

$$A = 1$$

$$B - 6A = 0$$

$$6A - 4B + C = 4$$

$$2B - 2C + E = 0.$$

Then $A = 1$, $B = 6$, $C = 22$, $E = 32$, and the general solution is

$$y = c_1 e^x + c_2 x e^x + x^3 + 6x^2 + 22x + 32.$$

43. Applying $D - 4$ to the differential equation we obtain

$$(D - 4)(D^2 - D - 12)y = (D - 4)^2(D + 3)y = 0.$$

Then

$$y = \underbrace{c_1 e^{4x} + c_2 e^{-3x}}_{y_c} + c_3 x e^{4x}$$

and $y_p = Axe^{4x}$. Substituting y_p into the differential equation yields $7Ae^{4x} = e^{4x}$. Equating coefficients gives $A = 1/7$. The general solution is

$$y = c_1 e^{4x} + c_2 e^{-3x} + \frac{1}{7}xe^{4x}.$$

44. Applying $D - 6$ to the differential equation we obtain

$$(D - 6)(D^2 + 2D + 2)y = 0.$$

Then

$$y = \underbrace{e^{-x}(c_1 \cos x + c_2 \sin x)}_{y_c} + c_3 e^{6x}$$

and $y_p = Ae^{6x}$. Substituting y_p into the differential equation yields $50Ae^{6x} = 5e^{6x}$. Equating coefficients gives $A = 1/10$. The general solution is

$$y = e^{-x}(c_1 \cos x + c_2 \sin x) + \frac{1}{10}e^{6x}.$$

45. Applying $D(D-1)$ to the differential equation we obtain

$$D(D-1)(D^2 - 2D - 3)y = D(D-1)(D+1)(D-3)y = 0.$$

Then

$$y = \underbrace{c_1 e^{3x} + c_2 e^{-x}}_{y_c} + c_3 e^x + c_4$$

and $y_p = Ae^x + B$. Substituting y_p into the differential equation yields $-4Ae^x - 3B = 4e^x - 9$. Equating coefficients gives $A = -1$ and $B = 3$. The general solution is

$$y = c_1 e^{3x} + c_2 e^{-x} - e^x + 3.$$

46. Applying $D^2(D+2)$ to the differential equation we obtain

$$D^2(D+2)(D^2 + 6D + 8)y = D^2(D+2)^2(D+4)y = 0.$$

Then

$$y = \underbrace{c_1 e^{-2x} + c_2 e^{-4x}}_{y_c} + c_3 xe^{-2x} + c_4 x + c_5$$

and $y_p = Axe^{-2x} + Bx + C$. Substituting y_p into the differential equation yields

$$2Ae^{-2x} + 8Bx + (6B + 8C) = 3e^{-2x} + 2x.$$

Equating coefficients gives

$$2A = 3$$

$$8B = 2$$

$$6B + 8C = 0.$$

Then $A = 3/2$, $B = 1/4$, $C = -3/16$, and the general solution is

$$y = c_1 e^{-2x} + c_2 e^{-4x} + \frac{3}{2}xe^{-2x} + \frac{1}{4}x - \frac{3}{16}.$$

47. Applying $D^2 + 1$ to the differential equation we obtain

$$(D^2 + 1)(D^2 + 25)y = 0.$$

Then

$$y = \underbrace{c_1 \cos 5x + c_2 \sin 5x}_{y_c} + c_3 \cos x + c_4 \sin x$$

and $y_p = A\cos x + B\sin x$. Substituting y_p into the differential equation yields

$$24A\cos x + 24B\sin x = 6\sin x.$$

Equating coefficients gives $A = 0$ and $B = 1/4$. The general solution is

$$y = c_1 \cos 5x + c_2 \sin 5x + \frac{1}{4}\sin x.$$

48. Applying $D(D^2 + 1)$ to the differential equation we obtain

$$D(D^2 + 1)(D^2 + 4)y = 0.$$

Then

$$y = \underbrace{c_1 \cos 2x + c_2 \sin 2x}_{y_c} + c_3 \cos x + c_4 \sin x + c_5$$

and $y_p = A\cos x + B\sin x + C$. Substituting y_p into the differential equation yields

$$3A\cos x + 3B\sin x + 4C = 4\cos x + 3\sin x - 8.$$

Equating coefficients gives $A = 4/3$, $B = 1$, and $C = -2$. The general solution is

$$y = c_1 \cos 2x + c_2 \sin 2x + \frac{4}{3}\cos x + \sin x - 2.$$

49. Applying $(D - 4)^2$ to the differential equation we obtain

$$(D - 4)^2(D^2 + 6D + 9)y = (D - 4)^2(D + 3)^2 y = 0.$$

Then

$$y = \underbrace{c_1 e^{-3x} + c_2 x e^{-3x}}_{y_c} + c_3 x e^{4x} + c_4 e^{4x}$$

and $y_p = Axe^{4x} + Be^{4x}$. Substituting y_p into the differential equation yields

$$49Axe^{4x} + (14A + 49B)e^{4x} = -xe^{4x}.$$

Equating coefficients gives

$$49A = -1$$

$$14A + 49B = 0.$$

Then $A = -1/49$, $B = 2/343$, and the general solution is

$$y = c_1 e^{-3x} + c_2 x e^{-3x} - \frac{1}{49}xe^{4x} + \frac{2}{343}e^{4x}.$$

50. Applying $D^2(D - 1)^2$ to the differential equation we obtain

$$D^2(D - 1)^2(D^2 + 3D - 10)y = D^2(D - 1)^2(D - 2)(D + 5)y = 0.$$

171

Then

$$y = \underbrace{c_1 e^{2x} + c_2 e^{-5x}}_{y_c} + c_3 x e^x + c_4 e^x + c_5 x + c_6$$

and $y_p = Axe^x + Be^x + Cx + E$. Substituting y_p into the differential equation yields

$$-6Axe^x + (5A - 6B)e^x - 10Cx + (3C - 10E) = xe^x + x.$$

Equating coefficients gives

$$-6A = 1$$

$$5A - 6B = 0$$

$$-10C = 1$$

$$3C - 10E = 0.$$

Then $A = -1/6$, $B = -5/36$, $C = -1/10$, $E = -3/100$, and the general solution is

$$y = c_1 e^{2x} + c_2 e^{-5x} - \frac{1}{6} x e^x - \frac{5}{36} e^x - \frac{1}{10} x - \frac{3}{100}.$$

51. Applying $D(D-1)^3$ to the differential equation we obtain

$$D(D-1)^3(D^2 - 1)y = D(D-1)^4(D+1)y = 0.$$

Then

$$y = \underbrace{c_1 e^x + c_2 e^{-x}}_{y_c} + c_3 x^3 e^x + c_4 x^2 e^x + c_5 x e^x + c_6$$

and $y_p = Ax^3 e^x + Bx^2 e^x + Cxe^x + E$. Substituting y_p into the differential equation yields

$$6Ax^2 e^x + (6A + 4B)xe^x + (2B + 2C)e^x - E = x^2 e^x + 5.$$

Equating coefficients gives

$$6A = 1$$

$$6A + 4B = 0$$

$$2B + 2C = 0$$

$$-E = 5.$$

Then $A = 1/6$, $B = -1/4$, $C = 1/4$, $E = -5$, and the general solution is

$$y = c_1 e^x + c_2 e^{-x} + \frac{1}{6} x^3 e^x - \frac{1}{4} x^2 e^x + \frac{1}{4} x e^x - 5.$$

52. Applying $(D+1)^3$ to the differential equation we obtain

$$(D+1)^3(D^2 + 2D + 1)y = (D+1)^5 y = 0.$$

172

Then

$$y = \underbrace{c_1 e^{-x} + c_2 x e^{-x}}_{y_c} + c_3 x^4 e^{-x} + c_4 x^3 e^{-x} + c_5 x^2 e^{-x}$$

and $y_p = A x^4 e^{-x} + B x^3 e^{-x} + C x^2 e^{-x}$. Substituting y_p into the differential equation yields

$$12 A x^2 e^{-x} + 6 B x e^{-x} + 2 C e^{-x} = x^2 e^{-x}.$$

Equating coefficients gives $A = \frac{1}{12}$, $B = 0$, and $C = 0$. The general solution is

$$y = c_1 e^{-x} + c_2 x e^{-x} + \frac{1}{12} x^4 e^{-x}.$$

53. Applying $D^2 - 2D + 2$ to the differential equation we obtain

$$(D^2 - 2D + 2)(D^2 - 2D + 5)y = 0.$$

Then

$$y = \underbrace{e^x (c_1 \cos 2x + c_2 \sin 2x)}_{y_c} + e^x (c_3 \cos x + c_4 \sin x)$$

and $y_p = A e^x \cos x + B e^x \sin x$. Substituting y_p into the differential equation yields

$$3 A e^x \cos x + 3 B e^x \sin x = e^x \sin x.$$

Equating coefficients gives $A = 0$ and $B = 1/3$. The general solution is

$$y = e^x (c_1 \cos 2x + c_2 \sin 2x) + \frac{1}{3} e^x \sin x.$$

54. Applying $D^2 - 2D + 10$ to the differential equation we obtain

$$(D^2 - 2D + 10)\left(D^2 + D + \frac{1}{4}\right) y = (D^2 - 2D + 10)\left(D + \frac{1}{2}\right)^2 y = 0.$$

Then

$$y = \underbrace{c_1 e^{-x/2} + c_2 x e^{-x/2}}_{y_c} + c_3 e^x \cos 3x + c_4 e^x \sin 3x$$

and $y_p = A e^x \cos 3x + B e^x \sin 3x$. Substituting y_p into the differential equation yields

$$(9B - 27A/4) e^x \cos 3x - (9A + 27B/4) e^x \sin 3x = -e^x \cos 3x + e^x \sin 3x.$$

Equating coefficients gives

$$-\frac{27}{4} A + 9B = -1$$

$$-9A - \frac{27}{4} B = 1.$$

Then $A = -4/225$, $B = -28/225$, and the general solution is

$$y = c_1 e^{-x/2} + c_2 x e^{-x/2} - \frac{4}{225} e^x \cos 3x - \frac{28}{225} e^x \sin 3x.$$

173

55. Applying $D^2 + 25$ to the differential equation we obtain
$$(D^2 + 25)(D^2 + 25) = (D^2 + 25)^2 = 0.$$

Then
$$y = \underbrace{c_1 \cos 5x + c_2 \sin 5x}_{y_c} + c_3 x \cos 5x + c_4 x \cos 5x$$

and $y_p = Ax \cos 5x + Bx \sin 5x$. Substituting y_p into the differential equation yields
$$10B \cos 5x - 10A \sin 5x = 20 \sin 5x.$$

Equating coefficients gives $A = -2$ and $B = 0$. The general solution is
$$y = c_1 \cos 5x + c_2 \sin 5x - 2x \cos 5x.$$

56. Applying $D^2 + 1$ to the differential equation we obtain
$$(D^2 + 1)(D^2 + 1) = (D^2 + 1)^2 = 0.$$

Then
$$y = \underbrace{c_1 \cos x + c_2 \sin x}_{y_c} + c_3 x \cos x + c_4 x \cos x$$

and $y_p = Ax \cos x + Bx \sin x$. Substituting y_p into the differential equation yields
$$2B \cos x - 2A \sin x = 4 \cos x - \sin x.$$

Equating coefficients gives $A = 1/2$ and $B = 2$. The general solution is
$$y = c_1 \cos x + c_2 \sin x + \frac{1}{2} x \cos x - 2x \sin x.$$

57. Applying $(D^2 + 1)^2$ to the differential equation we obtain
$$(D^2 + 1)^2 (D^2 + D + 1) = 0.$$

Then
$$y = e^{-x/2} \underbrace{\left[c_1 \cos \frac{\sqrt{3}}{2} x + c_2 \sin \frac{\sqrt{3}}{2} x \right]}_{y_c} + c_3 \cos x + c_4 \sin x + c_5 x \cos x + c_6 x \sin x$$

and $y_p = A \cos x + B \sin x + Cx \cos x + Ex \sin x$. Substituting y_p into the differential equation yields
$$(B + C + 2E) \cos x + Ex \cos x + (-A - 2C + E) \sin x - Cx \sin x = x \sin x.$$

Equating coefficients gives
$$B + C + 2E = 0$$
$$E = 0$$
$$-A - 2C + E = 0$$
$$-C = 1.$$

Then $A = 2$, $B = 1$, $C = -1$, and $E = 0$, and the general solution is

$$y = e^{-x/2}\left[c_1\cos\frac{\sqrt{3}}{2}x + c_2\sin\frac{\sqrt{3}}{2}x\right] + 2\cos x + \sin x - x\cos x.$$

58. Writing $\cos^2 x = \frac{1}{2}(1 + \cos 2x)$ and applying $D(D^2 + 4)$ to the differential equation we obtain

$$D(D^2 + 4)(D^2 + 4) = D(D^2 + 4)^2 = 0.$$

Then

$$y = \underbrace{c_1\cos 2x + c_2\sin 2x}_{y_c} + c_3 x\cos 2x + c_4 x\sin 2x + c_5$$

and $y_p = Ax\cos 2x + Bx\sin 2x + C$. Substituting y_p into the differential equation yields

$$-4A\sin 2x + 4B\cos 2x + 4C = \frac{1}{2} + \frac{1}{2}\cos 2x.$$

Equating coefficients gives $A = 0$, $B = 1/8$, and $C = 1/8$. The general solution is

$$y = c_1\cos 2x + c_2\sin 2x + \frac{1}{8}x\sin 2x + \frac{1}{8}.$$

59. Applying D^3 to the differential equation we obtain

$$D^3(D^3 + 8D^2) = D^5(D + 8) = 0.$$

Then

$$y = \underbrace{c_1 + c_2 x + c_3 e^{-8x}}_{y_c} + c_4 x^2 + c_5 x^3 + c_6 x^4$$

and $y_p = Ax^2 + Bx^3 + Cx^4$. Substituting y_p into the differential equation yields

$$16A + 6B + (48B + 24C)x + 96Cx^2 = 2 + 9x - 6x^2.$$

Equating coefficients gives

$$16A + 6B = 2$$

$$48B + 24C = 9$$

$$96C = -6.$$

Then $A = 11/256$, $B = 7/32$, and $C = -1/16$, and the general solution is

$$y = c_1 + c_2 x + c_3 e^{-8x} + \frac{11}{256}x^2 + \frac{7}{32}x^3 - \frac{1}{16}x^4.$$

60. Applying $D(D-1)^2(D+1)$ to the differential equation we obtain

$$D(D-1)^2(D+1)(D^3 - D^2 + D - 1) = D(D-1)^3(D+1)(D^2 + 1) = 0.$$

Then

$$y = \underbrace{c_1 e^x + c_2\cos x + c_3\sin x}_{y_c} + c_4 + c_5 e^{-x} + c_6 x e^x + c_7 x^2 e^x$$

and $y_p = A + Be^{-x} + Cxe^x + Ex^2e^x$. Substituting y_p into the differential equation yields

$$4Exe^x + (2C + 4E)e^x - 4Be^{-x} - A = xe^x - e^{-x} + 7.$$

Equating coefficients gives

$$4E = 1$$

$$2C + 4E = 0$$

$$-4B = -1$$

$$-A = 7.$$

Then $A = -7$, $B = 1/4$, $C = -1/2$, and $E = 1/4$, and the general solution is

$$y = c_1e^x + c_2 \cos x + c_3 \sin x - 7 + \frac{1}{4}e^{-x} - \frac{1}{2}xe^x + \frac{1}{4}x^2e^x.$$

61. Applying $D^2(D-1)$ to the differential equation we obtain

$$D^2(D-1)(D^3 - 3D^2 + 3D - 1) = D^2(D-1)^4 = 0.$$

Then

$$y = \underbrace{c_1e^x + c_2xe^x + c_3x^2e^x}_{y_c} + c_4 + c_5x + c_6x^3e^x$$

and $y_p = A + Bx + Cx^3e^x$. Substituting y_p into the differential equation yields

$$(-A + 3B) - Bx + 6Ce^x = 16 - x + e^x.$$

Equating coefficients gives

$$-A + 3B = 16$$

$$-B = -1$$

$$6C = 1.$$

Then $A = -13$, $B = 1$, and $C = 1/6$, and the general solution is

$$y = c_1e^x + c_2xe^x + c_3x^2e^x - 13 + x + \frac{1}{6}x^3e^x.$$

62. Writing $(e^x + e^{-x})^2 = 2 + e^{2x} + e^{-2x}$ and applying $D(D-2)(D+2)$ to the differential equation we obtain

$$D(D-2)(D+2)(2D^3 - 3D^2 - 3D + 2) = D(D-2)^2(D+2)(D+1)(2D-1) = 0.$$

Then

$$y = \underbrace{c_1e^{-x} + c_2e^{2x} + c_3e^{x/2}}_{y_c} + c_4 + c_5xe^{2x} + c_6e^{-2x}$$

and $y_p = A + Bxe^{2x} + Ce^{-2x}$. Substituting y_p into the differential equation yields

$$2A + 9Be^{2x} - 20Ce^{-2x} = 2 + e^{2x} + e^{-2x}.$$

Equating coefficients gives $A = 1$, $B = 1/9$, and $C = -1/20$. The general solution is

$$y = c_1 e^{-x} + c_2 e^{2x} + c_3 e^{x/2} + 1 + \frac{1}{9}xe^{2x} - \frac{1}{20}e^{-2x}.$$

63. Applying $D(D-1)$ to the differential equation we obtain

$$D(D-1)(D^4 - 2D^3 + D^2) = D^3(D-1)^3 = 0.$$

Then

$$y = \underbrace{c_1 + c_2 x + c_3 e^x + c_4 xe^x}_{y_c} + c_5 x^2 + c_6 x^2 e^x$$

and $y_p = Ax^2 + Bx^2 e^x$. Substituting y_p into the differential equation yields $2A + 2Be^x = 1 + e^x$. Equating coefficients gives $A = 1/2$ and $B = 1/2$. The general solution is

$$y = c_1 + c_2 x + c_3 e^x + c_4 xe^x + \frac{1}{2}x^2 + \frac{1}{2}x^2 e^x.$$

64. Applying $D^3(D-2)$ to the differential equation we obtain

$$D^3(D-2)(D^4 - 4D^2) = D^5(D-2)^2(D+2) = 0.$$

Then

$$y = \underbrace{c_1 + c_2 x + c_3 e^{2x} + c_4 e^{-2x}}_{y_c} + c_5 x^2 + c_6 x^3 + c_7 x^4 + c_8 xe^{2x}$$

and $y_p = Ax^2 + Bx^3 + Cx^4 + Exe^{2x}$. Substituting y_p into the differential equation yields

$$(-8A + 24C) - 24Bx - 48Cx^2 + 16Ee^{2x} = 5x^2 - e^{2x}.$$

Equating coefficients gives

$$-8A + 24C = 0$$

$$-24B = 0$$

$$-48C = 5$$

$$16E = -1.$$

Then $A = -5/16$, $B = 0$, $C = -5/48$, and $E = -1/16$, and the general solution is

$$y = c_1 + c_2 x + c_3 e^{2x} + c_4 e^{-2x} - \frac{5}{16}x^2 - \frac{5}{48}x^4 - \frac{1}{16}xe^{2x}.$$

65. The complementary function is $y_c = c_1 e^{8x} + c_2 e^{-8x}$. Using D to annihilate 16 we find $y_p = A$. Substituting y_p into the differential equation we obtain $-64A = 16$. Thus $A = -1/4$ and

$$y = c_1 e^{8x} + c_2 e^{-8x} - \frac{1}{4}$$

$$y' = 8c_1 e^{8x} - 8c_2 e^{-8x}.$$

The initial conditions imply

$$c_1 + c_2 = \frac{5}{4}$$

$$8c_1 - 8c_2 = 0.$$

Thus $c_1 = c_2 = 5/8$ and

$$y = \frac{5}{8} e^{8x} + \frac{5}{8} e^{-8x} - \frac{1}{4}.$$

66. The complementary function is $y_c = c_1 + c_2 e^{-x}$. Using D^2 to annihilate x we find $y_p = Ax + Bx^2$. Substituting y_p into the differential equation we obtain $(A + 2B) + 2Bx = x$. Thus $A = -1$ and $B = 1/2$, and

$$y = c_1 + c_2 e^{-x} - x + \frac{1}{2} x^2$$

$$y' = -c_2 e^{-x} - 1 + x.$$

The initial conditions imply

$$c_1 + c_2 = 1$$

$$-c_2 = 1.$$

Thus $c_1 = 2$ and $c_2 = -1$, and

$$y = 2 - e^{-x} - x + \frac{1}{2} x^2.$$

67. The complementary function is $y_c = c_1 + c_2 e^{5x}$. Using D^2 to annihilate $x - 2$ we find $y_p = Ax + Bx^2$. Substituting y_p into the differential equation we obtain $(-5A + 2B) - 10Bx = -2 + x$. Thus $A = 9/25$ and $B = -1/10$, and

$$y = c_1 + c_2 e^{5x} + \frac{9}{25} x - \frac{1}{10} x^2$$

$$y' = 5c_2 e^{5x} + \frac{9}{25} - \frac{1}{5} x.$$

The initial conditions imply

$$c_1 + c_2 = 0$$

$$c_2 = \frac{41}{125}.$$

Thus $c_1 = -41/125$ and $c_2 = 41/125$, and

$$y = -\frac{41}{125} + \frac{41}{125}e^{5x} + \frac{9}{25}x - \frac{1}{10}x^2.$$

68. The complementary function is $y_c = c_1 e^x + c_2 e^{-6x}$. Using $D - 2$ to annihilate $10e^{2x}$ we find $y_p = Ae^{2x}$. Substituting y_p into the differential equation we obtain $8Ae^{2x} = 10e^{2x}$. Thus $A = 5/4$ and

$$y = c_1 e^x + c_2 e^{-6x} + \frac{5}{4}e^{2x}$$

$$y' = c_1 e^x - 6c_2 e^{-6x} + \frac{5}{2}e^{2x}.$$

The initial conditions imply

$$c_1 + c_2 = -\frac{1}{4}$$

$$c_1 - 6c_2 = -\frac{3}{2}.$$

Thus $c_1 = -3/7$ and $c_2 = 5/28$, and

$$y = -\frac{3}{7}e^x + \frac{5}{28}e^{-6x} + \frac{5}{4}e^{2x}$$

69. The complementary function is $y_c = c_1 \cos x + c_2 \sin x$. Using $(D^2 + 1)(D^2 + 4)$ to annihilate $8 \cos 2x - 4 \sin x$ we find $y_p = Ax \cos x + Bx \sin x + C \cos 2x + E \sin 2x$. Substituting y_p into the differential equation we obtain $2B \cos x - 3C \cos 2x - 2A \sin x - 3E \sin 2x = 8 \cos 2x - 4 \sin x$. Thus $A = 2$, $B = 0$, $C = -8/3$, and $E = 0$, and

$$y = c_1 \cos x + c_2 \sin x + 2x \cos x - \frac{8}{3} \cos 2x$$

$$y' = -c_1 \sin x + c_2 \cos x + 2 \cos x - 2x \sin x + \frac{16}{3} \sin 2x.$$

The initial conditions imply

$$c_2 + \frac{8}{3} = -1$$

$$-c_1 - \pi = 0.$$

Thus $c_1 = -\pi$ and $c_2 = -11/3$, and

$$y = -\pi \cos x - \frac{11}{3} \sin x + 2x \cos x - \frac{8}{3} \cos 2x.$$

70. The complementary function is $y_c = c_1 + c_2 e^x + c_3 x e^x$. Using $D(D - 1)^2$ to annihilate $xe^x + 5$ we find $y_p = Ax + Bx^2 e^x + Cx^3 e^x$. Substituting y_p into the differential equation we obtain

$A + (2B + 6C)e^x + 6Cxe^x = xe^x + 5$. Thus $A = 5$, $B = -1/2$, and $C = 1/6$, and

$$y = c_1 + c_2 e^x + c_3 x e^x + 5x - \frac{1}{2}x^2 e^x + \frac{1}{6}x^3 e^x$$

$$y' = c_2 e^x + c_3(xe^x + e^x) + 5 - xe^x + \frac{1}{6}x^3 e^x$$

$$y'' = c_2 e^x + c_3(xe^x + 2e^x) - e^x - xe^x + \frac{1}{2}x^2 e^x + \frac{1}{6}x^3 e^x.$$

The initial conditions imply

$$c_1 + c_2 = 2$$

$$c_2 + c_3 + 5 = 2$$

$$c_2 + 2c_3 - 1 = -1.$$

Thus $c_1 = 8$, $c_2 = -6$, and $c_3 = 3$, and

$$y = 8 - 6e^x + 3xe^x + 5x - \frac{1}{2}x^2 e^x + \frac{1}{6}x^3 e^x.$$

71. The complementary function is $y_c = e^{2x}(c_1 \cos 2x + c_2 \sin 2x)$. Using D^4 to annihilate x^3 we find $y_p = A + Bx + Cx^2 + Ex^3$. Substituting y_p into the differential equation we obtain $(8A - 4B + 2C) + (8B - 8C + 6E)x + (8C - 12E)x^2 + 8Ex^3 = x^3$. Thus $A = 0$, $B = 3/32$, $C = 3/16$, and $E = 1/8$, and

$$y = e^{2x}(c_1 \cos 2x + c_2 \sin 2x) + \frac{3}{32}x + \frac{3}{16}x^2 + \frac{1}{8}x^3$$

$$y' = e^{2x}[c_1(2 \cos 2x - 2 \sin 2x) + c_2(2 \cos 2x + 2 \sin 2x)] + \frac{3}{32} + \frac{3}{8}x + \frac{3}{8}x^2.$$

The initial conditions imply

$$c_1 = 2$$

$$2c_1 + 2c_2 + \frac{3}{32} = 4.$$

Thus $c_1 = 2$, $c_2 = -3/64$, and

$$y = e^{2x}\left(2 \cos 2x - \frac{3}{64} \sin 2x\right) + \frac{3}{32}x + \frac{3}{16}x^2 + \frac{1}{8}x^3.$$

72. The complementary function is $y_c = c_1 + c_2 x + c_3 x^2 + c_4 e^x$. Using $D^2(D - 1)$ to annihilate $x + e^x$ we find $y_p = Ax^3 + Bx^4 + Cxe^x$. Substituting y_p into the differential equation we obtain

180

$(-6A + 24B) - 24Bx + Ce^x = x + e^x$. Thus $A = -1/6$, $B = -1/24$, and $C = 1$, and

$$y = c_1 + c_2 x + c_3 x^2 + c_4 e^x - \frac{1}{6}x^3 - \frac{1}{24}x^4 + xe^x$$

$$y' = c_2 + 2c_3 x + c_4 e^x - \frac{1}{2}x^2 - \frac{1}{6}x^3 + e^x + xe^x$$

$$y'' = 2c_3 + c_4 e^x - x - \frac{1}{2}x^2 + 2e^x + xe^x.$$

$$y''' = c_4 e^x - 1 - x + 3e^x + xe^x$$

The initial conditions imply

$$c_1 + c_4 = 0$$

$$c_2 + c_4 + 1 = 0$$

$$2c_3 + c_4 + 2 = 0$$

$$2 + c_4 = 0.$$

Thus $c_1 = 2$, $c_2 = 1$, $c_3 = 0$, and $c_4 = -2$, and

$$y = 2 + x - 2e^x - \frac{1}{6}x^3 - \frac{1}{24}x^4 + xe^x.$$

73. To see in this case that the factors of L do not commute consider the operators $(xD - 1)(D + 4)$ and $(D + 4)(xD - 1)$. Applying the operators to the function x we find

$$(xD - 1)(D + 4)x = (xD^2 + 4xD - D - 4)x$$

$$= xD^2 x + 4xDx - Dx - 4x$$

$$= x(0) + 4x(1) - 1 - 4x = -1$$

and

$$(D + 4)(xD - 1)x = (D + 4)(xDx - x)$$

$$= (D + 4)(x \cdot 1 - x) = 0.$$

Thus, the operators are not the same.

The particular solution, $y_p = u_1 y_1 + u_2 y_2$, in the following problems can take on a variety of forms, especially where trigonometric functions are involved. The validity of a particular form can best be checked by substituting it back into the differential equation.

1. The auxiliary equation is $m^2 + 1 = 0$, so $y_c = c_1 \cos x + c_2 \sin x$ and

$$W = \begin{vmatrix} \cos x & \sin x \\ -\sin x & \cos x \end{vmatrix} = 1.$$

Identifying $f(x) = \sec x$ we obtain

$$u_1' = -\frac{\sin x \sec x}{1} = -\tan x$$

$$u_2' = \frac{\cos x \sec x}{1} = 1.$$

Then $u_1 = \ln|\cos x|$, $u_2 = x$, and

$$y = c_1 \cos x + c_2 \sin x + \cos x \ln|\cos x| + x \sin x.$$

2. The auxiliary equation is $m^2 + 1 = 0$, so $y_c = c_1 \cos x + c_2 \sin x$ and

$$W = \begin{vmatrix} \cos x & \sin x \\ -\sin x & \cos x \end{vmatrix} = 1.$$

Identifying $f(x) = \tan x$ we obtain

$$u_1' = -\sin x \tan x = \frac{\cos^2 x - 1}{\cos x} = \cos x - \sec x$$

$$u_2' = \sin x.$$

Then $u_1 = \sin x - \ln|\sec x + \tan x|$, $u_2 = -\cos x$, and

$$y = c_1 \cos x + c_2 \sin x + \cos x \left(\sin x - \ln|\sec x + \tan x|\right) - \cos x \sin x$$

$$= c_1 \cos x + c_2 \sin x - \cos x \ln|\sec x + \tan x|.$$

3. The auxiliary equation is $m^2 + 1 = 0$, so $y_c = c_1 \cos x + c_2 \sin x$ and

$$W = \begin{vmatrix} \cos x & \sin x \\ -\sin x & \cos x \end{vmatrix} = 1.$$

Identifying $f(x) = \sin x$ we obtain

$$u_1' = -\sin^2 x$$

$$u_2' = \cos x \sin x.$$

Then

$$u_1 = \frac{1}{4}\sin 2x - \frac{1}{2}x = \frac{1}{2}\sin x \cos x - \frac{1}{2}x$$

$$u_2 = -\frac{1}{2}\cos^2 x.$$

and

$$y = c_1 \cos x + c_2 \sin x + \frac{1}{2}\sin x \cos^2 x - \frac{1}{2}x\cos x - \frac{1}{2}\cos^2 x \sin x$$

$$= c_1 \cos x + c_2 \sin x - \frac{1}{2}x\cos x.$$

4. The auxiliary equation is $m^2 + 1 = 0$, so $y_c = c_1 \cos x + c_2 \sin x$ and

$$W = \begin{vmatrix} \cos x & \sin x \\ -\sin x & \cos x \end{vmatrix} = 1.$$

Identifying $f(x) = \sec x \tan x$ we obtain

$$u_1' = -\sin x(\sec x \tan x) = -\tan^2 x = 1 - \sec^2 x$$

$$u_2' = \cos x(\sec x \tan x) = \tan x.$$

Then $u_1 = x - \tan x$, $u_2 = -\ln|\cos x|$, and

$$y = c_1 \cos x + c_2 \sin x + x\cos x - \sin x - \sin x \ln|\cos x|$$

$$= c_1 \cos x + c_3 \sin x + x\cos x - \sin x \ln|\cos x|.$$

5. The auxiliary equation is $m^2 + 1 = 0$, so $y_c = c_1 \cos x + c_2 \sin x$ and

$$W = \begin{vmatrix} \cos x & \sin x \\ -\sin x & \cos x \end{vmatrix} = 1.$$

Identifying $f(x) = \cos^2 x$ we obtain

$$u_1' = -\sin x \cos^2 x$$

$$u_2' = \cos^3 x = \cos x\left(1 - \sin^2 x\right).$$

Then $u_1 = \frac{1}{3}\cos^3 x$, $u_2 = \sin x - \frac{1}{3}\sin^3 x$, and

$$y = c_1\cos x + c_2\sin x + \frac{1}{3}\cos^4 x + \sin^2 x - \frac{1}{3}\sin^4 x$$

$$= c_1\cos x + c_2\sin x + \frac{1}{3}\left(\cos^2 x + \sin^2 x\right)\left(\cos^2 x - \sin^2 x\right) + \sin^2 x$$

$$= c_1\cos x + c_2\sin x + \frac{1}{3}\cos^2 x + \frac{2}{3}\sin^2 x$$

$$= c_1\cos x + c_2\sin x + \frac{1}{3} + \frac{1}{3}\sin^2 x.$$

6. The auxiliary equation is $m^2 + 1 = 0$, so $y_c = c_1\cos x + c_2\sin x$ and

$$W = \begin{vmatrix} \cos x & \sin x \\ -\sin x & \cos x \end{vmatrix} = 1.$$

Identifying $f(x) = \sec^2 x$ we obtain

$$u_1' = -\frac{\sin x}{\cos^2 x}$$

$$u_2' = \sec x.$$

Then

$$u_1 = -\frac{1}{\cos x} = -\sec x$$

$$u_2 = \ln|\sec x + \tan x|$$

and

$$y = c_1\cos x + c_2\sin x - \cos x\sec x + \sin x\ln|\sec x + \tan x|$$

$$= c_1\cos x + c_2\sin x - 1 + \sin x\ln|\sec x + \tan x|.$$

7. The auxiliary equation is $m^2 - 1 = 0$, so $y_c = c_1 e^x + c_2 e^{-x}$ and

$$W = \begin{vmatrix} e^x & e^{-x} \\ e^x & -e^{-x} \end{vmatrix} = -2.$$

Identifying $f(x) = \cosh x = \frac{1}{2}(e^{-x} + e^x)$ we obtain

$$u_1' = \frac{1}{4}e^{-2x} + \frac{1}{4}$$

$$u_2' = -\frac{1}{4} - \frac{1}{4}e^{2x}.$$

Then

$$u_1 = -\frac{1}{8}e^{-2x} + \frac{1}{4}x$$

$$u_2 = -\frac{1}{8}e^{2x} - \frac{1}{4}x$$

and

$$y = c_1 e^x + c_2 e^{-x} - \frac{1}{8}e^{-x} + \frac{1}{4}xe^x - \frac{1}{8}e^x - \frac{1}{4}xe^{-x}$$

$$= c_3 e^x + c_4 e^{-x} + \frac{1}{4}x(e^x - e^{-x})$$

$$= c_3 e^x + c_4 e^{-x} + \frac{1}{2}x \sinh x.$$

8. The auxiliary equation is $m^2 - 1 = 0$, so $y_c = c_1 e^x + c_2 e^{-x}$ and

$$W = \begin{vmatrix} e^x & e^{-x} \\ e^x & -e^{-x} \end{vmatrix} = -2.$$

Identifying $f(x) = \sinh 2x$ we obtain

$$u_1' = -\frac{1}{4}e^{-3x} + \frac{1}{4}e^x$$

$$u_2' = \frac{1}{4}e^{-x} - \frac{1}{4}e^{3x}.$$

Then

$$u_1 = \frac{1}{12}e^{-3x} + \frac{1}{4}e^x$$

$$u_2 = -\frac{1}{4}e^{-x} - \frac{1}{12}e^{3x}.$$

and

$$y = c_1 e^x + c_2 e^{-x} + \frac{1}{12}e^{-2x} + \frac{1}{4}e^{2x} - \frac{1}{4}e^{-2x} - \frac{1}{12}e^{2x}$$

$$= c_1 e^x + c_2 e^{-x} + \frac{1}{6}\left(e^{2x} - e^{-2x}\right)$$

$$= c_1 e^x + c_2 e^{-x} + \frac{1}{3}\sinh 2x.$$

9. The auxiliary equation is $m^2 - 4 = 0$, so $y_c = c_1 e^{2x} + c_2 e^{-2x}$ and

$$W = \begin{vmatrix} e^{2x} & e^{-2x} \\ 2e^{2x} & -2e^{-2x} \end{vmatrix} = -4.$$

Identifying $f(x) = e^{2x}/x$ we obtain $u_1' = 1/4x$ and $u_2' = -e^{4x}/4x$. Then

$$u_1 = \frac{1}{4}\ln|x|,$$

$$u_2 = -\frac{1}{4}\int_{x_0}^{x} \frac{e^{4t}}{t}\, dt$$

and

$$y = c_1 e^{2x} + c_2 e^{-2x} + \frac{1}{4}\left(e^{2x}\ln|x| - e^{-2x}\int_{x_0}^{x} \frac{e^{4t}}{t}\, dt\right), \qquad x_0 > 0.$$

10. The auxiliary equation is $m^2 - 9 = 0$, so $y_c = c_1 e^{3x} + c_2 e^{-3x}$ and

$$W = \begin{vmatrix} e^{3x} & e^{-3x} \\ 3e^{3x} & -3e^{-3x} \end{vmatrix} = -6.$$

Identifying $f(x) = 9x/e^{3x}$ we obtain $u_1' = \frac{3}{2}xe^{-6x}$ and $u_2' = -\frac{3}{2}x$. Then

$$u_1 = -\frac{1}{24}e^{-6x} - \frac{1}{4}xe^{-6x},$$

$$u_2 = -\frac{3}{4}x^2$$

and

$$y = c_1 e^{3x} + c_2 e^{-3x} - \frac{1}{24}e^{-3x} - \frac{1}{4}xe^{-3x} - \frac{3}{4}x^2 e^{-3x}$$

$$= c_1 e^{3x} + c_3 e^{-3x} - \frac{1}{4}xe^{-3x}(1 - 3x).$$

11. The auxiliary equation is $m^2 + 3m + 2 = (m+1)(m+2) = 0$, so $y_c = c_1 e^{-x} + c_2 e^{-2x}$ and

$$W = \begin{vmatrix} e^{-x} & e^{-2x} \\ -e^{-x} & -2e^{-2x} \end{vmatrix} = -e^{-3x}.$$

Identifying $f(x) = 1/(1 + e^x)$ we obtain

$$u_1' = \frac{e^x}{1 + e^x}$$

$$u_2' = -\frac{e^{2x}}{1 + e^x} = \frac{e^x}{1 + e^x} - e^x.$$

Then $u_1 = \ln(1 + e^x)$, $u_2 = \ln(1 + e^x) - e^x$, and

$$y = c_1 e^{-x} + c_2 e^{-2x} + e^{-x}\ln(1 + e^x) + e^{-2x}\ln(1 + e^x) - e^{-x}$$

$$= c_3 e^{-x} + c_2 e^{-2x} + (1 + e^{-x})e^{-x}\ln(1 + e^x).$$

12. The auxiliary equation is $m^2 - 2m + 1 = (m-1)^2 = 0$, so $y_c = c_1 e^x + c_2 xe^x$ and

$$W = \begin{vmatrix} e^x & xe^x \\ e^x & xe^x + e^x \end{vmatrix} = e^{2x}.$$

Identifying $f(x) = e^x/\left(1 + x^2\right)$ we obtain

$$u_1' = -\frac{xe^x e^x}{e^{2x}\left(1 + x^2\right)} = -\frac{x}{1 + x^2}$$

$$u_2' = \frac{e^x e^x}{e^{2x}\left(1 + x^2\right)} = \frac{1}{1 + x^2}.$$

Then $u_1 = -\frac{1}{2}\ln\left(1+x^2\right)$, $u_2 = \tan^{-1} x$, and

$$y = c_1 e^x + c_2 x e^x - \frac{1}{2} e^x \ln\left(1+x^2\right) + x e^x \tan^{-1} x.$$

13. The auxiliary equation is $m^2 + 3m + 2 = (m+1)(m+2) = 0$, so $y_c = c_1 e^{-x} + c_2 e^{-2x}$ and

$$W = \begin{vmatrix} e^{-x} & e^{-2x} \\ -e^{-x} & -2e^{-2x} \end{vmatrix} = -e^{-3x}.$$

Identifying $f(x) = \sin e^x$ we obtain

$$u_1' = \frac{e^{-2x}\sin e^x}{e^{-3x}} = e^x \sin e^x$$

$$u_2' = \frac{e^{-x}\sin e^x}{-e^{-3x}} = -e^{2x}\sin e^x.$$

Then $u_1 = -\cos e^x$, $u_2 = e^x \cos e^x - \sin e^x$, and

$$y = c_1 e^{-x} + c_2 e^{-2x} - e^{-x}\cos e^x + e^{-x}\cos e^x - e^{-2x}\sin e^x$$

$$= c_1 e^{-x} + c_2 e^{-2x} - e^{-2x}\sin e^x.$$

14. The auxiliary equation is $m^2 - 2m + 1 = (m-1)^2 = 0$, so $y_c = c_1 e^t + c_2 t e^t$ and

$$W = \begin{vmatrix} e^t & te^t \\ e^t & te^t + e^t \end{vmatrix} = e^{2t}.$$

Identifying $f(t) = e^t \tan^{-1} t$ we obtain

$$u_1' = -\frac{te^t e^t \tan^{-1} t}{e^{2t}} = -t\tan^{-1} t$$

$$u_2' = \frac{e^t e^t \tan^{-1} t}{e^{2t}} = \tan^{-1} t.$$

Then

$$u_1 = -\frac{1+t^2}{2}\tan^{-1} t + \frac{t}{2}$$

$$u_2 = t\tan^{-1} t - \frac{1}{2}\ln\left(1+t^2\right)$$

and

$$y = c_1 e^t + c_2 t e^t + \left(-\frac{1+t^2}{2}\tan^{-1} t + \frac{t}{2}\right)e^t + \left(t\tan^{-1} t - \frac{1}{2}\ln\left(1+t^2\right)\right)te^t$$

$$= c_1 e^t + c_3 t e^t + \frac{1}{2}e^t\left[\left(t^2 - 1\right)\tan^{-1} t - \ln\left(1+t^2\right)\right].$$

15. The auxiliary equation is $m^2 + 2m + 1 = (m+1)^2 = 0$, so $y_c = c_1 e^{-t} + c_2 t e^{-t}$ and

$$W = \begin{vmatrix} e^{-t} & te^{-t} \\ -e^{-t} & -te^{-t} + e^{-t} \end{vmatrix} = e^{-2t}.$$

Identifying $f(t) = e^{-t} \ln t$ we obtain

$$u_1' = -\frac{te^{-t}e^{-t}\ln t}{e^{-2t}} = -t \ln t$$

$$u_2' = \frac{e^{-t}e^{-t}\ln t}{e^{-2t}} = \ln t.$$

Then

$$u_1 = -\frac{1}{2}t^2 \ln t + \frac{1}{4}t^2$$

$$u_2 = t \ln t - t$$

and

$$y = c_1 e^{-t} + c_2 t e^{-t} - \frac{1}{2}t^2 e^{-t} \ln t + \frac{1}{4}t^2 e^{-t} + t^2 e^{-t} \ln t - t^2 e^{-t}$$

$$= c_1 e^{-t} + c_2 t e^{-t} + \frac{1}{2}t^2 e^{-t} \ln t - \frac{3}{4}t^2 e^{-t}.$$

16. The auxiliary equation is $2m^2 + 2m + 1 = 0$, so $y_c = e^{-x/2}[c_1 \cos(x/2) + c_2 \sin(x/2)]$ and

$$W = \begin{vmatrix} e^{-x/2}\cos\dfrac{x}{2} & e^{-x/2}\sin\dfrac{x}{2} \\ -\dfrac{1}{2}e^{-x/2}\cos\dfrac{x}{2} - \dfrac{1}{2}e^{-x/2}\sin\dfrac{x}{2} & \dfrac{1}{2}e^{-x/2}\cos\dfrac{x}{2} - \dfrac{1}{2}e^{x/2}\sin\dfrac{x}{2} \end{vmatrix} = \frac{1}{2}e^{-x}.$$

Identifying $f(x) = 2\sqrt{x}$ we obtain

$$u_1' = -\frac{e^{-x/2}\sin(x/2)2\sqrt{x}}{e^{-x/2}} = -4e^{x/2}\sqrt{x}\sin\frac{x}{2}$$

$$u_2' = -\frac{e^{-x/2}\cos(x/2)2\sqrt{x}}{e^{-x/2}} = 4e^{x/2}\sqrt{x}\cos\frac{x}{2}.$$

Then

$$u_1 = -4\int_{x_0}^{x} e^{t/2}\sqrt{t}\sin\frac{t}{2}\,dt$$

$$u_2 = 4\int_{x_0}^{x} e^{t/2}\sqrt{t}\cos\frac{t}{2}\,dt$$

and

$$y = e^{-x/2}\left(c_1 \cos\frac{x}{2} + c_2 \sin\frac{x}{2}\right) - 4e^{-x/2}\cos\frac{x}{2}\int_{x_0}^{x} e^{t/2}\sqrt{t}\sin\frac{t}{2}\,dt + 4e^{-x/2}\sin\frac{x}{2}\int_{x_0}^{x} e^{t/2}\sqrt{t}\cos\frac{t}{2}\,dt.$$

17. The auxiliary equation is $3m^2 - 6m + 6 = 0$, so $y_c = e^x(c_1 \cos x + c_2 \sin x)$ and

$$W = \begin{vmatrix} e^x \cos x & e^x \sin x \\ e^x \cos x - e^x \sin x & e^x \cos x + e^x \sin x \end{vmatrix} = e^{2x}.$$

Identifying $f(x) = \frac{1}{3}e^x \sec x$ we obtain

$$u_1' = -\frac{(e^x \sin x)(e^x \sec x)/3}{e^{2x}} = -\frac{1}{3}\tan x$$

$$u_2' = \frac{(e^x \cos x)(e^x \sec x)/3}{e^{2x}} = \frac{1}{3}.$$

Then $u_1 = \frac{1}{3}\ln(\cos x)$, $u_2 = \frac{1}{3}x$, and

$$y = c_1 e^x \cos x + c_2 e^x \sin x + \frac{1}{3}\ln(\cos x)e^x \cos x + \frac{1}{3}xe^x \sin x.$$

18. The auxiliary equation is $4m^2 - 4m + 1 = (2m - 1)^2 = 0$, so $y_c = c_1 e^{x/2} + c_2 xe^{x/2}$ and

$$W = \begin{vmatrix} e^{x/2} & xe^{x/2} \\ \frac{1}{2}e^{x/2} & \frac{1}{2}xe^{x/2} + e^{x/2} \end{vmatrix} = e^x.$$

Identifying $f(x) = \frac{1}{4}e^{x/2}\sqrt{1 - x^2}$ we obtain

$$u_1' = -\frac{xe^{x/2}e^{x/2}\sqrt{1 - x^2}}{4e^x} = -\frac{1}{4}x\sqrt{1 - x^2}$$

$$u_2' = \frac{e^{x/2}e^{x/2}\sqrt{1 - x^2}}{4e^x} = \frac{1}{4}\sqrt{1 - x^2}.$$

To find u_1 and u_2 we use the substitution $v = 1 - x^2$ and the trig substitution $x = \sin\theta$, respectively:

$$u_1 = \frac{1}{12}\left(1 - x^2\right)^{3/2}$$

$$u_2 = \frac{x}{8}\sqrt{1 - x^2} + \frac{1}{8}\sin^{-1} x.$$

Thus

$$y = c_1 e^{x/2} + c_2 xe^{x/2} + \frac{1}{12}e^{x/2}\left(1 - x^2\right)^{3/2} + \frac{1}{8}x^2 e^{x/2}\sqrt{1 - x^2} + \frac{1}{8}xe^{x/2}\sin^{-1} x.$$

19. The auxiliary equation is $4m^2 - 1 = (2m - 1)(2m + 1) = 0$, so $y_c = c_1 e^{x/2} + c_2 e^{-x/2}$ and

$$W = \begin{vmatrix} e^{x/2} & e^{-x/2} \\ \frac{1}{2}e^{x/2} & -\frac{1}{2}e^{-x/2} \end{vmatrix} = -1.$$

Identifying $f(x) = xe^{x/2}/4$ we obtain $u_1' = x/4$ and $u_2' = -xe^x/4$. Then $u_1 = x^2/8$ and $u_2 = -xe^x/4 + e^x/4$. Thus

$$y = c_1 e^{x/2} + c_2 e^{-x/2} + \frac{1}{8}x^2 e^{x/2} - \frac{1}{4}xe^{x/2} + \frac{1}{4}e^{x/2}$$

$$= c_3 e^{x/2} + c_2 e^{-x/2} + \frac{1}{8}x^2 e^{x/2} - \frac{1}{4}xe^{x/2}$$

and

$$y' = \frac{1}{2}c_3 e^{x/2} - \frac{1}{2}c_2 e^{-x/2} + \frac{1}{16}x^2 e^{x/2} + \frac{1}{8}xe^{x/2} - \frac{1}{4}e^{x/2}.$$

The initial conditions imply

$$c_3 + \; c_2 \quad\;\; = 1$$

$$\frac{1}{2}c_3 - \frac{1}{2}c_2 - \frac{1}{4} = 0.$$

Thus $c_3 = 3/4$ and $c_2 = 1/4$, and

$$y = \frac{3}{4}e^{x/2} + \frac{1}{4}e^{-x/2} + \frac{1}{8}x^2e^{x/2} - \frac{1}{4}xe^{x/2}.$$

20. The auxiliary equation is $2m^2 + m - 1 = (2m-1)(m+1) = 0$, so $y_c = c_1e^{x/2} + c_2e^{-x}$ and

$$W = \begin{vmatrix} e^{x/2} & e^{-x} \\ \frac{1}{2}e^{x/2} & -e^{-x} \end{vmatrix} = -\frac{3}{2}e^{-x/2}.$$

Identifying $f(x) = (x+1)/2$ we obtain

$$u_1' = \frac{1}{3}e^{-x/2}(x+1)$$

$$u_2' = -\frac{1}{3}e^x(x+1).$$

Then

$$u_1 = -e^{-x/2}\left(\frac{2}{3}x - 2\right)$$

$$u_2 = -\frac{1}{3}xe^x.$$

Thus

$$y = c_1e^{x/2} + c_2e^{-x} - x - 2$$

and

$$y' = \frac{1}{2}c_1e^{x/2} - c_2e^{-x} - 1.$$

The initial conditions imply

$$c_1 - c_2 - 2 = 1$$

$$\frac{1}{2}c_1 - c_2 - 1 = 0.$$

Thus $c_1 = 8/3$ and $c_2 = 1/3$, and

$$y = \frac{8}{3}e^{x/2} + \frac{1}{3}e^{-x} - x - 2.$$

21. The auxiliary equation is $m^2 + 2m - 8 = (m-2)(m+4) = 0$, so $y_c = c_1e^{2x} + c_2e^{-4x}$ and

$$W = \begin{vmatrix} e^{2x} & e^{-4x} \\ 2e^{2x} & -4e^{-4x} \end{vmatrix} = -6e^{-2x}.$$

Identifying $f(x) = 2e^{-2x} - e^{-x}$ we obtain

$$u_1' = \frac{1}{3}e^{-4x} - \frac{1}{6}e^{-3x}$$

$$u_2' = \frac{1}{6}e^{3x} - \frac{1}{3}e^{2x}.$$

Then

$$u_1 = -\frac{1}{12}e^{-4x} + \frac{1}{18}e^{-3x}$$

$$u_2 = \frac{1}{18}e^{3x} - \frac{1}{6}e^{2x}.$$

Thus

$$y = c_1 e^{2x} + c_2 e^{-4x} - \frac{1}{12}e^{-2x} + \frac{1}{18}e^{-x} + \frac{1}{18}e^{-x} - \frac{1}{6}e^{-2x}$$

$$= c_1 e^{2x} + c_2 e^{-4x} - \frac{1}{4}e^{-2x} + \frac{1}{9}e^{-x}$$

and

$$y' = 2c_1 e^{2x} - 4c_2 e^{-4x} + \frac{1}{2}e^{-2x} - \frac{1}{9}e^{-x}.$$

The initial conditions imply

$$c_1 + c_2 - \frac{5}{36} = 1$$

$$2c_1 - 4c_2 + \frac{7}{18} = 0.$$

Thus $c_1 = 25/36$ and $c_2 = 4/9$, and

$$y = \frac{25}{36}e^{2x} + \frac{4}{9}e^{-4x} - \frac{1}{4}e^{-2x} + \frac{1}{9}e^{-x}.$$

22. The auxiliary equation is $m^2 - 4m + 4 = (m-2)^2 = 0$, so $y_c = c_1 e^{2x} + c_2 x e^{2x}$ and

$$W = \begin{vmatrix} e^{2x} & xe^{2x} \\ 2e^{2x} & 2xe^{2x} + e^{2x} \end{vmatrix} = e^{4x}.$$

Identifying $f(x) = \left(12x^2 - 6x\right)e^{2x}$ we obtain

$$u_1' = 6x^2 - 12x^3$$

$$u_2' = 12x^2 - 6x.$$

Then

$$u_1 = 2x^3 - 3x^4$$

$$u_2 = 4x^3 - 3x^2.$$

Thus

$$y = c_1 e^{2x} + c_2 x e^{2x} + \left(2x^3 - 3x^4\right) e^{2x} + \left(4x^3 - 3x^2\right) x e^{2x}$$

$$= c_1 e^{2x} + c_2 x e^{2x} + e^{2x} \left(x^4 - x^3\right)$$

and

$$y' = 2c_1 e^{2x} + c_2 \left(2x e^{2x} + e^{2x}\right) + e^{2x} \left(4x^3 - 3x^2\right) + 2e^{2x} \left(x^4 - x^3\right).$$

The initial conditions imply

$$c_1 = 1$$

$$2c_1 + c_2 = 0.$$

Thus $c_1 = 1$ and $c_2 = -2$, and

$$y = e^{2x} - 2x e^{2x} + e^{2x} \left(x^4 - x^3\right) = e^{2x} \left(x^4 - x^3 - 2x + 1\right).$$

23. Write the equation in the form

$$y'' + \frac{1}{x}y' + \left(1 - \frac{1}{4x^2}\right)y = x^{-1/2}$$

and identify $f(x) = x^{-1/2}$. From $y_1 = x^{-1/2}\cos x$ and $y_2 = x^{-1/2}\sin x$ we compute

$$W(y_1, y_2) = \begin{vmatrix} x^{-1/2}\cos x & x^{-1/2}\sin x \\ -x^{-1/2}\sin x - \frac{1}{2}x^{-3/2}\cos x & x^{-1/2}\cos x - \frac{1}{2}x^{-3/2}\sin x \end{vmatrix} = \frac{1}{x}.$$

Now

$$u_1' = -\sin x \quad \text{so} \quad u_1 = \cos x,$$

and

$$u_2' = \cos x \quad \text{so} \quad u_2 = \sin x.$$

Thus a particular solution is

$$y_p = x^{-1/2}\cos^2 x + x^{-1/2}\sin^2 x,$$

and the general solution is

$$y = c_1 x^{-1/2}\cos x + c_2 x^{-1/2}\sin x + x^{-1/2}\cos^2 x + x^{-1/2}\sin^2 x$$

$$= c_1 x^{-1/2}\cos x + c_2 x^{-1/2}\sin x + x^{-1/2}.$$

24. Write the equation in the form

$$y'' + \frac{1}{x}y' + \frac{1}{x^2}y = \frac{\sec(\ln x)}{x^2}$$

and identify $f(x) = \sec(\ln x)/x^2$. From $y_1 = \cos(\ln x)$ and $y_2 = \sin(\ln x)$ we compute

$$W = \begin{vmatrix} \cos(\ln x) & \sin(\ln x) \\ -\dfrac{\sin(\ln x)}{x} & \dfrac{\cos(\ln x)}{x} \end{vmatrix} = \frac{1}{x}.$$

Now

$$u_1' = -\frac{\tan(\ln x)}{x} \quad \text{so} \quad u_1 = \ln|\cos(\ln x)|,$$

and

$$u_2' = \frac{1}{x} \quad \text{so} \quad u_2 = \ln x.$$

Thus, a particular solution is

$$y_p = \cos(\ln x)\ln|\cos(\ln x)| + (\ln x)\sin(\ln x),$$

and the general solution is

$$y = c_1\cos(\ln x) + c_2\sin(\ln x) + \cos(\ln x)\ln|\cos(\ln x)| + (\ln x)\sin(\ln x).$$

25. The auxiliary equation is $m^3 + m = m(m^2 + 1) = 0$, so $y_c = c_1 + c_2\cos x + c_3\sin x$ and

$$W = \begin{vmatrix} 1 & \cos x & \sin x \\ 0 & -\sin x & \cos x \\ 0 & -\cos x & -\sin x \end{vmatrix} = 1.$$

Identifying $f(x) = \tan x$ we obtain

$$u_1' = W_1 = \begin{vmatrix} 0 & \cos x & \sin x \\ 0 & -\sin x & \cos x \\ \tan x & -\cos x & -\sin x \end{vmatrix} = \tan x$$

$$u_2' = W_2 = \begin{vmatrix} 1 & 0 & \sin x \\ 0 & 0 & \cos x \\ 0 & \tan x & -\sin x \end{vmatrix} = -\sin x$$

$$u_3' = W_3 = \begin{vmatrix} 1 & \cos x & 0 \\ 0 & -\sin x & 0 \\ 0 & -\cos x & \tan x \end{vmatrix} = -\sin x\tan x = \frac{\cos^2 x - 1}{\cos x} = \cos x - \sec x.$$

Then

$$u_1 = -\ln|\cos x|$$

$$u_2 = \cos x$$

$$u_3 = \sin x - \ln|\sec x + \tan x|$$

and

$$y = c_1 + c_2\cos x + c_3\sin x - \ln|\cos x| + \cos^2 x$$

$$+ \sin^2 x - \sin x\ln|\sec x + \tan x|$$

$$= c_4 + c_2\cos x + c_3\sin x - \ln|\cos x| - \sin x\ln|\sec x + \tan x|$$

for $-\pi/2 < x < \pi/2$.

26. The auxiliary equation is $m^3 + 4m = m\left(m^2 + 4\right) = 0$, so $y_c = c_1 + c_2 \cos 2x + c_3 \sin 2x$ and

$$W = \begin{vmatrix} 1 & \cos 2x & \sin 2x \\ 0 & -2\sin 2x & 2\cos 2x \\ 0 & -4\cos 2x & -4\sin 2x \end{vmatrix} = 8.$$

Identifying $f(x) = \sec 2x$ we obtain

$$u_1' = \frac{1}{8}W_1 = \frac{1}{8}\begin{vmatrix} 0 & \cos 2x & \sin 2x \\ 0 & -2\sin 2x & 2\cos 2x \\ \sec 2x & -4\cos 2x & -4\sin 2x \end{vmatrix} = \frac{1}{4}\sec 2x$$

$$u_2' = \frac{1}{8}W_2 = \frac{1}{8}\begin{vmatrix} 1 & 0 & \sin 2x \\ 0 & 0 & 2\cos 2x \\ 0 & \sec 2x & -4\sin 2x \end{vmatrix} = -\frac{1}{4}$$

$$u_3' = \frac{1}{8}W_3 = \frac{1}{8}\begin{vmatrix} 1 & \cos 2x & 0 \\ 0 & -2\sin 2x & 0 \\ 0 & -4\cos 2x & \sec 2x \end{vmatrix} = -\frac{1}{4}\tan 2x.$$

Then

$$u_1 = \frac{1}{8}\ln|\sec 2x + \tan 2x|$$

$$u_2 = -\frac{1}{4}x$$

$$u_3 = \frac{1}{8}\ln|\cos 2x|$$

and

$$y = c_1 + c_2\cos 2x + c_3\sin 2x + \frac{1}{8}\ln|\sec 2x + \tan 2x| - \frac{1}{4}x\cos 2x + \frac{1}{8}\sin 2x\ln|\cos 2x|$$

for $-\pi/4 < x < \pi/4$.

27. The auxiliary equation is $3m^2 - 6m + 30 = 0$, which has roots $1 \pm 3i$, so $y_c = e^x(c_1\cos 3x + c_2\sin 3x)$. We consider first the differential equation $3y'' - 6y' + 30y = 15\sin x$, which can be solved using undetermined coefficients. Letting $y_{p_1} = A\cos x + B\sin x$ and substituting into the differential equation we get

$$(27A - 6B)\cos x + (6A + 27B)\sin x = 15\sin x.$$

Then

$$27A - 6B = 0 \quad \text{and} \quad 6A + 27B = 15,$$

so $A = \frac{2}{17}$ and $B = \frac{9}{17}$. Thus, $y_{p_1} = \frac{2}{17}\cos x + \frac{9}{17}\sin x$. Next, we consider the differential equation $3y'' - 6y' + 30y$, for which a particular solution y_{p_2} can be found using variation of parameters. The Wronskian is

$$W = \begin{vmatrix} e^x \cos 3x & e^x \sin 3x \\ e^x \cos 3x - 3e^x \sin 3x & 3e^x \cos 3x + e^x \sin 3x \end{vmatrix} = 3e^{2x}.$$

Identifying $f(x) = \frac{1}{3}e^x \tan x$ we obtain

$$u_1' = -\frac{1}{9}\sin 3x \tan 3x = -\frac{1}{9}\left(\frac{\sin^2 3x}{\cos 3x}\right) = -\frac{1}{9}\left(\frac{1 - \cos^2 3x}{\cos 3x}\right) = -\frac{1}{9}(\sec 3x - \cos 3x)$$

so

$$u_1 = -\frac{1}{27}\ln|\sec 3x + \tan 3x| + \frac{1}{27}\sin 3x.$$

Next

$$u_2' = \frac{1}{9}\sin 3x \quad \text{so} \quad u_2 = -\frac{1}{27}\cos 3x.$$

Thus

$$y_{p_2} = -\frac{1}{27}e^x \cos 3x(\ln|\sec 3x + \tan 3x| - \sin 3x) - \frac{1}{27}e^x \sin 3x \cos 3x$$

$$= -\frac{1}{27}e^x(\cos 3x)\ln|\sec 3x + \tan 3x|$$

and the general solution of the original differential equation is

$$y = e^x(c_1 \cos 3x + c_2 \sin 3x) + y_{p_1}(x) + y_{p_2}(x).$$

28. The auxiliary equation is $m^2 - 2m + 1 = (m-1)^2 = 0$, which has repeated root 1, so $y_c = c_1 e^x + c_2 x e^x$. We consider first the differential equation $y'' - 2y' + y = 4x^2 - 3$, which can be solved using undetermined coefficients. Letting $y_{p_1} = Ax^2 + Bx + C$ and substituting into the differential equation we get

$$Ax^2 + (-4A + B)x + (2A - 2B + C) = 4x^2 - 3.$$

Then

$$A = 4, \quad -4A + B = 0, \quad \text{and} \quad 2A - 2B + C = -3,$$

so $A = 4$, $B = 16$, and $C = 21$. Thus, $y_{p_1} = 4x^2 + 16x + 21$. Next we consider the differential equation $y'' - 2y' + y = x^{-1}e^x$, for which a particular solution y_{p_2} can be found using variation of parameters. The Wronskian is

$$W = \begin{vmatrix} e^x & xe^x \\ e^x & xe^x + e^x \end{vmatrix} = e^{2x}.$$

Identifying $f(x) = e^x/x$ we obtain $u_1' = -1$ and $u_2' = 1/x$. Then $u_1 = -x$ and $u_2 = \ln x$, so that

$$y_{p_2} = -xe^x + xe^x \ln x,$$

195

and the general solution of the original differential equation is

$$y = y_c + y_{p_1} + y_{p_2} = c_1 e^x + c_2 x e^x + 4x^2 + 16x + 21 - x e^x + x e^x \ln x$$
$$= c_1 e^x + c_3 x e^x + 4x^2 + 16x + 21 + x e^x \ln x$$

29. The interval of definition for Problem 1 is $(-\pi/2, \pi/2)$, for Problem 7 is $(-\infty, \infty)$, for Problem 9 is $(0, \infty)$, and for Problem 18 is $(-1, 1)$. In Problem 24 the general solution is

$$y = c_1 \cos(\ln x) + c_2 \sin(\ln x) + \cos(\ln x) \ln|\cos(\ln x)| + (\ln x) \sin(\ln x)$$

for $-\pi/2 < \ln x < \pi/2$ or $e^{-\pi/2} < x < e^{\pi/2}$. The bounds on $\ln x$ are due to the presence of $\sec(\ln x)$ in the differential equation.

30. We are given that $y_1 = x^2$ is a solution of $x^4 y'' + x^3 y' - 4x^2 y = 0$. To find a second solution we use reduction of order. Let $y = x^2 u(x)$. Then the product rule gives

$$y' = x^2 u' + 2xu \quad \text{and} \quad y'' = x^2 u'' + 4xu' + 2u,$$

so

$$x^4 y'' + x^3 y' - 4x^2 y = x^5 (x u'' + 5u') = 0.$$

Letting $w = u'$, this becomes $xw' + 5w = 0$. Separating variables and integrating we have

$$\frac{dw}{w} = -\frac{5}{x} dx \quad \text{and} \quad \ln|w| = -5 \ln x + c.$$

Thus, $w = x^{-5}$ and $u = -\frac{1}{4} x^{-4}$. A second solution is then $y_2 = x^2 x^{-4} = 1/x^2$, and the general solution of the homogeneous differential equation is $y_c = c_1 x^2 + c_2/x^2$. To find a particular solution, y_p, we use variation of parameters. The Wronskian is

$$W = \begin{vmatrix} x^2 & 1/x^2 \\ 2x & -2/x^3 \end{vmatrix} = -\frac{4}{x}.$$

Identifying $f(x) = 1/x^4$ we obtain $u_1' = \frac{1}{4} x^{-5}$ and $u_2' = -\frac{1}{4} x^{-1}$. Then $u_1 = -\frac{1}{16} x^{-4}$ and $u_2 = -\frac{1}{4} \ln x$, so

$$y_p = -\frac{1}{16} x^{-4} x^2 - \frac{1}{4} (\ln x) x^{-2} = -\frac{1}{16} x^{-2} - \frac{1}{4} x^{-2} \ln x.$$

The general solution is

$$y = c_1 x^2 + \frac{c_2}{x^2} - \frac{1}{16x^2} - \frac{1}{4x^2} \ln x.$$

31. Suppose $y_p(x) = u_1(x) y_1(x) + u_2(x) y_2(x)$, where u_1 and u_2 are defined by (5) of Section 4.6 in the

196

text. Then, for x and x_0 in I,

$$y_p(x) = y_1(x) \int_{x_0}^x \frac{-y_2(t)f(t)}{W(t)}\, dt + y_2(x) \int_{x_0}^x \frac{y_1(t)f(t)}{W(t)}\, dt$$

$$= \int_{x_0}^x \frac{-y_1(x)y_2(t)f(t)}{W(t)}\, dt + \int_{x_0}^x \frac{y_1(t)y_2(x)f(t)}{W(t)}\, dt$$

$$= \int_{x_0}^x \left[\frac{y_1(t)y_2(x)f(t)}{W(t)} + \frac{-y_1(x)y_2(t)f(t)}{W(t)} \right] dt$$

$$= \int_{x_0}^x \frac{y_1(t)y_2(x)f(t) - y_1(x)y_2(t)f(t)}{W(t)}\, dt$$

$$= \int_{x_0}^x \frac{y_1(t)y_2(x) - y_1(x)y_2(t)}{W(t)}\, f(t)dt$$

$$= \int_{x_0}^x G(x,t)f(t)\, dt.$$

32. In the solution of Example 3 in the text we saw that $y_1 = e^x$, $y_2 = e^{-x}$, $f(x) = 1/x$, and $W(y_1, y_2) = -2$. From (13) the Green's function for the differential equation is

$$G(x,t) = \frac{e^t e^{-x} - e^x e^{-t}}{-2} = \frac{e^{x-t} - e^{-(x-t)}}{2} = \sinh(x-t).$$

The general solution of the differential equation on any interval $[x_0, x]$ not containing the origin is then

$$y = c_1 e^x + c_2 e^{-x} + \int_{x_0}^x \frac{\sinh(x-t)}{t}\, dt.$$

33. We already know that $y_p(x)$ is a particular solution of the differential equation. We simply need to show that it satisfies the initial conditions. Certainly

$$y(x_0) = \int_{x_0}^{x_0} G(x,t)f(t)dt = 0.$$

Using Leibniz's rule for differentiation under an integral sign we have

$$y_p'(x) = \frac{d}{dx} \int_{x_0}^x G(x,t)f(t)dt = \int_{x_0}^x \frac{d}{dx} G(x,t)f(t)dt + f(t)G(x,x) \cdot 1 - f(t)G(x_0, x) \cdot 0.$$

From (13) in the text, $G(x,x) = 0$ so

$$y_p'(x) = \frac{d}{dx} \int_{x_0}^x G(x,t)f(t)dt$$

and

$$y_p'(x_0) = \frac{d}{dx} \int_{x_0}^{x_0} G(x,t)f(t)dt = 0.$$

197

34. From the solution of Problem 32 we have that a particular solution of the differential equation is

$$y_p(x) = \int_0^x G(x,t)e^{2t}dt,$$

where $G(x,t) = \sinh(x-t)$. Then

$$y_p(x) = \int_0^x e^{2t}\sinh(x-t)dt = \int_0^x e^{2t}\frac{e^{x-t} - e^{-(x-t)}}{2}dt$$

$$= \frac{1}{2}\int_0^x \left[e^{x+t} - e^{-x+3t}\right]dt = \frac{1}{2}\left[e^{x+t} - \frac{1}{3}e^{-x+3t}\right]\Big|_0^x$$

$$= \frac{1}{2}e^{2x} - \frac{1}{6}e^{2x} - \frac{1}{2}e^{x} + \frac{1}{6}e^{-x} = \frac{1}{3}e^{2x} - \frac{1}{2}e^{x} + \frac{1}{6}e^{-x}.$$

Exercises 4.7

Cauchy-Euler Equation

1. The auxiliary equation is $m^2 - m - 2 = (m+1)(m-2) = 0$ so that $y = c_1 x^{-1} + c_2 x^2$.

2. The auxiliary equation is $4m^2 - 4m + 1 = (2m-1)^2 = 0$ so that $y = c_1 x^{1/2} + c_2 x^{1/2}\ln x$.

3. The auxiliary equation is $m^2 = 0$ so that $y = c_1 + c_2\ln x$.

4. The auxiliary equation is $m^2 - 4m = m(m-4) = 0$ so that $y = c_1 + c_2 x^4$.

5. The auxiliary equation is $m^2 + 4 = 0$ so that $y = c_1\cos(2\ln x) + c_2\sin(2\ln x)$.

6. The auxiliary equation is $m^2 + 4m + 3 = (m+1)(m+3) = 0$ so that $y = c_1 x^{-1} + c_2 x^{-3}$.

7. The auxiliary equation is $m^2 - 4m - 2 = 0$ so that $y = c_1 x^{2-\sqrt{6}} + c_2 x^{2+\sqrt{6}}$.

8. The auxiliary equation is $m^2 + 2m - 4 = 0$ so that $y = c_1 x^{-1+\sqrt{5}} + c_2 x^{-1-\sqrt{5}}$.

9. The auxiliary equation is $25m^2 + 1 = 0$ so that $y = c_1\cos\left(\frac{1}{5}\ln x\right) + c_2\sin\left(\frac{1}{5}\ln x\right)$.

10. The auxiliary equation is $4m^2 - 1 = (2m-1)(2m+1) = 0$ so that $y = c_1 x^{1/2} + c_2 x^{-1/2}$.

11. The auxiliary equation is $m^2 + 4m + 4 = (m+2)^2 = 0$ so that $y = c_1 x^{-2} + c_2 x^{-2}\ln x$.

12. The auxiliary equation is $m^2 + 7m + 6 = (m+1)(m+6) = 0$ so that $y = c_1 x^{-1} + c_2 x^{-6}$.

13. The auxiliary equation is $3m^2 + 3m + 1 = 0$ so that

$$y = x^{-1/2}\left[c_1\cos\left(\frac{\sqrt{3}}{6}\ln x\right) + c_2\sin\left(\frac{\sqrt{3}}{6}\ln x\right)\right].$$

14. The auxiliary equation is $m^2 - 8m + 41 = 0$ so that $y = x^4\left[c_1\cos(5\ln x) + c_2\sin(5\ln x)\right]$.

15. Assuming that $y = x^m$ and substituting into the differential equation we obtain

$$m(m-1)(m-2) - 6 = m^3 - 3m^2 + 2m - 6 = (m-3)(m^2+2) = 0.$$

Thus

$$y = c_1 x^3 + c_2 \cos\left(\sqrt{2}\ln x\right) + c_3 \sin\left(\sqrt{2}\ln x\right).$$

16. Assuming that $y = x^m$ and substituting into the differential equation we obtain

$$m(m-1)(m-2) + m - 1 = m^3 - 3m^2 + 3m - 1 = (m-1)^3 = 0.$$

Thus

$$y = c_1 x + c_2 x \ln x + c_3 x (\ln x)^2.$$

17. Assuming that $y = x^m$ and substituting into the differential equation we obtain

$$m(m-1)(m-2)(m-3) + 6m(m-1)(m-2) = m^4 - 7m^2 + 6m = m(m-1)(m-2)(m+3) = 0.$$

Thus

$$y = c_1 + c_2 x + c_3 x^2 + c_4 x^{-3}.$$

18. Assuming that $y = x^m$ and substituting into the differential equation we obtain

$$m(m-1)(m-2)(m-3) + 6m(m-1)(m-2) + 9m(m-1) + 3m + 1 = m^4 + 2m^2 + 1 = (m^2+1)^2 = 0.$$

Thus

$$y = c_1 \cos(\ln x) + c_2 \sin(\ln x) + c_3 (\ln x) \cos(\ln x) + c_4 (\ln x) \sin(\ln x).$$

19. The auxiliary equation is $m^2 - 5m = m(m-5) = 0$ so that $y_c = c_1 + c_2 x^5$ and

$$W(1, x^5) = \begin{vmatrix} 1 & x^5 \\ 0 & 5x^4 \end{vmatrix} = 5x^4.$$

Identifying $f(x) = x^3$ we obtain $u_1' = -\frac{1}{5}x^4$ and $u_2' = 1/5x$. Then $u_1 = -\frac{1}{25}x^5$, $u_2 = \frac{1}{5}\ln x$, and

$$y = c_1 + c_2 x^5 - \frac{1}{25}x^5 + \frac{1}{5}x^5 \ln x = c_1 + c_3 x^5 + \frac{1}{5}x^5 \ln x.$$

20. The auxiliary equation is $2m^2 + 3m + 1 = (2m+1)(m+1) = 0$ so that $y_c = c_1 x^{-1} + c_2 x^{-1/2}$ and

$$W(x^{-1}, x^{-1/2}) = \begin{vmatrix} x^{-1} & x^{-1/2} \\ -x^{-2} & -\frac{1}{2}x^{-3/2} \end{vmatrix} = \frac{1}{2}x^{-5/2}.$$

Identifying $f(x) = \frac{1}{2} - \frac{1}{2x}$ we obtain $u_1' = x - x^2$ and $u_2' = x^{3/2} - x^{1/2}$. Then $u_1 = \frac{1}{2}x^2 - \frac{1}{3}x^3$, $u_2 = \frac{2}{5}x^{5/2} - \frac{2}{3}x^{3/2}$, and

$$y = c_1 x^{-1} + c_2 x^{-1/2} + \frac{1}{2}x - \frac{1}{3}x^2 + \frac{2}{5}x^2 - \frac{2}{3}x = c_1 x^{-1} + c_2 x^{-1/2} - \frac{1}{6}x + \frac{1}{15}x^2.$$

21. The auxiliary equation is $m^2 - 2m + 1 = (m - 1)^2 = 0$ so that $y_c = c_1 x + c_2 x \ln x$ and

$$W(x, x \ln x) = \begin{vmatrix} x & x \ln x \\ 1 & 1 + \ln x \end{vmatrix} = x.$$

Identifying $f(x) = 2/x$ we obtain $u_1' = -2\ln x/x$ and $u_2' = 2/x$. Then $u_1 = -(\ln x)^2$, $u_2 = 2\ln x$, and

$$y = c_1 x + c_2 x \ln x - x(\ln x)^2 + 2x(\ln x)^2$$

$$= c_1 x + c_2 x \ln x + x(\ln x)^2, \qquad x > 0.$$

22. The auxiliary equation is $m^2 - 3m + 2 = (m - 1)(m - 2) = 0$ so that $y_c = c_1 x + c_2 x^2$ and

$$W(x, x^2) = \begin{vmatrix} x & x^2 \\ 1 & 2x \end{vmatrix} = x^2.$$

Identifying $f(x) = x^2 e^x$ we obtain $u_1' = -x^2 e^x$ and $u_2' = xe^x$. Then $u_1 = -x^2 e^x + 2xe^x - 2e^x$, $u_2 = xe^x - e^x$, and

$$y = c_1 x + c_2 x^2 - x^3 e^x + 2x^2 e^x - 2xe^x + x^3 e^x - x^2 e^x$$

$$= c_1 x + c_2 x^2 + x^2 e^x - 2xe^x.$$

23. The auxiliary equation $m(m - 1) + m - 1 = m^2 - 1 = 0$ has roots $m_1 = -1$, $m_2 = 1$, so $y_c = c_1 x^{-1} + c_2 x$. With $y_1 = x^{-1}$, $y_2 = x$, and the identification $f(x) = \ln x/x^2$, we get

$$W = 2x^{-1}, \qquad W_1 = -\ln x/x, \qquad \text{and} \qquad W_2 = \ln x/x^3.$$

Then $u_1' = W_1/W = -(\ln x)/2$, $u_2' = W_2/W = (\ln x)/2x^2$, and integration by parts gives

$$u_1 = \frac{1}{2}x - \frac{1}{2}x \ln x$$

$$u_2 = -\frac{1}{2}x^{-1} \ln x - \frac{1}{2}x^{-1},$$

so

$$y_p = u_1 y_1 + u_2 y_2 = \left(\frac{1}{2}x - \frac{1}{2}x \ln x\right) x^{-1} + \left(-\frac{1}{2}x^{-1} \ln x - \frac{1}{2}x^{-1}\right) x = -\ln x$$

and

$$y = y_c + y_p = c_1 x^{-1} + c_2 x - \ln x, \qquad x > 0.$$

24. The auxiliary equation $m(m - 1) + m - 1 = m^2 - 1 = 0$ has roots $m_1 = -1$, $m_2 = 1$, so $y_c = c_1 x^{-1} + c_2 x$. With $y_1 = x^{-1}$, $y_2 = x$, and the identification $f(x) = 1/x^2(x + 1)$, we get

$$W = 2x^{-1}, \qquad W_1 = -1/x(x + 1), \qquad \text{and} \qquad W_2 = 1/x^3(x + 1).$$

Then $u_1' = W_1/W = -1/2(x+1)$, $u_2' = W_2/W = 1/2x^2(x+1)$, and integration (by partial fractions for u_2') gives

$$u_1 = -\frac{1}{2}\ln(x+1)$$

$$u_2 = -\frac{1}{2}x^{-1} - \frac{1}{2}\ln x + \frac{1}{2}\ln(x+1),$$

so

$$y_p = u_1 y_1 + u_2 y_2 = \left[-\frac{1}{2}\ln(x+1)\right]x^{-1} + \left[-\frac{1}{2}x^{-1} - \frac{1}{2}\ln x + \frac{1}{2}\ln(x+1)\right]x$$

$$= -\frac{1}{2} - \frac{1}{2}x\ln x + \frac{1}{2}x\ln(x+1) - \frac{\ln(x+1)}{2x} = -\frac{1}{2} + \frac{1}{2}x\ln\left(1+\frac{1}{x}\right) - \frac{\ln(x+1)}{2x}$$

and

$$y = y_c + y_p = c_1 x^{-1} + c_2 x - \frac{1}{2} + \frac{1}{2}x\ln\left(1+\frac{1}{x}\right) - \frac{\ln(x+1)}{2x}, \qquad x > 0.$$

25. The auxiliary equation is $m^2 + 2m = m(m+2) = 0$, so that $y = c_1 + c_2 x^{-2}$ and $y' = -2c_2 x^{-3}$. The initial conditions imply

$$c_1 + c_2 = 0$$

$$-2c_2 = 4.$$

Thus, $c_1 = 2$, $c_2 = -2$, and $y = 2 - 2x^{-2}$. The graph is given to the right.

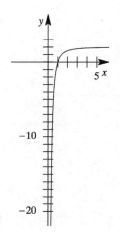

26. The auxiliary equation is $m^2 - 6m + 8 = (m-2)(m-4) = 0$, so that

$$y = c_1 x^2 + c_2 x^4 \quad \text{and} \quad y' = 2c_1 x + 4c_2 x^3.$$

The initial conditions imply

$$4c_1 + 16c_2 = 32$$

$$4c_1 + 32c_2 = 0.$$

Thus, $c_1 = 16$, $c_2 = -2$, and $y = 16x^2 - 2x^4$. The graph is given to the right.

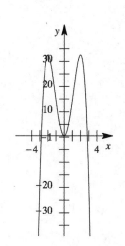

27. The auxiliary equation is $m^2 + 1 = 0$, so that

$$y = c_1 \cos(\ln x) + c_2 \sin(\ln x)$$

and

$$y' = -c_1 \frac{1}{x} \sin(\ln x) + c_2 \frac{1}{x} \cos(\ln x).$$

The initial conditions imply $c_1 = 1$ and $c_2 = 2$. Thus $y = \cos(\ln x) + 2\sin(\ln x)$. The graph is given to the right.

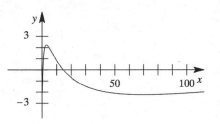

28. The auxiliary equation is $m^2 - 4m + 4 = (m-2)^2 = 0$, so that

$$y = c_1 x^2 + c_2 x^2 \ln x \quad \text{and} \quad y' = 2c_1 x + c_2(x + 2x \ln x).$$

The initial conditions imply $c_1 = 5$ and $c_2 + 10 = 3$. Thus $y = 5x^2 - 7x^2 \ln x$. The graph is given to the right.

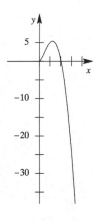

29. The auxiliary equation is $m^2 = 0$ so that $y_c = c_1 + c_2 \ln x$ and

$$W(1, \ln x) = \begin{vmatrix} 1 & \ln x \\ 0 & 1/x \end{vmatrix} = \frac{1}{x}.$$

Identifying $f(x) = 1$ we obtain $u_1' = -x \ln x$ and $u_2' = x$. Then $u_1 = \frac{1}{4}x^2 - \frac{1}{2}x^2 \ln x$, $u_2 = \frac{1}{2}x^2$, and

$$y = c_1 + c_2 \ln x + \frac{1}{4}x^2 - \frac{1}{2}x^2 \ln x + \frac{1}{2}x^2 \ln x = c_1 + c_2 \ln x + \frac{1}{4}x^2.$$

The initial conditions imply $c_1 + \frac{1}{4} = 1$ and $c_2 + \frac{1}{2} = -\frac{1}{2}$. Thus, $c_1 = \frac{3}{4}$, $c_2 = -1$, and $y = \frac{3}{4} - \ln x + \frac{1}{4}x^2$. The graph is given to the right.

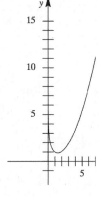

30. The auxiliary equation is $m^2 - 6m + 8 = (m-2)(m-4) = 0$, so that $y_c = c_1 x^2 + c_2 x^4$ and

$$W = \begin{vmatrix} x^2 & x^4 \\ 2x & 4x^3 \end{vmatrix} = 2x^5.$$

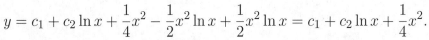

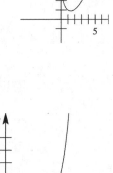

Identifying $f(x) = 8x^4$ we obtain $u_1' = -4x^3$ and $u_2' = 4x$. Then $u_1 = -x^4$, $u_2 = 2x^2$, and $y = c_1 x^2 + c_2 x^4 + x^6$. The initial conditions imply

$$\frac{1}{4}c_1 + \frac{1}{16}c_2 = -\frac{1}{64}$$

$$c_1 + \frac{1}{2}c_2 = -\frac{3}{16}.$$

Thus $c_1 = \frac{1}{16}$, $c_2 = -\frac{1}{2}$, and $y = \frac{1}{16}x^2 - \frac{1}{2}x^4 + x^6$. The graph is given above.

31. Substituting $x = e^t$ into the differential equation we obtain

$$\frac{d^2y}{dt^2} + 8\frac{dy}{dt} - 20y = 0.$$

The auxiliary equation is $m^2 + 8m - 20 = (m+10)(m-2) = 0$ so that

$$y = c_1 e^{-10t} + c_2 e^{2t} = c_1 x^{-10} + c_2 x^2.$$

32. Substituting $x = e^t$ into the differential equation we obtain

$$\frac{d^2y}{dt^2} - 10\frac{dy}{dt} + 25y = 0.$$

The auxiliary equation is $m^2 - 10m + 25 = (m-5)^2 = 0$ so that

$$y = c_1 e^{5t} + c_2 t e^{5t} = c_1 x^5 + c_2 x^5 \ln x.$$

33. Substituting $x = e^t$ into the differential equation we obtain

$$\frac{d^2y}{dt^2} + 9\frac{dy}{dt} + 8y = e^{2t}.$$

The auxiliary equation is $m^2 + 9m + 8 = (m+1)(m+8) = 0$ so that $y_c = c_1 e^{-t} + c_2 e^{-8t}$. Using undetermined coefficients we try $y_p = Ae^{2t}$. This leads to $30Ae^{2t} = e^{2t}$, so that $A = 1/30$ and

$$y = c_1 e^{-t} + c_2 e^{-8t} + \frac{1}{30}e^{2t} = c_1 x^{-1} + c_2 x^{-8} + \frac{1}{30}x^2.$$

34. Substituting $x = e^t$ into the differential equation we obtain

$$\frac{d^2y}{dt^2} - 5\frac{dy}{dt} + 6y = 2t.$$

The auxiliary equation is $m^2 - 5m + 6 = (m-2)(m-3) = 0$ so that $y_c = c_1 e^{2t} + c_2 e^{3t}$. Using undetermined coefficients we try $y_p = At + B$. This leads to $(-5A + 6B) + 6At = 2t$, so that $A = 1/3$, $B = 5/18$, and

$$y = c_1 e^{2t} + c_2 e^{3t} + \frac{1}{3}t + \frac{5}{18} = c_1 x^2 + c_2 x^3 + \frac{1}{3}\ln x + \frac{5}{18}.$$

35. Substituting $x = e^t$ into the differential equation we obtain

$$\frac{d^2y}{dt^2} - 4\frac{dy}{dt} + 13y = 4 + 3e^t.$$

The auxiliary equation is $m^2 - 4m + 13 = 0$ so that $y_c = e^{2t}(c_1 \cos 3t + c_2 \sin 3t)$. Using undetermined coefficients we try $y_p = A + Be^t$. This leads to $13A + 10Be^t = 4 + 3e^t$, so that $A = 4/13$, $B = 3/10$, and

$$y = e^{2t}(c_1 \cos 3t + c_2 \sin 3t) + \frac{4}{13} + \frac{3}{10}e^t$$

$$= x^2 \left[c_1 \cos(3\ln x) + c_2 \sin(3\ln x) \right] + \frac{4}{13} + \frac{3}{10}x.$$

36. From

$$\frac{d^2y}{dx^2} = \frac{1}{x^2}\left(\frac{d^2y}{dt^2} - \frac{dy}{dt} \right)$$

it follows that

$$\frac{d^3y}{dx^3} = \frac{1}{x^2}\frac{d}{dx}\left(\frac{d^2y}{dt^2} - \frac{dy}{dt} \right) - \frac{2}{x^3}\left(\frac{d^2y}{dt^2} - \frac{dy}{dt} \right)$$

$$= \frac{1}{x^2}\frac{d}{dx}\left(\frac{d^2y}{dt^2} \right) - \frac{1}{x^2}\frac{d}{dx}\left(\frac{dy}{dt} \right) - \frac{2}{x^3}\frac{d^2y}{dt^2} + \frac{2}{x^3}\frac{dy}{dt}$$

$$= \frac{1}{x^2}\frac{d^3y}{dt^3}\left(\frac{1}{x} \right) - \frac{1}{x^2}\frac{d^2y}{dt^2}\left(\frac{1}{x} \right) - \frac{2}{x^3}\frac{d^2y}{dt^2} + \frac{2}{x^3}\frac{dy}{dt}$$

$$= \frac{1}{x^3}\left(\frac{d^3y}{dt^3} - 3\frac{d^2y}{dt^2} + 2\frac{dy}{dt} \right).$$

Substituting into the differential equation we obtain

$$\frac{d^3y}{dt^3} - 3\frac{d^2y}{dt^2} + 2\frac{dy}{dt} - 3\left(\frac{d^2y}{dt^2} - \frac{dy}{dt} \right) + 6\frac{dy}{dt} - 6y = 3 + 3t$$

or

$$\frac{d^3y}{dt^3} - 6\frac{d^2y}{dt^2} + 11\frac{dy}{dt} - 6y = 3 + 3t.$$

The auxiliary equation is $m^3 - 6m^2 + 11m - 6 = (m-1)(m-2)(m-3) = 0$ so that $y_c = c_1 e^t + c_2 e^{2t} + c_3 e^{3t}$. Using undetermined coefficients we try $y_p = A + Bt$. This leads to $(11B - 6A) - 6Bt = 3 + 3t$, so that $A = -17/12$, $B = -1/2$, and

$$y = c_1 e^t + c_2 e^{2t} + c_3 e^{3t} - \frac{17}{12} - \frac{1}{2}t = c_1 x + c_2 x^2 + c_3 x^3 - \frac{17}{12} - \frac{1}{2}\ln x.$$

In the next two problems we use the substitution $t = -x$ since the initial conditions are on the interval $(-\infty, 0)$. In this case

$$\frac{dy}{dt} = \frac{dy}{dx}\frac{dx}{dt} = -\frac{dy}{dx}$$

and

$$\frac{d^2y}{dt^2} = \frac{d}{dt}\left(\frac{dy}{dt} \right) = \frac{d}{dt}\left(-\frac{dy}{dx} \right) = -\frac{d}{dt}(y') = -\frac{dy'}{dx}\frac{dx}{dt} = -\frac{d^2y}{dx^2}\frac{dx}{dt} = \frac{d^2y}{dx^2}.$$

37. The differential equation and initial conditions become

$$4t^2 \frac{d^2y}{dt^2} + y = 0; \quad y(t)\Big|_{t=1} = 2, \quad y'(t)\Big|_{t=1} = -4.$$

The auxiliary equation is $4m^2 - 4m + 1 = (2m-1)^2 = 0$, so that

$$y = c_1 t^{1/2} + c_2 t^{1/2} \ln t \quad \text{and} \quad y' = \frac{1}{2}c_1 t^{-1/2} + c_2 \left(t^{-1/2} + \frac{1}{2}t^{-1/2}\ln t \right).$$

The initial conditions imply $c_1 = 2$ and $1 + c_2 = -4$. Thus

$$y = 2t^{1/2} - 5t^{1/2}\ln t = 2(-x)^{1/2} - 5(-x)^{1/2}\ln(-x), \quad x < 0.$$

38. The differential equation and initial conditions become

$$t^2 \frac{d^2y}{dt^2} - 4t \frac{dy}{dt} + 6y = 0; \quad y(t)\Big|_{t=2} = 8, \quad y'(t)\Big|_{t=2} = 0.$$

The auxiliary equation is $m^2 - 5m + 6 = (m-2)(m-3) = 0$, so that

$$y = c_1 t^2 + c_2 t^3 \quad \text{and} \quad y' = 2c_1 t + 3c_2 t^2.$$

The initial conditions imply

$$4c_1 + 8c_2 = 8$$

$$4c_1 + 12c_2 = 0$$

from which we find $c_1 = 6$ and $c_2 = -2$. Thus

$$y = 6t^2 - 2t^3 = 6x^2 + 2x^3, \quad x < 0.$$

39. Letting $u = x + 2$ we obtain $dy/dx = dy/du$ and, using the Chain Rule,

$$\frac{d^2y}{dx^2} = \frac{d}{dx}\left(\frac{dy}{du}\right) = \frac{d^2y}{du^2}\frac{du}{dx} = \frac{d^2y}{du^2}(1) = \frac{d^2y}{du^2}.$$

Substituting into the differential equation we obtain

$$u^2 \frac{d^2y}{du^2} + u \frac{dy}{du} + y = 0.$$

The auxiliary equation is $m^2 + 1 = 0$ so that

$$y = c_1 \cos(\ln u) + c_2 \sin(\ln u) = c_1 \cos[\ln(x+2)] + c_2 \sin[\ln(x+2)].$$

40. If $1 - i$ is a root of the auxiliary equation then so is $1 + i$, and the auxiliary equation is

$$(m-2)[m - (1+i)][m - (1-i)] = m^3 - 4m^2 + 6m - 4 = 0.$$

We need $m^3 - 4m^2 + 6m - 4$ to have the form $m(m-1)(m-2) + bm(m-1) + cm + d$. Expanding this last expression and equating coefficients we get $b = -1$, $c = 3$, and $d = -4$. Thus, the differential equation is

$$x^3 y''' - x^2 y'' + 3xy' - 4y = 0.$$

41. For $x^2y'' = 0$ the auxiliary equation is $m(m-1) = 0$ and the general solution is $y = c_1 + c_2x$. The initial conditions imply $c_1 = y_0$ and $c_2 = y_1$, so $y = y_0 + y_1x$. The initial conditions are satisfied for all real values of y_0 and y_1.

For $x^2y'' - 2xy' + 2y = 0$ the auxiliary equation is $m^2 - 3m + 2 = (m-1)(m-2) = 0$ and the general solution is $y = c_1x + c_2x^2$. The initial condition $y(0) = y_0$ implies $0 = y_0$ and the condition $y'(0) = y_1$ implies $c_1 = y_1$. Thus, the initial conditions are satisfied for $y_0 = 0$ and for all real values of y_1.

For $x^2y'' - 4xy' + 6y = 0$ the auxiliary equation is $m^2 - 5m + 6 = (m-2)(m-3) = 0$ and the general solution is $y = c_1x^2 + c_2x^3$. The initial conditions imply $y(0) = 0 = y_0$ and $y'(0) = 0$. Thus, the initial conditions are satisfied only for $y_0 = y_1 = 0$.

42. The function $y(x) = -\sqrt{x}\cos(\ln x)$ is defined for $x > 0$ and has x-intercepts where $\ln x = \pi/2 + k\pi$ for k an integer or where $x = e^{\pi/2+k\pi}$. Solving $\pi/2 + k\pi = 0.5$ we get $k \approx -0.34$, so $e^{\pi/2+k\pi} < 0.5$ for all negative integers and the graph has infinitely many x-intercepts in the interval $(0, 0.5)$.

43. The auxiliary equation is $2m(m-1)(m-2) - 10.98m(m-1) + 8.5m + 1.3 = 0$, so that $m_1 = -0.053299$, $m_2 = 1.81164$, $m_3 = 6.73166$, and

$$y = c_1x^{-0.053299} + c_2x^{1.81164} + c_3x^{6.73166}.$$

44. The auxiliary equation is $m(m-1)(m-2) + 4m(m-1) + 5m - 9 = 0$, so that $m_1 = 1.40819$ and the two complex roots are $-1.20409 \pm 2.22291i$. The general solution of the differential equation is

$$y = c_1x^{1.40819} + x^{-1.20409}[c_2\cos(2.22291\ln x) + c_3\sin(2.22291\ln x)].$$

45. The auxiliary equation is $m(m-1)(m-2)(m-3) + 6m(m-1)(m-2) + 3m(m-1) - 3m + 4 = 0$, so that $m_1 = m_2 = \sqrt{2}$ and $m_3 = m_4 = -\sqrt{2}$. The general solution of the differential equation is

$$y = c_1x^{\sqrt{2}} + c_2x^{\sqrt{2}}\ln x + c_3x^{-\sqrt{2}} + c_4x^{-\sqrt{2}}\ln x.$$

46. The auxiliary equation is $m(m-1)(m-2)(m-3) - 6m(m-1)(m-2) + 33m(m-1) - 105m + 169 = 0$, so that $m_1 = m_2 = 3 + 2i$ and $m_3 = m_4 = 3 - 2i$. The general solution of the differential equation is

$$y = x^3[c_1\cos(2\ln x) + c_2\sin(2\ln x)] + x^3\ln x[c_3\cos(2\ln x) + c_4\sin(2\ln x)].$$

47. The auxiliary equation

$$m(m-1)(m-2) - m(m-1) - 2m + 6 = m^3 - 4m^2 + m + 6 = 0$$

has roots $m_1 = -1$, $m_2 = 2$, and $m_3 = 3$, so $y_c = c_1x^{-1} + c_2x^2 + c_3x^3$. With $y_1 = x^{-1}$, $y_2 = x^2$, $y_3 = x^3$, and the identification $f(x) = 1/x$, we get from (11) of Section 4.6 in the text

$$W_1 = x^3, \qquad W_2 = -4, \qquad W_3 = 3/x, \qquad \text{and} \qquad W = 12x.$$

Then $u_1' = W_1/W = x^2/12$, $u_2' = W_2/W = -1/3x$, $u_3' = 1/4x^2$, and integration gives

$$u_1 = \frac{x^3}{36}, \qquad u_2 = -\frac{1}{3}\ln x, \qquad \text{and} \qquad u_3 = -\frac{1}{4x},$$

so

$$y_p = u_1 y_1 + u_2 y_2 + u_3 y_3 = \frac{x^3}{36}x^{-1} + x^2\left(-\frac{1}{3}\ln x\right) + x^3\left(-\frac{1}{4x}\right) = -\frac{2}{9}x^2 - \frac{1}{3}x^2\ln x,$$

and

$$y = y_c + y_p = c_1 x^{-1} + c_2 x^2 + c_3 x^3 - \frac{2}{9}x^2 - \frac{1}{3}x^2\ln x, \qquad x > 0.$$

Exercises 4.8

Solving Systems of Linear DEs by Elimination

1. From $Dx = 2x - y$ and $Dy = x$ we obtain $y = 2x - Dx$, $Dy = 2Dx - D^2x$, and $(D^2 - 2D + 1)x = 0$. The solution is

$$x = c_1 e^t + c_2 t e^t$$

$$y = (c_1 - c_2)e^t + c_2 t e^t.$$

2. From $Dx = 4x + 7y$ and $Dy = x - 2y$ we obtain $y = \frac{1}{7}Dx - \frac{4}{7}x$, $Dy = \frac{1}{7}D^2x - \frac{4}{7}Dx$, and $(D^2 - 2D - 15)x = 0$. The solution is

$$x = c_1 e^{5t} + c_2 e^{-3t}$$

$$y = \frac{1}{7}c_1 e^{5t} - c_2 e^{-3t}.$$

3. From $Dx = -y + t$ and $Dy = x - t$ we obtain $y = t - Dx$, $Dy = 1 - D^2x$, and $(D^2 + 1)x = 1 + t$. The solution is

$$x = c_1 \cos t + c_2 \sin t + 1 + t$$

$$y = c_1 \sin t - c_2 \cos t + t - 1.$$

4. From $Dx - 4y = 1$ and $x + Dy = 2$ we obtain $y = \frac{1}{4}Dx - \frac{1}{4}$, $Dy = \frac{1}{4}D^2x$, and $(D^2 + 1)x = 2$. The solution is

$$x = c_1 \cos t + c_2 \sin t + 2$$

$$y = \frac{1}{4}c_2 \cos t - \frac{1}{4}c_1 \sin t - \frac{1}{4}.$$

5. From $(D^2+5)x - 2y = 0$ and $-2x + (D^2+2)y = 0$ we obtain $y = \frac{1}{2}(D^2+5)x$, $D^2y = \frac{1}{2}(D^4+5D^2)x$, and $(D^2+1)(D^2+6)x = 0$. The solution is

$$x = c_1 \cos t + c_2 \sin t + c_3 \cos \sqrt{6}\,t + c_4 \sin \sqrt{6}\,t$$

$$y = 2c_1 \cos t + 2c_2 \sin t - \frac{1}{2}c_3 \cos \sqrt{6}\,t - \frac{1}{2}c_4 \sin \sqrt{6}\,t.$$

6. From $(D+1)x + (D-1)y = 2$ and $3x + (D+2)y = -1$ we obtain $x = -\frac{1}{3} - \frac{1}{3}(D+2)y$, $Dx = -\frac{1}{3}(D^2+2D)y$, and $(D^2+5)y = -7$. The solution is

$$y = c_1 \cos \sqrt{5}\,t + c_2 \sin \sqrt{5}\,t - \frac{7}{5}$$

$$x = \left(-\frac{2}{3}c_1 - \frac{\sqrt{5}}{3}c_2\right) \cos \sqrt{5}\,t + \left(\frac{\sqrt{5}}{3}c_1 - \frac{2}{3}c_2\right) \sin \sqrt{5}\,t + \frac{3}{5}.$$

7. From $D^2x = 4y + e^t$ and $D^2y = 4x - e^t$ we obtain $y = \frac{1}{4}D^2x - \frac{1}{4}e^t$, $D^2y = \frac{1}{4}D^4x - \frac{1}{4}e^t$, and $(D^2+4)(D-2)(D+2)x = -3e^t$. The solution is

$$x = c_1 \cos 2t + c_2 \sin 2t + c_3 e^{2t} + c_4 e^{-2t} + \frac{1}{5}e^t$$

$$y = -c_1 \cos 2t - c_2 \sin 2t + c_3 e^{2t} + c_4 e^{-2t} - \frac{1}{5}e^t.$$

8. From $(D^2+5)x + Dy = 0$ and $(D+1)x + (D-4)y = 0$ we obtain $(D-5)(D^2+4)x = 0$ and $(D-5)(D^2+4)y = 0$. The solution is

$$x = c_1 e^{5t} + c_2 \cos 2t + c_3 \sin 2t$$

$$y = c_4 e^{5t} + c_5 \cos 2t + c_6 \sin 2t.$$

Substituting into $(D+1)x + (D-4)y = 0$ gives

$$(6c_1 + c_4)e^{5t} + (c_2 + 2c_3 - 4c_5 + 2c_6) \cos 2t + (-2c_2 + c_3 - 2c_5 - 4c_6) \sin 2t = 0$$

so that $c_4 = -6c_1$, $c_5 = \frac{1}{2}c_3$, $c_6 = -\frac{1}{2}c_2$, and

$$y = -6c_1 e^{5t} + \frac{1}{2}c_3 \cos 2t - \frac{1}{2}c_2 \sin 2t.$$

9. From $Dx + D^2y = e^{3t}$ and $(D+1)x + (D-1)y = 4e^{3t}$ we obtain $D(D^2+1)x = 34e^{3t}$ and $D(D^2+1)y = -8e^{3t}$. The solution is

$$y = c_1 + c_2 \sin t + c_3 \cos t - \frac{4}{15}e^{3t}$$

$$x = c_4 + c_5 \sin t + c_6 \cos t + \frac{17}{15}e^{3t}.$$

Substituting into $(D+1)x + (D-1)y = 4e^{3t}$ gives

$$(c_4 - c_1) + (c_5 - c_6 - c_3 - c_2) \sin t + (c_6 + c_5 + c_2 - c_3) \cos t = 0$$

so that $c_4 = c_1$, $c_5 = c_3$, $c_6 = -c_2$, and

$$x = c_1 - c_2 \cos t + c_3 \sin t + \frac{17}{15}e^{3t}.$$

10. From $D^2x - Dy = t$ and $(D+3)x + (D+3)y = 2$ we obtain $D(D+1)(D+3)x = 1 + 3t$ and $D(D+1)(D+3)y = -1 - 3t$. The solution is

$$x = c_1 + c_2e^{-t} + c_3e^{-3t} - t + \frac{1}{2}t^2$$

$$y = c_4 + c_5e^{-t} + c_6e^{-3t} + t - \frac{1}{2}t^2.$$

Substituting into $(D+3)x + (D+3)y = 2$ and $D^2x - Dy = t$ gives

$$3(c_1 + c_4) + 2(c_2 + c_5)e^{-t} = 2$$

and

$$(c_2 + c_5)e^{-t} + 3(3c_3 + c_6)e^{-3t} = 0$$

so that $c_4 = -c_1$, $c_5 = -c_2$, $c_6 = -3c_3$, and

$$y = -c_1 - c_2e^{-t} - 3c_3e^{-3t} + t - \frac{1}{2}t^2.$$

11. From $(D^2 - 1)x - y = 0$ and $(D-1)x + Dy = 0$ we obtain $y = (D^2 - 1)x$, $Dy = (D^3 - D)x$, and $(D-1)(D^2 + D + 1)x = 0$. The solution is

$$x = c_1e^t + e^{-t/2}\left[c_2 \cos \frac{\sqrt{3}}{2}t + c_3 \sin \frac{\sqrt{3}}{2}t\right]$$

$$y = \left(-\frac{3}{2}c_2 - \frac{\sqrt{3}}{2}c_3\right)e^{-t/2} \cos \frac{\sqrt{3}}{2}t + \left(\frac{\sqrt{3}}{2}c_2 - \frac{3}{2}c_3\right)e^{-t/2} \sin \frac{\sqrt{3}}{2}t.$$

12. From $(2D^2 - D - 1)x - (2D+1)y = 1$ and $(D-1)x + Dy = -1$ we obtain $(2D+1)(D-1)(D+1)x = -1$ and $(2D+1)(D+1)y = -2$. The solution is

$$x = c_1e^{-t/2} + c_2e^{-t} + c_3e^t + 1$$

$$y = c_4e^{-t/2} + c_5e^{-t} - 2.$$

Substituting into $(D-1)x + Dy = -1$ gives

$$\left(-\frac{3}{2}c_1 - \frac{1}{2}c_4\right)e^{-t/2} + (-2c_2 - c_5)e^{-t} = 0$$

so that $c_4 = -3c_1$, $c_5 = -2c_2$, and

$$y = -3c_1e^{-t/2} - 2c_2e^{-t} - 2.$$

13. From $(2D-5)x+Dy = e^t$ and $(D-1)x+Dy = 5e^t$ we obtain $Dy = (5-2D)x+e^t$ and $(4-D)x = 4e^t$. Then

$$x = c_1 e^{4t} + \frac{4}{3}e^t$$

and $Dy = -3c_1 e^{4t} + 5e^t$ so that

$$y = -\frac{3}{4}c_1 e^{4t} + c_2 + 5e^t.$$

14. From $Dx+Dy = e^t$ and $(-D^2+D+1)x+y = 0$ we obtain $y = (D^2-D-1)x$, $Dy = (D^3-D^2-D)x$, and $D^2(D-1)x = e^t$. The solution is

$$x = c_1 + c_2 t + c_3 e^t + te^t$$

$$y = -c_1 - c_2 - c_2 t - c_3 e^t - te^t + e^t.$$

15. Multiplying the first equation by $D+1$ and the second equation by D^2+1 and subtracting we obtain $(D^4 - D^2)x = 1$. Then

$$x = c_1 + c_2 t + c_3 e^t + c_4 e^{-t} - \frac{1}{2}t^2.$$

Multiplying the first equation by $D+1$ and subtracting we obtain $D^2(D+1)y = 1$. Then

$$y = c_5 + c_6 t + c_7 e^{-t} - \frac{1}{2}t^2.$$

Substituting into $(D-1)x + (D^2+1)y = 1$ gives

$$(-c_1 + c_2 + c_5 - 1) + (-2c_4 + 2c_7)e^{-t} + (-1 - c_2 + c_6)t = 1$$

so that $c_5 = c_1 - c_2 + 2$, $c_6 = c_2 + 1$, and $c_7 = c_4$. The solution of the system is

$$x = c_1 + c_2 t + c_3 e^t + c_4 e^{-t} - \frac{1}{2}t^2$$

$$y = (c_1 - c_2 + 2) + (c_2 + 1)t + c_4 e^{-t} - \frac{1}{2}t^2.$$

16. From $D^2 x - 2(D^2 + D)y = \sin t$ and $x + Dy = 0$ we obtain $x = -Dy$, $D^2 x = -D^3 y$, and $D(D^2 + 2D + 2)y = -\sin t$. The solution is

$$y = c_1 + c_2 e^{-t}\cos t + c_3 e^{-t}\sin t + \frac{1}{5}\cos t + \frac{2}{5}\sin t$$

$$x = (c_2 + c_3)e^{-t}\sin t + (c_2 - c_3)e^{-t}\cos t + \frac{1}{5}\sin t - \frac{2}{5}\cos t.$$

17. From $Dx = y$, $Dy = z$. and $Dz = x$ we obtain $x = D^2 y = D^3 x$ so that $(D-1)(D^2+D+1)x = 0$,

$$x = c_1 e^t + e^{-t/2}\left[c_2 \sin\frac{\sqrt{3}}{2}t + c_3 \cos\frac{\sqrt{3}}{2}t\right],$$

$$y = c_1 e^t + \left(-\frac{1}{2}c_2 - \frac{\sqrt{3}}{2}c_3\right) e^{-t/2} \sin \frac{\sqrt{3}}{2}t + \left(\frac{\sqrt{3}}{2}c_2 - \frac{1}{2}c_3\right) e^{-t/2} \cos \frac{\sqrt{3}}{2}t,$$

and

$$z = c_1 e^t + \left(-\frac{1}{2}c_2 + \frac{\sqrt{3}}{2}c_3\right) e^{-t/2} \sin \frac{\sqrt{3}}{2}t + \left(-\frac{\sqrt{3}}{2}c_2 - \frac{1}{2}c_3\right) e^{-t/2} \cos \frac{\sqrt{3}}{2}t.$$

18. From $Dx + z = e^t$, $(D-1)x + Dy + Dz = 0$, and $x + 2y + Dz = e^t$ we obtain $z = -Dx + e^t$, $Dz = -D^2 x + e^t$, and the system $(-D^2 + D - 1)x + Dy = -e^t$ and $(-D^2 + 1)x + 2y = 0$. Then $y = \frac{1}{2}(D^2 - 1)x$, $Dy = \frac{1}{2}D(D^2 - 1)x$, and $(D-2)(D^2 + 1)x = -2e^t$ so that the solution is

$$x = c_1 e^{2t} + c_2 \cos t + c_3 \sin t + e^t$$

$$y = \frac{3}{2}c_1 e^{2t} - c_2 \cos t - c_3 \sin t$$

$$z = -2c_1 e^{2t} - c_3 \cos t + c_2 \sin t.$$

19. Write the system in the form

$$Dx - 6y = 0$$

$$x - Dy + z = 0$$

$$x + y - Dz = 0.$$

Multiplying the second equation by D and adding to the third equation we obtain $(D+1)x - (D^2 - 1)y = 0$. Eliminating y between this equation and $Dx - 6y = 0$ we find

$$(D^3 - D - 6D - 6)x = (D+1)(D+2)(D-3)x = 0.$$

Thus

$$x = c_1 e^{-t} + c_2 e^{-2t} + c_3 e^{3t},$$

and, successively substituting into the first and second equations, we get

$$y = -\frac{1}{6}c_1 e^{-t} - \frac{1}{3}c_2 e^{-2t} + \frac{1}{2}c_3 e^{3t}$$

$$z = -\frac{5}{6}c_1 e^{-t} - \frac{1}{3}c_2 e^{-2t} + \frac{1}{2}c_3 e^{3t}.$$

20. Write the system in the form

$$(D+1)x - z = 0$$

$$(D+1)y - z = 0$$

$$x - y + Dz = 0.$$

Multiplying the third equation by $D+1$ and adding to the second equation we obtain $(D+1)x + (D^2 + D - 1)z = 0$. Eliminating z between this equation and $(D+1)x - z = 0$

we find $D(D+1)^2 x = 0$. Thus

$$x = c_1 + c_2 e^{-t} + c_3 t e^{-t},$$

and, successively substituting into the first and third equations, we get

$$y = c_1 + (c_2 - c_3)e^{-t} + c_3 t e^{-t}$$

$$z = c_1 + c_3 e^{-t}.$$

21. From $(D+5)x+y=0$ and $4x-(D+1)y=0$ we obtain $y = -(D+5)x$ so that $Dy = -(D^2+5D)x$. Then $4x + (D^2 + 5D)x + (D+5)x = 0$ and $(D+3)^2 x = 0$. Thus

$$x = c_1 e^{-3t} + c_2 t e^{-3t}$$

$$y = -(2c_1 + c_2)e^{-3t} - 2c_2 t e^{-3t}.$$

Using $x(1) = 0$ and $y(1) = 1$ we obtain

$$c_1 e^{-3} + c_2 e^{-3} = 0$$

$$-(2c_1 + c_2)e^{-3} - 2c_2 e^{-3} = 1$$

or

$$c_1 + c_2 = 0$$

$$2c_1 + 3c_2 = -e^3.$$

Thus $c_1 = e^3$ and $c_2 = -e^3$. The solution of the initial value problem is

$$x = e^{-3t+3} - te^{-3t+3}$$

$$y = -e^{-3t+3} + 2te^{-3t+3}.$$

22. From $Dx - y = -1$ and $3x + (D-2)y = 0$ we obtain $x = -\frac{1}{3}(D-2)y$ so that $Dx = -\frac{1}{3}(D^2 - 2D)y$. Then $-\frac{1}{3}(D^2 - 2D)y = y - 1$ and $(D^2 - 2D + 3)y = 3$. Thus

$$y = e^t \left(c_1 \cos \sqrt{2}\, t + c_2 \sin \sqrt{2}\, t \right) + 1$$

and

$$x = \frac{1}{3} e^t \left[\left(c_1 - \sqrt{2}\, c_2 \right) \cos \sqrt{2}\, t + \left(\sqrt{2}\, c_1 + c_2 \right) \sin \sqrt{2}\, t \right] + \frac{2}{3}.$$

Using $x(0) = y(0) = 0$ we obtain

$$c_1 + 1 = 0$$

$$\frac{1}{3} \left(c_1 - \sqrt{2}\, c_2 \right) + \frac{2}{3} = 0.$$

212

Thus $c_1 = -1$ and $c_2 = \sqrt{2}/2$. The solution of the initial value problem is

$$x = e^t \left(-\frac{2}{3} \cos \sqrt{2}\, t - \frac{\sqrt{2}}{6} \sin \sqrt{2}\, t \right) + \frac{2}{3}$$

$$y = e^t \left(- \cos \sqrt{2}\, t + \frac{\sqrt{2}}{2} \sin \sqrt{2}\, t \right) + 1.$$

23. Equating Newton's law with the net forces in the x- and y-directions gives $m\, d^2x/dt^2 = 0$ and $m\, d^2y/dt^2 = -mg$, respectively. From $mD^2x = 0$ we obtain $x(t) = c_1t + c_2$, and from $mD^2y = -mg$ or $D^2y = -g$ we obtain $y(t) = -\frac{1}{2}gt^2 + c_3t + c_4$.

24. From Newton's second law in the x-direction we have

$$m\frac{d^2x}{dt^2} = -k\cos\theta = -k\frac{1}{v}\frac{dx}{dt} = -|c|\frac{dx}{dt}.$$

In the y-direction we have

$$m\frac{d^2y}{dt^2} = -mg - k\sin\theta = -mg - k\frac{1}{v}\frac{dy}{dt} = -mg - |c|\frac{dy}{dt}.$$

From $mD^2x + |c|Dx = 0$ we have $D(mD + |c|)x = 0$ so that $(mD + |c|)x = c_1$ or $(D + |c|/m)x = c_2$. This is a linear first-order differential equation. An integrating factor is $e^{\int |c|dt/m} = e^{|c|t/m}$ so that

$$\frac{d}{dt}[e^{|c|t/m}x] = c_2 e^{|c|t/m}$$

and $e^{|c|t/m}x = (c_2m/|c|)e^{|c|t/m} + c_3$. The general solution of this equation is $x(t) = c_4 + c_3 e^{-|c|t/m}$.

From $(mD^2 + |c|D)y = -mg$ we have $D(mD + |c|)y = -mg$ so that $(mD + |c|)y = -mgt + c_1$ or $(D + |c|/m)y = -gt + c_2$. This is a linear first-order differential equation with integrating factor $e^{\int |c|dt/m} = e^{|c|t/m}$. Thus

$$\frac{d}{dt}[e^{|c|t/m}y] = (-gt + c_2)e^{|c|t/m}$$

$$e^{|c|t/m}y = -\frac{mg}{|c|}te^{|c|t/m} + \frac{m^2g}{c^2}e^{|c|t/m} + c_3e^{|c|t/m} + c_4$$

and

$$y(t) = -\frac{mg}{|c|}t + \frac{m^2g}{c^2} + c_3 + c_4e^{-|c|t/m}.$$

25. Multiplying the first equation by $D + 1$ and the second equation by D we obtain

$$D(D+1)x - 2D(D+1)y = 2t + t^2$$

$$D(D+1)x - 2D(D+1)y = 0.$$

This leads to $2t + t^2 = 0$, so the system has no solution.

26. The **FindRoot** application of *Mathematica* gives a solution of $x_1(t) = x_2(t)$ as approximately $t = 13.73$ minutes. So tank B contains more salt than tank A for $t > 13.73$ minutes.

27. (a) Separating variables in the first equation, we have $dx_1/x_1 = -dt/50$, so $x_1 = c_1 e^{-t/50}$. From $x_1(0) = 15$ we get $c_1 = 15$. The second differential equation then becomes

$$\frac{dx_2}{dt} = \frac{15}{50}e^{-t/50} - \frac{2}{75}x_2 \qquad \text{or} \qquad \frac{dx_2}{dt} + \frac{2}{75}x_2 = \frac{3}{10}e^{-t/50}.$$

This differential equation is linear and has the integrating factor $e^{\int 2\,dt/75} = e^{2t/75}$. Then

$$\frac{d}{dt}[e^{2t/75}x_2] = \frac{3}{10}e^{-t/50+2t/75} = \frac{3}{10}e^{t/150}$$

so

$$e^{2t/75}x_2 = 45e^{t/150} + c_2$$

and

$$x_2 = 45e^{-t/50} + c_2 e^{-2t/75}.$$

From $x_2(0) = 10$ we get $c_2 = -35$. The third differential equation then becomes

$$\frac{dx_3}{dt} = \frac{90}{75}e^{-t/50} - \frac{70}{75}e^{-2t/75} - \frac{1}{25}x_3$$

or

$$\frac{dx_3}{dt} + \frac{1}{25}x_3 = \frac{6}{5}e^{-t/50} - \frac{14}{15}e^{-2t/75}.$$

This differential equation is linear and has the integrating factor $e^{\int dt/25} = e^{t/25}$. Then

$$\frac{d}{dt}[e^{t/25}x_3] = \frac{6}{5}e^{-t/50+t/25} - \frac{14}{15}e^{-2t/75+t/25} = \frac{6}{5}e^{t/50} - \frac{14}{15}e^{t/75},$$

so

$$e^{t/25}x_3 = 60e^{t/50} - 70e^{t/75} + c_3$$

and

$$x_3 = 60e^{-t/50} - 70e^{-2t/75} + c_3 e^{-t/25}.$$

From $x_3(0) = 5$ we get $c_3 = 15$. The solution of the initial-value problem is

$$x_1(t) = 15e^{-t/50}$$

$$x_2(t) = 45e^{-t/50} - 35e^{-2t/75}$$

$$x_3(t) = 60e^{-t/50} - 70e^{-2t/75} + 15e^{-t/25}.$$

(b)

pounds salt

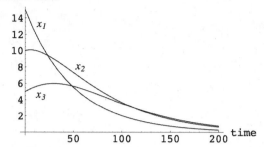

(c) Solving $x_1(t) = \frac{1}{2}$, $x_2(t) = \frac{1}{2}$, and $x_3(t) = \frac{1}{2}$, **FindRoot** gives, respectively, $t_1 = 170.06\,\text{min}$, $t_2 = 214.7\,\text{min}$, and $t_3 = 224.4\,\text{min}$. Thus, all three tanks will contain less than or equal to 0.5 pounds of salt after 224.4 minutes.

Exercises 4.9

Nonlinear Differential Equations

1. We have $y_1' = y_1'' = e^x$, so

$$(y_1'')^2 = (e^x)^2 = e^{2x} = y_1^2.$$

Also, $y_2' = -\sin x$ and $y_2'' = -\cos x$, so

$$(y_2'')^2 = (-\cos x)^2 = \cos^2 x = y_2^2.$$

However, if $y = c_1 y_1 + c_2 y_2$, we have $(y'')^2 = (c_1 e^x - c_2 \cos x)^2$ and $y^2 = (c_1 e^x + c_2 \cos x)^2$. Thus $(y'')^2 \neq y^2$.

2. We have $y_1' = y_1'' = 0$, so

$$y_1 y_1'' = 1 \cdot 0 = 0 = \frac{1}{2}(0)^2 = \frac{1}{2}(y_1')^2.$$

Also, $y_2' = 2x$ and $y_2'' = 2$, so

$$y_2 y_2'' = x^2(2) = 2x^2 = \frac{1}{2}(2x)^2 = \frac{1}{2}(y_2')^2.$$

However, if $y = c_1 y_1 + c_2 y_2$, we have $yy'' = (c_1 \cdot 1 + c_2 x^2)(c_1 \cdot 0 + 2c_2) = 2c_2(c_1 + c_2 x^2)$ and $\frac{1}{2}(y')^2 = \frac{1}{2}[c_1 \cdot 0 + c_2(2x)]^2 = 2c_2^2 x^2$. Thus $yy'' \neq \frac{1}{2}(y')^2$.

3. Let $u = y'$ so that $u' = y''$. The equation becomes $u' = -u^2 - 1$ which is separable. Thus

$$\frac{du}{u^2 + 1} = -dx \implies \tan^{-1} u = -x + c_1 \implies y' = \tan(c_1 - x) \implies y = \ln|\cos(c_1 - x)| + c_2.$$

4. Let $u = y'$ so that $u' = y''$. The equation becomes $u' = 1 + u^2$. Separating variables we obtain

$$\frac{du}{1 + u^2} = dx \implies \tan^{-1} u = x + c_1 \implies u = \tan(x + c_1) \implies y = -\ln|\cos(x + c_1)| + c_2.$$

5. Let $u = y'$ so that $u' = y''$. The equation becomes $x^2 u' + u^2 = 0$. Separating variables we obtain

$$\frac{du}{u^2} = -\frac{dx}{x^2} \implies -\frac{1}{u} = \frac{1}{x} + c_1 = \frac{c_1 x + 1}{x} \implies u = -\frac{1}{c_1}\left(\frac{x}{x + 1/c_1}\right) = \frac{1}{c_1}\left(\frac{1}{c_1 x + 1} - 1\right)$$

$$\implies y = \frac{1}{c_1^2}\ln|c_1 x + 1| - \frac{1}{c_1}x + c_2.$$

6. Let $u = y'$ so that $y'' = u\,du/dy$. The equation becomes $(y+1)u\,du/dy = u^2$. Separating variables we obtain

$$\frac{du}{u} = \frac{dy}{y+1} \implies \ln|u| = \ln|y+1| + \ln c_1 \implies u = c_1(y+1)$$

$$\implies \frac{dy}{dx} = c_1(y+1) \implies \frac{dy}{y+1} = c_1\,dx$$

$$\implies \ln|y+1| = c_1 x + c_2 \implies y + 1 = c_3 e^{c_1 x}.$$

7. Let $u = y'$ so that $y'' = u\,du/dy$. The equation becomes $u\,du/dy + 2yu^3 = 0$. Separating variables we obtain

$$\frac{du}{u^2} + 2y\,dy = 0 \implies -\frac{1}{u} + y^2 = c_1 \implies u = \frac{1}{y^2 - c_1} \implies y' = \frac{1}{y^2 - c_1}$$

$$\implies \left(y^2 - c_1\right)dy = dx \implies \frac{1}{3}y^3 - c_1 y = x + c_2.$$

8. Let $u = y'$ so that $y'' = u\,du/dy$. The equation becomes $y^2 u\,du/dy = u$. Separating variables we obtain

$$du = \frac{dy}{y^2} \implies u = -\frac{1}{y} + c_1 \implies y' = \frac{c_1 y - 1}{y} \implies \frac{y}{c_1 y - 1}\,dy = dx$$

$$\implies \frac{1}{c_1}\left(1 + \frac{1}{c_1 y - 1}\right)dy = dx \text{ (for } c_1 \neq 0) \implies \frac{1}{c_1}y + \frac{1}{c_1^2}\ln|y - 1| = x + c_2.$$

If $c_1 = 0$, then $y\,dy = -dx$ and another solution is $\frac{1}{2}y^2 = -x + c_2$.

9. (a)

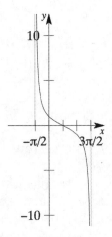

(b) Let $u = y'$ so that $y'' = u\,du/dy$. The equation becomes $u\,du/dy + yu = 0$. Separating variables we obtain

$$du = -y\,dy \implies u = -\frac{1}{2}y^2 + c_1 \implies y' = -\frac{1}{2}y^2 + c_1.$$

When $x = 0$, $y = 1$ and $y' = -1$ so $-1 = -1/2 + c_1$ and $c_1 = -1/2$. Then

$$\frac{dy}{dx} = -\frac{1}{2}y^2 - \frac{1}{2} \implies \frac{dy}{y^2+1} = -\frac{1}{2}\,dx \implies \tan^{-1} y = -\frac{1}{2}x + c_2$$

$$\implies y = \tan\left(-\frac{1}{2}x + c_2\right).$$

When $x = 0$, $y = 1$ so $1 = \tan c_2$ and $c_2 = \pi/4$. The solution of the initial-value problem is

$$y = \tan\left(\frac{\pi}{4} - \frac{1}{2}x\right).$$

The graph is shown in part (a).

(c) The interval of definition is $-\pi/2 < \pi/4 - x/2 < \pi/2$ or $-\pi/2 < x < 3\pi/2$.

10. Let $u = y'$ so that $u' = y''$. The equation becomes $(u')^2 + u^2 = 1$ which results in $u' = \pm\sqrt{1 - u^2}$. To solve $u' = \sqrt{1 - u^2}$ we separate variables:

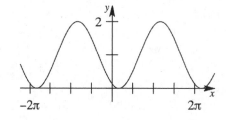

$$\frac{du}{\sqrt{1-u^2}} = dx \implies \sin^{-1} u = x + c_1 \implies u = \sin(x + c_1)$$

$$\implies y' = \sin(x + c_1).$$

When $x = \pi/2$, $y' = \sqrt{3}/2$, so $\sqrt{3}/2 = \sin(\pi/2 + c_1)$ and $c_1 = -\pi/6$. Thus

$$y' = \sin\left(x - \frac{\pi}{6}\right) \implies y = -\cos\left(x - \frac{\pi}{6}\right) + c_2.$$

When $x = \pi/2$, $y = 1/2$, so $1/2 = -\cos(\pi/2 - \pi/6) + c_2 = -1/2 + c_2$ and $c_2 = 1$. The solution of the initial-value problem is $y = 1 - \cos(x - \pi/6)$.

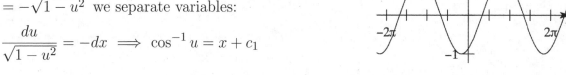

To solve $u' = -\sqrt{1 - u^2}$ we separate variables:

$$\frac{du}{\sqrt{1-u^2}} = -dx \implies \cos^{-1} u = x + c_1$$

$$\implies u = \cos(x + c_1) \implies y' = \cos(x + c_1).$$

When $x = \pi/2$, $y' = \sqrt{3}/2$, so $\sqrt{3}/2 = \cos(\pi/2 + c_1)$ and $c_1 = -\pi/3$. Thus

$$y' = \cos\left(x - \frac{\pi}{3}\right) \implies y = \sin\left(x - \frac{\pi}{3}\right) + c_2.$$

When $x = \pi/2$, $y = 1/2$, so $1/2 = \sin(\pi/2 - \pi/3) + c_2 = 1/2 + c_2$ and $c_2 = 0$. The solution of the initial-value problem is $y = \sin(x - \pi/3)$.

11. Let $u = y'$ so that $u' = y''$. The equation becomes $u' - (1/x)u = (1/x)u^3$, which is Bernoulli. Using $w = u^{-2}$ we obtain $dw/dx + (2/x)w = -2/x$. An integrating factor is x^2, so

$$\frac{d}{dx}[x^2 w] = -2x \implies x^2 w = -x^2 + c_1 \implies w = -1 + \frac{c_1}{x^2}$$

$$\implies u^{-2} = -1 + \frac{c_1}{x^2} \implies u = \frac{x}{\sqrt{c_1 - x^2}}$$

$$\implies \frac{dy}{dx} = \frac{x}{\sqrt{c_1 - x^2}} \implies y = -\sqrt{c_1 - x^2} + c_2$$

$$\implies c_1 - x^2 = (c_2 - y)^2 \implies x^2 + (c_2 - y)^2 = c_1.$$

12. Let $u = y'$ so that $u' = y''$. The equation becomes $u' - (1/x)u = u^2$, which is a Bernoulli differential equation. Using the substitution $w = u^{-1}$ we obtain $dw/dx + (1/x)w = -1$. An integrating factor is x, so

$$\frac{d}{dx}[xw] = -x \implies w = -\frac{1}{2}x + \frac{1}{x}c \implies \frac{1}{u} = \frac{c_1 - x^2}{2x} \implies u = \frac{2x}{c_1 - x^2} \implies y = -\ln\left|c_1 - x^2\right| + c_2.$$

In Problems 13-16 the thinner curve is obtained using a numerical solver, while the thicker curve is the graph of the Taylor polynomial.

13. We look for a solution of the form

$$y(x) = y(0) + y'(0)x + \frac{1}{2!}y''(0)x^2 + \frac{1}{3!}y'''(0)x^3 + \frac{1}{4!}y^{(4)}(0)x^4 + \frac{1}{5!}y^{(5)}(0)x^5.$$

From $y''(x) = x + y^2$ we compute

$$y'''(x) = 1 + 2yy'$$

$$y^{(4)}(x) = 2yy'' + 2(y')^2$$

$$y^{(5)}(x) = 2yy''' + 6y'y''.$$

Using $y(0) = 1$ and $y'(0) = 1$ we find

$$y''(0) = 1, \quad y'''(0) = 3, \quad y^{(4)}(0) = 4, \quad y^{(5)}(0) = 12.$$

An approximate solution is

$$y(x) = 1 + x + \frac{1}{2}x^2 + \frac{1}{2}x^3 + \frac{1}{6}x^4 + \frac{1}{10}x^5.$$

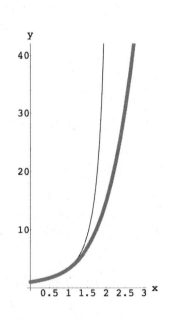

14. We look for a solution of the form

$$y(x) = y(0) + y'(0)x + \frac{1}{2!}y''(0)x^2 + \frac{1}{3!}y'''(0)x^3 + \frac{1}{4!}y^{(4)}(0)x^4 + \frac{1}{5!}y^{(5)}(0)x^5.$$

From $y''(x) = 1 - y^2$ we compute

$$y'''(x) = -2yy'$$

$$y^{(4)}(x) = -2yy'' - 2(y')^2$$

$$y^{(5)}(x) = -2yy''' - 6y'y''.$$

Using $y(0) = 2$ and $y'(0) = 3$ we find

$$y''(0) = -3, \quad y'''(0) = -12, \quad y^{(4)}(0) = -6, \quad y^{(5)}(0) = 102.$$

An approximate solution is

$$y(x) = 2 + 3x - \frac{3}{2}x^2 - 2x^3 - \frac{1}{4}x^4 + \frac{17}{20}x^5.$$

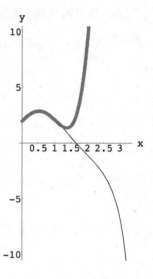

15. We look for a solution of the form

$$y(x) = y(0) + y'(0)x + \frac{1}{2!}y''(0)x^2 + \frac{1}{3!}y'''(0)x^3 + \frac{1}{4!}y^{(4)}(0)x^4 + \frac{1}{5!}y^{(5)}(0)x^5.$$

From $y''(x) = x^2 + y^2 - 2y'$ we compute

$$y'''(x) = 2x + 2yy' - 2y''$$

$$y^{(4)}(x) = 2 + 2(y')^2 + 2yy'' - 2y'''$$

$$y^{(5)}(x) = 6y'y'' + 2yy''' - 2y^{(4)}.$$

Using $y(0) = 1$ and $y'(0) = 1$ we find

$$y''(0) = -1, \quad y'''(0) = 4, \quad y^{(4)}(0) = -6, \quad y^{(5)}(0) = 14.$$

An approximate solution is

$$y(x) = 1 + x - \frac{1}{2}x^2 + \frac{2}{3}x^3 - \frac{1}{4}x^4 + \frac{7}{60}x^5.$$

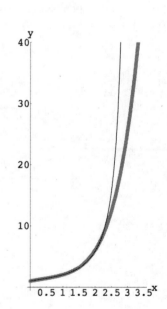

16. We look for a solution of the form

$$y(x) = y(0) + y'(0)x + \frac{1}{2!}y''(0)x^2 + \frac{1}{3!}y'''(0)x^3 + \frac{1}{4!}y^{(4)}(0)x^4$$

$$+ \frac{1}{5!}y^{(5)}(0)x^5 + \frac{1}{6!}y^{(6)}(0)x^6.$$

From $y''(x) = e^y$ we compute

$$y'''(x) = e^y y'$$

$$y^{(4)}(x) = e^y(y')^2 + e^y y''$$

$$y^{(5)}(x) = e^y(y')^3 + 3e^y y' y'' + e^y y'''$$

$$y^{(6)}(x) = e^y(y')^4 + 6e^y(y')^2 y'' + 3e^y(y'')^2 + 4e^y y' y''' + e^y y^{(4)}.$$

Using $y(0) = 0$ and $y'(0) = -1$ we find

$$y''(0) = 1, \quad y'''(0) = -1, \quad y^{(4)}(0) = 2, \quad y^{(5)}(0) = -5, \quad y^{(6)}(0) = 16.$$

An approximate solution is

$$y(x) = -x + \frac{1}{2}x^2 - \frac{1}{6}x^3 + \frac{1}{12}x^4 + \frac{1}{24}x^5 + \frac{1}{45}x^6.$$

17. We need to solve $[1 + (y')^2]^{3/2} = y''$. Let $u = y'$ so that $u' = y''$. The equation becomes $(1 + u^2)^{3/2} = u'$ or $(1 + u^2)^{3/2} = du/dx$. Separating variables and using the substitution $u = \tan\theta$ we have

$$\frac{du}{(1 + u^2)^{3/2}} = dx \implies \int \frac{\sec^2\theta}{\left(1 + \tan^2\theta\right)^{3/2}} d\theta = x \implies \int \frac{\sec^2\theta}{\sec^3\theta} d\theta = x$$

$$\implies \int \cos\theta \, d\theta = x \implies \sin\theta = x \implies \frac{u}{\sqrt{1 + u^2}} = x$$

$$\implies \frac{y'}{\sqrt{1 + (y')^2}} = x \implies (y')^2 = x^2\left[1 + (y')^2\right] = \frac{x^2}{1 - x^2}$$

$$\implies y' = \frac{x}{\sqrt{1 - x^2}} \quad \text{(for } x > 0\text{)} \implies y = -\sqrt{1 - x^2}.$$

18. When $y = \sin x$, $y' = \cos x$, $y'' = -\sin x$, and

$$(y'')^2 - y^2 = \sin^2 x - \sin^2 x = 0.$$

When $y = e^{-x}$, $y' = -e^{-x}$, $y'' = e^{-x}$, and

$$(y'')^2 - y^2 = e^{-2x} - e^{-2x} = 0.$$

From $(y'')^2 - y^2 = 0$ we have $y'' = \pm y$, which can be treated as two linear equations. Since linear combinations of solutions of linear homogeneous differential equations are also solutions, we see that $y = c_1 e^x + c_2 e^{-x}$ and $y = c_3 \cos x + c_4 \sin x$ must satisfy the differential equation. However, linear combinations that involve both exponential and trigonometric functions will not be solutions since the differential equation is not linear and each type of function satisfies a different linear differential equation that is part of the original differential equation.

19. Letting $u = y''$, separating variables, and integrating we have

$$\frac{du}{dx} = \sqrt{1 + u^2}, \qquad \frac{du}{\sqrt{1 + u^2}} = dx, \quad \text{and} \quad \sinh^{-1} u = x + c_1.$$

Then

$$u = y'' = \sinh(x + c_1), \quad y' = \cosh(x + c_1) + c_2, \quad \text{and} \quad y = \sinh(x + c_1) + c_2 x + c_3.$$

20. If the constant $-c_1^2$ is used instead of c_1^2, then, using partial fractions,

$$y = -\int \frac{dx}{x^2 - c_1^2} = -\frac{1}{2c_1} \int \left(\frac{1}{x - c_1} - \frac{1}{x + c_1} \right) dx = \frac{1}{2c_1} \ln \left| \frac{x + c_1}{x - c_1} \right| + c_2.$$

Alternatively, the inverse hyperbolic tangent can be used.

21. Let $u = dx/dt$ so that $d^2x/dt^2 = u\, du/dx$. The equation becomes $u\, du/dx = -k^2/x^2$. Separating variables we obtain

$$u\, du = -\frac{k^2}{x^2} dx \implies \frac{1}{2}u^2 = \frac{k^2}{x} + c \implies \frac{1}{2}v^2 = \frac{k^2}{x} + c.$$

When $t = 0$, $x = x_0$ and $v = 0$ so $0 = (k^2/x_0) + c$ and $c = -k^2/x_0$. Then

$$\frac{1}{2}v^2 = k^2 \left(\frac{1}{x} - \frac{1}{x_0} \right) \quad \text{and} \quad \frac{dx}{dt} = -k\sqrt{2} \sqrt{\frac{x_0 - x}{x x_0}}.$$

Separating variables we have

$$-\sqrt{\frac{x x_0}{x_0 - x}}\, dx = k\sqrt{2}\, dt \implies t = -\frac{1}{k}\sqrt{\frac{x_0}{2}} \int \sqrt{\frac{x}{x_0 - x}}\, dx.$$

Using *Mathematica* to integrate we obtain

$$t = -\frac{1}{k}\sqrt{\frac{x_0}{2}} \left[-\sqrt{x(x_0 - x)} - \frac{x_0}{2} \tan^{-1} \frac{(x_0 - 2x)}{2x} \sqrt{\frac{x}{x_0 - x}} \right]$$

$$= \frac{1}{k}\sqrt{\frac{x_0}{2}} \left[\sqrt{x(x_0 - x)} + \frac{x_0}{2} \tan^{-1} \frac{x_0 - 2x}{2\sqrt{x(x_0 - x)}} \right].$$

22.

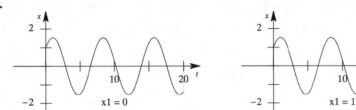

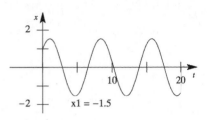

For $d^2x/dt^2 + \sin x = 0$ the motion appears to be periodic with amplitude 1 when $x_1 = 0$. The amplitude and period are larger for larger magnitudes of x_1.

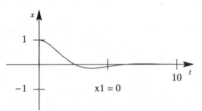

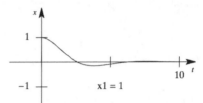

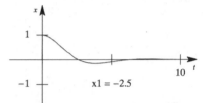

For $d^2x/dt^2 + dx/dt + \sin x = 0$ the motion appears to be periodic with decreasing amplitude. The dx/dt term could be said to have a damping effect.

Chapter 4 in Review

1. $y = 0$

2. Since $y_c = c_1 e^x + c_2 e^{-x}$, a particular solution for $y'' - y = 1 + e^x$ is $y_p = A + Bxe^x$.

3. It is not true unless the differential equation is homogeneous. For example, $y_1 = x$ is a solution of $y'' + y = x$, but $y_2 = 5x$ is not.

4. True

5. The set is linearly independent over $(-\infty, 0)$ and linearly dependent over $(0, \infty)$.

6. (a) Since $f_2(x) = 2 \ln x = 2f_1(x)$, the set of functions is linearly dependent.

   (b) Since x^{n+1} is not a constant multiple of x^n, the set of functions is linearly independent.

   (c) Since $x + 1$ is not a constant multiple of x, the set of functions is linearly independent.

   (d) Since $f_1(x) = \cos x \cos(\pi/2) - \sin x \sin(\pi/2) = -\sin x = -f_2(x)$, the set of functions is linearly dependent.

   (e) Since $f_1(x) = 0 \cdot f_2(x)$, the set of functions is linearly dependent.

   (f) Since $2x$ is not a constant multiple of 2, the set of functions is linearly independent.

(g) Since $3(x^2) + 2(1 - x^2) - (2 + x^2) = 0$, the set of functions is linearly dependent.

(h) Since $xe^{x+1} + 0(4x - 5)e^x - exe^x = 0$, the set of functions is linearly dependent.

7. (a) The general solution is

$$y = c_1 e^{3x} + c_2 e^{-5x} + c_3 x e^{-5x} + c_4 e^x + c_5 x e^x + c_6 x^2 e^x.$$

(b) The general solution is

$$y = c_1 x^3 + c_2 x^{-5} + c_3 x^{-5} \ln x + c_4 x + c_5 x \ln x + c_6 x (\ln x)^2.$$

8. Variation of parameters will work for all choices of $g(x)$, although the integral involved may not always be able to be expressed in terms of elementary functions. The method of undetermined coefficients will work for the functions in (b), (c), and (e).

9. From $m^2 - 2m - 2 = 0$ we obtain $m = 1 \pm \sqrt{3}$ so that

$$y = c_1 e^{(1+\sqrt{3})x} + c_2 e^{(1-\sqrt{3})x}.$$

10. From $2m^2 + 2m + 3 = 0$ we obtain $m = -1/2 \pm (\sqrt{5}/2)i$ so that

$$y = e^{-x/2} \left(c_1 \cos \frac{\sqrt{5}}{2} x + c_2 \sin \frac{\sqrt{5}}{2} x \right).$$

11. From $m^3 + 10m^2 + 25m = 0$ we obtain $m = 0$, $m = -5$, and $m = -5$ so that

$$y = c_1 + c_2 e^{-5x} + c_3 x e^{-5x}.$$

12. From $2m^3 + 9m^2 + 12m + 5 = 0$ we obtain $m = -1$, $m = -1$, and $m = -5/2$ so that

$$y = c_1 e^{-5x/2} + c_2 e^{-x} + c_3 x e^{-x}.$$

13. From $3m^3 + 10m^2 + 15m + 4 = 0$ we obtain $m = -1/3$ and $m = -3/2 \pm (\sqrt{7}/2)i$ so that

$$y = c_1 e^{-x/3} + e^{-3x/2} \left(c_2 \cos \frac{\sqrt{7}}{2} x + c_3 \sin \frac{\sqrt{7}}{2} x \right).$$

14. From $2m^4 + 3m^3 + 2m^2 + 6m - 4 = 0$ we obtain $m = 1/2$, $m = -2$, and $m = \pm\sqrt{2}\,i$ so that

$$y = c_1 e^{x/2} + c_2 e^{-2x} + c_3 \cos \sqrt{2}\, x + c_4 \sin \sqrt{2}\, x.$$

15. Applying D^4 to the differential equation we obtain $D^4(D^2 - 3D + 5) = 0$. Then

$$y = \underbrace{e^{3x/2} \left(c_1 \cos \frac{\sqrt{11}}{2} x + c_2 \sin \frac{\sqrt{11}}{2} x \right)}_{y_c} + c_3 + c_4 x + c_5 x^2 + c_6 x^3$$

and $y_p = A + Bx + Cx^2 + Dx^3$. Substituting y_p into the differential equation yields

$$(5A - 3B + 2C) + (5B - 6C + 6D)x + (5C - 9D)x^2 + 5Dx^3 = -2x + 4x^3.$$

Equating coefficients gives $A = -222/625$, $B = 46/125$, $C = 36/25$, and $D = 4/5$. The general solution is

$$y = e^{3x/2}\left(c_1\cos\frac{\sqrt{11}}{2}x + c_2\sin\frac{\sqrt{11}}{2}x\right) - \frac{222}{625} + \frac{46}{125}x + \frac{36}{25}x^2 + \frac{4}{5}x^3.$$

16. Applying $(D-1)^3$ to the differential equation we obtain $(D-1)^3(D-2D+1) = (D-1)^5 = 0$. Then

$$y = \underbrace{c_1e^x + c_2xe^x}_{y_c} + c_3x^2e^x + c_4x^3e^x + c_5x^4e^x$$

and $y_p = Ax^2e^x + Bx^3e^x + Cx^4e^x$. Substituting y_p into the differential equation yields

$$12Cx^2e^x + 6Bxe^x + 2Ae^x = x^2e^x.$$

Equating coefficients gives $A = 0$, $B = 0$, and $C = 1/12$. The general solution is

$$y = c_1e^x + c_2xe^x + \frac{1}{12}x^4e^x.$$

17. Applying $D(D^2+1)$ to the differential equation we obtain

$$D(D^2+1)(D^3 - 5D^2 + 6D) = D^2(D^2+1)(D-2)(D-3) = 0.$$

Then

$$y = \underbrace{c_1 + c_2e^{2x} + c_3e^{3x}}_{y_c} + c_4x + c_5\cos x + c_6\sin x$$

and $y_p = Ax + B\cos x + C\sin x$. Substituting y_p into the differential equation yields

$$6A + (5B + 5C)\cos x + (-5B + 5C)\sin x = 8 + 2\sin x.$$

Equating coefficients gives $A = 4/3$, $B = -1/5$, and $C = 1/5$. The general solution is

$$y = c_1 + c_2e^{2x} + c_3e^{3x} + \frac{4}{3}x - \frac{1}{5}\cos x + \frac{1}{5}\sin x.$$

18. Applying D to the differential equation we obtain $D(D^3 - D^2) = D^3(D-1) = 0$. Then

$$y = \underbrace{c_1 + c_2x + c_3e^x}_{y_c} + c_4x^2$$

and $y_p = Ax^2$. Substituting y_p into the differential equation yields $-2A = 6$. Equating coefficients gives $A = -3$. The general solution is

$$y = c_1 + c_2x + c_3e^x - 3x^2.$$

19. The auxiliary equation is $m^2 - 2m + 2 = [m - (1+i)][m - (1-i)] = 0$, so $y_c = c_1e^x\sin x + c_2e^x\cos x$ and

$$W = \begin{vmatrix} e^x\sin x & e^x\cos x \\ e^x\cos x + e^x\sin x & -e^x\sin x + e^x\cos x \end{vmatrix} = -e^{2x}.$$

Identifying $f(x) = e^x \tan x$ we obtain

$$u'_1 = -\frac{(e^x \cos x)(e^x \tan x)}{-e^{2x}} = \sin x$$

$$u'_2 = \frac{(e^x \sin x)(e^x \tan x)}{-e^{2x}} = -\frac{\sin^2 x}{\cos x} = \cos x - \sec x.$$

Then $u_1 = -\cos x$, $u_2 = \sin x - \ln|\sec x + \tan x|$, and

$$y = c_1 e^x \sin x + c_2 e^x \cos x - e^x \sin x \cos x + e^x \sin x \cos x - e^x \cos x \ln|\sec x + \tan x|$$

$$= c_1 e^x \sin x + c_2 e^x \cos x - e^x \cos x \ln|\sec x + \tan x|.$$

20. The auxiliary equation is $m^2 - 1 = 0$, so $y_c = c_1 e^x + c_2 e^{-x}$ and

$$W = \begin{vmatrix} e^x & e^{-x} \\ e^x & -e^{-x} \end{vmatrix} = -2.$$

Identifying $f(x) = 2e^x/(e^x + e^{-x})$ we obtain

$$u'_1 = \frac{1}{e^x + e^{-x}} = \frac{e^x}{1 + e^{2x}}$$

$$u'_2 = -\frac{e^{2x}}{e^x + e^{-x}} = -\frac{e^{3x}}{1 + e^{2x}} = -e^x + \frac{e^x}{1 + e^{2x}}.$$

Then $u_1 = \tan^{-1} e^x$, $u_2 = -e^x + \tan^{-1} e^x$, and

$$y = c_1 e^x + c_2 e^{-x} + e^x \tan^{-1} e^x - 1 + e^{-x} \tan^{-1} e^x.$$

21. The auxiliary equation is $6m^2 - m - 1 = 0$ so that

$$y = c_1 x^{1/2} + c_2 x^{-1/3}.$$

22. The auxiliary equation is $2m^3 + 13m^2 + 24m + 9 = (m+3)^2(m+1/2) = 0$ so that

$$y = c_1 x^{-3} + c_2 x^{-3} \ln x + c_3 x^{-1/2}.$$

23. The auxiliary equation is $m^2 - 5m + 6 = (m-2)(m-3) = 0$ and a particular solution is $y_p = x^4 - x^2 \ln x$ so that

$$y = c_1 x^2 + c_2 x^3 + x^4 - x^2 \ln x.$$

24. The auxiliary equation is $m^2 - 2m + 1 = (m-1)^2 = 0$ and a particular solution is $y_p = \frac{1}{4}x^3$ so that

$$y = c_1 x + c_2 x \ln x + \frac{1}{4}x^3.$$

25. (a) The auxiliary equation is $m^2 + \omega^2 = 0$, so $y_c = c_1 \cos \omega x + c_2 \sin \omega x$. When $\omega \neq \alpha$, $y_p = A \cos \alpha x + B \sin \alpha x$ and

$$y = c_1 \cos \omega x + c_2 \sin \omega x + A \cos \alpha x + B \sin \alpha x.$$

When $\omega = \alpha$, $y_p = Ax\cos\omega x + Bx\sin\omega x$ and

$$y = c_1\cos\omega x + c_2\sin\omega x + Ax\cos\omega x + Bx\sin\omega x.$$

(b) The auxiliary equation is $m^2 - \omega^2 = 0$, so $y_c = c_1 e^{\omega x} + c_2 e^{-\omega x}$. When $\omega \neq \alpha$, $y_p = Ae^{\alpha x}$ and

$$y = c_1 e^{\omega x} + c_2 e^{-\omega x} + Ae^{\alpha x}.$$

When $\omega = \alpha$, $y_p = Axe^{\omega x}$ and

$$y = c_1 e^{\omega x} + c_2 e^{-\omega x} + Axe^{\omega x}.$$

26. (a) If $y = \sin x$ is a solution then so is $y = \cos x$ and $m^2 + 1$ is a factor of the auxiliary equation $m^4 + 2m^3 + 11m^2 + 2m + 10 = 0$. Dividing by $m^2 + 1$ we get $m^2 + 2m + 10$, which has roots $-1 \pm 3i$. The general solution of the differential equation is

$$y = c_1\cos x + c_2\sin x + e^{-x}(c_3\cos 3x + c_4\sin 3x).$$

(b) The auxiliary equation is $m(m+1) = m^2 + m = 0$, so the associated homogeneous differential equation is $y'' + y' = 0$. Letting $y = c_1 + c_2 e^{-x} + \frac{1}{2}x^2 - x$ and computing $y'' + y'$ we get x. Thus, the differential equation is $y'' + y' = x$.

27. (a) The auxiliary equation is $m^4 - 2m^2 + 1 = (m^2 - 1)^2 = 0$, so the general solution of the differential equation is

$$y = c_1\sinh x + c_2\cosh x + c_3 x\sinh x + c_4 x\cosh x.$$

(b) Since both $\sinh x$ and $x\sinh x$ are solutions of the associated homogeneous differential equation, a particular solution of $y^{(4)} - 2y'' + y = \sinh x$ has the form $y_p = Ax^2\sinh x + Bx^2\cosh x$.

28. Since $y_1' = 1$ and $y_1'' = 0$, $x^2 y_1'' - (x^2 + 2x)y_1' + (x+2)y_1 = -x^2 - 2x + x^2 + 2x = 0$, and $y_1 = x$ is a solution of the associated homogeneous equation. Using the method of reduction of order, we let $y = ux$. Then $y' = xu' + u$ and $y'' = xu'' + 2u'$, so

$$x^2 y'' - (x^2 + 2x)y' + (x+2)y = x^3 u'' + 2x^2 u' - x^3 u' - 2x^2 u' - x^2 u - 2xu + x^2 u + 2xu$$

$$= x^3 u'' - x^3 u' = x^3(u'' - u').$$

To find a second solution of the homogeneous equation we note that $u = e^x$ is a solution of $u'' - u' = 0$. Thus, $y_c = c_1 x + c_2 x e^x$. To find a particular solution we set $x^3(u'' - u') = x^3$ so that $u'' - u' = 1$. This differential equation has a particular solution of the form Ax. Substituting, we find $A = -1$, so a particular solution of the original differential equation is $y_p = -x^2$ and the general solution is $y = c_1 x + c_2 x e^x - x^2$.

29. The auxiliary equation is $m^2 - 2m + 2 = 0$ so that $m = 1 \pm i$ and $y = e^x(c_1\cos x + c_2\sin x)$. Setting $y(\pi/2) = 0$ and $y(\pi) = -1$ we obtain $c_1 = e^{-\pi}$ and $c_2 = 0$. Thus, $y = e^{x-\pi}\cos x$.

30. The auxiliary equation is $m^2+2m+1 = (m+1)^2 = 0$, so that $y = c_1 e^{-x}+c_2 x e^{-x}$. Setting $y(-1) = 0$ and $y'(0) = 0$ we get $c_1 e - c_2 e = 0$ and $-c_1 + c_2 = 0$. Thus $c_1 = c_2$ and $y = c_1(e^{-x} + x e^{-x})$ is a solution of the boundary-value problem for any real number c_1.

31. The auxiliary equation is $m^2 - 1 = (m - 1)(m + 1) = 0$ so that $m = \pm 1$ and $y = c_1 e^x + c_2 e^{-x}$. Assuming $y_p = Ax + B + C \sin x$ and substituting into the differential equation we find $A = -1$, $B = 0$, and $C = -\frac{1}{2}$. Thus $y_p = -x - \frac{1}{2}\sin x$ and

$$y = c_1 e^x + c_2 e^{-x} - x - \frac{1}{2}\sin x.$$

Setting $y(0) = 2$ and $y'(0) = 3$ we obtain

$$c_1 + c_2 = 2$$

$$c_1 - c_2 - \frac{3}{2} = 3.$$

Solving this system we find $c_1 = \frac{13}{4}$ and $c_2 = -\frac{5}{4}$. The solution of the initial-value problem is

$$y = \frac{13}{4}e^x - \frac{5}{4}e^{-x} - x - \frac{1}{2}\sin x.$$

32. The auxiliary equation is $m^2 + 1 = 0$, so $y_c = c_1 \cos x + c_2 \sin x$ and

$$W = \begin{vmatrix} \cos x & \sin x \\ -\sin x & \cos x \end{vmatrix} = 1.$$

Identifying $f(x) = \sec^3 x$ we obtain

$$u_1' = -\sin x \sec^3 x = -\frac{\sin x}{\cos^3 x}$$

$$u_2' = \cos x \sec^3 x = \sec^2 x.$$

Then

$$u_1 = -\frac{1}{2}\frac{1}{\cos^2 x} = -\frac{1}{2}\sec^2 x$$

$$u_2 = \tan x.$$

Thus

$$y = c_1 \cos x + c_2 \sin x - \frac{1}{2}\cos x \sec^2 x + \sin x \tan x$$

$$= c_1 \cos x + c_2 \sin x - \frac{1}{2}\sec x + \frac{1 - \cos^2 x}{\cos x}$$

$$= c_3 \cos x + c_2 \sin x + \frac{1}{2}\sec x.$$

and

$$y' = -c_3 \sin x + c_2 \cos x + \frac{1}{2} \sec x \tan x.$$

The initial conditions imply

$$c_3 + \frac{1}{2} = 1$$

$$c_2 = \frac{1}{2}.$$

Thus $c_3 = c_2 = 1/2$ and

$$y = \frac{1}{2} \cos x + \frac{1}{2} \sin x + \frac{1}{2} \sec x.$$

33. Let $u = y'$ so that $u' = y''$. The equation becomes $u\, du/dx = 4x$. Separating variables we obtain

$$u\, du = 4x\, dx \implies \frac{1}{2} u^2 = 2x^2 + c_1 \implies u^2 = 4x^2 + c_2.$$

When $x = 1$, $y' = u = 2$, so $4 = 4 + c_2$ and $c_2 = 0$. Then

$$u^2 = 4x^2 \implies \frac{dy}{dx} = 2x \quad \text{or} \quad \frac{dy}{dx} = -2x$$

$$\implies y = x^2 + c_3 \quad \text{or} \quad y = -x^2 + c_4.$$

When $x = 1$, $y = 5$, so $5 = 1 + c_3$ and $5 = -1 + c_4$. Thus $c_3 = 4$ and $c_4 = 6$. We have $y = x^2 + 4$ and $y = -x^2 + 6$. Note however that when $y = -x^2 + 6$, $y' = -2x$ and $y'(1) = -2 \neq 2$. Thus, the solution of the initial-value problem is $y = x^2 + 4$.

34. Let $u = y'$ so that $y'' = u\, du/dy$. The equation becomes $2u\, du/dy = 3y^2$. Separating variables we obtain

$$2u\, du = 3y^2\, dy \implies u^2 = y^3 + c_1.$$

When $x = 0$, $y = 1$ and $y' = u = 1$ so $1 = 1 + c_1$ and $c_1 = 0$. Then

$$u^2 = y^3 \implies \left(\frac{dy}{dx}\right)^2 = y^3 \implies \frac{dy}{dx} = y^{3/2} \implies y^{-3/2}\, dy = dx$$

$$\implies -2y^{-1/2} = x + c_2 \implies y = \frac{4}{(x + c_2)^2}.$$

When $x = 0$, $y = 1$, so $1 = 4/c_2^2$ and $c_2 = \pm 2$. Thus, $y = 4/(x+2)^2$ and $y = 4/(x-2)^2$. Note, however, that when $y = 4/(x+2)^2$, $y' = -8/(x+2)^3$ and $y'(0) = -1 \neq 1$. Thus, the solution of the initial-value problem is $y = 4/(x-2)^2$.

35. (a) The auxiliary equation is $12m^4 + 64m^3 + 59m^2 - 23m - 12 = 0$ and has roots -4, $-\frac{3}{2}$, $-\frac{1}{3}$, and $\frac{1}{2}$. The general solution is

$$y = c_1 e^{-4x} + c_2 e^{-3x/2} + c_3 e^{-x/3} + c_4 e^{x/2}.$$

(b) The system of equations is

$$c_1 + c_2 + c_3 + c_4 = -1$$

$$-4c_1 - \frac{3}{2}c_2 - \frac{1}{3}c_3 + \frac{1}{2}c_4 = 2$$

$$16c_1 + \frac{9}{4}c_2 + \frac{1}{9}c_3 + \frac{1}{4}c_4 = 5$$

$$-64c_1 - \frac{27}{8}c_2 - \frac{1}{27}c_3 + \frac{1}{8}c_4 = 0.$$

Using a CAS we find $c_1 = -\frac{73}{495}$, $c_2 = \frac{109}{35}$, $c_3 = -\frac{3726}{385}$, and $c_4 = \frac{257}{45}$. The solution of the initial-value problem is

$$y = -\frac{73}{495}e^{-4x} + \frac{109}{35}e^{-3x/2} - \frac{3726}{385}e^{-x/3} + \frac{257}{45}e^{x/2}.$$

36. Consider $xy'' + y' = 0$ and look for a solution of the form $y = x^m$. Substituting into the differential equation we have

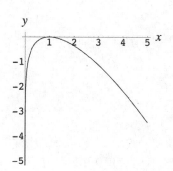

$$xy'' + y' = m(m-1)x^{m-1} + mx^{m-1} = m^2 x^{m-1}.$$

Thus, the general solution of $xy'' + y' = 0$ is $y_c = c_1 + c_2 \ln x$. To find a particular solution of $xy'' + y' = -\sqrt{x}$ we use variation of parameters. The Wronskian is

$$W = \begin{vmatrix} 1 & \ln x \\ 0 & 1/x \end{vmatrix} = \frac{1}{x}.$$

Identifying $f(x) = -x^{-1/2}$ we obtain

$$u_1' = \frac{x^{-1/2} \ln x}{1/x} = \sqrt{x} \ln x \quad \text{and} \quad u_2' = \frac{-x^{-1/2}}{1/x} = -\sqrt{x},$$

so that

$$u_1 = x^{3/2}\left(\frac{2}{3} \ln x - \frac{4}{9}\right) \quad \text{and} \quad u_2 = -\frac{2}{3}x^{3/2}.$$

Then

$$y_p = x^{3/2}\left(\frac{2}{3} \ln x - \frac{4}{9}\right) - \frac{2}{3}x^{3/2} \ln x = -\frac{4}{9}x^{3/2}$$

and the general solution of the differential equation is

$$y = c_1 + c_2 \ln x - \frac{4}{9}x^{3/2}.$$

The initial conditions are $y(1) = 0$ and $y'(1) = 0$. These imply that $c_1 = \frac{4}{9}$ and $c_2 = \frac{2}{3}$. The solution of the initial-value problem is

$$y = \frac{4}{9} + \frac{2}{3}\ln x - \frac{4}{9}x^{3/2}.$$

The graph is shown above.

37. From $(D-2)x + (D-2)y = 1$ and $Dx + (2D-1)y = 3$ we obtain $(D-1)(D-2)y = -6$ and $Dx = 3 - (2D-1)y$. Then

$$y = c_1 e^{2t} + c_2 e^t - 3 \quad \text{and} \quad x = -c_2 e^t - \frac{3}{2}c_1 e^{2t} + c_3.$$

Substituting into $(D-2)x + (D-2)y = 1$ gives $c_3 = \frac{5}{2}$ so that

$$x = -c_2 e^t - \frac{3}{2}c_1 e^{2t} + \frac{5}{2}.$$

38. From $(D-2)x - y = t - 2$ and $-3x + (D-4)y = -4t$ we obtain $(D-1)(D-5)x = 9 - 8t$. Then

$$x = c_1 e^t + c_2 e^{5t} - \frac{8}{5}t - \frac{3}{25}$$

and

$$y = (D-2)x - t + 2 = -c_1 e^t + 3c_2 e^{5t} + \frac{16}{25} + \frac{11}{25}t.$$

39. From $(D-2)x - y = -e^t$ and $-3x + (D-4)y = -7e^t$ we obtain $(D-1)(D-5)x = -4e^t$ so that

$$x = c_1 e^t + c_2 e^{5t} + te^t.$$

Then

$$y = (D-2)x + e^t = -c_1 e^t + 3c_2 e^{5t} - te^t + 2e^t.$$

40. From $(D+2)x + (D+1)y = \sin 2t$ and $5x + (D+3)y = \cos 2t$ we obtain $(D^2+5)y = 2\cos 2t - 7\sin 2t$. Then

$$y = c_1 \cos t + c_2 \sin t - \frac{2}{3}\cos 2t + \frac{7}{3}\sin 2t$$

and

$$x = -\frac{1}{5}(D+3)y + \frac{1}{5}\cos 2t$$

$$= \left(\frac{1}{5}c_1 - \frac{3}{5}c_2\right)\sin t + \left(-\frac{1}{5}c_2 - \frac{3}{5}c_1\right)\cos t - \frac{5}{3}\sin 2t - \frac{1}{3}\cos 2t.$$

5 Modeling with Higher-Order Differential Equations

Exercises 5.1 — Linear Models: Initial-Value Problems

1. From $\frac{1}{8}x'' + 16x = 0$ we obtain

$$x = c_1 \cos 8\sqrt{2}\,t + c_2 \sin 8\sqrt{2}\,t$$

so that the period of motion is $2\pi/8\sqrt{2} = \sqrt{2}\,\pi/8$ seconds.

2. From $20x'' + kx = 0$ we obtain

$$x = c_1 \cos \frac{1}{2}\sqrt{\frac{k}{5}}\,t + c_2 \sin \frac{1}{2}\sqrt{\frac{k}{5}}\,t$$

so that the frequency $2/\pi = \frac{1}{4}\sqrt{k/5}\,\pi$ and $k = 320$ N/m. If $80x'' + 320x = 0$ then

$$x = c_1 \cos 2t + c_2 \sin 2t$$

so that the frequency is $2/2\pi = 1/\pi$ cycles/s.

3. From $\frac{3}{4}x'' + 72x = 0$, $x(0) = -1/4$, and $x'(0) = 0$ we obtain $x = -\frac{1}{4}\cos 4\sqrt{6}\,t$.

4. From $\frac{3}{4}x'' + 72x = 0$, $x(0) = 0$, and $x'(0) = 2$ we obtain $x = \frac{\sqrt{6}}{12}\sin 4\sqrt{6}\,t$.

5. From $\frac{5}{8}x'' + 40x = 0$, $x(0) = 1/2$, and $x'(0) = 0$ we obtain $x = \frac{1}{2}\cos 8t$.

   (a) $x(\pi/12) = -1/4$, $x(\pi/8) = -1/2$, $x(\pi/6) = -1/4$, $x(\pi/4) = 1/2$, $x(9\pi/32) = \sqrt{2}/4$.

   (b) $x' = -4\sin 8t$ so that $x'(3\pi/16) = 4$ ft/s directed downward.

   (c) If $x = \frac{1}{2}\cos 8t = 0$ then $t = (2n+1)\pi/16$ for $n = 0, 1, 2, \ldots$.

6. From $50x'' + 200x = 0$, $x(0) = 0$, and $x'(0) = -10$ we obtain $x = -5\sin 2t$ and $x' = -10\cos 2t$.

7. From $20x'' + 20x = 0$, $x(0) = 0$, and $x'(0) = -10$ we obtain $x = -10\sin t$ and $x' = -10\cos t$.

   (a) The 20 kg mass has the larger amplitude.

   (b) 20 kg: $x'(\pi/4) = -5\sqrt{2}$ m/s, $x'(\pi/2) = 0$ m/s;   50 kg: $x'(\pi/4) = 0$ m/s, $x'(\pi/2) = 10$ m/s

   (c) If $-5\sin 2t = -10\sin t$ then $\sin t(\cos t - 1) = 0$ so that $t = n\pi$ for $n = 0, 1, 2, \ldots$, placing both masses at the equilibrium position. The 50 kg mass is moving upward; the 20 kg mass is moving upward when n is even and downward when n is odd.

8. From $x'' + 16x = 0$, $x(0) = -1$, and $x'(0) = -2$ we obtain

$$x = -\cos 4t - \frac{1}{2}\sin 4t = \frac{\sqrt{5}}{2}\sin(4t - 4.249).$$

The period is $\pi/2$ seconds and the amplitude is $\sqrt{5}/2$ feet. In 4π seconds it will make 8 complete cycles.

9. From $\frac{1}{4}x'' + x = 0$, $x(0) = 1/2$, and $x'(0) = 3/2$ we obtain

$$x = \frac{1}{2}\cos 2t + \frac{3}{4}\sin 2t = \frac{\sqrt{13}}{4}\sin(2t + 0.588).$$

10. From $1.6x'' + 40x = 0$, $x(0) = -1/3$, and $x'(0) = 5/4$ we obtain

$$x = -\frac{1}{3}\cos 5t + \frac{1}{4}\sin 5t = \frac{5}{12}\sin(5t - 0.927).$$

If $x = 5/24$ then $t = \frac{1}{5}\left(\frac{\pi}{6} + 0.927 + 2n\pi\right)$ and $t = \frac{1}{5}\left(\frac{5\pi}{6} + 0.927 + 2n\pi\right)$ for $n = 0, 1, 2, \ldots$.

11. From $2x'' + 200x = 0$, $x(0) = -2/3$, and $x'(0) = 5$ we obtain

  (a) $x = -\frac{2}{3}\cos 10t + \frac{1}{2}\sin 10t = \frac{5}{6}\sin(10t - 0.927)$.

  (b) The amplitude is $5/6$ ft and the period is $2\pi/10 = \pi/5$

  (c) $3\pi = \pi k/5$ and $k = 15$ cycles.

  (d) If $x = 0$ and the weight is moving downward for the second time, then $10t - 0.927 = 2\pi$ or $t = 0.721$ s.

  (e) If $x' = \frac{25}{3}\cos(10t - 0.927) = 0$ then $10t - 0.927 = \pi/2 + n\pi$ or $t = (2n+1)\pi/20 + 0.0927$ for $n = 0, 1, 2, \ldots$.

  (f) $x(3) = -0.597$ ft

  (g) $x'(3) = -5.814$ ft/s

  (h) $x''(3) = 59.702$ ft/s^2

  (i) If $x = 0$ then $t = \frac{1}{10}(0.927 + n\pi)$ for $n = 0, 1, 2, \ldots$. The velocity at these times is $x' = \pm 8.33$ ft/s.

  (j) If $x = 5/12$ then $t = \frac{1}{10}(\pi/6 + 0.927 + 2n\pi)$ and $t = \frac{1}{10}(5\pi/6 + 0.927 + 2n\pi)$ for $n = 0, 1, 2, \ldots$.

  (k) If $x = 5/12$ and $x' < 0$ then $t = \frac{1}{10}(5\pi/6 + 0.927 + 2n\pi)$ for $n = 0, 1, 2, \ldots$.

12. From $x'' + 9x = 0$, $x(0) = -1$, and $x'(0) = -\sqrt{3}$ we obtain

$$x = -\cos 3t - \frac{\sqrt{3}}{3}\sin 3t = \frac{2}{\sqrt{3}}\sin\left(3t + \frac{4\pi}{3}\right)$$

and $x' = 2\sqrt{3}\cos(3t + 4\pi/3)$. If $x' = 3$ then $t = -7\pi/18 + 2n\pi/3$ and $t = -\pi/2 + 2n\pi/3$ for $n = 1, 2, 3, \ldots$.

13. From $k_1 = 40$ and $k_2 = 120$ we compute the effective spring constant $k = 4(40)(120)/160 = 120$. Now, $m = 20/32$ so $k/m = 120(32)/20 = 192$ and $x'' + 192x = 0$. Using $x(0) = 0$ and $x'(0) = 2$ we obtain $x(t) = \frac{\sqrt{3}}{12}\sin 8\sqrt{3}\,t$.

14. Let m be the mass and k_1 and k_2 the spring constants. Then $k = 4k_1k_2/(k_1 + k_2)$ is the effective spring constant of the system. Since the initial mass stretches one spring $\frac{1}{3}$ foot and another spring $\frac{1}{2}$ foot, using $F = ks$, we have $\frac{1}{3}k_1 = \frac{1}{2}k_2$ or $2k_1 = 3k_2$. The given period of the combined system is $2\pi/\omega = \pi/15$, so $\omega = 30$. Since a mass weighing 8 pounds is $\frac{1}{4}$ slug, we have from $w^2 = k/m$

$$30^2 = \frac{k}{1/4} = 4k \quad \text{or} \quad k = 225.$$

We now have the system of equations

$$\frac{4k_1k_2}{k_1 + k_2} = 225$$

$$2k_1 = 3k_2.$$

Solving the second equation for k_1 and substituting in the first equation, we obtain

$$\frac{4(3k_2/2)k_2}{3k_2/2 + k_2} = \frac{12k_2^2}{5k_2} = \frac{12k_2}{5} = 225.$$

Thus, $k_2 = 375/4$ and $k_1 = 1125/8$. Finally, the weight of the first mass is

$$32m = \frac{k_1}{3} = \frac{1125/8}{3} = \frac{375}{8} \approx 46.88 \text{ lb}.$$

15. For large values of t the differential equation is approximated by $x'' = 0$. The solution of this equation is the linear function $x = c_1t + c_2$. Thus, for large time, the restoring force will have decayed to the point where the spring is incapable of returning the mass, and the spring will simply keep on stretching.

16. As t becomes larger the spring constant increases; that is, the spring is stiffening. It would seem that the oscillations would become periodic and the spring would oscillate more rapidly. It is likely that the amplitudes of the oscillations would decrease as t increases.

17. (a) above   **(b)** heading upward

18. (a) below   **(b)** from rest

19. (a) below   **(b)** heading upward

20. (a) above   **(b)** heading downward

21. From $\frac{1}{8}x'' + x' + 2x = 0$, $x(0) = -1$, and $x'(0) = 8$ we obtain $x = 4te^{-4t} - e^{-4t}$ and $x' = 8e^{-4t} - 16te^{-4t}$. If $x = 0$ then $t = 1/4$ second. If $x' = 0$ then $t = 1/2$ second and the extreme displacement is $x = e^{-2}$ feet.

22. From $\frac{1}{4}x'' + \sqrt{2}\,x' + 2x = 0$, $x(0) = 0$, and $x'(0) = 5$ we obtain $x = 5te^{-2\sqrt{2}t}$ and

$x' = 5e^{-2\sqrt{2}t}\left(1 - 2\sqrt{2}\,t\right)$. If $x' = 0$ then $t = \sqrt{2}/4$ second and the extreme displacement is $x = 5\sqrt{2}\,e^{-1}/4$ feet.

23. (a) From $x'' + 10x' + 16x = 0$, $x(0) = 1$, and $x'(0) = 0$ we obtain $x = \frac{4}{3}e^{-2t} - \frac{1}{3}e^{-8t}$.

(b) From $x'' + x' + 16x = 0$, $x(0) = 1$, and $x'(0) = -12$ then $x = -\frac{2}{3}e^{-2t} + \frac{5}{3}e^{-8t}$.

24. (a) $x = \frac{1}{3}e^{-8t}\left(4e^{6t} - 1\right)$ is not zero for $t \geq 0$; the extreme displacement is $x(0) = 1$ meter.

(b) $x = \frac{1}{3}e^{-8t}\left(5 - 2e^{6t}\right) = 0$ when $t = \frac{1}{6}\ln\frac{5}{2} \approx 0.153$ second; if $x' = \frac{4}{3}e^{-8t}\left(e^{6t} - 10\right) = 0$ then $t = \frac{1}{6}\ln 10 \approx 0.384$ second and the extreme displacement is $x = -0.232$ meter.

25. (a) From $0.1x'' + 0.4x' + 2x = 0$, $x(0) = -1$, and $x'(0) = 0$ we obtain $x = e^{-2t}\left[-\cos 4t - \frac{1}{2}\sin 4t\right]$.

(b) $x = \dfrac{\sqrt{5}}{2}e^{-2t}\sin(4t + 4.25)$

(c) If $x = 0$ then $4t + 4.25 = 2\pi,\ 3\pi,\ 4\pi,\ \ldots$ so that the first time heading upward is $t = 1.294$ seconds.

26. (a) From $\frac{1}{4}x'' + x' + 5x = 0$, $x(0) = 1/2$, and $x'(0) = 1$ we obtain $x = e^{-2t}\left(\frac{1}{2}\cos 4t + \frac{1}{2}\sin 4t\right)$.

(b) $x = \dfrac{1}{\sqrt{2}}e^{-2t}\sin\left(4t + \dfrac{\pi}{4}\right)$.

(c) If $x = 0$ then $4t + \pi/4 = \pi,\ 2\pi,\ 3\pi,\ \ldots$ so that the times heading downward are $t = (7 + 8n)\pi/16$ for $n = 0,\ 1,\ 2,\ \ldots$.

(d)

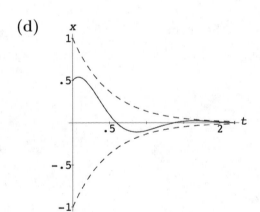

27. From $\frac{5}{16}x'' + \beta x' + 5x = 0$ we find that the roots of the auxiliary equation are $m = -\frac{8}{5}\beta \pm \frac{4}{5}\sqrt{4\beta^2 - 25}$.

(a) If $4\beta^2 - 25 > 0$ then $\beta > 5/2$.
(b) If $4\beta^2 - 25 = 0$ then $\beta = 5/2$.
(c) If $4\beta^2 - 25 < 0$ then $0 < \beta < 5/2$.

28. From $0.75x'' + \beta x' + 6x = 0$ and $\beta > 3\sqrt{2}$ we find that the roots of the auxiliary equation are

234

$m = -\frac{2}{3}\beta \pm \frac{2}{3}\sqrt{\beta^2 - 18}$ and

$$x = e^{-2\beta t/3}\left[c_1 \cosh\frac{2}{3}\sqrt{\beta^2 - 18}\,t + c_2 \sinh\frac{2}{3}\sqrt{\beta^2 - 18}\,t\right].$$

If $x(0) = 0$ and $x'(0) = -2$ then $c_1 = 0$ and $c_2 = -3/\sqrt{\beta^2 - 18}$.

29. If $\frac{1}{2}x'' + \frac{1}{2}x' + 6x = 10\cos 3t$, $x(0) = 2$, and $x'(0) = 0$ then

$$x_c = e^{-t/2}\left(c_1 \cos\frac{\sqrt{47}}{2}t + c_2 \sin\frac{\sqrt{47}}{2}t\right)$$

and $x_p = \frac{10}{3}(\cos 3t + \sin 3t)$ so that the equation of motion is

$$x = e^{-t/2}\left(-\frac{4}{3}\cos\frac{\sqrt{47}}{2}t - \frac{64}{3\sqrt{47}}\sin\frac{\sqrt{47}}{2}t\right) + \frac{10}{3}(\cos 3t + \sin 3t).$$

30. (a) If $x'' + 2x' + 5x = 12\cos 2t + 3\sin 2t$, $x(0) = 1$, and $x'(0) = 5$ then $x_c = e^{-t}(c_1 \cos 2t + c_2 \sin 2t)$ and $x_p = 3\sin 2t$ so that the equation of motion is

$$x = e^{-t}\cos 2t + 3\sin 2t.$$

(b)

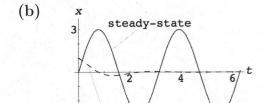

(c)

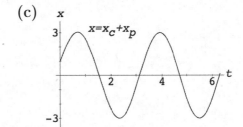

31. From $x'' + 8x' + 16x = 8\sin 4t$, $x(0) = 0$, and $x'(0) = 0$ we obtain $x_c = c_1 e^{-4t} + c_2 t e^{-4t}$ and $x_p = -\frac{1}{4}\cos 4t$ so that the equation of motion is

$$x = \frac{1}{4}e^{-4t} + te^{-4t} - \frac{1}{4}\cos 4t.$$

32. From $x'' + 8x' + 16x = e^{-t}\sin 4t$, $x(0) = 0$, and $x'(0) = 0$ we obtain $x_c = c_1 e^{-4t} + c_2 t e^{-4t}$ and $x_p = -\frac{24}{625}e^{-t}\cos 4t - \frac{7}{625}e^{-t}\sin 4t$ so that

$$x = \frac{1}{625}e^{-4t}(24 + 100t) - \frac{1}{625}e^{-t}(24\cos 4t + 7\sin 4t).$$

As $t \to \infty$ the displacement $x \to 0$.

33. From $2x'' + 32x = 68e^{-2t}\cos 4t$, $x(0) = 0$, and $x'(0) = 0$ we obtain $x_c = c_1 \cos 4t + c_2 \sin 4t$ and $x_p = \frac{1}{2}e^{-2t}\cos 4t - 2e^{-2t}\sin 4t$ so that

$$x = -\frac{1}{2}\cos 4t + \frac{9}{4}\sin 4t + \frac{1}{2}e^{-2t}\cos 4t - 2e^{-2t}\sin 4t.$$

34. Since $x = \frac{\sqrt{85}}{4}\sin(4t - 0.219) - \frac{\sqrt{17}}{2}e^{-2t}\sin(4t - 2.897)$, the amplitude approaches $\sqrt{85}/4$ as $t \to \infty$.

235

35. (a) By Hooke's law the external force is $F(t) = kh(t)$ so that $mx'' + \beta x' + kx = kh(t)$.

(b) From $\frac{1}{2}x'' + 2x' + 4x = 20\cos t$, $x(0) = 0$, and $x'(0) = 0$ we obtain $x_c = e^{-2t}(c_1\cos 2t + c_2\sin 2t)$ and $x_p = \frac{56}{13}\cos t + \frac{32}{13}\sin t$ so that

$$x = e^{-2t}\left(-\frac{56}{13}\cos 2t - \frac{72}{13}\sin 2t\right) + \frac{56}{13}\cos t + \frac{32}{13}\sin t.$$

36. (a) From $100x'' + 1600x = 1600\sin 8t$, $x(0) = 0$, and $x'(0) = 0$ we obtain $x_c = c_1\cos 4t + c_2\sin 4t$ and $x_p = -\frac{1}{3}\sin 8t$ so that by a trig identity

$$x = \frac{2}{3}\sin 4t - \frac{1}{3}\sin 8t = \frac{2}{3}\sin 4t - \frac{2}{3}\sin 4t\cos 4t.$$

(b) If $x = \frac{1}{3}\sin 4t(2 - 2\cos 4t) = 0$ then $t = n\pi/4$ for $n = 0, 1, 2, \ldots$.

(c) If $x' = \frac{8}{3}\cos 4t - \frac{8}{3}\cos 8t = \frac{8}{3}(1 - \cos 4t)(1 + 2\cos 4t) = 0$ then $t = \pi/3 + n\pi/2$ and $t = \pi/6 + n\pi/2$ for $n = 0, 1, 2, \ldots$ at the extreme values. *Note:* There are many other values of t for which $x' = 0$.

(d) $x(\pi/6 + n\pi/2) = \sqrt{3}/2$ cm and $x(\pi/3 + n\pi/2) = -\sqrt{3}/2$ cm

(e)

37. From $x'' + 4x = -5\sin 2t + 3\cos 2t$, $x(0) = -1$, and $x'(0) = 1$ we obtain $x_c = c_1\cos 2t + c_2\sin 2t$, $x_p = \frac{3}{4}t\sin 2t + \frac{5}{4}t\cos 2t$, and

$$x = -\cos 2t - \frac{1}{8}\sin 2t + \frac{3}{4}t\sin 2t + \frac{5}{4}t\cos 2t.$$

38. From $x'' + 9x = 5\sin 3t$, $x(0) = 2$, and $x'(0) = 0$ we obtain $x_c = c_1\cos 3t + c_2\sin 3t$, $x_p = -\frac{5}{6}t\cos 3t$, and

$$x = 2\cos 3t + \frac{5}{18}\sin 3t - \frac{5}{6}t\cos 3t.$$

39. (a) From $x'' + \omega^2 x = F_0\cos\gamma t$, $x(0) = 0$, and $x'(0) = 0$ we obtain $x_c = c_1\cos\omega t + c_2\sin\omega t$ and $x_p = (F_0\cos\gamma t)/(\omega^2 - \gamma^2)$ so that

$$x = -\frac{F_0}{\omega^2 - \gamma^2}\cos\omega t + \frac{F_0}{\omega^2 - \gamma^2}\cos\gamma t.$$

(b) $\displaystyle\lim_{\gamma\to\omega}\frac{F_0}{\omega^2 - \gamma^2}(\cos\gamma t - \cos\omega t) = \lim_{\gamma\to\omega}\frac{-F_0 t\sin\gamma t}{-2\gamma} = \frac{F_0}{2\omega}t\sin\omega t.$

40. From $x'' + \omega^2 x = F_0 \cos \omega t$, $x(0) = 0$, and $x'(0) = 0$ we obtain $x_c = c_1 \cos \omega t + c_2 \sin \omega t$ and $x_p = (F_0 t/2\omega) \sin \omega t$ so that $x = (F_0 t/2\omega) \sin \omega t$.

41. (a) From $\cos(u - v) = \cos u \cos v + \sin u \sin v$ and $\cos(u + v) = \cos u \cos v - \sin u \sin v$ we obtain $\sin u \sin v = \frac{1}{2}[\cos(u - v) - \cos(u + v)]$. Letting $u = \frac{1}{2}(\gamma - \omega)t$ and $v = \frac{1}{2}(\gamma + \omega)t$, the result follows.

(b) If $\epsilon = \frac{1}{2}(\gamma - \omega)$ then $\gamma \approx \omega$ so that $x = (F_0/2\epsilon\gamma) \sin \epsilon t \sin \gamma t$.

42. See the article "Distinguished Oscillations of a Forced Harmonic Oscillator" by T.G. Procter in *The College Mathematics Journal*, March, 1995. In this article the author illustrates that for $F_0 = 1$, $\lambda = 0.01$, $\gamma = 22/9$, and $\omega = 2$ the system exhibits beats oscillations on the interval $[0, 9\pi]$, but that this phenomenon is transient as $t \to \infty$.

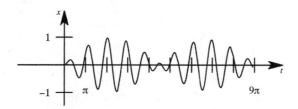

43. (a) The general solution of the homogeneous equation is

$$x_c(t) = c_1 e^{-\lambda t} \cos(\sqrt{\omega^2 - \lambda^2}\, t) + c_2 e^{-\lambda t} \sin(\sqrt{\omega^2 - \lambda^2}\, t)$$

$$= A e^{-\lambda t} \sin[\sqrt{\omega^2 - \lambda^2}\, t + \phi],$$

where $A = \sqrt{c_1^2 + c_2^2}$, $\sin \phi = c_1/A$, and $\cos \phi = c_2/A$. Now

$$x_p(t) = \frac{F_0(\omega^2 - \gamma^2)}{(\omega^2 - \gamma^2)^2 + 4\lambda^2\gamma^2} \sin \gamma t + \frac{F_0(-2\lambda\gamma)}{(\omega^2 - \gamma^2)^2 + 4\lambda^2\gamma^2} \cos \gamma t = A \sin(\gamma t + \theta),$$

where

$$\sin \theta = \frac{\dfrac{F_0(-2\lambda\gamma)}{(\omega^2 - \gamma^2)^2 + 4\lambda^2\gamma^2}}{\dfrac{F_0}{\sqrt{\omega^2 - \gamma^2 + 4\lambda^2\gamma^2}}} = \frac{-2\lambda\gamma}{\sqrt{(\omega^2 - \gamma^2)^2 + 4\lambda^2\gamma^2}}$$

and

$$\cos \theta = \frac{\dfrac{F_0(\omega^2 - \gamma^2)}{(\omega^2 - \gamma^2)^2 + 4\lambda^2\gamma^2}}{\dfrac{F_0}{\sqrt{\omega^2 - \gamma^2)^2 + 4\lambda^2\gamma^2}}} = \frac{\omega^2 - \gamma^2}{\sqrt{(\omega^2 - \gamma^2)^2 + 4\lambda^2\gamma^2}}.$$

(b) If $g'(\gamma) = 0$ then $\gamma\left(\gamma^2 + 2\lambda^2 - \omega^2\right) = 0$ so that $\gamma = 0$ or $\gamma = \sqrt{\omega^2 - 2\lambda^2}$. The first derivative test shows that g has a maximum value at $\gamma = \sqrt{\omega^2 - 2\lambda^2}$. The maximum value of g is

$$g\left(\sqrt{\omega^2 - 2\lambda^2}\right) = F_0/2\lambda\sqrt{\omega^2 - \lambda^2}.$$

(c) We identify $\omega^2 = k/m = 4$, $\lambda = \beta/2$, and $\gamma_1 = \sqrt{\omega^2 - 2\lambda^2} = \sqrt{4 - \beta^2/2}$. As $\beta \to 0$, $\gamma_1 \to 2$ and the resonance curve grows without bound at $\gamma_1 = 2$. That is, the system approaches pure resonance.

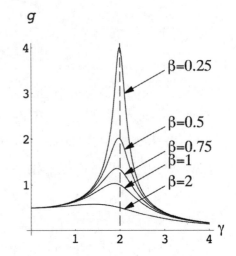

β	$\gamma1$	g
2.00	1.41	0.58
1.00	1.87	1.03
0.75	1.93	1.36
0.50	1.97	2.02
0.25	1.99	4.01

44. (a) For $n = 2$, $\sin^2 \gamma t = \frac{1}{2}(1 - \cos 2\gamma t)$. The system is in pure resonance when $2\gamma_1/2\pi = \omega/2\pi$, or when $\gamma_1 = \omega/2$.

(b) Note that

$$\sin^3 \gamma t = \sin \gamma t \sin^2 \gamma t = \frac{1}{2}[\sin \gamma t - \sin \gamma t \cos 2\gamma t].$$

Now

$$\sin(A + B) + \sin(A - B) = 2 \sin A \cos B$$

so

$$\sin \gamma t \cos 2\gamma t = \frac{1}{2}[\sin 3\gamma t - \sin \gamma t]$$

and

$$\sin^3 \gamma t = \frac{3}{4} \sin \gamma t - \frac{1}{4} \sin 3\gamma t.$$

Thus

$$x'' + \omega^2 x = \frac{3}{4} \sin \gamma t - \frac{1}{4} \sin 3\gamma t.$$

The frequency of free vibration is $\omega/2\pi$. Thus, when $\gamma_1/2\pi = \omega/2\pi$ or $\gamma_1 = \omega$, and when $3\gamma_2/2\pi = \omega/2\pi$ or $3\gamma_2 = \omega$ or $\gamma_3 = \omega/3$, the system will be in pure resonance.

(c)

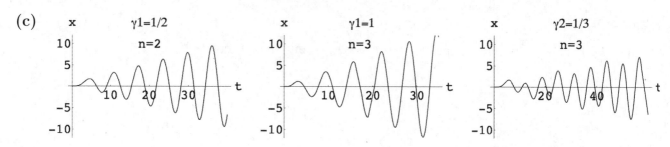

45. Solving $\frac{1}{20}q'' + 2q' + 100q = 0$ we obtain $q(t) = e^{-20t}(c_1 \cos 40t + c_2 \sin 40t)$. The initial conditions $q(0) = 5$ and $q'(0) = 0$ imply $c_1 = 5$ and $c_2 = 5/2$. Thus

$$q(t) = e^{-20t}\left(5\cos 40t + \frac{5}{2}\sin 40t\right) = \sqrt{25 + 25/4}\, e^{-20t}\sin(40t + 1.1071)$$

and $q(0.01) \approx 4.5676$ coulombs. The charge is zero for the first time when $40t + 1.1071 = \pi$ or $t \approx 0.0509$ second.

46. Solving $\frac{1}{4}q'' + 20q' + 300q = 0$ we obtain $q(t) = c_1 e^{-20t} + c_2 e^{-60t}$. The initial conditions $q(0) = 4$ and $q'(0) = 0$ imply $c_1 = 6$ and $c_2 = -2$. Thus

$$q(t) = 6e^{-20t} - 2e^{-60t}.$$

Setting $q = 0$ we find $e^{40t} = 1/3$ which implies $t < 0$. Therefore the charge is not 0 for $t \geq 0$.

47. Solving $\frac{5}{3}q'' + 10q' + 30q = 300$ we obtain $q(t) = e^{-3t}(c_1 \cos 3t + c_2 \sin 3t) + 10$. The initial conditions $q(0) = q'(0) = 0$ imply $c_1 = c_2 = -10$. Thus

$$q(t) = 10 - 10e^{-3t}(\cos 3t + \sin 3t) \quad \text{and} \quad i(t) = 60e^{-3t}\sin 3t.$$

Solving $i(t) = 0$ we see that the maximum charge occurs when $t = \pi/3$ and $q(\pi/3) \approx 10.432$.

48. Solving $q'' + 100q' + 2500q = 30$ we obtain $q(t) = c_1 e^{-50t} + c_2 te^{-50t} + 0.012$. The initial conditions $q(0) = 0$ and $q'(0) = 2$ imply $c_1 = -0.012$ and $c_2 = 1.4$. Thus, using $i(t) = q'(t)$ we get

$$q(t) = -0.012e^{-50t} + 1.4te^{-50t} + 0.012 \quad \text{and} \quad i(t) = 2e^{-50t} - 70te^{-50t}.$$

Solving $i(t) = 0$ we see that the maximum charge occurs when $t = 1/35$ second and $q(1/35) \approx 0.01871$ coulomb.

49. Solving $q'' + 2q' + 4q = 0$ we obtain $q_c = e^{-t}\left(\cos\sqrt{3}\,t + \sin\sqrt{3}\,t\right)$. The steady-state charge has the form $q_p = A\cos t + B\sin t$. Substituting into the differential equation we find

$$(3A + 2B)\cos t + (3B - 2A)\sin t = 50\cos t.$$

Thus, $A = 150/13$ and $B = 100/13$. The steady-state charge is

$$q_p(t) = \frac{150}{13}\cos t + \frac{100}{13}\sin t$$

and the steady-state current is

$$i_p(t) = -\frac{150}{13}\sin t + \frac{100}{13}\cos t.$$

50. From

$$i_p(t) = \frac{E_0}{Z}\left(\frac{R}{Z}\sin\gamma t - \frac{X}{Z}\cos\gamma t\right)$$

and $Z = \sqrt{X^2 + R^2}$ we see that the amplitude of $i_p(t)$ is

$$A = \sqrt{\frac{E_0^2 R^2}{Z^4} + \frac{E_0^2 X^2}{Z^4}} = \frac{E_0}{Z^2}\sqrt{R^2 + X^2} = \frac{E_0}{Z}.$$

51. The differential equation is $\frac{1}{2}q'' + 20q' + 1000q = 100\sin 60t$. To use Example 10 in the text we identify $E_0 = 100$ and $\gamma = 60$. Then

$$X = L\gamma - \frac{1}{c\gamma} = \frac{1}{2}(60) - \frac{1}{0.001(60)} \approx 13.3333,$$

$$Z = \sqrt{X^2 + R^2} = \sqrt{X^2 + 400} \approx 24.0370,$$

and

$$\frac{E_0}{Z} = \frac{100}{Z} \approx 4.1603.$$

From Problem 50, then

$$i_p(t) \approx 4.1603\sin(60t + \phi)$$

where $\sin\phi = -X/Z$ and $\cos\phi = R/Z$. Thus $\tan\phi = -X/R \approx -0.6667$ and ϕ is a fourth quadrant angle. Now $\phi \approx -0.5880$ and

$$i_p(t) = 4.1603\sin(60t - 0.5880).$$

52. Solving $\frac{1}{2}q'' + 20q' + 1000q = 0$ we obtain $q_c(t) = e^{-20t}(c_1\cos 40t + c_2\sin 40t)$. The steady-state charge has the form $q_p(t) = A\sin 60t + B\cos 60t + C\sin 40t + D\cos 40t$. Substituting into the differential equation we find

$$(-1600A - 2400B)\sin 60t + (2400A - 1600B)\cos 60t$$

$$+ (400C - 1600D)\sin 40t + (1600C + 400D)\cos 40t$$

$$= 200\sin 60t + 400\cos 40t.$$

Equating coefficients we obtain $A = -1/26$, $B = -3/52$, $C = 4/17$, and $D = 1/17$. The steady-state charge is

$$q_p(t) = -\frac{1}{26}\sin 60t - \frac{3}{52}\cos 60t + \frac{4}{17}\sin 40t + \frac{1}{17}\cos 40t$$

and the steady-state current is

$$i_p(t) = -\frac{30}{13}\cos 60t + \frac{45}{13}\sin 60t + \frac{160}{17}\cos 40t - \frac{40}{17}\sin 40t.$$

53. Solving $\frac{1}{2}q'' + 10q' + 100q = 150$ we obtain $q(t) = e^{-10t}(c_1\cos 10t + c_2\sin 10t) + 3/2$. The initial conditions $q(0) = 1$ and $q'(0) = 0$ imply $c_1 = c_2 = -1/2$. Thus

$$q(t) = -\frac{1}{2}e^{-10t}(\cos 10t + \sin 10t) + \frac{3}{2}.$$

As $t \to \infty$, $q(t) \to 3/2$.

54. In Problem 50 it is shown that the amplitude of the steady-state current is E_0/Z, where $Z = \sqrt{X^2 + R^2}$ and $X = L\gamma - 1/C\gamma$. Since E_0 is constant the amplitude will be a maximum when Z is a minimum. Since R is constant, Z will be a minimum when $X = 0$. Solving $L\gamma - 1/C\gamma = 0$ for γ we obtain $\gamma = 1/\sqrt{LC}$. The maximum amplitude will be E_0/R.

55. By Problem 50 the amplitude of the steady-state current is E_0/Z, where $Z = \sqrt{X^2 + R^2}$ and $X = L\gamma - 1/C\gamma$. Since E_0 is constant the amplitude will be a maximum when Z is a minimum. Since R is constant, Z will be a minimum when $X = 0$. Solving $L\gamma - 1/C\gamma = 0$ for C we obtain $C = 1/L\gamma^2$.

56. Solving $0.1q'' + 10q = 100\sin\gamma t$ we obtain

$$q(t) = c_1\cos 10t + c_2\sin 10t + q_p(t)$$

where $q_p(t) = A\sin\gamma t + B\cos\gamma t$. Substituting $q_p(t)$ into the differential equation we find

$$(100 - \gamma^2)A\sin\gamma t + (100 - \gamma^2)B\cos\gamma t = 100\sin\gamma t.$$

Equating coefficients we obtain $A = 100/(100 - \gamma^2)$ and $B = 0$. Thus, $q_p(t) = \dfrac{100}{100 - \gamma^2}\sin\gamma t$. The initial conditions $q(0) = q'(0) = 0$ imply $c_1 = 0$ and $c_2 = -10\gamma/(100 - \gamma^2)$. The charge is

$$q(t) = \frac{10}{100 - \gamma^2}(10\sin\gamma t - \gamma\sin 10t)$$

and the current is

$$i(t) = \frac{100\gamma}{100 - \gamma^2}(\cos\gamma t - \cos 10t).$$

57. In an LC-series circuit there is no resistor, so the differential equation is

$$L\frac{d^2q}{dt^2} + \frac{1}{C}q = E(t).$$

Then $q(t) = c_1\cos\left(t/\sqrt{LC}\right) + c_2\sin\left(t/\sqrt{LC}\right) + q_p(t)$ where $q_p(t) = A\sin\gamma t + B\cos\gamma t$. Substituting $q_p(t)$ into the differential equation we find

$$\left(\frac{1}{C} - L\gamma^2\right)A\sin\gamma t + \left(\frac{1}{C} - L\gamma^2\right)B\cos\gamma t = E_0\cos\gamma t.$$

Equating coefficients we obtain $A = 0$ and $B = E_0C/(1 - LC\gamma^2)$. Thus, the charge is

$$q(t) = c_1\cos\frac{1}{\sqrt{LC}}t + c_2\sin\frac{1}{\sqrt{LC}}t + \frac{E_0C}{1 - LC\gamma^2}\cos\gamma t.$$

The initial conditions $q(0) = q_0$ and $q'(0) = i_0$ imply $c_1 = q_0 - E_0C/(1 - LC\gamma^2)$ and $c_2 = i_0\sqrt{LC}$. The current is $i(t) = q'(t)$ or

$$i(t) = -\frac{c_1}{\sqrt{LC}}\sin\frac{1}{\sqrt{LC}}t + \frac{c_2}{\sqrt{LC}}\cos\frac{1}{\sqrt{LC}}t - \frac{E_0C\gamma}{1 - LC\gamma^2}\sin\gamma t$$

$$= i_0\cos\frac{1}{\sqrt{LC}}t - \frac{1}{\sqrt{LC}}\left(q_0 - \frac{E_0C}{1 - LC\gamma^2}\right)\sin\frac{1}{\sqrt{LC}}t - \frac{E_0C\gamma}{1 - LC\gamma^2}\sin\gamma t.$$

58. When the circuit is in resonance the form of $q_p(t)$ is $q_p(t) = At\cos kt + Bt\sin kt$ where $k = 1/\sqrt{LC}$. Substituting $q_p(t)$ into the differential equation we find

$$q_p'' + k^2 q_p = -2kA\sin kt + 2kB\cos kt = \frac{E_0}{L}\cos kt.$$

Equating coefficients we obtain $A = 0$ and $B = E_0/2kL$. The charge is

$$q(t) = c_1\cos kt + c_2\sin kt + \frac{E_0}{2kL}t\sin kt.$$

The initial conditions $q(0) = q_0$ and $q'(0) = i_0$ imply $c_1 = q_0$ and $c_2 = i_0/k$. The current is

$$i(t) = -c_1 k\sin kt + c_2 k\cos kt + \frac{E_0}{2kL}(kt\cos kt + \sin kt)$$

$$= \left(\frac{E_0}{2kL} - q_0 k\right)\sin kt + i_0\cos kt + \frac{E_0}{2L}t\cos kt.$$

Exercises 5.2   Linear Models: Boundary-Value Problems

1. (a) The general solution is

$$y(x) = c_1 + c_2 x + c_3 x^2 + c_4 x^3 + \frac{w_0}{24EI}x^4.$$

The boundary conditions are $y(0) = 0$, $y'(0) = 0$, $y''(L) = 0$, $y'''(L) = 0$. The first two conditions give $c_1 = 0$ and $c_2 = 0$. The conditions at $x = L$ give the system

$$2c_3 + 6c_4 L + \frac{w_0}{2EI}L^2 = 0$$

$$6c_4 + \frac{w_0}{EI}L = 0.$$

Solving, we obtain $c_3 = w_0 L^2/4EI$ and $c_4 = -w_0 L/6EI$. The deflection is

$$y(x) = \frac{w_0}{24EI}(6L^2 x^2 - 4Lx^3 + x^4).$$

(b)

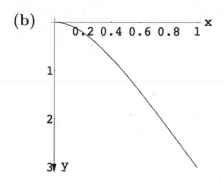

2. (a) The general solution is

$$y(x) = c_1 + c_2 x + c_3 x^2 + c_4 x^3 + \frac{w_0}{24EI} x^4.$$

The boundary conditions are $y(0) = 0$, $y''(0) = 0$, $y(L) = 0$, $y''(L) = 0$. The first two conditions give $c_1 = 0$ and $c_3 = 0$. The conditions at $x = L$ give the system

$$c_2 L + c_4 L^3 + \frac{w_0}{24EI} L^4 = 0$$

$$6 c_4 L + \frac{w_0}{2EI} L^2 = 0.$$

Solving, we obtain $c_2 = w_0 L^3 / 24EI$ and $c_4 = -w_0 L / 12EI$. The deflection is

$$y(x) = \frac{w_0}{24EI}(L^3 x - 2L x^3 + x^4).$$

(b)

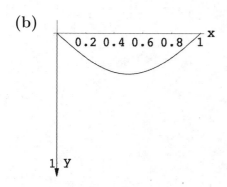

3. (a) The general solution is

$$y(x) = c_1 + c_2 x + c_3 x^2 + c_4 x^3 + \frac{w_0}{24EI} x^4.$$

The boundary conditions are $y(0) = 0$, $y'(0) = 0$, $y(L) = 0$, $y''(L) = 0$. The first two conditions give $c_1 = 0$ and $c_2 = 0$. The conditions at $x = L$ give the system

$$c_3 L^2 + c_4 L^3 + \frac{w_0}{24EI} L^4 = 0$$

$$2 c_3 + 6 c_4 L + \frac{w_0}{2EI} L^2 = 0.$$

Solving, we obtain $c_3 = w_0 L^2 / 16EI$ and $c_4 = -5 w_0 L / 48EI$. The deflection is

$$y(x) = \frac{w_0}{48EI}(3L^2 x^2 - 5L x^3 + 2x^4).$$

(b)

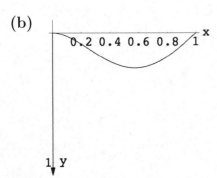

4. (a) The general solution is

$$y(x) = c_1 + c_2 x + c_3 x^2 + c_4 x^3 + \frac{w_0 L^4}{EI\pi^4} \sin \frac{\pi}{L} x.$$

The boundary conditions are $y(0) = 0$, $y'(0) = 0$, $y(L) = 0$, $y''(L) = 0$. The first two conditions give $c_1 = 0$ and $c_2 = -w_0 L^3 / EI\pi^3$. The conditions at $x = L$ give the system

$$c_3 L^2 + c_4 L^3 + \frac{w_0}{EI\pi^3} L^4 = 0$$

$$2c_3 + 6c_4 L = 0.$$

Solving, we obtain $c_3 = 3w_0 L^2 / 2EI\pi^3$ and $c_4 = -w_0 L / 2EI\pi^3$. The deflection is

$$y(x) = \frac{w_0 L}{2EI\pi^3} \left(-2L^2 x + 3Lx^2 - x^3 + \frac{2L^3}{\pi} \sin \frac{\pi}{L} x \right).$$

(b)

(c) Using a CAS we find the maximum deflection to be 0.270806 when $x = 0.572536$.

5. (a) The general solution is

$$y(x) = c_1 + c_2 x + c_3 x^2 + c_4 x^3 + \frac{w_0}{120EI} x^5.$$

The boundary conditions are $y(0) = 0$, $y''(0) = 0$, $y(L) = 0$, $y''(L) = 0$. The first two conditions give $c_1 = 0$ and $c_3 = 0$. The conditions at $x = L$ give the system

$$c_2 L + c_4 L^3 + \frac{w_0}{120EI} L^5 = 0$$

$$6c_4 L + \frac{w_0}{6EI} L^3 = 0.$$

Solving, we obtain $c_2 = 7w_0 L^4/360EI$ and $c_4 = -w_0 L^2/36EI$. The deflection is

$$y(x) = \frac{w_0}{360EI}(7L^4 x - 10L^2 x^3 + 3x^5).$$

(b)

(c) Using a CAS we find the maximum deflection to be 0.234799 when $x = 0.51933$.

6. (a) $y_{max} = y(L) = w_0 L^4/8EI$

(b) Replacing both L and x by $L/2$ in $y(x)$ we obtain $w_0 L^4/128EI$, which is $1/16$ of the maximum deflection when the length of the beam is L.

(c) $y_{max} = y(L/2) = 5w_0 L^4/384EI$

(d) The maximum deflection in Example 1 is $y(L/2) = (w_0/24EI)L^4/16 = w_0 L^4/384EI$, which is $1/5$ of the maximum displacement of the beam in part (c).

7. The general solution of the differential equation is

$$y = c_1 \cosh \sqrt{\frac{P}{EI}}\, x + c_2 \sinh \sqrt{\frac{P}{EI}}\, x + \frac{w_0}{2P} x^2 + \frac{w_0 EI}{P^2}.$$

Setting $y(0) = 0$ we obtain $c_1 = -w_0 EI/P^2$, so that

$$y = -\frac{w_0 EI}{P^2} \cosh \sqrt{\frac{P}{EI}}\, x + c_2 \sinh \sqrt{\frac{P}{EI}}\, x + \frac{w_0}{2P} x^2 + \frac{w_0 EI}{P^2}.$$

Setting $y'(L) = 0$ we find

$$c_2 = \left(\sqrt{\frac{P}{EI}}\, \frac{w_0 EI}{P^2} \sinh \sqrt{\frac{P}{EI}}\, L - \frac{w_0 L}{P} \right) \bigg/ \sqrt{\frac{P}{EI}} \cosh \sqrt{\frac{P}{EI}}\, L.$$

8. The general solution of the differential equation is

$$y = c_1 \cos \sqrt{\frac{P}{EI}}\, x + c_2 \sin \sqrt{\frac{P}{EI}}\, x + \frac{w_0}{2P} x^2 + \frac{w_0 EI}{P^2}.$$

Setting $y(0) = 0$ we obtain $c_1 = -w_0 EI/P^2$, so that

$$y = -\frac{w_0 EI}{P^2} \cos \sqrt{\frac{P}{EI}}\, x + c_2 \sin \sqrt{\frac{P}{EI}}\, x + \frac{w_0}{2P} x^2 + \frac{w_0 EI}{P^2}.$$

Setting $y'(L) = 0$ we find

$$c_2 = \left(-\sqrt{\frac{P}{EI}} \frac{w_0 EI}{P^2} \sin\sqrt{\frac{P}{EI}} L - \frac{w_0 L}{P} \right) \bigg/ \sqrt{\frac{P}{EI}} \cos\sqrt{\frac{P}{EI}} L.$$

9. This is Example 2 in the text with $L = \pi$. The eigenvalues are $\lambda_n = n^2\pi^2/\pi^2 = n^2$, $n = 1, 2, 3, \ldots$ and the corresponding eigenfunctions are $y_n = \sin(n\pi x/\pi) = \sin nx$, $n = 1, 2, 3, \ldots$.

10. This is Example 2 in the text with $L = \pi/4$. The eigenvalues are $\lambda_n = n^2\pi^2/(\pi/4)^2 = 16n^2$, $n = 1$, 2, 3, $\ldots$ and the eigenfunctions are $y_n = \sin(n\pi x/(\pi/4)) = \sin 4nx$, $n = 1, 2, 3, \ldots$.

11. For $\lambda \le 0$ the only solution of the boundary-value problem is $y = 0$. For $\lambda = \alpha^2 > 0$ we have

$$y = c_1 \cos \alpha x + c_2 \sin \alpha x.$$

Now

$$y'(x) = -c_1 \alpha \sin \alpha x + c_2 \alpha \cos \alpha x$$

and $y'(0) = 0$ implies $c_2 = 0$, so

$$y(L) = c_1 \cos \alpha L = 0$$

gives

$$\alpha L = \frac{(2n-1)\pi}{2} \quad \text{or} \quad \lambda = \alpha^2 = \frac{(2n-1)^2\pi^2}{4L^2}, \quad n = 1, 2, 3, \ldots .$$

The eigenvalues $(2n-1)^2\pi^2/4L^2$ correspond to the eigenfunctions $\cos\dfrac{(2n-1)\pi}{2L}x$ for $n = 1, 2, 3, \ldots$.

12. For $\lambda \le 0$ the only solution of the boundary-value problem is $y = 0$. For $\lambda = \alpha^2 > 0$ we have

$$y = c_1 \cos \alpha x + c_2 \sin \alpha x.$$

Since $y(0) = 0$ implies $c_1 = 0$, $y = c_2 \sin x\, dx$. Now

$$y'\left(\frac{\pi}{2}\right) = c_2 \alpha \cos \alpha\frac{\pi}{2} = 0$$

gives

$$\alpha\frac{\pi}{2} = \frac{(2n-1)\pi}{2} \quad \text{or} \quad \lambda = \alpha^2 = (2n-1)^2, \quad n = 1, 2, 3, \ldots .$$

The eigenvalues $\lambda_n = (2n-1)^2$ correspond to the eigenfunctions $y_n = \sin(2n-1)x$.

13. For $\lambda = -\alpha^2 < 0$ the only solution of the boundary-value problem is $y = 0$. For $\lambda = 0$ we have $y = c_1 x + c_2$. Now $y' = c_1$ and $y'(0) = 0$ implies $c_1 = 0$. Then $y = c_2$ and $y'(\pi) = 0$. Thus, $\lambda = 0$ is an eigenvalue with corresponding eigenfunction $y = 1$.

For $\lambda = \alpha^2 > 0$ we have

$$y = c_1 \cos \alpha x + c_2 \sin \alpha x.$$

Now

$$y'(x) = -c_1 \alpha \sin \alpha x + c_2 \alpha \cos \alpha x$$

and $y'(0) = 0$ implies $c_2 = 0$, so

$$y'(\pi) = -c_1 \alpha \sin \alpha \pi = 0$$

gives

$$\alpha \pi = n\pi \quad \text{or} \quad \lambda = \alpha^2 = n^2, \ n = 1, 2, 3, \ldots .$$

The eigenvalues n^2 correspond to the eigenfunctions $\cos nx$ for $n = 0, 1, 2, \ldots .$

14. For $\lambda \leq 0$ the only solution of the boundary-value problem is $y = 0$. For $\lambda = \alpha^2 > 0$ we have

$$y = c_1 \cos \alpha x + c_2 \sin \alpha x.$$

Now $y(-\pi) = y(\pi) = 0$ implies

$$c_1 \cos \alpha \pi - c_2 \sin \alpha \pi = 0$$

$$\tag{1}$$

$$c_1 \cos \alpha \pi + c_2 \sin \alpha \pi = 0.$$

This homogeneous system will have a nontrivial solution when

$$\begin{vmatrix} \cos \alpha \pi & -\sin \alpha \pi \\ \cos \alpha \pi & \sin \alpha \pi \end{vmatrix} = 2 \sin \alpha \pi \cos \alpha \pi = \sin 2\alpha \pi = 0.$$

Then

$$2\alpha \pi = n\pi \quad \text{or} \quad \lambda = \alpha^2 = \frac{n^2}{4}; \quad n = 1, 2, 3, \ldots .$$

When $n = 2k - 1$ is odd, the eigenvalues are $(2k - 1)^2/4$. Since $\cos(2k - 1)\pi/2 = 0$ and $\sin(2k - 1)\pi/2 \neq 0$, we see from either equation in (1) that $c_2 = 0$. Thus, the eigenfunctions corresponding to the eigenvalues $(2k - 1)^2/4$ are $y = \cos(2k - 1)x/2$ for $k = 1, 2, 3, \ldots .$ Similarly, when $n = 2k$ is even, the eigenvalues are k^2 with corresponding eigenfunctions $y = \sin kx$ for $k = 1, 2, 3, \ldots .$

15. The auxiliary equation has solutions

$$m = \frac{1}{2} \left(-2 \pm \sqrt{4 - 4(\lambda + 1)} \right) = -1 \pm \alpha.$$

For $\lambda = -\alpha^2 < 0$ we have

$$y = e^{-x} \left(c_1 \cosh \alpha x + c_2 \sinh \alpha x \right).$$

The boundary conditions imply

$$y(0) = c_1 = 0$$

$$y(5) = c_2 e^{-5} \sinh 5\alpha = 0$$

so $c_1 = c_2 = 0$ and the only solution of the boundary-value problem is $y = 0$.

For $\lambda = 0$ we have

$$y = c_1 e^{-x} + c_2 x e^{-x}$$

and the only solution of the boundary-value problem is $y = 0$.

For $\lambda = \alpha^2 > 0$ we have

$$y = e^{-x}\left(c_1 \cos \alpha x + c_2 \sin \alpha x\right).$$

Now $y(0) = 0$ implies $c_1 = 0$, so

$$y(5) = c_2 e^{-5} \sin 5\alpha = 0$$

gives

$$5\alpha = n\pi \quad \text{or} \quad \lambda = \alpha^2 = \frac{n^2 \pi^2}{25}, \quad n = 1, 2, 3, \ldots.$$

The eigenvalues $\lambda_n = \dfrac{n^2 \pi^2}{25}$ correspond to the eigenfunctions $y_n = e^{-x} \sin \dfrac{n\pi}{5} x$ for $n = 1, 2, 3, \ldots$.

16. For $\lambda < -1$ the only solution of the boundary-value problem is $y = 0$. For $\lambda = -1$ we have $y = c_1 x + c_2$. Now $y' = c_1$ and $y'(0) = 0$ implies $c_1 = 0$. Then $y = c_2$ and $y'(1) = 0$. Thus, $\lambda = -1$ is an eigenvalue with corresponding eigenfunction $y = 1$.

For $\lambda > -1$ or $\lambda + 1 = \alpha^2 > 0$ we have

$$y = c_1 \cos \alpha x + c_2 \sin \alpha x.$$

Now

$$y' = -c_1 \alpha \sin \alpha x + c_2 \alpha \cos \alpha x$$

and $y'(0) = 0$ implies $c_2 = 0$, so

$$y'(1) = -c_1 \alpha \sin \alpha = 0$$

gives

$$\alpha = n\pi, \quad \lambda + 1 = \alpha^2 = n^2 \pi^2, \quad \text{or} \quad \lambda = n^2 \pi^2 - 1, \ n = 1, 2, 3, \ldots.$$

The eigenvalues $n^2 \pi^2 - 1$ correspond to the eigenfunctions $\cos n\pi x$ for $n = 0, 1, 2, \ldots$.

17. For $\lambda = \alpha^2 > 0$ a general solution of the given differential equation is

$$y = c_1 \cos(\alpha \ln x) + c_2 \sin(\alpha \ln x).$$

Since $\ln 1 = 0$, the boundary condition $y(1) = 0$ implies $c_1 = 0$. Therefore

$$y = c_2 \sin(\alpha \ln x).$$

Using $\ln e^{\pi} = \pi$ we find that $y\left(e^{\pi}\right) = 0$ implies

$$c_2 \sin \alpha\pi = 0$$

or $\alpha\pi = n\pi$, $n = 1, 2, 3, \ldots$. The eigenvalues and eigenfunctions are, in turn,

$$\lambda = \alpha^2 = n^2, \quad n = 1, 2, 3, \ldots \quad \text{and} \quad y = \sin(n \ln x).$$

For $\lambda \leq 0$ the only solution of the boundary-value problem is $y = 0$.

18. For $\lambda = 0$ the general solution is $y = c_1 + c_2 \ln x$. Now $y' = c_2/x$, so $y'(e^{-1}) = c_2 e = 0$ implies $c_2 = 0$. Then $y = c_1$ and $y(1) = 0$ gives $c_1 = 0$. Thus $y(x) = 0$.

For $\lambda = -\alpha^2 < 0$, $y = c_1 x^{-\alpha} + c_2 x^{\alpha}$. The boundary conditions give $c_2 = c_1 e^{2\alpha}$ and $c_1 = 0$, so that $c_2 = 0$ and $y(x) = 0$.

For $\lambda = \alpha^2 > 0$, $y = c_1 \cos(\alpha \ln x) + c_2 \sin(\alpha \ln x)$. From $y(1) = 0$ we obtain $c_1 = 0$ and $y = c_2 \sin(\alpha \ln x)$. Now $y' = c_2(\alpha/x)\cos(\alpha \ln x)$, so $y'(e^{-1}) = c_2 e\alpha \cos\alpha = 0$ implies $\cos\alpha = 0$ or $\alpha = (2n-1)\pi/2$ and $\lambda = \alpha^2 = (2n-1)^2\pi^2/4$ for $n = 1, 2, 3, \ldots$. The corresponding eigenfunctions are

$$y_n = \sin\left(\frac{2n-1}{2}\pi \ln x\right).$$

19. For $\lambda = \alpha^4$, $\alpha > 0$, the general solution of the boundary-value problem

$$y^{(4)} - \lambda y = 0, \quad y(0) = 0, \ y''(0) = 0, \ y(1) = 0, \ y''(1) = 0$$

is

$$y = c_1 \cos\alpha x + c_2 \sin\alpha x + c_3 \cosh\alpha x + c_4 \sinh\alpha x.$$

The boundary conditions $y(0) = 0$, $y''(0) = 0$ give $c_1 + c_3 = 0$ and $-c_1\alpha^2 + c_3\alpha^2 = 0$, from which we conclude $c_1 = c_3 = 0$. Thus, $y = c_2 \sin\alpha x + c_4 \sinh\alpha x$. The boundary conditions $y(1) = 0$, $y''(1) = 0$ then give

$$c_2 \sin\alpha + c_4 \sinh\alpha = 0$$

$$-c_2\alpha^2 \sin\alpha + c_4\alpha^2 \sinh\alpha = 0.$$

In order to have nonzero solutions of this system, we must have the determinant of the coefficients equal zero, that is,

$$\begin{vmatrix} \sin\alpha & \sinh\alpha \\ -\alpha^2 \sin\alpha & \alpha^2 \sinh\alpha \end{vmatrix} = 0 \quad \text{or} \quad 2\alpha^2 \sinh\alpha \sin\alpha = 0.$$

But since $\alpha > 0$, the only way that this is satisfied is to have $\sin\alpha = 0$ or $\alpha = n\pi$. The system is then satisfied by choosing $c_2 \neq 0$, $c_4 = 0$, and $\alpha = n\pi$. The eigenvalues and corresponding eigenfunctions are then

$$\lambda_n = \alpha^4 = (n\pi)^4, \ n = 1, 2, 3, \ldots \quad \text{and} \quad y = \sin n\pi x.$$

20. For $\lambda = \alpha^4$, $\alpha > 0$, the general solution of the differential equation is

$$y = c_1 \cos\alpha x + c_2 \sin\alpha x + c_3 \cosh\alpha x + c_4 \sinh\alpha x.$$

The boundary conditions $y'(0) = 0$, $y'''(0) = 0$ give $c_2\alpha + c_4\alpha = 0$ and $-c_2\alpha^3 + c_4\alpha^3 = 0$ from which we conclude $c_2 = c_4 = 0$. Thus, $y = c_1 \cos\alpha x + c_3 \cosh\alpha x$. The boundary conditions $y(\pi) = 0$,

$y''(\pi) = 0$ then give

$$c_2 \cos \alpha\pi + c_4 \cosh \alpha\pi = 0$$

$$-c_2\lambda^2 \cos \alpha\pi + c_4\lambda^2 \cosh \alpha\pi = 0.$$

The determinant of the coefficients is $2\alpha^2 \cosh \alpha \cos \alpha = 0$. But since $\alpha > 0$, the only way that this is satisfied is to have $\cos \alpha\pi = 0$ or $\alpha = (2n-1)/2$, $n = 1, 2, 3, \ldots$. The eigenvalues and corresponding eigenfunctions are

$$\lambda_n = \alpha^4 = \left(\frac{2n-1}{2}\right)^4, \ n = 1, 2, 3, \ldots \quad \text{and} \quad y = \cos\left(\frac{2n-1}{2}\right)x.$$

21. If restraints are put on the column at $x = L/4$, $x = L/2$, and $x = 3L/4$, then the critical load will be P_4.

22. **(a)** The general solution of the differential equation is

$$y = c_1 \cos \sqrt{\frac{P}{EI}}\, x + c_2 \sin \sqrt{\frac{P}{EI}}\, x + \delta.$$

Since the column is embedded at $x = 0$, the boundary conditions are $y(0) = y'(0) = 0$. If $\delta = 0$ this implies that $c_1 = c_2 = 0$ and $y(x) = 0$. That is, there is no deflection.

(b) If $\delta \neq 0$, the boundary conditions give, in turn, $c_1 = -\delta$ and $c_2 = 0$. Then

$$y = \delta\left(1 - \cos \sqrt{\frac{P}{EI}}\, x\right).$$

In order to satisfy the boundary condition $y(L) = \delta$ we must have

$$\delta = \delta\left(1 - \cos \sqrt{\frac{P}{EI}}\, L\right) \quad \text{or} \quad \cos \sqrt{\frac{P}{EI}}\, L = 0.$$

This gives $\sqrt{P/EI}\, L = n\pi/2$ for $n = 1, 2, 3, \ldots$. The smallest value of P_n, the Euler load, is then

$$\sqrt{\frac{P_1}{EI}}\, L = \frac{\pi}{2} \quad \text{or} \quad P_1 = \frac{1}{4}\left(\frac{\pi^2 EI}{L^2}\right).$$

23. If $\lambda = \alpha^2 = P/EI$, then the solution of the differential equation is

$$y = c_1 \cos \alpha x + c_2 \sin \alpha x + c_3 x + c_4.$$

The conditions $y(0) = 0$, $y''(0) = 0$ yield, in turn, $c_1 + c_4 = 0$ and $c_1 = 0$. With $c_1 = 0$ and $c_4 = 0$ the solution is $y = c_2 \sin \alpha x + c_3 x$. The conditions $y(L) = 0$, $y''(L) = 0$, then yield

$$c_2 \sin \alpha L + c_3 L = 0 \quad \text{and} \quad c_2 \sin \alpha L = 0.$$

Hence, nontrivial solutions of the problem exist only if $\sin \alpha L = 0$. From this point on, the analysis is the same as in Example 3 in the text.

24. (a) The boundary-value problem is

$$\frac{d^4 y}{dx^4} + \lambda \frac{d^2 y}{dx^2} = 0, \quad y(0) = 0, y''(0) = 0, \; y(L) = 0, y'(L) = 0,$$

where $\lambda = \alpha^2 = P/EI$. The solution of the differential equation is $y = c_1 \cos \alpha x + c_2 \sin \alpha x + c_3 x + c_4$ and the conditions $y(0) = 0$, $y''(0) = 0$ yield $c_1 = 0$ and $c_4 = 0$. Next, by applying $y(L) = 0$, $y'(L) = 0$ to $y = c_2 \sin \alpha x + c_3 x$ we get the system of equations

$$c_2 \sin \alpha L + c_3 L = 0$$

$$\alpha c_2 \cos \alpha L + c_3 = 0.$$

To obtain nontrivial solutions c_2, c_3, we must have the determinant of the coefficients equal to zero:

$$\begin{vmatrix} \sin \alpha L & L \\ \alpha \cos \alpha L & 1 \end{vmatrix} = 0 \quad \text{or} \quad \tan \beta = \beta,$$

where $\beta = \alpha L$. If β_n denotes the positive roots of the last equation, then the eigenvalues are found from $\beta_n = \alpha_n L = \sqrt{\lambda_n}\, L$ or $\lambda_n = (\beta_n/L)^2$. From $\lambda = P/EI$ we see that the critical loads are $P_n = \beta_n^2 EI/L^2$. With the aid of a CAS we find that the first positive root of $\tan \beta = \beta$ is (approximately) $\beta_1 = 4.4934$, and so the Euler load is (approximately) $P_1 = 20.1907 EI/L^2$. Finally, if we use $c_3 = -c_2 \alpha \cos \alpha L$, then the deflection curves are

$$y_n(x) = c_2 \sin \alpha_n x + c_3 x = c_2 \left[\sin \left(\frac{\beta_n}{L} x \right) - \left(\frac{\beta_n}{L} \cos \beta_n \right) x \right].$$

(b) With $L = 1$ and c_2 appropriately chosen, the general shape of the first buckling mode,

$$y_1(x) = c_2 \left[\sin \left(\frac{4.4934}{L} x \right) - \left(\frac{4.4934}{L} \cos(4.4934) \right) x \right],$$

is shown below.

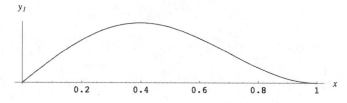

25. The general solution is

$$y = c_1 \cos \sqrt{\frac{\rho}{T}}\, \omega x + c_2 \sin \sqrt{\frac{\rho}{T}}\, \omega x.$$

From $y(0) = 0$ we obtain $c_1 = 0$. Setting $y(L) = 0$ we find $\sqrt{\rho/T}\,\omega L = n\pi$, $n = 1, 2, 3, \ldots$. Thus, critical speeds are $\omega_n = n\pi\sqrt{T}/L\sqrt{\rho}$, $n = 1, 2, 3, \ldots$. The corresponding deflection curves are

$$y(x) = c_2 \sin\frac{n\pi}{L}x, \quad n = 1, 2, 3, \ldots,$$

where $c_2 \neq 0$.

26. (a) When $T(x) = x^2$ the given differential equation is the Cauchy-Euler equation

$$x^2 y'' + 2xy' + \rho\omega^2 y = 0.$$

The solutions of the auxiliary equation

$$m(m-1) + 2m + \rho\omega^2 = m^2 + m + \rho\omega^2 = 0$$

are

$$m_1 = -\frac{1}{2} - \frac{1}{2}\sqrt{4\rho\omega^2 - 1}\,i, \quad m_2 = -\frac{1}{2} + \frac{1}{2}\sqrt{4\rho\omega^2 - 1}\,i$$

when $\rho\omega^2 > 0.25$. Thus

$$y = c_1 x^{-1/2}\cos(\lambda\ln x) + c_2 x^{-1/2}\sin(\lambda\ln x)$$

where $\lambda = \frac{1}{2}\sqrt{4\rho\omega^2 - 1}$. Applying $y(1) = 0$ gives $c_1 = 0$ and consequently

$$y = c_2 x^{-1/2}\sin(\lambda\ln x).$$

The condition $y(e) = 0$ requires $c_2 e^{-1/2}\sin\lambda = 0$. We obtain a nontrivial solution when $\lambda_n = n\pi$, $n = 1, 2, 3, \ldots$. But

$$\lambda_n = \frac{1}{2}\sqrt{4\rho\omega_n^2 - 1} = n\pi.$$

Solving for ω_n gives

$$\omega_n = \frac{1}{2}\sqrt{(4n^2\pi^2 + 1)/\rho}.$$

The corresponding solutions are

$$y_n(x) = c_2 x^{-1/2}\sin(n\pi\ln x).$$

(b)

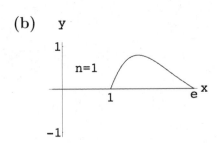

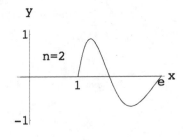

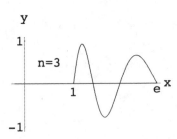

27. The auxiliary equation is $m^2 + m = m(m+1) = 0$ so that $u(r) = c_1 r^{-1} + c_2$. The boundary conditions $u(a) = u_0$ and $u(b) = u_1$ yield the system $c_1 a^{-1} + c_2 = u_0$, $c_1 b^{-1} + c_2 = u_1$. Solving gives

$$c_1 = \left(\frac{u_0 - u_1}{b - a}\right)ab \quad \text{and} \quad c_2 = \frac{u_1 b - u_0 a}{b - a}.$$

Thus

$$u(r) = \left(\frac{u_0 - u_1}{b - a}\right)\frac{ab}{r} + \frac{u_1 b - u_0 a}{b - a}.$$

28. The auxiliary equation is $m^2 = 0$ so that $u(r) = c_1 + c_2 \ln r$. The boundary conditions $u(a) = u_0$ and $u(b) = u_1$ yield the system $c_1 + c_2 \ln a = u_0$, $c_1 + c_2 \ln b = u_1$. Solving gives

$$c_1 = \frac{u_1 \ln a - u_0 \ln b}{\ln(a/b)} \quad \text{and} \quad c_2 = \frac{u_0 - u_1}{\ln(a/b)}.$$

Thus

$$u(r) = \frac{u_1 \ln a - u_0 \ln b}{\ln(a/b)} + \frac{u_0 - u_1}{\ln(a/b)}\ln r = \frac{u_0 \ln(r/b) - u_1 \ln(r/a)}{\ln(a/b)}.$$

29. The solution of the initial-value problem

$$x'' + \omega^2 x = 0, \quad x(0) = 0, \; x'(0) = v_0, \; \omega^2 = 10/m$$

is $x(t) = (v_0/\omega)\sin \omega t$. To satisfy the additional boundary condition $x(1) = 0$ we require that $\omega = n\pi$, $n = 1, 2, 3, \ldots$. The eigenvalues $\lambda = \omega^2 = n^2\pi^2$ and eigenfunctions of the problem are then $x(t) = (v_0/n\pi)\sin n\pi t$. Using $\omega^2 = 10/m$ we find that the *only* masses that can pass through the equilibrium position at $t = 1$ are $m_n = 10/n^2\pi^2$. Note for $n = 1$, the heaviest mass $m_1 = 10/\pi^2$ will *not* pass through the equilibrium position on the interval $0 < t < 1$ (the period of $x(t) = (v_0/\pi)\sin \pi t$ is $T = 2$, so on $0 \le t \le 1$ its graph passes through $x = 0$ only at $t = 0$ and $t = 1$). Whereas for $n > 1$, masses of lighter weight will pass through the equilibrium position $n - 1$ times prior to passing through at $t = 1$. For example, if $n = 2$, the period of $x(t) = (v_0/2\pi)\sin 2\pi t$ is $2\pi/2\pi = 1$, the mass will pass through $x = 0$ only *once* $(t = \frac{1}{2})$ prior to $t = 1$; if $n = 3$, the period of $x(t) = (v_0/3\pi)\sin 3\pi t$ is $\frac{2}{3}$, the mass will pass through $x = 0$ *twice* $(t = \frac{1}{3}$ and $t = \frac{2}{3})$ prior to $t = 1$; and so on.

30. The initial-value problem is

$$x'' + \frac{2}{m}x' + \frac{k}{m}x = 0, \quad x(0) = 0, \; x'(0) = v_0.$$

With $k = 10$, the auxiliary equation has roots $\gamma = -1/m \pm \sqrt{1 - 10m}/m$. Consider the three cases:

(*i*) $m = \frac{1}{10}$. The roots are $\gamma_1 = \gamma_2 = 10$ and the solution of the differential equation is $x(t) = c_1 e^{-10t} + c_2 t e^{-10t}$. The initial conditions imply $c_1 = 0$ and $c_2 = v_0$ and so $x(t) = v_0 t e^{-10t}$. The condition $x(1) = 0$ implies $v_0 e^{-10} = 0$ which is impossible because $v_0 \ne 0$.

(*ii*) $1 - 10m > 0$ or $0 < m < \frac{1}{10}$. The roots are

$$\gamma_1 = -\frac{1}{m} - \frac{1}{m}\sqrt{1 - 10m} \quad \text{and} \quad \gamma_2 = -\frac{1}{m} + \frac{1}{m}\sqrt{1 - 10m}$$

and the solution of the differential equation is $x(t) = c_1 e^{\gamma_1 t} + c_2 e^{\gamma_2 t}$. The initial conditions imply

$$c_1 + c_2 = 0$$

$$\gamma_1 c_1 + \gamma_2 c_2 = v_0$$

so $c_1 = v_0/(\gamma_1 - \gamma_2)$, $c_2 = -v_0/(\gamma_1 - \gamma_2)$, and

$$x(t) = \frac{v_0}{\gamma_1 - \gamma_2}(e^{\gamma_1 t} - e^{\gamma_2 t}).$$

Again, $x(1) = 0$ is impossible because $v_0 \neq 0$.

(*iii*) $1 - 10m < 0$ or $m > \frac{1}{10}$. The roots of the auxiliary equation are

$$\gamma_1 = -\frac{1}{m} - \frac{1}{m}\sqrt{10m-1}\,i \quad \text{and} \quad \gamma_2 = -\frac{1}{m} + \frac{1}{m}\sqrt{10m-1}\,i$$

and the solution of the differential equation is

$$x(t) = c_1 e^{-t/m}\cos\frac{1}{m}\sqrt{10m-1}\,t + c_2 e^{-t/m}\sin\frac{1}{m}\sqrt{10m-1}\,t.$$

The initial conditions imply $c_1 = 0$ and $c_2 = mv_0/\sqrt{10m-1}$, so that

$$x(t) = \frac{mv_0}{\sqrt{10m-1}}\,e^{-t/m}\sin\left(\frac{1}{m}\sqrt{10m-1}\,t\right),$$

The condition $x(1) = 0$ implies

$$\frac{mv_0}{\sqrt{10m-1}}e^{-1/m}\sin\frac{1}{m}\sqrt{10m-1} = 0$$

$$\sin\frac{1}{m}\sqrt{10m-1} = 0$$

$$\frac{1}{m}\sqrt{10m-1} = n\pi$$

$$\frac{10m-1}{m^2} = n^2\pi^2, \ \ n = 1, 2, 3, \ldots$$

$$(n^2\pi^2)m^2 - 10m + 1 = 0$$

$$m = \frac{10\sqrt{100 - 4n^2\pi^2}}{2n^2\pi^2} = \frac{5 \pm \sqrt{25 - n^2\pi^2}}{n^2\pi^2}.$$

Since m is real, $25 - n^2\pi^2 \geq 0$. If $25 - n^2\pi^2 = 0$, then $n^2 = 25/\pi^2$, and n is not an integer. Thus, $25 - n^2\pi^2 = (5 - n\pi)(5 + n\pi) > 0$ and since $n > 0$, $5 + n\pi > 0$, so $5 - n\pi > 0$ also. Then $n < 5/\pi$, and so $n = 1$. Therefore, the mass m will pass through the equilibrium position when $t = 1$ for

$$m_1 = \frac{5 + \sqrt{25 - \pi^2}}{\pi^2} \quad \text{and} \quad m_2 = \frac{5 - \sqrt{25 - \pi^2}}{\pi^2}.$$

31. (a) The general solution of the differential equation is $y = c_1\cos 4x + c_2\sin 4x$. From $y_0 = y(0) = c_1$ we see that $y = y_0\cos 4x + c_2\sin 4x$. From $y_1 = y(\pi/2) = y_0$ we see that any solution must satisfy $y_0 = y_1$. We also see that when $y_0 = y_1$, $y = y_0\cos 4x + c_2\sin 4x$ is a solution of the boundary-value problem for any choice of c_2. Thus, the boundary-value problem does not have a unique solution for any choice of y_0 and y_1.

(b) Whenever $y_0 = y_1$ there are infinitely many solutions.

(c) When $y_0 \neq y_1$ there will be no solutions.

(d) The boundary-value problem will have the trivial solution when $y_0 = y_1 = 0$. This solution will not be unique.

32. (a) The general solution of the differential equation is $y = c_1 \cos 4x + c_2 \sin 4x$. From $1 = y(0) = c_1$ we see that $y = \cos 4x + c_2 \sin 4x$. From $1 = y(L) = \cos 4L + c_2 \sin 4L$ we see that $c_2 = (1 - \cos 4L)/\sin 4L$. Thus,

$$y = \cos 4x + \left(\frac{1 - \cos 4L}{\sin 4L}\right) \sin 4x$$

will be a unique solution when $\sin 4L \neq 0$; that is, when $L \neq k\pi/4$ where $k = 1, 2, 3, \ldots$.

(b) There will be infinitely many solutions when $\sin 4L = 0$ and $1 - \cos 4L = 0$; that is, when $L = k\pi/2$ where $k = 1, 2, 3, \ldots$.

(c) There will be no solution when $\sin 4L \neq 0$ and $1 - \cos 4L \neq 0$; that is, when $L = k\pi/4$ where $k = 1, 3, 5, \ldots$.

(d) There can be no trivial solution since it would fail to satisfy the boundary conditions.

33. (a) A solution curve has the same y-coordinate at both ends of the interval $[-\pi, \pi]$ and the tangent lines at the endpoints of the interval are parallel.

(b) For $\lambda = 0$ the solution of $y'' = 0$ is $y = c_1 x + c_2$. From the first boundary condition we have

$$y(-\pi) = -c_1\pi + c_2 = y(\pi) = c_1\pi + c_2$$

or $2c_1\pi = 0$. Thus, $c_1 = 0$ and $y = c_2$. This constant solution is seen to satisfy the boundary-value problem.

For $\lambda = -\alpha^2 < 0$ we have $y = c_1 \cosh \alpha x + c_2 \sinh \alpha x$. In this case the first boundary condition gives

$$y(-\pi) = c_1 \cosh(-\alpha\pi) + c_2 \sinh(-\alpha\pi)$$

$$= c_1 \cosh \alpha\pi - c_2 \sinh \alpha\pi$$

$$= y(\pi) = c_1 \cosh \alpha\pi + c_2 \sinh \alpha\pi$$

or $2c_2 \sinh \alpha\pi = 0$. Thus $c_2 = 0$ and $y = c_1 \cosh \alpha x$. The second boundary condition implies in a similar fashion that $c_1 = 0$. Thus, for $\lambda < 0$, the only solution of the boundary-value problem is $y = 0$.

For $\lambda = \alpha^2 > 0$ we have $y = c_1 \cos \alpha x + c_2 \sin \alpha x$. The first boundary condition implies

$$y(-\pi) = c_1 \cos(-\alpha\pi) + c_2 \sin(-\alpha\pi)$$

$$= c_1 \cos \alpha\pi - c_2 \sin \alpha\pi$$

$$= y(\pi) = c_1 \cos \alpha\pi + c_2 \sin \alpha\pi$$

or $2c_2 \sin \alpha \pi = 0$. Similarly, the second boundary condition implies $2c_1 \alpha \sin \alpha \pi = 0$. If $c_1 = c_2 = 0$ the solution is $y = 0$. However, if $c_1 \neq 0$ or $c_2 \neq 0$, then $\sin \alpha \pi = 0$, which implies that α must be an integer, n. Therefore, for c_1 and c_2 not both 0, $y = c_1 \cos nx + c_2 \sin nx$ is a nontrivial solution of the boundary-value problem. Since $\cos(-nx) = \cos nx$ and $\sin(-nx) = -\sin nx$, we may assume without loss of generality that the eigenvalues are $\lambda_n = \alpha^2 = n^2$, for n a positive integer. The corresponding eigenfunctions are $y_n = \cos nx$ and $y_n = \sin nx$.

(c)

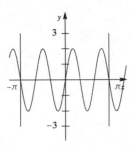   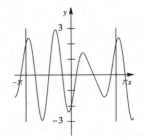

$$y = 2 \sin 3x \qquad\qquad y = \sin 4x - 2 \cos 3x$$

34. For $\lambda = \alpha^2 > 0$ the general solution is $y = c_1 \cos \sqrt{\alpha}\, x + c_2 \sin \sqrt{\alpha}\, x$. Setting $y(0) = 0$ we find $c_1 = 0$, so that $y = c_2 \sin \sqrt{\alpha}\, x$. The boundary condition $y(1) + y'(1) = 0$ implies

$$c_2 \sin \sqrt{\alpha} + c_2 \sqrt{\alpha} \cos \sqrt{\alpha} = 0.$$

Taking $c_2 \neq 0$, this equation is equivalent to $\tan \sqrt{\alpha} = -\sqrt{\alpha}$. Thus, the eigenvalues are $\lambda_n = \alpha_n^2 = x_n^2$, $n = 1, 2, 3, \ldots$, where the x_n are the consecutive positive roots of $\tan \sqrt{\alpha} = -\sqrt{\alpha}$.

35. We see from the graph that $\tan x = -x$ has infinitely many roots. Since $\lambda_n = \alpha_n^2$, there are no new eigenvalues when $\alpha_n < 0$. For $\lambda = 0$, the differential equation $y'' = 0$ has general solution $y = c_1 x + c_2$. The boundary conditions imply $c_1 = c_2 = 0$, so $y = 0$.

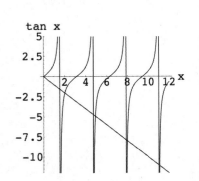

36. Using a CAS we find that the first four nonnegative roots of $\tan x = -x$ are approximately $2.02876, 4.91318, 7.97867$, and 11.0855. The corresponding eigenvalues are $4.11586, 24.1393$, 63.6591, and 122.889, with eigenfunctions $\sin(2.02876x), \sin(4.91318x), \sin(7.97867x)$, and $\sin(11.0855x)$.

37. In the case when $\lambda = -\alpha^2 < 0$, the solution of the differential equation is $y = c_1 \cosh \alpha x + c_2 \sinh \alpha x$. The condition $y(0) = 0$ gives $c_1 = 0$. The condition $y(1) - \frac{1}{2}y'(1) = 0$ applied to $y = c_2 \sinh \alpha x$ gives $c_2(\sinh \alpha - \frac{1}{2}\alpha \cosh \alpha) = 0$ or $\tanh \alpha = \frac{1}{2}\alpha$. As can be seen from the figure, the graphs of $y = \tanh x$ and $y = \frac{1}{2}x$ intersect at a single point with approximate x-coordinate $\alpha_1 = 1.915$. Thus, there is a single negative eigenvalue $\lambda_1 = -\alpha_1^2 \approx -3.667$ and the corresponding eigenfuntion is $y_1 = \sinh 1.915x$.

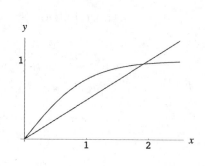

For $\lambda = 0$ the only solution of the boundary-value problem is $y = 0$.

For $\lambda = \alpha^2 > 0$ the solution of the differential equation is $y = c_1 \cos \alpha x + c_2 \sin \alpha x$. The condition $y(0) = 0$ gives $c_1 = 0$, so $y = c_2 \sin \alpha x$. The condition $y(1) - \frac{1}{2}y'(1) = 0$ gives $c_2(\sin \alpha - \frac{1}{2}\alpha \cos \alpha) = 0$, so the eigenvalues are $\lambda_n = \alpha_n^2$ when α_n, $n = 2, 3, 4, \ldots$, are the positive roots of $\tan \alpha = \frac{1}{2}\alpha$. Using a CAS we find that the first three values of α are $\alpha_2 = 4.27487$, $\alpha_3 = 7.59655$, and $\alpha_4 = 10.8127$. The first three eigenvalues are then $\lambda_2 = \alpha_2^2 = 18.2738$, $\lambda_3 = \alpha_3^2 = 57.7075$, and $\lambda_4 = \alpha_4^2 = 116.9139$ with corresponding eigenfunctions $y_2 = \sin 4.27487x$, $y_3 = \sin 7.59655x$, and $y_4 = \sin 10.8127x$.

38. For $\lambda = \alpha^4$, $\alpha > 0$, the solution of the differential equation is

$$y = c_1 \cos \alpha x + c_2 \sin \alpha x + c_3 \cosh \alpha x + c_4 \sinh \alpha x.$$

The boundary conditions $y(0) = 0$, $y'(0) = 0$, $y(1) = 0$, $y'(1) = 0$ give, in turn,

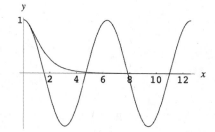

$$c_1 + c_3 = 0$$

$$\alpha c_2 + \alpha c_4 = 0,$$

$$c_1 \cos \alpha + c_2 \sin \alpha + c_3 \cosh \alpha + c_4 \sinh \alpha = 0$$

$$-c_1 \alpha \sin \alpha + c_2 \alpha \cos \alpha + c_3 \alpha \sinh \alpha + c_4 \alpha \cosh \alpha = 0.$$

The first two equations enable us to write

$$c_1(\cos \alpha - \cosh \alpha) + c_2(\sin \alpha - \sinh \alpha) = 0$$

$$c_1(-\sin \alpha - \sinh \alpha) + c_2(\cos \alpha - \cosh \alpha) = 0.$$

The determinant

$$\begin{vmatrix} \cos \alpha - \cosh \alpha & \sin \alpha - \sinh \alpha \\ -\sin \alpha - \sinh \alpha & \cos \alpha - \cosh \alpha \end{vmatrix} = 0$$

simplifies to $\cos \alpha \cosh \alpha = 1$. From the figure showing the graphs of $1/\cosh x$ and $\cos x$, we see that this equation has an infinite number of positive roots. With the aid of a CAS the first four roots are found to be $\alpha_1 = 4.73004$, $\alpha_2 = 7.8532$, $\alpha_3 = 10.9956$, and $\alpha_4 = 14.1372$, and the corresponding eigenvalues are $\lambda_1 = 500.5636$, $\lambda_2 = 3803.5281$, $\lambda_3 = 14{,}617.5885$, and $\lambda_4 = 39{,}944.1890$. Using the third equation in the system to eliminate c_2, we find that the eigenfunctions are

$$y_n = (-\sin\alpha_n + \sinh\alpha_n)(\cos\alpha_n x - \cosh\alpha_n x) + (\cos\alpha_n - \cosh\alpha_n)(\sin\alpha_n x - \sinh\alpha_n x).$$

Exercises 5.3 Nonlinear Models

1. The period corresponding to $x(0) = 1$, $x'(0) = 1$ is approximately 5.6. The period corresponding to $x(0) = 1/2$, $x'(0) = -1$ is approximately 6.2.

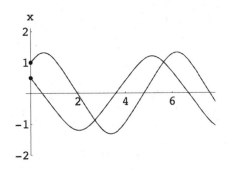

2. The solutions are not periodic.

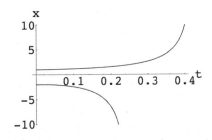

3. The period corresponding to $x(0) = 1$, $x'(0) = 1$ is approximately 5.8. The second initial-value problem does not have a periodic solution.

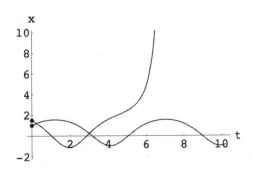

4. Both solutions have periods of approximately 6.3.

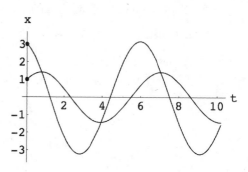

5. From the graph we see that $|x_1| \approx 1.2$.

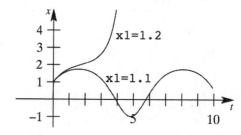

6. From the graphs we see that the interval is approximately $(-0.8, 1.1)$.

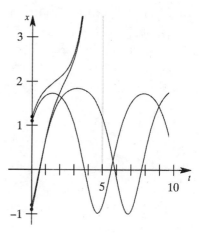

7. Since

$$xe^{0.01x} = x[1 + 0.01x + \frac{1}{2!}(0.01x)^2 + \cdots] \approx x$$

for small values of x, a linearization is $\dfrac{d^2x}{dt^2} + x = 0$.

8.

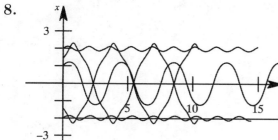

For $x(0) = 1$ and $x'(0) = 1$ the oscillations are symmetric about the line $x = 0$ with amplitude slightly greater than 1.

For $x(0) = -2$ and $x'(0) = 0.5$ the oscillations are symmetric about the line $x = -2$ with small amplitude.

For $x(0) = \sqrt{2}$ and $x'(0) = 1$ the oscillations are symmetric about the line $x = 0$ with amplitude a little greater than 2.

For $x(0) = 2$ and $x'(0) = 0.5$ the oscillations are symmetric about the line $x = 2$ with small amplitude.

For $x(0) = -2$ and $x'(0) = 0$ there is no oscillation; the solution is constant.

For $x(0) = -\sqrt{2}$ and $x'(0) = -1$ the oscillations are symmetric about the line $x = 0$ with amplitude a little greater than 2.

9. This is a damped hard spring, so x will approach 0 as t approaches ∞.

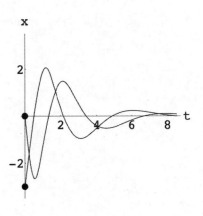

10. This is a damped soft spring, so we might expect no oscillatory solutions. However, if the initial conditions are sufficiently small the spring can oscillate.

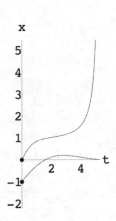

11.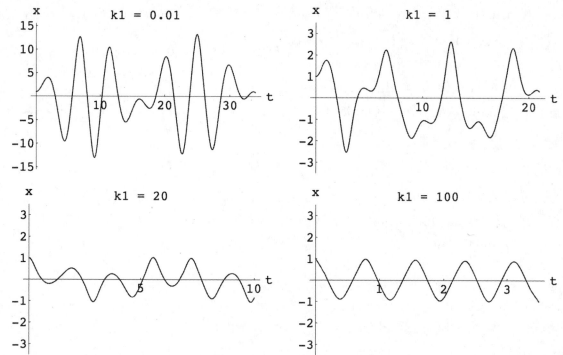

When k_1 is very small the effect of the nonlinearity is greatly diminished, and the system is close to pure resonance.

12. (a)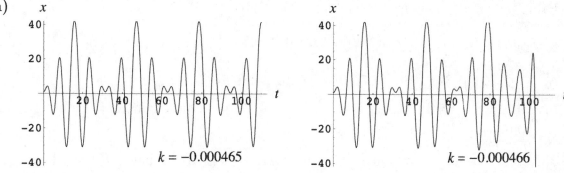

The system appears to be oscillatory for $-0.000465 \leq k_1 < 0$ and nonoscillatory for $k_1 \leq -0.000466$.

(b)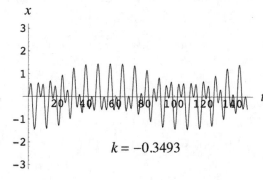

The system appears to be oscillatory for $-0.3493 \leq k_1 < 0$ and nonoscillatory for $k_1 \leq -0.3494$.

13. For $\lambda^2 - \omega^2 > 0$ we choose $\lambda = 2$ and $\omega = 1$ with $x(0) = 1$ and $x'(0) = 2$. For $\lambda^2 - \omega^2 < 0$ we choose $\lambda = 1/3$ and $\omega = 1$ with $x(0) = -2$ and $x'(0) = 4$. In both cases the motion corresponds to the overdamped and underdamped cases for spring/mass systems.

variable is θ not x

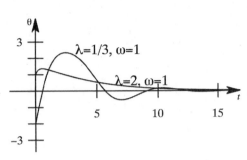

14. (a) Setting $dy/dt = v$, the differential equation in (13) becomes $dv/dt = -gR^2/y^2$. But, by the chain rule, $dv/dt = (dv/dy)(dy/dt) = v\,dv/dt$, so $v\,dv/dy = -gR^2/y^2$. Separating variables and integrating we obtain

$$v\,dv = -gR^2 \frac{dy}{y^2} \quad \text{and} \quad \frac{1}{2}v^2 = \frac{gR^2}{y} + c.$$

Setting $v = v_0$ and $y = R$ we find $c = -gR + \frac{1}{2}v_0^2$ and

$$v^2 = 2g\frac{R^2}{y} - 2gR + v_0^2.$$

(b) As $y \to \infty$ we assume that $v \to 0^+$. Then $v_0^2 = 2gR$ and $v_0 = \sqrt{2gR}$.

(c) Using $g = 32$ ft/s and $R = 4000(5280)$ ft we find

$$v_0 = \sqrt{2(32)(4000)(5280)} \approx 36765.2 \text{ ft/s} \approx 25067 \text{ mi/hr}.$$

(d) $v_0 = \sqrt{2(0.165)(32)(1080)} \approx 7760 \text{ ft/s} \approx 5291 \text{ mi/hr}$

15. (a) Intuitively, one might expect that only half of a 10-pound chain could be lifted by a 5-pound vertical force.

(b) Since $x = 0$ when $t = 0$, and $v = dx/dt = \sqrt{160 - 64x/3}$, we have $v(0) = \sqrt{160} \approx 12.65$ ft/s.

(c) Since x should always be positive, we solve $x(t) = 0$, getting $t = 0$ and $t = \frac{3}{2}\sqrt{5/2} \approx 2.3717$. Since the graph of $x(t)$ is a parabola, the maximum value occurs at $t_m = \frac{3}{4}\sqrt{5/2}$. (This can also be obtained by solving $x'(t) = 0$.) At this time the height of the chain is $x(t_m) \approx 7.5$ ft. This is higher than predicted because of the momentum generated by the force. When the chain is 5 feet high it still has a positive velocity of about 7.3 ft/s, which keeps it going higher for a while.

(d) As discussed in the solution to part (c) of this problem, the chain has momentum generated by the force applied to it that will cause it to go higher than expected. It will then fall back to below the expected maximum height, again due to momentum. This, in turn, will cause it to next go higher than expected, and so on.

16. (a) Setting $dx/dt = v$, the differential equation becomes $(L - x)dv/dt - v^2 = Lg$. But, by the Chain Rule, $dv/dt = (dv/dx)(dx/dt) = v\,dv/dx$, so $(L - x)v\,dv/dx - v^2 = Lg$. Separating variables and integrating we obtain

$$\frac{v}{v^2 + Lg}\,dv = \frac{1}{L - x}\,dx \quad \text{and} \quad \frac{1}{2}\ln(v^2 + Lg) = -\ln(L - x) + \ln c,$$

so $\sqrt{v^2 + Lg} = c/(L - x)$. When $x = 0$, $v = 0$, and $c = L\sqrt{Lg}$. Solving for v and simplifying we get

$$\frac{dx}{dt} = v(x) = \frac{\sqrt{Lg(2Lx - x^2)}}{L - x}.$$

Again, separating variables and integrating we obtain

$$\frac{L - x}{\sqrt{Lg(2Lx - x^2)}}dx = dt \quad \text{and} \quad \frac{\sqrt{2Lx - x^2}}{\sqrt{Lg}} = t + c_1.$$

Since $x(0) = 0$, we have $c_1 = 0$ and $\sqrt{2Lx - x^2}/\sqrt{Lg} = t$. Solving for x we get

$$x(t) = L - \sqrt{L^2 - Lgt^2} \quad \text{and} \quad v(t) = \frac{dx}{dt} = \frac{\sqrt{Lg}\,t}{\sqrt{L - gt^2}}.$$

(b) The chain will be completely on the ground when $x(t) = L$ or $t = \sqrt{L/g}$.

(c) The predicted velocity of the upper end of the chain when it hits the ground is infinity.

17. (a) Let (x, y) be the coordinates of S_2 on the curve C. The slope at (x, y) is then

$$dy/dx = (v_1 t - y)/(0 - x) = (y - v_1 t)/x \quad \text{or} \quad xy' - y = -v_1 t.$$

(b) Differentiating with respect to x and using $r = v_1/v_2$ gives

$$xy'' + y' - y' = -v_1 \frac{dt}{dx}$$

$$xy'' = -v_1 \frac{dt}{ds}\frac{ds}{dx}$$

$$xy'' = -v_1 \frac{1}{v_2}\left(-\sqrt{1 + (y')^2}\right)$$

$$xy'' = r\sqrt{1 + (y')^2}.$$

Letting $u = y'$ and separating variables, we obtain

$$x\frac{du}{dx} = r\sqrt{1+u^2}$$

$$\frac{du}{\sqrt{1+u^2}} = \frac{r}{x}\,dx$$

$$\sinh^{-1} u = r\ln x + \ln c = \ln(cx^r)$$

$$u = \sinh(\ln cx^r)$$

$$\frac{dy}{dx} = \frac{1}{2}\left(cx^r - \frac{1}{cx^r}\right).$$

At $t = 0$, $dy/dx = 0$ and $x = a$, so $0 = ca^r - 1/ca^r$. Thus $c = 1/a^r$ and

$$\frac{dy}{dx} = \frac{1}{2}\left[\left(\frac{x}{a}\right)^r - \left(\frac{a}{x}\right)^r\right] = \frac{1}{2}\left[\left(\frac{x}{a}\right)^r - \left(\frac{x}{a}\right)^{-r}\right].$$

If $r > 1$ or $r < 1$, integrating gives

$$y = \frac{a}{2}\left[\frac{1}{1+r}\left(\frac{x}{a}\right)^{1+r} - \frac{1}{1-r}\left(\frac{x}{a}\right)^{1-r}\right] + c_1.$$

When $t = 0$, $y = 0$ and $x = a$, so $0 = (a/2)[1/(1+r) - 1/(1-r)] + c_1$. Thus $c_1 = ar/(1-r^2)$ and

$$y = \frac{a}{2}\left[\frac{1}{1+r}\left(\frac{x}{a}\right)^{1+r} - \frac{1}{1-r}\left(\frac{x}{a}\right)^{1-r}\right] + \frac{ar}{1-r^2}.$$

(c) To see if the paths ever intersect we first note that if $r > 1$, then $v_1 > v_2$ and $y \to \infty$ as $x \to 0^+$. In other words, S_2 always lags behind S_1. Next, if $r < 1$, then $v_1 < v_2$ and $y = ar/(1-r^2)$ when $x = 0$. In other words, when the submarine's speed is greater than the ship's, their paths will intersect at the point $(0, ar/(1-r^2))$.

Finally, if $r = 1$, then integration gives

$$y = \frac{1}{2}\left[\frac{x^2}{2a} - \frac{1}{a}\ln x\right] + c_2.$$

When $t = 0$, $y = 0$ and $x = a$, so $0 = (1/2)[a/2 - (1/a)\ln a] + c_2$. Thus $c_2 = -(1/2)[a/2 - (1/a)\ln a]$ and

$$y = \frac{1}{2}\left[\frac{x^2}{2a} - \frac{1}{a}\ln x\right] - \frac{1}{2}\left[\frac{a}{2} - \frac{1}{a}\ln a\right] = \frac{1}{2}\left[\frac{1}{2a}(x^2 - a^2) + \frac{1}{a}\ln\frac{a}{x}\right].$$

Since $y \to \infty$ as $x \to 0^+$, S_2 will never catch up with S_1.

18. (a) Let (r, θ) denote the polar coordinates of the destroyer S_1. When S_1 travels the 6 miles from $(9, 0)$ to $(3, 0)$ it stands to reason, since S_2 travels half as fast as S_1, that the polar coordinates of S_2 are $(3, \theta_2)$, where θ_2 is unknown. In other words, the distances of the ships from $(0, 0)$

are the same and $r(t) = 15t$ then gives the radial distance of both ships. This is necessary if S_1 is to intercept S_2.

(b)   The differential of arc length in polar coordinates is $(ds)^2 = (r\,d\theta)^2 + (dr)^2$, so that

$$\left(\frac{ds}{dt}\right)^2 = r^2 \left(\frac{d\theta}{dt}\right)^2 + \left(\frac{dr}{dt}\right)^2.$$

Using $ds/dt = 30$ and $dr/dt = 15$ then gives

$$900 = 225t^2 \left(\frac{d\theta}{dt}\right)^2 + 225$$

$$675 = 225t^2 \left(\frac{d\theta}{dt}\right)^2$$

$$\frac{d\theta}{dt} = \frac{\sqrt{3}}{t}$$

$$\theta(t) = \sqrt{3}\ln t + c = \sqrt{3}\ln\frac{r}{15} + c.$$

When $r = 3$, $\theta = 0$, so $c = -\sqrt{3}\ln\frac{1}{5}$ and

$$\theta(t) = \sqrt{3}\left(\ln\frac{r}{15} - \ln\frac{1}{5}\right) = \sqrt{3}\ln\frac{r}{3}.$$

Thus $r = 3e^{\theta/\sqrt{3}}$, whose graph is a logarithmic spiral.

(c)   The time for S_1 to go from $(9,0)$ to $(3,0)$ = $\frac{1}{5}$ hour. Now S_1 must intercept the path of S_2 for some angle β, where $0 < \beta < 2\pi$. At the time of interception t_2 we have $15t_2 = 3e^{\beta/\sqrt{3}}$ or $t = \frac{1}{5}e^{\beta/\sqrt{3}}$. The total time is then

$$t = \frac{1}{5} + \frac{1}{5}e^{\beta/\sqrt{3}} < \frac{1}{5}(1 + e^{2\pi/\sqrt{3}}).$$

19. Since $(dx/dt)^2$ is always positive, it is necessary to use $|dx/dt|(dx/dt)$ in order to account for the fact that the motion is oscillatory and the velocity (or its square) should be negative when the spring is contracting.

20. (a) From the graph we see that the approximations appears to be quite good for $0 \le x \le 0.4$. Using an equation solver to solve $\sin x - x = 0.05$ and $\sin x - x = 0.005$, we find that the approximation is accurate to one decimal place for $\theta_1 = 0.67$ and to two decimal places for $\theta_1 = 0.31$.

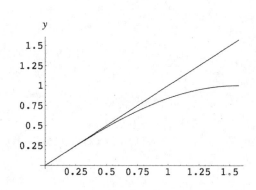

(b)

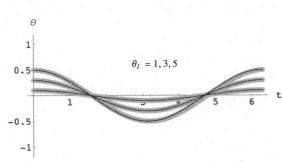

21. **(a)** Write the differential equation as

$$\frac{d^2\theta}{dt^2} + \omega^2 \sin\theta = 0,$$

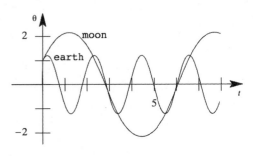

where $\omega^2 = g/l$. To test for differences between the earth and the moon we take $l = 3$, $\theta(0) = 1$, and $\theta'(0) = 2$. Using $g = 32$ on the earth and $g = 5.5$ on the moon we obtain the graphs shown in the figure. Comparing the apparent periods of the graphs, we see that the pendulum oscillates faster on the earth than on the moon.

(b) The amplitude is greater on the moon than on the earth.

(c) The linear model is

$$\frac{d^2\theta}{dt^2} + \omega^2\theta = 0,$$

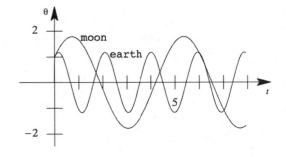

where $\omega^2 = g/l$. When $g = 32$, $l = 3$, $\theta(0) = 1$, and $\theta'(0) = 2$, the solution is

$$\theta(t) = \cos 3.266t + 0.612 \sin 3.266t.$$

When $g = 5.5$ the solution is

$$\theta(t) = \cos 1.354t + 1.477 \sin 1.354t.$$

As in the nonlinear case, the pendulum oscillates faster on the earth than on the moon and still has greater amplitude on the moon.

22. **(a)** The general solution of

$$\frac{d^2\theta}{dt^2} + \theta = 0$$

is $\theta(t) = c_1 \cos t + c_2 \sin t$. From $\theta(0) = \pi/12$ and $\theta'(0) = -1/3$ we find

$$\theta(t) = (\pi/12)\cos t - (1/3)\sin t.$$

Setting $\theta(t) = 0$ we have $\tan t = \pi/4$ which implies $t_1 = \tan^{-1}(\pi/4) \approx 0.66577$.

(b) We set $\theta(t) = \theta(0) + \theta'(0)t + \frac{1}{2}\theta''(0)t^2 + \frac{1}{6}\theta'''(0)t^3 + \cdots$ and use $\theta''(t) = -\sin\theta(t)$ together with $\theta(0) = \pi/12$ and $\theta'(0) = -1/3$. Then

$$\theta''(0) = -\sin(\pi/12) = -\sqrt{2}\,(\sqrt{3} - 1)/4$$

and

$$\theta'''(0) = -\cos\theta(0) \cdot \theta'(0) = -\cos(\pi/12)(-1/3) = \sqrt{2}\,(\sqrt{3} + 1)/12.$$

Thus

$$\theta(t) = \frac{\pi}{12} - \frac{1}{3}t - \frac{\sqrt{2}\,(\sqrt{3} - 1)}{8}t^2 + \frac{\sqrt{2}\,(\sqrt{3} + 1)}{72}t^3 + \cdots.$$

(c) Setting $\pi/12 - t/3 = 0$ we obtain $t_1 = \pi/4 \approx 0.785398$.

(d) Setting

$$\frac{\pi}{12} - \frac{1}{3}t - \frac{\sqrt{2}\,(\sqrt{3} - 1)}{8}t^2 = 0$$

and using the positive root we obtain $t_1 \approx 0.63088$.

(e) Setting

$$\frac{\pi}{12} - \frac{1}{3}t - \frac{\sqrt{2}\,(\sqrt{3} - 1)}{8}t^2 + \frac{\sqrt{2}\,(\sqrt{3} + 1)}{72}t^3 = 0$$

we find with the help of a CAS that $t_1 \approx 0.661973$ is the first positive root.

(f) From the output we see that $y(t)$ is an interpolating function on the interval $0 \le t \le 5$, whose graph is shown. The positive root of $y(t) = 0$ near $t = 1$ is $t_1 = 0.666404$.

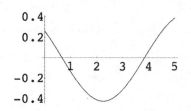

(g) To find the next two positive roots we change the interval used in **NDSolve** and **Plot** from $\{t,0,5\}$ to $\{t,0,10\}$. We see from the graph that the second and third positive roots are near 4 and 7, respectively. Replacing $\{t,1\}$ in **FindRoot** with $\{t,4\}$ and then $\{t,7\}$ we obtain $t_2 = 3.84411$ and $t_3 = 7.0218$.

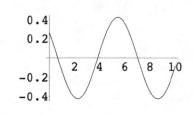

23. From the table below we see that the pendulum first passes the vertical position between 1.7 and 1.8 seconds. To refine our estimate of t_1 we estimate the solution of the differential equation on $[1.7, 1.8]$ using a step size of $h = 0.01$. From the resulting table we see that t_1 is between 1.76 and 1.77 seconds. Repeating the process with $h = 0.001$ we conclude that $t_1 \approx 1.767$. Then the period of the pendulum is approximately $4t_1 = 7.068$. The error when using $t_1 = 2\pi$ is $7.068 - 6.283 = 0.785$ and the percentage relative error is $(0.785/7.068)100 = 11.1$.

h=0.1		h=0.01	
t_n	θ_n	t_n	θ_n
0.00	0.78540	1.70	0.07706
0.10	0.78523	1.71	0.06572
0.20	0.78407	1.72	0.05428
0.30	0.78092	1.73	0.04275
0.40	0.77482	1.74	0.03111
0.50	0.76482	1.75	0.01938
0.60	0.75004	1.76	0.00755
0.70	0.72962	1.77	-0.00438
0.80	0.70275	1.78	-0.01641
0.90	0.66872	1.79	-0.02854
1.00	0.62687	1.80	-0.04076
1.10	0.57660		
1.20	0.51744	h=0.001	
1.30	0.44895	1.763	0.00398
1.40	0.37085	1.764	0.00279
1.50	0.28289	1.765	0.00160
1.60	0.18497	1.766	0.00040
1.70	0.07706	1.767	-0.00079
1.80	-0.04076	1.768	-0.00199
1.90	-0.16831	1.769	-0.00318
2.00	-0.30531	1.770	-0.00438

24. (*This is a Contributed Problem and the solution has been provided by the author of the problem.*)

(a) The auxiliary equation is $m^2 + g/\ell = 0$, so the general solution of the differential equation is

$$\theta(t) = c_1 \cos \sqrt{\frac{g}{\ell}}\, t + c_2 \sin \sqrt{\frac{g}{\ell}}\, t.$$

The initial condtion $\theta(0) = 0$ implies $c_1 = 0$ and $\theta'(0) = \omega_0$ implies $c_2 = \omega_0 \sqrt{\ell/g}$. Thus,

$$\theta(t) = \omega_0 \sqrt{\frac{\ell}{g}}\, \sin \sqrt{\frac{g}{\ell}}\, t.$$

(b) At $\theta_{\max}$, $\sin \sqrt{g/\ell}\, t = 1$, so

$$\theta_{\max} = \omega_0 \sqrt{\frac{\ell}{g}} = \frac{m_b}{m_w + m_b} \frac{v_b}{\ell} \sqrt{\frac{\ell}{g}} = \frac{m_b}{m_w + m_b} \frac{v_b}{\sqrt{\ell g}}$$

and

$$v_b = \frac{m_w + m_b}{m_b} \sqrt{\ell g}\, \theta_{\max}.$$

(c) We have $\cos \theta_{\max} = (\ell - h)/\ell = 1 - h/\ell$. Then

$$\cos \theta_{\max} \approx 1 - \frac{1}{2}\theta_{\max}^2 = 1 - \frac{h}{\ell}$$

and

$$\theta_{\max}^2 = \frac{2h}{\ell} \qquad \text{or} \qquad \theta_{\max} = \sqrt{\frac{2h}{\ell}}.$$

Thus

$$v_b = \frac{m_w + m_b}{m_b} \sqrt{\ell g} \sqrt{\frac{2h}{\ell}} = \frac{m_w + m_b}{m_b} \sqrt{2gh}\,.$$

(d) When $m_b = 5\,\text{g}$, $m_w = 1\,\text{kg}$, and $h = 6\,\text{cm}$, we have

$$v_b = \frac{1005}{5} \sqrt{2(980)(6)} \approx 21,797 \text{ cm/s}.$$

Chapter 5 in Review

1. 8 ft, since $k = 4$

2. $2\pi/5$, since $\frac{1}{4}x'' + 6.25x = 0$

3. $5/4$ m, since $x = -\cos 4t + \frac{3}{4}\sin 4t$

4. True

5. False; since an external force may exist

6. False; since the equation of motion in this case is $x(t) = e^{-\lambda t}(c_1 + c_2 t)$ and $x(t) = 0$ can have at most one real solution

7. overdamped

8. From $x(0) = (\sqrt{2}/2)\sin\phi = -1/2$ we see that $\sin\phi = -1/\sqrt{2}$, so ϕ is an angle in the third or fourth quadrant. Since $x'(t) = \sqrt{2}\cos(2t + \phi)$, $x'(0) = \sqrt{2}\cos\phi = 1$ and $\cos\phi > 0$. Thus ϕ is in the fourth quadrant and $\phi = -\pi/4$.

9. $y = 0$ because $\lambda = 8$ is not an eigenvalue

10. $y = \cos 6x$ because $\lambda = (6)^2 = 36$ is an eigenvalue

11. The period of a spring/mass system is given by $T = 2\pi/\omega$ where $\omega^2 = k/m = kg/W$, where k is the spring constant, W is the weight of the mass attached to the spring, and g is the acceleration due to gravity. Thus, the period of oscillation is $T = (2\pi/\sqrt{kg})\sqrt{W}$. If the weight of the original mass is W, then $(2\pi/\sqrt{kg})\sqrt{W} = 3$ and $(2\pi/\sqrt{kg})\sqrt{W-8} = 2$. Dividing, we get $\sqrt{W}/\sqrt{W-8} = 3/2$ or $W = \frac{9}{4}(W - 8)$. Solving for W we find that the weight of the original mass was 14.4 pounds.

12. (a) Solving $\frac{3}{8}x'' + 6x = 0$ subject to $x(0) = 1$ and $x'(0) = -4$ we obtain

$$x = \cos 4t - \sin 4t = \sqrt{2}\sin(4t + 3\pi/4)\,.$$

(b) The amplitude is $\sqrt{2}$, period is $\pi/2$, and frequency is $2/\pi$.

   (c) If $x = 1$ then $t = n\pi/2$ and $t = -\pi/8 + n\pi/2$ for $n = 1, 2, 3, \ldots$.

   (d) If $x = 0$ then $t = \pi/16 + n\pi/4$ for $n = 0, 1, 2, \ldots$. The motion is upward for n even and downward for n odd.

   (e) $x'(3\pi/16) = 0$

   (f) If $x' = 0$ then $4t + 3\pi/4 = \pi/2 + n\pi$ or $t = 3\pi/16 + n\pi$.

13. We assume that he spring is initially compressed by 4 inches and that the positive direction on the x-axis is in the direction of elongation of the spring. Then, from $\frac{1}{4}x'' + \frac{3}{2}x' + 2x = 0$, $x(0) = -1/3$, and $x'(0) = 0$ we obtain $x = -\frac{2}{3}e^{-2t} + \frac{1}{3}e^{-4t}$.

14. From $x'' + \beta x' + 64x = 0$ we see that oscillatory motion results if $\beta^2 - 256 < 0$ or $0 \le \beta < 16$.

15. From $mx'' + 4x' + 2x = 0$ we see that nonoscillatory motion results if $16 - 8m \ge 0$ or $0 < m \le 2$.

16. From $\frac{1}{4}x'' + x' + x = 0$, $x(0) = 4$, and $x'(0) = 2$ we obtain $x = 4e^{-2t} + 10te^{-2t}$. If $x'(t) = 0$, then $t = 1/10$, so that the maximum displacement is $x = 5e^{-0.2} \approx 4.094$.

17. Writing $\frac{1}{8}x'' + \frac{8}{3}x = \cos\gamma t + \sin\gamma t$ in the form $x'' + \frac{64}{3}x = 8\cos\gamma t + 8\sin\gamma t$ we identify $\omega^2 = \frac{64}{3}$. The system is in a state of pure resonance when $\gamma = \omega = \sqrt{64/3} = 8/\sqrt{3}$.

18. Clearly $x_p = A/\omega^2$ suffices.

19. From $\frac{1}{8}x'' + x' + 3x = e^{-t}$, $x(0) = 2$, and $x'(0) = 0$ we obtain $x_c = e^{-4t}\left(c_1 \cos 2\sqrt{2}\,t + c_2 \sin 2\sqrt{2}\,t\right)$, $x_p = \frac{8}{17}e^{-t}$, and

$$x = e^{-4t}\left(\frac{26}{17}\cos 2\sqrt{2}\,t + \frac{28\sqrt{2}}{17}\sin 2\sqrt{2}\,t\right) + \frac{8}{17}e^{-t}.$$

20. (a) Let k be the effective spring constant and x_1 and x_2 the elongation of springs k_1 and k_2. The restoring forces satisfy $k_1 x_1 = k_2 x_2$ so $x_2 = (k_1/k_2)x_1$. From $k(x_1 + x_2) = k_1 x_1$ we have

$$k\left(x_1 + \frac{k_1}{k_2}x_2\right) = k_1 x_1$$

$$k\left(\frac{k_2 + k_1}{k_2}\right) = k_1$$

$$k = \frac{k_1 k_2}{k_1 + k_2}$$

$$\frac{1}{k} = \frac{1}{k_1} + \frac{1}{k_2}.$$

   (b) From $k_1 = 2W$ and $k_2 = 4W$ we find $1/k = 1/2W + 1/4W = 3/4W$. Then $k = 4W/3 = 4mg/3$. The differential equation $mx'' + kx = 0$ then becomes $x'' + (4g/3)x = 0$. The solution is

$$x(t) = c_1 \cos 2\sqrt{\frac{g}{3}}\,t + c_2 \sin 2\sqrt{\frac{g}{3}}\,t.$$

The initial conditions $x(0) = 1$ and $x'(0) = 2/3$ imply $c_1 = 1$ and $c_2 = 1/\sqrt{3g}$.

(c) To compute the maximum speed of the mass we compute

$$x'(t) = 2\sqrt{\frac{g}{3}}\sin 2\sqrt{\frac{g}{3}}\,t + \frac{2}{3}\cos 2\sqrt{\frac{g}{3}}\,t \quad \text{and} \quad |x'(t)| = \sqrt{4\frac{g}{3} + \frac{4}{9}} = \frac{2}{3}\sqrt{3g+1}.$$

21. From $q'' + 10^4 q = 100\sin 50t$, $q(0) = 0$, and $q'(0) = 0$ we obtain $q_c = c_1\cos 100t + c_2\sin 100t$, $q_p = \frac{1}{75}\sin 50t$, and

(a) $q = -\frac{1}{150}\sin 100t + \frac{1}{75}\sin 50t$,

(b) $i = -\frac{2}{3}\cos 100t + \frac{2}{3}\cos 50t$, and

(c) $q = 0$ when $\sin 50t(1 - \cos 50t) = 0$ or $t = n\pi/50$ for $n = 0, 1, 2, \ldots$.

22. (a) By Kirchhoff's second law,

$$L\frac{d^2q}{dt^2} + R\frac{dq}{dt} + \frac{1}{C}q = E(t).$$

Using $q'(t) = i(t)$ we can write the differential equation in the form

$$L\frac{di}{dt} + Ri + \frac{1}{C}q = E(t).$$

Then differentiating we obtain

$$L\frac{d^2i}{dt^2} + R\frac{di}{dt} + \frac{1}{C}i = E'(t).$$

(b) From $Li'(t) + Ri(t) + (1/C)q(t) = E(t)$ we find

$$Li'(0) + Ri(0) + (1/C)q(0) = E(0)$$

or

$$Li'(0) + Ri_0 + (1/C)q_0 = E(0).$$

Solving for $i'(0)$ we get

$$i'(0) = \frac{1}{L}\left[E(0) - \frac{1}{C}q_0 - Ri_0\right].$$

23. For $\lambda = \alpha^2 > 0$ the general solution is $y = c_1\cos\alpha x + c_2\sin\alpha x$. Now

$$y(0) = c_1 \quad \text{and} \quad y(2\pi) = c_1\cos 2\pi\alpha + c_2\sin 2\pi\alpha,$$

so the condition $y(0) = y(2\pi)$ implies

$$c_1 = c_1\cos 2\pi\alpha + c_2\sin 2\pi\alpha$$

which is true when $\alpha = \sqrt{\lambda} = n$ or $\lambda = n^2$ for $n = 1, 2, 3, \ldots$. Since

$$y' = -\alpha c_1\sin\alpha x + \alpha c_2\cos\alpha x = -nc_1\sin nx + nc_2\cos nx,$$

we see that $y'(0) = nc_2 = y'(2\pi)$ for $n = 1, 2, 3, \ldots$. Thus, the eigenvalues are n^2 for $n = 1, 2, 3, \ldots$, with corresponding eigenfunctions $\cos nx$ and $\sin nx$. When $\lambda = 0$, the general solution is $y = c_1 x + c_2$ and the corresponding eigenfunction is $y = 1$.

For $\lambda = -\alpha^2 < 0$ the general solution is $y = c_1 \cosh \alpha x + c_2 \sinh \alpha x$. In this case $y(0) = c_1$ and $y(2\pi) = c_1 \cosh 2\pi\alpha + c_2 \sinh 2\pi\alpha$, so $y(0) = y(2\pi)$ can only be valid for $\alpha = 0$. Thus, there are no eigenvalues corresponding to $\lambda < 0$.

24. **(a)** The differential equation is $d^2 r/dt^2 - \omega^2 r = -g \sin \omega t$. The auxiliary equation is $m^2 - \omega^2 = 0$, so $r_c = c_1 e^{\omega t} + c_2 e^{-\omega t}$. A particular solution has the form $r_p = A \sin \omega t + B \cos \omega t$. Substituting into the differential equation we find $-2A\omega^2 \sin \omega t - 2B\omega^2 \cos \omega t = -g \sin \omega t$. Thus, $B = 0$, $A = g/2\omega^2$, and $r_p = (g/2\omega^2) \sin \omega t$. The general solution of the differential equation is $r(t) = c_1 e^{\omega t} + c_2 e^{-\omega t} + (g/2\omega^2) \sin \omega t$. The initial conditions imply $c_1 + c_2 = r_0$ and $g/2\omega - \omega c_1 + \omega c_2 = v_0$. Solving for c_1 and c_2 we get

$$c_1 = (2\omega^2 r_0 + 2\omega v_0 - g)/4\omega^2 \quad \text{and} \quad c_2 = (2\omega^2 r_0 - 2\omega v_0 + g)/4\omega^2,$$

so that

$$r(t) = \frac{2\omega^2 r_0 + 2\omega v_0 - g}{4\omega^2} e^{\omega t} + \frac{2\omega^2 r_0 - 2\omega v_0 + g}{4\omega^2} e^{-\omega t} + \frac{g}{2\omega^2} \sin \omega t.$$

(b) The bead will exhibit simple harmonic motion when the exponential terms are missing. Solving $c_1 = 0$, $c_2 = 0$ for r_0 and v_0 we find $r_0 = 0$ and $v_0 = g/2\omega$.

To find the minimum length of rod that will accommodate simple harmonic motion we determine the amplitude of $r(t)$ and double it. Thus $L = g/\omega^2$.

(c) As t increases, $e^{\omega t}$ approaches infinity and $e^{-\omega t}$ approaches 0. Since $\sin \omega t$ is bounded, the distance, $r(t)$, of the bead from the pivot point increases without bound and the distance of the bead from P will eventually exceed $L/2$.

(d)

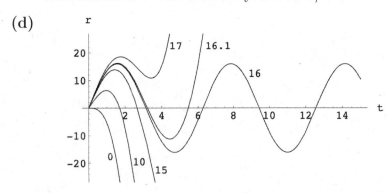

(e) For each v_0 we want to find the smallest value of t for which $r(t) = \pm 20$. Whether we look for $r(t) = -20$ or $r(t) = 20$ is determined by looking at the graphs in part (d). The total times that the bead stays on the rod is shown in the table below.

v_0	0	10	15		17
r	−20	−20	−20	20	20
t	1.55007	2.35494	3.43088	6.11627	4.22339

When $v_0 = 16$ the bead never leaves the rod.

25. Unlike the derivation given in (1) of Section 5.1 in the text, the weight mg of the mass m does not appear in the net force since the spring is not stretched by the weight of the mass when it is in the equilibrium position (i.e. there is no $mg - ks$ term in the net force). The only force acting on the mass when it is in motion is the restoring force of the spring. By Newton's second law,

$$m\frac{d^2x}{dt^2} = -kx \qquad \text{or} \qquad \frac{d^2x}{dt^2} + \frac{k}{m}x = 0.$$

26. The force of kinetic friction opposing the motion of the mass in μN, where μ is the coefficient of sliding friction and N is the normal component of the weight. Since friction is a force opposite to the direction of motion and since N is pointed directly downward (it is simply the weight of the mass), Newton's second law gives, for motion to the right $(x' > 0)$,

$$m\frac{d^2x}{dt^2} = -kx - \mu mg,$$

and for motion to the left $(x' < 0)$,

$$m\frac{d^2x}{dt^2} = -kx + \mu mg.$$

Traditionally, these two equations are written as one expression

$$m\frac{d^2x}{dt^2} + f_k\,\text{sgn}(x') + kx = 0,$$

where $f_k = \mu mg$ and

$$\text{sgn}(x') = \begin{cases} 1, & x' > 0 \\ -1, & x' < 0. \end{cases}$$

6 Series Solutions of Linear Equations

Exercises 6.1   Solutions About Ordinary Points

1. $\lim\limits_{n\to\infty}\left|\dfrac{2^{n+1}x^{n+1}/(n+1)}{2^n x^n/n}\right| = \lim\limits_{n\to\infty}\dfrac{2n}{n+1}|x| = 2|x|$

   The series is absolutely convergent for $2|x| < 1$ or $|x| < \frac{1}{2}$. The radius of convergence is $R = \frac{1}{2}$. At $x = -\frac{1}{2}$, the series $\sum_{n=1}^{\infty}(-1)^n/n$ converges by the alternating series test. At $x = \frac{1}{2}$, the series $\sum_{n=1}^{\infty}1/n$ is the harmonic series which diverges. Thus, the given series converges on $[-\frac{1}{2}, \frac{1}{2})$.

2. $\lim\limits_{n\to\infty}\left|\dfrac{100^{n+1}(x+7)^{n+1}/(n+1)!}{100^n(x+7)^n/n!}\right| = \lim\limits_{n\to\infty}\dfrac{100}{n+1}|x+7| = 0$

   The radius of convergence is $R = \infty$. The series is absolutely convergent on $(-\infty, \infty)$.

3. By the ratio test,

   $$\lim_{k\to\infty}\left|\dfrac{(x-5)^{k+1}/10^{k+1}}{(x-5)^k/10^k}\right| = \lim_{k\to\infty}\dfrac{1}{10}|x-5| = \dfrac{1}{10}|x-5|.$$

   The series is absolutely convergent for $\frac{1}{10}|x-5| < 1$, $|x-5| < 10$, or on $(-5, 15)$. The radius of convergence is $R = 10$. At $x = -5$, the series $\sum_{k=1}^{\infty}(-1)^k(-10)^k/10^k = \sum_{k=1}^{\infty}1$ diverges by the nth term test. At $x = 15$, the series $\sum_{k=1}^{\infty}(-1)^k 10^k/10^k = \sum_{k=1}^{\infty}(-1)^k$ diverges by the nth term test. Thus, the series converges on $(-5, 15)$.

4. $\lim\limits_{k\to\infty}\left|\dfrac{(k+1)!(x-1)^{k+1}}{k!(x-1)^k}\right| = \lim\limits_{k\to\infty}(k+1)|x-1| = \begin{cases} \infty, & x \neq 1 \\ 0, & x = 1 \end{cases}$

   The radius of convergence is $R = 0$ and the series converges only for $x = 1$.

5. $\sin x \cos x = \left(x - \dfrac{x^3}{6} + \dfrac{x^5}{120} - \dfrac{x^7}{5040} + \cdots\right)\left(1 - \dfrac{x^2}{2} + \dfrac{x^4}{24} - \dfrac{x^6}{720} + \cdots\right) = x - \dfrac{2x^3}{3} + \dfrac{2x^5}{15} - \dfrac{4x^7}{315} + \cdots$

6. $e^{-x}\cos x = \left(1 - x + \dfrac{x^2}{2} - \dfrac{x^3}{6} + \dfrac{x^4}{24} - \cdots\right)\left(1 - \dfrac{x^2}{2} + \dfrac{x^4}{24} - \cdots\right) = 1 - x + \dfrac{x^3}{3} - \dfrac{x^4}{6} + \cdots$

7. $\dfrac{1}{\cos x} = \dfrac{1}{1 - \dfrac{x^2}{2} + \dfrac{x^4}{4!} - \dfrac{x^6}{6!} + \cdots} = 1 + \dfrac{x^2}{2} + \dfrac{5x^4}{4!} + \dfrac{61x^6}{6!} + \cdots$

   Since $\cos(\pi/2) = \cos(-\pi/2) = 0$, the series converges on $(-\pi/2, \pi/2)$.

8. $\dfrac{1-x}{2+x} = \dfrac{1}{2} - \dfrac{3}{4}x + \dfrac{3}{8}x^2 - \dfrac{3}{16}x^3 + \cdots$

Since the function is undefined at $x = -2$, the series converges on $(-2, 2)$.

9. Let $k = n + 2$ so that $n = k - 2$ and

$$\sum_{n=1}^{\infty} n c_n x^{n+2} = \sum_{k=3}^{\infty} (k-2) c_{k-2} x^k.$$

10. Let $k = n - 3$ so that $n = k + 3$ and

$$\sum_{n=3}^{\infty} (2n-1) c_n x^{n-3} = \sum_{k=0}^{\infty} (2k+5) c_{k+3} x^k.$$

11. $\displaystyle\sum_{n=1}^{\infty} 2n c_n x^{n-1} + \sum_{n=0}^{\infty} 6 c_n x^{n+1} = 2 \cdot 1 \cdot c_1 x^0 + \underbrace{\sum_{n=2}^{\infty} 2n c_n x^{n-1}}_{k=n-1} + \underbrace{\sum_{n=0}^{\infty} 6 c_n x^{n+1}}_{k=n+1}$

$$= 2c_1 + \sum_{k=1}^{\infty} 2(k+1) c_{k+1} x^k + \sum_{k=1}^{\infty} 6 c_{k-1} x^k$$

$$= 2c_1 + \sum_{k=1}^{\infty} [2(k+1) c_{k+1} + 6 c_{k-1}] x^k$$

12. $\displaystyle\sum_{n=2}^{\infty} n(n-1) c_n x^n + 2 \sum_{n=2}^{\infty} n(n-1) c_n x^{n-2} + 3 \sum_{n=1}^{\infty} n c_n x^n$

$$= 2 \cdot 2 \cdot 1 c_2 x^0 + 2 \cdot 3 \cdot 2 c_3 x^1 + 3 \cdot 1 \cdot c_1 x^1 + \underbrace{\sum_{n=2}^{\infty} n(n-1) c_n x^n}_{k=n} + 2 \underbrace{\sum_{n=4}^{\infty} n(n-1) c_n x^{n-2}}_{k=n-2} + 3 \underbrace{\sum_{n=2}^{\infty} n c_n x^n}_{k=n}$$

$$= 4c_2 + (3c_1 + 12c_3)x + \sum_{k=2}^{\infty} k(k-1) c_k x^k + 2 \sum_{k=2}^{\infty} (k+2)(k+1) c_{k+2} x^k + 3 \sum_{k=2}^{\infty} k c_k x^k$$

$$= 4c_2 + (3c_1 + 12c_3)x + \sum_{k=2}^{\infty} [(k(k-1) + 3k) c_k + 2(k+2)(k+1) c_{k+2}] x^k$$

$$= 4c_2 + (3c_1 + 12c_3)x + \sum_{k=2}^{\infty} [k(k+2) c_k + 2(k+1)(k+2) c_{k+2}] x^k$$

13. $y' = \displaystyle\sum_{n=1}^{\infty} (-1)^{n+1} x^{n-1}, \qquad y'' = \sum_{n=2}^{\infty} (-1)^{n+1} (n-1) x^{n-2}$

275

$$(x+1)y'' + y' = (x+1)\sum_{n=2}^{\infty}(-1)^{n+1}(n-1)x^{n-2} + \sum_{n=1}^{\infty}(-1)^{n+1}x^{n-1}$$

$$= \sum_{n=2}^{\infty}(-1)^{n+1}(n-1)x^{n-1} + \sum_{n=2}^{\infty}(-1)^{n+1}(n-1)x^{n-2} + \sum_{n=1}^{\infty}(-1)^{n+1}x^{n-1}$$

$$= -x^0 + x^0 + \underbrace{\sum_{n=2}^{\infty}(-1)^{n+1}(n-1)x^{n-1}}_{k=n-1} + \underbrace{\sum_{n=3}^{\infty}(-1)^{n+1}(n-1)x^{n-2}}_{k=n-2} + \underbrace{\sum_{n=2}^{\infty}(-1)^{n+1}x^{n-1}}_{k=n-1}$$

$$= \sum_{k=1}^{\infty}(-1)^{k+2}kx^k + \sum_{k=1}^{\infty}(-1)^{k+3}(k+1)x^k + \sum_{k=1}^{\infty}(-1)^{k+2}x^k$$

$$= \sum_{k=1}^{\infty}\left[(-1)^{k+2}k - (-1)^{k+2}k - (-1)^{k+2} + (-1)^{k+2}\right]x^k = 0$$

14. $y' = \displaystyle\sum_{n=1}^{\infty}\frac{(-1)^n 2n}{2^{2n}(n!)^2}x^{2n-1}, \qquad y'' = \sum_{n=1}^{\infty}\frac{(-1)^n 2n(2n-1)}{2^{2n}(n!)^2}x^{2n-2}$

$$xy'' + y' + xy = \underbrace{\sum_{n=1}^{\infty}\frac{(-1)^n 2n(2n-1)}{2^{2n}(n!)^2}x^{2n-1}}_{k=n} + \underbrace{\sum_{n=1}^{\infty}\frac{(-1)^n 2n}{2^{2n}(n!)^2}x^{2n-1}}_{k=n} + \underbrace{\sum_{n=0}^{\infty}\frac{(-1)^n}{2^{2n}(n!)^2}x^{2n+1}}_{k=n+1}$$

$$= \sum_{k=1}^{\infty}\left[\frac{(-1)^k 2k(2k-1)}{2^{2k}(k!)^2} + \frac{(-1)^k 2k}{2^{2k}(k!)^2} + \frac{(-1)^{k-1}}{2^{2k-2}[(k-1)!]^2}\right]x^{2k-1}$$

$$= \sum_{k=1}^{\infty}\left[\frac{(-1)^k(2k)^2}{2^{2k}(k!)^2} - \frac{(-1)^k}{2^{2k-2}[(k-1)!]^2}\right]x^{2k-1}$$

$$= \sum_{k=1}^{\infty}(-1)^k\left[\frac{(2k)^2 - 2^2k^2}{2^{2k}(k!)^2}\right]x^{2k-1} = 0$$

15. The singular points of $(x^2 - 25)y'' + 2xy' + y = 0$ are -5 and 5. The distance from 0 to either of these points is 5. The distance from 1 to the closest of these points is 4.

16. The singular points of $(x^2 - 2x + 10)y'' + xy' - 4y = 0$ are $1 + 3i$ and $1 - 3i$. The distance from 0 to either of these points is $\sqrt{10}$. The distance from 1 to either of these points is 3.

17. Substituting $y = \sum_{n=0}^{\infty}c_n x^n$ into the differential equation we have

$$y'' - xy = \underbrace{\sum_{n=2}^{\infty}n(n-1)c_n x^{n-2}}_{k=n-2} - \underbrace{\sum_{n=0}^{\infty}c_n x^{n+1}}_{k=n+1} = \sum_{k=0}^{\infty}(k+2)(k+1)c_{k+2}x^k - \sum_{k=1}^{\infty}c_{k-1}x^k$$

$$= 2c_2 + \sum_{k=1}^{\infty}[(k+2)(k+1)c_{k+2} - c_{k-1}]x^k = 0.$$

Thus

$$c_2 = 0$$

$$(k+2)(k+1)c_{k+2} - c_{k-1} = 0$$

and

$$c_{k+2} = \frac{1}{(k+2)(k+1)} c_{k-1}, \quad k = 1, 2, 3, \ldots.$$

Choosing $c_0 = 1$ and $c_1 = 0$ we find

$$c_3 = \frac{1}{6}$$

$$c_4 = c_5 = 0$$

$$c_6 = \frac{1}{180}$$

and so on. For $c_0 = 0$ and $c_1 = 1$ we obtain

$$c_3 = 0$$

$$c_4 = \frac{1}{12}$$

$$c_5 = c_6 = 0$$

$$c_7 = \frac{1}{504}$$

and so on. Thus, two solutions are

$$y_1 = 1 + \frac{1}{6}x^3 + \frac{1}{180}x^6 + \cdots \qquad \text{and} \qquad y_2 = x + \frac{1}{12}x^4 + \frac{1}{504}x^7 + \cdots.$$

18. Substituting $y = \sum_{n=0}^{\infty} c_n x^n$ into the differential equation we have

$$y'' + x^2 y = \underbrace{\sum_{n=2}^{\infty} n(n-1)c_n x^{n-2}}_{k=n-2} + \underbrace{\sum_{n=0}^{\infty} c_n x^{n+2}}_{k=n+2} = \sum_{k=0}^{\infty} (k+2)(k+1)c_{k+2} x^k + \sum_{k=2}^{\infty} c_{k-2} x^k$$

$$= 2c_2 + 6c_3 x + \sum_{k=2}^{\infty} [(k+2)(k+1)c_{k+2} + c_{k-2}]x^k = 0.$$

Thus

$$c_2 = c_3 = 0$$

$$(k+2)(k+1)c_{k+2} + c_{k-2} = 0$$

and

$$c_{k+2} = -\frac{1}{(k+2)(k+1)} c_{k-2}, \quad k = 2, 3, 4, \ldots.$$

Exercises 6.1   Solutions About Ordinary Points

Choosing $c_0 = 1$ and $c_1 = 0$ we find

$$c_4 = -\frac{1}{12}$$

$$c_5 = c_6 = c_7 = 0$$

$$c_8 = \frac{1}{672}$$

and so on. For $c_0 = 0$ and $c_1 = 1$ we obtain

$$c_4 = 0$$

$$c_5 = -\frac{1}{20}$$

$$c_6 = c_7 = c_8 = 0$$

$$c_9 = \frac{1}{1440}$$

and so on. Thus, two solutions are

$$y_1 = 1 - \frac{1}{12}x^4 + \frac{1}{672}x^8 - \cdots \qquad \text{and} \qquad y_2 = x - \frac{1}{20}x^5 + \frac{1}{1440}x^9 - \cdots .$$

19. Substituting $y = \sum_{n=0}^{\infty} c_n x^n$ into the differential equation we have

$$y'' - 2xy' + y = \underbrace{\sum_{n=2}^{\infty} n(n-1)c_n x^{n-2}}_{k=n-2} - 2\underbrace{\sum_{n=1}^{\infty} nc_n x^n}_{k=n} + \underbrace{\sum_{n=0}^{\infty} c_n x^n}_{k=n}$$

$$= \sum_{k=0}^{\infty} (k+2)(k+1)c_{k+2}x^k - 2\sum_{k=1}^{\infty} kc_k x^k + \sum_{k=0}^{\infty} c_k x^k$$

$$= 2c_2 + c_0 + \sum_{k=1}^{\infty} [(k+2)(k+1)c_{k+2} - (2k-1)c_k]x^k = 0.$$

Thus

$$2c_2 + c_0 = 0$$

$$(k+2)(k+1)c_{k+2} - (2k-1)c_k = 0$$

and

$$c_2 = -\frac{1}{2}c_0$$

$$c_{k+2} = \frac{2k-1}{(k+2)(k+1)}c_k, \quad k = 1, 2, 3, \ldots .$$

Choosing $c_0 = 1$ and $c_1 = 0$ we find

$$c_2 = -\frac{1}{2}$$

$$c_3 = c_5 = c_7 = \cdots = 0$$

$$c_4 = -\frac{1}{8}$$

$$c_6 = -\frac{7}{240}$$

and so on. For $c_0 = 0$ and $c_1 = 1$ we obtain

$$c_2 = c_4 = c_6 = \cdots = 0$$

$$c_3 = \frac{1}{6}$$

$$c_5 = \frac{1}{24}$$

$$c_7 = \frac{1}{112}$$

and so on. Thus, two solutions are

$$y_1 = 1 - \frac{1}{2}x^2 - \frac{1}{8}x^4 - \frac{7}{240}x^6 - \cdots \qquad \text{and} \qquad y_2 = x + \frac{1}{6}x^3 + \frac{1}{24}x^5 + \frac{1}{112}x^7 + \cdots .$$

20. Substituting $y = \sum_{n=0}^{\infty} c_n x^n$ into the differential equation we have

$$y'' - xy' + 2y = \underbrace{\sum_{n=2}^{\infty} n(n-1)c_n x^{n-2}}_{k=n-2} - \underbrace{\sum_{n=1}^{\infty} nc_n x^n}_{k=n} + 2\underbrace{\sum_{n=0}^{\infty} c_n x^n}_{k=n}$$

$$= \sum_{k=0}^{\infty} (k+2)(k+1)c_{k+2}x^k - \sum_{k=1}^{\infty} kc_k x^k + 2\sum_{k=0}^{\infty} c_k x^k$$

$$= 2c_2 + 2c_0 + \sum_{k=1}^{\infty} [(k+2)(k+1)c_{k+2} - (k-2)c_k]x^k = 0.$$

Thus

$$2c_2 + 2c_0 = 0$$

$$(k+2)(k+1)c_{k+2} - (k-2)c_k = 0$$

and

$$c_2 = -c_0$$

$$c_{k+2} = \frac{k-2}{(k+2)(k+1)}c_k, \quad k = 1, 2, 3, \ldots .$$

Choosing $c_0 = 1$ and $c_1 = 0$ we find

$$c_2 = -1$$

$$c_3 = c_5 = c_7 = \cdots = 0$$

$$c_4 = 0$$

$$c_6 = c_8 = c_{10} = \cdots = 0.$$

For $c_0 = 0$ and $c_1 = 1$ we obtain

$$c_2 = c_4 = c_6 = \cdots = 0$$

$$c_3 = -\frac{1}{6}$$

$$c_5 = -\frac{1}{120}$$

and so on. Thus, two solutions are

$$y_1 = 1 - x^2 \qquad \text{and} \qquad y_2 = x - \frac{1}{6}x^3 - \frac{1}{120}x^5 - \cdots.$$

21. Substituting $y = \sum_{n=0}^{\infty} c_n x^n$ into the differential equation we have

$$y'' + x^2 y' + xy = \underbrace{\sum_{n=2}^{\infty} n(n-1)c_n x^{n-2}}_{k=n-2} + \underbrace{\sum_{n=1}^{\infty} nc_n x^{n+1}}_{k=n+1} + \underbrace{\sum_{n=0}^{\infty} c_n x^{n+1}}_{k=n+1}$$

$$= \sum_{k=0}^{\infty} (k+2)(k+1)c_{k+2} x^k + \sum_{k=2}^{\infty} (k-1)c_{k-1} x^k + \sum_{k=1}^{\infty} c_{k-1} x^k$$

$$= 2c_2 + (6c_3 + c_0)x + \sum_{k=2}^{\infty} [(k+2)(k+1)c_{k+2} + kc_{k-1}]x^k = 0.$$

Thus

$$c_2 = 0$$

$$6c_3 + c_0 = 0$$

$$(k+2)(k+1)c_{k+2} + kc_{k-1} = 0$$

and

$$c_2 = 0$$

$$c_3 = -\frac{1}{6}c_0$$

$$c_{k+2} = -\frac{k}{(k+2)(k+1)}c_{k-1}, \qquad k = 2, 3, 4, \ldots.$$

Choosing $c_0 = 1$ and $c_1 = 0$ we find

$$c_3 = -\frac{1}{6}$$

$$c_4 = c_5 = 0$$

$$c_6 = \frac{1}{45}$$

and so on. For $c_0 = 0$ and $c_1 = 1$ we obtain

$$c_3 = 0$$

$$c_4 = -\frac{1}{6}$$

$$c_5 = c_6 = 0$$

$$c_7 = \frac{5}{252}$$

and so on. Thus, two solutions are

$$y_1 = 1 - \frac{1}{6}x^3 + \frac{1}{45}x^6 - \cdots \qquad \text{and} \qquad y_2 = x - \frac{1}{6}x^4 + \frac{5}{252}x^7 - \cdots .$$

22. Substituting $y = \sum_{n=0}^{\infty} c_n x^n$ into the differential equation we have

$$y'' + 2xy' + 2y = \underbrace{\sum_{n=2}^{\infty} n(n-1)c_n x^{n-2}}_{k=n-2} + 2\underbrace{\sum_{n=1}^{\infty} nc_n x^n}_{k=n} + 2\underbrace{\sum_{n=0}^{\infty} c_n x^n}_{k=n}$$

$$= \sum_{k=0}^{\infty} (k+2)(k+1)c_{k+2}x^k + 2\sum_{k=1}^{\infty} kc_k x^k + 2\sum_{k=0}^{\infty} c_k x^k$$

$$= 2c_2 + 2c_0 + \sum_{k=1}^{\infty} [(k+2)(k+1)c_{k+2} + 2(k+1)c_k]x^k = 0.$$

Thus

$$2c_2 + 2c_0 = 0$$

$$(k+2)(k+1)c_{k+2} + 2(k+1)c_k = 0$$

and

$$c_2 = -c_0$$

$$c_{k+2} = -\frac{2}{k+2}c_k, \quad k = 1, 2, 3, \ldots .$$

Choosing $c_0 = 1$ and $c_1 = 0$ we find

$$c_2 = -1$$

$$c_3 = c_5 = c_7 = \cdots = 0$$

$$c_4 = \frac{1}{2}$$

$$c_6 = -\frac{1}{6}$$

and so on. For $c_0 = 0$ and $c_1 = 1$ we obtain

$$c_2 = c_4 = c_6 = \cdots = 0$$

$$c_3 = -\frac{2}{3}$$

$$c_5 = \frac{4}{15}$$

$$c_7 = -\frac{8}{105}$$

and so on. Thus, two solutions are

$$y_1 = 1 - x^2 + \frac{1}{2}x^4 - \frac{1}{6}x^6 + \cdots \quad \text{and} \quad y_2 = x - \frac{2}{3}x^3 + \frac{4}{15}x^5 - \frac{8}{105}x^7 + \cdots .$$

23. Substituting $y = \sum_{n=0}^{\infty} c_n x^n$ into the differential equation we have

$$(x-1)y'' + y' = \underbrace{\sum_{n=2}^{\infty} n(n-1)c_n x^{n-1}}_{k=n-1} - \underbrace{\sum_{n=2}^{\infty} n(n-1)c_n x^{n-2}}_{k=n-2} + \underbrace{\sum_{n=1}^{\infty} nc_n x^{n-1}}_{k=n-1}$$

$$= \sum_{k=1}^{\infty}(k+1)kc_{k+1}x^k - \sum_{k=0}^{\infty}(k+2)(k+1)c_{k+2}x^k + \sum_{k=0}^{\infty}(k+1)c_{k+1}x^k$$

$$= -2c_2 + c_1 + \sum_{k=1}^{\infty}[(k+1)kc_{k+1} - (k+2)(k+1)c_{k+2} + (k+1)c_{k+1}]x^k = 0.$$

Thus

$$-2c_2 + c_1 = 0$$

$$(k+1)^2 c_{k+1} - (k+2)(k+1)c_{k+2} = 0$$

and

$$c_2 = \frac{1}{2}c_1$$

$$c_{k+2} = \frac{k+1}{k+2}c_{k+1}, \quad k = 1, 2, 3, \ldots .$$

Choosing $c_0 = 1$ and $c_1 = 0$ we find $c_2 = c_3 = c_4 = \cdots = 0$. For $c_0 = 0$ and $c_1 = 1$ we obtain

$$c_2 = \frac{1}{2}, \qquad c_3 = \frac{1}{3}, \qquad c_4 = \frac{1}{4},$$

and so on. Thus, two solutions are

$$y_1 = 1 \qquad \text{and} \qquad y_2 = x + \frac{1}{2}x^2 + \frac{1}{3}x^3 + \frac{1}{4}x^4 + \cdots.$$

24. Substituting $y = \sum_{n=0}^{\infty} c_n x^n$ into the differential equation we have

$$(x+2)y'' + xy' - y = \underbrace{\sum_{n=2}^{\infty} n(n-1)c_n x^{n-1}}_{k=n-1} + \underbrace{\sum_{n=2}^{\infty} 2n(n-1)c_n x^{n-2}}_{k=n-2} + \underbrace{\sum_{n=1}^{\infty} nc_n x^n}_{k=n} - \underbrace{\sum_{n=0}^{\infty} c_n x^n}_{k=n}$$

$$= \sum_{k=1}^{\infty} (k+1)k c_{k+1} x^k + \sum_{k=0}^{\infty} 2(k+2)(k+1)c_{k+2} x^k + \sum_{k=1}^{\infty} k c_k x^k - \sum_{k=0}^{\infty} c_k x^k$$

$$= 4c_2 - c_0 + \sum_{k=1}^{\infty} [(k+1)k c_{k+1} + 2(k+2)(k+1)c_{k+2} + (k-1)c_k] x^k = 0.$$

Thus

$$4c_2 - c_0 = 0$$

$$(k+1)k c_{k+1} + 2(k+2)(k+1)c_{k+2} + (k-1)c_k = 0, \quad k = 1, 2, 3, \ldots$$

and

$$c_2 = \frac{1}{4}c_0$$

$$c_{k+2} = -\frac{(k+1)k c_{k+1} + (k-1)c_k}{2(k+2)(k+1)}, \quad k = 1, 2, 3, \ldots.$$

Choosing $c_0 = 1$ and $c_1 = 0$ we find

$$c_1 = 0, \qquad c_2 = \frac{1}{4}, \qquad c_3 = -\frac{1}{24}, \qquad c_4 = 0, \qquad c_5 = \frac{1}{480}$$

and so on. For $c_0 = 0$ and $c_1 = 1$ we obtain

$$c_2 = 0$$

$$c_3 = 0$$

$$c_4 = c_5 = c_6 = \cdots = 0.$$

Thus, two solutions are

$$y_1 = c_0 \left[1 + \frac{1}{4}x^2 - \frac{1}{24}x^3 + \frac{1}{480}x^5 + \cdots \right] \qquad \text{and} \qquad y_2 = c_1 x.$$

should these
be cap
C_1 & C_2
as in
probs 29 & 31?

or
delete
as in
#23?
& #25

25. Substituting $y = \sum_{n=0}^{\infty} c_n x^n$ into the differential equation we have

$$
y'' - (x+1)y' - y = \sum_{n=2}^{\infty} n(n-1)c_n x^{n-2} - \underbrace{\sum_{n=1}^{\infty} nc_n x^n}_{k=n} - \underbrace{\sum_{n=1}^{\infty} nc_n x^{n-1}}_{k=n-1} - \underbrace{\sum_{n=0}^{\infty} c_n x^n}_{k=n}
$$

$$
\underbrace{\phantom{\sum_{n=2}^{\infty} n(n-1)c_n x^{n-2}}}_{k=n-2}
$$

$$
= \sum_{k=0}^{\infty} (k+2)(k+1)c_{k+2} x^k - \sum_{k=1}^{\infty} kc_k x^k - \sum_{k=0}^{\infty} (k+1)c_{k+1} x^k - \sum_{k=0}^{\infty} c_k x^k
$$

$$
= 2c_2 - c_1 - c_0 + \sum_{k=1}^{\infty} [(k+2)(k+1)c_{k+2} - (k+1)c_{k+1} - (k+1)c_k]x^k = 0.
$$

Thus

$$
2c_2 - c_1 - c_0 = 0
$$

$$
(k+2)(k+1)c_{k+2} - (k+1)(c_{k+1} + c_k) = 0
$$

and

$$
c_2 = \frac{c_1 + c_0}{2}
$$

$$
c_{k+2} = \frac{c_{k+1} + c_k}{k+2}, \quad k = 1, 2, 3, \ldots .
$$

Choosing $c_0 = 1$ and $c_1 = 0$ we find

$$
c_2 = \frac{1}{2}, \qquad c_3 = \frac{1}{6}, \qquad c_4 = \frac{1}{6},
$$

and so on. For $c_0 = 0$ and $c_1 = 1$ we obtain

$$
c_2 = \frac{1}{2}, \qquad c_3 = \frac{1}{2}, \qquad c_4 = \frac{1}{4},
$$

and so on. Thus, two solutions are

$$
y_1 = 1 + \frac{1}{2}x^2 + \frac{1}{6}x^3 + \frac{1}{6}x^4 + \cdots \qquad \text{and} \qquad y_2 = x + \frac{1}{2}x^2 + \frac{1}{2}x^3 + \frac{1}{4}x^4 + \cdots .
$$

26. Substituting $y = \sum_{n=0}^{\infty} c_n x^n$ into the differential equation we have

$$
\left(x^2 + 1\right)y'' - 6y = \underbrace{\sum_{n=2}^{\infty} n(n-1)c_n x^n}_{k=n} + \underbrace{\sum_{n=2}^{\infty} n(n-1)c_n x^{n-2}}_{k=n-2} - 6\underbrace{\sum_{n=0}^{\infty} c_n x^n}_{k=n}
$$

$$
= \sum_{k=2}^{\infty} k(k-1)c_k x^k + \sum_{k=0}^{\infty} (k+2)(k+1)c_{k+2} x^k - 6\sum_{k=0}^{\infty} c_k x^k
$$

$$
= 2c_2 - 6c_0 + (6c_3 - 6c_1)x + \sum_{k=2}^{\infty} \left[\left(k^2 - k - 6\right)c_k + (k+2)(k+1)c_{k+2} \right] x^k = 0.
$$

Thus

$$2c_2 - 6c_0 = 0$$

$$6c_3 - 6c_1 = 0$$

$$(k-3)(k+2)c_k + (k+2)(k+1)c_{k+2} = 0$$

and

$$c_2 = 3c_0$$

$$c_3 = c_1$$

$$c_{k+2} = -\frac{k-3}{k+1}c_k, \quad k = 2, 3, 4, \ldots .$$

Choosing $c_0 = 1$ and $c_1 = 0$ we find

$$c_2 = 3$$

$$c_3 = c_5 = c_7 = \cdots = 0$$

$$c_4 = 1$$

$$c_6 = -\frac{1}{5}$$

and so on. For $c_0 = 0$ and $c_1 = 1$ we obtain

$$c_2 = c_4 = c_6 = \cdots = 0$$

$$c_3 = 1$$

$$c_5 = c_7 = c_9 = \cdots = 0.$$

Thus, two solutions are

$$y_1 = 1 + 3x^2 + x^4 - \frac{1}{5}x^6 + \cdots \quad \text{and} \quad y_2 = x + x^3.$$

27. Substituting $y = \sum_{n=0}^{\infty} c_n x^n$ into the differential equation we have

$$\left(x^2 + 2\right)y'' + 3xy' - y = \underbrace{\sum_{n=2}^{\infty} n(n-1)c_n x^n}_{k=n} + 2\underbrace{\sum_{n=2}^{\infty} n(n-1)c_n x^{n-2}}_{k=n-2} + 3\underbrace{\sum_{n=1}^{\infty} nc_n x^n}_{k=n} - \underbrace{\sum_{n=0}^{\infty} c_n x^n}_{k=n}$$

$$= \sum_{k=2}^{\infty} k(k-1)c_k x^k + 2\sum_{k=0}^{\infty} (k+2)(k+1)c_{k+2}x^k + 3\sum_{k=1}^{\infty} kc_k x^k - \sum_{k=0}^{\infty} c_k x^k$$

$$= (4c_2 - c_0) + (12c_3 + 2c_1)x + \sum_{k=2}^{\infty} \left[2(k+2)(k+1)c_{k+2} + \left(k^2 + 2k - 1\right)c_k\right]x^k = 0.$$

285

Thus

$$4c_2 - c_0 = 0$$

$$12c_3 + 2c_1 = 0$$

$$2(k+2)(k+1)c_{k+2} + \left(k^2 + 2k - 1\right)c_k = 0$$

and

$$c_2 = \frac{1}{4}c_0$$

$$c_3 = -\frac{1}{6}c_1$$

$$c_{k+2} = -\frac{k^2 + 2k - 1}{2(k+2)(k+1)}c_k, \quad k = 2, 3, 4, \dots .$$

Choosing $c_0 = 1$ and $c_1 = 0$ we find

$$c_2 = \frac{1}{4}$$

$$c_3 = c_5 = c_7 = \cdots = 0$$

$$c_4 = -\frac{7}{96}$$

and so on. For $c_0 = 0$ and $c_1 = 1$ we obtain

$$c_2 = c_4 = c_6 = \cdots = 0$$

$$c_3 = -\frac{1}{6}$$

$$c_5 = \frac{7}{120}$$

and so on. Thus, two solutions are

$$y_1 = 1 + \frac{1}{4}x^2 - \frac{7}{96}x^4 + \cdots \qquad \text{and} \qquad y_2 = x - \frac{1}{6}x^3 + \frac{7}{120}x^5 - \cdots .$$

28. Substituting $y = \sum_{n=0}^{\infty} c_n x^n$ into the differential equation we have

$$\left(x^2 - 1\right)y'' + xy' - y = \underbrace{\sum_{n=2}^{\infty} n(n-1)c_n x^n}_{k=n} - \underbrace{\sum_{n=2}^{\infty} n(n-1)c_n x^{n-2}}_{k=n-2} + \underbrace{\sum_{n=1}^{\infty} nc_n x^n}_{k=n} - \underbrace{\sum_{n=0}^{\infty} c_n x^n}_{k=n}$$

$$= \sum_{k=2}^{\infty} k(k-1)c_k x^k - \sum_{k=0}^{\infty} (k+2)(k+1)c_{k+2} x^k + \sum_{k=1}^{\infty} kc_k x^k - \sum_{k=0}^{\infty} c_k x^k$$

$$= (-2c_2 - c_0) - 6c_3 x + \sum_{k=2}^{\infty} \left[-(k+2)(k+1)c_{k+2} + \left(k^2 - 1\right)c_k\right]x^k = 0.$$

Thus

$$-2c_2 - c_0 = 0$$

$$-6c_3 = 0$$

$$-(k+2)(k+1)c_{k+2} + (k-1)(k+1)c_k = 0$$

and

$$c_2 = -\frac{1}{2}c_0$$

$$c_3 = 0$$

$$c_{k+2} = \frac{k-1}{k+2}c_k, \quad k = 2,3,4,\dots.$$

Choosing $c_0 = 1$ and $c_1 = 0$ we find

$$c_2 = -\frac{1}{2}$$

$$c_3 = c_5 = c_7 = \cdots = 0$$

$$c_4 = -\frac{1}{8}$$

and so on. For $c_0 = 0$ and $c_1 = 1$ we obtain

$$c_2 = c_4 = c_6 = \cdots = 0$$

$$c_3 = c_5 = c_7 = \cdots = 0.$$

Thus, two solutions are

$$y_1 = 1 - \frac{1}{2}x^2 - \frac{1}{8}x^4 - \cdots \qquad \text{and} \qquad y_2 = x.$$

29. Substituting $y = \sum_{n=0}^{\infty} c_n x^n$ into the differential equation we have

$$(x-1)y'' - xy' + y = \underbrace{\sum_{n=2}^{\infty} n(n-1)c_n x^{n-1}}_{k=n-1} - \underbrace{\sum_{n=2}^{\infty} n(n-1)c_n x^{n-2}}_{k=n-2} - \underbrace{\sum_{n=1}^{\infty} nc_n x^n}_{k=n} + \underbrace{\sum_{n=0}^{\infty} c_n x^n}_{k=n}$$

$$= \sum_{k=1}^{\infty}(k+1)kc_{k+1}x^k - \sum_{k=0}^{\infty}(k+2)(k+1)c_{k+2}x^k - \sum_{k=1}^{\infty} kc_k x^k + \sum_{k=0}^{\infty} c_k x^k$$

$$= -2c_2 + c_0 + \sum_{k=1}^{\infty}[-(k+2)(k+1)c_{k+2} + (k+1)kc_{k+1} - (k-1)c_k]x^k = 0.$$

Thus

$$-2c_2 + c_0 = 0$$

$$-(k+2)(k+1)c_{k+2} + (k+1)kc_{k+1} - (k-1)c_k = 0$$

and

$$c_2 = \frac{1}{2}c_0$$

$$c_{k+2} = \frac{kc_{k+1}}{k+2} - \frac{(k-1)c_k}{(k+2)(k+1)}, \quad k = 1, 2, 3, \ldots.$$

Choosing $c_0 = 1$ and $c_1 = 0$ we find

$$c_2 = \frac{1}{2}, \qquad c_3 = \frac{1}{6}, \qquad c_4 = \frac{1}{24},$$

and so on. For $c_0 = 0$ and $c_1 = 1$ we obtain $c_2 = c_3 = c_4 = \cdots = 0$. Thus,

$$y = C_1 \left(1 + \frac{1}{2}x^2 + \frac{1}{6}x^3 + \cdots\right) + C_2 x$$

and

$$y' = C_1 \left(x + \frac{1}{2}x^2 + \cdots\right) + C_2.$$

The initial conditions imply $C_1 = -2$ and $C_2 = 6$, so

$$y = -2 \left(1 + \frac{1}{2}x^2 + \frac{1}{6}x^3 + \cdots\right) + 6x = 8x - 2e^x.$$

30. Substituting $y = \sum_{n=0}^{\infty} c_n x^n$ into the differential equation we have

$$(x+1)y'' - (2-x)y' + y$$

$$= \sum_{n=2}^{\infty} n(n-1)c_n x^{n-1} + \sum_{n=2}^{\infty} n(n-1)c_n x^{n-2} - 2\sum_{n=1}^{\infty} nc_n x^{n-1} + \sum_{n=1}^{\infty} nc_n x^n + \sum_{n=0}^{\infty} c_n x^n$$

$$\underbrace{\phantom{\sum_{n=2}^{\infty} n(n-1)c_n x^{n-1}}}_{k=n-1} \quad \underbrace{\phantom{\sum_{n=2}^{\infty} n(n-1)c_n x^{n-2}}}_{k=n-2} \quad \underbrace{\phantom{2\sum nc_n x^{n-1}}}_{k=n-1} \quad \underbrace{}_{k=n} \quad \underbrace{}_{k=n}$$

$$= \sum_{k=1}^{\infty} (k+1)kc_{k+1}x^k + \sum_{k=0}^{\infty} (k+2)(k+1)c_{k+2}x^k - 2\sum_{k=0}^{\infty} (k+1)c_{k+1}x^k + \sum_{k=1}^{\infty} kc_k x^k + \sum_{k=0}^{\infty} c_k x^k$$

$$= 2c_2 - 2c_1 + c_0 + \sum_{k=1}^{\infty} [(k+2)(k+1)c_{k+2} - (k+1)c_{k+1} + (k+1)c_k]x^k = 0.$$

Thus

$$2c_2 - 2c_1 + c_0 = 0$$

$$(k+2)(k+1)c_{k+2} - (k+1)c_{k+1} + (k+1)c_k = 0$$

and

$$c_2 = c_1 - \frac{1}{2}c_0$$

$$c_{k+2} = \frac{1}{k+2}c_{k+1} - \frac{1}{k+2}c_k, \quad k = 1, 2, 3, \ldots.$$

Choosing $c_0 = 1$ and $c_1 = 0$ we find

$$c_2 = -\frac{1}{2}, \qquad c_3 = -\frac{1}{6}, \qquad c_4 = \frac{1}{12},$$

and so on. For $c_0 = 0$ and $c_1 = 1$ we obtain

$$c_2 = 1, \qquad c_3 = 0, \qquad c_4 = -\frac{1}{4},$$

and so on. Thus,

$$y = C_1 \left(1 - \frac{1}{2}x^2 - \frac{1}{6}x^3 + \frac{1}{12}x^4 + \cdots\right) + C_2 \left(x + x^2 - \frac{1}{4}x^4 + \cdots\right)$$

and

$$y' = C_1 \left(-x - \frac{1}{2}x^2 + \frac{1}{3}x^3 + \cdots\right) + C_2 \left(1 + 2x - x^3 + \cdots\right).$$

The initial conditions imply $C_1 = 2$ and $C_2 = -1$, so

$$y = 2\left(1 - \frac{1}{2}x^2 - \frac{1}{6}x^3 + \frac{1}{12}x^4 + \cdots\right) - \left(x + x^2 - \frac{1}{4}x^4 + \cdots\right)$$

$$= 2 - x - 2x^2 - \frac{1}{3}x^3 + \frac{5}{12}x^4 + \cdots.$$

31. Substituting $y = \sum_{n=0}^{\infty} c_n x^n$ into the differential equation we have

$$y'' - 2xy' + 8y = \underbrace{\sum_{n=2}^{\infty} n(n-1)c_n x^{n-2}}_{k=n-2} - 2\underbrace{\sum_{n=1}^{\infty} nc_n x^n}_{k=n} + 8\underbrace{\sum_{n=0}^{\infty} c_n x^n}_{k=n}$$

$$= \sum_{k=0}^{\infty} (k+2)(k+1)c_{k+2} x^k - 2\sum_{k=1}^{\infty} kc_k x^k + 8\sum_{k=0}^{\infty} c_k x^k$$

$$= 2c_2 + 8c_0 + \sum_{k=1}^{\infty} [(k+2)(k+1)c_{k+2} + (8-2k)c_k]x^k = 0.$$

Thus

$$2c_2 + 8c_0 = 0$$

$$(k+2)(k+1)c_{k+2} + (8-2k)c_k = 0$$

and

$$c_2 = -4c_0$$

$$c_{k+2} = \frac{2(k-4)}{(k+2)(k+1)} c_k, \quad k = 1, 2, 3, \ldots.$$

Choosing $c_0 = 1$ and $c_1 = 0$ we find

$$c_2 = -4$$

$$c_3 = c_5 = c_7 = \cdots = 0$$

$$c_4 = \frac{4}{3}$$

$$c_6 = c_8 = c_{10} = \cdots = 0.$$

For $c_0 = 0$ and $c_1 = 1$ we obtain

$$c_2 = c_4 = c_6 = \cdots = 0$$

$$c_3 = -1$$

$$c_5 = \frac{1}{10}$$

and so on. Thus,

$$y = C_1 \left(1 - 4x^2 + \frac{4}{3}x^4\right) + C_2 \left(x - x^3 + \frac{1}{10}x^5 + \cdots\right)$$

and

$$y' = C_1 \left(-8x + \frac{16}{3}x^3\right) + C_2 \left(1 - 3x^2 + \frac{1}{2}x^4 + \cdots\right).$$

The initial conditions imply $C_1 = 3$ and $C_2 = 0$, so

$$y = 3\left(1 - 4x^2 + \frac{4}{3}x^4\right) = 3 - 12x^2 + 4x^4.$$

32. Substituting $y = \sum_{n=0}^{\infty} c_n x^n$ into the differential equation we have

$$(x^2 + 1)y'' + 2xy' = \underbrace{\sum_{n=2}^{\infty} n(n-1)c_n x^n}_{k=n} + \underbrace{\sum_{n=2}^{\infty} n(n-1)c_n x^{n-2}}_{k=n-2} + \underbrace{\sum_{n=1}^{\infty} 2nc_n x^n}_{k=n}$$

$$= \sum_{k=2}^{\infty} k(k-1)c_k x^k + \sum_{k=0}^{\infty} (k+2)(k+1)c_{k+2}x^k + \sum_{k=1}^{\infty} 2kc_k x^k$$

$$= 2c_2 + (6c_3 + 2c_1)x + \sum_{k=2}^{\infty} [k(k+1)c_k + (k+2)(k+1)c_{k+2}]x^k = 0.$$

Thus

$$2c_2 = 0$$

$$6c_3 + 2c_1 = 0$$

$$k(k+1)c_k + (k+2)(k+1)c_{k+2} = 0$$

and

$$c_2 = 0$$

$$c_3 = -\frac{1}{3}c_1$$

$$c_{k+2} = -\frac{k}{k+2}c_k, \quad k = 2, 3, 4, \ldots .$$

Choosing $c_0 = 1$ and $c_1 = 0$ we find $c_3 = c_4 = c_5 = \cdots = 0$. For $c_0 = 0$ and $c_1 = 1$ we obtain

$$c_3 = -\frac{1}{3}$$

$$c_4 = c_6 = c_8 = \cdots = 0$$

$$c_5 = -\frac{1}{5}$$

$$c_7 = \frac{1}{7}$$

and so on. Thus

$$y = C_0 + C_1\left(x - \frac{1}{3}x^3 + \frac{1}{5}x^5 - \frac{1}{7}x^7 + \cdots\right)$$

and

$$y' = C_1\left(1 - x^2 + x^4 - x^6 + \cdots\right).$$

The initial conditions imply $c_0 = 0$ and $c_1 = 1$, so

$$y = x - \frac{1}{3}x^3 + \frac{1}{5}x^5 - \frac{1}{7}x^7 + \cdots .$$

33. Substituting $y = \sum_{n=0}^{\infty} c_n x^n$ into the differential equation we have

$$y'' + (\sin x)y = \sum_{n=2}^{\infty} n(n-1)c_n x^{n-2} + \left(x - \frac{1}{6}x^3 + \frac{1}{120}x^5 - \cdots\right)\left(c_0 + c_1 x + c_2 x^2 + \cdots\right)$$

$$= \left[2c_2 + 6c_3 x + 12c_4 x^2 + 20c_5 x^3 + \cdots\right] + \left[c_0 x + c_1 x^2 + \left(c_2 - \frac{1}{6}c_0\right)x^3 + \cdots\right]$$

$$= 2c_2 + (6c_3 + c_0)x + (12c_4 + c_1)x^2 + \left(20c_5 + c_2 - \frac{1}{6}c_0\right)x^3 + \cdots = 0.$$

Thus

$$2c_2 = 0$$

$$6c_3 + c_0 = 0$$

$$12c_4 + c_1 = 0$$

$$20c_5 + c_2 - \frac{1}{6}c_0 = 0$$

291

and

$$c_2 = 0$$

$$c_3 = -\frac{1}{6}c_0$$

$$c_4 = -\frac{1}{12}c_1$$

$$c_5 = -\frac{1}{20}c_2 + \frac{1}{120}c_0.$$

Choosing $c_0 = 1$ and $c_1 = 0$ we find

$$c_2 = 0, \qquad c_3 = -\frac{1}{6}, \qquad c_4 = 0, \qquad c_5 = \frac{1}{120}$$

and so on. For $c_0 = 0$ and $c_1 = 1$ we obtain

$$c_2 = 0, \qquad c_3 = 0, \qquad c_4 = -\frac{1}{12}, \qquad c_5 = 0$$

and so on. Thus, two solutions are

$$y_1 = 1 - \frac{1}{6}x^3 + \frac{1}{120}x^5 + \cdots \qquad \text{and} \qquad y_2 = x - \frac{1}{12}x^4 + \cdots .$$

34. Substituting $y = \sum_{n=0}^{\infty} c_n x^n$ into the differential equation we have

$$y'' + e^x y' - y = \sum_{n=2}^{\infty} n(n-1)c_n x^{n-2}$$

$$+ \left(1 + x + \frac{1}{2}x^2 + \frac{1}{6}x^3 + \cdots\right)\left(c_1 + 2c_2 x + 3c_3 x^2 + 4c_4 x^3 + \cdots\right) - \sum_{n=0}^{\infty} c_n x^n$$

$$= \left[2c_2 + 6c_3 x + 12c_4 x^2 + 20c_5 x^3 + \cdots\right]$$

$$+ \left[c_1 + (2c_2 + c_1)x + \left(3c_3 + 2c_2 + \frac{1}{2}c_1\right)x^2 + \cdots\right] - \left[c_0 + c_1 x + c_2 x^2 + \cdots\right]$$

$$= (2c_2 + c_1 - c_0) + (6c_3 + 2c_2)x + \left(12c_4 + 3c_3 + c_2 + \frac{1}{2}c_1\right)x^2 + \cdots = 0.$$

Thus

$$2c_2 + c_1 - c_0 = 0$$

$$6c_3 + 2c_2 = 0$$

$$12c_4 + 3c_3 + c_2 + \frac{1}{2}c_1 = 0$$

and

$$c_2 = \frac{1}{2}c_0 - \frac{1}{2}c_1$$

$$c_3 = -\frac{1}{3}c_2$$

$$c_4 = -\frac{1}{4}c_3 + \frac{1}{12}c_2 - \frac{1}{24}c_1.$$

Choosing $c_0 = 1$ and $c_1 = 0$ we find

$$c_2 = \frac{1}{2}, \qquad c_3 = -\frac{1}{6}, \qquad c_4 = 0$$

and so on. For $c_0 = 0$ and $c_1 = 1$ we obtain

$$c_2 = -\frac{1}{2}, \qquad c_3 = \frac{1}{6}, \qquad c_4 = -\frac{1}{24}$$

and so on. Thus, two solutions are

$$y_1 = 1 + \frac{1}{2}x^2 - \frac{1}{6}x^3 + \cdots \qquad \text{and} \qquad y_2 = x - \frac{1}{2}x^2 + \frac{1}{6}x^3 - \frac{1}{24}x^4 + \cdots .$$

35. The singular points of $(\cos x)y'' + y' + 5y = 0$ are odd integer multiples of $\pi/2$. The distance from 0 to either $\pm\pi/2$ is $\pi/2$. The singular point closest to 1 is $\pi/2$. The distance from 1 to the closest singular point is then $\pi/2 - 1$.

36. Substituting $y = \sum_{n=0}^{\infty} c_n x^n$ into the first differential equation leads to

$$y'' - xy = \underbrace{\sum_{n=2}^{\infty} n(n-1)c_n x^{n-2}}_{k=n-2} - \underbrace{\sum_{n=0}^{\infty} c_n x^{n+1}}_{k=n+1} = \sum_{k=0}^{\infty} (k+2)(k+1)c_{k+2}x^k - \sum_{k=1}^{\infty} c_{k-1}x^k$$

$$= 2c_2 + \sum_{k=1}^{\infty} [(k+2)(k+1)c_{k+2} - c_{k-1}]x^k = 1.$$

Thus

$$2c_2 = 1$$

$$(k+2)(k+1)c_{k+2} - c_{k-1} = 0$$

and

$$c_2 = \frac{1}{2}$$

$$c_{k+2} = \frac{c_{k-1}}{(k+2)(k+1)}, \qquad k = 1, 2, 3, \ldots .$$

Let c_0 and c_1 be arbitrary and iterate to find

$$c_2 = \frac{1}{2}$$

$$c_3 = \frac{1}{6}c_0$$

$$c_4 = \frac{1}{12}c_1$$

$$c_5 = \frac{1}{20}c_2 = \frac{1}{40}$$

and so on. The solution is

$$y = c_0 + c_1 x + \frac{1}{2}x^2 + \frac{1}{6}c_0 x^3 + \frac{1}{12}c_1 x^4 + \frac{1}{40}c_5 + \cdots$$

$$= c_0\left(1 + \frac{1}{6}x^3 + \cdots\right) + c_1\left(x + \frac{1}{12}x^4 + \cdots\right) + \frac{1}{2}x^2 + \frac{1}{40}x^5 + \cdots .$$

Substituting $y = \sum_{n=0}^{\infty} c_n x^n$ into the second differential equation leads to

$$y'' - 4xy' - 4y = \underbrace{\sum_{n=2}^{\infty} n(n-1)c_n x^{n-2}}_{k=n-2} - \underbrace{\sum_{n=1}^{\infty} 4nc_n x^n}_{k=n} - \underbrace{\sum_{n=0}^{\infty} 4c_n x^n}_{k=n}$$

$$= \sum_{k=0}^{\infty} (k+2)(k+1)c_{k+2}x^k - \sum_{k=1}^{\infty} 4kc_k x^k - \sum_{k=0}^{\infty} 4c_k x^k$$

$$= 2c_2 - 4c_0 + \sum_{k=1}^{\infty} [(k+2)(k+1)c_{k+2} - 4(k+1)c_k]x^k$$

$$= e^x = 1 + \sum_{k=1}^{\infty} \frac{1}{k!}x^k.$$

Thus

$$2c_2 - 4c_0 = 1$$

$$(k+2)(k+1)c_{k+2} - 4(k+1)c_k = \frac{1}{k!}$$

and

$$c_2 = \frac{1}{2} + 2c_0$$

$$c_{k+2} = \frac{1}{(k+2)!} + \frac{4}{k+2}c_k, \qquad k = 1, 2, 3, \ldots .$$

Let c_0 and c_1 be arbitrary and iterate to find

$$c_2 = \frac{1}{2} + 2c_0$$

$$c_3 = \frac{1}{3!} + \frac{4}{3}c_1 = \frac{1}{3!} + \frac{4}{3}c_1$$

$$c_4 = \frac{1}{4!} + \frac{4}{4}c_2 = \frac{1}{4!} + \frac{1}{2} + 2c_0 = \frac{13}{4!} + 2c_0$$

$$c_5 = \frac{1}{5!} + \frac{4}{5}c_3 = \frac{1}{5!} + \frac{4}{5 \cdot 3!} + \frac{16}{15}c_1 = \frac{17}{5!} + \frac{16}{15}c_1$$

$$c_6 = \frac{1}{6!} + \frac{4}{6}c_4 = \frac{1}{6!} + \frac{4 \cdot 13}{6 \cdot 4!} + \frac{8}{6}c_0 = \frac{261}{6!} + \frac{4}{3}c_0$$

$$c_7 = \frac{1}{7!} + \frac{4}{7}c_5 = \frac{1}{7!} + \frac{4 \cdot 17}{7 \cdot 5!} + \frac{64}{105}c_1 = \frac{409}{7!} + \frac{64}{105}c_1$$

and so on. The solution is

$$y = c_0 + c_1 x + \left(\frac{1}{2} + 2c_0\right)x^2 + \left(\frac{1}{3!} + \frac{4}{3}c_1\right)x^3 + \left(\frac{13}{4!} + 2c_0\right)x^4 + \left(\frac{17}{5!} + \frac{16}{15}c_1\right)x^5$$

$$+ \left(\frac{261}{6!} + \frac{4}{3}c_0\right)x^6 + \left(\frac{409}{7!} + \frac{64}{105}c_1\right)x^7 + \cdots$$

$$= c_0\left[1 + 2x^2 + 2x^4 + \frac{4}{3}x^6 + \cdots\right] + c_1\left[x + \frac{4}{3}x^3 + \frac{16}{15}x^5 + \frac{64}{105}x^7 + \cdots\right]$$

$$+ \frac{1}{2}x^2 + \frac{1}{3!}x^3 + \frac{13}{4!}x^4 + \frac{17}{5!}x^5 + \frac{261}{6!}x^6 + \frac{409}{7!}x^7 + \cdots .$$

37. We identify $P(x) = 0$ and $Q(x) = \sin x/x$. The Taylor series representation for $\sin x/x$ is $1 - x^2/3! + x^4/5! - \cdots$, for $|x| < \infty$. Thus, $Q(x)$ is analytic at $x = 0$ and $x = 0$ is an ordinary point of the differential equation.

38. If $x > 0$ and $y > 0$, then $y'' = -xy < 0$ and the graph of a solution curve is concave down. Thus, whatever portion of a solution curve lies in the first quadrant is concave down. When $x > 0$ and $y < 0$, $y'' = -xy > 0$, so whatever portion of a solution curve lies in the fourth quadrant is concave up.

39. (a) Substituting $y = \sum_{n=0}^{\infty} c_n x^n$ into the differential equation we have

$$y'' + xy' + y = \underbrace{\sum_{n=2}^{\infty} n(n-1)c_n x^{n-2}}_{k=n-2} + \underbrace{\sum_{n=1}^{\infty} nc_n x^n}_{k=n} + \underbrace{\sum_{n=0}^{\infty} c_n x^n}_{k=n}$$

$$= \sum_{k=0}^{\infty} (k+2)(k+1)c_{k+2} x^k + \sum_{k=1}^{\infty} kc_k x^k + \sum_{k=0}^{\infty} c_k x^k$$

$$= (2c_2 + c_0) + \sum_{k=1}^{\infty} [(k+2)(k+1)c_{k+2} + (k+1)c_k] x^k = 0.$$

Thus

$$2c_2 + c_0 = 0$$

$$(k+2)(k+1)c_{k+2} + (k+1)c_k = 0$$

and

$$c_2 = -\frac{1}{2}c_0$$

$$c_{k+2} = -\frac{1}{k+2}c_k, \quad k = 1, 2, 3, \ldots.$$

Choosing $c_0 = 1$ and $c_1 = 0$ we find

$$c_2 = -\frac{1}{2}$$

$$c_3 = c_5 = c_7 = \cdots = 0$$

$$c_4 = -\frac{1}{4}\left(-\frac{1}{2}\right) = \frac{1}{2^2 \cdot 2}$$

$$c_6 = -\frac{1}{6}\left(\frac{1}{2^2 \cdot 2}\right) = -\frac{1}{2^3 \cdot 3!}$$

and so on. For $c_0 = 0$ and $c_1 = 1$ we obtain

$$c_2 = c_4 = c_6 = \cdots = 0$$

$$c_3 = -\frac{1}{3} = -\frac{2}{3!}$$

$$c_5 = -\frac{1}{5}\left(-\frac{1}{3}\right) = \frac{1}{5 \cdot 3} = \frac{4 \cdot 2}{5!}$$

$$c_7 = -\frac{1}{7}\left(\frac{4 \cdot 2}{5!}\right) = -\frac{6 \cdot 4 \cdot 2}{7!}$$

and so on. Thus, two solutions are

$$y_1 = \sum_{k=0}^{\infty} \frac{(-1)^k}{2^k \cdot k!} x^{2k} \qquad \text{and} \qquad y_2 = \sum_{k=0}^{\infty} \frac{(-1)^k 2^k k!}{(2k+1)!} x^{2k+1}.$$

(b) For y_1, $S_3 = S_2$ and $S_5 = S_4$, so we plot S_2, S_4, S_6, S_8, and S_{10}.

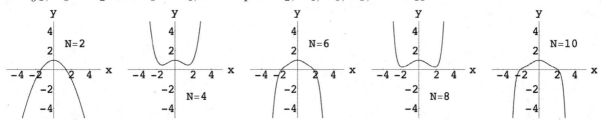

For y_2, $S_3 = S_4$ and $S_5 = S_6$, so we plot S_2, S_4, S_6, S_8, and S_{10}.

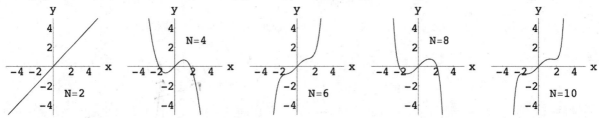

(c)

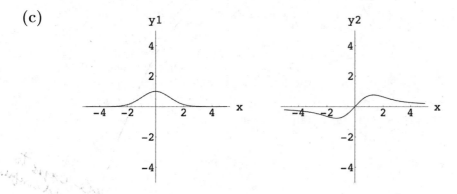

The graphs of y_1 and y_2 obtained from a numerical solver are shown. We see that the partial sum representations indicate the even and odd natures of the solution, but don't really give a very accurate representation of the true solution. Increasing N to about 20 gives a much more accurate representation on $[-4, 4]$.

(d) From $e^x = \sum_{k=0}^{\infty} x^k/k!$ we see that $e^{-x^2/2} = \sum_{k=0}^{\infty} (-x^2/2)^k/k! = \sum_{k=0}^{\infty} (-1)^k x^{2k}/2^k k!$. From (5) of Section 4.2 we have

$$y_2 = y_1 \int \frac{e^{-\int x\,dx}}{y_1^2}\,dx = e^{-x^2/2} \int \frac{e^{-x^2/2}}{(e^{-x^2/2})^2}\,dx = e^{-x^2/2} \int \frac{e^{-x^2/2}}{e^{-x^2}}\,dx = e^{-x^2/2}\int e^{x^2/2}\,dx$$

$$= \sum_{k=0}^{\infty} \frac{(-1)^k}{2^k k!} x^{2k} \int \sum_{k=0}^{\infty} \frac{1}{2^k k!} x^{2k}\,dx = \left(\sum_{k=0}^{\infty} \frac{(-1)^k}{2^k k!} x^{2k}\right)\left(\sum_{k=0}^{\infty} \int \frac{1}{2^k k!} x^{2k}\,dx\right)$$

$$= \left(\sum_{k=0}^{\infty} \frac{(-1)^k}{2^k k!} x^{2k}\right)\left(\sum_{k=0}^{\infty} \frac{1}{(2k+1)2^k k!} x^{2k+1}\right)$$

$$= \left(1 - \frac{1}{2}x^2 + \frac{1}{2^2 \cdot 2}x^4 - \frac{1}{2^3 \cdot 3!}x^6 + \cdots\right)\left(x + \frac{1}{3 \cdot 2}x^3 + \frac{1}{5 \cdot 2^2 \cdot 2}x^5 + \frac{1}{7 \cdot 2^3 \cdot 3!}x^7 + \cdots\right)$$

$$= x - \frac{2}{3!}x^3 + \frac{4 \cdot 2}{5!}x^5 - \frac{6 \cdot 4 \cdot 2}{7!}x^7 + \cdots = \sum_{k=0}^{\infty} \frac{(-1)^k 2^k k!}{(2k+1)!} x^{2k+1}.$$

40. (a) We have

$$y'' + (\cos x)y = 2c_2 + 6c_3 x + 12c_4 x^2 + 20c_5 x^3 + 30c_6 x^4 + 42c_7 x^5 + \cdots$$

$$+ \left(1 - \frac{x^2}{2!} + \frac{x^4}{4!} - \frac{x^6}{6!} + \cdots\right)(c_0 + c_1 x + c_2 x^2 + c_3 x^3 + c_4 x^4 + c_5 x^5 + \cdots)$$

$$= (2c_2 + c_0) + (6c_3 + c_1)x + \left(12c_4 + c_2 - \frac{1}{2}c_0\right)x^2 + \left(20c_5 + c_3 - \frac{1}{2}c_1\right)x^3$$

$$+ \left(30c_6 + c_4 + \frac{1}{24}c_0 - \frac{1}{2}c_2\right)x^4 + \left(42c_7 + c_5 + \frac{1}{24}c_1 - \frac{1}{2}c_3\right)x^5 + \cdots .$$

Then

$$30c_6 + c_4 + \frac{1}{24}c_0 - \frac{1}{2}c_2 = 0 \quad \text{and} \quad 42c_7 + c_5 + \frac{1}{24}c_1 - \frac{1}{2}c_3 = 0,$$

which gives $c_6 = -c_0/80$ and $c_7 = -19c_1/5040$. Thus

$$y_1(x) = 1 - \frac{1}{2}x^2 + \frac{1}{12}x^4 - \frac{1}{80}x^6 + \cdots$$

and

$$y_2(x) = x - \frac{1}{6}x^3 + \frac{1}{30}x^5 - \frac{19}{5040}x^7 + \cdots .$$

use caps to avoid confusion with $\int$ C_1 C_2

(b) From part (a) the general solution of the differential equation is $y = c_1 y_1 + c_2 y_2$. Then $y(0) = c_1 + c_2 \cdot 0 = c_1$ and $y'(0) = c_1 \cdot 0 + c_2 = c_2$, so the solution of the initial-value problem is

$$y = y_1 + y_2 = 1 + x - \frac{1}{2}x^2 - \frac{1}{6}x^3 + \frac{1}{12}x^4 + \frac{1}{30}x^5 - \frac{1}{80}x^6 - \frac{19}{5040}x^7 + \cdots .$$

298

(c)

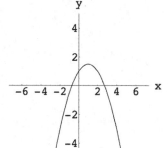
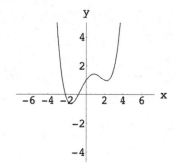
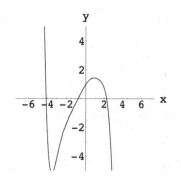

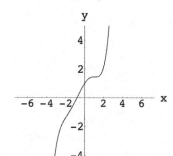
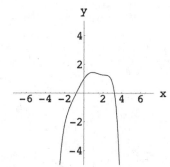
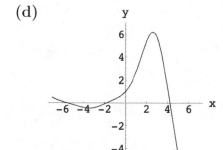

(d)

Exercises 6.2

Solutions About Singular Points

1. Irregular singular point: $x = 0$

2. Regular singular points: $x = 0, -3$

3. Irregular singular point: $x = 3$; regular singular point: $x = -3$

4. Irregular singular point: $x = 1$; regular singular point: $x = 0$

5. Regular singular points: $x = 0, \pm 2i$

6. Irregular singular point: $x = 5$; regular singular point: $x = 0$

299

7. Regular singular points: $x = -3, 2$

8. Regular singular points: $x = 0, \pm i$

9. Irregular singular point: $x = 0$; regular singular points: $x = 2, \pm 5$

10. Irregular singular point: $x = -1$; regular singular points: $x = 0, 3$

11. Writing the differential equation in the form

$$y'' + \frac{5}{x-1}y' + \frac{x}{x+1}y = 0$$

we see that $x_0 = 1$ and $x_0 = -1$ are regular singular points. For $x_0 = 1$ the differential equation can be put in the form

$$(x-1)^2 y'' + 5(x-1)y' + \frac{x(x-1)^2}{x+1}y = 0.$$

In this case $p(x) = 5$ and $q(x) = x(x-1)^2/(x+1)$. For $x_0 = -1$ the differential equation can be put in the form

$$(x+1)^2 y'' + 5(x+1)\frac{x+1}{x-1}y' + x(x+1)y = 0.$$

In this case $p(x) = 5(x+1)/(x-1)$ and $q(x) = x(x+1)$.

12. Writing the differential equation in the form

$$y'' + \frac{x+3}{x}y' + 7xy = 0$$

we see that $x_0 = 0$ is a regular singular point. Multiplying by x^2, the differential equation can be put in the form

$$x^2 y'' + x(x+3)y' + 7x^3 y = 0.$$

We identify $p(x) = x + 3$ and $q(x) = 7x^3$.

13. We identify $P(x) = 5/3x + 1$ and $Q(x) = -1/3x^2$, so that $p(x) = xP(x) = \frac{5}{3} + x$ and $q(x) = x^2 Q(x) = -\frac{1}{3}$. Then $a_0 = \frac{5}{3}$, $b_0 = -\frac{1}{3}$, and the indicial equation is

$$r(r-1) + \frac{5}{3}r - \frac{1}{3} = r^2 + \frac{2}{3}r - \frac{1}{3} = \frac{1}{3}(3r^2 + 2r - 1) = \frac{1}{3}(3r - 1)(r+1) = 0.$$

The indicial roots are $\frac{1}{3}$ and -1. Since these do not differ by an integer we expect to find two series solutions using the method of Frobenius.

14. We identify $P(x) = 1/x$ and $Q(x) = 10/x$, so that $p(x) = xP(x) = 1$ and $q(x) = x^2 Q(x) = 10x$. Then $a_0 = 1$, $b_0 = 0$, and the indicial equation is

$$r(r-1) + r = r^2 = 0.$$

The indicial roots are 0 and 0. Since these are equal, we expect the method of Frobenius to yield a single series solution.

15. Substituting $y = \sum_{n=0}^{\infty} c_n x^{n+r}$ into the differential equation and collecting terms, we obtain

$$2xy'' - y' + 2y = \left(2r^2 - 3r\right)c_0 x^{r-1} + \sum_{k=1}^{\infty} [2(k+r-1)(k+r)c_k - (k+r)c_k + 2c_{k-1}]x^{k+r-1} = 0,$$

which implies

$$2r^2 - 3r = r(2r - 3) = 0$$

and

$$(k+r)(2k+2r-3)c_k + 2c_{k-1} = 0.$$

The indicial roots are $r = 0$ and $r = 3/2$. For $r = 0$ the recurrence relation is

$$c_k = -\frac{2c_{k-1}}{k(2k-3)}, \quad k = 1, 2, 3, \ldots,$$

and

$$c_1 = 2c_0, \qquad c_2 = -2c_0, \qquad c_3 = \frac{4}{9}c_0,$$

and so on. For $r = 3/2$ the recurrence relation is

$$c_k = -\frac{2c_{k-1}}{(2k+3)k}, \quad k = 1, 2, 3, \ldots,$$

and

$$c_1 = -\frac{2}{5}c_0, \qquad c_2 = \frac{2}{35}c_0, \qquad c_3 = -\frac{4}{945}c_0,$$

and so on. The general solution on $(0, \infty)$ is

$$y = C_1\left(1 + 2x - 2x^2 + \frac{4}{9}x^3 + \cdots\right) + C_2 x^{3/2}\left(1 - \frac{2}{5}x + \frac{2}{35}x^2 - \frac{4}{945}x^3 + \cdots\right).$$

16. Substituting $y = \sum_{n=0}^{\infty} c_n x^{n+r}$ into the differential equation and collecting terms, we obtain

$$2xy'' + 5y' + xy = \left(2r^2 + 3r\right)c_0 x^{r-1} + \left(2r^2 + 7r + 5\right)c_1 x^r$$

$$+ \sum_{k=2}^{\infty} [2(k+r)(k+r-1)c_k + 5(k+r)c_k + c_{k-2}]x^{k+r-1}$$

$$= 0,$$

which implies

$$2r^2 + 3r = r(2r + 3) = 0,$$

$$\left(2r^2 + 7r + 5\right)c_1 = 0,$$

and

$$(k+r)(2k+2r+3)c_k + c_{k-2} = 0.$$

The indicial roots are $r = -3/2$ and $r = 0$, so $c_1 = 0$. For $r = -3/2$ the recurrence relation is

$$c_k = -\frac{c_{k-2}}{(2k-3)k}, \quad k = 2, 3, 4, \ldots,$$

and

$$c_2 = -\frac{1}{2}c_0, \qquad c_3 = 0, \qquad c_4 = \frac{1}{40}c_0,$$

and so on. For $r = 0$ the recurrence relation is

$$c_k = -\frac{c_{k-2}}{k(2k+3)}, \quad k = 2, 3, 4, \ldots,$$

and

$$c_2 = -\frac{1}{14}c_0, \qquad c_3 = 0, \qquad c_4 = \frac{1}{616}c_0,$$

and so on. The general solution on $(0, \infty)$ is

$$y = C_1 x^{-3/2} \left(1 - \frac{1}{2}x^2 + \frac{1}{40}x^4 + \cdots \right) + C_2 \left(1 - \frac{1}{14}x^2 + \frac{1}{616}x^4 + \cdots \right).$$

17. Substituting $y = \sum_{n=0}^{\infty} c_n x^{n+r}$ into the differential equation and collecting terms, we obtain

$$4xy'' + \frac{1}{2}y' + y = \left(4r^2 - \frac{7}{2}r \right) c_0 x^{r-1} + \sum_{k=1}^{\infty} \left[4(k+r)(k+r-1)c_k + \frac{1}{2}(k+r)c_k + c_{k-1} \right] x^{k+r-1}$$

$$= 0,$$

which implies

$$4r^2 - \frac{7}{2}r = r \left(4r - \frac{7}{2} \right) = 0$$

and

$$\frac{1}{2}(k+r)(8k + 8r - 7)c_k + c_{k-1} = 0.$$

The indicial roots are $r = 0$ and $r = 7/8$. For $r = 0$ the recurrence relation is

$$c_k = -\frac{2c_{k-1}}{k(8k-7)}, \quad k = 1, 2, 3, \ldots,$$

and

$$c_1 = -2c_0, \qquad c_2 = \frac{2}{9}c_0, \qquad c_3 = -\frac{4}{459}c_0,$$

and so on. For $r = 7/8$ the recurrence relation is

$$c_k = -\frac{2c_{k-1}}{(8k+7)k}, \quad k = 1, 2, 3, \ldots,$$

and

$$c_1 = -\frac{2}{15}c_0, \qquad c_2 = \frac{2}{345}c_0, \qquad c_3 = -\frac{4}{32{,}085}c_0,$$

and so on. The general solution on $(0, \infty)$ is

$$y = C_1 \left(1 - 2x + \frac{2}{9}x^2 - \frac{4}{459}x^3 + \cdots \right) + C_2 x^{7/8} \left(1 - \frac{2}{15}x + \frac{2}{345}x^2 - \frac{4}{32{,}085}x^3 + \cdots \right).$$

18. Substituting $y = \sum_{n=0}^{\infty} c_n x^{n+r}$ into the differential equation and collecting terms, we obtain

$$2x^2 y'' - xy' + \left(x^2 + 1 \right) y = \left(2r^2 - 3r + 1 \right) c_0 x^r + \left(2r^2 + r \right) c_1 x^{r+1}$$

$$+ \sum_{k=2}^{\infty} [2(k+r)(k+r-1)c_k - (k+r)c_k + c_k + c_{k-2}]x^{k+r}$$

$$= 0,$$

which implies

$$2r^2 - 3r + 1 = (2r - 1)(r - 1) = 0,$$

$$\left(2r^2 + r \right) c_1 = 0,$$

and

$$[(k+r)(2k+2r-3)+1]c_k + c_{k-2} = 0.$$

The indicial roots are $r = 1/2$ and $r = 1$, so $c_1 = 0$. For $r = 1/2$ the recurrence relation is

$$c_k = -\frac{c_{k-2}}{k(2k-1)}, \quad k = 2, 3, 4, \ldots,$$

and

$$c_2 = -\frac{1}{6}c_0, \qquad c_3 = 0, \qquad c_4 = \frac{1}{168}c_0,$$

and so on. For $r = 1$ the recurrence relation is

$$c_k = -\frac{c_{k-2}}{k(2k+1)}, \quad k = 2, 3, 4, \ldots,$$

and

$$c_2 = -\frac{1}{10}c_0, \qquad c_3 = 0, \qquad c_4 = \frac{1}{360}c_0,$$

and so on. The general solution on $(0, \infty)$ is

$$y = C_1 x^{1/2} \left(1 - \frac{1}{6}x^2 + \frac{1}{168}x^4 + \cdots \right) + C_2 x \left(1 - \frac{1}{10}x^2 + \frac{1}{360}x^4 + \cdots \right).$$

19. Substituting $y = \sum_{n=0}^{\infty} c_n x^{n+r}$ into the differential equation and collecting terms, we obtain

$$3xy'' + (2 - x)y' - y = \left(3r^2 - r \right) c_0 x^{r-1}$$

$$+ \sum_{k=1}^{\infty} [3(k+r-1)(k+r)c_k + 2(k+r)c_k - (k+r)c_{k-1}]x^{k+r-1}$$

$$= 0,$$

which implies

$$3r^2 - r = r(3r - 1) = 0$$

and

$$(k+r)(3k+3r-1)c_k - (k+r)c_{k-1} = 0.$$

The indicial roots are $r = 0$ and $r = 1/3$. For $r = 0$ the recurrence relation is

$$c_k = \frac{c_{k-1}}{3k-1}, \quad k = 1, 2, 3, \ldots,$$

and

$$c_1 = \frac{1}{2}c_0, \qquad c_2 = \frac{1}{10}c_0, \qquad c_3 = \frac{1}{80}c_0,$$

and so on. For $r = 1/3$ the recurrence relation is

$$c_k = \frac{c_{k-1}}{3k}, \quad k = 1, 2, 3, \ldots,$$

and

$$c_1 = \frac{1}{3}c_0, \qquad c_2 = \frac{1}{18}c_0, \qquad c_3 = \frac{1}{162}c_0,$$

and so on. The general solution on $(0, \infty)$ is

$$y = C_1\left(1 + \frac{1}{2}x + \frac{1}{10}x^2 + \frac{1}{80}x^3 + \cdots\right) + C_2 x^{1/3}\left(1 + \frac{1}{3}x + \frac{1}{18}x^2 + \frac{1}{162}x^3 + \cdots\right).$$

20. Substituting $y = \sum_{n=0}^{\infty} c_n x^{n+r}$ into the differential equation and collecting terms, we obtain

$$x^2 y'' - \left(x - \frac{2}{9}\right)y = \left(r^2 - r + \frac{2}{9}\right)c_0 x^r + \sum_{k=1}^{\infty}\left[(k+r)(k+r-1)c_k + \frac{2}{9}c_k - c_{k-1}\right]x^{k+r}$$

$$= 0,$$

which implies

$$r^2 - r + \frac{2}{9} = \left(r - \frac{2}{3}\right)\left(r - \frac{1}{3}\right) = 0$$

and

$$\left[(k+r)(k+r-1) + \frac{2}{9}\right]c_k - c_{k-1} = 0.$$

The indicial roots are $r = 2/3$ and $r = 1/3$. For $r = 2/3$ the recurrence relation is

$$c_k = \frac{3c_{k-1}}{3k^2 + k}, \quad k = 1, 2, 3, \ldots,$$

and

$$c_1 = \frac{3}{4}c_0, \qquad c_2 = \frac{9}{56}c_0, \qquad c_3 = \frac{9}{560}c_0,$$

and so on. For $r = 1/3$ the recurrence relation is

$$c_k = \frac{3c_{k-1}}{3k^2 - k}, \quad k = 1, 2, 3, \ldots,$$

304

and

$$c_1 = \frac{3}{2}c_0, \qquad c_2 = \frac{9}{20}c_0, \qquad c_3 = \frac{9}{160}c_0,$$

and so on. The general solution on $(0, \infty)$ is

$$y = C_1 x^{2/3}\left(1 + \frac{3}{4}x + \frac{9}{56}x^2 + \frac{9}{560}x^3 + \cdots\right) + C_2 x^{1/3}\left(1 + \frac{3}{2}x + \frac{9}{20}x^2 + \frac{9}{160}x^3 + \cdots\right).$$

21. Substituting $y = \sum_{n=0}^{\infty} c_n x^{n+r}$ into the differential equation and collecting terms, we obtain

$$2xy'' - (3+2x)y' + y = \left(2r^2 - 5r\right)c_0 x^{r-1} + \sum_{k=1}^{\infty}[2(k+r)(k+r-1)c_k$$

$$- 3(k+r)c_k - 2(k+r-1)c_{k-1} + c_{k-1}]x^{k+r-1}$$

$$= 0,$$

which implies

$$2r^2 - 5r = r(2r - 5) = 0$$

and

$$(k+r)(2k+2r-5)c_k - (2k+2r-3)c_{k-1} = 0.$$

The indicial roots are $r = 0$ and $r = 5/2$. For $r = 0$ the recurrence relation is

$$c_k = \frac{(2k-3)c_{k-1}}{k(2k-5)}, \qquad k = 1, 2, 3, \ldots,$$

and

$$c_1 = \frac{1}{3}c_0, \qquad c_2 = -\frac{1}{6}c_0, \qquad c_3 = -\frac{1}{6}c_0,$$

and so on. For $r = 5/2$ the recurrence relation is

$$c_k = \frac{2(k+1)c_{k-1}}{k(2k+5)}, \qquad k = 1, 2, 3, \ldots,$$

and

$$c_1 = \frac{4}{7}c_0, \qquad c_2 = \frac{4}{21}c_0, \qquad c_3 = \frac{32}{693}c_0,$$

and so on. The general solution on $(0, \infty)$ is

$$y = C_1\left(1 + \frac{1}{3}x - \frac{1}{6}x^2 - \frac{1}{6}x^3 + \cdots\right) + C_2 x^{5/2}\left(1 + \frac{4}{7}x + \frac{4}{21}x^2 + \frac{32}{693}x^3 + \cdots\right).$$

22. Substituting $y = \sum_{n=0}^{\infty} c_n x^{n+r}$ into the differential equation and collecting terms, we obtain

$$x^2 y'' + xy' + \left(x^2 - \frac{4}{9}\right)y = \left(r^2 - \frac{4}{9}\right)c_0 x^r + \left(r^2 + 2r + \frac{5}{9}\right)c_1 x^{r+1}$$

$$+ \sum_{k=2}^{\infty}\left[(k+r)(k+r-1)c_k + (k+r)c_k - \frac{4}{9}c_k + c_{k-2}\right]x^{k+r}$$

$$= 0,$$

Exercises 6.2 Solutions About Singular Points

which implies

$$r^2 - \frac{4}{9} = \left(r + \frac{2}{3}\right)\left(r - \frac{2}{3}\right) = 0,$$

$$\left(r^2 + 2r + \frac{5}{9}\right)c_1 = 0,$$

and

$$\left[(k+r)^2 - \frac{4}{9}\right]c_k + c_{k-2} = 0.$$

The indicial roots are $r = -2/3$ and $r = 2/3$, so $c_1 = 0$. For $r = -2/3$ the recurrence relation is

$$c_k = -\frac{9c_{k-2}}{3k(3k-4)}, \quad k = 2, 3, 4, \ldots,$$

and

$$c_2 = -\frac{3}{4}c_0, \qquad c_3 = 0, \qquad c_4 = \frac{9}{128}c_0,$$

and so on. For $r = 2/3$ the recurrence relation is

$$c_k = -\frac{9c_{k-2}}{3k(3k+4)}, \quad k = 2, 3, 4, \ldots,$$

and

$$c_2 = -\frac{3}{20}c_0, \qquad c_3 = 0, \qquad c_4 = \frac{9}{1,280}c_0,$$

and so on. The general solution on $(0, \infty)$ is

$$y = C_1 x^{-2/3}\left(1 - \frac{3}{4}x^2 + \frac{9}{128}x^4 + \cdots\right) + C_2 x^{2/3}\left(1 - \frac{3}{20}x^2 + \frac{9}{1,280}x^4 + \cdots\right).$$

23. Substituting $y = \sum_{n=0}^{\infty} c_n x^{n+r}$ into the differential equation and collecting terms, we obtain

$$9x^2y'' + 9x^2y' + 2y = \left(9r^2 - 9r + 2\right)c_0 x^r$$

$$+ \sum_{k=1}^{\infty}[9(k+r)(k+r-1)c_k + 2c_k + 9(k+r-1)c_{k-1}]x^{k+r}$$

$$= 0,$$

which implies

$$9r^2 - 9r + 2 = (3r-1)(3r-2) = 0$$

and

$$[9(k+r)(k+r-1) + 2]c_k + 9(k+r-1)c_{k-1} = 0.$$

The indicial roots are $r = 1/3$ and $r = 2/3$. For $r = 1/3$ the recurrence relation is

$$c_k = -\frac{(3k-2)c_{k-1}}{k(3k-1)}, \quad k = 1, 2, 3, \ldots,$$

306

and

$$c_1 = -\frac{1}{2}c_0, \qquad c_2 = \frac{1}{5}c_0, \qquad c_3 = -\frac{7}{120}c_0,$$

and so on. For $r = 2/3$ the recurrence relation is

$$c_k = -\frac{(3k-1)c_{k-1}}{k(3k+1)}, \qquad k = 1, 2, 3, \ldots,$$

and

$$c_1 = -\frac{1}{2}c_0, \qquad c_2 = \frac{5}{28}c_0, \qquad c_3 = -\frac{1}{21}c_0,$$

and so on. The general solution on $(0, \infty)$ is

$$y = C_1 x^{1/3}\left(1 - \frac{1}{2}x + \frac{1}{5}x^2 - \frac{7}{120}x^3 + \cdots\right) + C_2 x^{2/3}\left(1 - \frac{1}{2}x + \frac{5}{28}x^2 - \frac{1}{21}x^3 + \cdots\right).$$

24. Substituting $y = \sum_{n=0}^{\infty} c_n x^{n+r}$ into the differential equation and collecting terms, we obtain

$$2x^2 y'' + 3xy' + (2x - 1)y = \left(2r^2 + r - 1\right)c_0 x^r$$

$$+ \sum_{k=1}^{\infty}[2(k+r)(k+r-1)c_k + 3(k+r)c_k - c_k + 2c_{k-1}]x^{k+r}$$

$$= 0,$$

which implies

$$2r^2 + r - 1 = (2r-1)(r+1) = 0$$

and

$$[(k+r)(2k+2r+1) - 1]c_k + 2c_{k-1} = 0.$$

The indicial roots are $r = -1$ and $r = 1/2$. For $r = -1$ the recurrence relation is

$$c_k = -\frac{2c_{k-1}}{k(2k-3)}, \qquad k = 1, 2, 3, \ldots,$$

and

$$c_1 = 2c_0, \qquad c_2 = -2c_0, \qquad c_3 = \frac{4}{9}c_0,$$

and so on. For $r = 1/2$ the recurrence relation is

$$c_k = -\frac{2c_{k-1}}{k(2k+3)}, \qquad k = 1, 2, 3, \ldots,$$

and

$$c_1 = -\frac{2}{5}c_0, \qquad c_2 = \frac{2}{35}c_0, \qquad c_3 = -\frac{4}{945}c_0,$$

and so on. The general solution on $(0, \infty)$ is

$$y = C_1 x^{-1}\left(1 + 2x - 2x^2 + \frac{4}{9}x^3 + \cdots\right) + C_2 x^{1/2}\left(1 - \frac{2}{5}x + \frac{2}{35}x^2 - \frac{4}{945}x^3 + \cdots\right).$$

25. Substituting $y = \sum_{n=0}^{\infty} c_n x^{n+r}$ into the differential equation and collecting terms, we obtain

$$xy'' + 2y' - xy = \left(r^2 + r\right) c_0 x^{r-1} + \left(r^2 + 3r + 2\right) c_1 x^r$$

$$+ \sum_{k=2}^{\infty} [(k+r)(k+r-1)c_k + 2(k+r)c_k - c_{k-2}]x^{k+r-1}$$

$$= 0,$$

which implies

$$r^2 + r = r(r+1) = 0,$$

$$\left(r^2 + 3r + 2\right) c_1 = 0,$$

and

$$(k+r)(k+r+1)c_k - c_{k-2} = 0.$$

The indicial roots are $r_1 = 0$ and $r_2 = -1$, so $c_1 = 0$. For $r_1 = 0$ the recurrence relation is

$$c_k = \frac{c_{k-2}}{k(k+1)}, \quad k = 2, 3, 4, \ldots,$$

and

$$c_2 = \frac{1}{3!}c_0$$

$$c_3 = c_5 = c_7 = \cdots = 0$$

$$c_4 = \frac{1}{5!}c_0$$

$$c_{2n} = \frac{1}{(2n+1)!}c_0.$$

For $r_2 = -1$ the recurrence relation is

$$c_k = \frac{c_{k-2}}{k(k-1)}, \quad k = 2, 3, 4, \ldots,$$

and

$$c_2 = \frac{1}{2!}c_0$$

$$c_3 = c_5 = c_7 = \cdots = 0$$

$$c_4 = \frac{1}{4!}c_0$$

$$c_{2n} = \frac{1}{(2n)!}c_0.$$

The general solution on $(0, \infty)$ is

$$y = C_1 \sum_{n=0}^{\infty} \frac{1}{(2n+1)!} x^{2n} + C_2 x^{-1} \sum_{n=0}^{\infty} \frac{1}{(2n)!} x^{2n}$$

$$= \frac{1}{x} \left[C_1 \sum_{n=0}^{\infty} \frac{1}{(2n+1)!} x^{2n+1} + C_2 \sum_{n=0}^{\infty} \frac{1}{(2n)!} x^{2n} \right]$$

$$= \frac{1}{x} [C_1 \sinh x + C_2 \cosh x].$$

26. Substituting $y = \sum_{n=0}^{\infty} c_n x^{n+r}$ into the differential equation and collecting terms, we obtain

$$x^2 y'' + xy' + \left(x^2 - \frac{1}{4} \right) y = \left(r^2 - \frac{1}{4} \right) c_0 x^r + \left(r^2 + 2r + \frac{3}{4} \right) c_1 x^{r+1}$$

$$+ \sum_{k=2}^{\infty} \left[(k+r)(k+r-1)c_k + (k+r)c_k - \frac{1}{4} c_k + c_{k-2} \right] x^{k+r}$$

$$= 0,$$

which implies

$$r^2 - \frac{1}{4} = \left(r - \frac{1}{2} \right) \left(r + \frac{1}{2} \right) = 0,$$

$$\left(r^2 + 2r + \frac{3}{4} \right) c_1 = 0,$$

and

$$\left[(k+r)^2 - \frac{1}{4} \right] c_k + c_{k-2} = 0.$$

The indicial roots are $r_1 = 1/2$ and $r_2 = -1/2$, so $c_1 = 0$. For $r_1 = 1/2$ the recurrence relation is

$$c_k = -\frac{c_{k-2}}{k(k+1)}, \quad k = 2, 3, 4, \dots,$$

and

$$c_2 = -\frac{1}{3!} c_0$$

$$c_3 = c_5 = c_7 = \cdots = 0$$

$$c_4 = \frac{1}{5!} c_0$$

$$c_{2n} = \frac{(-1)^n}{(2n+1)!} c_0.$$

For $r_2 = -1/2$ the recurrence relation is

$$c_k = -\frac{c_{k-2}}{k(k-1)}, \quad k = 2, 3, 4, \dots,$$

and

$$c_2 = -\frac{1}{2!}c_0$$

$$c_3 = c_5 = c_7 = \cdots = 0$$

$$c_4 = \frac{1}{4!}c_0$$

$$c_{2n} = \frac{(-1)^n}{(2n)!}c_0.$$

The general solution on $(0, \infty)$ is

$$y = C_1 x^{1/2} \sum_{n=0}^{\infty} \frac{(-1)^n}{(2n+1)!} x^{2n} + C_2 x^{-1/2} \sum_{n=0}^{\infty} \frac{(-1)^n}{(2n)!} x^{2n}$$

$$= C_1 x^{-1/2} \sum_{n=0}^{\infty} \frac{(-1)^n}{(2n+1)!} x^{2n+1} + C_2 x^{-1/2} \sum_{n=0}^{\infty} \frac{(-1)^n}{(2n)!} x^{2n}$$

$$= x^{-1/2}[C_1 \sin x + C_2 \cos x].$$

27. Substituting $y = \sum_{n=0}^{\infty} c_n x^{n+r}$ into the differential equation and collecting terms, we obtain

$$xy'' - xy' + y = \left(r^2 - r\right) c_0 x^{r-1} + \sum_{k=0}^{\infty} [(k+r+1)(k+r)c_{k+1} - (k+r)c_k + c_k]x^{k+r} = 0$$

which implies

$$r^2 - r = r(r-1) = 0$$

and

$$(k+r+1)(k+r)c_{k+1} - (k+r-1)c_k = 0.$$

The indicial roots are $r_1 = 1$ and $r_2 = 0$. For $r_1 = 1$ the recurrence relation is

$$c_{k+1} = \frac{kc_k}{(k+2)(k+1)}, \quad k = 0, 1, 2, \ldots, \qquad \Rightarrow c_1 = c_2 = \cdots = 0,$$

and one solution is $y_1 = c_0 x$. A second solution is

$$y_2 = x \int \frac{e^{-\int (-1)dx}}{x^2} dx = x \int \frac{e^x}{x^2} dx = x \int \frac{1}{x^2}\left(1 + x + \frac{1}{2}x^2 + \frac{1}{3!}x^3 + \cdots\right) dx$$

$$= x \int \left(\frac{1}{x^2} + \frac{1}{x} + \frac{1}{2} + \frac{1}{3!}x + \frac{1}{4!}x^2 + \cdots\right) dx = x\left[-\frac{1}{x} + \ln x + \frac{1}{2}x + \frac{1}{12}x^2 + \frac{1}{72}x^3 + \cdots\right]$$

$$= x\ln x - 1 + \frac{1}{2}x^2 + \frac{1}{12}x^3 + \frac{1}{72}x^4 + \cdots.$$

The general solution on $(0, \infty)$ is

$$y = C_1 x + C_2 y_2(x).$$

28. Substituting $y = \sum_{n=0}^{\infty} c_n x^{n+r}$ into the differential equation and collecting terms, we obtain

$$y'' + \frac{3}{x}y' - 2y = \left(r^2 + 2r\right)c_0 x^{r-2} + \left(r^2 + 4r + 3\right)c_1 x^{r-1}$$

$$+ \sum_{k=2}^{\infty} [(k+r)(k+r-1)c_k + 3(k+r)c_k - 2c_{k-2}]x^{k+r-2}$$

$$= 0,$$

which implies

$$r^2 + 2r = r(r+2) = 0$$

$$\left(r^2 + 4r + 3\right)c_1 = 0$$

$$(k+r)(k+r+2)c_k - 2c_{k-2} = 0.$$

The indicial roots are $r_1 = 0$ and $r_2 = -2$, so $c_1 = 0$. For $r_1 = 0$ the recurrence relation is

$$c_k = \frac{2c_{k-2}}{k(k+2)}, \quad k = 2, 3, 4, \ldots,$$

and

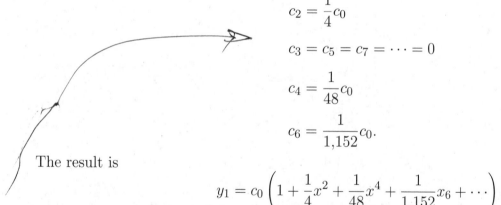

$$c_2 = \frac{1}{4}c_0$$

$$c_3 = c_5 = c_7 = \cdots = 0$$

$$c_4 = \frac{1}{48}c_0$$

$$c_6 = \frac{1}{1,152}c_0.$$

The result is

$$y_1 = c_0\left(1 + \frac{1}{4}x^2 + \frac{1}{48}x^4 + \frac{1}{1,152}x^6 + \cdots\right).$$

A second solution is

$$y_2 = y_1 \int \frac{e^{-\int (3/x)dx}}{y_1^2}\, dx = y_1 \int \frac{dx}{x^3\left(1 + \frac{1}{4}x^2 + \frac{1}{48}x^4 + \cdots\right)^2}$$

$$= y_1 \int \frac{dx}{x^3\left(1 + \frac{1}{2}x^2 + \frac{5}{48}x^4 + \frac{7}{576}x^6 + \cdots\right)} = y_1 \int \frac{1}{x^3}\left(1 - \frac{1}{2}x^2 + \frac{7}{48}x^4 + \frac{19}{576}x^6 + \cdots\right)dx$$

$$= y_1 \int \left(\frac{1}{x^3} - \frac{1}{2x} + \frac{7}{48}x - \frac{19}{576}x^3 + \cdots\right)dx = y_1\left[-\frac{1}{2x^2} - \frac{1}{2}\ln x + \frac{7}{96}x^2 - \frac{19}{2,304}x^4 + \cdots\right]$$

$$= -\frac{1}{2}y_1 \ln x + y\left[-\frac{1}{2x^2} + \frac{7}{96}x^2 - \frac{19}{2,304}x^4 + \cdots\right].$$

The general solution on $(0, \infty)$ is

$$y = C_1 y_1(x) + C_2 y_2(x).$$

29. Substituting $y = \sum_{n=0}^{\infty} c_n x^{n+r}$ into the differential equation and collecting terms, we obtain

$$xy'' + (1-x)y' - y = r^2 c_0 x^{r-1} + \sum_{k=1}^{\infty} [(k+r)(k+r-1)c_k + (k+r)c_k - (k+r)c_{k-1}]x^{k+r-1} = 0,$$

which implies $r^2 = 0$ and

$$(k+r)^2 c_k - (k+r)c_{k-1} = 0.$$

The indicial roots are $r_1 = r_2 = 0$ and the recurrence relation is

$$c_k = \frac{c_{k-1}}{k}, \quad k = 1, 2, 3, \ldots .$$

One solution is

$$y_1 = c_0 \left(1 + x + \frac{1}{2}x^2 + \frac{1}{3!}x^3 + \cdots \right) = c_0 e^x.$$

A second solution is

$$y_2 = y_1 \int \frac{e^{-\int (1/x - 1)dx}}{e^{2x}}\, dx = e^x \int \frac{e^x / x}{e^{2x}}\, dx = e^x \int \frac{1}{x} e^{-x}\, dx$$

$$= e^x \int \frac{1}{x}\left(1 - x + \frac{1}{2}x^2 - \frac{1}{3!}x^3 + \cdots \right) dx = e^x \int \left(\frac{1}{x} - 1 + \frac{1}{2}x - \frac{1}{3!}x^2 + \cdots \right) dx$$

$$= e^x \left[\ln x - x + \frac{1}{2 \cdot 2}x^2 - \frac{1}{3 \cdot 3!}x^3 + \cdots \right] = e^x \ln x - e^x \sum_{n=1}^{\infty} \frac{(-1)^{n+1}}{n \cdot n!}x^n.$$

The general solution on $(0, \infty)$ is

$$y = C_1 e^x + C_2 e^x \left(\ln x - \sum_{n=1}^{\infty} \frac{(-1)^{n+1}}{n \cdot n!}x^n \right).$$

30. Substituting $y = \sum_{n=0}^{\infty} c_n x^{n+r}$ into the differential equation and collecting terms, we obtain

$$xy'' + y' + y = r^2 c_0 x^{r-1} + \sum_{k=1}^{\infty} [(k+r)(k+r-1)c_k + (k+r)c_k + c_{k-1}]x^{k+r-1} = 0$$

which implies $r^2 = 0$ and

$$(k+r)^2 c_k + c_{k-1} = 0.$$

The indicial roots are $r_1 = r_2 = 0$ and the recurrence relation is

$$c_k = -\frac{c_{k-1}}{k^2}, \quad k = 1, 2, 3, \ldots .$$

One solution is

$$y_1 = c_0 \left(1 - x + \frac{1}{2^2}x^2 - \frac{1}{(3!)^2}x^3 + \frac{1}{(4!)^2}x^4 - \cdots \right) = c_0 \sum_{n=0}^{\infty} \frac{(-1)^n}{(n!)^2}x^n.$$

A second solution is

$$y_2 = y_1 \int \frac{e^{-\int (1/x)dx}}{y_1^2}\, dx = y_1 \int \frac{dx}{x\left(1 - x + \frac{1}{4}x^2 - \frac{1}{36}x^3 + \cdots\right)^2}$$

$$= y_1 \int \frac{dx}{x\left(1 - 2x + \frac{3}{2}x^2 - \frac{5}{9}x^3 + \frac{35}{288}x^4 - \cdots\right)}$$

$$= y_1 \int \frac{1}{x}\left(1 + 2x + \frac{5}{2}x^2 + \frac{23}{9}x^3 + \frac{677}{288}x^4 + \cdots\right)dx$$

$$= y_1 \int \left(\frac{1}{x} + 2 + \frac{5}{2}x + \frac{23}{9}x^2 + \frac{677}{288}x^3 + \cdots\right)dx$$

$$= y_1 \left[\ln x + 2x + \frac{5}{4}x^2 + \frac{23}{27}x^3 + \frac{677}{1,152}x^4 + \cdots\right]$$

$$= y_1 \ln x + y_1 \left(2x + \frac{5}{4}x^2 + \frac{23}{27}x^3 + \frac{677}{1,152}x^4 + \cdots\right).$$

The general solution on $(0, \infty)$ is

$$y = C_1 y_1(x) + C_2 y_2(x).$$

31. Substituting $y = \sum_{n=0}^{\infty} c_n x^{n+r}$ into the differential equation and collecting terms, we obtain

$$xy'' + (x - 6)y' - 3y = (r^2 - 7r)c_0 x^{r-1} + \sum_{k=1}^{\infty} [(k+r)(k+r-1)c_k + (k+r-1)c_{k-1}$$

$$- 6(k+r)c_k - 3c_{k-1}]x^{k+r-1} = 0,$$

which implies

$$r^2 - 7r = r(r - 7) = 0$$

and

$$(k+r)(k+r-7)c_k + (k+r-4)c_{k-1} = 0.$$

The indicial roots are $r_1 = 7$ and $r_2 = 0$. For $r_1 = 7$ the recurrence relation is

$$(k+7)kc_k + (k+3)c_{k-1} = 0, \quad k = 1, 2, 3, \ldots,$$

or

$$c_k = -\frac{k+3}{k(k+7)}c_{k-1}, \quad k = 1, 2, 3, \ldots.$$

Taking $c_0 \neq 0$ we obtain

$$c_1 = -\frac{1}{2}c_0$$

$$c_2 = \frac{5}{18}c_0$$

$$c_3 = -\frac{1}{6}c_0,$$

and so on. Thus, the indicial root $r_1 = 7$ yields a single solution. Now, for $r_2 = 0$ the recurrence relation is

$$k(k-7)c_k + (k-4)c_{k-1} = 0, \quad k = 1, 2, 3, \ldots .$$

Then

$$-6c_1 - 3c_0 = 0$$

$$-10c_2 - 2c_1 = 0$$

$$-12c_3 - c_2 = 0$$

$$-12c_4 + 0c_3 = 0 \implies c_4 = 0$$

$$-10c_5 + c_4 = 0 \implies c_5 = 0$$

$$-6c_6 + 2c_5 = 0 \implies c_6 = 0$$

$$0c_7 + 3c_6 = 0 \implies c_7 \text{ is arbitrary}$$

and

$$c_k = -\frac{k-4}{k(k-7)}c_{k-1}, \quad k = 8, 9, 10, \ldots .$$

Taking $c_0 \neq 0$ and $c_7 = 0$ we obtain

$$c_1 = -\frac{1}{2}c_0$$

$$c_2 = \frac{1}{10}c_0$$

$$c_3 = -\frac{1}{120}c_0$$

$$c_4 = c_5 = c_6 = \cdots = 0.$$

Taking $c_0 = 0$ and $c_7 \neq 0$ we obtain

$$c_1 = c_2 = c_3 = c_4 = c_5 = c_6 = 0$$

$$c_8 = -\frac{1}{2}c_7$$

$$c_9 = \frac{5}{36}c_7$$

$$c_{10} = -\frac{1}{36}c_7,$$

and so on. In this case we obtain the two solutions

$$y_1 = 1 - \frac{1}{2}x + \frac{1}{10}x^2 - \frac{1}{120}x^3 \quad \text{and} \quad y_2 = x^7 - \frac{1}{2}x^8 + \frac{5}{36}x^9 - \frac{1}{36}x^{10} + \cdots.$$

32. Substituting $y = \sum_{n=0}^{\infty} c_n x^{n+r}$ into the differential equation and collecting terms, we obtain

$$x(x-1)y'' + 3y' - 2y$$

$$= \left(4r - r^2\right)c_0 x^{r-1} + \sum_{k=1}^{\infty}[(k+r-1)(k+r-12)c_{k-1} - (k+r)(k+r-1)c_k$$

$$+ 3(k+r)c_k - 2c_{k-1}]x^{k+r-1}$$

$$= 0,$$

which implies

$$4r - r^2 = r(4 - r) = 0$$

and

$$-(k+r)(k+r-4)c_k + [(k+r-1)(k+r-2) - 2]c_{k-1} = 0.$$

The indicial roots are $r_1 = 4$ and $r_2 = 0$. For $r_1 = 4$ the recurrence relation is

$$-(k+4)kc_k + [(k+3)(k+2) - 2]c_{k-1} = 0$$

or

$$c_k = \frac{k+1}{k}c_{k-1}, \quad k = 1, 2, 3, \ldots.$$

Taking $c_0 \neq 0$ we obtain

$$c_1 = 2c_0$$

$$c_2 = 3c_0$$

$$c_3 = 4c_0,$$

and so on. Thus, the indicial root $r_1 = 4$ yields a single solution. For $r_2 = 0$ the recurrence relation is

$$-k(k-4)c_k + k(k-3)c_{k-1} = 0, \quad k = 1, 2, 3, \ldots,$$

or

$$-(k-4)c_k + (k-3)c_{k-1} = 0, \quad k = 1, 2, 3, \ldots.$$

Then

$$3c_1 - 2c_0 = 0$$

$$2c_2 - c_1 = 0$$

$$c_3 + 0c_2 = 0 \quad \Rightarrow \quad c_3 = 0$$

$$0c_4 + c_3 = 0 \quad \Rightarrow \quad c_4 \text{ is arbitrary}$$

and

$$c_k = \frac{(k-3)c_{k-1}}{k-4}, \quad k = 5, 6, 7, \ldots.$$

Taking $c_0 \neq 0$ and $c_4 = 0$ we obtain

$$c_1 = \frac{2}{3}c_0$$

$$c_2 = \frac{1}{3}c_0$$

$$c_3 = c_4 = c_5 = \cdots = 0.$$

Taking $c_0 = 0$ and $c_4 \neq 0$ we obtain

$$c_1 = c_2 = c_3 = 0$$

$$c_5 = 2c_4$$

$$c_6 = 3c_4$$

$$c_7 = 4c_4,$$

and so on. In this case we obtain the two solutions

$$y_1 = 1 + \frac{2}{3}x + \frac{1}{3}x^2 \quad \text{and} \quad y_2 = x^4 + 2x^5 + 3x^6 + 4x^7 + \cdots.$$

33. (a) From $t = 1/x$ we have $dt/dx = -1/x^2 = -t^2$. Then

$$\frac{dy}{dx} = \frac{dy}{dt}\frac{dt}{dx} = -t^2\frac{dy}{dt}$$

and

$$\frac{d^2y}{dx^2} = \frac{d}{dx}\left(\frac{dy}{dx}\right) = \frac{d}{dx}\left(-t^2\frac{dy}{dt}\right) = -t^2\frac{d^2y}{dt^2}\frac{dt}{dx} - \frac{dy}{dt}\left(2t\frac{dt}{dx}\right) = t^4\frac{d^2y}{dt^2} + 2t^3\frac{dy}{dt}.$$

Now

$$x^4 \frac{d^2y}{dx^2} + \lambda y = \frac{1}{t^4}\left(t^4 \frac{d^2y}{dt^2} + 2t^3 \frac{dy}{dt}\right) + \lambda y = \frac{d^2y}{dt^2} + \frac{2}{t}\frac{dy}{dt} + \lambda y = 0$$

becomes

$$t\frac{d^2y}{dt^2} + 2\frac{dy}{dt} + \lambda t y = 0.$$

(b) Substituting $y = \sum_{n=0}^{\infty} c_n t^{n+r}$ into the differential equation and collecting terms, we obtain

$$t\frac{d^2y}{dt^2} + 2\frac{dy}{dt} + \lambda t y = (r^2 + r)c_0 t^{r-1} + (r^2 + 3r + 2)c_1 t^r$$

$$+ \sum_{k=2}^{\infty}[(k+r)(k+r-1)c_k + 2(k+r)c_k + \lambda c_{k-2}]t^{k+r-1}$$

$$= 0,$$

which implies

$$r^2 + r = r(r+1) = 0,$$

$$\left(r^2 + 3r + 2\right)c_1 = 0,$$

and

$$(k+r)(k+r+1)c_k + \lambda c_{k-2} = 0.$$

The indicial roots are $r_1 = 0$ and $r_2 = -1$, so $c_1 = 0$. For $r_1 = 0$ the recurrence relation is

$$c_k = -\frac{\lambda c_{k-2}}{k(k+1)}, \quad k = 2, 3, 4, \ldots,$$

and

$$c_2 = -\frac{\lambda}{3!}c_0$$

$$c_3 = c_5 = c_7 = \cdots = 0$$

$$c_4 = \frac{\lambda^2}{5!}c_0$$

$$\vdots$$

$$c_{2n} = (-1)^n \frac{\lambda^n}{(2n+1)!}c_0.$$

For $r_2 = -1$ the recurrence relation is

$$c_k = -\frac{\lambda c_{k-2}}{k(k-1)}, \quad k = 2, 3, 4, \ldots,$$

317

and

$$C_2 = -\frac{\lambda}{2!}C_0$$

$$C_3 = C_5 = C_7 = \cdots = 0$$

$$C_4 = \frac{\lambda^2}{4!}C_0$$

$$\vdots$$

$$C_{2n} = (-1)^n \frac{\lambda^n}{(2n)!}C_0.$$

The general solution on $(0, \infty)$ is

$$y(t) = c_1 \sum_{n=0}^{\infty} \frac{(-1)^n}{(2n+1)!}(\sqrt{\lambda}\,t)^{2n} + c_2 t^{-1} \sum_{n=0}^{\infty} \frac{(-1)^n}{(2n)!}(\sqrt{\lambda}\,t)^{2n}$$

$$= \frac{1}{t}\left[C_1 \sum_{n=0}^{\infty} \frac{(-1)^n}{(2n+1)!}(\sqrt{\lambda}\,t)^{2n+1} + C_2 \sum_{n=0}^{\infty} \frac{(-1)^n}{(2n)!}(\sqrt{\lambda}\,t)^{2n} \right]$$

$$= \frac{1}{t}[C_1 \sin\sqrt{\lambda}\,t + C_2 \cos\sqrt{\lambda}\,t].$$

(c) Using $t = 1/x$, the solution of the original equation is

$$y(x) = C_1 x \sin\frac{\sqrt{\lambda}}{x} + C_2 x \cos\frac{\sqrt{\lambda}}{x}.$$

34. (a) From the boundary conditions $y(a) = 0$, $y(b) = 0$ we find

$$C_1 \sin\frac{\sqrt{\lambda}}{a} + C_2 \cos\frac{\sqrt{\lambda}}{a} = 0$$

$$C_1 \sin\frac{\sqrt{\lambda}}{b} + C_2 \cos\frac{\sqrt{\lambda}}{b} = 0.$$

Since this is a homogeneous system of linear equations, it will have nontrivial solutions for C_1 and C_2 if

$$\begin{vmatrix} \sin\dfrac{\sqrt{\lambda}}{a} & \cos\dfrac{\sqrt{\lambda}}{a} \\[2mm] \sin\dfrac{\sqrt{\lambda}}{b} & \cos\dfrac{\sqrt{\lambda}}{b} \end{vmatrix} = \sin\frac{\sqrt{\lambda}}{a}\cos\frac{\sqrt{\lambda}}{b} - \cos\frac{\sqrt{\lambda}}{a}\sin\frac{\sqrt{\lambda}}{b}$$

$$= \sin\left(\frac{\sqrt{\lambda}}{a} - \frac{\sqrt{\lambda}}{b}\right) = \sin\left(\sqrt{\lambda}\,\frac{b-a}{ab}\right) = 0.$$

This will be the case if

$$\sqrt{\lambda}\left(\frac{b-a}{ab}\right) = n\pi \qquad \text{or} \qquad \sqrt{\lambda} = \frac{n\pi ab}{b-a} = \frac{n\pi ab}{L}, \quad n = 1, 2, \ldots,$$

or, if

$$\lambda_n = \frac{n^2\pi^2 a^2 b^2}{L^2} = \frac{P_n b^4}{EI}.$$

The critical loads are then $P_n = n^2\pi^2(a/b)^2 EI_0/L^2$. Using $C_2 = -C_1 \sin(\sqrt{\lambda}/a)/\cos(\sqrt{\lambda}/a)$ we have

$$y = C_1 x\left[\sin\frac{\sqrt{\lambda}}{x} - \frac{\sin(\sqrt{\lambda}/a)}{\cos(\sqrt{\lambda}/a)}\cos\frac{\sqrt{\lambda}}{x}\right]$$

$$= C_3 x\left[\sin\frac{\sqrt{\lambda}}{x}\cos\frac{\sqrt{\lambda}}{a} - \cos\frac{\sqrt{\lambda}}{x}\sin\frac{\sqrt{\lambda}}{a}\right]$$

$$= C_3 x\sin\sqrt{\lambda}\left(\frac{1}{x} - \frac{1}{a}\right),$$

and

$$y_n(x) = C_3 x\sin\frac{n\pi ab}{L}\left(\frac{1}{x} - \frac{1}{a}\right) = C_3 x\sin\frac{n\pi ab}{La}\left(\frac{a}{x} - 1\right) = C_4 x\sin\frac{n\pi ab}{L}\left(1 - \frac{a}{x}\right).$$

(b) When $n = 1$, $b = 11$, and $a = 1$, we have, for $C_4 = 1$,

$$y_1(x) = x\sin 1.1\pi\left(1 - \frac{1}{x}\right).$$

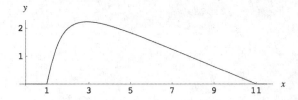

35. Express the differential equation in standard form:

$$y''' + P(x)y'' + Q(x)y' + R(x)y = 0.$$

Suppose x_0 is a singular point of the differential equation. Then we say that x_0 is a regular singular point if $(x - x_0)P(x)$, $(x - x_0)^2 Q(x)$, and $(x - x_0)^3 R(x)$ are analytic at $x = x_0$.

36. Substituting $y = \sum_{n=0}^{\infty} c_n x^{n+r}$ into the first differential equation and collecting terms, we obtain

$$x^3 y'' + y = c_0 x^r + \sum_{k=1}^{\infty}[c_k + (k+r-1)(k+r-2)c_{k-1}]x^{k+r} = 0.$$

It follows that $c_0 = 0$ and

$$c_k = -(k+r-1)(k+r-2)c_{k-1}.$$

The only solution we obtain is $y(x) = 0$.

Substituting $y = \sum_{n=0}^{\infty} c_n x^{n+r}$ into the second differential equation and collecting terms, we obtain

$$x^2 y'' + (3x-1)y' + y = -rc_0 + \sum_{k=0}^{\infty}[(k+r+1)^2 c_k - (k+r+1)c_{k+1}]x^{k+r} = 0,$$

which implies

$$-rc_0 = 0$$

$$(k+r+1)^2 c_k - (k+r+1)c_{k+1} = 0.$$

If $c_0 = 0$, then the solution of the differential equation is $y = 0$. Thus, we take $r = 0$, from which we obtain

$$c_{k+1} = (k+1)c_k, \quad k = 0, 1, 2, \ldots .$$

Letting $c_0 = 1$ we get $c_1 = 2$, $c_2 = 3!$, $c_3 = 4!$, and so on. The solution of the differential equation is then $y = \sum_{n=0}^{\infty} (n+1)! x^n$, which converges only at $x = 0$.

37. We write the differential equation in the form $x^2 y'' + (b/a)xy' + (c/a)y = 0$ and identify $a_0 = b/a$ and $b_0 = c/a$ as in (12) in the text. Then the indicial equation is

$$r(r-1) + \frac{b}{a} r + \frac{c}{a} = 0 \quad \text{or} \quad ar^2 + (b-a)r + c = 0,$$

which is also the auxiliary equation of $ax^2 y'' + bxy' + cy = 0$.

Exercises 6.3 Special Functions

1. Since $\nu^2 = 1/9$ the general solution is $y = c_1 J_{1/3}(x) + c_2 J_{-1/3}(x)$.

2. Since $\nu^2 = 1$ the general solution is $y = c_1 J_1(x) + c_2 Y_1(x)$.

3. Since $\nu^2 = 25/4$ the general solution is $y = c_1 J_{5/2}(x) + c_2 J_{-5/2}(x)$.

4. Since $\nu^2 = 1/16$ the general solution is $y = c_1 J_{1/4}(x) + c_2 J_{-1/4}(x)$.

5. Since $\nu^2 = 0$ the general solution is $y = c_1 J_0(x) + c_2 Y_0(x)$.

6. Since $\nu^2 = 4$ the general solution is $y = c_1 J_2(x) + c_2 Y_2(x)$.

7. We identify $\alpha = 3$ and $\nu = 2$. Then the general solution is $y = c_1 J_2(3x) + c_2 Y_2(3x)$.

8. We identify $\alpha = 6$ and $\nu = \frac{1}{2}$. Then the general solution is $y = c_1 J_{1/2}(6x) + c_2 J_{-1/2}(6x)$.

9. We identify $\alpha = 5$ and $\nu = \frac{2}{3}$. Then the general solution is $y = c_1 J_{2/3}(5x) + c_2 J_{-2/3}(5x)$.

10. We identify $\alpha = \sqrt{2}$ and $\nu = 8$. Then the general solution is $y = c_1 J_8(\sqrt{2}x) + c_2 Y_8(\sqrt{2}x)$.

11. If $y = x^{-1/2} v(x)$ then

$$y' = x^{-1/2} v'(x) - \frac{1}{2} x^{-3/2} v(x),$$

$$y'' = x^{-1/2} v''(x) - x^{-3/2} v'(x) + \frac{3}{4} x^{-5/2} v(x),$$

and

$$x^2 y'' + 2xy' + \alpha^2 x^2 y = x^{3/2} v''(x) + x^{1/2} v'(x) + \left(\alpha^2 x^{3/2} - \frac{1}{4} x^{-1/2}\right) v(x) = 0.$$

Multiplying by $x^{1/2}$ we obtain

$$x^2 v''(x) + x v'(x) + \left(\alpha^2 x^2 - \frac{1}{4}\right) v(x) = 0,$$

whose solution is $v = c_1 J_{1/2}(\alpha x) + c_2 J_{-1/2}(\alpha x)$. Then $y = c_1 x^{-1/2} J_{1/2}(\alpha x) + c_2 x^{-1/2} J_{-1/2}(\alpha x)$.

12. If $y = \sqrt{x}\, v(x)$ then

$$y' = x^{1/2} v'(x) + \frac{1}{2} x^{-1/2} v(x)$$

$$y'' = x^{1/2} v''(x) + x^{-1/2} v'(x) - \frac{1}{4} x^{-3/2} v(x)$$

and

$$x^2 y'' + \left(\alpha^2 x^2 - \nu^2 + \frac{1}{4}\right) y = x^{5/2} v''(x) + x^{3/2} v'(x) - \frac{1}{4} x^{1/2} v(x) + \left(\alpha^2 x^2 - \nu^2 + \frac{1}{4}\right) x^{1/2} v(x)$$

$$= x^{5/2} v''(x) + x^{3/2} v'(x) + (\alpha^2 x^{5/2} - \nu^2 x^{1/2}) v(x) = 0.$$

Multiplying by $x^{-1/2}$ we obtain

$$x^2 v''(x) + x v'(x) + (\alpha^2 x^2 - \nu^2) v(x) = 0,$$

whose solution is $v(x) = c_1 J_\nu(\alpha x) + c_2 Y_\nu(\alpha x)$. Then $y = c_1 \sqrt{x}\, J_\nu(\alpha x) + c_2 \sqrt{x}\, Y_\nu(\alpha x)$.

13. Write the differential equation in the form $y'' + (2/x)y' + (4/x)y = 0$. This is the form of (18) in the text with $a = -\frac{1}{2}$, $c = \frac{1}{2}$, $b = 4$, and $p = 1$, so, by (19) in the text, the general solution is

$$y = x^{-1/2}[c_1 J_1(4x^{1/2}) + c_2 Y_1(4x^{1/2})].$$

14. Write the differential equation in the form $y'' + (3/x)y' + y = 0$. This is the form of (18) in the text with $a = -1$, $c = 1$, $b = 1$, and $p = 1$, so, by (19) in the text, the general solution is

$$y = x^{-1}[c_1 J_1(x) + c_2 Y_1(x)].$$

15. Write the differential equation in the form $y'' - (1/x)y' + y = 0$. This is the form of (18) in the text with $a = 1$, $c = 1$, $b = 1$, and $p = 1$, so, by (19) in the text, the general solution is

$$y = x[c_1 J_1(x) + c_2 Y_1(x)].$$

16. Write the differential equation in the form $y'' - (5/x)y' + y = 0$. This is the form of (18) in the text with $a = 3$, $c = 1$, $b = 1$, and $p = 2$, so, by (19) in the text, the general solution is

$$y = x^3[c_1 J_3(x) + c_2 Y_3(x)].$$

321

17. Write the differential equation in the form $y'' + (1 - 2/x^2)y = 0$. This is the form of (18) in the text with $a = \frac{1}{2}$, $c = 1$, $b = 1$, and $p = \frac{3}{2}$, so, by (19) in the text, the general solution is

$$y = x^{1/2}[c_1 J_{3/2}(x) + c_2 Y_{3/2}(x)] = x^{1/2}[C_1 J_{3/2}(x) + C_2 J_{-3/2}(x)].$$

18. Write the differential equation in the form $y'' + (4 + 1/4x^2)y = 0$. This is the form of (18) in the text with $a = \frac{1}{2}$, $c = 1$, $b = 2$, and $p = 0$, so, by (19) in the text, the general solution is

$$y = x^{1/2}[c_1 J_0(2x) + c_2 Y_0(2x)].$$

19. Write the differential equation in the form $y'' + (3/x)y' + x^2 y = 0$. This is the form of (18) in the text with $a = -1$, $c = 2$, $b = \frac{1}{2}$, and $p = \frac{1}{2}$, so, by (19) in the text, the general solution is

$$y = x^{-1}\left[c_1 J_{1/2}\left(\frac{1}{2}x^2\right) + c_2 Y_{1/2}\left(\frac{1}{2}x^2\right)\right]$$

or

$$y = x^{-1}\left[C_1 J_{1/2}\left(\frac{1}{2}x^2\right) + C_2 J_{-1/2}\left(\frac{1}{2}x^2\right)\right].$$

20. Write the differential equation in the form $y'' + (1/x)y' + (\frac{1}{9}x^4 - 4/x^2)y = 0$. This is the form of (18) in the text with $a = 0$, $c = 3$, $b = \frac{1}{9}$, and $p = \frac{2}{3}$, so, by (19) in the text, the general solution is

$$y = c_1 J_{2/3}\left(\frac{1}{9}x^3\right) + c_2 Y_{2/3}\left(\frac{1}{9}x^3\right)$$

or

$$y = C_1 J_{2/3}\left(\frac{1}{9}x^3\right) + C_2 J_{-2/3}\left(\frac{1}{9}x^3\right).$$

21. Using the fact that $i^2 = -1$, along with the definition of $J_\nu(x)$ in (7) in the text, we have

$$I_\nu(x) = i^{-\nu} J_\nu(ix) = i^{-\nu} \sum_{n=0}^{\infty} \frac{(-1)^n}{n!\Gamma(1 + \nu + n)}\left(\frac{ix}{2}\right)^{2n+\nu}$$

$$= \sum_{n=0}^{\infty} \frac{(-1)^n}{n!\Gamma(1 + \nu + n)} i^{2n+\nu-\nu}\left(\frac{x}{2}\right)^{2n+\nu}$$

$$= \sum_{n=0}^{\infty} \frac{(-1)^n}{n!\Gamma(1 + \nu + n)} (i^2)^n\left(\frac{x}{2}\right)^{2n+\nu}$$

$$= \sum_{n=0}^{\infty} \frac{(-1)^{2n}}{n!\Gamma(1 + \nu + n)}\left(\frac{x}{2}\right)^{2n+\nu}$$

$$= \sum_{n=0}^{\infty} \frac{1}{n!\Gamma(1 + \nu + n)}\left(\frac{x}{2}\right)^{2n+\nu},$$

which is a real function.

22. (a) The differential equation has the form of (18) in the text with

$$1 - 2a = 0 \implies a = \frac{1}{2}$$

$$2c - 2 = 2 \implies c = 2$$

$$b^2 c^2 = -\beta^2 c^2 = -1 \implies \beta = \frac{1}{2} \quad \text{and} \quad b = \frac{1}{2}i$$

$$a^2 - p^2 c^2 = 0 \implies p = \frac{1}{4}.$$

Then, by (19) in the text,

$$y = x^{1/2} \left[c_1 J_{1/4} \left(\frac{1}{2} i x^2 \right) + c_2 J_{-1/4} \left(\frac{1}{2} i x^2 \right) \right].$$

In terms of real functions the general solution can be written

$$y = x^{1/2} \left[C_1 I_{1/4} \left(\frac{1}{2} x^2 \right) + C_2 K_{1/4} \left(\frac{1}{2} x^2 \right) \right].$$

(b) Write the differential equation in the form $y'' + (1/x)y' - 7x^2 y = 0$. This is the form of (18) in the text with

$$1 - 2a = 1 \implies a = 0$$

$$2c - 2 = 2 \implies c = 2$$

$$b^2 c^2 = -\beta^2 c^2 = -7 \implies \beta = \frac{1}{2}\sqrt{7} \quad \text{and} \quad b = \frac{1}{2}\sqrt{7}\,i$$

$$a^2 - p^2 c^2 = 0 \implies p = 0.$$

Then, by (19) in the text,

$$y = c_1 J_0 \left(\frac{1}{2}\sqrt{7}\,ix^2 \right) + c_2 Y_0 \left(\frac{1}{2}\sqrt{7}\,ix^2 \right).$$

In terms of real functions the general solution can be written

$$y = C_1 I_0 \left(\frac{1}{2}\sqrt{7}x^2 \right) + C_2 K_0 \left(\frac{1}{2}\sqrt{7}x^2 \right).$$

23. The differential equation has the form of (18) in the text with

$$1 - 2a = 0 \implies a = \frac{1}{2}$$

$$2c - 2 = 0 \implies c = 1$$

$$b^2 c^2 = 1 \implies b = 1$$

$$a^2 - p^2 c^2 = 0 \implies p = \frac{1}{2}.$$

Then, by (19) in the text,

$$y = x^{1/2}[c_1 J_{1/2}(x) + c_2 J_{-1/2}(x)] = x^{1/2}\left[c_1\sqrt{\frac{2}{\pi x}}\sin x + c_2\sqrt{\frac{2}{\pi x}}\cos x\right] = C_1\sin x + C_2\cos x.$$

24. Write the differential equation in the form $y'' + (4/x)y' + (1 + 2/x^2)y = 0$. This is the form of (18) in the text with

$$1 - 2a = 4 \implies a = -\frac{3}{2}$$

$$2c - 2 = 0 \implies c = 1$$

$$b^2 c^2 = 1 \implies b = 1$$

$$a^2 - p^2 c^2 = 2 \implies p = \frac{1}{2}.$$

Then, by (19), (23), and (24) in the text,

$$y = x^{-3/2}[c_1 J_{1/2}(x) + c_2 J_{-1/2}(x)] = x^{-3/2}\left[c_1\sqrt{\frac{2}{\pi x}}\sin x + c_2\sqrt{\frac{2}{\pi x}}\cos x\right]$$

$$= C_1\frac{1}{x^2}\sin x + C_2\frac{1}{x^2}\cos x.$$

25. Write the differential equation in the form $y'' + (2/x)y' + (\frac{1}{16}x^2 - 3/4x^2)y = 0$. This is the form of (18) in the text with

$$1 - 2a = 2 \implies a = -\frac{1}{2}$$

$$2c - 2 = 2 \implies c = 2$$

$$b^2 c^2 = \frac{1}{16} \implies b = \frac{1}{8}$$

$$a^2 - p^2 c^2 = -\frac{3}{4} \implies p = \frac{1}{2}.$$

Then, by (19) in the text,

$$y = x^{-1/2}\left[c_1 J_{1/2}\left(\frac{1}{8}x^2\right) + c_2 J_{-1/2}\left(\frac{1}{8}x^2\right)\right]$$

$$= x^{-1/2}\left[c_1\sqrt{\frac{16}{\pi x^2}}\sin\left(\frac{1}{8}x^2\right) + c_2\sqrt{\frac{16}{\pi x^2}}\cos\left(\frac{1}{8}x^2\right)\right]$$

$$= C_1 x^{-3/2}\sin\left(\frac{1}{8}x^2\right) + C_2 x^{-3/2}\cos\left(\frac{1}{8}x^2\right).$$

26. Write the differential equation in the form $y'' - (1/x)y' + (4 + 3/4x^2)y = 0$. This is the form of (18)

in the text with

$$1 - 2a = -1 \implies a = 1$$

$$2c - 2 = 0 \implies c = 1$$

$$b^2 c^2 = 4 \implies b = 2$$

$$a^2 - p^2 c^2 = \frac{3}{4} \implies p = \frac{1}{2}.$$

Then, by (19) in the text,

$$y = x[c_1 J_{1/2}(2x) + c_2 J_{-1/2}(2x)] = x\left[c_1 \sqrt{\frac{2}{\pi 2x}} \sin 2x + c_2 \sqrt{\frac{2}{\pi 2x}} \cos 2x\right]$$

$$= C_1 x^{1/2} \sin 2x + C_2 x^{1/2} \cos 2x.$$

27. (a) The recurrence relation follows from

$$-\nu J_\nu(x) + x J_{\nu-1}(x) = -\sum_{n=0}^{\infty} \frac{(-1)^n \nu}{n!\Gamma(1+\nu+n)} \left(\frac{x}{2}\right)^{2n+\nu} + x\sum_{n=0}^{\infty} \frac{(-1)^n}{n!\Gamma(\nu+n)} \left(\frac{x}{2}\right)^{2n+\nu-1}$$

$$= -\sum_{n=0}^{\infty} \frac{(-1)^n \nu}{n!\Gamma(1+\nu+n)} \left(\frac{x}{2}\right)^{2n+\nu} + \sum_{n=0}^{\infty} \frac{(-1)^n (\nu+n)}{n!\Gamma(1+\nu+n)} \cdot 2 \left(\frac{x}{2}\right)\left(\frac{x}{2}\right)^{2n+\nu-1}$$

$$= \sum_{n=0}^{\infty} \frac{(-1)^n (2n+\nu)}{n!\Gamma(1+\nu+n)} \left(\frac{x}{2}\right)^{2n+\nu} = x J_\nu'(x).$$

(b) The formula in part (a) is a linear first-order differential equation in $J_\nu(x)$. An integrating factor for this equation is x^ν, so

$$\frac{d}{dx}[x^\nu J_\nu(x)] = x^\nu J_{\nu-1}(x).$$

28. Subtracting the formula in part (a) of Problem 27 from the formula in Example 5 we obtain

$$0 = 2\nu J_\nu(x) - x J_{\nu+1}(x) - x J_{\nu-1}(x) \quad \text{or} \quad 2\nu J_\nu(x) = x J_{\nu+1}(x) + x J_{\nu-1}(x).$$

29. Letting $\nu = 1$ in (21) in the text we have

$$x J_0(x) = \frac{d}{dx}[x J_1(x)] \quad \text{so} \quad \int_0^x r J_0(r)\, dr = r J_1(r)\Big|_{r=0}^{r=x} = x J_1(x).$$

30. From (20) we obtain $J_0'(x) = -J_1(x)$, and from (21) we obtain $J_0'(x) = J_{-1}(x)$. Thus $J_0'(x) = J_{-1}(x) = -J_1(x)$.

31. Since $\Gamma(\frac{1}{2}) = \sqrt{\pi}$ and

$$\Gamma\left(1 - \frac{1}{2} + n\right) = \frac{(2n-1)!}{(n-1)! 2^{2n-1}} \sqrt{\pi} \quad n = 1, 2, 3, \ldots,$$

we obtain

$$J_{-1/2}(x) = \sum_{n=0}^{\infty} \frac{(-1)^n}{n!\,\Gamma(1 - \frac{1}{2} + n)}\left(\frac{x}{2}\right)^{2n-1/2} = \frac{1}{\Gamma(\frac{1}{2})}\left(\frac{x}{2}\right)^{-1/2} + \sum_{n=1}^{\infty} \frac{(-1)^n(n-1)!\,2^{2n-1}x^{2n-1/2}}{n!\,(2n-1)!\,2^{2n-1/2}\sqrt{\pi}}$$

$$= \frac{1}{\sqrt{\pi}}\sqrt{\frac{2}{x}} + \sum_{n=1}^{\infty} \frac{(-1)^n 2^{1/2}x^{-1/2}}{2n(2n-1)!\sqrt{\pi}}\,x^{2n} = \sqrt{\frac{2}{\pi x}} + \sqrt{\frac{2}{\pi x}}\sum_{n=1}^{\infty}\frac{(-1)^n}{(2n)!}\,x^{2n} = \sqrt{\frac{2}{\pi x}}\,\cos x.$$

32. (a) By Problem 28, with $\nu = 1/2$, we obtain $J_{1/2}(x) = xJ_{3/2}(x) + xJ_{-1/2}(x)$ so that

$$J_{3/2}(x) = \sqrt{\frac{2}{\pi x}}\left(\frac{\sin x}{x} - \cos x\right);$$

with $\nu = -1/2$ we obtain $-J_{-1/2}(x) = xJ_{1/2}(x) + xJ_{-3/2}(x)$ so that

$$J_{-3/2}(x) = -\sqrt{\frac{2}{\pi x}}\left(\frac{\cos x}{x} + \sin x\right);$$

and with $\nu = 3/2$ we obtain $3J_{3/2}(x) = xJ_{5/2}(x) + xJ_{1/2}(x)$ so that

$$J_{5/2}(x) = \sqrt{\frac{2}{\pi x}}\left(\frac{3\sin x}{x^2} - \frac{3\cos x}{x} - \sin x\right).$$

(b)

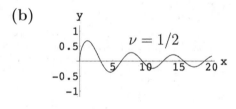 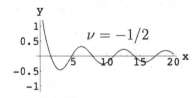 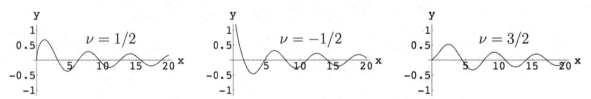

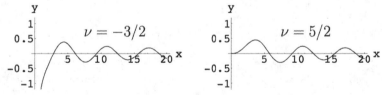 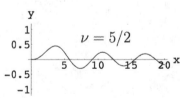

33. Letting

$$s = \frac{2}{\alpha}\sqrt{\frac{k}{m}}\,e^{-\alpha t/2},$$

we have

$$\frac{dx}{dt} = \frac{dx}{ds}\frac{ds}{dt} = \frac{dx}{dt}\left[\frac{2}{\alpha}\sqrt{\frac{k}{m}}\left(-\frac{\alpha}{2}\right)e^{-\alpha t/2}\right] = \frac{dx}{ds}\left(-\sqrt{\frac{k}{m}}\,e^{-\alpha t/2}\right)$$

and

$$\frac{d^2x}{dt^2} = \frac{d}{dt}\left(\frac{dx}{dt}\right) = \frac{dx}{ds}\left(\frac{\alpha}{2}\sqrt{\frac{k}{m}}\,e^{-\alpha t/2}\right) + \frac{d}{dt}\left(\frac{dx}{ds}\right)\left(-\sqrt{\frac{k}{m}}\,e^{-\alpha t/2}\right)$$

$$= \frac{dx}{ds}\left(\frac{\alpha}{2}\sqrt{\frac{k}{m}}\,e^{-\alpha t/2}\right) + \frac{d^2x}{ds^2}\frac{ds}{dt}\left(-\sqrt{\frac{k}{m}}\,e^{-\alpha t/2}\right)$$

$$= \frac{dx}{ds}\left(\frac{\alpha}{2}\sqrt{\frac{k}{m}}\,e^{-\alpha t/2}\right) + \frac{d^2x}{ds^2}\left(\frac{k}{m}\,e^{-\alpha t}\right).$$

Then

$$m\frac{d^2x}{dt^2} + ke^{-\alpha t}x = ke^{-\alpha t}\frac{d^2x}{ds^2} + \frac{m\alpha}{2}\sqrt{\frac{k}{m}}\,e^{-\alpha t/2}\frac{dx}{ds} + ke^{-\alpha t}x = 0.$$

Multiplying by $2^2/\alpha^2 m$ we have

$$\frac{2^2}{\alpha^2}\frac{k}{m}e^{-\alpha t}\frac{d^2x}{ds^2} + \frac{2}{\alpha}\sqrt{\frac{k}{m}}\,e^{-\alpha t/2}\frac{dx}{ds} + \frac{2^2}{\alpha^2}\frac{k}{m}e^{-\alpha t}x = 0$$

or, since $s = (2/\alpha)\sqrt{k/m}\,e^{-\alpha t/2}$,

$$s^2\frac{d^2x}{ds^2} + s\frac{dx}{ds} + s^2x = 0.$$

34. Differentiating $y = x^{1/2}w\left(\frac{2}{3}\alpha x^{3/2}\right)$ with respect to $\frac{2}{3}\alpha x^{3/2}$ we obtain

$$y' = x^{1/2}w'\left(\frac{2}{3}\alpha x^{3/2}\right)\alpha x^{1/2} + \frac{1}{2}x^{-1/2}w\left(\frac{2}{3}\alpha x^{3/2}\right)$$

and

$$y'' = \alpha x w''\left(\frac{2}{3}\alpha x^{3/2}\right)\alpha x^{1/2} + \alpha w'\left(\frac{2}{3}\alpha x^{3/2}\right)$$

$$+ \frac{1}{2}\alpha w'\left(\frac{2}{3}\alpha x^{3/2}\right) - \frac{1}{4}x^{-3/2}w\left(\frac{2}{3}\alpha x^{3/2}\right).$$

Then, after combining terms and simplifying, we have

$$y'' + \alpha^2 xy = \alpha\left[\alpha x^{3/2}w'' + \frac{3}{2}w' + \left(\alpha x^{3/2} - \frac{1}{4\alpha x^{3/2}}\right)w\right] = 0.$$

Letting $t = \frac{2}{3}\alpha x^{3/2}$ or $\alpha x^{3/2} = \frac{3}{2}t$ this differential equation becomes

$$\frac{3}{2}\frac{\alpha}{t}\left[t^2 w''(t) + tw'(t) + \left(t^2 - \frac{1}{9}\right)w(t)\right] = 0, \qquad t > 0.$$

35. (a) By Problem 34, a solution of Airy's equation is $y = x^{1/2}w(\frac{2}{3}\alpha x^{3/2})$, where

$$w(t) = c_1 J_{1/3}(t) + c_2 J_{-1/3}(t)$$

is a solution of Bessel's equation of order $\frac{1}{3}$. Thus, the general solution of Airy's equation for $x > 0$ is

$$y = x^{1/2}w\left(\frac{2}{3}\alpha x^{3/2}\right) = c_1 x^{1/2}J_{1/3}\left(\frac{2}{3}\alpha x^{3/2}\right) + c_2 x^{1/2}J_{-1/3}\left(\frac{2}{3}\alpha x^{3/2}\right).$$

(b) Airy's equation, $y'' + \alpha^2 xy = 0$, has the form of (18) in the text with

$$1 - 2a = 0 \implies a = \frac{1}{2}$$

$$2c - 2 = 1 \implies c = \frac{3}{2}$$

$$b^2 c^2 = \alpha^2 \implies b = \frac{2}{3}\alpha$$

$$a^2 - p^2 c^2 = 0 \implies p = \frac{1}{3}.$$

Then, by (19) in the text,

$$y = x^{1/2}\left[c_1 J_{1/3}\left(\frac{2}{3}\alpha x^{3/2}\right) + c_2 J_{-1/3}\left(\frac{2}{3}\alpha x^{3/2}\right)\right].$$

36. The general solution of the differential equation is

$$y(x) = c_1 J_0(\alpha x) + c_2 Y_0(\alpha x).$$

In order to satisfy the conditions that $\lim_{x \to 0^+} y(x)$ and $\lim_{x \to 0^+} y'(x)$ are finite we are forced to define $c_2 = 0$. Thus, $y(x) = c_1 J_0(\alpha x)$. The second boundary condition, $y(2) = 0$, implies $c_1 = 0$ or $J_0(2\alpha) = 0$. In order to have a nontrivial solution we require that $J_0(2\alpha) = 0$. From Table 6.1, the first three positive zeros of J_0 are found to be

$$2\alpha_1 = 2.4048, \quad 2\alpha_2 = 5.5201, \quad 2\alpha_3 = 8.6537$$

and so $\alpha_1 = 1.2024$, $\alpha_2 = 2.7601$, $\alpha_3 = 4.3269$. The eigenfunctions corresponding to the eigenvalues $\lambda_1 = \alpha_1^2$, $\lambda_2 = \alpha_2^2$, $\lambda_3 = \alpha_3^2$ are $J_0(1.2024x)$, $J_0(2.7601x)$, and $J_0(4.3269x)$.

37. (a) The differential equation $y'' + (\lambda/x)y = 0$ has the form of (18) in the text with

$$1 - 2a = 0 \implies a = \frac{1}{2}$$

$$2c - 2 = -1 \implies c = \frac{1}{2}$$

$$b^2 c^2 = \lambda \implies b = 2\sqrt{\lambda}$$

$$a^2 - p^2 c^2 = 0 \implies p = 1.$$

Then, by (19) in the text,

$$y = x^{1/2}[c_1 J_1(2\sqrt{\lambda x}) + c_2 Y_1(2\sqrt{\lambda x})].$$

(b) We first note that $y = J_1(t)$ is a solution of Bessel's equation, $t^2 y'' + ty' + (t^2 - 1)y = 0$, with $\nu = 1$. That is,

$$t^2 J_1''(t) + t J_1'(t) + (t^2 - 1)J_1(t) = 0,$$

or, letting $t = 2\sqrt{x}$,

$$4x J_1''(2\sqrt{x}) + 2\sqrt{x} J_1'(2\sqrt{x}) + (4x-1)J_1(2\sqrt{x}) = 0.$$

Now, if $y = \sqrt{x} J_1(2\sqrt{x})$, we have

$$y' = \sqrt{x}\, J_1'(2\sqrt{x})\frac{1}{\sqrt{x}} + \frac{1}{2\sqrt{x}}\, J_1(2\sqrt{x}) = J_1'(2\sqrt{x}) + \frac{1}{2}x^{-1/2}J_1(2\sqrt{x})$$

and

$$y'' = x^{-1/2}J_1''(2\sqrt{x}) + \frac{1}{2x}\, J_1'(2\sqrt{x}) - \frac{1}{4}x^{-3/2}J_1(2\sqrt{x}).$$

Then

$$xy'' + y = \sqrt{x}\, J_1''\, 2\sqrt{x} + \frac{1}{2}J_1'(2\sqrt{x}) - \frac{1}{4}x^{-1/2}J_1(2\sqrt{x}) + \sqrt{x}\, J(2\sqrt{x})$$

$$= \frac{1}{4\sqrt{x}}[4x J_1''(2\sqrt{x}) + 2\sqrt{x}\, J_1'(2\sqrt{x}) - J_1(2\sqrt{x}) + 4x J(2\sqrt{x})]$$

$$= 0,$$

and $y = \sqrt{x}\, J_1(2\sqrt{x})$ is a solution of Airy's differential equation.

38. We see from the graphs below that the graphs of the modified Bessel functions are not oscillatory, while those of the Bessel functions, shown in Figures 6.3.1 and 6.3.2 in the text, are oscillatory.

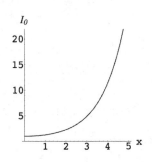 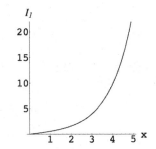 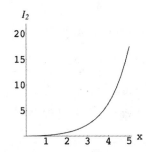

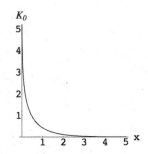 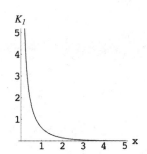 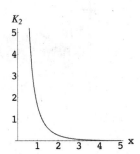

39. (a) We identify $m = 4$, $k = 1$, and $\alpha = 0.1$. Then

$$x(t) = c_1 J_0(10e^{-0.05t}) + c_2 Y_0(10e^{-0.05t})$$

and

$$x'(t) = -0.5c_1 J_0'(10e^{-0.05t}) - 0.5c_2 Y_0'(10e^{-0.05t}).$$

Now $x(0) = 1$ and $x'(0) = -1/2$ imply

$$c_1 J_0(10) + c_2 Y_0(10) = 1$$

$$c_1 J_0'(10) + c_2 Y_0'(10) = 1.$$

Using Cramer's rule we obtain

$$c_1 = \frac{Y_0'(10) - Y_0(10)}{J_0(10)Y_0'(10) - J_0'(10)Y_0(10)}$$

and

$$c_2 = \frac{J_0(10) - J_0'(10)}{J_0(10)Y_0'(10) - J_0'(10)Y_0(10)} .$$

Using $Y_0' = -Y_1$ and $J_0' = -J_1$ and Table 6.2 we find $c_1 = -4.7860$ and $c_2 = -3.1803$. Thus

$$x(t) = -4.7860 J_0(10e^{-0.05t}) - 3.1803 Y_0(10e^{-0.05t}).$$

(b)

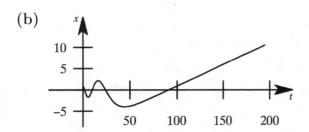

40. (a) Identifying $\alpha = \frac{1}{2}$, the general solution of $x'' + \frac{1}{4}tx = 0$ is

$$x(t) = c_1 x^{1/2} J_{1/3}\left(\frac{1}{3}x^{3/2}\right) + c_2 x^{1/2} J_{-1/3}\left(\frac{1}{3}x^{3/2}\right).$$

Solving the system $x(0.1) = 1$, $x'(0.1) = -\frac{1}{2}$ we find $c_1 = -0.809264$ and $c_2 = 0.782397$.

(b)

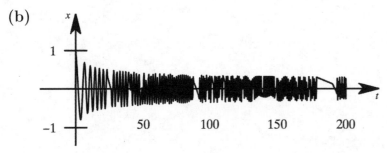

41. (a) Letting $t = L - x$, the boundary-value problem becomes

$$\frac{d^2\theta}{dt^2} + \alpha^2 t\theta = 0, \qquad \theta'(0) = 0, \quad \theta(L) = 0,$$

where $\alpha^2 = \delta g/EI$. This is Airy's differential equation, so by Problem 35 its solution is

$$y = c_1 t^{1/2} J_{1/3}\left(\frac{2}{3}\alpha t^{3/2}\right) + c_2 t^{1/2} J_{-1/3}\left(\frac{2}{3}\alpha t^{3/2}\right) = c_1\theta_1(t) + c_2\theta_2(t).$$

(b) Looking at the series forms of θ_1 and θ_2 we see that $\theta_1'(0) \neq 0$, while $\theta_2'(0) = 0$. Thus, the boundary condition $\theta'(0) = 0$ implies $c_1 = 0$, and so

$$\theta(t) = c_2 \sqrt{t}\, J_{-1/3}\left(\frac{2}{3}\alpha t^{3/2}\right).$$

From $\theta(L) = 0$ we have

$$c_2 \sqrt{L}\, J_{-1/3}\left(\frac{2}{3}\alpha L^{3/2}\right) = 0,$$

so either $c_2 = 0$, in which case $\theta(t) = 0$, or $J_{-1/3}(\frac{2}{3}\alpha L^{3/2}) = 0$. The column will just start to bend when L is the length corresponding to the smallest positive zero of $J_{-1/3}$.

(c) Using *Mathematica*, the first positive root of $J_{-1/3}(x)$ is $x_1 \approx 1.86635$. Thus $\frac{2}{3}\alpha L^{3/2} = 1.86635$ implies

$$L = \left(\frac{3(1.86635)}{2\alpha}\right)^{2/3} = \left[\frac{9EI}{4\delta g}(1.86635)^2\right]^{1/3}$$

$$= \left[\frac{9(2.6 \times 10^7)\pi(0.05)^4/4}{4(0.28)\pi(0.05)^2}(1.86635)^2\right]^{1/3} \approx 76.9 \text{ in.}$$

42. (a) Writing the differential equation in the form $xy'' + (PL/M)y = 0$, we identify $\lambda = PL/M$. From Problem 37 the solution of this differential equation is

$$y = c_1 \sqrt{x}\, J_1\left(2\sqrt{PLx/M}\right) + c_2 \sqrt{x}\, Y_1\left(2\sqrt{PLx/M}\right).$$

Now $J_1(0) = 0$, so $y(0) = 0$ implies $c_2 = 0$ and

$$y = c_1 \sqrt{x}\, J_1\left(2\sqrt{PLx/M}\right).$$

(b) From $y(L) = 0$ we have $y = J_1(2L\sqrt{PM}) = 0$. The first positive zero of J_1 is 3.8317 so, solving $2L\sqrt{P_1/M} = 3.8317$, we find $P_1 = 3.6705M/L^2$. Therefore,

$$y_1(x) = c_1 \sqrt{x}\, J_1\left(2\sqrt{\frac{3.6705x}{L}}\right) = c_1 \sqrt{x}\, J_1\left(\frac{3.8317}{\sqrt{L}}\sqrt{x}\right).$$

(c) For $c_1 = 1$ and $L = 1$ the graph of $y_1 = \sqrt{x}\, J_1(3.8317\sqrt{x})$ is shown.

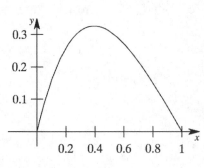

43. (a) Since $l' = v$, we integrate to obtain $l(t) = vt + c$. Now $l(0) = l_0$ implies $c = l_0$, so $l(t) = vt + l_0$.

Using $\sin\theta \approx \theta$ in $l\,d^2\theta/dt^2 + 2l'\,d\theta/dt + g\sin\theta = 0$ gives

$$(l_0 + vt)\frac{d^2\theta}{dt^2} + 2v\frac{d\theta}{dt} + g\theta = 0.$$

(b) Dividing by v, the differential equation in part (a) becomes

$$\frac{l_0 + vt}{v}\frac{d^2\theta}{dt^2} + 2\frac{d\theta}{dt} + \frac{g}{v}\theta = 0.$$

Letting $x = (l_0 + vt)/v = t + l_0/v$ we have $dx/dt = 1$, so

$$\frac{d\theta}{dt} = \frac{d\theta}{dx}\frac{dx}{dt} = \frac{d\theta}{dx}$$

and

$$\frac{d^2\theta}{dt^2} = \frac{d(d\theta/dt)}{dt} = \frac{d(d\theta/dx)}{dx}\frac{dx}{dt} = \frac{d^2\theta}{dx^2}.$$

Thus, the differential equation becomes

$$x\frac{d^2\theta}{dx^2} + 2\frac{d\theta}{dx} + \frac{g}{v}\theta = 0 \qquad \text{or} \qquad \frac{d^2\theta}{dx^2} + \frac{2}{x}\frac{d\theta}{dx} + \frac{g}{vx}\theta = 0.$$

(c) The differential equation in part (b) has the form of (18) in the text with

$$1 - 2a = 2 \implies a = -\frac{1}{2}$$

$$2c - 2 = -1 \implies c = \frac{1}{2}$$

$$b^2c^2 = \frac{g}{v} \implies b = 2\sqrt{\frac{g}{v}}$$

$$a^2 - p^2c^2 = 0 \implies p = 1.$$

Then, by (19) in the text,

$$\theta(x) = x^{-1/2}\left[c_1 J_1\left(2\sqrt{\frac{g}{v}}\,x^{1/2}\right) + c_2 Y_1\left(2\sqrt{\frac{g}{v}}\,x^{1/2}\right)\right]$$

or

$$\theta(t) = \sqrt{\frac{v}{l_0 + vt}}\left[c_1 J_1\left(\frac{2}{v}\sqrt{g(l_0 + vt)}\right) + c_2 Y_1\left(\frac{2}{v}\sqrt{g(l_0 + vt)}\right)\right].$$

(d) To simplify calculations, let

$$u = \frac{2}{v}\sqrt{g(l_0 + vt)} = 2\sqrt{\frac{g}{v}}\,x^{1/2},$$

and at $t = 0$ let $u_0 = 2\sqrt{gl_0}/v$. The general solution for $\theta(t)$ can then be written

$$\theta = C_1 u^{-1} J_1(u) + C_2 u^{-1} Y_1(u). \tag{1}$$

Before applying the initial conditions, note that

$$\frac{d\theta}{dt} = \frac{d\theta}{du}\frac{du}{dt}$$

so when $d\theta/dt = 0$ at $t = 0$ we have $d\theta/du = 0$ at $u = u_0$. Also,

$$\frac{d\theta}{du} = C_1 \frac{d}{du}[u^{-1}J_1(u)] + C_2 \frac{d}{du}[u^{-1}Y_1(u)]$$

which, in view of (20) in the text, is the same as

$$\frac{d\theta}{du} = -C_1 u^{-1}J_2(u) - C_2 u^{-1}Y_2(u). \tag{2}$$

Now at $t = 0$, or $u = u_0$, (1) and (2) give the system

$$C_1 u_0^{-1}J_1(u_0) + C_2 u_0^{-1}Y_1(u_0) = \theta_0$$

$$C_1 u_0^{-1}J_2(u_0) + C_2 u_0^{-1}Y_2(u_0) = 0$$

whose solution is easily obtained using Cramer's rule:

$$C_1 = \frac{u_0 \theta_0 Y_2(u_0)}{J_1(u_0)Y_2(u_0) - J_2(u_0)Y_1(u_0)}, \qquad C_2 = \frac{-u_0 \theta_0 J_2(u_0)}{J_1(u_0)Y_2(u_0) - J_2(u_0)Y_1(u_0)}.$$

In view of the given identity these results simplify to

$$C_1 = -\frac{\pi}{2}u_0^2 \theta_0 Y_2(u_0) \qquad \text{and} \qquad C_2 = \frac{\pi}{2}u_0^2 \theta_0 J_2(u_0).$$

The solution is then

$$\theta = \frac{\pi}{2}u_0^2 \theta_0 \left[-Y_2(u_0)\frac{J_1(u)}{u} + J_2(u_0)\frac{Y_1(u)}{u} \right].$$

Returning to $u = (2/v)\sqrt{g(l_0 + vt)}$ and $u_0 = (2/v)\sqrt{gl_0}$, we have

$$\theta(t) = \frac{\pi\sqrt{gl_0}\,\theta_0}{v} \left[-Y_2\left(\frac{2}{v}\sqrt{gl_0}\right)\frac{J_1\left(\frac{2}{v}\sqrt{g(l_0+vt)}\right)}{\sqrt{l_0+vt}} + J_2\left(\frac{2}{v}\sqrt{gl_0}\right)\frac{Y_1\left(\frac{2}{v}\sqrt{g(l_0+vt)}\right)}{\sqrt{l_0+vt}} \right].$$

(e) When $l_0 = 1$ ft, $\theta_0 = \frac{1}{10}$ radian, and $v = \frac{1}{60}$ ft/s, the above function is

$$\theta(t) = -1.69045\,\frac{J_1(480\sqrt{2}(1+t/60))}{\sqrt{1+t/60}} - 2.79381\,\frac{Y_1(480\sqrt{2}(1+t/60))}{\sqrt{1+t/60}}.$$

The plots of $\theta(t)$ on $[0, 10]$, $[0, 30]$, and $[0, 60]$ are

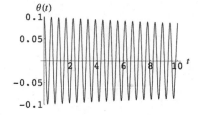 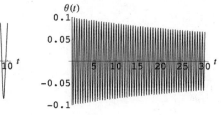 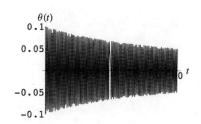

(f) The graphs indicate that $\theta(t)$ decreases as l increases. The graph of $\theta(t)$ on $[0, 300]$ is shown.

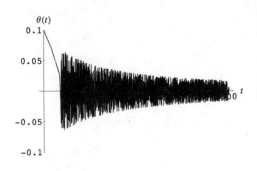

44. (a) From (26) in the text, we have

$$P_6(x) = c_0 \left(1 - \frac{6 \cdot 7}{2!} x^2 + \frac{4 \cdot 6 \cdot 7 \cdot 9}{4!} x^4 = \frac{2 \cdot 4 \cdot 6 \cdot 7 \cdot 9 \cdot 11}{6!} x^6 \right),$$

where

$$c_0 = (-1)^3 \frac{1 \cdot 3 \cdot 5}{2 \cdot 4 \cdot 6} = -\frac{5}{16}.$$

Thus,

$$P_6(x) = -\frac{5}{16} \left(1 - 21x^2 + 63x^4 - \frac{231}{5} x^6 \right) = \frac{1}{16} (231x^6 - 315x^4 + 105x^2 - 5).$$

Also, from (26) in the text we have

$$P_7(x) = c_1 \left(x - \frac{6 \cdot 9}{3!} x^3 + \frac{4 \cdot 6 \cdot 9 \cdot 11}{5!} x^5 - \frac{2 \cdot 4 \cdot 6 \cdot 9 \cdot 11 \cdot 13}{7!} x^7 \right)$$

where

$$c_1 = (-1)^3 \frac{1 \cdot 3 \cdot 5 \cdot 7}{2 \cdot 4 \cdot 6} = -\frac{35}{16}.$$

Thus

$$P_7(x) = -\frac{35}{16} \left(x - 9x^3 + \frac{99}{5} x^5 - \frac{429}{35} x^7 \right) = \frac{1}{16} (429x^7 - 693x^5 + 315x^3 - 35x).$$

(b) $P_6(x)$ satisfies $\left(1 - x^2 \right) y'' - 2xy' + 42y = 0$ and $P_7(x)$ satisfies $\left(1 - x^2 \right) y'' - 2xy' + 56y = 0$.

45. The recurrence relation can be written

$$P_{k+1}(x) = \frac{2k+1}{k+1} x P_k(x) - \frac{k}{k+1} P_{k-1}(x), \qquad k = 2, 3, 4, \ldots .$$

$k = 1$: $P_2(x) = \frac{3}{2} x^2 - \frac{1}{2}$

$k = 2$: $P_3(x) = \frac{5}{3} x \left(\frac{3}{2} x^2 - \frac{1}{2} \right) - \frac{2}{3} x = \frac{5}{2} x^3 - \frac{3}{2} x$

$k = 3$: $P_4(x) = \frac{7}{4} x \left(\frac{5}{2} x^3 - \frac{3}{2} x \right) - \frac{3}{4} \left(\frac{3}{2} x^2 - \frac{1}{2} \right) = \frac{35}{8} x^4 - \frac{30}{8} x^2 + \frac{3}{8}$

$k = 4$: $P_5(x) = \frac{9}{5} x \left(\frac{35}{8} x^4 - \frac{30}{8} x^2 + \frac{3}{8} \right) - \frac{4}{5} \left(\frac{5}{2} x^3 - \frac{3}{2} x \right) = \frac{63}{8} x^5 - \frac{35}{4} x^3 + \frac{15}{8} x$

$k = 5$: $P_6(x) = \dfrac{11}{6}x\left(\dfrac{63}{8}x^5 - \dfrac{35}{4}x^3 + \dfrac{15}{8}x\right) - \dfrac{5}{6}\left(\dfrac{35}{8}x^4 - \dfrac{30}{8}x^2 + \dfrac{3}{8}\right) = \dfrac{231}{16}x^6 - \dfrac{315}{16}x^4 + \dfrac{105}{16}x^2 - \dfrac{5}{16}$

$k = 6$: $P_7(x) = \dfrac{13}{7}x\left(\dfrac{231}{16}x^6 - \dfrac{315}{16}x^4 + \dfrac{105}{16}x^2 - \dfrac{5}{16}\right) - \dfrac{6}{7}\left(\dfrac{63}{8}x^5 - \dfrac{35}{4}x^3 + \dfrac{15}{8}x\right)$

$$= \dfrac{429}{16}x^7 - \dfrac{693}{16}x^5 + \dfrac{315}{16}x^3 - \dfrac{35}{16}x$$

46. If $x = \cos\theta$ then

$$\frac{dy}{d\theta} = -\sin\theta\,\frac{dy}{dx},$$

$$\frac{d^2y}{d\theta^2} = \sin^2\theta\,\frac{d^2y}{dx^2} - \cos\theta\,\frac{dy}{dx},$$

and

$$\sin\theta\,\frac{d^2y}{d\theta^2} + \cos\theta\,\frac{dy}{d\theta} + n(n+1)(\sin\theta)y = \sin\theta\left[\left(1 - \cos^2\theta\right)\frac{d^2y}{dx^2} - 2\cos\theta\,\frac{dy}{dx} + n(n+1)y\right] = 0.$$

That is,

$$\left(1 - x^2\right)\frac{d^2y}{dx^2} - 2x\frac{dy}{dx} + n(n+1)y = 0.$$

47. The only solutions bounded on $[-1, 1]$ are $y = cP_n(x)$, c a constant and $n = 0, 1, 2, \ldots$. By (iv) of the properties of the Legendre polynomials, $y(0) = 0$ or $P_n(0) = 0$ implies n must be odd. Thus the first three positive eigenvalues correspond to $n = 1, 3$, and 5 or $\lambda_1 = 1 \cdot 2$, $\lambda_2 = 3 \cdot 4 = 12$, and $\lambda_3 = 5 \cdot 6 = 30$. We can take the eigenfunctions to be $y_1 = P_1(x)$, $y_2 = P_3(x)$, and $y_3 = P_5(x)$.

48. Using a CAS we find

$$P_1(x) = \frac{1}{2}\frac{d}{dx}\left(x^2 - 1\right)^1 = x$$

$$P_2(x) = \frac{1}{2^2 2!}\frac{d^2}{dx^2}\left(x^2 - 1\right)^2 = \frac{1}{2}(3x^2 - 1)$$

$$P_3(x) = \frac{1}{2^3 3!}\frac{d^3}{dx^3}\left(x^2 - 1\right)^3 = \frac{1}{2}(5x^3 - 3x)$$

$$P_4(x) = \frac{1}{2^4 4!}\frac{d^4}{dx^4}\left(x^2 - 1\right)^4 = \frac{1}{8}(35x^4 - 30x^2 + 3)$$

$$P_5(x) = \frac{1}{2^5 5!}\frac{d^5}{dx^5}\left(x^2 - 1\right)^5 = \frac{1}{8}(63x^5 - 70x^3 + 15x)$$

$$P_6(x) = \frac{1}{2^6 6!}\frac{d^6}{dx^6}\left(x^2 - 1\right)^6 = \frac{1}{16}(231x^6 - 315x^4 + 105x^2 - 5)$$

$$P_7(x) = \frac{1}{2^7 7!}\frac{d^7}{dx^7}\left(x^2 - 1\right)^7 = \frac{1}{16}(429x^7 - 693x^5 + 315x^3 - 35x)$$

49.

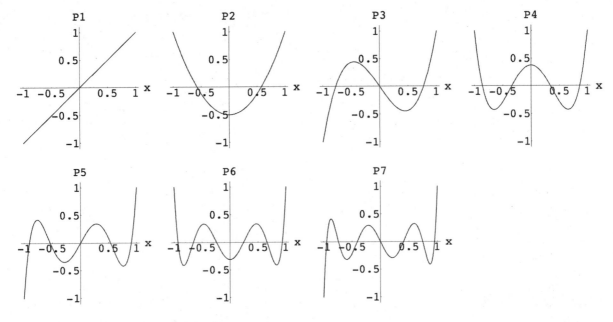

50. Zeros of Legendre polynomials for $n \geq 1$ are

$P_1(x)$: 0

$P_2(x)$: ± 0.57735

$P_3(x)$: 0, ± 0.77460

$P_4(x)$: ± 0.33998, ± 0.86115

$P_5(x)$: 0, ± 0.53847, ± 0.90618

$P_6(x)$: ± 0.23862, ± 0.66121, ± 0.93247

$P_7(x)$: 0, ± 0.40585, ± 0.74153 , ± 0.94911

$P_{10}(x)$: ± 0.14887, ± 0.43340, ± 0.67941, ± 0.86506, ± 0.097391

The zeros of any Legendre polynomial are in the interval $(-1, 1)$ and are symmetric with respect to 0.

Chapter 6 in Review

1. False; $J_1(x)$ and $J_{-1}(x)$ are not linearly independent when ν is a positive integer. (In this case $\nu = 1$). The general solution of $x^2 y'' + xy' + (x^2 - 1)y = 0$ is $y = c_1 J_1(x) + c_2 Y_1(x)$.

2. False; $y = x$ is a solution that is analytic at $x = 0$.

3. $x = -1$ is the nearest singular point to the ordinary point $x = 0$. Theorem 6.1.1 guarantees the existence of two power series solutions $y = \sum_{n=1}^{\infty} c_n x^n$ of the differential equation that converge at least for $-1 < x < 1$. Since $-\frac{1}{2} \le x \le \frac{1}{2}$ is properly contained in $-1 < x < 1$, both power series must converge for all points contained in $-\frac{1}{2} \le x \le \frac{1}{2}$.

4. The easiest way to solve the system

$$2c_2 + 2c_1 + c_0 = 0$$

$$6c_3 + 4c_2 + c_1 = 0$$

$$12c_4 + 6c_3 - \frac{1}{3}c_1 + c_2 = 0$$

$$20c_5 + 8c_4 - \frac{2}{3}c_2 + c_3 = 0$$

is to choose, in turn, $c_0 \ne 0$, $c_1 = 0$ and $c_0 = 0$, $c_1 \ne 0$. Assuming that $c_0 \ne 0$, $c_1 = 0$, we have

$$c_2 = -\frac{1}{2}c_0$$

$$c_3 = -\frac{2}{3}c_2 = \frac{1}{3}c_0$$

$$c_4 = -\frac{1}{2}c_3 - \frac{1}{12}c_2 = -\frac{1}{8}c_0$$

$$c_5 = -\frac{2}{5}c_4 + \frac{1}{30}c_2 - \frac{1}{20}c_3 = \frac{1}{60}c_0;$$

whereas the assumption that $c_0 = 0$, $c_1 \ne 0$ implies

$$c_2 = -c_1$$

$$c_3 = -\frac{2}{3}c_2 - \frac{1}{6}c_1 = \frac{1}{2}c_1$$

$$c_4 = -\frac{1}{2}c_3 + \frac{1}{36}c_1 - \frac{1}{12}c_2 = -\frac{5}{36}c_1$$

$$c_5 = -\frac{2}{5}c_4 + \frac{1}{30}c_2 - \frac{1}{20}c_3 = -\frac{1}{360}c_1.$$

five terms of two power series solutions are then

$$y_1(x) = c_0 \left[1 - \frac{1}{2}x^2 + \frac{1}{3}x^3 - \frac{1}{8}x^4 + \frac{1}{60}x^5 + \cdots \right]$$

and

$$y_2(x) = c_1 \left[x - x^2 + \frac{1}{2}x^3 - \frac{5}{36}x^4 - \frac{1}{360}x^5 + \cdots \right].$$

5. The interval of convergence is centered at 4. Since the series converges at -2, it converges at least on the interval $[-2, 10)$. Since it diverges at 13, it converges at most on the interval $[-5, 13)$. Thus, at -7 it does not converge, at 0 and 7 it does converge, and at 10 and 11 it might converge.

6. We have

$$f(x) = \frac{\sin x}{\cos x} = \frac{x - \dfrac{x^3}{6} + \dfrac{x^5}{120} - \cdots}{1 - \dfrac{x^2}{2} + \dfrac{x^4}{24} - \cdots} = x + \frac{x^3}{3} + \frac{2x^5}{15} + \cdots.$$

7. The differential equation $(x^3 - x^2)y'' + y' + y = 0$ has a regular singular point at $x = 1$ and an irregular singular point at $x = 0$.

8. The differential equation $(x - 1)(x + 3)y'' + y = 0$ has regular singular points at $x = 1$ and $x = -3$.

9. Substituting $y = \sum_{n=0}^{\infty} c_n x^{n+r}$ into the differential equation we obtain

$$2xy'' + y' + y = \left(2r^2 - r\right)c_0 x^{r-1} + \sum_{k=1}^{\infty} [2(k+r)(k+r-1)c_k + (k+r)c_k + c_{k-1}]x^{k+r-1} = 0$$

which implies

$$2r^2 - r = r(2r - 1) = 0$$

and

$$(k+r)(2k + 2r - 1)c_k + c_{k-1} = 0.$$

The indicial roots are $r = 0$ and $r = 1/2$. For $r = 0$ the recurrence relation is

$$c_k = -\frac{c_{k-1}}{k(2k-1)}, \quad k = 1, 2, 3, \ldots,$$

so

$$c_1 = -c_0, \qquad c_2 = \frac{1}{6}c_0, \qquad c_3 = -\frac{1}{90}c_0.$$

For $r = 1/2$ the recurrence relation is

$$c_k = -\frac{c_{k-1}}{k(2k+1)}, \quad k = 1, 2, 3, \ldots,$$

so

$$c_1 = -\frac{1}{3}c_0, \qquad c_2 = \frac{1}{30}c_0, \qquad c_3 = -\frac{1}{630}c_0.$$

Two linearly independent solutions are

$$y_1 = 1 - x + \frac{1}{6}x^2 - \frac{1}{90}x^3 + \cdots$$

and

$$y_2 = x^{1/2}\left(1 - \frac{1}{3}x + \frac{1}{30}x^2 - \frac{1}{630}x^3 + \cdots\right).$$

10. Substituting $y = \sum_{n=0}^{\infty} c_n x^n$ into the differential equation we have

$$y'' - xy' - y = \underbrace{\sum_{n=2}^{\infty} n(n-1)c_n x^{n-2}}_{k=n-2} - \underbrace{\sum_{n=1}^{\infty} nc_n x^n}_{k=n} - \underbrace{\sum_{n=0}^{\infty} c_n x^n}_{k=n}$$

$$= \sum_{k=0}^{\infty} (k+2)(k+1)c_{k+2} x^k - \sum_{k=1}^{\infty} kc_k x^k - \sum_{k=0}^{\infty} c_k x^k$$

$$= 2c_2 - c_0 + \sum_{k=1}^{\infty} [(k+2)(k+1)c_{k+2} - (k+1)c_k]x^k = 0.$$

Thus

$$2c_2 - c_0 = 0$$

$$(k+2)(k+1)c_{k+2} - (k+1)c_k = 0$$

and

$$c_2 = \frac{1}{2}c_0$$

$$c_{k+2} = \frac{1}{k+2} c_k, \quad k = 1, 2, 3, \dots.$$

Choosing $c_0 = 1$ and $c_1 = 0$ we find

$$c_2 = \frac{1}{2}$$

$$c_3 = c_5 = c_7 = \cdots = 0$$

$$c_4 = \frac{1}{8}$$

$$c_6 = \frac{1}{48}$$

and so on. For $c_0 = 0$ and $c_1 = 1$ we obtain

$$c_2 = c_4 = c_6 = \cdots = 0$$

$$c_3 = \frac{1}{3}$$

$$c_5 = \frac{1}{15}$$

$$c_7 = \frac{1}{105}$$

and so on. Thus, two solutions are

$$y_1 = 1 + \frac{1}{2}x^2 + \frac{1}{8}x^4 + \frac{1}{48}x^6 + \cdots$$

and

$$y_2 = x + \frac{1}{3}x^3 + \frac{1}{15}x^5 + \frac{1}{105}x^7 + \cdots.$$

11. Substituting $y = \sum_{n=0}^{\infty} c_n x^n$ into the differential equation we obtain

$$(x-1)y'' + 3y = (-2c_2 + 3c_0) + \sum_{k=1}^{\infty}[(k+1)kc_{k+1} - (k+2)(k+1)c_{k+2} + 3c_k]x^k = 0$$

which implies $c_2 = 3c_0/2$ and

$$c_{k+2} = \frac{(k+1)kc_{k+1} + 3c_k}{(k+2)(k+1)}, \quad k = 1, 2, 3, \ldots.$$

Choosing $c_0 = 1$ and $c_1 = 0$ we find

$$c_2 = \frac{3}{2}, \quad c_3 = \frac{1}{2}, \quad c_4 = \frac{5}{8}$$

and so on. For $c_0 = 0$ and $c_1 = 1$ we obtain

$$c_2 = 0, \quad c_3 = \frac{1}{2}, \quad c_4 = \frac{1}{4}$$

and so on. Thus, two solutions are

$$y_1 = 1 + \frac{3}{2}x^2 + \frac{1}{2}x^3 + \frac{5}{8}x^4 + \cdots$$

and

$$y_2 = x + \frac{1}{2}x^3 + \frac{1}{4}x^4 + \cdots.$$

12. Substituting $y = \sum_{n=0}^{\infty} c_n x^n$ into the differential equation we obtain

$$y'' - x^2 y' + xy = 2c_2 + (6c_3 + c_0)x + \sum_{k=1}^{\infty}[(k+3)(k+2)c_{k+3} - (k-1)c_k]x^{k+1} = 0$$

which implies $c_2 = 0$, $c_3 = -c_0/6$, and

$$c_{k+3} = \frac{k-1}{(k+3)(k+2)}c_k, \quad k = 1, 2, 3, \ldots.$$

Choosing $c_0 = 1$ and $c_1 = 0$ we find

$$c_3 = -\frac{1}{6}$$

$$c_4 = c_7 = c_{10} = \cdots = 0$$

$$c_5 = c_8 = c_{11} = \cdots = 0$$

$$c_6 = -\frac{1}{90}$$

and so on. For $c_0 = 0$ and $c_1 = 1$ we obtain

$$c_3 = c_6 = c_9 = \cdots = 0$$

$$c_4 = c_7 = c_{10} = \cdots = 0$$

$$c_5 = c_8 = c_{11} = \cdots = 0$$

and so on. Thus, two solutions are

$$y_1 = 1 - \frac{1}{6}x^3 - \frac{1}{90}x^6 - \cdots \quad \text{and} \quad y_2 = x.$$

13. Substituting $y = \sum_{n=0}^{\infty} c_n x^{n+r}$ into the differential equation, we obtain

$$xy'' - (x+2)y' + 2y = (r^2 - 3r)c_0 x^{r-1} + \sum_{k=1}^{\infty} [(k+r)(k+r-3)c_k$$

$$- (k+r-3)c_{k-1}]x^{k+r-1} = 0,$$

which implies

$$r^2 - 3r = r(r-3) = 0$$

and

$$(k+r)(k+r-3)c_k - (k+r-3)c_{k-1} = 0.$$

The indicial roots are $r_1 = 3$ and $r_2 = 0$. For $r_2 = 0$ the recurrence relation is

$$k(k-3)c_k - (k-3)c_{k-1} = 0, \qquad k = 1, 2, 3, \ldots.$$

Then

$$c_1 - c_0 = 0$$

$$2c_2 - c_1 = 0$$

$$0c_3 - 0c_2 = 0 \implies c_3 \text{ is arbitrary}$$

and

$$c_k = \frac{1}{k}c_{k-1}, \qquad k = 4, 5, 6, \ldots.$$

Taking $c_0 \neq 0$ and $c_3 = 0$ we obtain

$$c_1 = c_0$$

$$c_2 = \frac{1}{2}c_0$$

$$c_3 = c_4 = c_5 = \cdots = 0.$$

Taking $c_0 = 0$ and $c_3 \neq 0$ we obtain

$$c_0 = c_1 = c_2 = 0$$

$$c_4 = \frac{1}{4}c_3 = \frac{6}{4!}c_3$$

$$c_5 = \frac{1}{5 \cdot 4}c_3 = \frac{6}{5!}c_3$$

$$c_6 = \frac{1}{6 \cdot 5 \cdot 4}c_3 = \frac{6}{6!}c_3,$$

and so on. In this case we obtain the two solutions

$$y_1 = 1 + x + \frac{1}{2}x^2$$

and

$$y_2 = x^3 + \frac{6}{4!}x^4 + \frac{6}{5!}x^5 + \frac{6}{6!}x^6 + \cdots = 6e^x - 6\left(1 + x + \frac{1}{2}x^2\right).$$

14. Substituting $y = \sum_{n=0}^{\infty} c_n x^n$ into the differential equation we have

$$(\cos x)y'' + y = \left(1 - \frac{1}{2}x^2 + \frac{1}{24}x^4 - \frac{1}{720}x^6 + \cdots\right)\left(2c_2 + 6c_3 x + 12c_4 x^2 + 20c_5 x^3 + 30c_6 x^4 + \cdots\right)$$

$$+ \sum_{n=0}^{\infty} c_n x^n$$

$$= \left[2c_2 + 6c_3 x + (12c_4 - c_2)x^2 + (20c_5 - 3c_3)x^3 + \left(30c_6 - 6c_4 + \frac{1}{12}c_2\right)x^4 + \cdots\right]$$

$$+ \left[c_0 + c_1 x + c_2 x^2 + c_3 x^3 + c_4 x^4 + \cdots\right]$$

$$= (c_0 + 2c_2) + (c_1 + 6c_3)x + 12c_4 x^2 + (20c_5 - 2c_3)x^3 + \left(30c_6 - 5c_4 + \frac{1}{12}c_2\right)x^4 + \cdots$$

$$= 0.$$

Thus

$$c_0 + 2c_2 = 0$$

$$c_1 + 6c_3 = 0$$

$$12c_4 = 0$$

$$20c_5 - 2c_3 = 0$$

$$30c_6 - 5c_4 + \frac{1}{12}c_2 = 0$$

and

$$c_2 = -\frac{1}{2}c_0$$

$$c_3 = -\frac{1}{6}c_1$$

$$c_4 = 0$$

$$c_5 = \frac{1}{10}c_3$$

$$c_6 = \frac{1}{6}c_4 - \frac{1}{360}c_2.$$

Choosing $c_0 = 1$ and $c_1 = 0$ we find

$$c_2 = -\frac{1}{2}, \quad c_3 = 0, \quad c_4 = 0, \quad c_5 = 0, \quad c_6 = \frac{1}{720}$$

and so on. For $c_0 = 0$ and $c_1 = 1$ we find

$$c_2 = 0, \quad c_3 = -\frac{1}{6}, \quad c_4 = 0, \quad c_5 = -\frac{1}{60}, \quad c_6 = 0$$

and so on. Thus, two solutions are

$$y_1 = 1 - \frac{1}{2}x^2 + \frac{1}{720}x^6 + \cdots \quad \text{and} \quad y_2 = x - \frac{1}{6}x^3 - \frac{1}{60}x^5 + \cdots.$$

15.

$$y'' + xy' + 2y = \underbrace{\sum_{n=2}^{\infty} n(n-1)c_n x^{n-2}}_{k=n-2} + \underbrace{\sum_{n=1}^{\infty} nc_n x^n}_{k=n} + 2\underbrace{\sum_{n=0}^{\infty} c_n x^n}_{k=n}$$

$$= \sum_{k=0}^{\infty}(k+2)(k+1)c_{k+2}x^k + \sum_{k=1}^{\infty} kc_k x^k + 2\sum_{k=0}^{\infty} c_k x^k$$

$$= 2c_2 + 2c_0 + \sum_{k=1}^{\infty}[(k+2)(k+1)c_{k+2} + (k+2)c_k]x^k = 0.$$

Thus

$$2c_2 + 2c_0 = 0$$

$$(k+2)(k+1)c_{k+2} + (k+2)c_k = 0$$

and

$$c_2 = -c_0$$

$$c_{k+2} = -\frac{1}{k+1}c_k, \quad k = 1, 2, 3, \ldots.$$

Choosing $c_0 = 1$ and $c_1 = 0$ we find

$$c_2 = -1$$

$$c_3 = c_5 = c_7 = \cdots = 0$$

$$c_4 = \frac{1}{3}$$

$$c_6 = -\frac{1}{15}$$

and so on. For $c_0 = 0$ and $c_1 = 1$ we obtain

$$c_2 = c_4 = c_6 = \cdots = 0$$

$$c_3 = -\frac{1}{2}$$

$$c_5 = \frac{1}{8}$$

$$c_7 = -\frac{1}{48}$$

and so on. Thus, the general solution is

$$y = C_0 \left(1 - x^2 + \frac{1}{3}x^4 - \frac{1}{15}x^6 + \cdots\right) + C_1 \left(x - \frac{1}{2}x^3 + \frac{1}{8}x^5 - \frac{1}{48}x^7 + \cdots\right)$$

and

$$y' = C_0 \left(-2x + \frac{4}{3}x^3 - \frac{2}{5}x^5 + \cdots\right) + C_1 \left(1 - \frac{3}{2}x^2 + \frac{5}{8}x^4 - \frac{7}{48}x^6 + \cdots\right).$$

Setting $y(0) = 3$ and $y'(0) = -2$ we find $c_0 = 3$ and $c_1 = -2$. Therefore, the solution of the initial-value problem is

$$y = 3 - 2x - 3x^2 + x^3 + x^4 - \frac{1}{4}x^5 - \frac{1}{5}x^6 + \frac{1}{24}x^7 + \cdots.$$

16. Substituting $y = \sum_{n=0}^{\infty} c_n x^n$ into the differential equation we have

$$(x+2)y'' + 3y = \underbrace{\sum_{n=2}^{\infty} n(n-1)c_n x^{n-1}}_{k=n-1} + 2\underbrace{\sum_{n=2}^{\infty} n(n-1)c_n x^{n-2}}_{k=n-2} + 3\underbrace{\sum_{n=0}^{\infty} c_n x^n}_{k=n}$$

$$= \sum_{k=1}^{\infty}(k+1)k c_{k+1} x^k + 2\sum_{k=0}^{\infty}(k+2)(k+1)c_{k+2} x^k + 3\sum_{k=0}^{\infty} c_k x^k$$

$$= 4c_2 + 3c_0 + \sum_{k=1}^{\infty}[(k+1)k c_{k+1} + 2(k+2)(k+1)c_{k+2} + 3c_k]x^k = 0.$$

Thus

$$4c_2 + 3c_0 = 0$$

$$(k+1)k c_{k+1} + 2(k+2)(k+1)c_{k+2} + 3c_k = 0$$

and

$$c_2 = -\frac{3}{4}c_0$$

$$c_{k+2} = -\frac{k}{2(k+2)}c_{k+1} - \frac{3}{2(k+2)(k+1)}c_k, \quad k = 1, 2, 3, \ldots.$$

Choosing $c_0 = 1$ and $c_1 = 0$ we find

$$c_2 = -\frac{3}{4}$$

$$c_3 = \frac{1}{8}$$

$$c_4 = \frac{1}{16}$$

$$c_5 = -\frac{9}{320}$$

and so on. For $c_0 = 0$ and $c_1 = 1$ we obtain

$$c_2 = 0$$

$$c_3 = -\frac{1}{4}$$

$$c_4 = \frac{1}{16}$$

$$c_5 = 0$$

and so on. Thus, the general solution is

$$y = C_0 \left(1 - \frac{3}{4}x^2 + \frac{1}{8}x^3 + \frac{1}{16}x^4 - \frac{9}{320}x^5 + \cdots\right) + C_1 \left(x - \frac{1}{4}x^3 + \frac{1}{16}x^4 + \cdots\right)$$

and

$$y' = C_0 \left(-\frac{3}{2}x + \frac{3}{8}x^2 + \frac{1}{4}x^3 - \frac{9}{64}x^4 + \cdots\right) + C_1 \left(1 - \frac{3}{4}x^2 + \frac{1}{4}x^3 + \cdots\right).$$

Setting $y(0) = 0$ and $y'(0) = 1$ we find $c_0 = 0$ and $c_1 = 1$. Therefore, the solution of the initial-value problem is

$$y = x - \frac{1}{4}x^3 + \frac{1}{16}x^4 + \cdots.$$

17. The singular point of $(1 - 2\sin x)y'' + xy = 0$ closest to $x = 0$ is $\pi/6$. Hence a lower bound is $\pi/6$.

18. While we can find two solutions of the form

$$y_1 = c_0[1 + \cdots] \quad \text{and} \quad y_2 = c_1[x + \cdots],$$

the initial conditions at $x = 1$ give solutions for c_0 and c_1 in terms of infinite series. Letting $t = x - 1$ the initial-value problem becomes

$$\frac{d^2y}{dt^2} + (t+1)\frac{dy}{dt} + y = 0, \qquad y(0) = -6, \ y'(0) = 3.$$

Substituting $y = \sum_{n=0}^{\infty} c_n t^n$ into the differential equation, we have

$$\frac{d^2y}{dt^2} + (t+1)\frac{dy}{dt} + y = \underbrace{\sum_{n=2}^{\infty} n(n-1)c_n t^{n-2}}_{k=n-2} + \underbrace{\sum_{n=1}^{\infty} nc_n t^n}_{k=n} + \underbrace{\sum_{n=1}^{\infty} nc_n t^{n-1}}_{k=n-1} + \underbrace{\sum_{n=0}^{\infty} c_n t^n}_{k=n}$$

$$= \sum_{k=0}^{\infty}(k+2)(k+1)c_{k+2}t^k + \sum_{k=1}^{\infty} kc_k t^k + \sum_{k=0}^{\infty}(k+1)c_{k+1}t^k + \sum_{k=0}^{\infty} c_k t^k$$

$$= 2c_2 + c_1 + c_0 + \sum_{k=1}^{\infty}[(k+2)(k+1)c_{k+2} + (k+1)c_{k+1} + (k+1)c_k]t^k = 0.$$

Thus

$$2c_2 + c_1 + c_0 = 0$$

$$(k+2)(k+1)c_{k+2} + (k+1)c_{k+1} + (k+1)c_k = 0$$

and
$$c_2 = -\frac{c_1 + c_0}{2}$$

$$c_{k+2} = -\frac{c_{k+1} + c_k}{k+2}, \quad k = 1, 2, 3, \ldots.$$

Choosing $c_0 = 1$ and $c_1 = 0$ we find
$$c_2 = -\frac{1}{2}, \quad c_3 = \frac{1}{6}, \quad c_4 = \frac{1}{12},$$

and so on. For $c_0 = 0$ and $c_1 = 1$ we find
$$c_2 = -\frac{1}{2}, \quad c_3 = -\frac{1}{6}, \quad c_4 = \frac{1}{6},$$

and so on. Thus, the general solution is
$$y = c_0 \left[1 - \frac{1}{2}t^2 + \frac{1}{6}t^3 + \frac{1}{12}t^4 + \cdots \right] + c_1 \left[t - \frac{1}{2}t^2 - \frac{1}{6}t^3 + \frac{1}{6}t^4 + \cdots \right].$$

The initial conditions then imply $c_0 = -6$ and $c_1 = 3$. Thus the solution of the initial-value problem is
$$y = -6\left[1 - \frac{1}{2}(x-1)^2 + \frac{1}{6}(x-1)^3 + \frac{1}{12}(x-1)^4 + \cdots \right]$$
$$+ 3\left[(x-1) - \frac{1}{2}(x-1)^2 - \frac{1}{6}(x-1)^3 + \frac{1}{6}(x-1)^4 + \cdots \right].$$

19. Writing the differential equation in the form
$$y'' + \left(\frac{1 - \cos x}{x} \right) y' + xy = 0,$$

and noting that
$$\frac{1 - \cos x}{x} = \frac{x}{2} - \frac{x^3}{24} + \frac{x^5}{720} - \cdots$$

is analytic at $x = 0$, we conclude that $x = 0$ is an ordinary point of the differential equation.

20. Writing the differential equation in the form
$$y'' + \left(\frac{x}{e^x - 1 - x} \right) y = 0$$

and noting that
$$\frac{x}{e^x - 1 - x} = \frac{2}{x} - \frac{2}{3} + \frac{x}{18} + \frac{x^2}{270} - \cdots$$

we see that $x = 0$ is a singular point of the differential equation. Since
$$x^2 \left(\frac{x}{e^x - 1 - x} \right) = 2x - \frac{2x^2}{3} + \frac{x^3}{18} + \frac{x^4}{270} - \cdots,$$

we conclude that $x = 0$ is a regular singular point.

21. Substituting $y = \sum_{n=0}^{\infty} c_n x^n$ into the differential equation we have

$$y'' + x^2 y' + 2xy = \underbrace{\sum_{n=2}^{\infty} n(n-1)c_n x^{n-2}}_{k=n-2} + \underbrace{\sum_{n=1}^{\infty} nc_n x^{n+1}}_{k=n+1} + 2\underbrace{\sum_{n=0}^{\infty} c_n x^{n+1}}_{k=n+1}$$

$$= \sum_{k=0}^{\infty} (k+2)(k+1)c_{k+2}x^k + \sum_{k=2}^{\infty} (k-1)c_{k-1}x^k + 2\sum_{k=1}^{\infty} c_{k-1}x^k$$

$$= 2c_2 + (6c_3 + 2c_0)x + \sum_{k=2}^{\infty} [(k+2)(k+1)c_{k+2} + (k+1)c_{k-1}]x^k = 5 - 2x + 10x^3.$$

Thus, equating coefficients of like powers of x gives

$$2c_2 = 5$$

$$6c_3 + 2c_0 = -2$$

$$12c_4 + 3c_1 = 0$$

$$20c_5 + 4c_2 = 10$$

$$(k+2)(k+1)c_{k+2} + (k+1)c_{k-1} = 0, \quad k = 4, 5, 6, \dots,$$

and

$$c_2 = \frac{5}{2}$$

$$c_3 = -\frac{1}{3}c_0 - \frac{1}{3}$$

$$c_4 = -\frac{1}{4}c_1$$

$$c_5 = \frac{1}{2} - \frac{1}{5}c_2 = \frac{1}{2} - \frac{1}{5}\left(\frac{5}{2}\right) = 0$$

$$c_{k+2} = -\frac{1}{k+2}c_{k-1}.$$

Using the recurrence relation, we find

$$c_6 = -\frac{1}{6}c_3 = \frac{1}{3 \cdot 6}(c_0 + 1) = \frac{1}{3^2 \cdot 2!}c_0 + \frac{1}{3^2 \cdot 2!}$$

$$c_7 = -\frac{1}{7}c_4 = \frac{1}{4 \cdot 7}c_1$$

$$c_8 = c_{11} = c_{14} = \cdots = 0$$

$$c_9 = -\frac{1}{9}c_6 = -\frac{1}{3^3 \cdot 3!}c_0 - \frac{1}{3^3 \cdot 3!}$$

$$c_{10} = -\frac{1}{10}c_7 = -\frac{1}{4 \cdot 7 \cdot 10}c_1$$

$$c_{12} = -\frac{1}{12}c_9 = \frac{1}{3^4 \cdot 4!}c_0 + \frac{1}{3^4 \cdot 4!}$$

$$c_{13} = -\frac{1}{13}c_0 = \frac{1}{4 \cdot 7 \cdot 10 \cdot 13}c_1$$

and so on. Thus

$$y = c_0 \left[1 - \frac{1}{3}x^3 + \frac{1}{3^2 \cdot 2!}x^6 - \frac{1}{3^3 \cdot 3!}x^9 + \frac{1}{3^4 \cdot 4!}x^{12} - \cdots\right]$$

$$+ c_1 \left[x - \frac{1}{4}x^4 + \frac{1}{4 \cdot 7}x^7 - \frac{1}{4 \cdot 7 \cdot 10}x^{10} + \frac{1}{4 \cdot 7 \cdot 10 \cdot 13}x^{13} - \cdots\right]$$

$$+ \left[\frac{5}{2}x^2 - \frac{1}{3}x^3 + \frac{1}{3^2 \cdot 2!}x^6 - \frac{1}{3^3 \cdot 3!}x^9 + \frac{1}{3^4 \cdot 4!}x^{12} - \cdots\right].$$

22. (a) From $y = -\frac{1}{u}\frac{du}{dx}$ we obtain

$$\frac{dy}{dx} = -\frac{1}{u}\frac{d^2u}{dx^2} + \frac{1}{u^2}\left(\frac{du}{dx}\right)^2.$$

Then $dy/dx = x^2 + y^2$ becomes

$$-\frac{1}{u}\frac{d^2u}{dx^2} + \frac{1}{u^2}\left(\frac{du}{dx}\right)^2 = x^2 + \frac{1}{u^2}\left(\frac{du}{dx}\right)^2,$$

so $\quad \dfrac{d^2u}{dx^2} + x^2u = 0.$

(b) The differential equation $u'' + x^2u = 0$ has the form of (18) in Section 6.3 in the text with

$$1 - 2a = 0 \implies a = \frac{1}{2}$$

$$2c - 2 = 2 \implies c = 2$$

$$b^2c^2 = 1 \implies b = \frac{1}{2}$$

$$a^2 - p^2c^2 = 0 \implies p = \frac{1}{4}.$$

Then, by (19) of Section 6.3 in the text,

$$u = x^{1/2}\left[c_1 J_{1/4}\left(\frac{1}{2}x^2\right) + c_2 J_{-1/4}\left(\frac{1}{2}x^2\right)\right].$$

(c) We have

$$y = -\frac{1}{u}\frac{du}{dx} = -\frac{1}{x^{1/2}w(t)}\frac{d}{dx}x^{1/2}w(t)$$

$$= -\frac{1}{x^{1/2}w}\left[x^{1/2}\frac{dw}{dt}\frac{dt}{dx} + \frac{1}{2}x^{-1/2}w\right]$$

$$= -\frac{1}{x^{1/2}w}\left[x^{3/2}\frac{dw}{dt} + \frac{1}{2x^{1/2}}w\right]$$

$$= -\frac{1}{2xw}\left[2x^2\frac{dw}{dt} + w\right] = -\frac{1}{2xw}\left[4t\frac{dw}{dt} + w\right].$$

Now

$$4t\frac{dw}{dt} + w = 4t\frac{d}{dt}[c_1 J_{1/4}(t) + c_2 J_{-1/4}(t)] + c_1 J_{1/4}(t) + c_2 J_{-1/4}(t)$$

$$= 4t\left[c_1\left(J_{-3/4}(t) - \frac{1}{4t}J_{1/4}(t)\right) + c_2\left(-\frac{1}{4t}J_{-1/4}(t) - J_{3/4}(t)\right)\right]$$

$$\quad + c_1 J_{1/4}(t) + c_2 J_{-1/4}(t)$$

$$= 4c_1 t J_{-3/4}(t) - 4c_2 t J_{3/4}(t)$$

$$= 2c_1 x^2 J_{-3/4}\left(\frac{1}{2}x^2\right) - 2c_2 x^2 J_{3/4}\left(\frac{1}{2}x^2\right),$$

so

$$y = -\frac{2c_1 x^2 J_{-3/4}(\frac{1}{2}x^2) - 2c_2 x^2 J_{3/4}(\frac{1}{2}x^2)}{2x[c_1 J_{1/4}(\frac{1}{2}x^2) + c_2 J_{-1/4}(\frac{1}{2}x^2)]}$$

$$= x\frac{-c_1 J_{-3/4}(\frac{1}{2}x^2) + c_2 J_{3/4}(\frac{1}{2}x^2)}{c_1 J_{1/4}(\frac{1}{2}x^2) + c_2 J_{-1/4}(\frac{1}{2}x^2)}.$$

Letting $c = c_1/c_2$ we have

$$y = x\frac{J_{3/4}(\frac{1}{2}x^2) - cJ_{-3/4}(\frac{1}{2}x^2)}{cJ_{1/4}(\frac{1}{2}x^2) + J_{-1/4}(\frac{1}{2}x^2)}.$$

23. (a) Equations (10) and (24) of Section 6.3 in the text imply

$$Y_{1/2}(x) = \frac{\cos\frac{\pi}{2}J_{1/2}(x) - J_{-1/2}(x)}{\sin\frac{\pi}{2}} = -J_{-1/2}(x) = -\sqrt{\frac{2}{\pi x}}\cos x.$$

(b) From (15) of Section 6.3 in the text

$$I_{1/2}(x) = i^{-1/2} J_{1/2}(ix) \qquad \text{and} \qquad I_{-1/2}(x) = i^{1/2} J_{-1/2}(ix)$$

so

$$I_{1/2}(x) = \sqrt{\frac{2}{\pi x}} \sum_{n=0}^{\infty} \frac{1}{(2n+1)!} x^{2n+1} = \sqrt{\frac{2}{\pi x}} \sinh x$$

and

$$I_{-1/2}(x) = \sqrt{\frac{2}{\pi x}} \sum_{n=0}^{\infty} \frac{1}{(2n)!} x^{2n} = \sqrt{\frac{2}{\pi x}} \cosh x.$$

(c) Equation (16) of Section 6.3 in the text and part (b) imply

$$K_{1/2}(x) = \frac{\pi}{2} \frac{I_{-1/2}(x) - I_{1/2}(x)}{\sin \frac{\pi}{2}} = \frac{\pi}{2} \left[\sqrt{\frac{2}{\pi x}} \cosh x - \sqrt{\frac{2}{\pi x}} \sinh x \right]$$

$$= \sqrt{\frac{\pi}{2x}} \left[\frac{e^x + e^{-x}}{2} - \frac{e^x - e^{-x}}{2} \right] = \sqrt{\frac{\pi}{2x}} e^{-x}.$$

24. (a) Using formula (5) of Section 4.2 in the text, we find that a second solution of $(1-x^2)y'' - 2xy' = 0$ is

$$y_2(x) = 1 \cdot \int \frac{e^{\int 2x\, dx/(1-x^2)}}{1^2} \, dx = \int e^{-\ln(1-x^2)} \, dx$$

$$= \int \frac{dx}{1 - x^2} = \frac{1}{2} \ln \left(\frac{1+x}{1-x} \right),$$

where partial fractions was used to obtain the last integral.

(b) Using formula (5) of Section 4.2 in the text, we find that a second solution of $(1 - x^2)y'' - 2xy' + 2y = 0$ is

$$y_2(x) = x \cdot \int \frac{e^{\int 2x\, dx/(1-x^2)}}{x^2} \, dx = x \int \frac{e^{-\ln(1-x^2)}}{x^2} \, dx$$

$$= x \int \frac{dx}{x^2(1 - x^2)} \, dx = x \left[\frac{1}{2} \ln \left(\frac{1+x}{1-x} \right) - \frac{1}{x} \right]$$

$$= \frac{x}{2} \ln \left(\frac{1+x}{1-x} \right) - 1,$$

where partial fractions was used to obtain the last integral.

350

(c)

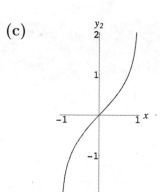 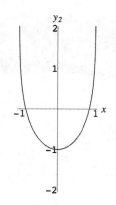

$$y_2(x) = \frac{1}{2}\ln\left(\frac{1+x}{1-x}\right) \qquad y_2 = \frac{x}{2}\ln\left(\frac{1+x}{1-x}\right) - 1$$

25. (a) By the binomial theorem we have

$$\left[1 + \left(t^2 - 2xt\right)\right]^{-1/2} = 1 - \frac{1}{2}\left(t^2 - 2xt\right) + \frac{(-1/2)(-3/2)}{2!}\left(t^2 - 2xt\right)^2$$

$$+ \frac{(-1/2)(-3/2)(-5/2)}{3!}\left(t^2 - 2xt\right)^3 + \cdots$$

$$= 1 - \frac{1}{2}(t^2 - 2xt) + \frac{3}{8}(t^2 - 2xt)^2 - \frac{5}{16}(t^2 - 2xt)^3 + \cdots$$

$$= 1 + xt + \frac{1}{2}(3x^2 - 1)t^2 + \frac{1}{2}(5x^3 - 3x)t^3 + \cdots$$

$$= \sum_{n=0}^{\infty} P_n(x)t^n.$$

(b) Letting $x = 1$ in $(1 - 2xt + t^2)^{-1/2}$, we have

$$(1 - 2t + t^2)^{-1/2} = (1 - t)^{-1} = \frac{1}{1-t} = 1 + t + t^2 + t^3 + \cdots \quad (|t| < 1)$$

$$= \sum_{n=0}^{\infty} t^n.$$

From part (a) we have

$$\sum_{n=0}^{\infty} P_n(1)t^n = (1 - 2t + t^2)^{-1/2} = \sum_{n=0}^{\infty} t^n.$$

Equating the coefficients of corresponding terms in the two series, we see that $P_n(1) = 1$. Similarly, letting $x = -1$ we have

$$(1 + 2t + t^2)^{-1/2} = (1 + t)^{-1} = \frac{1}{1+t} = 1 - t + t^2 - 3t^3 + \cdots \quad (|t| < 1)$$

$$= \sum_{n=0}^{\infty}(-1)^n t^n = \sum_{n=0}^{\infty} P_n(-1)t^n,$$

so that $P_n(-1) = (-1)^n$.

7 The Laplace Transform

Exercises 7.1   Definition of the Laplace Transform

1. $\mathscr{L}\{f(t)\} = \displaystyle\int_0^1 -e^{-st}\,dt + \int_1^\infty e^{-st}\,dt = \dfrac{1}{s}e^{-st}\,\Big|_0^1 - \dfrac{1}{s}e^{-st}\,\Big|_1^\infty$

$\qquad = \dfrac{1}{s}e^{-s} - \dfrac{1}{s} - \left(0 - \dfrac{1}{s}e^{-s}\right) = \dfrac{2}{s}e^{-s} - \dfrac{1}{s}, \quad s > 0$

2. $\mathscr{L}\{f(t)\} = \displaystyle\int_0^2 4e^{-st}\,dt = -\dfrac{4}{s}e^{-st}\,\Big|_0^2 = -\dfrac{4}{s}(e^{-2s} - 1), \quad s > 0$

3. $\mathscr{L}\{f(t)\} = \displaystyle\int_0^1 te^{-st}\,dt + \int_1^\infty e^{-st}\,dt = \left(-\dfrac{1}{s}te^{-st} - \dfrac{1}{s^2}e^{-st}\right)\Big|_0^1 - \dfrac{1}{s}e^{-st}\,\Big|_1^\infty$

$\qquad = \left(-\dfrac{1}{s}e^{-s} - \dfrac{1}{s^2}e^{-s}\right) - \left(0 - \dfrac{1}{s^2}\right) - \dfrac{1}{s}(0 - e^{-s}) = \dfrac{1}{s^2}(1 - e^{-s}), \quad s > 0$

4. $\mathscr{L}\{f(t)\} = \displaystyle\int_0^1 (2t + 1)e^{-st}\,dt = \left(-\dfrac{2}{s}te^{-st} - \dfrac{2}{s^2}e^{-st} - \dfrac{1}{s}e^{-st}\right)\Big|_0^1$

$\qquad = \left(-\dfrac{2}{s}e^{-s} - \dfrac{2}{s^2}e^{-s} - \dfrac{1}{s}e^{-s}\right) - \left(0 - \dfrac{2}{s^2} - \dfrac{1}{s}\right) = \dfrac{1}{s}(1 - 3e^{-s}) + \dfrac{2}{s^2}(1 - e^{-s}), \quad s > 0$

5. $\mathscr{L}\{f(t)\} = \displaystyle\int_0^\pi (\sin t)e^{-st}\,dt = \left(-\dfrac{s}{s^2 + 1}e^{-st}\sin t - \dfrac{1}{s^2 + 1}e^{-st}\cos t\right)\Big|_0^\pi$

$\qquad = \left(0 + \dfrac{1}{s^2 + 1}e^{-\pi s}\right) - \left(0 - \dfrac{1}{s^2 + 1}\right) = \dfrac{1}{s^2 + 1}(e^{-\pi s} + 1), \quad s > 0$

6. $\mathscr{L}\{f(t)\} = \displaystyle\int_{\pi/2}^\infty (\cos t)e^{-st}\,dt = \left(-\dfrac{s}{s^2 + 1}e^{-st}\cos t + \dfrac{1}{s^2 + 1}e^{-st}\sin t\right)\Big|_{\pi/2}^\infty$

$\qquad = 0 - \left(0 + \dfrac{1}{s^2 + 1}e^{-\pi s/2}\right) = -\dfrac{1}{s^2 + 1}e^{-\pi s/2}, \quad s > 0$

7. $f(t) = \begin{cases} 0, & 0 < t < 1 \\ t, & t > 1 \end{cases}$

$\qquad \mathscr{L}\{f(t)\} = \displaystyle\int_1^\infty te^{-st}\,dt = \left(-\dfrac{1}{s}te^{-st} - \dfrac{1}{s^2}e^{-st}\right)\Big|_1^\infty = \dfrac{1}{s}e^{-s} + \dfrac{1}{s^2}e^{-s}, \quad s > 0$

8. $f(t) = \begin{cases} 0, & 0 < t < 1 \\ 2t - 2, & t > 1 \end{cases}$

$$\mathscr{L}\{f(t)\} = 2\int_1^\infty (t-1)e^{-st}\,dt = 2\left(-\frac{1}{s}(t-1)e^{-st} - \frac{1}{s^2}e^{-st}\right)\bigg|_1^\infty = \frac{2}{s^2}e^{-s}, \quad s > 0$$

9. The function is $f(t) = \begin{cases} 1-t, & 0 < t < 1 \\ 0, & t > 1 \end{cases}$ so

$$\mathscr{L}\{f(t)\} = \int_0^1 (1-t)e^{-st}\,dt + \int_1^\infty 0e^{-st}\,dt = \int_0^1 (1-t)e^{-st}\,dt = \left(-\frac{1}{s}(1-t)e^{-st} + \frac{1}{s^2}e^{-st}\right)\bigg|_0^1$$

$$= \frac{1}{s^2}e^{-s} + \frac{1}{s} - \frac{1}{s^2}, \quad s > 0$$

10. $f(t) = \begin{cases} 0, & 0 < t < a \\ c, & a < t < b; \\ 0, & t > b \end{cases}$ $\mathscr{L}\{f(t)\} = \int_a^b ce^{-st}\,dt = -\frac{c}{s}e^{-st}\bigg|_a^b = \frac{c}{s}(e^{-sa} - e^{-sb}), \quad s > 0$

11. $\mathscr{L}\{f(t)\} = \int_0^\infty e^{t+7}e^{-st}\,dt = e^7\int_0^\infty e^{(1-s)t}\,dt = \dfrac{e^7}{1-s}e^{(1-s)t}\bigg|_0^\infty = 0 - \dfrac{e^7}{1-s} = \dfrac{e^7}{s-1}, \quad s > 1$

12. $\mathscr{L}\{f(t)\} = \int_0^\infty e^{-2t-5}e^{-st}\,dt = e^{-5}\int_0^\infty e^{-(s+2)t}\,dt = -\dfrac{e^{-5}}{s+2}e^{-(s+2)t}\bigg|_0^\infty = \dfrac{e^{-5}}{s+2}, \quad s > -2$

13. $\mathscr{L}\{f(t)\} = \int_0^\infty te^{4t}e^{-st}\,dt = \int_0^\infty te^{(4-s)t}\,dt = \left(\dfrac{1}{4-s}te^{(4-s)t} - \dfrac{1}{(4-s)^2}e^{(4-s)t}\right)\bigg|_0^\infty$

$$= \frac{1}{(4-s)^2}, \quad s > 4$$

14. $\mathscr{L}\{f(t)\} = \int_0^\infty t^2 e^{-2t}e^{-st}\,dt = \int_0^\infty t^2 e^{-(s+2)t}\,dt$

$$= \left(-\frac{1}{s+2}t^2 e^{-(s+2)t} - \frac{2}{(s+2)^2}te^{-(s+2)t} - \frac{2}{(s+2)^3}e^{-(s+2)t}\right)\bigg|_0^\infty = \frac{2}{(s+2)^3}, \quad s > -2$$

15. $\mathscr{L}\{f(t)\} = \int_0^\infty e^{-t}(\sin t)e^{-st}\,dt = \int_0^\infty (\sin t)e^{-(s+1)t}\,dt$

$$= \left(\frac{-(s+1)}{(s+1)^2+1}e^{-(s+1)t}\sin t - \frac{1}{(s+1)^2+1}e^{-(s+1)t}\cos t\right)\bigg|_0^\infty$$

$$= \frac{1}{(s+1)^2+1} = \frac{1}{s^2+2s+2}, \quad s > -1$$

16. $\mathscr{L}\{f(t)\} = \int_0^\infty e^t(\cos t)e^{-st}\,dt = \int_0^\infty (\cos t)e^{(1-s)t}\,dt$

$$= \left(\frac{1-s}{(1-s)^2+1}e^{(1-s)t}\cos t + \frac{1}{(1-s)^2+1}e^{(1-s)t}\sin t\right)\bigg|_0^\infty$$

$$= -\frac{1-s}{(1-s)^2+1} = \frac{s-1}{s^2-2s+2}, \quad s > 1$$

17. $\mathcal{L}\{f(t)\} = \displaystyle\int_0^\infty t(\cos t)e^{-st}\,dt$

$$= \left[\left(-\frac{st}{s^2+1} - \frac{s^2-1}{(s^2+1)^2}\right)(\cos t)e^{-st} + \left(\frac{t}{s^2+1} + \frac{2s}{(s^2+1)^2}\right)(\sin t)e^{-st}\right]_0^\infty$$

$$= \frac{s^2-1}{(s^2+1)^2}, \quad s > 0$$

18. $\mathcal{L}\{f(t)\} = \displaystyle\int_0^\infty t(\sin t)e^{-st}\,dt$

$$= \left[\left(-\frac{t}{s^2+1} - \frac{2s}{(s^2+1)^2}\right)(\cos t)e^{-st} - \left(\frac{st}{s^2+1} + \frac{s^2-1}{(s^2+1)^2}\right)(\sin t)e^{-st}\right]_0^\infty$$

$$= \frac{2s}{(s^2+1)^2}, \quad s > 0$$

19. $\mathcal{L}\{2t^4\} = 2\dfrac{4!}{s^5}$

20. $\mathcal{L}\{t^5\} = \dfrac{5!}{s^6}$

21. $\mathcal{L}\{4t - 10\} = \dfrac{4}{s^2} - \dfrac{10}{s}$

22. $\mathcal{L}\{7t + 3\} = \dfrac{7}{s^2} + \dfrac{3}{s}$

23. $\mathcal{L}\{t^2 + 6t - 3\} = \dfrac{2}{s^3} + \dfrac{6}{s^2} - \dfrac{3}{s}$

24. $\mathcal{L}\{-4t^2 + 16t + 9\} = -4\dfrac{2}{s^3} + \dfrac{16}{s^2} + \dfrac{9}{s}$

25. $\mathcal{L}\{t^3 + 3t^2 + 3t + 1\} = \dfrac{3!}{s^4} + 3\dfrac{2}{s^3} + 3\dfrac{1}{s^2} + \dfrac{1}{s}$

26. $\mathcal{L}\{8t^3 - 12t^2 + 6t - 1\} = 8\dfrac{3!}{s^4} - 12\dfrac{2}{s^3} + \dfrac{6}{s^2} - \dfrac{1}{s}$

27. $\mathcal{L}\{1 + e^{4t}\} = \dfrac{1}{s} + \dfrac{1}{s-4}$

28. $\mathcal{L}\{t^2 - e^{-9t} + 5\} = \dfrac{2}{s^3} - \dfrac{1}{s+9} + \dfrac{5}{s}$

29. $\mathcal{L}\{1 + 2e^{2t} + e^{4t}\} = \dfrac{1}{s} + \dfrac{2}{s-2} + \dfrac{1}{s-4}$

30. $\mathcal{L}\{e^{2t} - 2 + e^{-2t}\} = \dfrac{1}{s-2} - \dfrac{2}{s} + \dfrac{1}{s+2}$

31. $\mathcal{L}\{4t^2 - 5\sin 3t\} = 4\dfrac{2}{s^3} - 5\dfrac{3}{s^2+9}$

32. $\mathcal{L}\{\cos 5t + \sin 2t\} = \dfrac{s}{s^2+25} + \dfrac{2}{s^2+4}$

33. $\mathcal{L}\{\sinh kt\} = \dfrac{1}{2}\mathcal{L}\{e^{kt} - e^{-kt}\} = \dfrac{1}{2}\left[\dfrac{1}{s-k} - \dfrac{1}{s+k}\right] = \dfrac{k}{s^2-k^2}$

34. $\mathcal{L}\{\cosh kt\} = \dfrac{1}{2}\mathcal{L}\{e^{kt} + e^{kt}\} = \dfrac{s}{s^2-k^2}$

35. $\mathcal{L}\{e^t \sinh t\} = \mathcal{L}\left\{e^t \dfrac{e^t - e^{-t}}{2}\right\} = \mathcal{L}\left\{\dfrac{1}{2}e^{2t} - \dfrac{1}{2}\right\} = \dfrac{1}{2(s-2)} - \dfrac{1}{2s}$

36. $\mathcal{L}\{e^{-t} \cosh t\} = \mathcal{L}\left\{e^{-t} \dfrac{e^t + e^{-t}}{2}\right\} = \mathcal{L}\left\{\dfrac{1}{2} + \dfrac{1}{2}e^{-2t}\right\} = \dfrac{1}{2s} + \dfrac{1}{2(s+2)}$

37. $\mathscr{L}\{\sin 2t \cos 2t\} = \mathscr{L}\left\{\dfrac{1}{2}\sin 4t\right\} = \dfrac{2}{s^2 + 16}$

38. $\mathscr{L}\{\cos^2 t\} = \mathscr{L}\left\{\dfrac{1}{2} + \dfrac{1}{2}\cos 2t\right\} = \dfrac{1}{2s} + \dfrac{1}{2}\dfrac{s}{s^2 + 4}$

39. From the addition formula for the sine function, $\sin(4t + 5) = (\sin 4t)(\cos 5) + (\cos 4t)(\sin 5)$ so

$$\mathscr{L}\{\sin(4t + 5)\} = (\cos 5)\,\mathscr{L}\{\sin 4t\} + (\sin 5)\,\mathscr{L}\{\cos 4t\}$$

$$= (\cos 5)\,\dfrac{4}{s^2 + 16} + (\sin 5)\,\dfrac{s}{s^2 + 16}$$

$$= \dfrac{4\cos 5 + (\sin 5)s}{s^2 + 16}.$$

40. From the addition formula for the cosine function,

$$\cos\left(t - \dfrac{\pi}{6}\right) = \cos t \cos\dfrac{\pi}{6} + \sin t \sin\dfrac{\pi}{6} = \dfrac{\sqrt{3}}{2}\cos t + \dfrac{1}{2}\sin t$$

so

$$\mathscr{L}\left\{\cos\left(t - \dfrac{\pi}{6}\right)\right\} = \dfrac{\sqrt{3}}{2}\mathscr{L}\{\cos t\} + \dfrac{1}{2}\mathscr{L}\{\sin t\}$$

$$= \dfrac{\sqrt{3}}{2}\dfrac{s}{s^2 + 1} + \dfrac{1}{2}\dfrac{1}{s^2 + 1} = \dfrac{1}{2}\dfrac{\sqrt{3}\,s + 1}{s^2 + 1}.$$

41. (a) Using integration by parts for $\alpha > 0$,

$$\Gamma(\alpha + 1) = \int_0^\infty t^\alpha e^{-t}\,dt = -t^\alpha e^{-t}\,\Big|_0^\infty + \alpha\int_0^\infty t^{\alpha-1}e^{-t}\,dt = \alpha\Gamma(\alpha).$$

(b) Let $u = st$ so that $du = s\,dt$. Then

$$\mathscr{L}\{t^\alpha\} = \int_0^\infty e^{-st}t^\alpha dt = \int_0^\infty e^{-u}\left(\dfrac{u}{s}\right)^\alpha\dfrac{1}{s}\,du = \dfrac{1}{s^{\alpha+1}}\Gamma(\alpha + 1), \quad \alpha > -1.$$

42. (a) $\mathscr{L}\{t^{-1/2}\} = \dfrac{\Gamma(1/2)}{s^{1/2}} = \sqrt{\dfrac{\pi}{s}}$

(b) $\mathscr{L}\{t^{1/2}\} = \dfrac{\Gamma(3/2)}{s^{3/2}} = \dfrac{\sqrt{\pi}}{2s^{3/2}}$

(c) $\mathscr{L}\{t^{3/2}\} = \dfrac{\Gamma(5/2)}{s^{5/2}} = \dfrac{3\sqrt{\pi}}{4s^{5/2}}$

43. Let $F(t) = t^{1/3}$. Then $F(t)$ is of exponential order, but $f(t) = F'(t) = \frac{1}{3}t^{-2/3}$ is unbounded near $t = 0$ and hence is not of exponential order. Let

$$f(t) = 2te^{t^2}\cos e^{t^2} = \dfrac{d}{dt}\sin e^{t^2}.$$

355

This function is not of exponential order, but we can show that its Laplace transform exists. Using integration by parts we have

$$\mathscr{L}\{2te^{t^2}\cos e^{t^2}\} = \int_0^\infty e^{-st}\left(\frac{d}{dt}\sin e^{t^2}\right)dt = \lim_{a\to\infty}\left[e^{-st}\sin e^{t^2}\,\Big|_0^a + s\int_0^a e^{-st}\sin e^{t^2}\,dt\right]$$

$$= -\sin 1 + s\int_0^\infty e^{-st}\sin e^{t^2}\,dt = s\mathscr{L}\{\sin e^{t^2}\} - \sin 1.$$

Since $\sin e^{t^2}$ is continuous and of exponential order, $\mathscr{L}\{\sin e^{t^2}\}$ exists, and therefore $\mathscr{L}\{2te^{t^2}\cos e^{t^2}\}$ exists.

44. The relation will be valid when s is greater than the maximum of c_1 and c_2.

45. Since e^t is an increasing function and $t^2 > \ln M + ct$ for $M > 0$ we have $e^{t^2} > e^{\ln M + ct} = Me^{ct}$ for t sufficiently large and for any c. Thus, e^{t^2} is not of exponential order.

46. Assuming that (c) of Theorem 7.1.1 is applicable with a complex exponent, we have

$$\mathscr{L}\{e^{(a+ib)t}\} = \frac{1}{s-(a+ib)} = \frac{1}{(s-a)-ib}\frac{(s-a)+ib}{(s-a)+ib} = \frac{s-a+ib}{(s-a)^2+b^2}.$$

By Euler's formula, $e^{i\theta} = \cos\theta + i\sin\theta$, so

$$\mathscr{L}\{e^{(a+ib)t}\} = \mathscr{L}\{e^{at}e^{ibt}\} = \mathscr{L}\{e^{at}(\cos bt + i\sin bt)\}$$

$$= \mathscr{L}\{e^{at}\cos bt\} + i\mathscr{L}\{e^{at}\sin bt\}$$

$$= \frac{s-a}{(s-a)^2+b^2} + i\frac{b}{(s-a)^2+b^2}.$$

Equating real and imaginary parts we get

$$\mathscr{L}\{e^{at}\cos bt\} = \frac{s-a}{(s-a)^2+b^2} \quad\text{and}\quad \mathscr{L}\{e^{at}\sin bt\} = \frac{b}{(s-a)^2+b^2}.$$

47. We want $f(\alpha x + \beta y) = \alpha f(x) + \beta f(y)$ or

$$m(\alpha x + \beta y) + b = \alpha(mx + b) + \beta(my + b) = m(\alpha x + \beta y) + (\alpha + \beta)b$$

for all real numbers α and β. Taking $\alpha = \beta = 1$ we see that $b = 2b$, so $b = 0$. Thus, $f(x) = mx + b$ will be a linear transformation when $b = 0$.

48. Assume that $\mathscr{L}\{t^{n-1}\} = (n-1)!/s^n$. Then, using the definition of the Laplace transform and integration by parts, we have

$$\mathscr{L}\{t^n\} = \int_0^\infty e^{-st}t^n\,dt = -\frac{1}{s}e^{-st}t^n\,\Big|_0^\infty + \frac{n}{s}\int_0^\infty e^{-st}t^{n-1}\,dt$$

$$= 0 + \frac{n}{s}\mathscr{L}\{t^{n-1}\} = \frac{n}{s}\frac{(n-1)!}{s^n} = \frac{n!}{s^{n+1}}.$$

1. $\mathscr{L}^{-1}\left\{\dfrac{1}{s^3}\right\} = \dfrac{1}{2}\mathscr{L}^{-1}\left\{\dfrac{2}{s^3}\right\} = \dfrac{1}{2}t^2$

2. $\mathscr{L}^{-1}\left\{\dfrac{1}{s^4}\right\} = \dfrac{1}{6}\mathscr{L}^{-1}\left\{\dfrac{3!}{s^4}\right\} = \dfrac{1}{6}t^3$

3. $\mathscr{L}^{-1}\left\{\dfrac{1}{s^2} - \dfrac{48}{s^5}\right\} = \mathscr{L}^{-1}\left\{\dfrac{1}{s^2} - \dfrac{48}{24}\cdot\dfrac{4!}{s^5}\right\} = t - 2t^4$

4. $\mathscr{L}^{-1}\left\{\left(\dfrac{2}{s} - \dfrac{1}{s^3}\right)^2\right\} = \mathscr{L}^{-1}\left\{4\cdot\dfrac{1}{s^2} - \dfrac{4}{6}\cdot\dfrac{3!}{s^4} + \dfrac{1}{120}\cdot\dfrac{5!}{s^6}\right\} = 4t - \dfrac{2}{3}t^3 + \dfrac{1}{120}t^5$

5. $\mathscr{L}^{-1}\left\{\dfrac{(s+1)^3}{s^4}\right\} = \mathscr{L}^{-1}\left\{\dfrac{1}{s} + 3\cdot\dfrac{1}{s^2} + \dfrac{3}{2}\cdot\dfrac{2}{s^3} + \dfrac{1}{6}\cdot\dfrac{3!}{s^4}\right\} = 1 + 3t + \dfrac{3}{2}t^2 + \dfrac{1}{6}t^3$

6. $\mathscr{L}^{-1}\left\{\dfrac{(s+2)^2}{s^3}\right\} = \mathscr{L}^{-1}\left\{\dfrac{1}{s} + 4\cdot\dfrac{1}{s^2} + 2\cdot\dfrac{2}{s^3}\right\} = 1 + 4t + 2t^2$

7. $\mathscr{L}^{-1}\left\{\dfrac{1}{s^2} - \dfrac{1}{s} + \dfrac{1}{s-2}\right\} = t - 1 + e^{2t}$

8. $\mathscr{L}^{-1}\left\{\dfrac{4}{s} + \dfrac{6}{s^5} - \dfrac{1}{s+8}\right\} = \mathscr{L}^{-1}\left\{4\cdot\dfrac{1}{s} + \dfrac{1}{4}\cdot\dfrac{4!}{s^5} - \dfrac{1}{s+8}\right\} = 4 + \dfrac{1}{4}t^4 - e^{-8t}$

9. $\mathscr{L}^{-1}\left\{\dfrac{1}{4s+1}\right\} = \dfrac{1}{4}\mathscr{L}^{-1}\left\{\dfrac{1}{s+1/4}\right\} = \dfrac{1}{4}e^{-t/4}$

10. $\mathscr{L}^{-1}\left\{\dfrac{1}{5s-2}\right\} = \mathscr{L}^{-1}\left\{\dfrac{1}{5}\cdot\dfrac{1}{s-2/5}\right\} = \dfrac{1}{5}e^{2t/5}$

11. $\mathscr{L}^{-1}\left\{\dfrac{5}{s^2+49}\right\} = \mathscr{L}^{-1}\left\{\dfrac{5}{7}\cdot\dfrac{7}{s^2+49}\right\} = \dfrac{5}{7}\sin 7t$

12. $\mathscr{L}^{-1}\left\{\dfrac{10s}{s^2+16}\right\} = 10\cos 4t$

13. $\mathscr{L}^{-1}\left\{\dfrac{4s}{4s^2+1}\right\} = \mathscr{L}^{-1}\left\{\dfrac{s}{s^2+1/4}\right\} = \cos\dfrac{1}{2}t$

14. $\mathscr{L}^{-1}\left\{\dfrac{1}{4s^2+1}\right\} = \mathscr{L}^{-1}\left\{\dfrac{1}{2}\cdot\dfrac{1/2}{s^2+1/4}\right\} = \dfrac{1}{2}\sin\dfrac{1}{2}t$

15. $\mathscr{L}^{-1}\left\{\dfrac{2s-6}{s^2+9}\right\} = \mathscr{L}^{-1}\left\{2\cdot\dfrac{s}{s^2+9} - 2\cdot\dfrac{3}{s^2+9}\right\} = 2\cos 3t - 2\sin 3t$

16. $\mathscr{L}^{-1}\left\{\dfrac{s+1}{s^2+2}\right\} = \mathscr{L}^{-1}\left\{\dfrac{s}{s^2+2} + \dfrac{1}{\sqrt{2}}\dfrac{\sqrt{2}}{s^2+2}\right\} = \cos\sqrt{2}t + \dfrac{\sqrt{2}}{2}\sin\sqrt{2}\,t$

17. $\mathscr{L}^{-1}\left\{\dfrac{1}{s^2+3s}\right\} = \mathscr{L}^{-1}\left\{\dfrac{1}{3}\cdot\dfrac{1}{s} - \dfrac{1}{3}\cdot\dfrac{1}{s+3}\right\} = \dfrac{1}{3} - \dfrac{1}{3}e^{-3t}$

18. $\mathscr{L}^{-1}\left\{\dfrac{s+1}{s^2-4s}\right\} = \mathscr{L}^{-1}\left\{-\dfrac{1}{4}\cdot\dfrac{1}{s} + \dfrac{5}{4}\cdot\dfrac{1}{s-4}\right\} = -\dfrac{1}{4} + \dfrac{5}{4}e^{4t}$

19. $\mathscr{L}^{-1}\left\{\dfrac{s}{s^2+2s-3}\right\} = \mathscr{L}^{-1}\left\{\dfrac{1}{4}\cdot\dfrac{1}{s-1} + \dfrac{3}{4}\cdot\dfrac{1}{s+3}\right\} = \dfrac{1}{4}e^t + \dfrac{3}{4}e^{-3t}$

20. $\mathscr{L}^{-1}\left\{\dfrac{1}{s^2+s-20}\right\} = \mathscr{L}^{-1}\left\{\dfrac{1}{9}\cdot\dfrac{1}{s-4} - \dfrac{1}{9}\cdot\dfrac{1}{s+5}\right\} = \dfrac{1}{9}e^{4t} - \dfrac{1}{9}e^{-5t}$

21. $\mathscr{L}^{-1}\left\{\dfrac{0.9s}{(s-0.1)(s+0.2)}\right\} = \mathscr{L}^{-1}\left\{(0.3)\cdot\dfrac{1}{s-0.1} + (0.6)\cdot\dfrac{1}{s+0.2}\right\} = 0.3e^{0.1t} + 0.6e^{-0.2t}$

22. $\mathscr{L}^{-1}\left\{\dfrac{s-3}{(s-\sqrt{3})(s+\sqrt{3})}\right\} = \mathscr{L}^{-1}\left\{\dfrac{s}{s^2-3} - \sqrt{3}\cdot\dfrac{\sqrt{3}}{s^2-3}\right\} = \cosh\sqrt{3}t - \sqrt{3}\sinh\sqrt{3}t$

23. $\mathscr{L}^{-1}\left\{\dfrac{s}{(s-2)(s-3)(s-6)}\right\} = \mathscr{L}^{-1}\left\{\dfrac{1}{2}\cdot\dfrac{1}{s-2} - \dfrac{1}{s-3} + \dfrac{1}{2}\cdot\dfrac{1}{s-6}\right\} = \dfrac{1}{2}e^{2t} - e^{3t} + \dfrac{1}{2}e^{6t}$

24. $\mathscr{L}^{-1}\left\{\dfrac{s^2+1}{s(s-1)(s+1)(s-2)}\right\} = \mathscr{L}^{-1}\left\{\dfrac{1}{2}\cdot\dfrac{1}{s} - \dfrac{1}{s-1} - \dfrac{1}{3}\cdot\dfrac{1}{s+1} + \dfrac{5}{6}\cdot\dfrac{1}{s-2}\right\}$

$$= \dfrac{1}{2} - e^t - \dfrac{1}{3}e^{-t} + \dfrac{5}{6}e^{2t}$$

25. $\mathscr{L}^{-1}\left\{\dfrac{1}{s^3+5s}\right\} = \mathscr{L}^{-1}\left\{\dfrac{1}{s(s^2+5)}\right\} = \mathscr{L}^{-1}\left\{\dfrac{1}{5}\cdot\dfrac{1}{s} - \dfrac{1}{5}\dfrac{s}{s^2+5}\right\} = \dfrac{1}{5} - \dfrac{1}{5}\cos\sqrt{5}t$

26. $\mathscr{L}^{-1}\left\{\dfrac{s}{(s^2+4)(s+2)}\right\} = \mathscr{L}^{-1}\left\{\dfrac{1}{4}\cdot\dfrac{s}{s^2+4} + \dfrac{1}{4}\cdot\dfrac{2}{s^2+4} - \dfrac{1}{4}\cdot\dfrac{1}{s+2}\right\} = \dfrac{1}{4}\cos 2t + \dfrac{1}{4}\sin 2t - \dfrac{1}{4}e^{-2t}$

27. $\mathscr{L}^{-1}\left\{\dfrac{2s-4}{(s^2+s)(s^2+1)}\right\} = \mathscr{L}^{-1}\left\{\dfrac{2s-4}{s(s+1)(s^2+1)}\right\} = \mathscr{L}^{-1}\left\{-\dfrac{4}{s} + \dfrac{3}{s+1} + \dfrac{s}{s^2+1} + \dfrac{3}{s^2+1}\right\}$

$$= -4 + 3e^{-t} + \cos t + 3\sin t$$

28. $\mathscr{L}^{-1}\left\{\dfrac{1}{s^4-9}\right\} = \mathscr{L}^{-1}\left\{\dfrac{1}{6\sqrt{3}}\cdot\dfrac{\sqrt{3}}{s^2-3} - \dfrac{1}{6\sqrt{3}}\cdot\dfrac{\sqrt{3}}{s^2+3}\right\} = \dfrac{1}{6\sqrt{3}}\sinh\sqrt{3}t - \dfrac{1}{6\sqrt{3}}\sin\sqrt{3}t$

29. $\mathscr{L}^{-1}\left\{\dfrac{1}{(s^2+1)(s^2+4)}\right\} = \mathscr{L}^{-1}\left\{\dfrac{1}{3}\cdot\dfrac{1}{s^2+1} - \dfrac{1}{3}\cdot\dfrac{1}{s^2+4}\right\}$

$\qquad\qquad\qquad\qquad = \mathscr{L}^{-1}\left\{\dfrac{1}{3}\cdot\dfrac{1}{s^2+1} - \dfrac{1}{6}\cdot\dfrac{2}{s^2+4}\right\}$

$\qquad\qquad\qquad\qquad = \dfrac{1}{3}\sin t - \dfrac{1}{6}\sin 2t$

30. $\mathscr{L}^{-1}\left\{\dfrac{6s+3}{(s^2+1)(s^2+4)}\right\} = \mathscr{L}^{-1}\left\{2\cdot\dfrac{s}{s^2+1} + \dfrac{1}{s^2+1} - 2\cdot\dfrac{s}{s^2+4} - \dfrac{1}{2}\cdot\dfrac{2}{s^2+4}\right\}$

$\qquad\qquad\qquad\qquad = 2\cos t + \sin t - 2\cos 2t - \dfrac{1}{2}\sin 2t$

31. The Laplace transform of the initial-value problem is

$$s\mathscr{L}\{y\} - y(0) - \mathscr{L}\{y\} = \dfrac{1}{s}.$$

Solving for $\mathscr{L}\{y\}$ we obtain

$$\mathscr{L}\{y\} = -\dfrac{1}{s} + \dfrac{1}{s-1}.$$

Thus

$$y = -1 + e^t.$$

32. The Laplace transform of the initial-value problem is

$$2s\mathscr{L}\{y\} - 2y(0) + \mathscr{L}\{y\} = 0.$$

Solving for $\mathscr{L}\{y\}$ we obtain

$$\mathscr{L}\{y\} = \dfrac{6}{2s+1} = \dfrac{3}{s+1/2}.$$

Thus

$$y = 3e^{-t/2}.$$

33. The Laplace transform of the initial-value problem is

$$s\mathscr{L}\{y\} - y(0) + 6\mathscr{L}\{y\} = \dfrac{1}{s-4}.$$

Solving for $\mathscr{L}\{y\}$ we obtain

$$\mathscr{L}\{y\} = \dfrac{1}{(s-4)(s+6)} + \dfrac{2}{s+6} = \dfrac{1}{10}\cdot\dfrac{1}{s-4} + \dfrac{19}{10}\cdot\dfrac{1}{s+6}.$$

Thus

$$y = \dfrac{1}{10}e^{4t} + \dfrac{19}{10}e^{-6t}.$$

34. The Laplace transform of the initial-value problem is

$$s\mathscr{L}\{y\} - \mathscr{L}\{y\} = \dfrac{2s}{s^2+25}.$$

Solving for $\mathscr{L}\{y\}$ we obtain

$$\mathscr{L}\{y\} = \frac{2s}{(s-1)(s^2+25)} = \frac{1}{13} \cdot \frac{1}{s-1} - \frac{1}{13} \frac{s}{s^2+25} + \frac{5}{13} \cdot \frac{5}{s^2+25}.$$

Thus

$$y = \frac{1}{13}e^t - \frac{1}{13}\cos 5t + \frac{5}{13}\sin 5t.$$

35. The Laplace transform of the initial-value problem is

$$s^2 \mathscr{L}\{y\} - sy(0) - y'(0) + 5\left[s \mathscr{L}\{y\} - y(0)\right] + 4\mathscr{L}\{y\} = 0.$$

Solving for $\mathscr{L}\{y\}$ we obtain

$$\mathscr{L}\{y\} = \frac{s+5}{s^2+5s+4} = \frac{4}{3} \frac{1}{s+1} - \frac{1}{3} \frac{1}{s+4}.$$

Thus

$$y = \frac{4}{3}e^{-t} - \frac{1}{3}e^{-4t}.$$

36. The Laplace transform of the initial-value problem is

$$s^2 \mathscr{L}\{y\} - sy(0) - y'(0) - 4\left[s \mathscr{L}\{y\} - y(0)\right] = \frac{6}{s-3} - \frac{3}{s+1}.$$

Solving for $\mathscr{L}\{y\}$ we obtain

$$\mathscr{L}\{y\} = \frac{6}{(s-3)(s^2-4s)} - \frac{3}{(s+1)(s^2-4s)} + \frac{s-5}{s^2-4s}$$

$$= \frac{5}{2} \cdot \frac{1}{s} - \frac{2}{s-3} - \frac{3}{5} \cdot \frac{1}{s+1} + \frac{11}{10} \cdot \frac{1}{s-4}.$$

Thus

$$y = \frac{5}{2} - 2e^{3t} - \frac{3}{5}e^{-t} + \frac{11}{10}e^{4t}.$$

37. The Laplace transform of the initial-value problem is

$$s^2 \mathscr{L}\{y\} - sy(0) + \mathscr{L}\{y\} = \frac{2}{s^2+2}.$$

Solving for $\mathscr{L}\{y\}$ we obtain

$$\mathscr{L}\{y\} = \frac{2}{(s^2+1)(s^2+2)} + \frac{10s}{s^2+1} = \frac{10s}{s^2+1} + \frac{2}{s^2+1} - \frac{2}{s^2+2}.$$

Thus

$$y = 10\cos t + 2\sin t - \sqrt{2}\sin\sqrt{2}\,t.$$

38. The Laplace transform of the initial-value problem is

$$s^2 \mathscr{L}\{y\} + 9\mathscr{L}\{y\} = \frac{1}{s-1}.$$

Solving for $\mathscr{L}\{y\}$ we obtain

$$\mathscr{L}\{y\} = \frac{1}{(s-1)(s^2+9)} = \frac{1}{10} \cdot \frac{1}{s-1} - \frac{1}{10} \cdot \frac{1}{s^2+9} - \frac{1}{10} \cdot \frac{s}{s^2+9}.$$

Thus

$$y = \frac{1}{10}e^t - \frac{1}{30}\sin 3t - \frac{1}{10}\cos 3t.$$

39. The Laplace transform of the initial-value problem is

$$2\left[s^3\,\mathscr{L}\{y\} - s^2 y(0) - sy'(0) - y''(0)\right] + 3[s^2\,\mathscr{L}\{y\} - sy(0) - y'(0)] - 3[s\,\mathscr{L}\{y\} - y(0)] - 2\,\mathscr{L}\{y\} = \frac{1}{s+1}.$$

Solving for $\mathscr{L}\{y\}$ we obtain

$$\mathscr{L}\{y\} = \frac{2s+3}{(s+1)(s-1)(2s+1)(s+2)} = \frac{1}{2}\frac{1}{s+1} + \frac{5}{18}\frac{1}{s-1} - \frac{8}{9}\frac{1}{s+1/2} + \frac{1}{9}\frac{1}{s+2}.$$

Thus

$$y = \frac{1}{2}e^{-t} + \frac{5}{18}e^t - \frac{8}{9}e^{-t/2} + \frac{1}{9}e^{-2t}.$$

40. The Laplace transform of the initial-value problem is

$$s^3\,\mathscr{L}\{y\} - s^2(0) - sy'(0) - y''(0) + 2[s^2\,\mathscr{L}\{y\} - sy(0) - y'(0)] - [s\,\mathscr{L}\{y\} - y(0)] - 2\,\mathscr{L}\{y\} = \frac{3}{s^2+9}.$$

Solving for $\mathscr{L}\{y\}$ we obtain

$$\mathscr{L}\{y\} = \frac{s^2+12}{(s-1)(s+1)(s+2)(s^2+9)}$$

$$= \frac{13}{60}\frac{1}{s-1} - \frac{13}{20}\frac{1}{s+1} + \frac{16}{39}\frac{1}{s+2} + \frac{3}{130}\frac{s}{s^2+9} - \frac{1}{65}\frac{3}{s^2+9}.$$

Thus

$$y = \frac{13}{60}e^t - \frac{13}{20}e^{-t} + \frac{16}{39}e^{-2t} + \frac{3}{130}\cos 3t - \frac{1}{65}\sin 3t.$$

41. The Laplace transform of the initial-value problem is

$$s\,\mathscr{L}\{y\} + \mathscr{L}\{y\} = \frac{s+3}{s^2+6s+13}.$$

Solving for $\mathscr{L}\{y\}$ we obtain

$$\mathscr{L}\{y\} = \frac{s+3}{(s+1)(s^2+6s+13)} = \frac{1}{4} \cdot \frac{1}{s+1} - \frac{1}{4} \cdot \frac{s+1}{s^2+6s+13}$$

$$= \frac{1}{4} \cdot \frac{1}{s+1} - \frac{1}{4}\left(\frac{s+3}{(s+3)^2+4} - \frac{2}{(s+3)^2+4}\right).$$

Thus

$$y = \frac{1}{4}e^{-t} - \frac{1}{4}e^{-3t}\cos 2t + \frac{1}{4}e^{-3t}\sin 2t.$$

42. The Laplace transform of the initial-value problem is

$$s^2 \mathcal{L}\{y\} - s \cdot 1 - 3 - 2[s \mathcal{L}\{y\} - 1] + 5 \mathcal{L}\{y\} = (s^2 - 2s + 5) \mathcal{L}\{y\} - s - 1 = 0.$$

Solving for $\mathcal{L}\{y\}$ we obtain

$$\mathcal{L}\{y\} = \frac{s+1}{s^2 - 2s + 5} = \frac{s - 1 + 2}{(s-1)^2 + 2^2} = \frac{s-1}{(s-1)^2 + 2^2} + \frac{2}{(s-1)^2 + 2^2}.$$

Thus

$$y = e^t \cos 2t + e^t \sin 2t.$$

43. (a) Differentiating $f(t) = t e^{at}$ we get $f'(t) = at e^{at} + e^{at}$ so $\mathcal{L}\{at e^{at} + e^{at}\} = s \mathcal{L}\{t e^{at}\}$, where we have used $f(0) = 0$. Writing the equation as

$$a \mathcal{L}\{t e^{at}\} + \mathcal{L}\{e^{at}\} = s \mathcal{L}\{t e^{at}\}$$

and solving for $\mathcal{L}\{t e^{at}\}$ we get

$$\mathcal{L}\{t e^{at}\} = \frac{1}{s - a} \mathcal{L}\{e^{at}\} = \frac{1}{(s-a)^2}.$$

(b) Starting with $f(t) = t \sin kt$ we have

$$f'(t) = kt \cos kt + \sin kt$$

$$f''(t) = -k^2 t \sin kt + 2k \cos kt.$$

Then

$$\mathcal{L}\{-k^2 t \sin t + 2k \cos kt\} = s^2 \mathcal{L}\{t \sin kt\}$$

where we have used $f(0) = 0$ and $f'(0) = 0$. Writing the above equation as

$$-k^2 \mathcal{L}\{t \sin kt\} + 2k \mathcal{L}\{\cos kt\} = s^2 \mathcal{L}\{t \sin kt\}$$

and solving for $\mathcal{L}\{t \sin kt\}$ gives

$$\mathcal{L}\{t \sin kt\} = \frac{2k}{s^2 + k^2} \mathcal{L}\{\cos kt\} = \frac{2k}{s^2 + k^2} \frac{s}{s^2 + k^2} = \frac{2ks}{(s^2 + k^2)^2}.$$

44. Let $f_1(t) = 1$ and $f_2(t) = \begin{cases} 1, & t \geq 0, \ t \neq 1 \\ 0, & t = 1 \end{cases}$. Then $\mathcal{L}\{f_1(t)\} = \mathcal{L}\{f_2(t)\} = 1/s$, but $f_1(t) \neq f_2(t)$.

45. For $y'' - 4y' = 6e^{3t} - 3e^{-t}$ the transfer function is $W(s) = 1/(s^2 - 4s)$. The zero-input response is

$$y_0(t) = \mathcal{L}^{-1}\left\{\frac{s-5}{s^2 - 4s}\right\} = \mathcal{L}^{-1}\left\{\frac{5}{4} \cdot \frac{1}{s} - \frac{1}{4} \cdot \frac{1}{s-4}\right\} = \frac{5}{4} - \frac{1}{4}e^{4t},$$

and the zero-state response is

$$y_1(t) = \mathscr{L}^{-1}\left\{\frac{6}{(s-3)(s^2-4s)} - \frac{3}{(s+1)(s^2-4s)}\right\}$$

$$= \mathscr{L}^{-1}\left\{\frac{27}{20}\cdot\frac{1}{s-4} - \frac{2}{s-3} + \frac{5}{4}\cdot\frac{1}{s} - \frac{3}{5}\cdot\frac{1}{s+1}\right\}$$

$$= \frac{27}{20}e^{4t} - 2e^{3t} + \frac{5}{4} - \frac{3}{5}e^{-t}.$$

46. From Theorem 7.2.2, if f and f' are continuous and of exponential order, $\mathscr{L}\{f'(t)\} = sF(s) - f(0)$. From Theorem 7.1.3, $\lim_{s\to\infty}\mathscr{L}\{f'(t)\} = 0$ so

$$\lim_{s\to\infty}[sF(s) - f(0)] = 0 \quad\text{and}\quad \lim_{s\to\infty}F(s) = f(0).$$

For $f(t) = \cos kt$,

$$\lim_{s\to\infty}sF(s) = \lim_{s\to\infty}s\,\frac{s}{s^2+k^2} = 1 = f(0).$$

Exercises 7.3

Operational Properties I

1. $\mathscr{L}\{te^{10t}\} = \dfrac{1}{(s-10)^2}$

2. $\mathscr{L}\{te^{-6t}\} = \dfrac{1}{(s+6)^2}$

3. $\mathscr{L}\{t^3 e^{-2t}\} = \dfrac{3!}{(s+2)^4}$

4. $\mathscr{L}\{t^{10} e^{-7t}\} = \dfrac{10!}{(s+7)^{11}}$

5. $\mathscr{L}\left\{t\left(e^t + e^{2t}\right)^2\right\} = \mathscr{L}\left\{te^{2t} + 2te^{3t} + te^{4t}\right\} = \dfrac{1}{(s-2)^2} + \dfrac{2}{(s-3)^2} + \dfrac{1}{(s-4)^2}$

6. $\mathscr{L}\left\{e^{2t}(t-1)^2\right\} = \mathscr{L}\left\{t^2 e^{2t} - 2te^{2t} + e^{2t}\right\} = \dfrac{2}{(s-2)^3} - \dfrac{2}{(s-2)^2} + \dfrac{1}{s-2}$

7. $\mathscr{L}\{e^t \sin 3t\} = \dfrac{3}{(s-1)^2 + 9}$

8. $\mathscr{L}\{e^{-2t} \cos 4t\} = \dfrac{s+2}{(s+2)^2 + 16}$

9. $\mathscr{L}\{(1 - e^t + 3e^{-4t})\cos 5t\} = \mathscr{L}\{\cos 5t - e^t \cos 5t + 3e^{-4t}\cos 5t\}$

$$= \frac{s}{s^2 + 25} - \frac{s - 1}{(s - 1)^2 + 25} + \frac{3(s + 4)}{(s + 4)^2 + 25}$$

10. $\mathscr{L}\left\{e^{3t}\left(9 - 4t + 10\sin\frac{t}{2}\right)\right\} = \mathscr{L}\left\{9e^{3t} - 4te^{3t} + 10e^{3t}\sin\frac{t}{2}\right\} = \frac{9}{s - 3} - \frac{4}{(s - 3)^2} + \frac{5}{(s - 3)^2 + 1/4}$

11. $\mathscr{L}^{-1}\left\{\frac{1}{(s + 2)^3}\right\} = \mathscr{L}^{-1}\left\{\frac{1}{2}\frac{2}{(s + 2)^3}\right\} = \frac{1}{2}t^2 e^{-2t}$

12. $\mathscr{L}^{-1}\left\{\frac{1}{(s - 1)^4}\right\} = \frac{1}{6}\mathscr{L}^{-1}\left\{\frac{3!}{(s - 1)^4}\right\} = \frac{1}{6}t^3 e^t$

13. $\mathscr{L}^{-1}\left\{\frac{1}{s^2 - 6s + 10}\right\} = \mathscr{L}^{-1}\left\{\frac{1}{(s - 3)^2 + 1^2}\right\} = e^{3t}\sin t$

14. $\mathscr{L}^{-1}\left\{\frac{1}{s^2 + 2s + 5}\right\} = \mathscr{L}^{-1}\left\{\frac{1}{2}\frac{2}{(s + 1)^2 + 2^2}\right\} = \frac{1}{2}e^{-t}\sin 2t$

15. $\mathscr{L}^{-1}\left\{\frac{s}{s^2 + 4s + 5}\right\} = \mathscr{L}^{-1}\left\{\frac{s + 2}{(s + 2)^2 + 1^2} - 2\frac{1}{(s + 2)^2 + 1^2}\right\} = e^{-2t}\cos t - 2e^{-2t}\sin t$

16. $\mathscr{L}^{-1}\left\{\frac{2s + 5}{s^2 + 6s + 34}\right\} = \mathscr{L}^{-1}\left\{2\frac{(s + 3)}{(s + 3)^2 + 5^2} - \frac{1}{5}\frac{5}{(s + 3)^2 + 5^2}\right\} = 2e^{-3t}\cos 5t - \frac{1}{5}e^{-3t}\sin 5t$

17. $\mathscr{L}^{-1}\left\{\frac{s}{(s + 1)^2}\right\} = \mathscr{L}^{-1}\left\{\frac{s + 1 - 1}{(s + 1)^2}\right\} = \mathscr{L}^{-1}\left\{\frac{1}{s + 1} - \frac{1}{(s + 1)^2}\right\} = e^{-t} - te^{-t}$

18. $\mathscr{L}^{-1}\left\{\frac{5s}{(s - 2)^2}\right\} = \mathscr{L}^{-1}\left\{\frac{5(s - 2) + 10}{(s - 2)^2}\right\} = \mathscr{L}^{-1}\left\{\frac{5}{s - 2} + \frac{10}{(s - 2)^2}\right\} = 5e^{2t} + 10te^{2t}$

19. $\mathscr{L}^{-1}\left\{\frac{2s - 1}{s^2(s + 1)^3}\right\} = \mathscr{L}^{-1}\left\{\frac{5}{s} - \frac{1}{s^2} - \frac{5}{s + 1} - \frac{4}{(s + 1)^2} - \frac{3}{2}\frac{2}{(s + 1)^3}\right\} = 5 - t - 5e^{-t} - 4te^{-t} - \frac{3}{2}t^2 e^{-t}$

20. $\mathscr{L}^{-1}\left\{\frac{(s + 1)^2}{(s + 2)^4}\right\} = \mathscr{L}^{-1}\left\{\frac{1}{(s + 2)^2} - \frac{2}{(s + 2)^3} + \frac{1}{6}\frac{3!}{(s + 2)^4}\right\} = te^{-2t} - t^2 e^{-2t} + \frac{1}{6}t^3 e^{-2t}$

21. The Laplace transform of the differential equation is

$$s\mathscr{L}\{y\} - y(0) + 4\mathscr{L}\{y\} = \frac{1}{s + 4}.$$

Solving for $\mathscr{L}\{y\}$ we obtain

$$\mathscr{L}\{y\} = \frac{1}{(s + 4)^2} + \frac{2}{s + 4}.$$

Thus

$$y = te^{-4t} + 2e^{-4t}.$$

22. The Laplace transform of the differential equation is

$$s \mathscr{L}\{y\} - \mathscr{L}\{y\} = \frac{1}{s} + \frac{1}{(s-1)^2}.$$

Solving for $\mathscr{L}\{y\}$ we obtain

$$\mathscr{L}\{y\} = \frac{1}{s(s-1)} + \frac{1}{(s-1)^3} = -\frac{1}{s} + \frac{1}{s-1} + \frac{1}{(s-1)^3}.$$

Thus

$$y = -1 + e^t + \frac{1}{2}t^2 e^t.$$

23. The Laplace transform of the differential equation is

$$s^2 \mathscr{L}\{y\} - sy(0) - y'(0) + 2[s \mathscr{L}\{y\} - y(0)] + \mathscr{L}\{y\} = 0.$$

Solving for $\mathscr{L}\{y\}$ we obtain

$$\mathscr{L}\{y\} = \frac{s+3}{(s+1)^2} = \frac{1}{s+1} + \frac{2}{(s+1)^2}.$$

Thus

$$y = e^{-t} + 2te^{-t}.$$

24. The Laplace transform of the differential equation is

$$s^2 \mathscr{L}\{y\} - sy(0) - y'(0) - 4[s \mathscr{L}\{y\} - y(0)] + 4\mathscr{L}\{y\} = \frac{6}{(s-2)^4}.$$

Solving for $\mathscr{L}\{y\}$ we obtain $\mathscr{L}\{y\} = \frac{1}{20}\frac{5!}{(s-2)^6}$. Thus, $y = \frac{1}{20}t^5 e^{2t}$.

25. The Laplace transform of the differential equation is

$$s^2 \mathscr{L}\{y\} - sy(0) - y'(0) - 6[s \mathscr{L}\{y\} - y(0)] + 9\mathscr{L}\{y\} = \frac{1}{s^2}.$$

Solving for $\mathscr{L}\{y\}$ we obtain

$$\mathscr{L}\{y\} = \frac{1+s^2}{s^2(s-3)^2} = \frac{2}{27}\frac{1}{s} + \frac{1}{9}\frac{1}{s^2} - \frac{2}{27}\frac{1}{s-3} + \frac{10}{9}\frac{1}{(s-3)^2}.$$

Thus

$$y = \frac{2}{27} + \frac{1}{9}t - \frac{2}{27}e^{3t} + \frac{10}{9}te^{3t}.$$

26. The Laplace transform of the differential equation is

$$s^2 \mathscr{L}\{y\} - sy(0) - y'(0) - 4[s \mathscr{L}\{y\} - y(0)] + 4\mathscr{L}\{y\} = \frac{6}{s^4}.$$

Solving for $\mathscr{L}\{y\}$ we obtain

$$\mathscr{L}\{y\} = \frac{s^5 - 4s^4 + 6}{s^4(s-2)^2} = \frac{3}{4}\frac{1}{s} + \frac{9}{8}\frac{1}{s^2} + \frac{3}{4}\frac{2}{s^3} + \frac{1}{4}\frac{3!}{s^4} + \frac{1}{4}\frac{1}{s-2} - \frac{13}{8}\frac{1}{(s-2)^2}.$$

Thus

$$y = \frac{3}{4} + \frac{9}{8}t + \frac{3}{4}t^2 + \frac{1}{4}t^3 + \frac{1}{4}e^{2t} - \frac{13}{8}te^{2t}.$$

27. The Laplace transform of the differential equation is

$$s^2 \mathscr{L}\{y\} - sy(0) - y'(0) - 6[s\mathscr{L}\{y\} - y(0)] + 13\mathscr{L}\{y\} = 0.$$

Solving for $\mathscr{L}\{y\}$ we obtain

$$\mathscr{L}\{y\} = -\frac{3}{s^2 - 6s + 13} = -\frac{3}{2}\frac{2}{(s-3)^2 + 2^2}.$$

Thus

$$y = -\frac{3}{2}e^{3t}\sin 2t.$$

28. The Laplace transform of the differential equation is

$-y'(0)$

$$2[s^2\mathscr{L}\{y\} - sy(0)] + 20[s\mathscr{L}\{y\} - y(0)] + 51\mathscr{L}\{y\} = 0.$$

Solving for $\mathscr{L}\{y\}$ we obtain

$$\mathscr{L}\{y\} = \frac{4s + 40}{2s^2 + 20s + 51} = \frac{2s + 20}{(s+5)^2 + 1/2} = \frac{2(s+5)}{(s+5)^2 + 1/2} + \frac{10}{(s+5)^2 + 1/2}.$$

Thus

$$y = 2e^{-5t}\cos(t/\sqrt{2}) + 10\sqrt{2}\,e^{-5t}\sin(t/\sqrt{2}).$$

29. The Laplace transform of the differential equation is

$$s^2\mathscr{L}\{y\} - sy(0) - y'(0) - [s\mathscr{L}\{y\} - y(0)] = \frac{s-1}{(s-1)^2 + 1}.$$

Solving for $\mathscr{L}\{y\}$ we obtain

$$\mathscr{L}\{y\} = \frac{1}{s(s^2 - 2s + 2)} = \frac{1}{2}\frac{1}{s} - \frac{1}{2}\frac{s-1}{(s-1)^2 + 1} + \frac{1}{2}\frac{1}{(s-1)^2 + 1}.$$

Thus

$$y = \frac{1}{2} - \frac{1}{2}e^t\cos t + \frac{1}{2}e^t\sin t.$$

30. The Laplace transform of the differential equation is

$$s^2\mathscr{L}\{y\} - sy(0) - y'(0) - 2[s\mathscr{L}\{y\} - y(0)] + 5\mathscr{L}\{y\} = \frac{1}{s} + \frac{1}{s^2}.$$

Solving for $\mathscr{L}\{y\}$ we obtain

$$\mathscr{L}\{y\} = \frac{4s^2 + s + 1}{s^2(s^2 - 2s + 5)} = \frac{7}{25}\frac{1}{s} + \frac{1}{5}\frac{1}{s^2} + \frac{-7s/25 + 109/25}{s^2 - 2s + 5}$$

$$= \frac{7}{25}\frac{1}{s} + \frac{1}{5}\frac{1}{s^2} - \frac{7}{25}\frac{s-1}{(s-1)^2 + 2^2} + \frac{51}{25}\frac{2}{(s-1)^2 + 2^2}.$$

Thus

$$y = \frac{7}{25} + \frac{1}{5}t - \frac{7}{25}e^t \cos 2t + \frac{51}{25}e^t \sin 2t.$$

31. Taking the Laplace transform of both sides of the differential equation and letting $c = y(0)$ we obtain

$$\mathscr{L}\{y''\} + \mathscr{L}\{2y'\} + \mathscr{L}\{y\} = 0$$

$$s^2\mathscr{L}\{y\} - sy(0) - y'(0) + 2s\,\mathscr{L}\{y\} - 2y(0) + \mathscr{L}\{y\} = 0$$

$$s^2\mathscr{L}\{y\} - cs - 2 + 2s\,\mathscr{L}\{y\} - 2c + \mathscr{L}\{y\} = 0$$

$$\left(s^2 + 2s + 1\right)\mathscr{L}\{y\} = cs + 2c + 2$$

$$\mathscr{L}\{y\} = \frac{cs}{(s+1)^2} + \frac{2c+2}{(s+1)^2}$$

$$= c\frac{s+1-1}{(s+1)^2} + \frac{2c+2}{(s+1)^2}$$

$$= \frac{c}{s+1} + \frac{c+2}{(s+1)^2}.$$

Therefore,

$$y(t) = c\mathscr{L}^{-1}\left\{\frac{1}{s+1}\right\} + (c+2)\,\mathscr{L}^{-1}\left\{\frac{1}{(s+1)^2}\right\} = ce^{-t} + (c+2)te^{-t}.$$

To find c we let $y(1) = 2$. Then $2 = ce^{-1} + (c+2)e^{-1} = 2(c+1)e^{-1}$ and $c = e - 1$. Thus

$$y(t) = (e - 1)e^{-t} + (e + 1)te^{-t}.$$

32. Taking the Laplace transform of both sides of the differential equation and letting $c = y'(0)$ we obtain

$$\mathscr{L}\{y''\} + \mathscr{L}\{8y'\} + \mathscr{L}\{20y\} = 0$$

$$s^2\mathscr{L}\{y\} - y'(0) + 8s\,\mathscr{L}\{y\} + 20\,\mathscr{L}\{y\} = 0$$

$$s^2\mathscr{L}\{y\} - c + 8s\,\mathscr{L}\{y\} + 20\,\mathscr{L}\{y\} = 0$$

$$(s^2 + 8s + 20)\,\mathscr{L}\{y\} = c$$

$$\mathscr{L}\{y\} = \frac{c}{s^2 + 8s + 20} = \frac{c}{(s+4)^2 + 4}.$$

Therefore,

$$y(t) = \mathscr{L}^{-1}\left\{\frac{c}{(s+4)^2 + 4}\right\} = \frac{c}{2}e^{-4t}\sin 2t = c_1 e^{-4t}\sin 2t.$$

To find c we let $y'(\pi) = 0$. Then $0 = y'(\pi) = ce^{-4\pi}$ and $c = 0$. Thus, $y(t) = 0$. (Since the differential equation is homogeneous and both boundary conditions are 0, we can see immediately that $y(t) = 0$ is a solution. We have shown that it is the only solution.)

33. Recall from Section 5.1 that $mx'' = -kx - \beta x'$. Now $m = W/g = 4/32 = \frac{1}{8}$ slug, and $4 = 2k$ so that $k = 2$ lb/ft. Thus, the differential equation is $x'' + 7x' + 16x = 0$. The initial conditions are $x(0) = -3/2$ and $x'(0) = 0$. The Laplace transform of the differential equation is

$$s^2 \mathscr{L}\{x\} + \frac{3}{2}s + 7s\mathscr{L}\{x\} + \frac{21}{2} + 16\mathscr{L}\{x\} = 0.$$

Solving for $\mathscr{L}\{x\}$ we obtain

$$\mathscr{L}\{x\} = \frac{-3s/2 - 21/2}{s^2 + 7s + 16} = -\frac{3}{2}\frac{s + 7/2}{(s + 7/2)^2 + (\sqrt{15}/2)^2} - \frac{7\sqrt{15}}{10}\frac{\sqrt{15}/2}{(s + 7/2)^2 + (\sqrt{15}/2)^2}.$$

Thus

$$x = -\frac{3}{2}e^{-7t/2}\cos\frac{\sqrt{15}}{2}t - \frac{7\sqrt{15}}{10}e^{-7t/2}\sin\frac{\sqrt{15}}{2}t.$$

34. The differential equation is

$$\frac{d^2q}{dt^2} + 20\frac{dq}{dt} + 200q = 150, \quad q(0) = q'(0) = 0.$$

The Laplace transform of this equation is

$$s^2\mathscr{L}\{q\} + 20s\mathscr{L}\{q\} + 200\mathscr{L}\{q\} = \frac{150}{s}.$$

Solving for $\mathscr{L}\{q\}$ we obtain

$$\mathscr{L}\{q\} = \frac{150}{s(s^2 + 20s + 200)} = \frac{3}{4}\frac{1}{s} - \frac{3}{4}\frac{s + 10}{(s + 10)^2 + 10^2} - \frac{3}{4}\frac{10}{(s + 10)^2 + 10^2}.$$

Thus

$$q(t) = \frac{3}{4} - \frac{3}{4}e^{-10t}\cos 10t - \frac{3}{4}e^{-10t}\sin 10t$$

and

$$i(t) = q'(t) = 15e^{-10t}\sin 10t.$$

35. The differential equation is

$$\frac{d^2q}{dt^2} + 2\lambda\frac{dq}{dt} + \omega^2q = \frac{E_0}{L}, \quad q(0) = q'(0) = 0.$$

The Laplace transform of this equation is

$$s^2\mathscr{L}\{q\} + 2\lambda s\mathscr{L}\{q\} + \omega^2\mathscr{L}\{q\} = \frac{E_0}{L}\frac{1}{s}$$

or

$$\left(s^2 + 2\lambda s + \omega^2\right)\mathscr{L}\{q\} = \frac{E_0}{L}\frac{1}{s}.$$

Solving for $\mathscr{L}\{q\}$ and using partial fractions we obtain

$$\mathscr{L}\{q\} = \frac{E_0}{L}\left(\frac{1/\omega^2}{s} - \frac{(1/\omega^2)s + 2\lambda/\omega^2}{s^2 + 2\lambda s + \omega^2}\right) = \frac{E_0}{L\omega^2}\left(\frac{1}{s} - \frac{s + 2\lambda}{s^2 + 2\lambda s + \omega^2}\right).$$

For $\lambda > \omega$ we write $s^2 + 2\lambda s + \omega^2 = (s + \lambda)^2 - \left(\lambda^2 - \omega^2\right)$, so (recalling that $\omega^2 = 1/LC$)

$$\mathscr{L}\{q\} = E_0 C\left(\frac{1}{s} - \frac{s + \lambda}{(s + \lambda)^2 - (\lambda^2 - \omega^2)} - \frac{\lambda}{(s + \lambda)^2 - (\lambda^2 - \omega^2)}\right).$$

Thus for $\lambda > \omega$,

$$q(t) = E_0 C\left[1 - e^{-\lambda t}\left(\cosh\sqrt{\lambda^2 - \omega^2}\,t - \frac{\lambda}{\sqrt{\lambda^2 - \omega^2}}\sinh\sqrt{\lambda^2 - \omega^2}\,t\right)\right].$$

For $\lambda < \omega$ we write $s^2 + 2\lambda s + \omega^2 = (s + \lambda)^2 + \left(\omega^2 - \lambda^2\right)$, so

$$\mathscr{L}\{q\} = E_0 C\left(\frac{1}{s} - \frac{s + \lambda}{(s + \lambda)^2 + (\omega^2 - \lambda^2)} - \frac{\lambda}{(s + \lambda)^2 + (\omega^2 - \lambda^2)}\right).$$

Thus for $\lambda < \omega$,

$$q(t) = E_0 C\left[1 - e^{-\lambda t}\left(\cos\sqrt{\omega^2 - \lambda^2}\,t - \frac{\lambda}{\sqrt{\omega^2 - \lambda^2}}\sin\sqrt{\omega^2 - \lambda^2}\,t\right)\right].$$

For $\lambda = \omega$, $s^2 + 2\lambda + \omega^2 = (s + \lambda)^2$ and

$$\mathscr{L}\{q\} = \frac{E_0}{L}\frac{1}{s(s + \lambda)^2} = \frac{E_0}{L}\left(\frac{1/\lambda^2}{s} - \frac{1/\lambda^2}{s + \lambda} - \frac{1/\lambda}{(s + \lambda)^2}\right) = \frac{E_0}{L\lambda^2}\left(\frac{1}{s} - \frac{1}{s + \lambda} - \frac{\lambda}{(s + \lambda)^2}\right).$$

Thus for $\lambda = \omega$,

$$q(t) = E_0 C\left(1 - e^{-\lambda t} - \lambda t e^{-\lambda t}\right).$$

36. The differential equation is

$$R\frac{dq}{dt} + \frac{1}{C}q = E_0 e^{-kt}, \quad q(0) = 0.$$

The Laplace transform of this equation is

$$Rs\,\mathscr{L}\{q\} + \frac{1}{C}\mathscr{L}\{q\} = E_0\frac{1}{s + k}.$$

Solving for $\mathscr{L}\{q\}$ we obtain

$$\mathscr{L}\{q\} = \frac{E_0 C}{(s + k)(RCs + 1)} = \frac{E_0/R}{(s + k)(s + 1/RC)}.$$

When $1/RC \neq k$ we have by partial fractions

$$\mathscr{L}\{q\} = \frac{E_0}{R}\left(\frac{1/(1/RC - k)}{s + k} - \frac{1/(1/RC - k)}{s + 1/RC}\right) = \frac{E_0}{R}\frac{1}{1/RC - k}\left(\frac{1}{s + k} - \frac{1}{s + 1/RC}\right).$$

Thus

$$q(t) = \frac{E_0 C}{1 - kRC}\left(e^{-kt} - e^{-t/RC}\right).$$

When $1/RC = k$ we have

$$\mathscr{L}\{q\} = \frac{E_0}{R} \frac{1}{(s+k)^2}.$$

Thus

$$q(t) = \frac{E_0}{R} te^{-kt} = \frac{E_0}{R} te^{-t/RC}.$$

37. $\mathscr{L}\{(t-1)\,\mathscr{U}(t-1)\} = \dfrac{e^{-s}}{s^2}$

38. $\mathscr{L}\{e^{2-t}\,\mathscr{U}(t-2)\} = \mathscr{L}\left\{e^{-(t-2)}\,\mathscr{U}(t-2)\right\} = \dfrac{e^{-2s}}{s+1}$

39. $\mathscr{L}\{t\,\mathscr{U}(t-2)\} = \mathscr{L}\{(t-2)\,\mathscr{U}(t-2) + 2\,\mathscr{U}(t-2)\} = \dfrac{e^{-2s}}{s^2} + \dfrac{2e^{-2s}}{s}$

Alternatively, (16) of this section in the text could be used:

$$\mathscr{L}\{t\,\mathscr{U}(t-2)\} = e^{-2s}\,\mathscr{L}\{t+2\} = e^{-2s}\left(\frac{1}{s^2} + \frac{2}{s}\right).$$

40. $\mathscr{L}\{(3t+1)\,\mathscr{U}(t-1)\} = 3\,\mathscr{L}\{(t-1)\,\mathscr{U}(t-1)\} + 4\,\mathscr{L}\{\mathscr{U}(t-1)\} = \dfrac{3e^{-s}}{s^2} + \dfrac{4e^{-s}}{s}$

Alternatively, (16) of this section in the text could be used:

$$\mathscr{L}\{(3t+1)\,\mathscr{U}(t-1)\} = e^{-s}\,\mathscr{L}\{3t+4\} = e^{-s}\left(\frac{3}{s^2} + \frac{4}{s}\right).$$

41. $\mathscr{L}\{\cos 2t\,\mathscr{U}(t-\pi)\} = \mathscr{L}\{\cos 2(t-\pi)\,\mathscr{U}(t-\pi)\} = \dfrac{se^{-\pi s}}{s^2+4}$

Alternatively, (16) of this section in the text could be used:

$$\mathscr{L}\{\cos 2t\,\mathscr{U}(t-\pi)\} = e^{-\pi s}\,\mathscr{L}\{\cos 2(t+\pi)\} = e^{-\pi s}\,\mathscr{L}\{\cos 2t\} = e^{-\pi s}\frac{s}{s^2+4}.$$

42. $\mathscr{L}\left\{\sin t\,\mathscr{U}\left(t-\dfrac{\pi}{2}\right)\right\} = \mathscr{L}\left\{\cos\left(t-\dfrac{\pi}{2}\right)\mathscr{U}\left(t-\dfrac{\pi}{2}\right)\right\} = \dfrac{se^{-\pi s/2}}{s^2+1}$

Alternatively, (16) of this section in the text could be used:

$$\mathscr{L}\left\{\sin t\,\mathscr{U}\left(t-\frac{\pi}{2}\right)\right\} = e^{-\pi s/2}\,\mathscr{L}\left\{\sin\left(t+\frac{\pi}{2}\right)\right\} = e^{-\pi s/2}\,\mathscr{L}\{\cos t\} = e^{-\pi s/2}\frac{s}{s^2+1}.$$

43. $\mathscr{L}^{-1}\left\{\dfrac{e^{-2s}}{s^3}\right\} = \mathscr{L}^{-1}\left\{\dfrac{1}{2}\cdot\dfrac{2}{s^3}e^{-2s}\right\} = \dfrac{1}{2}(t-2)^2\,\mathscr{U}(t-2)$

44. $\mathscr{L}^{-1}\left\{\dfrac{(1+e^{-2s})^2}{s+2}\right\} = \mathscr{L}^{-1}\left\{\dfrac{1}{s+2} + \dfrac{2e^{-2s}}{s+2} + \dfrac{e^{-4s}}{s+2}\right\} = e^{-2t} + 2e^{-2(t-2)}\,\mathscr{U}(t-2) + e^{-2(t-4)}\,\mathscr{U}(t-4)$

45. $\mathscr{L}^{-1}\left\{\dfrac{e^{-\pi s}}{s^2+1}\right\} = \sin(t-\pi)\,\mathscr{U}(t-\pi) = -\sin t\,\mathscr{U}(t-\pi)$

46. $\mathcal{L}^{-1}\left\{\dfrac{se^{-\pi s/2}}{s^2+4}\right\} = \cos 2\left(t-\dfrac{\pi}{2}\right)\mathcal{U}\left(t-\dfrac{\pi}{2}\right) = -\cos 2t\,\mathcal{U}\left(t-\dfrac{\pi}{2}\right)$

47. $\mathcal{L}^{-1}\left\{\dfrac{e^{-s}}{s(s+1)}\right\} = \mathcal{L}^{-1}\left\{\dfrac{e^{-s}}{s}-\dfrac{e^{-s}}{s+1}\right\} = \mathcal{U}(t-1)-e^{-(t-1)}\mathcal{U}(t-1)$

48. $\mathcal{L}^{-1}\left\{\dfrac{e^{-2s}}{s^2(s-1)}\right\} = \mathcal{L}^{-1}\left\{-\dfrac{e^{-2s}}{s}-\dfrac{e^{-2s}}{s^2}+\dfrac{e^{-2s}}{s-1}\right\} = -\mathcal{U}(t-2)-(t-2)\mathcal{U}(t-2)+e^{t-2}\mathcal{U}(t-2)$

49. (c)   **50. (e)**   **51. (f)**   **52. (b)**   **53. (a)**   **54. (d)**

55. $\mathcal{L}\{2-4\mathcal{U}(t-3)\} = \dfrac{2}{s}-\dfrac{4}{s}e^{-3s}$

56. $\mathcal{L}\{1-\mathcal{U}(t-4)+\mathcal{U}(t-5)\} = \dfrac{1}{s}-\dfrac{e^{-4s}}{s}+\dfrac{e^{-5s}}{s}$

57. $\mathcal{L}\{t^2\mathcal{U}(t-1)\} = \mathcal{L}\left\{\left[(t-1)^2+2t-1\right]\mathcal{U}(t-1)\right\} = \mathcal{L}\left\{\left[(t-1)^2+2(t-1)+1\right]\mathcal{U}(t-1)\right\}$

$$= \left(\dfrac{2}{s^3}+\dfrac{2}{s^2}+\dfrac{1}{s}\right)e^{-s}$$

Alternatively, by (16) of this section in the text,

$$\mathcal{L}\{t^2\mathcal{U}(t-1)\} = e^{-s}\mathcal{L}\{t^2+2t+1\} = e^{-s}\left(\dfrac{2}{s^3}+\dfrac{2}{s^2}+\dfrac{1}{s}\right).$$

58. $\mathcal{L}\left\{\sin t\,\mathcal{U}\left(t-\dfrac{3\pi}{2}\right)\right\} = \mathcal{L}\left\{-\cos\left(t-\dfrac{3\pi}{2}\right)\mathcal{U}\left(t-\dfrac{3\pi}{2}\right)\right\} = -\dfrac{se^{-3\pi s/2}}{s^2+1}$

59. $\mathcal{L}\{t-t\,\mathcal{U}(t-2)\} = \mathcal{L}\{t-(t-2)\mathcal{U}(t-2)-2\mathcal{U}(t-2)\} = \dfrac{1}{s^2}-\dfrac{e^{-2s}}{s^2}-\dfrac{2e^{-2s}}{s}$

60. $\mathcal{L}\{\sin t-\sin t\,\mathcal{U}(t-2\pi)\} = \mathcal{L}\{\sin t-\sin(t-2\pi)\mathcal{U}(t-2\pi)\} = \dfrac{1}{s^2+1}-\dfrac{e^{-2\pi s}}{s^2+1}$

61. $\mathcal{L}\{f(t)\} = \mathcal{L}\{\mathcal{U}(t-a)-\mathcal{U}(t-b)\} = \dfrac{e^{-as}}{s}-\dfrac{e^{-bs}}{s}$

62. $\mathcal{L}\{f(t)\} = \mathcal{L}\{\mathcal{U}(t-1)+\mathcal{U}(t-2)+\mathcal{U}(t-3)+\cdots\} = \dfrac{e^{-s}}{s}+\dfrac{e^{-2s}}{s}+\dfrac{e^{-3s}}{s}+\cdots = \dfrac{1}{s}\dfrac{e^{-s}}{1-e^{-s}}$

63. The Laplace transform of the differential equation is

$$s\mathcal{L}\{y\}-y(0)+\mathcal{L}\{y\} = \dfrac{5}{s}e^{-s}.$$

Solving for $\mathcal{L}\{y\}$ we obtain

$$\mathcal{L}\{y\} = \dfrac{5e^{-s}}{s(s+1)} = 5e^{-s}\left[\dfrac{1}{s}-\dfrac{1}{s+1}\right].$$

Thus

$$y = 5\,\mathcal{U}(t-1) - 5e^{-(t-1)}\,\mathcal{U}(t-1).$$

64. The Laplace transform of the differential equation is

$$s\,\mathcal{L}\{y\} - y(0) + \mathcal{L}\{y\} = \frac{1}{s} - \frac{2}{s}e^{-s}.$$

Solving for $\mathcal{L}\{y\}$ we obtain

$$\mathcal{L}\{y\} = \frac{1}{s(s+1)} - \frac{2e^{-s}}{s(s+1)} = \frac{1}{s} - \frac{1}{s+1} - 2e^{-s}\left[\frac{1}{s} - \frac{1}{s+1}\right].$$

Thus

$$y = 1 - e^{-t} - 2\left[1 - e^{-(t-1)}\right]\mathcal{U}(t-1).$$

65. The Laplace transform of the differential equation is

$$s\,\mathcal{L}\{y\} - y(0) + 2\,\mathcal{L}\{y\} = \frac{1}{s^2} - e^{-s}\frac{s+1}{s^2}.$$

Solving for $\mathcal{L}\{y\}$ we obtain

$$\mathcal{L}\{y\} = \frac{1}{s^2(s+2)} - e^{-s}\frac{s+1}{s^2(s+2)} = -\frac{1}{4}\frac{1}{s} + \frac{1}{2}\frac{1}{s^2} + \frac{1}{4}\frac{1}{s+2} - e^{-s}\left[\frac{1}{4}\frac{1}{s} + \frac{1}{2}\frac{1}{s^2} - \frac{1}{4}\frac{1}{s+2}\right].$$

Thus

$$y = -\frac{1}{4} + \frac{1}{2}t + \frac{1}{4}e^{-2t} - \left[\frac{1}{4} + \frac{1}{2}(t-1) - \frac{1}{4}e^{-2(t-1)}\right]\mathcal{U}(t-1).$$

66. The Laplace transform of the differential equation is

$$s^2\,\mathcal{L}\{y\} - sy(0) - y'(0) + 4\,\mathcal{L}\{y\} = \frac{1}{s} - \frac{e^{-s}}{s}.$$

Solving for $\mathcal{L}\{y\}$ we obtain

$$\mathcal{L}\{y\} = \frac{1-s}{s(s^2+4)} - e^{-s}\frac{1}{s(s^2+4)} = \frac{1}{4}\frac{1}{s} - \frac{1}{4}\frac{s}{s^2+4} - \frac{1}{2}\frac{2}{s^2+4} - e^{-s}\left[\frac{1}{4}\frac{1}{s} - \frac{1}{4}\frac{s}{s^2+4}\right].$$

Thus

$$y = \frac{1}{4} - \frac{1}{4}\cos 2t - \frac{1}{2}\sin 2t - \left[\frac{1}{4} - \frac{1}{4}\cos 2(t-1)\right]\mathcal{U}(t-1).$$

67. The Laplace transform of the differential equation is

$$s^2\,\mathcal{L}\{y\} - sy(0) - y'(0) + 4\,\mathcal{L}\{y\} = e^{-2\pi s}\frac{1}{s^2+1}.$$

Solving for $\mathcal{L}\{y\}$ we obtain

$$\mathcal{L}\{y\} = \frac{s}{s^2+4} + e^{-2\pi s}\left[\frac{1}{3}\frac{1}{s^2+1} - \frac{1}{6}\frac{2}{s^2+4}\right].$$

Thus

$$y = \cos 2t + \left[\frac{1}{3}\sin(t-2\pi) - \frac{1}{6}\sin 2(t-2\pi)\right]\mathcal{U}(t-2\pi).$$

68. The Laplace transform of the differential equation is

$$s^2 \mathcal{L}\{y\} - sy(0) - y'(0) - 5\left[s\mathcal{L}\{y\} - y(0)\right] + 6\mathcal{L}\{y\} = \frac{e^{-s}}{s}.$$

Solving for $\mathcal{L}\{y\}$ we obtain

$$\mathcal{L}\{y\} = e^{-s}\frac{1}{s(s-2)(s-3)} + \frac{1}{(s-2)(s-3)}$$

$$= e^{-s}\left[\frac{1}{6}\frac{1}{s} - \frac{1}{2}\frac{1}{s-2} + \frac{1}{3}\frac{1}{s-3}\right] - \frac{1}{s-2} + \frac{1}{s-3}.$$

Thus

$$y = \left[\frac{1}{6} - \frac{1}{2}e^{2(t-1)} + \frac{1}{3}e^{3(t-1)}\right]\mathcal{U}(t-1) - e^{2t} + e^{3t}.$$

69. The Laplace transform of the differential equation is

$$s^2 \mathcal{L}\{y\} - sy(0) - y'(0) + \mathcal{L}\{y\} = \frac{e^{-\pi s}}{s} - \frac{e^{-2\pi s}}{s}.$$

Solving for $\mathcal{L}\{y\}$ we obtain

$$\mathcal{L}\{y\} = e^{-\pi s}\left[\frac{1}{s} - \frac{s}{s^2+1}\right] - e^{-2\pi s}\left[\frac{1}{s} - \frac{s}{s^2+1}\right] + \frac{1}{s^2+1}.$$

Thus

$$y = \left[1 - \cos(t-\pi)\right]\mathcal{U}(t-\pi) - \left[1 - \cos(t-2\pi)\right]\mathcal{U}(t-2\pi) + \sin t.$$

70. The Laplace transform of the differential equation is

$$s^2 \mathcal{L}\{y\} - sy(0) - y'(0) + 4[s\mathcal{L}\{y\} - y(0)] + 3\mathcal{L}\{y\} = \frac{1}{s} - \frac{e^{-2s}}{s} - \frac{e^{-4s}}{s} + \frac{e^{-6s}}{s}.$$

Solving for $\mathcal{L}\{y\}$ we obtain

$$\mathcal{L}\{y\} = \frac{1}{3}\frac{1}{s} - \frac{1}{2}\frac{1}{s+1} + \frac{1}{6}\frac{1}{s+3} - e^{-2s}\left[\frac{1}{3}\frac{1}{s} - \frac{1}{2}\frac{1}{s+1} + \frac{1}{6}\frac{1}{s+3}\right]$$

$$- e^{-4s}\left[\frac{1}{3}\frac{1}{s} - \frac{1}{2}\frac{1}{s+1} + \frac{1}{6}\frac{1}{s+3}\right] + e^{-6s}\left[\frac{1}{3}\frac{1}{s} - \frac{1}{2}\frac{1}{s+1} + \frac{1}{6}\frac{1}{s+3}\right].$$

Thus

$$y = \frac{1}{3} - \frac{1}{2}e^{-t} + \frac{1}{6}e^{-3t} - \left[\frac{1}{3} - \frac{1}{2}e^{-(t-2)} + \frac{1}{6}e^{-3(t-2)}\right]\mathcal{U}(t-2)$$

$$- \left[\frac{1}{3} - \frac{1}{2}e^{-(t-4)} + \frac{1}{6}e^{-3(t-4)}\right]\mathcal{U}(t-4) + \left[\frac{1}{3} - \frac{1}{2}e^{-(t-6)} + \frac{1}{6}e^{-3(t-6)}\right]\mathcal{U}(t-6).$$

71. Recall from Section 5.1 that $mx'' = -kx + f(t)$. Now $m = W/g = 32/32 = 1$ slug, and $32 = 2k$ so that $k = 16$ lb/ft. Thus, the differential equation is $x'' + 16x = f(t)$. The initial conditions are $x(0) = 0$, $x'(0) = 0$. Also, since

$$f(t) = \begin{cases} 20t, & 0 \le t < 5 \\ 0, & t \ge 5 \end{cases}$$

and $20t = 20(t-5) + 100$ we can write

$$f(t) = 20t - 20t\,\mathcal{U}(t-5) = 20t - 20(t-5)\,\mathcal{U}(t-5) - 100\,\mathcal{U}(t-5).$$

The Laplace transform of the differential equation is

$$s^2\mathcal{L}\{x\} + 16\,\mathcal{L}\{x\} = \frac{20}{s^2} - \frac{20}{s^2}e^{-5s} - \frac{100}{s}e^{-5s}.$$

Solving for $\mathcal{L}\{x\}$ we obtain

$$\mathcal{L}\{x\} = \frac{20}{s^2(s^2+16)} - \frac{20}{s^2(s^2+16)}e^{-5s} - \frac{100}{s(s^2+16)}e^{-5s}.$$

$$= \left(\frac{5}{4}\cdot\frac{1}{s^2} - \frac{5}{16}\cdot\frac{4}{s^2+16}\right)(1-e^{-5s}) - \left(\frac{25}{4}\cdot\frac{1}{s} - \frac{25}{4}\cdot\frac{s}{s^2+16}\right)e^{-5s}.$$

Thus

$$x(t) = \frac{5}{4}t - \frac{5}{16}\sin 4t - \left[\frac{5}{4}(t-5) - \frac{5}{16}\sin 4(t-5)\right]\mathcal{U}(t-5) - \left[\frac{25}{4} - \frac{25}{4}\cos 4(t-5)\right]\mathcal{U}(t-5)$$

$$= \frac{5}{4}t - \frac{5}{16}\sin 4t - \frac{5}{4}t\,\mathcal{U}(t-5) + \frac{5}{16}\sin 4(t-5)\,\mathcal{U}(t-5) + \frac{25}{4}\cos 4(t-5)\,\mathcal{U}(t-5).$$

72. Recall from Section 5.1 that $mx'' = -kx + f(t)$. Now $m = W/g = 32/32 = 1$ slug, and $32 = 2k$ so that $k = 16$ lb/ft. Thus, the differential equation is $x'' + 16x = f(t)$. The initial conditions are $x(0) = 0$, $x'(0) = 0$. Also, since

$$f(t) = \begin{cases} \sin t, & 0 \le t < 2\pi \\ 0, & t \ge 2\pi \end{cases}$$

and $\sin t = \sin(t - 2\pi)$ we can write

$$f(t) = \sin t - \sin(t-2\pi)\,\mathcal{U}(t-2\pi).$$

The Laplace transform of the differential equation is

$$s^2\mathcal{L}\{x\} + 16\,\mathcal{L}\{x\} = \frac{1}{s^2+1} - \frac{1}{s^2+1}e^{-2\pi s}.$$

Solving for $\mathcal{L}\{x\}$ we obtain

$$\mathcal{L}\{x\} = \frac{1}{(s^2+16)(s^2+1)} - \frac{1}{(s^2+16)(s^2+1)}e^{-2\pi s}$$

$$= \frac{-1/15}{s^2+16} + \frac{1/15}{s^2+1} - \left[\frac{-1/15}{s^2+16} + \frac{1/15}{s^2+1}\right]e^{-2\pi s}.$$

Thus

$$x(t) = -\frac{1}{60}\sin 4t + \frac{1}{15}\sin t + \frac{1}{60}\sin 4(t-2\pi)\,\mathcal{U}(t-2\pi) - \frac{1}{15}\sin(t-2\pi)\,\mathcal{U}(t-2\pi)$$

$$= \begin{cases} -\frac{1}{60}\sin 4t + \frac{1}{15}\sin t, & 0 \le t < 2\pi \\ 0, & t \ge 2\pi. \end{cases}$$

73. The differential equation is

$$2.5\frac{dq}{dt} + 12.5q = 5\,\mathcal{U}(t-3).$$

The Laplace transform of this equation is

$$s\mathcal{L}\{q\} + 5\mathcal{L}\{q\} = \frac{2}{s}e^{-3s}.$$

Solving for $\mathcal{L}\{q\}$ we obtain

$$\mathcal{L}\{q\} = \frac{2}{s(s+5)}e^{-3s} = \left(\frac{2}{5}\cdot\frac{1}{s} - \frac{2}{5}\cdot\frac{1}{s+5}\right)e^{-3s}.$$

Thus

$$q(t) = \frac{2}{5}\mathcal{U}(t-3) - \frac{2}{5}e^{-5(t-3)}\mathcal{U}(t-3).$$

74. The differential equation is

$$10\frac{dq}{dt} + 10q = 30e^{t} - 30e^{t}\,\mathcal{U}(t-1.5).$$

The Laplace transform of this equation is

$$s\mathcal{L}\{q\} - q_0 + \mathcal{L}\{q\} = \frac{3}{s-1} - \frac{3e^{1.5}}{s-1.5}e^{-1.5s}.$$

Solving for $\mathcal{L}\{q\}$ we obtain

$$\mathcal{L}\{q\} = \left(q_0 - \frac{3}{2}\right)\cdot\frac{1}{s+1} + \frac{3}{2}\cdot\frac{1}{s-1} - 3e^{1.5}\left(\frac{-2/5}{s+1} + \frac{2/5}{s-1.5}\right)e^{-1.5s}.$$

Thus

$$q(t) = \left(q_0 - \frac{3}{2}\right)e^{-t} + \frac{3}{2}e^{t} + \frac{6}{5}e^{1.5}\left(e^{-(t-1.5)} - e^{1.5(t-1.5)}\right)\mathcal{U}(t-1.5).$$

75. (a) The differential equation is

$$\frac{di}{dt} + 10i = \sin t + \cos\left(t - \frac{3\pi}{2}\right)\mathcal{U}\left(t - \frac{3\pi}{2}\right), \quad i(0) = 0.$$

The Laplace transform of this equation is

$$s\mathcal{L}\{i\} + 10\mathcal{L}\{i\} = \frac{1}{s^2+1} + \frac{se^{-3\pi s/2}}{s^2+1}.$$

Solving for $\mathcal{L}\{i\}$ we obtain

$$\mathcal{L}\{i\} = \frac{1}{(s^2+1)(s+10)} + \frac{s}{(s^2+1)(s+10)}e^{-3\pi s/2}$$

$$= \frac{1}{101}\left(\frac{1}{s+10} - \frac{s}{s^2+1} + \frac{10}{s^2+1}\right) + \frac{1}{101}\left(\frac{-10}{s+10} + \frac{10s}{s^2+1} + \frac{1}{s^2+1}\right)e^{-3\pi s/2}.$$

Thus

$$i(t) = \frac{1}{101}\left(e^{-10t} - \cos t + 10 \sin t\right)$$

$$+ \frac{1}{101}\left[-10e^{-10(t-3\pi/2)} + 10\cos\left(t - \frac{3\pi}{2}\right) + \sin\left(t - \frac{3\pi}{2}\right)\right]\mathcal{U}\left(t - \frac{3\pi}{2}\right).$$

(b)

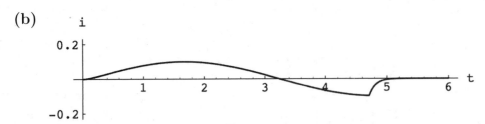

The maximum value of $i(t)$ is approximately 0.1 at $t = 1.7$, the minimum is approximately -0.1 at 4.7. [Using *Mathematica* we see that the maximum value is $i(t)$ is 0.0995037 at $t = 1.670465$, and the mininum value is $i(3\pi/2) \approx -0.0990099$ at $t = 3\pi/2$.]

76. (a) The differential equation is

$$50\frac{dq}{dt} + \frac{1}{0.01}q = E_0[\mathcal{U}(t-1) - \mathcal{U}(t-3)], \quad q(0) = 0$$

or

$$50\frac{dq}{dt} + 100q = E_0[\mathcal{U}(t-1) - \mathcal{U}(t-3)], \quad q(0) = 0.$$

The Laplace transform of this equation is

$$50s\,\mathscr{L}\{q\} + 100\,\mathscr{L}\{q\} = E_0\left(\frac{1}{s}e^{-s} - \frac{1}{s}e^{-3s}\right).$$

Solving for $\mathscr{L}\{q\}$ we obtain

$$\mathscr{L}\{q\} = \frac{E_0}{50}\left[\frac{e^{-s}}{s(s+2)} - \frac{e^{-3s}}{s(s+2)}\right] = \frac{E_0}{50}\left[\frac{1}{2}\left(\frac{1}{s} - \frac{1}{s+2}\right)e^{-s} - \frac{1}{2}\left(\frac{1}{s} - \frac{1}{s+2}\right)e^{-3s}\right].$$

Thus

$$q(t) = \frac{E_0}{100}\left[\left(1 - e^{-2(t-1)}\right)\mathcal{U}(t-1) - \left(1 - e^{-2(t-3)}\right)\mathcal{U}(t-3)\right].$$

(b)

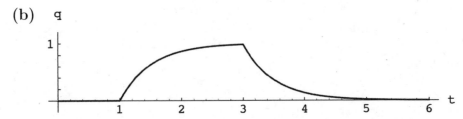

Assuming $E_0 = 100$, the maximum value of $q(t)$ is approximately 1 at $t = 3$. [Using *Mathematica* we see that the maximum value of $q(t)$ is 0.981684 at $t = 3$.]

77. The differential equation is

$$EI\frac{d^4y}{dx^4} = w_0[1 - \mathcal{U}(x - L/2)].$$

Taking the Laplace transform of both sides and using $y(0) = y'(0) = 0$ we obtain

$$s^4 \mathcal{L}\{y\} - sy''(0) - y'''(0) = \frac{w_0}{EI}\frac{1}{s}\left(1 - e^{-Ls/2}\right).$$

Letting $y''(0) = c_1$ and $y'''(0) = c_2$ we have

$$\mathcal{L}\{y\} = \frac{c_1}{s^3} + \frac{c_2}{s^4} + \frac{w_0}{EI}\frac{1}{s^5}\left(1 - e^{-Ls/2}\right)$$

so that

$$y(x) = \frac{1}{2}c_1 x^2 + \frac{1}{6}c_2 x^3 + \frac{1}{24}\frac{w_0}{EI}\left[x^4 - \left(x - \frac{L}{2}\right)^4 \mathcal{U}\left(x - \frac{L}{2}\right)\right].$$

To find c_1 and c_2 we compute

$$y''(x) = c_1 + c_2 x + \frac{1}{2}\frac{w_0}{EI}\left[x^2 - \left(x - \frac{L}{2}\right)^2 \mathcal{U}\left(x - \frac{L}{2}\right)\right]$$

and

$$y'''(x) = c_2 + \frac{w_0}{EI}\left[x - \left(x - \frac{L}{2}\right)\mathcal{U}\left(x - \frac{L}{2}\right)\right].$$

Then $y''(L) = y'''(L) = 0$ yields the system

$$c_1 + c_2 L + \frac{1}{2}\frac{w_0}{EI}\left[L^2 - \left(\frac{L}{2}\right)^2\right] = c_1 + c_2 L + \frac{3}{8}\frac{w_0 L^2}{EI} = 0$$

$$c_2 + \frac{w_0}{EI}\left(\frac{L}{2}\right) = c_2 + \frac{1}{2}\frac{w_0 L}{EI} = 0.$$

Solving for c_1 and c_2 we obtain $c_1 = \frac{1}{8}w_0 L^2/EI$ and $c_2 = -\frac{1}{2}w_0 L/EI$. Thus

$$y(x) = \frac{w_0}{EI}\left[\frac{1}{16}L^2 x^2 - \frac{1}{12}Lx^3 + \frac{1}{24}x^4 - \frac{1}{24}\left(x - \frac{L}{2}\right)^4 \mathcal{U}\left(x - \frac{L}{2}\right)\right].$$

78. The differential equation is

$$EI\frac{d^4y}{dx^4} = w_0[\mathcal{U}(x - L/3) - \mathcal{U}(x - 2L/3)].$$

Taking the Laplace transform of both sides and using $y(0) = y'(0) = 0$ we obtain

$$s^4 \mathcal{L}\{y\} - sy''(0) - y'''(0) = \frac{w_0}{EI}\frac{1}{s}\left(e^{-Ls/3} - e^{-2Ls/3}\right).$$

Letting $y''(0) = c_1$ and $y'''(0) = c_2$ we have

$$\mathcal{L}\{y\} = \frac{c_1}{s^3} + \frac{c_2}{s^4} + \frac{w_0}{EI}\frac{1}{s^5}\left(e^{-Ls/3} - e^{-2Ls/3}\right)$$

so that

$$y(x) = \frac{1}{2}c_1 x^2 + \frac{1}{6}c_2 x^3 + \frac{1}{24}\frac{w_0}{EI}\left[\left(x - \frac{L}{3}\right)^4 \mathcal{U}\left(x - \frac{L}{3}\right) - \left(x - \frac{2L}{3}\right)^4 \mathcal{U}\left(x - \frac{2L}{3}\right)\right].$$

To find c_1 and c_2 we compute

$$y''(x) = c_1 + c_2 x + \frac{1}{2}\frac{w_0}{EI}\left[\left(x - \frac{L}{3}\right)^2 \mathcal{U}\left(x - \frac{L}{3}\right) - \left(x - \frac{2L}{3}\right)^2 \mathcal{U}\left(x - \frac{2L}{3}\right)\right]$$

and

$$y'''(x) = c_2 + \frac{w_0}{EI}\left[\left(x - \frac{L}{3}\right)\mathcal{U}\left(x - \frac{L}{3}\right) - \left(x - \frac{2L}{3}\right)\mathcal{U}\left(x - \frac{2L}{3}\right)\right].$$

Then $y''(L) = y'''(L) = 0$ yields the system

$$c_1 + c_2 L + \frac{1}{2}\frac{w_0}{EI}\left[\left(\frac{2L}{3}\right)^2 - \left(\frac{L}{3}\right)^2\right] = c_1 + c_2 L + \frac{1}{6}\frac{w_0 L^2}{EI} = 0$$

$$c_2 + \frac{w_0}{EI}\left[\frac{2L}{3} - \frac{L}{3}\right] = c_2 + \frac{1}{3}\frac{w_0 L}{EI} = 0.$$

Solving for c_1 and c_2 we obtain $c_1 = \frac{1}{6}w_0 L^2/EI$ and $c_2 = -\frac{1}{3}w_0 L/EI$. Thus

$$y(x) = \frac{w_0}{EI}\left(\frac{1}{12}L^2 x^2 - \frac{1}{18}Lx^3 + \frac{1}{24}\left[\left(x - \frac{L}{3}\right)^4 \mathcal{U}\left(x - \frac{L}{3}\right) - \left(x - \frac{2L}{3}\right)^4 \mathcal{U}\left(x - \frac{2L}{3}\right)\right]\right).$$

79. The differential equation is

$$EI\frac{d^4 y}{dx^4} = \frac{2w_0}{L}\left[\frac{L}{2} - x + \left(x - \frac{L}{2}\right)\mathcal{U}\left(x - \frac{L}{2}\right)\right].$$

Taking the Laplace transform of both sides and using $y(0) = y'(0) = 0$ we obtain

$$s^4 \mathcal{L}\{y\} - sy''(0) - y'''(0) = \frac{2w_0}{EIL}\left[\frac{L}{2s} - \frac{1}{s^2} + \frac{1}{s^2}e^{-Ls/2}\right].$$

Letting $y''(0) = c_1$ and $y'''(0) = c_2$ we have

$$\mathcal{L}\{y\} = \frac{c_1}{s^3} + \frac{c_2}{s^4} + \frac{2w_0}{EIL}\left[\frac{L}{2s^5} - \frac{1}{s^6} + \frac{1}{s^6}e^{-Ls/2}\right]$$

so that

$$y(x) = \frac{1}{2}c_1 x^2 + \frac{1}{6}c_2 x^3 + \frac{2w_0}{EIL}\left[\frac{L}{48}x^4 - \frac{1}{120}x^5 + \frac{1}{120}\left(x - \frac{L}{2}\right)^5 \mathcal{U}\left(x - \frac{L}{2}\right)\right]$$

$$= \frac{1}{2}c_1 x^2 + \frac{1}{6}c_2 x^3 + \frac{w_0}{60EIL}\left[\frac{5L}{2}x^4 - x^5 + \left(x - \frac{L}{2}\right)^5 \mathcal{U}\left(x - \frac{L}{2}\right)\right].$$

To find c_1 and c_2 we compute

$$y''(x) = c_1 + c_2 x + \frac{w_0}{60EIL}\left[30Lx^2 - 20x^3 + 20\left(x - \frac{L}{2}\right)^3 \mathcal{U}\left(x - \frac{L}{2}\right)\right]$$

and

$$y'''(x) = c_2 + \frac{w_0}{60EIL}\left[60Lx - 60x^2 + 60\left(x - \frac{L}{2}\right)^2 \mathcal{U}\left(x - \frac{L}{2}\right)\right].$$

Then $y''(L) = y'''(L) = 0$ yields the system

$$c_1 + c_2 L + \frac{w_0}{60EIL}\left[30L^3 - 20L^3 + \frac{5}{2}L^3\right] = c_1 + c_2 L + \frac{5w_0 L^2}{24EI} = 0$$

$$c_2 + \frac{w_0}{60EIL}[60L^2 - 60L^2 + 15L^2] = c_2 + \frac{w_0 L}{4EI} = 0.$$

Solving for c_1 and c_2 we obtain $c_1 = w_0 L^2/24EI$ and $c_2 = -w_0 L/4EI$. Thus

$$y(x) = \frac{w_0 L^2}{48EI}x^2 - \frac{w_0 L}{24EI}x^3 + \frac{w_0}{60EIL}\left[\frac{5L}{2}x^4 - x^5 + \left(x - \frac{L}{2}\right)^5 \mathcal{U}\left(x - \frac{L}{2}\right)\right].$$

80. The differential equation is

$$EI\frac{d^4 y}{dx^4} = w_0[1 - \mathcal{U}(x - L/2)].$$

Taking the Laplace transform of both sides and using $y(0) = y'(0) = 0$ we obtain

$$s^4 \mathcal{L}\{y\} - sy''(0) - y'''(0) = \frac{w_0}{EI}\frac{1}{s}\left(1 - e^{-Ls/2}\right).$$

Letting $y''(0) = c_1$ and $y'''(0) = c_2$ we have

$$\mathcal{L}\{y\} = \frac{c_1}{s^3} + \frac{c_2}{s^4} + \frac{w_0}{EI}\frac{1}{s^5}\left(1 - e^{-Ls/2}\right)$$

so that

$$y(x) = \frac{1}{2}c_1 x^2 + \frac{1}{6}c_2 x^3 + \frac{1}{24}\frac{w_0}{EI}\left[x^4 - \left(x - \frac{L}{2}\right)^4 \mathcal{U}\left(x - \frac{L}{2}\right)\right].$$

To find c_1 and c_2 we compute

$$y''(x) = c_1 + c_2 x + \frac{1}{2}\frac{w_0}{EI}\left[x^2 - \left(x - \frac{L}{2}\right)^2 \mathcal{U}\left(x - \frac{L}{2}\right)\right].$$

Then $y(L) = y''(L) = 0$ yields the system

$$\frac{1}{2}c_1 L^2 + \frac{1}{6}c_2 L^3 + \frac{1}{24}\frac{w_0}{EI}\left[L^4 - \left(\frac{L}{2}\right)^4\right] = \frac{1}{2}c_1 L^2 + \frac{1}{6}c_2 L^3 + \frac{5w_0}{128EI}L^4 = 0$$

$$c_1 + c_2 L + \frac{1}{2}\frac{w_0}{EI}\left[L^2 - \left(\frac{L}{2}\right)^2\right] = c_1 + c_2 L + \frac{3w_0}{8EI}L^2 = 0.$$

Solving for c_1 and c_2 we obtain $c_1 = \frac{9}{128}w_0 L^2/EI$ and $c_2 = -\frac{57}{128}w_0 L/EI$. Thus

$$y(x) = \frac{w_0}{EI}\left[\frac{9}{256}L^2 x^2 - \frac{19}{256}Lx^3 + \frac{1}{24}x^4 - \frac{1}{24}\left(x - \frac{L}{2}\right)^4 \mathcal{U}\left(x - \frac{L}{2}\right)\right].$$

81. (a) The temperature T of the cake inside the oven is modeled by

$$\frac{dT}{dt} = k(T - T_m)$$

where T_m is the ambient temperature of the oven. For $0 \leq t \leq 4$, we have

$$T_m = 70 + \frac{300 - 70}{4 - 0}t = 70 + 57.5t.$$

Hence for $t \geq 0$,

$$T_m = \begin{cases} 70 + 57.5t, & 0 \leq t < 4 \\ 300, & t \geq 4. \end{cases}$$

In terms of the unit step function,

$$T_m = (70 + 57.5t)[1 - \mathcal{U}(t - 4)] + 300\,\mathcal{U}(t - 4) = 70 + 57.5t + (230 - 57.5t)\,\mathcal{U}(t - 4).$$

The initial-value problem is then

$$\frac{dT}{dt} = k[T - 70 - 57.5t - (230 - 57.5t)\,\mathcal{U}(t - 4)], \qquad T(0) = 70.$$

(b) Let $t(s) = \mathcal{L}\{T(t)\}$. Transforming the equation, using $230 - 57.5t = -57.5(t - 4)$ and Theorem 7.3.2, gives

$$st(s) - 70 = k\left(t(s) - \frac{70}{s} - \frac{57.5}{s^2} + \frac{57.5}{s^2}e^{-4s}\right)$$

or

$$t(s) = \frac{70}{s - k} - \frac{70k}{s(s - k)} - \frac{57.5k}{s^2(s - k)} + \frac{57.5k}{s^2(s - k)}e^{-4s}.$$

After using partial functions, the inverse transform is then

$$T(t) = 70 + 57.5\left(\frac{1}{k} + t - \frac{1}{k}e^{kt}\right) - 57.5\left(\frac{1}{k} + t - 4 - \frac{1}{k}e^{k(t-4)}\right)\mathcal{U}(t - 4).$$

Of course, the obvious question is: What is k? If the cake is supposed to bake for, say, 20 minutes, then $T(20) = 300$. That is,

$$300 = 70 + 57.5\left(\frac{1}{k} + 20 - \frac{1}{k}e^{20k}\right) - 57.5\left(\frac{1}{k} + 16 - \frac{1}{k}e^{16k}\right).$$

But this equation has no physically meaningful solution. This should be no surprise since the model predicts the asymptotic behavior $T(t) \to 300$ as t increases. Using $T(20) = 299$ instead, we find, with the help of a CAS, that $k \approx -0.3$.

82. We use the fact that Theorem 7.3.2 can be written as

$$\mathcal{L}\{f(t - a)\,\mathcal{U}(t - a)\} = e^{-as}\,\mathcal{L}\{f(t)\}.$$

(a) Indentifying $a = 1$ we have

$$\mathcal{L}\{(2t + 1)\,\mathcal{U}(t - 1)\} = \mathcal{L}\{[2(t - 1) + 3]\,\mathcal{U}(t - 1)\} = e^{-s}\,\mathcal{L}\{2t + 3\} = e^{-s}\left(\frac{2}{s^2} + \frac{3}{s}\right).$$

Using (16) in the text we have

$$\mathscr{L}\{(2t+1)\,\mathscr{U}(t-1)\} = e^{-s}\,\mathscr{L}\{2(t+1)+1\} = e^{-s}\,\mathscr{L}\{2t+3\} = e^{-s}\left(\frac{2}{s^2}+\frac{3}{s}\right).$$

(b) Indentifying $a = 5$ we have

$$\mathscr{L}\{e^t\,\mathscr{U}(t-5)\} = \mathscr{L}\,\{e^{t-5+5}\,\mathscr{U}(t-5)\} = e^5\,\mathscr{L}\{e^{t-5}\,\mathscr{U}(t-5)\} = e^5 e^{-5s}\,\mathscr{L}\{e^t\} = \frac{e^{-5(s-1)}}{s-1}.$$

Using (16) in the text we have

$$\mathscr{L}\{e^t\,\mathscr{U}(t-5)\} = e^{-5s}\,\mathscr{L}\{e^{t+5}\} = e^{-5s}e^5\,\mathscr{L}\{e^t\} = \frac{e^{-5(s-1)}}{s-1}.$$

(c) Indentifying $a = \pi$ we have

$$\mathscr{L}\{\cos t\,\mathscr{U}(t-\pi)\} = -\,\mathscr{L}\{\cos(t-\pi)\,\mathscr{U}(t-\pi)\} = -e^{-\pi s}\,\mathscr{L}\{\cos t\} = -\frac{se^{-\pi s}}{s^2+1}.$$

Using (16) in the text we have

$$\mathscr{L}\{\cos t\,\mathscr{U}(t-\pi)\} = e^{-\pi s}\,\mathscr{L}\{\cos(t+\pi)\} = -e^{-\pi s}\,\mathscr{L}\{\cos t\} = -\frac{se^{-\pi s}}{s^2+1}.$$

(d) Indentifying $a = 2$ we have

$$\mathscr{L}\{(t^2-3t)\,\mathscr{U}(t-2)\} = \mathscr{L}\,\{[(t-2)^2 + 4t - 4 - 3t]\,\mathscr{U}(t-2)\}$$

$$= \mathscr{L}\,\{[(t-2)^2 + (t-2) - 2]\,\mathscr{U}(t-2)\}$$

$$= e^{-2s}\,\mathscr{L}\{t^2 + t - 2\} = e^{-2s}\left(\frac{2}{s^3}+\frac{1}{s^2}-\frac{2}{s}\right).$$

Using (16) in the text we have

$$\mathscr{L}\{(t^2-3t)\,\mathscr{U}(t-2)\} = e^{-2s}\,\mathscr{L}\{(t+2)^2 - 3(t+2)\}$$

$$= e^{-2s}\,\mathscr{L}\{t^2 + t - 2\} = e^{-2s}\left(\frac{2}{s^3}+\frac{1}{s^2}-\frac{2}{s}\right).$$

83. (a) From Theorem 7.3.1 we have $\mathscr{L}\{te^{kti}\} = 1/(s-ki)^2$. Then, using Euler's formula,

$$\mathscr{L}\{te^{kti}\} = \mathscr{L}\,\{t\cos kt + it\sin kt\} = \mathscr{L}\,\{t\cos kt\} + i\,\mathscr{L}\{t\sin kt\}$$

$$= \frac{1}{(s-ki)^2} = \frac{(s+ki)^2}{(s^2+k^2)^2} = \frac{s^2-k^2}{(s^2+k^2)^2} + i\,\frac{2ks}{(s^2+k^2)^2}.$$

Equating real and imaginary parts we have

$$\mathscr{L}\{t\cos kt\} = \frac{s^2-k^2}{(s^2+k^2)^2} \quad \text{and} \quad \mathscr{L}\{t\sin kt\} = \frac{2ks}{(s^2+k^2)^2}.$$

(b) The Laplace transform of the differential equation is

$$s^2 \mathcal{L}\{x\} + \omega^2 \mathcal{L}\{x\} = \frac{s}{s^2 + \omega^2} \, .$$

Solving for $\mathcal{L}\{x\}$ we obtain $\mathcal{L}\{x\} = s/(s^2 + \omega^2)^2$. Thus $x = (1/2\omega)t \sin \omega t$.

Exercises 7.4   Operational Properties II

1. $\mathcal{L}\{te^{-10t}\} = -\dfrac{d}{ds}\left(\dfrac{1}{s+10}\right) = \dfrac{1}{(s+10)^2}$

2. $\mathcal{L}\{t^3 e^t\} = (-1)^3 \dfrac{d^3}{ds^3}\left(\dfrac{1}{s-1}\right) = \dfrac{6}{(s-1)^4}$

3. $\mathcal{L}\{t\cos 2t\} = -\dfrac{d}{ds}\left(\dfrac{s}{s^2+4}\right) = \dfrac{s^2 - 4}{(s^2+4)^2}$

4. $\mathcal{L}\{t\sinh 3t\} = -\dfrac{d}{ds}\left(\dfrac{3}{s^2-9}\right) = \dfrac{6s}{(s^2-9)^2}$

5. $\mathcal{L}\{t^2 \sinh t\} = \dfrac{d^2}{ds^2}\left(\dfrac{1}{s^2-1}\right) = \dfrac{6s^2 + 2}{(s^2-1)^3}$

6. $\mathcal{L}\{t^2 \cos t\} = \dfrac{d^2}{ds^2}\left(\dfrac{s}{s^2+1}\right) = \dfrac{d}{ds}\left(\dfrac{1-s^2}{(s^2+1)^2}\right) = \dfrac{2s\left(s^2-3\right)}{(s^2+1)^3}$

7. $\mathcal{L}\{te^{2t}\sin 6t\} = -\dfrac{d}{ds}\left(\dfrac{6}{(s-2)^2+36}\right) = \dfrac{12(s-2)}{[(s-2)^2+36]^2}$

8. $\mathcal{L}\{te^{-3t}\cos 3t\} = -\dfrac{d}{ds}\left(\dfrac{s+3}{(s+3)^2+9}\right) = \dfrac{(s+3)^2 - 9}{[(s+3)^2+9]^2}$

9. The Laplace transform of the differential equation is

$$s\mathcal{L}\{y\} + \mathcal{L}\{y\} = \frac{2s}{(s^2+1)^2} \, .$$

Solving for $\mathcal{L}\{y\}$ we obtain

$$\mathcal{L}\{y\} = \frac{2s}{(s+1)(s^2+1)^2} = -\frac{1}{2}\frac{1}{s+1} - \frac{1}{2}\frac{1}{s^2+1} + \frac{1}{2}\frac{s}{s^2+1} + \frac{1}{(s^2+1)^2} + \frac{s}{(s^2+1)^2} \, .$$

Thus

$$y(t) = -\frac{1}{2}e^{-t} - \frac{1}{2}\sin t + \frac{1}{2}\cos t + \frac{1}{2}(\sin t - t\cos t) + \frac{1}{2}t\sin t$$

$$= -\frac{1}{2}e^{-t} + \frac{1}{2}\cos t - \frac{1}{2}t\cos t + \frac{1}{2}t\sin t.$$

10. The Laplace transform of the differential equation is

$$s\,\mathscr{L}\{y\} - \mathscr{L}\{y\} = \frac{2(s-1)}{((s-1)^2+1)^2}.$$

Solving for $\mathscr{L}\{y\}$ we obtain

$$\mathscr{L}\{y\} = \frac{2}{((s-1)^2+1)^2}.$$

Thus

$$y = e^t\sin t - te^t\cos t.$$

11. The Laplace transform of the differential equation is

$$s^2\,\mathscr{L}\{y\} - sy(0) - y'(0) + 9\,\mathscr{L}\{y\} = \frac{s}{s^2+9}.$$

Letting $y(0) = 2$ and $y'(0) = 5$ and solving for $\mathscr{L}\{y\}$ we obtain

$$\mathscr{L}\{y\} = \frac{2s^3 + 5s^2 + 19s + 45}{(s^2+9)^2} = \frac{2s}{s^2+9} + \frac{5}{s^2+9} + \frac{s}{(s^2+9)^2}.$$

Thus

$$y = 2\cos 3t + \frac{5}{3}\sin 3t + \frac{1}{6}t\sin 3t.$$

12. The Laplace transform of the differential equation is

$$s^2\,\mathscr{L}\{y\} - sy(0) - y'(0) + \mathscr{L}\{y\} = \frac{1}{s^2+1}.$$

Solving for $\mathscr{L}\{y\}$ we obtain

$$\mathscr{L}\{y\} = \frac{s^3 - s^2 + s}{(s^2+1)^2} = \frac{s}{s^2+1} - \frac{1}{s^2+1} + \frac{1}{(s^2+1)^2}.$$

Thus

$$y = \cos t - \sin t + \left(\frac{1}{2}\sin t - \frac{1}{2}t\cos t\right) = \cos t - \frac{1}{2}\sin t - \frac{1}{2}t\cos t.$$

13. The Laplace transform of the differential equation is

$$s^2\,\mathscr{L}\{y\} - sy(0) - y'(0) + 16\,\mathscr{L}\{y\} = \mathscr{L}\{\cos 4t - \cos 4t\,\mathscr{U}(t-\pi)\}$$

or by (16) of Section 7.3,

$$(s^2+16)\,\mathscr{L}\{y\} = 1 + \frac{s}{s^2+16} - e^{-\pi s}\,\mathscr{L}\{\cos 4(t+\pi)\}$$

$$= 1 + \frac{s}{s^2+16} - e^{-\pi s}\,\mathscr{L}\{\cos 4t\}$$

$$= 1 + \frac{s}{s^2+16} - \frac{s}{s^2+16}e^{-\pi s}.$$

Thus

$$\mathscr{L}\{y\} = \frac{1}{s^2 + 16} + \frac{s}{(s^2 + 16)^2} - \frac{s}{(s^2 + 16)^2} e^{-\pi s}$$

and

$$y = \frac{1}{4}\sin 4t + \frac{1}{8}t\sin 4t - \frac{1}{8}(t - \pi)\sin 4(t - \pi)\mathscr{U}(t - \pi).$$

14. The Laplace transform of the differential equation is

$$s^2 \mathscr{L}\{y\} - sy(0) - y'(0) + \mathscr{L}\{y\} = \mathscr{L}\left\{1 - \mathscr{U}\left(t - \frac{\pi}{2}\right) + \sin t\, \mathscr{U}\left(t - \frac{\pi}{2}\right)\right\}$$

or

$$(s^2 + 1)\,\mathscr{L}\{y\} = s + \frac{1}{s} - \frac{1}{s}e^{-\pi s/2} + e^{-\pi s/2}\,\mathscr{L}\left\{\sin\left(t + \frac{\pi}{2}\right)\right\}$$

$$= s + \frac{1}{s} - \frac{1}{s}e^{-\pi s/2} + e^{-\pi s/2}\,\mathscr{L}\{\cos t\}$$

$$= s + \frac{1}{s} - \frac{1}{s}e^{-\pi s/2} + \frac{s}{s^2 + 1}e^{-\pi s/2}.$$

Thus

$$\mathscr{L}\{y\} = \frac{s}{s^2 + 1} + \frac{1}{s(s^2 + 1)} - \frac{1}{s(s^2 + 1)}e^{-\pi s/2} + \frac{s}{(s^2 + 1)^2}e^{-\pi s/2}$$

$$= \frac{s}{s^2 + 1} + \frac{1}{s} - \frac{s}{s^2 + 1} - \left(\frac{1}{s} - \frac{s}{s^2 + 1}\right)e^{-\pi s/2} + \frac{s}{(s^2 + 1)^2}e^{-\pi s/2}$$

$$= \frac{1}{s} - \left(\frac{1}{s} - \frac{s}{s^2 + 1}\right)e^{-\pi s/2} + \frac{s}{(s^2 + 1)^2}e^{-\pi s/2}$$

and

$$y = 1 - \left[1 - \cos\left(t - \frac{\pi}{2}\right)\right]\mathscr{U}\left(t - \frac{\pi}{2}\right) + \frac{1}{2}\left(t - \frac{\pi}{2}\right)\sin\left(t - \frac{\pi}{2}\right)\mathscr{U}\left(t - \frac{\pi}{2}\right)$$

$$= 1 - (1 - \sin t)\,\mathscr{U}\left(t - \frac{\pi}{2}\right) - \frac{1}{2}\left(t - \frac{\pi}{2}\right)\cos t\,\mathscr{U}\left(t - \frac{\pi}{2}\right).$$

15.

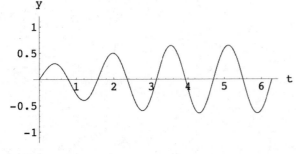

16.

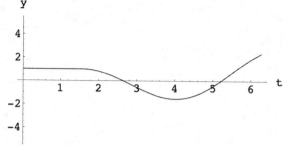

17. From (7) of Section 7.2 in the text along with Theorem 7.4.1,

$$\mathscr{L}\{ty''\} = -\frac{d}{ds}\mathscr{L}\{y''\} = -\frac{d}{ds}[s^2Y(s) - sy(0) - y'(0)] = -s^2\frac{dY}{ds} - 2sY + y(0),$$

so that the transform of the given second-order differential equation is the linear first-order differential equation in $Y(s)$:

$$s^2Y' + 3sY = -\frac{4}{s^3} \qquad \text{or} \qquad Y' + \frac{3}{s}Y = -\frac{4}{s^5}.$$

The solution of the latter equation is $Y(s) = 4/s^4 + c/s^3$, so

$$y(t) = \mathscr{L}^{-1}\{Y(s)\} = \frac{2}{3}t^3 + \frac{c}{2}t^2.$$

c_1 as in #32, p.367

18. From Theorem 7.4.1 in the text

$$\mathscr{L}\{ty'\} = -\frac{d}{ds}\mathscr{L}\{y'\} = -\frac{d}{ds}[sY(s) - y(0)] = -s\frac{dY}{ds} - Y$$

so that the transform of the given second-order differential equation is the linear first-order differential equation in $Y(s)$:

$$Y' + \left(\frac{3}{s} - 2s\right)Y = -\frac{10}{s}.$$

Using the integrating factor $s^3e^{-s^2}$, the last equation yields

$$Y(s) = \frac{5}{s^3} + \frac{c}{s^3}e^{s^2}.$$

But if $Y(s)$ is the Laplace transform of a piecewise-continuous function of exponential order, we must have, in view of Theorem 7.1.3, $\lim_{s\to\infty} Y(s) = 0$. In order to obtain this condition we require $c = 0$. Hence

$$y(t) = \mathscr{L}^{-1}\left\{\frac{5}{s^3}\right\} = \frac{5}{2}t^2.$$

19. $\mathscr{L}\{1 * t^3\} = \frac{1}{s}\frac{3!}{s^4} = \frac{6}{s^5}$

20. $\mathscr{L}\{t^2 * te^t\} = \frac{2}{s^3(s-1)^2}$

21. $\mathscr{L}\{e^{-t} * e^t\cos t\} = \frac{s-1}{(s+1)[(s-1)^2 + 1]}$

22. $\mathscr{L}\left\{e^{2t} * \sin t\right\} = \dfrac{1}{(s-2)(s^2+1)}$

23. $\mathscr{L}\left\{\displaystyle\int_0^t e^\tau \, d\tau\right\} = \dfrac{1}{s}\mathscr{L}\{e^t\} = \dfrac{1}{s(s-1)}$

24. $\mathscr{L}\left\{\displaystyle\int_0^t \cos\tau \, d\tau\right\} = \dfrac{1}{s}\mathscr{L}\{\cos t\} = \dfrac{s}{s(s^2+1)} = \dfrac{1}{s^2+1}$

25. $\mathscr{L}\left\{\displaystyle\int_0^t e^{-\tau}\cos\tau \, d\tau\right\} = \dfrac{1}{s}\,\mathscr{L}\left\{e^{-t}\cos t\right\} = \dfrac{1}{s}\dfrac{s+1}{(s+1)^2+1} = \dfrac{s+1}{s\,(s^2+2s+2)}$

26. $\mathscr{L}\left\{\displaystyle\int_0^t \tau\sin\tau \, d\tau\right\} = \dfrac{1}{s}\,\mathscr{L}\{t\sin t\} = \dfrac{1}{s}\left(-\dfrac{d}{ds}\dfrac{1}{s^2+1}\right) = -\dfrac{1}{s}\dfrac{-2s}{(s^2+1)^2} = \dfrac{2}{(s^2+1)^2}$

27. $\mathscr{L}\left\{\displaystyle\int_0^t \tau e^{t-\tau} \, d\tau\right\} = \mathscr{L}\{t\}\,\mathscr{L}\{e^t\} = \dfrac{1}{s^2(s-1)}$

28. $\mathscr{L}\left\{\displaystyle\int_0^t \sin\tau\cos(t-\tau) \, d\tau\right\} = \mathscr{L}\{\sin t\}\,\mathscr{L}\{\cos t\} = \dfrac{s}{(s^2+1)^2}$

29. $\mathscr{L}\left\{t\displaystyle\int_0^t \sin\tau \, d\tau\right\} = -\dfrac{d}{ds}\mathscr{L}\left\{\displaystyle\int_0^t \sin\tau \, d\tau\right\} = -\dfrac{d}{ds}\left(\dfrac{1}{s}\dfrac{1}{s^2+1}\right) = \dfrac{3s^2+1}{s^2\,(s^2+1)^2}$

30. $\mathscr{L}\left\{t\displaystyle\int_0^t \tau e^{-\tau}d\tau\right\} = -\dfrac{d}{ds}\mathscr{L}\left\{\displaystyle\int_0^t \tau e^{-\tau}d\tau\right\} = -\dfrac{d}{ds}\left(\dfrac{1}{s}\dfrac{1}{(s+1)^2}\right) = \dfrac{3s+1}{s^2(s+1)^3}$

31. $\mathscr{L}^{-1}\left\{\dfrac{1}{s(s-1)}\right\} = \mathscr{L}^{-1}\left\{\dfrac{1/(s-1)}{s}\right\} = \displaystyle\int_0^t e^\tau d\tau = e^t - 1$

32. $\mathscr{L}^{-1}\left\{\dfrac{1}{s^2(s-1)}\right\} = \mathscr{L}^{-1}\left\{\dfrac{1/s(s-1)}{s}\right\} = \displaystyle\int_0^t (e^\tau - 1)d\tau = e^t - t - 1$

33. $\mathscr{L}^{-1}\left\{\dfrac{1}{s^3(s-1)}\right\} = \mathscr{L}^{-1}\left\{\dfrac{1/s^2(s-1)}{s}\right\} = \displaystyle\int_0^t (e^\tau - \tau - 1)d\tau = e^t - \dfrac{1}{2}t^2 - t - 1$

34. Using $\mathscr{L}^{-1}\left\{\dfrac{1}{(s-a)^2}\right\} = te^{at}$, (8) in the text gives

$$\mathscr{L}^{-1}\left\{\dfrac{1}{s(s-a)^2}\right\} = \int_0^t \tau e^{a\tau}\, d\tau = \dfrac{1}{a^2}(ate^{at} - e^{at} + 1).$$

35. (a) The result in (4) in the text is $\mathscr{L}^{-1}\{F(s)G(s)\} = f * g$, so identify

$$F(s) = \dfrac{2k^3}{(s^2+k^2)^2} \qquad\text{and}\qquad G(s) = \dfrac{4s}{s^2+k^2}.$$

Then

$$f(t) = \sin kt - kt \cos kt \qquad \text{and} \qquad g(t) = 4 \cos kt$$

so

$$\mathscr{L}^{-1}\left\{\frac{8k^3 s}{(s^2 + k^2)^3}\right\} = \mathscr{L}^{-1}\{F(s)G(s)\} = f * g = 4 \int_0^t f(\tau)g(t - \tau)dt$$

$$= 4 \int_0^t (\sin k\tau - k\tau \cos k\tau) \cos k(t - \tau)d\tau.$$

Using a CAS to evaluate the integral we get

$$\mathscr{L}^{-1}\left\{\frac{8k^3 s}{(s^2 + k^2)^3}\right\} = t \sin kt - kt^2 \cos kt.$$

(b) Observe from part (a) that

$$\mathscr{L}\{t(\sin kt - kt \cos kt)\} = \frac{8k^3 s}{(s^2 + k^2)^3},$$

and from Theorem 7.4.1 that $\mathscr{L}\{tf(t)\} = -F'(s)$. We saw in (5) in the text that

$$\mathscr{L}\{\sin kt - kt \cos kt\} = 2k^3/(s^2 + k^2)^2,$$

so

$$\mathscr{L}\{t(\sin kt - kt \cos kt)\} = -\frac{d}{ds}\frac{2k^3}{(s^2 + k^2)^2} = \frac{8k^3 s}{(s^2 + k^2)^3}.$$

36. The Laplace transform of the differential equation is

$$s^2 \mathscr{L}\{y\} + \mathscr{L}\{y\} = \frac{1}{(s^2 + 1)} + \frac{2s}{(s^2 + 1)^2}.$$

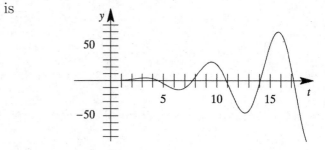

Thus

$$\mathscr{L}\{y\} = \frac{1}{(s^2 + 1)^2} + \frac{2s}{(s^2 + 1)^3}$$

and, using Problem 35 with $k = 1$,

$$y = \frac{1}{2}(\sin t - t \cos t) + \frac{1}{4}(t \sin t - t^2 \cos t).$$

37. The Laplace transform of the given equation is

$$\mathscr{L}\{f\} + \mathscr{L}\{t\}\mathscr{L}\{f\} = \mathscr{L}\{t\}.$$

Solving for $\mathscr{L}\{f\}$ we obtain $\mathscr{L}\{f\} = \dfrac{1}{s^2 + 1}$. Thus, $f(t) = \sin t$.

38. The Laplace transform of the given equation is

$$\mathscr{L}\{f\} = \mathscr{L}\{2t\} - 4\mathscr{L}\{\sin t\}\mathscr{L}\{f\}.$$

Solving for $\mathscr{L}\{f\}$ we obtain

$$\mathscr{L}\{f\} = \frac{2s^2 + 2}{s^2(s^2 + 5)} = \frac{2}{5}\frac{1}{s^2} + \frac{8}{5\sqrt{5}}\frac{\sqrt{5}}{s^2 + 5}.$$

Thus

$$f(t) = \frac{2}{5}t + \frac{8}{5\sqrt{5}}\sin\sqrt{5}\,t.$$

39. The Laplace transform of the given equation is

$$\mathscr{L}\{f\} = \mathscr{L}\{te^t\} + \mathscr{L}\{t\}\mathscr{L}\{f\}.$$

Solving for $\mathscr{L}\{f\}$ we obtain

$$\mathscr{L}\{f\} = \frac{s^2}{(s-1)^3(s+1)} = \frac{1}{8}\frac{1}{s-1} + \frac{3}{4}\frac{1}{(s-1)^2} + \frac{1}{4}\frac{2}{(s-1)^3} - \frac{1}{8}\frac{1}{s+1}.$$

Thus

$$f(t) = \frac{1}{8}e^t + \frac{3}{4}te^t + \frac{1}{4}t^2e^t - \frac{1}{8}e^{-t}$$

40. The Laplace transform of the given equation is

$$\mathscr{L}\{f\} + 2\mathscr{L}\{\cos t\}\mathscr{L}\{f\} = 4\mathscr{L}\{e^{-t}\} + \mathscr{L}\{\sin t\}.$$

Solving for $\mathscr{L}\{f\}$ we obtain

$$\mathscr{L}\{f\} = \frac{4s^2 + s + 5}{(s+1)^3} = \frac{4}{s+1} - \frac{7}{(s+1)^2} + 4\frac{2}{(s+1)^3}.$$

Thus

$$f(t) = 4e^{-t} - 7te^{-t} + 4t^2e^{-t}.$$

41. The Laplace transform of the given equation is

$$\mathscr{L}\{f\} + \mathscr{L}\{1\}\mathscr{L}\{f\} = \mathscr{L}\{1\}.$$

Solving for $\mathscr{L}\{f\}$ we obtain $\mathscr{L}\{f\} = \dfrac{1}{s+1}$. Thus, $f(t) = e^{-t}$.

42. The Laplace transform of the given equation is

$$\mathscr{L}\{f\} = \mathscr{L}\{\cos t\} + \mathscr{L}\{e^{-t}\}\mathscr{L}\{f\}.$$

Solving for $\mathscr{L}\{f\}$ we obtain

$$\mathscr{L}\{f\} = \frac{s}{s^2 + 1} + \frac{1}{s^2 + 1}.$$

Thus

$$f(t) = \cos t + \sin t.$$

43. The Laplace transform of the given equation is

$$\mathscr{L}\{f\} = \mathscr{L}\{1\} + \mathscr{L}\{t\} - \mathscr{L}\left\{\frac{8}{3}\int_0^t (t-\tau)^3 f(\tau)\,d\tau\right\}$$

$$= \frac{1}{s} + \frac{1}{s^2} + \frac{8}{3}\mathscr{L}\{t^3\}\mathscr{L}\{f\} = \frac{1}{s} + \frac{1}{s^2} + \frac{16}{s^4}\mathscr{L}\{f\}.$$

Solving for $\mathscr{L}\{f\}$ we obtain

$$\mathscr{L}\{f\} = \frac{s^2(s+1)}{s^4 - 16} = \frac{1}{8}\frac{1}{s+2} + \frac{3}{8}\frac{1}{s-2} + \frac{1}{4}\frac{2}{s^2+4} + \frac{1}{2}\frac{s}{s^2+4}.$$

Thus

$$f(t) = \frac{1}{8}e^{-2t} + \frac{3}{8}e^{2t} + \frac{1}{4}\sin 2t + \frac{1}{2}\cos 2t.$$

44. The Laplace transform of the given equation is

$$\mathscr{L}\{t\} - 2\mathscr{L}\{f\} = \mathscr{L}\left\{e^t - e^{-t}\right\}\mathscr{L}\{f\}.$$

Solving for $\mathscr{L}\{f\}$ we obtain

$$\mathscr{L}\{f\} = \frac{s^2-1}{2s^4} = \frac{1}{2}\frac{1}{s^2} - \frac{1}{12}\frac{3!}{s^4}.$$

Thus

$$f(t) = \frac{1}{2}t - \frac{1}{12}t^3.$$

45. The Laplace transform of the given equation is

$$s\mathscr{L}\{y\} - y(0) = \mathscr{L}\{1\} - \mathscr{L}\{\sin t\} - \mathscr{L}\{1\}\mathscr{L}\{y\}.$$

Solving for $\mathscr{L}\{f\}$ we obtain

$$\mathscr{L}\{y\} = \frac{s^2 - s + 1}{(s^2+1)^2} = \frac{1}{s^2+1} - \frac{1}{2}\frac{2s}{(s^2+1)^2}.$$

Thus

$$y = \sin t - \frac{1}{2}t\sin t.$$

46. The Laplace transform of the given equation is

$$s\mathscr{L}\{y\} - y(0) + 6\mathscr{L}\{y\} + 9\mathscr{L}\{1\}\mathscr{L}\{y\} = \mathscr{L}\{1\}.$$

Solving for $\mathscr{L}\{f\}$ we obtain $\mathscr{L}\{y\} = \dfrac{1}{(s+3)^2}$. Thus, $y = te^{-3t}$.

47. The differential equation is

$$0.1\frac{di}{dt} + 3i + \frac{1}{0.05}\int_0^t i(\tau)d\tau = 100[\mathcal{U}(t-1) - \mathcal{U}(t-2)]$$

or

$$\frac{di}{dt} + 30i + 200\int_0^t i(\tau)d\tau = 1000[\mathcal{U}(t-1) - \mathcal{U}(t-2)],$$

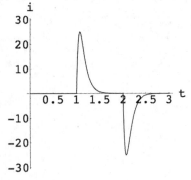

where $i(0) = 0$. The Laplace transform of the differential equation is

$$s\,\mathcal{L}\{i\} - y(0) + 30\,\mathcal{L}\{i\} + \frac{200}{s}\mathcal{L}\{i\} = \frac{1000}{s}(e^{-s} - e^{-2s}).$$

Solving for $\mathcal{L}\{i\}$ we obtain

$$\mathcal{L}\{i\} = \frac{1000e^{-s} - 1000e^{-2s}}{s^2 + 30s + 200} = \left(\frac{100}{s+10} - \frac{100}{s+20}\right)(e^{-s} - e^{-2s}).$$

Thus

$$i(t) = 100(e^{-10(t-1)} - e^{-20(t-1)})\,\mathcal{U}(t-1) - 100(e^{-10(t-2)} - e^{-20(t-2)})\,\mathcal{U}(t-2).$$

48. The differential equation is

$$0.005\frac{di}{dt} + i + \frac{1}{0.02}\int_0^t i(\tau)d\tau = 100[t - (t-1)\mathcal{U}(t-1)]$$

or

$$\frac{di}{dt} + 200i + 10{,}000\int_0^t i(\tau)d\tau = 20{,}000[t - (t-1)\mathcal{U}(t-1)],$$

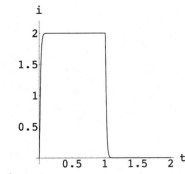

where $i(0) = 0$. The Laplace transform of the differential equation is

$$s\,\mathcal{L}\{i\} + 200\,\mathcal{L}\{i\} + \frac{10{,}000}{s}\mathcal{L}\{i\} = 20{,}000\left(\frac{1}{s^2} - \frac{1}{s^2}e^{-s}\right).$$

Solving for $\mathcal{L}\{i\}$ we obtain

$$\mathcal{L}\{i\} = \frac{20{,}000}{s(s+100)^2}(1 - e^{-s}) = \left[\frac{2}{s} - \frac{2}{s+100} - \frac{200}{(s+100)^2}\right](1 - e^{-s}).$$

Thus

$$i(t) = 2 - 2e^{-100t} - 200te^{-100t} - 2\,\mathcal{U}(t-1) + 2e^{-100(t-1)}\,\mathcal{U}(t-1) + 200(t-1)e^{-100(t-1)}\,\mathcal{U}(t-1).$$

49. $\mathcal{L}\{f(t)\} = \dfrac{1}{1 - e^{-2as}}\left[\displaystyle\int_0^a e^{-st}dt - \int_a^{2a} e^{-st}dt\right] = \dfrac{(1 - e^{-as})^2}{s(1 - e^{-2as})} = \dfrac{1 - e^{-as}}{s(1 + e^{-as})}$

50. $\mathcal{L}\{f(t)\} = \dfrac{1}{1 - e^{-2as}}\displaystyle\int_0^a e^{-st}dt = \dfrac{1}{s(1 + e^{-as})}$

51. Using integration by parts,

$$\mathcal{L}\{f(t)\} = \frac{1}{1 - e^{-bs}} \int_0^b \frac{a}{b} t e^{-st} \, dt = \frac{a}{s}\left(\frac{1}{bs} - \frac{1}{e^{bs} - 1}\right).$$

52. $\mathcal{L}\{f(t)\} = \dfrac{1}{1 - e^{-2s}} \left[\displaystyle\int_0^1 t e^{-st} \, dt + \int_1^2 (2 - t)e^{-st}\, dt\right] = \dfrac{1 - e^{-s}}{s^2(1 - e^{-2s})}$

53. $\mathcal{L}\{f(t)\} = \dfrac{1}{1 - e^{-\pi s}} \displaystyle\int_0^\pi e^{-st} \sin t \, dt = \dfrac{1}{s^2 + 1} \cdot \dfrac{e^{\pi s/2} + e^{-\pi s/2}}{e^{\pi s/2} - e^{-\pi s/2}} = \dfrac{1}{s^2 + 1} \coth \dfrac{\pi s}{2}$

54. $\mathcal{L}\{f(t)\} = \dfrac{1}{1 - e^{-2\pi s}} \displaystyle\int_0^\pi e^{-st} \sin t \, dt = \dfrac{1}{s^2 + 1} \cdot \dfrac{1}{1 - e^{-\pi s}}$

55. The differential equation is $L\, di/dt + Ri = E(t)$, where $i(0) = 0$. The Laplace transform of the equation is

$$Ls\, \mathcal{L}\{i\} + R\, \mathcal{L}\{i\} = \mathcal{L}\{E(t)\}.$$

From Problem 49 we have $\mathcal{L}\{E(t)\} = (1 - e^{-s})/s(1 + e^{-s})$. Thus

$$(Ls + R)\, \mathcal{L}\{i\} = \frac{1 - e^{-s}}{s(1 + e^{-s})}$$

and

$$\mathcal{L}\{i\} = \frac{1}{L} \frac{1 - e^{-s}}{s(s + R/L)(1 + e^{-s})} = \frac{1}{L} \frac{1 - e^{-s}}{s(s + R/L)} \frac{1}{1 + e^{-s}}$$

$$= \frac{1}{R}\left(\frac{1}{s} - \frac{1}{s + R/L}\right)(1 - e^{-s})(1 - e^{-s} + e^{-2s} - e^{-3s} + e^{-4s} - \cdots)$$

$$= \frac{1}{R}\left(\frac{1}{s} - \frac{1}{s + R/L}\right)(1 - 2e^{-s} + 2e^{-2s} - 2e^{-3s} + 2e^{-4s} - \cdots).$$

Therefore,

$$i(t) = \frac{1}{R}\left(1 - e^{-Rt/L}\right) - \frac{2}{R}\left(1 - e^{-R(t-1)/L}\right)\mathcal{U}(t - 1)$$

$$+ \frac{2}{R}\left(1 - e^{-R(t-2)/L}\right)\mathcal{U}(t - 2) - \frac{2}{R}\left(1 - e^{-R(t-3)/L}\right)\mathcal{U}(t - 3) + \cdots$$

$$= \frac{1}{R}\left(1 - e^{-Rt/L}\right) + \frac{2}{R}\sum_{n=1}^\infty (-1)^n \left(1 - e^{-R(t-n)/L}\right)\mathcal{U}(t - n).$$

The graph of $i(t)$ with $L = 1$ and $R = 1$ is shown below.

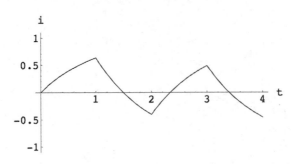

56. The differential equation is $L\, di/dt + Ri = E(t)$, where $i(0) = 0$. The Laplace transform of the equation is

$$Ls\,\mathscr{L}\{i\} + R\,\mathscr{L}\{i\} = \mathscr{L}\{E(t)\}.$$

From Problem 51 we have

$$\mathscr{L}\{E(t)\} = \frac{1}{s}\left(\frac{1}{s} - \frac{1}{e^s - 1}\right) = \frac{1}{s^2} - \frac{1}{s}\frac{1}{e^s - 1}.$$

Thus

$$(Ls + R)\,\mathscr{L}\{i\} = \frac{1}{s^2} - \frac{1}{s}\frac{1}{e^s - 1}$$

and

$$\mathscr{L}\{i\} = \frac{1}{L}\frac{1}{s^2(s + R/L)} - \frac{1}{L}\frac{1}{s(s + R/L)}\frac{1}{e^s - 1}$$

$$= \frac{1}{R}\left(\frac{1}{s^2} - \frac{L}{R}\frac{1}{s} + \frac{L}{R}\frac{1}{s + R/L}\right) - \frac{1}{R}\left(\frac{1}{s} - \frac{1}{s + R/L}\right)(e^{-s} + e^{-2s} + e^{-3s} + \cdots).$$

Therefore

$$i(t) = \frac{1}{R}\left(t - \frac{L}{R} + \frac{L}{R}e^{-Rt/L}\right) - \frac{1}{R}\left(1 - e^{-R(t-1)/L}\right)\mathscr{U}(t - 1)$$

$$- \frac{1}{R}\left(1 - e^{-R(t-2)/L}\right)\mathscr{U}(t - 2) - \frac{1}{R}\left(1 - e^{-R(t-3)/L}\right)\mathscr{U}(t - 3) - \cdots$$

$$= \frac{1}{R}\left(t - \frac{L}{R} + \frac{L}{R}e^{-Rt/L}\right) - \frac{1}{R}\sum_{n=1}^{\infty}\left(1 - e^{-R(t-n)/L}\right)\mathscr{U}(t - n).$$

The graph of $i(t)$ with $L = 1$ and $R = 1$ is shown below.

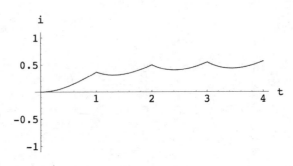

57. The differential equation is $x'' + 2x' + 10x = 20f(t)$, where $f(t)$ is the meander function in Problem 49 with $a = \pi$. Using the initial conditions $x(0) = x'(0) = 0$ and taking the Laplace transform we obtain

$$(s^2 + 2s + 10)\,\mathscr{L}\{x(t)\} = \frac{20}{s}(1 - e^{-\pi s})\frac{1}{1 + e^{-\pi s}}$$

$$= \frac{20}{s}(1 - e^{-\pi s})(1 - e^{-\pi s} + e^{-2\pi s} - e^{-3\pi s} + \cdots)$$

$$= \frac{20}{s}(1 - 2e^{-\pi s} + 2e^{-2\pi s} - 2e^{-3\pi s} + \cdots)$$

$$= \frac{20}{s} + \frac{40}{s}\sum_{n=1}^{\infty}(-1)^n e^{-n\pi s}.$$

Then

$$\mathscr{L}\{x(t)\} = \frac{20}{s(s^2 + 2s + 10)} + \frac{40}{s(s^2 + 2s + 10)}\sum_{n=1}^{\infty}(-1)^n e^{-n\pi s}$$

$$= \frac{2}{s} - \frac{2s + 4}{s^2 + 2s + 10} + \sum_{n=1}^{\infty}(-1)^n\left[\frac{4}{s} - \frac{4s + 8}{s^2 + 2s + 10}\right]e^{-n\pi s}$$

$$= \frac{2}{s} - \frac{2(s + 1) + 2}{(s + 1)^2 + 9} + 4\sum_{n=1}^{\infty}(-1)^n\left[\frac{1}{s} - \frac{(s + 1) + 1}{(s + 1)^2 + 9}\right]e^{-n\pi s}$$

and

$$x(t) = 2\left(1 - e^{-t}\cos 3t - \frac{1}{3}e^{-t}\sin 3t\right) + 4\sum_{n=1}^{\infty}(-1)^n\left[1 - e^{-(t - n\pi)}\cos 3(t - n\pi)\right.$$

$$\left. - \frac{1}{3}e^{-(t - n\pi)}\sin 3(t - n\pi)\right]\mathscr{U}(t - n\pi).$$

The graph of $x(t)$ on the interval $[0, 2\pi)$ is shown below.

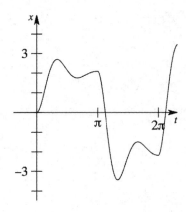

58. The differential equation is $x'' + 2x' + x = 5f(t)$, where $f(t)$ is the square wave function with $a = \pi$.

Using the initial conditions $x(0) = x'(0) = 0$ and taking the Laplace transform, we obtain

$$(s^2 + 2s + 1)\mathscr{L}\{x(t)\} = \frac{5}{s}\frac{1}{1 + e^{-\pi s}} = \frac{5}{s}(1 - e^{-\pi s} + e^{-2\pi s} - e^{-3\pi s} + e^{-4\pi s} - \cdots)$$

$$= \frac{5}{s}\sum_{n=0}^{\infty}(-1)^n e^{-n\pi s}.$$

Then

$$\mathscr{L}\{x(t)\} = \frac{5}{s(s+1)^2}\sum_{n=0}^{\infty}(-1)^n e^{-n\pi s} = 5\sum_{n=0}^{\infty}(-1)^n\left(\frac{1}{s} - \frac{1}{s+1} - \frac{1}{(s+1)^2}\right)e^{-n\pi s}$$

and

$$x(t) = 5\sum_{n=0}^{\infty}(-1)^n(1 - e^{-(t-n\pi)} - (t - n\pi)e^{-(t-n\pi)})\,\mathscr{U}(t - n\pi).$$

The graph of $x(t)$ on the interval $[0, 4\pi)$ is shown below.

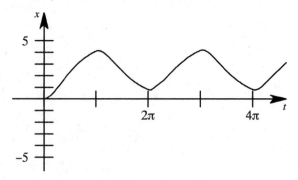

59. $f(t) = -\frac{1}{t}\mathscr{L}^{-1}\left\{\frac{d}{ds}[\ln(s-3) - \ln(s+1)]\right\} = -\frac{1}{t}\mathscr{L}^{-1}\left\{\frac{1}{s-3} - \frac{1}{s+1}\right\} = -\frac{1}{t}\left(e^{3t} - e^{-t}\right)$

60. The transform of Bessel's equation is

$$-\frac{d}{ds}[s^2 Y(s) - sy(0) - y'(0)] + sY(s) - y(0) - \frac{d}{ds}Y(s) = 0$$

or, after simplifying and using the initial condition, $(s^2 + 1)Y' + sY = 0$. This equation is both separable and linear. Solving gives $Y(s) = c/\sqrt{s^2 + 1}$. Now $Y(s) = \mathscr{L}\{J_0(t)\}$, where J_0 has a derivative that is continuous and of exponential order, implies by Problem 46 of Exercises 7.2 that

$$1 = J_0(0) = \lim_{s\to\infty}sY(s) = c\lim_{s\to\infty}\frac{s}{\sqrt{s^2 + k^2}} = c$$

so $c = 1$ and

$$Y(s) = \frac{1}{\sqrt{s^2 + 1}} \qquad \text{or} \qquad \mathscr{L}\{J_0(t)\} = \frac{1}{\sqrt{s^2 + 1}}.$$

61. (a) Using Theorem 7.4.1, the Laplace transform of the differential equation is

$$-\frac{d}{ds}[s^2 Y - sy(0) - y'(0)] + sY - y(0) + \frac{d}{ds}[sY - y(0)] + nY$$

$$= -\frac{d}{ds}[s^2 Y] + sY + \frac{d}{ds}[sY] + nY$$

$$= -s^2\left(\frac{dY}{ds}\right) - 2sY + sY + s\left(\frac{dY}{ds}\right) + Y + nY$$

$$= (s - s^2)\left(\frac{dY}{ds}\right) + (1 + n - s)Y = 0.$$

Separating variables, we find

$$\frac{dY}{Y} = \frac{1 + n - s}{s^2 - s}\,ds = \left(\frac{n}{s-1} - \frac{1+n}{s}\right)ds$$

$$\ln Y = n\ln(s - 1) - (1 + n)\ln s + c$$

$$Y = c_1 \frac{(s-1)^n}{s^{1+n}}.$$

Since the differential equation is homogeneous, any constant multiple of a solution will still be a solution, so for convenience we take $c_1 = 1$. The following polynomials are solutions of Laguerre's differential equation:

$$n = 0: \quad L_0(t) = \mathscr{L}^{-1}\left\{\frac{1}{s}\right\} = 1$$

$$n = 1: \quad L_1(t) = \mathscr{L}^{-1}\left\{\frac{s-1}{s^2}\right\} = \mathscr{L}^{-1}\left\{\frac{1}{s} - \frac{1}{s^2}\right\} = 1 - t$$

$$n = 2: \quad L_2(t) = \mathscr{L}^{-1}\left\{\frac{(s-1)^2}{s^3}\right\} = \mathscr{L}^{-1}\left\{\frac{1}{s} - \frac{2}{s^2} + \frac{1}{s^3}\right\} = 1 - 2t + \frac{1}{2}t^2$$

$$n = 3: \quad L_3(t) = \mathscr{L}^{-1}\left\{\frac{(s-1)^3}{s^4}\right\} = \mathscr{L}^{-1}\left\{\frac{1}{s} - \frac{3}{s^2} + \frac{3}{s^3} - \frac{1}{s^4}\right\} = 1 - 3t + \frac{3}{2}t^2 - \frac{1}{6}t^3$$

$$n = 4: \quad L_4(t) = \mathscr{L}^{-1}\left\{\frac{(s-1)^4}{s^5}\right\} = \mathscr{L}^{-1}\left\{\frac{1}{s} - \frac{4}{s^2} + \frac{6}{s^3} - \frac{4}{s^4} + \frac{1}{s^5}\right\}$$

$$= 1 - 4t + 3t^2 - \frac{2}{3}t^3 + \frac{1}{24}t^4.$$

(b) Letting $f(t) = t^n e^{-t}$ we note that $f^{(k)}(0) = 0$ for $k = 0, 1, 2, \ldots, n-1$ and $f^{(n)}(0) = n!$.

Now, by the first translation theorem,

$$\mathscr{L}\left\{\frac{e^t}{n!}\frac{d^n}{dt^n}\,t^n e^{-t}\right\} = \frac{1}{n!}\mathscr{L}\{e^t f^{(n)}(t)\} = \frac{1}{n!}\mathscr{L}\{f^{(n)}(t)\}\,|_{s\to s-1}$$

$$= \frac{1}{n!}\left[s^n\,\mathscr{L}\{t^n e^{-t}\} - s^{n-1}f(0) - s^{n-2}f'(0) - \cdots - f^{(n-1)}(0)\right]_{s\to s-1}$$

$$= \frac{1}{n!}\left[s^n\,\mathscr{L}\{t^n e^{-t}\}\right]_{s\to s-1}$$

$$= \frac{1}{n!}\left[s^n\,\frac{n!}{(s+1)^{n+1}}\right]_{s\to s-1} = \frac{(s-1)^n}{s^{n+1}} = Y,$$

where $Y = \mathscr{L}\{L_n(t)\}$. Thus

$$L_n(t) = \frac{e^t}{n!}\frac{d^n}{dt^n}(t^n e^{-t}), \quad n = 0, 1, 2, \ldots.$$

62. The output for the first three lines of the program are

$$9y[t] + 6y'[t] + y''[t] == t\,\sin[t]$$

$$1 - 2s + 9Y + s^2 Y + 6(-2 + sY) == \frac{2s}{(1+s^2)^2}$$

$$Y \to -\left(\frac{-11 - 4s - 22s^2 - 4s^3 - 11s^4 - 2s^5}{(1+s^2)^2(9 + 6s + s^2)}\right)$$

The fourth line is the same as the third line with $Y \to$ removed. The final line of output shows a solution involving complex coefficients of e^{it} and e^{-it}. To get the solution in more standard form write the last line as two lines:

euler={E^(It)−>Cos[t] + I Sin[t], E^(-It)−>Cos[t] - I Sin[t]}
InverseLaplaceTransform[Y, s, t]/.euler//Expand

We see that the solution is

$$y(t) = \left(\frac{487}{250} + \frac{247}{50}t\right)e^{-3t} + \frac{1}{250}(13\cos t - 15t\cos t - 9\sin t + 20t\sin t).$$

63. The solution is

$$y(t) = \frac{1}{6}e^t - \frac{1}{6}e^{-t/2}\cos\sqrt{15}\,t - \frac{\sqrt{3/5}}{6}e^{-t/2}\sin\sqrt{15}\,t.$$

64. The solution is

$$q(t) = 1 - \cos t + (6 - 6\cos t)\,\mathscr{U}(t - 3\pi) - (4 + 4\cos t)\,\mathscr{U}(t - \pi).$$

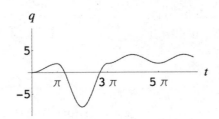

1. The Laplace transform of the differential equation yields

$$\mathcal{L}\{y\} = \frac{1}{s-3}e^{-2s}$$

so that

$$y = e^{3(t-2)}\,\mathcal{U}(t-2).$$

2. The Laplace transform of the differential equation yields

$$\mathcal{L}\{y\} = \frac{2}{s+1} + \frac{e^{-s}}{s+1}$$

so that

$$y = 2e^{-t} + e^{-(t-1)}\mathcal{U}(t-1).$$

3. The Laplace transform of the differential equation yields

$$\mathcal{L}\{y\} = \frac{1}{s^2+1}\left(1 + e^{-2\pi s}\right)$$

so that

$$y = \sin t + \sin t\,\mathcal{U}(t-2\pi).$$

4. The Laplace transform of the differential equation yields

$$\mathcal{L}\{y\} = \frac{1}{4}\frac{4}{s^2+16}e^{-2\pi s}$$

so that

$$y = \frac{1}{4}\sin 4(t-2\pi)\,\mathcal{U}(t-2\pi) = \frac{1}{4}\sin 4t\,\mathcal{U}(t-2\pi).$$

5. The Laplace transform of the differential equation yields

$$\mathcal{L}\{y\} = \frac{1}{s^2+1}\left(e^{-\pi s/2} + e^{-3\pi s/2}\right)$$

so that

$$y = \sin\left(t - \frac{\pi}{2}\right)\mathcal{U}\left(t - \frac{\pi}{2}\right) + \sin\left(t - \frac{3\pi}{2}\right)\mathcal{U}\left(t - \frac{3\pi}{2}\right)$$

$$= -\cos t\,\mathcal{U}\left(t - \frac{\pi}{2}\right) + \cos t\,\mathcal{U}\left(t - \frac{3\pi}{2}\right).$$

6. The Laplace transform of the differential equation yields

$$\mathcal{L}\{y\} = \frac{s}{s^2+1} + \frac{1}{s^2+1}(e^{-2\pi s} + e^{-4\pi s})$$

so that

$$y = \cos t + \sin t [\mathcal{U}(t - 2\pi) + \mathcal{U}(t - 4\pi)].$$

7. The Laplace transform of the differential equation yields

$$\mathcal{L}\{y\} = \frac{1}{s^2 + 2s}(1 + e^{-s}) = \left[\frac{1}{2}\frac{1}{s} - \frac{1}{2}\frac{1}{s+2}\right](1 + e^{-s})$$

so that

$$y = \frac{1}{2} - \frac{1}{2}e^{-2t} + \left[\frac{1}{2} - \frac{1}{2}e^{-2(t-1)}\right]\mathcal{U}(t - 1).$$

8. The Laplace transform of the differential equation yields

$$\mathcal{L}\{y\} = \frac{s + 1}{s^2(s - 2)} + \frac{1}{s(s - 2)}e^{-2s} = \frac{3}{4}\frac{1}{s - 2} - \frac{3}{4}\frac{1}{s} - \frac{1}{2}\frac{1}{s^2} + \left[\frac{1}{2}\frac{1}{s - 2} - \frac{1}{2}\frac{1}{s}\right]e^{-2s}$$

so that

$$y = \frac{3}{4}e^{2t} - \frac{3}{4} - \frac{1}{2}t + \left[\frac{1}{2}e^{2(t-2)} - \frac{1}{2}\right]\mathcal{U}(t - 2).$$

9. The Laplace transform of the differential equation yields

$$\mathcal{L}\{y\} = \frac{1}{(s + 2)^2 + 1}e^{-2\pi s}$$

so that

$$y = e^{-2(t-2\pi)}\sin t\, \mathcal{U}(t - 2\pi).$$

10. The Laplace transform of the differential equation yields

$$\mathcal{L}\{y\} = \frac{1}{(s + 1)^2}e^{-s}$$

so that

$$y = (t - 1)e^{-(t-1)}\mathcal{U}(t - 1).$$

11. The Laplace transform of the differential equation yields

$$\mathcal{L}\{y\} = \frac{4 + s}{s^2 + 4s + 13} + \frac{e^{-\pi s} + e^{-3\pi s}}{s^2 + 4s + 13}$$

$$= \frac{2}{3}\frac{3}{(s + 2)^2 + 3^2} + \frac{s + 2}{(s + 2)^2 + 3^2} + \frac{1}{3}\frac{3}{(s + 2)^2 + 3^2}\left(e^{-\pi s} + e^{-3\pi s}\right)$$

so that

$$y = \frac{2}{3}e^{-2t}\sin 3t + e^{-2t}\cos 3t + \frac{1}{3}e^{-2(t-\pi)}\sin 3(t - \pi)\mathcal{U}(t - \pi)$$

$$+ \frac{1}{3}e^{-2(t-3\pi)}\sin 3(t - 3\pi)\,\mathcal{U}(t - 3\pi).$$

12. The Laplace transform of the differential equation yields

$$\mathcal{L}\{y\} = \frac{1}{(s-1)^2(s-6)} + \frac{e^{-2s}+e^{-4s}}{(s-1)(s-6)}$$

$$= -\frac{1}{25}\frac{1}{s-1} - \frac{1}{5}\frac{1}{(s-1)^2} + \frac{1}{25}\frac{1}{s-6} + \left[-\frac{1}{5}\frac{1}{s-1} + \frac{1}{5}\frac{1}{s-6}\right]\left(e^{-2s}+e^{-4s}\right)$$

so that

$$y = -\frac{1}{25}e^t - \frac{1}{5}te^t + \frac{1}{25}e^{6t} + \left[-\frac{1}{5}e^{t-2} + \frac{1}{5}e^{6(t-2)}\right]\mathcal{U}(t-2)$$

$$+ \left[-\frac{1}{5}e^{t-4} + \frac{1}{5}e^{6(t-4)}\right]\mathcal{U}(t-4).$$

13. The Laplace transform of the differential equation yields

$$\mathcal{L}\{y\} = \frac{1}{2}\frac{2}{s^3}y''(0) + \frac{1}{6}\frac{3!}{s^4}y'''(0) + \frac{1}{6}\frac{P_0}{EI}\frac{3!}{s^4}e^{-Ls/2}$$

so that

$$y = \frac{1}{2}y''(0)x^2 + \frac{1}{6}y'''(0)x^3 + \frac{1}{6}\frac{P_0}{EI}\left(x-\frac{L}{2}\right)^3\mathcal{U}\left(x-\frac{L}{2}\right).$$

Using $y''(L)=0$ and $y'''(L)=0$ we obtain

$$y = \frac{1}{4}\frac{P_0L}{EI}x^2 - \frac{1}{6}\frac{P_0}{EI}x^3 + \frac{1}{6}\frac{P_0}{EI}\left(x-\frac{L}{2}\right)^3\mathcal{U}\left(x-\frac{L}{2}\right)$$

$$= \begin{cases} \frac{P_0}{EI}\left(\frac{L}{4}x^2 - \frac{1}{6}x^3\right), & 0 \le x < \frac{L}{2} \\ \frac{P_0L^2}{4EI}\left(\frac{1}{2}x - \frac{L}{12}\right), & \frac{L}{2} \le x \le L. \end{cases}$$

P_0 slb w$_0$

14. From Problem 13 we know that

$$y = \frac{1}{2}y''(0)x^2 + \frac{1}{6}y'''(0)x^3 + \frac{1}{6}\frac{P_0}{EI}\left(x-\frac{L}{2}\right)^3\mathcal{U}\left(x-\frac{L}{2}\right).$$

Using $y(L)=0$ and $y'(L)=0$ we obtain

$$y = \frac{1}{16}\frac{P_0L}{EI}x^2 - \frac{1}{12}\frac{P_0}{EI}x^3 + \frac{1}{6}\frac{P_0}{EI}\left(x-\frac{L}{2}\right)^3\mathcal{U}\left(x-\frac{L}{2}\right)$$

$$= \begin{cases} \frac{P_0}{EI}\left(\frac{L}{16}x^2 - \frac{1}{12}x^3\right), & 0 \le x < \frac{L}{2} \\ \frac{P_0}{EI}\left(\frac{L}{16}x^2 - \frac{1}{12}x^3\right) + \frac{1}{6}\frac{P_0}{EI}\left(x-\frac{L}{2}\right)^3, & \frac{L}{2} \le x \le L. \end{cases}$$

15. You should disagree. Although formal manipulations of the Laplace transform lead to $y(t) = \frac{1}{3}e^{-t}\sin 3t$ in both cases, this function does not satisfy the initial condition $y'(0)=0$ of the second initial-value problem.

Exercises 7.6

Systems of Linear Differential Equations

1. Taking the Laplace transform of the system gives

$$s\mathscr{L}\{x\} = -\mathscr{L}\{x\} + \mathscr{L}\{y\}$$

$$s\mathscr{L}\{y\} - 1 = 2\mathscr{L}\{x\}$$

so that

$$\mathscr{L}\{x\} = \frac{1}{(s-1)(s+2)} = \frac{1}{3}\frac{1}{s-1} - \frac{1}{3}\frac{1}{s+2}$$

and

$$\mathscr{L}\{y\} = \frac{1}{s} + \frac{2}{s(s-1)(s+2)} = \frac{2}{3}\frac{1}{s-1} + \frac{1}{3}\frac{1}{s+2}.$$

Then

$$x = \frac{1}{3}e^t - \frac{1}{3}e^{-2t} \qquad \text{and} \qquad y = \frac{2}{3}e^t + \frac{1}{3}e^{-2t}.$$

2. Taking the Laplace transform of the system gives

$$s\mathscr{L}\{x\} - 1 = 2\mathscr{L}\{y\} + \frac{1}{s-1}$$

$$s\mathscr{L}\{y\} - 1 = 8\mathscr{L}\{x\} - \frac{1}{s^2}$$

so that

$$\mathscr{L}\{y\} = \frac{s^3 + 7s^2 - s + 1}{s(s-1)(s^2-16)} = \frac{1}{16}\frac{1}{s} - \frac{8}{15}\frac{1}{s-1} + \frac{173}{96}\frac{1}{s-4} - \frac{53}{160}\frac{1}{s+4}$$

and

$$y = \frac{1}{16} - \frac{8}{15}e^t + \frac{173}{96}e^{4t} - \frac{53}{160}e^{-4t}.$$

Then

$$x = \frac{1}{8}y' + \frac{1}{8}t = \frac{1}{8}t - \frac{1}{15}e^t + \frac{173}{192}e^{4t} + \frac{53}{320}e^{-4t}.$$

3. Taking the Laplace transform of the system gives

$$s\mathscr{L}\{x\} + 1 = \mathscr{L}\{x\} - 2\mathscr{L}\{y\}$$

$$s\mathscr{L}\{y\} - 2 = 5\mathscr{L}\{x\} - \mathscr{L}\{y\}$$

so that

$$\mathscr{L}\{x\} = \frac{-s-5}{s^2+9} = -\frac{s}{s^2+9} - \frac{5}{3}\frac{3}{s^2+9}$$

and

$$x = -\cos 3t - \frac{5}{3}\sin 3t.$$

Then

$$y = \frac{1}{2}x - \frac{1}{2}x' = 2\cos 3t - \frac{7}{3}\sin 3t.$$

4. Taking the Laplace transform of the system gives

$$(s+3)\,\mathscr{L}\{x\} + s\,\mathscr{L}\{y\} = \frac{1}{s}$$

$$(s-1)\,\mathscr{L}\{x\} + (s-1)\,\mathscr{L}\{y\} = \frac{1}{s-1}$$

so that

$$\mathscr{L}\{y\} = \frac{5s-1}{3s(s-1)^2} = -\frac{1}{3}\frac{1}{s} + \frac{1}{3}\frac{1}{s-1} + \frac{4}{3}\frac{1}{(s-1)^2}$$

and

$$\mathscr{L}\{x\} = \frac{1-2s}{3s(s-1)^2} = \frac{1}{3}\frac{1}{s} - \frac{1}{3}\frac{1}{s-1} - \frac{1}{3}\frac{1}{(s-1)^2}.$$

Then

$$x = \frac{1}{3} - \frac{1}{3}e^t - \frac{1}{3}te^t \qquad \text{and} \qquad y = -\frac{1}{3} + \frac{1}{3}e^t + \frac{4}{3}te^t.$$

5. Taking the Laplace transform of the system gives

$$(2s-2)\,\mathscr{L}\{x\} + s\,\mathscr{L}\{y\} = \frac{1}{s}$$

$$(s-3)\,\mathscr{L}\{x\} + (s-3)\,\mathscr{L}\{y\} = \frac{2}{s}$$

so that

$$\mathscr{L}\{x\} = \frac{-s-3}{s(s-2)(s-3)} = -\frac{1}{2}\frac{1}{s} + \frac{5}{2}\frac{1}{s-2} - \frac{2}{s-3}$$

and

$$\mathscr{L}\{y\} = \frac{3s-1}{s(s-2)(s-3)} = -\frac{1}{6}\frac{1}{s} - \frac{5}{2}\frac{1}{s-2} + \frac{8}{3}\frac{1}{s-3}.$$

Then

$$x = -\frac{1}{2} + \frac{5}{2}e^{2t} - 2e^{3t} \qquad \text{and} \qquad y = -\frac{1}{6} - \frac{5}{2}e^{2t} + \frac{8}{3}e^{3t}.$$

6. Taking the Laplace transform of the system gives

$$(s+1)\,\mathscr{L}\{x\} - (s-1)\mathscr{L}\{y\} = -1$$

$$s\,\mathscr{L}\{x\} + (s+2)\,\mathscr{L}\{y\} = 1$$

so that

$$\mathscr{L}\{y\} = \frac{s+1/2}{s^2+s+1} = \frac{s+1/2}{(s+1/2)^2 + (\sqrt{3}/2)^2}$$

and

$$\mathcal{L}\{x\} = \frac{-3/2}{s^2+s+1} = -\sqrt{3}\frac{\sqrt{3}/2}{(s+1/2)^2+(\sqrt{3}/2)^2}.$$

Then

$$y = e^{-t/2}\cos\frac{\sqrt{3}}{2}t \quad\text{and}\quad x = -\sqrt{3}\,e^{-t/2}\sin\frac{\sqrt{3}}{2}t.$$

7. Taking the Laplace transform of the system gives

$$(s^2+1)\,\mathcal{L}\{x\} - \mathcal{L}\{y\} = -2$$

$$-\mathcal{L}\{x\} + (s^2+1)\,\mathcal{L}\{y\} = 1$$

so that

$$\mathcal{L}\{x\} = \frac{-2s^2-1}{s^4+2s^2} = -\frac{1}{2}\frac{1}{s^2} - \frac{3}{2}\frac{1}{s^2+2}$$

and

$$x = -\frac{1}{2}t - \frac{3}{2\sqrt{2}}\sin\sqrt{2}\,t.$$

Then

$$y = x'' + x = -\frac{1}{2}t + \frac{3}{2\sqrt{2}}\sin\sqrt{2}\,t.$$

8. Taking the Laplace transform of the system gives

$$(s+1)\,\mathcal{L}\{x\} + \mathcal{L}\{y\} = 1$$

$$4\mathcal{L}\{x\} - (s+1)\,\mathcal{L}\{y\} = 1$$

so that

$$\mathcal{L}\{x\} = \frac{s+2}{s^2+2s+5} = \frac{s+1}{(s+1)^2+2^2} + \frac{1}{2}\frac{2}{(s+1)^2+2^2}$$

and

$$\mathcal{L}\{y\} = \frac{-s+3}{s^2+2s+5} = -\frac{s+1}{(s+1)^2+2^2} + 2\frac{2}{(s+1)^2+2^2}.$$

Then

$$x = e^{-t}\cos 2t + \frac{1}{2}e^{-t}\sin 2t \quad\text{and}\quad y = -e^{-t}\cos 2t + 2e^{-t}\sin 2t.$$

9. Adding the equations and then subtracting them gives

$$\frac{d^2x}{dt^2} = \frac{1}{2}t^2 + 2t$$

$$\frac{d^2y}{dt^2} = \frac{1}{2}t^2 - 2t.$$

Taking the Laplace transform of the system gives

$$\mathcal{L}\{x\} = 8\frac{1}{s} + \frac{1}{24}\frac{4!}{s^5} + \frac{1}{3}\frac{3!}{s^4}$$

and

$$\mathscr{L}\{y\} = \frac{1}{24}\frac{4!}{s^5} - \frac{1}{3}\frac{3!}{s^4}$$

so that

$$x = 8 + \frac{1}{24}t^4 + \frac{1}{3}t^3 \quad \text{and} \quad y = \frac{1}{24}t^4 - \frac{1}{3}t^3.$$

10. Taking the Laplace transform of the system gives

$$(s - 4)\,\mathscr{L}\{x\} + s^3\,\mathscr{L}\{y\} = \frac{6}{s^2 + 1}$$

$$(s + 2)\,\mathscr{L}\{x\} - 2s^3\,\mathscr{L}\{y\} = 0$$

so that

$$\mathscr{L}\{x\} = \frac{4}{(s-2)(s^2+1)} = \frac{4}{5}\frac{1}{s-2} - \frac{4}{5}\frac{s}{s^2+1} - \frac{8}{5}\frac{1}{s^2+1}$$

and

$$\mathscr{L}\{y\} = \frac{2s+4}{s^3(s-2)(s^2+1)} = \frac{1}{s} - \frac{2}{s^2} - 2\frac{2}{s^3} + \frac{1}{5}\frac{1}{s-2} - \frac{6}{5}\frac{s}{s^2+1} + \frac{8}{5}\frac{1}{s^2+1}.$$

Then

$$x = \frac{4}{5}e^{2t} - \frac{4}{5}\cos t - \frac{8}{5}\sin t$$

and

$$y = 1 - 2t - 2t^2 + \frac{1}{5}e^{2t} - \frac{6}{5}\cos t + \frac{8}{5}\sin t.$$

11. Taking the Laplace transform of the system gives

$$s^2\mathscr{L}\{x\} + 3(s+1)\,\mathscr{L}\{y\} = 2$$

$$s^2\,\mathscr{L}\{x\} + 3\mathscr{L}\{y\} = \frac{1}{(s+1)^2}$$

so that

$$\mathscr{L}\{x\} = -\frac{2s+1}{s^3(s+1)} = \frac{1}{s} + \frac{1}{s^2} + \frac{1}{2}\frac{2}{s^3} - \frac{1}{s+1}.$$

Then

$$x = 1 + t + \frac{1}{2}t^2 - e^{-t}$$

and

$$y = \frac{1}{3}te^{-t} - \frac{1}{3}x'' = \frac{1}{3}te^{-t} + \frac{1}{3}e^{-t} - \frac{1}{3}.$$

12. Taking the Laplace transform of the system gives

$$(s-4)\,\mathscr{L}\{x\} + 2\mathscr{L}\{y\} = \frac{2e^{-s}}{s}$$

$$-3\,\mathscr{L}\{x\} + (s+1)\,\mathscr{L}\{y\} = \frac{1}{2} + \frac{e^{-s}}{s}$$

so that

$$\mathscr{L}\{x\} = \frac{-1/2}{(s-1)(s-2)} + e^{-s}\frac{1}{(s-1)(s-2)}$$

$$= \frac{1}{2}\frac{1}{s-1} - \frac{1}{2}\frac{1}{s-2} + e^{-s}\left[-\frac{1}{s-1} + \frac{1}{s-2}\right]$$

and

$$\mathscr{L}\{y\} = \frac{e^{-s}}{s} + \frac{s/4-1}{(s-1)(s-2)} + e^{-s}\frac{-s/2+2}{(s-1)(s-2)}$$

$$= \frac{3}{4}\frac{1}{s-1} - \frac{1}{2}\frac{1}{s-2} + e^{-s}\left[\frac{1}{s} - \frac{3}{2}\frac{1}{s-1} + \frac{1}{s-2}\right].$$

Then

$$x = \frac{1}{2}e^t - \frac{1}{2}e^{2t} + \left[-e^{t-1} + e^{2(t-1)}\right]\mathcal{U}(t-1)$$

and

$$y = \frac{3}{4}e^t - \frac{1}{2}e^{2t} + \left[1 - \frac{3}{2}e^{t-1} + e^{2(t-1)}\right]\mathcal{U}(t-1).$$

13. The system is

$$x_1'' = -3x_1 + 2(x_2 - x_1)$$

$$x_2'' = -2(x_2 - x_1)$$

$$x_1(0) = 0$$

$$x_1'(0) = 1$$

$$x_2(0) = 1$$

$$x_2'(0) = 0.$$

Taking the Laplace transform of the system gives

$$(s^2 + 5)\mathscr{L}\{x_1\} - 2\mathscr{L}\{x_2\} = 1$$

$$-2\mathscr{L}\{x_1\} + (s^2 + 2)\mathscr{L}\{x_2\} = s$$

so that

$$\mathscr{L}\{x_1\} = \frac{s^2 + 2s + 2}{s^4 + 7s^2 + 6} = \frac{2}{5}\frac{s}{s^2+1} + \frac{1}{5}\frac{1}{s^2+1} - \frac{2}{5}\frac{s}{s^2+6} + \frac{4}{5\sqrt{6}}\frac{\sqrt{6}}{s^2+6}$$

and

$$\mathscr{L}\{x_2\} = \frac{s^3 + 5s + 2}{(s^2+1)(s^2+6)} = \frac{4}{5}\frac{s}{s^2+1} + \frac{2}{5}\frac{1}{s^2+1} + \frac{1}{5}\frac{s}{s^2+6} - \frac{2}{5\sqrt{6}}\frac{\sqrt{6}}{s^2+6}.$$

Then

$$x_1 = \frac{2}{5}\cos t + \frac{1}{5}\sin t - \frac{2}{5}\cos\sqrt{6}\,t + \frac{4}{5\sqrt{6}}\sin\sqrt{6}\,t$$

and

$$x_2 = \frac{4}{5}\cos t + \frac{2}{5}\sin t + \frac{1}{5}\cos\sqrt{6}\,t - \frac{2}{5\sqrt{6}}\sin\sqrt{6}\,t.$$

14. In this system x_1 and x_2 represent displacements of masses m_1 and m_2 from their equilibrium positions. Since the net forces acting on m_1 and m_2 are

$$-k_1 x_1 + k_2(x_2 - x_1) \qquad \text{and} \qquad -k_2(x_2 - x_1) - k_3 x_2,$$

respectively, Newton's second law of motion gives

$$m_1 x_1'' = -k_1 x_1 + k_2(x_2 - x_1)$$

$$m_2 x_2'' = -k_2(x_2 - x_1) - k_3 x_2.$$

Using $k_1 = k_2 = k_3 = 1$, $m_1 = m_2 = 1$, $x_1(0) = 0$, $x_1(0) = -1$, $x_2(0) = 0$, and $x_2'(0) = 1$, and taking the Laplace transform of the system, we obtain

$$(2 + s^2)\mathscr{L}\{x_1\} - \mathscr{L}\{x_2\} = -1$$

$$\mathscr{L}\{x_1\} - (2 + s^2)\mathscr{L}\{x_2\} = -1$$

so that

$$\mathscr{L}\{x_1\} = -\frac{1}{s^2 + 3} \qquad \text{and} \qquad \mathscr{L}\{x_2\} = \frac{1}{s^2 + 3}.$$

Then

$$x_1 = -\frac{1}{\sqrt{3}}\sin\sqrt{3}\,t \qquad \text{and} \qquad x_2 = \frac{1}{\sqrt{3}}\sin\sqrt{3}\,t.$$

15. **(a)** By Kirchhoff's first law we have $i_1 = i_2 + i_3$. By Kirchhoff's second law, on each loop we have $E(t) = Ri_1 + L_1 i_2'$ and $E(t) = Ri_1 + L_2 i_3'$ or $L_1 i_2' + Ri_2 + Ri_3 = E(t)$ and $L_2 i_3' + Ri_2 + Ri_3 = E(t)$.

(b) Taking the Laplace transform of the system

$$0.01 i_2' + 5i_2 + 5i_3 = 100$$

$$0.0125 i_3' + 5i_2 + 5i_3 = 100$$

gives

$$(s + 500)\mathscr{L}\{i_2\} + 500\mathscr{L}\{i_3\} = \frac{10,000}{s}$$

$$400\mathscr{L}\{i_2\} + (s + 400)\mathscr{L}\{i_3\} = \frac{8,000}{s}$$

so that

$$\mathscr{L}\{i_3\} = \frac{8,000}{s^2 + 900s} = \frac{80}{9}\frac{1}{s} - \frac{80}{9}\frac{1}{s + 900}.$$

Then

$$i_3 = \frac{80}{9} - \frac{80}{9}e^{-900t} \qquad \text{and} \qquad i_2 = 20 - 0.0025 i_3' - i_3 = \frac{100}{9} - \frac{100}{9}e^{-900t}.$$

(c) $i_1 = i_2 + i_3 = 20 - 20e^{-900t}$

16. (a) Taking the Laplace transform of the system

$$i_2' + i_3' + 10i_2 = 120 - 120\,\mathcal{U}(t-2)$$

$$-10i_2' + 5i_3' + 5i_3 = 0$$

gives

$$(s+10)\,\mathcal{L}\{i_2\} + s\mathcal{L}\{i_3\} = \frac{120}{s}\left(1 - e^{-2s}\right)$$

$$-10s\mathcal{L}\{i_2\} + 5(s+1)\,\mathcal{L}\{i_3\} = 0$$

so that

$$\mathcal{L}\{i_2\} = \frac{120(s+1)}{(3s^2 + 11s + 10)s}\left(1 - e^{-2s}\right) = \left[\frac{48}{s+5/3} - \frac{60}{s+2} + \frac{12}{s}\right]\left(1 - e^{-2s}\right)$$

and

$$\mathcal{L}\{i_3\} = \frac{240}{3s^2 + 11s + 10}\left(1 - e^{-2s}\right) = \left[\frac{240}{s+5/3} - \frac{240}{s+2}\right]\left(1 - e^{-2s}\right).$$

Then

$$i_2 = 12 + 48e^{-5t/3} - 60e^{-2t} - \left[12 + 48e^{-5(t-2)/3} - 60e^{-2(t-2)}\right]\mathcal{U}(t-2)$$

and

$$i_3 = 240e^{-5t/3} - 240e^{-2t} - \left[240e^{-5(t-2)/3} - 240e^{-2(t-2)}\right]\mathcal{U}(t-2).$$

(b) $i_1 = i_2 + i_3 = 12 + 288e^{-5t/3} - 300e^{-2t} - \left[12 + 288e^{-5(t-2)/3} - 300e^{-2(t-2)}\right]\mathcal{U}(t-2)$

17. Taking the Laplace transform of the system

$$i_2' + 11i_2 + 6i_3 = 50\sin t$$

$$i_3' + 6i_2 + 6i_3 = 50\sin t$$

gives

$$(s+11)\,\mathcal{L}\{i_2\} + 6\mathcal{L}\{i_3\} = \frac{50}{s^2 + 1}$$

$$6\mathcal{L}\{i_2\} + (s+6)\,\mathcal{L}\{i_3\} = \frac{50}{s^2 + 1}$$

so that

$$\mathcal{L}\{i_2\} = \frac{50s}{(s+2)(s+15)(s^2+1)} = -\frac{20}{13}\frac{1}{s+2} + \frac{375}{1469}\frac{1}{s+15} + \frac{145}{113}\frac{s}{s^2+1} + \frac{85}{113}\frac{1}{s^2+1}.$$

Then

$$i_2 = -\frac{20}{13}e^{-2t} + \frac{375}{1469}e^{-15t} + \frac{145}{113}\cos t + \frac{85}{113}\sin t$$

and

$$i_3 = \frac{25}{3} \sin t - \frac{1}{6} i_2' - \frac{11}{6} i_2 = \frac{30}{13} e^{-2t} + \frac{250}{1469} e^{-15t} - \frac{280}{113} \cos t + \frac{810}{113} \sin t.$$

18. Taking the Laplace transform of the system

$$0.5i_1' + 50i_2 = 60$$

$$0.005i_2' + i_2 - i_1 = 0$$

gives

$$s \mathcal{L}\{i_1\} + 100 \mathcal{L}\{i_2\} = \frac{120}{s}$$

$$-200 \mathcal{L}\{i_1\} + (s + 200) \mathcal{L}\{i_2\} = 0$$

so that

$$\mathcal{L}\{i_2\} = \frac{24{,}000}{s(s^2 + 200s + 20{,}000)} = \frac{6}{5}\frac{1}{s} - \frac{6}{5}\frac{s+100}{(s+100)^2 + 100^2} - \frac{6}{5}\frac{100}{(s+100)^2 + 100^2}.$$

Then

$$i_2 = \frac{6}{5} - \frac{6}{5} e^{-100t} \cos 100t - \frac{6}{5} e^{-100t} \sin 100t$$

and

$$i_1 = 0.005i_2' + i_2 = \frac{6}{5} - \frac{6}{5} e^{-100t} \cos 100t.$$

19. Taking the Laplace transform of the system

$$2i_1' + 50i_2 = 60$$

$$0.005i_2' + i_2 - i_1 = 0$$

gives

$$2s \mathcal{L}\{i_1\} + 50 \mathcal{L}\{i_2\} = \frac{60}{s}$$

$$-200 \mathcal{L}\{i_1\} + (s + 200) \mathcal{L}\{i_2\} = 0$$

so that

$$\mathcal{L}\{i_2\} = \frac{6{,}000}{s(s^2 + 200s + 5{,}000)}$$

$$= \frac{6}{5}\frac{1}{s} - \frac{6}{5}\frac{s+100}{(s+100)^2 - (50\sqrt{2})^2} - \frac{6\sqrt{2}}{5}\frac{50\sqrt{2}}{(s+100)^2 - (50\sqrt{2})^2}.$$

Then

$$i_2 = \frac{6}{5} - \frac{6}{5} e^{-100t} \cosh 50\sqrt{2}\,t - \frac{6\sqrt{2}}{5} e^{-100t} \sinh 50\sqrt{2}\,t$$

and

$$i_1 = 0.005i_2' + i_2 = \frac{6}{5} - \frac{6}{5} e^{-100t} \cosh 50\sqrt{2}\,t - \frac{9\sqrt{2}}{10} e^{-100t} \sinh 50\sqrt{2}\,t.$$

Exercises 7.6   Systems of Linear Differential Equations

20. (a) Using Kirchhoff's first law we write $i_1 = i_2 + i_3$. Since $i_2 = dq/dt$ we have $i_1 - i_3 = dq/dt$. Using Kirchhoff's second law and summing the voltage drops across the shorter loop gives

$$E(t) = iR_1 + \frac{1}{C}q, \tag{1}$$

so that

$$i_1 = \frac{1}{R_1}E(t) - \frac{1}{R_1 C}q.$$

Then

$$\frac{dq}{dt} = i_1 - i_3 = \frac{1}{R_1}E(t) - \frac{1}{R_1 C}q - i_3$$

and

$$R_1\frac{dq}{dt} + \frac{1}{C}q + R_1 i_3 = E(t).$$

Summing the voltage drops across the longer loop gives

$$E(t) = i_1 R_1 + L\frac{di_3}{dt} + R_2 i_3.$$

Combining this with (1) we obtain

$$i_1 R_1 + L\frac{di_3}{dt} + R_2 i_3 = i_1 R_1 + \frac{1}{C}q$$

or

$$L\frac{di_3}{dt} + R_2 i_3 - \frac{1}{C}q = 0.$$

(b) Using $L = R_1 = R_2 = C = 1$, $E(t) = 50e^{-t}\,\mathcal{U}(t-1) = 50e^{-1}e^{-(t-1)}\,\mathcal{U}(t-1)$, $q(0) = i_3(0) = 0$, and taking the Laplace transform of the system we obtain

$$(s+1)\mathcal{L}\{q\} + \mathcal{L}\{i_3\} = \frac{50e^{-1}}{s+1}e^{-s}$$

$$(s+1)\mathcal{L}\{i_3\} - \mathcal{L}\{q\} = 0,$$

so that

$$\mathcal{L}\{q\} = \frac{50e^{-1}e^{-s}}{(s+1)^2 + 1}$$

and

$$q(t) = 50e^{-1}e^{-(t-1)}\sin(t-1)\,\mathcal{U}(t-1) = 50e^{-t}\sin(t-1)\,\mathcal{U}(t-1).$$

21. (a) Taking the Laplace transform of the system

$$4\theta_1'' + \theta_2'' + 8\theta_1 = 0$$

$$\theta_1'' + \theta_2'' + 2\theta_2 = 0$$

408

gives

$$4\left(s^2+2\right)\mathcal{L}\{\theta_1\}+s^2\mathcal{L}\{\theta_2\}=3s$$

$$s^2\mathcal{L}\{\theta_1\}+\left(s^2+2\right)\mathcal{L}\{\theta_2\}=0$$

so that

$$\left(3s^2+4\right)\left(s^2+4\right)\mathcal{L}\{\theta_2\}=-3s^3$$

or

$$\mathcal{L}\{\theta_2\}=\frac{1}{2}\frac{s}{s^2+4/3}-\frac{3}{2}\frac{s}{s^2+4}.$$

Then

$$\theta_2=\frac{1}{2}\cos\frac{2}{\sqrt{3}}t-\frac{3}{2}\cos 2t\qquad\text{and}\qquad \theta_1''=-\theta_2''-2\theta_2$$

so that

$$\theta_1=\frac{1}{4}\cos\frac{2}{\sqrt{3}}t+\frac{3}{4}\cos 2t.$$

(b)

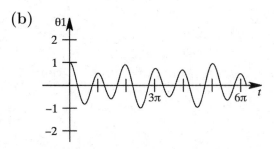 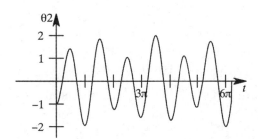

Mass m_2 has extreme displacements of greater magnitude. Mass m_1 first passes through its equilibrium position at about $t=0.87$, and mass m_2 first passes through its equilibrium position at about $t=0.66$. The motion of the pendulums is not periodic since $\cos(2t/\sqrt{3})$ has period $\sqrt{3}\,\pi$, $\cos 2t$ has period π, and the ratio of these periods is $\sqrt{3}$, which is not a rational number.

(c) The Lissajous curve is plotted for $0\le t\le 30$.

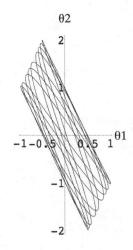

(d)

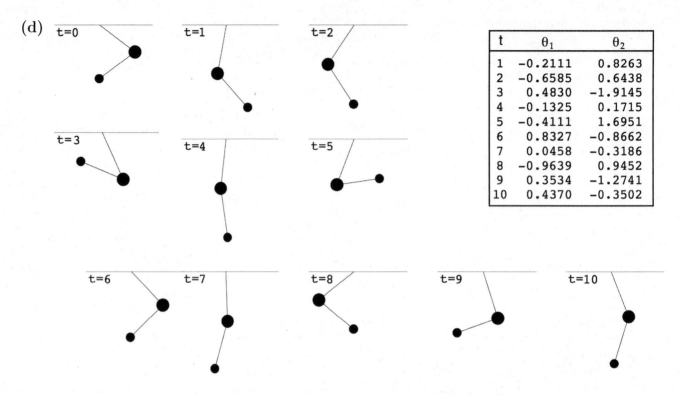

t	θ_1	θ_2
1	-0.2111	0.8263
2	-0.6585	0.6438
3	0.4830	-1.9145
4	-0.1325	0.1715
5	-0.4111	1.6951
6	0.8327	-0.8662
7	0.0458	-0.3186
8	-0.9639	0.9452
9	0.3534	-1.2741
10	0.4370	-0.3502

(e) Using a CAS to solve $\theta_1(t) = \theta_2(t)$ we see that $\theta_1 = \theta_2$ (so that the double pendulum is straight out) when t is about 0.75 seconds.

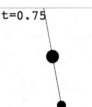

(f) To make a movie of the pendulum it is necessary to locate the mass in the plane as a function of time. Suppose that the upper arm is attached to the origin and that the equilibrium position lies along the negative y-axis. Then mass m_1 is at $(x, (t), y_1(t))$ and mass m_2 is at $(x_2(t), y_2(t))$, where

$$x_1(t) = 16 \sin \theta_1(t) \quad \text{and} \quad y_1(t) = -16 \cos \theta_1(t)$$

and

$$x_2(t) = x_1(t) + 16 \sin \theta_2(t) \quad \text{and} \quad y_2(t) = y_1(t) - 16 \cos \theta_2(t).$$

A reasonable movie can be constructed by letting t range from 0 to 10 in increments of 0.1 seconds.

Chapter 7 in Review

1. $\mathscr{L}\{f(t)\} = \int_0^1 te^{-st}\,dt + \int_1^{\infty} (2-t)e^{-st}\,dt = \dfrac{1}{s^2} - \dfrac{2}{s^2}e^{-s}$

2. $\mathscr{L}\{f(t)\} = \int_2^4 e^{-st}\,dt = \dfrac{1}{s}\left(e^{-2s} - e^{-4s}\right)$

3. False; consider $f(t) = t^{-1/2}$.

4. False, since $f(t) = (e^t)^{10} = e^{10t}$.

5. True, since $\lim_{s\to\infty} F(s) = 1 \neq 0$. (See Theorem 7.1.3 in the text.)

6. False; consider $f(t) = 1$ and $g(t) = 1$.

7. $\mathscr{L}\left\{e^{-7t}\right\} = \dfrac{1}{s+7}$

8. $\mathscr{L}\left\{te^{-7t}\right\} = \dfrac{1}{(s+7)^2}$

9. $\mathscr{L}\{\sin 2t\} = \dfrac{2}{s^2+4}$

10. $\mathscr{L}\left\{e^{-3t}\sin 2t\right\} = \dfrac{2}{(s+3)^2+4}$

11. $\mathscr{L}\{t\sin 2t\} = -\dfrac{d}{ds}\left[\dfrac{2}{s^2+4}\right] = \dfrac{4s}{(s^2+4)^2}$

12. $\mathscr{L}\{\sin 2t\,\mathscr{U}(t-\pi)\} = \mathscr{L}\{\sin 2(t-\pi)\,\mathscr{U}(t-\pi)\} = \dfrac{2}{s^2+4}e^{-\pi s}$

13. $\mathscr{L}^{-1}\left\{\dfrac{20}{s^6}\right\} = \mathscr{L}^{-1}\left\{\dfrac{1}{6}\dfrac{5!}{s^6}\right\} = \dfrac{1}{6}t^5$

14. $\mathscr{L}^{-1}\left\{\dfrac{1}{3s-1}\right\} = \mathscr{L}^{-1}\left\{\dfrac{1}{3}\dfrac{1}{s-1/3}\right\} = \dfrac{1}{3}e^{t/3}$

15. $\mathscr{L}^{-1}\left\{\dfrac{1}{(s-5)^3}\right\} = \dfrac{1}{2}\mathscr{L}^{-1}\left\{\dfrac{2}{(s-5)^3}\right\} = \dfrac{1}{2}t^2e^{5t}$

16. $\mathscr{L}^{-1}\left\{\dfrac{1}{s^2-5}\right\} = \mathscr{L}^{-1}\left\{-\dfrac{1}{2\sqrt{5}}\dfrac{1}{s+\sqrt{5}} + \dfrac{1}{2\sqrt{5}}\dfrac{1}{s-\sqrt{5}}\right\} = -\dfrac{1}{2\sqrt{5}}e^{-\sqrt{5}t} + \dfrac{1}{2\sqrt{5}}e^{\sqrt{5}t}$

17. $\mathcal{L}^{-1}\left\{\dfrac{s}{s^2 - 10s + 29}\right\} = \mathcal{L}^{-1}\left\{\dfrac{s-5}{(s-5)^2 + 2^2} + \dfrac{5}{2}\dfrac{2}{(s-5)^2 + 2^2}\right\} = e^{5t}\cos 2t + \dfrac{5}{2}e^{5t}\sin 2t$

18. $\mathcal{L}^{-1}\left\{\dfrac{1}{s^2}e^{-5s}\right\} = (t-5)\mathcal{U}(t-5)$

19. $\mathcal{L}^{-1}\left\{\dfrac{s+\pi}{s^2+\pi^2}e^{-s}\right\} = \mathcal{L}^{-1}\left\{\dfrac{s}{s^2+\pi^2}e^{-s} + \dfrac{\pi}{s^2+\pi^2}e^{-s}\right\}$

$$= \cos\pi(t-1)\mathcal{U}(t-1) + \sin\pi(t-1)\mathcal{U}(t-1)$$

20. $\mathcal{L}^{-1}\left\{\dfrac{1}{L^2 s^2 + n^2\pi^2}\right\} = \dfrac{1}{L^2}\dfrac{L}{n\pi}\mathcal{L}^{-1}\left\{\dfrac{n\pi/L}{s^2 + (n^2\pi^2)/L^2}\right\} = \dfrac{1}{Ln\pi}\sin\dfrac{n\pi}{L}t$

21. $\mathcal{L}\left\{e^{-5t}\right\}$ exists for $s > -5$.

22. $\mathcal{L}\left\{te^{8t}f(t)\right\} = -\dfrac{d}{ds}F(s-8)$.

23. $\mathcal{L}\left\{e^{at}f(t-k)\,\mathcal{U}(t-k)\right\} = e^{-ks}\,\mathcal{L}\left\{e^{a(t+k)}f(t)\right\} = e^{-ks}e^{ak}\,\mathcal{L}\left\{e^{at}f(t)\right\} = e^{-k(s-a)}F(s-a)$

24. $\mathcal{L}\left\{\displaystyle\int_0^t e^{a\tau}f(\tau)\,d\tau\right\} = \dfrac{1}{s}\mathcal{L}\left\{e^{at}f(t)\right\} = \dfrac{F(s-a)}{s}$, whereas

$$\mathcal{L}\left\{e^{at}\int_0^t f(\tau)\,d\tau\right\} = \mathcal{L}\left\{\int_0^t f(\tau)\,d\tau\right\}\Bigg|_{s\to s-a} = \dfrac{F(s)}{s}\Bigg|_{s\to s-a} = \dfrac{F(s-a)}{s-a}.$$

25. $f(t)\mathcal{U}(t-t_0)$

26. $f(t) - f(t)\mathcal{U}(t-t_0)$

27. $f(t-t_0)\mathcal{U}(t-t_0)$

28. $f(t) - f(t)\mathcal{U}(t-t_0) + f(t)\mathcal{U}(t-t_1)$

29. $f(t) = t - [(t-1)+1]\mathcal{U}(t-1) + \mathcal{U}(t-1) - \mathcal{U}(t-4) = t - (t-1)\mathcal{U}(t-1) - \mathcal{U}(t-4)$

$$\mathcal{L}\{f(t)\} = \dfrac{1}{s^2} - \dfrac{1}{s^2}e^{-s} - \dfrac{1}{s}e^{-4s}$$

$$\mathcal{L}\{e^t f(t)\} = \dfrac{1}{(s-1)^2} - \dfrac{1}{(s-1)^2}e^{-(s-1)} - \dfrac{1}{s-1}e^{-4(s-1)}$$

30. $f(t) = \sin t\,\mathcal{U}(t-\pi) - \sin t\,\mathcal{U}(t-3\pi) = -\sin(t-\pi)\mathcal{U}(t-\pi) + \sin(t-3\pi)\mathcal{U}(t-3\pi)$

$$\mathcal{L}\{f(t)\} = -\dfrac{1}{s^2+1}e^{-\pi s} + \dfrac{1}{s^2+1}e^{-3\pi s}$$

$$\mathcal{L}\{e^t f(t)\} = -\dfrac{1}{(s-1)^2+1}e^{-\pi(s-1)} + \dfrac{1}{(s-1)^2+1}e^{-3\pi(s-1)}$$

31. $f(t) = 2 - 2\mathcal{U}(t-2) + [(t-2)+2]\mathcal{U}(t-2) = 2 + (t-2)\mathcal{U}(t-2)$

$$\mathscr{L}\{f(t)\} = \frac{2}{s} + \frac{1}{s^2}e^{-2s}$$

$$\mathscr{L}\{e^t f(t)\} = \frac{2}{s-1} + \frac{1}{(s-1)^2}e^{-2(s-1)}$$

32. $f(t) = t - t\,\mathscr{U}(t-1) + (2-t)\mathscr{U}(t-1) - (2-t)\mathscr{U}(t-2) = t - 2(t-1)\mathscr{U}(t-1) + (t-2)\mathscr{U}(t-2)$

$$\mathscr{L}\{f(t)\} = \frac{1}{s^2} - \frac{2}{s^2}e^{-s} + \frac{1}{s^2}e^{-2s}$$

$$\mathscr{L}\{e^t f(t)\} = \frac{1}{(s-1)^2} - \frac{2}{(s-1)^2}e^{-(s-1)} + \frac{1}{(s-1)^2}e^{-2(s-1)}$$

33. Taking the Laplace transform of the differential equation we obtain

$$\mathscr{L}\{y\} = \frac{5}{(s-1)^2} + \frac{1}{2}\frac{2}{(s-1)^3}$$

so that

$$y = 5te^t + \frac{1}{2}t^2 e^t.$$

34. Taking the Laplace transform of the differential equation we obtain

$$\mathscr{L}\{y\} = \frac{1}{(s-1)^2(s^2 - 8s + 20)}$$

$$= \frac{6}{169}\frac{1}{s-1} + \frac{1}{13}\frac{1}{(s-1)^2} - \frac{6}{169}\frac{s-4}{(s-4)^2 + 2^2} + \frac{5}{338}\frac{2}{(s-4)^2 + 2^2}$$

so that

$$y = \frac{6}{169}e^t + \frac{1}{13}te^t - \frac{6}{169}e^{4t}\cos 2t + \frac{5}{338}e^{4t}\sin 2t.$$

35. Taking the Laplace transform of the given differential equation we obtain

$$\mathscr{L}\{y\} = \frac{s^3 + 6s^2 + 1}{s^2(s+1)(s+5)} - \frac{1}{s^2(s+1)(s+5)}e^{-2s} - \frac{2}{s(s+1)(s+5)}e^{-2s}$$

$$= -\frac{6}{25}\cdot\frac{1}{s} + \frac{1}{5}\cdot\frac{1}{s^2} + \frac{3}{2}\cdot\frac{1}{s+1} - \frac{13}{50}\cdot\frac{1}{s+5}$$

$$- \left(-\frac{6}{25}\cdot\frac{1}{s} + \frac{1}{5}\cdot\frac{1}{s^2} + \frac{1}{4}\cdot\frac{1}{s+1} - \frac{1}{100}\cdot\frac{1}{s+5}\right)e^{-2s}$$

$$- \left(\frac{2}{5}\cdot\frac{1}{s} - \frac{1}{2}\cdot\frac{1}{s+1} + \frac{1}{10}\cdot\frac{1}{s+5}\right)e^{-2s}$$

so that

$$y = -\frac{6}{25} + \frac{1}{5}t + \frac{3}{2}e^{-t} - \frac{13}{50}e^{-5t} - \frac{4}{25}\mathscr{U}(t-2) - \frac{1}{5}(t-2)\mathscr{U}(t-2)$$

$$+ \frac{1}{4}e^{-(t-2)}\mathscr{U}(t-2) - \frac{9}{100}e^{-5(t-2)}\mathscr{U}(t-2).$$

36. Taking the Laplace transform of the differential equation we obtain

$$\mathscr{L}\{y\} = \frac{s^3 + 2}{s^3(s-5)} - \frac{2 + 2s + s^2}{s^3(s-5)}e^{-s}$$

$$= -\frac{2}{125}\frac{1}{s} - \frac{2}{25}\frac{1}{s^2} - \frac{1}{5}\frac{2}{s^3} + \frac{127}{125}\frac{1}{s-5} - \left[-\frac{37}{125}\frac{1}{s} - \frac{12}{25}\frac{1}{s^2} - \frac{1}{5}\frac{2}{s^3} + \frac{37}{125}\frac{1}{s-5}\right]e^{-s}$$

so that

$$y = -\frac{2}{125} - \frac{2}{25}t - \frac{1}{5}t^2 + \frac{127}{125}e^{5t} - \left[-\frac{37}{125} - \frac{12}{25}(t-1) - \frac{1}{5}(t-1)^2 + \frac{37}{125}e^{5(t-1)}\right]\mathscr{U}(t-1).$$

37. Taking the Laplace transform of the integral equation we obtain

$$\mathscr{L}\{y\} = \frac{1}{s} + \frac{1}{s^2} + \frac{1}{2}\frac{2}{s^3}$$

so that

$$y(t) = 1 + t + \frac{1}{2}t^2.$$

38. Taking the Laplace transform of the integral equation we obtain

$$(\mathscr{L}\{f\})^2 = 6 \cdot \frac{6}{s^4} \quad \text{or} \quad \mathscr{L}\{f\} = \pm 6 \cdot \frac{1}{s^2}$$

so that $f(t) = \pm 6t$.

39. Taking the Laplace transform of the system gives

$$s\mathscr{L}\{x\} + \mathscr{L}\{y\} = \frac{1}{s^2} + 1$$

$$4\mathscr{L}\{x\} + s\mathscr{L}\{y\} = 2$$

so that

$$\mathscr{L}\{x\} = \frac{s^2 - 2s + 1}{s(s-2)(s+2)} = -\frac{1}{4}\frac{1}{s} + \frac{1}{8}\frac{1}{s-2} + \frac{9}{8}\frac{1}{s+2}.$$

Then

$$x = -\frac{1}{4} + \frac{1}{8}e^{2t} + \frac{9}{8}e^{-2t} \quad \text{and} \quad y = -x' + t = \frac{9}{4}e^{-2t} - \frac{1}{4}e^{2t} + t.$$

40. Taking the Laplace transform of the system gives

$$s^2\mathscr{L}\{x\} + s^2\mathscr{L}\{y\} = \frac{1}{s-2}$$

$$2s\mathscr{L}\{x\} + s^2\mathscr{L}\{y\} = -\frac{1}{s-2}$$

so that

$$\mathscr{L}\{x\} = \frac{2}{s(s-2)^2} = \frac{1}{2}\frac{1}{s} - \frac{1}{2}\frac{1}{s-2} + \frac{1}{(s-2)^2}$$

and

$$\mathcal{L}\{y\} = \frac{-s-2}{s^2(s-2)^2} = -\frac{3}{4}\frac{1}{s} - \frac{1}{2}\frac{1}{s^2} + \frac{3}{4}\frac{1}{s-2} - \frac{1}{(s-2)^2}.$$

Then

$$x = \frac{1}{2} - \frac{1}{2}e^{2t} + te^{2t} \quad \text{and} \quad y = -\frac{3}{4} - \frac{1}{2}t + \frac{3}{4}e^{2t} - te^{2t}.$$

41. The integral equation is

43

$$10i + 2\int_0^t i(\tau)\,d\tau = 2t^2 + 2t.$$

Taking the Laplace transform we obtain

$$\mathcal{L}\{i\} = \left(\frac{4}{s^3} + \frac{2}{s^2}\right)\frac{s}{10s+2} = \frac{s+2}{s^2(5s+2)} = -\frac{9}{s} + \frac{2}{s^2} + \frac{45}{5s+1} = -\frac{9}{s} + \frac{2}{s^2} + \frac{9}{s+1/5}.$$

Thus

$$i(t) = -9 + 2t + 9e^{-t/5}.$$

42. The differential equation is

44

$$\frac{1}{2}\frac{d^2q}{dt^2} + 10\frac{dq}{dt} + 100q = 10 - 10\,\mathcal{U}(t-5).$$

Taking the Laplace transform we obtain

$$\mathcal{L}\{q\} = \frac{20}{s(s^2+20s+200)}\left(1 - e^{-5s}\right)$$

$$= \left[\frac{1}{10}\frac{1}{s} - \frac{1}{10}\frac{s+10}{(s+10)^2+10^2} - \frac{1}{10}\frac{10}{(s+10)^2+10^2}\right]\left(1 - e^{-5s}\right)$$

so that

$$q(t) = \frac{1}{10} - \frac{1}{10}e^{-10t}\cos 10t - \frac{1}{10}e^{-10t}\sin 10t$$

$$- \left[\frac{1}{10} - \frac{1}{10}e^{-10(t-5)}\cos 10(t-5) - \frac{1}{10}e^{-10(t-5)}\sin 10(t-5)\right]\mathcal{U}(t-5).$$

43. Taking the Laplace transform of the given differential equation we obtain

45

$$\mathcal{L}\{y\} = \frac{2w_0}{EIL}\left(\frac{L}{48}\cdot\frac{4!}{s^5} - \frac{1}{120}\cdot\frac{5!}{s^6} + \frac{1}{120}\cdot\frac{5!}{s^6}e^{-sL/2}\right) + \frac{c_1}{2}\cdot\frac{2!}{s^3} + \frac{c_2}{6}\cdot\frac{3!}{s^4}$$

so that

$$y = \frac{2w_0}{EIL}\left[\frac{L}{48}x^4 - \frac{1}{120}x^5 + \frac{1}{120}\left(x - \frac{L}{2}\right)^5\mathcal{U}\left(x - \frac{L}{2}\right) + \frac{c_1}{2}x^2 + \frac{c_2}{6}x^3\right]$$

where $y''(0) = c_1$ and $y'''(0) = c_2$. Using $y''(L) = 0$ and $y'''(L) = 0$ we find

$$c_1 = w_0L^2/24EI, \qquad c_2 = -w_0L/4EI.$$

Hence

$$y = \frac{w_0}{12EIL}\left[-\frac{1}{5}x^5 + \frac{L}{2}x^4 - \frac{L^2}{2}x^3 + \frac{L^3}{4}x^2 + \frac{1}{5}\left(x - \frac{L}{2}\right)^5\mathcal{U}\left(x - \frac{L}{2}\right)\right].$$

415

44. (a) In this case the boundary conditions are $y(0) = y''(0) = 0$ and $y(\pi) = y''(\pi) = 0$. If we let $c_1 = y'(0)$ and $c_2 = y'''(0)$ then

$$s^4 \mathscr{L}\{y\} - s^3 y(0) - s^2 y'(0) - sy(0) - y'''(0) + 4\mathscr{L}\{y\} = \mathscr{L}\{w_0/EI\}$$

and

$$\mathscr{L}\{y\} = \frac{c_1}{2} \cdot \frac{2s^2}{s^4 + 4} + \frac{c_2}{4} \cdot \frac{4}{s^4 + 4} + \frac{w_0}{8EI}\left(\frac{2}{s} - \frac{s-1}{(s-1)^2 + 1} - \frac{s+1}{(s+1)^2 + 1}\right).$$

From the table of transforms we get

$$y = \frac{c_1}{2}(\sin x \cosh x + \cos x \sinh x) + \frac{c_2}{4}(\sin x \cosh x - \cos x \sinh x) + \frac{w_0}{4EI}(1 - \cos x \cosh x)$$

Using $y(\pi) = 0$ and $y''(\pi) = 0$ we find

$$c_1 = \frac{w_0}{4EI}(1 + \cosh \pi)\operatorname{csch} \pi, \qquad c_2 = -\frac{w_0}{2EI}(1 + \cosh \pi)\operatorname{csch} \pi.$$

Hence

$$y = \frac{w_0}{8EI}(1 + \cosh \pi)\operatorname{csch} \pi(\sin x \cosh x + \cos x \sinh x)$$

$$- \frac{w_0}{8EI}(1 + \cosh \pi)\operatorname{csch} \pi(\sin x \cosh x - \cos x \sinh x) + \frac{w_0}{4EI}(1 - \cos x \cosh x).$$

(b) In this case the boundary conditions are $y(0) = y'(0) = 0$ and $y(\pi) = y'(\pi) = 0$. If we let $c_1 = y''(0)$ and $c_2 = y'''(0)$ then

$$s^4 \mathscr{L}\{y\} - s^3 y(0) - s^2 y'(0) - sy(0) - y'''(0) + 4\mathscr{L}\{y\} = \mathscr{L}\{\delta(t - \pi/2)\}$$

and

$$\mathscr{L}\{y\} = \frac{c_1}{2} \cdot \frac{2s}{s^4 + 4} + \frac{c_2}{4} \cdot \frac{4}{s^4 + 4} + \frac{w_0}{4EI} \cdot \frac{4}{s^4 + 4} e^{-s\pi/2}.$$

From the table of transforms we get

$$y = \frac{c_1}{2} \sin x \sinh x + \frac{c_2}{4}(\sin x \cosh x - \cos x \sinh x)$$

$$+ \frac{w_0}{4EI}\left[\sin\left(x - \frac{\pi}{2}\right)\cosh\left(x - \frac{\pi}{2}\right) - \cos\left(x - \frac{\pi}{2}\right)\sinh\left(x - \frac{\pi}{2}\right)\right]\mathscr{U}\left(x - \frac{\pi}{2}\right)$$

Using $y(\pi) = 0$ and $y'(\pi) = 0$ we find

$$c_1 = \frac{w_0}{EI} \frac{\sinh\frac{\pi}{2}}{\sinh \pi}, \qquad c_2 = -\frac{w_0}{EI} \frac{\cosh\frac{\pi}{2}}{\sinh \pi}.$$

Hence

$$y = \frac{w_0}{2EI} \frac{\sinh\frac{\pi}{2}}{\sinh \pi} \sin x \sinh x - \frac{w_0}{4EI} \frac{\cosh\frac{\pi}{2}}{\sinh \pi}(\sin x \cosh x - \cos x \sinh x)$$

$$+ \frac{w_0}{4EI}\left[\sin\left(x - \frac{\pi}{2}\right)\cosh\left(x - \frac{\pi}{2}\right) - \cos\left(x - \frac{\pi}{2}\right)\sinh\left(x - \frac{\pi}{2}\right)\right]\mathscr{U}\left(x - \frac{\pi}{2}\right).$$

45. **(a)** With $\omega^2 = g/l$ and $K = k/m$ the system of differential equations is

$$\theta_1'' + \omega^2\theta_1 = -K(\theta_1 - \theta_2)$$

$$\theta_2'' + \omega^2\theta_2 = K(\theta_1 - \theta_2).$$

Denoting the Laplace transform of $\theta(t)$ by $\Theta(s)$ we have that the Laplace transform of the system is

$$(s^2 + \omega^2)\Theta_1(s) = -K\Theta_1(s) + K\Theta_2(s) + s\theta_0$$

$$(s^2 + \omega^2)\Theta_2(s) = K\Theta_1(s) - K\Theta_2(s) + s\psi_0.$$

If we add the two equations, we get

$$\Theta_1(s) + \Theta_2(s) = (\theta_0 + \psi_0)\frac{s}{s^2 + \omega^2}$$

which implies

$$\theta_1(t) + \theta_2(t) = (\theta_0 + \psi_0)\cos\omega t.$$

This enables us to solve for first, say, $\theta_1(t)$ and then find $\theta_2(t)$ from

$$\theta_2(t) = -\theta_1(t) + (\theta_0 + \psi_0)\cos\omega t.$$

Now solving

$$(s^2 + \omega^2 + K)\Theta_1(s) - K\Theta_2(s) = s\theta_0$$

$$-k\Theta_1(s) + (s^2 + \omega^2 + K)\Theta_2(s) = s\psi_0$$

gives

$$[(s^2 + \omega^2 + K)^2 - K^2]\Theta_1(s) = s(s^2 + \omega^2 + K)\theta_0 + Ks\psi_0.$$

Factoring the difference of two squares and using partial fractions we get

$$\Theta_1(s) = \frac{s(s^2 + \omega^2 + K)\theta_0 + Ks\psi_0}{(s^2 + \omega^2)(s^2 + \omega^2 + 2K)} = \frac{\theta_0 + \psi_0}{2}\frac{s}{s^2 + \omega^2} + \frac{\theta_0 - \psi_0}{2}\frac{s}{s^2 + \omega^2 + 2K},$$

so

$$\theta_1(t) = \frac{\theta_0 + \psi_0}{2}\cos\omega t + \frac{\theta_0 - \psi_0}{2}\cos\sqrt{\omega^2 + 2K}\,t.$$

Then from $\theta_2(t) = -\theta_1(t) + (\theta_0 + \psi_0)\cos\omega t$ we get

$$\theta_2(t) = \frac{\theta_0 + \psi_0}{2}\cos\omega t - \frac{\theta_0 - \psi_0}{2}\cos\sqrt{\omega^2 + 2K}\,t.$$

(b) With the initial conditions $\theta_1(0) = \theta_0$, $\theta_1'(0) = 0$, $\theta_2(0) = \theta_0$, $\theta_2'(0) = 0$ we have

$$\theta_1(t) = \theta_0\cos\omega t, \qquad \theta_2(t) = \theta_0\cos\omega t.$$

Physically this means that both pendulums swing in the same direction as if they were free since the spring exerts no influence on the motion ($\theta_1(t)$ and $\theta_2(t)$ are free of K).

With the initial conditions $\theta_1(0) = \theta_0$, $\theta_1'(0) = 0$, $\theta_2(0) = -\theta_0$, $\theta_2'(0) = 0$ we have

$$\theta_1(t) = \theta_0 \cos \sqrt{\omega^2 + 2K}\, t, \qquad \theta_2(t) = -\theta_0 \cos \sqrt{\omega^2 + 2K}\, t.$$

Physically this means that both pendulums swing in the opposite directions, stretching and compressing the spring. The amplitude of both displacements is $|\theta_0|$. Moreover, $\theta_1(t) = \theta_0$ and $\theta_2(t) = -\theta_0$ at precisely the same times. At these times the spring is stretched to its maximum.

48-50
new

8 Systems of Linear First-Order Differential Equations

1. Let $\mathbf{X} = \begin{pmatrix} x \\ y \end{pmatrix}$. Then $\mathbf{X}' = \begin{pmatrix} 3 & -5 \\ 4 & 8 \end{pmatrix} \mathbf{X}$.

2. Let $\mathbf{X} = \begin{pmatrix} x \\ y \end{pmatrix}$. Then $\mathbf{X}' = \begin{pmatrix} 4 & -7 \\ 5 & 0 \end{pmatrix} \mathbf{X}$.

3. Let $\mathbf{X} = \begin{pmatrix} x \\ y \\ z \end{pmatrix}$. Then $\mathbf{X}' = \begin{pmatrix} -3 & 4 & -9 \\ 6 & -1 & 0 \\ 10 & 4 & 3 \end{pmatrix} \mathbf{X}$.

4. Let $\mathbf{X} = \begin{pmatrix} x \\ y \\ z \end{pmatrix}$. Then $\mathbf{X}' = \begin{pmatrix} 1 & -1 & 0 \\ 1 & 0 & 2 \\ -1 & 0 & 1 \end{pmatrix} \mathbf{X}$.

5. Let $\mathbf{X} = \begin{pmatrix} x \\ y \\ z \end{pmatrix}$. Then $\mathbf{X}' = \begin{pmatrix} 1 & -1 & 1 \\ 2 & 1 & -1 \\ 1 & 1 & 1 \end{pmatrix} \mathbf{X} + \begin{pmatrix} 0 \\ -3t^2 \\ t^2 \end{pmatrix} + \begin{pmatrix} t \\ 0 \\ -t \end{pmatrix} + \begin{pmatrix} -1 \\ 0 \\ 2 \end{pmatrix}$.

6. Let $\mathbf{X} = \begin{pmatrix} x \\ y \\ z \end{pmatrix}$. Then $\mathbf{X}' = \begin{pmatrix} -3 & 4 & 0 \\ 5 & 9 & 0 \\ 0 & 1 & 6 \end{pmatrix} \mathbf{X} + \begin{pmatrix} e^{-t} \sin 2t \\ 4e^{-t} \cos 2t \\ -e^{-t} \end{pmatrix}$.

7. $\dfrac{dx}{dt} = 4x + 2y + e^t; \quad \dfrac{dy}{dt} = -x + 3y - e^t$

8. $\dfrac{dx}{dt} = 7x + 5y - 9z - 8e^{-2t}; \quad \dfrac{dy}{dt} = 4x + y + z + 2e^{5t}; \quad \dfrac{dz}{dt} = -2y + 3z + e^{5t} - 3e^{-2t}$

9. $\dfrac{dx}{dt} = x - y + 2z + e^{-t} - 3t; \quad \dfrac{dy}{dt} = 3x - 4y + z + 2e^{-t} + t; \quad \dfrac{dz}{dt} = -2x + 5y + 6z + 2e^{-t} - t$

10. $\dfrac{dx}{dt} = 3x - 7y + 4\sin t + (t - 4)e^{4t}; \quad \dfrac{dy}{dt} = x + y + 8\sin t + (2t + 1)e^{4t}$

11. Since

$$\mathbf{X}' = \begin{pmatrix} -5 \\ -10 \end{pmatrix} e^{-5t} \quad \text{and} \quad \begin{pmatrix} 3 & -4 \\ 4 & -7 \end{pmatrix} \mathbf{X} = \begin{pmatrix} -5 \\ -10 \end{pmatrix} e^{-5t}$$

we see that

$$\mathbf{X}' = \begin{pmatrix} 3 & -4 \\ 4 & -7 \end{pmatrix} \mathbf{X}.$$

12. Since

$$\mathbf{X}' = \begin{pmatrix} 5\cos t - 5\sin t \\ 2\cos t - 4\sin t \end{pmatrix} e^t \quad \text{and} \quad \begin{pmatrix} -2 & 5 \\ -2 & 4 \end{pmatrix} \mathbf{X} = \begin{pmatrix} 5\cos t - 5\sin t \\ 2\cos t - 4\sin t \end{pmatrix} e^t$$

we see that

$$\mathbf{X}' = \begin{pmatrix} -2 & 5 \\ -2 & 4 \end{pmatrix} \mathbf{X}.$$

13. Since

$$\mathbf{X}' = \begin{pmatrix} 3/2 \\ -3 \end{pmatrix} e^{-3t/2} \quad \text{and} \quad \begin{pmatrix} -1 & 1/4 \\ 1 & -1 \end{pmatrix} \mathbf{X} = \begin{pmatrix} 3/2 \\ -3 \end{pmatrix} e^{-3t/2}$$

we see that

$$\mathbf{X}' = \begin{pmatrix} -1 & 1/4 \\ 1 & -1 \end{pmatrix} \mathbf{X}.$$

14. Since

$$\mathbf{X}' = \begin{pmatrix} 5 \\ -1 \end{pmatrix} e^t + \begin{pmatrix} 4 \\ -4 \end{pmatrix} te^t \quad \text{and} \quad \begin{pmatrix} 2 & 1 \\ -1 & 0 \end{pmatrix} \mathbf{X} = \begin{pmatrix} 5 \\ -1 \end{pmatrix} e^t + \begin{pmatrix} 4 \\ -4 \end{pmatrix} te^t$$

we see that

$$\mathbf{X}' = \begin{pmatrix} 2 & 1 \\ -1 & 0 \end{pmatrix} \mathbf{X}.$$

15. Since

$$\mathbf{X}' = \begin{pmatrix} 0 \\ 0 \\ 0 \end{pmatrix} \quad \text{and} \quad \begin{pmatrix} 1 & 2 & 1 \\ 6 & -1 & 0 \\ -1 & -2 & -1 \end{pmatrix} \mathbf{X} = \begin{pmatrix} 0 \\ 0 \\ 0 \end{pmatrix}$$

we see that

$$\mathbf{X}' = \begin{pmatrix} 1 & 2 & 1 \\ 6 & -1 & 0 \\ -1 & -2 & -1 \end{pmatrix} \mathbf{X}.$$

16. Since

$$\mathbf{X}' = \begin{pmatrix} \cos t \\ \frac{1}{2}\sin t - \frac{1}{2}\cos t \\ -\cos t - \sin t \end{pmatrix} \quad \text{and} \quad \begin{pmatrix} 1 & 0 & 1 \\ 1 & 1 & 0 \\ -2 & 0 & -1 \end{pmatrix} \mathbf{X} = \begin{pmatrix} \cos t \\ \frac{1}{2}\sin t - \frac{1}{2}\cos t \\ -\cos t - \sin t \end{pmatrix}$$

we see that

$$\mathbf{X}' = \begin{pmatrix} 1 & 0 & 1 \\ 1 & 1 & 0 \\ -2 & 0 & -1 \end{pmatrix} \mathbf{X}.$$

17. Yes, since $W(\mathbf{X}_1, \mathbf{X}_2) = -2e^{-8t} \neq 0$ the set $\mathbf{X}_1$, $\mathbf{X}_2$ is linearly independent on $-\infty < t < \infty$.

18. Yes, since $W(\mathbf{X}_1, \mathbf{X}_2) = 8e^{2t} \neq 0$ the set $\mathbf{X}_1$, $\mathbf{X}_2$ is linearly independent on $-\infty < t < \infty$.

19. No, since $W(\mathbf{X}_1, \mathbf{X}_2, \mathbf{X}_3) = 0$ the set $\mathbf{X}_1$, $\mathbf{X}_2$, $\mathbf{X}_3$ is linearly dependent on $-\infty < t < \infty$.

20. Yes, since $W(\mathbf{X}_1, \mathbf{X}_2, \mathbf{X}_3) = -84e^{-t} \neq 0$ the set $\mathbf{X}_1$, $\mathbf{X}_2$, $\mathbf{X}_3$ is linearly independent on $-\infty < t < \infty$.

21. Since

$$\mathbf{X}'_p = \begin{pmatrix} 2 \\ -1 \end{pmatrix} \quad \text{and} \quad \begin{pmatrix} 1 & 4 \\ 3 & 2 \end{pmatrix} \mathbf{X}_p + \begin{pmatrix} 2 \\ -4 \end{pmatrix} t + \begin{pmatrix} -7 \\ -18 \end{pmatrix} = \begin{pmatrix} 2 \\ -1 \end{pmatrix}$$

we see that

$$\mathbf{X}'_p = \begin{pmatrix} 1 & 4 \\ 3 & 2 \end{pmatrix} \mathbf{X}_p + \begin{pmatrix} 2 \\ -4 \end{pmatrix} t + \begin{pmatrix} -7 \\ -18 \end{pmatrix}.$$

22. Since

$$\mathbf{X}'_p = \begin{pmatrix} 0 \\ 0 \end{pmatrix} \quad \text{and} \quad \begin{pmatrix} 2 & 1 \\ 1 & -1 \end{pmatrix} \mathbf{X}_p + \begin{pmatrix} -5 \\ 2 \end{pmatrix} = \begin{pmatrix} 0 \\ 0 \end{pmatrix}$$

we see that

$$\mathbf{X}'_p = \begin{pmatrix} 2 & 1 \\ 1 & -1 \end{pmatrix} \mathbf{X}_p + \begin{pmatrix} -5 \\ 2 \end{pmatrix}.$$

23. Since

$$\mathbf{X}'_p = \begin{pmatrix} 2 \\ 0 \end{pmatrix} e^t + \begin{pmatrix} 1 \\ -1 \end{pmatrix} te^t \quad \text{and} \quad \begin{pmatrix} 2 & 1 \\ 3 & 4 \end{pmatrix} \mathbf{X}_p - \begin{pmatrix} 1 \\ 7 \end{pmatrix} e^t = \begin{pmatrix} 2 \\ 0 \end{pmatrix} e^t + \begin{pmatrix} 1 \\ -1 \end{pmatrix} te^t$$

we see that

$$\mathbf{X}'_p = \begin{pmatrix} 2 & 1 \\ 3 & 4 \end{pmatrix} \mathbf{X}_p - \begin{pmatrix} 1 \\ 7 \end{pmatrix} e^t.$$

24. Since

$$\mathbf{X}'_p = \begin{pmatrix} 3\cos 3t \\ 0 \\ -3\sin 3t \end{pmatrix} \quad \text{and} \quad \begin{pmatrix} 1 & 2 & 3 \\ -4 & 2 & 0 \\ -6 & 1 & 0 \end{pmatrix} \mathbf{X}_p + \begin{pmatrix} -1 \\ 4 \\ 3 \end{pmatrix} \sin 3t = \begin{pmatrix} 3\cos 3t \\ 0 \\ -3\sin 3t \end{pmatrix}$$

we see that

$$\mathbf{X}'_p = \begin{pmatrix} 1 & 2 & 3 \\ -4 & 2 & 0 \\ -6 & 1 & 0 \end{pmatrix} \mathbf{X}_p + \begin{pmatrix} -1 \\ 4 \\ 3 \end{pmatrix} \sin 3t.$$

25. Let

$$\mathbf{X}_1 = \begin{pmatrix} 6 \\ -1 \\ -5 \end{pmatrix} e^{-t}, \quad \mathbf{X}_2 = \begin{pmatrix} -3 \\ 1 \\ 1 \end{pmatrix} e^{-2t}, \quad \mathbf{X}_3 = \begin{pmatrix} 2 \\ 1 \\ 1 \end{pmatrix} e^{3t}, \quad \text{and} \quad \mathbf{A} = \begin{pmatrix} 0 & 6 & 0 \\ 1 & 0 & 1 \\ 1 & 1 & 0 \end{pmatrix}.$$

Then

$$\mathbf{X}_1' = \begin{pmatrix} -6 \\ 1 \\ 5 \end{pmatrix} e^{-t} = \mathbf{A}\mathbf{X}_1,$$

$$\mathbf{X}_2' = \begin{pmatrix} 6 \\ -2 \\ -2 \end{pmatrix} e^{-2t} = \mathbf{A}\mathbf{X}_2,$$

$$\mathbf{X}_3' = \begin{pmatrix} 6 \\ 3 \\ 3 \end{pmatrix} e^{3t} = \mathbf{A}\mathbf{X}_3,$$

and $W(\mathbf{X}_1, \mathbf{X}_2, \mathbf{X}_3) = 20 \neq 0$ so that $\mathbf{X}_1$, $\mathbf{X}_2$, and $\mathbf{X}_3$ form a fundamental set for $\mathbf{X}' = \mathbf{A}\mathbf{X}$ on $-\infty < t < \infty$.

26. Let

$$\mathbf{X}_1 = \begin{pmatrix} 1 \\ -1 - \sqrt{2} \end{pmatrix} e^{\sqrt{2}t},$$

$$\mathbf{X}_2 = \begin{pmatrix} 1 \\ -1 + \sqrt{2} \end{pmatrix} e^{-\sqrt{2}t},$$

$$\mathbf{X}_p = \begin{pmatrix} 1 \\ 0 \end{pmatrix} t^2 + \begin{pmatrix} -2 \\ 4 \end{pmatrix} t + \begin{pmatrix} 1 \\ 0 \end{pmatrix},$$

and

$$\mathbf{A} = \begin{pmatrix} -1 & -1 \\ -1 & 1 \end{pmatrix}.$$

Then

$$\mathbf{X}_1' = \begin{pmatrix} \sqrt{2} \\ -2 - \sqrt{2} \end{pmatrix} e^{\sqrt{2}t} = \mathbf{A}\mathbf{X}_1,$$

$$\mathbf{X}_2' = \begin{pmatrix} -\sqrt{2} \\ -2 + \sqrt{2} \end{pmatrix} e^{-\sqrt{2}t} = \mathbf{A}\mathbf{X}_2,$$

$$\mathbf{X}_p' = \begin{pmatrix} 2 \\ 0 \end{pmatrix} t + \begin{pmatrix} -2 \\ 4 \end{pmatrix} = \mathbf{A}\mathbf{X}_p + \begin{pmatrix} 1 \\ 1 \end{pmatrix} t^2 + \begin{pmatrix} 4 \\ -6 \end{pmatrix} t + \begin{pmatrix} -1 \\ 5 \end{pmatrix},$$

and $W(\mathbf{X}_1, \mathbf{X}_2) = 2\sqrt{2} \neq 0$ so that $\mathbf{X}_p$ is a particular solution and $\mathbf{X}_1$ and $\mathbf{X}_2$ form a fundamental set on $-\infty < t < \infty$.

Exercises 8.2

1. The system is

$$\mathbf{X}' = \begin{pmatrix} 1 & 2 \\ 4 & 3 \end{pmatrix} \mathbf{X}$$

and $\det(\mathbf{A} - \lambda \mathbf{I}) = (\lambda - 5)(\lambda + 1) = 0$. For $\lambda_1 = 5$ we obtain

$$\begin{pmatrix} -4 & 2 & | & 0 \\ 4 & -2 & | & 0 \end{pmatrix} \Longrightarrow \begin{pmatrix} 1 & -1/2 & | & 0 \\ 0 & 0 & | & 0 \end{pmatrix} \quad \text{so that} \quad \mathbf{K}_1 = \begin{pmatrix} 1 \\ 2 \end{pmatrix}.$$

For $\lambda_2 = -1$ we obtain

$$\begin{pmatrix} 2 & 2 & | & 0 \\ 4 & 4 & | & 0 \end{pmatrix} \Longrightarrow \begin{pmatrix} 1 & 1 & | & 0 \\ 0 & 0 & | & 0 \end{pmatrix} \quad \text{so that} \quad \mathbf{K}_2 = \begin{pmatrix} -1 \\ 1 \end{pmatrix}.$$

Then

$$\mathbf{X} = c_1 \begin{pmatrix} 1 \\ 2 \end{pmatrix} e^{5t} + c_2 \begin{pmatrix} -1 \\ 1 \end{pmatrix} e^{-t}.$$

2. The system is

$$\mathbf{X}' = \begin{pmatrix} 2 & 2 \\ 1 & 3 \end{pmatrix} \mathbf{X}$$

and $\det(\mathbf{A} - \lambda \mathbf{I}) = (\lambda - 1)(\lambda - 4) = 0$. For $\lambda_1 = 1$ we obtain

$$\begin{pmatrix} 1 & 2 & | & 0 \\ 1 & 2 & | & 0 \end{pmatrix} \Longrightarrow \begin{pmatrix} 1 & 2 & | & 0 \\ 0 & 0 & | & 0 \end{pmatrix} \quad \text{so that} \quad \mathbf{K}_1 = \begin{pmatrix} -2 \\ 1 \end{pmatrix}.$$

For $\lambda_2 = 4$ we obtain

$$\begin{pmatrix} -2 & 2 & | & 0 \\ 1 & -1 & | & 0 \end{pmatrix} \Longrightarrow \begin{pmatrix} -1 & 1 & | & 0 \\ 0 & 0 & | & 0 \end{pmatrix} \quad \text{so that} \quad \mathbf{K}_2 = \begin{pmatrix} 1 \\ 1 \end{pmatrix}.$$

Then

$$\mathbf{X} = c_1 \begin{pmatrix} -2 \\ 1 \end{pmatrix} e^{t} + c_2 \begin{pmatrix} 1 \\ 1 \end{pmatrix} e^{4t}.$$

3. The system is

$$\mathbf{X}' = \begin{pmatrix} -4 & 2 \\ -5/2 & 2 \end{pmatrix} \mathbf{X}$$

and $\det(\mathbf{A} - \lambda \mathbf{I}) = (\lambda - 1)(\lambda + 3) = 0$. For $\lambda_1 = 1$ we obtain

$$\begin{pmatrix} -5 & 2 & | & 0 \\ -5/2 & 1 & | & 0 \end{pmatrix} \Longrightarrow \begin{pmatrix} -5 & 2 & | & 0 \\ 0 & 0 & | & 0 \end{pmatrix} \quad \text{so that} \quad \mathbf{K}_1 = \begin{pmatrix} 2 \\ 5 \end{pmatrix}.$$

For $\lambda_2 = -3$ we obtain

$$\begin{pmatrix} -1 & 2 & | & 0 \\ -5/2 & 5 & | & 0 \end{pmatrix} \implies \begin{pmatrix} -1 & 2 & | & 0 \\ 0 & 0 & | & 0 \end{pmatrix} \quad \text{so that} \quad \mathbf{K}_2 = \begin{pmatrix} 2 \\ 1 \end{pmatrix}.$$

Then

$$\mathbf{X} = c_1 \begin{pmatrix} 2 \\ 5 \end{pmatrix} e^t + c_2 \begin{pmatrix} 2 \\ 1 \end{pmatrix} e^{-3t}.$$

4. The system is

$$\mathbf{X}' = \begin{pmatrix} -5/2 & 2 \\ 3/4 & -2 \end{pmatrix} \mathbf{X}$$

and $\det(\mathbf{A} - \lambda\mathbf{I}) = \frac{1}{2}(\lambda + 1)(2\lambda + 7) = 0$. For $\lambda_1 = -7/2$ we obtain

$$\begin{pmatrix} 1 & 2 & | & 0 \\ 3/4 & 3/2 & | & 0 \end{pmatrix} \implies \begin{pmatrix} 1 & 2 & | & 0 \\ 0 & 0 & | & 0 \end{pmatrix} \quad \text{so that} \quad \mathbf{K}_1 = \begin{pmatrix} -2 \\ 1 \end{pmatrix}.$$

For $\lambda_2 = -1$ we obtain

$$\begin{pmatrix} -3/2 & 2 & | & 0 \\ 3/4 & -1 & | & 0 \end{pmatrix} \implies \begin{pmatrix} -3 & 4 & | & 0 \\ 0 & 0 & | & 0 \end{pmatrix} \quad \text{so that} \quad \mathbf{K}_2 = \begin{pmatrix} 4 \\ 3 \end{pmatrix}.$$

Then

$$\mathbf{X} = c_1 \begin{pmatrix} -2 \\ 1 \end{pmatrix} e^{-7t/2} + c_2 \begin{pmatrix} 4 \\ 3 \end{pmatrix} e^{-t}.$$

5. The system is

$$\mathbf{X}' = \begin{pmatrix} 10 & -5 \\ 8 & -12 \end{pmatrix} \mathbf{X}$$

and $\det(\mathbf{A} - \lambda\mathbf{I}) = (\lambda - 8)(\lambda + 10) = 0$. For $\lambda_1 = 8$ we obtain

$$\begin{pmatrix} 2 & -5 & | & 0 \\ 8 & -20 & | & 0 \end{pmatrix} \implies \begin{pmatrix} 1 & -5/2 & | & 0 \\ 0 & 0 & | & 0 \end{pmatrix} \quad \text{so that} \quad \mathbf{K}_1 = \begin{pmatrix} 5 \\ 2 \end{pmatrix}.$$

For $\lambda_2 = -10$ we obtain

$$\begin{pmatrix} 20 & -5 & | & 0 \\ 8 & -2 & | & 0 \end{pmatrix} \implies \begin{pmatrix} 1 & -1/4 & | & 0 \\ 0 & 0 & | & 0 \end{pmatrix} \quad \text{so that} \quad \mathbf{K}_2 = \begin{pmatrix} 1 \\ 4 \end{pmatrix}.$$

Then

$$\mathbf{X} = c_1 \begin{pmatrix} 5 \\ 2 \end{pmatrix} e^{8t} + c_2 \begin{pmatrix} 1 \\ 4 \end{pmatrix} e^{-10t}.$$

6. The system is

$$\mathbf{X}' = \begin{pmatrix} -6 & 2 \\ -3 & 1 \end{pmatrix} \mathbf{X}$$

and $\det(\mathbf{A} - \lambda\mathbf{I}) = \lambda(\lambda + 5) = 0$. For $\lambda_1 = 0$ we obtain

$$\begin{pmatrix} -6 & 2 & | & 0 \\ -3 & 1 & | & 0 \end{pmatrix} \implies \begin{pmatrix} 1 & -1/3 & | & 0 \\ 0 & 0 & | & 0 \end{pmatrix} \quad \text{so that} \quad \mathbf{K}_1 = \begin{pmatrix} 1 \\ 3 \end{pmatrix}.$$

For $\lambda_2 = -5$ we obtain

$$\begin{pmatrix} -1 & 2 & | & 0 \\ -3 & 6 & | & 0 \end{pmatrix} \implies \begin{pmatrix} 1 & -2 & | & 0 \\ 0 & 0 & | & 0 \end{pmatrix} \quad \text{so that} \quad \mathbf{K}_2 = \begin{pmatrix} 2 \\ 1 \end{pmatrix}.$$

Then

$$\mathbf{X} = c_1 \begin{pmatrix} 1 \\ 3 \end{pmatrix} + c_2 \begin{pmatrix} 2 \\ 1 \end{pmatrix} e^{-5t}.$$

7. The system is

$$\mathbf{X}' = \begin{pmatrix} 1 & 1 & -1 \\ 0 & 2 & 0 \\ 0 & 1 & -1 \end{pmatrix} \mathbf{X}$$

and $\det(\mathbf{A} - \lambda\mathbf{I}) = (\lambda - 1)(2 - \lambda)(\lambda + 1) = 0$. For $\lambda_1 = 1$, $\lambda_2 = 2$, and $\lambda_3 = -1$ we obtain

$$\mathbf{K}_1 = \begin{pmatrix} 1 \\ 0 \\ 0 \end{pmatrix}, \quad \mathbf{K}_2 = \begin{pmatrix} 2 \\ 3 \\ 1 \end{pmatrix}, \quad \text{and} \quad \mathbf{K}_3 = \begin{pmatrix} 1 \\ 0 \\ 2 \end{pmatrix},$$

so that

$$\mathbf{X} = c_1 \begin{pmatrix} 1 \\ 0 \\ 0 \end{pmatrix} e^t + c_2 \begin{pmatrix} 2 \\ 3 \\ 1 \end{pmatrix} e^{2t} + c_3 \begin{pmatrix} 1 \\ 0 \\ 2 \end{pmatrix} e^{-t}.$$

8. The system is

$$\mathbf{X}' = \begin{pmatrix} 2 & -7 & 0 \\ 5 & 10 & 4 \\ 0 & 5 & 2 \end{pmatrix} \mathbf{X}$$

and $\det(\mathbf{A} - \lambda\mathbf{I}) = (2 - \lambda)(\lambda - 5)(\lambda - 7) = 0$. For $\lambda_1 = 2$, $\lambda_2 = 5$, and $\lambda_3 = 7$ we obtain

$$\mathbf{K}_1 = \begin{pmatrix} 4 \\ 0 \\ -5 \end{pmatrix}, \quad \mathbf{K}_2 = \begin{pmatrix} -7 \\ 3 \\ 5 \end{pmatrix}, \quad \text{and} \quad \mathbf{K}_3 = \begin{pmatrix} -7 \\ 5 \\ 5 \end{pmatrix},$$

so that

$$\mathbf{X} = c_1 \begin{pmatrix} 4 \\ 0 \\ -5 \end{pmatrix} e^{2t} + c_2 \begin{pmatrix} -7 \\ 3 \\ 5 \end{pmatrix} e^{5t} + c_3 \begin{pmatrix} -7 \\ 5 \\ 5 \end{pmatrix} e^{7t}.$$

9. We have $\det(\mathbf{A} - \lambda\mathbf{I}) = -(\lambda + 1)(\lambda - 3)(\lambda + 2) = 0$. For $\lambda_1 = -1$, $\lambda_2 = 3$, and $\lambda_3 = -2$ we obtain

$$\mathbf{K}_1 = \begin{pmatrix} -1 \\ 0 \\ 1 \end{pmatrix}, \quad \mathbf{K}_2 = \begin{pmatrix} 1 \\ 4 \\ 3 \end{pmatrix}, \quad \text{and} \quad \mathbf{K}_3 = \begin{pmatrix} 1 \\ -1 \\ 3 \end{pmatrix},$$

so that

$$\mathbf{X} = c_1 \begin{pmatrix} -1 \\ 0 \\ 1 \end{pmatrix} e^{-t} + c_2 \begin{pmatrix} 1 \\ 4 \\ 3 \end{pmatrix} e^{3t} + c_3 \begin{pmatrix} 1 \\ -1 \\ 3 \end{pmatrix} e^{-2t}.$$

10. We have $\det(\mathbf{A} - \lambda\mathbf{I}) = -\lambda(\lambda - 1)(\lambda - 2) = 0$. For $\lambda_1 = 0$, $\lambda_2 = 1$, and $\lambda_3 = 2$ we obtain

$$\mathbf{K}_1 = \begin{pmatrix} 1 \\ 0 \\ -1 \end{pmatrix}, \quad \mathbf{K}_2 = \begin{pmatrix} 0 \\ 1 \\ 0 \end{pmatrix}, \quad \text{and} \quad \mathbf{K}_3 = \begin{pmatrix} 1 \\ 0 \\ 1 \end{pmatrix},$$

so that

$$\mathbf{X} = c_1 \begin{pmatrix} 1 \\ 0 \\ -1 \end{pmatrix} + c_2 \begin{pmatrix} 0 \\ 1 \\ 0 \end{pmatrix} e^t + c_3 \begin{pmatrix} 1 \\ 0 \\ 1 \end{pmatrix} e^{2t}.$$

11. We have $\det(\mathbf{A} - \lambda\mathbf{I}) = -(\lambda + 1)(\lambda + 1/2)(\lambda + 3/2) = 0$. For $\lambda_1 = -1$, $\lambda_2 = -1/2$, and $\lambda_3 = -3/2$ we obtain

$$\mathbf{K}_1 = \begin{pmatrix} 4 \\ 0 \\ -1 \end{pmatrix}, \quad \mathbf{K}_2 = \begin{pmatrix} -12 \\ 6 \\ 5 \end{pmatrix}, \quad \text{and} \quad \mathbf{K}_3 = \begin{pmatrix} 4 \\ 2 \\ -1 \end{pmatrix},$$

so that

$$\mathbf{X} = c_1 \begin{pmatrix} 4 \\ 0 \\ -1 \end{pmatrix} e^{-t} + c_2 \begin{pmatrix} -12 \\ 6 \\ 5 \end{pmatrix} e^{-t/2} + c_3 \begin{pmatrix} 4 \\ 2 \\ -1 \end{pmatrix} e^{-3t/2}.$$

12. We have $\det(\mathbf{A} - \lambda\mathbf{I}) = (\lambda - 3)(\lambda + 5)(6 - \lambda) = 0$. For $\lambda_1 = 3$, $\lambda_2 = -5$, and $\lambda_3 = 6$ we obtain

$$\mathbf{K}_1 = \begin{pmatrix} 1 \\ 1 \\ 0 \end{pmatrix}, \quad \mathbf{K}_2 = \begin{pmatrix} 1 \\ -1 \\ 0 \end{pmatrix}, \quad \text{and} \quad \mathbf{K}_3 = \begin{pmatrix} 2 \\ -2 \\ 11 \end{pmatrix},$$

so that

$$\mathbf{X} = c_1 \begin{pmatrix} 1 \\ 1 \\ 0 \end{pmatrix} e^{3t} + c_2 \begin{pmatrix} 1 \\ -1 \\ 0 \end{pmatrix} e^{-5t} + c_3 \begin{pmatrix} 2 \\ -2 \\ 11 \end{pmatrix} e^{6t}.$$

13. We have $\det(\mathbf{A} - \lambda\mathbf{I}) = (\lambda + 1/2)(\lambda - 1/2) = 0$. For $\lambda_1 = -1/2$ and $\lambda_2 = 1/2$ we obtain

$$\mathbf{K}_1 = \begin{pmatrix} 0 \\ 1 \end{pmatrix} \quad \text{and} \quad \mathbf{K}_2 = \begin{pmatrix} 1 \\ 1 \end{pmatrix},$$

so that

$$\mathbf{X} = c_1 \begin{pmatrix} 0 \\ 1 \end{pmatrix} e^{-t/2} + c_2 \begin{pmatrix} 1 \\ 1 \end{pmatrix} e^{t/2}.$$

If

$$\mathbf{X}(0) = \begin{pmatrix} 3 \\ 5 \end{pmatrix}$$

then $c_1 = 2$ and $c_2 = 3$.

14. We have $\det(\mathbf{A} - \lambda\mathbf{I}) = (2 - \lambda)(\lambda - 3)(\lambda + 1) = 0$. For $\lambda_1 = 2$, $\lambda_2 = 3$, and $\lambda_3 = -1$ we obtain

$$\mathbf{K}_1 = \begin{pmatrix} 5 \\ -3 \\ 2 \end{pmatrix}, \quad \mathbf{K}_2 = \begin{pmatrix} 2 \\ 0 \\ 1 \end{pmatrix}, \quad \text{and} \quad \mathbf{K}_3 = \begin{pmatrix} -2 \\ 0 \\ 1 \end{pmatrix},$$

so that

$$\mathbf{X} = c_1 \begin{pmatrix} 5 \\ -3 \\ 2 \end{pmatrix} e^{2t} + c_2 \begin{pmatrix} 2 \\ 0 \\ 1 \end{pmatrix} e^{3t} + c_3 \begin{pmatrix} -2 \\ 0 \\ 1 \end{pmatrix} e^{-t}.$$

If

$$\mathbf{X}(0) = \begin{pmatrix} 1 \\ 3 \\ 0 \end{pmatrix}$$

then $c_1 = -1$, $c_2 = 5/2$, and $c_3 = -1/2$.

15. $\mathbf{X} = c_1 \begin{pmatrix} 0.382175 \\ 0.851161 \\ 0.359815 \end{pmatrix} e^{8.58979t} + c_2 \begin{pmatrix} 0.405188 \\ -0.676043 \\ 0.615458 \end{pmatrix} e^{2.25684t} + c_3 \begin{pmatrix} -0.923562 \\ -0.132174 \\ 0.35995 \end{pmatrix} e^{-0.0466321t}$

16. $\mathbf{X} = c_1 \begin{pmatrix} 0.0312209 \\ 0.949058 \\ 0.239535 \\ 0.195825 \\ 0.0508861 \end{pmatrix} e^{5.05452t} + c_2 \begin{pmatrix} -0.280232 \\ -0.836611 \\ -0.275304 \\ 0.176045 \\ 0.338775 \end{pmatrix} e^{4.09561t} + c_3 \begin{pmatrix} 0.262219 \\ -0.162664 \\ -0.826218 \\ -0.346439 \\ 0.31957 \end{pmatrix} e^{-2.92362t}$

$+ c_4 \begin{pmatrix} 0.313235 \\ 0.64181 \\ 0.31754 \\ 0.173787 \\ -0.599108 \end{pmatrix} e^{2.02882t} + c_5 \begin{pmatrix} -0.301294 \\ 0.466599 \\ 0.222136 \\ 0.0534311 \\ -0.799567 \end{pmatrix} e^{-0.155338t}$

17. (a)

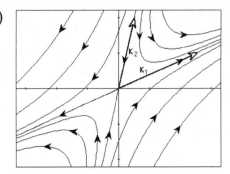

(b) Letting $c_1 = 1$ and $c_2 = 0$ we get $x = 5e^{8t}$, $y = 2e^{8t}$. Eliminating the parameter we find $y = \frac{2}{5}x$, $x > 0$. When $c_1 = -1$ and $c_2 = 0$ we find $y = \frac{2}{5}x$, $x < 0$. Letting $c_1 = 0$ and $c_2 = 1$ we get $x = e^{-10t}$, $y = 4e^{-10t}$. Eliminating the parameter we find $y = 4x$, $x > 0$. Letting $c_1 = 0$ and $c_2 = -1$ we find $y = 4x$, $x < 0$.

(c) The eigenvectors $\mathbf{K}_1 = (5, 2)$ and $\mathbf{K}_2 = (1, 4)$ are shown in the figure in part (a).

18. In Problem 2, letting $c_1 = 1$ and $c_2 = 0$ we get $x = -2e^t$, $y = e^t$. Eliminating the parameter we find $y = -\frac{1}{2}x$, $x < 0$. When $c_1 = -1$ and $c_2 = 0$ we find $y = -\frac{1}{2}x$, $x > 0$. Letting $c_1 = 0$ and $c_2 = 1$ we get $x = e^{4t}$, $y = e^{4t}$. Eliminating the parameter we find $y = x$, $x > 0$. When $c_1 = 0$ and $c_2 = -1$ we find $y = x$, $x < 0$.

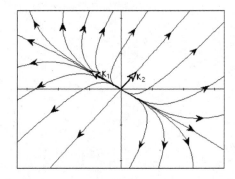

In Problem 4, letting $c_1 = 1$ and $c_2 = 0$ we get $x = -2e^{-7t/2}$, $y = e^{-7t/2}$. Eliminating the parameter we find $y = -\frac{1}{2}x$, $x < 0$. When $c_1 = -1$ and $c_2 = 0$ we find $y = -\frac{1}{2}x$, $x > 0$. Letting $c_1 = 0$ and $c_2 = 1$ we get $x = 4e^{-t}$, $y = 3e^{-t}$. Eliminating the parameter we find $y = \frac{3}{4}x$, $x > 0$. When $c_1 = 0$ and $c_2 = -1$ we find $y = \frac{3}{4}x$, $x < 0$.

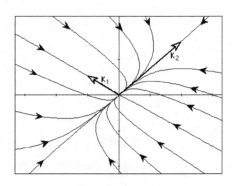

19. We have $\det(\mathbf{A} - \lambda\mathbf{I}) = \lambda^2 = 0$. For $\lambda_1 = 0$ we obtain

$$\mathbf{K} = \begin{pmatrix} 1 \\ 3 \end{pmatrix}.$$

A solution of $(\mathbf{A} - \lambda_1\mathbf{I})\mathbf{P} = \mathbf{K}$ is

$$\mathbf{P} = \begin{pmatrix} 1 \\ 2 \end{pmatrix}$$

so that

$$\mathbf{X} = c_1 \begin{pmatrix} 1 \\ 3 \end{pmatrix} + c_2 \left[\begin{pmatrix} 1 \\ 3 \end{pmatrix} t + \begin{pmatrix} 1 \\ 2 \end{pmatrix} \right].$$

20. We have $\det(\mathbf{A} - \lambda\mathbf{I}) = (\lambda + 1)^2 = 0$. For $\lambda_1 = -1$ we obtain

$$\mathbf{K} = \begin{pmatrix} 1 \\ 1 \end{pmatrix}.$$

A solution of $(\mathbf{A} - \lambda_1\mathbf{I})\mathbf{P} = \mathbf{K}$ is

$$\mathbf{P} = \begin{pmatrix} 0 \\ 1/5 \end{pmatrix}$$

so that

$$\mathbf{X} = c_1 \begin{pmatrix} 1 \\ 1 \end{pmatrix} e^{-t} + c_2 \left[\begin{pmatrix} 1 \\ 1 \end{pmatrix} te^{-t} + \begin{pmatrix} 0 \\ 1/5 \end{pmatrix} e^{-t} \right].$$

21. We have $\det(\mathbf{A} - \lambda\mathbf{I}) = (\lambda - 2)^2 = 0$. For $\lambda_1 = 2$ we obtain

$$\mathbf{K} = \begin{pmatrix} 1 \\ 1 \end{pmatrix}.$$

A solution of $(\mathbf{A} - \lambda_1\mathbf{I})\mathbf{P} = \mathbf{K}$ is

$$\mathbf{P} = \begin{pmatrix} -1/3 \\ 0 \end{pmatrix}$$

so that

$$\mathbf{X} = c_1 \begin{pmatrix} 1 \\ 1 \end{pmatrix} e^{2t} + c_2 \left[\begin{pmatrix} 1 \\ 1 \end{pmatrix} te^{2t} + \begin{pmatrix} -1/3 \\ 0 \end{pmatrix} e^{2t} \right].$$

22. We have $\det(\mathbf{A} - \lambda\mathbf{I}) = (\lambda - 6)^2 = 0$. For $\lambda_1 = 6$ we obtain

$$\mathbf{K} = \begin{pmatrix} 3 \\ 2 \end{pmatrix}.$$

A solution of $(\mathbf{A} - \lambda_1\mathbf{I})\mathbf{P} = \mathbf{K}$ is

$$\mathbf{P} = \begin{pmatrix} 1/2 \\ 0 \end{pmatrix}$$

so that

$$\mathbf{X} = c_1 \begin{pmatrix} 3 \\ 2 \end{pmatrix} e^{6t} + c_2 \left[\begin{pmatrix} 3 \\ 2 \end{pmatrix} te^{6t} + \begin{pmatrix} 1/2 \\ 0 \end{pmatrix} e^{6t} \right].$$

23. We have $\det(\mathbf{A} - \lambda\mathbf{I}) = (1 - \lambda)(\lambda - 2)^2 = 0$. For $\lambda_1 = 1$ we obtain

$$\mathbf{K}_1 = \begin{pmatrix} 1 \\ 1 \\ 1 \end{pmatrix}.$$

For $\lambda_2 = 2$ we obtain

$$\mathbf{K}_2 = \begin{pmatrix} 1 \\ 0 \\ 1 \end{pmatrix} \quad \text{and} \quad \mathbf{K}_3 = \begin{pmatrix} 1 \\ 1 \\ 0 \end{pmatrix}.$$

Then

$$\mathbf{X} = c_1 \begin{pmatrix} 1 \\ 1 \\ 1 \end{pmatrix} e^t + c_2 \begin{pmatrix} 1 \\ 0 \\ 1 \end{pmatrix} e^{2t} + c_3 \begin{pmatrix} 1 \\ 1 \\ 0 \end{pmatrix} e^{2t}.$$

24. We have $\det(\mathbf{A} - \lambda\mathbf{I}) = (\lambda - 8)(\lambda + 1)^2 = 0$. For $\lambda_1 = 8$ we obtain

$$\mathbf{K}_1 = \begin{pmatrix} 2 \\ 1 \\ 2 \end{pmatrix}.$$

For $\lambda_2 = -1$ we obtain

$$\mathbf{K}_2 = \begin{pmatrix} 0 \\ -2 \\ 1 \end{pmatrix} \quad \text{and} \quad \mathbf{K}_3 = \begin{pmatrix} 1 \\ -2 \\ 0 \end{pmatrix}.$$

Then

$$\mathbf{X} = c_1 \begin{pmatrix} 2 \\ 1 \\ 2 \end{pmatrix} e^{8t} + c_2 \begin{pmatrix} 0 \\ -2 \\ 1 \end{pmatrix} e^{-t} + c_3 \begin{pmatrix} 1 \\ -2 \\ 0 \end{pmatrix} e^{-t}.$$

25. We have $\det(\mathbf{A} - \lambda\mathbf{I}) = -\lambda(5 - \lambda)^2 = 0$. For $\lambda_1 = 0$ we obtain

$$\mathbf{K}_1 = \begin{pmatrix} -4 \\ -5 \\ 2 \end{pmatrix}.$$

For $\lambda_2 = 5$ we obtain

$$\mathbf{K} = \begin{pmatrix} -2 \\ 0 \\ 1 \end{pmatrix}.$$

A solution of $(\mathbf{A} - \lambda_2\mathbf{I})\mathbf{P} = \mathbf{K}$ is

$$\mathbf{P} = \begin{pmatrix} 5/2 \\ 1/2 \\ 0 \end{pmatrix}$$

so that

$$\mathbf{X} = c_1 \begin{pmatrix} -4 \\ -5 \\ 2 \end{pmatrix} + c_2 \begin{pmatrix} -2 \\ 0 \\ 1 \end{pmatrix} e^{5t} + c_3 \left[\begin{pmatrix} -2 \\ 0 \\ 1 \end{pmatrix} te^{5t} + \begin{pmatrix} 5/2 \\ 1/2 \\ 0 \end{pmatrix} e^{5t} \right].$$

26. We have $\det(\mathbf{A} - \lambda\mathbf{I}) = (1 - \lambda)(\lambda - 2)^2 = 0$. For $\lambda_1 = 1$ we obtain

$$\mathbf{K}_1 = \begin{pmatrix} 1 \\ 0 \\ 0 \end{pmatrix}.$$

For $\lambda_2 = 2$ we obtain

$$\mathbf{K} = \begin{pmatrix} 0 \\ -1 \\ 1 \end{pmatrix}.$$

A solution of $(\mathbf{A} - \lambda_2 \mathbf{I})\mathbf{P} = \mathbf{K}$ is

$$\mathbf{P} = \begin{pmatrix} 0 \\ -1 \\ 0 \end{pmatrix}$$

so that

$$\mathbf{X} = c_1 \begin{pmatrix} 1 \\ 0 \\ 0 \end{pmatrix} e^t + c_2 \begin{pmatrix} 0 \\ -1 \\ 1 \end{pmatrix} e^{2t} + c_3 \left[\begin{pmatrix} 0 \\ -1 \\ 1 \end{pmatrix} te^{2t} + \begin{pmatrix} 0 \\ -1 \\ 0 \end{pmatrix} e^{2t} \right].$$

27. We have $\det(\mathbf{A} - \lambda\mathbf{I}) = -(\lambda - 1)^3 = 0$. For $\lambda_1 = 1$ we obtain

$$\mathbf{K} = \begin{pmatrix} 0 \\ 1 \\ 1 \end{pmatrix}.$$

Solutions of $(\mathbf{A} - \lambda_1 \mathbf{I})\mathbf{P} = \mathbf{K}$ and $(\mathbf{A} - \lambda_1 \mathbf{I})\mathbf{Q} = \mathbf{P}$ are

$$\mathbf{P} = \begin{pmatrix} 0 \\ 1 \\ 0 \end{pmatrix} \quad \text{and} \quad \mathbf{Q} = \begin{pmatrix} 1/2 \\ 0 \\ 0 \end{pmatrix}$$

so that

$$\mathbf{X} = c_1 \begin{pmatrix} 0 \\ 1 \\ 1 \end{pmatrix} e^t + c_2 \left[\begin{pmatrix} 0 \\ 1 \\ 1 \end{pmatrix} te^t + \begin{pmatrix} 0 \\ 1 \\ 0 \end{pmatrix} e^t \right] + c_3 \left[\begin{pmatrix} 0 \\ 1 \\ 1 \end{pmatrix} \frac{t^2}{2} e^t + \begin{pmatrix} 0 \\ 1 \\ 0 \end{pmatrix} te^t + \begin{pmatrix} 1/2 \\ 0 \\ 0 \end{pmatrix} e^t \right].$$

28. We have $\det(\mathbf{A} - \lambda\mathbf{I}) = (\lambda - 4)^3 = 0$. For $\lambda_1 = 4$ we obtain

$$\mathbf{K} = \begin{pmatrix} 1 \\ 0 \\ 0 \end{pmatrix}.$$

Solutions of $(\mathbf{A} - \lambda_1 \mathbf{I})\mathbf{P} = \mathbf{K}$ and $(\mathbf{A} - \lambda_1 \mathbf{I})\mathbf{Q} = \mathbf{P}$ are

$$\mathbf{P} = \begin{pmatrix} 0 \\ 1 \\ 0 \end{pmatrix} \quad \text{and} \quad \mathbf{Q} = \begin{pmatrix} 0 \\ 0 \\ 1 \end{pmatrix}$$

so that

$$\mathbf{X} = c_1 \begin{pmatrix} 1 \\ 0 \\ 0 \end{pmatrix} e^{4t} + c_2 \left[\begin{pmatrix} 1 \\ 0 \\ 0 \end{pmatrix} te^{4t} + \begin{pmatrix} 0 \\ 1 \\ 0 \end{pmatrix} e^{4t} \right] + c_3 \left[\begin{pmatrix} 1 \\ 0 \\ 0 \end{pmatrix} \frac{t^2}{2} e^{4t} + \begin{pmatrix} 0 \\ 1 \\ 0 \end{pmatrix} te^{4t} + \begin{pmatrix} 0 \\ 0 \\ 1 \end{pmatrix} e^{4t} \right].$$

29. We have $\det(\mathbf{A} - \lambda\mathbf{I}) = (\lambda - 4)^2 = 0$. For $\lambda_1 = 4$ we obtain

$$\mathbf{K} = \begin{pmatrix} 2 \\ 1 \end{pmatrix}.$$

A solution of $(\mathbf{A} - \lambda_1\mathbf{I})\mathbf{P} = \mathbf{K}$ is

$$\mathbf{P} = \begin{pmatrix} 1 \\ 1 \end{pmatrix}$$

so that

$$\mathbf{X} = c_1 \begin{pmatrix} 2 \\ 1 \end{pmatrix} e^{4t} + c_2 \left[\begin{pmatrix} 2 \\ 1 \end{pmatrix} te^{4t} + \begin{pmatrix} 1 \\ 1 \end{pmatrix} e^{4t} \right].$$

If

$$\mathbf{X}(0) = \begin{pmatrix} -1 \\ 6 \end{pmatrix}$$

then $c_1 = -7$ and $c_2 = 13$.

30. We have $\det(\mathbf{A} - \lambda\mathbf{I}) = -(\lambda + 1)(\lambda - 1)^2 = 0$. For $\lambda_1 = -1$ we obtain

$$\mathbf{K}_1 = \begin{pmatrix} -1 \\ 0 \\ 1 \end{pmatrix}.$$

For $\lambda_2 = 1$ we obtain

$$\mathbf{K}_2 = \begin{pmatrix} 1 \\ 0 \\ 1 \end{pmatrix} \quad \text{and} \quad \mathbf{K}_3 = \begin{pmatrix} 0 \\ 1 \\ 0 \end{pmatrix}$$

so that

$$\mathbf{X} = c_1 \begin{pmatrix} -1 \\ 0 \\ 1 \end{pmatrix} e^{-t} + c_2 \begin{pmatrix} 1 \\ 0 \\ 1 \end{pmatrix} e^{t} + c_3 \begin{pmatrix} 0 \\ 1 \\ 0 \end{pmatrix} e^{t}.$$

If

$$\mathbf{X}(0) = \begin{pmatrix} 1 \\ 2 \\ 5 \end{pmatrix}$$

then $c_1 = 2$, $c_2 = 3$, and $c_3 = 2$.

31. In this case $\det(\mathbf{A} - \lambda\mathbf{I}) = (2 - \lambda)^5$, and $\lambda_1 = 2$ is an eigenvalue of multiplicity 5. Linearly independent eigenvectors are

$$\mathbf{K}_1 = \begin{pmatrix} 1 \\ 0 \\ 0 \\ 0 \\ 0 \end{pmatrix}, \quad \mathbf{K}_2 = \begin{pmatrix} 0 \\ 0 \\ 1 \\ 0 \\ 0 \end{pmatrix}, \quad \text{and} \quad \mathbf{K}_3 = \begin{pmatrix} 0 \\ 0 \\ 0 \\ 1 \\ 0 \end{pmatrix}.$$

32. In Problem 20 letting $c_1 = 1$ and $c_2 = 0$ we get $x = e^t$, $y = e^t$. Eliminating the parameter we find $y = x$, $x > 0$. When $c_1 = -1$ and $c_2 = 0$ we find $y = x$, $x < 0$.

In Problem 21 letting $c_1 = 1$ and $c_2 = 0$ we get $x = e^{2t}$, $y = e^{2t}$. Eliminating the parameter we find $y = x$, $x > 0$. When $c_1 = -1$ and $c_2 = 0$ we find $y = x$, $x < 0$.

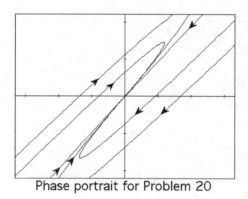
Phase portrait for Problem 20

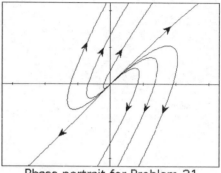
Phase portrait for Problem 21

In Problems 33-46 the form of the answer will vary according to the choice of eigenvector. For example, in Problem 33, if $\mathbf{K}_1$ is chosen to be $\begin{pmatrix} 1 \\ 2 - i \end{pmatrix}$ the solution has the form

$$\mathbf{X} = c_1 \begin{pmatrix} \cos t \\ 2\cos t + \sin t \end{pmatrix} e^{4t} + c_2 \begin{pmatrix} \sin t \\ 2\sin t - \cos t \end{pmatrix} e^{4t}.$$

33. We have $\det(\mathbf{A} - \lambda\mathbf{I}) = \lambda^2 - 8\lambda + 17 = 0$. For $\lambda_1 = 4 + i$ we obtain

$$\mathbf{K}_1 = \begin{pmatrix} 2 + i \\ 5 \end{pmatrix}$$

so that

$$\mathbf{X}_1 = \begin{pmatrix} 2 + i \\ 5 \end{pmatrix} e^{(4+i)t} = \begin{pmatrix} 2\cos t - \sin t \\ 5\cos t \end{pmatrix} e^{4t} + i \begin{pmatrix} \cos t + 2\sin t \\ 5\sin t \end{pmatrix} e^{4t}.$$

Then

$$\mathbf{X} = c_1 \begin{pmatrix} 2\cos t - \sin t \\ 5\cos t \end{pmatrix} e^{4t} + c_2 \begin{pmatrix} \cos t + 2\sin t \\ 5\sin t \end{pmatrix} e^{4t}.$$

34. We have $\det(\mathbf{A} - \lambda\mathbf{I}) = \lambda^2 + 1 = 0$. For $\lambda_1 = i$ we obtain

$$\mathbf{K}_1 = \begin{pmatrix} -1 - i \\ 2 \end{pmatrix}$$

so that

$$\mathbf{X}_1 = \begin{pmatrix} -1 - i \\ 2 \end{pmatrix} e^{it} = \begin{pmatrix} \sin t - \cos t \\ 2\cos t \end{pmatrix} + i \begin{pmatrix} -\cos t - \sin t \\ 2\sin t \end{pmatrix}.$$

Then
$$\mathbf{X} = c_1 \begin{pmatrix} \sin t - \cos t \\ 2\cos t \end{pmatrix} + c_2 \begin{pmatrix} -\cos t - \sin t \\ 2\sin t \end{pmatrix}.$$

35. We have $\det(\mathbf{A} - \lambda\mathbf{I}) = \lambda^2 - 8\lambda + 17 = 0$. For $\lambda_1 = 4 + i$ we obtain
$$\mathbf{K}_1 = \begin{pmatrix} -1 - i \\ 2 \end{pmatrix}$$

so that
$$\mathbf{X}_1 = \begin{pmatrix} -1 - i \\ 2 \end{pmatrix} e^{(4+i)t} = \begin{pmatrix} \sin t - \cos t \\ 2\cos t \end{pmatrix} e^{4t} + i \begin{pmatrix} -\sin t - \cos t \\ 2\sin t \end{pmatrix} e^{4t}.$$

Then
$$\mathbf{X} = c_1 \begin{pmatrix} \sin t - \cos t \\ 2\cos t \end{pmatrix} e^{4t} + c_2 \begin{pmatrix} -\sin t - \cos t \\ 2\sin t \end{pmatrix} e^{4t}.$$

36. We have $\det(\mathbf{A} - \lambda\mathbf{I}) = \lambda^2 - 10\lambda + 34 = 0$. For $\lambda_1 = 5 + 3i$ we obtain
$$\mathbf{K}_1 = \begin{pmatrix} 1 - 3i \\ 2 \end{pmatrix}$$

so that
$$\mathbf{X}_1 = \begin{pmatrix} 1 - 3i \\ 2 \end{pmatrix} e^{(5+3i)t} = \begin{pmatrix} \cos 3t + 3\sin 3t \\ 2\cos 3t \end{pmatrix} e^{5t} + i \begin{pmatrix} \sin 3t - 3\cos 3t \\ 2\sin 3t \end{pmatrix} e^{5t}.$$

Then
$$\mathbf{X} = c_1 \begin{pmatrix} \cos 3t + 3\sin 3t \\ 2\cos 3t \end{pmatrix} e^{5t} + c_2 \begin{pmatrix} \sin 3t - 3\cos 3t \\ 2\sin 3t \end{pmatrix} e^{5t}.$$

37. We have $\det(\mathbf{A} - \lambda\mathbf{I}) = \lambda^2 + 9 = 0$. For $\lambda_1 = 3i$ we obtain
$$\mathbf{K}_1 = \begin{pmatrix} 4 + 3i \\ 5 \end{pmatrix}$$

so that
$$\mathbf{X}_1 = \begin{pmatrix} 4 + 3i \\ 5 \end{pmatrix} e^{3it} = \begin{pmatrix} 4\cos 3t - 3\sin 3t \\ 5\cos 3t \end{pmatrix} + i \begin{pmatrix} 4\sin 3t + 3\cos 3t \\ 5\sin 3t \end{pmatrix}.$$

Then
$$\mathbf{X} = c_1 \begin{pmatrix} 4\cos 3t - 3\sin 3t \\ 5\cos 3t \end{pmatrix} + c_2 \begin{pmatrix} 4\sin 3t + 3\cos 3t \\ 5\sin 3t \end{pmatrix}.$$

38. We have $\det(\mathbf{A} - \lambda\mathbf{I}) = \lambda^2 + 2\lambda + 5 = 0$. For $\lambda_1 = -1 + 2i$ we obtain
$$\mathbf{K}_1 = \begin{pmatrix} 2 + 2i \\ 1 \end{pmatrix}$$

so that

$$\mathbf{X}_1 = \begin{pmatrix} 2 + 2i \\ 1 \end{pmatrix} e^{(-1+2i)t}$$

$$= \begin{pmatrix} 2\cos 2t - 2\sin 2t \\ \cos 2t \end{pmatrix} e^{-t} + i \begin{pmatrix} 2\cos 2t + 2\sin 2t \\ \sin 2t \end{pmatrix} e^{-t}.$$

Then

$$\mathbf{X} = c_1 \begin{pmatrix} 2\cos 2t - 2\sin 2t \\ \cos 2t \end{pmatrix} e^{-t} + c_2 \begin{pmatrix} 2\cos 2t + 2\sin 2t \\ \sin 2t \end{pmatrix} e^{-t}.$$

39. We have $\det(\mathbf{A} - \lambda\mathbf{I}) = -\lambda\left(\lambda^2 + 1\right) = 0$. For $\lambda_1 = 0$ we obtain

$$\mathbf{K}_1 = \begin{pmatrix} 1 \\ 0 \\ 0 \end{pmatrix}.$$

For $\lambda_2 = i$ we obtain

$$\mathbf{K}_2 = \begin{pmatrix} -i \\ i \\ 1 \end{pmatrix}$$

so that

$$\mathbf{X}_2 = \begin{pmatrix} -i \\ i \\ 1 \end{pmatrix} e^{it} = \begin{pmatrix} \sin t \\ -\sin t \\ \cos t \end{pmatrix} + i \begin{pmatrix} -\cos t \\ \cos t \\ \sin t \end{pmatrix}.$$

Then

$$\mathbf{X} = c_1 \begin{pmatrix} 1 \\ 0 \\ 0 \end{pmatrix} + c_2 \begin{pmatrix} \sin t \\ -\sin t \\ \cos t \end{pmatrix} + c_3 \begin{pmatrix} -\cos t \\ \cos t \\ \sin t \end{pmatrix}.$$

40. We have $\det(\mathbf{A} - \lambda\mathbf{I}) = -(\lambda+3)(\lambda^2 - 2\lambda + 5) = 0$. For $\lambda_1 = -3$ we obtain

$$\mathbf{K}_1 = \begin{pmatrix} 0 \\ -2 \\ 1 \end{pmatrix}.$$

For $\lambda_2 = 1 + 2i$ we obtain

$$\mathbf{K}_2 = \begin{pmatrix} -2 - i \\ -3i \\ 2 \end{pmatrix}$$

so that

$$\mathbf{X}_2 = \begin{pmatrix} -2\cos 2t + \sin 2t \\ 3\sin 2t \\ 2\cos 2t \end{pmatrix} e^t + i \begin{pmatrix} -\cos 2t - 2\sin 2t \\ -3\cos 2t \\ 2\sin 2t \end{pmatrix} e^t.$$

Then

$$\mathbf{X} = c_1 \begin{pmatrix} 0 \\ -2 \\ 1 \end{pmatrix} e^{-3t} + c_2 \begin{pmatrix} -2\cos 2t + \sin 2t \\ 3\sin 2t \\ 2\cos 2t \end{pmatrix} e^t + c_3 \begin{pmatrix} -\cos 2t - 2\sin 2t \\ -3\cos 2t \\ 2\sin 2t \end{pmatrix} e^t.$$

41. We have $\det(\mathbf{A} - \lambda\mathbf{I}) = (1 - \lambda)(\lambda^2 - 2\lambda + 2) = 0$. For $\lambda_1 = 1$ we obtain

$$\mathbf{K}_1 = \begin{pmatrix} 0 \\ 2 \\ 1 \end{pmatrix}.$$

For $\lambda_2 = 1 + i$ we obtain

$$\mathbf{K}_2 = \begin{pmatrix} 1 \\ i \\ i \end{pmatrix}$$

so that

$$\mathbf{X}_2 = \begin{pmatrix} 1 \\ i \\ i \end{pmatrix} e^{(1+i)t} = \begin{pmatrix} \cos t \\ -\sin t \\ -\sin t \end{pmatrix} e^t + i \begin{pmatrix} \sin t \\ \cos t \\ \cos t \end{pmatrix} e^t.$$

Then

$$\mathbf{X} = c_1 \begin{pmatrix} 0 \\ 2 \\ 1 \end{pmatrix} e^t + c_2 \begin{pmatrix} \cos t \\ -\sin t \\ -\sin t \end{pmatrix} e^t + c_3 \begin{pmatrix} \sin t \\ \cos t \\ \cos t \end{pmatrix} e^t.$$

42. We have $\det(\mathbf{A} - \lambda\mathbf{I}) = -(\lambda - 6)(\lambda^2 - 8\lambda + 20) = 0$. For $\lambda_1 = 6$ we obtain

$$\mathbf{K}_1 = \begin{pmatrix} 0 \\ 1 \\ 0 \end{pmatrix}.$$

For $\lambda_2 = 4 + 2i$ we obtain

$$\mathbf{K}_2 = \begin{pmatrix} -i \\ 0 \\ 2 \end{pmatrix}$$

so that

$$\mathbf{X}_2 = \begin{pmatrix} -i \\ 0 \\ 2 \end{pmatrix} e^{(4+2i)t} = \begin{pmatrix} \sin 2t \\ 0 \\ 2\cos 2t \end{pmatrix} e^{4t} + i \begin{pmatrix} -\cos 2t \\ 0 \\ 2\sin 2t \end{pmatrix} e^{4t}.$$

Then

$$\mathbf{X} = c_1 \begin{pmatrix} 0 \\ 1 \\ 0 \end{pmatrix} e^{6t} + c_2 \begin{pmatrix} \sin 2t \\ 0 \\ 2\cos 2t \end{pmatrix} e^{4t} + c_3 \begin{pmatrix} -\cos 2t \\ 0 \\ 2\sin 2t \end{pmatrix} e^{4t}.$$

43. We have $\det(\mathbf{A} - \lambda\mathbf{I}) = (2 - \lambda)(\lambda^2 + 4\lambda + 13) = 0$. For $\lambda_1 = 2$ we obtain

$$\mathbf{K}_1 = \begin{pmatrix} 28 \\ -5 \\ 25 \end{pmatrix}.$$

For $\lambda_2 = -2 + 3i$ we obtain

$$\mathbf{K}_2 = \begin{pmatrix} 4 + 3i \\ -5 \\ 0 \end{pmatrix}$$

so that

$$\mathbf{X}_2 = \begin{pmatrix} 4 + 3i \\ -5 \\ 0 \end{pmatrix} e^{(-2+3i)t} = \begin{pmatrix} 4\cos 3t - 3\sin 3t \\ -5\cos 3t \\ 0 \end{pmatrix} e^{-2t} + i \begin{pmatrix} 4\sin 3t + 3\cos 3t \\ -5\sin 3t \\ 0 \end{pmatrix} e^{-2t}.$$

Then

$$\mathbf{X} = c_1 \begin{pmatrix} 28 \\ -5 \\ 25 \end{pmatrix} e^{2t} + c_2 \begin{pmatrix} 4\cos 3t - 3\sin 3t \\ -5\cos 3t \\ 0 \end{pmatrix} e^{-2t} + c_3 \begin{pmatrix} 4\sin 3t + 3\cos 3t \\ -5\sin 3t \\ 0 \end{pmatrix} e^{-2t}.$$

44. We have $\det(\mathbf{A} - \lambda\mathbf{I}) = -(\lambda + 2)(\lambda^2 + 4) = 0$. For $\lambda_1 = -2$ we obtain

$$\mathbf{K}_1 = \begin{pmatrix} 0 \\ -1 \\ 1 \end{pmatrix}.$$

For $\lambda_2 = 2i$ we obtain

$$\mathbf{K}_2 = \begin{pmatrix} -2 - 2i \\ 1 \\ 1 \end{pmatrix}$$

so that

$$\mathbf{X}_2 = \begin{pmatrix} -2 - 2i \\ 1 \\ 1 \end{pmatrix} e^{2it} = \begin{pmatrix} -2\cos 2t + 2\sin 2t \\ \cos 2t \\ \cos 2t \end{pmatrix} + i \begin{pmatrix} -2\cos 2t - 2\sin 2t \\ \sin 2t \\ \sin 2t \end{pmatrix}.$$

Then

$$\mathbf{X} = c_1 \begin{pmatrix} 0 \\ -1 \\ 1 \end{pmatrix} e^{-2t} + c_2 \begin{pmatrix} -2\cos 2t + 2\sin 2t \\ \cos 2t \\ \cos 2t \end{pmatrix} + c_3 \begin{pmatrix} -2\cos 2t - 2\sin 2t \\ \sin 2t \\ \sin 2t \end{pmatrix}.$$

45. We have $\det(\mathbf{A} - \lambda\mathbf{I}) = (1 - \lambda)(\lambda^2 + 25) = 0$. For $\lambda_1 = 1$ we obtain

$$\mathbf{K}_1 = \begin{pmatrix} 25 \\ -7 \\ 6 \end{pmatrix}.$$

For $\lambda_2 = 5i$ we obtain

$$\mathbf{K}_2 = \begin{pmatrix} 1 + 5i \\ 1 \\ 1 \end{pmatrix}$$

so that

$$\mathbf{X}_2 = \begin{pmatrix} 1 + 5i \\ 1 \\ 1 \end{pmatrix} e^{5it} = \begin{pmatrix} \cos 5t - 5\sin 5t \\ \cos 5t \\ \cos 5t \end{pmatrix} + i \begin{pmatrix} \sin 5t + 5\cos 5t \\ \sin 5t \\ \sin 5t \end{pmatrix}.$$

Then

$$\mathbf{X} = c_1 \begin{pmatrix} 25 \\ -7 \\ 6 \end{pmatrix} e^t + c_2 \begin{pmatrix} \cos 5t - 5\sin 5t \\ \cos 5t \\ \cos 5t \end{pmatrix} + c_3 \begin{pmatrix} \sin 5t + 5\cos 5t \\ \sin 5t \\ \sin 5t \end{pmatrix}.$$

If

$$\mathbf{X}(0) = \begin{pmatrix} 4 \\ 6 \\ -7 \end{pmatrix}$$

then $c_1 = c_2 = -1$ and $c_3 = 6$.

46. We have $\det(\mathbf{A} - \lambda\mathbf{I}) = \lambda^2 - 10\lambda + 29 = 0$. For $\lambda_1 = 5 + 2i$ we obtain

$$\mathbf{K}_1 = \begin{pmatrix} 1 \\ 1 - 2i \end{pmatrix}$$

so that

$$\mathbf{X}_1 = \begin{pmatrix} 1 \\ 1 - 2i \end{pmatrix} e^{(5+2i)t} = \begin{pmatrix} \cos 2t \\ \cos 2t + 2\sin 2t \end{pmatrix} e^{5t} + i \begin{pmatrix} \sin 2t \\ \sin 2t - 2\cos 2t \end{pmatrix} e^{5t}.$$

and

$$\mathbf{X} = c_1 \begin{pmatrix} \cos 2t \\ \cos 2t + 2\sin 2t \end{pmatrix} e^{5t} + c_3 \begin{pmatrix} \sin 2t \\ \sin 2t - 2\cos 2t \end{pmatrix} e^{5t}.$$

If $\mathbf{X}(0) = \begin{pmatrix} -2 \\ 8 \end{pmatrix}$, then $c_1 = -2$ and $c_2 = 5$.

47.

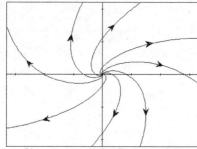

Phase portrait for Problem 36

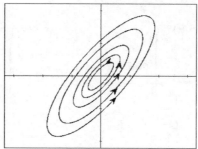

Phase portrait for Problem 37

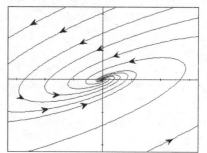

Phase portrait for Problem 38

48. (a) Letting $x_1 = y_1$, $x'_1 = y_2$, $x_2 = y_3$, and $x'_2 = y_4$ we have

$$y'_2 = x''_1 = -10x_1 + 4x_2 = -10y_1 + 4y_3$$

$$y'_4 = x''_2 = 4x_1 - 4x_2 = 4y_1 - 4y_3.$$

The corresponding linear system is

$$y'_1 = y_2$$

$$y'_2 = -10y_1 + 4y_3$$

$$y'_3 = y_4$$

$$y'_4 = 4y_1 - 4y_3$$

or

$$\mathbf{Y}' = \begin{pmatrix} 0 & 1 & 0 & 0 \\ -10 & 0 & 4 & 0 \\ 0 & 0 & 0 & 1 \\ 4 & 0 & -4 & 0 \end{pmatrix} \mathbf{Y}.$$

Using a CAS, we find eigenvalues $\pm\sqrt{2}i$ and $\pm2\sqrt{3}i$ with corresponding eigenvectors

$$\begin{pmatrix} \mp\sqrt{2}i/4 \\ 1/2 \\ \mp\sqrt{2}i/2 \\ 1 \end{pmatrix} = \begin{pmatrix} 0 \\ 1/2 \\ 0 \\ 1 \end{pmatrix} + i \begin{pmatrix} \mp\sqrt{2}/4 \\ 0 \\ \mp\sqrt{2}/2 \\ 0 \end{pmatrix}$$

and

$$\begin{pmatrix} \pm\sqrt{3}i/3 \\ -2 \\ \mp\sqrt{3}i/6 \\ 1 \end{pmatrix} = \begin{pmatrix} 0 \\ -2 \\ 0 \\ 1 \end{pmatrix} + i \begin{pmatrix} \pm\sqrt{3}/3 \\ 0 \\ \mp\sqrt{3}/6 \\ 0 \end{pmatrix}.$$

439

Thus

$$\mathbf{Y}(t) = c_1 \left[\begin{pmatrix} 0 \\ 1/2 \\ 0 \\ 1 \end{pmatrix} \cos \sqrt{2}t - \begin{pmatrix} -\sqrt{2}/4 \\ 0 \\ -\sqrt{2}/2 \\ 0 \end{pmatrix} \sin \sqrt{2}t \right]$$

$$+ c_2 \left[\begin{pmatrix} -\sqrt{2}/4 \\ 0 \\ -\sqrt{2}/2 \\ 0 \end{pmatrix} \cos \sqrt{2}t + \begin{pmatrix} 0 \\ 1/2 \\ 0 \\ 1 \end{pmatrix} \sin \sqrt{2}t \right]$$

$$+ c_3 \left[\begin{pmatrix} 0 \\ -2 \\ 0 \\ 1 \end{pmatrix} \cos 2\sqrt{3}t - \begin{pmatrix} \sqrt{3}/3 \\ 0 \\ -\sqrt{3}/6 \\ 0 \end{pmatrix} \sin 2\sqrt{3}t \right]$$

$$+ c_4 \left[\begin{pmatrix} \sqrt{3}/3 \\ 0 \\ -\sqrt{3}/6 \\ 0 \end{pmatrix} \cos 2\sqrt{3}t + \begin{pmatrix} 0 \\ -2 \\ 0 \\ 1 \end{pmatrix} \sin 2\sqrt{3}t \right].$$

The initial conditions $y_1(0) = 0$, $y_2(0) = 1$, $y_3(0) = 0$, and $y_4(0) = -1$ imply $c_1 = -\frac{2}{5}$, $c_2 = 0$, $c_3 = -\frac{3}{5}$, and $c_4 = 0$. Thus,

$$x_1(t) = y_1(t) = -\frac{\sqrt{2}}{10} \sin \sqrt{2}t + \frac{\sqrt{3}}{5} \sin 2\sqrt{3}t$$

$$x_2(t) = y_3(t) = -\frac{\sqrt{2}}{5} \sin \sqrt{2}t - \frac{\sqrt{3}}{10} \sin 2\sqrt{3}t.$$

(b) The second-order system is

$$x_1'' = -10x_1 + 4x_2$$

$$x_2'' = 4x_1 - 4x_2$$

or

$$\mathbf{X}'' = \begin{pmatrix} -10 & 4 \\ 4 & -4 \end{pmatrix} \mathbf{X}.$$

We assume solutions of the form $\mathbf{X} = \mathbf{V} \cos \omega t$ and $\mathbf{X} = \mathbf{V} \sin \omega t$. Since the eigenvalues are -2 and -12, $\omega_1 = \sqrt{-(-2)} = \sqrt{2}$ and $\omega_2 = \sqrt{-(-12)} = 2\sqrt{3}$. The corresponding eigenvectors are

$$\mathbf{V}_1 = \begin{pmatrix} 1 \\ 2 \end{pmatrix} \quad \text{and} \quad \mathbf{V}_2 = \begin{pmatrix} -2 \\ 1 \end{pmatrix}.$$

Then, the general solution of the system is

$$\mathbf{X} = c_1 \begin{pmatrix} 1 \\ 2 \end{pmatrix} \cos \sqrt{2}t + c_2 \begin{pmatrix} 1 \\ 2 \end{pmatrix} \sin \sqrt{2}t + c_3 \begin{pmatrix} -2 \\ 1 \end{pmatrix} \cos 2\sqrt{3}t + c_4 \begin{pmatrix} -2 \\ 1 \end{pmatrix} \sin 2\sqrt{3}t.$$

The initial conditions

$$\mathbf{X}(0) = \begin{pmatrix} 0 \\ 0 \end{pmatrix} \qquad \text{and} \qquad \mathbf{X}'(0) = \begin{pmatrix} 1 \\ -1 \end{pmatrix}$$

imply $c_1 = 0$, $c_2 = -\sqrt{2}/10$, $c_3 = 0$, and $c_4 = -\sqrt{3}/10$. Thus

$$x_1(t) = -\frac{\sqrt{2}}{10} \sin \sqrt{2}t + \frac{\sqrt{3}}{5} \sin 2\sqrt{3}t$$

$$x_2(t) = -\frac{\sqrt{2}}{5} \sin \sqrt{2}t - \frac{\sqrt{3}}{10} \sin 2\sqrt{3}t.$$

49. (a) From $\det(\mathbf{A} - \lambda\mathbf{I}) = \lambda(\lambda - 2) = 0$ we get $\lambda_1 = 0$ and $\lambda_2 = 2$. For $\lambda_1 = 0$ we obtain

$$\begin{pmatrix} 1 & 1 & | & 0 \\ 1 & 1 & | & 0 \end{pmatrix} \implies \begin{pmatrix} 1 & 1 & | & 0 \\ 0 & 0 & | & 0 \end{pmatrix} \qquad \text{so that} \quad \mathbf{K}_1 = \begin{pmatrix} -1 \\ 1 \end{pmatrix}.$$

For $\lambda_2 = 2$ we obtain

$$\begin{pmatrix} -1 & 1 & | & 0 \\ 1 & -1 & | & 0 \end{pmatrix} \implies \begin{pmatrix} -1 & 1 & | & 0 \\ 0 & 0 & | & 0 \end{pmatrix} \qquad \text{so that} \quad \mathbf{K}_2 = \begin{pmatrix} 1 \\ 1 \end{pmatrix}.$$

Then

$$\mathbf{X} = c_1 \begin{pmatrix} -1 \\ 1 \end{pmatrix} + c_2 \begin{pmatrix} 1 \\ 1 \end{pmatrix} e^{2t}.$$

The line $y = -x$ is not a trajectory of the system. Trajectories are $x = -c_1 + c_2 e^{2t}$, $y = c_1 + c_2 e^{2t}$ or $y = x + 2c_1$. This is a family of lines perpendicular to the line $y = -x$. All of the constant solutions of the system do, however, lie on the line $y = -x$.

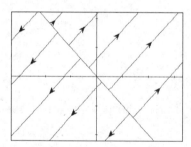

(b) From $\det(\mathbf{A} - \lambda\mathbf{I}) = \lambda^2 = 0$ we get $\lambda_1 = 0$ and

$$\mathbf{K} = \begin{pmatrix} -1 \\ 1 \end{pmatrix}.$$

A solution of $(\mathbf{A} - \lambda_1\mathbf{I})\mathbf{P} = \mathbf{K}$ is

$$\mathbf{P} = \begin{pmatrix} -1 \\ 0 \end{pmatrix}$$

so that

$$\mathbf{X} = c_1 \begin{pmatrix} -1 \\ 1 \end{pmatrix} + c_2 \left[\begin{pmatrix} -1 \\ 1 \end{pmatrix} t + \begin{pmatrix} -1 \\ 0 \end{pmatrix} \right].$$

All trajectories are parallel to $y = -x$, but $y = -x$ is not a trajectory. There are constant solutions of the system, however, that do lie on the line $y = -x$.

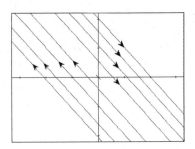

50. The system of differential equations is

$$x_1' = 2x_1 + x_2$$

$$x_2' = 2x_2$$

$$x_3' = 2x_3$$

$$x_4' = 2x_4 + x_5$$

$$x_5' = 2x_5.$$

We see immediately that $x_2 = c_2 e^{2t}$, $x_3 = c_3 e^{2t}$, and $x_5 = c_5 e^{2t}$. Then

$$x_1' = 2x_1 + c_2 e^{2t} \qquad \text{so} \qquad x_1 = c_2 t e^{2t} + c_1 e^{2t},$$

and

$$x_4' = 2x_4 + c_5 e^{2t} \qquad \text{so} \qquad x_4 = c_5 t e^{2t} + c_4 e^{2t}.$$

The general solution of the system is

$$\mathbf{X} = \begin{pmatrix} c_2 t e^{2t} + c_1 e^{2t} \\ c_2 e^{2t} \\ c_3 e^{2t} \\ c_5 t e^{2t} + c_4 e^{2t} \\ c_5 e^{2t} \end{pmatrix}$$

$$
= c_1 \begin{pmatrix} 1 \\ 0 \\ 0 \\ 0 \\ 0 \end{pmatrix} e^{2t} + c_2 \left[\begin{pmatrix} 1 \\ 0 \\ 0 \\ 0 \\ 0 \end{pmatrix} te^{2t} + \begin{pmatrix} 0 \\ 1 \\ 0 \\ 0 \\ 0 \end{pmatrix} e^{2t} \right]
$$

$$
+ c_3 \begin{pmatrix} 0 \\ 0 \\ 1 \\ 0 \\ 0 \end{pmatrix} e^{2t} + c_4 \begin{pmatrix} 0 \\ 0 \\ 0 \\ 1 \\ 0 \end{pmatrix} e^{2t} + c_5 \left[\begin{pmatrix} 0 \\ 0 \\ 0 \\ 1 \\ 0 \end{pmatrix} te^{2t} + \begin{pmatrix} 0 \\ 0 \\ 0 \\ 0 \\ 1 \end{pmatrix} e^{2t} \right]
$$

$$
= c_1 \mathbf{K}_1 e^{2t} + c_2 \left[\mathbf{K}_1 te^{2t} + \begin{pmatrix} 0 \\ 1 \\ 0 \\ 0 \\ 0 \end{pmatrix} e^{2t} \right]
$$

$$
+ c_3 \mathbf{K}_2 e^{2t} + c_4 \mathbf{K}_3 e^{2t} + c_5 \left[\mathbf{K}_3 te^{2t} + \begin{pmatrix} 0 \\ 0 \\ 0 \\ 0 \\ 1 \end{pmatrix} e^{2t} \right].
$$

There are three solutions of the form $\mathbf{X} = \mathbf{K}e^{2t}$, where $\mathbf{K}$ is an eigenvector, and two solutions of the form $\mathbf{X} = \mathbf{K}te^{2t} + \mathbf{P}e^{2t}$. See (12) in the text. From (13) and (14) in the text

$$
(\mathbf{A} - 2\mathbf{I})\mathbf{K}_1 = \mathbf{0}
$$

and

$$
(\mathbf{A} - 2\mathbf{I})\mathbf{K}_2 = \mathbf{K}_1.
$$

This implies

$$
\begin{pmatrix} 0 & 1 & 0 & 0 & 0 \\ 0 & 0 & 0 & 0 & 0 \\ 0 & 0 & 0 & 0 & 0 \\ 0 & 0 & 0 & 0 & 1 \\ 0 & 0 & 0 & 0 & 0 \end{pmatrix} \begin{pmatrix} p_1 \\ p_2 \\ p_3 \\ p_4 \\ p_5 \end{pmatrix} = \begin{pmatrix} 1 \\ 0 \\ 0 \\ 0 \\ 0 \end{pmatrix},
$$

so $p_2 = 1$ and $p_5 = 0$, while p_1, p_3, and p_4 are arbitrary. Choosing $p_1 = p_3 = p_4 = 0$ we have

$$\mathbf{P} = \begin{pmatrix} 1 \\ 0 \\ 0 \\ 0 \\ 0 \end{pmatrix}.$$

Therefore a solution is

$$\mathbf{X} = \begin{pmatrix} 1 \\ 0 \\ 0 \\ 0 \\ 0 \end{pmatrix} te^{2t} + \begin{pmatrix} 1 \\ 0 \\ 0 \\ 0 \\ 0 \end{pmatrix} e^{2t}.$$

Repeating for $\mathbf{K}_3$ we find

$$\mathbf{P} = \begin{pmatrix} 0 \\ 0 \\ 0 \\ 0 \\ 1 \end{pmatrix},$$

so another solution is

$$\mathbf{X} = \begin{pmatrix} 0 \\ 0 \\ 0 \\ 1 \\ 0 \end{pmatrix} te^{2t} + \begin{pmatrix} 0 \\ 0 \\ 0 \\ 0 \\ 1 \end{pmatrix} e^{2t}.$$

51. From $x = 2\cos 2t - 2\sin 2t$, $y = -\cos 2t$ we find $x + 2y = -2\sin 2t$. Then

$$(x + 2y)^2 = 4\sin^2 2t = 4(1 - \cos^2 2t) = 4 - 4\cos^2 2t = 4 - 4y^2$$

and

$$x^2 + 4xy + 4y^2 = 4 - 4y^2 \qquad \text{or} \qquad x^2 + 4xy + 8y^2 = 4.$$

This is a rotated conic section and, from the discriminant $b^2 - 4ac = 16 - 32 < 0$, we see that the curve is an ellipse.

52. Suppose the eigenvalues are $\alpha \pm i\beta$, $\beta > 0$. In Problem 36 the eigenvalues are $5 \pm 3i$, in Problem 37 they are $\pm 3i$, and in Problem 38 they are $-1 \pm 2i$. From Problem 47 we deduce that the phase portrait will consist of a family of closed curves when $\alpha = 0$ and spirals when $\alpha \neq 0$. The origin will be a repellor when $\alpha > 0$, and an attractor when $\alpha < 0$.

1. Solving

$$\det(\mathbf{A} - \lambda\mathbf{I}) = \begin{vmatrix} 2-\lambda & 3 \\ -1 & -2-\lambda \end{vmatrix} = \lambda^2 - 1 = (\lambda - 1)(\lambda + 1) = 0$$

we obtain eigenvalues $\lambda_1 = -1$ and $\lambda_2 = 1$. Corresponding eigenvectors are

$$\mathbf{K}_1 = \begin{pmatrix} -1 \\ 1 \end{pmatrix} \quad \text{and} \quad \mathbf{K}_2 = \begin{pmatrix} -3 \\ 1 \end{pmatrix}.$$

Thus

$$\mathbf{X}_c = c_1 \begin{pmatrix} -1 \\ 1 \end{pmatrix} e^{-t} + c_2 \begin{pmatrix} -3 \\ 1 \end{pmatrix} e^t.$$

Substituting

$$\mathbf{X}_p = \begin{pmatrix} a_1 \\ b_1 \end{pmatrix}$$

into the system yields

$$2a_1 + 3b_1 = 7$$

$$-a_1 - 2b_1 = -5,$$

from which we obtain $a_1 = -1$ and $b_1 = 3$. Then

$$\mathbf{X}(t) = c_1 \begin{pmatrix} -1 \\ 1 \end{pmatrix} e^{-t} + c_2 \begin{pmatrix} -3 \\ 1 \end{pmatrix} e^t + \begin{pmatrix} -1 \\ 3 \end{pmatrix}.$$

2. Solving

$$\det(\mathbf{A} - \lambda\mathbf{I}) = \begin{vmatrix} 5-\lambda & 9 \\ -1 & 11-\lambda \end{vmatrix} = \lambda^2 - 16\lambda + 64 = (\lambda - 8)^2 = 0$$

we obtain the eigenvalue $\lambda = 8$. A corresponding eigenvector is

$$\mathbf{K} = \begin{pmatrix} 3 \\ 1 \end{pmatrix}.$$

Solving $(\mathbf{A} - 8\mathbf{I})\mathbf{P} = \mathbf{K}$ we obtain

$$\mathbf{P} = \begin{pmatrix} 2 \\ 1 \end{pmatrix}.$$

Thus

$$\mathbf{X}_c = c_1 \begin{pmatrix} 3 \\ 1 \end{pmatrix} e^{8t} + c_2 \left[\begin{pmatrix} 3 \\ 1 \end{pmatrix} te^{8t} + \begin{pmatrix} 2 \\ 1 \end{pmatrix} e^{8t} \right].$$

Substituting

$$\mathbf{X}_p = \begin{pmatrix} a_1 \\ b_1 \end{pmatrix}$$

into the system yields

$$5a_1 + 9b_1 = -2$$

$$-a_1 + 11b_1 = -6,$$

from which we obtain $a_1 = 1/2$ and $b_1 = -1/2$. Then

$$\mathbf{X}(t) = c_1 \begin{pmatrix} 3 \\ 1 \end{pmatrix} e^{8t} + c_2 \left[\begin{pmatrix} 3 \\ 1 \end{pmatrix} t e^{8t} + \begin{pmatrix} 2 \\ 1 \end{pmatrix} e^{8t} \right] + \begin{pmatrix} 1/2 \\ -1/2 \end{pmatrix}.$$

3. Solving

$$\det(\mathbf{A} - \lambda \mathbf{I}) = \begin{vmatrix} 1 - \lambda & 3 \\ 3 & 1 - \lambda \end{vmatrix} = \lambda^2 - 2\lambda - 8 = (\lambda - 4)(\lambda + 2) = 0$$

we obtain eigenvalues $\lambda_1 = -2$ and $\lambda_2 = 4$. Corresponding eigenvectors are

$$\mathbf{K}_1 = \begin{pmatrix} 1 \\ -1 \end{pmatrix} \quad \text{and} \quad \mathbf{K}_2 = \begin{pmatrix} 1 \\ 1 \end{pmatrix}.$$

Thus

$$\mathbf{X}_c = c_1 \begin{pmatrix} 1 \\ -1 \end{pmatrix} e^{-2t} + c_2 \begin{pmatrix} 1 \\ 1 \end{pmatrix} e^{4t}.$$

Substituting

$$\mathbf{X}_p = \begin{pmatrix} a_3 \\ b_3 \end{pmatrix} t^2 + \begin{pmatrix} a_2 \\ b_2 \end{pmatrix} t + \begin{pmatrix} a_1 \\ b_1 \end{pmatrix}$$

into the system yields

$$a_3 + 3b_3 = 2 \qquad a_2 + 3b_2 = 2a_3 \qquad a_1 + 3b_1 = a_2$$

$$3a_3 + b_3 = 0 \qquad 3a_2 + b_2 + 1 = 2b_3 \qquad 3a_1 + b_1 + 5 = b_2$$

from which we obtain $a_3 = -1/4$, $b_3 = 3/4$, $a_2 = 1/4$, $b_2 = -1/4$, $a_1 = -2$, and $b_1 = 3/4$. Then

$$\mathbf{X}(t) = c_1 \begin{pmatrix} 1 \\ -1 \end{pmatrix} e^{-2t} + c_2 \begin{pmatrix} 1 \\ 1 \end{pmatrix} e^{4t} + \begin{pmatrix} -1/4 \\ 3/4 \end{pmatrix} t^2 + \begin{pmatrix} 1/4 \\ -1/4 \end{pmatrix} t + \begin{pmatrix} -2 \\ 3/4 \end{pmatrix}.$$

4. Solving

$$\det(\mathbf{A} - \lambda \mathbf{I}) = \begin{vmatrix} 1 - \lambda & -4 \\ 4 & 1 - \lambda \end{vmatrix} = \lambda^2 - 2\lambda + 17 = 0$$

we obtain eigenvalues $\lambda_1 = 1 + 4i$ and $\lambda_2 = 1 - 4i$. Corresponding eigenvectors are

$$\mathbf{K}_1 = \begin{pmatrix} i \\ 1 \end{pmatrix} \quad \text{and} \quad \mathbf{K}_2 = \begin{pmatrix} -i \\ 1 \end{pmatrix}.$$

Thus

$$\mathbf{X}_c = c_1 \left[\begin{pmatrix} 0 \\ 1 \end{pmatrix} \cos 4t + \begin{pmatrix} -1 \\ 0 \end{pmatrix} \sin 4t \right] e^t + c_2 \left[\begin{pmatrix} -1 \\ 0 \end{pmatrix} \cos 4t - \begin{pmatrix} 0 \\ 1 \end{pmatrix} \sin 4t \right] e^t$$

$$= c_1 \begin{pmatrix} -\sin 4t \\ \cos 4t \end{pmatrix} e^t + c_2 \begin{pmatrix} -\cos 4t \\ -\sin 4t \end{pmatrix} e^t.$$

Substituting

$$\mathbf{X}_p = \begin{pmatrix} a_3 \\ b_3 \end{pmatrix} t + \begin{pmatrix} a_2 \\ b_2 \end{pmatrix} + \begin{pmatrix} a_1 \\ b_1 \end{pmatrix} e^{6t}$$

into the system yields

$$a_3 - 4b_3 = -4 \qquad a_2 - 4b_2 = a_3 \qquad -5a_1 - 4b_1 = -9$$

$$4a_3 + b_3 = 1 \qquad 4a_2 + b_2 = b_3 \qquad 4a_1 - 5b_1 = -1$$

from which we obtain $a_3 = 0$, $b_3 = 1$, $a_2 = 4/17$, $b_2 = 1/17$, $a_1 = 1$, and $b_1 = 1$. Then

$$\mathbf{X}(t) = c_1 \begin{pmatrix} -\sin 4t \\ \cos 4t \end{pmatrix} e^t + c_2 \begin{pmatrix} -\cos 4t \\ -\sin 4t \end{pmatrix} e^t + \begin{pmatrix} 0 \\ 1 \end{pmatrix} t + \begin{pmatrix} 4/17 \\ 1/17 \end{pmatrix} + \begin{pmatrix} 1 \\ 1 \end{pmatrix} e^{6t}.$$

5. Solving

$$\det(\mathbf{A} - \lambda\mathbf{I}) = \begin{vmatrix} 4 - \lambda & 1/3 \\ 9 & 6 - \lambda \end{vmatrix} = \lambda^2 - 10\lambda + 21 = (\lambda - 3)(\lambda - 7) = 0$$

we obtain the eigenvalues $\lambda_1 = 3$ and $\lambda_2 = 7$. Corresponding eigenvectors are

$$\mathbf{K}_1 = \begin{pmatrix} 1 \\ -3 \end{pmatrix} \quad \text{and} \quad \mathbf{K}_2 = \begin{pmatrix} 1 \\ 9 \end{pmatrix}.$$

Thus

$$\mathbf{X}_c = c_1 \begin{pmatrix} 1 \\ -3 \end{pmatrix} e^{3t} + c_2 \begin{pmatrix} 1 \\ 9 \end{pmatrix} e^{7t}.$$

Substituting

$$\mathbf{X}_p = \begin{pmatrix} a_1 \\ b_1 \end{pmatrix} e^t$$

into the system yields

$$3a_1 + \frac{1}{3}b_1 = 3$$

$$9a_1 + 5b_1 = -10$$

from which we obtain $a_1 = 55/36$ and $b_1 = -19/4$. Then

$$\mathbf{X}(t) = c_1 \begin{pmatrix} 1 \\ -3 \end{pmatrix} e^{3t} + c_2 \begin{pmatrix} 1 \\ 9 \end{pmatrix} e^{7t} + \begin{pmatrix} 55/36 \\ -19/4 \end{pmatrix} e^t.$$

6. Solving

$$\det(\mathbf{A} - \lambda\mathbf{I}) = \begin{vmatrix} -1-\lambda & 5 \\ -1 & 1-\lambda \end{vmatrix} = \lambda^2 + 4 = 0$$

we obtain the eigenvalues $\lambda_1 = 2i$ and $\lambda_2 = -2i$. Corresponding eigenvectors are

$$\mathbf{K}_1 = \begin{pmatrix} 5 \\ 1+2i \end{pmatrix} \quad \text{and} \quad \mathbf{K}_2 = \begin{pmatrix} 5 \\ 1-2i \end{pmatrix}.$$

Thus

$$\mathbf{X}_c = c_1 \begin{pmatrix} 5\cos 2t \\ \cos 2t - 2\sin 2t \end{pmatrix} + c_2 \begin{pmatrix} 5\sin 2t \\ 2\cos 2t + \sin 2t \end{pmatrix}.$$

Substituting

$$\mathbf{X}_p = \begin{pmatrix} a_2 \\ b_2 \end{pmatrix}\cos t + \begin{pmatrix} a_1 \\ b_1 \end{pmatrix}\sin t$$

into the system yields

$$-a_2 + 5b_2 - a_1 = 0$$

$$-a_2 + b_2 - b_1 - 2 = 0$$

$$-a_1 + 5b_1 + a_2 + 1 = 0$$

$$-a_1 + b_1 + b_2 = 0$$

from which we obtain $a_2 = -3$, $b_2 = -2/3$, $a_1 = -1/3$, and $b_1 = 1/3$. Then

$$\mathbf{X}(t) = c_1 \begin{pmatrix} 5\cos 2t \\ \cos 2t - 2\sin 2t \end{pmatrix} + c_2 \begin{pmatrix} 5\sin 2t \\ 2\cos 2t + \sin 2t \end{pmatrix} + \begin{pmatrix} -3 \\ -2/3 \end{pmatrix}\cos t + \begin{pmatrix} -1/3 \\ 1/3 \end{pmatrix}\sin t.$$

7. Solving

$$\det(\mathbf{A} - \lambda\mathbf{I}) = \begin{vmatrix} 1-\lambda & 1 & 1 \\ 0 & 2-\lambda & 3 \\ 0 & 0 & 5-\lambda \end{vmatrix} = (1-\lambda)(2-\lambda)(5-\lambda) = 0$$

we obtain the eigenvalues $\lambda_1 = 1$, $\lambda_2 = 2$, and $\lambda_3 = 5$. Corresponding eigenvectors are

$$\mathbf{K}_1 = \begin{pmatrix} 1 \\ 0 \\ 0 \end{pmatrix}, \quad \mathbf{K}_2 = \begin{pmatrix} 1 \\ 1 \\ 0 \end{pmatrix} \quad \text{and} \quad \mathbf{K}_3 = \begin{pmatrix} 1 \\ 2 \\ 2 \end{pmatrix}.$$

Thus

$$\mathbf{X}_c = C_1 \begin{pmatrix} 1 \\ 0 \\ 0 \end{pmatrix} e^t + C_2 \begin{pmatrix} 1 \\ 1 \\ 0 \end{pmatrix} e^{2t} + C_3 \begin{pmatrix} 1 \\ 2 \\ 2 \end{pmatrix} e^{5t}.$$

Substituting

$$\mathbf{X}_p = \begin{pmatrix} a_1 \\ b_1 \\ c_1 \end{pmatrix} e^{4t}$$

into the system yields

$$-3a_1 + b_1 + c_1 = -1$$

$$-2b_1 + 3c_1 = 1$$

$$c_1 = -2$$

from which we obtain $c_1 = -2$, $b_1 = -7/2$, and $a_1 = -3/2$. Then

$$\mathbf{X}(t) = C_1 \begin{pmatrix} 1 \\ 0 \\ 0 \end{pmatrix} e^t + C_2 \begin{pmatrix} 1 \\ 1 \\ 0 \end{pmatrix} e^{2t} + C_3 \begin{pmatrix} 1 \\ 2 \\ 2 \end{pmatrix} e^{5t} + \begin{pmatrix} -3/2 \\ -7/2 \\ -2 \end{pmatrix} e^{4t}.$$

8. Solving

$$\det(\mathbf{A} - \lambda\mathbf{I}) = \begin{vmatrix} -\lambda & 0 & 5 \\ 0 & 5-\lambda & 0 \\ 5 & 0 & -\lambda \end{vmatrix} = -(\lambda - 5)^2(\lambda + 5) = 0$$

we obtain the eigenvalues $\lambda_1 = 5$, $\lambda_2 = 5$, and $\lambda_3 = -5$. Corresponding eigenvectors are

$$\mathbf{K}_1 = \begin{pmatrix} 1 \\ 0 \\ 0 \end{pmatrix}, \quad \mathbf{K}_2 = \begin{pmatrix} 1 \\ 1 \\ 1 \end{pmatrix} \quad \text{and} \quad \mathbf{K}_3 = \begin{pmatrix} 1 \\ 0 \\ -1 \end{pmatrix}.$$

Thus

$$\mathbf{X}_c = C_1 \begin{pmatrix} 1 \\ 0 \\ 1 \end{pmatrix} e^{5t} + C_2 \begin{pmatrix} 1 \\ 1 \\ 1 \end{pmatrix} e^{5t} + C_3 \begin{pmatrix} 1 \\ 0 \\ -1 \end{pmatrix} e^{-5t}.$$

Substituting

$$\mathbf{X}_p = \begin{pmatrix} a_1 \\ b_1 \\ c_1 \end{pmatrix}$$

into the system yields

$$5c_1 = -5$$

$$5b_1 = 10$$

$$5a_1 = -40$$

from which we obtain $c_1 = -1$, $b_1 = 2$, and $a_1 = -8$. Then

$$\mathbf{X}(t) = C_1 \begin{pmatrix} 1 \\ 0 \\ 1 \end{pmatrix} e^{5t} + C_2 \begin{pmatrix} 1 \\ 1 \\ 1 \end{pmatrix} e^{5t} + C_3 \begin{pmatrix} 1 \\ 0 \\ -1 \end{pmatrix} e^{-5t} + \begin{pmatrix} -8 \\ 2 \\ -1 \end{pmatrix}.$$

9. Solving

$$\det(\mathbf{A} - \lambda \mathbf{I}) = \begin{vmatrix} -1 - \lambda & -2 \\ 3 & 4 - \lambda \end{vmatrix} = \lambda^2 - 3\lambda + 2 = (\lambda - 1)(\lambda - 2) = 0$$

we obtain the eigenvalues $\lambda_1 = 1$ and $\lambda_2 = 2$. Corresponding eigenvectors are

$$\mathbf{K}_1 = \begin{pmatrix} 1 \\ -1 \end{pmatrix} \quad \text{and} \quad \mathbf{K}_2 = \begin{pmatrix} -4 \\ 6 \end{pmatrix}.$$

Thus

$$\mathbf{X}_c = c_1 \begin{pmatrix} 1 \\ -1 \end{pmatrix} e^t + c_2 \begin{pmatrix} -4 \\ 6 \end{pmatrix} e^{2t}.$$

Substituting

$$\mathbf{X}_p = \begin{pmatrix} a_1 \\ b_1 \end{pmatrix}$$

into the system yields

$$-a_1 - 2b_1 = -3$$

$$3a_1 + 4b_1 = -3$$

from which we obtain $a_1 = -9$ and $b_1 = 6$. Then

$$\mathbf{X}(t) = c_1 \begin{pmatrix} 1 \\ -1 \end{pmatrix} e^t + c_2 \begin{pmatrix} -4 \\ 6 \end{pmatrix} e^{2t} + \begin{pmatrix} -9 \\ 6 \end{pmatrix}.$$

Setting

$$\mathbf{X}(0) = \begin{pmatrix} -4 \\ 5 \end{pmatrix}$$

we obtain

$$c_1 - 4c_2 - 9 = -4$$

$$-c_1 + 6c_2 + 6 = 5.$$

Then $c_1 = 13$ and $c_2 = 2$ so

$$\mathbf{X}(t) = 13 \begin{pmatrix} 1 \\ -1 \end{pmatrix} e^t + 2 \begin{pmatrix} -4 \\ 6 \end{pmatrix} e^{2t} + \begin{pmatrix} -9 \\ 6 \end{pmatrix}.$$

10. (a) Let $\mathbf{I} = \begin{pmatrix} i_2 \\ i_3 \end{pmatrix}$ so that

$$\mathbf{I}' = \begin{pmatrix} -2 & -2 \\ -2 & -5 \end{pmatrix} \mathbf{I} + \begin{pmatrix} 60 \\ 60 \end{pmatrix}$$

and

$$\mathbf{I}_c = c_1 \begin{pmatrix} 2 \\ -1 \end{pmatrix} e^{-t} + c_2 \begin{pmatrix} 1 \\ 2 \end{pmatrix} e^{-6t}.$$

If $\mathbf{I}_p = \begin{pmatrix} a_1 \\ b_1 \end{pmatrix}$ then $\mathbf{I}_p = \begin{pmatrix} 30 \\ 0 \end{pmatrix}$ so that

$$\mathbf{I} = c_1 \begin{pmatrix} 2 \\ -1 \end{pmatrix} e^{-t} + c_2 \begin{pmatrix} 1 \\ 2 \end{pmatrix} e^{-6t} + \begin{pmatrix} 30 \\ 0 \end{pmatrix}.$$

For $\mathbf{I}(0) = \begin{pmatrix} 0 \\ 0 \end{pmatrix}$ we find $c_1 = -12$ and $c_2 = -6$.

(b) $i_1(t) = i_2(t) + i_3(t) = -12e^{-t} - 18e^{-6t} + 30.$

11. From

$$\mathbf{X}' = \begin{pmatrix} 3 & -3 \\ 2 & -2 \end{pmatrix} \mathbf{X} + \begin{pmatrix} 4 \\ -1 \end{pmatrix}$$

we obtain

$$\mathbf{X}_c = c_1 \begin{pmatrix} 1 \\ 1 \end{pmatrix} + c_2 \begin{pmatrix} 3 \\ 2 \end{pmatrix} e^t.$$

Then

$$\mathbf{\Phi} = \begin{pmatrix} 1 & 3e^t \\ 1 & 2e^t \end{pmatrix} \quad \text{and} \quad \mathbf{\Phi}^{-1} = \begin{pmatrix} -2 & 3 \\ e^{-t} & -e^{-t} \end{pmatrix}$$

so that

$$\mathbf{U} = \int \mathbf{\Phi}^{-1}\mathbf{F}\, dt = \int \begin{pmatrix} -11 \\ 5e^{-t} \end{pmatrix} dt = \begin{pmatrix} -11t \\ -5e^{-t} \end{pmatrix}$$

and

$$\mathbf{X}_p = \mathbf{\Phi}\mathbf{U} = \begin{pmatrix} -11 \\ -11 \end{pmatrix} t + \begin{pmatrix} -15 \\ -10 \end{pmatrix}.$$

12. From

$$\mathbf{X}' = \begin{pmatrix} 2 & -1 \\ 3 & -2 \end{pmatrix} \mathbf{X} + \begin{pmatrix} 0 \\ 4 \end{pmatrix} t$$

we obtain

$$\mathbf{X}_c = c_1 \begin{pmatrix} 1 \\ 1 \end{pmatrix} e^t + c_2 \begin{pmatrix} 1 \\ 3 \end{pmatrix} e^{-t}.$$

Then

$$\mathbf{\Phi} = \begin{pmatrix} e^t & e^{-t} \\ e^t & 3e^{-t} \end{pmatrix} \quad \text{and} \quad \mathbf{\Phi}^{-1} = \begin{pmatrix} \frac{3}{2}e^{-t} & -\frac{1}{2}e^{-t} \\ -\frac{1}{2}e^t & \frac{1}{2}e^t \end{pmatrix}$$

so that

$$\mathbf{U} = \int \mathbf{\Phi}^{-1}\mathbf{F}\, dt = \int \begin{pmatrix} -2te^{-t} \\ 2te^t \end{pmatrix} dt = \begin{pmatrix} 2te^{-t} + 2e^{-t} \\ 2te^t - 2e^t \end{pmatrix}$$

and

$$\mathbf{X}_p = \mathbf{\Phi}\mathbf{U} = \begin{pmatrix} 4 \\ 8 \end{pmatrix} t + \begin{pmatrix} 0 \\ -4 \end{pmatrix}.$$

13. From

$$\mathbf{X}' = \begin{pmatrix} 3 & -5 \\ 3/4 & -1 \end{pmatrix} \mathbf{X} + \begin{pmatrix} 1 \\ -1 \end{pmatrix} e^{t/2}$$

we obtain

$$\mathbf{X}_c = c_1 \begin{pmatrix} 10 \\ 3 \end{pmatrix} e^{3t/2} + c_2 \begin{pmatrix} 2 \\ 1 \end{pmatrix} e^{t/2}.$$

Then

$$\mathbf{\Phi} = \begin{pmatrix} 10e^{3t/2} & 2e^{t/2} \\ 3e^{3t/2} & e^{t/2} \end{pmatrix} \quad \text{and} \quad \mathbf{\Phi}^{-1} = \begin{pmatrix} \frac{1}{4}e^{-3t/2} & -\frac{1}{2}e^{-3t/2} \\ -\frac{3}{4}e^{-t/2} & \frac{5}{2}e^{-t/2} \end{pmatrix}$$

so that

$$\mathbf{U} = \int \mathbf{\Phi}^{-1}\mathbf{F}\,dt = \int \begin{pmatrix} \frac{3}{4}e^{-t} \\ -\frac{13}{4} \end{pmatrix} dt = \begin{pmatrix} -\frac{3}{4}e^{-t} \\ -\frac{13}{4}t \end{pmatrix}$$

and

$$\mathbf{X}_p = \mathbf{\Phi}\mathbf{U} = \begin{pmatrix} -13/2 \\ -13/4 \end{pmatrix} te^{t/2} + \begin{pmatrix} -15/2 \\ -9/4 \end{pmatrix} e^{t/2}.$$

14. From

$$\mathbf{X}' = \begin{pmatrix} 2 & -1 \\ 4 & 2 \end{pmatrix} \mathbf{X} + \begin{pmatrix} \sin 2t \\ 2\cos 2t \end{pmatrix}$$

we obtain

$$\mathbf{X}_c = c_1 \begin{pmatrix} -\sin 2t \\ 2\cos 2t \end{pmatrix} e^{2t} + c_2 \begin{pmatrix} \cos 2t \\ 2\sin 2t \end{pmatrix} e^{2t}.$$

Then

$$\mathbf{\Phi} = \begin{pmatrix} -e^{2t}\sin 2t & e^{2t}\cos 2t \\ 2e^{2t}\cos 2t & 2e^{2t}\sin 2t \end{pmatrix} \quad \text{and} \quad \mathbf{\Phi}^{-1} = \begin{pmatrix} -\frac{1}{2}e^{-2t}\sin 2t & \frac{1}{4}e^{-2t}\cos 2t \\ \frac{1}{2}e^{-2t}\cos 2t & \frac{1}{4}e^{-2t}\sin 2t \end{pmatrix}$$

so that

$$\mathbf{U} = \int \mathbf{\Phi}^{-1}\mathbf{F}\,dt = \int \begin{pmatrix} \frac{1}{2}\cos 4t \\ \frac{1}{2}\sin 4t \end{pmatrix} dt = \begin{pmatrix} \frac{1}{8}\sin 4t \\ -\frac{1}{8}\cos 4t \end{pmatrix}$$

and

$$\mathbf{X}_p = \mathbf{\Phi}\mathbf{U} = \begin{pmatrix} -\frac{1}{8}\sin 2t\cos 4t - \frac{1}{8}\cos 2t\cos 4t \\ \frac{1}{4}\cos 2t\sin 4t - \frac{1}{4}\sin 2t\cos 4t \end{pmatrix} e^{2t}.$$

15. From

$$\mathbf{X}' = \begin{pmatrix} 0 & 2 \\ -1 & 3 \end{pmatrix} \mathbf{X} + \begin{pmatrix} 1 \\ -1 \end{pmatrix} e^t$$

we obtain

$$\mathbf{X}_c = c_1 \begin{pmatrix} 2 \\ 1 \end{pmatrix} e^t + c_2 \begin{pmatrix} 1 \\ 1 \end{pmatrix} e^{2t}.$$

Then

$$\mathbf{\Phi} = \begin{pmatrix} 2e^t & e^{2t} \\ e^t & e^{2t} \end{pmatrix} \quad \text{and} \quad \mathbf{\Phi}^{-1} = \begin{pmatrix} e^{-t} & -e^{-t} \\ -e^{-2t} & 2e^{-2t} \end{pmatrix}$$

so that

$$\mathbf{U} = \int \mathbf{\Phi}^{-1}\mathbf{F}\,dt = \int \begin{pmatrix} 2 \\ -3e^{-t} \end{pmatrix} dt = \begin{pmatrix} 2t \\ 3e^{-t} \end{pmatrix}$$

and

$$\mathbf{X}_p = \mathbf{\Phi}\mathbf{U} = \begin{pmatrix} 4 \\ 2 \end{pmatrix} te^t + \begin{pmatrix} 3 \\ 3 \end{pmatrix} e^t.$$

16. From

$$\mathbf{X}' = \begin{pmatrix} 0 & 2 \\ -1 & 3 \end{pmatrix} \mathbf{X} + \begin{pmatrix} 2 \\ e^{-3t} \end{pmatrix}$$

we obtain

$$\mathbf{X}_c = c_1 \begin{pmatrix} 2 \\ 1 \end{pmatrix} e^t + c_2 \begin{pmatrix} 1 \\ 1 \end{pmatrix} e^{2t}.$$

Then

$$\mathbf{\Phi} = \begin{pmatrix} 2e^t & e^{2t} \\ e^t & e^{2t} \end{pmatrix} \quad \text{and} \quad \mathbf{\Phi}^{-1} = \begin{pmatrix} e^{-t} & -e^{-t} \\ -e^{-2t} & 2e^{-2t} \end{pmatrix}$$

so that

$$\mathbf{U} = \int \mathbf{\Phi}^{-1}\mathbf{F}\, dt = \int \begin{pmatrix} 2e^{-t} - e^{-4t} \\ -2e^{-2t} + 2e^{-5t} \end{pmatrix} dt = \begin{pmatrix} -2e^{-t} + \frac{1}{4}e^{-4t} \\ e^{-2t} - \frac{2}{5}e^{-5t} \end{pmatrix}$$

and

$$\mathbf{X}_p = \mathbf{\Phi}\mathbf{U} = \begin{pmatrix} \frac{1}{10}e^{-3t} - 3 \\ -\frac{3}{20}e^{-3t} - 1 \end{pmatrix}.$$

17. From

$$\mathbf{X}' = \begin{pmatrix} 1 & 8 \\ 1 & -1 \end{pmatrix} \mathbf{X} + \begin{pmatrix} 12 \\ 12 \end{pmatrix} t$$

we obtain

$$\mathbf{X}_c = c_1 \begin{pmatrix} 4 \\ 1 \end{pmatrix} e^{3t} + c_2 \begin{pmatrix} -2 \\ 1 \end{pmatrix} e^{-3t}.$$

Then

$$\mathbf{\Phi} = \begin{pmatrix} 4e^{3t} & -2e^{-3t} \\ e^{3t} & e^{-3t} \end{pmatrix} \quad \text{and} \quad \mathbf{\Phi}^{-1} = \begin{pmatrix} \frac{1}{6}e^{-3t} & \frac{1}{3}e^{-3t} \\ -\frac{1}{6}e^{3t} & \frac{2}{3}e^{3t} \end{pmatrix}$$

so that

$$\mathbf{U} = \int \mathbf{\Phi}^{-1}\mathbf{F}\, dt = \int \begin{pmatrix} 6te^{-3t} \\ 6te^{3t} \end{pmatrix} dt = \begin{pmatrix} -2te^{-3t} - \frac{2}{3}e^{-3t} \\ 2te^{3t} - \frac{2}{3}e^{3t} \end{pmatrix}$$

and

$$\mathbf{X}_p = \mathbf{\Phi}\mathbf{U} = \begin{pmatrix} -12 \\ 0 \end{pmatrix} t + \begin{pmatrix} -4/3 \\ -4/3 \end{pmatrix}.$$

18. From

$$\mathbf{X}' = \begin{pmatrix} 1 & 8 \\ 1 & -1 \end{pmatrix} \mathbf{X} + \begin{pmatrix} e^{-t} \\ te^t \end{pmatrix}$$

we obtain

$$\mathbf{X}_c = c_1 \begin{pmatrix} 4 \\ 1 \end{pmatrix} e^{3t} + c_2 \begin{pmatrix} -2 \\ 1 \end{pmatrix} e^{-3t}.$$

Then

$$\Phi = \begin{pmatrix} 4e^{3t} & -2e^{3t} \\ e^{3t} & e^{-3t} \end{pmatrix} \quad \text{and} \quad \Phi^{-1} = \begin{pmatrix} \frac{1}{6}e^{-3t} & \frac{1}{3}e^{-3t} \\ -\frac{1}{6}e^{3t} & \frac{2}{3}e^{3t} \end{pmatrix}$$

so that

$$\mathbf{U} = \int \Phi^{-1}\mathbf{F}\, dt = \int \begin{pmatrix} \frac{1}{6}e^{-4t} + \frac{1}{3}te^{-2t} \\ -\frac{1}{6}e^{2t} + \frac{2}{3}te^{4t} \end{pmatrix} dt = \begin{pmatrix} -\frac{1}{24}e^{-4t} - \frac{1}{6}te^{-2t} - \frac{1}{12}e^{-2t} \\ -\frac{1}{12}e^{2t} + \frac{1}{6}te^{4t} - \frac{1}{24}e^{4t} \end{pmatrix}$$

and

$$\mathbf{X}_p = \Phi\mathbf{U} = \begin{pmatrix} -te^t - \frac{1}{4}e^t \\ -\frac{1}{8}e^{-t} - \frac{1}{8}e^t \end{pmatrix}.$$

19. From

$$\mathbf{X}' = \begin{pmatrix} 3 & 2 \\ -2 & -1 \end{pmatrix}\mathbf{X} + \begin{pmatrix} 2 \\ 1 \end{pmatrix}e^{-t}$$

we obtain

$$\mathbf{X}_c = c_1 \begin{pmatrix} 1 \\ -1 \end{pmatrix}e^t + c_2\left[\begin{pmatrix} 1 \\ -1 \end{pmatrix}te^t + \begin{pmatrix} 0 \\ 1/2 \end{pmatrix}e^t\right].$$

Then

$$\Phi = \begin{pmatrix} e^t & te^t \\ -e^t & \frac{1}{2}e^t - te^t \end{pmatrix} \quad \text{and} \quad \Phi^{-1} = \begin{pmatrix} e^{-t} - 2te^{-t} & -2te^{-t} \\ 2e^{-t} & 2e^{-t} \end{pmatrix}$$

so that

$$\mathbf{U} = \int \Phi^{-1}\mathbf{F}\, dt = \int \begin{pmatrix} 2e^{-2t} - 6te^{-2t} \\ 6e^{-2t} \end{pmatrix} dt = \begin{pmatrix} \frac{1}{2}e^{-2t} + 3te^{-2t} \\ -3e^{-2t} \end{pmatrix}$$

and

$$\mathbf{X}_p = \Phi\mathbf{U} = \begin{pmatrix} 1/2 \\ -2 \end{pmatrix}e^{-t}.$$

20. From

$$\mathbf{X}' = \begin{pmatrix} 3 & 2 \\ -2 & -1 \end{pmatrix}\mathbf{X} + \begin{pmatrix} 1 \\ 1 \end{pmatrix}$$

we obtain

$$\mathbf{X}_c = c_1 \begin{pmatrix} 1 \\ -1 \end{pmatrix}e^t + c_2\left[\begin{pmatrix} 1 \\ -1 \end{pmatrix}te^t + \begin{pmatrix} 0 \\ 1/2 \end{pmatrix}e^t\right].$$

Then

$$\Phi = \begin{pmatrix} e^t & te^t \\ -e^t & \frac{1}{2}e^t - te^t \end{pmatrix} \quad \text{and} \quad \Phi^{-1} = \begin{pmatrix} e^{-t} - 2te^{-t} & -2te^{-t} \\ 2e^{-t} & 2e^{-t} \end{pmatrix}$$

so that

$$\mathbf{U} = \int \Phi^{-1}\mathbf{F}\, dt = \int \begin{pmatrix} e^{-t} - 4te^{-t} \\ 2e^{-t} \end{pmatrix} dt = \begin{pmatrix} 3e^{-t} + 4te^{-t} \\ -2e^{-t} \end{pmatrix}$$

and

$$\mathbf{X}_p = \Phi\mathbf{U} = \begin{pmatrix} 3 \\ -5 \end{pmatrix}.$$

21. From

$$\mathbf{X}' = \begin{pmatrix} 0 & -1 \\ 1 & 0 \end{pmatrix} \mathbf{X} + \begin{pmatrix} \sec t \\ 0 \end{pmatrix}$$

we obtain

$$\mathbf{X}_c = c_1 \begin{pmatrix} \cos t \\ \sin t \end{pmatrix} + c_2 \begin{pmatrix} \sin t \\ -\cos t \end{pmatrix}.$$

Then

$$\mathbf{\Phi} = \begin{pmatrix} \cos t & \sin t \\ \sin t & -\cos t \end{pmatrix} \quad \text{and} \quad \mathbf{\Phi}^{-1} = \begin{pmatrix} \cos t & \sin t \\ \sin t & -\cos t \end{pmatrix}$$

so that

$$\mathbf{U} = \int \mathbf{\Phi}^{-1}\mathbf{F}\, dt = \int \begin{pmatrix} 1 \\ \tan t \end{pmatrix} dt = \begin{pmatrix} t \\ -\ln|\cos t| \end{pmatrix}$$

and

$$\mathbf{X}_p = \mathbf{\Phi}\mathbf{U} = \begin{pmatrix} t\cos t - \sin t \ln|\cos t| \\ t\sin t + \cos t \ln|\cos t| \end{pmatrix}.$$

22. From

$$\mathbf{X}' = \begin{pmatrix} 1 & -1 \\ 1 & 1 \end{pmatrix} \mathbf{X} + \begin{pmatrix} 3 \\ 3 \end{pmatrix} e^t$$

we obtain

$$\mathbf{X}_c = c_1 \begin{pmatrix} -\sin t \\ \cos t \end{pmatrix} e^t + c_2 \begin{pmatrix} \cos t \\ \sin t \end{pmatrix} e^t.$$

Then

$$\mathbf{\Phi} = \begin{pmatrix} -\sin t & \cos t \\ \cos t & \sin t \end{pmatrix} e^t \quad \text{and} \quad \mathbf{\Phi}^{-1} = \begin{pmatrix} -\sin t & \cos t \\ \cos t & \sin t \end{pmatrix} e^{-t}$$

so that

$$\mathbf{U} = \int \mathbf{\Phi}^{-1}\mathbf{F}\, dt = \int \begin{pmatrix} -3\sin t + 3\cos t \\ 3\cos t + 3\sin t \end{pmatrix} dt = \begin{pmatrix} 3\cos t + 3\sin t \\ 3\sin t - 3\cos t \end{pmatrix}$$

and

$$\mathbf{X}_p = \mathbf{\Phi}\mathbf{U} = \begin{pmatrix} -3 \\ 3 \end{pmatrix} e^t.$$

23. From

$$\mathbf{X}' = \begin{pmatrix} 1 & -1 \\ 1 & 1 \end{pmatrix} \mathbf{X} + \begin{pmatrix} \cos t \\ \sin t \end{pmatrix} e^t$$

we obtain

$$\mathbf{X}_c = c_1 \begin{pmatrix} -\sin t \\ \cos t \end{pmatrix} e^t + c_2 \begin{pmatrix} \cos t \\ \sin t \end{pmatrix} e^t.$$

Then

$$\mathbf{\Phi} = \begin{pmatrix} -\sin t & \cos t \\ \cos t & \sin t \end{pmatrix} e^t \quad \text{and} \quad \mathbf{\Phi}^{-1} = \begin{pmatrix} -\sin t & \cos t \\ \cos t & \sin t \end{pmatrix} e^{-t}$$

455

so that

$$\mathbf{U} = \int \mathbf{\Phi}^{-1}\mathbf{F}\, dt = \int \begin{pmatrix} 0 \\ 1 \end{pmatrix} dt = \begin{pmatrix} 0 \\ t \end{pmatrix}$$

and

$$\mathbf{X}_p = \mathbf{\Phi U} = \begin{pmatrix} \cos t \\ \sin t \end{pmatrix} te^t.$$

24. From

$$\mathbf{X}' = \begin{pmatrix} 2 & -2 \\ 8 & -6 \end{pmatrix} \mathbf{X} + \begin{pmatrix} 1 \\ 3 \end{pmatrix} \frac{1}{t} e^{-2t}$$

we obtain

$$\mathbf{X}_c = c_1 \begin{pmatrix} 1 \\ 2 \end{pmatrix} e^{-2t} + c_2 \left[\begin{pmatrix} 1 \\ 2 \end{pmatrix} te^{-2t} + \begin{pmatrix} 1/2 \\ 1/2 \end{pmatrix} e^{-2t} \right].$$

Then

$$\mathbf{\Phi} = \begin{pmatrix} 1 & t + \frac{1}{2} \\ 2 & 2t + \frac{1}{2} \end{pmatrix} e^{-2t} \quad \text{and} \quad \mathbf{\Phi}^{-1} = \begin{pmatrix} -4t-1 & 2t+1 \\ 4 & -2 \end{pmatrix} e^{2t}$$

so that

$$\mathbf{U} = \int \mathbf{\Phi}^{-1}\mathbf{F}\, dt = \int \begin{pmatrix} 2 + 2/t \\ -2/t \end{pmatrix} dt = \begin{pmatrix} 2t + 2\ln t \\ -2\ln t \end{pmatrix}$$

and

$$\mathbf{X}_p = \mathbf{\Phi U} = \begin{pmatrix} 2t + \ln t - 2t\ln t \\ 4t + 3\ln t - 4t\ln t \end{pmatrix} e^{-2t}.$$

25. From

$$\mathbf{X}' = \begin{pmatrix} 0 & 1 \\ -1 & 0 \end{pmatrix} \mathbf{X} + \begin{pmatrix} 0 \\ \sec t \tan t \end{pmatrix}$$

we obtain

$$\mathbf{X}_c = c_1 \begin{pmatrix} \cos t \\ -\sin t \end{pmatrix} + c_2 \begin{pmatrix} \sin t \\ \cos t \end{pmatrix}.$$

Then

$$\mathbf{\Phi} = \begin{pmatrix} \cos t & \sin t \\ -\sin t & \cos t \end{pmatrix} t \quad \text{and} \quad \mathbf{\Phi}^{-1} = \begin{pmatrix} \cos t & -\sin t \\ \sin t & \cos t \end{pmatrix}$$

so that

$$\mathbf{U} = \int \mathbf{\Phi}^{-1}\mathbf{F}\, dt = \int \begin{pmatrix} -\tan^2 t \\ \tan t \end{pmatrix} dt = \begin{pmatrix} t - \tan t \\ -\ln|\cos t| \end{pmatrix}$$

and

$$\mathbf{X}_p = \mathbf{\Phi U} = \begin{pmatrix} \cos t \\ -\sin t \end{pmatrix} t + \begin{pmatrix} -\sin t \\ \sin t \tan t \end{pmatrix} - \begin{pmatrix} \sin t \\ \cos t \end{pmatrix} \ln|\cos t|.$$

26. From

$$\mathbf{X}' = \begin{pmatrix} 0 & 1 \\ -1 & 0 \end{pmatrix} \mathbf{X} + \begin{pmatrix} 1 \\ \cot t \end{pmatrix}$$

we obtain

$$\mathbf{X}_c = c_1 \begin{pmatrix} \cos t \\ -\sin t \end{pmatrix} + c_2 \begin{pmatrix} \sin t \\ \cos t \end{pmatrix}.$$

Then

$$\boldsymbol{\Phi} = \begin{pmatrix} \cos t & \sin t \\ -\sin t & \cos t \end{pmatrix} \quad \text{and} \quad \boldsymbol{\Phi}^{-1} = \begin{pmatrix} \cos t & -\sin t \\ \sin t & \cos t \end{pmatrix}$$

so that

$$\mathbf{U} = \int \boldsymbol{\Phi}^{-1} \mathbf{F} \, dt = \int \begin{pmatrix} 0 \\ \csc t \end{pmatrix} dt = \begin{pmatrix} 0 \\ \ln|\csc t - \cot t| \end{pmatrix}$$

and

$$\mathbf{X}_p = \boldsymbol{\Phi}\mathbf{U} = \begin{pmatrix} \sin t \ln|\csc t - \cot t| \\ \cos t \ln|\csc t - \cot t| \end{pmatrix}.$$

27. From

$$\mathbf{X}' = \begin{pmatrix} 1 & 2 \\ -1/2 & 1 \end{pmatrix} \mathbf{X} + \begin{pmatrix} \csc t \\ \sec t \end{pmatrix} e^t$$

we obtain

$$\mathbf{X}_c = c_1 \begin{pmatrix} 2\sin t \\ \cos t \end{pmatrix} e^t + c_2 \begin{pmatrix} 2\cos t \\ -\sin t \end{pmatrix} e^t.$$

Then

$$\boldsymbol{\Phi} = \begin{pmatrix} 2\sin t & 2\cos t \\ \cos t & -\sin t \end{pmatrix} e^t \quad \text{and} \quad \boldsymbol{\Phi}^{-1} = \begin{pmatrix} \frac{1}{2}\sin t & \cos t \\ \frac{1}{2}\cos t & -\sin t \end{pmatrix} e^{-t}$$

so that

$$\mathbf{U} = \int \boldsymbol{\Phi}^{-1} \mathbf{F} \, dt = \int \begin{pmatrix} \frac{3}{2} \\ \frac{1}{2}\cot t - \tan t \end{pmatrix} dt = \begin{pmatrix} \frac{3}{2}t \\ \frac{1}{2}\ln|\sin t| + \ln|\cos t| \end{pmatrix}$$

and

$$\mathbf{X}_p = \boldsymbol{\Phi}\mathbf{U} = \begin{pmatrix} 3\sin t \\ \frac{3}{2}\cos t \end{pmatrix} te^t + \begin{pmatrix} \cos t \\ -\frac{1}{2}\sin t \end{pmatrix} e^t \ln|\sin t| + \begin{pmatrix} 2\cos t \\ -\sin t \end{pmatrix} e^t \ln|\cos t|.$$

28. From

$$\mathbf{X}' = \begin{pmatrix} 1 & -2 \\ 1 & -1 \end{pmatrix} \mathbf{X} + \begin{pmatrix} \tan t \\ 1 \end{pmatrix}$$

we obtain

$$\mathbf{X}_c = c_1 \begin{pmatrix} \cos t - \sin t \\ \cos t \end{pmatrix} + c_2 \begin{pmatrix} \cos t + \sin t \\ \sin t \end{pmatrix}.$$

Then

$$\boldsymbol{\Phi} = \begin{pmatrix} \cos t - \sin t & \cos t + \sin t \\ \cos t & \sin t \end{pmatrix} \quad \text{and} \quad \boldsymbol{\Phi}^{-1} = \begin{pmatrix} -\sin t & \cos t + \sin t \\ \cos t & \sin t - \cos t \end{pmatrix}$$

so that

$$\mathbf{U} = \int \boldsymbol{\Phi}^{-1} \mathbf{F} \, dt = \int \begin{pmatrix} 2\cos t + \sin t - \sec t \\ 2\sin t - \cos t \end{pmatrix} dt = \begin{pmatrix} 2\sin t - \cos t - \ln|\sec t + \tan t| \\ -2\cos t - \sin t \end{pmatrix}$$

and

$$\mathbf{X}_p = \mathbf{\Phi}\mathbf{U} = \begin{pmatrix} 3\sin t\cos t - \cos^2 t - 2\sin^2 t + (\sin t - \cos t)\ln|\sec t + \tan t| \\ \sin^2 t - \cos^2 t - \cos t(\ln|\sec t + \tan t|) \end{pmatrix}.$$

29. From

$$\mathbf{X}' = \begin{pmatrix} 1 & 1 & 0 \\ 1 & 1 & 0 \\ 0 & 0 & 3 \end{pmatrix}\mathbf{X} + \begin{pmatrix} e^t \\ e^{2t} \\ te^{3t} \end{pmatrix}$$

we obtain

$$\mathbf{X}_c = c_1 \begin{pmatrix} 1 \\ -1 \\ 0 \end{pmatrix} + c_2 \begin{pmatrix} 1 \\ 1 \\ 0 \end{pmatrix}e^{2t} + c_3 \begin{pmatrix} 0 \\ 0 \\ 1 \end{pmatrix}e^{3t}.$$

Then

$$\mathbf{\Phi} = \begin{pmatrix} 1 & e^{2t} & 0 \\ -1 & e^{2t} & 0 \\ 0 & 0 & e^{3t} \end{pmatrix} \quad \text{and} \quad \mathbf{\Phi}^{-1} = \begin{pmatrix} \frac{1}{2} & -\frac{1}{2} & 0 \\ \frac{1}{2}e^{-2t} & \frac{1}{2}e^{-2t} & 0 \\ 0 & 0 & e^{-3t} \end{pmatrix}$$

so that

$$\mathbf{U} = \int \mathbf{\Phi}^{-1}\mathbf{F}\,dt = \int \begin{pmatrix} \frac{1}{2}e^t - \frac{1}{2}e^{2t} \\ \frac{1}{2}e^{-t} + \frac{1}{2} \\ t \end{pmatrix} dt = \begin{pmatrix} \frac{1}{2}e^t - \frac{1}{4}e^{2t} \\ -\frac{1}{2}e^{-t} + \frac{1}{2}t \\ \frac{1}{2}t^2 \end{pmatrix}$$

and

$$\mathbf{X}_p = \mathbf{\Phi}\mathbf{U} = \begin{pmatrix} -\frac{1}{4}e^{2t} + \frac{1}{2}te^{2t} \\ -e^t + \frac{1}{4}e^{2t} + \frac{1}{2}te^{2t} \\ \frac{1}{2}t^2 e^{3t} \end{pmatrix}.$$

30. From

$$\mathbf{X}' = \begin{pmatrix} 3 & -1 & -1 \\ 1 & 1 & -1 \\ 1 & -1 & 1 \end{pmatrix}\mathbf{X} + \begin{pmatrix} 0 \\ t \\ 2e^t \end{pmatrix}$$

we obtain

$$\mathbf{X}_c = c_1 \begin{pmatrix} 1 \\ 1 \\ 1 \end{pmatrix}e^t + c_2 \begin{pmatrix} 1 \\ 1 \\ 0 \end{pmatrix}e^{2t} + c_3 \begin{pmatrix} 1 \\ 0 \\ 1 \end{pmatrix}e^{2t}.$$

Then

$$\mathbf{\Phi} = \begin{pmatrix} e^t & e^{2t} & e^{2t} \\ e^t & e^{2t} & 0 \\ e^t & 0 & e^{2t} \end{pmatrix} \quad \text{and} \quad \mathbf{\Phi}^{-1} = \begin{pmatrix} -e^{-t} & e^{-t} & e^{-t} \\ e^{-2t} & 0 & -e^{-2t} \\ e^{-2t} & -e^{-2t} & 0 \end{pmatrix}$$

so that

$$\mathbf{U} = \int \mathbf{\Phi}^{-1}\mathbf{F}\,dt = \int \begin{pmatrix} te^{-t} + 2 \\ -2e^{-t} \\ -te^{-2t} \end{pmatrix} dt = \begin{pmatrix} -te^{-t} - e^{-t} + 2t \\ 2e^{-t} \\ \frac{1}{2}te^{-2t} + \frac{1}{4}e^{-2t} \end{pmatrix}$$

and

$$\mathbf{X}_p = \mathbf{\Phi}\mathbf{U} = \begin{pmatrix} -1/2 \\ -1 \\ -1/2 \end{pmatrix} t + \begin{pmatrix} -3/4 \\ -1 \\ -3/4 \end{pmatrix} + \begin{pmatrix} 2 \\ 2 \\ 0 \end{pmatrix} e^t + \begin{pmatrix} 2 \\ 2 \\ 2 \end{pmatrix} te^t.$$

31. From

$$\mathbf{X}' = \begin{pmatrix} 3 & -1 \\ -1 & 3 \end{pmatrix} \mathbf{X} + \begin{pmatrix} 4e^{2t} \\ 4e^{4t} \end{pmatrix}$$

we obtain

$$\mathbf{\Phi} = \begin{pmatrix} -e^{4t} & e^{2t} \\ e^{4t} & e^{2t} \end{pmatrix}, \quad \mathbf{\Phi}^{-1} = \begin{pmatrix} -\frac{1}{2}e^{-4t} & \frac{1}{2}e^{-4t} \\ \frac{1}{2}e^{-2t} & \frac{1}{2}e^{-2t} \end{pmatrix},$$

and

$$\mathbf{X} = \mathbf{\Phi}\mathbf{\Phi}^{-1}(0)\mathbf{X}(0) + \mathbf{\Phi}\int_0^t \mathbf{\Phi}^{-1}\mathbf{F}\,ds = \mathbf{\Phi} \cdot \begin{pmatrix} 0 \\ 1 \end{pmatrix} + \mathbf{\Phi} \cdot \begin{pmatrix} e^{-2t} + 2t - 1 \\ e^{2t} + 2t - 1 \end{pmatrix}$$

$$= \begin{pmatrix} 2 \\ 2 \end{pmatrix} te^{2t} + \begin{pmatrix} -1 \\ 1 \end{pmatrix} e^{2t} + \begin{pmatrix} -2 \\ 2 \end{pmatrix} te^{4t} + \begin{pmatrix} 2 \\ 0 \end{pmatrix} e^{4t}.$$

32. From

$$\mathbf{X}' = \begin{pmatrix} 1 & -1 \\ 1 & -1 \end{pmatrix} \mathbf{X} + \begin{pmatrix} 1/t \\ 1/t \end{pmatrix}$$

we obtain

$$\mathbf{\Phi} = \begin{pmatrix} 1 & 1+t \\ 1 & t \end{pmatrix}, \quad \mathbf{\Phi}^{-1} = \begin{pmatrix} -t & 1+t \\ 1 & -1 \end{pmatrix},$$

and

$$\mathbf{X} = \mathbf{\Phi}\mathbf{\Phi}^{-1}(1)\mathbf{X}(1) + \mathbf{\Phi}\int_1^t \mathbf{\Phi}^{-1}\mathbf{F}\,ds = \mathbf{\Phi} \cdot \begin{pmatrix} -4 \\ 3 \end{pmatrix} + \mathbf{\Phi} \cdot \begin{pmatrix} \ln t \\ 0 \end{pmatrix} = \begin{pmatrix} 3 \\ 3 \end{pmatrix} t - \begin{pmatrix} 1 \\ 4 \end{pmatrix} + \begin{pmatrix} 1 \\ 1 \end{pmatrix} \ln t.$$

33. Let $\mathbf{I} = \begin{pmatrix} i_1 \\ i_2 \end{pmatrix}$ so that

$$\mathbf{I}' = \begin{pmatrix} -11 & 3 \\ 3 & -3 \end{pmatrix} \mathbf{I} + \begin{pmatrix} 100\sin t \\ 0 \end{pmatrix}$$

and

$$\mathbf{I}_c = c_1 \begin{pmatrix} 1 \\ 3 \end{pmatrix} e^{-2t} + c_2 \begin{pmatrix} 3 \\ -1 \end{pmatrix} e^{-12t}.$$

Then

$$\Phi = \begin{pmatrix} e^{-2t} & 3e^{-12t} \\ 3e^{-2t} & -e^{-12t} \end{pmatrix}, \qquad \Phi^{-1} = \begin{pmatrix} \frac{1}{10}e^{2t} & \frac{3}{10}e^{2t} \\ \frac{3}{10}e^{12t} & -\frac{1}{10}e^{12t} \end{pmatrix},$$

$$\mathbf{U} = \int \Phi^{-1}\mathbf{F}\, dt = \int \begin{pmatrix} 10e^{2t}\sin t \\ 30e^{12t}\sin t \end{pmatrix} dt = \begin{pmatrix} 2e^{2t}(2\sin t - \cos t) \\ \frac{6}{29}e^{12t}(12\sin t - \cos t) \end{pmatrix},$$

and

$$\mathbf{I}_p = \Phi\mathbf{U} = \begin{pmatrix} \frac{332}{29}\sin t - \frac{76}{29}\cos t \\ \frac{276}{29}\sin t - \frac{168}{29}\cos t \end{pmatrix}$$

so that

$$\mathbf{I} = c_1 \begin{pmatrix} 1 \\ 3 \end{pmatrix} e^{-2t} + c_2 \begin{pmatrix} 3 \\ -1 \end{pmatrix} e^{-12t} + \mathbf{I}_p.$$

If $\mathbf{I}(0) = \begin{pmatrix} 0 \\ 0 \end{pmatrix}$ then $c_1 = 2$ and $c_2 = \frac{6}{29}$.

34. Write the differential equation as a system

$$\begin{aligned} y' &= v \\ v' &= -Qy - Pv + f \end{aligned} \qquad \text{or} \qquad \begin{pmatrix} y \\ v \end{pmatrix}' = \begin{pmatrix} 0 & 1 \\ -Q & -P \end{pmatrix}\begin{pmatrix} y \\ v \end{pmatrix} + \begin{pmatrix} 0 \\ f \end{pmatrix}.$$

From (9) in the text of this section, a particular solution is then $\mathbf{X}_p = \Phi(x)\int \Phi^{-1}(x)\mathbf{F}(x)\, dx$ where

$$\Phi(x) = \begin{pmatrix} y_1 & y_2 \\ y_1' & y_2' \end{pmatrix} \qquad \text{and} \qquad \mathbf{X}_p = \begin{pmatrix} u_1 \\ u_2 \end{pmatrix}.$$

Then

$$\Phi^{-1}(x) = \frac{1}{y_1 y_2' - y_2 y_1'}\begin{pmatrix} y_2' & -y_2 \\ -y_1' & y_1 \end{pmatrix},$$

so

$$\mathbf{X}_p = \int \frac{1}{W}\begin{pmatrix} y_2' & -y_2 \\ -y_1' & y_1 \end{pmatrix}\begin{pmatrix} 0 \\ f \end{pmatrix} dx$$

and $W = y_1 y_2' - y_2 y_1'$. Thus

$$u_1 = \int \frac{-y_2 f(x)}{W}\, dx \qquad \text{and} \qquad u_2 = \int \frac{y_1 f(x)}{W}\, dx,$$

which are the antiderivative forms of the equations in (5) of Section 4.6 in the text.

35. (a) The eigenvalues are 0, 1, 3, and 4, with corresponding eigenvectors

$$\begin{pmatrix} -6 \\ -4 \\ 1 \\ 2 \end{pmatrix}, \qquad \begin{pmatrix} 2 \\ 1 \\ 0 \\ 0 \end{pmatrix}, \qquad \begin{pmatrix} 3 \\ 1 \\ 2 \\ 1 \end{pmatrix}, \qquad \text{and} \qquad \begin{pmatrix} -1 \\ 1 \\ 0 \\ 0 \end{pmatrix}.$$

(b) $\Phi = \begin{pmatrix} -6 & 2e^t & 3e^{3t} & -e^{4t} \\ -4 & e^t & e^{3t} & e^{4t} \\ 1 & 0 & 2e^{3t} & 0 \\ 2 & 0 & e^{3t} & 0 \end{pmatrix}$, $\Phi^{-1} = \begin{pmatrix} 0 & 0 & -\frac{1}{3} & \frac{2}{3} \\ \frac{1}{3}e^{-t} & \frac{1}{3}e^{-t} & -2e^{-t} & \frac{8}{3}e^{-t} \\ 0 & 0 & \frac{2}{3}e^{-3t} & -\frac{1}{3}e^{-3t} \\ -\frac{1}{3}e^{-4t} & \frac{2}{3}e^{-4t} & 0 & \frac{1}{3}e^{-4t} \end{pmatrix}$

(c) $\Phi^{-1}(t)\mathbf{F}(t) = \begin{pmatrix} \frac{2}{3} - \frac{1}{3}e^{2t} \\ \frac{1}{3}e^{-2t} + \frac{8}{3}e^{-t} - 2e^t + \frac{1}{3}t \\ -\frac{1}{3}e^{-3t} + \frac{2}{3}e^{-t} \\ \frac{2}{3}e^{-5t} + \frac{1}{3}e^{-4t} - \frac{1}{3}te^{-3t} \end{pmatrix}$,

$\displaystyle \int \Phi^{-1}(t)\mathbf{F}(t)\,dt = \begin{pmatrix} -\frac{1}{6}e^{2t} + \frac{2}{3}t \\ -\frac{1}{6}e^{-2t} - \frac{8}{3}e^{-t} - 2e^t + \frac{1}{6}t^2 \\ \frac{1}{9}e^{-3t} - \frac{2}{3}e^{-t} \\ -\frac{2}{15}e^{-5t} - \frac{1}{12}e^{-4t} + \frac{1}{27}e^{-3t} + \frac{1}{9}te^{-3t} \end{pmatrix}$,

$\displaystyle \mathbf{X}_p(t) = \Phi(t)\int \Phi^{-1}(t)\mathbf{F}(t)\,dt = \begin{pmatrix} -5e^{2t} - \frac{1}{5}e^{-t} - \frac{1}{27}e^t - \frac{1}{9}te^t + \frac{1}{3}t^2e^t - 4t - \frac{59}{12} \\ -2e^{2t} - \frac{3}{10}e^{-t} + \frac{1}{27}e^t + \frac{1}{9}te^t + \frac{1}{6}t^2e^t - \frac{8}{3}t - \frac{95}{36} \\ -\frac{3}{2}e^{2t} + \frac{2}{3}t + \frac{2}{9} \\ -e^{2t} + \frac{4}{3}t - \frac{1}{9} \end{pmatrix}$,

$\displaystyle \mathbf{X}_c(t) = \Phi(t)\mathbf{C} = \begin{pmatrix} -6c_1 + 2c_2e^t + 3c_3e^{3t} - c_4e^{4t} \\ -4c_1 + c_2e^t + c_3e^{3t} + c_4e^{4t} \\ c_1 + 2c_3e^{3t} \\ 2c_1 + c_3e^{3t} \end{pmatrix}$,

$\mathbf{X}(t) = \Phi(t)\mathbf{C} + \Phi(t)\int \Phi^{-1}(t)\mathbf{F}(t)\,dt$

$\displaystyle = \begin{pmatrix} -6c_1 + 2c_2e^t + 3c_3e^{3t} - c_4e^{4t} \\ -4c_1 + c_2e^t + c_3e^{3t} + c_4e^{4t} \\ c_1 + 2c_3e^{3t} \\ 2c_1 + c_3e^{3t} \end{pmatrix} + \begin{pmatrix} -5e^{2t} - \frac{1}{5}e^{-t} - \frac{1}{27}e^t - \frac{1}{9}te^t + \frac{1}{3}t^2e^t - 4t - \frac{59}{12} \\ -2e^{2t} - \frac{3}{10}e^{-t} + \frac{1}{27}e^t + \frac{1}{9}te^t + \frac{1}{6}t^2e^t - \frac{8}{3}t - \frac{95}{36} \\ -\frac{3}{2}e^{2t} + \frac{2}{3}t + \frac{2}{9} \\ -e^{2t} + \frac{4}{3}t - \frac{1}{9} \end{pmatrix}$

(d) $\mathbf{X}(t) = c_1 \begin{pmatrix} -6 \\ -4 \\ 1 \\ 2 \end{pmatrix} + c_2 \begin{pmatrix} 2 \\ 1 \\ 0 \\ 0 \end{pmatrix} e^t + c_3 \begin{pmatrix} 3 \\ 1 \\ 2 \\ 1 \end{pmatrix} e^{3t} + c_4 \begin{pmatrix} -1 \\ 1 \\ 0 \\ 0 \end{pmatrix} e^{4t}$

$\displaystyle + \begin{pmatrix} -5e^{2t} - \frac{1}{5}e^{-t} - \frac{1}{27}e^t - \frac{1}{9}te^t + \frac{1}{3}t^2e^t - 4t - \frac{59}{12} \\ -2e^{2t} - \frac{3}{10}e^{-t} + \frac{1}{27}e^t + \frac{1}{9}te^t + \frac{1}{6}t^2e^t - \frac{8}{3}t - \frac{95}{36} \\ -\frac{3}{2}e^{2t} + \frac{2}{3}t + \frac{2}{9} \\ -e^{2t} + \frac{4}{3}t - \frac{1}{9} \end{pmatrix}$

Exercises 8.4   Matrix Exponential

1. For $\mathbf{A} = \begin{pmatrix} 1 & 0 \\ 0 & 2 \end{pmatrix}$ we have

$$\mathbf{A}^2 = \begin{pmatrix} 1 & 0 \\ 0 & 2 \end{pmatrix}\begin{pmatrix} 1 & 0 \\ 0 & 2 \end{pmatrix} = \begin{pmatrix} 1 & 0 \\ 0 & 4 \end{pmatrix},$$

$$\mathbf{A}^3 = \mathbf{A}\mathbf{A}^2 = \begin{pmatrix} 1 & 0 \\ 0 & 2 \end{pmatrix}\begin{pmatrix} 1 & 0 \\ 0 & 4 \end{pmatrix} = \begin{pmatrix} 1 & 0 \\ 0 & 8 \end{pmatrix},$$

$$\mathbf{A}^4 = \mathbf{A}\mathbf{A}^3 = \begin{pmatrix} 1 & 0 \\ 0 & 2 \end{pmatrix}\begin{pmatrix} 1 & 0 \\ 0 & 8 \end{pmatrix} = \begin{pmatrix} 1 & 0 \\ 0 & 16 \end{pmatrix},$$

and so on. In general

$$\mathbf{A}^k = \begin{pmatrix} 1 & 0 \\ 0 & 2^k \end{pmatrix} \quad \text{for} \quad k = 1, 2, 3, \ldots.$$

Thus

$$e^{\mathbf{A}t} = \mathbf{I} + \frac{\mathbf{A}}{1!}t + \frac{\mathbf{A}^2}{2!}t^2 + \frac{\mathbf{A}^3}{3!}t^3 + \cdots$$

$$= \begin{pmatrix} 1 & 0 \\ 0 & 1 \end{pmatrix} + \frac{1}{1!}\begin{pmatrix} 1 & 0 \\ 0 & 2 \end{pmatrix}t + \frac{1}{2!}\begin{pmatrix} 1 & 0 \\ 0 & 4 \end{pmatrix}t^2 + \frac{1}{3!}\begin{pmatrix} 1 & 0 \\ 0 & 8 \end{pmatrix}t^3 + \cdots$$

$$= \begin{pmatrix} 1 + t + \dfrac{t^2}{2!} + \dfrac{t^3}{3!} + \cdots & 0 \\ 0 & 1 + 2t + \dfrac{(2t)^2}{2!} + \dfrac{(2t)^3}{3!} + \cdots \end{pmatrix} = \begin{pmatrix} e^t & 0 \\ 0 & e^{2t} \end{pmatrix}$$

and

$$e^{-\mathbf{A}t} = \begin{pmatrix} e^{-t} & 0 \\ 0 & e^{-2t} \end{pmatrix}.$$

2. For $\mathbf{A} = \begin{pmatrix} 0 & 1 \\ 1 & 0 \end{pmatrix}$ we have

$$\mathbf{A}^2 = \begin{pmatrix} 0 & 1 \\ 1 & 0 \end{pmatrix}\begin{pmatrix} 0 & 1 \\ 1 & 0 \end{pmatrix} = \begin{pmatrix} 1 & 0 \\ 0 & 1 \end{pmatrix} = \mathbf{I}$$

$$\mathbf{A}^3 = \mathbf{A}\mathbf{A}^2 = \begin{pmatrix} 0 & 1 \\ 1 & 0 \end{pmatrix}\mathbf{I} = \begin{pmatrix} 0 & 1 \\ 1 & 0 \end{pmatrix} = \mathbf{A}$$

$$\mathbf{A}^4 = (\mathbf{A}^2)^2 = \mathbf{I}$$

$$\mathbf{A}^5 = \mathbf{A}\mathbf{A}^4 = \mathbf{A}\mathbf{I} = \mathbf{A},$$

and so on. In general,

$$\mathbf{A}^k = \begin{cases} \mathbf{A}, & k = 1,\, 3,\, 5,\, \ldots \\ \mathbf{I}, & k = 2,\, 4,\, 6,\, \ldots . \end{cases}$$

Thus

$$e^{\mathbf{A}t} = \mathbf{I} + \frac{\mathbf{A}}{1!}t + \frac{\mathbf{A}^2}{2!}t^2 + \frac{\mathbf{A}^3}{3!}t^3 + \cdots$$

$$= \mathbf{I} + \mathbf{A}t + \frac{1}{2!}\mathbf{I}t^2 + \frac{1}{3!}\mathbf{A}t^3 + \cdots$$

$$= \mathbf{I}\left(1 + \frac{1}{2!}t^2 + \frac{1}{4!}t^4 + \cdots\right) + \mathbf{A}\left(t + \frac{1}{3!}t^3 + \frac{1}{5!}t^5 + \cdots\right)$$

$$= \mathbf{I}\cosh t + \mathbf{A}\sinh t = \begin{pmatrix} \cosh t & \sinh t \\ \sinh t & \cosh t \end{pmatrix}$$

and

$$e^{-\mathbf{A}t} = \begin{pmatrix} \cosh(-t) & \sinh(-t) \\ \sinh(-t) & \cosh(-t) \end{pmatrix} = \begin{pmatrix} \cosh t & -\sinh t \\ -\sinh t & \cosh t \end{pmatrix}.$$

3. For

$$\mathbf{A} = \begin{pmatrix} 1 & 1 & 1 \\ 1 & 1 & 1 \\ -2 & -2 & -2 \end{pmatrix}$$

we have

$$\mathbf{A}^2 = \begin{pmatrix} 1 & 1 & 1 \\ 1 & 1 & 1 \\ -2 & -2 & -2 \end{pmatrix}\begin{pmatrix} 1 & 1 & 1 \\ 1 & 1 & 1 \\ -2 & -2 & -2 \end{pmatrix} = \begin{pmatrix} 0 & 0 & 0 \\ 0 & 0 & 0 \\ 0 & 0 & 0 \end{pmatrix}.$$

Thus, $\mathbf{A}^3 = \mathbf{A}^4 = \mathbf{A}^5 = \cdots = \mathbf{0}$ and

$$e^{\mathbf{A}t} = \mathbf{I} + \mathbf{A}t = \begin{pmatrix} 1 & 0 & 0 \\ 0 & 1 & 0 \\ 0 & 0 & 1 \end{pmatrix} + \begin{pmatrix} t & t & t \\ t & t & t \\ -2t & -2t & -2t \end{pmatrix} = \begin{pmatrix} t+1 & t & t \\ t & t+1 & t \\ -2t & -2t & -2t+1 \end{pmatrix}.$$

4. For

$$\mathbf{A} = \begin{pmatrix} 0 & 0 & 0 \\ 3 & 0 & 0 \\ 5 & 1 & 0 \end{pmatrix}$$

we have

$$A^2 = \begin{pmatrix} 0 & 0 & 0 \\ 3 & 0 & 0 \\ 5 & 1 & 0 \end{pmatrix} \begin{pmatrix} 0 & 0 & 0 \\ 3 & 0 & 0 \\ 5 & 1 & 0 \end{pmatrix} = \begin{pmatrix} 0 & 0 & 0 \\ 0 & 0 & 0 \\ 3 & 0 & 0 \end{pmatrix}$$

$$A^3 = AA^2 = \begin{pmatrix} 0 & 0 & 0 \\ 3 & 0 & 0 \\ 5 & 1 & 0 \end{pmatrix} \begin{pmatrix} 0 & 0 & 0 \\ 0 & 0 & 0 \\ 3 & 0 & 0 \end{pmatrix} = \begin{pmatrix} 0 & 0 & 0 \\ 0 & 0 & 0 \\ 0 & 0 & 0 \end{pmatrix}.$$

Thus, $A^4 = A^5 = A^6 = \cdots = 0$ and

$$e^{At} = I + At + \frac{1}{2}A^2t^2$$

$$= \begin{pmatrix} 1 & 0 & 0 \\ 0 & 1 & 0 \\ 0 & 0 & 1 \end{pmatrix} + \begin{pmatrix} 0 & 0 & 0 \\ 3t & 0 & 0 \\ 5t & t & 0 \end{pmatrix} + \begin{pmatrix} 0 & 0 & 0 \\ 0 & 0 & 0 \\ \frac{3}{2}t^2 & 0 & 0 \end{pmatrix} = \begin{pmatrix} 1 & 0 & 0 \\ 3t & 1 & 0 \\ \frac{3}{2}t^2 + 5t & t & 1 \end{pmatrix}.$$

5. Using the result of Problem 1,

$$X = \begin{pmatrix} e^t & 0 \\ 0 & e^{2t} \end{pmatrix} \begin{pmatrix} c_1 \\ c_2 \end{pmatrix} = c_1 \begin{pmatrix} e^t \\ 0 \end{pmatrix} + c_2 \begin{pmatrix} 0 \\ e^t \end{pmatrix}.$$

6. Using the result of Problem 2,

$$X = \begin{pmatrix} \cosh t & \sinh t \\ \sinh t & \cosh t \end{pmatrix} \begin{pmatrix} c_1 \\ c_2 \end{pmatrix} = c_1 \begin{pmatrix} \cosh t \\ \sinh t \end{pmatrix} + c_2 \begin{pmatrix} \sinh t \\ \cosh t \end{pmatrix}.$$

7. Using the result of Problem 3,

$$X = \begin{pmatrix} t+1 & t & t \\ t & t+1 & t \\ -2t & -2t & -2t+1 \end{pmatrix} \begin{pmatrix} c_1 \\ c_2 \\ c_3 \end{pmatrix} = c_1 \begin{pmatrix} t+1 \\ t \\ -2t \end{pmatrix} + c_2 \begin{pmatrix} t \\ t+1 \\ -2t \end{pmatrix} + c_3 \begin{pmatrix} t \\ t \\ -2t+1 \end{pmatrix}.$$

8. Using the result of Problem 4,

$$X = \begin{pmatrix} 1 & 0 & 0 \\ 3t & 1 & 0 \\ \frac{3}{2}t^2 + 5t & t & 1 \end{pmatrix} \begin{pmatrix} c_1 \\ c_2 \\ c_3 \end{pmatrix} = c_1 \begin{pmatrix} 1 \\ 3t \\ \frac{3}{2}t^2 + 5t \end{pmatrix} + c_2 \begin{pmatrix} 0 \\ 1 \\ t \end{pmatrix} + c_3 \begin{pmatrix} 0 \\ 0 \\ 1 \end{pmatrix}.$$

9. To solve

$$X' = \begin{pmatrix} 1 & 0 \\ 0 & 2 \end{pmatrix} X + \begin{pmatrix} 3 \\ -1 \end{pmatrix}$$

we identify $t_0 = 0$, $\mathbf{F}(t) = \begin{pmatrix} 3 \\ -1 \end{pmatrix}$, and use the results of Problem 1 and equation (5) in the text.

$$\mathbf{X}(t) = e^{\mathbf{A}t}\mathbf{C} + e^{\mathbf{A}t}\int_{t_0}^{t} e^{-\mathbf{A}s}\mathbf{F}(s)\,ds$$

$$= \begin{pmatrix} e^t & 0 \\ 0 & e^{2t} \end{pmatrix}\begin{pmatrix} c_1 \\ c_2 \end{pmatrix} + \begin{pmatrix} e^t & 0 \\ 0 & e^{2t} \end{pmatrix}\int_0^t \begin{pmatrix} e^{-s} & 0 \\ 0 & e^{-2s} \end{pmatrix}\begin{pmatrix} 3 \\ -1 \end{pmatrix}\,ds$$

$$= \begin{pmatrix} c_1 e^t \\ c_2 e^{2t} \end{pmatrix} + \begin{pmatrix} e^t & 0 \\ 0 & e^{2t} \end{pmatrix}\int_0^t \begin{pmatrix} 3e^{-s} \\ -e^{-2s} \end{pmatrix}\,ds$$

$$= \begin{pmatrix} c_1 e^t \\ c_2 e^{2t} \end{pmatrix} + \begin{pmatrix} e^t & 0 \\ 0 & e^{2t} \end{pmatrix}\begin{pmatrix} -3e^{-s} \\ \frac{1}{2}e^{-2s} \end{pmatrix}\Big|_0^t$$

$$= \begin{pmatrix} c_1 e^t \\ c_2 e^{2t} \end{pmatrix} + \begin{pmatrix} e^t & 0 \\ 0 & e^{2t} \end{pmatrix}\begin{pmatrix} -3e^{-t} + 3 \\ \frac{1}{2}e^{-2t} - \frac{1}{2} \end{pmatrix}$$

$$= \begin{pmatrix} c_1 e^t \\ c_2 e^{2t} \end{pmatrix} + \begin{pmatrix} -3 + 3e^t \\ \frac{1}{2} - \frac{1}{2}e^{2t} \end{pmatrix} = c_3 \begin{pmatrix} 1 \\ 0 \end{pmatrix}e^t + c_4 \begin{pmatrix} 0 \\ 1 \end{pmatrix}e^{2t} + \begin{pmatrix} -3 \\ \frac{1}{2} \end{pmatrix}.$$

10. To solve

$$\mathbf{X}' = \begin{pmatrix} 1 & 0 \\ 0 & 2 \end{pmatrix}\mathbf{X} + \begin{pmatrix} t \\ e^{4t} \end{pmatrix}$$

we identify $t_0 = 0$, $\mathbf{F}(t) = \begin{pmatrix} t \\ e^{4t} \end{pmatrix}$, and use the results of Problem 1 and equation (5) in the text.

$$\mathbf{X}(t) = e^{\mathbf{A}t}\mathbf{C} + e^{\mathbf{A}t}\int_{t_0}^{t} e^{-\mathbf{A}s}\mathbf{F}(s)\,ds$$

$$= \begin{pmatrix} e^t & 0 \\ 0 & e^{2t} \end{pmatrix}\begin{pmatrix} c_1 \\ c_2 \end{pmatrix} + \begin{pmatrix} e^t & 0 \\ 0 & e^{2t} \end{pmatrix}\int_0^t \begin{pmatrix} e^{-s} & 0 \\ 0 & e^{-2s} \end{pmatrix}\begin{pmatrix} s \\ e^{4s} \end{pmatrix}\,ds$$

$$= \begin{pmatrix} c_1 e^t \\ c_2 e^{2t} \end{pmatrix} + \begin{pmatrix} e^t & 0 \\ 0 & e^{2t} \end{pmatrix}\int_0^t \begin{pmatrix} se^{-s} \\ e^{2s} \end{pmatrix}\,ds$$

$$= \begin{pmatrix} c_1 e^t \\ c_2 e^{2t} \end{pmatrix} + \begin{pmatrix} e^t & 0 \\ 0 & e^{2t} \end{pmatrix}\begin{pmatrix} -se^{-s} - e^{-s} \\ \frac{1}{2}e^{2s} \end{pmatrix}\Big|_0^t$$

$$= \begin{pmatrix} c_1 e^t \\ c_2 e^{2t} \end{pmatrix} + \begin{pmatrix} e^t & 0 \\ 0 & e^{2t} \end{pmatrix}\begin{pmatrix} -te^{-t} - e^{-t} + 1 \\ \frac{1}{2}e^{2t} - \frac{1}{2} \end{pmatrix}$$

$$= \begin{pmatrix} c_1 e^t \\ c_2 e^{2t} \end{pmatrix} + \begin{pmatrix} -t - 1 + e^t \\ \frac{1}{2}e^{4t} - \frac{1}{2}e^{2t} \end{pmatrix} = c_3 \begin{pmatrix} 1 \\ 0 \end{pmatrix}e^t + c_4 \begin{pmatrix} 0 \\ 1 \end{pmatrix}e^{2t} + \begin{pmatrix} -t - 1 \\ \frac{1}{2}e^{4t} \end{pmatrix}.$$

11. To solve

$$\mathbf{X}' = \begin{pmatrix} 0 & 1 \\ 1 & 0 \end{pmatrix} \mathbf{X} + \begin{pmatrix} 1 \\ 1 \end{pmatrix}$$

we identify $t_0 = 0$, $\mathbf{F}(t) = \begin{pmatrix} 1 \\ 1 \end{pmatrix}$, and use the results of Problem 2 and equation (5) in the text.

$$\mathbf{X}(t) = e^{\mathbf{A}t}\mathbf{C} + e^{\mathbf{A}t} \int_{t_0}^{t} e^{-\mathbf{A}s}\mathbf{F}(s)\,ds$$

$$= \begin{pmatrix} \cosh t & \sinh t \\ \sinh t & \cosh t \end{pmatrix}\begin{pmatrix} c_1 \\ c_2 \end{pmatrix} + \begin{pmatrix} \cosh t & \sinh t \\ \sinh t & \cosh t \end{pmatrix}\int_0^t \begin{pmatrix} \cosh s & -\sinh s \\ -\sinh s & \cosh s \end{pmatrix}\begin{pmatrix} 1 \\ 1 \end{pmatrix}\,ds$$

$$= \begin{pmatrix} c_1\cosh t + c_2\sinh t \\ c_1\sinh t + c_2\cosh t \end{pmatrix} + \begin{pmatrix} \cosh t & \sinh t \\ \sinh t & \cosh t \end{pmatrix}\int_0^t \begin{pmatrix} \cosh s - \sinh s \\ -\sinh s + \cosh s \end{pmatrix}\,ds$$

$$= \begin{pmatrix} c_1\cosh t + c_2\sinh t \\ c_1\sinh t + c_2\cosh t \end{pmatrix} + \begin{pmatrix} \cosh t & \sinh t \\ \sinh t & \cosh t \end{pmatrix}\begin{pmatrix} \sinh s - \cosh s \\ -\cosh s + \sinh s \end{pmatrix}\Big|_0^t$$

$$= \begin{pmatrix} c_1\cosh t + c_2\sinh t \\ c_1\sinh t + c_2\cosh t \end{pmatrix} + \begin{pmatrix} \cosh t & \sinh t \\ \sinh t & \cosh t \end{pmatrix}\begin{pmatrix} \sinh t - \cosh t + 1 \\ -\cosh t + \sinh t + 1 \end{pmatrix}$$

$$= \begin{pmatrix} c_1\cosh t + c_2\sinh t \\ c_1\sinh t + c_2\cosh t \end{pmatrix} + \begin{pmatrix} \sinh^2 t - \cosh^2 t + \cosh t + \sinh t \\ \sinh^2 t - \cosh^2 t + \sinh t + \cosh t \end{pmatrix}$$

$$= c_1\begin{pmatrix} \cosh t \\ \sinh t \end{pmatrix} + c_2\begin{pmatrix} \sinh t \\ \cosh t \end{pmatrix} + \begin{pmatrix} \cosh t \\ \sinh t \end{pmatrix} + \begin{pmatrix} \sinh t \\ \cosh t \end{pmatrix} - \begin{pmatrix} 1 \\ 1 \end{pmatrix}$$

$$= c_3\begin{pmatrix} \cosh t \\ \sinh t \end{pmatrix} + c_4\begin{pmatrix} \sinh t \\ \cosh t \end{pmatrix} - \begin{pmatrix} 1 \\ 1 \end{pmatrix}.$$

12. To solve

$$\mathbf{X}' = \begin{pmatrix} 0 & 1 \\ 1 & 0 \end{pmatrix} \mathbf{X} + \begin{pmatrix} \cosh t \\ \sinh t \end{pmatrix}$$

we identify $t_0 = 0$, $\mathbf{F}(t) = \begin{pmatrix} \cosh t \\ \sinh t \end{pmatrix}$, and use the results of Problem 2 and equation (5) in the text.

$$\mathbf{X}(t) = e^{\mathbf{A}t}\mathbf{C} + e^{\mathbf{A}t} \int_{t_0}^{t} e^{-\mathbf{A}s}\mathbf{F}(s)\,ds$$

$$= \begin{pmatrix} \cosh t & \sinh t \\ \sinh t & \cosh t \end{pmatrix}\begin{pmatrix} c_1 \\ c_2 \end{pmatrix} + \begin{pmatrix} \cosh t & \sinh t \\ \sinh t & \cosh t \end{pmatrix}\int_0^t \begin{pmatrix} \cosh s & -\sinh s \\ -\sinh s & \cosh s \end{pmatrix}\begin{pmatrix} \cosh s \\ \sinh s \end{pmatrix}\,ds$$

$$= \begin{pmatrix} c_1 \cosh t + c_2 \sinh t \\ c_1 \sinh t + c_2 \cosh t \end{pmatrix} + \begin{pmatrix} \cosh t & \sinh t \\ \sinh t & \cosh t \end{pmatrix} \int_0^t \begin{pmatrix} 1 \\ 0 \end{pmatrix} ds$$

$$= \begin{pmatrix} c_1 \cosh t + c_2 \sinh t \\ c_1 \sinh t + c_2 \cosh t \end{pmatrix} + \begin{pmatrix} \cosh t & \sinh t \\ \sinh t & \cosh t \end{pmatrix} \begin{pmatrix} s \\ 0 \end{pmatrix} \Big|_0^t$$

$$= \begin{pmatrix} c_1 \cosh t + c_2 \sinh t \\ c_1 \sinh t + c_2 \cosh t \end{pmatrix} + \begin{pmatrix} \cosh t & \sinh t \\ \sinh t & \cosh t \end{pmatrix} \begin{pmatrix} t \\ 0 \end{pmatrix}$$

$$= \begin{pmatrix} c_1 \cosh t + c_2 \sinh t \\ c_1 \sinh t + c_2 \cosh t \end{pmatrix} + \begin{pmatrix} t \cosh t \\ t \sinh t \end{pmatrix} = c_1 \begin{pmatrix} \cosh t \\ \sinh t \end{pmatrix} + c_2 \begin{pmatrix} \sinh t \\ \cosh t \end{pmatrix} + t \begin{pmatrix} \cosh t \\ \sinh t \end{pmatrix}.$$

13. We have

$$\mathbf{X}(0) = c_1 \begin{pmatrix} 1 \\ 0 \\ 0 \end{pmatrix} + c_2 \begin{pmatrix} 0 \\ 1 \\ 0 \end{pmatrix} + c_3 \begin{pmatrix} 0 \\ 0 \\ 1 \end{pmatrix} = \begin{pmatrix} c_1 \\ c_2 \\ c_3 \end{pmatrix} = \begin{pmatrix} 1 \\ -4 \\ 6 \end{pmatrix}.$$

Thus, the solution of the initial-value problem is

$$\mathbf{X} = \begin{pmatrix} t+1 \\ t \\ -2t \end{pmatrix} - 4 \begin{pmatrix} t \\ t+1 \\ -2t \end{pmatrix} + 6 \begin{pmatrix} t \\ t \\ -2t+1 \end{pmatrix}.$$

14. We have

$$\mathbf{X}(0) = c_3 \begin{pmatrix} 1 \\ 0 \end{pmatrix} + c_4 \begin{pmatrix} 0 \\ 1 \end{pmatrix} + \begin{pmatrix} -3 \\ \frac{1}{2} \end{pmatrix} = \begin{pmatrix} c_3 - 3 \\ c_4 + \frac{1}{2} \end{pmatrix} = \begin{pmatrix} 4 \\ 3 \end{pmatrix}.$$

Thus, $c_3 = 7$ and $c_4 = \frac{5}{2}$, so

$$\mathbf{X} = 7 \begin{pmatrix} 1 \\ 0 \end{pmatrix} e^t + \frac{5}{2} \begin{pmatrix} 0 \\ 1 \end{pmatrix} e^{2t} + \begin{pmatrix} -3 \\ \frac{1}{2} \end{pmatrix}.$$

15. From $s\mathbf{I} - \mathbf{A} = \begin{pmatrix} s-4 & -3 \\ 4 & s+4 \end{pmatrix}$ we find

$$(s\mathbf{I} - \mathbf{A})^{-1} = \begin{pmatrix} \dfrac{3/2}{s-2} - \dfrac{1/2}{s+2} & \dfrac{3/4}{s-2} - \dfrac{3/4}{s+2} \\ \dfrac{-1}{s-2} + \dfrac{1}{s+2} & \dfrac{-1/2}{s-2} + \dfrac{3/2}{s+2} \end{pmatrix}$$

and

$$e^{\mathbf{A}t} = \begin{pmatrix} \frac{3}{2}e^{2t} - \frac{1}{2}e^{-2t} & \frac{3}{4}e^{2t} - \frac{3}{4}e^{-2t} \\ -e^{2t} + e^{-2t} & -\frac{1}{2}e^{2t} + \frac{3}{2}e^{-2t} \end{pmatrix}.$$

The general solution of the system is then

$$X = e^{At}C = \begin{pmatrix} \frac{3}{2}e^{2t} - \frac{1}{2}e^{-2t} & \frac{3}{4}e^{2t} - \frac{3}{4}e^{-2t} \\ -e^{2t} + e^{-2t} & -\frac{1}{2}e^{2t} + \frac{3}{2}e^{-2t} \end{pmatrix} \begin{pmatrix} c_1 \\ c_2 \end{pmatrix}$$

$$= c_1 \begin{pmatrix} 3/2 \\ -1 \end{pmatrix} e^{2t} + c_1 \begin{pmatrix} -1/2 \\ 1 \end{pmatrix} e^{-2t} + c_2 \begin{pmatrix} 3/4 \\ -1/2 \end{pmatrix} e^{2t} + c_2 \begin{pmatrix} -3/4 \\ 3/2 \end{pmatrix} e^{-2t}$$

$$= \left(\frac{1}{2}c_1 + \frac{1}{4}c_2\right) \begin{pmatrix} 3 \\ -2 \end{pmatrix} e^{2t} + \left(-\frac{1}{2}c_1 - \frac{3}{4}c_2\right) \begin{pmatrix} 1 \\ -2 \end{pmatrix} e^{-2t}$$

$$= c_3 \begin{pmatrix} 3 \\ -2 \end{pmatrix} e^{2t} + c_4 \begin{pmatrix} 1 \\ -2 \end{pmatrix} e^{-2t}.$$

16. From $sI - A = \begin{pmatrix} s - 4 & 2 \\ -1 & s - 1 \end{pmatrix}$ we find

$$(sI - A)^{-1} = \begin{pmatrix} \dfrac{2}{s-3} - \dfrac{1}{s-2} & -\dfrac{2}{s-3} + \dfrac{2}{s-2} \\ \dfrac{1}{s-3} - \dfrac{1}{s-2} & \dfrac{-1}{s-3} + \dfrac{2}{s-2} \end{pmatrix}$$

and

$$e^{At} = \begin{pmatrix} 2e^{3t} - e^{2t} & -2e^{3t} + 2e^{2t} \\ e^{3t} - e^{2t} & -e^{3t} + 2e^{2t} \end{pmatrix}.$$

The general solution of the system is then

$$X = e^{At}C = \begin{pmatrix} 2e^{3t} - e^{2t} & -2e^{3t} + 2e^{2t} \\ e^{3t} - e^{2t} & -e^{3t} + 2e^{2t} \end{pmatrix} \begin{pmatrix} c_1 \\ c_2 \end{pmatrix}$$

$$= c_1 \begin{pmatrix} 2 \\ 1 \end{pmatrix} e^{3t} + c_1 \begin{pmatrix} -1 \\ -1 \end{pmatrix} e^{2t} + c_2 \begin{pmatrix} -2 \\ -1 \end{pmatrix} e^{3t} + c_2 \begin{pmatrix} 2 \\ 2 \end{pmatrix} e^{2t}$$

$$= (c_1 - c_2) \begin{pmatrix} 2 \\ 1 \end{pmatrix} e^{3t} + (-c_1 + 2c_2) \begin{pmatrix} 1 \\ 1 \end{pmatrix} e^{2t}$$

$$= c_3 \begin{pmatrix} 2 \\ 1 \end{pmatrix} e^{3t} + c_4 \begin{pmatrix} 1 \\ 1 \end{pmatrix} e^{2t}.$$

17. From $sI - A = \begin{pmatrix} s - 5 & 9 \\ -1 & s + 1 \end{pmatrix}$ we find

$$(sI - A)^{-1} = \begin{pmatrix} \dfrac{1}{s-2} + \dfrac{3}{(s-2)^2} & -\dfrac{9}{(s-2)^2} \\ \dfrac{1}{(s-2)^2} & \dfrac{1}{s-2} - \dfrac{3}{(s-2)^2} \end{pmatrix}$$

and

$$e^{\mathbf{A}t} = \begin{pmatrix} e^{2t} + 3te^{2t} & -9te^{2t} \\ te^{2t} & e^{2t} - 3te^{2t} \end{pmatrix}.$$

The general solution of the system is then

$$\mathbf{X} = e^{\mathbf{A}t}\mathbf{C} = \begin{pmatrix} e^{2t} + 3te^{2t} & -9te^{2t} \\ te^{2t} & e^{2t} - 3te^{2t} \end{pmatrix} \begin{pmatrix} c_1 \\ c_2 \end{pmatrix}$$

$$= c_1 \begin{pmatrix} 1 \\ 0 \end{pmatrix} e^{2t} + c_1 \begin{pmatrix} 3 \\ 1 \end{pmatrix} te^{2t} + c_2 \begin{pmatrix} 0 \\ 1 \end{pmatrix} e^{2t} + c_2 \begin{pmatrix} -9 \\ -3 \end{pmatrix} te^{2t}$$

$$= c_1 \begin{pmatrix} 1 + 3t \\ t \end{pmatrix} e^{2t} + c_2 \begin{pmatrix} -9t \\ 1 - 3t \end{pmatrix} e^{2t}.$$

18. From $s\mathbf{I} - \mathbf{A} = \begin{pmatrix} s & -1 \\ 2 & s+2 \end{pmatrix}$ we find

$$(s\mathbf{I} - \mathbf{A})^{-1} = \begin{pmatrix} \dfrac{s+1+1}{(s+1)^2+1} & \dfrac{1}{(s+1)^2+1} \\ \dfrac{-2}{(s+1)^2+1} & \dfrac{s+1-1}{(s+1)^2+1} \end{pmatrix}$$

and

$$e^{\mathbf{A}t} = \begin{pmatrix} e^{-t}\cos t + e^{-t}\sin t & e^{-t}\sin t \\ -2e^{-t}\sin t & e^{-t}\cos t - e^{-t}\sin t \end{pmatrix}.$$

The general solution of the system is then

$$\mathbf{X} = e^{\mathbf{A}t}\mathbf{C} = \begin{pmatrix} e^{-t}\cos t + e^{-t}\sin t & e^{-t}\sin t \\ -2e^{-t}\sin t & e^{-t}\cos t - e^{-t}\sin t \end{pmatrix} \begin{pmatrix} c_1 \\ c_2 \end{pmatrix}$$

$$= c_1 \begin{pmatrix} 1 \\ 0 \end{pmatrix} e^{-t}\cos t + c_1 \begin{pmatrix} 1 \\ -2 \end{pmatrix} e^{-t}\sin t + c_2 \begin{pmatrix} 0 \\ 1 \end{pmatrix} e^{-t}\cos t + c_2 \begin{pmatrix} 1 \\ -1 \end{pmatrix} e^{-t}\sin t$$

$$= c_1 \begin{pmatrix} \cos t + \sin t \\ -2\sin t \end{pmatrix} e^{-t} + c_2 \begin{pmatrix} \sin t \\ \cos t - \sin t \end{pmatrix} e^{-t}.$$

19. Solving

$$\det(\mathbf{A} - \lambda\mathbf{I}) = \begin{vmatrix} 2 - \lambda & 1 \\ -3 & 6 - \lambda \end{vmatrix} = \lambda^2 - 8\lambda + 15 = (\lambda - 3)(\lambda - 5) = 0$$

we find eigenvalues $\lambda_1 = 3$ and $\lambda_2 = 5$. Corresponding eigenvectors are

$$\mathbf{K}_1 = \begin{pmatrix} 1 \\ 1 \end{pmatrix} \quad \text{and} \quad \mathbf{K}_2 = \begin{pmatrix} 1 \\ 3 \end{pmatrix}.$$

Then

$$\mathbf{P} = \begin{pmatrix} 1 & 1 \\ 1 & 3 \end{pmatrix}, \quad \mathbf{P}^{-1} = \begin{pmatrix} 3/2 & -1/2 \\ -1/2 & 1/2 \end{pmatrix}, \quad \text{and} \quad \mathbf{D} = \begin{pmatrix} 3 & 0 \\ 0 & 5 \end{pmatrix},$$

so that

$$\mathbf{PDP}^{-1} = \begin{pmatrix} 2 & 1 \\ -3 & 6 \end{pmatrix}.$$

20. Solving

$$\det(\mathbf{A} - \lambda\mathbf{I}) = \begin{vmatrix} 2 - \lambda & 1 \\ 1 & 2 - \lambda \end{vmatrix} = \lambda^2 - 4\lambda + 3 = (\lambda - 1)(\lambda - 3) = 0$$

we find eigenvalues $\lambda_1 = 1$ and $\lambda_2 = 3$. Corresponding eigenvectors are

$$\mathbf{K}_1 = \begin{pmatrix} -1 \\ 1 \end{pmatrix} \quad \text{and} \quad \mathbf{K}_2 = \begin{pmatrix} 1 \\ 1 \end{pmatrix}.$$

Then

$$\mathbf{P} = \begin{pmatrix} -1 & 1 \\ 1 & 1 \end{pmatrix}, \quad \mathbf{P}^{-1} = \begin{pmatrix} -1/2 & 1/2 \\ 1/2 & 1/2 \end{pmatrix}, \quad \text{and} \quad \mathbf{D} = \begin{pmatrix} 1 & 0 \\ 0 & 3 \end{pmatrix}$$

so that

$$\mathbf{PDP}^{-1} = \begin{pmatrix} 2 & 1 \\ 1 & 2 \end{pmatrix}.$$

21. From equation (3) in the text

$$e^{t\mathbf{A}} = e^{t\mathbf{PDP}^{-1}} = \mathbf{I} + t(\mathbf{PDP}^{-1}) + \frac{1}{2!}t^2(\mathbf{PDP}^{-1})^2 + \frac{1}{3!}t^3(\mathbf{PDP}^{-1})^3 + \cdots$$

$$= \mathbf{P}\left[\mathbf{I} + t\mathbf{D} + \frac{1}{2!}(t\mathbf{D})^2 + \frac{1}{3!}(t\mathbf{D})^3 + \cdots\right]\mathbf{P}^{-1} = \mathbf{P}e^{t\mathbf{D}}\mathbf{P}^{-1}.$$

22. From equation (3) in the text

$$e^{t\mathbf{D}} = \begin{pmatrix} 1 & 0 & \cdots & 0 \\ 0 & 1 & \cdots & 0 \\ \vdots & \vdots & \ddots & \vdots \\ 0 & 0 & \cdots & 1 \end{pmatrix} + t\begin{pmatrix} \lambda_1 & 0 & \cdots & 0 \\ 0 & \lambda_2 & \cdots & 0 \\ \vdots & \vdots & \ddots & \vdots \\ 0 & 0 & \cdots & \lambda_n \end{pmatrix} + \frac{1}{2!}t^2\begin{pmatrix} \lambda_1^2 & 0 & \cdots & 0 \\ 0 & \lambda_2^2 & \cdots & 0 \\ \vdots & \vdots & \ddots & \vdots \\ 0 & 0 & \cdots & \lambda_n^2 \end{pmatrix}$$

$$+ \frac{1}{3!}t^3\begin{pmatrix} \lambda_1^3 & 0 & \cdots & 0 \\ 0 & \lambda_2^3 & \cdots & 0 \\ \vdots & \vdots & \ddots & \vdots \\ 0 & 0 & \cdots & \lambda_n^3 \end{pmatrix} + \cdots$$

$$= \begin{pmatrix} 1 + \lambda_1 t + \frac{1}{2!}(\lambda_1 t)^2 + \cdots & 0 & \cdots & 0 \\ 0 & 1 + \lambda_2 t + \frac{1}{2!}(\lambda_2 t)^2 + \cdots & \cdots & 0 \\ \vdots & \vdots & \ddots & \vdots \\ 0 & 0 & \cdots & 1 + \lambda_n t + \frac{1}{2!}(\lambda_n t)^2 + \cdots \end{pmatrix}$$

$$= \begin{pmatrix} e^{\lambda_1 t} & 0 & \cdots & 0 \\ 0 & e^{\lambda_2 t} & \cdots & 0 \\ \vdots & \vdots & \ddots & \vdots \\ 0 & 0 & \cdots & e^{\lambda_n t} \end{pmatrix}.$$

23. From Problems 19, 21, and 22, and equation (1) in the text

$$\mathbf{X} = e^{t\mathbf{A}}\mathbf{C} = \mathbf{P}e^{t\mathbf{D}}\mathbf{P}^{-1}\mathbf{C}$$

$$= \begin{pmatrix} e^{3t} & e^{5t} \\ e^{3t} & 3e^{5t} \end{pmatrix} \begin{pmatrix} e^{3t} & 0 \\ 0 & e^{5t} \end{pmatrix} \begin{pmatrix} \frac{3}{2}e^{-3t} & -\frac{1}{2}e^{-3t} \\ -\frac{1}{2}e^{-5t} & \frac{1}{2}e^{-5t} \end{pmatrix} \begin{pmatrix} c_1 \\ c_2 \end{pmatrix}$$

$$= \begin{pmatrix} \frac{3}{2}e^{3t} - \frac{1}{2}e^{5t} & -\frac{1}{2}e^{3t} + \frac{1}{2}e^{5t} \\ \frac{3}{2}e^{3t} - \frac{3}{2}e^{5t} & -\frac{1}{2}e^{3t} + \frac{3}{2}e^{5t} \end{pmatrix} \begin{pmatrix} c_1 \\ c_2 \end{pmatrix}.$$

24. From Problems 20-22 and equation (1) in the text

$$\mathbf{X} = e^{t\mathbf{A}}\mathbf{C} = \mathbf{P}e^{t\mathbf{D}}\mathbf{P}^{-1}\mathbf{C}$$

$$= \begin{pmatrix} -e^t & e^{3t} \\ e^t & e^{3t} \end{pmatrix} \begin{pmatrix} e^t & 0 \\ 0 & e^{3t} \end{pmatrix} \begin{pmatrix} -\frac{1}{2}e^{-t} & \frac{1}{2}e^{-t} \\ \frac{1}{2}e^{3t} & \frac{1}{2}e^{-3t} \end{pmatrix} \begin{pmatrix} c_1 \\ c_2 \end{pmatrix}$$

$$= \begin{pmatrix} \frac{1}{2}e^t + \frac{1}{2}e^{9t} & -\frac{1}{2}e^t + \frac{1}{2}e^{3t} \\ -\frac{1}{2}e^t + \frac{1}{2}e^{9t} & \frac{1}{2}e^t + \frac{1}{2}e^{3t} \end{pmatrix} \begin{pmatrix} c_1 \\ c_2 \end{pmatrix}.$$

25. If $\det(s\mathbf{I} - \mathbf{A}) = 0$, then s is an eigenvalue of $\mathbf{A}$. Thus $s\mathbf{I} - \mathbf{A}$ has an inverse if s is not an eigenvalue of $\mathbf{A}$. For the purposes of the discussion in this section, we take s to be larger than the largest eigenvalue of $\mathbf{A}$. Under this condition $s\mathbf{I} - \mathbf{A}$ has an inverse.

26. Since $\mathbf{A}^3 = \mathbf{0}$, $\mathbf{A}$ is nilpotent. Since

$$e^{\mathbf{A}t} = \mathbf{I} + \mathbf{A}t + \mathbf{A}^2 \frac{t^2}{2!} + \cdots + \mathbf{A}^k \frac{t^k}{k!} + \cdots,$$

if $\mathbf{A}$ is nilpotent and $\mathbf{A}^m = \mathbf{0}$, then $\mathbf{A}^k = \mathbf{0}$ for $k \geq m$ and

$$e^{\mathbf{A}t} = \mathbf{I} + \mathbf{A}t + \mathbf{A}^2 \frac{t^2}{2!} + \cdots + \mathbf{A}^{m-1} \frac{t^{m-1}}{(m-1)!}.$$

In this problem $\mathbf{A}^3 = \mathbf{0}$, so

$$e^{\mathbf{A}t} = \mathbf{I} + \mathbf{A}t + \mathbf{A}^2 \frac{t^2}{2} = \begin{pmatrix} 1 & 0 & 0 \\ 0 & 1 & 0 \\ 0 & 0 & 1 \end{pmatrix} + \begin{pmatrix} -1 & 1 & 1 \\ -1 & 0 & 1 \\ -1 & 1 & 1 \end{pmatrix} t + \begin{pmatrix} -1 & 0 & 1 \\ 0 & 0 & 0 \\ -1 & 0 & 1 \end{pmatrix} \frac{t^2}{2}$$

$$= \begin{pmatrix} 1 - t - t^2/2 & t & t + t^2/2 \\ -t & 1 & t \\ -t - t^2/2 & t & 1 + t + t^2/2 \end{pmatrix}$$

and the solution of $\mathbf{X}' = \mathbf{A}\mathbf{X}$ is

$$\mathbf{X}(t) = e^{\mathbf{A}t}\mathbf{C} = e^{\mathbf{A}t} \begin{pmatrix} c_1 \\ c_2 \\ c_3 \end{pmatrix} = \begin{pmatrix} c_1(1 - t - t^2/2) + c_2 t + c_3(t + t^2/2) \\ -c_1 t + c_2 + c_3 t \\ c_1(-t - t^2/2) + c_2 t + c_3(1 + t + t^2/2) \end{pmatrix}.$$

27. **(a)** The following commands can be used in *Mathematica*:

> A={{4, 2},{3, 3}};
> c={c1, c2};
> m=MatrixExp[A t];
> sol=Expand[m.c]
> Collect[sol, {c1, c2}]//MatrixForm

The output gives

$$x(t) = c_1 \left(\frac{2}{5}e^t + \frac{3}{5}e^{6t} \right) + c_2 \left(-\frac{2}{5}e^t + \frac{2}{5}e^{6t} \right)$$

$$y(t) = c_1 \left(-\frac{3}{5}e^t + \frac{3}{5}e^{6t} \right) + c_2 \left(\frac{3}{5}e^t + \frac{2}{5}e^{6t} \right).$$

The eigenvalues are 1 and 6 with corresponding eigenvectors

$$\begin{pmatrix} -2 \\ 3 \end{pmatrix} \quad \text{and} \quad \begin{pmatrix} 1 \\ 1 \end{pmatrix},$$

so the solution of the system is

$$\mathbf{X}(t) = b_1 \begin{pmatrix} -2 \\ 3 \end{pmatrix} e^t + b_2 \begin{pmatrix} 1 \\ 1 \end{pmatrix} e^{6t}$$

or

$$x(t) = -2b_1 e^t + b_2 e^{6t}$$

$$y(t) = 3b_1 e^t + b_2 e^{6t}.$$

If we replace b_1 with $-\frac{1}{5}c_1 + \frac{1}{5}c_2$ and b_2 with $\frac{3}{5}c_1 + \frac{2}{5}c_2$, we obtain the solution found using the matrix exponential.

(b) $x(t) = c_1 e^{-2t} \cos t - (c_1 + c_2)e^{-2t} \sin t$

$\quad\quad y(t) = c_2 e^{-2t} \cos t + (2c_1 + c_2)e^{-2t} \sin t$

28. $x(t) = c_1(3e^{-2t} - 2e^{-t}) + c_3(-6e^{-2t} + 6e^{-t})$

$\quad y(t) = c_2(4e^{-2t} - 3e^{-t}) + c_4(4e^{-2t} - 4e^{-t})$

$\quad z(t) = c_1(e^{-2t} - e^{-t}) + c_3(-2e^{-2t} + 3e^{-t})$

$\quad w(t) = c_2(-3e^{-2t} + 3e^{-t}) + c_4(-3e^{-2t} + 4e^{-t})$

Chapter 8 in Review

1. If $\mathbf{X} = k \begin{pmatrix} 4 \\ 5 \end{pmatrix}$, then $\mathbf{X}' = \mathbf{0}$ and

$$k \begin{pmatrix} 1 & 4 \\ 2 & -1 \end{pmatrix} \begin{pmatrix} 4 \\ 5 \end{pmatrix} - \begin{pmatrix} 8 \\ 1 \end{pmatrix} = k \begin{pmatrix} 24 \\ 3 \end{pmatrix} - \begin{pmatrix} 8 \\ 1 \end{pmatrix} = \begin{pmatrix} 0 \\ 0 \end{pmatrix}.$$

We see that $k = \frac{1}{3}$.

2. Solving for c_1 and c_2 we find $c_1 = -\frac{3}{4}$ and $c_2 = \frac{1}{4}$.

3. Since

$$\begin{pmatrix} 4 & 6 & 6 \\ 1 & 3 & 2 \\ -1 & -4 & -3 \end{pmatrix} \begin{pmatrix} 3 \\ 1 \\ -1 \end{pmatrix} = \begin{pmatrix} 12 \\ 4 \\ -4 \end{pmatrix} = 4 \begin{pmatrix} 3 \\ 1 \\ -1 \end{pmatrix},$$

we see that $\lambda = 4$ is an eigenvalue with eigenvector $\mathbf{K}_3$. The corresponding solution is $\mathbf{X}_3 = \mathbf{K}_3 e^{4t}$.

4. The other eigenvalue is $\lambda_2 = 1 - 2i$ with corresponding eigenvector $\mathbf{K}_2 = \begin{pmatrix} 1 \\ -i \end{pmatrix}$. The general solution is

$$\mathbf{X}(t) = c_1 \begin{pmatrix} \cos 2t \\ -\sin 2t \end{pmatrix} e^t + c_2 \begin{pmatrix} \sin 2t \\ \cos 2t \end{pmatrix} e^t.$$

5. We have $\det(\mathbf{A} - \lambda\mathbf{I}) = (\lambda - 1)^2 = 0$ and $\mathbf{K} = \begin{pmatrix} 1 \\ -1 \end{pmatrix}$. A solution to $(\mathbf{A} - \lambda\mathbf{I})\mathbf{P} = \mathbf{K}$ is $\mathbf{P} = \begin{pmatrix} 0 \\ 1 \end{pmatrix}$ so that

$$\mathbf{X} = c_1 \begin{pmatrix} 1 \\ -1 \end{pmatrix} e^t + c_2 \left[\begin{pmatrix} 1 \\ -1 \end{pmatrix} te^t + \begin{pmatrix} 0 \\ 1 \end{pmatrix} e^t \right].$$

6. We have $\det(\mathbf{A} - \lambda\mathbf{I}) = (\lambda + 6)(\lambda + 2) = 0$ so that

$$\mathbf{X} = c_1 \begin{pmatrix} 1 \\ -1 \end{pmatrix} e^{-6t} + c_2 \begin{pmatrix} 1 \\ 1 \end{pmatrix} e^{-2t}.$$

7. We have $\det(\mathbf{A} - \lambda\mathbf{I}) = \lambda^2 - 2\lambda + 5 = 0$. For $\lambda = 1 + 2i$ we obtain $\mathbf{K}_1 = \begin{pmatrix} 1 \\ i \end{pmatrix}$ and

$$\mathbf{X}_1 = \begin{pmatrix} 1 \\ i \end{pmatrix} e^{(1+2i)t} = \begin{pmatrix} \cos 2t \\ -\sin 2t \end{pmatrix} e^t + i \begin{pmatrix} \sin 2t \\ \cos 2t \end{pmatrix} e^t.$$

Then

$$\mathbf{X} = c_1 \begin{pmatrix} \cos 2t \\ -\sin 2t \end{pmatrix} e^t + c_2 \begin{pmatrix} \sin 2t \\ \cos 2t \end{pmatrix} e^t.$$

8. We have $\det(\mathbf{A} - \lambda\mathbf{I}) = \lambda^2 - 2\lambda + 2 = 0$. For $\lambda = 1 + i$ we obtain $\mathbf{K}_1 = \begin{pmatrix} 3 - i \\ 2 \end{pmatrix}$ and

$$\mathbf{X}_1 = \begin{pmatrix} 3 - i \\ 2 \end{pmatrix} e^{(1+i)t} = \begin{pmatrix} 3\cos t + \sin t \\ 2\cos t \end{pmatrix} e^t + i \begin{pmatrix} -\cos t + 3\sin t \\ 2\sin t \end{pmatrix} e^t.$$

Then

$$\mathbf{X} = c_1 \begin{pmatrix} 3\cos t + \sin t \\ 2\cos t \end{pmatrix} e^t + c_2 \begin{pmatrix} -\cos t + 3\sin t \\ 2\sin t \end{pmatrix} e^t.$$

9. We have $\det(\mathbf{A} - \lambda\mathbf{I}) = -(\lambda - 2)(\lambda - 4)(\lambda + 3) = 0$ so that

$$\mathbf{X} = c_1 \begin{pmatrix} -2 \\ 3 \\ 1 \end{pmatrix} e^{2t} + c_2 \begin{pmatrix} 0 \\ 1 \\ 1 \end{pmatrix} e^{4t} + c_3 \begin{pmatrix} 7 \\ 12 \\ -16 \end{pmatrix} e^{-3t}.$$

10. We have $\det(\mathbf{A} - \lambda\mathbf{I}) = -(\lambda + 2)(\lambda^2 - 2\lambda + 3) = 0$. The eigenvalues are $\lambda_1 = -2$, $\lambda_2 = 1 + \sqrt{2}i$, and $\lambda_2 = 1 - \sqrt{2}i$, with eigenvectors

$$\mathbf{K}_1 = \begin{pmatrix} -7 \\ 5 \\ 4 \end{pmatrix}, \quad \mathbf{K}_2 = \begin{pmatrix} 1 \\ \sqrt{2}i/2 \\ 1 \end{pmatrix}, \quad \text{and} \quad \mathbf{K}_3 = \begin{pmatrix} 1 \\ -\sqrt{2}i/2 \\ 1 \end{pmatrix}.$$

Thus

$$\mathbf{X} = c_1 \begin{pmatrix} -7 \\ 5 \\ 4 \end{pmatrix} e^{-2t} + c_2 \left[\begin{pmatrix} 1 \\ 0 \\ 1 \end{pmatrix} \cos\sqrt{2}t - \begin{pmatrix} 0 \\ \sqrt{2}/2 \\ 0 \end{pmatrix} \sin\sqrt{2}t \right] e^t$$

$$+ c_3 \left[\begin{pmatrix} 0 \\ \sqrt{2}/2 \\ 0 \end{pmatrix} \cos\sqrt{2}t + \begin{pmatrix} 1 \\ 0 \\ 1 \end{pmatrix} \sin\sqrt{2}t \right] e^t$$

$$= c_1 \begin{pmatrix} -7 \\ 5 \\ 4 \end{pmatrix} e^{-2t} + c_2 \begin{pmatrix} \cos\sqrt{2}t \\ -\frac{1}{2}\sqrt{2}\sin\sqrt{2}t \\ \cos\sqrt{2}t \end{pmatrix} e^t + c_3 \begin{pmatrix} \sin\sqrt{2}t \\ \frac{1}{2}\sqrt{2}\cos\sqrt{2}t \\ \sin\sqrt{2}t \end{pmatrix} e^t.$$

11. We have

$$\mathbf{X}_c = c_1 \begin{pmatrix} 1 \\ 0 \end{pmatrix} e^{2t} + c_2 \begin{pmatrix} 4 \\ 1 \end{pmatrix} e^{4t}.$$

Then

$$\boldsymbol{\Phi} = \begin{pmatrix} e^{2t} & 4e^{4t} \\ 0 & e^{4t} \end{pmatrix}, \quad \boldsymbol{\Phi}^{-1} = \begin{pmatrix} e^{-2t} & -4e^{-2t} \\ 0 & e^{-4t} \end{pmatrix},$$

and

$$\mathbf{U} = \int \boldsymbol{\Phi}^{-1}\mathbf{F}\,dt = \int \begin{pmatrix} 2e^{-2t} - 64te^{-2t} \\ 16te^{-4t} \end{pmatrix} dt = \begin{pmatrix} 15e^{-2t} + 32te^{-2t} \\ -e^{-4t} - 4te^{-4t} \end{pmatrix},$$

so that

$$\mathbf{X}_p = \boldsymbol{\Phi}\mathbf{U} = \begin{pmatrix} 11 + 16t \\ -1 - 4t \end{pmatrix}.$$

12. We have

$$\mathbf{X}_c = c_1 \begin{pmatrix} 2\cos t \\ -\sin t \end{pmatrix} e^t + c_2 \begin{pmatrix} 2\sin t \\ \cos t \end{pmatrix} e^t.$$

Then

$$\boldsymbol{\Phi} = \begin{pmatrix} 2\cos t & 2\sin t \\ -\sin t & \cos t \end{pmatrix} e^t, \quad \boldsymbol{\Phi}^{-1} = \begin{pmatrix} \frac{1}{2}\cos t & -\sin t \\ \frac{1}{2}\sin t & \cos t \end{pmatrix} e^{-t},$$

and

$$\mathbf{U} = \int \boldsymbol{\Phi}^{-1}\mathbf{F}\,dt = \int \begin{pmatrix} \cos t - \sec t \\ \sin t \end{pmatrix} dt = \begin{pmatrix} \sin t - \ln|\sec t + \tan t| \\ -\cos t \end{pmatrix},$$

so that

$$\mathbf{X}_p = \boldsymbol{\Phi}\mathbf{U} = \begin{pmatrix} -2\cos t \ln|\sec t + \tan t| \\ -1 + \sin t \ln|\sec t + \tan t| \end{pmatrix} e^t.$$

13. We have

$$\mathbf{X}_c = c_1 \begin{pmatrix} \cos t + \sin t \\ 2\cos t \end{pmatrix} + c_2 \begin{pmatrix} \sin t - \cos t \\ 2\sin t \end{pmatrix}.$$

Then

$$\boldsymbol{\Phi} = \begin{pmatrix} \cos t + \sin t & \sin t - \cos t \\ 2\cos t & 2\sin t \end{pmatrix}, \quad \boldsymbol{\Phi}^{-1} = \begin{pmatrix} \sin t & \frac{1}{2}\cos t - \frac{1}{2}\sin t \\ -\cos t & \frac{1}{2}\cos t + \frac{1}{2}\sin t \end{pmatrix},$$

and

$$\mathbf{U} = \int \boldsymbol{\Phi}^{-1}\mathbf{F}\,dt = \int \begin{pmatrix} \frac{1}{2}\sin t - \frac{1}{2}\cos t + \frac{1}{2}\csc t \\ -\frac{1}{2}\sin t - \frac{1}{2}\cos t + \frac{1}{2}\csc t \end{pmatrix} dt$$

$$= \begin{pmatrix} -\frac{1}{2}\cos t - \frac{1}{2}\sin t + \frac{1}{2}\ln|\csc t - \cot t| \\ \frac{1}{2}\cos t - \frac{1}{2}\sin t + \frac{1}{2}\ln|\csc t - \cot t| \end{pmatrix},$$

so that

$$\mathbf{X}_p = \mathbf{\Phi U} = \begin{pmatrix} -1 \\ -1 \end{pmatrix} + \begin{pmatrix} \sin t \\ \sin t + \cos t \end{pmatrix} \ln|\csc t - \cot t|.$$

14. We have

$$\mathbf{X}_c = c_1 \begin{pmatrix} 1 \\ -1 \end{pmatrix} e^{2t} + c_2 \left[\begin{pmatrix} 1 \\ -1 \end{pmatrix} t e^{2t} + \begin{pmatrix} 1 \\ 0 \end{pmatrix} e^{2t} \right].$$

Then

$$\mathbf{\Phi} = \begin{pmatrix} e^{2t} & te^{2t} + e^{2t} \\ -e^{2t} & -te^{2t} \end{pmatrix}, \qquad \mathbf{\Phi}^{-1} = \begin{pmatrix} -te^{-2t} & -te^{-2t} - e^{-2t} \\ e^{-2t} & e^{-2t} \end{pmatrix},$$

and

$$\mathbf{U} = \int \mathbf{\Phi}^{-1}\mathbf{F}\, dt = \int \begin{pmatrix} t - 1 \\ -1 \end{pmatrix} dt = \begin{pmatrix} \frac{1}{2}t^2 - t \\ -t \end{pmatrix},$$

so that

$$\mathbf{X}_p = \mathbf{\Phi U} = \begin{pmatrix} -1/2 \\ 1/2 \end{pmatrix} t^2 e^{2t} + \begin{pmatrix} -2 \\ 1 \end{pmatrix} te^{2t}.$$

15. (a) Letting

$$\mathbf{K} = \begin{pmatrix} k_1 \\ k_2 \\ k_3 \end{pmatrix}$$

we note that $(\mathbf{A} - 2\mathbf{I})\mathbf{K} = \mathbf{0}$ implies that $3k_1 + 3k_2 + 3k_3 = 0$, so $k_1 = -(k_2 + k_3)$. Choosing $k_2 = 0$, $k_3 = 1$ and then $k_2 = 1$, $k_3 = 0$ we get

$$\mathbf{K}_1 = \begin{pmatrix} -1 \\ 0 \\ 1 \end{pmatrix} \quad \text{and} \quad \mathbf{K}_2 = \begin{pmatrix} -1 \\ 1 \\ 0 \end{pmatrix},$$

respectively. Thus,

$$\mathbf{X}_1 = \begin{pmatrix} -1 \\ 0 \\ 1 \end{pmatrix} e^{2t} \quad \text{and} \quad \mathbf{X}_2 = \begin{pmatrix} -1 \\ 1 \\ 0 \end{pmatrix} e^{2t}$$

are two solutions.

(b) From $\det(\mathbf{A} - \lambda\mathbf{I}) = \lambda^2(3 - \lambda) = 0$ we see that $\lambda_1 = 3$, and 0 is an eigenvalue of multiplicity two. Letting

$$\mathbf{K} = \begin{pmatrix} k_1 \\ k_2 \\ k_3 \end{pmatrix},$$

as in part (**a**), we note that $(\mathbf{A} - 0\mathbf{I})\mathbf{K} = \mathbf{AK} = \mathbf{0}$ implies that $k_1 + k_2 + k_3 = 0$, so $k_1 = -(k_2 + k_3)$. Choosing $k_2 = 0$, $k_3 = 1$, and then $k_2 = 1$, $k_3 = 0$ we get

$$\mathbf{K}_2 = \begin{pmatrix} -1 \\ 0 \\ 1 \end{pmatrix} \quad \text{and} \quad \mathbf{K}_3 = \begin{pmatrix} -1 \\ 1 \\ 0 \end{pmatrix},$$

respectively. Since the eigenvector corresponding to $\lambda_1 = 3$ is

$$\mathbf{K}_1 = \begin{pmatrix} 1 \\ 1 \\ 1 \end{pmatrix},$$

the general solution of the system is

$$\mathbf{X} = c_1 \begin{pmatrix} 1 \\ 1 \\ 1 \end{pmatrix} e^{3t} + c_2 \begin{pmatrix} -1 \\ 0 \\ 1 \end{pmatrix} + c_3 \begin{pmatrix} -1 \\ 1 \\ 0 \end{pmatrix}.$$

16. For $\mathbf{X} = \begin{pmatrix} c_1 \\ c_2 \end{pmatrix} e^t$ we have $\mathbf{X}' = \mathbf{X} = \mathbf{IX}$.

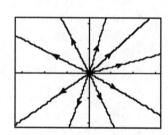

9 Numerical Solutions of Ordinary Differential Equations

1.

h=0.1			h=0.05	
x_n	y_n		x_n	y_n
1.00	5.0000		1.00	5.0000
1.10	3.9900		1.05	4.4475
1.20	3.2546		1.10	3.9763
1.30	2.7236		1.15	3.5751
1.40	2.3451		1.20	3.2342
1.50	2.0801		1.25	2.9452
			1.30	2.7009
			1.35	2.4952
			1.40	2.3226
			1.45	2.1786
			1.50	2.0592

2.

h=0.1			h=0.05	
x_n	y_n		x_n	y_n
0.00	2.0000		0.00	2.0000
0.10	1.6600		0.05	1.8150
0.20	1.4172		0.10	1.6571
0.30	1.2541		0.15	1.5237
0.40	1.1564		0.20	1.4124
0.50	1.1122		0.25	1.3212
			0.30	1.2482
			0.35	1.1916
			0.40	1.1499
			0.45	1.1217
			0.50	1.1056

3.

h=0.1			h=0.05	
x_n	y_n		x_n	y_n
0.00	0.0000		0.00	0.0000
0.10	0.1005		0.05	0.0501
0.20	0.2030		0.10	0.1004
0.30	0.3098		0.15	0.1512
0.40	0.4234		0.20	0.2028
0.50	0.5470		0.25	0.2554
			0.30	0.3095
			0.35	0.3652
			0.40	0.4230
			0.45	0.4832
			0.50	0.5465

4.

h=0.1			h=0.05	
x_n	y_n		x_n	y_n
0.00	1.0000		0.00	1.0000
0.10	1.1110		0.05	1.0526
0.20	1.2515		0.10	1.1113
0.30	1.4361		0.15	1.1775
0.40	1.6880		0.20	1.2526
0.50	2.0488		0.25	1.3388
			0.30	1.4387
			0.35	1.5556
			0.40	1.6939
			0.45	1.8598
			0.50	2.0619

5. $h=0.1$ $h=0.05$

x_n	y_n
0.00	0.0000
0.10	0.0952
0.20	0.1822
0.30	0.2622
0.40	0.3363
0.50	0.4053

x_n	y_n
0.00	0.0000
0.05	0.0488
0.10	0.0953
0.15	0.1397
0.20	0.1823
0.25	0.2231
0.30	0.2623
0.35	0.3001
0.40	0.3364
0.45	0.3715
0.50	0.4054

6. $h=0.1$ $h=0.05$

x_n	y_n
0.00	0.0000
0.10	0.0050
0.20	0.0200
0.30	0.0451
0.40	0.0805
0.50	0.1266

x_n	y_n
0.00	0.0000
0.05	0.0013
0.10	0.0050
0.15	0.0113
0.20	0.0200
0.25	0.0313
0.30	0.0451
0.35	0.0615
0.40	0.0805
0.45	0.1022
0.50	0.1266

7. $h=0.1$ $h=0.05$

x_n	y_n
0.00	0.5000
0.10	0.5215
0.20	0.5362
0.30	0.5449
0.40	0.5490
0.50	0.5503

x_n	y_n
0.00	0.5000
0.05	0.5116
0.10	0.5214
0.15	0.5294
0.20	0.5359
0.25	0.5408
0.30	0.5444
0.35	0.5469
0.40	0.5484
0.45	0.5492
0.50	0.5495

8. $h=0.1$ $h=0.05$

x_n	y_n
0.00	1.0000
0.10	1.1079
0.20	1.2337
0.30	1.3806
0.40	1.5529
0.50	1.7557

x_n	y_n
0.00	1.0000
0.05	1.0519
0.10	1.1079
0.15	1.1684
0.20	1.2337
0.25	1.3043
0.30	1.3807
0.35	1.4634
0.40	1.5530
0.45	1.6503
0.50	1.7560

9. $h=0.1$ $h=0.05$

x_n	y_n
1.00	1.0000
1.10	1.0095
1.20	1.0404
1.30	1.0967
1.40	1.1866
1.50	1.3260

x_n	y_n
1.00	1.0000
1.05	1.0024
1.10	1.0100
1.15	1.0228
1.20	1.0414
1.25	1.0663
1.30	1.0984
1.35	1.1389
1.40	1.1895
1.45	1.2526
1.50	1.3315

10. $h=0.1$ $h=0.05$

x_n	y_n
0.00	0.5000
0.10	0.5250
0.20	0.5498
0.30	0.5744
0.40	0.5986
0.50	0.6224

x_n	y_n
0.00	0.5000
0.05	0.5125
0.10	0.5250
0.15	0.5374
0.20	0.5498
0.25	0.5622
0.30	0.5744
0.35	0.5866
0.40	0.5987
0.45	0.6106
0.50	0.6224

11. To obtain the analytic solution use the substitution $u = x+y-1$. The resulting differential equation in $u(x)$ will be separable.

$h=0.1$

x_n	y_n	Actual Value
0.00	2.0000	2.0000
0.10	2.1220	2.1230
0.20	2.3049	2.3085
0.30	2.5858	2.5958
0.40	3.0378	3.0650
0.50	3.8254	3.9082

$h=0.05$

x_n	y_n	Actual Value
0.00	2.0000	2.0000
0.05	2.0553	2.1230
0.10	2.1228	2.3085
0.15	2.2056	2.5958
0.20	2.3075	3.0650
0.25	2.4342	3.9082
0.30	2.5931	2.5958
0.35	2.7953	2.7997
0.40	3.0574	3.0650
0.45	3.4057	3.4189
0.50	3.8840	3.9082

12. (a)

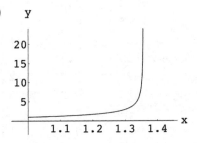

(b)

x_n	Euler	Imp. Euler
1.00	1.0000	1.0000
1.10	1.2000	1.2469
1.20	1.4938	1.6430
1.30	1.9711	2.4042
1.40	2.9060	4.5085

13. (a) Using Euler's method we obtain $y(0.1) \approx y_1 = 1.2$.

(b) Using $y'' = 4e^{2x}$ we see that the local truncation error is

$$y''(c)\frac{h^2}{2} = 4e^{2c}\frac{(0.1)^2}{2} = 0.02e^{2c}.$$

Since e^{2x} is an increasing function, $e^{2c} \le e^{2(0.1)} = e^{0.2}$ for $0 \le c \le 0.1$. Thus an upper bound for the local truncation error is $0.02e^{0.2} = 0.0244$.

(c) Since $y(0.1) = e^{0.2} = 1.2214$, the actual error is $y(0.1) - y_1 = 0.0214$, which is less than 0.0244.

(d) Using Euler's method with $h = 0.05$ we obtain $y(0.1) \approx y_2 = 1.21$.

(e) The error in (d) is $1.2214 - 1.21 = 0.0114$. With global truncation error $O(h)$, when the step size is halved we expect the error for $h = 0.05$ to be one-half the error when $h = 0.1$. Comparing 0.0114 with 0.0214 we see that this is the case.

14. (a) Using the improved Euler's method we obtain $y(0.1) \approx y_1 = 1.22$.

(b) Using $y''' = 8e^{2x}$ we see that the local truncation error is

$$y'''(c)\frac{h^3}{6} = 8e^{2c}\frac{(0.1)^3}{6} = 0.001333e^{2c}.$$

480

Since e^{2x} is an increasing function, $e^{2c} \leq e^{2(0.1)} = e^{0.2}$ for $0 \leq c \leq 0.1$. Thus an upper bound for the local truncation error is $0.001333e^{0.2} = 0.001628$.

(c) Since $y(0.1) = e^{0.2} = 1.221403$, the actual error is $y(0.1) - y_1 = 0.001403$ which is less than 0.001628.

(d) Using the improved Euler's method with $h = 0.05$ we obtain $y(0.1) \approx y_2 = 1.221025$.

(e) The error in (d) is $1.221403 - 1.221025 = 0.000378$. With global truncation error $O(h^2)$, when the step size is halved we expect the error for $h = 0.05$ to be one-fourth the error for $h = 0.1$. Comparing 0.000378 with 0.001403 we see that this is the case.

15. (a) Using Euler's method we obtain $y(0.1) \approx y_1 = 0.8$.

   (b) Using $y'' = 5e^{-2x}$ we see that the local truncation error is

$$5e^{-2c} \frac{(0.1)^2}{2} = 0.025e^{-2c}.$$

   Since e^{-2x} is a decreasing function, $e^{-2c} \leq e^0 = 1$ for $0 \leq c \leq 0.1$. Thus an upper bound for the local truncation error is $0.025(1) = 0.025$.

   (c) Since $y(0.1) = 0.8234$, the actual error is $y(0.1) - y_1 = 0.0234$, which is less than 0.025.

   (d) Using Euler's method with $h = 0.05$ we obtain $y(0.1) \approx y_2 = 0.8125$.

   (e) The error in (d) is $0.8234 - 0.8125 = 0.0109$. With global truncation error $O(h)$, when the step size is halved we expect the error for $h = 0.05$ to be one-half the error when $h = 0.1$. Comparing 0.0109 with 0.0234 we see that this is the case.

16. (a) Using the improved Euler's method we obtain $y(0.1) \approx y_1 = 0.825$.

   (b) Using $y''' = -10e^{-2x}$ we see that the local truncation error is

$$10e^{-2c} \frac{(0.1)^3}{6} = 0.001667e^{-2c}.$$

   Since e^{-2x} is a decreasing function, $e^{-2c} \leq e^0 = 1$ for $0 \leq c \leq 0.1$. Thus an upper bound for the local truncation error is $0.001667(1) = 0.001667$.

   (c) Since $y(0.1) = 0.823413$, the actual error is $y(0.1) - y_1 = 0.001587$, which is less than 0.001667.

   (d) Using the improved Euler's method with $h = 0.05$ we obtain $y(0.1) \approx y_2 = 0.823781$.

   (e) The error in (d) is $|0.823413 - 0.8237181| = 0.000305$. With global truncation error $O(h^2)$, when the step size is halved we expect the error for $h = 0.05$ to be one-fourth the error when $h = 0.1$. Comparing 0.000305 with 0.001587 we see that this is the case.

17. (a) Using $y'' = 38e^{-3(x-1)}$ we see that the local truncation error is

$$y''(c) \frac{h^2}{2} = 38e^{-3(c-1)} \frac{h^2}{2} = 19h^2 e^{-3(c-1)}.$$

(b) Since $e^{-3(x-1)}$ is a decreasing function for $1 \leq x \leq 1.5$, $e^{-3(c-1)} \leq e^{-3(1-1)} = 1$ for $1 \leq c \leq 1.5$ and

$$y''(c)\frac{h^2}{2} \leq 19(0.1)^2(1) = 0.19.$$

(c) Using Euler's method with $h = 0.1$ we obtain $y(1.5) \approx 1.8207$. With $h = 0.05$ we obtain $y(1.5) \approx 1.9424$.

(d) Since $y(1.5) = 2.0532$, the error for $h = 0.1$ is $E_{0.1} = 0.2325$, while the error for $h = 0.05$ is $E_{0.05} = 0.1109$. With global truncation error $O(h)$ we expect $E_{0.1}/E_{0.05} \approx 2$. We actually have $E_{0.1}/E_{0.05} = 2.10$.

18. (a) Using $y''' = -114e^{-3(x-1)}$ we see that the local truncation error is

$$\left| y'''(c)\frac{h^3}{6} \right| = 114e^{-3(x-1)}\frac{h^3}{6} = 19h^3 e^{-3(c-1)}.$$

(b) Since $e^{-3(x-1)}$ is a decreasing function for $1 \leq x \leq 1.5$, $e^{-3(c-1)} \leq e^{-3(1-1)} = 1$ for $1 \leq c \leq 1.5$ and

$$\left| y'''(c)\frac{h^3}{6} \right| \leq 19(0.1)^3(1) = 0.019.$$

(c) Using the improved Euler's method with $h = 0.1$ we obtain $y(1.5) \approx 2.080108$. With $h = 0.05$ we obtain $y(1.5) \approx 2.059166$.

(d) Since $y(1.5) = 2.053216$, the error for $h = 0.1$ is $E_{0.1} = 0.026892$, while the error for $h = 0.05$ is $E_{0.05} = 0.005950$. With global truncation error $O(h^2)$ we expect $E_{0.1}/E_{0.05} \approx 4$. We actually have $E_{0.1}/E_{0.05} = 4.52$.

19. (a) Using $y'' = -1/(x+1)^2$ we see that the local truncation error is

$$\left| y''(c)\frac{h^2}{2} \right| = \frac{1}{(c+1)^2}\frac{h^2}{2}.$$

(b) Since $1/(x+1)^2$ is a decreasing function for $0 \leq x \leq 0.5$, $1/(c+1)^2 \leq 1/(0+1)^2 = 1$ for $0 \leq c \leq 0.5$ and

$$\left| y''(c)\frac{h^2}{2} \right| \leq (1)\frac{(0.1)^2}{2} = 0.005.$$

(c) Using Euler's method with $h = 0.1$ we obtain $y(0.5) \approx 0.4198$. With $h = 0.05$ we obtain $y(0.5) \approx 0.4124$.

(d) Since $y(0.5) = 0.4055$, the error for $h = 0.1$ is $E_{0.1} = 0.0143$, while the error for $h = 0.05$ is $E_{0.05} = 0.0069$. With global truncation error $O(h)$ we expect $E_{0.1}/E_{0.05} \approx 2$. We actually have $E_{0.1}/E_{0.05} = 2.06$.

20. (a) Using $y''' = 2/(x+1)^3$ we see that the local truncation error is

$$y'''(c)\frac{h^3}{6} = \frac{1}{(c+1)^3}\frac{h^3}{3}.$$

(b) Since $1/(x+1)^3$ is a decreasing function for $0 \le x \le 0.5$, $1/(c+1)^3 \le 1/(0+1)^3 = 1$ for $0 \le c \le 0.5$ and

$$y'''(c)\frac{h^3}{6} \le (1)\frac{(0.1)^3}{3} = 0.000333.$$

(c) Using the improved Euler's method with $h = 0.1$ we obtain $y(0.5) \approx 0.405281$. With $h = 0.05$ we obtain $y(0.5) \approx 0.405419$.

(d) Since $y(0.5) = 0.405465$, the error for $h = 0.1$ is $E_{0.1} = 0.000184$, while the error for $h = 0.05$ is $E_{0.05} = 0.000046$. With global truncation error $O(h^2)$ we expect $E_{0.1}/E_{0.05} \approx 4$. We actually have $E_{0.1}/E_{0.05} = 3.98$.

21. Because y_{n+1}^* depends on y_n and is used to determine y_{n+1}, all of the y_n^* cannot be computed at one time independently of the corresponding y_n values. For example, the computation of y_4^* involves the value of y_3.

Exercises 9.2

Runge-Kutta Methods

1.

x_n	y_n	Actual Value
0.00	2.0000	2.0000
0.10	2.1230	2.1230
0.20	2.3085	2.3085
0.30	2.5958	2.5958
0.40	3.0649	3.0650
0.50	3.9078	3.9082

2. In this problem we use $h = 0.1$. Substituting $w_2 = \frac{3}{4}$ into the equations in (4) in the text, we obtain

$$w_1 = 1 - w_2 = \frac{1}{4}, \quad \alpha = \frac{1}{2w_2} = \frac{2}{3}, \quad \text{and} \quad \beta = \frac{1}{2w_2} = \frac{2}{3}.$$

x_n	Second–Order Runge–Kutta	Improved Euler
0.00	2.0000	2.0000
0.10	2.1213	2.1220
0.20	2.3030	2.3049
0.30	2.5814	2.5858
0.40	3.0277	3.0378
0.50	3.8002	3.8254

The resulting second-order Runge-Kutta method is

$$y_{n+1} = y_n + h\left(\frac{1}{4}k_1 + \frac{3}{4}k_2\right) = y_n + \frac{h}{4}(k_1 + 3k_2)$$

where

$$k_1 = f(x_n, y_n)$$

$$k_2 = f\left(x_n + \frac{2}{3}h, y_n + \frac{2}{3}hk_1\right).$$

The table compares the values obtained using this second-order Runge-Kutta method with the values obtained using the improved Euler's method.

3.

x_n	y_n
1.00	5.0000
1.10	3.9724
1.20	3.2284
1.30	2.6945
1.40	2.3163
1.50	2.0533

4.

x_n	y_n
0.00	2.0000
0.10	1.6562
0.20	1.4110
0.30	1.2465
0.40	1.1480
0.50	1.1037

5.

x_n	y_n
0.00	0.0000
0.10	0.1003
0.20	0.2027
0.30	0.3093
0.40	0.4228
0.50	0.5463

6.

x_n	y_n
0.00	1.0000
0.10	1.1115
0.20	1.2530
0.30	1.4397
0.40	1.6961
0.50	2.0670

7.

x_n	y_n
0.00	0.0000
0.10	0.0953
0.20	0.1823
0.30	0.2624
0.40	0.3365
0.50	0.4055

8.

x_n	y_n
0.00	0.0000
0.10	0.0050
0.20	0.0200
0.30	0.0451
0.40	0.0805
0.50	0.1266

9.

x_n	y_n
0.00	0.5000
0.10	0.5213
0.20	0.5358
0.30	0.5443
0.40	0.5482
0.50	0.5493

10.

x_n	y_n
0.00	1.0000
0.10	1.1079
0.20	1.2337
0.30	1.3807
0.40	1.5531
0.50	1.7561

11.

x_n	y_n
1.00	1.0000
1.10	1.0101
1.20	1.0417
1.30	1.0989
1.40	1.1905
1.50	1.3333

12.

x_n	y_n
0.00	0.5000
0.10	0.5250
0.20	0.5498
0.30	0.5744
0.40	0.5987
0.50	0.6225

13. (a) Write the equation in the form

$$\frac{dv}{dt} = 32 - 0.025v^2 = f(t, v).$$

t_n	v_n
0.0	0.0000
1.0	25.2570
2.0	32.9390
3.0	34.9770
4.0	35.5500
5.0	35.7130

(b)

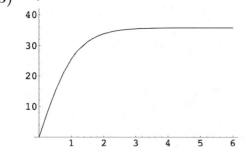

(c) Separating variables and using partial fractions we have

$$\frac{1}{2\sqrt{32}} \left(\frac{1}{\sqrt{32} - \sqrt{0.125}\,v} + \frac{1}{\sqrt{32} + \sqrt{0.125}\,v} \right) dv = dt$$

and

$$\frac{1}{2\sqrt{32}\sqrt{0.125}} \left(\ln|\sqrt{32} + \sqrt{0.125}\,v| - \ln|\sqrt{32} - \sqrt{0.125}\,v| \right) = t + c.$$

Since $v(0) = 0$ we find $c = 0$. Solving for v we obtain

$$v(t) = \frac{16\sqrt{5}\,(e^{\sqrt{3.2}\,t} - 1)}{e^{\sqrt{3.2}\,t} + 1}$$

and $v(5) \approx 35.7678$. Alternatively, the solution can be expressed as

$$v(t) = \sqrt{\frac{mg}{k}} \, \tanh \sqrt{\frac{kg}{m}} \, t.$$

14. (a)

t (days)	1	2	3	4	5
A (observed)	2.78	13.53	36.30	47.50	49.40
A (approximated)	1.93	12.50	36.46	47.23	49.00

(b) From the graph we estimate $A(1) \approx 1.68$, $A(2) \approx 13.2$, $A(3) \approx 36.8$, $A(4) \approx 46.9$, and $A(5) \approx 48.9$.

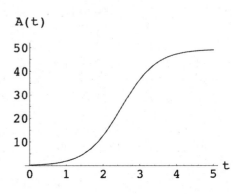

(c) Let $\alpha = 2.128$ and $\beta = 0.0432$. Separating variables we obtain

$$\frac{dA}{A(\alpha - \beta A)} = dt$$

$$\frac{1}{\alpha} \left(\frac{1}{A} + \frac{\beta}{\alpha - \beta A} \right) dA = dt$$

$$\frac{1}{\alpha} [\ln A - \ln(\alpha - \beta A)] = t + c$$

$$\ln \frac{A}{\alpha - \beta A} = \alpha(t + c)$$

$$\frac{A}{\alpha - \beta A} = e^{\alpha(t+c)}$$

$$A = \alpha e^{\alpha(t+c)} - \beta A e^{\alpha(t+c)}$$

$$\left[1 + \beta e^{\alpha(t+c)} \right] A = \alpha e^{\alpha(t+c)}.$$

Thus

$$A(t) = \frac{\alpha e^{\alpha(t+c)}}{1 + \beta e^{\alpha(t+c)}} = \frac{\alpha}{\beta + e^{-\alpha(t+c)}} = \frac{\alpha}{\beta + e^{-\alpha c} e^{-\alpha t}} \, .$$

From $A(0) = 0.24$ we obtain

$$0.24 = \frac{\alpha}{\beta + e^{-\alpha c}}$$

so that $e^{-\alpha c} = \alpha/0.24 - \beta \approx 8.8235$ and

$$A(t) \approx \frac{2.128}{0.0432 + 8.8235e^{-2.128t}}.$$

t (days)	1	2	3	4	5
A (observed)	2.78	13.53	36.30	47.50	49.40
A (actual)	1.93	12.50	36.46	47.23	49.00

15. (a)

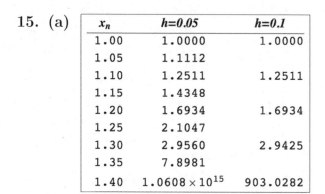

(b)

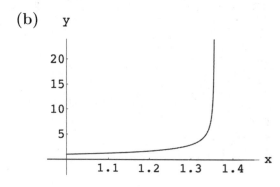

16. (a) Using the RK4 method we obtain $y(0.1) \approx y_1 = 1.2214$.

(b) Using $y^{(5)}(x) = 32e^{2x}$ we see that the local truncation error is

$$y^{(5)}(c)\frac{h^5}{120} = 32e^{2c}\frac{(0.1)^5}{120} = 0.000002667e^{2c}.$$

Since e^{2x} is an increasing function, $e^{2c} \le e^{2(0.1)} = e^{0.2}$ for $0 \le c \le 0.1$. Thus an upper bound for the local truncation error is $0.000002667e^{0.2} = 0.000003257$.

(c) Since $y(0.1) = e^{0.2} = 1.221402758$, the actual error is $y(0.1) - y_1 = 0.000002758$ which is less than 0.000003257.

(d) Using the RK4 formula with $h = 0.05$ we obtain $y(0.1) \approx y_2 = 1.221402571$.

(e) The error in (d) is $1.221402758 - 1.221402571 = 0.000000187$. With global truncation error $O(h^4)$, when the step size is halved we expect the error for $h = 0.05$ to be one-sixteenth the error for $h = 0.1$. Comparing 0.000000187 with 0.000002758 we see that this is the case.

17. (a) Using the RK4 method we obtain $y(0.1) \approx y_1 = 0.823416667$.

(b) Using $y^{(5)}(x) = -40e^{-2x}$ we see that the local truncation error is

$$40e^{-2c}\frac{(0.1)^5}{120} = 0.000003333.$$

Since e^{-2x} is a decreasing function, $e^{-2c} \le e^0 = 1$ for $0 \le c \le 0.1$. Thus an upper bound for the local truncation error is $0.000003333(1) = 0.000003333$.

(c) Since $y(0.1) = 0.823413441$, the actual error is $|y(0.1) - y_1| = 0.000003225$, which is less than 0.000003333.

(d) Using the RK4 method with $h = 0.05$ we obtain $y(0.1) \approx y_2 = 0.823413627$.

(e) The error in (d) is $|0.823413441 - 0.823413627| = 0.000000185$. With global truncation error $O(h^4)$, when the step size is halved we expect the error for $h = 0.05$ to be one-sixteenth the error when $h = 0.1$. Comparing 0.000000185 with 0.000003225 we see that this is the case.

18. (a) Using $y^{(5)} = -1026e^{-3(x-1)}$ we see that the local truncation error is

$$\left| y^{(5)}(c) \frac{h^5}{120} \right| = 8.55h^5 e^{-3(c-1)}.$$

(b) Since $e^{-3(x-1)}$ is a decreasing function for $1 \le x \le 1.5$, $e^{-3(c-1)} \le e^{-3(1-1)} = 1$ for $1 \le c \le 1.5$ and

$$y^{(5)}(c) \frac{h^5}{120} \le 8.55(0.1)^5(1) = 0.0000855.$$

(c) Using the RK4 method with $h = 0.1$ we obtain $y(1.5) \approx 2.053338827$. With $h = 0.05$ we obtain $y(1.5) \approx 2.053222989$.

19. (a) Using $y^{(5)} = 24/(x+1)^5$ we see that the local truncation error is

$$y^{(5)}(c) \frac{h^5}{120} = \frac{1}{(c+1)^5} \frac{h^5}{5}.$$

(b) Since $1/(x+1)^5$ is a decreasing function for $0 \le x \le 0.5$, $1/(c+1)^5 \le 1/(0+1)^5 = 1$ for $0 \le c \le 0.5$ and

$$y^{(5)}(c) \frac{h^5}{5} \le (1) \frac{(0.1)^5}{5} = 0.000002.$$

(c) Using the RK4 method with $h = 0.1$ we obtain $y(0.5) \approx 0.405465168$. With $h = 0.05$ we obtain $y(0.5) \approx 0.405465111$.

20. Each step of Euler's method requires only 1 function evaluation, while each step of the improved Euler's method requires 2 function evaluations – once at (x_n, y_n) and again at (x_{n+1}, y_{n+1}^*). The second-order Runge-Kutta methods require 2 function evaluations per step, while the RK4 method requires 4 function evaluations per step. To compare the methods we approximate the solution of $y' = (x + y - 1)^2$, $y(0) = 2$, at $x = 0.2$ using $h = 0.1$ for the Runge-Kutta method, $h = 0.05$ for the improved Euler's method, and $h = 0.025$ for Euler's method. For each method a total of 8 function evaluations is required. By comparing with the exact solution we see that the RK4 method appears to still give the most accurate result.

x_n	Euler h=0.025	Imp. Euler h=0.05	RK4 h=0.1	Actual
0.000	2.0000	2.0000	2.0000	2.0000
0.025	2.0250			2.0263
0.050	2.0526	2.0553		2.0554
0.075	2.0830			2.0875
0.100	2.1165	2.1228	2.1230	2.1230
0.125	2.1535			2.1624
0.150	2.1943	2.2056		2.2061
0.175	2.2395			2.2546
0.200	2.2895	2.3075	2.3085	2.3085

21. (a) For $y' + y = 10 \sin 3x$ an integrating factor is e^x so that

$$\frac{d}{dx}[e^x y] = 10 e^x \sin 3x \implies e^x y = e^x \sin 3x - 3 e^x \cos 3x + c$$

$$\implies y = \sin 3x - 3 \cos 3x + c e^{-x}.$$

When $x = 0$, $y = 0$, so $0 = -3 + c$ and $c = 3$. The solution is

$$y = \sin 3x - 3 \cos 3x + 3 e^{-x}.$$

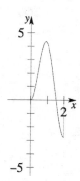

Using Newton's method we find that $x = 1.53235$ is the only positive root in $[0, 2]$.

(b) Using the RK4 method with $h = 0.1$ we obtain the table of values shown. These values are used to obtain an interpolating function in *Mathematica*. The graph of the interpolating function is shown. Using *Mathematica*'s root finding capability we see that the only positive root in $[0, 2]$ is $x = 1.53236$.

x_n	y_n	x_n	y_n
0.0	0.0000	1.0	4.2147
0.1	0.1440	1.1	3.8033
0.2	0.5448	1.2	3.1513
0.3	1.1409	1.3	2.3076
0.4	1.8559	1.4	1.3390
0.5	2.6049	1.5	0.3243
0.6	3.3019	1.6	-0.6530
0.7	3.8675	1.7	-1.5117
0.8	4.2356	1.8	-2.1809
0.9	4.3593	1.9	-2.6061
1.0	4.2147	2.0	-2.7539

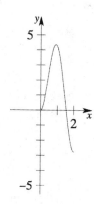

22. (*This is a Contributed Problem and the solution has been provided by the author of the problem.*)

The answers shown here pertain to the case $F \neq 0$, i.e. answers to question (**h**). Answers to questions (**a**) - (**g**) are obtained by setting $F = 0$.

(**a**) Divide both sides of the equation given in the text by the quantity $(M/2)$ to obtain

$$\left(\frac{dx}{dt}\right)^2 + \omega^2 x^2 + (2F/M)x = C,$$

where $\omega = \sqrt{k/M}$.

(**b**) Set $C = 1$ to obtain

$$\left(\frac{dx}{dt}\right)^2 + \omega^2 x^2 + (2F/M)x = 1.$$

Upon completing the square in the above equation we have

$$\left(\frac{dx}{dt}\right)^2 = -\left(\omega x + \frac{F}{M\omega}\right)^2 + \frac{F^2 + M^2\omega^2}{M^2\omega^2}.$$

If we let $u = \omega x + F/(M\omega)$ then this equation reduces to

$$\frac{du}{dt} = \frac{\sqrt{F^2 + M^2\omega^2}}{M}\sqrt{1 - \left(\frac{M^2\omega^2}{F^2 + M^2\omega^2}\right)u^2}. \tag{1}$$

Finally, with $y = M\omega^2/\sqrt{F^2 + M^2\omega^2}\, u$, equation (1) reduces to

$$\frac{dy}{dt} = \omega\sqrt{1 - y^2}, \quad \text{with} \quad y(0) = \frac{F}{\sqrt{M^2\omega^2 + F^2}}. \tag{2}$$

(**c**) Use Euler's method with $F = 10, k = 48,$ and $M = 3$ to solve

$$\frac{dy}{dt} = \omega\sqrt{1 - y^2}, \quad y(0) = \frac{F}{\sqrt{M^2\omega^2 + F^2}} \quad \text{or} \quad \frac{dy}{dt} = 4\sqrt{1 - y^2}, \quad y(0) = \frac{5}{\sqrt{61}}.$$

(**d**) Graphically, we observe (this can also be shown analytically) that the solution $y(t)$ starts at the intitial point $y_0 = y(0)$, increases almost linearly until it reaches 1 at time

$$t^* = \frac{\pi/2 - \arcsin y_0}{\omega}$$

and remains at 1 afterwards. The numerical solution is described by

$$y(t) = \begin{cases} \sin(\omega t + \arcsin y_0) & \text{if } 0 \leq t \leq t^*; \\ 1 & \text{if } t > t^*. \end{cases}$$

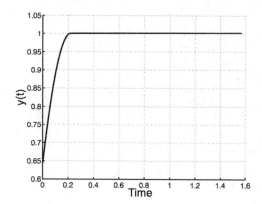

Figure 1: Plot of y(t) versus time for N=5000

Therefore, the numerical solution does not seem to capture the physics involved after $t > t^*$ since there are no oscillations. Note that the constant solution $y = 1$ is a solution to the initial-value problem. However, the solution is not physical.

(e) First separate variables and integrate

$$\int \frac{dy}{\sqrt{1-y^2}} = \int \omega \, dt$$

to obtain

$$\arcsin y = \omega t + C_0.$$

Upon using the initial condition, we find

$$y(t) = \sin(\omega t + \arcsin y_0).$$

The analytic solution does capture the oscillations of the spring.

(f) Differentiate both sides of equation (2) with respect to time to obtain

$$\frac{d^2 y}{dt^2} = \omega \left(-y \frac{dy}{dt} \right) \frac{1}{\sqrt{1-y^2}},$$

and then use the fact that $dy/dt = \omega\sqrt{1-y^2}$.

From equation (2), we have $y(0) = y_0$ and from equation 2 again, we have

$$y'(0) = \omega\sqrt{1-y_0^2}.$$

(g) First create the following function file (name it spring2.m)

```
function out=spring2(t,y);
omega=4;
out(1)=y(2);
```

out(2)=-$\omega^2 * y(1)$;

out=out';

---

then in the Matlab window, type the following commands:

$>>$ M $= 3; k = 48; \omega = \sqrt{k/M}; F = 10 :$

$>>$ y0 $= F/\sqrt{(M^2\omega^2 + F^2)}$

$>>$ $y_1 = \omega\sqrt{1 - y_0^2}$

$>>$ $[t, y] = ode45('spring2', [0, pi/2], [y_0, y_1] :$

$>>$ plot(t,y(:,1))

---

where $y_1 = dy/dt$ at $t = 0$. The resulting plot is shown in figure 2. The graph is consistent with the analytical solution $y(t) = \sin(\omega t + \arcsin y_0)$ from part (**e**).

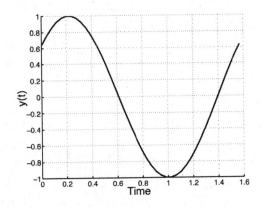

Figure 2: Plot of y(t) versus time using ODE45

The second-order differential equation has constant coefficients. The analytic solution can easily be obtained,

$$x(t) = \frac{\sqrt{F^2 + M^2\omega^2}}{M\omega^2}\left(y_0 \cos(\omega t) + \frac{y_1}{\omega} \sin(\omega t)\right) - \frac{F}{M\omega^2} .$$

In the tables in this section "ABM" stands for Adams-Bashforth-Moulton.

1. Writing the differential equation in the form $y' - y = x - 1$ we see that an integrating factor is $e^{-\int dx} = e^{-x}$, so that

$$\frac{d}{dx}[e^{-x}y] = (x-1)e^{-x}$$

and

$$y = e^x(-xe^{-x} + c) = -x + ce^x.$$

From $y(0) = 1$ we find $c = 1$, so the solution of the initial-value problem is $y = -x + e^x$. Actual values of the analytic solution above are compared with the approximated values in the table.

x_n	y_n	Actual	
0.0	1.00000000	1.00000000	init. cond.
0.2	1.02140000	1.02140276	RK4
0.4	1.09181796	1.09182470	RK4
0.6	1.22210646	1.22211880	RK4
0.8	1.42552788	1.42554093	ABM

2. The following program is written in *Mathematica*. It uses the Adams-Bashforth-Moulton method to approximate the solution of the initial-value problem $y' = x + y - 1$, $y(0) = 1$, on the interval $[0, 1]$.

```
Clear[f, x, y, h, a, b, y0];
f[x_, y_]:= x + y - 1;        (* define the differential equation *)
h = 0.2;                      (* set the step size *)
a = 0; y0 = 1; b = 1;         (* set the initial condition and the interval *)
f[x, y]                       (* display the DE *)

Clear[k1, k2, k3, k4, x, y, u, v]
x = u[0] = a;
y = v[0] = y0;
n = 0;
While[x < a + 3h,             (* use RK4 to compute the first 3 values after y(0) *)
    n = n + 1;
```

$$k1 = f[x, y];$$
$$k2 = f[x + h/2, y + h\ k1/2];$$
$$k3 = f[x + h/2, y + h\ k2/2];$$
$$k4 = f[x + h, y + h\ k3];$$
$$x = x + h;$$
$$y = y + (h/6)(k1 + 2k2 + 2k3 + k4);$$
$$u[n] = x;$$
$$v[n] = y];$$

While[x ≤ b,   (* use Adams-Bashforth-Moulton *)
$$p3 = f[u[n - 3], v[n - 3]];$$
$$p2 = f[u[n - 2], v[n - 2]];$$
$$p1 = f[u[n - 1], v[n - 1]];$$
$$p0 = f[u[n], v[n]];$$
pred = y + (h/24)(55p0 - 59p1 + 37p2 - 9p3);   (* predictor *)
$$x = x + h;$$
$$p4 = f[x, pred];$$
y = y + (h/24)(9p4 + 19p0 - 5p1 + p2);   (* corrector *)
$$n = n + 1;$$
$$u[n] = x;$$
$$v[n] = y]$$

(*display the table *)
TableForm[Prepend[Table[{u[n], v[n]}, {n, 0, (b-a)/h}], {"x(n)", "y(n)"}]];

3. The first predictor is $y_4^* = 0.73318477$.

x_n	y_n	
0.0	1.00000000	init. cond.
0.2	0.73280000	RK4
0.4	0.64608032	RK4
0.6	0.65851653	RK4
0.8	0.72319464	ABM

4. The first predictor is $y_4^* = 1.21092217$.

x_n	y_n	
0.0	2.00000000	init. cond.
0.2	1.41120000	RK4
0.4	1.14830848	RK4
0.6	1.10390600	RK4
0.8	1.20486982	ABM

5. The first predictor for $h = 0.2$ is $y_4^* = 1.02343488$.

x_n	h=0.2		h=0.1	
0.0	0.00000000	init. cond.	0.00000000	init. cond.
0.1			0.10033459	RK4
0.2	0.20270741	RK4	0.20270988	RK4
0.3			0.30933604	RK4
0.4	0.42278899	RK4	0.42279808	ABM
0.5			0.54631491	ABM
0.6	0.68413340	RK4	0.68416105	ABM
0.7			0.84233188	ABM
0.8	1.02969040	ABM	1.02971420	ABM
0.9			1.26028800	ABM
1.0	1.55685960	ABM	1.55762558	ABM

6. The first predictor for $h = 0.2$ is $y_4^* = 3.34828434$.

x_n	h=0.2		h=0.1	
0.0	1.00000000	init. cond.	1.00000000	init. cond.
0.1			1.21017082	RK4
0.2	1.44139950	RK4	1.44140511	RK4
0.3			1.69487942	RK4
0.4	1.97190167	RK4	1.97191536	ABM
0.5			2.27400341	ABM
0.6	2.60280694	RK4	2.60283209	ABM
0.7			2.96031780	ABM
0.8	3.34860927	ABM	3.34863769	ABM
0.9			3.77026548	ABM
1.0	4.22797875	ABM	4.22801028	ABM

7. The first predictor for $h = 0.2$ is $y_4^* = 0.13618654$.

x_n	h=0.2		h=0.1	
0.0	0.00000000	init. cond.	0.00000000	init. cond.
0.1			0.00033209	RK4
0.2	0.00262739	RK4	0.00262486	RK4
0.3			0.00868768	RK4
0.4	0.02005764	RK4	0.02004821	ABM
0.5			0.03787884	ABM
0.6	0.06296284	RK4	0.06294717	ABM
0.7			0.09563116	ABM
0.8	0.13598600	ABM	0.13596515	ABM
0.9			0.18370712	ABM
1.0	0.23854783	ABM	0.23841344	ABM

8. The first predictor for $h = 0.2$ is $y_4^* = 2.61796154$.

x_n	h=0.2		h=0.1	
0.0	1.00000000	init. cond.	1.00000000	init. cond.
0.1			1.10793839	RK4
0.2	1.23369623	RK4	1.23369772	RK4
0.3			1.38068454	RK4
0.4	1.55308554	RK4	1.55309381	ABM
0.5			1.75610064	ABM
0.6	1.99610329	RK4	1.99612995	ABM
0.7			2.28119129	ABM
0.8	2.62136177	ABM	2.62131818	ABM
0.9			3.02914333	ABM
1.0	3.52079042	ABM	3.52065536	ABM

Exercises 9.4

Higher-Order Equations and Systems

1. The substitution $y' = u$ leads to the iteration formulas

$$y_{n+1} = y_n + hu_n, \qquad u_{n+1} = u_n + h(4u_n - 4y_n).$$

The initial conditions are $y_0 = -2$ and $u_0 = 1$. Then

$$y_1 = y_0 + 0.1u_0 = -2 + 0.1(1) = -1.9$$

$$u_1 = u_0 + 0.1(4u_0 - 4y_0) = 1 + 0.1(4 + 8) = 2.2$$

$$y_2 = y_1 + 0.1u_1 = -1.9 + 0.1(2.2) = -1.68.$$

The general solution of the differential equation is $y = c_1 e^{2x} + c_2 x e^{2x}$. From the initial conditions we find $c_1 = -2$ and $c_2 = 5$. Thus $y = -2e^{2x} + 5xe^{2x}$ and $y(0.2) \approx -1.4918$.

2. The substitution $y' = u$ leads to the iteration formulas

$$y_{n+1} = y_n + hu_n, \qquad u_{n+1} = u_n + h\left(\frac{2}{x}u_n - \frac{2}{x^2}y_n\right).$$

The initial conditions are $y_0 = 4$ and $u_0 = 9$. Then

$$y_1 = y_0 + 0.1u_0 = 4 + 0.1(9) = 4.9$$

$$u_1 = u_0 + 0.1\left(\frac{2}{1}u_0 - \frac{2}{1}y_0\right) = 9 + 0.1[2(9) - 2(4)] = 10$$

$$y_2 = y_1 + 0.1u_1 = 4.9 + 0.1(10) = 5.9.$$

The general solution of the Cauchy-Euler differential equation is $y = c_1 x + c_2 x^2$. From the initial conditions we find $c_1 = -1$ and $c_2 = 5$. Thus $y = -x + 5x^2$ and $y(1.2) = 6$.

3. The substitution $y' = u$ leads to the system

$$y' = u, \qquad u' = 4u - 4y.$$

Using formula (4) in the text with x corresponding to t, y corresponding to x, and u corresponding to y, we obtain the table shown.

x_n	h=0.2 y_n	h=0.2 u_n	h=0.1 y_n	h=0.1 u_n
0.0	-2.0000	1.0000	-2.0000	1.0000
0.1			-1.8321	2.4427
0.2	-1.4928	4.4731	-1.4919	4.4753

4. The substitution $y' = u$ leads to the system

$$y' = u, \qquad u' = \frac{2}{x}u - \frac{2}{x^2}y.$$

Using formula (4) in the text with x corresponding to t, y corresponding to x, and u corresponding to y, we obtain the table shown.

x_n	h=0.2 y_n	h=0.2 u_n	h=0.1 y_n	h=0.1 u_n
1.0	4.0000	9.0000	4.0000	9.0000
1.1			4.9500	10.0000
1.2	6.0001	11.0002	6.0000	11.0000

5. The substitution $y' = u$ leads to the system

$$y' = u, \qquad u' = 2u - 2y + e^t \cos t.$$

Using formula (4) in the text with y corresponding to x and u corresponding to y, we obtain the table shown.

x_n	h=0.2 y_n	h=0.2 u_n	h=0.1 y_n	h=0.1 u_n
0.0	1.0000	2.0000	1.0000	2.0000
0.1			1.2155	2.3150
0.2	1.4640	2.6594	1.4640	2.6594

6. Using $h = 0.1$, the RK4 method for a system, and a numerical solver, we obtain

t_n	h=0.2 i_{1n}	h=0.2 i_{3n}
0.0	0.0000	0.0000
0.1	2.5000	3.7500
0.2	2.8125	5.7813
0.3	2.0703	7.4023
0.4	0.6104	9.1919
0.5	-1.5619	11.4877

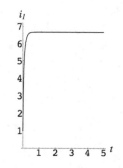
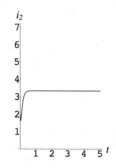

7.

t_n	h=0.2 x_n	h=0.2 y_n	h=0.1 x_n	h=0.1 y_n
0.0	6.0000	2.0000	6.0000	2.0000
0.1			7.0731	2.6524
0.2	8.3055	3.4199	8.3055	3.4199

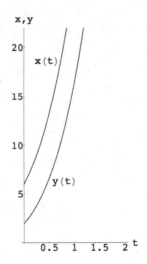

8.

t_n	h=0.2 x_n	h=0.2 y_n	h=0.1 x_n	h=0.1 y_n
0.0	1.0000	1.0000	1.0000	1.0000
0.1			1.4006	1.8963
0.2	2.0785	3.3382	2.0845	3.3502

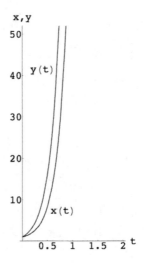

9.

t_n	h=0.2 x_n	h=0.2 y_n	h=0.1 x_n	h=0.1 y_n
0.0	-3.0000	5.0000	-3.0000	5.0000
0.1			-3.4790	4.6707
0.2	-3.9123	4.2857	-3.9123	4.2857

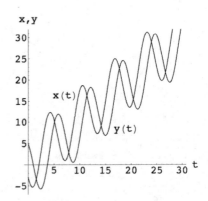

10.

t_n	h=0.2 x_n	h=0.2 y_n	h=0.1 x_n	h=0.1 y_n
0.0	0.5000	0.2000	0.5000	0.2000
0.1			1.0207	1.0115
0.2	2.1589	2.3279	2.1904	2.3592

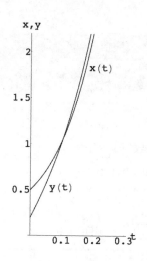

11. Solving for x' and y' we obtain the system

$$x' = -2x + y + 5t$$

$$y' = 2x + y - 2t.$$

t_n	h=0.2 x_n	h=0.2 y_n	h=0.1 x_n	h=0.1 y_n
0.0	1.0000	-2.0000	1.0000	-2.0000
0.1			0.6594	-2.0476
0.2	0.4179	-2.1824	0.4173	-2.1821

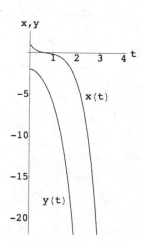

12. Solving for x' and y' we obtain the system

$$x' = \frac{1}{2}y - 3t^2 + 2t - 5$$

$$y' = -\frac{1}{2}y + 3t^2 + 2t + 5.$$

t_n	h=0.2 x_n	h=0.2 y_n	h=0.1 x_n	h=0.1 y_n
0.0	3.0000	-1.0000	3.0000	-1.0000
0.1			2.4727	-0.4527
0.2	1.9867	0.0933	1.9867	0.0933

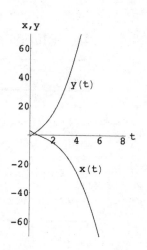

1. We identify $P(x) = 0$, $Q(x) = 9$, $f(x) = 0$, and $h = (2-0)/4 = 0.5$. Then the finite difference equation is

$$y_{i+1} + 0.25y_i + y_{i-1} = 0.$$

The solution of the corresponding linear system gives

x	0.0	0.5	1.0	1.5	2.0
y	4.0000	-5.6774	-2.5807	6.3226	1.0000

2. We identify $P(x) = 0$, $Q(x) = -1$, $f(x) = x^2$, and $h = (1-0)/4 = 0.25$. Then the finite difference equation is

$$y_{i+1} - 2.0625y_i + y_{i-1} = 0.0625x_i^2.$$

The solution of the corresponding linear system gives

x	0.00	0.25	0.50	0.75	1.00
y	0.0000	-0.0172	-0.0316	-0.0324	0.0000

3. We identify $P(x) = 2$, $Q(x) = 1$, $f(x) = 5x$, and $h = (1-0)/5 = 0.2$. Then the finite difference equation is

$$1.2y_{i+1} - 1.96y_i + 0.8y_{i-1} = 0.04(5x_i).$$

The solution of the corresponding linear system gives

x	0.0	0.2	0.4	0.6	0.8	1.0
y	0.0000	-0.2259	-0.3356	-0.3308	-0.2167	0.0000

4. We identify $P(x) = -10$, $Q(x) = 25$, $f(x) = 1$, and $h = (1-0)/5 = 0.2$. Then the finite difference equation is

$$-y_i + 2y_{i-1} = 0.04.$$

The solution of the corresponding linear system gives

x	0.0	0.2	0.4	0.6	0.8	1.0
y	1.0000	1.9600	3.8800	7.7200	15.4000	0.0000

5. We identify $P(x) = -4$, $Q(x) = 4$, $f(x) = (1+x)e^{2x}$, and $h = (1-0)/6 = 0.1667$. Then the finite difference equation is

$$0.6667y_{i+1} - 1.8889y_i + 1.3333y_{i-1} = 0.2778(1+x_i)e^{2x_i}.$$

The solution of the corresponding linear system gives

x	0.0000	0.1667	0.3333	0.5000	0.6667	0.8333	1.0000
y	3.0000	3.3751	3.6306	3.6448	3.2355	2.1411	0.0000

6. We identify $P(x) = 5$, $Q(x) = 0$, $f(x) = 4\sqrt{x}$, and $h = (2-1)/6 = 0.1667$. Then the finite difference equation is

$$1.4167 y_{i+1} - 2y_i + 0.5833 y_{i-1} = 0.2778(4\sqrt{x_i}).$$

The solution of the corresponding linear system gives

x	1.0000	1.1667	1.3333	1.5000	1.6667	1.8333	2.0000
y	1.0000	-0.5918	-1.1626	-1.3070	-1.2704	-1.1541	-1.0000

7. We identify $P(x) = 3/x$, $Q(x) = 3/x^2$, $f(x) = 0$, and $h = (2-1)/8 = 0.125$. Then the finite difference equation is

$$\left(1 + \frac{0.1875}{x_i}\right) y_{i+1} + \left(-2 + \frac{0.0469}{x_i^2}\right) y_i + \left(1 - \frac{0.1875}{x_i}\right) y_{i-1} = 0.$$

The solution of the corresponding linear system gives

x	1.000	1.125	1.250	1.375	1.500	1.625	1.750	1.875	2.000
y	5.0000	3.8842	2.9640	2.2064	1.5826	1.0681	0.6430	0.2913	0.0000

8. We identify $P(x) = -1/x$, $Q(x) = x^{-2}$, $f(x) = \ln x/x^2$, and $h = (2-1)/8 = 0.125$. Then the finite difference equation is

$$\left(1 - \frac{0.0625}{x_i}\right) y_{i+1} + \left(-2 + \frac{0.0156}{x_i^2}\right) y_i + \left(1 + \frac{0.0625}{x_i}\right) y_{i-1} = 0.0156 \ln x_i.$$

The solution of the corresponding linear system gives

x	1.000	1.125	1.250	1.375	1.500	1.625	1.750	1.875	2.000
y	0.0000	-0.1988	-0.4168	-0.6510	-0.8992	-1.1594	-1.4304	-1.7109	-2.0000

9. We identify $P(x) = 1 - x$, $Q(x) = x$, $f(x) = x$, and $h = (1-0)/10 = 0.1$. Then the finite difference equation is

$$[1 + 0.05(1 - x_i)]y_{i+1} + [-2 + 0.01x_i]y_i + [1 - 0.05(1 - x_i)]y_{i-1} = 0.01x_i.$$

The solution of the corresponding linear system gives

x	0.0	0.1	0.2	0.3	0.4	0.5	0.6
y	0.0000	0.2660	0.5097	0.7357	0.9471	1.1465	1.3353

0.7	0.8	0.9	1.0
1.5149	1.6855	1.8474	2.0000

10. We identify $P(x) = x$, $Q(x) = 1$, $f(x) = x$, and $h = (1-0)/10 = 0.1$. Then the finite difference equation is

$$(1 + 0.05x_i)y_{i+1} - 1.99y_i + (1 - 0.05x_i)y_{i-1} = 0.01x_i.$$

The solution of the corresponding linear system gives

x	0.0	0.1	0.2	0.3	0.4	0.5	0.6
y	1.0000	0.8929	0.7789	0.6615	0.5440	0.4296	0.3216

	0.7	0.8	0.9	1.0
	0.2225	0.1347	0.0601	0.0000

11. We identify $P(x) = 0$, $Q(x) = -4$, $f(x) = 0$, and $h = (1-0)/8 = 0.125$. Then the finite difference equation is

$$y_{i+1} - 2.0625 y_i + y_{i-1} = 0.$$

The solution of the corresponding linear system gives

x	0.000	0.125	0.250	0.375	0.500	0.625	0.750	0.875	1.000
y	0.0000	0.3492	0.7202	1.1363	1.6233	2.2118	2.9386	3.8490	5.0000

12. We identify $P(r) = 2/r$, $Q(r) = 0$, $f(r) = 0$, and $h = (4-1)/6 = 0.5$. Then the finite difference equation is

$$\left(1 + \frac{0.5}{r_i}\right) u_{i+1} - 2 u_i + \left(1 - \frac{0.5}{r_i}\right) u_{i-1} = 0.$$

The solution of the corresponding linear system gives

r	1.0	1.5	2.0	2.5	3.0	3.5	4.0
u	50.0000	72.2222	83.3333	90.0000	94.4444	97.6190	100.0000

13. **(a)** The difference equation

$$\left(1 + \frac{h}{2} P_i\right) y_{i+1} + (-2 + h^2 Q_i) y_i + \left(1 - \frac{h}{2} P_i\right) y_{i-1} = h^2 f_i$$

is the same as equation (8) in the text. The equations are the same because the derivation was based only on the differential equation, not the boundary conditions. If we allow i to range from 0 to $n-1$ we obtain n equations in the $n+1$ unknowns $y_{-1}, y_0, y_1, \ldots, y_{n-1}$. Since y_n is one of the given boundary conditions, it is not an unknown.

(b) Identifying $y_0 = y(0)$, $y_{-1} = y(0-h)$, and $y_1 = y(0+h)$ we have from equation (5) in the text

$$\frac{1}{2h}[y_1 - y_{-1}] = y'(0) = 1 \quad \text{or} \quad y_1 - y_{-1} = 2h.$$

The difference equation corresponding to $i = 0$,

$$\left(1 + \frac{h}{2} P_0\right) y_1 + (-2 + h^2 Q_0) y_0 + \left(1 - \frac{h}{2} P_0\right) y_{-1} = h^2 f_0$$

becomes, with $y_{-1} = y_1 - 2h$,

$$\left(1 + \frac{h}{2} P_0\right) y_1 + (-2 + h^2 Q_0) y_0 + \left(1 - \frac{h}{2} P_0\right)(y_1 - 2h) = h^2 f_0$$

502

or

$$2y_1 + (-2 + h^2 Q_0)y_0 = h^2 f_0 + 2h - P_0.$$

Alternatively, we may simply add the equation $y_1 - y_{-1} = 2h$ to the list of n difference equations obtaining $n + 1$ equations in the $n + 1$ unknowns $y_{-1}, y_0, y_1, \ldots, y_{n-1}$.

(c) Using $n = 5$ we obtain

x	0.0	0.2	0.4	0.6	0.8	1.0
y	-2.2755	-2.0755	-1.8589	-1.6126	-1.3275	-1.0000

14. Using $h = 0.1$ and, after shooting a few times, $y'(0) = 0.43535$ we obtain the following table with the RK4 method.

x	0.0	0.1	0.2	0.3	0.4	0.5	0.6
y	1.00000	1.04561	1.09492	1.14714	1.20131	1.25633	1.31096

	0.7	0.8	0.9	1.0
	1.36392	1.41388	1.45962	1.50003

Chapter 9 in Review

1.

x_n	Euler h=0.1	Euler h=0.05	Imp. Euler h=0.1	Imp. Euler h=0.05	RK4 h=0.1	RK4 h=0.05
1.00	2.0000	2.0000	2.0000	2.0000	2.0000	2.0000
1.05		2.0693		2.0735		2.0736
1.10	2.1386	2.1469	2.1549	2.1554	2.1556	2.1556
1.15		2.2328		2.2459		2.2462
1.20	2.3097	2.3272	2.3439	2.3450	2.3454	2.3454
1.25		2.4299		2.4527		2.4532
1.30	2.5136	2.5409	2.5672	2.5689	2.5695	2.5695
1.35		2.6604		2.6937		2.6944
1.40	2.7504	2.7883	2.8246	2.8269	2.8278	2.8278
1.45		2.9245		2.9686		2.9696
1.50	3.0201	3.0690	3.1157	3.1187	3.1197	3.1197

2.

x_n	Euler h=0.1	Euler h=0.05	Imp. Euler h=0.1	Imp. Euler h=0.05	RK4 h=0.1	RK4 h=0.05
0.00	0.0000	0.0000	0.0000	0.0000	0.0000	0.0000
0.05		0.0500		0.0501		0.0500
0.10	0.1000	0.1001	0.1005	0.1004	0.1003	0.1003
0.15		0.1506		0.1512		0.1511
0.20	0.2010	0.2017	0.2030	0.2027	0.2026	0.2026
0.25		0.2537		0.2552		0.2551
0.30	0.3049	0.3067	0.3092	0.3088	0.3087	0.3087
0.35		0.3610		0.3638		0.3637
0.40	0.4135	0.4167	0.4207	0.4202	0.4201	0.4201
0.45		0.4739		0.4782		0.4781
0.50	0.5279	0.5327	0.5382	0.5378	0.5376	0.5376

3.

x_n	Euler h=0.1	Euler h=0.05	Imp. Euler h=0.1	Imp. Euler h=0.05	RK4 h=0.1	RK4 h=0.05
0.50	0.5000	0.5000	0.5000	0.5000	0.5000	0.5000
0.55		0.5500		0.5512		0.5512
0.60	0.6000	0.6024	0.6048	0.6049	0.6049	0.6049
0.65		0.6573		0.6609		0.6610
0.70	0.7095	0.7144	0.7191	0.7193	0.7194	0.7194
0.75		0.7739		0.7800		0.7801
0.80	0.8283	0.8356	0.8427	0.8430	0.8431	0.8431
0.85		0.8996		0.9082		0.9083
0.90	0.9559	0.9657	0.9752	0.9755	0.9757	0.9757
0.95		1.0340		1.0451		1.0452
1.00	1.0921	1.1044	1.1163	1.1168	1.1169	1.1169

4.

x_n	Euler h=0.1	Euler h=0.05	Imp. Euler h=0.1	Imp. Euler h=0.05	RK4 h=0.1	RK4 h=0.05
1.00	1.0000	1.0000	1.0000	1.0000	1.0000	1.0000
1.05		1.1000		1.1091		1.1095
1.10	1.2000	1.2183	1.2380	1.2405	1.2415	1.2415
1.15		1.3595		1.4010		1.4029
1.20	1.4760	1.5300	1.5910	1.6001	1.6036	1.6036
1.25		1.7389		1.8523		1.8586
1.30	1.8710	1.9988	2.1524	2.1799	2.1909	2.1911
1.35		2.3284		2.6197		2.6401
1.40	2.4643	2.7567	3.1458	3.2360	3.2745	3.2755
1.45		3.3296		4.1528		4.2363
1.50	3.4165	4.1253	5.2510	5.6404	5.8338	5.8446

5. Using

$$y_{n+1} = y_n + hu_n, \qquad\qquad y_0 = 3$$

$$u_{n+1} = u_n + h(2x_n + 1)y_n, \qquad u_0 = 1$$

504

we obtain (when $h = 0.2$) $y_1 = y(0.2) = y_0 + hu_0 = 3 + (0.2)1 = 3.2$. When $h = 0.1$ we have

$$y_1 = y_0 + 0.1u_0 = 3 + (0.1)1 = 3.1$$

$$u_1 = u_0 + 0.1(2x_0 + 1)y_0 = 1 + 0.1(1)3 = 1.3$$

$$y_2 = y_1 + 0.1u_1 = 3.1 + 0.1(1.3) = 3.23.$$

6. The first predictor is $y_3^* = 1.14822731$.

x_n	y_n	
0.0	2.00000000	init. cond.
0.1	1.65620000	RK4
0.2	1.41097281	RK4
0.3	1.24645047	RK4
0.4	1.14796764	ABM

7. Using $x_0 = 1$, $y_0 = 2$, and $h = 0.1$ we have

$$x_1 = x_0 + h(x_0 + y_0) = 1 + 0.1(1 + 2) = 1.3$$

$$y_1 = y_0 + h(x_0 - y_0) = 2 + 0.1(1 - 2) = 1.9$$

and

$$x_2 = x_1 + h(x_1 + y_1) = 1.3 + 0.1(1.3 + 1.9) = 1.62$$

$$y_2 = y_1 + h(x_1 - y_1) = 1.9 + 0.1(1.3 - 1.9) = 1.84.$$

Thus, $x(0.2) \approx 1.62$ and $y(0.2) \approx 1.84$.

8. We identify $P(x) = 0$, $Q(x) = 6.55(1 + x)$, $f(x) = 1$, and $h = (1 - 0)/10 = 0.1$. Then the finite difference equation is

$$y_{i+1} + [-2 + 0.0655(1 + x_i)]y_i + y_{i-1} = 0.001$$

or

$$y_{i+1} + (0.0655x_i - 1.9345)y_i + y_{i-1} = 0.001.$$

The solution of the corresponding linear system gives

x	0.0	0.1	0.2	0.3	0.4	0.5	0.6
y	0.0000	4.1987	8.1049	11.3840	13.7038	14.7770	14.4083

	0.7	0.8	0.9	1.0
	12.5396	9.2847	4.9450	0.0000

10 Plane Autonomous Systems

1. The corresponding plane autonomous system is

$$x' = y, \quad y' = -9\sin x.$$

If (x, y) is a critical point, $y = 0$ and $-9\sin x = 0$. Therefore $x = \pm n\pi$ and so the critical points are $(\pm n\pi, 0)$ for $n = 0, 1, 2, \ldots$.

2. The corresponding plane autonomous system is

$$x' = y, \quad y' = -2x - y^2.$$

If (x, y) is a critical point, then $y = 0$ and so $-2x - y^2 = -2x = 0$. Therefore $(0, 0)$ is the sole critical point.

3. The corresponding plane autonomous system is

$$x' = y, \quad y' = x^2 - y(1 - x^3).$$

If (x, y) is a critical point, $y = 0$ and so $x^2 - y(1 - x^3) = x^2 = 0$. Therefore $(0, 0)$ is the sole critical point.

4. The corresponding plane autonomous system is

$$x' = y, \quad y' = -4\frac{x}{1 + x^2} - 2y.$$

If (x, y) is a critical point, $y = 0$ and so $-4x/(1 + x^2) - 2(0) = 0$. Therefore $x = 0$ and so $(0, 0)$ is the sole critical point.

5. The corresponding plane autonomous system is

$$x' = y, \quad y' = -x + \epsilon x^3.$$

If (x, y) is a critical point, $y = 0$ and $-x + \epsilon x^3 = 0$. Hence $x(-1 + \epsilon x^2) = 0$ and so $x = 0, \sqrt{1/\epsilon},$ $-\sqrt{1/\epsilon}$. The critical points are $(0, 0)$, $(\sqrt{1/\epsilon}, 0)$ and $(-\sqrt{1/\epsilon}, 0)$.

6. The corresponding plane autonomous system is

$$x' = y, \quad y' = -x + \epsilon x|x|.$$

506

If (x, y) is a critical point, $y = 0$ and $-x + \epsilon x|x| = x(-1 + \epsilon|x|) = 0$. Hence $x = 0, 1/\epsilon, -1/\epsilon$. The critical points are $(0, 0)$, $(1/\epsilon, 0)$ and $(-1/\epsilon, 0)$.

7. From $x + xy = 0$ we have $x(1 + y) = 0$. Therefore $x = 0$ or $y = -1$. If $x = 0$, then, substituting into $-y - xy = 0$, we obtain $y = 0$, Likewise, if $y = -1$, $1 + x = 0$ or $x = -1$. We can conclude that $(0, 0)$ and $(-1, -1)$ are critical points of the system.

8. From $y^2 - x = 0$ we have $x = y^2$. Substituting into $x^2 - y = 0$, we obtain $y^4 - y = 0$ or $y(y^3 - 1) = 0$. It follows that $y = 0, 1$ and so $(0, 0)$ and $(1, 1)$ are the critical points of the system.

9. From $x - y = 0$ we have $y = x$. Substituting into $3x^2 - 4y = 0$ we obtain $3x^2 - 4x = x(3x - 4) = 0$. It follows that $(0, 0)$ and $(4/3, 4/3)$ are the critical points of the system.

10. From $x^3 - y = 0$ we have $y = x^3$. Substituting into $x - y^3 = 0$ we obtain $x - x^9 = 0$ or $x(1 - x^8)$. Therefore $x = 0, 1, -1$ and so the critical points of the system are $(0, 0)$, $(1, 1)$, and $(-1, -1)$.

11. From $x(10 - x - \frac{1}{2}y) = 0$ we obtain $x = 0$ or $x + \frac{1}{2}y = 10$. Likewise $y(16 - y - x) = 0$ implies that $y = 0$ or $x + y = 16$. We therefore have four cases. If $x = 0$, $y = 0$ or $y = 16$. If $x + \frac{1}{2}y = 10$, we can conclude that $y(-\frac{1}{2}y + 6) = 0$ and so $y = 0, 12$. Therefore the critical points of the system are $(0, 0)$, $(0, 16)$, $(10, 0)$, and $(4, 12)$.

12. Adding the two equations we obtain $10 - 15y/(y + 5) = 0$. It follows that $y = 10$, and from $-2x + y + 10 = 0$ we can conclude that $x = 10$. Therefore $(10, 10)$ is the sole critical point of the system.

13. From $x^2 e^y = 0$ we have $x = 0$. Since $e^x - 1 = e^0 - 1 = 0$, the second equation is satisfied for an arbitrary value of y. Therefore any point of the form $(0, y)$ is a critical point.

14. From $\sin y = 0$ we have $y = \pm n\pi$. From $e^{x-y} = 1$, we can conclude that $x - y = 0$ or $x = y$. The critical points of the system are therefore $(\pm n\pi, \pm n\pi)$ for $n = 0, 1, 2, \dots$.

15. From $x(1 - x^2 - 3y^2) = 0$ we have $x = 0$ or $x^2 + 3y^2 = 1$. If $x = 0$, then substituting into $y(3 - x^2 - 3y^2)$ gives $y(3 - 3y^2) = 0$. Therefore $y = 0, 1, -1$. Likewise $x^2 = 1 - 3y^2$ yields $2y = 0$ so that $y = 0$ and $x^2 = 1 - 3(0)^2 = 1$. The critical points of the system are therefore $(0, 0)$, $(0, 1)$, $(0, -1)$, $(1, 0)$, and $(-1, 0)$.

16. From $-x(4 - y^2) = 0$ we obtain $x = 0$, $y = 2$, or $y = -2$. If $x = 0$, then substituting into $4y(1 - x^2)$ yields $y = 0$. Likewise $y = 2$ gives $8(1 - x^2) = 0$ or $x = 1, -1$. Finally $y = -2$ yields $-8(1 - x^2) = 0$ or $x = 1, -1$. The critical points of the system are therefore $(0, 0)$, $(1, 2)$, $(-1, 2)$, $(1, -2)$, and $(-1, -2)$.

17. (a) From Exercises 8.2, Problem 1, $x = c_1 e^{5t} - c_2 e^{-t}$ and $y = 2c_1 e^{5t} + c_2 e^{-t}$, which are not periodic.

    (b) From $\mathbf{X}(0) = (-2, 2)$ it follows that $c_1 = 0$ and $c_2 = 2$. Therefore $x = -2e^{-t}$ and $y = 2e^{-t}$.

(c)

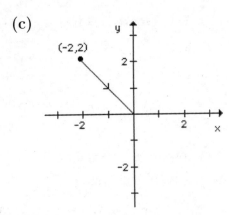

18. **(a)** From Exercises 8.2, Problem 6, $x = c_1 + 2c_2e^{-5t}$ and $y = 3c_1 + c_2e^{-5t}$, which is not periodic.

   (b) From $\mathbf{X}(0) = (3,4)$ it follows that $c_1 = c_2 = 1$. Therefore $x = 1 + 2e^{-5t}$ and $y = 3 + e^{-5t}$ gives $y = \frac{1}{2}(x-1) + 3$.

   (c)

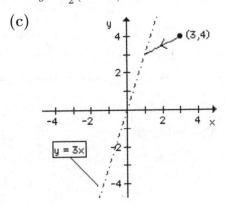

19. **(a)** From Exercises 8.2, Problem 37, $x = c_1(4\cos 3t - 3\sin 3t) + c_2(4\sin 3t + 3\cos 3t)$ and $y = c_1(5\cos 3t) + c_2(5\sin 3t)$. All solutions are one periodic with $p = 2\pi/3$.

   (b) From $\mathbf{X}(0) = (4,5)$ it follows that $c_1 = 1$ and $c_2 = 0$. Therefore $x = 4\cos 3t - 3\sin 3t$ and $y = 5\cos 3t$.

   (c)

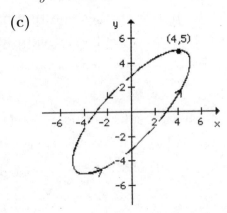

20. **(a)** From Exercises 8.2, Problem 34, $x = c_1(\sin t - \cos t) + c_2(-\cos t - \sin t)$ and $y = 2c_1\cos t + 2c_2\sin t$. All solutions are periodic with $p = 2\pi$.

(b) From $\mathbf{X}(0) = (-2, 2)$ it follows that $c_1 = c_2 = 1$. Therefore $x = -2\cos t$ and $y = 2\cos t + 2\sin t$.

(c)

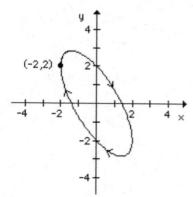

21. (a) From Exercises 8.2, Problem 35, $x = c_1(\sin t - \cos t)e^{4t} + c_2(-\sin t - \cos t)e^{4t}$ and $y = 2c_1(\cos t)e^{4t} + 2c_2(\sin t)e^{4t}$. Because of the presence of e^{4t}, there are no periodic solutions.

(b) From $\mathbf{X}(0) = (-1, 2)$ it follows that $c_1 = 1$ and $c_2 = 0$. Therefore $x = (\sin t - \cos t)e^{4t}$ and $y = 2(\cos t)e^{4t}$.

(c)

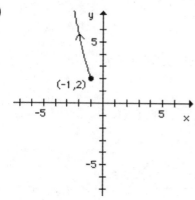

22. (a) From Exercises 8.2, Problem 38, $x = c_1 e^{-t}(2\cos 2t - 2\sin 2t) + c_2 e^{-t}(2\cos 2t + 2\sin 2t)$ and $y = c_1 e^{-t}\cos 2t + c_2 e^{-t}\sin 2t$. Because of the presence of e^{-t}, there are no periodic solutions.

(b) From $\mathbf{X}(0) = (2, 1)$ it follows that $c_1 = 1$ and $c_2 = 0$. Therefore $x = e^{-t}(2\cos 2t - 2\sin 2t)$ and $y = e^{-t}\cos 2t$.

(c)

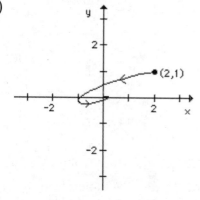

23. Switching to polar coordinates,

$$\frac{dr}{dt} = \frac{1}{r}\left(x\frac{dx}{dt} + y\frac{dy}{dt}\right) = \frac{1}{r}(-xy - x^2r^4 + xy - y^2r^4) = -r^5$$

$$\frac{d\theta}{dt} = \frac{1}{r^2}\left(-y\frac{dx}{dt} + x\frac{dy}{dt}\right) = \frac{1}{r^2}(y^2 + xyr^4 + x^2 - xyr^4) = 1.$$

If we use separation of variables on $\dfrac{dr}{dt} = -r^5$ we obtain

$$r = \left(\frac{1}{4t + c_1}\right)^{1/4} \quad \text{and} \quad \theta = t + c_2.$$

Since $\mathbf{X}(0) = (4, 0)$, $r = 4$ and $\theta = 0$ when $t = 0$. It follows that $c_2 = 0$ and $c_1 = \frac{1}{256}$. The final solution can be written as

$$r = \frac{4}{\sqrt[4]{1024t + 1}}, \qquad \theta = t$$

and so the solution spirals toward the origin as t increases.

24. Switching to polar coordinates,

$$\frac{dr}{dt} = \frac{1}{r}\left(x\frac{dx}{dt} + y\frac{dy}{dt}\right) = \frac{1}{r}(xy - x^2r^2 - xy + y^2r^2) = r^3$$

$$\frac{d\theta}{dt} = \frac{1}{r^2}\left(-y\frac{dx}{dt} + x\frac{dy}{dt}\right) = \frac{1}{r^2}(-y^2 - xyr^2 - x^2 + xyr^2) = -1.$$

If we use separation of variables, it follows that

$$r = \frac{1}{\sqrt{-2t + c_1}} \quad \text{and} \quad \theta = -t + c_2.$$

Since $\mathbf{X}(0) = (4, 0)$, $r = 4$ and $\theta = 0$ when $t = 0$. It follows that $c_2 = 0$ and $c_1 = \frac{1}{16}$. The final solution can be written as

$$r = \frac{4}{\sqrt{1 - 32t}}, \qquad \theta = -t.$$

Note that $r \to \infty$ as $t \to \left(\frac{1}{32}\right)$. Because $0 \le t \le \frac{1}{32}$, the curve is not a spiral.

25. Switching to polar coordinates,

$$\frac{dr}{dt} = \frac{1}{r}\left(x\frac{dx}{dt} + y\frac{dy}{dt}\right) = \frac{1}{r}[-xy + x^2(1 - r^2) + xy + y^2(1 - r^2)] = r(1 - r^2)$$

$$\frac{d\theta}{dt} = \frac{1}{r^2}\left(-y\frac{dx}{dt} + x\frac{dy}{dt}\right) = \frac{1}{r^2}[y^2 - xy(1 - r^2) + x^2 + xy(1 - r^2)] = 1.$$

Now $dr/dt = r - r^3$ or $(dr/dt) - r = -r^3$ is a Bernoulli differential equation. Following the procedure in Section 2.5 of the text, we let $w = r^{-2}$ so that $w' = -2r^{-3}(dr/dt)$. Therefore $w' + 2w = 2$, a linear first order differential equation. It follows that $w = 1 + c_1 e^{-2t}$ and so $r^2 = 1/(1 + c_1 e^{-2t})$. The general solution can be written as

$$r = \frac{1}{\sqrt{1 + c_1 e^{-2t}}}, \qquad \theta = t + c_2.$$

If $\mathbf{X}(0) = (1, 0)$, $r = 1$ and $\theta = 0$ when $t = 0$. Therefore $c_1 = 0 = c_2$ and so $x = r \cos t = \cos t$ and $y = r \sin t = \sin t$. This solution generates the circle $r = 1$. If $\mathbf{X}(0) = (2, 0)$, $r = 2$ and $\theta = 0$ when $t = 0$. Therefore $c_1 = -3/4$, $c_2 = 0$ and so

$$r = \frac{1}{\sqrt{1 - \frac{3}{4} e^{-2t}}}, \qquad \theta = t.$$

This solution spirals toward the circle $r = 1$ as t increases.

26. Switching to polar coordinates,

$$\frac{dr}{dt} = \frac{1}{r}\left(x \frac{dx}{dt} + y \frac{dy}{dt}\right) = \frac{1}{r}\left[xy - \frac{x^2}{r}(4 - r^2) - xy - \frac{y^2}{r}(4 - r^2)\right] = r^2 - 4$$

$$\frac{d\theta}{dt} = \frac{1}{r^2}\left(-y \frac{dx}{dt} + x \frac{dy}{dt}\right) = \frac{1}{r^2}\left[-y^2 + \frac{xy}{r}(4 - r^2) - x^2 - \frac{xy}{r}(4 - r^2)\right] = -1.$$

From Example 3, Section 2.2,

$$r = 2 \frac{1 + c_1 e^{4t}}{1 - c_1 e^{4t}} \quad \text{and} \quad \theta = -t + c_2.$$

If $\mathbf{X}(0) = (1, 0)$, $r = 1$ and $\theta = 0$ when $t = 0$. It follows that $c_2 = 0$ and $c_1 = -\frac{1}{3}$. Therefore

$$r = 2 \frac{1 - \frac{1}{3} e^{4t}}{1 + \frac{1}{3} e^{4t}} \quad \text{and} \quad \theta = -t.$$

Note that $r = 0$ when $e^{4t} = 3$ or $t = (\ln 3)/4$ and $r \to -2$ as $t \to \infty$. The solution therefore approaches the circle $r = 2$. If $\mathbf{X}(0) = (2, 0)$, it follows that $c_1 = c_2 = 0$. Therefore $r = 2$ and $\theta = -t$ so that the solution generates the circle $r = 2$ traversed in the clockwise direction. Note also that the original system is not defined at $(0, 0)$ but the corresponding polar system is defined for $r = 0$. If the Runge-Kutta method is applied to the original system, the solution corresponding to $\mathbf{X}(0) = (1, 0)$ will stall at the origin.

27. The system has no critical points, so there are no periodic solutions.

28. From $x(6y - 1) = 0$ and $y(2 - 8x) = 0$ we see that $(0, 0)$ and $(1/4, 1/6)$ are critical points. From the graph we see that there are periodic solutions around $(1/4, 1/6)$.

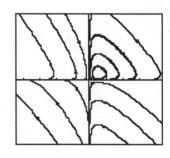

29. The only critical point is $(0,0)$. There appears to be a single periodic solution around $(0,0)$.

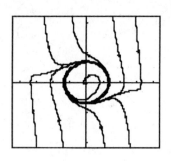

30. The system has no critical points, so there are no periodic solutions.

Exercises 10.2   Stability of Linear Systems

1. (a) If $\mathbf{X}(0) = \mathbf{X}_0$ lies on the line $y = 2x$, then $\mathbf{X}(t)$ approaches $(0,0)$ along this line. For all other initial conditions, $\mathbf{X}(t)$ approaches $(0,0)$ from the direction determined by the line $y = -x/2$.

(b)

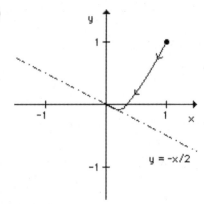

2. (a) If $\mathbf{X}(0) = \mathbf{X}_0$ lies on the line $y = -x$, then $\mathbf{X}(t)$ becomes unbounded along this line. For all other initial conditions, $\mathbf{X}(t)$ becomes unbounded and $y = -3x/2$ serves as an asymptote.

(b)

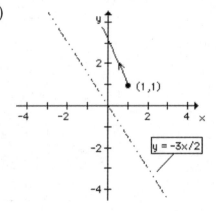

3. **(a)** All solutions are unstable spirals which become unbounded as t increases.

(b)

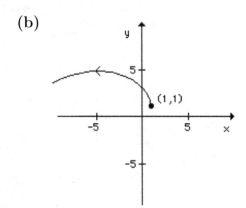

4. **(a)** All solutions are spirals which approach the origin.

(b)

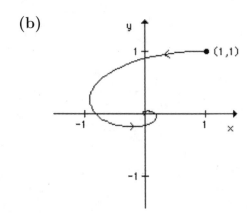

5. **(a)** All solutions approach $(0,0)$ from the direction specified by the line $y = x$.

(b)

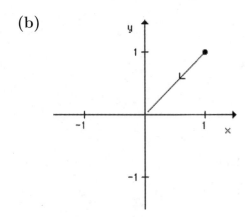

6. **(a)** All solutions become unbounded and $y = x/2$ serves as the asymptote.

(b)

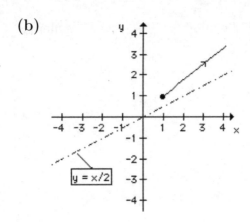

7. (a) If $\mathbf{X}(0) = \mathbf{X}_0$ lies on the line $y = 3x$, then $\mathbf{X}(t)$ approaches $(0,0)$ along this line. For all other initial conditions, $\mathbf{X}(t)$ becomes unbounded and $y = x$ serves as the asymptote.

(b)

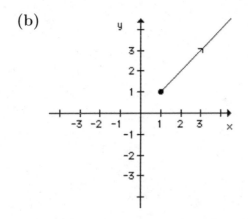

8. (a) The solutions are ellipses which encircle the origin.

(b)

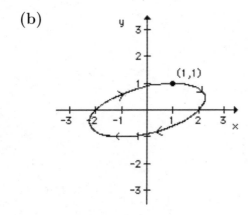

9. Since $\Delta = -41 < 0$, we can conclude from Figure 10.2.12 that $(0,0)$ is a saddle point.

10. Since $\Delta = 29$ and $\tau = -12$, $\tau^2 - 4\Delta > 0$ and so from Figure 10.2.12, $(0,0)$ is a stable node.

11. Since $\Delta = -19 < 0$, we can conclude from Figure 10.2.12 that $(0,0)$ is a saddle point.

12. Since $\Delta = 1$ and $\tau = -1$, $\tau^2 - 4\Delta = -3$ and so from Figure 10.2.12, $(0,0)$ is a stable spiral point.

13. Since $\Delta = 1$ and $\tau = -2$, $\tau^2 - 4\Delta = 0$ and so from Figure 10.2.12, $(0,0)$ is a degenerate stable node.

14. Since $\Delta = 1$ and $\tau = 2$, $\tau^2 - 4\Delta = 0$ and so from Figure 10.2.12, $(0,0)$ is a degenerate unstable node.

15. Since $\Delta = 0.01$ and $\tau = -0.03$, $\tau^2 - 4\Delta < 0$ and so from Figure 10.2.12, $(0,0)$ is a stable spiral point.

16. Since $\Delta = 0.0016$ and $\tau = 0.08$, $\tau^2 - 4\Delta = 0$ and so from Figure 10.2.12, $(0,0)$ is a degenerate unstable node.

17. $\Delta = 1 - \mu^2$, $\tau = 0$, and so we need $\Delta = 1 - \mu^2 > 0$ for $(0,0)$ to be a center. Therefore $|\mu| < 1$.

18. Note that $\Delta = 1$ and $\tau = \mu$. Therefore we need both $\tau = \mu < 0$ and $\tau^2 - 4\Delta = \mu^2 - 4 < 0$ for $(0,0)$ to be a stable spiral point. These two conditions can be written as $-2 < \mu < 0$.

19. Note that $\Delta = \mu + 1$ and $\tau = \mu + 1$ and so $\tau^2 - 4\Delta = (\mu+1)^2 - 4(\mu+1) = (\mu+1)(\mu-3)$. It follows that $\tau^2 - 4\Delta < 0$ if and only if $-1 < \mu < 3$. We can conclude that $(0,0)$ will be a saddle point when $\mu < -1$. Likewise $(0,0)$ will be an unstable spiral point when $\tau = \mu + 1 > 0$ and $\tau^2 - 4\Delta < 0$. This condition reduces to $-1 < \mu < 3$.

20. $\tau = 2\alpha$, $\Delta = \alpha^2 + \beta^2 > 0$, and $\tau^2 - 4\Delta = -4\beta < 0$. If $\alpha < 0$, $(0,0)$ is a stable spiral point. If $\alpha > 0$, $(0,0)$ is an unstable spiral point. Therefore $(0,0)$ cannot be a node or saddle point.

21. $\mathbf{AX}_1 + \mathbf{F} = \mathbf{0}$ implies that $\mathbf{AX}_1 = -\mathbf{F}$ or $\mathbf{X}_1 = -\mathbf{A}^{-1}\mathbf{F}$. Since $\mathbf{X}_p(t) = -\mathbf{A}^{-1}\mathbf{F}$ is a particular solution, it follows from Theorem 8.1.6 that $\mathbf{X}(t) = \mathbf{X}_c(t) + \mathbf{X}_1$ is the general solution to $\mathbf{X}' = \mathbf{AX} + \mathbf{F}$. If $\tau < 0$ and $\Delta > 0$ then $\mathbf{X}_c(t)$ approaches $(0,0)$ by Theorem 10.1(a). It follows that $\mathbf{X}(t)$ approaches $\mathbf{X}_1$ as $t \to \infty$.

22. If $bc < 1$, $\Delta = ad\hat{x}\hat{y}(1 - bc) > 0$ and $\tau^2 - 4\Delta = (a\hat{x} - d\hat{y})^2 + 4abcd\hat{x}\hat{y} > 0$. Therefore $(0,0)$ is a stable node.

23. **(a)** The critical point is $\mathbf{X}_1 = (-3, 4)$.

    (b) From the graph, $\mathbf{X}_1$ appears to be an unstable node or a saddle point.

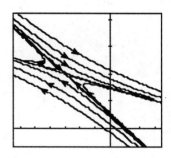

    (c) Since $\Delta = -1$, $(0,0)$ is a saddle point.

24. (a) The critical point is $\mathbf{X}_1 = (-1, -2)$.

(b) From the graph, $\mathbf{X}_1$ appears to be a stable node or a degenerate stable node.

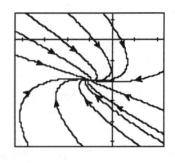

(c) Since $\tau = -16$, $\Delta = 64$, and $\tau^2 - 4\Delta = 0$, $(0,0)$ is a degenerate stable node.

25. (a) The critical point is $\mathbf{X}_1 = (0.5, 2)$.

(b) From the graph, $\mathbf{X}_1$ appears to be an unstable spiral point.

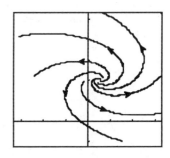

(c) Since $\tau = 0.2$, $\Delta = 0.03$, and $\tau 2 - 4\Delta = -0.08$, $(0,0)$ is an unstable spiral point.

26. (a) The critical point is $\mathbf{X}_1 = (1, 1)$.

(b) From the graph, $\mathbf{X}_1$ appears to be a center.

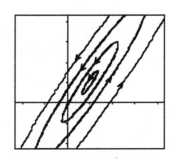

(c) Since $\tau = 0$ and $\Delta = 1$, $(0,0)$ is a center.

Exercises 10.3   Linearization and Local Stability

1. Switching to polar coordinates,

$$\frac{dr}{dt} = \frac{1}{r}\left(x\frac{dx}{dt} + y\frac{dy}{dt}\right) = \frac{1}{r}(\alpha x^2 - \beta xy + xy^2 + \beta xy + \alpha y^2 - xy^2) = \frac{1}{r}\alpha r^2 = \alpha r.$$

Therefore $r = ce^{\alpha t}$ and so $r \to 0$ if and only if $\alpha < 0$.

2. The differential equation $dr/dt = \alpha r(5 - r)$ is a logistic differential equation. [See Section 3.2, (4) and (5).] It follows that

$$r = \frac{5}{1 + c_1 e^{-5\alpha t}} \quad \text{and} \quad \theta = -t + c_2.$$

If $\alpha > 0$, $r \to 5$ as $t \to +\infty$ and so the critical point $(0, 0)$ is unstable. If $\alpha < 0$, $r \to 0$ as $t \to +\infty$ and so $(0, 0)$ is asymptotically stable.

3. The critical points are $x = 0$ and $x = n + 1$. Since $g'(x) = k(n+1) - 2kx$, $g'(0) = k(n+1) > 0$ and $g'(n+1) = -k(n+1) < 0$. Therefore $x = 0$ is unstable while $x = n + 1$ is asymptotically stable. See Theorem 10.2.

4. Note that $x = k$ is the only critical point since $\ln(x/k)$ is not defined at $x = 0$. Since $g'(x) = -k - k\ln(x/k)$, $g'(k) = -k < 0$. Therefore $x = k$ is an asymptotically stable critical point by Theorem 10.3.1.

5. The only critical point is $T = T_0$. Since $g'(T) = k$, $g'(T_0) = k > 0$. Therefore $T = T_0$ is unstable by Theorem 10.3.1.

6. The only critical point is $v = mg/k$. Now $g(v) = g - (k/m)v$ and so $g'(v) = -k/m < 0$. Therefore $v = mg/k$ is an asymptotically stable critical point by Theorem 10.3.1.

7. Critical points occur at $x = \alpha, \beta$. Since $g'(x) = k(-\alpha - \beta + 2x)$, $g'(\alpha) = k(\alpha - \beta)$ and $g'(\beta) = k(\beta - \alpha)$. Since $\alpha > \beta$, $g'(\alpha) > 0$ and so $x = \alpha$ is unstable. Likewise $x = \beta$ is asymptotically stable.

8. Critical points occur at $x = \alpha, \beta, \gamma$. Since

$$g'(x) = k(\alpha - x)(-\beta - \gamma - 2x) + k(\beta - x)(\gamma - x)(-1),$$

$g'(\alpha) = -k(\beta - \alpha)(\gamma - \alpha) < 0$ since $\alpha > \beta > \gamma$. Therefore $x = \alpha$ is asymptotically stable. Similarly $g'(\beta) > 0$ and $g'(\gamma) < 0$. Therefore $x = \beta$ is unstable while $x = \gamma$ is asymptotically stable.

9. Critical points occur at $P = a/b$, c but not at $P = 0$. Since $g'(P) = (a - bP) + (P - c)(-b)$,

$$g'(a/b) = (a/b - c)(-b) = -a + bc \quad \text{and} \quad g'(c) = a - bc.$$

Since $a < bc$, $-a + bc > 0$ and $a - bc < 0$. Therefore $P = a/b$ is unstable while $P = c$ is asymptotically stable.

10. Since $A > 0$, the only critical point is $A = K^2$. Since $g'(A) = \frac{1}{2}kKA^{-1/2} - k$, $g'(K^2) = -k/2 < 0$. Therefore $A = K^2$ is asymptotically stable.

11. The sole critical point is $(1/2, 1)$ and

$$g'(\mathbf{X}) = \begin{pmatrix} -2y & -2x \\ 2y & 2x - 1 \end{pmatrix}.$$

Computing $\mathbf{g}'((1/2, 1))$ we find that $\tau = -2$ and $\Delta = 2$ so that $\tau^2 - 4\Delta = -4 < 0$. Therefore $(1/2, 1)$ is a stable spiral point.

12. Critical points are $(1, 0)$ and $(-1, 0)$, and

$$\mathbf{g}'(\mathbf{X}) = \begin{pmatrix} 2x & -2y \\ 0 & 2 \end{pmatrix}.$$

At $\mathbf{X} = (1, 0)$, $\tau = 4$, $\Delta = 4$, and so $\tau^2 - 4\Delta = 0$. We can conclude that $(1, 0)$ is unstable but we are unable to classify this critical point any further. At $\mathbf{X} = (-1, 0)$, $\Delta = -4 < 0$ and so $(-1, 0)$ is a saddle point.

13. $y' = 2xy - y = y(2x - 1)$. Therefore if (x, y) is a critical point, either $x = 1/2$ or $y = 0$. The case $x = 1/2$ and $y - x^2 + 2 = 0$ implies that $(x, y) = (1/2, -7/4)$. The case $y = 0$ leads to the critical points $(\sqrt{2}, 0)$ and $(-\sqrt{2}, 0)$. We next use the Jacobian matrix

$$\mathbf{g}'(\mathbf{X}) = \begin{pmatrix} -2x & 1 \\ 2y & 2x - 1 \end{pmatrix}$$

to classify these three critical points. For $\mathbf{X} = (\sqrt{2}, 0)$ or $(-\sqrt{2}, 0)$, $\tau = -1$ and $\Delta < 0$. Therefore both critical points are saddle points. For $\mathbf{X} = (1/2, -7/4)$, $\tau = -1$, $\Delta = 7/2$ and so $\tau^2 - 4\Delta = -13 < 0$. Therefore $(1/2, -7/4)$ is a stable spiral point.

14. $y' = -y + xy = y(-1 + x)$. Therefore if (x, y) is a critical point, either $y = 0$ or $x = 1$. The case $y = 0$ and $2x - y^2 = 0$ implies that $(x, y) = (0, 0)$. The case $x = 1$ leads to the critical points $(1, \sqrt{2})$ and $(1, -\sqrt{2})$. We next use the Jacobian matrix

$$\mathbf{g}'(\mathbf{X}) = \begin{pmatrix} 2 & -2y \\ y & x - 1 \end{pmatrix}$$

to classify these critical points. For $\mathbf{X} = (0, 0)$, $\Delta = -2 < 0$ and so $(0, 0)$ is a saddle point. For either $(1, \sqrt{2})$ or $(1, -\sqrt{2})$, $\tau = 2$, $\Delta = 4$, and so $\tau^2 - 4\Delta = -12$. Therefore $(1, \sqrt{2})$ and $(1, -\sqrt{2})$ are unstable spiral points.

15. Since $x^2 - y^2 = 0$, $y^2 = x^2$ and so $x^2 - 3x + 2 = (x - 1)(x - 2) = 0$. It follows that the critical points are $(1, 1)$, $(1, -1)$, $(2, 2)$, and $(2, -2)$. We next use the Jacobian

$$\mathbf{g}'(\mathbf{X}) = \begin{pmatrix} -3 & 2y \\ 2x & -2y \end{pmatrix}$$

to classify these four critical points. For $\mathbf{X} = (1, 1)$, $\tau = -5$, $\Delta = 2$, and so $\tau^2 - 4\Delta = 17 > 0$. Therefore $(1, 1)$ is a stable node. For $\mathbf{X} = (1, -1)$, $\Delta = -2 < 0$ and so $(1, -1)$ is a saddle point. For $\mathbf{X} = (2, 2)$, $\Delta = -4 < 0$ and so we have another saddle point. Finally, if $\mathbf{X} = (2, -2)$, $\tau = 1$, $\Delta = 4$, and so $\tau^2 - 4\Delta = -15 < 0$. Therefore $(2, -2)$ is an unstable spiral point.

16. From $y^2 - x^2 = 0$, $y = x$ or $y = -x$. The case $y = x$ leads to $(4, 4)$ and $(-1, 1)$ but the case $y = -x$ leads to $x^2 - 3x + 4 = 0$ which has no real solutions. Therefore $(4, 4)$ and $(-1, 1)$ are the only

critical points. We next use the Jacobian matrix

$$\mathbf{g}'(\mathbf{X}) = \begin{pmatrix} y & x-3 \\ -2x & 2y \end{pmatrix}$$

to classify these two critical points. For $\mathbf{X} = (4,4)$, $\tau = 12$, $\Delta = 40$, and so $\tau^2 - 4\Delta < 0$. Therefore $(4,4)$ is an unstable spiral point. For $\mathbf{X} = (-1,1)$, $\tau = -3$, $\Delta = 10$, and so $x^2 - 4\Delta < 0$. It follows that $(-1,-1)$ is a stable spiral point.

17. Since $x' = -2xy = 0$, either $x = 0$ or $y = 0$. If $x = 0$, $y(1-y^2) = 0$ and so $(0,0)$, $(0,1)$, and $(0,-1)$ are critical points. The case $y = 0$ leads to $x = 0$. We next use the Jacobian matrix

$$\mathbf{g}'(\mathbf{X}) = \begin{pmatrix} -2y & -2x \\ -1+y & 1+x-3y^2 \end{pmatrix}$$

to classify these three critical points. For $\mathbf{X} = (0,0)$, $\tau = 1$ and $\Delta = 0$ and so the test is inconclusive. For $\mathbf{X} = (0,1)$, $\tau = -4$, $\Delta = 4$ and so $\tau^2 - 4\Delta = 0$. We can conclude that $(0,1)$ is a stable critical point but we are unable to classify this critical point further in this borderline case. For $\mathbf{X} = (0,-1)$, $\Delta = -4 < 0$ and so $(0,-1)$ is a saddle point.

18. We found that $(0,0)$, $(0,1)$, $(0,-1)$, $(1,0)$ and $(-1,0)$ were the critical points in Exercise 15, Section 10.1. The Jacobian is

$$\mathbf{g}'(\mathbf{X}) = \begin{pmatrix} 1 - 3x^2 - 3y^2 & -6xy \\ -2xy & 3 - x^2 - 9y^2 \end{pmatrix}.$$

For $\mathbf{X} = (0,0)$, $\tau = 4$, $\Delta = 3$ and so $\tau^2 - 4\Delta = 4 > 0$. Therefore $(0,0)$ is an unstable node. Both $(0,1)$ and $(0,-1)$ give $\tau = -8$, $\Delta = 12$, and $\tau^2 - 4\Delta = 16 > 0$. These two critical points are therefore stable nodes. For $\mathbf{X} = (1,0)$ or $(-1,0)$, $\Delta = -4 < 0$ and so saddle points occur.

19. We found the critical points $(0,0)$, $(10,0)$, $(0,16)$ and $(4,12)$ in Exercise 11, Section 10.1. Since the Jacobian is

$$\mathbf{g}'(\mathbf{X}) = \begin{pmatrix} 10 - 2x - \frac{1}{2}y & -\frac{1}{2}x \\ -y & 16 - 2y - x \end{pmatrix}$$

we can classify the critical points as follows:

$\mathbf{X}$	τ	Δ	$\tau^2 - 4\Delta$	Conclusion
$(0,0)$	26	160	36	unstable node
$(10,0)$	-4	-60	$-$	saddle point
$(0,16)$	-14	-32	$-$	saddle point
$(4,12)$	-16	24	160	stable node

20. We found the sole critical point $(10,10)$ in Exercise 12, Section 10.1. The Jacobian is

$$\mathbf{g}'(\mathbf{X}) = \begin{pmatrix} -2 & 1 \\ 2 & -1 - \dfrac{15}{(y+5)^2} \end{pmatrix},$$

$\mathbf{g}'((10, 10))$ has trace $\tau = -46/15$, $\Delta = 2/15$, and $\tau^2 - 4\Delta > 0$. Therefore $(0, 0)$ is a stable node.

21. The corresponding plane autonomous system is

$$\theta' = y, \quad y' = (\cos\theta - \frac{1}{2})\sin\theta.$$

Since $|\theta| < \pi$, it follows that critical points are $(0, 0)$, $(\pi/3, 0)$ and $(-\pi/3, 0)$. The Jacobian matrix is

$$\mathbf{g}'(\mathbf{X}) = \begin{pmatrix} 0 & 1 \\ \cos 2\theta - \frac{1}{2}\cos\theta & 0 \end{pmatrix}$$

and so at $(0, 0)$, $\tau = 0$ and $\Delta = -1/2$. Therefore $(0, 0)$ is a saddle point. For $\mathbf{X} = (\pm\pi/3, 0)$, $\tau = 0$ and $\Delta = 3/4$. It is not possible to classify either critical point in this borderline case.

22. The corresponding plane autonomous system is

$$x' = y, \quad y' = -x + \left(\frac{1}{2} - 3y^2\right)y - x^2.$$

If (x, y) is a critical point, $y = 0$ and so $-x - x^2 = -x(1 + x) = 0$. Therefore $(0, 0)$ and $(-1, 0)$ are the only two critical points. We next use the Jacobian matrix

$$\mathbf{g}'(\mathbf{X}) = \begin{pmatrix} 0 & 1 \\ -1 - 2x & \frac{1}{2} - 9y^2 \end{pmatrix}$$

to classify these critical points. For $\mathbf{X} = (0, 0)$, $\tau = 1/2$, $\Delta = 1$, and $\tau^2 - 4\Delta < 0$. Therefore $(0, 0)$ is an unstable spiral point. For $\mathbf{X} = (-1, 0)$, $\tau = 1/2$, $\Delta = -1$ and so $(-1, 0)$ is a saddle point.

23. The corresponding plane autonomous system is

$$x' = y, \quad y' = x^2 - y(1 - x^3)$$

and the only critical point is $(0, 0)$. Since the Jacobian matrix is

$$\mathbf{g}'(\mathbf{X}) = \begin{pmatrix} 0 & 1 \\ 2x + 3x^2y & x^3 - 1 \end{pmatrix},$$

$\tau = -1$ and $\Delta = 0$, and we are unable to classify the critical point in this borderline case.

24. The corresponding plane autonomous system is

$$x' = y, \quad y' = -\frac{4x}{1 + x^2} - 2y$$

and the only critical point is $(0, 0)$. Since the Jacobian matrix is

$$\mathbf{g}'(\mathbf{X}) = \begin{pmatrix} 0 & 1 \\ -4\dfrac{1 - x^2}{(1 + x^2)^2} & -2 \end{pmatrix},$$

$\tau = -2$, $\Delta = 4$, $\tau^2 - 4\Delta = -12$, and so $(0, 0)$ is a stable spiral point.

25. In Exercise 5, Section 10.1, we showed that $(0,0)$, $(\sqrt{1/\epsilon},0)$ and $(-\sqrt{1/\epsilon},0)$ are the critical points. We will use the Jacobian matrix

$$\mathbf{g}'(\mathbf{X}) = \begin{pmatrix} 0 & 1 \\ -1 + 3\epsilon x^2 & 0 \end{pmatrix}$$

to classify these three critical points. For $\mathbf{X} = (0,0)$, $\tau = 0$ and $\Delta = 1$ and we are unable to classify this critical point. For $(\pm\sqrt{1/\epsilon},0)$, $\tau = 0$ and $\Delta = -2$ and so both of these critical points are saddle points.

26. In Exercise 6, Section 10.1, we showed that $(0,0)$, $(1/\epsilon,0)$, and $(-1/\epsilon,0)$ are the critical points. Since $D_x x|x| = 2|x|$, the Jacobian matrix is

$$\mathbf{g}'(\mathbf{X}) = \begin{pmatrix} 0 & 1 \\ 2\epsilon|x| - 1 & 0 \end{pmatrix}.$$

For $\mathbf{X} = (0,0)$, $\tau = 0$, $\Delta = 1$ and we are unable to classify this critical point. For $(\pm1/\epsilon,0)$, $\tau = 0$, $\Delta = -1$, and so both of these critical points are saddle points.

27. The corresponding plane autonomous system is

$$x' = y, \quad y' = -\frac{(\beta + \alpha^2 y^2)x}{1 + \alpha^2 x^2}$$

and the Jacobian matrix is

$$\mathbf{g}'(\mathbf{X}) = \begin{pmatrix} 0 & 1 \\ \dfrac{(\beta + \alpha y^2)(\alpha^2 x^2 - 1)}{(1 + \alpha^2 x^2)^2} & \dfrac{-2\alpha^2 yx}{1 + \alpha^2 x^2} \end{pmatrix}.$$

For $\mathbf{X} = (0,0)$, $\tau = 0$ and $\Delta = \beta$. Since $\beta < 0$, we can conclude that $(0,0)$ is a saddle point.

28. From $x' = -\alpha x + xy = x(-\alpha + y) = 0$, either $x = 0$ or $y = \alpha$. If $x = 0$, then $1 - \beta y = 0$ and so $y = 1/\beta$. The case $y = \alpha$ implies that $1 - \beta\alpha - x^2 = 0$ or $x^2 = 1 - \alpha\beta$. Since $\alpha\beta > 1$, this equation has no real solutions. It follows that $(0, 1/\beta)$ is the unique critical point. Since the Jacobian matrix is

$$\mathbf{g}'(\mathbf{X}) = \begin{pmatrix} -\alpha + y & x \\ -2x & -\beta \end{pmatrix},$$

$\tau = -\alpha - \beta + \dfrac{1}{\beta} = -\beta + \dfrac{1 - \alpha\beta}{\beta} < 0$ and $\Delta = \alpha\beta - 1 > 0$. Therefore $(0, 1/\beta)$ is a stable critical point.

29. (a) The graphs of $-x + y - x^3 = 0$ and $-x - y + y^2 = 0$ are shown in the figure. The Jacobian matrix is

$$\mathbf{g}'(\mathbf{X}) = \begin{pmatrix} -1 - 3x^2 & 1 \\ -1 & -1 + 2y \end{pmatrix}.$$

For $\mathbf{X} = (0,0)$, $\tau = -2$, $\Delta = 2$, $\tau^2 - 4\Delta = -4$, and so $(0,0)$ is a stable spiral point.

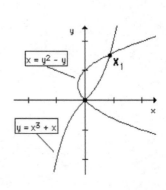

(b) For X_1, $\Delta = -6.07 < 0$ and so a saddle point occurs at X_1.

30. (a) The corresponding plane autonomous system is

$$x' = y, \quad y' = \epsilon\left(y - \frac{1}{3}y^3\right) - x$$

and so the only critical point is $(0,0)$. Since the Jacobian matrix is

$$g'(X) = \begin{pmatrix} 0 & 1 \\ -1 & \epsilon(1 - y^2) \end{pmatrix},$$

$\tau = \epsilon$, $\Delta = 1$, and so $\tau^2 - 4\Delta = \epsilon^2 - 4$ at the critical point $(0,0)$.

(b) When $\tau = \epsilon > 0$, $(0,0)$ is an unstable critical point.

(c) When $\epsilon < 0$ and $\tau^2 - 4\Delta = \epsilon^2 - 4 < 0$, $(0,0)$ is a stable spiral point. These two requirements can be written as $-2 < \epsilon < 0$.

(d) When $\epsilon = 0$, $x'' + x = 0$ and so $x = c_1 \cos t + c_2 \sin t$. Therefore all solutions are periodic (with period 2π) and so $(0,0)$ is a center.

31. The differential equation $dy/dx = y'/x' = -2x^3/y$ can be solved by separating variables. It follows that $y^2 + x^4 = c$. If $X(0) = (x_0, 0)$ where $x_0 > 0$, then $c = x_0^4$ so that $y^2 = x_0^4 - x^4$. Therefore if $-x_0 < x < x_0$, $y^2 > 0$ and so there are two values of y corresponding to each value of x. Therefore the solution $X(t)$ with $X(0) = (x_0, 0)$ is periodic and so $(0,0)$ is a center.

32. The differential equation $dy/dx = y'/x' = (x^2 - 2x)/y$ can be solved by separating variables. It follows that $y^2/2 = (x^3/3) - x^2 + c$ and since $X(0) = (x(0), x'(0)) = (1,0)$, $c = \frac{2}{3}$. Therefore

$$\frac{y^2}{2} = \frac{x^3 - 3x^2 + 2}{3} = \frac{(x-1)(x^2 - 2x - 2)}{3}.$$

But $(x-1)(x^2 - 2x - 2) > 0$ for $1 - \sqrt{3} < x < 1$ and so each x in this interval has 2 corresponding values of y. therefore $X(t)$ is a periodic solution.

33. (a) $x' = 2xy = 0$ implies that either $x = 0$ or $y = 0$. If $x = 0$, then from $1 - x^2 + y^2 = 0$, $y^2 = -1$ and there are no real solutions. If $y = 0$, $1 - x^2 = 0$ and so $(1,0)$ and $(-1,0)$ are critical points. The Jacobian matrix is

$$g'(X) = \begin{pmatrix} 2y & 2x \\ -2x & 2y \end{pmatrix}$$

and so $\tau = 0$ and $\Delta = 4$ at either $X = (1,0)$ or $(-1,0)$. We obtain no information about these critical points in this borderline case.

(b) The differential equation is

$$\frac{dy}{dx} = \frac{y'}{x'} = \frac{1 - x^2 + y^2}{2xy}$$

or

$$2xy\frac{dy}{dx} = 1 - x^2 + y^2.$$

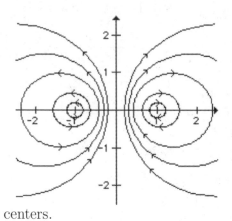

Letting $\mu = y^2/x$, it follows that $d\mu/dx = (1/x^2) - 1$ and so $\mu = -(1/x) - x + 2c$. Therefore $y^2/x = -(1/x) - x + 2c$ which can be put in the form $(x - c)^2 + y^2 = c^2 - 1$. The solution curves are shown and so both $(1, 0)$ and $(-1, 0)$ are centers.

34. (a) The differential equation is $dy/dx = y'/x' = (-x - y^2)/y = -(x/y) - y$ and so $dy/dx + y = -xy^{-1}$.

(b) Let $w = y^{1-n} = y^2$. It follows that $dw/dx + 2w = -2x$, a linear first order differential equation whose solution is $y^2 = w = ce^{-2x} + \left(\frac{1}{2} - x\right)$. Since $x(0) = \frac{1}{2}$ and $y(0) = x'(0) = 0$, $0 = c$ and so $y^2 = \frac{1}{2} - x$, a parabola with vertex at $(1/2, 0)$. Therefore the solution $\mathbf{X}(t)$ with $\mathbf{X}(0) = (1/2, 0)$ is not periodic.

35. The differential equation is $dy/dx = y'/x' = (x^3 - x)/y$ and so $y^2/2 = x^4/4 - x^2/2 + c$ or $y^2 = x^4/2 - x^2 + c_1$. Since $x(0) = 0$ and $y(0) = x'(0) = v_0$, it follows that $c_1 = v_0^2$ and so

$$y^2 = \frac{1}{2}x^4 - x^2 + v_0^2 = \frac{(x^2 - 1)^2 + 2v_0^2 - 1}{2}.$$

The x-intercepts on this graph satisfy

$$x^2 = 1 \pm \sqrt{1 - 2v_0^2}$$

and so we must require that $1 - 2v_0^2 \geq 0$ (or $|v_0| \leq \frac{1}{2}\sqrt{2}$) for real solutions to exist. If $x_0^2 = 1 - \sqrt{1 - 2v_0^2}$ and $-x_0 < x < x_0$, then $(x^2 - 1)^2 + 2v_0^2 - 1 > 0$ and so there are two corresponding values of y. Therefore $\mathbf{X}(t)$ with $\mathbf{X}(0) = (0, v_0)$ is periodic provided that $|v_0| \leq \frac{1}{2}\sqrt{2}$.

36. The corresponding plane autonomous system is

$$x' = y, \quad y' = \epsilon x^2 - x + 1$$

and so the critical points must satisfy $y = 0$ and

$$x = \frac{1 \pm \sqrt{1 - 4\epsilon}}{2\epsilon}.$$

Therefore we must require that $\epsilon \leq \frac{1}{4}$ for real solutions to exist. We will use the Jacobian matrix

$$\mathbf{g}'(\mathbf{X}) = \begin{pmatrix} 0 & 1 \\ 2\epsilon x - 1 & 0 \end{pmatrix}$$

to attempt to classify $((1 \pm \sqrt{1-4\epsilon})/2\epsilon, 0)$ when $\epsilon \le 1/4$. Note that $\tau = 0$ and $\Delta = \mp\sqrt{1-4\epsilon}$. For $\mathbf{X} = ((1 + \sqrt{1-4\epsilon})/2\epsilon, 0)$ and $\epsilon < 1/4$, $\Delta < 0$ and so a saddle point occurs. For $\mathbf{X} = ((1 - \sqrt{1-4\epsilon})/2\epsilon, 0)$, $\Delta \ge 0$ and we are not able to classify this critical point using linearization.

37. The corresponding plane autonomous system is

$$x' = y, \quad y' = -\frac{\alpha}{L}x - \frac{\beta}{L}x^3 - \frac{R}{L}y$$

where $x = q$ and $y = q'$. If $\mathbf{X} = (x, y)$ is a critical point, $y = 0$ and $-\alpha x - \beta x^3 = -x(\alpha + \beta x^2) = 0$. If $\beta > 0$, $\alpha + \beta x^2 = 0$ has no real solutions and so $(0,0)$ is the only critical point. Since

$$\mathbf{g}'(\mathbf{X}) = \begin{pmatrix} 0 & 1 \\ \dfrac{-\alpha - 3\beta x^2}{L} & -\dfrac{R}{L} \end{pmatrix},$$

$\tau = -R/L < 0$ and $\Delta = \alpha/L > 0$. Therefore $(0,0)$ is a stable critical point. If $\beta < 0$, $(0,0)$ and $(\pm\hat{x}, 0)$, where $\hat{x}^2 = -\alpha/\beta$ are critical points. At $\mathbf{X}(\pm\hat{x}, 0)$, $\tau = -R/L < 0$ and $\Delta = -2\alpha/L < 0$. Therefore both critical points are saddles.

38. If we let $dx/dt = y$, then $dy/dt = -x^3 - x$. From this we obtain the first-order differential equation

$$\frac{dy}{dx} = \frac{dy/dt}{dx/dt} = -\frac{x^3 + x}{y}.$$

Separating variables and integrating we obtain

$$\int y\, dy = -\int (x^3 + x)\, dx$$

and

$$\frac{1}{2}y^2 = -\frac{1}{4}x^4 - \frac{1}{2}x^2 + c_1.$$

Completing the square we can write the solution as $y^2 = -\frac{1}{2}(x^2 + 1)^2 + c_2$. If $\mathbf{X}(0) = (x_0, 0)$, then $c_2 = \frac{1}{2}(x_0^2 + 1)^2$ and so

$$y^2 = -\frac{1}{2}(x^2 + 1)^2 + \frac{1}{2}(x_0^2 + 1)^2 = \frac{x_0^4 + 2x_0^2 + 1 - x^4 - 2x^2 - 1}{2}$$

$$= \frac{(x_0^2 + x^2)(x_0^2 - x^2) + 2(x_0^2 - x^2)}{2} = \frac{(x_0^2 + x^2 + 2)(x_0^2 - x^2)}{2}.$$

Note that $y = 0$ when $x = -x_0$. In addition, the right-hand side is positive for $-x_0 < x < x_0$, and so there are two corresponding values of y for each x between $-x_0$ and x_0. The solution $\mathbf{X} = \mathbf{X}(t)$ that satisfies $\mathbf{X}(0) = (x_0, 0)$ is therefore periodic, and so $(0,0)$ is a center.

39. (a) Letting $x = \theta$ and $y = x'$ we obtain the system $x' = y$ and $y' = 1/2 - \sin x$. Since $\sin \pi/6 = \sin 5\pi/6 = 1/2$ we see that $(\pi/6, 0)$ and $(5\pi/6, 0)$ are critical points of the system.

(b) The Jacobian matrix is

$$\mathbf{g}'(\mathbf{X}) = \begin{pmatrix} 0 & 1 \\ -\cos x & 0 \end{pmatrix}$$

and so

$$\mathbf{A}_1 = \mathbf{g}' = ((\pi/6, 0)) = \begin{pmatrix} 0 & 1 \\ -\sqrt{3}/2 & 0 \end{pmatrix} \quad \text{and} \quad \mathbf{A}_2 = \mathbf{g}' = ((5\pi/6, 0)) = \begin{pmatrix} 0 & 1 \\ \sqrt{3}/2 & 0 \end{pmatrix}.$$

Since $\det \mathbf{A}_1 > 0$ and the trace of $\mathbf{A}_1$ is 0, no conclusion can be drawn regarding the critical point $(\pi/6, 0)$. Since $\det \mathbf{A}_2 < 0$, we see that $(5\pi/6, 0)$ is a saddle point.

(c) From the system in part (a) we obtain the first-order differential equation

$$\frac{dy}{dx} = \frac{1/2 - \sin x}{y}.$$

Separating variables and integrating we obtain

$$\int y \, dy = \int \left(\frac{1}{2} - \sin x \right) dx$$

and

$$\frac{1}{2}y^2 = \frac{1}{2}x + \cos x + c_1$$

or

$$y^2 = x + 2\cos x + c_2.$$

For x_0 near $\pi/6$, if $\mathbf{X}(0) = (x_0, 0)$ then $c_2 = -x_0 - 2\cos x_0$ and $y^2 = x + 2\cos x - x_0 - 2\cos x_0$. Thus, there are two values of y for each x in a sufficiently small interval around $\pi/6$. Therefore $(\pi/6, 0)$ is a center.

40. (a) Writing the system as $x' = x(x^3 - 2y^3)$ and $y' = y(2x^3 - y^3)$ we see that $(0, 0)$ is a critical point. Setting $x^3 - 2y^3 = 0$ we have $x^3 = 2y^3$ and $2x^3 - y^3 = 4y^3 - y^3 = 3y^3$. Thus, $(0, 0)$ is the only critical point of the system.

(b) From the system we obtain the first-order differential equation

$$\frac{dy}{dx} = \frac{2x^3y - y^4}{x^4 - 2xy^3}$$

or

$$(2x^3y - y^4) \, dx + (2xy^3 - x^4) \, dy = 0$$

which is homogeneous. If we let $y = ux$ it follows that

$$(2x^4u - x^4u^4) \, dx + (2x^4u^3 - x^4)(u \, dx + x \, du) = 0$$

$$x^4u(1 + u^3) \, dx + x^5(2u^3 - 1) \, du = 0$$

$$\frac{1}{x} \, dx + \frac{2u^3 - 1}{u(u^3 + 1)} \, du = 0$$

$$\frac{1}{x} \, dx + \left(\frac{1}{u+1} - \frac{1}{u} + \frac{2u - 1}{u^2 - u + 1} \right) du = 0.$$

Integrating gives

$$\ln |x| + \ln |u + 1| - \ln |u| + \ln |u^2 - u + 1| = c_1$$

525

or

$$x\left(\frac{u+1}{u}\right)\left(u^2 - u + 1\right) = c_2$$

$$x\left(\frac{y+x}{y}\right)\left(\frac{y^2}{x^2} - \frac{y}{x} + 1\right) = c_2$$

$$(xy + x^2)(y^2 - xy + x^2) = c_2 x^2 y$$

$$xy^3 + x^4 = c_2 x^2 y$$

$$x^3 + y^2 = 3c_3 xy.$$

(c) We see from the graph that $(0,0)$ is unstable. It is not possible to classify the critical point as a node, saddle, center, or spiral point.

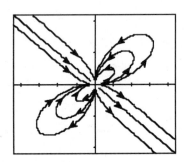

Exercises 10.4

Autonomous Systems as Mathematical Models

1. We are given that $x(0) = \theta(0) = \pi/3$ and $y(0) = \theta'(0) = w_0$. Since $y^2 = (2g/l)\cos x + c$, $w_0^2 = (2g/l)\cos(\pi/3) + c = g/l + c$ and so $c = w_0^2 - g/l$. Therefore

$$y^2 = \frac{2g}{l}\left(\cos x - \frac{1}{2} + \frac{l}{2g}w_0^2\right)$$

and the x-intercepts occur where $\cos x = 1/2 - (l/2g)w_0^2$ and so $1/2 - (l/2g)w_0^2$ must be greater than -1 for solutions to exist. This condition is equivalent to $|w_0| < \sqrt{3g/l}$.

2. (a) Since $y^2 = (2g/l)\cos x + c$, $x(0) = \theta(0) = \theta_0$ and $y(0) = \theta'(0) = 0$, $c = -(2g/l)\cos\theta_0$ and so $y^2 = 2g(\cos\theta - \cos\theta_0)/l$. When $\theta = -\theta_0$, $y^2 = 2g[\cos(-\theta_0) - \cos\theta_0]/l = 0$. Therefore $y = d\theta/dt = 0$ when $\theta = \theta_0$.

(b) Since $y = d\theta/dt$ and θ is decreasing between the time when $\theta = \theta_0$, $t = 0$, and $\theta = -\theta_0$, that is, $t = T$,

$$\frac{d\theta}{dt} = -\sqrt{\frac{2g}{l}}\sqrt{\cos\theta - \cos\theta_0}.$$

Therefore

$$\frac{dt}{d\theta} = -\sqrt{\frac{l}{2g}} \frac{1}{\sqrt{\cos\theta - \cos\theta_0}}$$

and so

$$T = -\sqrt{\frac{l}{2g}} \int_{\theta=\theta_0}^{\theta=-\theta_0} \frac{1}{\sqrt{\cos\theta - \cos\theta_0}} \, d\theta = \sqrt{\frac{l}{2g}} \int_{-\theta_0}^{\theta_0} \frac{1}{\sqrt{\cos\theta - \cos\theta_0}} \, d\theta.$$

3. The corresponding plane autonomous system is

$$x' = y, \quad y' = -g\frac{f'(x)}{1 + [f'(x)]^2} - \frac{\beta}{m}y$$

and

$$\frac{\partial}{\partial x}\left(-g\frac{f'(x)}{1 + [f'(x)]^2} - \frac{\beta}{m}y\right) = -g\frac{(1 + [f'(x)]^2)f''(x) - f'(x)2f'(x)f''(x)}{(1 + [f'(x)]^2)^2}.$$

If $\mathbf{X}_1 = (x_1, y_1)$ is a critical point, $y_1 = 0$ and $f'(x_1) = 0$. The Jacobian at this critical point is therefore

$$\mathbf{g}'(\mathbf{X}_1) = \begin{pmatrix} 0 & 1 \\ -gf''(x_1) & -\dfrac{\beta}{m} \end{pmatrix}.$$

4. When $\beta = 0$ the Jacobian matrix is

$$\begin{pmatrix} 0 & 1 \\ -gf''(x_1) & 0 \end{pmatrix}$$

which has complex eigenvalues $\lambda = \pm\sqrt{gf''(x_1)}\,i$. The approximating linear system with $x'(0) = 0$ has solution

$$x(t) = x(0)\cos\sqrt{gf''(x_1)}\,t$$

and period $2\pi/\sqrt{gf''(x_1)}$. Therefore $p \approx 2\pi/\sqrt{gf''(x_1)}$ for the actual solution.

5. **(a)** If $f(x) = x^2/2$, $f'(x) = x$ and so

$$\frac{dy}{dx} = \frac{y'}{x'} = -g\frac{x}{1+x^2}\frac{1}{y}.$$

We can separate variables to show that $y^2 = -g\ln(1+x^2) + c$. But $x(0) = x_0$ and $y(0) = x'(0) = v_0$. Therefore $c = v_0^2 + g\ln(1+x_0^2)$ and so

$$y^2 = v_0^2 - g\ln\left(\frac{1+x^2}{1+x_0^2}\right).$$

Now

$$v_0^2 - g\ln\left(\frac{1+x^2}{1+x_0^2}\right) \geq 0 \quad \text{if and only if} \quad x^2 \leq e^{v_0^2/g}(1+x_0^2) - 1.$$

Therefore, if $|x| \leq [e^{v_0^2/g}(1+x_0^2) - 1]^{1/2}$, there are two values of y for a given value of x and so the solution is periodic.

527

(b) Since $z = x^2/2$, the maximum height occurs at the largest value of x on the cycle. From (a), $x_{max} = [e^{v_0^2/g}(1 + x_0^2) - 1]^{1/2}$ and so

$$z_{max} = \frac{x_{max}^2}{2} = \frac{1}{2}[e^{v_0^2/g}(1 + x_0^2) - 1].$$

6. (a) If $f(x) = \cosh x$, $f'(x) = \sinh x$ and $[f'(x)]^2 + 1 = \sinh^2 x + 1 = \cosh^2 x$. Therefore

$$\frac{dy}{dx} = \frac{y'}{x'} = -g\frac{\sinh x}{\cosh^2 x}\frac{1}{y}.$$

We can separate variables to show that $y^2 = 2g/\cosh x + c$. But $x(0) = x_0$ and $y(0) = x'(0) = v_0$. Therefore $c = v_0^2 - (2g/\cosh x_0)$ and so

$$y^2 = \frac{2g}{\cosh x} - \frac{2g}{\cosh x_0} + v_0^2.$$

Now

$$\frac{2g}{\cosh x} - \frac{2g}{\cosh x_0} + v_0^2 \geq 0 \quad \text{if and only if} \quad \cosh x \leq \frac{2g\cosh x_0}{2g - v_0^2\cosh x_0}$$

and the solution to this inequality is an interval $[-a, a]$. Therefore each x in $(-a, a)$ has two corresponding values of y and so the solution is periodic.

(b) Since $z = \cosh x$, the maximum height occurs at the largest value of x on the cycle. From (a), $x_{max} = a$ where $\cosh a = 2g\cosh x_0/(2g - v_0^2\cosh x_0)$. Therefore

$$z_{max} = \frac{2g\cosh x_0}{2g - v_0^2\cosh x_0}.$$

7. If $x_m < x_1 < x_n$, then $F(x_1) > F(x_m) = F(x_n)$. Letting $x = x_1$,

$$G(y) = \frac{c_0}{F(x_1)} = \frac{F(x_m)G(a/b)}{F(x_1)} < G(a/b).$$

Therefore from Property (ii) on page 391 in this section of the text, $G(y) = c_0/F(x_1)$ has two solutions y_1 and y_2 that satisfy $y_1 < a/b < y_2$.

8. From Property (i) on page 391 in this section of the text, when $y = a/b$, x_n is taken on at some time t. From Property (iii) on page 391 in this section of the text, if $x > x_n$ there is no corresponding value of y. Therefore the maximum number of predators is x_n and x_n occurs when $y = a/b$.

9. (a) In the Lotka-Volterra Model the average number of predators is d/c and the average number of prey is a/b. But

$$x' = -ax + bxy - \epsilon_1 x = -(a + \epsilon_1)x + bxy$$

$$y' = -cxy + dy - \epsilon_2 y = -cxy + (d - \epsilon_2)y$$

and so the new critical point in the first quadrant is $(d/c - \epsilon_2/c, a/b + \epsilon_1/b)$.

528

(b) The average number of predators $d/c - \epsilon_2/c$ has decreased while the average number of prey $a/b + \epsilon_1/b$ has increased. The fishery science model is consistent with Volterra's principle.

10. (a) Solving

$$x(-0.1 + 0.02y) = 0$$

$$y(0.2 - 0.025x) = 0$$

in the first quadrant we obtain the critical point $(8,5)$. The graphs are plotted using $x(0) = 7$ and $y(0) = 4$.

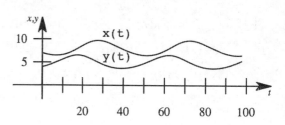

(b) The graph in part (a) was obtained using **NDSolve** in *Mathematica*. We see that the period is around 40. Since $x(0) = 7$, we use the **FindRoot** equation solver in *Mathematica* to approximate the solution of $x(t) = 7$ for t near 40. From this we see that the period is more closely approximated by $t = 44.65$.

11. Solving

$$x(20 - 0.4x - 0.3y) = 0$$

$$y(10 - 0.1y - 0.3x) = 0$$

we see that critical points are $(0,0)$, $(0,100)$, $(50,0)$, and $(20,40)$. The Jacobian matrix is

$$\mathbf{g}'(\mathbf{X}) = \begin{pmatrix} 0.08(20 - 0.8x - 0.3y) & -0.024x \\ -0.018y & 0.06(10 - 0.2y - 0.3x) \end{pmatrix}$$

and so

$$\mathbf{A}_1 = \mathbf{g}'((0,0)) = \begin{pmatrix} 1.6 & 0 \\ 0 & 0.6 \end{pmatrix} \qquad \mathbf{A}_2 = \mathbf{g}'((0,100)) = \begin{pmatrix} -0.8 & 0 \\ -1.8 & -0.6 \end{pmatrix}$$

$$\mathbf{A}_3 = \mathbf{g}'((50,0)) = \begin{pmatrix} -1.6 & -1.2 \\ 0 & -0.3 \end{pmatrix} \qquad \mathbf{A}_4 = \mathbf{g}'((20,40)) = \begin{pmatrix} -0.64 & -0.48 \\ -0.72 & -0.24 \end{pmatrix}.$$

Since $\det(\mathbf{A}_1) = \Delta_1 = 0.96 > 0$, $\tau = 2.2 > 0$, and $\tau_1^2 - 4\Delta_1 = 1 > 0$, we see that $(0,0)$ is an unstable node. Since $\det(\mathbf{A}_2) = \Delta_2 = 0.48 > 0$, $\tau = -1.4 < 0$, and $\tau_2^2 - 4\Delta_2 = 0.04 > 0$, we see that $(0,100)$ is a stable node. Since $\det(\mathbf{A}_3) = \Delta_3 = 0.48 > 0$, $\tau = -1.9 < 0$, and $\tau_3^2 - 4\Delta_3 = 1.69 > 0$, we see that $(50,0)$ is a stable node. Since $\det(\mathbf{A}_4) = -0.192 < 0$ we see that $(20,40)$ is a saddle point.

12. $\Delta = r_1 r_2$, $\tau = r_1 + r_2$ and $\tau^2 - 4\Delta = (r_1 + r_2)^2 - 4r_1 r_2 = (r_1 - r_2)^2$. Therefore when $r_1 \neq r_2$, $(0,0)$ is an unstable node.

13. For $\mathbf{X} = (K_1, 0)$, $\tau = -r_1 + r_2[1 - (K_1\alpha_{21}/K_2)]$ and $\Delta = -r_1 r_2[1 - (K_1\alpha_{21}/K_2)]$. If we let $c = 1 - K_1\alpha_{21}/K_2$, $\tau^2 - 4\Delta = (cr_2 + r_1)^2 > 0$. Now if $k_1 > K_2/\alpha_{21}$, $c < 0$ and so $\tau < 0$, $\Delta > 0$.

Therefore $(K_1, 0)$ is a stable node. If $K_1 < K_2/\alpha_{21}$, $c > 0$ and so $\Delta < 0$. In this case $(K_1, 0)$ is a saddle point.

14. $(\hat{x}, \hat{y})$ is a stable node if and only if $K_1/\alpha_{12} > K_2$ and $K_2/\alpha_{21} > K_1$. [See Figure 10.4.11(a) in the text.] From Problem 12, (0.0) is an unstable node and from Problem 13, since $K_1 < K_2/\alpha_{21}$, $(K_1, 0)$ is a saddle point. Finally, when $K_2 < K_1/\alpha_{12}$, $(0, K_2)$ is a saddle point. This is Problem 12 with the roles of 1 and 2 interchanged. Therefore $(0, 0)$, $(K_1, 0)$, and $(0, K_2)$ are unstable.

15. $K_1/\alpha_{12} < K_2 < K_1\alpha_{21}$ and so $\alpha_{12}\alpha_{21} > 1$. Therefore $\Delta = (1 - \alpha_{12}\alpha_{21})\hat{x}\hat{y}\,r_1r_2/K_1K_2 < 0$ and so $(\hat{x}, \hat{y})$ is a saddle point.

16. **(a)** The corresponding plane autonomous system is

$$x' = y, \quad y' = \frac{-g}{l}\sin x - \frac{\beta}{ml}y$$

and so critical points must satisfy both $y = 0$ and $\sin x = 0$. Therefore $(\pm n\pi, 0)$ are critical points.

(b) The Jacobian matrix

$$\begin{pmatrix} 0 & 1 \\ -\dfrac{g}{l}\cos x & -\dfrac{\beta}{ml} \end{pmatrix}$$

has trace $\tau = -\beta/ml$ and determinant $\Delta = g/l > 0$ at $(0, 0)$. Therefore

$$\tau^2 - 4\Delta = \frac{\beta^2}{m^2l^2} - 4\frac{g}{l} = \frac{\beta^2 - 4glm^2}{m^2l^2}.$$

We can conclude that $(0, 0)$ is a stable spiral point provided $\beta^2 - 4glm^2 < 0$ or $\beta < 2m\sqrt{gl}$.

17. **(a)** The corresponding plane autonomous system is

$$x = y, \quad y' = -\frac{\beta}{m}y|y| - \frac{k}{m}x$$

and so a critical point must satisfy both $y = 0$ and $x = 0$. Therefore $(0, 0)$ is the unique critical point.

(b) The Jacobian matrix is

$$\begin{pmatrix} 0 & 1 \\ -\dfrac{k}{m} & -\dfrac{\beta}{m}2|y| \end{pmatrix}$$

and so $\tau = 0$ and $\Delta = k/m > 0$. Therefore $(0, 0)$ is a center, stable spiral point, or an unstable spiral point. Physical considerations suggest that $(0, 0)$ must be asymptotically stable and so $(0, 0)$ must be a stable spiral point.

18. (a) The magnitude of the frictional force between the bead and the wire is $\mu(mg\cos\theta)$ for some $\mu > 0$. The component of this frictional force in the x-direction is

$$(\mu mg\cos\theta)\cos\theta = \mu mg\cos^2\theta.$$

But

$$\cos\theta = \frac{1}{\sqrt{1 + [f'(x)]^2}} \quad \text{and so} \quad \mu mg\cos^2\theta = \frac{\mu mg}{1 + [f'(x)]^2}.$$

It follows from Newton's Second Law that

$$mx'' = -mg\frac{f'(x)}{1 + [f'(x)]^2} - \beta x' + mg\frac{\mu}{1 + [f'(x)]^2}$$

and so

$$x'' = g\frac{\mu - f'(x)}{1 + [f'(x)]^2} - \frac{\beta}{m}x'.$$

(b) A critical point (x, y) must satisfy $y = 0$ and $f'(x) = \mu$. Therefore critical points occur at $(x_1, 0)$ where $f'(x_1) = \mu$. The Jacobian matrix of the plane autonomous system is

$$\mathbf{g}'(\mathbf{X}) = \begin{pmatrix} 0 & 1 \\ g\dfrac{(1 + [f'(x)]^2)(-f''(x)) - (\mu - f'(x))2f'(x)f''(x)}{(1 + [f'(x)]^2)^2} & -\dfrac{\beta}{m} \end{pmatrix}$$

and so at a critical point $\mathbf{X}_1$,

$$\mathbf{g}'(\mathbf{X}) = \begin{pmatrix} 0 & 1 \\ \dfrac{-gf''(x_1)}{1 + \mu^2} & -\dfrac{\beta}{m} \end{pmatrix}.$$

Therefore $\tau = -\beta/m < 0$ and $\Delta = gf''(x_1)/(1 + \mu^2)$. When $f''(x_1) < 0$, $\Delta < 0$ and so a saddle point occurs. When $f''(x_1) > 0$ and

$$\tau^2 - 4\Delta = \frac{\beta^2}{m^2} - 4g\frac{f''(x_1)}{1 + \mu^2} < 0,$$

$(x_1, 0)$ is a stable spiral point. This condition can also be written as

$$\beta^2 < 4gm^2\frac{f''(x_1)}{1 + \mu^2}.$$

19. We have $dy/dx = y'/x' = -f(x)/y$ and so, using separation of variables,

$$\frac{y^2}{2} = -\int_0^x f(\mu)\,d\mu + c \quad \text{or} \quad y^2 + 2F(x) = c.$$

We can conclude that for a given value of x there are at most two corresponding values of y. If $(0, 0)$ were a stable spiral point there would exist an x with more than two corresponding values of y. Note that the condition $f(0) = 0$ is required for $(0, 0)$ to be a critical point of the corresponding plane autonomous system $x' = y$, $y' = -f(x)$.

20. (a) $x' = x(-a + by) = 0$ implies that $x = 0$ or $y = a/b$. If $x = 0$, then, from

$$-cxy + \frac{r}{K} y(K - y) = 0,$$

$y = 0$ or K. Therefore $(0, 0)$ and $(0, K)$ are critical points. If $\hat{y} = a/b$, then

$$\hat{y}\left[-cx + \frac{r}{K}(K - \hat{y})\right] = 0.$$

The corresponding value of x, $x = \hat{x}$, therefore satisfies the equation $c\hat{x} = r(K - \hat{y})/K$.

(b) The Jacobian matrix is

$$\mathbf{g}'(\mathbf{X}) = \begin{pmatrix} -a + by & bx \\ -cy & -cx + \frac{r}{K}(K - 2y) \end{pmatrix}$$

and so at $\mathbf{X}_1 = (0, 0)$, $\Delta = -ar < 0$. For $\mathbf{X}_1 = (0, K)$, $\Delta = n(Kb - a) = -rb(K - a/b)$. Since we are given that $K > a/b$, $\Delta < 0$ in this case. Therefore $(0, 0)$ and $(0, K)$ are each saddle points. For $\mathbf{X}_1 = (\hat{x}, \hat{y})$ where $\hat{y} = a/b$ and $c\hat{x} = r(K - \hat{y})/K$, we can write the Jacobian matrix as

$$\mathbf{g}'((\hat{x}, \hat{y})) = \begin{pmatrix} 0 & b\hat{x} \\ -c\hat{y} & -\frac{r}{K}\hat{y} \end{pmatrix}$$

and so $\tau = -r\hat{y}/K < 0$ and $\Delta = bc\hat{x}\hat{y} > 0$. Therefore $(\hat{x}, \hat{y})$ is a stable critical point and so it is either a stable node (perhaps degenerate) or a stable spiral point.

(c) Write

$$\tau^2 - 4\Delta = \frac{r^2}{K^2}\hat{y}^2 - 4bc\hat{x}\hat{y} = \hat{y}\left[\frac{r^2}{K^2}\hat{y} - 4bc\hat{x}\right] = \hat{y}\left[\frac{r^2}{K^2}\hat{y} - 4b\frac{r}{K}(K - \hat{y})\right]$$

using

$$c\hat{x} = \frac{r}{K}(K - \hat{y}) = \frac{r}{K}\hat{y}\left[\left(\frac{r}{K} + 4b\right)\hat{y} - 4bK\right].$$

Therefore $\tau^2 - 4\Delta < 0$ if and only if

$$\hat{y} < \frac{4bK}{\frac{r}{K} + 4b} = \frac{4bK^2}{r + 4bK}.$$

Note that

$$\frac{4bK^2}{r + 4bK} = \frac{4bK}{r + 4bK} \cdot K \approx K$$

where K is large, and $\hat{y} = a/b < K$. Therefore $\tau^2 - 4\Delta < 0$ when K is large and a stable spiral point will result.

21. The equation

$$x' = \alpha\frac{y}{1 + y}x - x = x\left(\frac{\alpha y}{1 + y} - 1\right) = 0$$

implies that $x = 0$ or $y = 1/(\alpha - 1)$. When $\alpha > 0$, $\hat{y} = 1/(\alpha - 1) > 0$. If $x = 0$, then from the differential equation for y', $y = \beta$. On the other hand, if $\hat{y} = 1/(\alpha - 1)$, $\hat{y}/(1 + \hat{y}) = 1/\alpha$ and so $\hat{x}/\alpha - 1/(\alpha - 1) + \beta = 0$. It follows that

$$\hat{x} = \alpha\left(\beta - \frac{1}{\alpha - 1}\right) = \frac{\alpha}{\alpha - 1}[(\alpha - 1)\beta - 1]$$

and if $\beta(\alpha - 1) > 1$, $\hat{x} > 0$. Therefore $(\hat{x}, \hat{y})$ is the unique critical point in the first quadrant. The Jacobian matrix is

$$\mathbf{g}'(\mathbf{X}) = \begin{pmatrix} \alpha \dfrac{y}{y + 1} - 1 & \dfrac{\alpha x}{(1 + y)^2} \\ -\dfrac{y}{1 + y} & \dfrac{-x}{(1 + y)^2} - 1 \end{pmatrix}$$

and for $\mathbf{X} = (\hat{x}, \hat{y})$, the Jacobian can be written in the form

$$\mathbf{g}'((\hat{x}, \hat{y})) = \begin{pmatrix} 0 & \dfrac{(\alpha - 1)^2}{\alpha}\hat{x} \\ -\dfrac{1}{\alpha} & -\dfrac{(\alpha - 1)^2}{\alpha^2} - 1 \end{pmatrix}.$$

It follows that

$$\tau = -\left[\frac{(\alpha - 1)^2}{\alpha^2}\hat{x} + 1\right] < 0, \quad \Delta = \frac{(\alpha - 1)^2}{\alpha^2}\hat{x}$$

and so $\tau = -(\Delta + 1)$. Therefore $\tau^2 - 4\Delta = (\Delta + 1)^2 - 4\Delta = (\Delta - 1)^2 > 0$. Therefore $(\hat{x}, \hat{y})$ is a stable node.

22. Letting $y = x'$ we obtain the plane autonomous system

$$x' = y$$

$$y' = -8x + 6x^3 - x^5.$$

Solving $x^5 - 6x^3 + 8x = x(x^2 - 4)(x^2 - 2) = 0$ we see that critical points are $(0, 0)$, $(0, -2)$, $(0, 2)$, $(0, -\sqrt{2})$, and $(0, \sqrt{2})$. The Jacobian matrix is

$$\mathbf{g}'(\mathbf{X}) = \begin{pmatrix} 0 & 1 \\ -8 + 18x^2 - 5x^4 & 0 \end{pmatrix}$$

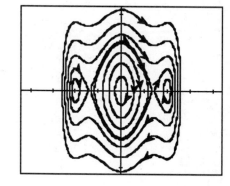

and we see that $\det(\mathbf{g}'(\mathbf{X})) = 5x^4 - 18x^2 + 8$ and the trace of $\mathbf{g}'(\mathbf{X})$ is 0. Since $\det(g'((\pm\sqrt{2}, 0))) = -8 < 0$, $(\pm\sqrt{2}, 0)$ are saddle points. For the other critical points the determinant is positive and linearization discloses no information. The graph of the phase plane suggests that $(0, 0)$ and $(\pm 2, 0)$ are centers.

Chapter 10 in Review

1. True

2. True

3. a center or a saddle point

4. complex with negative real parts

5. False; there are initial conditions for which $\lim\limits_{t\to\infty} \mathbf{X}(t) = (0,0)$.

6. True

7. False; this is a borderline case. See Figure 10.3.7 in the text.

8. False; see Figure 10.4.2 in the text.

9. The system is linear and we identify $\Delta = -\alpha$ and $\tau = \alpha + 1$. Since a critical point will be a center when $\Delta > 0$ and $\tau = 0$ we see that for $\alpha = -1$ critical points will be centers and solutions will be periodic. Note also that when $\alpha = -1$ the system is

$$x' = -x - 2y$$

$$y' = x + y,$$

which does have an isolated critical point at $(0,0)$.

10. We identify $g(x) = \sin x$ in Theorem 10.3.1. Then $x_1 = n\pi$ is a critical point for n an integer and $g'(n\pi) = \cos n\pi < 0$ when n is an odd integer. Thus, $n\pi$ is an asymptotically stable critical point when n is an odd integer.

11. Switching to polar coordinates,

$$\frac{dr}{dt} = \frac{1}{r}\left(x\frac{dx}{dt} + y\frac{dy}{dt}\right) = \frac{1}{r}(-xy - x^2r^3 + xy - y^2r^3) = -r^4$$

$$\frac{d\theta}{dt} = \frac{1}{r^2}\left(-y\frac{dx}{dt} + x\frac{dy}{dt}\right) = \frac{1}{r^2}(y^2 + xyr^3 + x^2 - xyr^3) = 1.$$

Using separation of variables it follows that $r = 1/\sqrt[3]{3t + c_1}$ and $\theta = t + c_2$. Since $\mathbf{X}(0) = (1,0)$, $r = 1$ and $\theta = 0$. It follows that $c_1 = 1$, $c_2 = 0$, and so

$$r = \frac{1}{\sqrt[3]{3t + 1}}, \quad \theta = t.$$

As $t \to \infty$, $r \to 0$ and the solution spirals toward the origin.

12. **(a)** If $\mathbf{X}(0) = \mathbf{X}_0$ lies on the line $y = -2x$, then $\mathbf{X}(t)$ approaches $(0,0)$ along this line. For all other initial conditions, $\mathbf{X}(t)$ approaches $(0,0)$ from the direction determined by the line $y = x$.

   (b) If $\mathbf{X}(0) = \mathbf{X}_0$ lies on the line $y = -x$, then $\mathbf{X}(t)$ approaches $(0,0)$ along this line. For all other initial conditions, $\mathbf{X}(t)$ becomes unbounded and $y = 2x$ serves as an asymptote.

13. **(a)** $\tau = 0$, $\Delta = 11 > 0$ and so $(0,0)$ is a center.

   (b) $\tau = -2$, $\Delta = 1$, $\tau^2 - 4\Delta = 0$ and so $(0,0)$ is a degenerate stable node.

14. From $x' = x(1 + y - 3x) = 0$, either $x = 0$ or $1 + y - 3x = 0$. If $x = 0$, then, from $y(4 - 2x - y) = 0$ we obtain $y(4 - y) = 0$. It follows that $(0,0)$ and $(0,4)$ are critical points. If $1 + y - 3x = 0$, then $y(5 - 5x) = 0$. Therefore $(1/3, 0)$ and $(1, 2)$ are the remaining critical points. We will use the Jacobian matrix

$$\mathbf{g}'(\mathbf{X}) = \begin{pmatrix} 1 + y - 6x & x \\ -2y & 4 - 2x - 2y \end{pmatrix}$$

to classify these four critical points. The results are as follows:

$\mathbf{X}$	τ	Δ	$\tau^2 - 4\Delta$	Conclusion
$(0,0)$	5	4	9	unstable node
$(0,4)$	–	-20	–	saddle point
$(\frac{1}{3}, 0)$	–	$-\frac{10}{3}$	–	saddle point
$(1,2)$	-5	10	-15	stable spiral point

15. From $x = r\cos\theta$, $y = r\sin\theta$ we have

$$\frac{dx}{dt} = -r\sin\theta\,\frac{d\theta}{dt} + \frac{dr}{dt}\cos\theta$$

$$\frac{dy}{dt} = r\cos\theta\,\frac{d\theta}{dt} + \frac{dr}{dt}\sin\theta.$$

Then $r' = \alpha r$, $\theta' = 1$ gives

$$\frac{dx}{dt} = -r\sin\theta + \alpha r\cos\theta$$

$$\frac{dy}{dt} = r\cos\theta + \alpha r\sin\theta.$$

We see that $r = 0$, which corresponds to $\mathbf{X} = (0,0)$, is a critical point. Solving $r' = \alpha r$ we have $r = c_1 e^{\alpha t}$. Thus, when $\alpha < 0$, $\lim_{t \to \infty} r(t) = 0$ and $(0,0)$ is a stable critical point. When $\alpha = 0$, $r' = 0$ and $r = c_1$. In this case $(0,0)$ is a center, which is stable. Therefore, $(0,0)$ is a stable critical point for the system when $\alpha \leq 0$.

16. The corresponding plane autonomous system is $x' = y$, $y' = \mu(1 - x^2) - x$ and so the Jacobian at the critical point $(0,0)$ is

$$\mathbf{g}'((0,0)) = \begin{pmatrix} 0 & 1 \\ -1 & \mu \end{pmatrix}.$$

Therefore $\tau = \mu$, $\Delta = 1$ and $\tau^2 - 4\Delta = \mu^2 - 4$. Now $\mu^2 - 4 < 0$ if and only if $-2 < \mu < 2$. We can therefore conclude that $(0,0)$ is a stable node for $\mu < -2$, a stable spiral point for $-2 < \mu < 0$, an unstable spiral point for $0 < \mu < 2$, and an unstable node for $\mu > 2$.

17. Critical points occur at $x = \pm 1$. Since

$$g'(x) = -\frac{1}{2}\, e^{-x/2}(x^2 - 4x - 1),$$

$g'(1) > 0$ and $g'(-1) < 0$. Therefore $x = 1$ is unstable and $x = -1$ is asymptotically stable.

18. Using the phase-plane method we obtain

$$\frac{dy}{dx} = \frac{y'}{x'} = \frac{-2x\sqrt{y^2 + 1}}{y}.$$

We can separate variables to show that $\sqrt{y^2 + 1} = -x^2 + c$. But $x(0) = x_0$ and $y(0) = x'(0) = 0$. It follows that $c = 1 + x_0^2$ so that $y^2 = (1 + x_0^2 - x^2)^2 - 1$. Note that $1 + x_0^2 - x^2 > 1$ for $-x_0 < x < x_0$ and $y = 0$ for $x = \pm x_0$. Each x with $-x_0 < x < x_0$ has two corresponding values of y and so the solution $\mathbf{X}(t)$ with $\mathbf{X}(0) = (x_0, 0)$ is periodic.

19. The corresponding plane autonomous system is

$$x' = y, \quad y' = -\frac{\beta}{m}y - \frac{k}{m}(s + x)^3 + g$$

and so the Jacobian is

$$\mathbf{g}'(\mathbf{X}) = \begin{pmatrix} 0 & 1 \\ -\dfrac{3k}{m}(s + x)^2 & -\dfrac{\beta}{m} \end{pmatrix}.$$

For $\mathbf{X} = (0,0)$, $\tau = -\beta/m < 0$, $\Delta = 3ks^2/m > 0$. Therefore

$$\tau^2 - 4\Delta = \frac{\beta^2}{m^2} - \frac{12k}{m}s^2 = \frac{1}{m^2}(\beta^2 - 12kms^2).$$

Therefore $(0,0)$ is a stable node if $\beta^2 > 12kms^2$ and a stable spiral point provided $\beta^2 < 12kms^2$, where $ks^3 = mg$.

20. (a) If (x,y) is a critical point, $y = 0$ and so $\sin x(\omega^2 \cos x - g/l) = 0$. Either $\sin x = 0$ (in which case $x = 0$) of $\cos x = g/\omega^2 l$. But if $\omega^2 < g/l$, $g/\omega^2 l > 1$ and so the latter equation has no real solutions. Therefore $(0,0)$ is the only critical point if $\omega^2 < g/l$. The Jacobian matrix is

$$\mathbf{g}'(\mathbf{X}) = \begin{pmatrix} 0 & 1 \\ \omega^2 \cos 2x - \frac{g}{l}\cos x & -\frac{\beta}{ml} \end{pmatrix}$$

and so $\tau = -\beta/ml < 0$ and $\Delta = g/l - \omega^2 > 0$ for $\mathbf{X} = (0,0)$. It follows that $(0,0)$ is asymptotically stable and so after a small displacement, the pendulum will return to $\theta = 0$, $\theta' = 0$.

(b) If $\omega^2 > g/l$, $\cos x = g/\omega^2 l$ will have two solutions $x = \pm\hat{x}$ that satisfy $-\pi < x < \pi$. Therefore $(\pm\hat{x}, 0)$ are two additional critical points. If $\mathbf{X}_1 = (0, 0)$, $\Delta = g/l - \omega^2 < 0$ and so $(0, 0)$ is a saddle point. If $\mathbf{X}_1 = (\pm\hat{x}, 0)$, $\tau = -\beta/ml < 0$ and

$$\Delta = \frac{g}{l}\cos\hat{x} - \omega^2\cos 2\hat{x} = \frac{g^2}{\omega^2 l^2} - \omega^2\left(2\frac{g^2}{\omega^4 l^2} - 1\right) = \omega^2 - \frac{g^2}{\omega^2 l^2} > 0.$$

Therefore $(\hat{x}, 0)$ and $(-\hat{x}, 0)$ are each stable. When $\theta(0) = \theta_0$, $\theta'(0) = 0$ and θ_0 is small we expect the pendulum to reach one of these two stable equilibrium positions.

11 Orthogonal Functions and Fourier Series

Exercises 11.1

Orthogonal Functions

1. $\displaystyle\int_{-2}^{2} x x^2 \, dx = \frac{1}{4} x^4 \Big|_{-2}^{2} = 0$

2. $\displaystyle\int_{-1}^{1} x^3 (x^2 + 1) \, dx = \frac{1}{6} x^6 \Big|_{-1}^{1} + \frac{1}{4} x^4 \Big|_{-1}^{1} = 0$

3. $\displaystyle\int_{0}^{2} e^x (x e^{-x} - e^{-x}) \, dx = \int_{0}^{2} (x - 1) \, dx = \left(\frac{1}{2} x^2 - x \right) \Big|_{0}^{2} = 0$

4. $\displaystyle\int_{0}^{\pi} \cos x \sin^2 x \, dx = \frac{1}{3} \sin^3 x \Big|_{0}^{\pi} = 0$

5. $\displaystyle\int_{-\pi/2}^{\pi/2} x \cos 2x \, dx = \frac{1}{2} \left(\frac{1}{2} \cos 2x + x \sin 2x \right) \Big|_{-\pi/2}^{\pi/2} = 0$

6. $\displaystyle\int_{\pi/4}^{5\pi/4} e^x \sin x \, dx = \left(\frac{1}{2} e^x \sin x - \frac{1}{2} e^x \cos x \right) \Big|_{\pi/4}^{5\pi/4} = 0$

7. For $m \neq n$

$$\int_{0}^{\pi/2} \sin(2n + 1)x \, \sin(2m + 1)x \, dx$$

$$= \frac{1}{2} \int_{0}^{\pi/2} \Big(\cos 2(n - m)x - \cos 2(n + m + 1)x \Big) \, dx$$

$$= \frac{1}{4(n - m)} \sin 2(n - m)x \Big|_{0}^{\pi/2} - \frac{1}{4(n + m + 1)} \sin 2(n + m + 1)x \Big|_{0}^{\pi/2}$$

$$= 0.$$

For $m = n$

$$\int_0^{\pi/2} \sin^2(2n+1)x \, dx = \int_0^{\pi/2} \left(\frac{1}{2} - \frac{1}{2}\cos 2(2n+1)x\right) dx$$

$$= \frac{1}{2}x \Big|_0^{\pi/2} - \frac{1}{4(2n+1)}\sin 2(2n+1)x \Big|_0^{\pi/2}$$

$$= \frac{\pi}{4}$$

so that

$$\|\sin(2n+1)x\| = \frac{1}{2}\sqrt{\pi}.$$

8. For $m \neq n$

$$\int_0^{\pi/2} \cos(2n+1)x \, \cos(2m+1)x \, dx$$

$$= \frac{1}{2}\int_0^{\pi/2}\left(\cos 2(n-m)x + \cos 2(n+m+1)x\right) dx$$

$$= \frac{1}{4(n-m)}\sin 2(n-m)x \Big|_0^{\pi/2} + \frac{1}{4(n+m+1)}\sin 2(n+m+1)x \Big|_0^{\pi/2}$$

$$= 0.$$

For $m = n$

$$\int_0^{\pi/2} \cos^2(2n+1)x \, dx = \int_0^{\pi/2}\left(\frac{1}{2} + \frac{1}{2}\cos 2(2n+1)x\right) dx$$

$$= \frac{1}{2}x \Big|_0^{\pi/2} + \frac{1}{4(2n+1)}\sin 2(2n+1)x \Big|_0^{\pi/2}$$

$$= \frac{\pi}{4}$$

so that

$$\|\cos(2n+1)x\| = \frac{1}{2}\sqrt{\pi}.$$

9. For $m \neq n$

$$\int_0^{\pi} \sin nx \sin mx \, dx = \frac{1}{2}\int_0^{\pi}\left(\cos(n-m)x - \cos(n+m)x\right) dx$$

$$= \frac{1}{2(n-m)}\sin(n-m)x \Big|_0^{\pi} - \frac{1}{2(n+m)}\sin(n+m)x \Big|_0^{\pi}$$

$$= 0.$$

For $m = n$

$$\int_0^{\pi} \sin^2 nx \, dx = \int_0^{\pi}\left(\frac{1}{2} - \frac{1}{2}\cos 2nx\right) dx = \frac{1}{2}x \Big|_0^{\pi} - \frac{1}{4n}\sin 2nx \Big|_0^{\pi} = \frac{\pi}{2}$$

so that

$$\| \sin nx \| = \sqrt{\frac{\pi}{2}} \, .$$

10. For $m \neq n$

$$\int_0^p \sin \frac{n\pi}{p} x \sin \frac{m\pi}{p} x \, dx = \frac{1}{2} \int_0^p \left(\cos \frac{(n-m)\pi}{p} x - \cos \frac{(n+m)\pi}{p} x \right) dx$$

$$= \frac{p}{2(n-m)\pi} \sin \frac{(n-m)\pi}{p} x \, \Big|_0^p - \frac{p}{2(n+m)\pi} \sin \frac{(n+m)\pi}{p} x \, \Big|_0^p$$

$$= 0.$$

For $m = n$

$$\int_0^p \sin^2 \frac{n\pi}{p} x \, dx = \int_0^p \left(\frac{1}{2} - \frac{1}{2} \cos \frac{2n\pi}{p} x \right) dx = \frac{1}{2} x \, \Big|_0^p - \frac{p}{4n\pi} \sin \frac{2n\pi}{p} x \, \Big|_0^p = \frac{p}{2}$$

so that

$$\left\| \sin \frac{n\pi}{p} x \right\| = \sqrt{\frac{p}{2}} \, .$$

11. For $m \neq n$

$$\int_0^p \cos \frac{n\pi}{p} x \cos \frac{m\pi}{p} x \, dx = \frac{1}{2} \int_0^p \left(\cos \frac{(n-m)\pi}{p} x + \cos \frac{(n+m)\pi}{p} x \right) dx$$

$$= \frac{p}{2(n-m)\pi} \sin \frac{(n-m)\pi}{p} x \, \Big|_0^p + \frac{p}{2(n+m)\pi} \sin \frac{(n+m)\pi}{p} x \, \Big|_0^p$$

$$= 0.$$

For $m = n$

$$\int_0^p \cos^2 \frac{n\pi}{p} x \, dx = \int_0^p \left(\frac{1}{2} + \frac{1}{2} \cos \frac{2n\pi}{p} x \right) dx = \frac{1}{2} x \, \Big|_0^p + \frac{p}{4n\pi} \sin \frac{2n\pi}{p} x \, \Big|_0^p = \frac{p}{2} \, .$$

Also

$$\int_0^p 1 \cdot \cos \frac{n\pi}{p} x \, dx = \frac{p}{n\pi} \sin \frac{n\pi}{p} x \, \Big|_0^p = 0 \quad \text{and} \quad \int_0^p 1^2 dx = p$$

so that

$$\| 1 \| = \sqrt{p} \quad \text{and} \quad \left\| \cos \frac{n\pi}{p} x \right\| = \sqrt{\frac{p}{2}} \, .$$

12. For $m \neq n$, we use Problems 11 and 10:

$$\int_{-p}^p \cos \frac{n\pi}{p} x \cos \frac{m\pi}{p} x \, dx = 2 \int_0^p \cos \frac{n\pi}{p} x \cos \frac{m\pi}{p} x \, dx = 0$$

$$\int_{-p}^p \sin \frac{n\pi}{p} x \sin \frac{m\pi}{p} x \, dx = 2 \int_0^p \sin \frac{n\pi}{p} x \sin \frac{m\pi}{p} x \, dx = 0.$$

Also

$$\int_{-p}^{p} \sin \frac{n\pi}{p} x \, \cos \frac{m\pi}{p} x \, dx = \frac{1}{2} \int_{-p}^{p} \left(\sin \frac{(n-m)\pi}{p} x + \sin \frac{(n+m)\pi}{p} x \right) dx = 0,$$

$$\int_{-p}^{p} 1 \cdot \cos \frac{n\pi}{p} x \, dx = \frac{p}{n\pi} \sin \frac{n\pi}{p} x \, \Big|_{-p}^{p} = 0,$$

$$\int_{-p}^{p} 1 \cdot \sin \frac{n\pi}{p} x \, dx = -\frac{p}{n\pi} \cos \frac{n\pi}{p} x \, \Big|_{-p}^{p} = 0,$$

and

$$\int_{-p}^{p} \sin \frac{n\pi}{p} x \, \cos \frac{n\pi}{p} x \, dx = \int_{-p}^{p} \frac{1}{2} \sin \frac{2n\pi}{p} x \, dx = -\frac{p}{4n\pi} \cos \frac{2n\pi}{p} x \, \Big|_{-p}^{p} = 0.$$

For $m = n$

$$\int_{-p}^{p} \cos^2 \frac{n\pi}{p} x \, dx = \int_{-p}^{p} \left(\frac{1}{2} + \frac{1}{2} \cos \frac{2n\pi}{p} x \right) dx = p,$$

$$\int_{-p}^{p} \sin^2 \frac{n\pi}{p} x \, dx = \int_{-p}^{p} \left(\frac{1}{2} - \frac{1}{2} \cos \frac{2n\pi}{p} x \right) dx = p,$$

and

$$\int_{-p}^{p} 1^2 dx = 2p$$

so that

$$\|1\| = \sqrt{2p}, \quad \left\| \cos \frac{n\pi}{p} x \right\| = \sqrt{p}, \quad \text{and} \quad \left\| \sin \frac{n\pi}{p} x \right\| = \sqrt{p}.$$

13. Since

$$\int_{-\infty}^{\infty} e^{-x^2} \cdot 1 \cdot 2x \, dx = -e^{-x^2} \, \Big|_{-\infty}^{0} - e^{-x^2} \, \Big|_{0}^{\infty} = 0,$$

$$\int_{-\infty}^{\infty} e^{-x^2} \cdot 1 \cdot (4x^2 - 2) \, dx = 2 \int_{-\infty}^{\infty} x \left(2xe^{-x^2} \right) dx - 2 \int_{-\infty}^{\infty} e^{-x^2} dx$$

$$= 2 \left(-xe^{-x^2} \, \Big|_{-\infty}^{\infty} + \int_{-\infty}^{\infty} e^{-x^2} dx \right) - 2 \int_{-\infty}^{\infty} e^{-x^2} dx$$

$$= 2 \left(-xe^{-x^2} \, \Big|_{-\infty}^{0} - xe^{-x^2} \, \Big|_{0}^{\infty} \right) = 0,$$

and

$$\int_{-\infty}^{\infty} e^{-x^2} \cdot 2x \cdot (4x^2 - 2)\, dx = 4\int_{-\infty}^{\infty} x^2 \left(2xe^{-x^2}\right) dx - 4\int_{-\infty}^{\infty} xe^{-x^2}\, dx$$

$$= 4\left(-x^2 e^{-x^2}\,\Big|_{-\infty}^{\infty} + 2\int_{-\infty}^{\infty} xe^{-x^2}\, dx\right) - 4\int_{-\infty}^{\infty} xe^{-x^2}\, dx$$

$$= 4\left(-x^2 e^{-x^2}\,\Big|_{-\infty}^{0} - x^2 e^{-x^2}\,\Big|_{0}^{\infty}\right) + 2\int_{-\infty}^{\infty} 2xe^{-x^2}\, dx = 0,$$

the functions are orthogonal.

14. Since

$$\int_{0}^{\infty} e^{-x} \cdot 1(1-x)\, dx = (x-1)e^{-x}\,\Big|_{0}^{\infty} - \int_{0}^{\infty} e^{-x}dx = 0,$$

$$\int_{0}^{\infty} e^{-x} \cdot 1 \cdot \left(\frac{1}{2}x^2 - 2x + 1\right) dx = \left(2x - 1 - \frac{1}{2}x^2\right)e^{-x}\,\Big|_{0}^{\infty} + \int_{0}^{\infty} e^{-x}(x-2)\, dx$$

$$= 1 + (2-x)e^{-x}\,\Big|_{0}^{\infty} + \int_{0}^{\infty} e^{-x}dx = 0,$$

and

$$\int_{0}^{\infty} e^{-x} \cdot (1-x)\left(\frac{1}{2}x^2 - 2x + 1\right) dx$$

$$= \int_{0}^{\infty} e^{-x}\left(-\frac{1}{2}x^3 + \frac{5}{2}x^2 - 3x + 1\right) dx$$

$$= e^{-x}\left(\frac{1}{2}x^3 - \frac{5}{2}x^2 + 3x - 1\right)\Big|_{0}^{\infty} + \int_{0}^{\infty} e^{-x}\left(-\frac{3}{2}x^2 + 5x - 3\right) dx$$

$$= 1 + e^{-x}\left(\frac{3}{2}x^2 - 5x + 3\right)\Big|_{0}^{\infty} + \int_{0}^{\infty} e^{-x}(5 - 3x)\, dx$$

$$= 1 - 3 + e^{-x}(3x - 5)\Big|_{0}^{\infty} - 3\int_{0}^{\infty} e^{-x}dx = 0,$$

the functions are orthogonal.

15. By orthogonality $\int_{a}^{b} \phi_0(x)\phi_n(x)dx = 0$ for $n = 1, 2, 3, \ldots$; that is, $\int_{a}^{b} \phi_n(x)dx = 0$ for $n = 1, 2, 3, \ldots$.

16. Using the facts that ϕ_0 and ϕ_1 are orthogonal to ϕ_n for $n > 1$, we have

$$\int_a^b (\alpha x + \beta)\phi_n(x)\, dx = \alpha \int_a^b x\phi_n(x)\, dx + \beta \int_a^b 1 \cdot \phi_n(x)\, dx$$

$$= \alpha \int_a^b \phi_1(x)\phi_n(x)\, dx + \beta \int_a^b \phi_0(x)\phi_n(x)\, dx$$

$$= \alpha \cdot 0 + \beta \cdot 0 = 0$$

for $n = 2, 3, 4, \ldots$.

17. Using the fact that ϕ_n and ϕ_m are orthogonal for $n \neq m$ we have

$$\|\phi_m(x) + \phi_n(x)\|^2 = \int_a^b [\phi_m(x) + \phi_n(x)]^2 dx = \int_a^b \left[\phi_m^2(x) + 2\phi_m(x)\phi_n(x) + \phi_n^2(x)\right] dx$$

$$= \int_a^b \phi_m^2(x)\, dx + 2\int_a^b \phi_m(x)\phi_n(x)\, dx + \int_a^b \phi_n^2(x)\, dx$$

$$= \|\phi_m(x)\|^2 + \|\phi_n(x)\|^2 .$$

18. Setting

$$0 = \int_{-2}^2 f_3(x)f_1(x)\, dx = \int_{-2}^2 \left(x^2 + c_1 x^3 + c_2 x^4\right) dx = \frac{16}{3} + \frac{64}{5}c_2$$

and

$$0 = \int_{-2}^2 f_3(x)f_2(x)\, dx = \int_{-2}^2 \left(x^3 + c_1 x^4 + c_2 x^5\right) dx = \frac{64}{5}c_1$$

we obtain $c_1 = 0$ and $c_2 = -5/12$.

19. Since $\sin nx$ is an odd function on $[-\pi, \pi]$,

$$(1, \sin nx) = \int_{-\pi}^{\pi} \sin nx\, dx = 0$$

and $f(x) = 1$ is orthogonal to every member of $\{\sin nx\}$. Thus $\{\sin nx\}$ is not complete.

20. $(f_1 + f_2, f_3) = \int_a^b [f_1(x) + f_2(x)]f_3(x)\, dx = \int_a^b f_1(x)f_3(x)\, dx + \int_a^b f_2(x)f_3(x)\, dx = (f_1, f_3) + (f_2, f_3)$

21. (a) The fundamental period is $2\pi/2\pi = 1$.

(b) The fundamental period is $2\pi/(4/L) = \frac{1}{2}\pi L$.

(c) The fundamental period of $\sin x + \sin 2x$ is 2π.

(d) The fundamental period of $\sin 2x + \cos 4x$ is $2\pi/2 = \pi$.

(e) The fundamental period of $\sin 3x + \cos 4x$ is 2π since the smallest integer multiples of $2\pi/3$ and $2\pi/4 = \pi/2$ that are equal are 3 and 4, respectively.

(f) The fundamental period of $f(x)$ is $2\pi/(\pi/p) = 2p$.

1. $a_0 = \dfrac{1}{\pi}\displaystyle\int_{-\pi}^{\pi} f(x)\,dx = \dfrac{1}{\pi}\int_{0}^{\pi} 1\,dx = 1$

$a_n = \dfrac{1}{\pi}\displaystyle\int_{-\pi}^{\pi} f(x)\cos\dfrac{n\pi}{\pi}x\,dx = \dfrac{1}{\pi}\int_{0}^{\pi}\cos nx\,dx = 0$

$b_n = \dfrac{1}{\pi}\displaystyle\int_{-\pi}^{\pi} f(x)\sin\dfrac{n\pi}{\pi}x\,dx = \dfrac{1}{\pi}\int_{0}^{\pi}\sin nx\,dx = \dfrac{1}{n\pi}(1-\cos n\pi) = \dfrac{1}{n\pi}[1-(-1)^n]$

$f(x) = \dfrac{1}{2} + \dfrac{1}{\pi}\displaystyle\sum_{n=1}^{\infty}\dfrac{1-(-1)^n}{n}\sin nx$

2. $a_0 = \dfrac{1}{\pi}\displaystyle\int_{-\pi}^{\pi} f(x)\,dx = \dfrac{1}{\pi}\int_{-\pi}^{0}(-1)\,dx + \dfrac{1}{\pi}\int_{0}^{\pi} 2\,dx = 1$

$a_n = \dfrac{1}{\pi}\displaystyle\int_{-\pi}^{\pi} f(x)\cos nx\,dx = \dfrac{1}{\pi}\int_{-\pi}^{0} -\cos nx\,dx + \dfrac{1}{\pi}\int_{0}^{\pi} 2\cos nx\,dx = 0$

$b_n = \dfrac{1}{\pi}\displaystyle\int_{-\pi}^{\pi} f(x)\sin nx\,dx = \dfrac{1}{\pi}\int_{-\pi}^{0} -\sin nx\,dx + \dfrac{1}{\pi}\int_{0}^{\pi} 2\sin nx\,dx = \dfrac{3}{n\pi}[1-(-1)^n]$

$f(x) = \dfrac{1}{2} + \dfrac{3}{\pi}\displaystyle\sum_{n=1}^{\infty}\dfrac{1-(-1)^n}{n}\sin nx$

3. $a_0 = \displaystyle\int_{-1}^{1} f(x)\,dx = \int_{-1}^{0} 1\,dx + \int_{0}^{1} x\,dx = \dfrac{3}{2}$

$a_n = \displaystyle\int_{-1}^{1} f(x)\cos n\pi x\,dx = \int_{-1}^{0}\cos n\pi x\,dx + \int_{0}^{1} x\cos n\pi x\,dx = \dfrac{1}{n^2\pi^2}[(-1)^n - 1]$

$b_n = \displaystyle\int_{-1}^{1} f(x)\sin n\pi x\,dx = \int_{-1}^{0}\sin n\pi x\,dx + \int_{0}^{1} x\sin n\pi x\,dx = -\dfrac{1}{n\pi}$

$f(x) = \dfrac{3}{4} + \displaystyle\sum_{n=1}^{\infty}\left[\dfrac{(-1)^n - 1}{n^2\pi^2}\cos n\pi x - \dfrac{1}{n\pi}\sin n\pi x\right]$

4. $a_0 = \displaystyle\int_{-1}^{1} f(x)\,dx = \int_{0}^{1} x\,dx = \dfrac{1}{2}$

$a_n = \displaystyle\int_{-1}^{1} f(x)\cos n\pi x\,dx = \int_{0}^{1} x\cos n\pi x\,dx = \dfrac{1}{n^2\pi^2}[(-1)^n - 1]$

$b_n = \displaystyle\int_{-1}^{1} f(x)\sin n\pi x\,dx = \int_{0}^{1} x\sin n\pi x\,dx = \dfrac{(-1)^{n+1}}{n\pi}$

$$f(x) = \frac{1}{4} + \sum_{n=1}^{\infty} \left[\frac{(-1)^n - 1}{n^2 \pi^2} \cos n\pi x + \frac{(-1)^{n+1}}{n\pi} \sin n\pi x \right]$$

5. $a_0 = \dfrac{1}{\pi} \displaystyle\int_{-\pi}^{\pi} f(x)\, dx = \dfrac{1}{\pi} \displaystyle\int_{0}^{\pi} x^2\, dx = \dfrac{1}{3}\pi^2$

$a_n = \dfrac{1}{\pi} \displaystyle\int_{-\pi}^{\pi} f(x) \cos nx\, dx = \dfrac{1}{\pi} \displaystyle\int_{0}^{\pi} x^2 \cos nx\, dx = \dfrac{1}{\pi} \left(\dfrac{x^2}{\pi} \sin nx \Big|_0^{\pi} - \dfrac{2}{n} \displaystyle\int_0^{\pi} x \sin nx\, dx \right) = \dfrac{2(-1)^n}{n^2}$

$b_n = \dfrac{1}{\pi} \displaystyle\int_0^{\pi} x^2 \sin nx\, dx = \dfrac{1}{\pi} \left(-\dfrac{x^2}{n} \cos nx \Big|_0^{\pi} + \dfrac{2}{n} \displaystyle\int_0^{\pi} x \cos nx\, dx \right) = \dfrac{\pi}{n}(-1)^{n+1} + \dfrac{2}{n^3\pi}[(-1)^n - 1]$

$f(x) = \dfrac{\pi^2}{6} + \displaystyle\sum_{n=1}^{\infty} \left[\dfrac{2(-1)^n}{n^2} \cos nx + \left(\dfrac{\pi}{n}(-1)^{n+1} + \dfrac{2[(-1)^n - 1]}{n^3\pi} \right) \sin nx \right]$

6. $a_0 = \dfrac{1}{\pi} \displaystyle\int_{-\pi}^{\pi} f(x)\, dx = \dfrac{1}{\pi} \displaystyle\int_{-\pi}^{0} \pi^2\, dx + \dfrac{1}{\pi} \displaystyle\int_0^{\pi} \left(\pi^2 - x^2 \right) dx = \dfrac{5}{3}\pi^2$

$a_n = \dfrac{1}{\pi} \displaystyle\int_{-\pi}^{\pi} f(x) \cos nx\, dx = \dfrac{1}{\pi} \displaystyle\int_{-\pi}^{0} \pi^2 \cos nx\, dx + \dfrac{1}{\pi} \displaystyle\int_0^{\pi} \left(\pi^2 - x^2 \right) \cos nx\, dx$

$= \dfrac{1}{\pi} \left(\dfrac{\pi^2 - x^2}{n} \sin nx \Big|_0^{\pi} + \dfrac{2}{n} \displaystyle\int_0^{\pi} x \sin nx\, dx \right) = \dfrac{2}{n^2}(-1)^{n+1}$

$b_n = \dfrac{1}{\pi} \displaystyle\int_{-\pi}^{\pi} f(x) \sin nx\, dx = \dfrac{1}{\pi} \displaystyle\int_{-\pi}^{0} \pi^2 \sin nx\, dx + \dfrac{1}{\pi} \displaystyle\int_0^{\pi} \left(\pi^2 - x^2 \right) \sin nx\, dx$

$= \dfrac{\pi}{n}[(-1)^n - 1] + \dfrac{1}{\pi} \left(\dfrac{x^2 - \pi^2}{n} \cos nx \Big|_0^{\pi} - \dfrac{2}{n} \displaystyle\int_0^{\pi} x \cos nx\, dx \right) = \dfrac{\pi}{n}(-1)^n + \dfrac{2}{n^3\pi}[1 - (-1)^n]$

$f(x) = \dfrac{5\pi^2}{6} + \displaystyle\sum_{n=1}^{\infty} \left[\dfrac{2}{n^2}(-1)^{n+1} \cos nx + \left(\dfrac{\pi}{n}(-1)^n + \dfrac{2[1 - (-1)^n]}{n^3\pi} \right) \sin nx \right]$

7. $a_0 = \dfrac{1}{\pi} \displaystyle\int_{-\pi}^{\pi} f(x)\, dx = \dfrac{1}{\pi} \displaystyle\int_{-\pi}^{\pi} (x + \pi)\, dx = 2\pi$

$a_n = \dfrac{1}{\pi} \displaystyle\int_{-\pi}^{\pi} f(x) \cos nx\, dx = \dfrac{1}{\pi} \displaystyle\int_{-\pi}^{\pi} (x + \pi) \cos nx\, dx = 0$

$b_n = \dfrac{1}{\pi} \displaystyle\int_{-\pi}^{\pi} f(x) \sin nx\, dx = \dfrac{2}{n}(-1)^{n+1}$

$f(x) = \pi + \displaystyle\sum_{n=1}^{\infty} \dfrac{2}{n}(-1)^{n+1} \sin nx$

8. $a_0 = \dfrac{1}{\pi} \displaystyle\int_{-\pi}^{\pi} f(x)\, dx = \dfrac{1}{\pi} \displaystyle\int_{-\pi}^{\pi} (3 - 2x)\, dx = 6$

$a_n = \dfrac{1}{\pi} \displaystyle\int_{-\pi}^{\pi} f(x) \cos nx\, dx = \dfrac{1}{\pi} \displaystyle\int_{-\pi}^{\pi} (3 - 2x) \cos nx\, dx = 0$

$$b_n = \frac{1}{\pi} \int_{-\pi}^{\pi} (3 - 2x) \sin nx \, dx = \frac{4}{n} (-1)^n$$

$$f(x) = 3 + 4 \sum_{n=1}^{\infty} \frac{(-1)^n}{n} \sin nx$$

9. $a_0 = \dfrac{1}{\pi} \int_{-\pi}^{\pi} f(x) \, dx = \dfrac{1}{\pi} \int_{0}^{\pi} \sin x \, dx = \dfrac{2}{\pi}$

$$a_n = \frac{1}{\pi} \int_{-\pi}^{\pi} f(x) \cos nx \, dx = \frac{1}{\pi} \int_{0}^{\pi} \sin x \cos nx \, dx = \frac{1}{2\pi} \int_{0}^{\pi} \Big(\sin(1+n)x + \sin(1-n)x \Big) \, dx$$

$$= \frac{1 + (-1)^n}{\pi(1 - n^2)} \quad \text{for } n = 2, 3, 4, \ldots$$

$$a_1 = \frac{1}{2\pi} \int_{0}^{\pi} \sin 2x \, dx = 0$$

$$b_n = \frac{1}{\pi} \int_{-\pi}^{\pi} f(x) \sin nx \, dx = \frac{1}{\pi} \int_{0}^{\pi} \sin x \sin nx \, dx$$

$$= \frac{1}{2\pi} \int_{0}^{\pi} \Big(\cos(1-n)x - \cos(1+n)x \Big) \, dx = 0 \quad \text{for } n = 2, 3, 4, \ldots$$

$$b_1 = \frac{1}{2\pi} \int_{0}^{\pi} (1 - \cos 2x) \, dx = \frac{1}{2}$$

$$f(x) = \frac{1}{\pi} + \frac{1}{2} \sin x + \sum_{n=2}^{\infty} \frac{1 + (-1)^n}{\pi(1 - n^2)} \cos nx$$

10. $a_0 = \dfrac{2}{\pi} \int_{-\pi/2}^{\pi/2} f(x) \, dx = \dfrac{2}{\pi} \int_{0}^{\pi/2} \cos x \, dx = \dfrac{2}{\pi}$

$$a_n = \frac{2}{\pi} \int_{-\pi/2}^{\pi/2} f(x) \cos 2nx \, dx = \frac{2}{\pi} \int_{0}^{\pi/2} \cos x \cos 2nx \, dx = \frac{1}{\pi} \int_{0}^{\pi/2} \Big(\cos(2n - 1)x + \cos(2n + 1)x \Big) \, dx$$

$$= \frac{2(-1)^{n+1}}{\pi(4n^2 - 1)}$$

$$b_n = \frac{2}{\pi} \int_{-\pi/2}^{\pi/2} f(x) \sin 2nx \, dx = \frac{2}{\pi} \int_{0}^{\pi/2} \cos x \sin 2nx \, dx = \frac{1}{\pi} \int_{0}^{\pi/2} \Big(\sin(2n - 1)x + \sin(2n + 1)x \Big) \, dx$$

$$= \frac{4n}{\pi(4n^2 - 1)}$$

$$f(x) = \frac{1}{\pi} + \sum_{n=1}^{\infty} \left[\frac{2(-1)^{n+1}}{\pi(4n^2 - 1)} \cos 2nx + \frac{4n}{\pi(4n^2 - 1)} \sin 2nx \right]$$

11. $a_0 = \dfrac{1}{2} \displaystyle\int_{-2}^{2} f(x) \, dx = \dfrac{1}{2} \left(\int_{-1}^{0} -2 \, dx + \int_{0}^{1} 1 \, dx \right) = -\dfrac{1}{2}$

$$a_n = \frac{1}{2}\int_{-2}^{2} f(x)\cos\frac{n\pi}{2}x\,dx = \frac{1}{2}\left(\int_{-1}^{0}(-2)\cos\frac{n\pi}{2}x\,dx + \int_{0}^{1}\cos\frac{n\pi}{2}x\,dx\right) = -\frac{1}{n\pi}\sin\frac{n\pi}{2}$$

$$b_n = \frac{1}{2}\int_{-2}^{2} f(x)\sin\frac{n\pi}{2}x\,dx = \frac{1}{2}\left(\int_{-1}^{0}(-2)\sin\frac{n\pi}{2}x\,dx + \int_{0}^{1}\sin\frac{n\pi}{2}x\,dx\right) = \frac{3}{n\pi}\left(1-\cos\frac{n\pi}{2}\right)$$

$$f(x) = -\frac{1}{4} + \sum_{n=1}^{\infty}\left[-\frac{1}{n\pi}\sin\frac{n\pi}{2}\cos\frac{n\pi}{2}x + \frac{3}{n\pi}\left(1-\cos\frac{n\pi}{2}\right)\sin\frac{n\pi}{2}x\right]$$

12. $\quad a_0 = \dfrac{1}{2}\displaystyle\int_{-2}^{2} f(x)\,dx = \dfrac{1}{2}\left(\int_{0}^{1} x\,dx + \int_{1}^{2} 1\,dx\right) = \dfrac{3}{4}$

$$a_n = \frac{1}{2}\int_{-2}^{2} f(x)\cos\frac{n\pi}{2}x\,dx = \frac{1}{2}\left(\int_{0}^{1} x\cos\frac{n\pi}{2}x\,dx + \int_{1}^{2}\cos\frac{n\pi}{2}x\,dx\right) = \frac{2}{n^2\pi^2}\left(\cos\frac{n\pi}{2}-1\right)$$

$$b_n = \frac{1}{2}\int_{-2}^{2} f(x)\sin\frac{n\pi}{2}x\,dx = \frac{1}{2}\left(\int_{0}^{1} x\sin\frac{n\pi}{2}x\,dx + \int_{1}^{2}\sin\frac{n\pi}{2}x\,dx\right)$$

$$= \frac{2}{n^2\pi^2}\left(\sin\frac{n\pi}{2} + \frac{n\pi}{2}(-1)^{n+1}\right)$$

$$f(x) = \frac{3}{8} + \sum_{n=1}^{\infty}\left[\frac{2}{n^2\pi^2}\left(\cos\frac{n\pi}{2}-1\right)\cos\frac{n\pi}{2}x + \frac{2}{n^2\pi^2}\left(\sin\frac{n\pi}{2}+\frac{n\pi}{2}(-1)^{n+1}\right)\sin\frac{n\pi}{2}x\right]$$

13. $\quad a_0 = \dfrac{1}{5}\displaystyle\int_{-5}^{5} f(x)\,dx = \dfrac{1}{5}\left(\int_{-5}^{0} 1\,dx + \int_{0}^{5}(1+x)\,dx\right) = \dfrac{9}{2}$

$$a_n = \frac{1}{5}\int_{-5}^{5} f(x)\cos\frac{n\pi}{5}x\,dx = \frac{1}{5}\left(\int_{-5}^{0}\cos\frac{n\pi}{5}x\,dx + \int_{0}^{5}(1+x)\cos\frac{n\pi}{5}x\,dx\right) = \frac{5}{n^2\pi^2}[(-1)^n - 1]$$

$$b_n = \frac{1}{5}\int_{-5}^{5} f(x)\sin\frac{n\pi}{5}x\,dx = \frac{1}{5}\left(\int_{-5}^{0}\sin\frac{n\pi}{5}x\,dx + \int_{0}^{5}(1+x)\cos\frac{n\pi}{5}x\,dx\right) = \frac{5}{n\pi}(-1)^{n+1}$$

$$f(x) = \frac{9}{4} + \sum_{n=1}^{\infty}\left[\frac{5}{n^2\pi^2}[(-1)^n - 1]\cos\frac{n\pi}{5}x + \frac{5}{n\pi}(-1)^{n+1}\sin\frac{n\pi}{5}x\right]$$

14. $\quad a_0 = \dfrac{1}{2}\displaystyle\int_{-2}^{2} f(x)\,dx = \dfrac{1}{2}\left(\int_{-2}^{0}(2+x)\,dx + \int_{0}^{2} 2\,dx\right) = 3$

$$a_n = \frac{1}{2}\int_{-2}^{2} f(x)\cos\frac{n\pi}{2}x\,dx = \frac{1}{2}\left(\int_{-2}^{0}(2+x)\cos\frac{n\pi}{2}x\,dx + \int_{0}^{2} 2\cos\frac{n\pi}{2}x\,dx\right) = \frac{2}{n^2\pi^2}[1-(-1)^n]$$

$$b_n = \frac{1}{2}\int_{-2}^{2} f(x)\sin\frac{n\pi}{2}x\,dx = \frac{1}{2}\left(\int_{-2}^{0}(2+x)\sin\frac{n\pi}{2}x\,dx + \int_{0}^{2} 2\sin\frac{n\pi}{2}x\,dx\right) = \frac{2}{n\pi}(-1)^{n+1}$$

$$f(x) = \frac{3}{2} + \sum_{n=1}^{\infty}\left[\frac{2}{n^2\pi^2}[1-(-1)^n]\cos\frac{n\pi}{2}x + \frac{2}{n\pi}(-1)^{n+1}\sin\frac{n\pi}{2}x\right]$$

15. $a_0 = \dfrac{1}{\pi} \displaystyle\int_{-\pi}^{\pi} f(x)\, dx = \dfrac{1}{\pi} \int_{-\pi}^{\pi} e^x\, dx = \dfrac{1}{\pi}(e^\pi - e^{-\pi})$

$a_n = \dfrac{1}{\pi} \displaystyle\int_{-\pi}^{\pi} f(x) \cos nx\, dx = \dfrac{(-1)^n(e^\pi - e^{-\pi})}{\pi(1+n^2)}$

$b_n = \dfrac{1}{\pi} \displaystyle\int_{-\pi}^{\pi} f(x) \sin nx\, dx = \dfrac{1}{\pi} \int_{-\pi}^{\pi} e^x \sin nx\, dx = \dfrac{(-1)^n n(e^{-\pi} - e^\pi)}{\pi(1+n^2)}$

$f(x) = \dfrac{e^\pi - e^{-\pi}}{2\pi} + \displaystyle\sum_{n=1}^{\infty} \left[\dfrac{(-1)^n(e^\pi - e^{-\pi})}{\pi(1+n^2)} \cos nx + \dfrac{(-1)^n n(e^{-\pi} - e^\pi)}{\pi(1+n^2)} \sin nx \right]$

16. $a_0 = \dfrac{1}{\pi} \displaystyle\int_{-\pi}^{\pi} f(x)\, dx = \dfrac{1}{\pi} \int_0^\pi (e^x - 1)\, dx = \dfrac{1}{\pi}(e^\pi - \pi - 1)$

$a_n = \dfrac{1}{\pi} \displaystyle\int_{-\pi}^{\pi} f(x) \cos nx\, dx = \dfrac{1}{\pi} \int_0^\pi (e^x - 1) \cos nx\, dx = \dfrac{[e^\pi(-1)^n - 1]}{\pi(1+n^2)}$

$b_n = \dfrac{1}{\pi} \displaystyle\int_{-\pi}^{\pi} f(x) \sin nx\, dx = \dfrac{1}{\pi} \int_0^\pi (e^x - 1) \sin nx\, dx = \dfrac{1}{\pi}\left(\dfrac{ne^\pi(-1)^{n+1}}{1+n^2} + \dfrac{n}{1+n^2} + \dfrac{(-1)^n}{n} - \dfrac{1}{n} \right)$

$f(x) = \dfrac{e^\pi - \pi - 1}{2\pi} + \displaystyle\sum_{n=1}^{\infty} \left[\dfrac{e^\pi(-1)^n - 1}{\pi(1+n^2)} \cos nx + \left(\dfrac{n}{1+n^2}\left[e^\pi(-1)^{n+1} + 1\right] + \dfrac{(-1)^n - 1}{n} \right) \sin nx \right]$

17. The function in Problem 5 is discontinuous at $x = \pi$, so the corresponding Fourier series converges to $\pi^2/2$ at $x = \pi$. That is,

$$\dfrac{\pi^2}{2} = \dfrac{\pi^2}{6} + \sum_{n=1}^{\infty} \left[\dfrac{2(-1)^n}{n^2} \cos n\pi + \left(\dfrac{\pi}{n}(-1)^{n+1} + \dfrac{2[(-1)^n - 1]}{n^3 \pi} \right) \sin n\pi \right]$$

$$= \dfrac{\pi^2}{6} + \sum_{n=1}^{\infty} \dfrac{2(-1)^n}{n^2}(-1)^n = \dfrac{\pi^2}{6} + \sum_{n=1}^{\infty} \dfrac{2}{n^2} = \dfrac{\pi^2}{6} + 2\left(1 + \dfrac{1}{2^2} + \dfrac{1}{3^2} + \cdots\right)$$

and

$$\dfrac{\pi^2}{6} = \dfrac{1}{2}\left(\dfrac{\pi^2}{2} - \dfrac{\pi^2}{6} \right) = 1 + \dfrac{1}{2^2} + \dfrac{1}{3^2} + \cdots.$$

At $x = 0$ the series converges to 0 and

$$0 = \dfrac{\pi^2}{6} + \sum_{n=1}^{\infty} \dfrac{2(-1)^n}{n^2} = \dfrac{\pi^2}{6} + 2\left(-1 + \dfrac{1}{2^2} - \dfrac{1}{3^2} + \dfrac{1}{4^2} - \cdots \right)$$

so

$$\dfrac{\pi^2}{12} = 1 - \dfrac{1}{2^2} + \dfrac{1}{3^2} - \dfrac{1}{4^2} + \cdots.$$

18. From Problem 17

$$\dfrac{\pi^2}{8} = \dfrac{1}{2}\left(\dfrac{\pi^2}{6} + \dfrac{\pi^2}{12} \right) = \dfrac{1}{2}\left(2 + \dfrac{2}{3^2} + \dfrac{2}{5^2} + \cdots \right) = 1 + \dfrac{1}{3^2} + \dfrac{1}{5^2} + \cdots.$$

19. The function in Problem 7 is continuous at $x = \pi/2$ so

$$\frac{3\pi}{2} = f\left(\frac{\pi}{2}\right) = \pi + \sum_{n=1}^{\infty} \frac{2}{n}(-1)^{n+1} \sin \frac{n\pi}{2} = \pi + 2\left(1 - \frac{1}{3} + \frac{1}{5} - \frac{1}{7} + \cdots\right)$$

and

$$\frac{\pi}{4} = 1 - \frac{1}{3} + \frac{1}{5} - \frac{1}{7} + \cdots.$$

20. The function in Problem 9 is continuous at $x = \pi/2$ so

$$1 = f\left(\frac{\pi}{2}\right) = \frac{1}{\pi} + \frac{1}{2} + \sum_{n=2}^{\infty} \frac{1 + (-1)^n}{\pi(1 - n^2)} \cos \frac{n\pi}{2}$$

$$1 = \frac{1}{\pi} + \frac{1}{2} + \frac{2}{3\pi} - \frac{2}{3 \cdot 5\pi} + \frac{2}{5 \cdot 7\pi} - \cdots$$

and

$$\pi = 1 + \frac{\pi}{2} + \frac{2}{3} - \frac{2}{3 \cdot 5} + \frac{2}{5 \cdot 7} - \cdots$$

or

$$\frac{\pi}{4} = \frac{1}{2} + \frac{1}{1 \cdot 3} - \frac{1}{3 \cdot 5} + \frac{1}{5 \cdot 7} - \cdots.$$

21. (a) Letting $c_0 = a_0/2$, $c_n = (a_n - ib_n)/2$, and $c_{-n} = (a_n + ib_n)/2$ we have

$$f(x) = \frac{a_0}{2} + \sum_{n=1}^{\infty} \left(a_n \cos \frac{n\pi}{p}x + b_n \sin \frac{n\pi}{p}x\right)$$

$$= c_0 + \sum_{n=1}^{\infty} \left(a_n \frac{e^{in\pi x/p} + e^{-in\pi x/p}}{2} + b_n \frac{e^{in\pi x/p} - e^{-in\pi x/p}}{2i}\right)$$

$$= c_0 + \sum_{n=1}^{\infty} \left(a_n \frac{e^{in\pi x/p} + e^{-in\pi x/p}}{2} - b_n \frac{ie^{in\pi x/p} - ie^{-in\pi x/p}}{2}\right)$$

$$= c_0 + \sum_{n=1}^{\infty} \left(\frac{a_n - ib_n}{2} e^{in\pi x/p} + \frac{a_n + ib_n}{2} e^{-in\pi x/p}\right)$$

$$= c_0 + \sum_{n=1}^{\infty} \left(c_n e^{in\pi x/p} + c_{-n} e^{i(-n)\pi x/p}\right) = \sum_{n=-\infty}^{\infty} c_n e^{in\pi x/p}.$$

(b) From part (a) we have

$$c_n = \frac{1}{2}(a_n - ib_n) = \frac{1}{2p} \int_{-p}^{p} f(x)\left(\cos \frac{n\pi}{p}x - i\sin \frac{n\pi}{p}x\right) dx = \frac{1}{2p} \int_{-p}^{p} f(x) e^{-in\pi x/p} dx$$

and

$$c_{-n} = \frac{1}{2}(a_n + ib_n) = \frac{1}{2p} \int_{-p}^{p} f(x)\left(\cos \frac{n\pi}{p}x + i\sin \frac{n\pi}{p}x\right) dx = \frac{1}{2p} \int_{-p}^{p} f(x) e^{in\pi x/p} dx$$

for $n = 1, 2, 3, \ldots$. Thus, for $n = \pm 1, \pm 2, \pm 3, \ldots$,

$$c_n = \frac{1}{2p} \int_{-p}^{p} f(x) e^{-in\pi x/p} dx.$$

When $n = 0$ the above formula gives

$$c_0 = \frac{1}{2p} \int_{-p}^{p} f(x) dx,$$

which is $a_0/2$ where a_0 is (9) in the text. Therefore

$$c_n = \frac{1}{2p} \int_{-p}^{p} f(x) e^{-in\pi x/p} dx, \quad n = 0, \pm 1, \pm 2, \ldots.$$

22. Identifying $f(x) = e^{-x}$ and $p = \pi$, we have

$$c_n = \frac{1}{2\pi} \int_{-\pi}^{\pi} e^{-x} e^{-inx} dx = \frac{1}{2\pi} \int_{-\pi}^{\pi} e^{-(in+1)x} dx$$

$$= -\frac{1}{2(in+1)\pi} e^{-(in+1)x} \Big|_{-\pi}^{\pi}$$

$$= -\frac{1}{2(in+1)\pi} \left[e^{-(in+1)\pi} - e^{(in+1)\pi} \right]$$

$$= \frac{e^{(in+1)\pi} - e^{-(in+1)\pi}}{2(in+1)\pi}$$

$$= \frac{e^{\pi}(\cos n\pi + i \sin n\pi) - e^{-\pi}(\cos n\pi - i \sin n\pi)}{2(in+1)\pi}$$

$$= \frac{(e^{\pi} - e^{-\pi}) \cos n\pi}{2(in+1)\pi} = \frac{(e^{\pi} - e^{-\pi})(-1)^n}{2(in+1)\pi}.$$

Thus

$$f(x) = \sum_{n=-\infty}^{\infty} (-1)^n \frac{e^{\pi} - e^{-\pi}}{2(in+1)\pi} e^{inx}.$$

Exercises 11.3   Fourier Cosine and Sine Series

1. Since $f(-x) = \sin(-3x) = -\sin 3x = -f(x)$, $f(x)$ is an odd function.

2. Since $f(-x) = -x \cos(-x) = -x \cos x = -f(x)$, $f(x)$ is an odd function.

3. Since $f(-x) = (-x)^2 - x = x^2 - x$, $f(x)$ is neither even nor odd.

4. Since $f(-x) = (-x)^3 + 4x = -(x^3 - 4x) = -f(x)$, $f(x)$ is an odd function.

5. Since $f(-x) = e^{|-x|} = e^{|x|} = f(x)$, $f(x)$ is an even function.

6. Since $f(-x) = e^{-x} - e^x = -f(x)$, $f(x)$ is an odd function.

7. For $0 < x < 1$, $f(-x) = (-x)^2 = x^2 = -f(x)$, $f(x)$ is an odd function.

8. For $0 \leq x < 2$, $f(-x) = -x + 5 = f(x)$, $f(x)$ is an even function.

9. Since $f(x)$ is not defined for $x < 0$, it is neither even nor odd.

10. Since $f(-x) = \left|(-x)^5\right| = \left|x^5\right| = f(x)$, $f(x)$ is an even function.

11. Since $f(x)$ is an odd function, we expand in a sine series:

$$b_n = \frac{2}{\pi} \int_0^\pi 1 \cdot \sin nx \, dx = \frac{2}{n\pi}[1 - (-1)^n].$$

Thus

$$f(x) = \sum_{n=1}^\infty \frac{2}{n\pi}[1 - (-1)^n] \sin nx.$$

12. Since $f(x)$ is an even function, we expand in a cosine series:

$$a_0 = \int_1^2 1 \, dx = 1$$

$$a_n = \int_1^2 \cos \frac{n\pi}{2} x \, dx = -\frac{2}{n\pi} \sin \frac{n\pi}{2}.$$

Thus

$$f(x) = \frac{1}{2} + \sum_{n=1}^\infty \frac{-2}{n\pi} \sin \frac{n\pi}{2} \cos \frac{n\pi}{2} x.$$

13. Since $f(x)$ is an even function, we expand in a cosine series:

$$a_0 = \frac{2}{\pi} \int_0^\pi x \, dx = \pi$$

$$a_n = \frac{2}{\pi} \int_0^\pi x \cos nx \, dx = \frac{2}{n^2\pi}[(-1)^n - 1].$$

Thus

$$f(x) = \frac{\pi}{2} + \sum_{n=1}^\infty \frac{2}{n^2\pi}[(-1)^n - 1] \cos nx.$$

14. Since $f(x)$ is an odd function, we expand in a sine series:

$$b_n = \frac{2}{\pi} \int_0^\pi x \sin nx \, dx = \frac{2}{n}(-1)^{n+1}.$$

Thus

$$f(x) = \sum_{n=1}^\infty \frac{2}{n}(-1)^{n+1} \sin nx.$$

15. Since $f(x)$ is an even function, we expand in a cosine series:

$$a_0 = 2 \int_0^1 x^2 \, dx = \frac{2}{3}$$

$$a_n = 2 \int_0^1 x^2 \cos n\pi x \, dx = 2 \left(\frac{x^2}{n\pi} \sin n\pi x \, \Big|_0^1 - \frac{2}{n\pi} \int_0^1 x \sin n\pi x \, dx \right) = \frac{4}{n^2 \pi^2} (-1)^n.$$

Thus

$$f(x) = \frac{1}{3} + \sum_{n=1}^{\infty} \frac{4}{n^2 \pi^2} (-1)^n \cos n\pi x.$$

16. Since $f(x)$ is an odd function, we expand in a sine series:

$$b_n = 2 \int_0^1 x^2 \sin n\pi x \, dx = 2 \left(-\frac{x^2}{n\pi} \cos n\pi x \, \Big|_0^1 + \frac{2}{n\pi} \int_0^1 x \cos n\pi x \, dx \right)$$

$$= \frac{2(-1)^{n+1}}{n\pi} + \frac{4}{n^3 \pi^3} [(-1)^n - 1].$$

Thus

$$f(x) = \sum_{n=1}^{\infty} \left(\frac{2(-1)^{n+1}}{n\pi} + \frac{4}{n^3 \pi^3} [(-1)^n - 1] \right) \sin n\pi x.$$

17. Since $f(x)$ is an even function, we expand in a cosine series:

$$a_0 = \frac{2}{\pi} \int_0^{\pi} (\pi^2 - x^2) \, dx = \frac{4}{3} \pi^2$$

$$a_n = \frac{2}{\pi} \int_0^{\pi} (\pi^2 - x^2) \cos nx \, dx = \frac{2}{\pi} \left(\frac{\pi^2 - x^2}{n} \sin nx \, \Big|_0^{\pi} + \frac{2}{n} \int_0^{\pi} x \sin nx \, dx \right) = \frac{4}{n^2} (-1)^{n+1}.$$

Thus

$$f(x) = \frac{2}{3} \pi^2 + \sum_{n=1}^{\infty} \frac{4}{n^2} (-1)^{n+1} \cos nx.$$

18. Since $f(x)$ is an odd function, we expand in a sine series:

$$b_n = \frac{2}{\pi} \int_0^{\pi} x^3 \sin nx \, dx = \frac{2}{\pi} \left(-\frac{x^3}{n} \cos nx \, \Big|_0^{\pi} + \frac{3}{n} \int_0^{\pi} x^2 \cos nx \, dx \right)$$

$$= \frac{2\pi^2}{n} (-1)^{n+1} - \frac{12}{n^2 \pi} \int_0^{\pi} x \sin nx \, dx$$

$$= \frac{2\pi^2}{n} (-1)^{n+1} - \frac{12}{n^2 \pi} \left(-\frac{x}{n} \cos nx \, \Big|_0^{\pi} + \frac{1}{n} \int_0^{\pi} \cos nx \, dx \right) = \frac{2\pi^2}{n} (-1)^{n+1} + \frac{12}{n^3} (-1)^n.$$

Thus

$$f(x) = \sum_{n=1}^{\infty} \left(\frac{2\pi^2}{n} (-1)^{n+1} + \frac{12}{n^3} (-1)^n \right) \sin nx.$$

19. Since $f(x)$ is an odd function, we expand in a sine series:

$$b_n = \frac{2}{\pi} \int_0^\pi (x+1) \sin nx \, dx = \frac{2(\pi+1)}{n\pi}(-1)^{n+1} + \frac{2}{n\pi}.$$

Thus

$$f(x) = \sum_{n=1}^\infty \left(\frac{2(\pi+1)}{n\pi}(-1)^{n+1} + \frac{2}{n\pi} \right) \sin nx.$$

20. Since $f(x)$ is an odd function, we expand in a sine series:

$$b_n = 2 \int_0^1 (x-1) \sin n\pi x \, dx = 2 \left[\int_0^1 x \sin n\pi x \, dx - \int_0^1 \sin n\pi x \, dx \right]$$

$$= 2 \left[\frac{1}{n^2\pi^2} \sin n\pi x - \frac{x}{n\pi} \cos n\pi x + \frac{1}{n\pi} \cos n\pi x \right]_0^1 = -\frac{2}{n\pi}.$$

Thus

$$f(x) = -\sum_{n=1}^\infty \frac{2}{n\pi} \sin n\pi x.$$

21. Since $f(x)$ is an even function, we expand in a cosine series:

$$a_0 = \int_0^1 x \, dx + \int_1^2 1 \, dx = \frac{3}{2}$$

$$a_n = \int_0^1 x \cos \frac{n\pi}{2} x \, dx + \int_1^2 \cos \frac{n\pi}{2} x \, dx = \frac{4}{n^2\pi^2} \left(\cos \frac{n\pi}{2} - 1 \right).$$

Thus

$$f(x) = \frac{3}{4} + \sum_{n=1}^\infty \frac{4}{n^2\pi^2} \left(\cos \frac{n\pi}{2} - 1 \right) \cos \frac{n\pi}{2} x.$$

22. Since $f(x)$ is an odd function, we expand in a sine series:

$$b_n = \frac{1}{\pi} \int_0^\pi x \sin \frac{n}{2} x \, dx + \int_\pi^{2\pi} \pi \sin \frac{n}{2} x \, dx = \frac{4}{n^2\pi} \sin \frac{n\pi}{2} + \frac{2}{n}(-1)^{n+1}.$$

Thus

$$f(x) = \sum_{n=1}^\infty \left(\frac{4}{n^2\pi} \sin \frac{n\pi}{2} + \frac{2}{n}(-1)^{n+1} \right) \sin \frac{n}{2} x.$$

23. Since $f(x)$ is an even function, we expand in a cosine series:

$$a_0 = \frac{2}{\pi} \int_0^\pi \sin x \, dx = \frac{4}{\pi}$$

$$a_n = \frac{2}{\pi} \int_0^\pi \sin x \cos nx \, dx = \frac{1}{\pi} \int_0^\pi \left(\sin(1+n)x + \sin(1-n)x \right) dx$$

$$= \frac{2}{\pi(1-n^2)} (1 + (-1)^n) \quad \text{for } n = 2, 3, 4, \dots$$

$$a_1 = \frac{1}{\pi} \int_0^\pi \sin 2x \, dx = 0.$$

Thus

$$f(x) = \frac{2}{\pi} + \sum_{n=2}^{\infty} \frac{2[1 + (-1)^n]}{\pi(1 - n^2)} \cos nx.$$

24. Since $f(x)$ is an even function, we expand in a cosine series. [See the solution of Problem 10 in Exercise 11.2 for the computation of the integrals.]

$$a_0 = \frac{2}{\pi/2} \int_0^{\pi/2} \cos x \, dx = \frac{4}{\pi}$$

$$a_n = \frac{2}{\pi/2} \int_0^{\pi/2} \cos x \cos \frac{n\pi}{\pi/2} x \, dx = \frac{4(-1)^{n+1}}{\pi(4n^2 - 1)}$$

Thus

$$f(x) = \frac{2}{\pi} + \sum_{n=1}^{\infty} \frac{4(-1)^{n+1}}{\pi(4n^2 - 1)} \cos 2nx.$$

25. $a_0 = 2 \int_0^{1/2} 1 \, dx = 1$

$$a_n = 2 \int_0^{1/2} 1 \cdot \cos n\pi x \, dx = \frac{2}{n\pi} \sin \frac{n\pi}{2}$$

$$b_n = 2 \int_0^{1/2} 1 \cdot \sin n\pi x \, dx = \frac{2}{n\pi} \left(1 - \cos \frac{n\pi}{2} \right)$$

$$f(x) = \frac{1}{2} + \sum_{n=1}^{\infty} \frac{2}{n\pi} \sin \frac{n\pi}{2} \cos n\pi x$$

$$f(x) = \sum_{n=1}^{\infty} \frac{2}{n\pi} \left(1 - \cos \frac{n\pi}{2} \right) \sin n\pi x$$

26. $a_0 = 2 \int_{1/2}^{1} 1 \, dx = 1$

$$a_n = 2 \int_{1/2}^{1} 1 \cdot \cos n\pi x \, dx = -\frac{2}{n\pi} \sin \frac{n\pi}{2}$$

$$b_n = 2 \int_{1/2}^{1} 1 \cdot \sin n\pi x \, dx = \frac{2}{n\pi} \left(\cos \frac{n\pi}{2} + (-1)^{n+1} \right)$$

$$f(x) = \frac{1}{2} + \sum_{n=1}^{\infty} \left(-\frac{2}{n\pi} \sin \frac{n\pi}{2} \right) \cos n\pi x$$

$$f(x) = \sum_{n=1}^{\infty} \frac{2}{n\pi} \left(\cos \frac{n\pi}{2} + (-1)^{n+1} \right) \sin n\pi x$$

27. $a_0 = \frac{4}{\pi} \int_0^{\pi/2} \cos x \, dx = \frac{4}{\pi}$

$$a_n = \frac{4}{\pi} \int_0^{\pi/2} \cos x \cos 2nx \, dx = \frac{2}{\pi} \int_0^{\pi/2} [\cos(2n+1)x + \cos(2n-1)x] \, dx = \frac{4(-1)^n}{\pi(1-4n^2)}$$

$$b_n = \frac{4}{\pi} \int_0^{\pi/2} \cos x \sin 2nx \, dx = \frac{2}{\pi} \int_0^{\pi/2} [\sin(2n+1)x + \sin(2n-1)x] \, dx = \frac{8n}{\pi(4n^2-1)}$$

$$f(x) = \frac{2}{\pi} + \sum_{n=1}^{\infty} \frac{4(-1)^n}{\pi(1-4n^2)} \cos 2nx$$

$$f(x) = \sum_{n=1}^{\infty} \frac{8n}{\pi(4n^2-1)} \sin 2nx$$

28. $a_0 = \dfrac{2}{\pi} \displaystyle\int_0^{\pi} \sin x \, dx = \dfrac{4}{\pi}$

$$a_n = \frac{2}{\pi} \int_0^{\pi} \sin x \cos nx \, dx = \frac{1}{\pi} \int_0^{\pi} [\sin(n+1)x - \sin(n-1)x] \, dx = \frac{2[(-1)^n + 1]}{\pi(1-n^2)} \quad \text{for } n = 2, 3, 4, \ldots$$

$$b_n = \frac{2}{\pi} \int_0^{\pi} \sin x \sin nx \, dx = \frac{1}{\pi} \int_0^{\pi} [\cos(n-1)x - \cos(n+1)x] \, dx = 0 \quad \text{for } n = 2, 3, 4, \ldots$$

$$a_1 = \frac{1}{\pi} \int_0^{\pi} \sin 2x \, dx = 0$$

$$b_1 = \frac{2}{\pi} \int_0^{\pi} \sin^2 x \, dx = 1$$

$$f(x) = \sin x$$

$$f(x) = \frac{2}{\pi} + \frac{2}{\pi} \sum_{n=2}^{\infty} \frac{(-1)^n + 1}{1 - n^2} \cos nx$$

29. $a_0 = \dfrac{2}{\pi} \left(\displaystyle\int_0^{\pi/2} x \, dx + \int_{\pi/2}^{\pi} (\pi - x) \, dx \right) = \dfrac{\pi}{2}$

$$a_n = \frac{2}{\pi} \left(\int_0^{\pi/2} x \cos nx \, dx + \int_{\pi/2}^{\pi} (\pi - x) \cos nx \, dx \right) = \frac{2}{n^2 \pi} \left(2 \cos \frac{n\pi}{2} + (-1)^{n+1} - 1 \right)$$

$$b_n = \frac{2}{\pi} \left(\int_0^{\pi/2} x \sin nx \, dx + \int_{\pi/2}^{\pi} (\pi - x) \sin nx \, dx \right) = \frac{4}{n^2 \pi} \sin \frac{n\pi}{2}$$

$$f(x) = \frac{\pi}{4} + \sum_{n=1}^{\infty} \frac{2}{n^2 \pi} \left(2 \cos \frac{n\pi}{2} + (-1)^{n+1} - 1 \right) \cos nx$$

$$f(x) = \sum_{n=1}^{\infty} \frac{4}{n^2 \pi} \sin \frac{n\pi}{2} \sin nx$$

30. $a_0 = \dfrac{1}{\pi} \displaystyle\int_{\pi}^{2\pi} (x - \pi) \, dx = \dfrac{\pi}{2}$

$$a_n = \frac{1}{\pi} \int_{\pi}^{2\pi} (x - \pi) \cos \frac{n}{2} x \, dx = \frac{4}{n^2 \pi} \left((-1)^n - \cos \frac{n\pi}{2} \right)$$

$$b_n = \frac{1}{\pi} \int_\pi^{2\pi} (x - \pi) \sin \frac{n}{2} x \, dx = \frac{2}{n}(-1)^{n+1} - \frac{4}{n^2\pi} \sin \frac{n\pi}{2}$$

$$f(x) = \frac{\pi}{4} + \sum_{n=1}^{\infty} \frac{4}{n^2\pi} \left((-1)^n - \cos \frac{n\pi}{2} \right) \cos \frac{n}{2} x$$

$$f(x) = \sum_{n=1}^{\infty} \left(\frac{2}{n}(-1)^{n+1} - \frac{4}{n^2\pi} \sin \frac{n\pi}{2} \right) \sin \frac{n}{2} x$$

31. $a_0 = \int_0^1 x \, dx + \int_1^2 1 \, dx = \frac{3}{2}$

$$a_n = \int_0^1 x \cos \frac{n\pi}{2} x \, dx = \frac{4}{n^2\pi^2} \left(\cos \frac{n\pi}{2} - 1 \right)$$

$$b_n = \int_0^1 x \sin \frac{n\pi}{2} x \, dx + \int_1^2 1 \cdot \sin \frac{n\pi}{2} x \, dx = \frac{4}{n^2\pi^2} \sin \frac{n\pi}{2} + \frac{2}{n\pi}(-1)^{n+1}$$

$$f(x) = \frac{3}{4} + \sum_{n=1}^{\infty} \frac{4}{n^2\pi^2} \left(\cos \frac{n\pi}{2} - 1 \right) \cos \frac{n\pi}{2} x$$

$$f(x) = \sum_{n=1}^{\infty} \left(\frac{4}{n^2\pi^2} \sin \frac{n\pi}{2} + \frac{2}{n\pi}(-1)^{n+1} \right) \sin \frac{n\pi}{2} x$$

32. $a_0 = \int_0^1 1 \, dx + \int_1^2 (2 - x) \, dx = \frac{3}{2}$

$$a_n = \int_0^1 1 \cdot \cos \frac{n\pi}{2} x \, dx + \int_1^2 (2 - x) \cos \frac{n\pi}{2} x \, dx = \frac{4}{n^2\pi^2} \left(\cos \frac{n\pi}{2} + (-1)^{n+1} \right)$$

$$b_n = \int_0^1 1 \cdot \sin \frac{n\pi}{2} x \, dx + \int_1^2 (2 - x) \sin \frac{n\pi}{2} x \, dx = \frac{2}{n\pi} + \frac{4}{n^2\pi^2} \sin \frac{n\pi}{2}$$

$$f(x) = \frac{3}{4} + \sum_{n=1}^{\infty} \frac{4}{n^2\pi^2} \left(\cos \frac{n\pi}{2} + (-1)^{n+1} \right) \cos \frac{n\pi}{2} x$$

$$f(x) = \sum_{n=1}^{\infty} \left(\frac{2}{n\pi} + \frac{4}{n^2\pi^2} \sin \frac{n\pi}{2} \right) \sin \frac{n\pi}{2} x$$

33. $a_0 = 2 \int_0^1 (x^2 + x) \, dx = \frac{5}{3}$

$$a_n = 2 \int_0^1 (x^2 + x) \cos n\pi x \, dx = \frac{2(x^2 + x)}{n\pi} \sin n\pi x \Big|_0^1 - \frac{2}{n\pi} \int_0^1 (2x + 1) \sin n\pi x \, dx = \frac{2}{n^2\pi^2} [3(-1)^n - 1]$$

$$b_n = 2 \int_0^1 (x^2 + x) \sin n\pi x \, dx = -\frac{2(x^2 + x)}{n\pi} \cos n\pi x \Big|_0^1 + \frac{2}{n\pi} \int_0^1 (2x + 1) \cos n\pi x \, dx$$

$$= \frac{4}{n\pi}(-1)^{n+1} + \frac{4}{n^3\pi^3} [(-1)^n - 1]$$

$$f(x) = \frac{5}{6} + \sum_{n=1}^{\infty} \frac{2}{n^2\pi^2}[3(-1)^n - 1]\cos n\pi x$$

$$f(x) = \sum_{n=1}^{\infty} \left(\frac{4}{n\pi}(-1)^{n+1} + \frac{4}{n^3\pi^3}[(-1)^n - 1] \right) \sin n\pi x$$

34. $a_0 = \displaystyle\int_0^2 (2x - x^2)\, dx = \frac{4}{3}$

$a_n = \displaystyle\int_0^2 (2x - x^2)\cos\frac{n\pi}{2}x\, dx = \frac{8}{n^2\pi^2}[(-1)^{n+1} - 1]$

$b_n = \displaystyle\int_0^2 (2x - x^2)\sin\frac{n\pi}{2}x\, dx = \frac{16}{n^3\pi^3}[1 - (-1)^n]$

$f(x) = \dfrac{2}{3} + \displaystyle\sum_{n=1}^{\infty} \frac{8}{n^2\pi^2}[(-1)^{n+1} - 1]\cos\frac{n\pi}{2}x$

$f(x) = \displaystyle\sum_{n=1}^{\infty} \frac{16}{n^3\pi^3}[1 - (-1)^n]\sin\frac{n\pi}{2}x$

35. $a_0 = \dfrac{1}{\pi}\displaystyle\int_0^{2\pi} x^2\, dx = \frac{8}{3}\pi^2$

$a_n = \dfrac{1}{\pi}\displaystyle\int_0^{2\pi} x^2\cos nx\, dx = \frac{4}{n^2}$

$b_n = \dfrac{1}{\pi}\displaystyle\int_0^{2\pi} x^2\sin nx\, dx = -\frac{4\pi}{n}$

$f(x) = \dfrac{4}{3}\pi^2 + \displaystyle\sum_{n=1}^{\infty} \left(\frac{4}{n^2}\cos nx - \frac{4\pi}{n}\sin nx \right)$

36. $a_0 = \dfrac{2}{\pi}\displaystyle\int_0^{\pi} x\, dx = \pi$

$a_n = \dfrac{2}{\pi}\displaystyle\int_0^{\pi} x\cos 2nx\, dx = 0$

$b_n = \dfrac{2}{\pi}\displaystyle\int_0^{\pi} x\sin 2nx\, dx = -\frac{1}{n}$

$f(x) = \dfrac{\pi}{2} - \displaystyle\sum_{n=1}^{\infty} \frac{1}{n}\sin 2nx$

37. $a_0 = 2\displaystyle\int_0^1 (x+1)\, dx = 3$

$a_n = 2\displaystyle\int_0^1 (x+1)\cos 2n\pi x\, dx = 0$

$b_n = 2\displaystyle\int_0^1 (x+1)\sin 2n\pi x\, dx = -\frac{1}{n\pi}$

$$f(x) = \frac{3}{2} - \sum_{n=1}^{\infty} \frac{1}{n\pi} \sin 2n\pi x$$

38. $a_0 = \dfrac{2}{2} \displaystyle\int_0^2 (2-x)\,dx = 2$

$a_n = \dfrac{2}{2} \displaystyle\int_0^2 (2-x)\cos n\pi x\,dx = 0$

$b_n = \dfrac{2}{2} \displaystyle\int_0^2 (2-x)\sin n\pi x\,dx = \dfrac{2}{n\pi}$

$f(x) = 1 + \displaystyle\sum_{n=1}^{\infty} \frac{2}{n\pi} \sin n\pi x$

39. We have

$$b_n = \frac{2}{\pi} \int_0^{\pi} 5\sin nt\,dt = \frac{10}{n\pi}[1 - (-1)^n]$$

so that

$$f(t) = \sum_{n=1}^{\infty} \frac{10[1 - (-1)^n]}{n\pi} \sin nt.$$

Substituting the assumption $x_p(t) = \sum_{n=1}^{\infty} B_n \sin nt$ into the differential equation then gives

$$x_p'' + 10x_p = \sum_{n=1}^{\infty} B_n(10 - n^2)\sin nt = \sum_{n=1}^{\infty} \frac{10[1 - (-1)^n]}{n\pi} \sin nt$$

and so $B_n = 10[1 - (-1)^n]/n\pi(10 - n^2)$. Thus

$$x_p(t) = \frac{10}{\pi} \sum_{n=1}^{\infty} \frac{1 - (-1)^n}{n(10 - n^2)} \sin nt.$$

40. We have

$$b_n = \frac{2}{\pi} \int_0^1 (1 - t)\sin n\pi t\,dt = \frac{2}{n\pi}$$

so that

$$f(t) = \sum_{n=1}^{\infty} \frac{2}{n\pi} \sin n\pi t.$$

Substituting the assumption $x_p(t) = \sum_{n=1}^{\infty} B_n \sin n\pi t$ into the differential equation then gives

$$x_p'' + 10x_p = \sum_{n=1}^{\infty} B_n(10 - n^2\pi^2)\sin n\pi t = \sum_{n=1}^{\infty} \frac{2}{n\pi} \sin n\pi t$$

and so $B_n = 2/n\pi(10 - n^2\pi^2)$. Thus

$$x_p(t) = \frac{2}{\pi} \sum_{n=1}^{\infty} \frac{1}{n(10 - n^2\pi^2)} \sin n\pi t.$$

41. We have

$$a_0 = \frac{2}{\pi} \int_0^\pi (2\pi t - t^2)\, dt = \frac{4}{3}\pi^2$$

$$a_n = \frac{2}{\pi} \int_0^\pi (2\pi t - t^2) \cos nt\, dt = -\frac{4}{n^2}$$

so that

$$f(t) = \frac{2\pi^2}{3} - \sum_{n=1}^\infty \frac{4}{n^2} \cos nt.$$

Substituting the assumption

$$x_p(t) = \frac{A_0}{2} + \sum_{n=1}^\infty A_n \cos nt$$

into the differential equation then gives

$$\frac{1}{4} x_p'' + 12 x_p = 6A_0 + \sum_{n=1}^\infty A_n \left(-\frac{1}{4}n^2 + 12\right) \cos nt = \frac{2\pi^2}{3} - \sum_{n=1}^\infty \frac{4}{n^2} \cos nt$$

and $A_0 = \pi^2/9$, $A_n = 16/n^2(n^2 - 48)$. Thus

$$x_p(t) = \frac{\pi^2}{18} + 16 \sum_{n=1}^\infty \frac{1}{n^2(n^2 - 48)} \cos nt.$$

42. We have

$$a_0 = \frac{2}{1/2} \int_0^{1/2} t\, dt = \frac{1}{2}$$

$$a_n = \frac{2}{1/2} \int_0^{1/2} t \cos 2n\pi t\, dt = \frac{1}{n^2\pi^2} \left[(-1)^n - 1\right]$$

so that

$$f(t) = \frac{1}{4} + \sum_{n=1}^\infty \frac{(-1)^n - 1}{n^2\pi^2} \cos 2n\pi t.$$

Substituting the assumption

$$x_p(t) = \frac{A_0}{2} + \sum_{n=1}^\infty A_n \cos 2n\pi t$$

into the differential equation then gives

$$\frac{1}{4} x_p'' + 12 x_p = 6A_0 + \sum_{n=1}^\infty A_n (12 - n^2\pi^2) \cos 2n\pi t = \frac{1}{4} + \sum_{n=1}^\infty \frac{(-1)^n - 1}{n^2\pi^2} \cos 2n\pi t$$

and $A_0 = 1/24$, $A_n = [(-1)^n - 1]/n^2\pi^2(12 - n^2\pi^2)$. Thus

$$x_p(t) = \frac{1}{48} + \frac{1}{\pi^2} \sum_{n=1}^\infty \frac{(-1)^n - 1}{n^2(12 - n^2\pi^2)} \cos 2n\pi t.$$

43. (a) The general solution is $x(t) = c_1 \cos \sqrt{10}t + c_2 \sin \sqrt{10}t + x_p(t)$, where

$$x_p(t) = \frac{10}{\pi} \sum_{n=1}^{\infty} \frac{1 - (-1)^n}{n(10 - n^2)} \sin nt.$$

The initial condition $x(0) = 0$ implies $c_1 + x_p(0) = 0$. Since $x_p(0) = 0$, we have $c_1 = 0$ and $x(t) = c_2 \sin \sqrt{10}t + x_p(t)$. Then $x'(t) = c_2 \sqrt{10} \cos \sqrt{10}t + x_p'(t)$ and $x'(0) = 0$ implies

$$c_2 \sqrt{10} + \frac{10}{\pi} \sum_{n=1}^{\infty} \frac{1 - (-1)^n}{10 - n^2} \cos 0 = 0.$$

Thus

$$c_2 = -\frac{\sqrt{10}}{\pi} \sum_{n=1}^{\infty} \frac{1 - (-1)^n}{10 - n^2}$$

and

$$x(t) = \frac{10}{\pi} \sum_{n=1}^{\infty} \frac{1 - (-1)^n}{10 - n^2} \left[\frac{1}{n} \sin nt - \frac{1}{\sqrt{10}} \sin \sqrt{10}t \right].$$

(b) The graph is plotted using eight nonzero terms in the series expansion of $x(t)$.

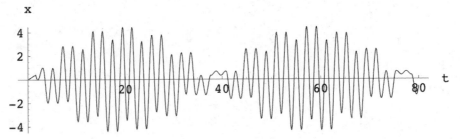

44. (a) The general solution is $x(t) = c_1 \cos 4\sqrt{3}t + c_2 \sin 4\sqrt{3}t + x_p(t)$, where

$$x_p(t) = \frac{\pi^2}{18} + 16 \sum_{n=1}^{\infty} \frac{1}{n^2(n^2 - 48)} \cos nt.$$

The initial condition $x(0) = 0$ implies $c_1 + x_p(0) = 1$ or

$$c_1 = 1 - x_p(0) = 1 - \frac{\pi^2}{18} - 16 \sum_{n=1}^{\infty} \frac{1}{n^2(n^2 - 48)}.$$

Now $x'(t) = -4\sqrt{3}c_1 \sin 4\sqrt{3}t + 4\sqrt{3}c_2 \cos 4\sqrt{3}t + x_p'(t)$, so $x'(0) = 0$ implies $4\sqrt{3}c_2 + x_p'(0) = 0$. Since $x_p'(0) = 0$, we have $c_2 = 0$ and

$$x(t) = \left(1 - \frac{\pi^2}{18} - 16 \sum_{n=1}^{\infty} \frac{1}{n^2(n^2 - 48)} \right) \cos 4\sqrt{3}t + \frac{\pi^2}{18} + 16 \sum_{n=1}^{\infty} \frac{1}{n^2(n^2 - 48)} \cos nt$$

$$= \frac{\pi^2}{18} + \left(1 - \frac{\pi^2}{18} \right) \cos 4\sqrt{3}t + 16 \sum_{n=1}^{\infty} \frac{1}{n^2(n^2 - 48)} [\cos nt - \cos 4\sqrt{3}t].$$

(b) The graph is plotted using five nonzero terms in the series expansion of $x(t)$.

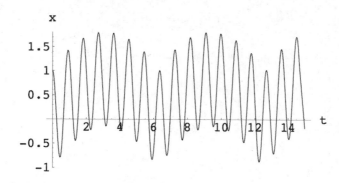

45. (a) We have

$$b_n = \frac{2}{L} \int_0^L \frac{w_0 x}{L} \sin \frac{n\pi}{L} x \, dx = \frac{2w_0}{n\pi} (-1)^{n+1}$$

so that

$$w(x) = \sum_{n=1}^{\infty} \frac{2w_0}{n\pi} (-1)^{n+1} \sin \frac{n\pi}{L} x.$$

(b) If we assume $y_p(x) = \sum_{n=1}^{\infty} B_n \sin(n\pi x/L)$ then

$$y_p^{(4)} = \sum_{n=1}^{\infty} \frac{n^4 \pi^4}{L^4} B_n \sin \frac{n\pi}{L} x$$

and so the differential equation $EIy_p^{(4)} = w(x)$ gives

$$B_n = \frac{2w_0(-1)^{n+1} L^4}{EI n^5 \pi^5}.$$

Thus

$$y_p(x) = \frac{2w_0 L^4}{EI \pi^5} \sum_{n=1}^{\infty} \frac{(-1)^{n+1}}{n^5} \sin \frac{n\pi}{L} x.$$

46. We have

$$b_n = \frac{2}{L} \int_{L/3}^{2L/3} w_0 \sin \frac{n\pi}{L} x \, dx = \frac{2w_0}{n\pi} \left(\cos \frac{n\pi}{3} - \cos \frac{2n\pi}{3} \right)$$

so that

$$w(x) = \sum_{n=1}^{\infty} \frac{2w_0}{n\pi} \left(\cos \frac{n\pi}{3} - \cos \frac{2n\pi}{3} \right) \sin \frac{n\pi}{L} x.$$

If we assume $y_p(x) = \sum_{n=1}^{\infty} B_n \sin(n\pi x/L)$ then

$$y_p^{(4)}(x) = \sum_{n=1}^{\infty} \frac{n^4 \pi^4}{L^4} B_n \sin \frac{n\pi}{L} x$$

and so the differential equation $EIy_p^{(4)}(x) = w(x)$ gives

$$B_n = 2w_0 L^4 \frac{\cos \frac{n\pi}{3} - \cos \frac{2n\pi}{3}}{EI n^5 \pi^5}.$$

Thus

$$y_p(x) = \frac{2w_0 L^4}{EI\pi^5} \sum_{n=1}^{\infty} \frac{\cos \frac{n\pi}{3} - \cos \frac{2n\pi}{3}}{n^5} \sin \frac{n\pi}{L} x.$$

47. We note that $w(x)$ is 2π-periodic and even. With $p = \pi$ we find the cosine expansion of

$$f(x) = \begin{cases} w_0 & 0 < x < \pi/2 \\ 0, & \pi/2 < x < \pi. \end{cases}$$

We have

$$a_0 = \frac{2}{\pi} \int_0^{\pi} f(x)\,dx = \frac{2}{\pi} \int_0^{\pi/2} w_0\,dx = w_0$$

$$a_n = \frac{2}{\pi} \int_0^{\pi} f(x) \cos nx\,dx = \frac{2}{\pi} \int_0^{\pi/2} w_0 \cos nx\,dx = \frac{2w_0}{n\pi} \sin \frac{n\pi}{2}.$$

Thus,

$$w(x) = \frac{w_0}{2} + \frac{2w_0}{\pi} \sum_{n=1}^{\infty} \frac{1}{n} \sin \frac{n\pi}{2} \cos nx.$$

Now we assume a particular solution of the form $y_p(x) = A_0/2 + \sum_{n=1}^{\infty} A_n \cos nx$. Then $y_p^{(4)}(x) = \sum_{n=1}^{\infty} A_n n^4 \cos nx$ and substituting into the differential equation, we obtain

$$EI y_p^{(4)}(x) + k y_p(x) = \frac{kA_0}{2} + \sum_{n=1}^{\infty} A_n(EIn^4 + k) \cos nx$$

$$= \frac{w_0}{2} + \frac{2w_0}{\pi} \sum_{n=1}^{\infty} \frac{1}{n} \sin \frac{n\pi}{2} \cos nx.$$

Thus

$$A_0 = \frac{w_0}{k} \qquad \text{and} \qquad A_n = \frac{2w_0}{\pi} \frac{\sin(n\pi/2)}{n(EIn^4 + k)},$$

and

$$y_p(x) = \frac{w_0}{2k} + \frac{2w_0}{\pi} \sum_{n=1}^{\infty} \frac{\sin(n\pi/2)}{n(EIn^4 + k)} \cos nx.$$

48. (a) If f and g are even and $h(x) = f(x)g(x)$ then

$$h(-x) = f(-x)g(-x) = f(x)g(x) = h(x)$$

and h is even.

(c) If f is even and g is odd and $h(x) = f(x)g(x)$ then

$$h(-x) = f(-x)g(-x) = f(x)[-g(x)] = -h(x)$$

and h is odd.

(d) Let $h(x) = f(x) \pm g(x)$ where f and g are even. Then

$$h(-x) = f(-x) \pm g(-x) = f(x) \pm g(x) = h(x),$$

and so h is an even function.

(f) If f is even then

$$\int_{-a}^{a} f(x)\,dx = -\int_{a}^{0} f(-u)\,du + \int_{0}^{a} f(x)\,dx = \int_{0}^{a} f(u)\,du + \int_{0}^{a} f(x)\,dx = 2\int_{0}^{a} f(x)\,dx.$$

(g) If f is odd then

$$\int_{-a}^{a} f(x)\,dx = -\int_{-a}^{0} f(-x)\,dx + \int_{0}^{a} f(x)\,dx = \int_{a}^{0} f(u)\,du + \int_{0}^{a} f(x)\,dx$$

$$= -\int_{0}^{a} f(u)\,du + \int_{0}^{a} f(x)\,dx = 0.$$

49. If $f(x)$ is even then $f(-x) = f(x)$. If $f(x)$ is odd then $f(-x) = -f(x)$. Thus, if $f(x)$ is both even and odd, $f(x) = f(-x) = -f(x)$, and $f(x) = 0$.

50. For $EIy^{(4)} + ky = 0$ the roots of the auxiliary equation are $m_1 = \alpha + \alpha i$, $m_2 = \alpha - \alpha i$, $m_3 = -\alpha + \alpha i$, and $m_4 = -\alpha - \alpha i$, where $\alpha = (k/EI)^{1/4}/\sqrt{2}$. Thus

$$y_c = e^{\alpha x}(c_1 \cos \alpha x + c_2 \sin \alpha x) + e^{-\alpha x}(c_3 \cos \alpha x + c_4 \sin \alpha x).$$

We expect $y(x)$ to be bounded as $x \to \infty$, so we must have $c_1 = c_2 = 0$. We also expect $y(x)$ to be bounded as $x \to -\infty$, so we must have $c_3 = c_4 = 0$. Thus, $y_c = 0$ and the solution of the differential equation in Problem 47 is $y_p(x)$.

51. The graph is obtained by summing the series from $n = 1$ to 20. It appears that

$$f(x) = \begin{cases} x, & 0 < x < \pi \\ -\pi, & \pi < x < 2\pi. \end{cases}$$

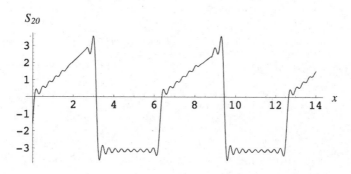

52. The graph is obtained by summing the series from $n = 1$ to 10. It appears that

$$f(x) = \begin{cases} 1 - x, & 0 < x < 1 \\ 0, & 1 < x < 2. \end{cases}$$

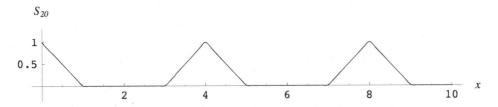

53. (a) The function in Problem 51 is not unique; it could also be defined as

$$f(x) = \begin{cases} x, & 0 < x < \pi \\ 1, & x = \pi \\ -\pi, & \pi < x < 2\pi. \end{cases}$$

(b) The function in Problem 52 is not unique; it could also be defined as

$$f(x) = \begin{cases} 0, & -2 < x < -1 \\ x + 1, & -1 < x < 0 \\ -x + 1, & 0 < x < 1 \\ 0, & 1 < x < 2. \end{cases}$$

Exercises 11.4

Sturm-Liouville Problem

1. For $\lambda \leq 0$ the only solution of the boundary-value problem is $y = 0$. For $\lambda = \alpha^2 > 0$ we have

$$y = c_1 \cos \alpha x + c_2 \sin \alpha x.$$

Now

$$y'(x) = -c_1 \alpha \sin \alpha x + c_2 \alpha \cos \alpha x$$

and $y'(0) = 0$ implies $c_2 = 0$, so

$$y(1) + y'(1) = c_1(\cos \alpha - \alpha \sin \alpha) = 0 \quad \text{or} \quad \cot \alpha = \alpha.$$

The eigenvalues are $\lambda_n = \alpha_n^2$ where $\alpha_1, \alpha_2, \alpha_3, \ldots$ are the consecutive positive solutions of $\cot \alpha = \alpha$. The corresponding eigenfunctions are $\cos \alpha_n x$ for $n = 1, 2, 3, \ldots$. Using a CAS we find that the first four eigenvalues are approximately 0.7402, 11.7349, 41.4388, and 90.8082 with corresponding approximate eigenfunctions $\cos 0.8603x$, $\cos 3.4256x$, $\cos 6.4373x$, and $\cos 9.5293x$.

2. For $\lambda < 0$ the only solution of the boundary-value problem is $y = 0$. For $\lambda = 0$ we have $y = c_1x + c_2$. Now $y' = c_1$ and the boundary conditions both imply $c_1 + c_2 = 0$. Thus, $\lambda = 0$ is an eigenvalue with corresponding eigenfunction $y_0 = x - 1$.

For $\lambda = \alpha^2 > 0$ we have

$$y = c_1 \cos \alpha x + c_2 \sin \alpha x$$

and

$$y'(x) = -c_1\alpha \sin \alpha x + c_2\alpha \cos \alpha x.$$

The boundary conditions imply

$$c_1 + c_2\alpha = 0$$

$$c_1 \cos \alpha + c_2 \sin \alpha = 0$$

which gives

$$-c_2\alpha \cos \alpha + c_2 \sin \alpha = 0 \qquad \text{or} \qquad \tan \alpha = \alpha.$$

The eigenvalues are $\lambda_n = \alpha_n^2$ where $\alpha_1, \alpha_2, \alpha_3, \ldots$ are the consecutive positive solutions of $\tan \alpha = \alpha$. The corresponding eigenfunctions are $\alpha \cos \alpha x - \sin \alpha x$ (obtained by taking $c_2 = -1$ in the first equation of the system.) Using a CAS we find that the first four positive eigenvalues are 20.1907, 59.6795, 118.9000, and 197.858 with corresponding eigenfunctions $4.4934 \cos 4.4934x - \sin 4.4934x$, $7.7253 \cos 7.7253x - \sin 7.7253x$, $10.9041 \cos 10.9041x - \sin 10.9041x$, and $14.0662 \cos 14.0662x - \sin 14.0662x$.

3. For $\lambda = 0$ the solution of $y'' = 0$ is $y = c_1x + c_2$. The condition $y'(0) = 0$ implies $c_1 = 0$, so $\lambda = 0$ is an eigenvalue with corresponding eigenfunction 1.

For $\lambda = -\alpha^2 < 0$ we have $y = c_1 \cosh \alpha x + c_2 \sinh \alpha x$ and $y' = c_1\alpha \sinh \alpha x + c_2\alpha \cosh \alpha x$. The condition $y'(0) = 0$ implies $c_2 = 0$ and so $y = c_1 \cosh \alpha x$. Now the condition $y'(L) = 0$ implies $c_1 = 0$. Thus $y = 0$ and there are no negative eigenvalues.

For $\lambda = \alpha^2 > 0$ we have $y = c_1 \cos \alpha x + c_2 \sin \alpha x$ and $y' = -c_1\alpha \sin \alpha x + c_2\alpha \cos \alpha x$. The condition $y'(0) = 0$ implies $c_2 = 0$ and so $y = c_1 \cos \alpha x$. Now the condition $y'(L) = 0$ implies $-c_1\alpha \sin \alpha L = 0$. For $c_1 \neq 0$ this condition will hold when $\alpha L = n\pi$ or $\lambda = \alpha^2 = n^2\pi^2/L^2$, where $n = 1, 2, 3, \ldots$. These are the positive eigenvalues with corresponding eigenfunctions $\cos(n\pi x/L)$, $n = 1, 2, 3, \ldots$.

4. For $\lambda = -\alpha^2 < 0$ we have

$$y = c_1 \cosh \alpha x + c_2 \sinh \alpha x$$

$$y' = c_1\alpha \sinh \alpha x + c_2\alpha \cosh \alpha x.$$

Using the fact that $\cosh x$ is an even function and $\sinh x$ is odd we have

$$y(-L) = c_1 \cosh(-\alpha L) + c_2 \sinh(-\alpha L)$$

$$= c_1 \cosh \alpha L - c_2 \sinh \alpha L$$

and

$$y'(-L) = c_1 \alpha \sinh(-\alpha L) + c_2 \alpha \cosh(-\alpha L)$$

$$= -c_1 \alpha \sinh \alpha L + c_2 \alpha \cosh \alpha L.$$

The boundary conditions imply

$$c_1 \cosh \alpha L - c_2 \sinh \alpha L = c_1 \cosh \alpha L + c_2 \sinh \alpha L$$

or

$$2c_2 \sinh \alpha L = 0$$

and

$$-c_1 \alpha \sinh \alpha L + c_2 \alpha \cosh \alpha L = c_1 \alpha \sinh \alpha L + c_2 \alpha \cosh \alpha L$$

or

$$2c_1 \alpha \sinh \alpha L = 0.$$

Since $\alpha L \neq 0$, $c_1 = c_2 = 0$ and the only solution of the boundary-value problem in this case is $y = 0$.

For $\lambda = 0$ we have

$$y = c_1 x + c_2$$

$$y' = c_1.$$

From $y(-L) = y(L)$ we obtain

$$-c_1 L + c_2 = c_1 L + c_2.$$

Then $c_1 = 0$ and $y = 1$ is an eigenfunction corresponding to the eigenvalue $\lambda = 0$.

For $\lambda = \alpha^2 > 0$ we have

$$y = c_1 \cos \alpha x + c_2 \sin \alpha x$$

$$y' = -c_1 \alpha \sin \alpha x + c_2 \alpha \cos \alpha x.$$

The first boundary condition implies

$$c_1 \cos \alpha L - c_2 \sin \alpha L = c_1 \cos \alpha L + c_2 \sin \alpha L$$

or

$$2c_2 \sin \alpha L = 0.$$

566

Thus, if $c_1 = 0$ and $c_2 \neq 0$,

$$\alpha L = n\pi \quad \text{or} \quad \lambda = \alpha^2 = \frac{n^2 \pi^2}{L^2}, \quad n = 1, 2, 3, \ldots .$$

The corresponding eigenfunctions are $\sin(n\pi x/L)$, for $n = 1,\ 2,\ 3,\ \ldots$. Similarly, the second boundary condition implies

$$2c_1 \alpha \sin \alpha L = 0.$$

If $c_1 \neq 0$ and $c_2 = 0$,

$$\alpha L = n\pi \quad \text{or} \quad \lambda = \alpha^2 = \frac{n^2 \pi^2}{L^2}, \quad n = 1, 2, 3, \ldots,$$

and the corresponding eigenfunctions are $\cos(n\pi x/L)$, for $n = 1,\ 2,\ 3,\ \ldots$.

5. The eigenfunctions are $\cos \alpha_n x$ where $\cot \alpha_n = \alpha_n$. Thus

$$\| \cos \alpha_n x \|^2 = \int_0^1 \cos^2 \alpha_n x \, dx = \frac{1}{2} \int_0^1 (1 + \cos 2\alpha_n x) \, dx$$

$$= \frac{1}{2} \left(x + \frac{1}{2\alpha_n} \sin 2\alpha_n x \right) \Big|_0^1 = \frac{1}{2} \left(1 + \frac{1}{2\alpha_n} \sin 2\alpha_n \right)$$

$$= \frac{1}{2} \left[1 + \frac{1}{2\alpha_n} (2 \sin \alpha_n \cos \alpha_n) \right]$$

$$= \frac{1}{2} \left[1 + \frac{1}{\alpha_n} \sin \alpha_n \cot \alpha_n \sin \alpha_n \right]$$

$$= \frac{1}{2} \left[1 + \frac{1}{\alpha_n} (\sin \alpha_n) \alpha_n (\sin \alpha_n) \right] = \frac{1}{2} \left(1 + \sin^2 \alpha_n \right).$$

6. The eigenfunctions are $\sin \alpha_n x$ where $\tan \alpha_n = -\alpha_n$. Thus

$$\| \sin \alpha_n x \|^2 = \int_0^1 \sin^2 \alpha_n x \, dx = \frac{1}{2} \int_0^1 (1 - \cos 2\alpha_n x) \, dx$$

$$= \frac{1}{2} \left(x - \frac{1}{2\alpha_n} \sin 2\alpha_n x \right) \Big|_0^1 = \frac{1}{2} \left(1 - \frac{1}{2\alpha_n} \sin 2\alpha_n \right)$$

$$= \frac{1}{2} \left[1 - \frac{1}{2\alpha_n} (2 \sin \alpha_n \cos \alpha_n) \right]$$

$$= \frac{1}{2} \left[1 - \frac{1}{\alpha_n} \tan \alpha_n \cos \alpha_n \cos \alpha_n \right]$$

$$= \frac{1}{2} \left[1 - \frac{1}{\alpha_n} \left(-\alpha_n \cos^2 \alpha_n \right) \right] = \frac{1}{2} \left(1 + \cos^2 \alpha_n \right).$$

7. (a) If $\lambda \leq 0$ the initial conditions imply $y = 0$. For $\lambda = \alpha^2 > 0$ the general solution of the Cauchy-Euler differential equation is $y = c_1 \cos(\alpha \ln x) + c_2 \sin(\alpha \ln x)$. The condition $y(1) = 0$ implies

Exercises 11.4   Sturm-Liouville Problem

$c_1 = 0$, so that $y = c_2 \sin(\alpha \ln x)$. The condition $y(5) = 0$ implies $\alpha \ln 5 = n\pi$, $n = 1, 2, 3, \ldots$. Thus, the eigenvalues are $n^2\pi^2/(\ln 5)^2$ for $n = 1, 2, 3, \ldots$, with corresponding eigenfunctions $\sin[(n\pi/\ln 5)\ln x]$.

(b) The self-adjoint form is

$$\frac{d}{dx}[xy'] + \frac{\lambda}{x}y = 0.$$

(c) An orthogonality relation is

$$\int_1^5 \frac{1}{x}\sin\left(\frac{m\pi}{\ln 5}\ln x\right)\sin\left(\frac{n\pi}{\ln 5}\ln x\right)dx = 0, \quad m \neq n.$$

8. (a) The roots of the auxiliary equation $m^2 + m + \lambda = 0$ are $\frac{1}{2}(-1 \pm \sqrt{1-4\lambda})$. When $\lambda = 0$ the general solution of the differential equation is $c_1 + c_2 e^{-x}$. The boundary conditions imply $c_1 + c_2 = 0$ and $c_1 + c_2 e^{-2} = 0$. Since the determinant of the coefficients is not 0, the only solution of this homogeneous system is $c_1 = c_2 = 0$, in which case $y = 0$. When $\lambda = \frac{1}{4}$, the general solution of the differential equation is $c_1 e^{-x/2} + c_2 x e^{-x/2}$. The boundary conditions imply $c_1 = 0$ and $c_1 + 2c_2 = 0$, so $c_1 = c_2 = 0$ and $y = 0$. Similarly, if $0 < \lambda < \frac{1}{4}$, the general solution is

$$y = c_1 e^{\frac{1}{2}(-1+\sqrt{1-4\lambda})x} + c_2 e^{\frac{1}{2}(-1-\sqrt{1-4\lambda})x}.$$

In this case the boundary conditions again imply $c_1 = c_2 = 0$, and so $y = 0$. Now, for $\lambda > \frac{1}{4}$, the general solution of the differential equation is

$$y = c_1 e^{-x/2}\cos\sqrt{4\lambda - 1}\,x + c_2 e^{-x/2}\sin\sqrt{4\lambda - 1}\,x.$$

The condition $y(0) = 0$ implies $c_1 = 0$ so $y = c_2 e^{-x/2}\sin\sqrt{4\lambda - 1}\,x$. From

$$y(2) = c_2 e^{-1}\sin 2\sqrt{4\lambda - 1} = 0$$

we see that the eigenvalues are determined by $2\sqrt{4\lambda - 1} = n\pi$ for $n = 1, 2, 3, \ldots$. Thus, the eigenvalues are $n^2\pi^2/4^2 + 1/4$ for $n = 1, 2, 3, \ldots$, with corresponding eigenfunctions $e^{-x/2}\sin(n\pi x/2)$.

(b) The self-adjoint form is

$$\frac{d}{dx}[e^x y'] + \lambda e^x y = 0.$$

(c) An orthogonality relation is

$$\int_0^2 e^x\left(e^{-x/2}\sin\frac{m\pi}{2}x\right)\left(e^{-x/2}\cos\frac{n\pi}{2}x\right)dx = \int_0^2 \sin\frac{m\pi}{2}x\,\cos\frac{n\pi}{2}x\,dx = 0.$$

9. To obtain the self-adjoint form we note that an integrating factor is $(1/x)e^{\int (1-x)dx/x} = e^{-x}$. Thus, the differential equation is

$$xe^{-x}y'' + (1-x)e^{-x}y' + ne^{-x}y = 0$$

568

and the self-adjoint form is

$$\frac{d}{dx}\left[xe^{-x}y'\right] + ne^{-x}y = 0.$$

Identifying the weight function $p(x) = e^{-x}$ and noting that since $r(x) = xe^{-x}$, $r(0) = 0$ and $\lim_{x\to\infty} r(x) = 0$, we have the orthogonality relation

$$\int_0^\infty e^{-x}L_m(x)L_n(x)\,dx = 0,\ m \neq n.$$

10. To obtain the self-adjoint form we note that an integrating factor is $e^{\int -2x\,dx} = e^{-x^2}$. Thus, the differential equation is

$$e^{-x^2}y'' - 2xe^{-x^2}y' + 2ne^{-x^2}y = 0$$

and the self-adjoint form is

$$\frac{d}{dx}\left[e^{-x^2}y'\right] + 2ne^{-x^2}y = 0.$$

Identifying the weight function $p(x) = e^{-x^2}$ and noting that since $r(x) = e^{-x^2}$, $\lim_{x\to-\infty} r(x) = \lim_{x\to\infty} r(x) = 0$, we have the orthogonality relation

$$\int_{-\infty}^\infty e^{-x^2}H_m(x)H_n(x)\,dx = 0,\ m \neq n.$$

11. (a) The differential equation is

$$(1+x^2)y'' + 2xy' + \frac{\lambda}{1+x^2}y = 0.$$

Letting $x = \tan\theta$ we have $\theta = \tan^{-1}x$ and

$$\frac{dy}{dx} = \frac{dy}{d\theta}\frac{d\theta}{dx} = \frac{1}{1+x^2}\frac{dy}{d\theta}$$

$$\frac{d^2y}{dx^2} = \frac{d}{dx}\left[\frac{1}{1+x^2}\frac{dy}{d\theta}\right] = \frac{1}{1+x^2}\left(\frac{d^2y}{d\theta^2}\frac{d\theta}{dx}\right) - \frac{2x}{(1+x^2)^2}\frac{dy}{d\theta}$$

$$= \frac{1}{(1+x^2)^2}\frac{d^2y}{d\theta^2} - \frac{2x}{(1+x^2)^2}\frac{dy}{d\theta}.$$

The differential equation can then be written in terms of $y(\theta)$ as

$$(1+x^2)\left[\frac{1}{(1+x^2)^2}\frac{d^2y}{d\theta^2} - \frac{2x}{(1+x^2)^2}\frac{dy}{d\theta}\right] + 2x\left[\frac{1}{1+x^2}\frac{dy}{d\theta}\right] + \frac{\lambda}{1+x^2}y$$

$$= \frac{1}{1+x^2}\frac{d^2y}{d\theta^2} + \frac{\lambda}{1+x^2}y = 0$$

or

$$\frac{d^2y}{d\theta^2} + \lambda y = 0.$$

The boundary conditions become $y(0) = y(\pi/4) = 0$. For $\lambda \le 0$ the only solution of the boundary-value problem is $y = 0$. For $\lambda = \alpha^2 > 0$ the general solution of the differential equation is $y = c_1 \cos \alpha\theta + c_2 \sin \alpha\theta$. The condition $y(0) = 0$ implies $c_1 = 0$ so $y = c_2 \sin \alpha\theta$. Now the condition $y(\pi/4) = 0$ implies $c_2 \sin \alpha\pi/4 = 0$. For $c_2 \ne 0$ this condition will hold when $\alpha\pi/4 = n\pi$ or $\lambda = \alpha^2 = 16n^2$, where $n = 1, 2, 3, \ldots$. These are the eigenvalues with corresponding eigenfunctions $\sin 4n\theta = \sin(4n \tan^{-1} x)$, for $n = 1, 2, 3, \ldots$.

(b) An orthogonality relation is

$$\int_0^1 \frac{1}{x^2 + 1} \sin(4m \tan^{-1} x) \sin(4n \tan^{-1} x)\, dx = 0, \quad m \ne n.$$

12. (a) Letting $\lambda = \alpha^2$ the differential equation becomes $x^2 y'' + x y' + (\alpha^2 x^2 - 1)y = 0$. This is the parametric Bessel equation with $\nu = 1$. The general solution is

$$y = c_1 J_1(\alpha x) + c_2 Y_1(\alpha x).$$

Since Y is unbounded at 0 we must have $c_2 = 0$, so that $y = c_1 J_1(\alpha x)$. The condition $J_1(3\alpha) = 0$ defines the eigenvalues $\lambda_n = \alpha_n^2$ for $n = 1, 2, 3, \ldots$. The corresponding eigenfunctions are $J_1(\alpha_n x)$.

(b) Using a CAS or Table 6.1 in Section 6.3 of the text to solve $J_1(3\alpha) = 0$ we find $3\alpha_1 = 3.8317$, $3\alpha_2 = 7.0156$, $3\alpha_3 = 10.1735$, and $3\alpha_4 = 13.3237$. The corresponding eigenvalues are $\lambda_1 = \alpha_1^2 = 1.6313$, $\lambda_2 = \alpha_2^2 = 5.4687$, $\lambda_3 = \alpha_3^2 = 11.4999$, and $\lambda_4 = \alpha_4^2 = 19.7245$.

13. When $\lambda = 0$ the differential equation is $r(x)y'' + r'(x)y' = 0$. By inspection we see that $y = 1$ is a solution of the boundary-value problem. Thus, $\lambda = 0$ is an eigenvalue.

14. (a) An orthogonality relation is

$$\int_0^1 \cos x_m x \cos x_n x\, dx = 0$$

where $x_m \ne x_n$ are positive solutions of $\cot x = x$.

(b) Referring to Problem 1 we use a CAS to compute

$$\int_0^1 (\cos 0.8603 x)(\cos 3.4256 x)\, dx = -1.8771 \times 10^{-6} \approx 0.$$

15. (a) An orthogonality relation is

$$\int_0^1 (x_m \cos x_m x - \sin x_m x)(x_n \cos x_n x - \sin x_n x)\, dx = 0$$

where $x_m \ne x_n$ are positive solutions of $\tan x = x$.

(b) Referring to Problem 2 we use a CAS to compute

$$\int_0^1 (4.4934 \cos 4.4934 x - \sin 4.4934 x)(7.7253 \cos 7.7253 x - \sin 7.7253 x)\, dx = -2.5650 \times 10^{-4} \approx 0.$$

1. Identifying $b = 3$, we have $\alpha_1 = 1.2772$, $\alpha_2 = 2.3385$, $\alpha_3 = 3.3912$, and $\alpha_4 = 4.4412$.

2. By (6) in the text $J_0'(2\alpha) = -J_1(2\alpha)$. Thus, $J_0'(2\alpha) = 0$ is equivalent to $J_1(2\alpha)$. Then $\alpha_1 = 1.9159$, $\alpha_2 = 3.5078$, $\alpha_3 = 5.0867$, and $\alpha_4 = 6.6618$.

3. The boundary condition indicates that we use (15) and (16) in the text. With $b = 2$ we obtain

$$c_i = \frac{2}{4J_1^2(2\alpha_i)} \int_0^2 x J_0(\alpha_i x)\, dx$$

$$\boxed{t = \alpha_i x \qquad dt = \alpha_i\, dx}$$

$$= \frac{1}{2J_1^2(2\alpha_i)} \cdot \frac{1}{\alpha_i^2} \int_0^{2\alpha_i} t J_0(t)\, dt$$

$$= \frac{1}{2\alpha_i^2 J_1^2(2\alpha_i)} \int_0^{2\alpha_i} \frac{d}{dt}[t J_1(t)]\, dt \qquad \text{[From (5) in the text]}$$

$$= \frac{1}{2\alpha_i^2 J_1^2(2\alpha_i)} t J_1(t)\Big|_0^{2\alpha_i}$$

$$= \frac{1}{\alpha_i J_1(2\alpha_i)}.$$

Thus

$$f(x) = \sum_{i=1}^{\infty} \frac{1}{\alpha_i J_1(2\alpha_i)} J_0(\alpha_i x).$$

4. The boundary condition indicates that we use (19) and (20) in the text. With $b = 2$ we obtain

$$c_1 = \frac{2}{4} \int_0^2 x\, dx = \frac{2}{4}\frac{x^2}{2}\Big|_0^2 = 1,$$

$$c_i = \frac{2}{4J_0^2(2\alpha_i)} \int_0^2 x J_0(\alpha_i x)\, dx$$

$$\boxed{t = \alpha_i x \qquad dt = \alpha_i\, dx}$$

$$= \frac{1}{2J_0^2(2\alpha_i)} \cdot \frac{1}{\alpha_i^2} \int_0^{2\alpha_i} t J_0(t)\, dt$$

$$= \frac{1}{2\alpha_i^2 J_0^2(2\alpha_i)} \int_0^{2\alpha_i} \frac{d}{dt}[t J_1(t)]\, dt \qquad \text{[From (5) in the text]}$$

$$= \frac{1}{2\alpha_i^2 J_0^2(2\alpha_i)} t J_1(t)\Big|_0^{2\alpha_i}$$

$$= \frac{J_1(2\alpha_i)}{\alpha_i J_0^2(2\alpha_i)}.$$

Now since $J_0'(2\alpha_i) = 0$ is equivalent to $J_1(2\alpha_i) = 0$ we conclude $c_i = 0$ for $i = 2, 3, 4, \ldots$. Thus the expansion of f on $0 < x < 2$ consists of a series with one nontrivial term:

$$f(x) = c_1 = 1.$$

5. The boundary condition indicates that we use (17) and (18) in the text. With $b = 2$ and $h = 1$ we obtain

$$c_i = \frac{2\alpha_i^2}{(4\alpha_i^2 + 1)J_0^2(2\alpha_i)} \int_0^2 x J_0(\alpha_i x)\, dx$$

$$\boxed{t = \alpha_i x \qquad dt = \alpha_i\, dx}$$

$$= \frac{2\alpha_i^2}{(4\alpha_i^2 + 1)J_0^2(2\alpha_i)} \cdot \frac{1}{\alpha_i^2} \int_0^{2\alpha_i} t J_0(t)\, dt$$

$$= \frac{2}{(4\alpha_i^2 + 1)J_0^2(2\alpha_i)} \int_0^{2\alpha_i} \frac{d}{dt}[t J_1(t)]\, dt \qquad \text{[From (5) in the text]}$$

$$= \frac{2}{(4\alpha_i^2 + 1)J_0^2(2\alpha_i)} t J_1(t)\Big|_0^{2\alpha_i}$$

$$= \frac{4\alpha_i J_1(2\alpha_i)}{(4\alpha_i^2 + 1)J_0^2(2\alpha_i)}.$$

Thus

$$f(x) = 4\sum_{i=1}^{\infty} \frac{\alpha_i J_1(2\alpha_i)}{(4\alpha_i^2 + 1)J_0^2(2\alpha_i)} J_0(\alpha_i x).$$

6. Writing the boundary condition in the form

$$2J_0(2\alpha) + 2\alpha J_0'(2\alpha) = 0$$

we identify $b = 2$ and $h = 2$. Using (17) and (18) in the text we obtain

$$c_i = \frac{2\alpha_i^2}{(4\alpha_i^2 + 4)J_0^2(2\alpha_i)} \int_0^2 x J_0(\alpha_i x)\, dx$$

$$\boxed{t = \alpha_i x \qquad dt = \alpha_i\, dx}$$

$$= \frac{\alpha_i^2}{2(\alpha_i^2 + 1)J_0^2(2\alpha_i)} \cdot \frac{1}{\alpha_i^2} \int_0^{2\alpha_i} t J_0(t)\, dt$$

$$= \frac{1}{2(\alpha_i^2 + 1)J_0^2(2\alpha_i)} \int_0^{2\alpha_i} \frac{d}{dt}[t J_1(t)]\, dt \qquad \text{[From (5) in the text]}$$

$$= \frac{1}{2(\alpha_i^2 + 1)J_0^2(2\alpha_i)} t J_1(t)\Big|_0^{2\alpha_i}$$

$$= \frac{\alpha_i J_1(2\alpha_i)}{(\alpha_i^2 + 1)J_0^2(2\alpha_i)}.$$

Thus

$$f(x) = \sum_{i=1}^{\infty} \frac{\alpha_i J_1(2\alpha_i)}{(\alpha_i^2 + 1)J_0^2(2\alpha_i)} J_0(\alpha_i x).$$

7. The boundary condition indicates that we use (17) and (18) in the text. With $n = 1$, $b = 4$, and $h = 3$ we obtain

$$c_i = \frac{2\alpha_i^2}{(16\alpha_i^2 - 1 + 9)J_1^2(4\alpha_i)} \int_0^4 x J_1(\alpha_i x) 5x \, dx$$

$$\boxed{t = \alpha_i x \qquad dt = \alpha_i \, dx}$$

$$= \frac{5\alpha_i^2}{4(2\alpha_i^2 + 1)J_1^2(4\alpha_i)} \cdot \frac{1}{\alpha_i^3} \int_0^{4\alpha_i} t^2 J_1(t) \, dt$$

$$= \frac{5}{4\alpha_i(2\alpha_i^2 + 1)J_1^2(4\alpha_i)} \int_0^{4\alpha_i} \frac{d}{dt}[t^2 J_2(t)] \, dt \qquad \text{[From (5) in the text]}$$

$$= \frac{5}{4\alpha_i(2\alpha_i^2 + 1)J_1^2(4\alpha_i)} t^2 J_2(t) \Big|_0^{4\alpha_i}$$

$$= \frac{20\alpha_i J_2(4\alpha_i)}{(2\alpha_i^2 + 1)J_1^2(4\alpha_i)} \cdot$$

Thus

$$f(x) = 20 \sum_{i=1}^{\infty} \frac{\alpha_i J_2(4\alpha_i)}{(2\alpha_i^2 + 1)J_1^2(4\alpha_i)} J_1(\alpha_i x).$$

8. The boundary condition indicates that we use (15) and (16) in the text. With $n = 2$ and $b = 1$ we obtain

$$c_1 = \frac{2}{J_3^2(\alpha_i)} \int_0^1 x J_2(\alpha_i x) x^2 \, dx$$

$$\boxed{t = \alpha_i x \qquad dt = \alpha_i \, dx}$$

$$= \frac{2}{J_3^2(\alpha_i)} \cdot \frac{1}{\alpha_i^4} \int_0^{\alpha_i} t^3 J_2(t) \, dt$$

$$= \frac{2}{\alpha_i^4 J_3^2(\alpha_i)} \int_0^{\alpha_i} \frac{d}{dt}[t^3 J_3(t)] \, dt \qquad \text{[From (5) in the text]}$$

$$= \frac{2}{\alpha_i^4 J_3^2(\alpha_i)} t^3 J_3(t) \Big|_0^{\alpha_i}$$

$$= \frac{2}{\alpha_i J_3(\alpha_i)} \cdot$$

Thus

$$f(x) = 2 \sum_{i=1}^{\infty} \frac{1}{\alpha_i J_3(\alpha_i)} J_2(\alpha_i x).$$

9. The boundary condition indicates that we use (19) and (20) in the text. With $b = 3$ we obtain

$$c_1 = \frac{2}{9} \int_0^3 x x^2 \, dx = \frac{2}{9} \frac{x^4}{4} \Big|_0^3 = \frac{9}{2},$$

$$c_i = \frac{2}{9 J_0^2(3\alpha_i)} \int_0^3 x J_0(\alpha_i x) x^2 \, dx$$

$$\boxed{t = \alpha_i x \qquad dt = \alpha_i \, dx}$$

$$= \frac{2}{9 J_0^2(3\alpha_i)} \cdot \frac{1}{\alpha_i^4} \int_0^{3\alpha_i} t^3 J_0(t) \, dt$$

$$= \frac{2}{9\alpha_i^4 J_0^2(3\alpha_i)} \int_0^{3\alpha_i} t^2 \frac{d}{dt} [t J_1(t)] \, dt$$

$$\boxed{\begin{array}{ll} u = t^2 & dv = \frac{d}{dt}[t J_1(t)] \, dt \\ du = 2t \, dt & v = t J_1(t) \end{array}}$$

$$= \frac{2}{9\alpha_i^4 J_0^2(3\alpha_i)} \left(t^3 J_1(t) \Big|_0^{3\alpha_i} - 2 \int_0^{3\alpha_i} t^2 J_1(t) \, dt \right).$$

With $n = 0$ in equation (6) in the text we have $J_0'(x) = -J_1(x)$, so the boundary condition $J_0'(3\alpha_i) = 0$ implies $J_1(3\alpha_i) = 0$. Then

$$c_i = \frac{2}{9\alpha_i^4 J_0^2(3\alpha_i)} \left(-2 \int_0^{3\alpha_i} \frac{d}{dt} [t^2 J_2(t)] \, dt \right) = \frac{2}{9\alpha_i^4 J_0^2(3\alpha_i)} \left(-2t^2 J_2(t) \Big|_0^{3\alpha_i} \right)$$

$$= \frac{2}{9\alpha_i^4 J_0^2(3\alpha_i)} \left[-18\alpha_i^2 J_2(3\alpha_i) \right] = \frac{-4 J_2(3\alpha_i)}{\alpha_i^2 J_0^2(3\alpha_i)}.$$

Thus

$$f(x) = \frac{9}{2} - 4 \sum_{i=1}^{\infty} \frac{J_2(3\alpha_i)}{\alpha_i^2 J_0^2(3\alpha_i)} J_0(\alpha_i x).$$

10. The boundary condition indicates that we use (15) and (16) in the text. With $b = 1$ it follows that

$$c_i = \frac{2}{J_1^2(\alpha_i)} \int_0^1 x \left(1 - x^2 \right) J_0(\alpha_i x) \, dx$$

$$= \frac{2}{J_1^2(\alpha_i)} \left[\int_0^1 x J_0(\alpha_i x) \, dx - \int_0^1 x^3 J_0(\alpha_i x) \, dx \right]$$

$$\boxed{t = \alpha_i x \qquad dt = \alpha_i \, dx}$$

$$= \frac{2}{J_1^2(\alpha_i)} \left[\frac{1}{\alpha_i^2} \int_0^{\alpha_i} t J_0(t) \, dt - \frac{1}{\alpha_i^4} \int_0^{\alpha_i} t^3 J_0(t) \, dt \right]$$

$$= \frac{2}{J_1^2(\alpha_i)} \left[\frac{1}{\alpha_i^2} \int_0^{\alpha_i} \frac{d}{dt}[t J_1(t)] \, dt - \frac{1}{\alpha_i^4} \int_0^{\alpha_i} t^2 \frac{d}{dt}[t J_1(t)] \, dt \right]$$

$$\boxed{\begin{array}{ll} u = t^2 & dv = \frac{d}{dt}[t J_1(t)] \, dt \\ du = 2t \, dt & v = t J_1(t) \end{array}}$$

$$= \frac{2}{J_1^2(\alpha_i)} \left[\frac{1}{\alpha_i^2} t J_1(t) \Big|_0^{\alpha_i} - \frac{1}{\alpha_i^4} \left(t^3 J_1(t) \Big|_0^{\alpha_i} - 2 \int_0^{\alpha_i} t^2 J_1(t) \, dt \right) \right]$$

$$= \frac{2}{J_1^2(\alpha_i)} \left[\frac{J_1(\alpha_i)}{\alpha_i} - \frac{J_1(\alpha_i)}{\alpha_i} + \frac{2}{\alpha_i^4} \int_0^{\alpha_i} \frac{d}{dt}[t^2 J_2(t)] \, dt \right]$$

$$= \frac{2}{J_1^2(\alpha_i)} \left[\frac{2}{\alpha_i^4} t^2 J_2(t) \Big|_0^{\alpha_i} \right] = \frac{4 J_2(\alpha_i)}{\alpha_i^2 J_1^2(\alpha_i)}.$$

Thus

$$f(x) = 4 \sum_{i=1}^{\infty} \frac{J_2(\alpha_i)}{\alpha_i^2 J_1^2(\alpha_i)} J_0(\alpha_i x).$$

11. (a)

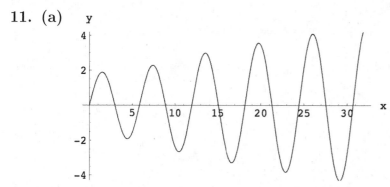

(b) Using **FindRoot** in *Mathematica* we find the roots $x_1 = 2.9496$, $x_2 = 5.8411$, $x_3 = 8.8727$, $x_4 = 11.9561$, and $x_5 = 15.0624$.

(c) Dividing the roots in part (b) by 4 we find the eigenvalues $\alpha_1 = 0.7374$, $\alpha_2 = 1.4603$, $\alpha_3 = 2.2182$, $\alpha_4 = 2.9890$, and $\alpha_5 = 3.7656$.

(d) The next five eigenvalues are $\alpha_6 = 4.5451$, $\alpha_7 = 5.3263$, $\alpha_8 = 6.1085$, $\alpha_9 = 6.8915$, and $\alpha_{10} = 7.6749$.

12. (a) From Problem 7, the coefficients of the Fourier-Bessel series are

$$c_i = \frac{20 \alpha_i J_2(4\alpha_i)}{(2\alpha_i^2 + 1) J_1^2(4\alpha_i)}.$$

Using a CAS we find $c_1 = 26.7896$, $c_2 = -12.4624$, $c_3 = 7.1404$, $c_4 = -4.68705$, and $c_5 = 3.35619$.

(b)

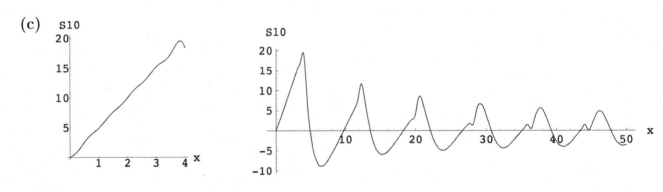

(c)

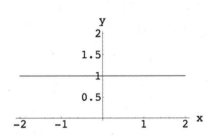

13. Since f is expanded as a series of Bessel functions, $J_1(\alpha_i x)$ and J_1 is an odd function, the series should represent an odd function.

14. (a) Since J_0 is an even function, a series expansion of a function defined on $(0, 2)$ would converge to the even extension of the function on $(-2, 0)$.

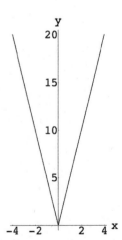

(b) In Section 6.3 we saw that $J_2'(x) = 2J_2(x)/x - J_3(x)$. Since J_2 is even and J_3 is odd we see that

$$J_2'(-x) = 2J_2(-x)/(-x) - J_3(-x)$$

$$= -2J_2(x)/x + J_3(x) = -J_2'(x),$$

so that J_2' is an odd function. Now, if $f(x) = 3J_2(x) + 2xJ_2'(x)$, we see that

$$f(-x) = 3J_2(-x) - 2xJ_2'(-x)$$

$$= 3J_2(x) + 2xJ_2'(x) = f(x),$$

so that f is an even function. Thus, a series expansion of a function defined on $(0, 4)$ would converge to the even extension of the function on $(-4, 0)$.

15. We compute

$$c_0 = \frac{1}{2}\int_0^1 xP_0(x)\,dx = \frac{1}{2}\int_0^1 x\,dx = \frac{1}{4}$$

$$c_1 = \frac{3}{2}\int_0^1 xP_1(x)\,dx = \frac{3}{2}\int_0^1 x^2\,dx = \frac{1}{2}$$

$$c_2 = \frac{5}{2}\int_0^1 xP_2(x)\,dx = \frac{5}{2}\int_0^1 \frac{1}{2}(3x^3 - x)\,dx = \frac{5}{16}$$

$$c_3 = \frac{7}{2}\int_0^1 xP_3(x)\,dx = \frac{7}{2}\int_0^1 \frac{1}{2}(5x^4 - 3x^2)\,dx = 0$$

$$c_4 = \frac{9}{2}\int_0^1 xP_4(x)\,dx = \frac{9}{2}\int_0^1 \frac{1}{8}(35x^5 - 30x^3 + 3x)\,dx = -\frac{3}{32}$$

$$c_5 = \frac{11}{2}\int_0^1 xP_5(x)\,dx = \frac{11}{2}\int_0^1 \frac{1}{8}(63x^6 - 70x^4 + 15x^2)\,dx = 0$$

$$c_6 = \frac{13}{2}\int_0^1 xP_6(x)\,dx = \frac{13}{2}\int_0^1 \frac{1}{16}(231x^7 - 315x^5 + 105x^3 - 5x)\,dx = \frac{13}{256}.$$

Thus

$$f(x) = \frac{1}{4}P_0(x) + \frac{1}{2}P_1(x) + \frac{5}{16}P_2(x) - \frac{3}{32}P_4(x) + \frac{13}{256}P_6(x) + \cdots.$$

The figure above is the graph of $S_5(x) = \frac{1}{4}P_0(x) + \frac{1}{2}P_1(x) + \frac{5}{16}P_2(x) - \frac{3}{32}P_4(x) + \frac{13}{256}P_6(x)$.

16. We compute

$$c_0 = \frac{1}{2}\int_{-1}^1 e^x P_0(x)\,dx = \frac{1}{2}\int_{-1}^1 e^x\,dx = \frac{1}{2}(e - e^{-1})$$

$$c_1 = \frac{3}{2}\int_{-1}^1 e^x P_1(x)\,dx = \frac{3}{2}\int_{-1}^1 xe^x\,dx = 3e^{-1}$$

$$c_2 = \frac{5}{2}\int_{-1}^1 e^x P_2(x)\,dx = \frac{5}{2}\int_{-1}^1 \frac{1}{2}(3x^2 e^x - e^x)\,dx$$

$$= \frac{5}{2}(e - 7e^{-1})$$

$$c_3 = \frac{7}{2}\int_{-1}^1 e^x P_3(x)\,dx = \frac{7}{2}\int_{-1}^1 \frac{1}{2}(5x^3 e^x - 3xe^x)\,dx = \frac{7}{2}(-5e + 37e^{-1})$$

$$c_4 = \frac{9}{2}\int_{-1}^1 e^x P_4(x)\,dx = \frac{9}{2}\int_{-1}^1 \frac{1}{8}(35x^4 e^x - 30x^2 e^x + 3e^x)\,dx = \frac{9}{2}(36e - 266e^{-1}).$$

Thus

$$f(x) = \frac{1}{2}(e - e^{-1})P_0(x) + 3e^{-1}P_1(x) + \frac{5}{2}(e - 7e^{-1})P_2(x)$$

$$+ \frac{7}{2}(-5e + 37e^{-1})P_3(x) + \frac{9}{2}(36e - 266e^{-1})P_4(x) + \cdots.$$

The figure above is the graph of $S_5(x)$.

17. Using $\cos^2 \theta = \frac{1}{2}(\cos 2\theta + 1)$ we have

$$P_2(\cos \theta) = \frac{1}{2}(3\cos^2 \theta - 1) = \frac{3}{2}\cos^2 \theta - \frac{1}{2}$$

$$= \frac{3}{4}(\cos 2\theta + 1) - \frac{1}{2} = \frac{3}{4}\cos 2\theta + \frac{1}{4} = \frac{1}{4}(3\cos 2\theta + 1).$$

18. From Problem 17 we have

$$P_2(\cos \theta) = \frac{1}{4}(3\cos 2\theta + 1) \qquad \text{or} \qquad \cos 2\theta = \frac{4}{3}P_2(\cos \theta) - \frac{1}{3}.$$

Then, using $P_0(\cos \theta) = 1$,

$$F(\theta) = 1 - \cos 2\theta = 1 - \left[\frac{4}{3}P_2(\cos \theta) - \frac{1}{3}\right]$$

$$= \frac{4}{3} - \frac{4}{3}P_2(\cos \theta) = \frac{4}{3}P_0(\cos \theta) - \frac{4}{3}P_2(\cos \theta).$$

19. If f is an even function on $(-1, 1)$ then

$$\int_{-1}^{1} f(x)P_{2n}(x)\,dx = 2\int_{0}^{1} f(x)P_{2n}(x)\,dx$$

and

$$\int_{-1}^{1} f(x)P_{2n+1}(x)\,dx = 0.$$

Thus

$$c_{2n} = \frac{2(2n) + 1}{2}\int_{-1}^{1} f(x)P_{2n}(x)\,dx = \frac{4n + 1}{2}\left(2\int_{0}^{1} f(x)P_{2n}(x)\,dx\right)$$

$$= (4n + 1)\int_{0}^{1} f(x)P_{2n}(x)\,dx,$$

$c_{2n+1} = 0$, and

$$f(x) = \sum_{n=0}^{\infty} c_{2n}P_{2n}(x).$$

20. If f is an odd function on $(-1, 1)$ then

$$\int_{-1}^{1} f(x)P_{2n}(x)\,dx = 0$$

and

$$\int_{-1}^{1} f(x)P_{2n+1}(x)\,dx = 2\int_{0}^{1} f(x)P_{2n+1}(x)\,dx.$$

Thus

$$c_{2n+1} = \frac{2(2n+1)+1}{2}\int_{-1}^{1} f(x)P_{2n+1}(x)\,dx = \frac{4n+3}{2}\left(2\int_{0}^{1} f(x)P_{2n+1}(x)\,dx\right)$$

$$= (4n+3)\int_{0}^{1} f(x)P_{2n+1}(x)\,dx,$$

$c_{2n} = 0$, and

$$f(x) = \sum_{n=0}^{\infty} c_{2n+1}P_{2n+1}(x).$$

21. From (26) in Problem 19 in the text we find

$$c_0 = \int_{0}^{1} xP_0(x)\,dx = \int_{0}^{1} x\,dx = \frac{1}{2},$$

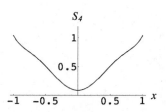

$$c_2 = 5\int_{0}^{1} xP_2(x)\,dx = 5\int_{0}^{1} \frac{1}{2}(3x^3 - x)\,dx = \frac{5}{8},$$

$$c_4 = 9\int_{0}^{1} xP_4(x)\,dx = 9\int_{0}^{1} \frac{1}{8}(35x^5 - 30x^3 + 3x)\,dx = -\frac{3}{16},$$

and

$$c_6 = 13\int_{0}^{1} xP_6(x)\,dx = 13\int_{0}^{1} \frac{1}{16}(231x^7 - 315x^5 + 105x^3 - 5x)\,dx = \frac{13}{128}.$$

Hence, from (25) in the text,

$$f(x) = \frac{1}{2}P_0(x) + \frac{5}{8}P_2(x) - \frac{3}{16}P_4(x) + \frac{13}{128}P_6 + \cdots.$$

On the interval $-1 < x < 1$ this series represents the function $f(x) = |x|$.

22. From (28) in Problem 20 in the text we find

$$c_1 = 3\int_{0}^{1} P_1(x)\,dx = 3\int_{0}^{1} x\,dx = \frac{3}{2},$$

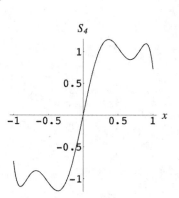

$$c_3 = 7\int_{0}^{1} P_3(x)\,dx = 7\int_{0}^{1} \frac{1}{2}\left(5x^3 - 3x\right)\,dx = -\frac{7}{8},$$

$$c_5 = 11\int_{0}^{1} P_5(x)\,dx = 11\int_{0}^{1} \frac{1}{8}\left(63x^5 - 70x^3 + 15x\right)\,dx = \frac{11}{16}$$

and

$$c_7 = 15\int_{0}^{1} P_7(x)\,dx = 15\int_{0}^{1} \frac{1}{16}\left(429x^7 - 693x^5 + 315x^3 - 35x\right)\,dx = -\frac{75}{128}.$$

Hence, from (27) in the text,

$$f(x) = \frac{3}{2}P_1(x) - \frac{7}{8}P_3(x) + \frac{11}{16}P_5(x) - \frac{75}{128}P_7(x) + \cdots.$$

On the interval $-1 < x < 1$ this series represents the odd function

$$f(x) = \begin{cases} -1, & -1 < x < 0 \\ 1, & 0 < x < 1. \end{cases}$$

23. Since there is a Legendre polynomial of any specified degree, every polynomial can be represented as a finite linear combination of Legendre polynomials.

24. For $f(x) = x^2$ we have

$$c_0 = \frac{1}{2}\int_{-1}^{1} x^2 \cdot 1 \, dx = \frac{1}{3}$$

$$c_1 = \frac{3}{2}\int_{-1}^{1} x^2 \cdot x \, dx = 0$$

$$c_2 = \frac{5}{2}\int_{-1}^{1} x^2 \cdot \frac{1}{2}(3x^2 - 1)dx = \frac{2}{3},$$

so

$$x^2 = \frac{1}{3}P_0(x) + \frac{2}{3}P_2(x).$$

For $f(x) = x^3$ we have

$$c_0 = \frac{1}{2}\int_{-1}^{1} x^3 \cdot 1 \, dx = 0$$

$$c_1 = \frac{3}{2}\int_{-1}^{1} x^3 \cdot x \, dx = \frac{3}{5}$$

$$c_2 = \frac{5}{2}\int_{-1}^{1} x^3 \cdot \frac{1}{2}(3x^2 - 1) \, dx = 0$$

$$c_3 = \frac{7}{2}\int_{-1}^{1} x^3 \cdot \frac{1}{2}(5x^3 - 3x) \, dx = \frac{2}{5},$$

so

$$x^3 = \frac{3}{5}P_1(x) + \frac{2}{5}P_3(x).$$

1. True, since $\int_{-\pi}^{\pi}(x^2-1)x^5\,dx = 0$

2. Even, since if f and g are odd then $h(-x) = f(-x)g(-x) = -f(x)[-g(x)] = f(x)g(x) = h(x)$

3. Cosine, since f is even

4. True

5. False; the Sturm-Liouville problem,

$$\frac{d}{dx}[r(x)y'] + \lambda p(x)y = 0, \qquad y'(a) = 0, \quad y'(b) = 0,$$

on the interval $[a, b]$, has eigenvalue $\lambda = 0$.

6. Periodically extending the function we see that at $x = -1$ the function converges to $\frac{1}{2}(-1+0) = -\frac{1}{2}$; at $x = 0$ it converges to $\frac{1}{2}(0 + 1) = \frac{1}{2}$, and at $x = 1$ it converges to $\frac{1}{2}(-1 + 0) = -\frac{1}{2}$.

7. The Fourier series will converge to 1, the cosine series to 1, and the sine series to 0 at $x = 0$. Respectively, this is because the rule $(x^2 + 1)$ defining $f(x)$ determines a continuous function on $(-3, 3)$, the even extension of f to $(-3, 0)$ is continuous at 0, and the odd extension of f to $(-3, 0)$ approaches -1 as x approaches 0 from the left.

8. $\cos 5x$, since the general solution is $y = c_1 \cos \alpha x + c_2 \sin \alpha x$ and $y'(0) = 0$ implies $c_2 = 0$.

9. Since the coefficient of y in the differential equation is n^2, the weight function is the integrating factor

$$\frac{1}{a(x)}e^{\int(b/a)dx} = \frac{1}{1-x^2}e^{\int -\frac{x}{1-x^2}\,dx} = \frac{1}{1-x^2}e^{\frac{1}{2}\ln(1-x^2)} = \frac{\sqrt{1-x^2}}{1-x^2} = \frac{1}{\sqrt{1-x^2}}$$

on the interval $(-1, 1)$. The orthogonality relation is

$$\int_{-1}^{1} \frac{1}{\sqrt{1-x^2}} T_m(x)T_n(x)\,dx = 0, \quad m \neq n.$$

10. Since $P_n(x)$ is orthogonal to $P_0(x) = 1$ for $n > 0$,

$$\int_{-1}^{1} P_n(x)\,dx = \int_{-1}^{1} P_0(x)P_n(x)\,dx = 0.$$

11. We know from a half-angle formula in trigonometry that $\cos^2 x = \frac{1}{2} + \frac{1}{2}\cos 2x$, which is a cosine series.

12. (a) For $m \neq n$

$$\int_{0}^{L} \sin\frac{(2n+1)\pi}{2L}x \sin\frac{(2m+1)\pi}{2L}x\,dx = \frac{1}{2}\int_{0}^{L}\left(\cos\frac{n-m}{L}\pi x - \cos\frac{n+m+1}{L}\pi x\right)dx = 0.$$

(b) From

$$\int_0^L \sin^2 \frac{(2n+1)\pi}{2L} x \, dx = \int_0^L \left(\frac{1}{2} - \frac{1}{2} \cos \frac{(2n+1)\pi}{L} x \right) dx = \frac{L}{2}$$

we see that

$$\left\| \sin \frac{(2n+1)\pi}{2L} x \right\| = \sqrt{\frac{L}{2}}.$$

13. Since

$$a_0 = \int_{-1}^0 (-2x) \, dx = 1,$$

$$a_n = \int_{-1}^0 (-2x) \cos n\pi x \, dx = \frac{2}{n^2\pi^2}[(-1)^n - 1],$$

and

$$b_n = \int_{-1}^0 (-2x) \sin n\pi x \, dx = \frac{4}{n\pi}(-1)^n$$

for $n = 1, 2, 3, \ldots$ we have

$$f(x) = \frac{1}{2} + \sum_{n=1}^\infty \left(\frac{2}{n^2\pi^2}[(-1)^n - 1] \cos n\pi x + \frac{4}{n\pi}(-1)^n \sin n\pi x \right).$$

14. Since

$$a_0 = \int_{-1}^1 (2x^2 - 1) \, dx = -\frac{2}{3},$$

$$a_n = \int_{-1}^1 (2x^2 - 1) \cos n\pi x \, dx = \frac{8}{n^2\pi^2}(-1)^n,$$

and

$$b_n = \int_{-1}^1 (2x^2 - 1) \sin n\pi x \, dx = 0$$

for $n = 1, 2, 3, \ldots$ we have

$$f(x) = -\frac{1}{3} + \sum_{n=1}^\infty \frac{8}{n^2\pi^2}(-1)^n \cos n\pi x.$$

15. (a) Since

$$a_0 = 2\int_0^1 e^{-x} dx = 2(1 - e^{-1})$$

and

$$a_n = 2\int_{-1}^1 e^{-x} \cos n\pi x \, dx = \frac{2}{1 + n^2\pi^2}[1 - (-1)^n e^{-1}]$$

for $n = 1, 2, 3, \ldots$ we have

$$f(x) = 1 - e^{-1} + 2\sum_{n=1}^\infty \frac{1 - (-1)^n e^{-1}}{1 + n^2\pi^2} \cos n\pi x.$$

(b) Since

$$b_n = 2\int_0^1 e^{-x}\sin n\pi x\,dx = \frac{2n\pi}{1+n^2\pi^2}[1-(-1)^n e^{-1}]$$

for $n = 1, 2, 3, \ldots$ we have

$$f(x) = \sum_{n=1}^{\infty}\frac{2n\pi}{1+n^2\pi^2}[1-(-1)^n e^{-1}]\sin n\pi x.$$

16.

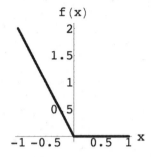
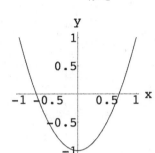
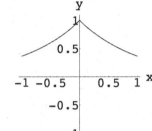

$$f(x) = |x| - x \qquad f(x) = 2x^2 - 1 \qquad f(x) = e^{-|x|} \qquad f(x) = \begin{cases} e^{-x}, & 0 < x < 1 \\ 0, & x = 0 \\ -e^x, & -1 < x < 0 \end{cases}$$

17. The cosine series of f in Problem 15 converges to $F(x)$ on the interval $-1 < x < 1$ since F is the even extension of f to the interval.

18. Expanding in a full Fourier series we have

$$a_0 = \frac{1}{2}\left(\int_0^2 x\,dx + \int_2^4 2\,dx\right) = 3$$

$$a_n = \frac{1}{2}\left(\int_0^2 x\cos\frac{n\pi x}{2}\,dx + \int_2^4 2\cos\frac{n\pi x}{2}\,dx\right) = 2\frac{(-1)^n-1}{n^2\pi^2}$$

$$b_n = \frac{1}{2}\left(\int_0^2 x\sin\frac{n\pi x}{2}\,dx + \int_2^4 2\sin\frac{n\pi x}{2}\,dx\right) = 4\frac{-1}{n\pi}$$

so

$$f(x) = \frac{3}{2} + 2\sum_{n=1}^{\infty}\left(\frac{(-1)^n-1}{n^2\pi^2}\cos\frac{n\pi x}{2} - \frac{2}{n\pi}\sin\frac{n\pi x}{2}\right).$$

19. For $\lambda = \alpha^2 > 0$ a general solution of the given differential equation is

$$y = c_1\cos(3\alpha\ln x) + c_2\sin(3\alpha\ln x)$$

and

$$y' = -\frac{3c_1\alpha}{x}\sin(3\alpha\ln x) + \frac{3c_2\alpha}{x}\cos(3\alpha\ln x).$$

Since $\ln 1 = 0$, the boundary condition $y'(1) = 0$ implies $c_2 = 0$. Therefore

$$y = c_1\cos(3\alpha\ln x).$$

Using $\ln e = 1$ we find that $y(e) = 0$ implies $c_1 \cos 3\alpha = 0$ or $3\alpha = (2n-1)\pi/2$, for $n = 1, 2, 3, \ldots$. The eigenvalues are $\lambda = \alpha^2 = (2n-1)^2\pi^2/36$ with corresponding eigenfunctions $\cos[(2n-1)\pi(\ln x)/2]$ for $n = 1, 2, 3, \ldots$.

20. To obtain the self-adjoint form of the differential equation in Problem 19 we note that an integrating factor is $(1/x^2)e^{\int dx/x} = 1/x$. Thus the weight function is $1/x$ and an orthogonality relation is

$$\int_1^e \frac{1}{x} \cos\left(\frac{2n-1}{2}\pi \ln x\right) \cos\left(\frac{2m-1}{2}\pi \ln x\right) dx = 0, \ m \neq n.$$

21. The boundary condition indicates that we use (15) and (16) of Section 11.5 in the text. With $b = 4$ we obtain

$$c_i = \frac{2}{16 J_1^2(4\alpha_i)} \int_0^4 x J_0(\alpha_i x) f(x)\, dx$$

$$= \frac{1}{8 J_1^2(4\alpha_i)} \int_0^2 x J_0(\alpha_i x)\, dx$$

$$\boxed{t = \alpha_i x \qquad dt = \alpha_i\, dx}$$

$$= \frac{1}{8 J_1^2(4\alpha_i)} \cdot \frac{1}{\alpha_i^2} \int_0^{2\alpha_i} t J_0(t)\, dt$$

$$= \frac{1}{8 J_1^2(4\alpha_i)} \int_0^{2\alpha_i} \frac{d}{dt}[t J_1(t)]\, dt \qquad \text{[From (5) in 11.5 in the text]}$$

$$= \frac{1}{8 J_1^2(4\alpha_i)} t J_1(t)\Big|_0^{2\alpha_i} = \frac{J_1(2\alpha_i)}{4\alpha_i J_1^2(4\alpha_i)}.$$

Thus

$$f(x) = \frac{1}{4} \sum_{i=1}^\infty \frac{J_1(2\alpha_i)}{\alpha_i J_1^2(4\alpha_i)} J_0(\alpha_i x).$$

22. Since $f(x) = x^4$ is a polynomial in x, an expansion of f in Legendre polynomials in x must terminate with the term having the same degree as f. Using the fact that $x^4 P_1(x)$ and $x^4 P_3(x)$ are odd functions, we see immediately that $c_1 = c_3 = 0$. Now

$$c_0 = \frac{1}{2} \int_{-1}^1 x^4 P_0(x)\, dx = \frac{1}{2} \int_{-1}^1 x^4 dx = \frac{1}{5}$$

$$c_2 = \frac{5}{2} \int_{-1}^1 x^4 P_2(x)\, dx = \frac{5}{2} \int_{-1}^1 \frac{1}{2}(3x^6 - x^4)\, dx = \frac{4}{7}$$

$$c_4 = \frac{9}{2} \int_{-1}^1 x^4 P_4(x)\, dx = \frac{9}{2} \int_{-1}^1 \frac{1}{8}(35x^8 - 30x^6 + 3x^4)\, dx = \frac{8}{35}.$$

Thus

$$f(x) = \frac{1}{5} P_0(x) + \frac{4}{7} P_2(x) + \frac{8}{35} P_4(x).$$

23. (a) $f_e(x) + f_o(x) = \dfrac{f(x) + f(-x)}{2} + \dfrac{f(x) - f(-x)}{2} = \dfrac{2f(x)}{2} = f(x)$

(b) $f_e(-x) = \dfrac{f(-x) + f(-(-x))}{2} = \dfrac{f(x) + f(-x)}{2} = f_e(x)$

$f_o(-x) = \dfrac{f(-x) - f(-(-x))}{2} = -\dfrac{f(x) - f(-x)}{2} = -f_o(x)$

24. Identifying

$$f_e(x) = \frac{f(x) + f(-x)}{2} = \frac{e^x + e^{-x}}{2} \quad \text{and} \quad f_o(x) = \frac{f(x) - f(-x)}{2} = \frac{e^x - e^{-x}}{2},$$

we have

$$e^x = \frac{e^x + e^{-x}}{2} + \frac{e^x - e^{-x}}{2}.$$

25. The details of the proof are a little complicated, but follow easily from the graph of a periodic function with a assumed to be between 0 and $2p$. It is then seen that this restriction on a is not used in the proof. From the substitution $u = x + 2p$, we have

$$\int_0^a f(x)\,dx = \int_{2p}^{a+2p} f(u - 2p)\,du.$$

But, since f is $2p$-periodic, $f(u - 2p) = f(u)$ and

$$\int_0^a f(x)\,dx = \int_{2p}^{a+2p} f(u)\,du = \int_{2p}^{a+2p} f(x)\,dx.$$

Then

$$\int_a^{a+2p} f(x)\,dx = \int_a^{2p} f(x)\,dx + \int_{2p}^{a+2p} f(x)\,dx = \int_a^{2p} f(x)\,dx + \int_0^a f(x)\,dx = \int_0^{2p} f(x)\,dx.$$

12 Boundary-Value Problems in Rectangular Coordinates

Exercises 12.1 — Separable Partial Differential Equations

1. Substituting $u(x, y) = X(x)Y(y)$ into the partial differential equation yields $X'Y = XY'$. Separating variables and using the separation constant $-\lambda$, where $\lambda \neq 0$, we obtain

$$\frac{X'}{X} = \frac{Y'}{Y} = -\lambda.$$

When $\lambda \neq 0$

$$X' + \lambda X = 0 \quad \text{and} \quad Y' + \lambda Y = 0$$

so that

$$X = c_1 e^{-\lambda x} \quad \text{and} \quad Y = c_2 e^{-\lambda y}.$$

A particular product solution of the partial differential equation is

$$u = XY = c_3 e^{-\lambda(x+y)}, \quad \lambda \neq 0.$$

When $\lambda = 0$ the differential equations become $X' = 0$ and $Y' = 0$, so in this case $X = c_4$, $Y = c_5$, and $u = XY = c_6$.

2. Substituting $u(x, y) = X(x)Y(y)$ into the partial differential equation yields $X'Y + 3XY' = 0$. Separating variables and using the separation constant $-\lambda$ we obtain

$$\frac{X'}{-3X} = \frac{Y'}{Y} = -\lambda.$$

When $\lambda \neq 0$

$$X' - 3\lambda X = 0 \quad \text{and} \quad Y' + \lambda Y = 0$$

so that

$$X = c_1 e^{3\lambda x} \quad \text{and} \quad Y = c_2 e^{-\lambda y}.$$

A particular product solution of the partial differential equation is

$$u = XY = c_3 e^{\lambda(3x-y)}.$$

When $\lambda = 0$ the differential equations become $X' = 0$ and $Y' = 0$, so in this case $X = c_4$, $Y = c_5$, and $u = XY = c_6$.

3. Substituting $u(x, y) = X(x)Y(y)$ into the partial differential equation yields $X'Y + XY' = XY$. Separating variables and using the separation constant $-\lambda$ we obtain

$$\frac{X'}{X} = \frac{Y - Y'}{Y} = -\lambda.$$

Then

$$X' + \lambda X = 0 \qquad \text{and} \qquad Y' - (1 + \lambda)Y = 0$$

so that

$$X = c_1 e^{-\lambda x} \qquad \text{and} \qquad Y = c_2 e^{(1+\lambda)y}.$$

A particular product solution of the partial differential equation is

$$u = XY = c_3 e^{y + \lambda(y - x)}.$$

4. Substituting $u(x, y) = X(x)Y(y)$ into the partial differential equation yields $X'Y = XY' + XY$. Separating variables and using the separation constant $-\lambda$ we obtain

$$\frac{X'}{X} = \frac{Y + Y'}{Y} = -\lambda.$$

Then

$$X' + \lambda X = 0 \qquad \text{and} \qquad y' + (1 + \lambda)Y = 0$$

so that

$$X = c_1 e^{-\lambda x} \qquad \text{and} \qquad Y = c_2 e^{-(1+\lambda)y} = 0.$$

A particular product solution of the partial differential equation is

$$u = XY = c_3 e^{-y - \lambda(x + y)}.$$

5. Substituting $u(x, y) = X(x)Y(y)$ into the partial differential equation yields $xX'Y = yXY'$. Separating variables and using the separation constant $-\lambda$ we obtain

$$\frac{xX'}{X} = \frac{yY'}{Y} = -\lambda.$$

When $\lambda \neq 0$

$$xX' + \lambda X = 0 \qquad \text{and} \qquad yY' + \lambda Y = 0$$

so that

$$X = c_1 x^{-\lambda} \qquad \text{and} \qquad Y = c_2 y^{-\lambda}.$$

A particular product solution of the partial differential equation is

$$u = XY = c_3 (xy)^{-\lambda}.$$

When $\lambda = 0$ the differential equations become $X' = 0$ and $Y' = 0$, so in this case $X = c_4$, $Y = c_5$, and $u = XY = c_6$.

6. Substituting $u(x, y) = X(x)Y(y)$ into the partial differential equation yields $yX'Y + xXY' = 0$. Separating variables and using the separation constant $-\lambda$ we obtain

$$\frac{X'}{xX} = -\frac{Y'}{yY} = -\lambda.$$

When $\lambda \neq 0$

$$X' + \lambda x X = 0 \qquad \text{and} \qquad Y' - \lambda y Y = 0$$

so that

$$X = c_1 e^{\lambda x^2/2} \qquad \text{and} \qquad Y = c_2 e^{-\lambda y^2/2}.$$

A particular product solution of the partial differential equation is

$$u = XY = c_3 e^{\lambda(x^2 - y^2)/2}.$$

When $\lambda = 0$ the differential equations become $X' = 0$ and $Y' = 0$, so in this case $X = c_4$, $Y = c_5$, and $u = XY = c_6$.

7. Substituting $u(x, y) = X(x)Y(y)$ into the partial differential equation yields $X''Y + X'Y' + XY'' = 0'$, which is not separable.

8. Substituting $u(x, y) = X(x)Y(y)$ into the partial differential equation yields $yX'Y' + XY = 0$. Separating variables and using the separation constant $-\lambda$ we obtain

$$\frac{X'}{X} = -\frac{Y}{yY'} = -\lambda.$$

When $\lambda \neq 0$

$$X' + \lambda X = 0 \qquad \text{and} \qquad \lambda y Y' - Y = 0$$

so that

$$X = c_1 e^{-\lambda x} \qquad \text{and} \qquad Y = c_2 y^{1/\lambda}.$$

A particular product solution of the partial differential equation is

$$u = XY = c_3 e^{-\lambda x} y^{1/\lambda}.$$

In this case $\lambda = 0$ yields no solution.

9. Substituting $u(x, t) = X(x)T(t)$ into the partial differential equation yields $kX''T - XT = XT'$. Separating variables and using the separation constant $-\lambda$ we obtain

$$\frac{kX'' - X}{X} = \frac{T'}{T} = -\lambda.$$

Then

$$X'' + \frac{\lambda - 1}{k}X = 0 \qquad \text{and} \qquad T' + \lambda T = 0.$$

The second differential equation implies $T(t) = c_1 e^{-\lambda t}$. For the first differential equation we consider three cases:

I. If $(\lambda - 1)/k = 0$ then $\lambda = 1$, $X'' = 0$, and $X(x) = c_2 x + c_3$, so

$$u = XT = e^{-t}(A_1 x + A_2).$$

II. If $(\lambda - 1)/k = -\alpha^2 < 0$, then $\lambda = 1 - k\alpha^2$, $X'' - \alpha^2 X = 0$, and $X(x) = c_4 \cosh \alpha x + c_5 \sinh \alpha x$, so

$$u = XT = (A_3 \cosh \alpha x + A_4 \sinh \alpha x)e^{-(1-k\alpha^2)t}.$$

III. If $(\lambda - 1)/k = \alpha^2 > 0$, then $\lambda = 1 + \lambda\alpha^2$, $X'' + \alpha^2 X = 0$, and $X(x) = c_6 \cos \alpha x + c_7 \sin \alpha x$, so

$$u = XT = (A_5 \cos \alpha x + A_6 \sin \alpha x)e^{-(1+\lambda\alpha^2)t}.$$

10. Substituting $u(x,t) = X(x)T(t)$ into the partial differential equation yields $kX''T = XT'$. Separating variables and using the separation constant $-\lambda$ we obtain

$$\frac{X''}{X} = \frac{T'}{kT} = -\lambda.$$

Then

$$X'' + \lambda X = 0 \quad \text{and} \quad T' + \lambda kT = 0.$$

The second differential equation implies $T(t) = c_1 e^{-\lambda kt}$. For the first differential equation we consider three cases:

I. If $\lambda = 0$ then $X'' = 0$ and $X(x) = c_2 x + c_3$, so

$$u = XT = A_1 x + A_2.$$

II. If $\lambda = -\alpha^2 < 0$, then $X'' - \alpha^2 X = 0$, and $X(x) = c_4 \cosh \alpha x + c_5 \sinh \alpha x$, so

$$u = XT = (A_3 \cosh \alpha x + A_4 \sinh \alpha x)e^{k\alpha^2 t}.$$

III. If $\lambda = \alpha^2 > 0$, then $X'' + \alpha^2 X = 0$, and $X(x) = c_6 \cos \alpha x + c_7 \sin \alpha x$, so

$$u = XT = (A_5 \cos \alpha x + A_6 \sin \alpha x)e^{-k\alpha^2 t}.$$

11. Substituting $u(x,t) = X(x)T(t)$ into the partial differential equation yields $a^2 X''T = XT''$. Separating variables and using the separation constant $-\lambda$ we obtain

$$\frac{X''}{X} = \frac{T''}{a^2 T} = -\lambda.$$

Then

$$X'' + \lambda X = 0 \quad \text{and} \quad T'' + a^2 \lambda T = 0.$$

We consider three cases:

I. If $\lambda = 0$ then $X'' = 0$ and $X(x) = c_1x + c_2$. Also, $T'' = 0$ and $T(t) = c_3t + c_4$, so

$$u = XT = (c_1x + c_2)(c_3t + c_4).$$

II. If $\lambda = -\alpha^2 < 0$, then $X'' - \alpha^2 X = 0$, and $X(x) = c_5 \cosh \alpha x + c_6 \sinh \alpha x$. Also, $T'' - \alpha^2 a^2 T = 0$ and $T(t) = c_7 \cosh \alpha at + c_8 \sinh \alpha at$, so

$$u = XT = (c_5 \cosh \alpha x + c_6 \sinh \alpha x)(c_7 \cosh \alpha at + c_8 \sinh \alpha at).$$

III. If $\lambda = \alpha^2 > 0$, then $X'' + \alpha^2 X = 0$, and $X(x) = c_9 \cos \alpha x + c_{10} \sin \alpha x$. Also, $T'' + \alpha^2 a^2 T = 0$ and $T(t) = c_{11} \cos \alpha at + c_{12} \sin \alpha at$, so

$$u = XT = (c_9 \cos \alpha x + c_{10} \sin \alpha x)(c_{11} \cos \alpha at + c_{12} \sin \alpha at).$$

12. Substituting $u(x, t) = X(x)T(t)$ into the partial differential equation yields $a^2 X''T = XT'' + 2kXT'$. Separating variables and using the separation constant $-\lambda$ we obtain

$$\frac{X''}{X} = \frac{T'' + 2kT'}{a^2 T} = -\lambda.$$

Then

$$X'' + \lambda X = 0 \qquad \text{and} \qquad T'' + 2kT' + a^2\lambda T = 0.$$

We consider three cases:

I. If $\lambda = 0$ then $X'' = 0$ and $X(x) = c_1x + c_2$. Also, $T'' + 2kT' = 0$ and $T(t) = c_3 + c_4 e^{-2kt}$, so

$$u = XT = (c_1x + c_2)(c_3 + c_4 e^{-2kt}).$$

II. If $\lambda = -\alpha^2 < 0$, then $X'' - \alpha^2 X = 0$, and $X(x) = c_5 \cosh \alpha x + c_6 \sinh \alpha x$. The auxiliary equation of $T'' + 2kT' - \alpha^2 a^2 T = 0$ is $m^2 + 2km - \alpha^2 a^2 = 0$. Solving for m we obtain $m = -k \pm \sqrt{k^2 + \alpha^2 a^2}$, so $T(t) = c_7 e^{(-k+\sqrt{k^2+\alpha^2 a^2})t} + c_8 e^{(-k-\sqrt{k^2+\alpha^2 a^2})t}$. Then

$$u = XT = (c_5 \cosh \alpha x + c_6 \sinh \alpha x)(c_7 e^{(-k+\sqrt{k^2+\alpha^2 a^2})t} + c_8 e^{(-k-\sqrt{k^2+\alpha^2 a^2})t}.$$

III. If $\lambda = \alpha^2 > 0$, then $X'' + \alpha^2 X = 0$, and $X(x) = c_9 \cos \alpha x + c_{10} \sin \alpha x$. The auxiliary equation of $T'' + 2kT' + \alpha^2 a^2 T = 0$ is $m^2 + 2km + \alpha^2 a^2 = 0$. Solving for m we obtain $m = -k \pm \sqrt{k^2 - \alpha^2 a^2}$. We consider three possibilities for the discriminant $k^2 - \alpha^2 a^2$:

(i) If $k^2 - \alpha^2 a^2 = 0$ then $T(t) = c_{11}e^{-kt} + c_{12}te^{-kt}$ and

$$u = XT = (c_9 \cos \alpha x + c_{10} \sin \alpha x)(c_{11}e^{-kt} + c_{12}te^{-kt}).$$

From $k^2 - \alpha^2 a^2 = 0$ we have $\alpha = k/a$ so the solution can be written

$$u = XT = (c_9 \cos kx/a + c_{10} \sin kx/a)(c_{11}e^{-kt} + c_{12}te^{-kt}).$$

(ii) If $k^2 - \alpha^2 a^2 < 0$ then $T(t) = e^{-kt}\left(c_{13} \cos \sqrt{\alpha^2 a^2 - k^2}\, t + c_{14} \sin \sqrt{\alpha^2 a^2 - k^2}\, t\right)$ and

$$u = XT = (c_9 \cos \alpha x + c_{10} \sin \alpha x)e^{-kt}\left(c_{13} \cos \sqrt{\alpha^2 a^2 - k^2}\, t + c_{14} \sin \sqrt{\alpha^2 a^2 - k^2}\, t\right).$$

(iii) If $k^2 - \alpha^2 a^2 > 0$ then $T(t) = c_{15}e^{(-k+\sqrt{k^2-\alpha^2 a^2})t} + c_{16}e^{(-k-\sqrt{k^2-\alpha^2 a^2})t}$ and

$$u = XT = (c_9 \cos \alpha x + c_{10} \sin \alpha x)\left(c_{15}e^{(-k+\sqrt{k^2-\alpha^2 a^2})t} + c_{16}e^{(-k-\sqrt{k^2-\alpha^2 a^2})t}\right).$$

13. Substituting $u(x, y) = X(x)Y(y)$ into the partial differential equation yields $X''Y + XY'' = 0$. Separating variables and using the separation constant $-\lambda$ we obtain

$$-\frac{X''}{X} = \frac{Y''}{Y} = -\lambda.$$

Then

$$X'' - \lambda X = 0 \qquad \text{and} \qquad Y'' + \lambda Y = 0.$$

We consider three cases:

I. If $\lambda = 0$ then $X'' = 0$ and $X(x) = c_1 x + c_2$. Also, $Y'' = 0$ and $Y(y) = c_3 y + c_4$ so

$$u = XY = (c_1 x + c_2)(c_3 y + c_4).$$

II. If $\lambda = -\alpha^2 < 0$ then $X'' + \alpha^2 X = 0$ and $X(x) = c_5 \cos \alpha x + c_6 \sin \alpha x$. Also, $Y'' - \alpha^2 Y = 0$ and $Y(y) = c_7 \cosh \alpha y + c_8 \sinh \alpha y$ so

$$u = XY = (c_5 \cos \alpha x + c_6 \sin \alpha x)(c_7 \cosh \alpha y + c_8 \sinh \alpha y).$$

III. If $\lambda = \alpha^2 > 0$ then $X'' - \alpha^2 X = 0$ and $X(x) = c_9 \cosh \alpha x + c_{10} \sinh \alpha x$. Also, $Y'' + \alpha^2 Y = 0$ and $Y(y) = c_{11} \cos \alpha y + c_{12} \sin \alpha y$ so

$$u = XY = (c_9 \cosh \alpha x + c_{10} \sinh \alpha x)(c_{11} \cos \alpha y + c_{12} \sin \alpha y).$$

14. Substituting $u(x, y) = X(x)Y(y)$ into the partial differential equation yields $x^2 X''Y + XY'' = 0$. Separating variables and using the separation constant $-\lambda$ we obtain

$$-\frac{x^2 X''}{X} = \frac{Y''}{Y} = -\lambda.$$

Then

$$x^2 X'' - \lambda X = 0 \qquad \text{and} \qquad Y'' + \lambda Y = 0.$$

We consider three cases:

I. If $\lambda = 0$ then $x^2 X'' = 0$ and $X(x) = c_1 x + c_2$. Also, $Y'' = 0$ and $Y(y) = c_3 y + c_4$ so

$$u = XY = (c_1 x + c_2)(c_3 y + c_4).$$

II. If $\lambda = -\alpha^2 < 0$ then $x^2 X'' + \alpha^2 X = 0$ and $Y'' - \alpha^2 Y = 0$. The solution of the second differential equation is $Y(y) = c_5 \cosh \alpha y + c_6 \sinh \alpha y$. The first equation is Cauchy-Euler with auxiliary equation $m^2 - m + \alpha^2 = 0$. Solving for m we obtain $m = \frac{1}{2} \pm \frac{1}{2}\sqrt{1 - 4\alpha^2}$. We consider three possibilities for the discriminant $1 - 4\alpha^2$.

(i) If $1 - 4\alpha^2 = 0$ then $X(x) = c_7 x^{1/2} + c_8 x^{1/2} \ln x$ and

$$u = XY = x^{1/2}(c_7 + c_8 \ln x)(c_5 \cosh \alpha y + c_6 \sinh \alpha y).$$

(ii) If $1 - 4\alpha^2 < 0$ then $X(x) = x^{1/2}\left[c_9 \cos\left(\sqrt{4\alpha^2 - 1}\, \ln x\right) + c_{10} \sin\left(\sqrt{4\alpha^2 - 1}\, \ln x\right)\right]$ and

$$u = XY = x^{1/2}\left[c_9 \cos\left(\sqrt{4\alpha^2 - 1}\, \ln x\right) + c_{10} \sin\left(\sqrt{4\alpha^2 - 1}\, \ln x\right)\right](c_5 \cosh \alpha y + c_6 \sinh \alpha y).$$

(iii) If $1 - 4\alpha^2 > 0$ then $X(x) = x^{1/2}\left(c_{11} x^{\sqrt{1-4\alpha^2}/2} + c_{12} x^{-\sqrt{1-4\alpha^2}/2}\right)$ and

$$u = XY = x^{1/2}\left(c_{11} x^{\sqrt{1-4\alpha^2}/2} + c_{12} x^{-\sqrt{1-4\alpha^2}/2}\right)(c_5 \cosh \alpha y + c_6 \sinh \alpha y).$$

III. If $\lambda = \alpha^2 > 0$ then $x^2 X'' - \alpha^2 X = 0$ and $Y'' + \alpha^2 Y = 0$. The solution of the second differential equation is $Y(y) = c_{13} \cos \alpha y + c_{14} \sin \alpha y$. The first equation is Cauchy-Euler with auxiliary equation $m^2 - m - \alpha^2 = 0$. Solving for m we obtain $m = \frac{1}{2} \pm \frac{1}{2}\sqrt{1 + 4\alpha^2}$. In this case the discriminant is always positive so the solution of the differential equation is $X(x) = x^{1/2}\left(c_{15} x^{\sqrt{1+4\alpha^2}/2} + c_{16} x^{-\sqrt{1+4\alpha^2}/2}\right)$ and

$$u = XY = x^{1/2}\left(c_{15} x^{\sqrt{1+4\alpha^2}/2} + c_{16} x^{-\sqrt{1+4\alpha^2}/2}\right)(c_{13} \cos \alpha y + c_{14} \sin \alpha y).$$

15. Substituting $u(x, y) = X(x)Y(y)$ into the partial differential equation yields $X''Y + XY'' = XY$. Separating variables and using the separation constant $-\lambda$ we obtain

$$\frac{X''}{X} = \frac{Y - Y''}{Y} = -\lambda.$$

Then

$$X'' + \lambda X = 0 \quad \text{and} \quad Y'' - (1 + \lambda)Y = 0.$$

We consider three cases:

I. If $\lambda = 0$ then $X'' = 0$ and $X(x) = c_1 x + c_2$. Also $Y'' - Y = 0$ and $Y(y) = c_3 \cosh y + c_4 \sinh y$ so

$$u = XY = (c_1 x + c_2)(c_3 \cosh y + c_4 \sinh y).$$

II. If $\lambda = -\alpha^2 < 0$ then $X'' - \alpha^2 X = 0$ and $Y'' + (\alpha^2 - 1)Y = 0$. The solution of the first differential equation is $X(x) = c_5 \cosh \alpha x + c_6 \sinh \alpha x$. The solution of the second differential equation depends on the nature of $\alpha^2 - 1$. We consider three cases:

(i) If $\alpha^2 - 1 = 0$, or $\alpha^2 = 1$, then $Y(y) = c_7 y + c_8$ nad

$$u = XY = (c_5 \cosh x + c_6 \sinh x)(c_7 y + c_8).$$

(ii) If $\alpha^2 - 1 < 0$, or $0 < \alpha^2 < 1$, then $Y(y) = c_9 \cosh \sqrt{1 - \alpha^2}\, y + c_{10} \sinh \sqrt{1 - \alpha^2}\, y$ and

$$u = XY = (c_5 \cosh \alpha x + c_6 \sinh \alpha x)\left(c_9 \cosh \sqrt{1 - \alpha^2}\, y + c_{10} \sinh \sqrt{1 - \alpha^2}\, y\right).$$

(iii) If $\alpha^2 - 1 > 0$, or $\alpha^2 > 1$, then $Y(y) = c_{11} \cos \sqrt{\alpha^2 - 1}\, y + c_{12} \sin \sqrt{\alpha^2 - 1}\, y$ and

$$u = XY = (c_5 \cosh \alpha x + c_6 \sinh \alpha x) \left(c_{11} \cos \sqrt{\alpha^2 - 1}\, y + c_{12} \sin \sqrt{\alpha^2 - 1}\, y \right).$$

III. If $\lambda = \alpha^2 > 0$, then $X'' + \alpha^2 X = 0$ and $X(x) = c_{13} \cos \alpha x + c_{14} \sin \alpha x$. Also,

$Y'' - (1 + \alpha^2)Y = 0$ and $Y(y) = c_{15} \cosh \sqrt{1 + \alpha^2}\, y + c_{16} \sinh \sqrt{1 + \alpha^2}\, y$ so

$$u = XY = (c_{13} \cos \alpha x + c_{14} \sin \alpha x) \left(c_{15} \cosh \sqrt{1 + \alpha^2}\, y + c_{16} \sinh \sqrt{1 + \alpha^2}\, y \right).$$

16. Substituting $u(x,t) = X(x)T(t)$ into the partial differential equation yields $a^2 X'' T - g = XT''$, which is not separable.

17. Identifying $A = B = C = 1$, we compute $B^2 - 4AC = -3 < 0$. The equation is elliptic.

18. Identifying $A = 3$, $B = 5$, and $C = 1$, we compute $B^2 - 4AC = 13 > 0$. The equation is hyperbolic.

19. Identifying $A = 1$, $B = 6$, and $C = 9$, we compute $B^2 - 4AC = 0$. The equation is parabolic.

20. Identifying $A = 1$, $B = -1$, and $C = -3$, we compute $B^2 - 4AC = 13 > 0$. The equation is hyperbolic.

21. Identifying $A = 1$, $B = -9$, and $C = 0$, we compute $B^2 - 4AC = 81 > 0$. The equation is hyperbolic.

22. Identifying $A = 0$, $B = 1$, and $C = 0$, we compute $B^2 - 4AC = 1 > 0$. The equation is hyperbolic.

23. Identifying $A = 1$, $B = 2$, and $C = 1$, we compute $B^2 - 4AC = 0$. The equation is parabolic.

24. Identifying $A = 1$, $B = 0$, and $C = 1$, we compute $B^2 - 4AC = -4 < 0$. The equation is elliptic.

25. Identifying $A = a^2$, $B = 0$, and $C = -1$, we compute $B^2 - 4AC = 4a^2 > 0$. The equation is hyperbolic.

26. Identifying $A = k > 0$, $B = 0$, and $C = 0$, we compute $B^2 - 4AC = -4k < 0$. The equation is elliptic.

27. Substituting $u(r,t) = R(r)T(t)$ into the partial differential equation yields

$$k \left(R'' T + \frac{1}{r} R' T \right) = RT'.$$

Separating variables and using the separation constant $-\lambda$ we obtain

$$\frac{rR'' + R'}{rR} = \frac{T'}{kT} = -\lambda.$$

Then

$$rR'' + R' + \lambda r R = 0 \qquad \text{and} \qquad T' + \lambda k T = 0.$$

Letting $\lambda = \alpha^2$ and writing the first equation as $r^2 R'' + rR' = \alpha^2 r^2 R = 0$ we see that it is a parametric Bessel equation of order 0. As discussed in Chapter 6 of the text, it has solution

$R(r) = c_1 J_0(\alpha r) + c_2 Y_0(\alpha r)$. Since a solution of $T' + \alpha^2 kT$ is $T(t) = e^{-k\alpha^2 t}$, we see that a solution of the partial differential equation is

$$u = RT = e^{-k\alpha^2 t}[c_1 J_0(\alpha r) + c_2 Y_0(\alpha r)].$$

28. Substituting $u(r, \theta) = R(r)\Theta(\theta)$ into the partial differential equation yields

$$R''\Theta + \frac{1}{r} R'\Theta + \frac{1}{r^2} R\Theta'' = 0.$$

Separating variables and using the separation constant $-\lambda$ we obtain

$$\frac{r^2 R'' + rR'}{R} = -\frac{\Theta''}{\Theta} = -\lambda.$$

Then

$$r^2 R'' + rR' + \lambda R = 0 \quad \text{and} \quad \Theta'' - \lambda\Theta = 0.$$

Letting $\lambda = -\alpha^2$ we have the Cauchy-Euler equation $r^2 R'' + rR' - \alpha^2 R = 0$ whose solution is $R(r) = c_3 r^\alpha + c_4 r^{-\alpha}$. Since the solution of $\Theta'' + \alpha^2\Theta = 0$ is $\Theta(\theta) = c_1 \cos\alpha\theta + c_2 \sin\alpha\theta$ we see that a solution of the partial differential equation is

$$u = R\Theta = (c_1 \cos\alpha\theta + c_2 \sin\alpha\theta)(c_3 r^\alpha + c_4 r^{-\alpha}).$$

29. For $u = A_1 + B_1 x$ we compute $\partial^2 u/\partial x^2 = 0 = \partial u/\partial y$. Then $\partial^2 u/\partial x^2 = 4\,\partial u/\partial y$.

For $u = A_2 e^{\alpha^2 y} \cos 2\alpha x + B_2 e^{\alpha^2 y} \sin 2\alpha x$ we compute

$$\frac{\partial u}{\partial x} = 2\alpha A_2 e^{\alpha^2 y} \sinh 2\alpha x + 2\alpha B_2 e^{\alpha^2 y} \cosh 2\alpha x$$

$$\frac{\partial^2 u}{\partial x^2} = 4\alpha^2 A_2 e^{\alpha^2 y} \cosh 2\alpha x + 4\alpha^2 B_2 e^{\alpha^2 y} \sinh 2\alpha x$$

and

$$\frac{\partial u}{\partial y} = \alpha^2 A_2 e^{\alpha^2 y} \cosh 2\alpha x + \alpha^2 B_2 e^{\alpha^2 y} \sinh 2\alpha x.$$

Then $\partial^2 u/\partial x^2 = 4\,\partial u/\partial y$.

For $u = A_3 e^{-\alpha^2 y} \cosh 2\alpha x + B_3 e^{-\alpha^2 y} \sinh 2\alpha x$ we compute

$$\frac{\partial u}{\partial x} = -2\alpha A_3 e^{-\alpha^2 y} \sin 2\alpha x + 2\alpha B_3 e^{-\alpha^2 y} \cos 2\alpha x$$

$$\frac{\partial^2 u}{\partial x^2} = -4\alpha^2 A_3 e^{-\alpha^2 y} \cos 2\alpha x - 4\alpha^2 B_3 e^{-\alpha^2 y} \sin 2\alpha x$$

and

$$\frac{\partial u}{\partial y} = -\alpha^2 A_3 e^{-\alpha^2 y} \cos 2\alpha x - \alpha^2 B_3 e^{-\alpha^2 y} \sin 2\alpha x.$$

Then $\partial^2 u/\partial x^2 = 4\,\partial u/\partial y$.

30. We identify $A = xy + 1$, $B = x + 2y$, and $C = 1$. Then $B^2 - 4AC = x^2 + 4y^2 - 4$. The equation $x^2 + 4y^2 = 4$ defines an ellipse. The partial differential equation is hyperbolic outside the ellipse, parabolic on the ellipse, and elliptic inside the ellipse.

Exercises 12.2   Classical PDEs and Boundary-Value Problems

1. $k\dfrac{\partial^2 u}{\partial x^2} = \dfrac{\partial u}{\partial t}$, $0 < x < L,\ t > 0$

$u(0, t) = 0, \quad \dfrac{\partial u}{\partial x}\bigg|_{x=L} = 0, \quad t > 0$

$u(x, 0) = f(x), \quad 0 < x < L$

2. $k\dfrac{\partial^2 u}{\partial x^2} = \dfrac{\partial u}{\partial t}$, $0 < x < L,\ t > 0$

$u(0, t) = u_0, \quad u(L, t) = u_1, \quad t > 0$

$u(x, 0) = 0, \quad 0 < x < L$

3. $k\dfrac{\partial^2 u}{\partial x^2} = \dfrac{\partial u}{\partial t}$, $0 < x < L,\ t > 0$

$u(0, t) = 100, \quad \dfrac{\partial u}{\partial x}\bigg|_{x=L} = -hu(L, t), \quad t > 0$

$u(x, 0) = f(x), \quad 0 < x < L$

4. $k\dfrac{\partial^2 u}{\partial x^2} + h(u - 50) = \dfrac{\partial u}{\partial t}$, $0 < x < L,\ t > 0$

$\dfrac{\partial u}{\partial x}\bigg|_{x=0} = 0, \quad \dfrac{\partial u}{\partial x}\bigg|_{x=L} = 0, \quad t > 0$

$u(x, 0) = 100, \quad 0 < x < L$

5. $a^2\dfrac{\partial^2 u}{\partial x^2} = \dfrac{\partial^2 u}{\partial t^2}$, $0 < x < L,\ t > 0$

$u(0, t) = 0, \quad u(L, t) = 0, \quad t > 0$

$u(x, 0) = x(L - x), \quad \dfrac{\partial u}{\partial t}\bigg|_{t=0} = 0, \quad 0 < x < L$

6. $a^2\dfrac{\partial^2 u}{\partial x^2} = \dfrac{\partial^2 u}{\partial t^2}$, $0 < x < L,\ t > 0$

$$u(0,t) = 0, \quad u(L,t) = 0, \quad t > 0$$

$$u(x,0) = 0, \quad \frac{\partial u}{\partial t}\bigg|_{t=0} = \sin\frac{\pi x}{L}, \quad 0 < x < L$$

7. $a^2\dfrac{\partial^2 u}{\partial x^2} - 2\beta\dfrac{\partial u}{\partial t} = \dfrac{\partial^2 u}{\partial t^2}, \quad 0 < x < L, \ t > 0$

$$u(0,t) = 0, \quad u(L,t) = \sin\pi t, \quad t > 0$$

$$u(x,0) = f(x), \quad \frac{\partial u}{\partial t}\bigg|_{t=0} = 0, \quad 0 < x < L$$

8. $a^2\dfrac{\partial^2 u}{\partial x^2} + Ax = \dfrac{\partial^2 u}{\partial t^2}, \quad 0 < x < L, \ t > 0, \ A \text{ a constant}$

$$u(0,t) = 0, \quad u(L,t) = 0, \quad t > 0$$

$$u(x,0) = 0, \quad \frac{\partial u}{\partial t}\bigg|_{t=0} = 0, \quad 0 < x < L$$

9. $\dfrac{\partial^2 u}{\partial x^2} + \dfrac{\partial^2 u}{\partial y^2} = 0, \quad 0 < x < 4, \ 0 < y < 2$

$$\frac{\partial u}{\partial x}\bigg|_{x=0} = 0, \quad u(4,y) = f(y), \quad 0 < y < 2$$

$$\frac{\partial u}{\partial y}\bigg|_{y=0} = 0, \quad u(x,2) = 0, \quad 0 < x < 4$$

10. $\dfrac{\partial^2 u}{\partial x^2} + \dfrac{\partial^2 u}{\partial y^2} = 0, \quad 0 < x < \pi, \ y > 0$

$$u(0,y) = e^{-y}, \quad u(\pi,y) = \begin{cases} 100, & 0 < y \le 1 \\ 0, & y > 1 \end{cases}$$

$$u(x,0) = f(x), \quad 0 < x < \pi$$

Exercises 12.3

Heat Equation

1. Using $u = XT$ and $-\lambda$ as a separation constant we obtain

$$X'' + \lambda X = 0,$$

$$X(0) = 0,$$

$$X(L) = 0,$$

and

$$T' + k\lambda T = 0.$$

This leads to

$$X = c_1 \sin \frac{n\pi}{L} x \qquad \text{and} \qquad T = c_2 e^{-kn^2\pi^2 t/L^2}$$

for $n = 1, 2, 3, \ldots$ so that

$$u = \sum_{n=1}^{\infty} A_n e^{-kn^2\pi^2 t/L^2} \sin \frac{n\pi}{L} x.$$

Imposing

$$u(x, 0) = \sum_{n=1}^{\infty} A_n \sin \frac{n\pi}{L} x$$

gives

$$A_n = \frac{2}{L} \int_0^{L/2} \sin \frac{n\pi}{L} x \, dx = \frac{2}{n\pi} \left(1 - \cos \frac{n\pi}{2} \right)$$

for $n = 1, 2, 3, \ldots$ so that

$$u(x, t) = \frac{2}{\pi} \sum_{n=1}^{\infty} \frac{1 - \cos \frac{n\pi}{2}}{n} e^{-kn^2\pi^2 t/L^2} \sin \frac{n\pi}{L} x.$$

2. Using $u = XT$ and $-\lambda$ as a separation constant we obtain

$$X'' + \lambda X = 0,$$

$$X(0) = 0,$$

$$X(L) = 0,$$

and

$$T' + k\lambda T = 0.$$

This leads to

$$X = c_1 \sin \frac{n\pi}{L} x \qquad \text{and} \qquad T = c_2 e^{-kn^2\pi^2 t/L^2}$$

for $n = 1, 2, 3, \ldots$ so that

$$u = \sum_{n=1}^{\infty} A_n e^{-kn^2\pi^2 t/L^2} \sin \frac{n\pi}{L} x.$$

Imposing

$$u(x, 0) = \sum_{n=1}^{\infty} A_n \sin \frac{n\pi}{L} x$$

gives

$$A_n = \frac{2}{L} \int_0^{L} x(L - x) \sin \frac{n\pi}{L} x \, dx = \frac{4L^2}{n^3\pi^3} [1 - (-1)^n]$$

for $n = 1, 2, 3, \ldots$ so that

$$u(x, t) = \frac{4L^2}{\pi^3} \sum_{n=1}^{\infty} \frac{1 - (-1)^n}{n^3} e^{-kn^2\pi^2 t/L^2} \sin \frac{n\pi}{L} x.$$

3. Using $u = XT$ and $-\lambda$ as a separation constant we obtain

$$X'' + \lambda X = 0,$$

$$X'(0) = 0,$$

$$X'(L) = 0,$$

and

$$T' + k\lambda T = 0.$$

This leads to

$$X = c_1 \cos \frac{n\pi}{L} x \quad \text{and} \quad T = c_2 e^{-kn^2\pi^2 t/L^2}$$

for $n = 0, 1, 2, \ldots$ ($\lambda = 0$ is an eigenvalue in this case) so that

$$u = \sum_{n=0}^{\infty} A_n e^{-kn^2\pi^2 t/L^2} \cos \frac{n\pi}{L} x.$$

Imposing

$$u(x,0) = f(x) = A_0 + \sum_{n=1}^{\infty} A_n \cos \frac{n\pi}{L} x$$

gives

$$u(x,t) = \frac{1}{L} \int_0^L f(x)\, dx + \frac{2}{L} \sum_{n=1}^{\infty} \left(\int_0^L f(x) \cos \frac{n\pi}{L} x\, dx \right) e^{-kn^2\pi^2 t/L^2} \cos \frac{n\pi}{L} x.$$

4. If $L = 2$ and $f(x)$ is x for $0 < x < 1$ and $f(x)$ is 0 for $1 < x < 2$ then

$$u(x,t) = \frac{1}{4} + 4 \sum_{n=1}^{\infty} \left[\frac{1}{2n\pi} \sin \frac{n\pi}{2} + \frac{1}{n^2\pi^2} \left(\cos \frac{n\pi}{2} - 1 \right) \right] e^{-kn^2\pi^2 t/4} \cos \frac{n\pi}{2} x.$$

5. Using $u = XT$ and $-\lambda$ as a separation constant leads to

$$X'' + \lambda X = 0,$$

$$X'(0) = 0,$$

$$X'(L) = 0,$$

and

$$T' + (h + k\lambda)T = 0.$$

Then

$$X = c_1 \cos \frac{n\pi}{L} x \quad \text{and} \quad T = c_2 e^{-ht - kn^2\pi^2 t/L^2}$$

for $n = 0, 1, 2, \ldots$ ($\lambda = 0$ is an eigenvalue in this case) so that

$$u = A_0 e^{-ht} + e^{-ht} \sum_{n=1}^{\infty} A_n e^{-kn^2\pi^2 t/L^2} \cos \frac{n\pi}{L} x.$$

Imposing

$$u(x,0) = f(x) = \sum_{n=0}^{\infty} A_n \cos \frac{n\pi}{L}x$$

gives

$$u(x,t) = \frac{e^{-ht}}{L} \int_0^L f(x)\,dx + \frac{2e^{-ht}}{L} \sum_{n=1}^{\infty} \left(\int_0^L f(x) \cos \frac{n\pi}{L}x\,dx \right) e^{-kn^2\pi^2 t/L^2} \cos \frac{n\pi}{L}x.$$

6. In Problem 5 we instead find that $X(0) = 0$ and $X(L) = 0$ so that

$$X = c_1 \sin \frac{n\pi}{L}x$$

and

$$u = \frac{2e^{-ht}}{L} \sum_{n=1}^{\infty} \left(\int_0^L f(x) \sin \frac{n\pi}{L}x\,dx \right) e^{-kn^2\pi^2 t/L^2} \sin \frac{n\pi}{L}x.$$

7.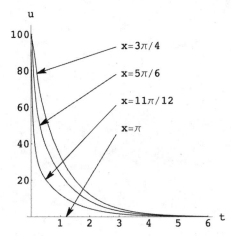

8. In this case we have from (13) of this section in the text that

$$u_n(x,0) = A_n \sin \frac{n\pi}{L}x = f(x) = 10 \sin \frac{5\pi x}{L},$$

so we can take $n = 5$ and $A_5 = 0.$ All other values of A_n are 0. Therefore, we can take the solution of the boundary-value problem to be

$$u(x,t) = 10 e^{-k(25\pi^2/L^2)t} \sin \frac{5\pi}{L}x.$$

9. (a) The solution is

$$u(x,t) = \sum_{n=1}^{\infty} A_n e^{-kn^2\pi^2 t/100^2} \sin \frac{n\pi}{100}x,$$

where

$$A_n = \frac{2}{100} \left[\int_0^{50} 0.8x \sin \frac{n\pi}{100}x\,dx + \int_{50}^{100} 0.8(100-x) \sin \frac{n\pi}{100}x\,dx \right] = \frac{320}{n^2\pi^2} \sin \frac{n\pi}{2}.$$

Thus,

$$u(x,t) = \frac{320}{\pi^2} \sum_{n=1}^{\infty} \frac{1}{n^2} \left(\sin \frac{n\pi}{2}\right) e^{-kn^2\pi^2 t/100^2} \sin \frac{n\pi}{100} x.$$

(b) Since $A_n = 0$ for n even, the first five nonzero terms correspond to $n = 1, 3, 5, 7, 9$. In this case $\sin(n\pi/2) = \sin(2p-1)/2 = (-1)^{p+1}$ for $p = 1, 2, 3, 4, 5$, and

$$u(x,t) = \frac{320}{\pi^2} \sum_{p=1}^{\infty} \frac{(-1)^{p+1}}{(2p-1)^2} e^{(-1.6352(2p-1)^2\pi^2/100^2)t} \sin \frac{(2p-1)\pi}{100} x.$$

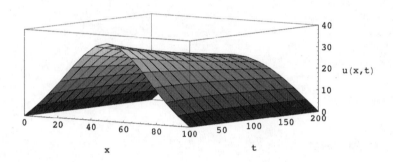

Exercises 12.4

Wave Equation

1. Using $u = XT$ and $-\lambda$ as a separation constant we obtain

$$X'' + \lambda X = 0,$$

$$X(0) = 0,$$

$$X(L) = 0,$$

and

$$T'' + \lambda a^2 T = 0,$$

$$T'(0) = 0.$$

Solving the differential equations we get

$$X = c_1 \sin \frac{n\pi}{L} x + c_2 \cos \frac{n\pi}{L} x \qquad \text{and} \qquad T = c_3 \cos \frac{n\pi a}{L} t + c_4 \sin \frac{n\pi a}{L} t$$

for $n = 1, 2, 3, \ldots$. The boundary and initial conditions give

$$u = \sum_{n=1}^{\infty} A_n \cos \frac{n\pi a}{L} t \sin \frac{n\pi}{L} x.$$

Imposing

$$u(x,0) = \frac{1}{4}x(L-x) = \sum_{n=1}^{\infty} A_n \sin \frac{n\pi}{L}x$$

gives

$$A_n = \frac{L^2}{n^3\pi^3}[1-(-1)^n]$$

for $n = 1, 2, 3, \ldots$ so that

$$u(x,t) = \frac{L^2}{\pi^3}\sum_{n=1}^{\infty} \frac{1-(-1)^n}{n^3} \cos \frac{n\pi a}{L}t \sin \frac{n\pi}{L}x.$$

2. Using $u = XT$ and $-\lambda$ as a separation constant we obtain

$$X'' + \lambda X = 0,$$

$$X(0) = 0,$$

$$X(L) = 0,$$

and

$$T'' + \lambda a^2 T = 0,$$

$$T(0) = 0.$$

Solving the differential equations we get

$$X = c_1 \sin \frac{n\pi}{L}x + c_2 \cos \frac{n\pi}{L}x \quad \text{and} \quad T = c_3 \cos \frac{n\pi a}{L}t + c_4 \sin \frac{n\pi a}{L}t$$

for $n = 1, 2, 3, \ldots$. The boundary and initial conditions give

$$u = \sum_{n=1}^{\infty} B_n \sin \frac{n\pi a}{L}t \sin \frac{n\pi}{L}x.$$

Imposing

$$u_t(x,0) = x(L-x) = \sum_{n=1}^{\infty} B_n \frac{n\pi a}{L} \sin \frac{n\pi}{L}x$$

gives

$$B_n \frac{n\pi a}{L} = \frac{4L^2}{n^3\pi^3}[1-(-1)^n]$$

for $n = 1, 2, 3, \ldots$ so that

$$u(x,t) = \frac{4L^3}{a\pi^4}\sum_{n=1}^{\infty} \frac{1-(-1)^n}{n^4} \sin \frac{n\pi a}{L}t \sin \frac{n\pi}{L}x.$$

3. Using $u = XT$ and $-\lambda$ as a separation constant we obtain

$$X'' + \lambda X = 0,$$

$$X(0) = 0,$$

$$X(L) = 0,$$

and

$$T'' + \lambda a^2 T = 0,$$

$$T'(0) = 0.$$

Solving the differential equations we get

$$X = c_1 \sin \frac{n\pi}{L} x + c_2 \cos \frac{n\pi}{L} x \qquad \text{and} \qquad T = c_3 \cos \frac{n\pi a}{L} t + c_4 \sin \frac{n\pi a}{L} t$$

for $n = 1, 2, 3, \ldots$. The boundary and initial conditions give

$$u = \sum_{n=1}^{\infty} A_n \cos \frac{n\pi a}{L} t \, \sin \frac{n\pi}{L} x.$$

Imposing

$$u(x,0) = \sum_{n=1}^{\infty} A_n \sin \frac{n\pi}{L} x$$

gives

$$A_n = \frac{2}{L} \left(\int_0^{L/3} \frac{3}{L} x \sin \frac{n\pi}{L} x \, dx + \int_{L/3}^{2L/3} \sin \frac{n\pi}{L} x \, dx + \int_{2L/3}^{L} \left(3 - \frac{3}{L} x \right) \sin \frac{n\pi}{L} x \, dx \right)$$

so that

$$A_1 = \frac{6\sqrt{3}}{\pi^2},$$

$$A_2 = A_3 = A_4 = 0,$$

$$A_5 = -\frac{6\sqrt{3}}{5^2 \pi^2},$$

$$A_6 = 0,$$

$$A_7 = \frac{6\sqrt{3}}{7^2 \pi^2}$$

$$\vdots$$

and

$$u(x,t) = \frac{6\sqrt{3}}{\pi^2} \left(\cos \frac{\pi a}{L} t \, \sin \frac{\pi}{L} x - \frac{1}{5^2} \cos \frac{5\pi a}{L} t \, \sin \frac{5\pi}{L} x + \frac{1}{7^2} \cos \frac{7\pi a}{L} t \, \sin \frac{7\pi}{L} x - \cdots \right).$$

4. Using $u = XT$ and $-\lambda$ as a separation constant we obtain

$$X'' + \lambda X = 0,$$

$$X(0) = 0,$$

$$X(\pi) = 0,$$

and

$$T'' + \lambda a^2 T = 0,$$

$$T'(0) = 0.$$

Solving the differential equations we get

$$X = c_1 \sin nx + c_2 \cos nx \qquad \text{and} \qquad T = c_3 \cos nat + c_4 \sin nat$$

for $n = 1, 2, 3, \ldots$. The boundary and initial conditions give

$$u = \sum_{n=1}^{\infty} A_n \cos nt \, \sin nx.$$

Imposing

$$u(x, 0) = \frac{1}{6} x(\pi^2 - x^2) = \sum_{n=1}^{\infty} A_n \sin nx \qquad \text{and} \qquad u_t(x, 0) = 0$$

gives

$$A_n = \frac{2}{n^3}(-1)^{n+1}$$

for $n = 1, 2, 3, \ldots$ so that

$$u(x, t) = 2 \sum_{n=1}^{\infty} \frac{(-1)^{n+1}}{n^3} \cos nat \, \sin nx.$$

5. Using $u = XT$ and $-\lambda$ as a separation constant we obtain

$$X'' + \lambda X = 0,$$

$$X(0) = 0,$$

$$X(\pi) = 0,$$

and

$$T'' + \lambda a^2 T = 0,$$

$$T'(0) = 0.$$

Solving the differential equations we get

$$X = c_1 \sin nx + c_2 \cos nx \qquad \text{and} \qquad T = c_3 \cos nat + c_4 \sin nat$$

for $n = 1, 2, 3, \ldots$. The boundary and initial conditions give

$$u = \sum_{n=1}^{\infty} A_n \cos nt \, \sin nx.$$

Imposing

$$u_t(x, 0) = \sin x = \sum_{n=1}^{\infty} B_n na \sin nx$$

603

gives

$$B_1 = \frac{1}{a^2}, \quad \text{and} \quad B_n = 0$$

for $n = 2, 3, 4, \ldots$ so that

$$u(x,t) = \frac{1}{a} \sin at \, \sin x.$$

6. Using $u = XT$ and $-\lambda$ as a separation constant we obtain

$$X'' + \lambda X = 0,$$

$$X(0) = 0,$$

$$X(1) = 0,$$

and

$$T'' + \lambda a^2 T = 0,$$

$$T'(0) = 0.$$

Solving the differential equations we get

$$X = c_1 \sin n\pi x + c_2 \cos n\pi x \quad \text{and} \quad T = c_3 \cos n\pi at + c_4 \sin n\pi at$$

for $n = 1, 2, 3, \ldots$. The boundary and initial conditions give

$$u = \sum_{n=1}^{\infty} A_n \cos nt \, \sin nx.$$

Imposing

$$u(x,0) = 0.01 \sin 3\pi x = \sum_{n=1}^{\infty} A_n \sin n\pi x$$

gives $A_3 = 0.01$, and $A_n = 0$ for $n = 1, 2, 4, 5, 6, \ldots$ so that

$$u(x,t) = 0.01 \sin 3\pi x \, \cos 3\pi at.$$

7. Using $u = XT$ and $-\lambda$ as a separation constant we obtain

$$X'' + \lambda X = 0,$$

$$X(0) = 0,$$

$$X(L) = 0,$$

and

$$T'' + \lambda a^2 T = 0,$$

$$T'(0) = 0.$$

Solving the differential equations we get

$$X = c_1 \sin \frac{n\pi}{L} x + c_2 \cos \frac{n\pi}{L} x \quad \text{and} \quad T = c_3 \cos \frac{n\pi a}{L} t + c_4 \sin \frac{n\pi a}{L} t$$

for $n = 1, 2, 3, \ldots$. The boundary and initial conditions give

$$u = \sum_{n=1}^{\infty} A_n \cos \frac{n\pi a}{L} t \, \sin \frac{n\pi}{L} x.$$

Imposing

$$u(x, 0) = \sum_{n=1}^{\infty} A_n \sin \frac{n\pi}{L} x$$

gives

$$A_n = \frac{8h}{n^2 \pi^2} \sin \frac{n\pi}{2}$$

for $n = 1, 2, 3, \ldots$ so that

$$u(x, t) = \frac{8h}{\pi^2} \sum_{n=1}^{\infty} \frac{1}{n^2} \sin \frac{n\pi}{2} \, \cos \frac{n\pi a}{L} t \, \sin \frac{n\pi}{L} x.$$

8. Using $u = XT$ and $-\lambda$ as a separation constant we obtain

$$X'' + \lambda X = 0,$$

$$X'(0) = 0,$$

$$X'(L) = 0,$$

and

$$T'' + \lambda a^2 T = 0,$$

$$T'(0) = 0.$$

Solving the differential equations we get

$$X = c_1 \sin \frac{n\pi}{L} x + c_2 \cos \frac{n\pi}{L} x \qquad \text{and} \qquad T = c_3 \cos \frac{n\pi a}{L} t + c_4 \sin \frac{n\pi a}{L} t$$

for $n = 1, 2, 3, \ldots$. The boundary and initial conditions, together with the fact that $\lambda = 0$ is an eigenvalue with eigenfunction $X(x) = 1$, give

$$u = A_0 + \sum_{n=1}^{\infty} A_n \cos \frac{n\pi a}{L} t \, \sin \frac{n\pi}{L} x.$$

Imposing

$$u(x, 0) = x = A_0 + \sum_{n=1}^{\infty} A_n \cos \frac{n\pi}{L} x$$

gives

$$A_0 = \frac{1}{L} \int_0^L x \, dx = \frac{L}{2}$$

and

$$A_n = \frac{2}{L} \int_0^L x \cos \frac{n\pi}{L} x \, dx = \frac{2L}{n^2 \pi^2} [(-1)^n - 1]$$

for $n = 1, 2, 3, \ldots$, so that

$$u(x,t) = \frac{L}{2} + \frac{2L}{\pi^2} \sum_{n=1}^{\infty} \frac{(-1)^n - 1}{n^2} \cos \frac{n\pi a}{L} t \cos \frac{n\pi}{L} x.$$

9. Using $u = XT$ and $-\lambda$ as a separation constant we obtain

$$X'' + \lambda X = 0,$$

$$X(0) = 0,$$

$$X(\pi) = 0,$$

and

$$T'' + 2\beta T' + \lambda T = 0,$$

$$T'(0) = 0.$$

Solving the differential equations we get

$$X = c_1 \sin nx + c_2 \cos nx \qquad \text{and} \qquad T = e^{-\beta t} \left(c_3 \cos \sqrt{n^2 - \beta^2}\, t + c_4 \sin \sqrt{n^2 - \beta^2}\, t \right)$$

The boundary conditions on X imply $c_2 = 0$ so

$$X = c_1 \sin nx \qquad \text{and} \qquad T = e^{-\beta t} \left(c_3 \cos \sqrt{n^2 - \beta^2}\, t + c_4 \sin \sqrt{n^2 - \beta^2}\, t \right)$$

and

$$u = \sum_{n=1}^{\infty} e^{-\beta t} \left(A_n \cos \sqrt{n^2 - \beta^2}\, t + B_n \sin \sqrt{n^2 - \beta^2}\, t \right) \sin nx.$$

Imposing

$$u(x,0) = f(x) = \sum_{n=1}^{\infty} A_n \sin nx$$

and

$$u_t(x,0) = 0 = \sum_{n=1}^{\infty} \left(B_n \sqrt{n^2 - \beta^2} - \beta A_n \right) \sin nx$$

gives

$$u(x,t) = e^{-\beta t} \sum_{n=1}^{\infty} A_n \left(\cos \sqrt{n^2 - \beta^2}\, t + \frac{\beta}{\sqrt{n^2 - \beta^2}} \sin \sqrt{n^2 - \beta^2}\, t \right) \sin nx,$$

where

$$A_n = \frac{2}{\pi} \int_0^{\pi} f(x) \sin nx \, dx.$$

10. Using $u = XT$ and $-\lambda =$ as a separation constant leads to $X'' + \alpha^2 X = 0$, $X(0) = 0$, $X(\pi) = 0$ and $T'' + (1 + \alpha^2)T = 0$, $T'(0) = 0$. Then $X = c_2 \sin nx$ and $T = c_3 \cos \sqrt{n^2 + 1}\, t$ for $n = 1, 2, 3, \ldots$ so that

$$u = \sum_{n1} B_n \cos \sqrt{n^2 + 1}\, t \sin nx.$$

Imposing $u(x,0) = \sum_{n=1}^{\infty} B_n \sin nx$ gives

$$B_n = \frac{2}{\pi} \int_0^{\pi/2} x \sin nx \, dx + \frac{2}{\pi} \int_{\pi/2}^{\pi} (\pi - x) \sin nx \, dx = \frac{4}{\pi n^2} \sin \frac{n\pi}{2}$$

$$= \begin{cases} 0, & n \text{ even}, \\ \frac{4}{\pi n^2}(-1)^{(n+3)/2}, & n = 2k-1, \ k = 1, 2, 3, \dots. \end{cases}$$

Thus with $n = 2k - 1$,

$$u(x,t) = \frac{4}{\pi} \sum_{n=1}^{\infty} \frac{\sin \frac{n\pi}{2}}{n^2} \cos \sqrt{n^2 + 1}\, t \sin nx = \frac{4}{\pi} \sum_{k=1}^{\infty} \frac{(-1)^{k+1}}{(2k-1)^2} \cos \sqrt{(2k-1)^2 + 1}\, t \sin(2k-1)x.$$

11. Separating variables in the partial differential equation and using the separation constant $-\lambda = \alpha^4$ gives

$$\frac{X^{(4)}}{X} = -\frac{T''}{a^2 T} = \alpha^4$$

so that

$$X^{(4)} - \alpha^4 X = 0$$

$$T'' + a^2 \alpha^4 T = 0$$

and

$$X = c_1 \cosh \alpha x + c_2 \sinh \alpha x + c_3 \cos \alpha x + c_4 \sin \alpha x$$

$$T = c_5 \cos a\alpha^2 t + c_6 \sin a\alpha^2 t.$$

The boundary conditions translate into $X(0) = X(L) = 0$ and $X''(0) = X''(L) = 0$. From $X(0) = X''(0) = 0$ we find $c_1 = c_3 = 0$. From

$$X(L) = c_2 \sinh \alpha L + c_4 \sin \alpha L = 0$$

$$X''(L) = \alpha^2 c_2 \sinh \alpha L - \alpha^2 c_4 \sin \alpha L = 0$$

we see by subtraction that $c_4 \sin \alpha L = 0$. This equation yields the eigenvalues $\alpha = n\pi L$ for $n = 1$, 2, 3, ... The corresponding eigenfunctions are

$$X = c_4 \sin \frac{n\pi}{L} x.$$

Thus

$$u(x,t) = \sum_{n=1}^{\infty} \left(A_n \cos \frac{n^2 \pi^2}{L^2} at + B_n \sin \frac{n^2 \pi^2}{L^2} at \right) \sin \frac{n\pi}{L} x.$$

From

$$u(x,0) = f(x) = \sum_{n=1}^{\infty} A_n \sin \frac{n\pi}{L} x$$

we obtain

$$A_n = \frac{2}{L} \int_0^L f(x) \sin \frac{n\pi}{L} x \, dx.$$

From

$$\frac{\partial u}{\partial t} = \sum_{n=1}^{\infty} \left(-A_n \frac{n^2 \pi^2 a}{L^2} \sin \frac{n^2 \pi^2}{L^2} at + B_n \frac{n^2 \pi^2 a}{L^2} \cos \frac{n^2 \pi^2}{L^2} at \right) \sin \frac{n\pi}{L} x$$

and

$$\frac{\partial u}{\partial t} \bigg|_{t=0} = g(x) = \sum_{n=1}^{\infty} B_n \frac{n^2 \pi^2 a}{L^2} \sin \frac{n\pi}{L} x$$

we obtain

$$B_n \frac{n^2 \pi^2 a}{L^2} = \frac{2}{L} \int_0^L g(x) \sin \frac{n\pi}{L} x \, dx$$

and

$$B_n = \frac{2L}{n^2 \pi^2 a} \int_0^L g(x) \sin \frac{n\pi}{L} x \, dx.$$

12. (a) Write the differential equation in X from Problem 11 as $X^{(4)} - \alpha^4 X = 0$ where the eigenvalues are $\lambda = \alpha^2$. Then

$$X = c_1 \cosh \alpha x + c_2 \sinh \alpha x + c_3 \cos \alpha x + c_4 \sin \alpha x$$

and using $X(0) = 0$ and $X'(0) = 0$ we find $c_3 = -c_1$ and $c_4 = -c_2$. The conditions $X(L) = 0$ and $X'(L) = 0$ yield the system of equations

$$c_1(\cosh \alpha L - \cos \alpha L) + c_2(\sinh \alpha L - \sin \alpha L) = 0$$

$$c_1(\alpha \sinh \alpha L + \alpha \sin \alpha L) + c_2(\alpha \cosh \alpha L - \alpha \cos \alpha L) = 0.$$

In order for this system to have nontrivial solutions the determinant of the coefficients must be zero. That is,

$$\alpha(\cosh \alpha L - \cos \alpha L)^2 - \alpha(\sinh^2 \alpha L - \sin^2 \alpha L) = 0.$$

Since $\alpha = 0$ leads to $X = 0$, $\lambda = \alpha^2 = 0^2 = 0$ is not an eigenvalue. Then, dividing the above equation by α, we have

$$(\cosh \alpha L - \cos \alpha L)^2 - (\sinh^2 \alpha L - \sin^2 \alpha L)$$

$$= \cosh^2 \alpha L - 2 \cosh \alpha L \cos \alpha L + \cos^2 \alpha L - \sinh^2 \alpha L + \sin^2 \alpha L$$

$$= -2 \cosh \alpha L \cos \alpha L + 2 = 0$$

or $\cosh \alpha L \cos \alpha L = 1$. Letting $x = \alpha L$ we see that the eigenvalues are $\lambda_n = \alpha_n^2 = x_n^2 / L^2$ where x_n, $n = 1, 2, 3, \ldots$, are the positive roots of the equation $\cosh x \cos x = 1$.

(b) The equation $\cosh x \cos x = 1$ is the same as $\cos x = \operatorname{sech} x$.
The figure indicates that the equation has an infinite number of roots.

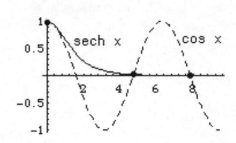

(c) Using a CAS we find the first four positive roots of $\cosh x \cos x = 1$ to be $x_1 = 4.7300$, $x_2 = 7.8532$, $x_3 = 10.9956$, and $x_4 = 14.1372$. Thus the first four eigenvalues are $\lambda_1 = x_1^2/L = 22.3733/L$, $\lambda_2 = x_2^2/L = 61.6728/L$, $\lambda_3 = x_3^2/L = 120.9034/L$, and $\lambda_4 = x_4^2/L = 199.8594/L$.

13. From (8) in the text we have

$$u(x,t) = \sum_{n=1}^{\infty} \left(A_n \cos \frac{n\pi a}{L}t + B_n \sin \frac{n\pi a}{L}t \right) \sin \frac{n\pi}{L}x.$$

Since $u_t(x,0) = g(x) = 0$ we have $B_n = 0$ and

$$u(x,t) = \sum_{n=1}^{\infty} A_n \cos \frac{n\pi a}{L}t \sin \frac{n\pi}{L}x$$

$$= \sum_{n=1}^{\infty} A_n \frac{1}{2} \left[\sin\left(\frac{n\pi}{L}x + \frac{n\pi a}{L}t\right) + \sin\left(\frac{n\pi}{L}x - \frac{n\pi a}{L}t\right) \right]$$

$$= \frac{1}{2} \sum_{n=1}^{\infty} A_n \left[\sin \frac{n\pi}{L}(x + at) + \sin \frac{n\pi}{L}(x - at) \right].$$

From

$$u(x,0) = f(x) = \sum_{n=1}^{\infty} A_n \sin \frac{n\pi}{L}x$$

we identify

$$f(x + at) = \sum_{n=1}^{\infty} A_n \sin \frac{n\pi}{L}(x + at)$$

and

$$f(x - at) = \sum_{n=1}^{\infty} A_n \sin \frac{n\pi}{L}(x - at),$$

so that

$$u(x,t) = \frac{1}{2}[f(x + at) + f(x - at)].$$

14. (a) We note that $\xi_x = \eta_x = 1$, $\xi_t = a$, and $\eta_t = -a$. Then

$$\frac{\partial u}{\partial x} = \frac{\partial u}{\partial \xi}\frac{\partial \xi}{\partial x} + \frac{\partial u}{\partial \eta}\frac{\partial \eta}{\partial x} = u_\xi + u_\eta$$

and

$$\frac{\partial^2 u}{\partial x^2} = \frac{\partial}{\partial x}(u_\xi + u_\eta) = \frac{\partial u_\xi}{\partial \xi}\frac{\partial \xi}{\partial x} + \frac{\partial u_\xi}{\partial \eta}\frac{\partial \eta}{\partial x} + \frac{\partial u_\eta}{\partial \xi}\frac{\partial \xi}{\partial x} + \frac{\partial u_\eta}{\partial \eta}\frac{\partial \eta}{\partial x}$$

$$= u_{\xi\xi} + 2u_{\xi\eta} + u_{\eta\eta}.$$

Similarly

$$\frac{\partial^2 u}{\partial t^2} = a^2(u_{\xi\xi} - 2u_{\xi\eta} + u_{\eta\eta}).$$

Thus

$$a^2\frac{\partial^2 u}{\partial x^2} = \frac{\partial^2 u}{\partial t^2} \quad \text{becomes} \quad \frac{\partial^2 u}{\partial \xi \partial \eta} = 0.$$

(b) Integrating

$$\frac{\partial^2 u}{\partial \xi \partial \eta} = \frac{\partial}{\partial \eta}u_\xi = 0$$

we obtain

$$\int \frac{\partial}{\partial \eta}u_\xi \, d\eta = \int 0 \, d\eta$$

$$u_\xi = f(\xi).$$

Integrating this result with respect to ξ we obtain

$$\int \frac{\partial u}{\partial \xi} \, d\xi = \int f(\xi) \, d\xi$$

$$u = F(\xi) + G(\eta).$$

Since $\xi = x + at$ and $\eta = x - at$, we then have

$$u = F(\xi) + G(\eta) = F(x + at) + G(x - at).$$

Next, we have

$$u(x,t) = F(x + at) + G(x - at)$$

$$u(x,0) = F(x) + G(x) = f(x)$$

$$u_t(x,0) = aF'(x) - aG'(x) = g(x)$$

Integrating the last equation with respect to x gives

$$F(x) - G(x) = \frac{1}{a}\int_{x_0}^{x} g(s) \, ds + c_1.$$

Substituting $G(x) = f(x) - F(x)$ we obtain

$$F(x) = \frac{1}{2}f(x) + \frac{1}{2a}\int_{x_0}^{x} g(s) \, ds + c$$

where $c = c_1/2$. Thus

$$G(x) = \frac{1}{2}f(x) - \frac{1}{2a}\int_{x_0}^{x} g(s) \, ds - c.$$

(c) From the expressions for F and G,

$$F(x + at) = \frac{1}{2}f(x + at) + \frac{1}{2a}\int_{x_0}^{x+at} g(s)\,ds + c$$

$$G(x - at) = \frac{1}{2}f(x - at) - \frac{1}{2a}\int_{x_0}^{x-at} g(s)\,ds - c.$$

Thus,

$$u(x, t) = F(x + at) + G(x - at) = \frac{1}{2}[f(x + at) + f(x - at)] + \frac{1}{2a}\int_{x-at}^{x+at} g(s)\,ds.$$

Here we have used $-\int_{x_0}^{x-at} g(s)\,ds = \int_{x-at}^{x_0} g(s)\,ds.$

15. $u(x, t) = \dfrac{1}{2}[\sin(x + at) + \sin(x - at)] + \dfrac{1}{2a}\displaystyle\int_{x-at}^{x+at} ds$

$= \dfrac{1}{2}[\sin x \cos at + \cos x \sin at + \sin x \cos at - \cos x \sin at] + \dfrac{1}{2a}s\,\Big|_{x-at}^{x+at} = \sin x \cos at + t$

16. $u(x, t) = \dfrac{1}{2}\sin(x + at) + \sin(x - at)] + \dfrac{1}{2a}\displaystyle\int_{x-at}^{x+at} \cos s\,ds$

$= \sin x \cos at + \dfrac{1}{2a}[\sin(x + at) - \sin(x - at)] = \sin x \cos at + \dfrac{1}{a}\cos x \sin at$

17. $u(x, t) = 0 + \dfrac{1}{2a}\displaystyle\int_{x-at}^{x+at} \sin 2s\,ds = \dfrac{1}{2a}\left[\dfrac{-\cos(2x + 2at) + \cos(2x - 2at)}{2}\right]$

$= \dfrac{1}{4a}[-\cos 2x \cos 2at + \sin 2x \sin 2at + \cos 2x \cos 2at + \sin 2x \sin 2at] = \dfrac{1}{2a}\sin 2x \sin 2at$

18. $u(x, t) = \dfrac{1}{2}\left[e^{-(x+at)^2} + e^{-(x-at)^2}\right]$

$= \dfrac{1}{2}\left[e^{-(x^2 + 2axt + a^2 t^2)} + e^{-(x^2 - 2axt + a^2 t^2)}\right]$

$= e^{-(x^2 + a^2 t^2)}\left[\dfrac{e^{-2axt} + e^{2axt}}{2}\right] = e^{-(x^2 + a^2 t^2)}\cosh 2axt$

19. (a)

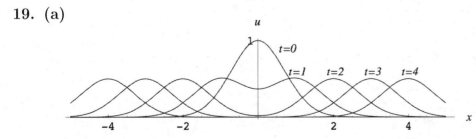

(b)

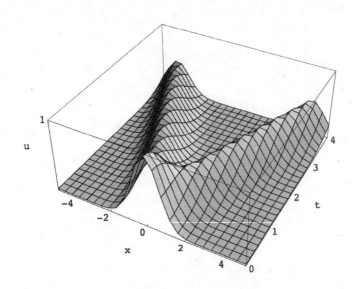

20. (a)

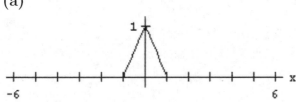

(b) Since $g(x) = 0$, d'Alembert's solution with $a = 1$ is

$$u(x,t) = \frac{1}{2}[f(x+t) + f(x-t)].$$

Sample plots are shown below.

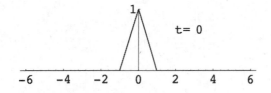

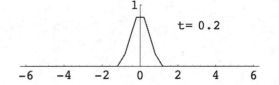

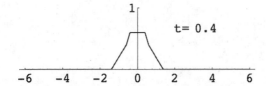

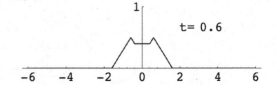

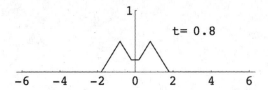

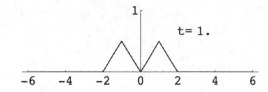

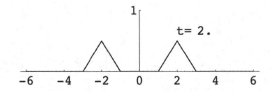

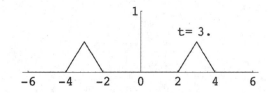

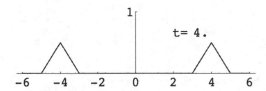

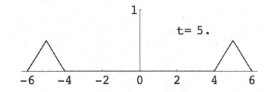

(c) The single peaked wave disolves into two peaks moving outward.

21. (a) With $a = 1$, d'Alembert's solution is

$$u(x,t) = \frac{1}{2} \int_{x-t}^{x+t} g(s)\,ds \qquad \text{where} \qquad g(s) = \begin{cases} 1, & |s| \le 0.1 \\ 0, & |s| > 0.1 \end{cases}.$$

Sample plots are shown below.

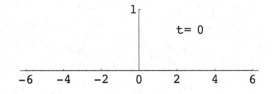

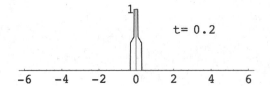

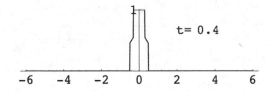

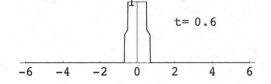

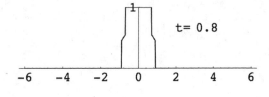

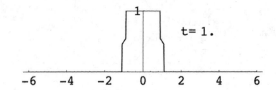

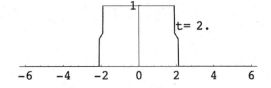

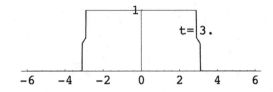

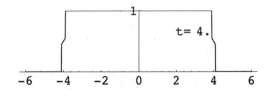

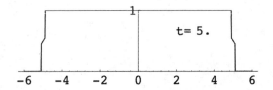

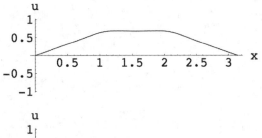

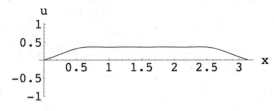

(b) Some frames of the movie are shown in part (a), The string has a roughly rectangular shape with the base on the x-axis increasing in length.

22. (a) and (b) With the given parameters, the solution is

$$u(x,t) = \frac{8}{\pi^2} \sum_{n=1}^{\infty} \frac{1}{n^2} \sin \frac{n\pi}{2} \cos nt \sin nx.$$

For n even, $\sin(n\pi/2) = 0$, so the first six nonzero terms correspond to $n = 1, 3, 5, 7, 9, 11$. In this case $\sin(n\pi/2) = \sin(2p-1)/2 = (-1)^{p+1}$ for $p = 1, 2, 3, 4, 5, 6$, and

$$u(x,t) = \frac{8}{\pi^2} \sum_{p=1}^{\infty} \frac{(-1)^{p+1}}{(2p-1)^2} \cos(2p-1)t \sin(2p-1)x.$$

Frames of the movie corresponding to $t = 0.5, 1, 1.5,$ and 2 are shown.

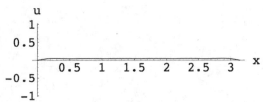

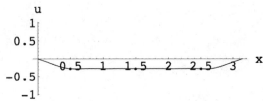

1. Using $u = XY$ and $-\lambda$ as a separation constant we obtain

$$X'' + \lambda X = 0,$$

$$X(0) = 0,$$

$$X(a) = 0,$$

and

$$Y'' - \lambda Y = 0,$$

$$Y(0) = 0.$$

With $\lambda = \alpha^2 > 0$ the solutions of the differential equations are

$$X = c_1 \cos \alpha x + c_2 \sin \alpha x \quad \text{and} \quad Y = c_3 \cosh \alpha y + c_4 \sinh \alpha y$$

The boundary and initial conditions imply

$$X = c_2 \sin \frac{n\pi}{a} x \quad \text{and} \quad Y = c_4 \sinh \frac{n\pi}{a} y$$

for $n = 1, 2, 3, \ldots$ so that

$$u = \sum_{n=1}^{\infty} A_n \sin \frac{n\pi}{a} x \sinh \frac{n\pi}{a} y.$$

Imposing

$$u(x, b) = f(x) = \sum_{n=1}^{\infty} A_n \sinh \frac{n\pi b}{a} \sin \frac{n\pi}{a} x$$

gives

$$A_n \sinh \frac{n\pi b}{a} = \frac{2}{a} \int_0^a f(x) \sin \frac{n\pi}{a} x \, dx$$

so that

$$u(x, y) = \sum_{n=1}^{\infty} A_n \sin \frac{n\pi}{a} x \sinh \frac{n\pi}{a} y$$

where

$$A_n = \frac{2}{a} \operatorname{csch} \frac{n\pi b}{a} \int_0^a f(x) \sin \frac{n\pi}{a} x \, dx.$$

2. Using $u = XY$ and $-\lambda$ as a separation constant we obtain

$$X'' + \lambda X = 0,$$

$$X(0) = 0,$$

$$X(a) = 0,$$

and

$$Y'' - \lambda Y = 0,$$

$$Y'(0) = 0.$$

With $\lambda = \alpha^2 > 0$ the solutions of the differential equations are

$$X = c_1 \cos \alpha x + c_2 \sin \alpha x \quad \text{and} \quad Y = c_3 \cosh \alpha y + c_4 \sinh \alpha y$$

The boundary and initial conditions imply

$$X = c_2 \sin \frac{n\pi}{a} x \quad \text{and} \quad Y = c_3 \cosh \frac{n\pi}{a} y$$

for $n = 1, 2, 3, \ldots$ so that

$$u = \sum_{n=1}^{\infty} A_n \sin \frac{n\pi}{a} x \cosh \frac{n\pi}{a} y.$$

Imposing

$$u(x, b) = f(x) = \sum_{n=1}^{\infty} A_n \cosh \frac{n\pi b}{a} \sin \frac{n\pi}{a} x$$

gives

$$A_n \cosh \frac{n\pi b}{a} = \frac{2}{a} \int_0^a f(x) \sin \frac{n\pi}{a} x \, dx$$

so that

$$u(x, y) = \sum_{n=1}^{\infty} A_n \sin \frac{n\pi}{a} x \cosh \frac{n\pi}{a} y$$

where

$$A_n = \frac{2}{a} \operatorname{sech} \frac{n\pi b}{a} \int_0^a f(x) \sin \frac{n\pi}{a} x \, dx.$$

3. Using $u = XY$ and $-\lambda$ as a separation constant we obtain

$$X'' + \lambda X = 0,$$

$$X(0) = 0,$$

$$X(a) = 0,$$

and

$$Y'' - \lambda Y = 0,$$

$$Y(b) = 0.$$

With $\lambda = \alpha^2 > 0$ the solutions of the differential equations are

$$X = c_1 \cos \alpha x + c_2 \sin \alpha x \quad \text{and} \quad Y = c_3 \cosh \alpha y + c_4 \sinh \alpha y$$

The boundary and initial conditions imply

$$X = c_2 \sin \frac{n\pi}{a} x \quad \text{and} \quad Y = c_2 \cosh \frac{n\pi}{a} y - c_2 \frac{\cosh \frac{n\pi b}{a}}{\sinh \frac{n\pi b}{a}} \sinh \frac{n\pi}{a} y$$

for $n = 1, 2, 3, \ldots$ so that

$$u = \sum_{n=1}^{\infty} A_n \left(\cosh \frac{n\pi}{a} y - \frac{\cosh \frac{n\pi b}{a}}{\sinh \frac{n\pi b}{a}} \sinh \frac{n\pi}{a} y \right) \sin \frac{n\pi}{a} x.$$

Imposing

$$u(x, 0) = f(x) = \sum_{n=1}^{\infty} A_n \sin \frac{n\pi}{a} x$$

gives

$$A_n = \frac{2}{a} \int_0^a f(x) \sin \frac{n\pi}{a} x \, dx$$

so that

$$u(x, y) = \frac{2}{a} \sum_{n=1}^{\infty} \left(\int_0^a f(x) \sin \frac{n\pi}{a} x \, dx \right) \left(\cosh \frac{n\pi}{a} y - \frac{\cosh \frac{n\pi b}{a}}{\sinh \frac{n\pi b}{a}} \sinh \frac{n\pi}{a} y \right) \sin \frac{n\pi}{a} x.$$

4. Using $u = XY$ and $-\lambda$ as a separation constant we obtain

$$X'' + \lambda X = 0,$$

$$X'(0) = 0,$$

$$X'(a) = 0,$$

and

$$Y'' - \lambda Y = 0,$$

$$Y(b) = 0.$$

With $\lambda = \alpha^2 > 0$ the solutions of the differential equations are

$$X = c_1 \cos \alpha x + c_2 \sin \alpha x \quad \text{and} \quad Y = c_3 \cosh \alpha y + c_4 \sinh \alpha y$$

The boundary and initial conditions imply

$$X = c_1 \cos \frac{n\pi}{a} x \quad \text{and} \quad Y = c_3 \cosh \frac{n\pi}{a} y - c_3 \frac{\cosh \frac{n\pi b}{a}}{\sinh \frac{n\pi b}{a}} \sinh \frac{n\pi}{a} y$$

for $n = 1, 2, 3, \ldots$. Since $\lambda = 0$ is an eigenvalue for both differential equations with corresponding eigenfunctions 1 and $y - b$, respectively we have

$$u = A_0(y - b) + \sum_{n=1}^{\infty} A_n \cos \frac{n\pi}{a} x \left(\cosh \frac{n\pi}{a} y - \frac{\cosh \frac{n\pi b}{a}}{\sinh \frac{n\pi b}{a}} \sinh \frac{n\pi}{a} y \right).$$

Imposing

$$u(x, 0) = x = -A_0 b + \sum_{n=1}^{\infty} A_n \cos \frac{n\pi}{a} x$$

617

gives

$$-A_0 b = \frac{1}{a} \int_0^a x\, dx = \frac{1}{2}a$$

and

$$A_n = \frac{2}{a} \int_0^a x \cos \frac{n\pi}{a} x\, dx = \frac{2a}{n^2\pi^2}[(-1)^n - 1]$$

so that

$$u(x,y) = \frac{a}{2b}(b-y) + \frac{2a}{\pi^2} \sum_{n=1}^{\infty} \frac{(-1)^n - 1}{n^2} \cos \frac{n\pi}{a} x \left(\cosh \frac{n\pi}{a} y - \frac{\cosh \frac{n\pi b}{a}}{\sinh \frac{n\pi b}{a}} \sinh \frac{n\pi}{a} y \right).$$

5. Using $u = XY$ and $-\lambda$ as a separation constant we obtain

$$X'' + \lambda X = 0,$$

$$X'(0) = 0,$$

$$X'(a) = 0,$$

and

$$Y'' - \lambda Y = 0,$$

$$Y(b) = 0.$$

With $\lambda = -\alpha^2 < 0$ the solutions of the differential equations are

$$X = c_1 \cosh \alpha x + c_2 \sinh \alpha x \quad \text{and} \quad Y = c_3 \cos \alpha y + c_4 \sin \alpha y$$

for $n = 1, 2, 3 \ldots$. The boundary and initial conditions imply

$$X = c_2 \sinh n\pi x \quad \text{and} \quad Y = c_3 \cos n\pi y$$

for $n = 1, 2, 3, \ldots$. Since $\lambda = 0$ is an eigenvalue for the differential equation in X with corresponding eigenfunction x we have

$$u = A_0 x + \sum_{n=1}^{\infty} A_n \sinh n\pi x \cos n\pi y.$$

Imposing

$$u(1,y) = 1 - y = A_0 + \sum_{n=1}^{\infty} A_n \sinh n\pi \cos n\pi y$$

gives

$$A_0 = \int_0^1 (1-y)\, dy$$

and

$$A_n \sinh n\pi = 2 \int_0^1 (1-y) \cos n\pi y = \frac{2[1 - (-1)^n]}{n^2\pi^2}$$

for $n = 1, 2, 3, \ldots$ so that

$$u(x,y) = \frac{1}{2}x + \frac{2}{\pi^2} \sum_{n=1}^{\infty} \frac{1 - (-1)^n}{n^2 \sinh n\pi} \sinh n\pi x \cos n\pi y.$$

6. Using $u = XY$ and $-\lambda$ as a separation constant we obtain

$$X'' + \lambda X = 0,$$

$$X'(1) = 0$$

and

$$Y'' - \lambda Y = 0,$$

$$Y'(0) = 0,$$

$$Y'(\pi) = 0.$$

With $\lambda = \alpha^2 < 0$ the solutions of the differential equations are

$$X = c_1 \cosh \alpha x + c_2 \sinh \alpha x \quad \text{and} \quad Y = c_3 \cos \alpha y + c_4 \sin \alpha y$$

The boundary and initial conditions imply

$$X = c_1 \cosh nx - c_1 \frac{\sinh n}{\cosh n} \sinh nx \quad \text{and} \quad Y = c_3 \cos ny$$

for $n = 1, 2, 3, \ldots$. Since $\lambda = 0$ is an eigenvalue for both differential equations with corresponding eigenfunctions 1 and 1 we have

$$u = A_0 + \sum_{n=1}^{\infty} A_n \left(\cosh nx - \frac{\sinh n}{\cosh n} \sinh nx \right) \cos ny.$$

Imposing

$$u(0, y) = g(y) = A_0 + \sum_{n=1}^{\infty} A_n \cos ny$$

gives

$$A_0 = \frac{1}{\pi} \int_0^{\pi} g(y) \, dy \quad \text{and} \quad A_n = \frac{2}{\pi} \int_0^{\pi} g(y) \cos ny \, dy$$

for $n = 1, 2, 3, \ldots$ so that

$$u(x, y) = \frac{1}{\pi} \int_0^{\pi} g(y) \, dy + \sum_{n=1}^{\infty} \left(\frac{2}{\pi} \int_0^{\pi} g(y) \cos ny \, dy \right) \left(\cosh nx - \frac{\sinh n}{\cosh n} \sinh nx \right) \cos ny.$$

7. Using $u = XY$ and $-\lambda$ as a separation constant we obtain

$$X'' + \lambda X = 0,$$

$$X'(0) = X(0)$$

and

$$Y'' - \lambda Y = 0,$$

$$Y(0) = 0,$$

$$Y(\pi) = 0.$$

619

With $\lambda = \alpha^2 < 0$ the solutions of the differential equations are

$$X = c_1 \cosh \alpha x + c_2 \sinh \alpha x \qquad \text{and} \qquad Y = c_3 \cos \alpha y + c_4 \sin \alpha y$$

The boundary and initial conditions imply

$$Y = c_4 \sin ny \qquad \text{and} \qquad X = c_2(n \cosh nx + \sinh nx)$$

for $n = 1, 2, 3, \ldots$ so that

$$u = \sum_{n=1}^{\infty} A_n(n \cosh nx + \sinh nx) \sin ny.$$

Imposing

$$u(\pi, y) = 1 = \sum_{n=1}^{\infty} A_n(n \cosh n\pi + \sinh n\pi) \sin ny$$

gives

$$A_n(n \cosh n\pi + \sinh n\pi) = \frac{2}{\pi} \int_0^{\pi} \sin ny \, dy = \frac{2[1 - (-1)^n]}{n\pi}$$

for $n = 1, 2, 3, \ldots$ so that

$$u(x, y) = \frac{2}{\pi} \sum_{n=1}^{\infty} \frac{1 - (-1)^n}{n} \frac{n \cosh nx + \sinh nx}{n \cosh n\pi + \sinh n\pi} \sin ny.$$

8. Using $u = XY$ and $-\lambda$ as a separation constant we obtain

$$X'' + \lambda X = 0,$$

$$X(0) = 0,$$

$$X(1) = 0,$$

and

$$Y'' - \lambda Y = 0,$$

$$Y'(0) = Y(0).$$

With $\lambda = \alpha^2 > 0$ the solutions of the differential equations are

$$X = c_1 \cos \alpha x + c_2 \sin \alpha x \qquad \text{and} \qquad Y = c_3 \cosh \alpha y + c_4 \sinh \alpha y$$

The boundary and initial conditions imply

$$X = c_2 \sin n\pi x \qquad \text{and} \qquad Y = c_4(n \cosh n\pi y + \sinh n\pi y)$$

for $n = 1, 2, 3, \ldots$ so that

$$u = \sum_{n=1}^{\infty} A_n(n \cosh n\pi y + \sinh n\pi y) \sin n\pi x.$$

Imposing

$$u(x, 1) = f(x) = \sum_{n=1}^{\infty} A_n(n \cosh n\pi + \sinh n\pi) \sin n\pi x$$

gives

$$A_n(n \cosh n\pi + \sinh n\pi) = \frac{2}{\pi} \int_0^{\pi} f(x) \sin n\pi x \, dx$$

for $n = 1, 2, 3, \ldots$ so that

$$u(x, y) = \sum_{n=1}^{\infty} A_n(n \cosh n\pi y + \sinh n\pi y) \sin n\pi x$$

where

$$A_n = \frac{2}{n\pi \cosh n\pi + \pi \sinh n\pi} \int_0^1 f(x) \sin n\pi x \, dx.$$

9. This boundary-value problem has the form of Problem 1 in this section, with $a = b = 1$, $f(x) = 100$, and $g(x) = 200$. The solution, then, is

$$u(x, y) = \sum_{n=1}^{\infty} (A_n \cosh n\pi y + B_n \sinh n\pi y) \sin n\pi x,$$

where

$$A_n = 2 \int_0^1 100 \sin n\pi x \, dx = 200 \left(\frac{1 - (-1)^n}{n\pi} \right)$$

and

$$B_n = \frac{1}{\sinh n\pi} \left[2 \int_0^1 200 \sin n\pi x \, dx - A_n \cosh n\pi \right]$$

$$= \frac{1}{\sinh n\pi} \left[400 \left(\frac{1 - (-1)^n}{n\pi} \right) - 200 \left(\frac{1 - (-1)^n}{n\pi} \right) \cosh n\pi \right]$$

$$= 200 \left[\frac{1 - (-1)^n}{n\pi} \right] [2 \operatorname{csch} n\pi - \coth n\pi].$$

10. This boundary-value problem has the form of Problem 2 in this section, with $a = 1$ and $b = 1$. Thus, the solution has the form

$$u(x, y) = \sum_{n=1}^{\infty} (A_n \cosh n\pi x + B_n \sinh n\pi x) \sin n\pi y.$$

The boundary condition $u(0, y) = 10y$ implies

$$10y = \sum_{n=1}^{\infty} A_n \sin n\pi y$$

and

$$A_n = \frac{2}{1} \int_0^1 10y \sin n\pi y \, dy = \frac{20}{n\pi} (-1)^{n+1}.$$

The boundary condition $u_x(1, y) = -1$ implies

$$-1 = \sum_{n=1}^{\infty} (n\pi A_n \sinh n\pi + n\pi B_n \cosh n\pi) \sin n\pi y$$

and

$$n\pi A_n \sinh n\pi + n\pi B_n \cosh n\pi = \frac{2}{1} \int_0^1 (-\sin n\pi y)\,dy$$

$$A_n \sinh n\pi + B_n \cos n\pi = -\frac{2}{n\pi}[1 - (-1)^n]$$

$$B_n = \frac{2}{n\pi}[(-1)^n - 1]\operatorname{sech} n\pi - \frac{20}{n\pi}(-1)^{n+1}\tanh n\pi.$$

11. Using $u = XY$ and $-\lambda$ as a separation constant we obtain

$$X'' + \lambda X = 0,$$

$$X(0) = 0,$$

$$X(\pi) = 0,$$

and

$$Y'' - \lambda Y = 0.$$

With $\lambda = \alpha^2 > 0$ the solutions of the differential equations are

$$X = c_1 \cos \alpha x + c_2 \sin \alpha x \qquad \text{and} \qquad Y = c_3 e^{\alpha y} + c_4 e^{-\alpha y}$$

Then the boundedness of u as $y \to \infty$ implies $c_3 = 0$, so $Y = c_4 e^{-ny}$. The boundary conditions at $x = 0$ and $x = \pi$ imply $c_1 = 0$ so $X = c_2 \sin nx$ for $n = 1, 2, 3, \ldots$ and

$$u = \sum_{n=1}^{\infty} A_n e^{-ny} \sin nx.$$

Imposing

$$u(x, 0) = f(x) = \sum_{n=1}^{\infty} A_n \sin nx$$

gives

$$A_n = \frac{2}{\pi} \int_0^{\pi} f(x) \sin nx\,dx$$

so that

$$u(x, y) = \sum_{n=1}^{\infty} \left(\frac{2}{\pi} \int_0^{\pi} f(x) \sin nx\,dx\right) e^{-ny} \sin nx.$$

12. Using $u = XY$ and $-\lambda$ as a separation constant we obtain

$$X'' + \lambda X = 0,$$

$$X'(0) = 0,$$

622

$$X'(\pi) = 0,$$

and

$$Y'' - \lambda Y = 0.$$

With $\lambda = \alpha^2 > 0$ the solutions of the differential equations are

$$X = c_1 \cos \alpha x + c_2 \sin \alpha x \quad \text{and} \quad Y = c_3 e^{\alpha y} + c_4 e^{-\alpha y}$$

The boundary conditions at $x = 0$ and $x = \pi$ imply $c_2 = 0$ so $X = c_1 \cos nx$ for $n = 1, 2, 3, \ldots$. Now the boundedness of u as $y \to \infty$ implies $c_3 = 0$, so $Y = c_4 e^{-ny}$. In this problem $\lambda = 0$ is also an eigenvalue with corresponding eigenfunction 1 so that

$$u = A_0 + \sum_{n=1}^{\infty} A_n e^{-ny} \cos nx.$$

Imposing

$$u(x, 0) = f(x) = A_0 + \sum_{n=1}^{\infty} A_n \cos nx$$

gives

$$A_0 = \frac{1}{\pi} \int_0^{\pi} f(x)\, dx \quad \text{and} \quad A_n = \frac{2}{\pi} \int_0^{\pi} f(x) \cos nx\, dx$$

so that

$$u(x, y) = \frac{1}{\pi} \int_0^{\pi} f(x)\, dx + \sum_{n=1}^{\infty} \left(\frac{2}{\pi} \int_0^{\pi} f(x) \cos nx\, dx \right) e^{-ny} \cos nx.$$

13. Since the boundary conditions at $y = 0$ and $y = b$ are functions of x we choose to separate Laplace's equation as

$$\frac{X''}{X} = -\frac{Y''}{Y} = -\lambda$$

so that

$$X'' + \lambda X = 0$$

$$Y'' - \lambda Y = 0.$$

Then with $\lambda = \alpha^2$ we have

$$X(x) = c_1 \cos \alpha x + c_2 \sin \alpha x$$

$$Y(y) = c_3 \cosh \alpha y + c_4 \sinh \alpha y.$$

Now $X(0) = 0$ gives $c_1 = 0$ and $X(a) = 0$ implies $\sin \alpha a = 0$ or $\alpha = n\pi/a$ for $n = 1, 2, 3, \ldots$. Thus

$$u_n(x, y) = XY = \left(A_n \cosh \frac{n\pi}{a} y + B_n \sinh \frac{n\pi}{a} y \right) \sin \frac{n\pi}{a} x$$

and

$$u(x, y) = \sum_{n=1}^{\infty} \left(A_n \cosh \frac{n\pi}{a} y + B_n \sinh \frac{n\pi}{a} y \right) \sin \frac{n\pi}{a} x. \tag{1}$$

At $y = 0$ we then have

$$f(x) = \sum_{n=1}^{\infty} A_n \sin \frac{n\pi}{a} x$$

and consequently

$$A_n = \frac{2}{a} \int_0^a f(x) \sin \frac{n\pi}{a} x \, dx. \tag{2}$$

At $y = b$,

$$g(y) = \sum_{n=1}^{\infty} \left(A_n \cosh \frac{n\pi}{a} b + B_n \sinh \frac{n\pi}{b} a \right) \sin \frac{n\pi}{a} x$$

indicates that the entire expression in the parentheses is given by

$$A_n \cosh \frac{n\pi}{a} b + B_n \sinh \frac{n\pi}{a} b = \frac{2}{a} \int_0^a g(x) \sin \frac{n\pi}{a} x \, dx.$$

We can now solve for B_n:

$$B_n \sinh \frac{n\pi}{a} b = \frac{2}{a} \int_0^a g(x) \sin \frac{n\pi}{a} x \, dx - A_n \cosh \frac{n\pi}{a} b$$

$$B_n = \frac{1}{\sinh \frac{n\pi}{a} b} \left(\frac{2}{a} \int_0^a g(x) \sin \frac{n\pi}{a} x \, dx - A_n \cosh \frac{n\pi}{a} b \right). \tag{3}$$

A solution to the given boundary-value problem consists of the series (1) with coefficients A_n and B_n given in (2) and (3), respectively.

14. Since the boundary conditions at $x = 0$ and $x = a$ are functions of y we choose to separate Laplace's equation as

$$\frac{X''}{X} = -\frac{Y''}{Y} = -\lambda$$

so that

$$X'' + \lambda X = 0$$

$$Y'' - \lambda Y = 0.$$

Then with $\lambda = -\alpha^2$ we have

$$X(x) = c_1 \cosh \alpha x + c_2 \sinh \alpha x$$

$$Y(y) = c_3 \cos \alpha y + c_4 \sin \alpha y.$$

Now $Y(0) = 0$ gives $c_3 = 0$ and $Y(b) = 0$ implies $\sin \alpha b = 0$ or $\alpha = n\pi/b$ for $n = 1, 2, 3, \ldots$. Thus

$$u_n(x, y) = XY = \left(A_n \cosh \frac{n\pi}{b} x + B_n \sinh \frac{n\pi}{b} x \right) \sin \frac{n\pi}{b} y$$

and

$$u(x, y) = \sum_{n=1}^{\infty} \left(A_n \cosh \frac{n\pi}{b} x + B_n \sinh \frac{n\pi}{b} x \right) \sin \frac{n\pi}{b} y. \tag{4}$$

At $x = 0$ we then have

$$F(y) = \sum_{n=1}^{\infty} A_n \sin \frac{n\pi}{b} y$$

and consequently

$$A_n = \frac{2}{b} \int_0^b F(y) \sin \frac{n\pi}{b} y \, dy. \tag{5}$$

At $x = a$,

$$G(y) = \sum_{n=1}^{\infty} \left(A_n \cosh \frac{n\pi}{b} a + B_n \sinh \frac{n\pi}{b} a \right) \sin \frac{n\pi}{b} y$$

indicates that the entire expression in the parentheses is given by

$$A_n \cosh \frac{n\pi}{b} a + B_n \sinh \frac{n\pi}{b} a = \frac{2}{b} \int_0^b G(y) \sin \frac{n\pi}{b} y \, dy.$$

We can now solve for B_n:

$$B_n \sinh \frac{n\pi}{b} a = \frac{2}{b} \int_0^b G(y) \sin \frac{n\pi}{b} y \, dy - A_n \cosh \frac{n\pi}{b} a$$

$$B_n = \frac{1}{\sinh \frac{n\pi}{b} a} \left(\frac{2}{b} \int_0^b G(y) \sin \frac{n\pi}{b} y \, dy - A_n \cosh \frac{n\pi}{b} a \right). \tag{6}$$

A solution to the given boundary-value problem consists of the series (4) with coefficients A_n and B_n given in (5) and (6), respectively.

15. Referring to the discussion in this section of the text we identify $a = b = \pi$, $f(x) = 0$, $g(x) = 1$, $F(y) = 1$, and $G(y) = 1$. Then $A_n = 0$ and

$$u_1(x, y) = \sum_{n=1}^{\infty} B_n \sinh ny \sin nx$$

where

$$B_n = \frac{2}{\pi \sinh n\pi} \int_0^{\pi} \sin nx \, dx = \frac{2[1 - (-1)^n]}{n\pi \sinh n\pi}.$$

Next

$$u_2(x, y) = \sum_{n=1}^{\infty} (A_n \cosh nx + B_n \sinh nx) \sin ny$$

where

$$A_n = \frac{2}{\pi} \int_0^{\pi} \sin ny \, dy = \frac{2[1 - (-1)^n]}{n\pi}$$

and

$$B_n = \frac{1}{\sinh n\pi} \left(\frac{2}{\pi} \int_0^{\pi} \sin ny \, dy - A_n \cosh n\pi \right)$$

$$= \frac{1}{\sinh n\pi} \left(\frac{2[1 - (-1)^n]}{n\pi} - \frac{2[1 - (-1)^n]}{n\pi} \cosh n\pi \right)$$

$$= \frac{2[1 - (-1)^n]}{n\pi \sinh n\pi} (1 - \cosh n\pi).$$

625

Now

$$A_n \cosh nx + B_n \sinh nx = \frac{2[1-(-1)^n]}{n\pi}\left[\cosh nx + \frac{\sinh nx}{\sinh n\pi}(1-\cosh n\pi)\right]$$

$$= \frac{2[1-(-1)^n]}{n\pi \sinh n\pi}[\cosh nx \sinh n\pi + \sinh nx - \sinh nx \cosh n\pi]$$

$$= \frac{2[1-(-1)^n]}{n\pi \sinh n\pi}[\sinh nx + \sinh n(\pi - x)]$$

and

$$u(x,y) = u_1 + u_2 = \frac{2}{\pi}\sum_{n=1}^{\infty}\frac{1-(-1)^n}{n \sinh n\pi}\sinh ny \sin nx$$

$$+ \frac{2}{\pi}\sum_{n=1}^{\infty}\frac{[1-(-1)^n][\sinh nx + \sinh n(\pi-x)]}{n \sinh n\pi}\sin ny.$$

16. Referring to the discussion in this section of the text we identify $a = b = 2$, $f(x) = 0$,

$$g(x) = \begin{cases} x, & 0 < x < 1 \\ 2-x, & 1 < x < 2, \end{cases}$$

$F(y) = 0$, and $G(y) = y(2-y)$. Then $A_n = 0$ and

$$u_1(x,y) = \sum_{n=1}^{\infty} B_n \sinh \frac{n\pi}{2} y \sin \frac{n\pi}{2} x$$

where

$$B_n = \frac{1}{\sinh n\pi}\int_0^2 g(x)\sin\frac{n\pi}{2}x\,dx$$

$$= \frac{1}{\sinh n\pi}\left(\int_0^1 x\sin\frac{n\pi}{2}x\,dx + \int_1^2 (2-x)\sin\frac{n\pi}{2}x\,dx\right)$$

$$= \frac{8\sin\frac{n\pi}{2}}{n^2\pi^2\sinh n\pi}.$$

Next, since $A_n = 0$ in u_2, we have

$$u_2(x,y) = \sum_{n=1}^{\infty} B_n \sinh \frac{n\pi}{2}x \sin\frac{n\pi}{2}$$

where

$$B_n = \frac{1}{\sinh n\pi}\int_0^b y(2-y)\sin\frac{n\pi}{2}y\,dy = \frac{16[1-(-1)^n]}{n^3\pi^3\sinh n\pi}.$$

Thus

$$u(x,y) = u_1 + u_2 = \frac{8}{\pi^2}\sum_{n=1}^{\infty}\frac{\sin\frac{n\pi}{2}}{n^2\sinh n\pi}\sinh\frac{n\pi}{2}y\sin\frac{n\pi}{2}x$$

$$+ \frac{16}{\pi^3}\sum_{n=1}^{\infty}\frac{[1-(-1)^n]}{n^3\sinh n\pi}\sinh\frac{n\pi}{2}x\sin\frac{n\pi}{2}y.$$

17. (a)

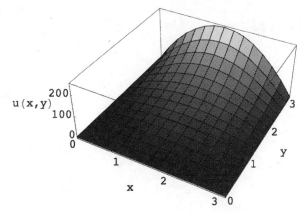

(b) The maximum value occurs at $(\pi/2, \pi)$ and is $f(\pi/2) = 25\pi^2$.

(c) The coefficients are

$$A_n = \frac{2}{\pi} \operatorname{csch} n\pi \int_0^\pi 100 x(\pi - x) \sin nx \, dx$$

$$= \frac{200 \operatorname{csch} n\pi}{\pi} \left[\frac{200}{n^3}\left(1 - (-1)^n\right) \right] = \frac{400}{n^3 \pi}\left[1 - (-1)^n\right] \operatorname{csch} n\pi.$$

See part (a) for the graph.

18. From the figure showing the boundary conditions we see that the maximum value of the temperature is 1 at $(1, 2)$ and $(2, 1)$.

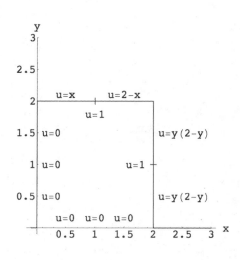

19. (a)

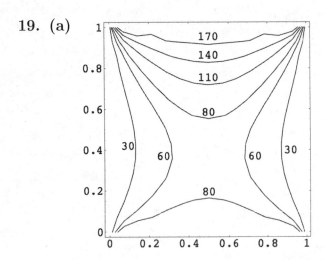

(b)

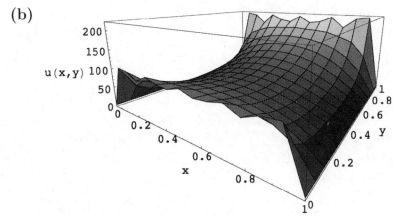

20.

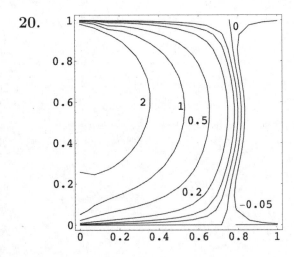

1. Using $v(x,t) = u(x,t) - 100$ we wish to solve $kv_{xx} = v_t$ subject to $v(0,t) = 0$, $v(1,t) = 0$, and $v(x,0) = -100$. Let $v = XT$ and use $-\lambda$ as a separation constant so that

$$X'' + \lambda X = 0,$$

$$X(0) = 0,$$

$$X(1) = 0,$$

and

$$T' + \lambda kT = 0.$$

This leads to

$$X = c_2 \sin(n\pi x) \qquad \text{and} \qquad T = c_3 e^{-kn^2\pi^2 t}$$

for $n = 1, 2, 3, \ldots$ so that

$$v = \sum_{n=1}^{\infty} A_n e^{-kn^2\pi^2 t} \sin n\pi x.$$

Imposing

$$v(x,0) = -100 = \sum_{n=1}^{\infty} A_n \sin n\pi x$$

gives

$$A_n = 2 \int_0^1 (-100) \sin n\pi x \, dx = \frac{-200}{n\pi}[1 - (-1)^n]$$

so that

$$u(x,t) = v(x,t) + 100 = 100 + \frac{200}{\pi} \sum_{n=1}^{\infty} \frac{(-1)^n - 1}{n} e^{-kn^2\pi^2 t} \sin n\pi x.$$

2. Letting $u(x,t) = v(x,t) + \psi(x)$ and proceeding as in Example 1 in the text we find $\psi(x) = u_0 - u_0 x$. Then $v(x,t) = u(x,t) + u_0 x - u_0$ and we wish to solve $kv_{xx} = v_t$ subject to $v(0,t) = 0$, $v(1,t) = 0$, and $v(x,0) = f(x) + u_0 x - u_0$. Let $v = XT$ and use $-\lambda$ as a separation constant so that

$$X'' + \lambda X = 0,$$

$$X(0) = 0,$$

$$X(1) = 0,$$

and

$$T' + \lambda kT = 0.$$

Then

$$X = c_2 \sin n\pi x \qquad \text{and} \qquad T = c_3 e^{-kn^2\pi^2 t}$$

for $n = 1, 2, 3, \ldots$ so that

$$v = \sum_{n=1}^{\infty} A_n e^{-kn^2\pi^2 t} \sin n\pi x.$$

Imposing

$$v(x,0) = f(x) + u_0 x - u_0 = \sum_{n=1}^{\infty} A_n \sin n\pi x$$

gives

$$A_n = 2 \int_0^1 (f(x) + u_0 x - u_0) \sin n\pi x \, dx$$

so that

$$u(x,t) = v(x,t) + u_0 - u_0 x = u_0 - u_0 x + \sum_{n=1}^{\infty} A_n e^{-kn^2\pi^2 t} \sin n\pi x.$$

3. If we let $u(x,t) = v(x,t) + \psi(x)$, then we obtain as in Example 1 in the text

$$k\psi'' + r = 0$$

or

$$\psi(x) = -\frac{r}{2k}x^2 + c_1 x + c_2.$$

The boundary conditions become

$$u(0,t) = v(0,t) + \psi(0) = u_0$$

$$u(1,t) = v(1,t) + \psi(1) = u_0.$$

Letting $\psi(0) = \psi(1) = u_0$ we obtain homogeneous boundary conditions in v:

$$v(0,t) = 0 \qquad \text{and} \qquad v(1,t) = 0.$$

Now $\psi(0) = \psi(1) = u_0$ implies $c_2 = u_0$ and $c_1 = r/2k$. Thus

$$\psi(x) = -\frac{r}{2k}x^2 + \frac{r}{2k}x + u_0 = u_0 - \frac{r}{2k}x(x-1).$$

To determine $v(x,t)$ we solve

$$k\frac{\partial^2 v}{\partial x^2} = \frac{\partial v}{dt}, \quad 0 < x < 1, \ t > 0$$

$$v(0,t) = 0, \quad v(1,t) = 0,$$

$$v(x,0) = \frac{r}{2k}x(x-1) - u_0.$$

Separating variables, we find

$$v(x,t) = \sum_{n=1}^{\infty} A_n e^{-kn^2\pi^2 t} \sin n\pi x,$$

where

$$A_n = 2 \int_0^1 \left[\frac{r}{2k} x(x-1) - u_0 \right] \sin n\pi x \, dx = 2 \left[\frac{u_0}{n\pi} + \frac{r}{kn^3\pi^3} \right] [(-1)^n - 1]. \tag{1}$$

Hence, a solution of the original problem is

$$u(x,t) = \psi(x) + v(x,t)$$

$$= u_0 - \frac{r}{2k} x(x-1) + \sum_{n=1}^{\infty} A_n e^{-kn^2\pi^2 t} \sin n\pi x,$$

where A_n is defined in (1).

4. If we let $u(x,t) = v(x,t) + \psi(x)$, then we obtain as in Example 1 in the text

$$k\psi'' + r = 0.$$

Integrating gives

$$\psi(x) = -\frac{r}{2k} x^2 + c_1 x + c_2.$$

The boundary conditions become

$$u(0,t) = v(0,t) + \psi(0) = u_0$$

$$u(1,t) = v(1,t) + \psi(1) = u_1.$$

Letting $\psi(0) = u_0$ and $\psi(1) = u_1$ we obtain homogeneous boundary conditions in v:

$$v(0,t) = 0 \quad \text{and} \quad v(1,t) = 0.$$

Now $\psi(0) = u_0$ and $\psi(1) = u_1$ imply $c_2 = u_0$ and $c_1 = u_1 - u_0 + r/2k$. Thus

$$\psi(x) = -\frac{r}{2k} x^2 + \left(u_1 - u_0 + \frac{r}{2k} \right) x + u_0.$$

To determine $v(x,t)$ we solve

$$k \frac{\partial^2 v}{\partial x^2} = \frac{\partial v}{dt}, \quad 0 < x < 1, \ t > 0$$

$$v(0,t) = 0, \quad v(1,t) = 0,$$

$$v(x,0) = f(x) - \psi(x).$$

Separating variables, we find

$$v(x,t) = \sum_{n=1}^{\infty} A_n e^{-kn^2\pi^2 t} \sin n\pi x,$$

where

$$A_n = 2 \int_0^1 [f(x) - \psi(x)] \sin n\pi x \, dx. \tag{2}$$

Hence, a solution of the original problem is

$$u(x,t) = \psi(x) + v(x,t)$$

$$= -\frac{r}{2k}x^2 + \left(u_1 - u_0 + \frac{r}{2k}\right)x + u_0 + \sum_{n=1}^{\infty} A_n e^{-kn^2\pi^2 t}\sin n\pi x,$$

where A_n is defined in (2).

5. Substituting $u(x,t) = v(x,t) + \psi(x)$ into the partial differential equation gives

$$k\frac{\partial^2 v}{\partial x^2} + k\psi'' + Ae^{-\beta x} = \frac{\partial v}{\partial t}.$$

This equation will be homogeneous provided ψ satisfies

$$k\psi'' + Ae^{-\beta x} = 0.$$

The solution of this differential equation is obtained by successive integrations:

$$\psi(x) = -\frac{A}{\beta^2 k}e^{-\beta x} + c_1 x + c_2.$$

From $\psi(0) = 0$ and $\psi(1) = 0$ we find

$$c_1 = \frac{A}{\beta^2 k}(e^{-\beta} - 1) \qquad \text{and} \qquad c_2 = \frac{A}{\beta^2 k}.$$

Hence

$$\psi(x) = -\frac{A}{\beta^2 k}e^{-\beta x} + \frac{A}{\beta^2 k}(e^{-\beta} - 1)x + \frac{A}{\beta^2 k}$$

$$= \frac{A}{\beta^2 k}\left[1 - e^{-\beta x} + (e^{-\beta} - 1)x\right].$$

Now the new problem is

$$k\frac{\partial^2 v}{\partial x^2} = \frac{\partial v}{\partial t} \quad , \quad 0 < x < 1, \quad t > 0,$$

$$v(0,t) = 0, \quad v(1,t) = 0, \quad t > 0,$$

$$v(x,0) = f(x) - \psi(x), \quad 0 < x < 1.$$

Identifying this as the heat equation solved in Section 12.3 in the text with $L = 1$ we obtain

$$v(x,t) = \sum_{n=1}^{\infty} A_n e^{-kn^2\pi^2 t}\sin n\pi x$$

where

$$A_n = 2\int_0^1 [f(x) - \psi(x)]\sin n\pi x\,dx.$$

Thus

$$u(x,t) = \frac{A}{\beta^2 k}\left[1 - e^{-\beta x} + (e^{-\beta} - 1)x\right] + \sum_{n=1}^{\infty} A_n e^{-kn^2\pi^2 t}\sin n\pi x.$$

6. Substituting $u(x,t) = v(x,t) + \psi(x)$ into the partial differential equation gives

$$k\frac{\partial^2 v}{\partial x^2} + k\psi'' - hv - h\psi = \frac{\partial v}{\partial t}.$$

This equation will be homogeneous provided ψ satisfies

$$k\psi'' - h\psi = 0.$$

Since k and h are positive, the general solution of this latter linear second-order equation is

$$\psi(x) = c_1 \cosh\sqrt{\frac{h}{k}}\, x + c_2 \sinh\sqrt{\frac{h}{k}}\, x.$$

From $\psi(0) = 0$ and $\psi(\pi) = u_0$ we find $c_1 = 0$ and $c_2 = u_0/\sinh\sqrt{h/k}\,\pi$. Hence

$$\psi(x) = u_0 \frac{\sinh\sqrt{h/k}\, x}{\sinh\sqrt{h/k}\,\pi}.$$

Now the new problem is

$$k\frac{\partial^2 v}{\partial x^2} - hv = \frac{\partial v}{\partial t}, \quad 0 < x < \pi,\ t > 0$$

$$v(0,t) = 0, \quad v(\pi,t) = 0, \quad t > 0$$

$$v(x,0) = -\psi(x), \quad 0 < x < \pi.$$

If we let $v = XT$ then

$$\frac{X''}{X} = \frac{T' + hT}{kT} = -\lambda.$$

With $\lambda = \alpha^2 > 0$, the separated differential equations

$$X'' + \alpha^2 X = 0 \quad \text{and} \quad T' + \left(h + k\alpha^2\right)T = 0.$$

have the respective solutions

$$X(x) = c_3 \cos\alpha x + c_4 \sin\alpha x$$

$$T(t) = c_5 e^{-\left(h + k\alpha^2\right)t}.$$

From $X(0) = 0$ we get $c_3 = 0$ and from $X(\pi) = 0$ we find $\alpha = n$ for $n = 1, 2, 3, \ldots$. Consequently, it follows that

$$v(x,t) = \sum_{n=1}^{\infty} A_n e^{-\left(h + kn^2\right)t}\sin nx$$

where

$$A_n = -\frac{2}{\pi}\int_0^{\pi} \psi(x)\sin nx\, dx.$$

Hence a solution of the original problem is

$$u(x,t) = u_0 \frac{\sinh\sqrt{h/k}\, x}{\sinh\sqrt{h/k}\,\pi} + e^{-ht}\sum_{n=1}^{\infty} A_n e^{-kn^2 t}\sin nx$$

where

$$A_n = -\frac{2}{\pi} \int_0^\pi u_0 \frac{\sinh \sqrt{h/k}\, x}{\sinh \sqrt{h/k}\, \pi} \sin nx\, dx.$$

Using the exponential definition of the hyperbolic sine and integration by parts we find

$$A_n = \frac{2u_0 nk(-1)^n}{\pi\left(h + kn^2\right)}.$$

7. Substituting $u(x, t) = v(x, t) + \psi(x)$ into the partial differential equation gives

$$k\frac{\partial^2 v}{\partial x^2} + k\psi'' - hv - h\psi + hu_0 = \frac{\partial v}{\partial t}.$$

This equation will be homogeneous provided ψ satisfies

$$k\psi'' - h\psi + hu_0 = 0 \qquad \text{or} \qquad k\psi'' - h\psi = -hu_0.$$

This non-homogeneous, linear, second-order, differential equation has solution

$$\psi(x) = c_1 \cosh \sqrt{\frac{h}{k}}\, x + c_2 \sinh \sqrt{\frac{h}{k}}\, x + u_0,$$

where we assume $h > 0$ and $k > 0$. From $\psi(0) = u_0$ and $\psi(1) = 0$ we find $c_1 = 0$ and $c_2 = -u_0/\sinh\sqrt{h/k}$. Thus, the steady-state solution is

$$\psi(x) = -\frac{u_0}{\sinh\sqrt{\frac{h}{k}}} \sinh \sqrt{\frac{h}{k}}\, x + u_0 = u_0\left(1 - \frac{\sinh\sqrt{\frac{h}{k}}\, x}{\sinh\sqrt{\frac{h}{k}}}\right).$$

8. The partial differential equation is

$$k\frac{\partial^2 u}{\partial x^2} - hu = \frac{\partial u}{\partial t}.$$

Substituting $u(x, t) = v(x, t) + \psi(x)$ gives

$$k\frac{\partial^2 v}{\partial x^2} + k\psi'' - hv - h\psi = \frac{\partial v}{\partial t}.$$

This equation will be homogeneous provided ψ satisfies

$$k\psi'' - h\psi = 0.$$

Assuming $h > 0$ and $k > 0$, we have

$$\psi = c_1 e^{\sqrt{h/k}\, x} + c_2 e^{-\sqrt{h/k}\, x},$$

where we have used the exponential form of the solution since the rod is infinite. Now, in order that the steady-state temperature $\psi(x)$ be bounded as $x \to \infty$, we require $c_1 = 0$. Then

$$\psi(x) = c_2 e^{-\sqrt{h/k}\, x}$$

and $\psi(0) = u_0$ implies $c_2 = u_0$. Thus

$$\psi(x) = u_0 e^{-\sqrt{h/k}\,x}.$$

9. Substituting $u(x,t) = v(x,t) + \psi(x)$ into the partial differential equation gives

$$a^2 \frac{\partial^2 v}{\partial x^2} + a^2 \psi'' + Ax = \frac{\partial^2 v}{\partial t^2}.$$

This equation will be homogeneous provided ψ satisfies

$$a^2 \psi'' + Ax = 0.$$

The solution of this differential equation is

$$\psi(x) = -\frac{A}{6a^2} x^3 + c_1 x + c_2.$$

From $\psi(0) = 0$ we obtain $c_2 = 0$, and from $\psi(1) = 0$ we obtain $c_1 = A/6a^2$. Hence

$$\psi(x) = \frac{A}{6a^2}(x - x^3).$$

Now the new problem is

$$a^2 \frac{\partial^2 v}{\partial x^2} = \frac{\partial^2 v}{\partial t^2}$$

$$v(0,t) = 0, \quad v(1,t) = 0, \quad t > 0,$$

$$v(x,0) = -\psi(x), \quad v_t(x,0) = 0, \quad 0 < x < 1.$$

Identifying this as the wave equation solved in Section 12.4 in the text with $L = 1$, $f(x) = -\psi(x)$, and $g(x) = 0$ we obtain

$$v(x,t) = \sum_{n=1}^{\infty} A_n \cos n\pi a t \sin n\pi x$$

where

$$A_n = 2 \int_0^1 [-\psi(x)] \sin n\pi x \, dx = \frac{A}{3a^2} \int_0^1 (x^3 - x) \sin n\pi x \, dx = \frac{2A(-1)^n}{a^2 \pi^3 n^3}.$$

Thus

$$u(x,t) = \frac{A}{6a^2}(x - x^3) + \frac{2A}{a^2 \pi^3} \sum_{n=1}^{\infty} \frac{(-1)^n}{n^3} \cos n\pi a t \sin n\pi x.$$

10. We solve

$$a^2 \frac{\partial^2 u}{\partial x^2} - g = \frac{\partial^2 u}{\partial t^2}, \quad 0 < x < 1, \, t > 0$$

$$u(0,t) = 0, \quad u(1,t) = 0, \quad t > 0$$

$$u(x,0) = 0, \quad \left.\frac{\partial u}{\partial t}\right|_{t=0} = 0, \quad 0 < x < 1.$$

The partial differential equation is nonhomogeneous. The substitution $u(x, t) = v(x, t) + \psi(x)$ yields a homogeneous partial differential equation provided ψ satisfies

$$a^2 \psi'' - g = 0.$$

By integrating twice we find

$$\psi(x) = \frac{g}{2a^2} x^2 + c_1 x + c_2.$$

The imposed conditions $\psi(0) = 0$ and $\psi(1) = 0$ then lead to $c_2 = 0$ and $c_1 = -g/2a^2$. Hence

$$\psi(x) = \frac{g}{2a^2} \left(x^2 - x \right).$$

The new problem is now

$$a^2 \frac{\partial^2 v}{\partial x^2} = \frac{\partial^2 v}{\partial t^2}, \quad 0 < x < 1, \ t > 0$$

$$v(0, t) = 0, \quad v(1, t) = 0$$

$$v(x, 0) = \frac{g}{2a^2} \left(x - x^2 \right), \quad \left. \frac{\partial v}{\partial t} \right|_{t=0} = 0.$$

Substituting $v = XT$ we find in the usual manner

$$X'' + \alpha^2 X = 0$$

$$T'' + a^2 \alpha^2 T = 0$$

with solutions

$$X(x) = c_3 \cos \alpha x + c_4 \sin \alpha x$$

$$T(t) = c_5 \cos a\alpha t + c_6 \sin a\alpha t.$$

The conditions $X(0) = 0$ and $X(1) = 0$ imply in turn that $c_3 = 0$ and $\alpha = n\pi$ for $n = 1, 2, 3, \ldots$. The condition $T'(0) = 0$ implies $c_6 = 0$. Hence, by the Superposition Principle

$$v(x, t) = \sum_{n=1}^{\infty} A_n \cos(an\pi t) \sin(n\pi x).$$

At $t = 0$,

$$\frac{g}{2a^2} \left(x - x^2 \right) = \sum_{n=1}^{\infty} A_n \sin(n\pi x)$$

and so

$$A_n = \frac{g}{a^2} \int_0^1 \left(x - x^2 \right) \sin(n\pi x) \, dx = \frac{2g}{a^2 n^3 \pi^3} \left[1 - (-1)^n \right].$$

Thus the solution to the original problem is

$$u(x, t) = \psi(x) + v(x, t) = \frac{g}{2a^2} \left(x^2 - x \right) + \frac{2g}{a^2 \pi^3} \sum_{n=1}^{\infty} \frac{1 - (-1)^n}{n^3} \cos(an\pi t) \sin(n\pi x).$$

11. Substituting $u(x, y) = v(x, y) + \psi(y)$ into Laplace's equation we obtain

$$\frac{\partial^2 v}{\partial x^2} + \frac{\partial^2 v}{\partial y^2} + \psi''(y) = 0.$$

This equation will be homogeneous provided ψ satisfies $\psi(y) = c_1 y + c_2$. Considering

$$u(x, 0) = v(x, 0) + \psi(0) = u_1$$

$$u(x, 1) = v(x, 1) + \psi(1) = u_0$$

$$u(0, y) = v(0, y) + \psi(y) = 0$$

we require that $\psi(0) = u_1$, $\psi_1 = u_0$ and $v(0, y) = -\psi(y)$. Then $c_1 = u_0 - u_1$ and $c_2 = u_1$. The new boundary-value problem is

$$\frac{\partial^2 v}{\partial x^2} + \frac{\partial^2 v}{\partial y^2} = 0$$

$$v(x, 0) = 0, \quad v(x, 1) = 0,$$

$$v(0, y) = -\psi(y), \quad 0 < y < 1,$$

where $v(x, y)$ is bounded at $x \to \infty$. This problem is similar to Problem 11 in Section 12.5. The solution is

$$v(x, y) = \sum_{n=1}^{\infty} \left(2 \int_0^1 [-\psi(y) \sin n\pi y] \, dy \right) e^{-n\pi x} \sin n\pi y$$

$$= 2 \sum_{n=1}^{\infty} \left[(u_1 - u_0) \int_0^1 y \sin n\pi y \, dy - u_1 \int_0^1 \sin n\pi y \, dy \right] e^{-n\pi x} \sin n\pi y$$

$$= \frac{2}{\pi} \sum_{n=1}^{\infty} \frac{u_0 (-1)^n - u_1}{n} e^{-n\pi x} \sin n\pi y.$$

Thus

$$u(x, y) = v(x, y) + \psi(y)$$

$$= (u_0 - u_1) y + u_1 + \frac{2}{\pi} \sum_{n=1}^{\infty} \frac{u_0 (-1)^n - u_1}{n} e^{-n\pi x} \sin n\pi y.$$

12. Substituting $u(x, y) = v(x, y) + \psi(x)$ into Poisson's equation we obtain

$$\frac{\partial^2 v}{\partial x^2} + \psi''(x) + h + \frac{\partial^2 v}{\partial y^2} = 0.$$

The equation will be homogeneous provided ψ satisfies $\psi''(x) + h = 0$ or $\psi(x) = -\frac{h}{2} x^2 + c_1 x + c_2$. From $\psi(0) = 0$ we obtain $c_2 = 0$. From $\psi(\pi) = 1$ we obtain

$$c_1 = \frac{1}{\pi} + \frac{h\pi}{2}.$$

Then

$$\psi(x) = \left(\frac{1}{\pi} + \frac{h\pi}{2}\right)x - \frac{h}{2}x^2.$$

The new boundary-value problem is

$$\frac{\partial^2 v}{\partial x^2} + \frac{\partial^2 v}{\partial y^2} = 0$$

$$v(0, y) = 0, \quad v(\pi, y) = 0,$$

$$v(x, 0) = -\psi(x), \quad 0 < x < \pi.$$

This is Problem 11 in Section 12.5. The solution is

$$v(x, y) = \sum_{n=1}^{\infty} A_n e^{-ny} \sin nx$$

where

$$A_n = \frac{2}{\pi} \int_0^{\pi} [-\psi(x) \sin nx]\, dx$$

$$= \frac{2(-1)^n}{m}\left(\frac{1}{\pi} + \frac{h\pi}{2}\right) - h(-1)^n\left(\frac{\pi}{n} + \frac{2}{n^2}\right).$$

Thus

$$u(x, y) = v(x, y) + \psi(x) = \left(\frac{1}{\pi} + \frac{h\pi}{2}\right)x - \frac{h}{2}x^2 + \sum_{n=1}^{\infty} A_n e^{-ny} \sin nx.$$

13. With $k = 1$ and $L = \pi$ in Method 2 the eigenfunctions of $X'' + \lambda X = 0$, $X(0) = 0$, $X(\pi) = 0$ are $\sin nx$, $n = 1, 2, 3, \ldots$. Assuming that $u(x, t) = \sum_{n=1}^{\infty} u_n(t) \sin nx$, the formal partial derivatives of u are

$$\frac{\partial^2 u}{\partial x^2} = \sum_{n=1}^{\infty} u_n(t)(-n^2) \sin nx \qquad \text{and} \qquad \frac{\partial u}{\partial t} = \sum_{n=1}^{\infty} u_n'(t) \sin nx.$$

Assuming that $xe^{-3t} = \sum_{n=1}^{\infty} F_n(t) \sin nx$ we have

$$F_n(t) = \frac{2}{\pi} \int_0^{\pi} xe^{-3t} \sin nx\, dx = \frac{2e^{-3t}}{\pi} \int_0^{\pi} x \sin nx\, dx = \frac{2e^{-3t}(-1)^{n+1}}{n}.$$

Then

$$xe^{-3t} = \sum_{n=1}^{\infty} \frac{2e^{-3t}(-1)^{n+1}}{n} \sin nx$$

and

$$u_t - u_{xx} = \sum_{n=1}^{\infty} [u_n'(t) + n^2 u_n(t)] \sin nx = xe^{-3t} = \sum_{n=1}^{\infty} \frac{2e^{-3t}(-1)^{n+1}}{n} \sin nx.$$

Equating coefficients we obtain

$$u_n'(t) + n^2 u_n(t) = \frac{2e^{-3t}(-1)^{n+1}}{n}.$$

This is a linear first-order differential equation whose solution is

$$u_n(t) = \frac{2(-1)^{n+1}}{n(n^2 - 3)}e^{-3t} + C_n e^{-n^2 t}.$$

Thus

$$u(x,t) = \sum_{n=1}^{\infty} \frac{2(-1)^{n+1}}{n(n^2 - 3)}e^{-3t}\sin nx + \sum_{n=1}^{\infty} C_n e^{-n^2 t}\sin nx$$

and $u(x,0) = 0$ implies

$$\sum_{n=1}^{\infty} \frac{2(-1)^{n+1}}{n(n^2 - 3)}\sin nx + \sum_{n=1}^{\infty} C_n \sin nx = 0$$

so that $C_n = 2(-1)^n/n(n^2 - 3)$. Therefore

$$u(x,t) = 2\sum_{n=1}^{\infty} \frac{(-1)^{n+1}}{n(n^2 - 3)}e^{-3t}\sin nx + 2\sum_{n=1}^{\infty} \frac{(-1)^n}{n(n^2 - 3)}e^{-n^2 t}\sin nx.$$

14. With $k = 1$ and $L = \pi$ in Method 2 the eigenfunctions of $X'' + \lambda X = 0$, $X(0) = 0$, $X'(\pi) = 0$ are $1, \cos nx$, $n = 1, 2, 3, \ldots$. Assuming that $u(x,t) = \frac{1}{2}u_0(t) + \sum_{n=1}^{\infty} u_n(t)\cos nx$, the formal partial derivatives of u are

$$\frac{\partial^2 u}{\partial x^2} = \sum_{n=1}^{\infty} u_n(t)(-n^2)\cos nx \qquad \text{and} \qquad \frac{\partial u}{\partial t} = \frac{1}{2}u_0' + \sum_{n=1}^{\infty} u_n'(t)\cos nx.$$

Assuming that $xe^{-3t} = \frac{1}{2}F_0(t) + \sum_{n=1}^{\infty} F_n(t)\cos nx$ we have

$$F_0(t) = \frac{2e^{-3t}}{\pi}\int_0^\pi x\,dx = \pi e^{-3t}$$

and

$$F_n(t) = \frac{2e^{-3t}}{\pi}\int_0^\pi x\cos nx\,dx = \frac{2e^{-3t}[(-1)^n - 1]}{\pi n^2}.$$

Then

$$xe^{-3t} = \frac{\pi}{2}e^{-3t} + \sum_{n=1}^{\infty} \frac{2e^{-3t}[(-1)^n - 1]}{\pi n^2}\cos nx$$

and

$$u_t - u_{xx} = \frac{1}{2}u_0'(t) + \sum_{n=1}^{\infty} [u_n'(t) + n^2 u_n(t)]\cos nx$$

$$= xe^{-3t} = \frac{\pi}{2}e^{-3t} + \sum_{n=1}^{\infty} \frac{2e^{-3t}[(-1)^n - 1]}{\pi n^2}\cos nx.$$

Equating coefficients, we obtain

$$u_0'(t) = \pi e^{-3t} \qquad \text{and} \qquad u_n'(t) + n^2 u_n(t) = \frac{2e^{-3t}[(-1)^n - 1]}{\pi n^2}\cos nx.$$

The first equation yields $u_0(t) = -(\pi/3)e^{-3t} + C_0$ and the second equation, which is a linear first-order differential equation, yields

$$u_n(t) = \frac{2[(-1)^n - 1]}{\pi n^2(n^2 - 3)} e^{-3t} + C_n e^{-n^2 t}.$$

Thus

$$u(x,t) = -\frac{\pi}{3}e^{-3t} + C_0 + \sum_{n=1}^{\infty} \frac{2[(-1)^n - 1]}{\pi n^2(n^2 - 3)} e^{-3t} \cos nx + \sum_{n=1}^{\infty} C_n e^{-n^2 t} \cos nx$$

and $u(x, 0) = 0$ implies

$$-\frac{\pi}{3} + C_0 + \sum_{n=1}^{\infty} \frac{2[(-1)^n - 1]}{\pi n^2(n^2 - 3)} \cos nx + \sum_{n=1}^{\infty} C_n \cos nx = 0$$

so that $C_0 = \pi/3$ and $C_n = 2[(-1)^n - 1]/\pi n^2(n^2 - 3)$. Therefore

$$u(x,t) = \frac{\pi}{3}(1 - e^{-3t}) + \frac{2}{\pi}\sum_{n=1}^{\infty} \frac{(-1)^n - 1}{n^2(n^2 - 3)} e^{-3t} \cos nx + \frac{2}{\pi}\sum_{n=1}^{\infty} \frac{1 - (-1)^n}{n^2(n^2 - 3)} e^{-n^2 t} \cos nx.$$

15. With $k = 1$ and $L = 1$ in Method 2 the eigenfunctions of $X'' + \lambda X = 0$, $X(0) = 0$, $X(1) = 0$ are $\sin n\pi x$, $n = 1, 2, 3, \ldots$. Assuming that $u(x,t) = \sum_{n=1}^{\infty} u_n(t) \sin n\pi x$, the formal partial derivatives of u are

$$\frac{\partial^2 u}{\partial x^2} = \sum_{n=1}^{\infty} u_n(t)(-n^2\pi^2) \sin n\pi x \qquad \text{and} \qquad \frac{\partial u}{\partial t} = \sum_{n=1}^{\infty} u_n'(t) \sin n\pi x.$$

Assuming that $-1 + x - x\cos t = \sum_{n=1}^{\infty} F_n(t) \sin n\pi x$ we have

$$F_n(t) = \frac{2}{1} \int_0^1 (-1 + x - x\cos t) \sin n\pi x \, dx = \frac{2[-1 + (-1)^n \cos t]}{n\pi}.$$

Then

$$-1 + x - x\cos t = \frac{2}{\pi}\sum_{n=1}^{\infty} \frac{-1 + (-1)^n \cos t}{n} \sin n\pi x$$

and

$$u_t - u_{xx} = \sum_{n=1}^{\infty} [u_n'(t) + n^2\pi^2 u_n(t)] \sin n\pi x$$

$$= -1 + x - x\cos t = \frac{2}{\pi}\sum_{n=1}^{\infty} \frac{-1 + (-1)^n \cos t}{n} \sin n\pi x.$$

Equating coefficients we obtain

$$u_n'(t) + n^2\pi^2 u_n(t) = \frac{2[-1 + (-1)^n \cos t]}{n\pi}.$$

This is a linear first-order differential equation whose solution is

$$u_n(t) = \frac{2}{n\pi}\left[-\frac{1}{n^2\pi^2} + (-1)^n \frac{n^2\pi^2 \cos t + \sin t}{n^4\pi^4 + 1}\right] + C_n e^{-n^2\pi^2 t}.$$

Thus

$$u(x,t) = \sum_{n=1}^{\infty} \frac{2}{n\pi} \left[-\frac{1}{n^2\pi^2} + (-1)^n \frac{n^2\pi^2 \cos t + \sin t}{n^4\pi^4 + 1} \right] \sin n\pi x + \sum_{n=1}^{\infty} C_n e^{-n^2\pi^2 t} \sin n\pi x$$

and $u(x,0) = x(1-x)$ implies

$$\sum_{n=1}^{\infty} \frac{2}{n\pi} \left[-\frac{1}{n^2\pi^2} + (-1)^n \frac{n^2\pi^2}{n^4\pi^4 + 1} + C_n \right] \sin n\pi x = x(1-x).$$

Hence

$$\frac{2}{n\pi} \left[-\frac{1}{n^2\pi^2} + (-1)^n \frac{n^2\pi^2}{n^4\pi^4 + 1} + C_n \right] = \frac{2}{1} \int_0^1 x(1-x) \sin n\pi x \, dx = 2 \left[\frac{1 - (-1)^n}{n^3\pi^3} \right]$$

and

$$C_n = \frac{4 - 2(-1)^n}{n^3\pi^3} - (-1)^n \frac{2n\pi}{n^4\pi^4 + 1}.$$

Therefore

$$u(x,t) = \sum_{n=1}^{\infty} \frac{2}{n\pi} \left[-\frac{1}{n^2\pi^2} + (-1)^n \frac{n^2\pi^2 \cos t + \sin t}{n^4\pi^4 + 1} \right] \sin n\pi x$$

$$+ \sum_{n=1}^{\infty} \left[\frac{4 - 2(-1)^n}{n^3\pi^3} - (-1)^n \frac{2n\pi}{n^4\pi^4 + 1} \right] e^{-n^2\pi^2 t} \sin n\pi x.$$

16. With $k = 1$ and $L = \pi$ in Method 2 the eigenfunctions of $X'' + \lambda X = 0$, $X(0) = 0$, $X(\pi) = 0$ are $\sin nx$, $n = 1, 2, 3, \ldots$. Assuming that $u(x,t) = \sum_{n=1}^{\infty} u_n(t) \sin nx$, the formal partial derivatives of u are

$$\frac{\partial^2 u}{\partial x^2} = \sum_{n=1}^{\infty} u_n(t)(-n^2) \sin nx \qquad \text{and} \qquad \frac{\partial^2 u}{\partial t^2} = \sum_{n=1}^{\infty} u_n''(t) \sin nx.$$

Then

$$u_{tt} - u_{xx} = \sum_{n=1}^{\infty} [u_n''(t) + n^2 u_n(t)] \sin nx = \cos t \sin x.$$

Equating coefficients, we obtain $u_1''(t) + u_1(t) \cos t$ and $u_n''(t) + n^2 u_n(t) = 0$ for $n = 2, 3, 4, \ldots$. Solving the first differential equation we obtain $u_1(t) = A_1 \cos t + B_1 \sin t + \frac{1}{2} t \sin t$. From the second differential equation we obtain $u_n(t) = A_n \cos nt + B_n \sin nt$ for $n = 2, 3, 4, \ldots$. Thus

$$u(x,t) = \left(A_1 \cos t + B_1 \sin t + \frac{1}{2} t \sin t \right) \sin x + \sum_{n=2}^{\infty} (A_n \cos nt + B_n \sin nt) \sin nx.$$

From

$$u(x,0) = A_1 \sin x + \sum_{n=2}^{\infty} A_n \sin nx = 0$$

we see that $A_n = 0$ for $n = 1, 2, 3, \ldots$. Thus

$$u(x,t) = \left(B_1 \sin t + \frac{1}{2} t \sin t \right) \sin x + \sum_{n=2}^{\infty} B_n \sin nt \sin nx$$

and

$$\frac{\partial u}{\partial t} = \left(B_1 \cos t + \frac{1}{2} t \cos t + \frac{1}{2} \sin t \right) \sin x + \sum_{n=2}^{\infty} n B_n \cos nt \sin nx,$$

so

$$\left. \frac{\partial u}{\partial t} \right|_{t=0} = B_1 \sin x + \sum_{n=2}^{\infty} n B_n \sin nx = 0.$$

We see that $B_n = 0$ for all n so $u(x,t) = \frac{1}{2} t \sin t \sin x$.

17. (*This is a Contributed Problem and the solution has been provided by the author of the problem.*)

Part a. We start with the homogeneous beam equation

$$\rho \frac{\partial^2 u}{\partial t^2} + \frac{\partial^2}{\partial x^2} \left(EI \frac{\partial^2 u}{\partial x^2} \right) = 0$$

with the boundary conditions

$$\frac{\partial^2 u}{\partial x^2}(0,t) = \frac{\partial^2 u}{\partial x^2}(L,t) = 0, \quad \frac{\partial^3 u}{\partial x^3}(0,t) = \frac{\partial^3 u}{\partial x^3}(L,t) = 0.$$

Plugging in $u(t,x) = T(t)X(x)$ into the equation, we have

$$\rho \ddot{T}(t)X(x) + EIT(t)X''''(x) = 0$$

$$\frac{\rho \ddot{T}}{EIT(t)} + \frac{X''''(x)}{X(x)} = 0$$

$$\frac{\rho \ddot{T}}{EIT(t)} = -\frac{X''''(x)}{X(x)}.$$

Noting that a function of x and a function of t can only be equal as functions if they are both constant functions, we have

$$\frac{\rho \ddot{T}}{EIT(t)} = -\frac{X''''(x)}{X(x)} = -\beta^4.$$

Part a.i. Considering the $T(t)$ part of this equation

$$\frac{\rho \ddot{T}}{EIT(t)} = -\beta^4$$

$$\frac{\rho}{EI} \ddot{T}(t) = -\beta^4 T(t)$$

$$\ddot{T}(t) + \frac{EI\beta^4}{\rho} T(t) = 0.$$

Setting $\omega = \sqrt{EI\beta^4 / \rho}$, we have $\ddot{T}(t) + \omega^2 T(t) = 0$, leading to

$$T(t) = P\cos(\omega t) + Q\sin(\omega t).$$

Part a.ii.

Considering the $X(x)$ part of the separation of variables result, we have

$$\frac{X''''(x)}{X(x)} = \beta^4$$

$$X''''(x) - \beta^4 X(x) = 0.$$

We see then that the solution of the characteristic equation

$$r^4 - \beta^4 = 0$$

$$r = \pm\beta, \ \pm i\beta.$$

so that, using the techniques of Chapter 4, we have

$$X(x) = Ae^{\beta x} + Be^{-\beta x} + C\cos(\beta x) + D\sin(\beta t).$$

Part a.iii.

We now apply the boundary conditions to determine the values of β. The free-free conditions lead to the equations

$$0 = X''(0) = A\beta^2 + B\beta^2 - C\beta^2$$

$$0 = X'''(0) = A\beta^3 - B\beta^3 - D\beta^3$$

$$0 = X''(L) = A\beta^2 e^{\beta L} + B\beta^2 e^{-\beta L} - C\beta^2 \cos(\beta L) - D\beta^2 \sin(\beta L)$$

$$0 = X'''(L) = A\beta^3 e^{\beta L} - B\beta^3 e^{-\beta L} + C\beta^3 \sin(\beta L) - D\beta^3 \cos(\beta L).$$

Simplifying, we have

$$0 = X''(0) = A + B - C$$

$$0 = X'''(0) = A - B - D$$

$$0 = X''(L) = Ae^{\beta L} + Be^{-\beta L} - C\cos(\beta L) - D\sin(\beta L)$$

$$0 = X'''(L) = Ae^{\beta L} - Be^{-\beta L} + C\sin(\beta L) - D\cos(\beta L).$$

The matrix form of this equation system is

$$\begin{bmatrix} 1 & 1 & -1 & 0 \\ 1 & -1 & 0 & -1 \\ e^{\beta L} & e^{-\beta L} & -\cos(\beta L) & -\sin(\beta L) \\ e^{\beta L} & -e^{-\beta L} & \sin(\beta L) & -\cos(\beta L) \end{bmatrix} \begin{bmatrix} A \\ B \\ C \\ D \end{bmatrix} = \begin{bmatrix} 0 \\ 0 \\ 0 \\ 0 \end{bmatrix}.$$

Part a. iv. The determinant of this matrix is $f(a) = -2 + \left(e^{-a} + e^a\right)\cos(a)$, where $a = \beta L$.

First, we note that a=0 is a root of f=0. Thus $\beta_1 = 0$. Below, we plot the determinant function over a sequence of intervals to help identify the roots of f=0.

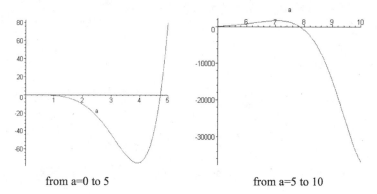

from a=0 to 5   from a=5 to 10

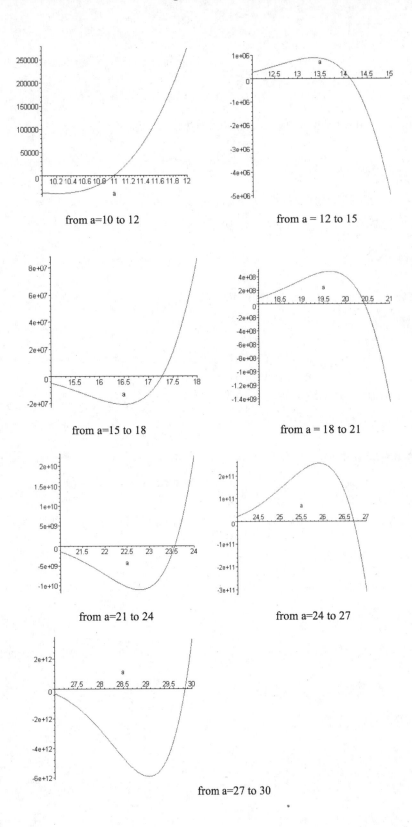

from a=10 to 12

from a = 12 to 15

from a=15 to 18

from a = 18 to 21

from a=21 to 24

from a=24 to 27

from a=27 to 30

The table below gives the first 10 values of $a = \beta L$. To obtain β, divide the tabulated values of a by L.

root number	root
1	0
2	4.730041
3	7.853205
4	10.995608
5	14.137165
6	17.278760
7	20.420352
8	23.561945
9	26.703538
10	29.845130

Part a.v. The equations for the coefficients of the modes $X(x)$ are

$$0 = A + B - C$$
$$0 = A - B - D$$
$$0 = Ae^{\beta L} + Be^{-\beta L} - C\cos(\beta L) - D\sin(\beta L)$$
$$0 = Ae^{\beta L} - Be^{-\beta L} + C\sin(\beta L) - D\cos(\beta L).$$

We begin by writing C and D in terms of A and B:

$$A + B = C$$
$$A - B = D.$$

Plugging these two equations in the third and fourth equation leads to

$$0 = Ae^{\beta L} + Be^{-\beta L} - (A + B)\cos(\beta L) - (A - B)\sin(\beta L)$$
$$0 = Ae^{\beta L} - Be^{-\beta L} + (A + B)\sin(\beta L) - (A - B)\cos(\beta L).$$

Factor to get things in terms of A and B separately:

$$0 = A(e^{\beta L} - \cos(\beta L) - \sin(\beta L)) + B(e^{-\beta L} - \cos(\beta L) + \sin(\beta L)).$$

Since the matrix determinant must be 0, we know that the system has multiple solutions. If we arbitrarily set A= -1, then

$$B = \frac{e^{\beta L} - \cos(\beta L) - \sin(\beta L)}{e^{-\beta L} - \cos(\beta L) + \sin(\beta L)}$$

$$C = -1 + \frac{e^{\beta L} - \cos(\beta L) - \sin(\beta L)}{e^{-\beta L} - \cos(\beta L) + \sin(\beta L)}$$

$$D = -1 - \frac{e^{\beta L} - \cos(\beta L) - \sin(\beta L)}{e^{-\beta L} - \cos(\beta L) + \sin(\beta L)}.$$

The mode shapes are determined using these coefficients:

$$X(x) = Ae^{\beta x} + Be^{-\beta x} + C\cos(\beta x) + D\sin(\beta t).$$

Of course, the first mode shape corresponds to $\beta = 0$, yielding $X_1(x) = 1$

Part b. Mode shape 1 is constant. The mode shapes 2 through 10 are graphed below.

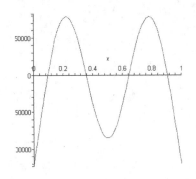

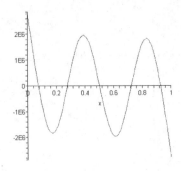

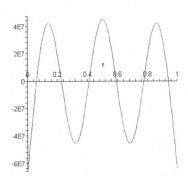

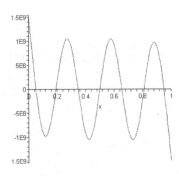

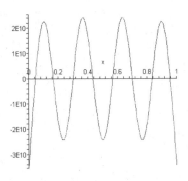

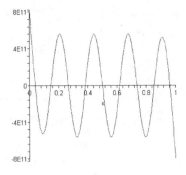

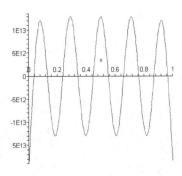

Part c. The forced equation has a solution of the form

$$u(t,x) = \sum_{i=1}^{\infty} T_i(t) X_i(x)$$

in which the mode shapes $X_i(x)$ are as in the Parts a and b. The differential equations for the time problem are

$$\ddot{T}_n(t) + \omega_n^2 T_n(t) = f_n(t)$$

$$f_n(t) = \frac{\displaystyle\int_0^L f(t,x) X_n(x)\,dx}{\displaystyle\int_0^L (X_n(x))^2\,dx} = \frac{F_0 X_n(L/2)}{\displaystyle\int_0^L (X_n(x))^2\,dx}\sin(\alpha t) = F_n \sin(\alpha t)$$

The amplitude F_n is given in the table below.

Term number	Forcing Amplitude Fn
1	F0/L
2	0.01063554295*F0/L
3	0
4	-0.00002386035567*F0/L
5	0
6	00000004429312018*F0/L
7	0
8	-(8.273648032E-11)*F0/L
9	0
10	(1.545038954E-13)*F0/L

With zero initial displacement and velocity, the ordinary differential equation $\ddot{T}_n(t) + \omega_n^2 T_n(t) = F_n \sin(\alpha t)$ has as its solution (see Chapter 4 in the text)

$$T_n(t) = \frac{F_n \alpha}{\omega_n(\alpha^2 - \omega_n^2)}\sin(\omega_n t) - \frac{F_n}{(\alpha^2 - \omega_n^2)}\sin(\alpha t).$$

Thus, the solution of the forced Euler-Bernoulli equation is given by the series

$$u(t,x) = \sum_{n=1}^{\infty} T_n(t) X_n(x)$$

in which the T's are as above and the X's are the mode shapes above. Recall that $\omega_n = \sqrt{EI\beta_n^4/\rho}$ and that the β's, found through the mode shape and boundary conditions, are tabulated above.

Part d. See the figure below. The first 10 modes are used to generate the plot, and the forcing function has a frequency of 5 hz.

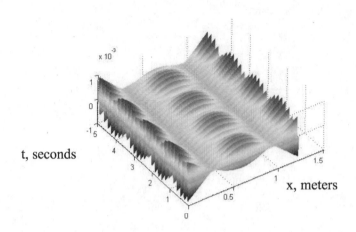

Part e. In the data, we placed an accelerometer at the x=0 end of the beam. We excited the beam at 5hz, 15hz, and 25hz. Each experiment lasted 4 seconds. The graphs below compare the data to the solution of the forced Euler-Bernoulli equation.

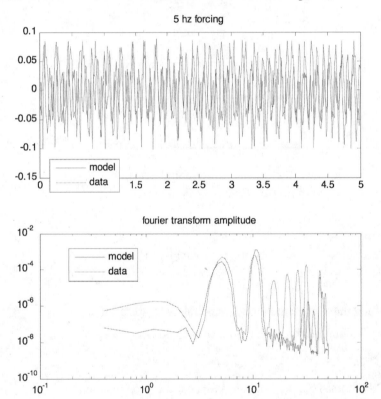

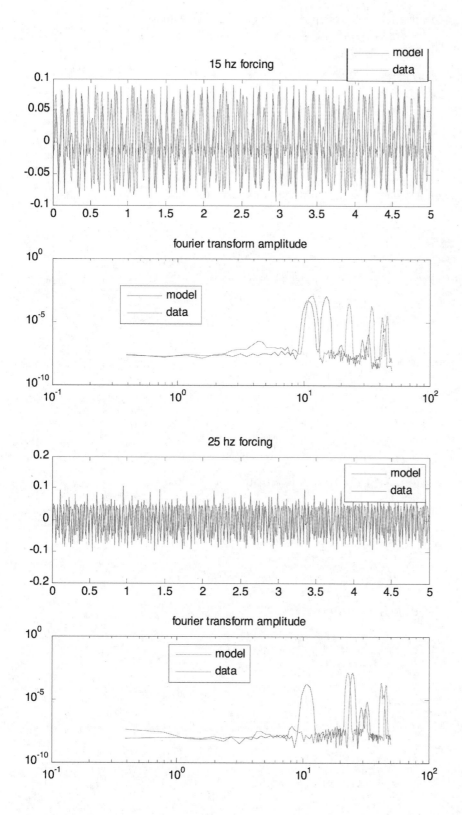

Part f. The Dirac function is an idealization of the clamped piston oscillator. A more appropriate function would be $f(t,x) = \begin{cases} F_0 \sin(\alpha t), & -r < x < r \\ 0, & \text{otherwise.} \end{cases}$.

Using the same techniques applied in Part c, we have

$$\ddot{T}_n(t) + \omega_n^2 T_n(t) = f_n(t)$$

$$f_n(t) = \frac{\int_0^L f(t,x) X_n(x)\,dx}{\int_0^L \left(X_n(x)\right)^2 dx} = \frac{F_0 \int_{-r}^{r} X_n(x)}{\int_0^L \left(X_n(x)\right)^2 dx} \sin(\alpha t) = F_n \sin(\alpha t).$$

With slightly modified amplitudes, the same solution formula as in Part c applies.

Exercises 12.7 — Orthogonal Series Expansions

1. Referring to Example 1 in the text we have

$$X(x) = c_1 \cos \alpha x + c_2 \sin \alpha x$$

and

$$T(t) = c_3 e^{-k\alpha^2 t}.$$

From $X'(0) = 0$ (since the left end of the rod is insulated), we find $c_2 = 0$. Then $X(x) = c_1 \cos \alpha x$ and the other boundary condition $X'(1) = -hX(1)$ implies

$$-\alpha \sin \alpha + h \cos \alpha = 0 \qquad \text{or} \qquad \cot \alpha = \frac{\alpha}{h}.$$

Denoting the consecutive positive roots of this latter equation by α_n for $n = 1, 2, 3, \ldots$, we have

$$u(x,t) = \sum_{n=1}^{\infty} A_n e^{-k\alpha_n^2 t} \cos \alpha_n x.$$

From the initial condition $u(x,0) = 1$ we obtain

$$1 = \sum_{n=1}^{\infty} A_n \cos \alpha_n x$$

and

$$A_n = \frac{\int_0^1 \cos \alpha_n x\, dx}{\int_0^1 \cos^2 \alpha_n x\, dx} = \frac{\sin \alpha_n / \alpha_n}{\frac{1}{2}\left[1 + \frac{1}{2\alpha_n} \sin 2\alpha_n\right]}$$

$$= \frac{2 \sin \alpha_n}{\alpha_n \left[1 + \frac{1}{\alpha_n} \sin \alpha_n \cos \alpha_n\right]} = \frac{2 \sin \alpha_n}{\alpha_n \left[1 + \frac{1}{h\alpha_n} \sin \alpha_n (\alpha_n \sin \alpha_n)\right]}$$

$$= \frac{2h \sin \alpha_n}{\alpha_n [h + \sin^2 \alpha_n]}.$$

The solution is

$$u(x,t) = 2h \sum_{n=1}^{\infty} \frac{\sin \alpha_n}{\alpha_n (h + \sin^2 \alpha_n)} e^{-k\alpha_n^2 t} \cos \alpha_n x.$$

2. Substituting $u(x,t) = v(x,t) + \psi(x)$ into the partial differential equation gives

$$k \frac{\partial^2 v}{\partial x^2} + k\psi'' = \frac{\partial v}{\partial t}.$$

This equation will be homogeneous if $\psi''(x) = 0$ or $\psi(x) = c_1 x + c_2$. The boundary condition $u(0,t) = 0$ implies $\psi(0) = 0$ which implies $c_2 = 0$. Thus $\psi(x) = c_1 x$. Using the second boundary condition we obtain

$$-\left(\frac{\partial v}{\partial x} + \psi' \right) \bigg|_{x=1} = -h[v(1,t) + \psi(1) - u_0],$$

which will be homogeneous when

$$-\psi'(1) = -h\psi(1) + hu_0.$$

Since $\psi(1) = \psi'(1) = c_1$ we have $-c_1 = -hc_1 + hu_0$ and $c_1 = hu_0/(h-1)$. Thus

$$\psi(x) = \frac{hu_0}{h-1} x.$$

The new boundary-value problem is

$$k \frac{\partial^2 v}{\partial x^2} = \frac{\partial v}{\partial t}, \quad 0 < x < 1, \quad t > 0$$

$$v(0,t) = 0, \quad \frac{\partial v}{\partial x} \bigg|_{x=1} = -hv(1,t), \quad h > 0, \quad t > 0$$

$$v(x,0) = f(x) - \frac{hu_0}{h-1} x, \quad 0 < x < 1.$$

Referring to Example 1 in the text we see that

$$v(x,t) = \sum_{n=1}^{\infty} A_n e^{-k\alpha_n^2 t} \sin \alpha_n x$$

and

$$u(x,t) = v(x,t) + \psi(x) = \frac{hu_0}{h-1} x + \sum_{n=1}^{\infty} A_n e^{-k\alpha_n^2 t} \sin \alpha_n x$$

where

$$f(x) - \frac{hu_0}{h-1} x = \sum_{n=1}^{\infty} A_n \sin \alpha_n x$$

651

and α_n is a solution of $\alpha_n \cos \alpha_n = -h \sin \alpha_n$. The coefficients are

$$A_n = \frac{\int_0^1 [f(x) - hu_0 x/(h-1)] \sin \alpha_n x \, dx}{\int_0^1 \sin^2 \alpha_n x \, dx}$$

$$= \frac{\int_0^1 [f(x) - hu_0 x/(h-1)] \sin \alpha_n x \, dx}{\frac{1}{2}\left[1 - \frac{1}{2\alpha_n} \sin 2\alpha_n\right]}$$

$$= \frac{2\int_0^1 [f(x) - hu_0 x/(h-1)] \sin \alpha_n x \, dx}{1 - \frac{1}{\alpha_n} \sin \alpha_n \cos \alpha_n}$$

$$= \frac{2\int_0^1 [f(x) - hu_0 x/(h-1)] \sin \alpha_n x \, dx}{1 - \frac{1}{h\alpha_n}(h \sin \alpha_n) \cos \alpha_n}$$

$$= \frac{2\int_0^1 [f(x) - hu_0 x/(h-1)] \sin \alpha_n x \, dx}{1 - \frac{1}{h\alpha_n}(-\alpha_n \cos \alpha_n) \cos \alpha_n}$$

$$= \frac{2h}{h + \cos^2 \alpha_n} \int_0^1 \left[f(x) - \frac{hu_0}{h-1}x\right] \sin \alpha_n x \, dx.$$

3. Separating variables in Laplace's equation gives

$$X'' + \alpha^2 X = 0$$

$$Y'' - \alpha^2 Y = 0$$

and

$$X(x) = c_1 \cos \alpha x + c_2 \sin \alpha x$$

$$Y(y) = c_3 \cosh \alpha y + c_4 \sinh \alpha y.$$

From $u(0, y) = 0$ we obtain $X(0) = 0$ and $c_1 = 0$. From $u_x(a, y) = -hu(a, y)$ we obtain $X'(a) = -hX(a)$ and

$$\alpha \cos \alpha a = -h \sin \alpha a \qquad \text{or} \qquad \tan \alpha a = -\frac{\alpha}{h}.$$

Let α_n, where $n = 1, 2, 3, \ldots$, be the consecutive positive roots of this equation. From $u(x, 0) = 0$ we obtain $Y(0) = 0$ and $c_3 = 0$. Thus

$$u(x, y) = \sum_{n=1}^{\infty} A_n \sinh \alpha_n y \sin \alpha_n x.$$

Now

$$f(x) = \sum_{n=1}^{\infty} A_n \sinh \alpha_n b \sin \alpha_n x$$

and

$$A_n \sinh \alpha_n b = \frac{\int_0^a f(x) \sin \alpha_n x \, dx}{\int_0^a \sin^2 \alpha_n x \, dx}.$$

Since

$$\int_0^a \sin^2 \alpha_n x \, dx = \frac{1}{2}\left[a - \frac{1}{2\alpha_n}\sin 2\alpha_n a\right] = \frac{1}{2}\left[a - \frac{1}{\alpha_n}\sin \alpha_n a \cos \alpha_n a\right]$$

$$= \frac{1}{2}\left[a - \frac{1}{h\alpha_n}(h\sin\alpha_n a)\cos\alpha_n a\right]$$

$$= \frac{1}{2}\left[a - \frac{1}{h\alpha_n}(-\alpha_n\cos\alpha_n a)\cos\alpha_n a\right] = \frac{1}{2h}\left[ah + \cos^2\alpha_n a\right],$$

we have

$$A_n = \frac{2h}{\sinh\alpha_n b[ah + \cos^2\alpha_n a]}\int_0^a f(x)\sin\alpha_n x\, dx.$$

4. Letting $u(x,y) = X(x)Y(y)$ and separating variables gives

$$X''Y + XY'' = 0.$$

The boundary conditions

$$\frac{\partial u}{\partial y}\bigg|_{y=0} = 0 \qquad \text{and} \qquad \frac{\partial u}{\partial y}\bigg|_{y=1} = -hu(x,1)$$

correspond to

$$X(x)Y'(0) = 0 \qquad \text{and} \qquad X(x)Y'(1) = -hX(x)Y(1)$$

or

$$Y'(0) = 0 \qquad \text{and} \qquad Y'(1) = -hY(1).$$

Since these homogeneous boundary conditions are in terms of Y, we separate the differential equation as

$$\frac{X''}{X} = -\frac{Y''}{Y} = \alpha^2.$$

Then

$$Y'' + \alpha^2 Y = 0$$

and

$$X'' - \alpha^2 X = 0$$

have solutions

$$Y(y) = c_1\cos\alpha y + c_2\sin\alpha y$$

and

$$X(x) = c_3 e^{-\alpha x} + c_4 e^{\alpha x}.$$

We use exponential functions in the solution of $X(x)$ since the interval over which X is defined is infinite. (See the informal rule given in Section 11.4 of the text that discusses when to use the exponential form and when to use the hyperbolic form of the solution of $y'' - \alpha^2 y = 0$.) Now,

Exercises 12.7 Orthogonal Series Expansions

$Y'(0) = 0$ implies $c_2 = 0$, so $Y(y) = c_1 \cos \alpha y$. Since $Y'(y) = -c_1 \alpha \sin \alpha y$, the boundary condition $Y'(1) = -hY(1)$ implies

$$-c_1 \alpha \sin \alpha = -hc_1 \cos \alpha \qquad \text{or} \qquad \cot \alpha = \frac{\alpha}{h}.$$

Consideration of the graphs of $f(\alpha) = \cot \alpha$ and $g(\alpha) = \alpha/h$ show that $\cos \alpha = \alpha h$ has an infinite number of roots. The consecutive positive roots α_n for $n = 1, 2, 3, \ldots$, are the eigenvalues of the problem. The corresponding eigenfunctions are $Y_n(y) = c_1 \cos \alpha_n y$. The condition $\lim_{x \to \infty} u(x, y) = 0$ is equivalent to $\lim_{x \to \infty} X(x) = 0$. Thus $c_4 = 0$ and $X(x) = c_3 e^{-\alpha x}$. Therefore

$$u_n(x, y) = X_n(x)Y_n(x) = A_n e^{-\alpha_n x} \cos \alpha_n y$$

and by the Superposition Principle

$$u(x, y) = \sum_{n=1}^{\infty} A_n e^{-\alpha_n x} \cos \alpha_n y.$$

[It is easily shown that there are no eigenvalues corresponding to $\alpha = 0$.]   Finally, the condition $u(0, y) = u_0$ implies

$$u_0 = \sum_{n=1}^{\infty} A_n \cos \alpha_n y.$$

This is not a Fourier cosine series since the coefficients α_n of y are not integer multiples of π/p, where $p = 1$ in this problem. The functions $\cos \alpha_n y$ are however orthogonal since they are eigenfunctions of the Sturm-Lionville problem

$$Y'' + \alpha^2 Y = 0,$$

$$Y'(0) = 0$$

$$Y'(1) + hY(1) = 0,$$

with weight function $p(x) = 1$. Thus we find

$$A_n = \frac{\int_0^1 u_0 \cos \alpha_n y \, dy}{\int_0^1 \cos^2 \alpha_n y \, dy}.$$

Now

$$\int_0^1 u_0 \cos \alpha_n y \, dy = \frac{u_0}{\alpha_n} \sin \alpha_n y \Big|_0^1 = \frac{u_0}{\alpha_n} \sin \alpha_n$$

and

$$\int_0^1 \cos^2 \alpha_n y \, dy = \frac{1}{2} \int_0^1 (1 + \cos 2\alpha_n y) \, dy = \frac{1}{2} \left[y + \frac{1}{2\alpha_n} \sin 2\alpha_n y \right]_0^1$$

$$= \frac{1}{2} \left[1 + \frac{1}{2\alpha_n} \sin 2\alpha_n \right] = \frac{1}{2} \left[1 + \frac{1}{\alpha_n} \sin \alpha_n \cos \alpha_n \right].$$

Since $\cot \alpha = \alpha/h$,

$$\frac{\cos \alpha}{\alpha} = \frac{\sin \alpha}{h}$$

and

$$\int_0^1 \cos^2 \alpha_n y \, dy = \frac{1}{2}\left[1 + \frac{\sin^2 \alpha_n}{h}\right].$$

Then

$$A_n = \frac{\frac{u_0}{\alpha_n}\sin \alpha_n}{\frac{1}{2}\left[1 + \frac{1}{h}\sin^2 \alpha_n\right]} = \frac{2hu_0 \sin \alpha_n}{\alpha_n\left(h + \sin^2 \alpha_n\right)}$$

and

$$u(x,y) = 2hu_0 \sum_{n=1}^{\infty} \frac{\sin \alpha_n}{\alpha_n\left(h + \sin^2 \alpha_n\right)} e^{-\alpha_n x}\cos \alpha_n y$$

where α_n for $n = 1, 2, 3, \ldots$ are the consecutive positive roots of $\cot \alpha = \alpha/h$.

5. The boundary-value problem is

$$k\frac{\partial^2 u}{\partial x^2} = \frac{\partial u}{\partial t}, \quad 0 < x < L, \quad t > 0,$$

$$u(0,t) = 0, \quad \frac{\partial u}{\partial x}\bigg|_{x=L} = 0, \quad t > 0,$$

$$u(x,0) = f(x), \quad 0 < x < L.$$

Separation of variables leads to

$$X'' + \alpha^2 X = 0$$

$$T' + k\alpha^2 T = 0$$

and

$$X(x) = c_1 \cos \alpha x + c_2 \sin \alpha x$$

$$T(t) = c_3 e^{-k\alpha^2 t}.$$

From $X(0) = 0$ we find $c_1 = 0$. From $X'(L) = 0$ we obtain $\cos \alpha L = 0$ and

$$\alpha = \frac{\pi(2n-1)}{2L}, \quad n = 1, 2, 3, \ldots .$$

Thus

$$u(x,t) = \sum_{n=1}^{\infty} A_n e^{-k(2n-1)^2 \pi^2 t/4L^2} \sin\left(\frac{2n-1}{2L}\right)\pi x$$

where

$$A_n = \frac{\int_0^L f(x)\sin\left(\frac{2n-1}{2L}\right)\pi x \, dx}{\int_0^L \sin^2\left(\frac{2n-1}{2L}\right)\pi x \, dx} = \frac{2}{L}\int_0^L f(x)\sin\left(\frac{2n-1}{2L}\right)\pi x \, dx.$$

6. Substituting $u(x,t) = v(x,t) + \psi(x)$ into the partial differential equation gives

$$a^2 \frac{\partial^2 v}{\partial x^2} + \psi''(x) = \frac{\partial^2 v}{\partial t^2}.$$

This equation will be homogeneous if $\psi''(x) = 0$ or $\psi(x) = c_1 x + c_2$. The boundary condition $u(0,t) = 0$ implies $\psi(0) = 0$ which implies $c_2 = 0$. Thus $\psi(x) = c_1 x$. Using the second boundary condition, we obtain

$$E\left(\frac{\partial v}{\partial x} + \psi'\right)\bigg|_{x=L} = F_0,$$

which will be homogeneous when

$$E\psi'(L) = F_0.$$

Since $\psi'(x) = c_1$ we conclude that $c_1 = F_0/E$ and

$$\psi(x) = \frac{F_0}{E}x.$$

The new boundary-value problem is

$$a^2 \frac{\partial^2 v}{\partial x^2} = \frac{\partial^2 v}{\partial t^2}, \quad 0 < x < L, \quad t > 0$$

$$v(0,t) = 0, \quad \frac{\partial v}{\partial x}\bigg|_{x=L} = 0, \quad t > 0,$$

$$v(x,0) = -\frac{F_0}{E}x, \quad \frac{\partial v}{\partial t}\bigg|_{t=0} = 0, \quad 0 < x < L.$$

Referring to Example 2 in the text we see that

$$v(x,t) = \sum_{n=1}^{\infty} A_n \cos a\left(\frac{2n-1}{2L}\right)\pi t \sin\left(\frac{2n-1}{2L}\right)\pi x$$

where

$$-\frac{F_0}{E}x = \sum_{n=1}^{\infty} A_n \sin\left(\frac{2n-1}{2L}\right)\pi x$$

and

$$A_n = \frac{-F_0 \int_0^L x \sin\left(\frac{2n-1}{2L}\right)\pi x \, dx}{E \int_0^L \sin^2\left(\frac{2n-1}{2L}\right)\pi x \, dx} = \frac{8F_0 L(-1)^n}{E\pi^2(2n-1)^2}.$$

Thus

$$u(x,t) = v(x,t) + \psi(x)$$

$$= \frac{F_0}{E}x + \frac{8F_0 L}{E\pi^2}\sum_{n=1}^{\infty} \frac{(-1)^n}{(2n-1)^2}\cos a\left(\frac{2n-1}{2L}\right)\pi t \sin\left(\frac{2n-1}{2L}\right)\pi x.$$

7. Separation of variables leads to

$$Y'' + \alpha^2 Y = 0$$

$$X'' - \alpha^2 X = 0$$

and

$$Y(y) = c_1 \cos \alpha y + c_2 \sin \alpha y$$

$$X(x) = c_3 \cosh \alpha x + c_4 \sinh \alpha x.$$

From $Y(0) = 0$ we find $c_1 = 0$. From $Y'(1) = 0$ we obtain $\cos \alpha = 0$ and

$$\alpha = \frac{\pi(2n-1)}{2}, \quad n = 1, 2, 3, \ldots .$$

Thus

$$Y(y) = c_2 \sin \left(\frac{2n-1}{2} \right) \pi y.$$

From $X'(0) = 0$ we find $c_4 = 0$. Then

$$u(x, y) = \sum_{n=1}^{\infty} A_n \cosh \left(\frac{2n-1}{2} \right) \pi x \, \sin \left(\frac{2n-1}{2} \right) \pi y$$

where

$$u_0 = u(1, y) = \sum_{n=1}^{\infty} A_n \cosh \left(\frac{2n-1}{2} \right) \pi \, \sin \left(\frac{2n-1}{2} \right) \pi y$$

and

$$A_n \cosh \left(\frac{2n-1}{2} \right) \pi = \frac{\int_0^1 u_0 \sin \left(\frac{2n-1}{2} \right) \pi y \, dy}{\int_0^1 \sin^2 \left(\frac{2n-1}{2} \right) \pi y \, dy} = \frac{4u_0}{(2n-1)\pi} .$$

Thus

$$u(x, y) = \frac{4u_0}{\pi} \sum_{n=1}^{\infty} \frac{1}{(2n-1) \cosh \left(\frac{2n-1}{2} \right) \pi} \cosh \left(\frac{2n-1}{2} \right) \pi x \, \sin \left(\frac{2n-1}{2} \right) \pi y.$$

8. The boundary-value problem is

$$k \frac{\partial^2 u}{\partial x^2} = \frac{\partial u}{\partial t}, \quad 0 < x < 1, \quad t > 0$$

$$\left. \frac{\partial u}{\partial x} \right|_{x=0} = hu(0, t), \quad \left. \frac{\partial u}{\partial x} \right|_{x=1} = -hu(1, t), \quad h > 0, \quad t > 0,$$

$$u(x, 0) = f(x), \quad 0 < x < 1.$$

Referring to Example 1 in the text we have

$$X(x) = c_1 \cos \alpha x + c_2 \sin \alpha x$$

and

$$T(t) = c_3 e^{-k\alpha^2 t}.$$

Applying the boundary conditions, we obtain

$$X'(0) = hX(0)$$

$$X'(1) = -hX(1)$$

or

$$\alpha c_2 = hc_1$$

$$-\alpha c_1 \sin \alpha + \alpha c_2 \cos \alpha = -hc_1 \cos \alpha - hc_2 \sin \alpha.$$

Choosing $c_1 = \alpha$ and $c_2 = h$ (to satisfy the first equation above) we obtain

$$-\alpha^2 \sin \alpha + h\alpha \cos \alpha = -h\alpha \cos \alpha - h^2 \sin \alpha$$

$$2h\alpha \cos \alpha = (\alpha^2 - h^2) \sin \alpha.$$

The eigenvalues α_n are the consecutive positive roots of

$$\tan \alpha = \frac{2h\alpha}{\alpha^2 - h^2}.$$

Then

$$u(x,t) = \sum_{n=1}^{\infty} A_n e^{-k\alpha_n^2 t}(\alpha_n \cos \alpha_n x + h \sin \alpha_n x)$$

where

$$f(x) = u(x,0) = \sum_{n=1}^{\infty} A_n(\alpha_n \cos \alpha_n x + h \sin \alpha_n x)$$

and

$$A_n = \frac{\int_0^1 f(x)(\alpha_n \cos \alpha_n x + h \sin \alpha_n x)dx}{\int_0^1 (\alpha_n \cos \alpha_n x + h \sin \alpha_n x)^2 dx}$$

$$= \frac{2}{\alpha_n^2 + 2h + h^2} \int_0^1 f(x)(\alpha_n \cos \alpha_n x + h \sin \alpha_n x)dx.$$

[Note: the evaluation and simplification of the integral in the denominator requires the use of the relationship $(\alpha^2 - h^2) \sin \alpha = 2h\alpha \cos \alpha$.]

9. The eigenfunctions of the associated homogeneous boundary-value problem are $\sin \alpha_n x$, $n = 1, 2, 3, \ldots$, where the α_n are the consecutive positive roots of $\tan \alpha = -\alpha$. We assume that

$$u(x,t) = \sum_{n=1}^{\infty} u_n(t) \sin \alpha_n x \qquad \text{and} \qquad xe^{-2t} = \sum_{n=1}^{\infty} F_n(t) \sin \alpha_n x.$$

Then

$$F_n(t) = \frac{e^{-2t} \int_0^1 x \sin \alpha_n x \, dx}{\int_0^1 \sin^2 \alpha_n x \, dx}.$$

Since $\alpha_n \cos \alpha_n = -\sin \alpha_n$ and

$$\int_0^1 \sin^2 \alpha_n x \, dx = \frac{1}{2}\left[1 - \frac{1}{2\alpha_n} \sin 2\alpha_n\right],$$

we have

$$e^{-2t} \int_0^1 x \sin \alpha_n x \, dx = e^{-2t}\left(\frac{\sin \alpha_n - \alpha_n \cos \alpha_n}{\alpha_n^2}\right) = \frac{2 \sin \alpha_n}{\alpha_n^2} e^{-2t}$$

$$\int_0^1 \sin^2 \alpha_n x \, dx = \frac{1}{2}[1 + \cos^2 \alpha_n]$$

and so

$$F_n(t) = \frac{4 \sin \alpha_n}{\alpha_n^2 (1 + \cos^2 \alpha_n)} e^{-2t}.$$

Substituting the assumptions into $u_t - k u_{xx} = x e^{-2t}$ and equating coefficients leads to the linear first-order differential equation

$$u_n'(t) + k \alpha_n^2 u(t) = \frac{4 \sin \alpha_n}{\alpha_n^2 (1 + \cos^2 \alpha_n)} e^{-2t}$$

whose solution is

$$u_n(t) = \frac{4 \sin \alpha_n}{\alpha_n^2 (1 + \cos^2 \alpha_n)(k\alpha_n^2 - 2)} e^{-2t} + C_n e^{-k\alpha_n^2 t}.$$

From

$$u(x,t) = \sum_{n=1}^{\infty} \left[\frac{4 \sin \alpha_n}{\alpha_n^2 (1 + \cos^2 \alpha_n)(k\alpha_n^2 - 2)} e^{-2t} + C_n e^{-k\alpha_n^2 t} \right] \sin \alpha_n x$$

and the initial condition $u(x, 0) = 0$ we see

$$C_n = -\frac{4 \sin \alpha_n}{\alpha_n^2 (1 + \cos^2 \alpha_n)(k\alpha_n^2 - 2)}.$$

The formal solution of the original problem is then

$$u(x,t) = \sum_{n=1}^{\infty} \frac{4 \sin \alpha_n}{\alpha_n^2 (1 + \cos^2 \alpha_n)(k\alpha_n^2 - 2)} (e^{-2t} - e^{-k\alpha_n^2 t}) \sin \alpha_n x.$$

10. Using $u = XT$ and separation constant $-\lambda = \alpha^4$ we find

$$X^{(4)} - \alpha^4 X = 0$$

and

$$X(x) = c_1 \cos \alpha x + c_2 \sin \alpha x + c_3 \cosh \alpha x + c_4 \sinh \alpha x.$$

Since $u = XT$ the boundary conditions become

$$X(0) = 0, \quad X'(0) = 0, \quad X''(1) = 0, \quad X'''(1) = 0.$$

Now $X(0) = 0$ implies $c_1 + c_3 = 0$, while $X'(0) = 0$ implies $c_2 + c_4 = 0$. Thus

$$X(x) = c_1 \cos \alpha x + c_2 \sin \alpha x - c_1 \cosh \alpha x - c_2 \sinh \alpha x.$$

The boundary condition $X''(1) = 0$ implies

$$-c_1 \cos \alpha - c_2 \sin \alpha - c_1 \cosh \alpha - c_2 \sinh \alpha = 0$$

while the boundary condition $X'''(1) = 0$ implies

$$c_1 \sin \alpha - c_2 \cos \alpha - c_1 \sinh \alpha - c_2 \cosh \alpha = 0.$$

659

We then have the system of two equations in two unknowns

$$(\cos\alpha + \cosh\alpha)c_1 + (\sin\alpha + \sinh\alpha)c_2 = 0$$

$$(\sin\alpha - \sinh\alpha)c_1 - (\cos\alpha + \cosh\alpha)c_2 = 0.$$

This homogeneous system will have nontrivial solutions for c_1 and c_2 provided

$$\begin{vmatrix} \cos\alpha + \cosh\alpha & \sin\alpha + \sinh\alpha \\ \sin\alpha - \sinh\alpha & -\cos\alpha - \cosh\alpha \end{vmatrix} = 0$$

or

$$-2 - 2\cos\alpha\cosh\alpha = 0.$$

Thus, the eigenvalues are determined by the equation $\cos\alpha\cosh\alpha = -1$.

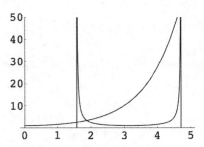

Using a computer to graph $\cosh\alpha$ and $-1/\cos\alpha = -\sec\alpha$ we see that the first two positive eigenvalues occur near 1.9 and 4.7. Applying Newton's method with these initial values we find that the eigenvalues are $\alpha_1 = 1.8751$ and $\alpha_2 = 4.6941$.

11. **(a)** In this case the boundary conditions are

$$u(0,t) = 0, \qquad \frac{\partial u}{\partial x}\bigg|_{x=0} = 0$$

$$u(1,t) = 0, \qquad \frac{\partial u}{\partial x}\bigg|_{x=1} = 0.$$

Separating variables leads to

$$X(x) = c_1\cos\alpha x + c_2\sin\alpha x + c_3\cosh\alpha x + c_4\sinh\alpha x$$

subject to

$$X(0) = 0, \quad X'(0) = 0, \quad X(1) = 0, \quad \text{and} \quad X'(1) = 0.$$

Now $X(0) = 0$ implies $c_1 + c_3 = 0$ while $X'(0) = 0$ implies $c_2 + c_4 = 0$. Thus

$$X(x) = c_1\cos\alpha x + c_2\sin\alpha x - c_1\cosh\alpha x - c_2\sinh\alpha x.$$

The boundary condition $X(1) = 0$ implies

$$c_1\cos\alpha + c_2\sin\alpha - c_1\cosh\alpha - c_2\sinh\alpha = 0$$

while the boundary condition $X'(1) = 0$ implies

$$-c_1\sin\alpha + c_2\cos\alpha - c_1\sinh\alpha - c_2\cosh\alpha = 0.$$

We then have the system of two equations in two unknowns

$$(\cos \alpha - \cosh \alpha)c_1 + (\sin \alpha - \sinh \alpha)c_2 = 0$$

$$-(\sin \alpha + \sinh \alpha)c_1 + (\cos \alpha - \cosh \alpha)c_2 = 0.$$

This homogeneous system will have nontrivial solutions for c_1 and c_2 provided

$$\begin{vmatrix} \cos \alpha - \cosh \alpha & \sin \alpha - \sinh \alpha \\ -\sin \alpha - \sinh \alpha & \cos \alpha - \cosh \alpha \end{vmatrix} = 0$$

or

$$2 - 2\cos \alpha \cosh \alpha = 0.$$

Thus, the eigenvalues are determined by the equation $\cos \alpha \cosh \alpha = 1$.

(b) Using a computer to graph $\cosh \alpha$ and $1/\cos \alpha = \sec \alpha$ we see that the first two positive eigenvalues occur near the vertical asymptotes of $\sec \alpha$, at $3\pi/2$ and $5\pi/2$. Applying Newton's method with these initial values we find that the eigenvalues are $\alpha_1 = 4.7300$ and $\alpha_2 = 7.8532$.

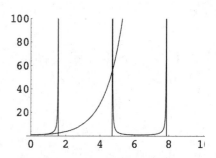

Exercises 12.8 Higher-Dimensional Problems

1. This boundary-value problem was solved in Example 1 in the text. Identifying $b = c = \pi$ and $f(x, y) = u_0$ we have

$$u(x, y, t) = \sum_{m=1}^{\infty} \sum_{n=1}^{\infty} A_{mn} e^{-k(m^2 + n^2)t} \sin mx \sin ny$$

where

$$A_{mn} = \frac{4}{\pi^2} \int_0^\pi \int_0^\pi u_0 \sin mx \sin ny \, dx \, dy$$

$$= \frac{4u_0}{\pi^2} \int_0^\pi \sin mx \, dx \int_0^\pi \sin ny \, dy$$

$$= \frac{4u_0}{mn\pi^2}[1 - (-1)^m][1 - (-1)^n].$$

2. As shown in Example 1 in the text, separation of variables leads to

$$X(x) = c_1 \cos \alpha x + c_2 \sin \alpha x$$

$$Y(y) = c_3 \cos \beta y + c_4 \sin \beta y$$

and

$$T(t) + c_5 e^{-k(\alpha^2 + \beta^2)t}.$$

The boundary conditions

$$\left.\begin{array}{ll} u_x(0, y, t) = 0, & u_x(1, y, t) = 0 \\[2mm] u_y(x, 0, t) = 0, & u_y(x, 1, t) = 0 \end{array}\right\} \quad \text{imply} \quad \left\{\begin{array}{ll} X'(0) = 0, & X'(1) = 0 \\[2mm] Y'(0) = 0, & Y'(1) = 0. \end{array}\right.$$

Applying these conditions to

$$X'(x) = -\alpha c_1 \sin \alpha x + \alpha c_2 \cos \alpha x$$

and

$$Y'(y) = -\beta c_3 \sin \beta y + \beta c_4 \cos \beta y$$

gives $c_2 = c_4 = 0$ and $\sin \alpha = \sin \beta = 0$. Then

$$\alpha = m\pi, \ m = 0, 1, 2, \ldots \quad \text{and} \quad \beta = n\pi, \ n = 0, 1, 2, \ldots .$$

By the Superposition Principle

$$u(x, y, t) = A_{00} + \sum_{m=1}^{\infty} A_{m0} e^{-km^2\pi^2 t} \cos m\pi x + \sum_{n=1}^{\infty} A_{0n} e^{-kn^2\pi^2 t} \cos n\pi y$$

$$+ \sum_{m=1}^{\infty} \sum_{n=1}^{\infty} A_{mn} e^{-k(m^2 + n^2)\pi^2 t} \cos m\pi x \cos n\pi y.$$

We now compute the coefficients of the double cosine series: Identifying $b = c = 1$ and $f(x, y) = xy$ we have

$$A_{00} = \int_0^1 \int_0^1 xy \, dx \, dy = \int_0^1 \frac{1}{2} x^2 y \Big|_0^1 dy = \frac{1}{2} \int_0^1 y \, dy = \frac{1}{4},$$

$$A_{m0} = 2 \int_0^1 \int_0^1 xy \cos m\pi x \, dx \, dy = 2 \int_0^1 \frac{1}{m^2\pi^2} (\cos m\pi x + m\pi x \sin m\pi x) \Big|_0^1 y \, dy$$

$$= 2 \int_0^1 \frac{\cos m\pi - 1}{m^2\pi^2} y \, dy = \frac{\cos m\pi - 1}{m^2\pi^2} = \frac{(-1)^m - 1}{m^2\pi^2},$$

$$A_{0n} = 2 \int_0^1 \int_0^1 xy \cos n\pi y \, dx \, dy = \frac{(-1)^n - 1}{n^2\pi^2},$$

and

$$A_{mn} = 4 \int_0^1 \int_0^1 xy \cos m\pi x \cos n\pi y \, dx \, dy = 4 \int_0^1 x \cos m\pi x \, dx \int_0^1 y \cos n\pi y \, dy$$

$$= 4 \left(\frac{(-1)^m - 1}{m^2\pi^2} \right) \left(\frac{(-1)^n - 1}{n^2\pi^2} \right).$$

In Problems 3 and 4 we need to solve the partial differential equation

$$a^2 \left(\frac{\partial^2 u}{\partial x^2} + \frac{\partial^2 u}{\partial y^2} \right) = \frac{\partial^2 u}{\partial t^2}.$$

To separate this equation we try $u(x, y, t) = X(x)Y(y)T(t)$:

$$a^2(X''YT + XY''T) = XYT''$$

$$\frac{X''}{X} = -\frac{Y''}{Y} + \frac{T''}{a^2 T} = -\alpha^2.$$

Then

$$X'' + \alpha^2 X = 0 \tag{1}$$

$$\frac{Y''}{Y} = \frac{T''}{a^2 T} + \alpha^2 = -\beta^2$$

$$Y'' + \beta^2 Y = 0 \tag{2}$$

$$T'' + a^2 \left(\alpha^2 + \beta^2 \right) T = 0. \tag{3}$$

The general solutions of equations (1), (2), and (3) are, respectively,

$$X(x) = c_1 \cos \alpha x + c_2 \sin \alpha x$$

$$Y(y) = c_3 \cos \beta y + c_4 \sin \beta y$$

$$T(t) = c_5 \cos a\sqrt{\alpha^2 + \beta^2}\, t + c_6 \sin a\sqrt{\alpha^2 + \beta^2}\, t.$$

3. The conditions $X(0) = 0$ and $Y(0) = 0$ give $c_1 = 0$ and $c_3 = 0$. The conditions $X(\pi) = 0$ and $Y(\pi) = 0$ yield two sets of eigenvalues:

$$\alpha = m, \; m = 1, 2, 3, \dots \quad \text{and} \quad \beta = n, \; n = 1, 2, 3, \dots.$$

A product solution of the partial differential equation that satisfies the boundary conditions is

$$u_{mn}(x, y, t) = \left(A_{mn} \cos a\sqrt{m^2 + n^2}\, t + B_{mn} \sin a\sqrt{m^2 + n^2}\, t \right) \sin mx \sin ny.$$

To satisfy the initial conditions we use the Superposition Principle:

$$u(x, y, t) = \sum_{m=1}^{\infty} \sum_{n=1}^{\infty} \left(A_{mn} \cos a\sqrt{m^2 + n^2}\, t + B_{mn} \sin a\sqrt{m^2 + n^2}\, t \right) \sin mx \sin ny.$$

The initial condition $u_t(x, y, 0) = 0$ implies $B_{mn} = 0$ and

$$u(x, y, t) = \sum_{m=1}^{\infty} \sum_{n=1}^{\infty} A_{mn} \cos a\sqrt{m^2 + n^2}\, t \sin mx \sin ny.$$

At $t = 0$ we have

$$xy(x - \pi)(y - \pi) = \sum_{m=1}^{\infty} \sum_{n=1}^{\infty} A_{mn} \sin mx \sin ny.$$

Using (11) and (12) in the text, it follows that

$$A_{mn} = \frac{4}{\pi^2} \int_0^\pi \int_0^\pi xy(x - \pi)(y - \pi) \sin mx \sin ny \, dx \, dy$$

$$= \frac{4}{\pi^2} \int_0^\pi x(x - \pi) \sin mx \, dx \int_0^\pi y(y - \pi) \sin ny \, dy$$

$$= \frac{16}{m^3 n^3 \pi^2} [(-1)^m - 1][(-1)^n - 1].$$

4. The conditions $X(0) = 0$ and $Y(0) = 0$ give $c_1 = 0$ and $c_3 = 0$. The conditions $X(b) = 0$ and $Y(c) = 0$ yield two sets of eigenvalues

$$\alpha = m\pi/b, \ m = 1, 2, 3, \ldots \quad \text{and} \quad \beta = n\pi/c, \ n = 1, 2, 3, \ldots.$$

A product solution of the partial differential equation that satisfies the boundary conditions is

$$u_{mn}(x, y, t) = (A_{mn} \cos a\omega_{mn}t + B_{mn} \sin a\omega_{mn}t) \sin \left(\frac{m\pi}{b}x\right) \sin \left(\frac{n\pi}{c}y\right),$$

where $\omega_{mn} = \sqrt{(m\pi/b)^2 + (n\pi/c)^2}$. To satisfy the initial conditions we use the Superposition Principle:

$$u(x, y, t) = \sum_{m=1}^{\infty} \sum_{n=1}^{\infty} (A_{mn} \cos a\omega_{mn}t + B_{mn} \sin a\omega_{mn}t) \sin \left(\frac{m\pi}{b}x\right) \sin \left(\frac{n\pi}{c}y\right).$$

At $t = 0$ we have

$$f(x, y) = \sum_{m=1}^{\infty} \sum_{n=1}^{\infty} A_{mn} \sin \left(\frac{m\pi}{b}x\right) \sin \left(\frac{n\pi}{c}y\right)$$

and

$$g(x, y) = \sum_{m=1}^{\infty} \sum_{n=1}^{\infty} B_{mn} a\omega_{mn} \sin \left(\frac{m\pi}{b}x\right) \sin \left(\frac{n\pi}{c}y\right).$$

Using (11) and (12) in the text, it follows that

$$A_{mn} = \frac{4}{bc} \int_0^c \int_0^b f(x, y) \sin \left(\frac{m\pi}{b}x\right) \sin \left(\frac{n\pi}{c}y\right) dx \, dy$$

$$B_{mn} = \frac{4}{abc\omega_{mn}} \int_0^c \int_0^b g(x, y) \sin \left(\frac{m\pi}{b}x\right) \sin \left(\frac{n\pi}{c}y\right) dx \, dy.$$

Note: In Problems 5 and 6 we try $u(x, y, z) = X(x)Y(y)Z(z)$ to separate Laplace's equation in three dimensions :

$$X''YZ + XY''Z + XYZ'' = 0$$

$$\frac{X''}{X} = -\frac{Y''}{Y} - \frac{Z''}{Z} = -\alpha^2.$$

Then

$$X'' + \alpha^2 X = 0 \tag{4}$$

$$\frac{Y''}{Y} = -\frac{Z''}{Z} + \alpha^2 = -\beta^2$$

$$Y'' + \beta^2 Y = 0 \tag{5}$$

$$Z'' - (\alpha^2 + \beta^2)Z = 0. \tag{6}$$

The general solutions of equations (4), (5), and (6) are, respectively

$$X(x) = c_1 \cos \alpha x + c_2 \sin \alpha x$$

$$Y(y) = c_3 \cos \beta y + c_4 \sin \beta y$$

$$Z(z) = c_5 \cosh \sqrt{\alpha^2 + \beta^2}\, z + c_6 \sinh \sqrt{\alpha^2 + \beta^2}\, z.$$

5. The boundary and initial conditions are

$$u(0, y, z) = 0, \quad u(a, y, z) = 0$$

$$u(x, 0, z) = 0, \quad u(x, b, z) = 0$$

$$u(x, y, 0) = 0, \quad u(x, y, c) = f(x, y).$$

The conditions $X(0) = Y(0) = Z(0) = 0$ give $c_1 = c_3 = c_5 = 0$. The conditions $X(a) = 0$ and $Y(b) = 0$ yield two sets of eigenvalues:

$$\alpha = \frac{m\pi}{a}, \ m = 1, 2, 3, \ldots \qquad \text{and} \qquad \beta = \frac{n\pi}{b}, \ n = 1, 2, 3, \ldots .$$

By the Superposition Principle

$$u(x, y, t) = \sum_{m=1}^{\infty} \sum_{n=1}^{\infty} A_{mn} \sinh \omega_{mn} z \sin \frac{m\pi}{a} x \sin \frac{n\pi}{b} y$$

where

$$\omega_{mn}^2 = \frac{m^2 \pi^2}{a^2} + \frac{n^2 \pi^2}{b^2}$$

and

$$A_{mn} = \frac{4}{ab \sinh \omega_{mn} c} \int_0^b \int_0^a f(x, y) \sin \frac{m\pi}{a} x \sin \frac{n\pi}{b} y \, dx \, dy.$$

6. The boundary and initial conditions are

$$u(0, y, z) = 0, \qquad u(a, y, z) = 0,$$

$$u(x, 0, z) = 0, \qquad u(x, b, z) = 0,$$

$$u(x, y, 0) = f(x, y), \qquad u(x, y, c) = 0.$$

665

The conditions $X(0) = Y(0) = 0$ give $c_1 = c_3 = 0$. The conditions $X(a) = Y(b) = 0$ yield two sets of eigenvalues:

$$\alpha = \frac{m\pi}{a}, \ m = 1, 2, 3, \ldots \qquad \text{and} \qquad \beta = \frac{n\pi}{b}, \ n = 1, 2, 3, \ldots .$$

Let

$$\omega_{mn}^2 = \frac{m^2\pi^2}{a^2} + \frac{n^2\pi^2}{b^2}.$$

Then the boundary condition $Z(c) = 0$ gives

$$c_5 \cosh c\omega_{mn} + c_6 \sinh c\omega_{mn} = 0$$

from which we obtain

$$Z(z) = c_5 \left(\cosh \omega_{mn} z - \frac{\cosh c\omega_{mn}}{\sinh c\omega_{mn}} \sinh \omega z \right)$$

$$= \frac{c_5}{\sinh c\omega_{mn}} (\sinh c\omega_{mn} \cosh \omega_{mn} z - \cosh c\omega_{mn} \sinh \omega_{mn} z) = c_{mn} \sinh \omega_{mn}(c - z).$$

By the Superposition Principle

$$u(x, y, t) = \sum_{m=1}^{\infty} \sum_{n=1}^{\infty} A_{mn} \sinh \omega_{mn}(c - z) \sin \frac{m\pi}{a} x \sin \frac{n\pi}{b} y$$

where

$$A_{mn} = \frac{4}{ab \sinh c\omega_{mn}} \int_0^b \int_0^a f(x, y) \sin \frac{m\pi}{a} x \sin \frac{n\pi}{b} y \, dx \, dy.$$

Chapter 12 in Review

1. Letting $u(x, y) = X(x) + Y(y)$ we have $X'Y' = XY$ and

$$\frac{X'}{X} = \frac{Y}{Y'} = -\lambda.$$

If $\lambda = 0$ then $X' = 0$ and $X(x) = c_1$. also $Y(y) = 0$ so $u = 0$.

If $\lambda \neq 0$ then $X' + \lambda X = 0$ and $Y + (1/\lambda)Y = 0$. Thus $X(x) = c_1 e^{-\lambda x}$ and $Y(y) = c_2 e^{-y/\lambda}$ so

$$u(x, y) = A e^{(-\lambda x - y/\lambda)}.$$

2. Letting $u = XY$ we have $X''Y + XY'' + 2X'Y + 2XY' = 0$ so that $(X'' + 2X')Y + X(Y'' + 2Y') = 0$. Separating variables and using the separation constant $-\lambda$ we obtain

$$\frac{X'' + 2X'}{-X} = \frac{Y'' + 2Y'}{Y} = -\lambda$$

so that

$$X'' + 2X' - \lambda X = 0 \qquad \text{and} \qquad Y'' + 2Y' + \lambda Y = 0.$$

The corresponding auxiliary equations are $m^2 + 2m - \lambda = 0$ and $m^2 + 2m + \lambda$ with solutions $m = -1 \pm \sqrt{1+\lambda}$ and $m = -1 \pm \sqrt{1-\lambda}$, respectively. We consider five cases:

I. $\lambda = -1$: In this case $X = c_1 e^{-x} + c_2 x e^{-x}$ and $Y = c_3 e^{(-1+\sqrt{2})y} + c_4 e^{(-1-\sqrt{2})y}$ so that

$$u = \left(c_1 e^x + c_2 x e^{-x} \right) \left(c_3 e^{(-1+\sqrt{2})y} + c_4 e^{(-1-\sqrt{2})y} \right).$$

II. $\lambda = 1$: In this case $X = c_5 e^{(-1+\sqrt{2})x} + c_6 e^{(-1-\sqrt{2})y}$ and $Y = c_7 e^{-y} + c_8 y e^{-y}$ so that

$$u = \left(c_5 e^{(-1+\sqrt{2})x} + c_6 e^{(-1-\sqrt{2})x} \right) (c_7 e^{-y} + c_8 y e^{-y}).$$

III. $-1 < \lambda < 1$: Here both $1 + \lambda$ and $1 - \lambda$ are positive so

$$u = \left(c_9 e^{(-1+\sqrt{1+\lambda})x} + c_{10} e^{(-1-\sqrt{1+\lambda})x} \right) \left(c_{11} e^{(-1+\sqrt{1-\lambda})y} + c_{12} e^{(-1-\sqrt{1-\lambda})y} \right).$$

IV. $\lambda < -1$: Here $1 + \lambda < 0$ and $1 - \lambda > 0$ so

$$u = e^{-x}(c_{13} \cos \sqrt{-1-\lambda}\, x + c_{14} \sin \sqrt{-1-\lambda}\, x) + \left(c_{15} e^{(-1+\sqrt{1-\lambda})y} + c_{16} e^{(-1-\sqrt{1-\lambda})y} \right).$$

V. $\lambda > 1$: Here $1 + \lambda > 0$ and $1 - \lambda < 0$ so

$$u = \left(c_{17} e^{(-1+\sqrt{1+\lambda})x} + c_{18} e^{(-1-\sqrt{1+\lambda})x} \right) + e^{-x}(c_{19} \cos \sqrt{\lambda-1}\, y + c_{20} \sin \sqrt{\lambda-1}\, y).$$

We see from the above that it is not possible to choose λ so that both X and Y are oscillatory.

3. Substituting $u(x,t) = v(x,t) + \psi(x)$ into the partial differential equation we obtain

$$k \frac{\partial^2 v}{\partial x^2} + k\psi''(x) = \frac{\partial v}{\partial t}.$$

This equation will be homogeneous provided ψ satisfies

$$k\psi'' = 0 \qquad \text{or} \qquad \psi = c_1 x + c_2.$$

Considering

$$u(0,t) = v(0,t) + \psi(0) = u_0$$

we set $\psi(0) = u_0$ so that $\psi(x) = c_1 x + u_0$. Now

$$-\frac{\partial u}{\partial x} \bigg|_{x=\pi} = -\frac{\partial v}{\partial x} \bigg|_{x=\pi} - \psi'(x) = v(\pi, t) + \psi(\pi) - u_1$$

is equivalent to

$$\frac{\partial v}{\partial x} \bigg|_{x=\pi} + v(\pi, t) = u_1 - \psi'(x) - \psi(\pi) = u_1 - c_1 - (c_1\pi + u_0),$$

which will be homogeneous when

$$u_1 - c_1 - c_1\pi - u_0 = 0 \qquad \text{or} \qquad c_1 = \frac{u_1 - u_0}{1 + \pi}.$$

The steady-state solution is

$$\psi(x) = \left(\frac{u_1 - u_0}{1 + \pi}\right) x + u_0.$$

4. The solution of the problem represents the heat of a thin rod of length π. The left boundary $x = 0$ is kept at constant temperature u_0 for $t > 0$. Heat is lost from the right end of the rod by being in contact with a medium that is held at constant temperature u_1.

5. The boundary-value problem is

$$a^2 \frac{\partial^2 u}{\partial x^2} = \frac{\partial^2 u}{\partial t^2}, \qquad 0 < x < 1, \quad t > 0,$$

$$u(0, t) = 0, \quad u = (1, t) = 0, \quad t > 0,$$

$$u(x, 0) = 0, \quad \frac{\partial u}{\partial t}\bigg|_{t=0} = g(x), \quad 0 < x < 1.$$

From Section 12.4 in the text we see that $A_n = 0$,

$$B_n = \frac{2}{n\pi a} \int_0^1 g(x) \sin n\pi x \, dx = \frac{2}{n\pi a} \int_{1/4}^{3/4} h \sin n\pi x \, dx$$

$$= \frac{2h}{n\pi a}\left(-\frac{1}{n\pi} \cos n\pi x\right)\bigg|_{1/4}^{3/4} = \frac{2h}{n^2 \pi^2 a}\left(\cos \frac{n\pi}{4} - \cos \frac{3n\pi}{4}\right)$$

and

$$u(x, t) = \sum_{n=1}^{\infty} B_n \sin n\pi a t \sin n\pi x.$$

6. The boundary-value problem is

$$\frac{\partial^2 u}{\partial x^2} + x^2 = \frac{\partial^2 u}{\partial t^2}, \qquad 0 < x < 1, \quad t > 0,$$

$$u(0, t) = 1, \quad u(1, t) = 0, \quad t > 0,$$

$$u(x, 0) = f(x), \quad u_t(x, 0) = 0, \quad 0 < x < 1.$$

Substituting $u(x, t) = v(x, t) + \psi(x)$ into the partial differential equation gives

$$\frac{\partial^2 v}{\partial x^2} + \psi''(x) + x^2 = \frac{\partial^2 v}{\partial t^2}.$$

This equation will be homogeneous provided $\psi''(x) + x^2 = 0$ or

$$\psi(x) = -\frac{1}{12}x^4 + c_1 x + c_2.$$

From $\psi(0) = 1$ and $\psi(1) = 0$ we obtain $c_1 = -11/12$ and $c_2 = 1$. The new problem is

$$\frac{\partial^2 v}{\partial x^2} = \frac{\partial^2 v}{\partial t^2}, \quad 0 < x < 1, \quad t > 0,$$

$$v(0, t) = 0, \quad v(1, t) = 0, \quad t > 0,$$

$$v(x, 0) = f(x) - \psi(x), \quad v_t(x, 0) = 0, \quad 0 < x < 1.$$

From Section 12.4 in the text we see that $B_n = 0$,

$$A_n = 2 \int_0^1 [f(x) - \psi(x)] \sin n\pi x \, dx = 2 \int_0^1 \left[f(x) + \frac{1}{12} x^4 + \frac{11}{12} x - 1 \right] \sin n\pi x \, dx,$$

and

$$v(x, t) = \sum_{n=1}^{\infty} A_n \cos n\pi t \sin n\pi x.$$

Thus

$$u(x, t) = v(x, t) + \psi(x) = -\frac{1}{12} x^4 - \frac{11}{12} x + 1 + \sum_{n=1}^{\infty} A_n \cos n\pi t \sin n\pi x.$$

7. Using $u = XY$ and $-\lambda$ as a separation constant leads to

$$X'' - \lambda X = 0,$$

$$X(0) = 0,$$

and

$$Y'' + \lambda Y = 0,$$

$$Y(0) = 0,$$

$$Y(\pi) = 0.$$

This leads to

$$Y = c_4 \sin ny \quad \text{and} \quad X = c_2 \sinh nx$$

for $n = 1, 2, 3, \ldots$ so that

$$u = \sum_{n=1}^{\infty} A_n \sinh nx \sin ny.$$

Imposing

$$u(\pi, y) = 50 = \sum_{n=1}^{\infty} A_n \sinh n\pi \sin ny$$

gives

$$A_n = \frac{100}{n\pi} \frac{1 - (-1)^n}{\sinh n\pi}$$

so that

$$u(x, y) = \frac{100}{\pi} \sum_{n=1}^{\infty} \frac{1 - (-1)^n}{n \sinh n\pi} \sinh nx \sin ny.$$

8. Using $u = XY$ and $-\lambda$ as a separation constant leads to

$$X'' - \lambda X = 0,$$

and

$$Y'' + \lambda Y = 0,$$

$$Y'(0) = 0,$$

$$Y'(\pi) = 0.$$

This leads to

$$Y = c_3 \cos ny \qquad \text{and} \qquad X = c_2 e^{-nx}$$

for $n = 1, 2, 3, \ldots$. In this problem we also have $\lambda = 0$ is an eigenvalue with corresponding eigenfunctions 1 and 1. Thus

$$u = A_0 + \sum_{n=1}^{\infty} A_n e^{-nx} \cos ny.$$

Imposing

$$u(0, y) = 50 = A_0 + \sum_{n=1}^{\infty} A_n \cos ny$$

gives

$$A_0 = \frac{1}{\pi} \int_0^{\pi} 50 \, dy = 50$$

and

$$A_n = \frac{2}{\pi} \int_0^{\pi} 50 \cos ny \, dy = 0$$

for $n = 1, 2, 3, \ldots$ so that

$$u(x, y) = 50.$$

9. Using $u = XY$ and $-\lambda$ as a separation constant leads to

$$X'' - \lambda X = 0,$$

and

$$Y'' + \lambda Y = 0,$$

$$Y(0) = 0,$$

$$Y(\pi) = 0.$$

Then

$$X = c_1 e^{nx} + c_2 e^{-nx} \qquad \text{and} \qquad Y = c_3 \cos ny + c_4 \sin ny$$

for $n = 1, 2, 3, \ldots$. Since u must be bounded as $x \to \infty$) we define $c_1 = 0$. Also $Y(0) = 0$ implies $c_3 = 0$ so

$$u = \sum_{n=1}^{\infty} A_n e^{-nx} \sin ny.$$

Imposing

$$u(0, y) = 50 = \sum_{n=1}^{\infty} A_n \sin ny$$

gives

$$A_n = \frac{2}{\pi} \int_0^\pi 50 \sin ny \, dy = \frac{100}{n\pi}[1 - (-1)^n]$$

so that

$$u(x, y) = \sum_{n=1}^{\infty} \frac{100}{n\pi}[1 - (-1)^n]e^{-nx} \sin ny.$$

10. The boundary-value problem is

$$k\frac{\partial^2 u}{\partial x^2} = \frac{\partial u}{\partial t}, \quad -L < x < L, \quad t > 0,$$

$$u(-L, t) = 0, \quad u(L, t) = 0, \quad t > 0,$$

$$u(x, 0) = u_0, \quad -L < x < L.$$

Referring to Section 12.3 in the text we have

$$X(x) = c_1 \cos \alpha x + c_2 \sin \alpha x$$

and

$$T(t) = c_3 e^{-k\alpha^2 t}.$$

Using the boundary conditions $u(-L, 0) = X(-L)T(0) = 0$ and $u(L, 0) = X(L)T(0) = 0$ we obtain $X(-L) = 0$ and $X(L) = 0$. Thus

$$c_1 \cos(-\alpha L) + c_2 \sin(-\alpha L) = 0$$

$$c_1 \cos \alpha L + c_2 \sin \alpha L = 0$$

or

$$c_1 \cos \alpha L - c_2 \sin \alpha L = 0$$

$$c_1 \cos \alpha L + c_2 \sin \alpha L = 0.$$

Adding, we find $\cos \alpha L = 0$ which gives the eigenvalues

$$\alpha = \frac{2n-1}{2L}\pi, \quad n = 1, 2, 3, \ldots .$$

Thus

$$u(x, t) = \sum_{n=1}^{\infty} A_n e^{-\left(\frac{2n-1}{2L}\pi\right)^2 kt} \cos\left(\frac{2n-1}{2L}\right)\pi x.$$

From

$$u(x,0) = u_0 = \sum_{n=1}^{\infty} A_n \cos\left(\frac{2n-1}{2L}\right)\pi x$$

we find

$$A_n = \frac{2\int_0^L u_0 \cos\left(\frac{2n-1}{2L}\right)\pi x \, dx}{2\int_0^L \cos^2\left(\frac{2n-1}{2L}\right)\pi x \, dx} = \frac{u_0(-1)^{n+1}2L/\pi(2n-1)}{L/2} = \frac{4u_0(-1)^{n+1}}{\pi(2n-1)}.$$

11. The coefficients of the series

$$u(x,0) = \sum_{n=1}^{\infty} B_n \sin nx$$

are

$$B_n = \frac{2}{\pi}\int_0^{\pi} \sin x \sin nx \, dx = \frac{2}{\pi}\int_0^{\pi} \frac{1}{2}[\cos(1-n)x - \cos(1+n)x]\, dx$$

$$= \frac{1}{\pi}\left[\frac{\sin(1-n)x}{1-n}\bigg|_0^{\pi} - \frac{\sin(1+n)x}{1+n}\bigg|_0^{\pi}\right] = 0 \text{ for } n \neq 1.$$

For $n = 1$,

$$B_1 = \frac{2}{\pi}\int_0^{\pi} \sin^2 x \, dx = \frac{1}{\pi}\int_0^{\pi}(1 - \cos 2x)\, dx = 1.$$

Thus

$$u(x,t) = \sum_{n=1}^{\infty} B_n e^{-n^2 t}\sin nx$$

reduces to $u(x,t) = e^{-t}\sin x$ for $n = 1$.

12. Substituting $u(x,t) = v(x,t) + \psi(x)$ into the partial differential equation results in $\psi'' = -\sin x$ and $\psi(x) = c_1 x + c_2 + \sin x$. The boundary conditions $\psi(0) = 400$ and $\psi(\pi) = 200$ imply $c_1 = -200/\pi$ and $c_2 = 400$ so

$$\psi(x) = -\frac{200}{\pi}x + 400 + \sin x.$$

Solving

$$\frac{\partial^2 v}{\partial x^2} = \frac{\partial v}{\partial t}, \qquad 0 < x < \pi, \quad t > 0$$

$$v(0,t) = 0, \qquad v(\pi, t) = 0, \quad t > 0$$

$$u(x,0) = 400 + \sin x - \left(-\frac{200}{\pi}x + 400 + \sin x\right) = \frac{200}{\pi}x, \qquad 0 < x < \pi$$

using separation of variables with separation constant $-\lambda$, where $\lambda = \alpha^2$, gives

$$X'' + \alpha^2 X = 0 \qquad \text{and} \qquad T' + \alpha^2 T = 0.$$

Using $X(0) = 0$ and $X(\pi) = 0$ we determine $\alpha^2 = n^2$, $X(x) = c_2 \sin nx$, and $T(t) = c_3 e^{-n^2 t}$. Then

$$v(x,t) = \sum_{n=1}^{\infty} A_n e^{-n^2 t}\sin nx$$

and

$$v(x,0) = \frac{200}{\pi}x = \sum_{n=1}^{\infty} A_n \sin nx$$

so

$$A_n = \frac{400}{\pi^2} \int_0^\pi x \sin nx \, dx = \frac{400}{n\pi}(-1)^{n+1}.$$

Thus

$$u(x,t) = -\frac{200}{\pi}x + 400 + \sin x + \frac{400}{\pi}\sum_{n=1}^{\infty} \frac{(-1)^{n+1}}{n}e^{-n^2 t}\sin nx.$$

13. Using $u = XT$ and $-\lambda$, where $\lambda = \alpha^2$, as a separation constant we find

$$X'' + 2X' + \alpha^2 X = 0 \quad \text{and} \quad T'' + 2T' + (1+\alpha^2)T = 0.$$

Thus for $\alpha > 1$

$$X = c_1 e^{-x}\cos\sqrt{\alpha^2-1}\,x + c_2 e^{-x}\sin\sqrt{\alpha^2-1}\,x$$

$$T = c_3 e^{-t}\cos\alpha t + c_4 e^{-t}\sin\alpha t.$$

For $0 \le \alpha \le 1$ we only obtain $X = 0$. Now the boundary conditions $X(0) = 0$ and $X(\pi) = 0$ give, in turn, $c_1 = 0$ and $\sqrt{\alpha^2-1}\,\pi = n\pi$ or $\alpha^2 = n^2 + 1$, $n = 1, 2, 3, \ldots$. The corresponding solutions are $X = c_2 e^{-x}\sin nx$. The initial condition $T'(0) = 0$ implies $c_3 = \alpha c_4$ and so

$$T = c_4 e^{-t}\left[\sqrt{n^2+1}\,\cos\sqrt{n^2+1}\,t + \sin\sqrt{n^2+1}\,t\right].$$

Using $u = XT$ and the Superposition Principle, a formal series solution is

$$u(x,t) = e^{-(x+t)}\sum_{n=1}^{\infty} A_n\left[\sqrt{n^2+1}\,\cos\sqrt{n^2+1}\,t + \sin\sqrt{n^2+1}\,t\right]\sin nx.$$

14. Letting $c = XT$ and separating variables we obtain

$$\frac{kX'' - hX'}{X} = \frac{T'}{T} \quad \text{or} \quad \frac{X'' - aX'}{X} = \frac{T'}{kT} = -\lambda$$

where $a = h/k$. Setting $\lambda = \alpha^2$ leads to the separated differential equations

$$X'' - aX' + \alpha^2 X = 0 \quad \text{and} \quad T' + k\alpha^2 T = 0.$$

The solution of the second equation is

$$T(t) = c_3 e^{-k\alpha^2 t}.$$

For the first equation we have $m = \frac{1}{2}(a \pm \sqrt{a^2 - 4\alpha^2})$, and we consider three cases using the boundary conditions $X(0) = X(1) = 0$:

$\boxed{a^2 > 4\alpha^2}$ The solution is $X = c_1 e^{m_1 x} + c_2 e^{m_2 x}$, where the boundary conditions imply $c_1 = c_2 = 0$, so $X = 0$. (Note in this case that if $\alpha = 0$, the solution is $X = c_1 + c_2 e^{ax}$ and the boundary conditions again imply $c_1 = c_2 = 0$, so $X = 0$.)

$\boxed{a^2 = 4\alpha^2}$ The solution is $X = c_1 e^{m_1 x} + c_2 x e^{m_1 x}$, where the boundary conditions imply $c_1 = c_2 = 0$, so $X = 0$.

$\boxed{a^2 < 4\alpha^2}$ The solution is

$$X(x) = c_1 e^{ax/2} \cos \frac{\sqrt{4\alpha^2 - a^2}}{2} x + c_2 e^{ax/2} \sin \frac{\sqrt{4\alpha^2 - a^2}}{2} x.$$

From $X(0) = 0$ we see that $c_1 = 0$. From $X(1) = 0$ we find

$$\frac{1}{2}\sqrt{4\alpha^2 - a^2} = n\pi \qquad \text{or} \qquad \alpha^2 = \frac{1}{4}(4n^2\pi^2 + a^2).$$

Thus

$$X(x) = c_2 e^{ax/2} \sin n\pi x,$$

and

$$c(x, t) = \sum_{n=1}^{\infty} A_n e^{ax/2} e^{-k(4n^2\pi^2 + a^2)t/4} \sin \pi x.$$

The initial condition $c(x, 0) = c_0$ implies

$$c_0 = \sum_{n=1}^{\infty} A_n e^{ax/2} \sin n\pi x. \tag{1}$$

From the self-adjoint form

$$\frac{d}{dx}[e^{-ax} X'] + \alpha^2 e^{-ax} X = 0$$

the eigenfunctions are orthogonal on $[0, 1]$ with weight function e^{-ax}. That is

$$\int_0^1 e^{-ax}(e^{ax/2} \sin n\pi x)(e^{ax/2} \sin m\pi x)\,dx = 0, \quad n \neq m.$$

Multiplying (1) by $e^{-ax}e^{ax/2}\sin m\pi x$ and integrating we obtain

$$\int_0^1 c_0 e^{-ax} e^{ax/2} \sin m\pi x\,dx = \sum_{n=1}^{\infty} A_n \int_0^1 e^{-ax}e^{ax/2}(\sin m\pi x)e^{ax/2} \sin n\pi x\,dx$$

$$c_0 \int_0^1 e^{-ax/2} \sin n\pi x\,dx = A_n \int_0^1 \sin^2 n\pi x\,dx = \frac{1}{2} A_n$$

and

$$A_n = 2c_0 \int_0^1 e^{-ax/2} \sin n\pi x\,dx = \frac{4c_0[2e^{a/2}n\pi - 2n\pi(-1)^n]}{e^{a/2}(a^2 + 4n^2\pi^2)} = \frac{8n\pi c_0[e^{a/2} - (-1)^n]}{e^{a/2}(a^2 + 4n^2\pi^2)}.$$

13 Boundary-Value Problems in Other Coordinate Systems

1. We have

$$A_0 = \frac{1}{2\pi} \int_0^\pi u_0 \, d\theta = \frac{u_0}{2}$$

$$A_n = \frac{1}{\pi} \int_0^\pi u_0 \cos n\theta \, d\theta = 0$$

$$B_n = \frac{1}{\pi} \int_0^\pi u_0 \sin n\theta \, d\theta = \frac{u_0}{n\pi}[1 - (-1)^n]$$

and so

$$u(r, \theta) = \frac{u_0}{2} + \frac{u_0}{\pi} \sum_{n=1}^\infty \frac{1 - (-1)^n}{n} r^n \sin n\theta.$$

2. We have

$$A_0 = \frac{1}{2\pi} \int_0^\pi \theta \, d\theta + \frac{1}{2\pi} \int_\pi^{2\pi} (\pi - \theta) \, d\theta = 0$$

$$A_n = \frac{1}{\pi} \int_0^\pi \theta \cos n\theta \, d\theta + \frac{1}{\pi} \int_\pi^{2\pi} (\pi - \theta) \cos n\theta \, d\theta = \frac{2}{n^2 \pi}[(-1)^n - 1]$$

$$B_n = \frac{1}{\pi} \int_0^\pi \theta \sin n\theta \, d\theta + \frac{1}{\pi} \int_\pi^{2\pi} (\pi - \theta) \sin n\theta \, d\theta = \frac{1}{n}[1 - (-1)^n]$$

and so

$$u(r, \theta) = \sum_{n=1}^\infty r^n \left[\frac{(-1)^n - 1}{n^2 \pi} \cos n\theta + \frac{1 - (-1)^n}{n} \sin n\theta \right].$$

3. We have

$$A_0 = \frac{1}{2\pi} \int_0^{2\pi} (2\pi\theta - \theta^2) \, d\theta = \frac{2\pi^2}{3}$$

$$A_n = \frac{1}{\pi} \int_0^{2\pi} (2\pi\theta - \theta^2) \cos n\theta \, d\theta = -\frac{4}{n^2}$$

$$B_n = \frac{1}{\pi} \int_0^{2\pi} (2\pi\theta - \theta^2) \sin n\theta \, d\theta = 0$$

and so

$$u(r,\theta) = \frac{2\pi^2}{3} - 4 \sum_{n=1}^{\infty} \frac{r^n}{n^2} \cos n\theta.$$

4. We have

$$A_0 = \frac{1}{2\pi} \int_0^{2\pi} \theta \, d\theta = \pi$$

$$A_n = \frac{1}{\pi} \int_0^{2\pi} \theta \cos n\theta \, d\theta = 0$$

$$B_n = \frac{1}{\pi} \int_0^{2\pi} \theta \sin n\theta \, d\theta = -\frac{2}{n}$$

and so

$$u(r,\theta) = \pi - 2 \sum_{n=1}^{\infty} \frac{r^n}{n} \sin n\theta.$$

5. As in Example 1 in the text we have $R(r) = c_3 r^n + c_4 r^{-n}$. In order that the solution be bounded as $r \to \infty$ we must define $c_3 = 0$. Hence

$$u(r,\theta) = A_0 + \sum_{n=1}^{\infty} r^{-n}(A_n \cos n\theta + B_n \sin n\theta)$$

where

$$A_0 = \frac{1}{2\pi} \int_0^{2\pi} f(\theta) \, d\theta$$

$$A_n = \frac{c^n}{\pi} \int_0^{2\pi} f(\theta) \cos n\theta \, d\theta$$

$$B_n = \frac{c^n}{\pi} \int_0^{2\pi} f(\theta) \sin n\theta \, d\theta.$$

6. We solve

$$\frac{\partial^2 u}{\partial r^2} + \frac{1}{r}\frac{\partial u}{\partial r} + \frac{1}{r^2}\frac{\partial^2 u}{\partial \theta^2} = 0, \quad 0 < \theta < \frac{\pi}{2}, \quad 0 < r < c,$$

$$u(c,\theta) = f(\theta), \quad 0 < \theta < \frac{\pi}{2},$$

$$u(r,0) = 0, \quad u(r,\pi/2) = 0, \quad 0 < r < c.$$

Proceeding as in Example 1 in the text we obtain the separated differential equations

$$r^2 R'' + r R' - \lambda R = 0$$

$$\Theta'' + \lambda \Theta = 0.$$

Taking $\lambda = \alpha^2$ the solutions are

$$\Theta(\theta) = c_1 \cos \alpha\theta + c_2 \sin \alpha\theta$$

$$R(r) = c_3 r^\alpha + c_4 r^{-\alpha}.$$

Since we want $R(r)$ to be bounded as $r \to 0$ we require $c_4 = 0$. Applying the boundary conditions $\Theta(0) = 0$ and $\Theta(\pi/2) = 0$ we find that $c_1 = 0$ and $\alpha = 2n$ for $n = 1, 2, 3, \ldots$. Therefore

$$u(r, \theta) = \sum_{n=1}^{\infty} A_n r^{2n} \sin 2n\theta.$$

From

$$u(c, \theta) = f(\theta) = \sum_{n=1}^{\infty} A_n c^n \sin 2n\theta$$

we find

$$A_n = \frac{4}{\pi c^{2n}} \int_0^{\pi/2} f(\theta) \sin 2n\theta \, d\theta.$$

7. Referring to the solution of Problem 6 above we have

$$\Theta(\theta) = c_1 \cos \alpha\theta + c_2 \sin \alpha\theta$$

$$R(r) = c_3 r^\alpha.$$

Applying the boundary conditions $\Theta'(0) = 0$ and $\Theta'(\pi/2) = 0$ we find that $c_2 = 0$ and $\alpha = 2n$ for $n = 0, 1, 2, \ldots$. Therefore

$$u(r, \theta) = A_0 + \sum_{n=1}^{\infty} A_n r^{2n} \cos 2n\theta.$$

From

$$u(c, \theta) = \begin{cases} 1, & 0 < \theta < \pi/4 \\ 0, & \pi/4 < \theta < \pi/2 \end{cases} = A_0 + \sum_{n=1}^{\infty} A_n c^{2n} \cos 2n\theta$$

we find

$$A_0 = \frac{1}{\pi/2} \int_0^{\pi/4} d\theta = \frac{1}{2}$$

and

$$c^{2n} A_n = \frac{2}{\pi/2} \int_0^{\pi/4} \cos 2n\theta \, d\theta = \frac{2}{n\pi} \sin \frac{n\pi}{2}.$$

Thus

$$u(r, \theta) = \frac{1}{2} + \frac{2}{\pi} \sum_{n=1}^{\infty} \frac{1}{n} \sin \frac{n\pi}{2} \left(\frac{r}{c}\right)^{2n} \cos 2n\theta.$$

8. We solve

$$\frac{\partial^2 u}{\partial r^2} + \frac{1}{r} \frac{\partial u}{\partial r} + \frac{1}{r^2} \frac{\partial^2 u}{\partial \theta^2} = 0, \quad 0 < \theta < \pi/4, \quad r > 0$$

$$u(r, 0) = 0, \quad r > 0$$

$$u(r, \pi/4) = 30, \quad r > 0.$$

Proceeding as in Example 1 in the text we find the separated ordinary differential equations to be

$$r^2 R'' + rR' - \lambda R = 0$$

$$\Theta'' + \lambda\Theta = 0.$$

With $\lambda = \alpha^2 > 0$ the corresponding general solutions are

$$R(r) = c_1 r^\alpha + c_2 r^{-\alpha}$$

$$\Theta(\theta) = c_3 \cos\alpha\theta + c_4 \sin\alpha\theta.$$

The condition $\Theta(0) = 0$ implies $c_3 = 0$ so that $\Theta = c_4 \sin\alpha\theta$. Now, in order that the temperature be bounded as $r \to \infty$ we define $c_1 = 0$. Similarly, in order that the temperature be bounded as $r \to 0$ we are forced to define $c_2 = 0$. Thus $R(r) = 0$ and so no nontrivial solution exists for $\lambda > 0$. For $\lambda = 0$ the separated differential equations are

$$r^2 R'' + rR' = 0 \qquad \text{and} \qquad \Theta'' = 0.$$

Solutions of these latter equations are

$$R(r) = c_1 + c_2 \ln r \qquad \text{and} \qquad \Theta(\theta) = c_3\theta + c_4.$$

$\Theta(0) = 0$ still implies $c_4 = 0$, whereas boundedness as $r \to 0$ demands $c_2 = 0$. Thus, a product solution is

$$u = c_1 c_3 \theta = A\theta.$$

From $u(r, \pi/4) = 0$ we obtain $A = 120/\pi$. Thus, a solution to the problem is

$$u(r, \theta) = \frac{120}{\pi}\theta.$$

9. Proceeding as in Example 1 in the text and again using the periodicity of $u(r, \theta)$, we have

$$\Theta(\theta) = c_1 \cos\alpha\theta + c_2 \sin\alpha\theta$$

where $\alpha = n$ for $n = 0, 1, 2, \ldots$. Then

$$R(r) = c_3 r^n + c_4 r^{-n}.$$

[We do not have $c_4 = 0$ in this case since $0 < a \le r$.] Since $u(b, \theta) = 0$ we have

$$u(r, \theta) = A_0 \ln\frac{r}{b} + \sum_{n=1}^{\infty} \left[\left(\frac{b}{r}\right)^n - \left(\frac{r}{b}\right)^n\right] [A_n \cos n\theta + B_n \sin n\theta].$$

From

$$u(a, \theta) = f(\theta) = A_0 \ln\frac{a}{b} + \sum_{n=1}^{\infty} \left[\left(\frac{b}{a}\right)^n - \left(\frac{a}{b}\right)^n\right] [A_n \cos n\theta + B_n \sin n\theta]$$

we find

$$A_0 \ln \frac{a}{b} = \frac{1}{2\pi} \int_0^{2\pi} f(\theta)\, d\theta,$$

$$\left[\left(\frac{b}{a}\right)^n - \left(\frac{a}{b}\right)^n \right] A_n = \frac{1}{\pi} \int_0^{2\pi} f(\theta) \cos n\theta\, d\theta,$$

and

$$\left[\left(\frac{b}{a}\right)^n - \left(\frac{a}{b}\right)^n \right] B_n = \frac{1}{\pi} \int_0^{2\pi} f(\theta) \sin n\theta\, d\theta.$$

10. Substituting $u(r,\theta) = v(r,\theta) + \psi(r)$ into the partial differential equation we obtain

$$\frac{\partial^2 v}{\partial r^2} + \psi''(r) + \frac{1}{r}\left[\frac{\partial v}{\partial r} + \psi'(r)\right] + \frac{1}{r^2}\frac{\partial^2 v}{\partial \theta^2} = 0.$$

This equation will be homogeneous provided

$$\psi''(r) + \frac{1}{r}\psi'(r) = 0 \qquad \text{or} \qquad r^2\psi''(r) + r\psi'(r) = 0.$$

The general solution of this Cauchy-Euler differential equation is

$$\psi(r) = c_1 + c_2 \ln r.$$

From

$$u_0 = u(a,\theta) = v(a,\theta) + \psi(a) \qquad \text{and} \qquad u_1 = u(b,\theta) = v(b,\theta) + \psi(b)$$

we see that in order for the boundary values $v(a,\theta)$ and $v(b,\theta)$ to be 0 we need $\psi(a) = u_0$ and $\psi(b) = u_1$. From this we have

$$\psi(a) = c_1 + c_2 \ln a = u_0$$

$$\psi(b) = c_1 + c_2 \ln b = u_1.$$

Solving for c_1 and c_2 we obtain

$$c_1 = \frac{u_1 \ln a - u_0 \ln b}{\ln(a/b)} \qquad \text{and} \qquad c_2 = \frac{u_0 - u_1}{\ln(a/b)}.$$

Then

$$\psi(r) = \frac{u_1 \ln a - u_0 \ln b}{\ln(a/b)} + \frac{u_0 - u_1}{\ln(a/b)} \ln r = \frac{u_0 \ln(r/b) - u_1 \ln(r/a)}{\ln(a/b)}.$$

From Problem 9 with $f(\theta) = 0$ we see that the solution of

$$\frac{\partial^2 v}{\partial r^2} + \frac{1}{r}\frac{\partial v}{\partial r} + \frac{1}{r^2}\frac{\partial^2 v}{\partial \theta^2} = 0, \quad 0 < \theta < 2\pi, \quad a < r < b,$$

$$v(a,\theta) = 0, \quad v(b,\theta) = 0, \quad 0 < \theta < 2\pi$$

is $v(r,\theta) = 0$. Thus the steady-state temperature of the ring is

$$u(r,\theta) = v(r,\theta) + \psi(r) = \frac{u_0 \ln(r/b) - u_1 \ln(r/a)}{\ln(a/b)}.$$

11 new

~~11.~~ We solve

12

$$\frac{\partial^2 u}{\partial r^2} + \frac{1}{r}\frac{\partial u}{\partial r} + \frac{1}{r^2}\frac{\partial^2 u}{\partial \theta^2} = 0, \quad 0 < \theta < \pi, \quad a < r < b,$$

$$u(a, \theta) = \theta(\pi - \theta), \quad u(b, \theta) = 0, \quad 0 < \theta < \pi,$$

$$u(r, 0) = 0, \quad u(r, \pi) = 0, \quad a < r < b.$$

Proceeding as in Example 1 in the text we obtain the separated differential equations

$$r^2 R'' + rR' - \lambda R = 0$$

$$\Theta'' + \lambda \Theta = 0.$$

Taking $\lambda = \alpha^2$ the solutions are

$$\Theta(\theta) = c_1 \cos \alpha \theta + c_2 \sin \alpha \theta$$

$$R(r) = c_3 r^\alpha + c_4 r^{-\alpha}.$$

Applying the boundary conditions $\Theta(0) = 0$ and $\Theta(\pi) = 0$ we find that $c_1 = 0$ and $\alpha = n$ for $n = 1, 2, 3, \ldots$. The boundary condition $R(b) = 0$ gives

$$c_3 b^n + c_4 b^{-n} = 0 \quad \text{and} \quad c_4 = -c_3 b^{2n}.$$

Then

$$R(r) = c_3 \left(r^n - \frac{b^{2n}}{r^n} \right) = c_3 \left(\frac{r^{2n} - b^{2n}}{r^n} \right)$$

and

$$u(r, \theta) = \sum_{n=1}^{\infty} A_n \left(\frac{r^{2n} - b^{2n}}{r^n} \right) \sin n\theta.$$

From

$$u(a, \theta) = \theta(\pi - \theta) = \sum_{n=1}^{\infty} A_n \left(\frac{a^{2n} - b^{2n}}{a^n} \right) \sin n\theta$$

we find

$$A_n \left(\frac{a^{2n} - b^{2n}}{a^n} \right) = \frac{2}{\pi} \int_0^\pi (\theta \pi - \theta^2) \sin n\theta \, d\theta = \frac{4}{n^3 \pi}[1 - (-1)^n].$$

Thus

$$u(r, \theta) = \frac{4}{\pi} \sum_{n=1}^{\infty} \frac{1 - (-1)^n}{n^3} \frac{r^{2n} - b^{2n}}{a^{2n} - b^{2n}} \left(\frac{a}{r} \right)^n \sin n\theta.$$

13 new

12. Letting $u(r, \theta) = v(r, \theta) + \psi(\theta)$ we obtain $\psi''(\theta) = 0$ and so $\psi(\theta) = c_1 \theta + c_2$. From $\psi(0) = 0$ and $\psi(\pi) = u_0$ we find, in turn, $c_2 = 0$ and $c_1 = u_0/\pi$. Therefore $\psi(\theta) = \frac{u_0}{\pi}\theta$.

Now $u(1, \theta) = v(1, \theta) + \psi(\theta)$ so that $v(1, \theta) = u_0 - \frac{u_0}{\pi}\theta$. From

$$v(r, \theta) = \sum_{n=1}^{\infty} A_n r^n \sin n\theta \quad \text{and} \quad v(1, \theta) = \sum_{n=1}^{\infty} A_n \sin n\theta$$

680

we obtain

$$A_n = \frac{2}{\pi} \int_0^\pi \left(u_0 - \frac{u_0}{\pi} \theta \right) \sin n\theta \, d\theta = \frac{2u_0}{\pi n} \, .$$

Thus

$$u(r, \theta) = \frac{u_0}{\pi} \theta + \frac{2u_0}{\pi} \sum_{n=1}^\infty \frac{r^n}{n} \sin n\theta.$$

13. We solve

$$\frac{\partial^2 u}{\partial r^2} + \frac{1}{r} \frac{\partial u}{\partial r} + \frac{1}{r^2} \frac{\partial^2 u}{\partial \theta^2} = 0, \quad 0 < \theta < \pi, \quad 0 < r < 2,$$

$$u(2, \theta) = \begin{cases} u_0, & 0 < \theta < \pi/2 \\ 0, & \pi/2 < \theta < \pi \end{cases}$$

$$\frac{\partial u}{\partial \theta} \bigg|_{\theta=0} = 0, \quad \frac{\partial u}{\partial \theta} \bigg|_{\theta=\pi} = 0, \quad 0 < r < 2.$$

Proceeding as in Example 1 in the text we obtain the separated differential equations

$$r^2 R'' + r R' - \lambda R = 0$$

$$\Theta'' + \lambda \Theta = 0.$$

Taking $\lambda = \alpha^2$ the solutions are

$$\Theta(\theta) = c_1 \cos \alpha\theta + c_2 \sin \alpha\theta$$

$$R(r) = c_3 r^\alpha + c_4 r^{-\alpha}.$$

Applying the boundary conditions $\Theta'(0) = 0$ and $\Theta'(\pi) = 0$ we find that $c_2 = 0$ and $\alpha = n$ for $n = 0, 1, 2, \dots$. Since we want $R(r)$ to be bounded as $r \to 0$ we require $c_4 = 0$. Thus

$$u(r, \theta) = A_0 + \sum_{n=1}^\infty A_n r^n \cos n\theta.$$

From

$$u(2, \theta) = \begin{cases} u_0, & 0 < \theta < \pi/2 \\ 0, & \pi/2 < \theta < \pi \end{cases} = A_0 + \sum_{n=1}^\infty A_n 2^n \cos n\theta$$

we find

$$A_0 = \frac{1}{2} \frac{2}{\pi} \int_0^{\pi/2} u_0 \, d\theta = \frac{u_0}{2}$$

and

$$2^n A_n = \frac{2u_0}{\pi} \int_0^{\pi/2} \cos n\theta \, d\theta = \frac{2u_0}{\pi} \frac{\sin n\pi/2}{n} \, .$$

Therefore

$$u(r, \theta) = \frac{u_0}{2} + \frac{2u_0}{\pi} \sum_{n=1}^\infty \frac{1}{n} \left(\sin \frac{n\pi}{2} \right) \left(\frac{r}{2} \right)^n \cos n\theta.$$

14. Separating variables we get $\Theta(\theta) = c_1 \cos \alpha\theta + c_2 \sin \alpha\theta$, so $\Theta(0) = 0$ and $c_1 = 0$. Now $\Theta(\theta) = c_2 \sin \alpha\theta$ and $\Theta(\pi/4) = 0$ implies $\sin(\alpha\pi/4) = 0$ or $\alpha = 4n$. Then $\lambda = (4n)^2$ and $\Theta(\theta) = c_2 \sin 4n\theta$. Now $R(r) = c_3 r^{4n} + c_4 r^{-4n}$, so $R(a) = 0$ implies $c_3 a^{4n} + c_4 a^{-4n} = 0$ and $c_4 = -a^{4n}/a^{-4n}$. Thus

$$R(r) = c_3 \frac{(r/a)^{4n} - (a/r)^{4n}}{a^{4n}}$$

$$u(r,\theta) = \sum_{n=1}^{\infty} A_n \frac{(r/a)^{4n} - (a/r)^{4n}}{a^{4n}} \sin 4n\theta$$

$$u(b,\theta) = 100 = \sum_{n=1}^{\infty} A_n \frac{(b/a)^{4n} - (a/b)^{4n}}{a^{4n}} \sin 4n\theta$$

$$\frac{(b/a)^{4n} - (a/b)^{4n}}{a^{4n}} A_n = \frac{8}{\pi} \int_0^{\pi/4} 100 \sin 4n\theta \, d\theta = \frac{800}{\pi} \frac{1 - (-1)^n}{4n}$$

and
$$u(r,\theta) = \frac{200}{\pi} \sum_{n=1}^{\infty} \frac{(r/a)^{4n} - (a/r)^{4n}}{(b/a)^{4n} - (a/b)^{4n}} \frac{1 - (-1)^n}{n} \sin 4n\theta.$$

15. Let u_1 be the solution of the boundary-value problem

$$\frac{\partial^2 u_1}{\partial r^2} + \frac{1}{r} \frac{\partial u_1}{\partial r} + \frac{1}{r^2} \frac{\partial^2 u_1}{\partial \theta^2} = 0, \quad 0 < \theta < 2\pi, \quad a < r < b$$

$$u_1(a,\theta) = f(\theta), \quad 0 < \theta < 2\pi$$

$$u_1(b,\theta) = 0, \quad 0 < \theta < 2\pi,$$

and let u_2 be the solution of the boundary-value problem

$$\frac{\partial^2 u_2}{\partial r^2} + \frac{1}{r} \frac{\partial u_2}{\partial r} + \frac{1}{r^2} \frac{\partial^2 u_2}{\partial \theta^2} = 0, \quad 0 < \theta < 2\pi, \quad a < r < b$$

$$u_2(a,\theta) = 0, \quad 0 < \theta < 2\pi$$

$$u_2(b,\theta) = g(\theta), \quad 0 < \theta < 2\pi.$$

Each of these problems can be solved using the methods shown in Problem 9 of this section. Now if $u(r,\theta) = u_1(r,\theta) + u_2(r,\theta)$, then

$$u(a,\theta) = u_1(a,\theta) + u_2(a,\theta) = f(\theta)$$

$$u(b,\theta) = u_1(b,\theta) + u_2(b,\theta) = g(\theta)$$

and $u(r,\theta)$ will be the steady-state temperature of the circular ring with boundary conditions $u(a,\theta) = f(\theta)$ and $u(b,\theta) = g(\theta)$.

16. Referring to Problem 15 above we solve boundary-value problems for $u_1(r,\theta)$ and $u_2(r,\theta)$. Using the answer to Problem 9 we find $A_0 = -100/\ln 2$, $A_1 = 100/3$, $A_n = 0$, $n > 1$, and $B_n = 0$ for all n. Then

$$u_1(r,\theta) = -100 \frac{\ln r}{\ln 2} + \frac{100}{3}(r^{-1} - r) \cos \theta.$$

Using the answer to Problem 10 we find

$$u_2(r, \theta) = 200 \frac{\ln 2r}{\ln 2} = 200 \left(1 + \frac{\ln r}{\ln 2} \right)$$

so

$$u(r, \theta) = u_1(r, \theta) + u_2(r, \theta) = 200 + 100 \frac{\ln r}{\ln 2} + \frac{100}{3}(r^{-1} - r) \cos \theta.$$

19 new

17. **(a)** From Problem 1 in this section, with $u_0 = 100$,

20

$$u(r, \theta) = 50 + \frac{100}{\pi} \sum_{n=1}^{\infty} \frac{1 - (-1)^n}{n} r^n \sin n\theta.$$

(b)

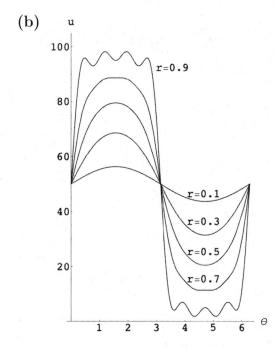

(c) We could use S_5 from part (b) of this problem to compute the approximations, but in a CAS it is just as easy to compute the sum with a much larger number of terms, thereby getting greater accuracy. In this case we use partial sums including the term with r^{99} to find

$$u(0.9, 1.3) \approx 96.5268 \qquad u(0.9, 2\pi - 1.3) \approx 3.4731$$
$$u(0.7, 2) \approx 87.871 \qquad u(0.7, 2\pi - 2) \approx 12.129$$
$$u(0.5, 3.5) \approx 36.0744 \qquad u(0.5, 2\pi - 3.5) \approx 63.9256$$
$$u(0.3, 4) \approx 35.2674 \qquad u(0.3, 2\pi - 4) \approx 64.7326$$
$$u(0.1, 5.5) \approx 45.4934 \qquad u(0.1, 2\pi - 5.5) \approx 54.5066$$

(d) At the center of the plate $u(0, 0) = 50$. From the graphs in part (b) we observe that the solution curves are symmetric about the point $(\pi, 50)$. In part (c) we observe that the horizontal pairs add up to 100, and hence average 50. This is consistent with the observation about part (b), so it is appropriate to say the average temperature in the plate is $50°$.

Exercises 13.2   Polar and Cylindrical Coordinates

1. Referring to the solution of Example 1 in the text we have

$$R(r) = c_1 J_0(\alpha_n r) \qquad \text{and} \qquad T(t) = c_3 \cos a\alpha_n t + c_4 \sin a\alpha_n t$$

where the α_n are the positive roots of $J_0(\alpha c) = 0$. Now, the initial condition $u(r,0) = R(r)T(0) = 0$ implies $T(0) = 0$ and so $c_3 = 0$. Thus

$$u(r,t) = \sum_{n=1}^{\infty} A_n \sin a\alpha_n t\, J_0(\alpha_n r) \qquad \text{and} \qquad \frac{\partial u}{\partial t} = \sum_{n=1}^{\infty} a\alpha_n A_n \cos a\alpha_n t\, J_0(\alpha_n r).$$

From

$$\left. \frac{\partial u}{\partial t} \right|_{t=0} = 1 = \sum_{n=1}^{\infty} a\alpha_n A_n J_0(\alpha_n r)$$

we find

$$a\alpha_n A_n = \frac{2}{c^2 J_1^2(\alpha_n c)} \int_0^c r J_0(\alpha_n r)\, dr \qquad \boxed{x = \alpha_n r, \ \ dx = \alpha_n\, dr}$$

$$= \frac{2}{c^2 J_1^2(\alpha_n c)} \int_0^{\alpha_n c} \frac{1}{\alpha_n^2} x J_0(x)\, dx$$

$$= \frac{2}{c^2 J_1^2(\alpha_n c)} \int_0^{\alpha_n c} \frac{1}{\alpha_n^2} \frac{d}{dx}[x J_1(x)]\, dx \qquad \boxed{\text{see (5) of Section 11.5 in text}}$$

$$= \frac{2}{c^2 \alpha_n^2 J_1^2(\alpha_n c)} \left. x J_1(x) \right|_0^{\alpha_n c} = \frac{2}{c\alpha_n J_1(\alpha_n c)}.$$

Then

$$A_n = \frac{2}{ac\alpha_n^2 J_1(\alpha_n c)}$$

and

$$u(r,t) = \frac{2}{ac} \sum_{n=1}^{\infty} \frac{J_0(\alpha_n r)}{\alpha_n^2 J_1(\alpha_n c)} \sin a\alpha_n t.$$

2. From Example 1 in the text we have $B_n = 0$ and

$$A_n = \frac{2}{J_1^2(\alpha_n)} \int_0^1 r(1 - r^2) J_0(\alpha_n r)\, dr.$$

From Problem 10, Exercises 11.5 we obtained $A_n = \dfrac{4J_2(\alpha_n)}{\alpha_n^2 J_1^2(\alpha_n)}$. Thus

$$u(r,t) = 4 \sum_{n=1}^{\infty} \frac{J_2(\alpha_n)}{J_1^2(\alpha_n)} \cos a\alpha_n t\, J_0(\alpha_n r).$$

3. Referring to Example 2 in the text we have

$$R(r) = c_1 J_0(\alpha r) + c_2 Y_0(\alpha r)$$

$$Z(z) = c_3 \cosh \alpha z + c_4 \sinh \alpha z$$

where $c_2 = 0$ and $J_0(2\alpha) = 0$ defines the positive eigenvalues $\lambda_n = \alpha_n^2$. From $Z(4) = 0$ we obtain

$$c_3 \cosh 4\alpha_n + c_4 \sinh 4\alpha_n = 0 \qquad \text{or} \qquad c_4 = -c_3 \frac{\cosh 4\alpha_n}{\sinh 4\alpha_n}.$$

Then

$$Z(z) = c_3 \left[\cosh \alpha_n z - \frac{\cosh 4\alpha_n}{\sinh 4\alpha_n} \sinh \alpha_n z \right] = c_3 \frac{\sinh 4\alpha_n \cosh \alpha_n z - \cosh 4\alpha_n \sinh \alpha_n z}{\sinh 4\alpha_n}$$

$$= c_3 \frac{\sinh \alpha_n (4 - z)}{\sinh 4\alpha_n}$$

and

$$u(r, z) = \sum_{n=1}^{\infty} A_n \frac{\sinh \alpha_n (4 - z)}{\sinh 4\alpha_n} J_0(\alpha_n r).$$

From

$$u(r, 0) = u_0 = \sum_{n=1}^{\infty} A_n J_0(\alpha_n r)$$

we obtain

$$A_n = \frac{2u_0}{4 J_1^2(2\alpha_n)} \int_0^2 r J_0(\alpha_n r)\, dr = \frac{u_0}{\alpha_n J_1(2\alpha_n)}.$$

Thus the temperature in the cylinder is

$$u(r, z) = u_0 \sum_{n=1}^{\infty} \frac{\sinh \alpha_n (4 - z) J_0(\alpha_n r)}{\alpha_n \sinh 4\alpha_n J_1(2\alpha_n)}.$$

4. (a) The boundary condition $u_r(2, z) = 0$ implies $R'(2) = 0$ or $J_0'(2\alpha) = 0$. Thus $\alpha = 0$ is also an eigenvalue and the separated equations are in this case $rR'' + R' = 0$ and $z'' = 0$. The solutions of these equations are then

$$R(r) = c_1 + c_2 \ln r, \qquad Z(z) = c_3 z + c_4.$$

Now $Z(0) = 0$ yields $c_4 = 0$ and the implicit condition that the temperature is bounded as $r \to 0$ demands that we define $c_2 = 0$. Thus we have

$$u(r, z) = A_1 z + \sum_{n=2}^{\infty} A_n \sinh \alpha_n z J_0(\alpha_n r). \tag{1}$$

At $z = 4$ we obtain

$$f(r) = 4A_1 + \sum_{n=2}^{\infty} A_n \sinh 4\alpha_n J_0(\alpha_n r).$$

Thus from (17) and (18) of Section 11.5 in the text we can write with $b = 2$,

$$A_1 = \frac{1}{8} \int_0^2 r f(r) \, dr \tag{2}$$

$$A_n = \frac{1}{2 \sinh 4\alpha_n J_0^2(2\alpha_n)} \int_0^2 r f(r) J_0(\alpha_n r) \, dr. \tag{3}$$

A solution of the problem consists of the series (1) with coefficients A_1 and A_n defined in (2) and (3), respectively.

(b) When $f(r) = u_0$ we get $A_1 = u_0/4$ and

$$A_n = \frac{u_0 J_1(2\alpha_n)}{\alpha_n \sinh 4\alpha_n J_0^2(2\alpha_n)} = 0$$

since $J_0'(2\alpha) = 0$ is equivalent to $J_1(2\alpha) = 0$. A solution of the problem is then $u(r, z) = \dfrac{u_0}{4} z$.

5. Letting the separation constant be $\lambda = \alpha^2$ and referring to Example 2 in Section 13.2 in the text we have

$$R(r) = c_1 J_0(\alpha r) + c_2 Y_0(\alpha r)$$

$$Z(z) = c_3 \cosh \alpha z + c_4 \sinh \alpha z$$

where $c_2 = 0$ and the positive eigenvalues λ_n are determined by $J_0(2\alpha) = 0$. From $Z'(0) = 0$ we obtain $c_4 = 0$. Then

$$u(r, z) = \sum_{n=1}^{\infty} A_n \cosh \alpha_n z \, J_0(\alpha_n r).$$

From

$$u(r, 4) = 50 = \sum_{n=1}^{\infty} A_n \cosh 4\alpha_n J_0(\alpha_n r)$$

we obtain (as in Example 1 of Section 13.1)

$$A_n \cosh 4\alpha_n = \frac{2(50)}{4 J_1^2(2\alpha_n)} \int_0^2 r J_0(\alpha_n r) \, dr = \frac{50}{\alpha_n J_1(2\alpha_n)}.$$

Thus the temperature in the cylinder is

$$u(r, z) = 50 \sum_{n=1}^{\infty} \frac{\cosh \alpha_n z \, J_0(\alpha_n r)}{\alpha_n \cosh 4\alpha_n J_1(2\alpha_n)}.$$

6. The boundary-value problem in this case is

$$\frac{\partial^2 u}{\partial r^2} + \frac{1}{r} \frac{\partial u}{\partial r} + \frac{\partial^2 u}{\partial z^2} = 0, \quad 0 < r < 2, \quad 0 < z < 4$$

$$u(2, z) = 50, \quad 0 < z < 4$$

$$\left. \frac{\partial u}{\partial z} \right|_{z=0} = 0, \quad \left. \frac{\partial u}{\partial z} \right|_{z=4} = 0, \quad 0 < r < 2.$$

We have $u(r, z) = v(r, z) + 50$, so

$$\frac{\partial^2 v}{\partial r^2} + \frac{1}{r}\frac{\partial v}{\partial r} + \frac{\partial^2 v}{\partial z^2} = 0, \quad 0 < r < 2, \quad 0 < z < 4$$

$$v(2, z) = 0, \quad 0 < z < 4$$

$$\frac{\partial v}{\partial z}\bigg|_{z=0} = 0, \quad \frac{\partial v}{\partial z}\bigg|_{z=4} = 0, \quad 0 < r < 2,$$

which implies that $v(r, z) = 0$ and so $u(r, z) = 50$.

7. Letting $u(r, t) = R(r)T(t)$ and separating variables we obtain

$$\frac{R'' + \frac{1}{r}R'}{R} = \frac{T'}{kT} = -\lambda \quad \text{and} \quad R'' + \frac{1}{r}R' + \lambda R = 0, \quad T' + \lambda kT = 0.$$

From the last equation we find $T(t) = e^{-\lambda kt}$. If $\lambda < 0$, $T(t)$ increases without bound as $t \to \infty$. Thus we assume $\lambda = \alpha^2 > 0$. Now

$$R'' + \frac{1}{r}R' + \alpha^2 R = 0$$

is a parametric Bessel equation with solution

$$R(r) = c_1 J_0(\alpha r) + c_2 Y_0(\alpha r).$$

Since Y_0 is unbounded as $r \to 0$ we take $c_2 = 0$. Then $R(r) = c_1 J_0(\alpha r)$ and the boundary condition $u(c, t) = R(c)T(t) = 0$ implies $J_0(\alpha c) = 0$. This latter equation defines the positive eigenvalues $\lambda_n = \alpha_n^2$. Thus

$$u(r, t) = \sum_{n=1}^{\infty} A_n J_0(\alpha_n r) e^{-\alpha_n^2 kt}.$$

From

$$u(r, 0) = f(r) = \sum_{n=1}^{\infty} A_n J_0(\alpha_n r)$$

we find

$$A_n = \frac{2}{c^2 J_1^2(\alpha_n c)} \int_0^c r J_0(\alpha_n r) f(r)\, dr, \quad n = 1, 2, 3, \dots.$$

8. If the edge $r = c$ is insulated we have the boundary condition $u_r(c, t) = 0$. Referring to the solution of Problem 7 above we have

$$R'(c) = \alpha c_1 J_0'(\alpha c) = 0$$

which defines an eigenvalue $\lambda = \alpha^2 = 0$ and positive eigenvalues $\lambda_n = \alpha_n^2$. Thus

$$u(r, t) = A_0 + \sum_{n=1}^{\infty} A_n J_0(\alpha_n r) e^{-\alpha_n^2 kt}.$$

From

$$u(r, 0) = f(r) = A_0 + \sum_{n=1}^{\infty} A_n J_0(\alpha_n r)$$

687

we find

$$A_0 = \frac{2}{c^2} \int_0^c r f(r) \, dr$$

$$A_n = \frac{2}{c^2 J_0^2(\alpha_n c)} \int_0^c r J_0(\alpha_n r) f(r) \, dr.$$

9. Referring to Problem 7 above we have $T(t) = e^{-\lambda k t}$ and $R(r) = c_1 J_0(\alpha r)$. The boundary condition $hu(1,t) + u_r(1,t) = 0$ implies $h J_0(\alpha) + \alpha J_0'(\alpha) = 0$ which defines positive eigenvalues $\lambda_n = \alpha_n^2$. Now

$$u(r,t) = \sum_{n=1}^{\infty} A_n J_0(\alpha_n r) e^{-\alpha_n^2 k t}$$

where

$$A_n = \frac{2\alpha_n^2}{(\alpha_n^2 + h^2) J_0^2(\alpha_n)} \int_0^1 r J_0(\alpha_n r) f(r) \, dr.$$

10. We solve

$$\frac{\partial^2 u}{\partial r^2} + \frac{1}{r} \frac{\partial u}{\partial r} + \frac{\partial^2 u}{\partial z^2} = 0, \quad 0 < r < 1, \quad z > 0$$

$$\left. \frac{\partial u}{\partial r} \right|_{r=1} = -hu(1,z), \quad z > 0$$

$$u(r,0) = u_0, \quad 0 < r < 1.$$

assuming $u = RZ$ we get

$$\frac{R'' + \frac{1}{r} R'}{R} = -\frac{Z''}{Z} = -\lambda$$

and so

$$r R'' + R' + \lambda^2 r R = 0 \quad \text{and} \quad Z'' - \lambda Z = 0.$$

Letting $\lambda = \alpha^2$ we then have

$$R(r) = c_1 J_0(\alpha r) + c_2 Y_0(\alpha r) \quad \text{and} \quad Z(z) = c_3 e^{-\alpha z} + c_4 e^{\alpha z}.$$

We use the exponential form of the solution of $Z'' - \alpha^2 Z = 0$ since the domain of the variable z is a semi-infinite interval. As usual we define $c_2 = 0$ since the temperature is surely bounded as $r \to 0$. Hence $R(r) = c_1 J_0(\alpha r)$. Now the boundary-condition $u_r(1,z) + hu(1,z) = 0$ is equivalent to

$$\alpha J_0'(\alpha) + h J_0(\alpha) = 0. \tag{4}$$

The eigenvalues α_n are the positive roots of (4) above. Finally, we must now define $c_4 = 0$ since the temperature is also expected to be bounded as $z \to \infty$. A product solution of the partial differential equation that satisfies the first boundary condition is given by

$$u_n(r,z) = A_n e^{-\alpha_n z} J_0(\alpha_n r).$$

Therefore

$$u(r, z) = \sum_{n=1}^{\infty} A_n e^{-\alpha_n z} J_0(\alpha_n r)$$

is another formal solution. At $z = 0$ we have $u_0 = A_n J_0(\alpha_n r)$. In view of (4) above we use equations (17) and (18) of Section 11.5 in the text with the identification $b = 1$:

$$A_n = \frac{2\alpha_n^2}{(\alpha_n^2 + h^2) J_0^2(\alpha_n)} \int_0^1 r J_0(\alpha_n r) u_0 \, dr$$

$$= \frac{2\alpha_n^2 u_0}{(\alpha_n^2 + h^2) J_0^2(\alpha_n) \alpha_n^2} t J_1(t) \Big|_0^{\alpha_n} = \frac{2\alpha_n u_0 J_1(\alpha_n)}{(\alpha_n^2 + h^2) J_0^2(\alpha_n)}. \tag{5}$$

Since $J_0' = -J_1$ [see equation (6) of Section 11.5 in the text] it follows from (4) above that $\alpha_n J_1(\alpha_n) = h J_0(\alpha_n)$. Thus (5) above simplifies to

$$A_n = \frac{2u_0 h}{(\alpha_n^2 + h^2) J_0(\alpha_n)}.$$

A solution to the boundary-value problem is then

$$u(r, z) = 2u_0 h \sum_{n=1}^{\infty} \frac{e^{-\alpha_n z}}{(\alpha_n^2 + h^2) J_0(\alpha_n)} J_0(\alpha_n r).$$

11. Substituting $u(r, t) = v(r, t) + \psi(r)$ into the partial differential equation gives

$$\frac{\partial^2 v}{\partial r^2} + \frac{1}{r} \frac{\partial v}{\partial r} + \psi'' + \frac{1}{r} \psi' = \frac{\partial v}{\partial t}.$$

This equation will be homogeneous provided $\psi'' + \frac{1}{r} \psi' = 0$ or $\psi(r) = c_1 \ln r + c_2$. Since $\ln r$ is unbounded as $r \to 0$ we take $c_1 = 0$. Then $\psi(r) = c_2$ and using $u(2, t) = v(2, t) + \psi(2) = 100$ we set $c_2 = \psi(2) = 100$. Therefore $\psi(r) = 100$. Referring to Problem 7 above, the solution of the boundary-value problem

$$\frac{\partial^2 v}{\partial r^2} + \frac{1}{r} \frac{\partial v}{\partial r} = \frac{\partial v}{\partial t}, \quad 0 < r < 2, \ t > 0,$$

$$v(2, t) = 0, \quad t > 0,$$

$$v(r, 0) = u(r, 0) - \psi(r)$$

is

$$v(r, t) = \sum_{n=1}^{\infty} A_n J_0(\alpha_n r) e^{-\alpha_n^2 t}$$

where

$$A_n = \frac{2}{2^2 J_1^2(2\alpha_n)} \int_0^2 r J_0(\alpha_n r)[u(r,0) - \psi(r)]\, dr$$

$$= \frac{1}{2 J_1^2(2\alpha_n)} \left[\int_0^1 r J_0(\alpha_n r)[200 - 100]\, dr + \int_1^2 r J_0(\alpha_n r)[100 - 100]\, dr \right]$$

$$= \frac{50}{J_1^2(2\alpha_n)} \int_0^1 r J_0(\alpha_n r)\, dr \qquad \boxed{x = \alpha_n r, \ dx = \alpha_n\, dr}$$

$$= \frac{50}{J_1^2(2\alpha_n)} \int_0^{\alpha_n} \frac{1}{\alpha_n^2} x J_0(x)\, dx$$

$$= \frac{50}{\alpha_n^2 J_1^2(2\alpha_n)} \int_0^{\alpha_n} \frac{d}{dx}[x J_1(x)]\, dx \qquad \boxed{\text{see (5) of Section 11.5 in text}}$$

$$= \frac{50}{\alpha_n^2 J_1^2(2\alpha_n)} (x J_1(x)) \Big|_0^{\alpha_n} = \frac{50 J_1(\alpha_n)}{\alpha_n J_1^2(2\alpha_n)}.$$

Thus

$$u(r,t) = v(r,t) + \psi(r) = 100 + 50 \sum_{n=1}^{\infty} \frac{J_1(\alpha_n) J_0(\alpha_n r)}{\alpha_n J_1^2(2\alpha_n)} e^{-\alpha_n^2 t}.$$

12. Letting $u(r,t) = u(r,t) + \psi(r)$ we obtain $r\psi'' + \psi' = -\beta r$. The general solution of this nonhomogeneous Cauchy-Euler equation is found with the aid of variation of parameters: $\psi = c_1 + c_2 \ln r - \beta r^2/4$. In order that this solution be bounded as $r \to 0$ we define $c_2 = 0$. Using $\psi(1) = 0$ then gives $c_1 = \beta/4$ and so $\psi(r) = \beta(1 - r^2)/4$. Using $v = RT$ we find that a solution of

$$\frac{\partial^2 v}{\partial r^2} + \frac{1}{r}\frac{\partial v}{\partial r} = \frac{\partial v}{\partial t}, \qquad 0 < r < 1, \quad t > 0$$

$$v(1,t) = 0, \qquad t > 0$$

$$v(r,0) = -\psi(r), \qquad 0 < r < 1$$

is

$$v(r,t) = \sum_{n=1}^{\infty} A_n e^{-\alpha_n^2 t} J_0(\alpha_n r)$$

where

$$A_n = -\frac{\beta}{4} \frac{2}{J_1^2(\alpha_n)} \int_0^1 r(1 - r^2) J_0(\alpha_n r)\, dr$$

and the α_n are defined by $J_0(\alpha) = 0$. From the result of Problem 10, Exercises 11.5 (see also Problem 2 of this exercise set) we get

$$A_n = -\frac{\beta J_2(\alpha_n)}{\alpha_n^2 J_1^2(\alpha_n)}.$$

690

Thus from $u = v + \psi(r)$ it follows that

$$u(r,t) = \frac{\beta}{4}(1 - r^2) - \beta \sum_{n=1}^{\infty} \frac{J_2(\alpha_n)}{\alpha_n^2 J_1^2(\alpha_n)} e^{-\alpha_n^2 t} J_0(\alpha_n r).$$

13. (a) Writing the partial differential equation in the form

$$g\left(x \frac{\partial^2 u}{\partial x^2} + \frac{\partial u}{\partial x}\right) = \frac{\partial^2 u}{\partial t^2}$$

and separating variables we obtain

$$\frac{x X'' + X'}{X} = \frac{T''}{gT} = -\lambda.$$

Letting $\lambda = \alpha^2$ we obtain

$$x X'' + X' + \alpha^2 X = 0 \qquad \text{and} \qquad T'' + g\alpha^2 T = 0.$$

Letting $x = \tau^2/4$ in the first equation we obtain $dx/d\tau = \tau/2$ or $d\tau/dx = 2\tau$. Then

$$\frac{dX}{dx} = \frac{dX}{d\tau} \frac{d\tau}{dx} = \frac{2}{\tau} \frac{dX}{d\tau}$$

and

$$\frac{d^2 X}{dx^2} = \frac{d}{dx}\left(\frac{2}{\tau} \frac{dX}{d\tau}\right) = \frac{2}{\tau} \frac{d}{dx}\left(\frac{dX}{d\tau}\right) + \frac{dX}{d\tau} \frac{d}{dx}\left(\frac{2}{\tau}\right)$$

$$= \frac{2}{\tau} \frac{d}{d\tau}\left(\frac{dX}{d\tau}\right) \frac{d\tau}{dx} + \frac{dX}{d\tau} \frac{d}{d\tau}\left(\frac{2}{\tau}\right) \frac{d\tau}{dx} = \frac{4}{\tau^2} \frac{d^2 X}{d\tau^2} - \frac{4}{\tau^3} \frac{dX}{d\tau}.$$

Thus

$$x X'' + X' + \alpha^2 X = \frac{\tau^2}{4}\left(\frac{4}{\tau^2} \frac{d^2 X}{d\tau^2} - \frac{4}{\tau^3} \frac{dX}{d\tau}\right) + \frac{2}{\tau} \frac{dX}{d\tau} + \alpha^2 X = \frac{d^2 X}{d\tau^2} + \frac{1}{\tau} \frac{dX}{d\tau} + \alpha^2 X = 0.$$

This is a parametric Bessel equation with solution

$$X(\tau) = c_1 J_0(\alpha\tau) + c_2 Y_0(\alpha\tau).$$

(b) To insure a finite solution at $x = 0$ (and thus $\tau = 0$) we set $c_2 = 0$. The condition $u(L,t) = X(L)T(t) = 0$ implies $X|_{x=L} = X|_{\tau=2\sqrt{L}} = c_1 J_0(2\alpha\sqrt{L}) = 0$, which defines positive eigenvalues $\lambda_n = \alpha_n^2$. The solution of $T'' + g\alpha^2 T = 0$ is

$$T(t) = c_3 \cos(\alpha_n \sqrt{g}\, t) + c_4 \sin(\alpha_n \sqrt{g}\, t).$$

The boundary condition $u_t(x,0) = X(x)T'(0) = 0$ implies $c_4 = 0$. Thus

$$u(\tau,t) = \sum_{n=1}^{\infty} A_n \cos(\alpha_n \sqrt{g}\, t) J_0(\alpha_n \tau).$$

From

$$u(\tau,0) = f(\tau^2/4) = \sum_{n=1}^{\infty} A_n J_0(\alpha_n \tau)$$

we find

$$A_n = \frac{2}{(2\sqrt{L})^2 J_1^2(2\alpha_n\sqrt{L})} \int_0^{2\sqrt{L}} \tau J_0(\alpha_n\tau) f(\tau^2/4)\, d\tau \qquad \boxed{v = \tau/2, \; dv = d\tau/2}$$

$$= \frac{1}{2LJ_1^2(2\alpha_n\sqrt{L})} \int_0^{\sqrt{L}} 2v\, J_0(2\alpha_n v) f(v^2) 2\, dv$$

$$= \frac{2}{LJ_1^2(2\alpha_n\sqrt{L})} \int_0^{\sqrt{L}} v\, J_0(2\alpha_n v) f(v^2)\, dv.$$

The solution of the boundary-value problem is

$$u(x,t) = \sum_{n=1}^{\infty} A_n \cos(\alpha_n\sqrt{g}\, t) J_0(2\alpha_n\sqrt{x}\,).$$

14. (a) First we see that

$$\frac{R''\Theta + \frac{1}{r}R'\Theta + \frac{1}{r^2}R\Theta''}{R\Theta} = \frac{T''}{a^2 T} = -\lambda.$$

This gives $T'' + a^2\lambda T = 0$ and from

$$\frac{R'' + \frac{1}{r}R' + \lambda R}{-R/r^2} = \frac{\Theta''}{\Theta} = -\nu$$

we get $\Theta'' + \nu\Theta = 0$ and $r^2 R'' + rR' + (\lambda r^2 - \nu)R = 0$.

(b) With $\lambda = \alpha^2$ and $\nu = \beta^2$ the general solutions of the differential equations in part (a) are

$$T = c_1\cos a\alpha t + c_2\sin a\alpha t$$

$$\Theta = c_3\cos\beta\theta + c_4\cos\beta\theta$$

$$R = c_5 J_\beta(\alpha r) + c_6 Y_\beta(\alpha r).$$

(c) Implicitly we expect $u(r,\theta,t) = u(r,\theta + 2\pi, t)$ and so Θ must be 2π-periodic. Therefore $\beta = n$, $n = 0, 1, 2, \ldots$. The corresponding eigenfunctions are 1, $\cos\theta$, $\cos 2\theta$, $\ldots$, $\sin\theta$, $\sin 2\theta$, $\ldots$. Arguing that $u(r,\theta,t)$ is bounded as $r \to 0$ we then define $c_6 = 0$ and so $R = c_3 J_n(\alpha r)$. But $R(c) = 0$ gives $J_n(\alpha c) = 0$; this equation defines the eigenvalues $\lambda_n = \alpha_n^2$. For each n, $\alpha_{ni} = x_{ni}/c$, $i = 1, 2, 3, \ldots$.

(d) $u(r,\theta,t) = \sum_{i=1}^{n} (A_{0i}\cos a\alpha_{0i} t + B_{0i}\sin a\alpha_{0i} t) J_0(\alpha_{0i} r)$

$$+ \sum_{n=1}^{\infty}\sum_{i=1}^{\infty} \Big[(A_{ni}\cos a\alpha_{ni} t + B_{ni}\sin a\alpha_{ni} t)\cos n\theta$$

$$+ (C_{ni}\cos a\alpha_{ni} t + D_{ni}\sin a\alpha_{ni} t)\sin n\theta\Big] J_n(\alpha_{ni} r)$$

15. (a) The boundary-value problem is

$$a^2\left(\frac{\partial^2 u}{\partial r^2} + \frac{1}{r}\frac{\partial u}{\partial r}\right) = \frac{\partial^2 u}{\partial t^2}, \quad 0 < r < 1, \ t > 0$$

$$u(1, t) = 0, \quad t > 0$$

$$u(r, 0) = 0, \quad \frac{\partial u}{\partial t}\Big|_{t=0} = \begin{cases} -v_0, & 0 \le r < b \\ 0, & b \le r < 1 \end{cases}, \quad 0 < r < 1,$$

and the solution is

$$u(r, t) = \sum_{n=1}^{\infty}(A_n \cos a\alpha_n t + B_n \sin a\alpha_n t)J_0(\alpha_n r),$$

where the eigenvalues $\lambda_n = \alpha_n^2$ are defined by $J_0(\alpha) = 0$ and $A_n = 0$ since $f(r) = 0$. The coefficients B_n are given by

$$B_n = \frac{2}{a\alpha_n J_1^2(\alpha_n)}\int_0^b r J_0(\alpha_n r)g(r)\,dr = -\frac{2v_0}{a\alpha_n J_1^2(\alpha_n)}\int_0^b r J_0(\alpha_n r)\,dr$$

$$\boxed{\text{let } x = \alpha_n r}$$

$$= -\frac{2v_0}{a\alpha_n J_1^2(\alpha_n)}\int_0^{\alpha_n b}\frac{x}{\alpha_n}J_0(x)\frac{1}{\alpha_n}\,dx = -\frac{2v_0}{a\alpha_n^3 J_1^2(\alpha_n)}\int_0^{\alpha_n b}x J_0(x)\,dx$$

$$= -\frac{2v_0}{a\alpha_n^3 J_1^2(\alpha_n)}(x J_1(x))\Big|_0^{\alpha_n b} = -\frac{2v_0}{a\alpha_n^3 J_1(\alpha_n)}(\alpha_n b J_1(\alpha_n b))$$

$$= -\frac{2v_0 b}{a\alpha_n^2}\frac{J_1(\alpha_n b)}{J_1^2(\alpha_n)}.$$

Thus,

$$u(r, t) = \frac{-2v_0 b}{a}\sum_{n=1}^{\infty}\frac{1}{\alpha_n^2}\frac{J_1(\alpha_n b)}{J_1^2(\alpha_n)}\sin(a\alpha_n t)J_0(\alpha_n r).$$

(b) The standing wave $u_n(r, t)$ is given by $u_n(r, t) = B_n \sin(a\alpha_n t)J_0(\alpha_n r)$, which has frequency $f_n = a\alpha_n/2\pi$, where α_n is the nth positive zero of $J_0(x)$. The fundamental frequency is $f_1 = a\alpha_1/2\pi$. The next two frequencies are

$$f_2 = \frac{a\alpha_2}{2\pi} = \frac{\alpha_2}{\alpha_1}\left(\frac{a\alpha_1}{2\pi}\right) = \frac{5.520}{2.405}f_1 = 2.295f_1$$

and

$$f_3 = \frac{a\alpha_3}{2\pi} = \frac{\alpha_3}{\alpha_1}\left(\frac{a\alpha_1}{2\pi}\right) = \frac{8.654}{2.405}f_1 = 3.598f_1.$$

(c) With $a = 1$, $b = \frac{1}{4}$, and $v_0 = 1$, the solution becomes

$$u(r, t) = -\frac{1}{2}\sum_{n=1}^{\infty}\frac{1}{\alpha_n^2}\frac{J_1^2(\alpha_n/4)}{J_1^2(\alpha_n)}\sin(\alpha_n t)J_0(\alpha_n r).$$

The graphs of $S_5(r, t)$ for $t = 1, 2, 3, 4, 5, 6$ are shown below.

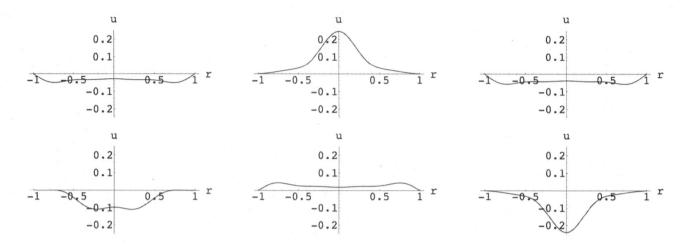

(d) Three frames from the movie are shown.

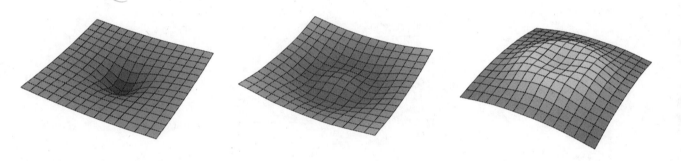

16. (a) With $c = 10$ in Example 1 in the text the eigenvalues are $\lambda_n = \alpha_n^2 = x_n^2/100$ where x_n is a positive root of $J_0(x) = 0$. From a CAS we find that $x_1 = 2.4048$, $x_2 = 5.5201$, and $x_3 = 8.6537$, so that the first three eigenvalues are $\lambda_1 = 0.0578$, $\lambda_2 = 0.3047$, and $\lambda_3 = 0.7489$. The corresponding coefficients are

$$A_1 = \frac{2}{100 J_1^2(x_1)} \int_0^{10} r J_0(x_1 r/10)(1 - r/10)\, dr = 0.7845,$$

$$A_2 = \frac{2}{100 J_1^2(x_2)} \int_0^{10} r J_0(x_2 r/10)(1 - r/10)\, dr = 0.0687,$$

and

$$A_3 = \frac{2}{100 J_1^2(x_3)} \int_0^{10} r J_0(x_3 r/10)(1 - r/10)\, dr = 0.0531.$$

Since $g(r) = 0$, $B_n = 0$, $n = 1, 2, 3, \ldots$, and the third partial sum of the series solution is

$$S_3(r, t) = \sum_{n=1}^{\infty} A_n \cos(x_n t / 10) J_0(x_n r / 10)$$

$$= 0.7845 \cos(0.2405t) J_0(0.2405r) + 0.0687 \cos(0.5520t) J_0(0.5520r)$$

$$+ 0.0531 \cos(0.8654t) J_0(0.8654r).$$

(b)

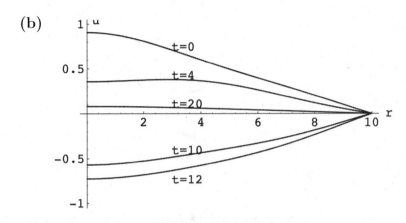

17. Because of the nonhomogeneous boundary condition $u(c, t) = 200$ we use the substitution $u(r, t) = v(r, t) + \psi(r)$. This gives

$$k \left(\frac{\partial^2 v}{\partial r^2} + \frac{1}{r} \frac{\partial v}{\partial r} + \psi'' + \frac{1}{r} \psi' \right) = \frac{\partial v}{\partial t}.$$

This equation will be homogeneous provided $\psi'' + (1/r)\psi' = 0$ or $\psi(r) = c_1 \ln r + c_2$. Since $\ln r$ is unbounded as $r \to 0$ we take $c_1 = 0$. Then $\psi(r) = c_2$ and using $u(c, t) = v(c, t) + c_2 = 200$ we set $c_2 = 200$, giving $v(c, t) = 0$. Referring to Problem 7 in this section, the solution of the boundary-value problem

$$k \left(\frac{\partial^2 v}{\partial r^2} + \frac{1}{r} \frac{\partial v}{\partial r} \right) = \frac{\partial v}{\partial t}, \quad 0 < r < c, \ t > 0$$

$$v(c, t) = 0, \quad t > 0$$

$$v(r, 0) = -200, \quad 0 < r < c$$

is

$$v(r, t) = \sum_{n=1}^{\infty} A_n J_0(\alpha_n r) e^{-\alpha_n^2 k t},$$

where the separation constant is $-\lambda = -\alpha^2$. The eigenvalues are $\lambda_n = \alpha^2 = x_n^2 / c^2$ where x_n is a positive root of $J_0(x) = 0$ and the coefficients A_n are

$$A_n = \frac{2}{c^2 J_1^2(\alpha_n c)} \int_0^c r J_0(\alpha_n r)(-200) dr = -\frac{400}{c^2 J_1^2(\alpha_n c)} \int_0^c r J_0(\alpha_n r) dr.$$

695

Taking $c = 10$ and $k = 0.1$ we have

$$u(r,t) = v(r,t) + 200 = 200 + \sum_{n=1}^{\infty} A_n J_0(x_n r/10) e^{-0.01 x_n^2 t/100}$$

where

$$A_n = -\frac{4}{J_1^2(x_n)} \int_0^{10} r J_0(x_n r/10)\, dr.$$

Using a CAS we find that the first five values of x_n are
$x_1 = 2.4048$, $x_2 = 5.5201$, $x_3 = 8.6537$, $x_4 = 11.7915$,
and $x_5 = 14.9309$, with corresponding eigenvalues
$\lambda_1 = 0.0578$, $\lambda_2 = 0.3047$, $\lambda_3 = 0.7489$, $\lambda_4 = 1.3904$,
and $\lambda_5 = 2.2293$. The first five values of A_n are
$A_1 = -320.4$, $A_2 = 213.0$, $A_3 = -170.3$, $A_4 = 145.9$,
and $A_5 = -129.7$. Using a root finding application in
a CAS we find that $u(5,t) = 100$ when $t \approx 1331$ and

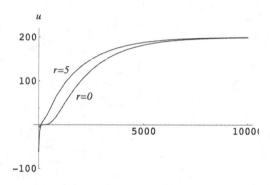

$u(0,t) = 100$ when $t \approx 2005$. Since $u = 200$ is an asymptote for the graphs of $u(0,t)$ and $u(5,t)$ we solve $u(5,t) = 199.9$ and $u(0,t) = 199.9$. This gives $t \approx 13{,}265$ and $t \approx 13{,}958$, respectively.

Exercises 13.3   Spherical Coordinates

1. To compute

$$A_n = \frac{2n+1}{2c^n} \int_0^\pi f(\theta) P_n(\cos\theta) \sin\theta\, d\theta$$

we substitute $x = \cos\theta$ and $dx = -\sin\theta\, d\theta$. Then

$$A_n = \frac{2n+1}{2c^n} \int_1^{-1} F(x) P_n(x)(-dx) = \frac{2n+1}{2c^n} \int_{-1}^{1} F(x) P_n(x)\, dx$$

where

$$F(x) = \begin{cases} 0, & -1 < x < 0 \\ 50, & 0 < x < 1 \end{cases} = 50 \begin{cases} 0, & -1 < x < 0 \\ 1, & 0 < x < 1 \end{cases}.$$

The coefficients A_n are computed in Example 3 of Section 11.5. Thus

$$u(r,\theta) = \sum_{n=0}^{\infty} A_n r^n P_n(\cos\theta)$$

$$= 50 \left[\frac{1}{2} P_0(\cos\theta) + \frac{3}{4} \left(\frac{r}{c}\right) P_1(\cos\theta) - \frac{7}{16} \left(\frac{r}{c}\right)^3 P_3(\cos\theta) + \frac{11}{32} \left(\frac{r}{c}\right)^5 P_5(\cos\theta) + \cdots \right].$$

2. In the solution of the Cauchy-Euler equation,

$$R(r) = c_1 r^n + c_2 r^{-(n+1)},$$

we define $c_1 = 0$ since we expect the potential u to be bounded as $r \to \infty$. Hence

$$u_n(r, \theta) = A_n r^{-(n+1)} P_n(\cos \theta)$$

$$u(r, \theta) = \sum_{n=0}^{\infty} A_n r^{-(n+1)} P_n(\cos \theta).$$

When $r = c$ we have

$$f(\theta) = \sum_{n=0}^{\infty} A_n c^{-(n+1)} P_n(\cos \theta)$$

so that

$$A_n = c^{n+1} \frac{(2n+1)}{2} \int_0^\pi f(\theta) P_n(\cos \theta) \sin \theta \, d\theta.$$

The solution of the problem is then

$$u(r, \theta) = \sum_{n=0}^{\infty} \left(\frac{2n+1}{2} \int_0^\pi f(\theta) P_n(\cos \theta) \sin \theta d\theta \right) \left(\frac{c}{r} \right)^{n+1} P_n(\cos \theta).$$

3. The coefficients are given by

$$A_n = \frac{2n+1}{2c^n} \int_0^\pi \cos \theta P_n(\cos \theta) \sin \theta \, d\theta = \frac{2n+1}{2c^n} \int_0^\pi P_1(\cos \theta) P_n(\cos \theta) \sin \theta \, d\theta$$

$$\boxed{x = \cos \theta, \ dx = -\sin \theta \, d\theta}$$

$$= \frac{2n+1}{2c^n} \int_{-1}^1 P_1(x) P_n(x) \, dx.$$

Since $P_n(x)$ and $P_m(x)$ are orthogonal for $m \neq n$, $A_n = 0$ for $n \neq 1$ and

$$A_1 = \frac{2(1)+1}{2c^1} \int_{-1}^1 P_1(x) P_1(x) \, dx = \frac{3}{2c} \int_{-1}^1 x^2 dx = \frac{1}{c}.$$

Thus

$$u(r, \theta) = \frac{r}{c} P_1(\cos \theta) = \frac{r}{c} \cos \theta.$$

4. The coefficients are given by

$$A_n = \frac{2n+1}{2c^n} \int_0^\pi (1 - \cos 2\theta) P_n(\cos \theta) \sin \theta \, d\theta.$$

These were computed in Problem 18 of Section 11.5. Thus

$$u(r, \theta) = \frac{4}{3} P_0(\cos \theta) - \frac{4}{3} \left(\frac{r}{c} \right)^2 P_2(\cos \theta).$$

5. Referring to Example 1 in the text we have

$$\Theta = P_n(\cos \theta) \quad \text{and} \quad R = c_1 r^n + c_2 r^{-(n+1)}.$$

Since $u(b, \theta) = R(b)\Theta(\theta) = 0$,

$$c_1 b^n + c_2 b^{-(n+1)} = 0 \qquad \text{or} \qquad c_1 = -c_2 b^{-2n-1},$$

and

$$R(r) = -c_2 b^{-2n-1} r^n + c_2 r^{-(n+1)} = c_2 \left(\frac{b^{2n+1} - r^{2n+1}}{b^{2n+1} r^{n+1}} \right).$$

Then

$$u(r, \theta) = \sum_{n=0}^{\infty} A_n \frac{b^{2n+1} - r^{2n+1}}{b^{2n+1} r^{n+1}} P_n(\cos \theta)$$

where

$$\frac{b^{2n+1} - a^{2n+1}}{b^{2n+1} a^{n+1}} A_n = \frac{2n+1}{2} \int_0^\pi f(\theta) P_n(\cos \theta) \sin \theta \, d\theta.$$

6. Referring to Example 1 in the text we have

$$R(r) = c_1 r^n \qquad \text{and} \qquad \Theta(\theta) = P_n(\cos \theta).$$

Now $\Theta(\pi/2) = 0$ implies that n is odd, so

$$u(r, \theta) = \sum_{n=0}^{\infty} A_{2n+1} r^{2n+1} P_{2n+1}(\cos \theta).$$

From

$$u(c, \theta) = f(\theta) = \sum_{n=0}^{\infty} A_{2n+1} c^{2n+1} P_{2n+1}(\cos \theta)$$

we see that

$$A_{2n+1} c^{2n+1} = (4n + 3) \int_0^{\pi/2} f(\theta) \sin \theta \, P_{2n+1}(\cos \theta) \, d\theta.$$

Thus

$$u(r, \theta) = \sum_{n=0}^{\infty} A_{2n+1} r^{2n+1} P_{2n+1}(\cos \theta)$$

where

$$A_{2n+1} = \frac{4n + 3}{c^{2n+1}} \int_0^{\pi/2} f(\theta) \sin \theta \, P_{2n+1}(\cos \theta) \, d\theta.$$

7. Referring to Example 1 in the text we have

$$r^2 R'' + 2r R' - \lambda R = 0$$

$$\sin \theta \, \Theta'' + \cos \theta \, \Theta' + \lambda \sin \theta \, \Theta = 0.$$

Substituting $x = \cos \theta$, $0 \leq \theta \leq \pi/2$, the latter equation becomes

$$(1 - x^2) \frac{d^2 \Theta}{dx^2} - 2x \frac{d\Theta}{dx} + \lambda \Theta = 0, \quad 0 \leq x \leq 1.$$

698

Taking the solutions of this equation to be the Legendre polynomials $P_n(x)$ corresponding to $\lambda = n(n+1)$ for $n = 1, 2, 3, \ldots$, we have $\Theta = P_n(\cos \theta)$. Since

$$\frac{\partial u}{\partial \theta}\bigg|_{\theta = \pi/2} = \Theta'(\pi/2)R(r) = 0$$

we have

$$\Theta'(\pi/2) = -(\sin \pi/2)P_n'(\cos \pi/2) = -P_n'(0) = 0.$$

As noted in the hint, $P_n'(0) = 0$ only if n is even. Thus $\Theta = P_n(\cos \theta)$, $n = 0, 2, 4, \ldots$. As in Example 1, $R(r) = c_1 r^n$. Hence

$$u(r, \theta) = \sum_{n=0}^{\infty} A_{2n} r^{2n} P_{2n}(\cos \theta).$$

At $r = c$,

$$f(\theta) = \sum_{n=0}^{\infty} A_{2n} c^{2n} P_{2n}(\cos \theta).$$

Using Problem 19 in Section 11.5, we obtain

$$c^{2n} A_{2n} = (4n+1) \int_{\pi/2}^{0} f(\theta) P_{2n}(\cos \theta)(-\sin \theta)\, d\theta$$

and

$$A_{2n} = \frac{4n+1}{c^{2n}} \int_{0}^{\pi/2} f(\theta) \sin \theta \, P_{2n}(\cos \theta)\, d\theta.$$

8. Referring to Example 1 in the text we have

$$R(r) = c_1 r^n + c_2 r^{-(n-1)} \qquad \text{and} \qquad \Theta(\theta) = P_n(\cos \theta).$$

Since we expect $u(r, \theta)$ to be bounded as $r \to \infty$, we define $c_1 = 0$. Also $\Theta(\pi/2) = 0$ implies that n is odd, so

$$u(r, \theta) = \sum_{n=0}^{\infty} A_{2n+1} r^{-2(n+1)} P_{2n+1}(\cos \theta).$$

From

$$u(c, \theta) = f(\theta) = \sum_{n=0}^{\infty} A_{2n+1} c^{-2(n+1)} P_{2n+1}(\cos \theta)$$

we see that

$$A_{2n+1} c^{-2(n+1)} = (4n+3) \int_{0}^{\pi/2} f(\theta) \sin \theta \, P_{2n+1}(\cos \theta)\, d\theta.$$

Thus

$$u(r, \theta) = \sum_{n=0}^{\infty} A_{2n+1} r^{-2(n+1)} P_{2n+1}(\cos \theta)$$

where

$$A_{2n+1} = (4n+3)c^{2(n+1)} \int_{0}^{\pi/2} f(\theta) \sin \theta \, P_{2n+1}(\cos \theta)\, d\theta.$$

Exercises 13.3 Spherical Coordinates

9. Checking the hint, we find

$$\frac{1}{r}\frac{\partial^2}{\partial r^2}(ru) = \frac{1}{r}\frac{\partial}{\partial r}\left[r\frac{\partial u}{\partial r} + u\right] = \frac{1}{r}\left[r\frac{\partial^2 u}{\partial r^2} + \frac{\partial u}{\partial r} + \frac{\partial u}{\partial r}\right] = \frac{\partial^2 u}{\partial r^2} + \frac{2}{r}\frac{\partial u}{\partial r}.$$

The partial differential equation then becomes

$$\frac{\partial^2}{\partial r^2}(ru) = r\frac{\partial u}{\partial t}.$$

Now, letting $ru(r,t) = v(r,t) + \psi(r)$, since the boundary condition is nonhomogeneous, we obtain

$$\frac{\partial^2}{\partial r^2}[v(r,t) + \psi(r)] = r\frac{\partial}{\partial t}\left[\frac{1}{r}v(r,t) + \psi(r)\right]$$

or

$$\frac{\partial^2 v}{\partial r^2} + \psi''(r) = \frac{\partial v}{\partial t}.$$

This differential equation will be homogeneous if $\psi''(r) = 0$ or $\psi(r) = c_1 r + c_2$. Now

$$u(r,t) = \frac{1}{r}v(r,t) + \frac{1}{r}\psi(r) \qquad \text{and} \qquad \frac{1}{r}\psi(r) = c_1 + \frac{c_2}{r}.$$

Since we want $u(r,t)$ to be bounded as r approaches 0, we require $c_2 = 0$. Then $\psi(r) = c_1 r$. When $r = 1$

$$u(1,t) = v(1,t) + \psi(1) = v(1,t) + c_1 = 100,$$

and we will have the homogeneous boundary condition $v(1,t) = 0$ when $c_1 = 100$. Consequently, $\psi(r) = 100r$. The initial condition

$$u(r,0) = \frac{1}{r}v(r,0) + \frac{1}{r}\psi(r) = \frac{1}{r}v(r,0) + 100 = 0$$

implies $v(r,0) = -100r$. We are thus led to solve the new boundary-value problem

$$\frac{\partial^2 v}{\partial r^2} = \frac{\partial v}{\partial t}, \quad 0 < r < 1, \quad t > 0,$$

$$v(1,t) = 0, \quad \lim_{r \to 0}\frac{1}{r}v(r,t) < \infty,$$

$$v(r,0) = -100r.$$

Letting $v(r,t) = R(r)T(t)$ and using the separation constant $-\lambda$ we obtain

$$R'' + \lambda R = 0 \qquad \text{and} \qquad T' + \lambda T = 0.$$

Using $\lambda = \alpha^2 > 0$ we then have

$$R(r) = c_3 \cos \alpha r + c_4 \sin \alpha r \qquad \text{and} \qquad T(t) = c_5 e^{-\alpha^2 t}.$$

The boundary conditions are equivalent to $R(1) = 0$ and $\lim_{r \to 0} R(r)/r < \infty$. Since

$$\lim_{r \to 0}\frac{\cos \alpha r}{r}$$

700

does not exist we must have $c_3 = 0$. Then $R(r) = c_4 \sin \alpha r$, and $R(1) = 0$ implies $\alpha = n\pi$ for $n = 1$, $2, 3, \ldots$. Thus

$$v_n(r, t) = A_n e^{-n^2 \pi^2 t} \sin n\pi r$$

for $n = 1, 2, 3, \ldots$. Using the condition $\lim_{r \to 0} R(r)/r < \infty$ it is easily shown that there are no eigenvalues for $\lambda = 0$, nor does setting the common constant to $-\lambda = \alpha^2$ when separating variables lead to any solutions. Now, by the Superposition Principle,

$$v(r, t) = \sum_{n=1}^{\infty} A_n e^{-n^2 \pi^2 t} \sin n\pi r.$$

The initial condition $v(r, 0) = -100r$ implies

$$-100r = \sum_{n=1}^{\infty} A_n \sin n\pi r.$$

This is a Fourier sine series and so

$$A_n = 2 \int_0^1 (-100r \sin n\pi r)\, dr = -200 \left[-\frac{r}{n\pi} \cos n\pi r \Big|_0^1 + \int_0^1 \frac{1}{n\pi} \cos n\pi r\, dr \right]$$

$$= -200 \left[-\frac{\cos n\pi}{n\pi} + \frac{1}{n^2 \pi^2} \sin n\pi r \Big|_0^1 \right] = -200 \left[-\frac{(-1)^n}{n\pi} \right] = \frac{(-1)^n 200}{n\pi}.$$

A solution of the problem is thus

$$u(r, t) = \frac{1}{r} v(r, t) + \frac{1}{r} \psi(r) = \frac{1}{r} \sum_{n=1}^{\infty} (-1)^n \frac{20}{n\pi} e^{-n^2 \pi^2 t} \sin n\pi r + \frac{1}{r}(100r)$$

$$= \frac{200}{\pi r} \sum_{n=1}^{\infty} \frac{(-1)^n}{n} e^{-n^2 \pi^2 t} \sin n\pi r + 100.$$

10. Referring to Problem 9 we have

$$\frac{\partial^2 v}{\partial r^2} + \psi''(r) = \frac{\partial v}{\partial t}$$

where $\psi(r) = c_1 r$. Since

$$u(r, t) = \frac{1}{r} v(r, t) + \frac{1}{r} \psi(r) = \frac{1}{r} v(r, t) + c_1$$

we have

$$\frac{\partial u}{\partial r} = \frac{1}{r} v_r(r, t) - \frac{1}{r^2} v(r, t).$$

When $r = 1$,

$$\frac{\partial u}{\partial r} \bigg|_{r=1} = v_r(1, t) - v(1, t)$$

and

$$\frac{\partial u}{\partial r} \bigg|_{r=1} + hu(1, t) = v_r(1, t) - v(1, t) + h[v(1, t) + \psi(1)] = v_r(1, t) + (h - 1)v(1, t) + hc_1.$$

Thus the boundary condition

$$\frac{\partial u}{\partial r}\bigg|_{r=1} + hu(1,t) = hu_1$$

will be homogeneous when $hc_1 = hu_1$ or $c_1 = u_1$. Consequently $\psi(r) = u_1 r$. The initial condition

$$u(r,0) = \frac{1}{r}v(r,0) + \frac{1}{r}\psi(r) = \frac{1}{r}v(r,0) + u_1 = u_0$$

implies $v(r,0) = (u_0 - u_1)r$. We are thus led to solve the new boundary-value problem

$$\frac{\partial^2 v}{\partial r^2} = \frac{\partial v}{\partial t}, \quad 0 < r < 1, \quad t > 0,$$

$$v_r(1,t) + (h-1)v(1,t) = 0, \quad t > 0,$$

$$\lim_{r \to 0} \frac{1}{r}v(r,t) < \infty,$$

$$v(r,0) = (u_0 - u_1)r.$$

Separating variables as in Problem 9 leads to

$$R(r) = c_3 \cos \alpha r + c_4 \sin \alpha r \quad \text{and} \quad T(t) = c_5 e^{-\alpha^2 t}.$$

The boundary conditions are equivalent to

$$R'(1) + (h-1)R(1) = 0 \quad \text{and} \quad \lim_{r \to 0} \frac{1}{r}R(r) < \infty.$$

As in Problem 6 we use the second condition to determine that $c_3 = 0$ and $R(r) = c_4 \sin \alpha r$. Then

$$R'(1) + (h-1)R(1) = c_4 \alpha \cos \alpha + c_4(h-1)\sin \alpha = 0$$

and the α_n are the consecutive nonnegative roots of $\tan \alpha = \alpha/(1-h)$. Now

$$v(r,t) = \sum_{n=1}^{\infty} A_n e^{-\alpha_n^2 t} \sin \alpha_n r.$$

From

$$v(r,0) = (u_0 - u_1)r = \sum_{n=1}^{\infty} A_n \sin \alpha_n r$$

we obtain

$$A_n = \frac{\int_0^1 (u_0 - u_1)r \sin \alpha_n r\, dr}{\int_0^1 \sin^2 \alpha_n r\, dr}.$$

We compute the integrals

$$\int_0^1 r \sin \alpha_n r\, dr = \left(\frac{1}{\alpha_n^2}\sin \alpha_n r - \frac{1}{\alpha_n}\cos \alpha_n r\right)\bigg|_0^1 = \frac{1}{\alpha_n^2}\sin \alpha_n - \frac{1}{\alpha_n}\cos \alpha_n$$

and

$$\int_0^1 \sin^2 \alpha_n r\, dr = \left(\frac{1}{2}r - \frac{1}{4\alpha_n}\sin 2\alpha_n r\right)\bigg|_0^1 = \frac{1}{2} - \frac{1}{4\alpha_n}\sin 2\alpha_n.$$

Using $\alpha_n \cos \alpha_n = -(h-1)\sin \alpha_n$ we then have

$$A_n = (u_0 - u_1)\frac{\frac{1}{\alpha_n^2}\sin \alpha_n - \frac{1}{\alpha_n}\cos \alpha_n}{\frac{1}{2} - \frac{1}{4\alpha_n}\sin 2\alpha_n} = (u_0 - u_1)\frac{4\sin \alpha_n - 4\alpha_n \cos \alpha_n}{2\alpha_n^2 - \alpha_n 2\sin \alpha_n \cos \alpha_n}$$

$$= 2(u_0 - u_1)\frac{\sin \alpha_n + (h-1)\sin \alpha_n}{\alpha_n^2 + (h-1)\sin \alpha_n \sin \alpha_n} = 2(u_0 - u_1)h\frac{\sin \alpha_n}{\alpha_n^2 + (h-1)\sin^2 \alpha_n}.$$

Therefore

$$u(r,t) = \frac{1}{r}v(r,t) + \frac{1}{r}\psi(r) = u_1 + 2(u_0 - u_1)h\sum_{n=1}^{\infty}\frac{\sin \alpha_n \sin \alpha_n r}{r[\alpha_n^2 + (h-1)\sin^2 \alpha_n]}e^{-\alpha_n^2 t}.$$

11. We write the differential equation in the form

$$a^2\frac{1}{r}\frac{\partial^2}{\partial r^2}(ru) = \frac{\partial^2 u}{\partial t^2} \qquad \text{or} \qquad a^2\frac{\partial^2}{\partial r^2}(ru) = r\frac{\partial^2 u}{\partial t^2},$$

and then let $v(r,t) = ru(r,t)$. The new boundary-value problem is

$$a^2\frac{\partial^2 v}{\partial r^2} = \frac{\partial^2 v}{\partial t^2}, \qquad 0 < r < c, \quad t > 0$$

$$v(c,t) = 0, \quad t > 0$$

$$v(r,0) = rf(r), \quad \frac{\partial v}{\partial t}\bigg|_{t=0} = rg(r).$$

Letting $v(r,t) = R(r)T(t)$ and using the separation constant $-\lambda = -\alpha^2$ we obtain

$$R'' + \alpha^2 R = 0$$

$$T'' + a^2\alpha^2 T = 0$$

and

$$R(r) = c_1 \cos \alpha r + c_2 \sin \alpha r$$

$$T(t) = c_3 \cos a\alpha t + c_4 \sin a\alpha t.$$

Since $u(r,t) = v(r,t)/r$, in order to insure boundedness at $r = 0$ we define $c_1 = 0$. Then $R(r) = c_2 \sin \alpha r$ and the condition $R(c) = 0$ implies $\alpha = n\pi/c$. Thus

$$v(r,t) = \sum_{n=1}^{\infty}\left(A_n \cos \frac{n\pi a}{c}t + B_n \sin \frac{n\pi a}{c}t\right)\sin \frac{n\pi}{c}r.$$

From

$$v(r,0) = rf(r) = \sum_{n=1}^{\infty}A_n \sin \frac{n\pi}{c}r$$

we see that

$$A_n = \frac{2}{c}\int_0^c rf(r)\sin \frac{n\pi}{c}r\,dr.$$

From

$$\frac{\partial v}{\partial t}\Big|_{t=0} = rg(r) = \sum_{n=1}^{\infty}\left(B_n\frac{n\pi a}{c}\right)\sin\frac{n\pi}{c}r$$

we see that

$$B_n = \frac{c}{n\pi a}\cdot\frac{2}{c}\int_0^c rg(r)\sin\frac{n\pi}{c}r\,dr = \frac{2}{n\pi a}\int_0^c rg(r)\sin\frac{n\pi}{c}r\,dr.$$

The solution is

$$u(r,t) = \frac{1}{r}\sum_{n=1}^{\infty}\left(A_n\cos\frac{n\pi a}{c}t + B_n\sin\frac{n\pi a}{c}t\right)\sin\frac{n\pi}{c}r,$$

where A_n and B_n are given above.

12. Proceeding as in Example 1 we obtain

$$\Theta(\theta) = P_n(\cos\theta) \qquad \text{and} \qquad R(r) = c_1 r^n + c_2 r^{-(n+1)}$$

so that

$$u(r,\theta) = \sum_{n=0}^{\infty}(A_n r^n + B_n r^{-(n+1)})P_n(\cos\theta).$$

To satisfy $\lim_{r\to\infty} u(r,\theta) = -Er\cos\theta$ we must have $A_n = 0$ for $n = 2, 3, 4, \ldots$. Then

$$\lim_{r\to\infty} u(r,\theta) = -Er\cos\theta = A_0\cdot 1 + A_1 r\cos\theta,$$

so $A_0 = 0$ and $A_1 = -E$. Thus

$$u(r,\theta) = -Er\cos\theta + \sum_{n=0}^{\infty}B_n r^{-(n+1)}P_n(\cos\theta).$$

Now

$$u(c,\theta) = 0 = -Ec\cos\theta + \sum_{n=0}^{\infty}B_n c^{-(n+1)}P_n(\cos\theta)$$

so

$$\sum_{n=0}^{\infty}B_n c^{-(n+1)}P_n(\cos\theta) = Ec\cos\theta$$

and

$$B_n c^{-(n+1)} = \frac{2n+1}{2}\int_0^{\pi} Ec\cos\theta\,P_n(\cos\theta)\sin\theta\,d\theta.$$

Now $\cos\theta = P_1(\cos\theta)$ so, for $n\neq 1$,

$$\int_0^{\pi}\cos\theta\,P_n(\cos\theta)\sin\theta\,d\theta = 0$$

by orthogonality. Thus $B_n = 0$ for $n\neq 1$ and

$$B_1 = \frac{3}{2}Ec^3\int_0^{\pi}\cos^2\theta\sin\theta\,d\theta = Ec^3.$$

Therefore,

$$u(r,\theta) = -Er\cos\theta + Ec^3 r^{-2}\cos\theta.$$

1. We have

$$A_0 = \frac{1}{2\pi} \int_0^\pi u_0 \, d\theta + \frac{1}{2\pi} \int_\pi^{2\pi} (-u_0) \, d\theta = 0$$

$$A_n = \frac{1}{c^n \pi} \int_0^\pi u_0 \cos n\theta \, d\theta + \frac{1}{c^n \pi} \int_\pi^{2\pi} (-u_0) \cos n\theta \, d\theta = 0$$

$$B_n = \frac{1}{c^n \pi} \int_0^\pi u_0 \sin n\theta \, d\theta + \frac{1}{c^n \pi} \int_\pi^{2\pi} (-u_0) \sin n\theta \, d\theta = \frac{2u_0}{c^n n\pi}[1 - (-1)^n]$$

and so

$$u(r, \theta) = \frac{2u_0}{\pi} \sum_{n=1}^\infty \frac{1 - (-1)^n}{n} \left(\frac{r}{c}\right)^n \sin n\theta.$$

2. We have

$$A_0 = \frac{1}{2\pi} \int_0^{\pi/2} d\theta + \frac{1}{2\pi} \int_{3\pi/2}^{2\pi} d\theta = \frac{1}{2}$$

$$A_n = \frac{1}{c^n \pi} \int_0^{\pi/2} \cos n\theta \, d\theta + \frac{1}{c^n \pi} \int_{3\pi/2}^{2\pi} \cos n\theta \, d\theta = \frac{1}{c^n n\pi}\left[\sin \frac{n\pi}{2} - \sin \frac{3n\pi}{2}\right]$$

$$B_n = \frac{1}{c^n \pi} \int_0^{\pi/2} \sin n\theta \, d\theta + \frac{1}{c^n \pi} \int_{3\pi/2}^{2\pi} \sin n\theta \, d\theta = \frac{1}{c^n n\pi}\left[\cos \frac{3n\pi}{2} - \cos \frac{n\pi}{2}\right]$$

and so

$$u(r, \theta) = \frac{1}{2} + \frac{1}{\pi} \sum_{n=1}^\infty \left(\frac{r}{c}\right)^n \left[\frac{\sin \frac{n\pi}{2} - \sin \frac{3n\pi}{2}}{n} \cos n\theta + \frac{\cos \frac{3n\pi}{2} - \cos \frac{n\pi}{2}}{n} \sin n\theta\right].$$

3. The conditions $\Theta(0) = 0$ and $\Theta(\pi) = 0$ applied to $\Theta = c_1 \cos \alpha\theta + c_2 \sin \alpha\theta$ give $c_1 = 0$ and $\alpha = n$, $n = 1, 2, 3, \ldots$, respectively. Thus we have the Fourier sine-series coefficients

$$A_n = \frac{2}{\pi} \int_0^\pi u_0(\pi\theta - \theta^2) \sin n\theta \, d\theta = \frac{4u_0}{n^3\pi}[1 - (-1)^n].$$

Thus

$$u(r, \theta) = \frac{4u_0}{\pi} \sum_{n=1}^\infty \frac{1 - (-1)^n}{n^3} r^n \sin n\theta.$$

4. In this case

$$A_n = \frac{2}{\pi} \int_0^\pi \sin \theta \sin n\theta \, d\theta = \frac{1}{\pi} \int_0^\pi [\cos(1-n)\theta - \cos(1+n)\theta] \, d\theta = 0, \quad n \neq 1.$$

For $n = 1$,

$$A_1 = \frac{2}{\pi} \int_0^\pi \sin^2 \theta \, d\theta = \frac{1}{\pi} \int_0^\pi (1 - \cos 2\theta) \, d\theta = 1.$$

Thus

$$u(r, \theta) = \sum_{n=1}^\infty A_n r^n \sin n\theta$$

reduces to

$$u(r, \theta) = r \sin \theta.$$

5. We solve

$$\frac{\partial^2 u}{\partial r^2} + \frac{1}{r} \frac{\partial u}{\partial r} + \frac{1}{r^2} \frac{\partial^2 u}{\partial \theta^2} = 0, \quad 0 < \theta < \frac{\pi}{4}, \quad \frac{1}{2} < r < 1,$$

$$u(r, 0) = 0, \quad u(r, \pi/4) = 0, \quad \frac{1}{2} < r < 1,$$

$$u(1/2, \theta) = u_0, \quad u_r(1, \theta) = 0, \quad 0 < \theta < \frac{\pi}{4}.$$

Proceeding as in Example 1 in Section 13.1 using the separation constant $\lambda = \alpha^2$ we obtain

$$r^2 R'' + r R' - \lambda R = 0$$

$$\Theta'' + \lambda \Theta = 0$$

with solutions

$$\Theta(\theta) = c_1 \cos \alpha\theta + c_2 \sin \alpha\theta$$

$$R(r) = c_3 r^\alpha + c_4 r^{-\alpha}.$$

Applying the boundary conditions $\Theta(0) = 0$ and $\Theta(\pi/4) = 0$ gives $c_1 = 0$ and $\alpha = 4n$ for $n = 1, 2, 3, \ldots$. From $R_r(1) = 0$ we obtain $c_3 = c_4$. Therefore

$$u(r, \theta) = \sum_{n=1}^\infty A_n \left(r^{4n} + r^{-4n} \right) \sin 4n\theta.$$

From

$$u(1/2, \theta) = u_0 = \sum_{n=1}^\infty A_n \left(\frac{1}{2^{4n}} + \frac{1}{2^{-4n}} \right) \sin 4n\theta$$

we find

$$A_n \left(\frac{1}{2^{4n}} + \frac{1}{2^{-4n}} \right) = \frac{2}{\pi/4} \int_0^{\pi/4} u_0 \sin 4n\theta \, d\theta = \frac{2u_0}{n\pi} [1 - (-1)^n]$$

or

$$A_n = \frac{2u_0}{n\pi(2^{4n} + 2^{-4n})} [1 - (-1)^n].$$

Thus the steady-state temperature in the plate is

$$u(r, \theta) = \frac{2u_0}{\pi} \sum_{n=1}^\infty \frac{[r^{4n} + r^{-4n}][1 - (-1)^n]}{n[2^{4n} + 2^{-4n}]} \sin 4n\theta.$$

6. We solve

$$\frac{\partial^2 u}{\partial r^2} + \frac{1}{r}\frac{\partial u}{\partial r} + \frac{1}{r^2}\frac{\partial^2 u}{\partial \theta^2} = 0, \quad r > 1, \quad 0 < \theta < \pi,$$

$$u(r, 0) = 0, \quad u(r, \pi) = 0, \quad r > 1,$$

$$u(1, \theta) = f(\theta), \quad 0 < \theta < \pi.$$

Separating variables we obtain

$$\Theta(\theta) = c_1 \cos \alpha\theta + c_2 \sin \alpha\theta$$

$$R(r) = c_3 r^\alpha + c_4 r^{-\alpha}.$$

Applying the boundary conditions $\Theta(0) = 0$, and $\Theta(\pi) = 0$ gives $c_1 = 0$ and $\alpha = n$ for $n = 1, 2, 3, \ldots$. Assuming $f(\theta)$ to be bounded, we expect the solution $u(r, \theta)$ to also be bounded as $r \to \infty$. This requires that $c_3 = 0$. Therefore

$$u(r, \theta) = \sum_{n=1}^{\infty} A_n r^{-n} \sin n\theta.$$

From

$$u(1, \theta) = f(\theta) = \sum_{n=1}^{\infty} A_n \sin n\theta$$

we obtain

$$A_n = \frac{2}{\pi} \int_0^{\pi} f(\theta) \sin n\theta \, d\theta.$$

7. Letting $u(r, t) = R(r)T(t)$ and separating variables we obtain

$$\frac{R'' + \frac{1}{r}R' - hR}{R} = \frac{T'}{T} = \lambda$$

so

$$R'' + \frac{1}{r}R' - (\lambda + h)R = 0 \quad \text{and} \quad T' - \lambda T = 0.$$

From the second equation we find $T(t) = c_1 e^{\lambda t}$. If $\lambda > 0$, $T(t)$ increases without bound as $t \to \infty$. Thus we assume $\lambda = -\alpha^2 < 0$. Since $h > 0$ we can take $\mu = -\alpha^2 - h$. Then

$$R'' + \frac{1}{r}R' + \alpha^2 R = 0$$

is a parametric Bessel equation with solution

$$R(r) = c_1 J_0(\alpha r) + c_2 Y_0(\alpha r).$$

Since Y_0 is unbounded as $r \to 0$ we take $c_2 = 0$. Then $R(r) = c_1 J_0(\alpha r)$ and the boundary condition $u(1, t) = R(1)T(t) = 0$ implies $J_0(\alpha) = 0$. This latter equation defines the positive eigenvalues λ_n. Thus

$$u(r, t) = \sum_{n=1}^{\infty} A_n J_0(\alpha_n r) e^{(-\alpha_n^2 - h)t}.$$

From

$$u(r,0) = 1 = \sum_{n=1}^{\infty} A_n J_0(\alpha_n r)$$

we find

$$A_n = \frac{2}{J_1^2(\alpha_n)} \int_0^1 r J_0(\alpha_n r)\, dr \qquad \boxed{x = \alpha_n r, \;\; dx = \alpha_n\, dr}$$

$$= \frac{2}{J_1^2(\alpha_n)} \int_0^{\alpha_n} \frac{1}{\alpha_n^2} x J_0(x)\, dx.$$

From recurrence relation (5) in Section 11.5 of the text we have

$$x J_0(x) = \frac{d}{dx}[x J_1(x)].$$

Then

$$A_n = \frac{2}{\alpha_n^2 J_1^2(\alpha_n)} \int_0^{\alpha_n} \frac{d}{dx}[x J_1(x)]\, dx = \frac{2}{\alpha_n^2 J_1^2(\alpha_n)} \left(x J_1(x) \right) \Big|_0^{\alpha_n} = \frac{2\alpha_1 J_1(\alpha_n)}{\alpha_n^2 J_1^2(\alpha_n)} = \frac{2}{\alpha_n J_1(\alpha_n)}$$

and

$$u(r,t) = 2e^{-ht} \sum_{n=1}^{\infty} \frac{J_0(\alpha_n r)}{\alpha_n J_1(\alpha_n)} e^{-\alpha_n^2 t}$$

8. Letting $\lambda = \alpha^2 > 0$ and proceeding in the usual manner we find

$$u(r,t) = \sum_{n=1}^{\infty} A_n \cos a\alpha_n t\, J_0(\alpha_n r)$$

where the eigenvalues $\lambda_n = \alpha_n^2$ are determined by $J_0(\alpha) = 0$. Then the initial condition gives

$$u_0 J_0(x_k r) = \sum_{n=1}^{\infty} A_n J_0(\alpha_n r)$$

and so

$$A_n = \frac{2}{J_1^2(\alpha_n)} \int_0^1 r\, (u_0 J_0(x_k r))\, J_0(\alpha_n r)\, dr.$$

But $J_0(\alpha) = 0$ implies that the eigenvalues are the positive zeros of J_0, that is, $\alpha_n = x_n$ for $n = 1, 2, 3, \dots$. Therefore

$$A_n = \frac{2u_0}{J_1^2(\alpha_n)} \int_0^1 r J_0(\alpha_k r) J_0(\alpha_n r)\, dr = 0, \quad n \neq k$$

by orthogonality. For $n = k$,

$$A_k = \frac{2u_0}{J_1^2(\alpha_k)} \int_0^1 r J_0^2(\alpha_k)\, dr = u_0$$

by (7) of Section 11.5. Thus the solution $u(r,t)$ reduces to one term when $n = k$, and

$$u(r,t) = u_0 \cos a\alpha_k t\, J_0(\alpha_k r) = u_0 \cos a x_k t\, J_0(x_k r).$$

9. The boundary-value problem is

$$\frac{\partial^2 u}{\partial r^2} + \frac{1}{r}\frac{\partial u}{\partial r} + \frac{\partial^2 u}{\partial z^2} = 0, \quad 0 < r < 2, \quad 0 < z < 4$$

$$u(2, z) = 50, \quad 0 < z < 4$$

$$\left.\frac{\partial u}{\partial z}\right|_{z=0} = 0, \quad u(r, 4) = 0, \quad 0 < r < 2.$$

We have $u(r, z) = v(r, z) + \psi(r)$ so $\psi'' + (1/r)\psi' = 0$ and $\psi = c_1 \ln r + c_2$. Now, boundedness at $r = 0$ implies $c_2 = 0$. Also, $\psi = c_2$ and $u(r, z) = v(r, z) + \psi(r)$, which implies

$$u(2, z) = v(2, z) + \psi(2) = 50 = c_2,$$

and $\psi(r) = 50$, all of which implies

$$\frac{\partial^2 v}{\partial r^2} + \frac{1}{r}\frac{\partial v}{\partial r} + \frac{\partial^2 v}{\partial z^2} = 0, \quad 0 < r < 2, \quad 0 < z < 4$$

$$v(2, z) = 0, \quad 0 < z < 4$$

$$\left.\frac{\partial v}{\partial z}\right|_{z=0} = 0, \quad v(r, 4) = -50, \quad 0 < r < 2.$$

(See the solution of Problem 5 in Exercises 13.2.) From

$$v(r, z) = -50 \sum_{n=1}^{\infty} \frac{\cosh(\alpha_n z)}{\alpha_n \cosh(4\alpha_n) J_1(2\alpha_n)} J_0(\alpha_n r)$$

we get

$$u(r, z) = 50 - 50 \sum_{n=1}^{\infty} \frac{\cosh(\alpha_n z)}{\alpha_n \cosh(4\alpha_n) J_1(2\alpha_n)} J_0(\alpha_n r).$$

10. Using $u = RZ$ and $-\lambda$ as a separation constant and then letting $\lambda = \alpha^2 > 0$ leads to

$$r^2 R'' + r R' + \alpha^2 r^2 R = 0, \quad R'(1) = 0, \quad \text{and} \quad Z'' - \alpha^2 Z = 0.$$

Thus

$$R(r) = c_1 J_0(\alpha r) + c_2 Y_0(\alpha r)$$

$$Z(z) = c_3 \cosh \alpha z + c_4 \sinh \alpha z$$

for $\alpha > 0$. Arguing that $u(r, z)$ is bounded as $r \to 0$ we define $c_2 = 0$. Since the eigenvalues are defined by $J_0'(\alpha) = 0$ we know that $\lambda = \alpha = 0$ is an eigenvalue. The solutions are then

$$R(r) = c_1 + c_2 \ln r \quad \text{and} \quad Z(z) = c_3 z + c_4$$

where boundedness again dictates that $c_2 = 0$. Thus,

$$u(r, z) = A_0 z + B_0 + \sum_{n=1}^{\infty} (A_n \sinh \alpha_n z + B_n \cosh \alpha_n z) J_0(\alpha_n r).$$

Finally, the specified conditions $z = 0$ and $z = 1$ give, in turn,

$$B_0 = 2 \int_0^1 r f(r)\, dr$$

$$B_n = \frac{2}{J_0^2(\alpha_n)} \int_0^1 r f(r) J_0(\alpha_n r)\, dr$$

$$A_0 = -B_0 + 2 \int_0^1 r g(r)\, dr$$

$$A_n = \frac{1}{\sinh \alpha_n} \left[-B_n \cosh \alpha_n + \frac{2}{J_0^2(\alpha_n)} \int_0^1 r g(r) J_0(\alpha_n r)\, dr \right].$$

11. Referring to Example 1 in Section 13.3 of the text we have

$$u(r, \theta) = \sum_{n=0}^{\infty} A_n r^n P_n(\cos \theta).$$

For $x = \cos \theta$

$$u(1, \theta) = \begin{cases} 100 & 0 < \theta < \pi/2 \\ -100 & \pi/2 < \theta < \pi \end{cases} = 100 \begin{cases} -1, & -1 < x < 0 \\ 1, & 0 < x < 1 \end{cases} = g(x).$$

From Problem 22 in Exercise 11.5 we have

$$u(r, \theta) = 100 \left[\frac{3}{2} r P_1(\cos \theta) - \frac{7}{8} r^3 P_3(\cos \theta) + \frac{11}{16} r^5 P_5(\cos \theta) + \cdots \right].$$

12. Since

$$\frac{1}{r} \frac{\partial^2}{\partial r^2}(ru) = \frac{1}{r} \frac{\partial}{\partial r}\left[r \frac{\partial u}{\partial r} + u \right] = \frac{1}{r}\left[r \frac{\partial^2 u}{\partial r^2} + \frac{\partial u}{\partial r} + \frac{\partial u}{\partial r} \right] = \frac{\partial^2 u}{\partial r^2} + \frac{2}{r} \frac{\partial u}{\partial r} \;\checkmark$$

the differential equation becomes

$$\frac{1}{r} \frac{\partial^2}{\partial r^2}(ru) = \frac{\partial^2 u}{\partial t^2} \quad \text{or} \quad \frac{\partial^2}{\partial r^2}(ru) = r \frac{\partial^2 u}{\partial t^2}.$$

s(b:

$$\partial r$$

Letting $v(r, t) = r u(r, t)$ we obtain the boundary-value problem

$$\frac{\partial^2 v}{\partial r^2} = \frac{\partial^2 v}{\partial t^2}, \quad 0 < r < 1, \quad t > 0$$

$$\left. \frac{\partial v}{\partial r} \right|_{r=1} - v(1, t) = 0, \quad t > 0$$

$$v(r, 0) = r f(r), \quad \left. \frac{\partial v}{\partial t} \right|_{t=0} = r g(r), \quad 0 < r < 1.$$

If we separate variables using $v(r, t) = R(r)T(t)$ and separation constant $-\lambda$ then we obtain

$$\frac{R''}{R} = \frac{T''}{T} = -\lambda$$

so that

$$R'' + \lambda R = 0$$

$$T'' + \lambda T = 0.$$

Letting $\lambda = \alpha^2 > 0$ and solving the differential equations we get

$$R(r) = c_1 \cos \alpha r + c_2 \sin \alpha r$$

$$T(t) = c_3 \cos \alpha t + c_4 \sin \alpha t.$$

Since $u(r,t) = v(r,t)/r$, in order to insure boundedness at $r = 0$ we define $c_1 = 0$. Then $R(r) = c_2 \sin \alpha r$. Now the boundary condition $R'(1) - R(1) = 0$ implies $\alpha \cos \alpha - \sin \alpha = 0$. Thus, the eigenvalues λ_n are determined by the positive solutions of $\tan \alpha = \alpha$. We now have

$$v_n(r,t) = (A_n \cos \alpha_n t + B_n \sin \alpha_n t) \sin \alpha_n r.$$

For the eigenvalue $\lambda = 0$,

$$R(r) = c_1 r + c_2 \qquad \text{and} \qquad T(t) = c_3 t + c_4,$$

and boundedness at $r = 0$ implies $c_2 = 0$. We then take

$$v_0(r,t) = A_0 t r + B_0 r$$

so that

$$v(r,t) = A_0 t r + B_0 r + \sum_{n=1}^{\infty} (a_n \cos \alpha_n t + B_n \sin \alpha_n t) \sin \alpha_n r.$$

Now

$$v(r,0) = r f(r) = B_0 r + \sum_{n=1}^{\infty} A_n \sin \alpha_n r.$$

Since $\{r, \sin \alpha_n r\}$ is an orthogonal set on $[0,1]$,

$$\int_0^1 r \sin \alpha_n r \, dr = 0 \qquad \text{and} \qquad \int_0^1 \sin \alpha_n r \sin \alpha_n r \, dr = 0$$

for $m \neq n$. Therefore

$$\int_0^1 r^2 f(r) \, dr = B_0 \int_0^1 r^2 \, dr = \frac{1}{3} B_0$$

and

$$B_0 = 3 \int_0^1 r^2 f(r) \, dr.$$

Also

$$\int_0^1 r f(r) \sin \alpha_n r \, dr = A_n \int_0^1 \sin^2 \alpha_n r \, dr$$

and

$$A_n = \frac{\int_0^1 r f(r) \sin \alpha_n r \, dr}{\int_0^1 \sin^2 \alpha_n r \, dr}.$$

Now

$$\int_0^1 \sin^2 \alpha_n r \, dr = \frac{1}{2} \int_0^1 (1 - \cos 2\alpha_n r) \, dr = \frac{1}{2} \left[1 - \frac{\sin 2\alpha_n}{2\alpha_n} \right] = \frac{1}{2} [1 - \cos^2 \alpha_n].$$

Since $\tan \alpha_n = \alpha_n$,

$$1 + \alpha_n^2 = 1 + \tan^2 \alpha_n = \sec^2 \alpha_n = \frac{1}{\cos^2 \alpha_n}$$

and

$$\cos^2 \alpha_n = \frac{1}{1 + \alpha_n^2}.$$

Then

$$\int_0^1 \sin^2 \alpha_n r \, dr = \frac{1}{2} \left[1 - \frac{1}{1 + \alpha_n^2} \right] = \frac{\alpha_n^2}{2(1 + \alpha_n^2)}$$

and

$$A_n = \frac{2(1 + \alpha_n^2)}{\alpha_n^2} \int_0^1 r f(r) \sin \alpha_n r \, dr.$$

Similarly, setting

$$\frac{\partial v}{\partial t} \bigg|_{t=0} = rg(r) = A_0 r + \sum_{n=1}^{\infty} B_n \alpha_n \sin \alpha_n r$$

we obtain

$$A_0 = 3 \int_0^1 r^2 g(r) \, dr$$

and

$$B_n = \frac{2(1 + \alpha_n^2)}{\alpha_n^3} \int_0^1 rg(r) \sin \alpha_n r \, dr.$$

Therefore, since $v(r, t) = ru(r, t)$ we have

$$u(r, t) = A_0 t + B_0 + \sum_{n=1}^{\infty} (A_n \cos \alpha_n t + B_n \sin \alpha_n t) \frac{\sin \alpha_n r}{r},$$

where the α_n are solutions of $\tan \alpha = \alpha$ and

$$A_0 = 3 \int_0^1 r^2 g(r) \, dr$$

$$B_0 = 3 \int_0^1 r^2 f(r) \, dr$$

$$A_n = \frac{2(1 + \alpha_n^2)}{\alpha_n^2} \int_0^1 r f(r) \sin \alpha_n r \, dr$$

$$B_n = \frac{2(1 + \alpha_n^2)}{\alpha_n^3} \int_0^1 rg(r) \sin \alpha_n r \, dr$$

for $n = 1, 2, 3, \ldots$.

13. We note that the differential equation can be expressed in the form

$$\frac{d}{dx}[xu'] = -\alpha^2 xu.$$

Thus

$$u_n \frac{d}{dx}[xu'_m] = -\alpha_m^2 x u_m u_n$$

and

$$u_m \frac{d}{dx}[xu'_n] = -\alpha_n^2 x u_n u_m.$$

Subtracting we obtain

$$u_n \frac{d}{dx}[xu'_m] - u_m \frac{d}{dx}[xu'_n] = (\alpha_n^2 - \alpha_m^2) x u_m u_n$$

and

$$\int_a^b u_n \frac{d}{dx}[xu'_m]\, dx - \int_a^b u_m \frac{d}{dx}[xu'_n] = (\alpha_n^2 - \alpha_m^2)\int_a^b x u_m u_n \, dx.$$

Using integration by parts this becomes

$$u_n x u'_m \Big|_a^b - \int_a^b x u'_m u'_n \, dx - u_m x u'_n \Big|_a^b + \int_a^b x u'_n u'_m \, dx$$

$$= b[u_n(b)u'_m(b) - u_m(b)u'_n(b)] - a[u_n(a)u'_m(a) - u_m(a)u'_n(a)]$$

$$= (\alpha_n^2 - \alpha_m^2)\int_a^b x u_m u_n \, dx.$$

Since

$$u(x) = Y_0(\alpha a) J_0(\alpha x) - J_0(\alpha a) Y_0(\alpha x)$$

we have

$$u_n(b) = Y_0(\alpha_n a) J_0(\alpha_n b) - J_0(\alpha_n a) Y_0(\alpha_n b) = 0$$

by the definition of the α_n. Similarly $u_m(b) = 0$. Also

$$u_n(a) = Y_0(\alpha a) J_0(\alpha_n a) - J_0(\alpha_n a) Y_0(\alpha_n a) = 0$$

and $u_m(a) = 0$. Therefore

$$\int_a^b x u_m u_n \, dx = \frac{1}{\alpha_n^2 - \alpha_m^2}\left(b[u_n(b)u'_m(b) - u_m(b)u'_n(b)] - a[u_n(a)u'_m(a) - u_m(a)u'_n(a)]\right) = 0$$

and the $u_n(x)$ are orthogonal with respect to the weight function x.

14. Letting $u(r,t) = R(r)T(t)$ and the separation constant be $-\lambda = -\alpha^2$ we obtain

$$rR'' + R' + \alpha^2 rR = 0$$

$$T' + \alpha^2 T = 0,$$

Chapter 13 in Review

with solutions

$$R(r) = c_1 J_0(\alpha r) + c_2 Y_0(\alpha r)$$

$$T(t) = c_3 e^{-\alpha^2 t}.$$

Now the boundary conditions imply

$$R(a) = 0 = c_1 J_0(\alpha a) + c_2 Y_0(\alpha a)$$

$$R(b) = 0 = c_1 J_0(\alpha b) + c_2 Y_0(\alpha b)$$

so that

$$c_2 = -\frac{c_1 J_0(\alpha a)}{Y_0(\alpha a)}$$

and

$$c_1 J_0(\alpha b) - \frac{c_1 J_0(\alpha a)}{Y_0(\alpha a)} Y_0(\alpha b) = 0$$

or

$$Y_0(\alpha a) J_0(\alpha b) - J_0(\alpha a) Y_0(\alpha b) = 0.$$

This equation defines α_n for $n = 1, 2, 3, \dots$. Now

$$R(r) = c_1 J_0(\alpha r) - c_1 \frac{J_0(\alpha a)}{Y_0(\alpha a)} Y_0(\alpha r) = \frac{c_1}{Y_0(\alpha a)} \Big[Y_0(\alpha a) J_0(\alpha r) - J_0(\alpha a) Y_0(\alpha r) \Big]$$

and

$$u_n(r, t) = A_n \Big[Y_0(\alpha_n a) J_0(\alpha_n r) - J_0(\alpha_n a) Y_0(\alpha_n r) \Big] e^{-\alpha_n^2 t} = A_n u_n(r) e^{-\alpha_n^2 t}.$$

Thus

$$u(r, t) = \sum_{n=1}^{\infty} A_n u_n(r) e^{-\alpha_n^2 t}.$$

From the initial condition

$$u(r, 0) = f(r) = \sum_{n=1}^{\infty} A_n u_n(r)$$

we obtain

$$A_n = \frac{\int_a^b r f(r) u_n(r)\, dr}{\int_a^b r u_n^2(r)\, dr}.$$

15. We use the Superposition Principle for Laplace's equation discussed in Section 12.5 and shown schematically in Figure 12.5.3 in the text. That is,

Solution $u =$ Solution u_1 of Problem $1 +$ Solution u_2 of Problem 2,

where in Problem 1 the boundary condition on the top and bottom of the cylinder is $u = 0$, while on the lateral surface $r = c$ it is $u = h(z)$, and in Problem 2 the boundary condition on the top of the cylinder $z = L$ is $u = f(r)$, on the bottom $z = 0$ it is $u = g(r)$, and on the lateral surface $r = c$ it is $u = 0$.

Solution for $u_1(r, z)$

Using λ as a separation constant we have

$$\frac{R'' + \frac{1}{r}R'}{R} = -\frac{Z''}{Z} = \lambda,$$

so

$$rR'' + R' - \lambda r R = 0 \qquad \text{and} \qquad Z'' + \lambda Z = 0.$$

The differential equation in Z, together with the boundary conditions $Z(0) = 0$ and $Z(L) = 0$ is a Sturm-Liouville problem. Letting $\lambda = \alpha^2 > 0$ we note that the above differential equation in R is a modified parametric Bessel equation which is discussed in Section 6.3 in the text. Also, we have $Z(z) = c_1 \cos \alpha z + c_2 \sin \alpha z$. The boundary conditions imply $c_1 = 0$ and $\sin \alpha L = 0$. Thus, $\alpha_n = n\pi/L$, $n = 1, 2, 3, \ldots$, so $\lambda_n = n^2\pi^2/L^2$ and

$$R(r) = c_3 I_0\left(\frac{n\pi}{L}r\right) + c_4 K_0\left(\frac{n\pi}{L}r\right).$$

Now boundedness at $r = 0$ implies $c_4 = 0$, so $R(r) = c_3 I_0(n\pi r/L)$ and

$$u_1(r, z) = \sum_{n=1}^{\infty} A_n I_0\left(\frac{n\pi}{L}r\right) \sin\left(\frac{n\pi}{L}z\right).$$

At $r = c$ for $0 < z < L$ we have

$$h(z) = u_1(c, z) = \sum_{n=1}^{\infty} A_n I_0\left(\frac{n\pi}{L}c\right) \sin\left(\frac{n\pi}{L}z\right)$$

which gives

$$A_n = \frac{2}{L I_0(n\pi c/L)} \int_0^L h(z) \sin\left(\frac{n\pi}{L}z\right) dz.$$

Solution for $u_2(r, z)$

In this case we use $-\lambda$ as a separation constant which leads to

$$\frac{R'' + \frac{1}{r}R'}{R} = -\frac{Z''}{Z} = -\lambda,$$

so

$$rR'' + R' + \lambda r R = 0 \qquad \text{and} \qquad Z'' - \lambda Z = 0.$$

The differential equation in R is a parametric Bessel equation. Using $\lambda = \alpha^2$ we find $R(r) = c_1 J_0(\alpha r) + c_2 Y_0(\alpha r)$. Boundedness at $r = 0$ implies $c_2 = 0$ so $R(r) = c_3 J_0(\alpha r)$. The boundary condition $R(c) = 0$ then gives the defining equation for the eigenvalues: $J_0(\alpha c) = 0$. Let $\lambda_n = \alpha_n^2$ where $\alpha_n c = x_n$ are the roots. The solution of the differential equation in Z is $Z(z) = c_4 \cosh \alpha_n z + c_5 \sinh \alpha_n z$, so

$$u_2(r, z) = \sum_{n=1}^{\infty} (B_n \cosh \alpha_n z + C_n \sinh \alpha_n z) J_0(\alpha_n r).$$

At $z = 0$, for $0 < r < c$, we have

$$f(r) = u_2(r, 0) = \sum_{n=1}^{\infty} B_n J_0(\alpha_n r),$$

so

$$B_n = \frac{2}{c^2 J_1^2(\alpha_n c)} \int_0^c r f(r) J_0(\alpha_n r) \, dr.$$

At $z = L$, for $0 < r < c$, we have

$$g(r) = u_2(r, L) = \sum_{n=1}^{\infty} (B_n \cosh \alpha_n L + C_n \sinh \alpha_n L) J_0(\alpha_n r),$$

so

$$B_n \cosh \alpha_n L + C_n \sinh \alpha_n L = \frac{2}{c^2 J_1^2(\alpha_n c)} \int_0^c r g(r) J_0(\alpha_n r) \, dr$$

and

$$C_n = -B_n \frac{\cosh \alpha_n L}{\sinh \alpha_n L} + \frac{2}{c^2 (\sinh \alpha_n L) J_1^2(\alpha_n c)} \int_0^c r g(r) J_0(\alpha_n r) \, dr.$$

By the Superposition Principle the solution of the original problem is

$$u(r, z) = u_1(r, z) + u_2(r, z).$$

14 Integral Transforms

Exercises 14.1 Error Function

1. (a) The result follows by letting $\tau = u^2$ or $u = \sqrt{\tau}$ in $\operatorname{erf}(\sqrt{t}) = \dfrac{2}{\sqrt{\pi}} \displaystyle\int_0^{\sqrt{t}} e^{-u^2} du$.

(b) Using $\mathscr{L}\{t^{-1/2}\} = \dfrac{\sqrt{\pi}}{s^{1/2}}$ and the first translation theorem, it follows from the convolution theorem that

$$\mathscr{L}\left\{\operatorname{erf}(\sqrt{t})\right\} = \frac{1}{\sqrt{\pi}} \mathscr{L}\left\{\int_0^t \frac{e^{-\tau}}{\sqrt{\tau}}\, d\tau\right\} = \frac{1}{\sqrt{\pi}} \mathscr{L}\{1\}\, \mathscr{L}\left\{t^{-1/2} e^{-t}\right\} = \frac{1}{\sqrt{\pi}} \frac{1}{s} \mathscr{L}\left\{t^{-1/2}\right\}\Big|_{s \to s+1}$$

$$= \frac{1}{\sqrt{\pi}} \frac{1}{s} \frac{\sqrt{\pi}}{\sqrt{s+1}} = \frac{1}{s\sqrt{s+1}}.$$

2. Since $\operatorname{erfc}(\sqrt{t}) = 1 - \operatorname{erf}(\sqrt{t})$ we have

$$\mathscr{L}\left\{\operatorname{erfc}(\sqrt{t})\right\} = \mathscr{L}\{1\} - \mathscr{L}\left\{\operatorname{erf}(\sqrt{t})\right\} = \frac{1}{s} - \frac{1}{s\sqrt{s+1}} = \frac{1}{s}\left[1 - \frac{1}{\sqrt{s+1}}\right].$$

3. By the first translation theorem,

$$\mathscr{L}\left\{e^t \operatorname{erf}(\sqrt{t})\right\} = \mathscr{L}\left\{\operatorname{erf}(\sqrt{t})\right\}\Big|_{s \to s-1} = \frac{1}{s\sqrt{s+1}}\Big|_{s \to s-1} = \frac{1}{\sqrt{s}\,(s-1)}.$$

4. By the first translation theorem and the result of Problem 2,

$$\mathscr{L}\left\{e^t \operatorname{erfc}(\sqrt{t})\right\} = \mathscr{L}\left\{\operatorname{erfc}(\sqrt{t})\right\}\Big|_{s \to s-1} = \left(\frac{1}{s} - \frac{1}{s\sqrt{s+1}}\right)\Big|_{s \to s-1} = \frac{1}{s-1} - \frac{1}{\sqrt{s}\,(s-1)}$$

$$= \frac{\sqrt{s}-1}{\sqrt{s}\,(s-1)} = \frac{\sqrt{s}-1}{\sqrt{s}\,(\sqrt{s}+1)(\sqrt{s}-1)} = \frac{1}{\sqrt{s}\,(\sqrt{s}+1)}.$$

Exercises 14.1 Error Function

5. From entry 3 in Table 14.1 and the first translation theorem we have

$$\mathscr{L}\left\{e^{-Gt/C}\mathrm{erf}\left(\frac{x}{2}\sqrt{\frac{RC}{t}}\right)\right\} = \mathscr{L}\left\{e^{-Gt/C}\left[1 - \mathrm{erfc}\left(\frac{x}{2}\sqrt{\frac{RC}{t}}\right)\right]\right\}$$

$$= \mathscr{L}\left\{e^{-Gt/C}\right\} - \mathscr{L}\left\{e^{-Gt/C}\,\mathrm{erfc}\left(\frac{x}{2}\sqrt{\frac{RC}{t}}\right)\right\}$$

$$= \frac{1}{s+G/C} - \frac{e^{-x\sqrt{RC}\,\sqrt{s}}}{s}\bigg|_{s\to s+G/C}$$

$$= \frac{1}{s+G/C} - \frac{e^{-x\sqrt{RC}\,\sqrt{s+G/C}}}{s+G/C} = \frac{C}{Cs+G}\left(1 - e^{x\sqrt{RCs+RG}}\right).$$

minus sign missing in super.

6. We first compute

$$\frac{\sinh a\sqrt{s}}{s\sinh\sqrt{s}} = \frac{e^{a\sqrt{s}} - e^{-a\sqrt{s}}}{s(e^{\sqrt{s}} - e^{-\sqrt{s}})} = \frac{e^{(a-1)\sqrt{s}} - e^{-(a+1)\sqrt{s}}}{s(1 - e^{-2\sqrt{s}})}$$

$$= \frac{e^{(a-1)\sqrt{s}}}{s}\left[1 + e^{-2\sqrt{s}} + e^{-4\sqrt{s}} + \cdots\right] - \frac{e^{-(a+1)\sqrt{s}}}{s}\left[1 + e^{-2\sqrt{s}} + e^{-4\sqrt{s}} + \cdots\right]$$

$$= \left[\frac{e^{-(1-a)\sqrt{s}}}{s} + \frac{e^{-(3-a)\sqrt{s}}}{s} + \frac{e^{-(5-a)\sqrt{s}}}{s} + \cdots\right]$$

$$\quad - \left[\frac{e^{-(1+a)\sqrt{s}}}{s} + \frac{e^{-(3+a)\sqrt{s}}}{s} + \frac{e^{-(5+a)\sqrt{s}}}{s} + \cdots\right]$$

$$= \sum_{n=0}^{\infty}\left[\frac{e^{-(2n+1-a)\sqrt{s}}}{s} - \frac{e^{-(2n+1+a)\sqrt{s}}}{s}\right].$$

Then

$$\mathscr{L}^{-1}\left\{\frac{\sinh a\sqrt{s}}{s\sinh\sqrt{s}}\right\} = \sum_{n=0}^{\infty}\left[\mathscr{L}^{-1}\left\{\frac{e^{-(2n+1-a)\sqrt{s}}}{s}\right\} - \mathscr{L}^{-1}\left\{-\frac{e^{-(2n+1+a)\sqrt{s}}}{s}\right\}\right]$$

$$= \sum_{n=0}^{\infty}\left[\mathrm{erfc}\left(\frac{2n+1-a}{2\sqrt{t}}\right) - \mathrm{erfc}\left(\frac{2n+1+a}{2\sqrt{t}}\right)\right]$$

$$= \sum_{n=0}^{\infty}\left(\left[1 - \mathrm{erf}\left(\frac{2n+1-a}{2\sqrt{t}}\right)\right] - \left[1 - \mathrm{erf}\left(\frac{2n+1+a}{2\sqrt{t}}\right)\right]\right)$$

$$= \sum_{n=0}^{\infty}\left[\mathrm{erf}\left(\frac{2n+1+a}{2\sqrt{t}}\right) - \mathrm{erf}\left(\frac{2n+1-a}{2\sqrt{t}}\right)\right].$$

718

7. Taking the Laplace transform of both sides of the equation we obtain

$$\mathscr{L}\{y(t)\} = \mathscr{L}\{1\} - \mathscr{L}\left\{\int_0^t \frac{y(\tau)}{\sqrt{t-\tau}}\,d\tau\right\}$$

$$Y(s) = \frac{1}{s} - Y(s)\frac{\sqrt{\pi}}{\sqrt{s}}$$

$$\frac{\sqrt{s}+\sqrt{\pi}}{\sqrt{s}}Y(s) = \frac{1}{s}$$

$$Y(s) = \frac{1}{\sqrt{s}\left(\sqrt{s}+\sqrt{\pi}\right)}.$$

Thus

$$y(t) = \mathscr{L}^{-1}\left\{\frac{1}{\sqrt{s}\left(\sqrt{s}+\sqrt{\pi}\right)}\right\} = e^{\pi t}\,\mathrm{erfc}(\sqrt{\pi t}). \qquad \boxed{\text{By entry 5 in Table 14.1}}$$

8. Using entries 3 and 5 in Table 14.1, we have

$$\mathscr{L}\left\{-e^{ab}e^{b^2 t}\,\mathrm{erfc}\left(b\sqrt{t}+\frac{a}{2\sqrt{t}}\right) + \mathrm{erfc}\left(\frac{a}{2\sqrt{t}}\right)\right\}$$

$$= -\mathscr{L}\left\{e^{ab}e^{b^2 t}\,\mathrm{erfc}\left(b\sqrt{t}+\frac{a}{2\sqrt{t}}\right)\right\} + \mathscr{L}\left\{\mathrm{erfc}\left(\frac{a}{2\sqrt{t}}\right)\right\}$$

$$= -\frac{e^{-a\sqrt{s}}}{\sqrt{s}\left(\sqrt{s}+b\right)} + \frac{e^{-a\sqrt{s}}}{s}$$

$$= e^{-a\sqrt{s}}\left[\frac{1}{s} - \frac{1}{\sqrt{s}\left(\sqrt{s}+b\right)}\right] = e^{-a\sqrt{s}}\left[\frac{1}{s} - \frac{\sqrt{s}}{s\left(\sqrt{s}+b\right)}\right]$$

$$= e^{-a\sqrt{s}}\left[\frac{\sqrt{s}+b-\sqrt{s}}{s\left(\sqrt{s}+b\right)}\right] = \frac{be^{-a\sqrt{s}}}{s\left(\sqrt{s}+b\right)}.$$

9. $\displaystyle\int_a^b e^{-u^2}\,du = \int_a^0 e^{-u^2}\,du + \int_0^b e^{-u^2}\,du = \int_0^b e^{-u^2}\,du - \int_0^a e^{-u^2}\,du$

$$= \frac{\sqrt{\pi}}{2}\,\mathrm{erf}(b) - \frac{\sqrt{\pi}}{2}\,\mathrm{erf}(a) = \frac{\sqrt{\pi}}{2}[\mathrm{erf}(b) - \mathrm{erf}(a)]$$

10. Since $f(x) = e^{-x^2}$ is an even function,

$$\int_{-a}^a e^{-u^2}\,du = 2\int_0^a e^{-u^2}\,du.$$

Therefore,

$$\int_{-a}^a e^{-u^2}\,du = \sqrt{\pi}\,\mathrm{erf}(a).$$

11. The function erf (x) is symmetric with respect to the origin, while erfc(x) appears to be symmetric with respect to the point $(0, 1)$. From the graph it appears that $\lim_{x \to -\infty} \text{erf}\,(x) = -1$ and $\lim_{x \to -\infty} \text{erfc}(x) = 2$.

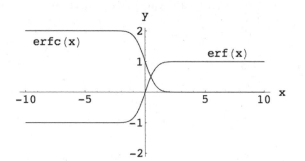

Exercises 14.2

Laplace Transform

1. The boundary-value problem is

$$a^2 \frac{\partial^2 u}{\partial x^2} = \frac{\partial^2 u}{\partial t^2}, \quad 0 < x < L, \quad t > 0,$$

$$u(0, t) = 0, \quad u(L, t) = 0, \quad t > 0,$$

$$u(x, 0) = A \sin \frac{\pi}{L}x, \quad \frac{\partial u}{\partial t}\bigg|_{t=0} = 0.$$

Transforming the partial differential equation gives

$$\frac{d^2 U}{dx^2} - \left(\frac{s}{a}\right)^2 U = -\frac{s}{a^2} A \sin \frac{\pi}{L}x.$$

Using undetermined coefficients we obtain

$$U(x, s) = c_1 \cosh \frac{s}{a}x + c_2 \sinh \frac{s}{a}x + \frac{As}{s^2 + a^2\pi^2/L^2} \sin \frac{\pi}{L}x.$$

The transformed boundary conditions, $U(0, s) = 0$, $U(L, s) = 0$ give in turn $c_1 = 0$ and $c_2 = 0$. Therefore

$$U(x, s) = \frac{As}{s^2 + a^2\pi^2/L^2} \sin \frac{\pi}{L}x$$

and

$$u(x, t) = A\mathscr{L}^{-1}\left\{\frac{s}{s^2 + a^2\pi^2/L^2}\right\} \sin \frac{\pi}{L}x = A\cos \frac{a\pi}{L}t \sin \frac{\pi}{L}x.$$

2. The transformed equation is

$$\frac{d^2 U}{dx^2} - s^2 U = -2 \sin \pi x - 4 \sin 3\pi x$$

and so

$$U(x, s) = c_1 \cosh sx + c_2 \sinh sx + \frac{2}{s^2 + \pi^2} \sin \pi x + \frac{4}{s^2 + 9\pi^2} \sin 3\pi x.$$

The transformed boundary conditions, $U(0, s) = 0$ and $U(1, s) = 0$ give $c_1 = 0$ and $c_2 = 0$. Thus

$$U(x, s) = \frac{2}{s^2 + \pi^2} \sin \pi x + \frac{4}{s^2 + 9\pi^2} \sin 3\pi x$$

and

$$u(x, t) = 2\mathcal{L}^{-1}\left\{\frac{1}{s^2 + \pi^2}\right\} \sin \pi x + 4\mathcal{L}^{-1}\left\{\frac{1}{s^2 + 9\pi^2}\right\} \sin 3\pi x$$

$$= \frac{2}{\pi} \sin \pi t \sin \pi x + \frac{4}{3\pi} \sin 3\pi t \sin 3\pi x.$$

3. The solution of

$$a^2 \frac{d^2 U}{dx^2} - s^2 U = 0$$

is in this case

$$U(x, s) = c_1 e^{-(x/a)s} + c_2 e^{(x/a)s}.$$

Since $\lim_{x \to \infty} u(x, t) = 0$ we have $\lim_{x \to \infty} U(x, s) = 0$. Thus $c_2 = 0$ and

$$U(x, s) = c_1 e^{-(x/a)s}.$$

If $\mathcal{L}\{u(0, t)\} = \mathcal{L}\{f(t)\} = F(s)$ then $U(0, s) = F(s)$. From this we have $c_1 = F(s)$ and

$$U(x, s) = F(s) e^{-(x/a)s}.$$

Hence, by the second translation theorem,

$$u(x, t) = f\left(t - \frac{x}{a}\right) \mathcal{U}\left(t - \frac{x}{a}\right).$$

4. Expressing $f(t)$ in the form $(\sin \pi t)[1 - \mathcal{U}(t - 1)]$ and using the result of Problem 3 we find

$$u(x, t) = f\left(t - \frac{x}{a}\right) \mathcal{U}\left(t - \frac{x}{a}\right)$$

$$= \sin \pi \left(t - \frac{x}{a}\right)\left[1 - \mathcal{U}\left(t - \frac{x}{a} - 1\right)\right] \mathcal{U}\left(t - \frac{x}{a}\right)$$

$$= \sin \pi \left(t - \frac{x}{a}\right)\left[\mathcal{U}\left(t - \frac{x}{a}\right) - \mathcal{U}\left(t - \frac{x}{a}\right)\mathcal{U}\left(t - \frac{x}{a} - 1\right)\right]$$

$$= \sin \pi \left(t - \frac{x}{a}\right)\left[\mathcal{U}\left(t - \frac{x}{a}\right) - \mathcal{U}\left(t - \frac{x}{a} - 1\right)\right]$$

Now

$$\mathcal{U}\left(t - \frac{x}{a}\right) - \mathcal{U}\left(t - \frac{x}{a} - 1\right) = \begin{cases} 0, & 0 \leq t < x/a \\ 1, & x/a \leq t \leq x/a + 1 \\ 0, & t > x/a + 1 \end{cases}$$

$$= \begin{cases} 0, & x < a(t - 1) \text{ or } x > at \\ 1, & a(t - 1) \leq x \leq at \end{cases}$$

so

$$u(x,t) = \begin{cases} 0, & x < a(t-1) \text{ or } x > at \\ \sin \pi(t - x/a), & a(t-1) \le x \le at. \end{cases}$$

The graph is shown for $t > 1$.

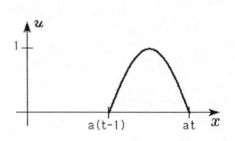

5. We use

$$U(x,s) = c_1 e^{-(x/a)s} - \frac{g}{s^3}.$$

Now

$$\mathcal{L}\{u(0,t)\} = U(0,s) = \frac{A\omega}{s^2 + \omega^2}$$

and so

$$U(0,s) = c_1 - \frac{g}{s^3} = \frac{A\omega}{s^2 + \omega^2} \quad\text{or}\quad c_1 = \frac{g}{s^3} + \frac{A\omega}{s^2 + \omega^2}.$$

Therefore

$$U(x,s) = \frac{A\omega}{s^2 + \omega^2} e^{-(x/a)s} + \frac{g}{s^3} e^{-(x/a)s} - \frac{g}{s^3}$$

and

$$u(x,t) = A\mathcal{L}^{-1}\left\{\frac{\omega e^{-(x/a)s}}{s^2 + \omega^2}\right\} + g\mathcal{L}^{-1}\left\{\frac{e^{-(x/a)s}}{s^3}\right\} - g\mathcal{L}^{-1}\left\{\frac{1}{s^3}\right\}$$

$$= A\sin\omega\left(t - \frac{x}{a}\right)\mathcal{U}\left(t - \frac{x}{a}\right) + \frac{1}{2}g\left(t - \frac{x}{a}\right)^2 \mathcal{U}\left(t - \frac{x}{a}\right) - \frac{1}{2}gt^2.$$

6. Transforming the partial differential equation gives

$$\frac{d^2U}{dx^2} - s^2 U = -\frac{\omega}{s^2 + \omega^2}\sin\pi x.$$

Using undetermined coefficients we obtain

$$U(x,s) = c_1\cosh sx + c_2\sinh sx + \frac{\omega}{(s^2 + \pi^2)(s^2 + \omega^2)}\sin\pi x.$$

The transformed boundary conditions $U(0,s) = 0$ and $U(1,s) = 0$ give, in turn, $c_1 = 0$ and $c_2 = 0$. Therefore

$$U(x,s) = \frac{\omega}{(s^2 + \pi^2)(s^2 + \omega^2)}\sin\pi x$$

and

$$u(x,t) = \omega \sin \pi x \, \mathcal{L}^{-1}\left\{\frac{1}{(s^2+\pi^2)(s^2+\omega^2)}\right\}$$

$$= \frac{\omega}{\omega^2 - \pi^2} \sin \pi x \, \mathcal{L}^{-1}\left\{\frac{1}{\pi}\frac{\pi}{s^2+\pi^2} - \frac{1}{\omega}\frac{\omega}{s^2+\omega^2}\right\}$$

$$= \frac{\omega}{\pi(\omega^2 - \pi^2)} \sin \pi t \sin \pi x - \frac{1}{\omega^2 - \pi^2} \sin \omega t \sin \pi x.$$

7. We use

$$U(x,s) = c_1 \cosh \frac{s}{a} x + c_2 \sinh \frac{s}{a} x.$$

Now $U(0,s) = 0$ implies $c_1 = 0$, so $U(x,s) = c_2 \sinh(s/a)x$. The condition $E \, dU/dx \,|_{x=L} = F_0$ then yields $c_2 = F_0 a / Es \cosh(s/a)L$ and so

$$U(x,s) = \frac{aF_0}{Es}\frac{\sinh(s/a)x}{\cosh(s/a)L} = \frac{aF_0}{Es}\frac{e^{(s/a)x} - e^{-(s/a)x}}{e^{(s/a)L} + e^{-(s/a)L}}$$

$$= \frac{aF_0}{Es}\frac{e^{(s/a)(x-L)} - e^{-(s/a)(x+L)}}{1 + e^{-2sL/a}}$$

$$= \frac{aF_0}{E}\left[\frac{e^{-(s/a)(L-x)}}{s} - \frac{e^{-(s/a)(3L-x)}}{s} + \frac{e^{-(s/a)(5L-x)}}{s} - \cdots\right]$$

$$\quad - \frac{aF_0}{E}\left[\frac{e^{-(s/a)(L+x)}}{s} - \frac{e^{-(s/a)(3L+x)}}{s} + \frac{e^{-(s/a)(5L+x)}}{s} - \cdots\right]$$

$$= \frac{aF_0}{E}\sum_{n=0}^{\infty}(-1)^n\left[\frac{e^{-(s/a)(2nL+L-x)}}{s} - \frac{e^{-(s/a)(2nL+L+x)}}{s}\right]$$

and

$$u(x,t) = \frac{aF_0}{E}\sum_{n=0}^{\infty}(-1)^n\left[\mathcal{L}^{-1}\left\{\frac{e^{-(s/a)(2nL+L-x)}}{s}\right\} - \mathcal{L}^{-1}\left\{\frac{e^{-(s/a)(2nL+L+x)}}{s}\right\}\right]$$

$$= \frac{aF_0}{E}\sum_{n=0}^{\infty}(-1)^n\left[\left(t - \frac{2nL+L-x}{a}\right)\mathcal{U}\left(t - \frac{2nL+L-x}{a}\right)\right.$$

$$\left. - \left(t - \frac{2nL+L+x}{a}\right)\mathcal{U}\left(t - \frac{2nL+L+x}{a}\right)\right].$$

8. We use

$$U(x,s) = c_1 e^{-(x/a)s} + c_2 e^{(x/a)s} - \frac{v_0}{s^2}.$$

Now $\lim_{x\to\infty} dU/dx = 0$ implies $c_2 = 0$, and $U(0,s) = 0$ then gives $c_1 = v_0/s^2$. Hence

$$U(x,s) = \frac{v_0}{s^2}e^{-(x/a)s} - \frac{v_0}{s^2}$$

and

$$u(x,t) = v_0 \left(t - \frac{x}{a}\right)\mathscr{U}\left(t - \frac{x}{a}\right) - v_0 t.$$

9. Transforming the partial differential equation gives

$$\frac{d^2 U}{dx^2} - s^2 U = -sxe^{-x}.$$

Using undetermined coefficients we obtain

$$U(x,s) = c_1 e^{-sx} + c_2 e^{sx} - \frac{2s}{(s^2 - 1)^2} e^{-x} + \frac{s}{s^2 - 1} xe^{-x}.$$

The transformed boundary conditions $\lim_{x\to\infty} U(x,s) = 0$ and $U(0,s) = 0$ give, in turn, $c_2 = 0$ and $c_1 = 2s/(s^2 - 1)^2$. Therefore

$$U(x,s) = \frac{2s}{(s^2 - 1)^2} e^{-sx} - \frac{2s}{(s^2 - 1)^2} e^{-x} + \frac{s}{s^2 - 1} xe^{-x}.$$

From entries (13) and (26) in the Table of Laplace transforms we obtain

$$u(x,t) = \mathscr{L}^{-1}\left\{\frac{2s}{(s^2 - 1)^2} e^{-sx} - \frac{2s}{(s^2 - 1)^2} e^{-x} + \frac{s}{s^2 - 1} xe^{-x}\right\}$$

$$= 2(t - x)\sinh(t - x)\,\mathscr{U}(t - x) - te^{-x}\sinh t + xe^{-x}\cosh t.$$

10. We use

$$U(x,s) = c_1 e^{-xs} + c_2 e^{xs} + \frac{s}{s^2 - 1} e^{-x}.$$

Now $\lim_{x\to\infty} u(x,t) = 0$ implies $\lim_{x\to\infty} U(x,s) = 0$, so we define $c_2 = 0$. Then

$$U(x,s) = c_1 e^{-xs} + \frac{s}{s^2 - 1} e^{-x}.$$

Finally, $U(0,s) = 1/s$ gives $c_1 = 1/s - s/(s^2 - 1)$. Thus

$$U(x,s) = \frac{1}{s} - \frac{s}{s^2 - 1} e^{-xs} + \frac{s}{s^2 - 1} e^{-x}$$

and

$$u(x,t) = -\mathscr{L}^{-1}\left\{\frac{s}{s^2 - 1} e^{-(x/a)s}\right\} + \mathscr{L}^{-1}\left\{\frac{s}{s^2 - 1}\right\} e^{-x}$$

$$= -\cosh\left(t - \frac{x}{a}\right)\mathscr{U}\left(t - \frac{x}{a}\right) + e^{-x}\cosh t.$$

11. We use

$$U(x,s) = c_1 e^{-\sqrt{s}\,x} + c_2 e^{\sqrt{s}\,x} + \frac{u_1}{s}.$$

The condition $\lim_{x\to\infty} u(x,t) = u_1$ implies $\lim_{x\to\infty} U(x,s) = u_1/s$, so we define $c_2 = 0$. Then

$$U(x,s) = c_1 e^{-\sqrt{s}\,x} + \frac{u_1}{s}.$$

From $U(0, s) = u_0/s$ we obtain $c_1 = (u_0 - u_1)/s$. Thus

$$U(x, s) = (u_0 - u_1)\frac{e^{-\sqrt{s}\, x}}{s} + \frac{u_1}{s}$$

and

$$u(x, t) = (u_0 - u_1)\,\mathscr{L}^{-1}\left\{\frac{e^{-x\sqrt{s}}}{s}\right\} + u_1\,\mathscr{L}^{-1}\left\{\frac{1}{s}\right\} = (u_0 - u_1)\operatorname{erfc}\left(\frac{x}{2\sqrt{t}}\right) + u_1.$$

12. We use

$$U(x, s) = c_1 e^{-\sqrt{s}\, x} + c_2 e^{\sqrt{s}\, x} + \frac{u_1 x}{s}.$$

The condition $\lim_{x\to\infty} u(x, t)/x = u_1$ implies $\lim_{x\to\infty} U(x, s)/x = u_1/s$, so we define $c_2 = 0$. Then

$$U(x, s) = c_1 e^{-\sqrt{s}\, x} + \frac{u_1 x}{s}.$$

From $U(0, s) = u_0/s$ we obtain $c_1 = u_0/s$. Hence

$$U(x, s) = u_0\frac{e^{-\sqrt{s}\, x}}{s} + \frac{u_1 x}{s}$$

and

$$u(x, t) = u_0\,\mathscr{L}^{-1}\left\{\frac{e^{-x\sqrt{s}}}{s}\right\} + u_1 x\,\mathscr{L}^{-1}\left\{\frac{1}{s}\right\} = u_0\operatorname{erfc}\left(\frac{x}{2\sqrt{t}}\right) + u_1 x.$$

13. We use

$$U(x, s) = c_1 e^{-\sqrt{s}\, x} + c_2 e^{\sqrt{s}\, x} + \frac{u_0}{s}.$$

The condition $\lim_{x\to\infty} u(x, t) = u_0$ implies $\lim_{x\to\infty} U(x, s) = u_0/s$, so we define $c_2 = 0$. Then

$$U(x, s) = c_1 e^{-\sqrt{s}\, x} + \frac{u_0}{s}.$$

The transform of the remaining boundary conditions gives

$$\left.\frac{dU}{dx}\right|_{x=0} = U(0, s).$$

This condition yields $c_1 = -u_0/s(\sqrt{s} + 1)$. Thus

$$U(x, s) = -u_0\frac{e^{-\sqrt{s}\, x}}{s(\sqrt{s} + 1)} + \frac{u_0}{s}$$

and

$$u(x, t) = -u_0\,\mathscr{L}^{-1}\left\{\frac{e^{-x\sqrt{s}}}{s(\sqrt{s} + 1)}\right\} + u_0\,\mathscr{L}^{-1}\left\{\frac{1}{s}\right\}$$

$$= u_0 e^{x+t}\operatorname{erfc}\left(\sqrt{t} + \frac{x}{2\sqrt{t}}\right) - u_0\operatorname{erfc}\left(\frac{x}{2\sqrt{t}}\right) + u_0 \qquad \boxed{\text{By entry (6) in Table 14.1}}$$

14. We use

$$U(x, s) = c_1 e^{-\sqrt{s}\, x} + c_2 e^{\sqrt{s}\, x}.$$

The condition $\lim_{x\to\infty} u(x,t) = 0$ implies $\lim_{x\to\infty} U(x,s) = 0$, so we define $c_2 = 0$. Hence

$$U(x,s) = c_1 e^{-\sqrt{s}\,x}.$$

The remaining boundary condition transforms into

$$\frac{dU}{dx}\bigg|_{x=0} = U(0,s) - \frac{50}{s}.$$

This condition gives $c_1 = 50/s(\sqrt{s}+1)$. Therefore

$$U(x,s) = 50\,\frac{e^{-\sqrt{s}\,x}}{s(\sqrt{s}+1)}$$

and

$$u(x,t) = 50\,\mathscr{L}^{-1}\left\{\frac{e^{-x\sqrt{s}}}{s(\sqrt{s}+1)}\right\} = -50 e^{x+t}\operatorname{erfc}\left(\sqrt{t}+\frac{x}{2\sqrt{t}}\right) + 50\operatorname{erfc}\left(\frac{x}{2\sqrt{t}}\right).$$

15. We use

$$U(x,s) = c_1 e^{-\sqrt{s}\,x} + c_2 e^{\sqrt{s}\,x}.$$

The condition $\lim_{x\to\infty} u(x,t) = 0$ implies $\lim_{x\to\infty} U(x,s) = 0$, so we define $c_2 = 0$. Hence

$$U(x,s) = c_1 e^{-\sqrt{s}\,x}.$$

The transform of $u(0,t) = f(t)$ is $U(0,s) = F(s)$. Therefore

$$U(x,s) = F(s)e^{-\sqrt{s}\,x}$$

and

$$u(x,t) = \mathscr{L}^{-1}\left\{F(s)e^{-x\sqrt{s}}\right\} = \frac{x}{2\sqrt{\pi}}\int_0^t \frac{f(t-\tau)e^{-x^2/4\tau}}{\tau^{3/2}}\,d\tau.$$

16. We use

$$U(x,s) = c_1 e^{-\sqrt{s}\,x} + c_2 e^{\sqrt{s}\,x}.$$

The condition $\lim_{x\to\infty} u(x,t) = 0$ implies $\lim_{x\to\infty} U(x,s) = 0$, so we define $c_2 = 0$. Then $U(x,s) = c_1 e^{-\sqrt{s}\,x}$. The transform of the remaining boundary condition gives

$$\frac{dU}{dx}\bigg|_{x=0} = -F(s)$$

where $F(s) = \mathscr{L}\{f(t)\}$. This condition yields $c_1 = F(s)/\sqrt{s}$. Thus

$$U(x,s) = F(s)\,\frac{e^{-\sqrt{s}\,x}}{\sqrt{s}}.$$

Using entry (44) in the Table of Laplace transforms and the convolution theorem we obtain

$$u(x,t) = \mathscr{L}^{-1}\left\{F(s)\cdot\frac{e^{-\sqrt{s}\,x}}{\sqrt{s}}\right\} = \frac{1}{\sqrt{\pi}}\int_0^t f(\tau)\frac{e^{-x^2/4(t-\tau)}}{\sqrt{t-\tau}}\,d\tau.$$

17. Transforming the partial differential equation gives

$$\frac{d^2U}{dx^2} - sU = -60.$$

Using undetermied coefficients we obtain

$$U(x, s) = c_1 e^{-\sqrt{s}\,x} + c_2 e^{\sqrt{s}\,x} + \frac{60}{s}.$$

The condition $\lim_{x\to\infty} u(x, t) = 60$ implies $\lim_{x\to\infty} U(x, s) = 60/s$, so we define $c_2 = 0$. The transform of the remaining boundary condition gives

$$U(0, s) = \frac{60}{s} + \frac{40}{s} e^{-2s}.$$

This condition yields $c_1 = \frac{40}{s} e^{-2s}$. Thus

$$U(x, s) = \frac{60}{s} + 40 e^{-2s} \frac{e^{-\sqrt{s}\,x}}{s}.$$

Using entry (46) in the Table of Laplace transforms and the second translation theorem we obtain

$$u(x, t) = \mathscr{L}^{-1}\left\{ \frac{60}{s} + 40 e^{-2s} \frac{e^{-\sqrt{s}\,x}}{s} \right\} = 60 + 40\,\mathrm{erfc}\left(\frac{x}{2\sqrt{t-2}} \right) \mathscr{U}(t - 2).$$

18. The solution of the transformed equation

$$\frac{d^2U}{dx^2} - sU = -100$$

by undetermined coefficients is

$$U(x, s) = c_1 e^{\sqrt{s}\,x} + c_2 e^{-\sqrt{s}\,x} + \frac{100}{s}.$$

From the fact that $\lim_{x\to\infty} U(x, s) = 100/s$ we see that $c_1 = 0$. Thus

$$U(x, s) = c_2 e^{-\sqrt{s}\,x} + \frac{100}{s}. \tag{1}$$

Now the transform of the boundary condition at $x = 0$ is

$$U(0, s) = 20\left[\frac{1}{s} - \frac{1}{s} e^{-s} \right].$$

It follows from (1) that

$$\frac{20}{s} - \frac{20}{s} e^{-s} = c_2 + \frac{100}{s} \qquad \text{or} \qquad c_2 = -\frac{80}{s} - \frac{20}{s} e^{-s}$$

and so

$$U(x, s) = \left(-\frac{80}{s} - \frac{20}{s} e^{-s} \right) e^{-\sqrt{s}\,x} + \frac{100}{s}$$

$$= \frac{100}{s} - \frac{80}{s} e^{-\sqrt{s}\,x} - \frac{20}{s} e^{-\sqrt{s}\,x} e^{-s}.$$

727

Thus

$$u(x,t) = 100\,\mathscr{L}^{-1}\left\{\frac{1}{s}\right\} - 80\,\mathscr{L}^{-1}\left\{\frac{e^{-\sqrt{s}\,x}}{s}\right\} - 20\,\mathscr{L}^{-1}\left\{\frac{e^{-\sqrt{s}\,x}}{s}e^{-s}\right\}$$

$$= 100 - 80\,\mathrm{erfc}\left(x/2\sqrt{t}\right) - 20\,\mathrm{erfc}\left(x/2\sqrt{t-1}\right)\mathscr{U}(t-1).$$

19. Transforming the partial differential equation gives

$$\frac{d^2U}{dx^2} - sU = 0$$

and so

$$U(x,s) = c_1 e^{-\sqrt{s}\,x} + c_2 e^{\sqrt{s}\,x}.$$

The condition $\lim_{x\to-\infty} u(x,t) = 0$ implies $\lim_{x\to-\infty} U(x,s) = 0$, so we define $c_1 = 0$. The transform of the remaining boundary condition gives

$$\left.\frac{dU}{dx}\right|_{x=1} = \frac{100}{s} - U(1,s).$$

This condition yields

$$c_2\sqrt{s}\,e^{\sqrt{s}} = \frac{100}{s} - c_2 e^{\sqrt{s}}$$

from which it follows that

$$c_2 = \frac{100}{s(\sqrt{s}+1)}e^{-\sqrt{s}}.$$

Thus

$$U(x,s) = 100\,\frac{e^{-(1-x)\sqrt{s}}}{s(\sqrt{s}+1)}.$$

Using entry (49) in the Table of Laplace transforms we obtain

$$u(x,t) = 100\,\mathscr{L}^{-1}\left\{\frac{e^{-(1-x)\sqrt{s}}}{s(\sqrt{s}+1)}\right\} = 100\left[-e^{1-x+t}\,\mathrm{erfc}\left(\sqrt{t}+\frac{1-x}{\sqrt{t}}\right) + \mathrm{erfc}\left(\frac{1-x}{2\sqrt{t}}\right)\right].$$

20. Transforming the partial differential equation gives

$$k\frac{d^2U}{dx^2} - sU = -\frac{r}{s}.$$

Using undetermined coefficients we obtain

$$U(x,s) = c_1 e^{-\sqrt{s/k}\,x} + c_2 e^{\sqrt{s/k}\,x} + \frac{r}{s^2}.$$

The condition $\lim_{x\to\infty} \partial u/\partial x = 0$ implies $\lim_{x\to\infty} dU/dx = 0$, so we define $c_2 = 0$. The transform of the remaining boundary condition gives $U(0,s) = 0$. This condition yields $c_1 = -r/s^2$. Thus

$$U(x,s) = r\left[\frac{1}{s^2} - \frac{e^{-\sqrt{s/k}\,x}}{s^2}\right].$$

Using entries (2) and (46) in the Table of Laplace transforms and the convolution theorem we obtain

$$u(x,t) = r\,\mathcal{L}^{-1}\left\{\frac{1}{s^2} - \frac{1}{s}\cdot\frac{e^{-\sqrt{s/k}\,x}}{s}\right\} = rt - r\int_0^t \mathrm{erfc}\left(\frac{x}{2\sqrt{k\tau}}\right)\,d\tau.$$

21. The solution of

$$\frac{d^2U}{dx^2} - sU = -u_0 - u_0\sin\frac{\pi}{L}x$$

is

$$U(x,s) = c_1\cosh(\sqrt{s}\,x) + c_2\sinh(\sqrt{s}\,x) + \frac{u_0}{s} + \frac{u_0}{s+\pi^2/L^2}\sin\frac{\pi}{L}x.$$

The transformed boundary conditions $U(0,s) = u_0/s$ and $U(L,s) = u_0/s$ give, in turn, $c_1 = 0$ and $c_2 = 0$. Therefore

$$U(x,s) = \frac{u_0}{s} + \frac{u_0}{s+\pi^2/L^2}\sin\frac{\pi}{L}x$$

and

$$u(x,t) = u_0\,\mathcal{L}^{-1}\left\{\frac{1}{s}\right\} + u_0\,\mathcal{L}^{-1}\left\{\frac{1}{s+\pi^2/L^2}\right\}\sin\frac{\pi}{L}x = u_0 + u_0 e^{-\pi^2 t/L^2}\sin\frac{\pi}{L}x.$$

22. The transform of the partial differential equation is

$$k\frac{d^2U}{dx^2} - hU + h\frac{u_m}{s} = sU - u_0$$

or

$$k\frac{d^2U}{dx^2} - (h+s)U = -h\frac{u_m}{s} - u_0.$$

By undetermined coefficients we find

$$U(x,s) = c_1 e^{\sqrt{(h+s)/k}\,x} + c_2 e^{-\sqrt{(h+s)/k}\,x} + \frac{hu_m + u_0 s}{s(s+h)}.$$

The transformed boundary conditions are $U'(0,s) = 0$ and $U'(L,s) = 0$. These conditions imply $c_1 = 0$ and $c_2 = 0$. By partial fractions we then get

$$U(x,s) = \frac{hu_m + u_0 s}{s(s+h)} = \frac{u_m}{s} - \frac{u_m}{s+h} + \frac{u_0}{s+h}.$$

Therefore,

$$u(x,t) = u_m\,\mathcal{L}^{-1}\left\{\frac{1}{s}\right\} - u_m\,\mathcal{L}^{-1}\left\{\frac{1}{s+h}\right\} + u_0\,\mathcal{L}^{-1}\left\{\frac{1}{s+h}\right\} = u_m - u_m e^{-ht} + u_0 e^{-ht}.$$

23. We use

$$U(x,s) = c_1\cosh\sqrt{\frac{s}{k}}\,x + c_2\sinh\sqrt{\frac{s}{k}}\,x + \frac{u_0}{s}.$$

The transformed boundary conditions $dU/dx\,|_{x=0} = 0$ and $U(1,s) = 0$ give, in turn, $c_2 = 0$ and

$c_1 = -u_0/s \cosh \sqrt{s/k}$. Therefore

$$U(x,s) = \frac{u_0}{s} - \frac{u_0 \cosh \sqrt{s/k}\, x}{s \cosh \sqrt{s/k}} = \frac{u_0}{s} - u_0 \frac{e^{\sqrt{s/k}\, x} + e^{-\sqrt{s/k}\, x}}{s(e^{\sqrt{s/k}} + e^{-\sqrt{s/k}})}$$

$$= \frac{u_0}{s} - u_0 \frac{e^{\sqrt{s/k}\,(x-1)} + e^{-\sqrt{s/k}\,(x+1)}}{s(1 + e^{-2\sqrt{s/k}})}$$

$$= \frac{u_0}{s} - u_0 \left[\frac{e^{-\sqrt{s/k}\,(1-x)}}{s} - \frac{e^{-\sqrt{s/k}\,(3-x)}}{s} + \frac{e^{-\sqrt{s/k}\,(5-x)}}{s} - \cdots \right]$$

$$- u_0 \left[\frac{e^{-\sqrt{s/k}\,(1+x)}}{s} - \frac{e^{-\sqrt{s/k}\,(3+x)}}{s} + \frac{e^{-\sqrt{s/k}\,(5+x)}}{s} - \cdots \right]$$

$$= \frac{u_0}{s} - u_0 \sum_{n=0}^{\infty} (-1)^n \left[\frac{e^{-(2n+1-x)\sqrt{s}/\sqrt{k}}}{s} + \frac{e^{-(2n+1+x)\sqrt{s}/\sqrt{k}}}{s} \right]$$

and

$$u(x,t) = u_0 \mathcal{L}^{-1}\left\{ \frac{1}{s} \right\} - u_0 \sum_{n=0}^{\infty} (-1)^n \left[\mathcal{L}^{-1}\left\{ \frac{e^{-(2n+1-x)\sqrt{s}/\sqrt{k}}}{s} \right\} - \mathcal{L}^{-1}\left\{ \frac{e^{-(2n+1+x)\sqrt{s}/\sqrt{k}}}{s} \right\} \right]$$

$$= u_0 - u_0 \sum_{n=0}^{\infty} (-1)^n \left[\mathrm{erfc}\left(\frac{2n+1-x}{2\sqrt{kt}} \right) - \mathrm{erfc}\left(\frac{2n+1+x}{2\sqrt{kt}} \right) \right].$$

24. We use $\mathcal{L}\{c(x,t)\} = C(x,s)$ *(see problem 30)*

s/b cap "cee" (3×)

$$C(x,s) = c_1 \cosh \sqrt{\frac{s}{D}}\, x + c_2 \sinh \sqrt{\frac{s}{D}}\, x.$$

The transform of the two boundary conditions are $C(0,s) = c_0/s$ and $C(1,s) = c_0/s$. From these conditions we obtain $c_1 = c_0/s$ and

$$c_2 = c_0(1 - \cosh \sqrt{s/D})/s \sinh \sqrt{s/D}.$$

Therefore

$$c(x,s) = c_0\left[\frac{\cosh\sqrt{s/D}\,x}{s} + \frac{(1-\cosh\sqrt{s/D}\,)}{s\sinh\sqrt{s/D}}\sinh\sqrt{s/D}\,x\right]$$

$$= c_0\left[\frac{\sinh\sqrt{s/D}\,(1-x)}{s\sinh\sqrt{s/D}} + \frac{\sin\sqrt{s/D}\,x}{s\sinh\sqrt{s/D}}\right]$$

$$= c_0\left[\frac{e^{\sqrt{s/D}\,(1-x)} - e^{-\sqrt{s/D}\,(1-x)}}{s(e^{\sqrt{s/D}} - e^{-\sqrt{s/D}})} + \frac{e^{\sqrt{s/D}\,x} - e^{-\sqrt{s/D}\,x}}{s(e^{\sqrt{s/D}} - e^{-\sqrt{s/D}})}\right]$$

$$= c_0\left[\frac{e^{-\sqrt{s/D}\,x} - e^{-\sqrt{s/D}\,(2-x)}}{s(1-e^{-2\sqrt{s/D}})} + \frac{e^{\sqrt{s/D}\,(x-1)} - e^{-\sqrt{s/D}\,(x+1)}}{s(1-e^{-2\sqrt{s/D}})}\right]$$

$$= c_0\frac{(e^{-\sqrt{s/D}\,x} - e^{-\sqrt{s/D}\,(2-x)})}{s}\left(1 + e^{-2\sqrt{s/D}} + e^{-4\sqrt{s/D}} + \cdots\right)$$

$$+ c_0\frac{(e^{\sqrt{s/D}\,(x-1)} - e^{-\sqrt{s/D}\,(x+1)})}{s}\left(1 + e^{-2\sqrt{s/D}} + e^{-4\sqrt{s/D}} + \cdots\right)$$

$$= c_0\sum_{n=0}^{\infty}\left[\frac{e^{-(2n+x)\sqrt{s/D}}}{s} - \frac{e^{-(2n+2-x)\sqrt{s/D}}}{s}\right]$$

$$+ c_0\sum_{n=0}^{\infty}\left[\frac{e^{-(2n+1-x)\sqrt{s/D}}}{s} - \frac{e^{-(2n+1+x)\sqrt{s/D}}}{s}\right]$$

and so

$$c(x,t) = c_0\sum_{n=0}^{\infty}\left[\mathcal{L}^{-1}\left\{\frac{e^{-\frac{(2n+x)}{\sqrt{D}}\sqrt{s}}}{s}\right\} - \mathcal{L}^{-1}\left\{\frac{e^{-\frac{(2n+2-x)}{\sqrt{D}}\sqrt{s}}}{s}\right\}\right]$$

$$+ c_0\sum_{n=0}^{\infty}\left[\mathcal{L}^{-1}\left\{\frac{e^{-\frac{(2n+1-x)}{\sqrt{D}}\sqrt{s}}}{s}\right\} - \mathcal{L}^{-1}\left\{\frac{e^{-\frac{(2n+1+x)}{\sqrt{D}}\sqrt{s}}}{s}\right\}\right]$$

$$= c_0\sum_{n=0}^{\infty}\left[\operatorname{erfc}\left(\frac{2n+x}{2\sqrt{Dt}}\right) - \operatorname{erfc}\left(\frac{2n+2-x}{2\sqrt{Dt}}\right)\right]$$

$$+ c_0\sum_{n=0}^{\infty}\left[\operatorname{erfc}\left(\frac{2n+1-x}{2\sqrt{Dt}}\right) - \operatorname{erfc}\left(\frac{2n+1+x}{2\sqrt{Dt}}\right)\right].$$

Now using $\operatorname{erfc}(x) = 1 - \operatorname{erf}(x)$ we get

$$c(x,t) = c_0 \sum_{n=0}^{\infty} \left[\operatorname{erf}\left(\frac{2n+2-x}{2\sqrt{Dt}} \right) - \operatorname{erf}\left(\frac{2n+x}{2\sqrt{Dt}} \right) \right]$$

$$+ c_0 \sum_{n=0}^{\infty} \left[\operatorname{erf}\left(\frac{2n+1+x}{2\sqrt{Dt}} \right) - \operatorname{erf}\left(\frac{2n+1-x}{2\sqrt{Dt}} \right) \right].$$

25. We use

$$U(x,s) = c_1 e^{-\sqrt{RCs+RG}\,x} + c_2 e^{\sqrt{RCs+RG}} + \frac{Cu_0}{Cs+G}.$$

The condition $\lim_{x\to\infty} \partial u / \partial x = 0$ implies $\lim_{x\to\infty} dU/dx = 0$, so we define $c_2 = 0$. Applying $U(0,s) = 0$ to

$$U(x,s) = c_1 e^{-\sqrt{RCsRG}\,x} + \frac{Cu_0}{Cs+G} \qquad \textit{(insert + sign in superscript)}$$

gives $c_1 = -Cu_0/(Cs+G)$. Therefore

$$U(x,s) = -Cu_0 \frac{e^{-\sqrt{RCs+RG}\,x}}{Cs+G} + \frac{Cu_0}{Cs+G}$$

and

$$u(x,t) = u_0 \mathcal{L}^{-1}\left\{ \frac{1}{s+G/C} \right\} - u_0 \mathcal{L}^{-1}\left\{ \frac{e^{-x\sqrt{RC}\sqrt{s+G/C}}}{s+G/C} \right\}$$

$$= u_0 e^{-Gt/C} - u_0 e^{-Gt/C} \operatorname{erfc}\left(\frac{x\sqrt{RC}}{2\sqrt{t}} \right)$$

$$= u_0 e^{-Gt/C}\left[1 - \operatorname{erfc}\left(\frac{x}{2}\sqrt{\frac{RC}{t}} \right) \right]$$

$$= u_0 e^{-Gt/C} \operatorname{erf}\left(\frac{x}{2}\sqrt{\frac{RC}{t}} \right).$$

26. We use

$$U(x,s) = c_1 e^{-\sqrt{s+h}\,x} + c_2 e^{\sqrt{s+h}\,x}.$$

The condition $\lim_{x\to\infty} u(x,t) = 0$ implies $\lim_{x\to\infty} U(x,s) = 0$, so we take $c_2 = 0$. Therefore

$$U(x,s) = c_1 e^{-\sqrt{s+h}\,x}.$$

The Laplace transform of $u(0,t) = u_0$ is $U(0,s) = u_0/s$ and so

$$U(x,s) = u_0 \frac{e^{-\sqrt{s+h}\,x}}{s}$$

and

$$u(x,t) = u_0 \mathcal{L}^{-1}\left\{ \frac{e^{-\sqrt{s+h}\,x}}{s} \right\} = u_0 \mathcal{L}^{-1}\left\{ \frac{1}{s} e^{-\sqrt{s+h}\,x} \right\}.$$

From the first translation theorem,

$$\mathscr{L}^{-1}\left\{e^{-\sqrt{s+h}\,x}\right\} = e^{-ht}\,\mathscr{L}^{-1}\left\{e^{-x\sqrt{s}}\right\} = e^{-ht}\frac{x}{2\sqrt{\pi t^3}}\,e^{-x^2/4t}.$$

Thus, from the convolution theorem we obtain

$$u(x, s) = \frac{u_0 x}{2\sqrt{\pi}}\int_0^t \frac{e^{-h\tau - x^2/4\tau}}{\tau^{3/2}}\,d\tau.$$

27. (a) We use

$$U(x, s) = c_1 e^{-(s/a)x} + c_2 e^{(s/a)x} + \frac{v_0^2 F_0}{(a^2 - v_0^2)s^2}\,e^{-(s/v_0)x}.$$

The condition $\lim_{x\to\infty} u(x, t) = 0$ implies $\lim_{x\to\infty} U(x, s) = 0$, so we must define $c_2 = 0$. Consequently

$$U(x, s) = c_1 e^{-(s/a)x} + \frac{v_0^2 F_0}{(a^2 - v_0^2)s^2}\,e^{-(s/v_0)x}.$$

The remaining boundary condition transforms into $U(0, s) = 0$. From this we find

$$c_1 = -v_0^2 F_0 / (a^2 - v_0^2)s^2.$$

Therefore, by the second translation theorem

$$U(x, s) = -\frac{v_0^2 F_0}{(a^2 - v_0^2)s^2}\,e^{-(s/a)x} + \frac{v_0^2 F_0}{(a^2 - v_0^2)s^2}\,e^{-(s/v_0)x}$$

and

$$u(x, t) = \frac{v_0^2 F_0}{a^2 - v_0^2}\left[\mathscr{L}^{-1}\left\{\frac{e^{-(x/v_0)s}}{s^2}\right\} - \mathscr{L}^{-1}\left\{\frac{e^{-(x/a)s}}{s^2}\right\}\right]$$

$$= \frac{v_0^2 F_0}{a^2 - v_0^2}\left[\left(t - \frac{x}{v_0}\right)\mathscr{U}\left(t - \frac{x}{v_0}\right) - \left(t - \frac{x}{a}\right)\mathscr{U}\left(t - \frac{x}{a}\right)\right].$$

(b) In the case when $v_0 = a$ the solution of the transformed equation is

$$U(x, s) = c_1 e^{-(s/a)x} + c_2 e^{(s/a)x} - \frac{F_0}{2as}\,xe^{-(s/a)x}.$$

The usual analysis then leads to $c_1 = 0$ and $c_2 = 0$. Therefore

$$U(x, s) = -\frac{F_0}{2as}\,xe^{-(s/a)x}$$

and

$$u(x, t) = -\frac{xF_0}{2a}\,\mathscr{L}^{-1}\left\{\frac{e^{-(x/a)s}}{s}\right\} = -\frac{xF_0}{2a}\,\mathscr{U}\left(t - \frac{x}{a}\right).$$

28. (a) We use

$$U(x, s) = c_1 e^{-\sqrt{s/k}\,x} + c_2 e^{\sqrt{s/k}\,x}.$$

Now $\lim_{x\to\infty} u(x,t) = 0$ implies $\lim_{x\to\infty} U(x,s) = 0$, so we define $c_2 = 0$. Then

$$U(x,s) = c_1 e^{-\sqrt{s/k}\,x}.$$

Finally, from $U(0,s) = u_0/s$ we obtain $c_1 = u_0/s$. Thus

$$U(x,s) = u_0\,\frac{e^{-\sqrt{s/k}\,x}}{s}$$

and

$$u(x,t) = u_0\,\mathscr{L}^{-1}\left\{\frac{e^{-\sqrt{s/k}\,x}}{s}\right\} = u_0\,\mathscr{L}^{-1}\left\{\frac{e^{-(x/\sqrt{k})\sqrt{s}}}{s}\right\} = u_0\,\mathrm{erfc}\left(\frac{x}{2\sqrt{kt}}\right).$$

Since $\mathrm{erfc}(0) = 1$,

$$\lim_{t\to\infty} u(x,t) = \lim_{t\to\infty} u_0\,\mathrm{erfc}(x/2\sqrt{kt}\,) = u_0.$$

(b)

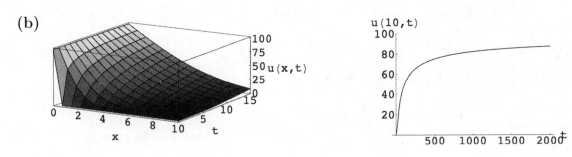

29. (a) Transforming the partial differential equation and using the initial condition gives

$$k\,\frac{d^2U}{dx^2} - sU = 0.$$

Since the domain of the variable x is an infinite interval we write the general solution of this differential equation as

$$U(x,s) = c_1 e^{-\sqrt{s/k}\,x} + c_2 e^{\sqrt{s/k}\,x}.$$

Transforming the boundary conditions gives $U'(0,s) = -A/s$ and $\lim_{x\to\infty} U(x,s) = 0$. Hence we find $c_2 = 0$ and $c_1 = A\sqrt{k}/s\sqrt{s}$. From

$$U(x,s) = A\sqrt{k}\,\frac{e^{-\sqrt{s/k}\,x}}{s\sqrt{s}}$$

we see that

$$u(x,t) = A\sqrt{k}\,\mathscr{L}^{-1}\left\{\frac{e^{-\sqrt{s/k}\,x}}{s\sqrt{s}}\right\}.$$

With the identification $a = x/\sqrt{k}$ it follows from (47) in the Table of Laplace transforms that

$$u(x,t) = A\sqrt{k}\left\{2\sqrt{\frac{t}{\pi}}\,e^{-x^2/4kt} - \frac{x}{\sqrt{k}}\,\text{erfc}\left(\frac{x}{2\sqrt{kt}}\right)\right\}$$

$$= 2A\sqrt{\frac{kt}{\pi}}\,e^{-x^2/4kt} - Ax\,\text{erfc}\left(x/2\sqrt{kt}\right).$$

Since $\text{erfc}(0) = 1$,

$$\lim_{t\to\infty} u(x,t) = \lim_{t\to\infty}\left(2A\sqrt{\frac{kt}{\pi}}\,e^{-x^2/4kt} - Ax\,\text{erfc}\left(\frac{x}{2\sqrt{kt}}\right)\right) = \infty.$$

(b)

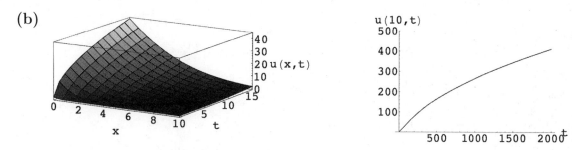

30. **(a)** Letting $C(x,s) = \mathscr{L}\{c(x,t)\}$ we obtain

$$\frac{d^2C}{dx^2} - \frac{s}{k}C = 0 \qquad \text{subject to} \qquad \left.\frac{dC}{dx}\right|_{x=0} = -A.$$

The solution of this initial-value problem is

$$C(x,s) = A\sqrt{k}\,\frac{e^{-(x/\sqrt{k})\sqrt{s}}}{\sqrt{s}},$$

so that

$$c(x,t) = A\sqrt{\frac{k}{\pi t}}\,e^{-x^2/4kt}.$$

(b) c(x,t)

t=0.1

t=0.5

t=1
t=2
t=5

x

(c) $\displaystyle\int_0^\infty c(x,t)\,dx = Ak\,\text{erf}\left(\frac{x}{2\sqrt{kt}}\right)\Big|_0^\infty = Ak(1-0) = Ak$

1. From formulas (5) and (6) in the text,

$$A(\alpha) = \int_{-1}^{0} (-1) \cos \alpha x \, dx + \int_{0}^{1} (2) \cos \alpha x \, dx = -\frac{\sin \alpha}{\alpha} + 2\frac{\sin \alpha}{\alpha} = \frac{\sin \alpha}{\alpha}$$

and

$$B(\alpha) = \int_{-1}^{0} (-1) \sin \alpha x \, dx + \int_{0}^{1} (2) \sin \alpha x \, dx$$

$$= \frac{1 - \cos \alpha}{\alpha} - 2\frac{\cos \alpha - 1}{\alpha} = \frac{3(1 - \cos \alpha)}{\alpha}.$$

Hence

$$f(x) = \frac{1}{\pi} \int_{0}^{\infty} \frac{\sin \alpha \cos \alpha x + 3(1 - \cos \alpha) \sin \alpha x}{\alpha} \, d\alpha.$$

2. From formulas (5) and (6) in the text,

$$A(\alpha) = \int_{\pi}^{2\pi} 4 \cos \alpha x \, dx = 4\frac{\sin 2\pi\alpha - \sin \pi\alpha}{\alpha}$$

and

$$B(\alpha) = \int_{\pi}^{2\pi} 4 \sin \alpha x \, dx = 4\frac{\cos \pi\alpha - \cos 2\pi\alpha}{\alpha}.$$

Hence

$$f(x) = \frac{4}{\pi} \int_{0}^{\infty} \frac{(\sin 2\pi\alpha - \sin \pi\alpha) \cos \alpha x + (\cos \pi\alpha - \cos 2\pi\alpha) \sin \alpha x}{\alpha} \, d\alpha$$

$$= \frac{4}{\pi} \int_{0}^{\infty} \frac{\sin 2\pi\alpha \cos \alpha x - \cos 2\pi\alpha \sin \alpha x - \sin \pi\alpha \cos \alpha x + \cos \pi\alpha \sin \alpha x}{\alpha} \, d\alpha$$

$$= \frac{4}{\pi} \int_{0}^{\infty} \frac{\sin \alpha(2\pi - x) - \sin \alpha(\pi - x)}{\alpha} \, d\alpha.$$

3. From formulas (5) and (6) in the text,

$$A(\alpha) = \int_{0}^{3} x \cos \alpha x \, dx = \frac{x \sin \alpha x}{\alpha} \bigg|_{0}^{3} - \frac{1}{\alpha} \int_{0}^{3} \sin \alpha x \, dx$$

$$= \frac{3 \sin 3\alpha}{\alpha} + \frac{\cos \alpha x}{\alpha^2} \bigg|_{0}^{3} = \frac{3\alpha \sin 3\alpha + \cos 3\alpha - 1}{\alpha^2}$$

and

$$B(\alpha) = \int_{0}^{3} x \sin \alpha x \, dx = -\frac{x \cos \alpha x}{\alpha} \bigg|_{0}^{3} + \frac{1}{\alpha} \int_{0}^{3} \cos \alpha x \, dx$$

$$= -\frac{3 \cos 3\alpha}{\alpha} + \frac{\sin \alpha x}{\alpha^2} \bigg|_{0}^{3} = \frac{\sin 3\alpha - 3\alpha \cos 3\alpha}{\alpha^2}.$$

Hence

$$f(x) = \frac{1}{\pi} \int_0^\infty \frac{(3\alpha \sin 3\alpha + \cos 3\alpha - 1)\cos \alpha x + (\sin 3\alpha - 3\alpha \cos 3\alpha)\sin \alpha x}{\alpha^2} \, d\alpha$$

$$= \frac{1}{\pi} \int_0^\infty \frac{3\alpha(\sin 3\alpha \cos \alpha x - \cos 3\alpha \sin \alpha x) + \cos 3\alpha \cos \alpha x + \sin 3\alpha \sin \alpha x - \cos \alpha x}{\alpha^2} \, d\alpha$$

$$= \frac{1}{\pi} \int_0^\infty \frac{3\alpha \sin \alpha(3 - x) + \cos \alpha(3 - x) - \cos \alpha x}{\alpha^2} \, d\alpha.$$

4. From formulas (5) and (6) in the text,

$$A(\alpha) = \int_{-\infty}^\infty f(x) \cos \alpha x \, dx$$

$$= \int_{-\infty}^0 0 \cdot \cos \alpha x \, dx + \int_0^\pi \sin x \cos \alpha x \, dx + \int_\pi^\infty 0 \cdot \cos \alpha x \, dx$$

$$= \frac{1}{2} \int_0^\pi [\sin(1 + \alpha)x + \sin(1 - \alpha)x] \, dx$$

$$= \frac{1}{2} \left[-\frac{\cos(1 + \alpha)x}{1 + \alpha} - \frac{\cos(1 - \alpha)x}{1 - \alpha} \right]_0^\pi$$

$$= -\frac{1}{2} \left[\frac{\cos(1 + \alpha)\pi - 1}{1 + \alpha} + \frac{\cos(1 - \alpha)\pi - 1}{1 - \alpha} \right]$$

$$= -\frac{1}{2} \left[\frac{\cos(1 + \alpha)\pi - \alpha\cos(1 + \alpha)\pi + \cos(1 - \alpha)\pi + \alpha\cos(1 - \alpha)\pi - 2}{1 - \alpha^2} \right]$$

$$= \frac{1 + \cos \alpha \pi}{1 - \alpha^2},$$

and

$$B(\alpha) = \int_0^\pi \sin x \sin \alpha x \, dx = \frac{1}{2} \int_0^\pi [\cos(1 - \alpha)x - \cos(1 + \alpha)] \, dx$$

$$= \frac{1}{2} \left[\frac{\sin(1 - \alpha)\pi}{1 - \alpha} - \frac{\sin(1 + \alpha)\pi}{1 + \alpha} \right] = \frac{\sin \alpha \pi}{1 - \alpha^2}.$$

Hence

$$f(x) = \frac{1}{\pi} \int_0^\infty \frac{\cos \alpha x + \cos \alpha x \cos \alpha \pi + \sin \alpha x \sin \alpha \pi}{1 - \alpha^2} \, d\alpha$$

$$= \frac{1}{\pi} \int_0^\infty \frac{\cos \alpha x + \cos \alpha(x - \pi)}{1 - \alpha^2} \, d\alpha.$$

5. From formula (5) in the text,

$$A(\alpha) = \int_0^\infty e^{-x} \cos \alpha x \, dx.$$

Recall $\mathcal{L}\{\cos kt\} = s/(s^2 + k^2)$. If we set $s = 1$ and $k = \alpha$ we obtain

$$A(\alpha) = \frac{1}{1 + \alpha^2}.$$

Now

$$B(\alpha) = \int_0^\infty e^{-x} \sin \alpha x \, dx.$$

Recall $\mathcal{L}\{\sin kt\} = k/(s^2 + k^2)$. If we set $s = 1$ and $k = \alpha$ we obtain

$$B(\alpha) = \frac{\alpha}{1 + \alpha^2}.$$

Hence

$$f(x) = \frac{1}{\pi} \int_0^\infty \frac{\cos \alpha x + \alpha \sin \alpha x}{1 + \alpha^2} \, d\alpha.$$

6. From formulas (5) and (6) in the text,

$$A(\alpha) = \int_{-1}^1 e^x \cos \alpha x \, dx$$

$$= \frac{e(\cos \alpha + \alpha \sin \alpha) - e^{-1}(\cos \alpha - \alpha \sin \alpha)}{1 + \alpha^2}$$

$$= \frac{2(\sinh 1) \cos \alpha - 2\alpha(\cosh 1) \sin \alpha}{1 + \alpha^2}$$

and

$$B(\alpha) = \int_{-1}^1 e^x \sin \alpha x \, dx$$

$$= \frac{e(\sin \alpha - \alpha \cos \alpha) - e^{-1}(-\sin \alpha - \alpha \cos \alpha)}{1 + \alpha^2}$$

$$= \frac{2(\cosh 1) \sin \alpha - 2\alpha(\sinh 1) \cos \alpha}{1 + \alpha^2}.$$

Hence

$$f(x) = \frac{1}{\pi} \int_0^\infty [A(\alpha) \cos \alpha x + B(\alpha) \sin \alpha x] \, d\alpha.$$

7. The function is odd. Thus from formula (11) in the text

$$B(\alpha) = 5 \int_0^1 \sin \alpha x \, dx = \frac{5(1 - \cos \alpha)}{\alpha}.$$

Hence from formula (10) in the text,

$$f(x) = \frac{10}{\pi} \int_0^\infty \frac{(1 - \cos \alpha) \sin \alpha x}{\alpha} \, d\alpha.$$

8. The function is even. Thus from formula (9) in the text

$$A(\alpha) = \pi \int_1^2 \cos \alpha x \, dx = \pi \left(\frac{\sin 2\alpha - \sin \alpha}{\alpha} \right).$$

Hence from formula (8) in the text,

$$f(x) = 2 \int_0^\infty \frac{(\sin 2\alpha - \sin \alpha) \cos \alpha x}{\alpha} \, d\alpha.$$

9. The function is even. Thus from formula (9) in the text

$$A(\alpha) = \int_0^\pi x \cos \alpha x \, dx = \frac{x \sin \alpha x}{\alpha} \bigg|_0^\pi - \frac{1}{\alpha} \int_0^\pi \sin \alpha x \, dx$$

$$= \frac{\pi \sin \pi \alpha}{\alpha} + \frac{1}{\alpha^2} \cos \alpha x \bigg|_0^\pi = \frac{\pi \alpha \sin \pi \alpha + \cos \pi \alpha - 1}{\alpha^2}.$$

Hence from formula (8) in the text

$$f(x) = \frac{2}{\pi} \int_0^\infty \frac{(\pi \alpha \sin \pi \alpha + \cos \pi \alpha - 1) \cos \alpha x}{\alpha^2} \, d\alpha.$$

10. The function is odd. Thus from formula (11) in the text

$$B(\alpha) = \int_0^\pi x \sin \alpha x \, dx = -\frac{x \cos \alpha x}{\alpha} \bigg|_0^\pi + \frac{1}{\alpha} \int_0^\pi \cos \alpha x \, dx$$

$$= -\frac{\pi \cos \pi \alpha}{\alpha} + \frac{1}{\alpha^2} \sin \alpha x \bigg|_0^\pi = \frac{-\pi \alpha \cos \pi \alpha + \sin \pi \alpha}{\alpha^2}.$$

Hence from formula (10) in the text,

$$f(x) = \frac{2}{\pi} \int_0^\infty \frac{(-\pi \alpha \cos \pi \alpha + \sin \pi \alpha) \sin \alpha x}{\alpha^2} \, d\alpha.$$

11. The function is odd. Thus from formula (11) in the text

$$B(\alpha) = \int_0^\infty (e^{-x} \sin x) \sin \alpha x \, dx$$

$$= \frac{1}{2} \int_0^\infty e^{-x} [\cos(1 - \alpha)x - \cos(1 + \alpha)x] \, dx$$

$$= \frac{1}{2} \int_0^\infty e^{-x} \cos(1 - \alpha)x \, dx - \frac{1}{2} \int_0^\infty e^{-x} \cos(1 + \alpha)x, dx.$$

Now recall

$$\mathscr{L}\{\cos kt\} = \int_0^\infty e^{-st} \cos kt \, dt = s/(s^2 + k^2).$$

If we set $s = 1$, and in turn, $k = 1 - \alpha$ and then $k = 1 + \alpha$, we obtain

$$B(\alpha) = \frac{1}{2} \frac{1}{1 + (1 - \alpha)^2} - \frac{1}{2} \frac{1}{1 + (1 + \alpha)^2} = \frac{1}{2} \frac{(1 + \alpha)^2 - (1 - \alpha)^2}{[1 + (1 - \alpha)^2][1 + (1 + \alpha)^2]}.$$

Simplifying the last expression gives

$$B(\alpha) = \frac{2\alpha}{4 + \alpha^4}.$$

Hence from formula (10) in the text

$$f(x) = \frac{4}{\pi} \int_0^\infty \frac{\alpha \sin \alpha x}{4 + \alpha^4} \, d\alpha.$$

739

12. The function is odd. Thus from formula (11) in the text

$$B(\alpha) = \int_0^\infty x e^{-x} \sin \alpha x \, dx.$$

Now recall

$$\mathscr{L}\{t \sin kt\} = -\frac{d}{ds} \mathscr{L}\{\sin kt\} = 2ks/(s^2 + k^2)^2.$$

If we set $s = 1$ and $k = \alpha$ we obtain

$$B(\alpha) = \frac{2\alpha}{(1 + \alpha^2)^2}.$$

Hence from formula (10) in the text

$$f(x) = \frac{4}{\pi} \int_0^\infty \frac{\alpha \sin \alpha x}{(1 + \alpha^2)^2} \, d\alpha.$$

13. For the cosine integral,

$$A(\alpha) = \int_0^\infty e^{-kx} \cos \alpha x \, dx = \frac{k}{k^2 + \alpha^2}.$$

Hence

$$f(x) = \frac{2}{\pi} \int_0^\infty \frac{k \cos \alpha x}{k^2 + \alpha^2} \, d\alpha = \frac{2k}{\pi} \int_0^\infty \frac{\cos \alpha x}{k^2 + \alpha^2} \, d\alpha.$$

For the sine integral,

$$B(\alpha) = \int_0^\infty e^{-kx} \sin \alpha x \, dx = \frac{\alpha}{k^2 + \alpha^2}.$$

Hence

$$f(x) = \frac{2}{\pi} \int_0^\infty \frac{\alpha \sin \alpha x}{k^2 + \alpha^2} \, d\alpha.$$

14. From Problem 13 the cosine and sine integral representations of e^{-kx}, $k > 0$, are respectively,

$$e^{-kx} = \frac{2k}{\pi} \int_0^\infty \frac{\cos \alpha x}{k^2 + \alpha^2} \, d\alpha \quad \text{and} \quad e^{-kx} = \frac{2}{\pi} \int_0^\infty \frac{\alpha \sin \alpha x}{k^2 + \alpha^2} \, d\alpha.$$

Hence, the cosine integral representation of $f(x) = e^{-x} - e^{-3x}$ is

$$e^{-x} - e^{-3x} = \frac{2}{\pi} \int_0^\infty \frac{\cos \alpha x}{1 + \alpha^2} \, d\alpha - \frac{2(3)}{\pi} \int_0^\infty \frac{\cos \alpha x}{9 + \alpha^2} \, d\alpha = \frac{4}{\pi} \int_0^\infty \frac{3 - \alpha^2}{(1 + \alpha^2)(9 + \alpha^2)} \cos \alpha x \, d\alpha.$$

The sine integral representation of f is

$$e^{-x} - e^{-3x} = \frac{2}{\pi} \int_0^\infty \frac{\alpha \sin \alpha x}{1 + \alpha^2} \, d\alpha - \frac{2}{\pi} \int_0^\infty \frac{\alpha \sin \alpha x}{9 + \alpha^2} \, d\alpha = \frac{16}{\pi} \int_0^\infty \frac{\alpha \sin \alpha x}{(1 + \alpha^2)(9 + \alpha^2)} \, d\alpha.$$

15. For the cosine integral,

$$A(\alpha) = \int_0^\infty x e^{-2x} \cos \alpha x \, dx.$$

But we know

$$\mathscr{L}\{t \cos kt\} = -\frac{d}{ds} \frac{s}{(s^2 + k^2)} = \frac{(s^2 - k^2)}{(s^2 + k^2)^2}.$$

If we set $s = 2$ and $k = \alpha$ we obtain

$$A(\alpha) = \frac{4 - \alpha^2}{(4 + \alpha^2)^2}.$$

Hence

$$f(x) = \frac{2}{\pi} \int_0^\infty \frac{(4 - \alpha^2) \cos \alpha x}{(4 + \alpha^2)^2} \, d\alpha.$$

For the sine integral,

$$B(\alpha) = \int_0^\infty x e^{-2x} \sin \alpha x \, dx.$$

From Problem 12, we know

$$\mathscr{L}\{t \sin kt\} = \frac{2ks}{(s^2 + k^2)^2}.$$

If we set $s = 2$ and $k = \alpha$ we obtain

$$B(\alpha) = \frac{4\alpha}{(4 + \alpha^2)^2}.$$

Hence

$$f(x) = \frac{8}{\pi} \int_0^\infty \frac{\alpha \sin \alpha x}{(4 + \alpha^2)^2} \, d\alpha.$$

16. For the cosine integral,

$$A(\alpha) = \int_0^\infty e^{-x} \cos x \cos \alpha x \, dx$$

$$= \frac{1}{2} \int_0^\infty e^{-x} [\cos(1 + \alpha)x + \cos(1 - \alpha)x] \, dx$$

$$= \frac{1}{2} \frac{1}{1 + (1 + \alpha)^2} + \frac{1}{2} \frac{1}{1 + (1 - \alpha)^2}$$

$$= \frac{1}{2} \frac{1 + (1 - \alpha)^2 + 1 + (1 + \alpha)^2}{[1 + (1 + \alpha)^2][1 + (1 - \alpha)^2]}$$

$$= \frac{2 + \alpha^2}{4 + \alpha^4}.$$

Hence

$$f(x) = \frac{2}{\pi} \int_0^\infty \frac{(2 + \alpha^2) \cos \alpha x}{4 + \alpha^4} \, d\alpha.$$

For the sine integral,

$$B(\alpha) = \int_0^\infty e^{-x} \cos x \sin \alpha x \, dx$$

$$= \frac{1}{2} \int_0^\infty e^{-x} [\sin(1+\alpha)x - \sin(1-\alpha)x] \, dx$$

$$= \frac{1}{2} \frac{1+\alpha}{1+(1+\alpha)^2} - \frac{1}{2} \frac{1-\alpha}{1+(1-\alpha)^2}$$

$$= \frac{1}{2} \left[\frac{(1+\alpha)[1+(1-\alpha)^2] - (1-\alpha)[1+(1+\alpha)^2]}{[1+(1+\alpha)^2][1+(1-\alpha)^2]} \right]$$

$$= \frac{\alpha^3}{4+\alpha^4}.$$

Hence

$$f(x) = \frac{2}{\pi} \int_0^\infty \frac{\alpha^3 \sin \alpha x}{4+\alpha^4} \, d\alpha.$$

17. By formula (8) in the text

$$f(x) = 2\pi \int_0^\infty e^{-\alpha} \cos \alpha x \, d\alpha = \frac{2}{\pi} \frac{1}{1+x^2}, \quad x > 0.$$

18. From the formula for sine integral of $f(x)$ we have

$$f(x) = \frac{2}{\pi} \int_0^\infty \left(\int_0^\infty f(x) \sin \alpha x \, dx \right) \sin \alpha x \, dx$$

$$= \frac{2}{\pi} \left[\int_0^1 1 \cdot \sin \alpha x \, d\alpha + \int_1^\infty 0 \cdot \sin \alpha x \, d\alpha \right]$$

$$= \frac{2}{\pi} \frac{(-\cos \alpha x)}{x} \Big|_0^1 = \frac{2}{\pi} \frac{1-\cos x}{x}.$$

19. (a) From formula (7) in the text with $x = 2$, we have

$$\frac{1}{2} = \frac{2}{\pi} \int_0^\infty \frac{\sin \alpha \cos \alpha}{\alpha} \, d\alpha = \frac{1}{\pi} \int_0^\infty \frac{\sin 2\alpha}{\alpha} \, d\alpha.$$

If we let $\alpha = x$ we obtain

$$\int_0^\infty \frac{\sin 2x}{x} \, dx = \frac{\pi}{2}.$$

(b) If we now let $2x = kt$ where $k > 0$, then $dx = (k/2)dt$ and the integral in part (a) becomes

$$\int_0^\infty \frac{\sin kt}{kt/2} (k/2) \, dt = \int_0^\infty \frac{\sin kt}{t} \, dt = \frac{\pi}{2}.$$

20. With $f(x) = e^{-|x|}$, formula (20) in the text is

$$C(\alpha) = \int_{-\infty}^\infty e^{-|x|} e^{i\alpha x} dx = \int_{-\infty}^\infty e^{-|x|} \cos \alpha x \, dx + i \int_{-\infty}^\infty e^{-|x|} \sin \alpha x \, dx.$$

The imaginary part in the last line is zero since the integrand is an odd function of x. Therefore,

$$C(\alpha) = \int_{-\infty}^{\infty} e^{-|x|} \cos \alpha x \, dx = 2 \int_{0}^{\infty} e^{-x} \cos \alpha x \, dx = \frac{2}{1 + \alpha^2}$$

and so from formula (19) in the text,

$$f(x) = \frac{1}{\pi} \int_{-\infty}^{\infty} \frac{\cos \alpha x}{1 + \alpha^2} \, d\alpha = \frac{2}{\pi} \int_{0}^{\infty} \frac{\cos \alpha x}{1 + \alpha^2} \, d\alpha.$$

This is the same result obtained from formulas (8) and (9) in the text.

21. (a) From the identity

$$\sin A \cos B = \frac{1}{2}[\sin(A + B) + \sin(A - B)]$$

we have

$$\sin \alpha \cos \alpha x = \frac{1}{2}[\sin(\alpha + \alpha x) + \sin(\alpha - \alpha x)]$$

$$= \frac{1}{2}[\sin \alpha(1 + x) + \sin \alpha(1 - x)]$$

$$= \frac{1}{2}[\sin \alpha(x + 1) - \sin \alpha(x - 1)].$$

Then

$$\frac{2}{\pi} \int_{0}^{\infty} \frac{\sin \alpha \cos \alpha x}{\alpha} \, d\alpha = \frac{1}{\pi} \int_{0}^{\infty} \frac{\sin \alpha(x + 1) - \sin \alpha(x - 1)}{\alpha} \, d\alpha.$$

(b) Noting that

$$F_b = \frac{1}{\pi} \int_{0}^{b} \frac{\sin \alpha(x + 1) - \sin \alpha(x - 1)}{\alpha} \, d\alpha$$

$$= \frac{1}{\pi} \left[\int_{0}^{b} \frac{\sin \alpha(x + 1)}{\alpha} \, d\alpha - \int_{0}^{b} \frac{\sin \alpha(x - 1)}{\alpha} \, d\alpha \right]$$

and letting $t = \alpha(x + 1)$ so that $dt = (x + 1) \, d\alpha$ in the first integral and $t = \alpha(x - 1)$ so that $dt = (x - 1) \, d\alpha$ in the second integral we have

$$F_b = \frac{1}{\pi} \left[\int_{0}^{b(x+1)} \frac{\sin t}{t} \, dt - \int_{0}^{b(x-1)} \frac{\sin t}{t} \, dt \right].$$

Since $\mathrm{Si}\,(x) = \int_{0}^{x}[(\sin t)/t] \, dt$, this becomes

$$F_b = \frac{1}{\pi}[\mathrm{Si}\,(b(x + 1)) - \mathrm{Si}\,(b(x - 1))].$$

(c) In *Mathematica* we define **f[b_] := (1/Pi)(SinIntegral[b(x + 1)] − SinIntegral[b(x − 1)]**.
Graphs of $F_b(x)$ for $b = 4, 6, 15$, and 75 are shown below.

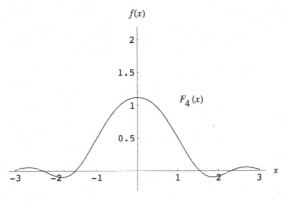

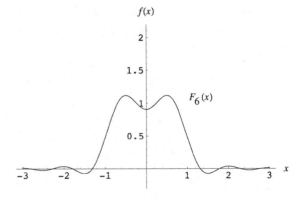

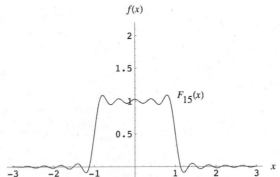

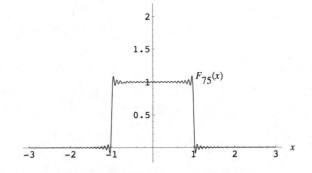

Exercises 14.4

Fourier Transforms

For the boundary-value problems in this section it is sometimes useful to note that the identities

$$e^{i\alpha} = \cos\alpha + i\sin\alpha \quad and \quad e^{-i\alpha} = \cos\alpha - i\sin\alpha$$

imply

$$e^{i\alpha} + e^{-i\alpha} = 2\cos\alpha \quad and \quad e^{i\alpha} - e^{-i\alpha} = 2i\sin\alpha.$$

1. Using the Fourier transform, the partial differential equation becomes

$$\frac{dU}{dt} + k\alpha^2 U = 0 \quad \text{and so} \quad U(\alpha, t) = ce^{-k\alpha^2 t}.$$

Now

$$\mathscr{F}\{u(x, 0)\} = U(\alpha, 0) = \mathscr{F}\left\{e^{-|x|}\right\}.$$

We have

$$\mathscr{F}\left\{e^{-|x|}\right\} = \int_{-\infty}^{\infty} e^{-|x|}e^{i\alpha x}\,dx = \int_{-\infty}^{\infty} e^{-|x|}(\cos\alpha x + i\sin\alpha x)\,dx = \int_{-\infty}^{\infty} e^{-|x|}\cos\alpha x\,dx.$$

The integral

$$\int_{-\infty}^{\infty} e^{-|x|}\sin\alpha x\,dx = 0$$

since the integrand is an odd function of x. Continuing we obtain

$$\mathscr{F}\left\{e^{-|x|}\right\} = 2\int_{0}^{\infty} e^{-x}\cos\alpha x\,dx = \frac{2}{1+\alpha^2}.$$

But $U(\alpha, 0) = c = 2/(1+\alpha^2)$ gives

$$U(\alpha, t) = \frac{2e^{-k\alpha^2 t}}{1+\alpha^2}$$

and so

$$u(x,t) = \frac{2}{2\pi}\int_{-\infty}^{\infty} \frac{e^{-k\alpha^2 t}e^{-i\alpha x}}{1+\alpha^2}\,d\alpha = \frac{1}{\pi}\int_{-\infty}^{\infty} \frac{e^{-k\alpha^2 t}}{1+\alpha^2}(\cos\alpha x - i\sin\alpha x)\,d\alpha$$

$$= \frac{1}{\pi}\int_{-\infty}^{\infty} \frac{e^{-k\alpha^2 t}\cos\alpha x}{1+\alpha^2}\,d\alpha = \frac{2}{\pi}\int_{0}^{\infty} \frac{e^{-k\alpha^2 t}\cos\alpha x}{1+\alpha^2}\,d\alpha.$$

2. Using the Fourier sine transform we find $U(\alpha, t) = ce^{-k\alpha^2 t}$. The Fourier sine transform of the initial condition is

$$\mathscr{F}_S\{u(x,0)\} = \int_{0}^{\infty} u(x,0)\sin\alpha x\,dx = \int_{0}^{1} 100\sin\alpha x\,dx = \frac{100}{\alpha}(1-\cos\alpha).$$

Thus $U(\alpha, 0) = (100/\alpha)(1-\cos\alpha)$ and since $c = U(\alpha, 0)$, we have

$$U(\alpha, t) = \frac{100}{\alpha}(1-\cos\alpha)e^{-k\alpha^2 t}.$$

Applying the inverse Fourier transform we obtain

$$u(x,t) = \mathscr{F}_S^{-1}\{U(\alpha, t)\} = \frac{2}{\pi}\int_{0}^{\infty} \frac{100}{\alpha}(1-\cos\alpha)e^{-k\alpha^2 t}\sin\alpha x\,d\alpha$$

$$= \frac{200}{\pi}\int_{0}^{\infty} \frac{1-\cos\alpha}{\alpha}e^{-k\alpha^2 t}\sin\alpha x\,dx.$$

3. Using the Fourier sine transform, the partial differential equation becomes

$$\frac{dU}{dt} + k\alpha^2 U = k\alpha u_0.$$

The general solution of this linear equation is

$$U(\alpha, t) = ce^{-k\alpha^2 t} + \frac{u_0}{\alpha}.$$

But $U(\alpha, 0) = 0$ implies $c = -u_0/\alpha$ and so

$$U(\alpha, t) = u_0\frac{1-e^{-k\alpha^2 t}}{\alpha}.$$

and

$$u(x,t) = \frac{2u_0}{\pi} \int_0^\infty \frac{1 - e^{-k\alpha^2 t}}{\alpha} \sin \alpha x \, d\alpha.$$

4. The solution of Problem 3 can be written

$$u(x,t) = \frac{2u_0}{\pi} \int_0^\infty \frac{\sin \alpha x}{\alpha} \, d\alpha - \frac{2u_0}{\pi} \int_0^\infty \frac{\sin \alpha x}{\alpha} e^{-k\alpha^2 t} \, d\alpha.$$

Using $\int_0^\infty \frac{\sin \alpha x}{\alpha} \, d\alpha = \pi/2$ the last line becomes

$$u(x,t) = u_0 - \frac{2u_0}{\pi} \int_0^\infty \frac{\sin \alpha x}{\alpha} e^{-k\alpha^2 t} \, d\alpha.$$

5. Using the Fourier sine transform we find

$$U(\alpha, t) = c e^{-k\alpha^2 t}.$$

Now

$$\mathscr{F}_S\{u(x,0)\} = U(\alpha, 0) = \int_0^1 \sin \alpha x \, dx = \frac{1 - \cos \alpha}{\alpha}.$$

From this we find $c = (1 - \cos \alpha)/\alpha$ and so

$$U(\alpha, t) = \frac{1 - \cos \alpha}{\alpha} e^{-k\alpha^2 t}$$

and

$$u(x,t) = \frac{2}{\pi} \int_0^\infty \frac{1 - \cos \alpha}{\alpha} e^{-k\alpha^2 t} \sin \alpha x \, d\alpha.$$

6. Since the domain of x is $(0, \infty)$ and the condition at $x = 0$ involves $\partial u/\partial x$ we use the Fourier cosine transform:

$$-k\alpha^2 U(\alpha, t) - k u_x(0, t) = \frac{dU}{dt}$$

$$\frac{dU}{dt} + k\alpha^2 U = kA$$

$$U(\alpha, t) = c e^{-k\alpha^2 t} + \frac{A}{\alpha^2}.$$

Since

subscript missing

$$\mathscr{F}_C\{u(x,0)\} = U(\alpha, 0) = 0$$

we find $c = -A/\alpha^2$, so that

$$U(\alpha, t) = A \frac{1 - e^{-k\alpha^2 t}}{\alpha^2}.$$

Applying the inverse Fourier cosine transform we obtain

$$u(x,t) = \mathscr{F}_C^{-1}\{U(\alpha, t)\} = \frac{2A}{\pi} \int_0^\infty \frac{1 - e^{-k\alpha^2 t}}{\alpha^2} \cos \alpha x \, d\alpha.$$

7. Using the Fourier cosine transform we find

$$U(\alpha, t) = ce^{-k\alpha^2 t}.$$

Now

$$\mathscr{F}_C\{u(x,0)\} = \int_0^1 \cos \alpha x \, dx = \frac{\sin \alpha}{\alpha} = U(\alpha, 0).$$

From this we obtain $c = (\sin \alpha)/\alpha$ and so

$$U(\alpha, t) = \frac{\sin \alpha}{\alpha} e^{-k\alpha^2 t}$$

and

$$u(x, t) = \frac{2}{\pi} \int_0^\infty \frac{\sin \alpha}{\alpha} e^{-k\alpha^2 t} \cos \alpha x \, d\alpha.$$

8. Using the Fourier sine transform we find

$$U(\alpha, t) = ce^{-k\alpha^2 t} + \frac{1}{\alpha}.$$

Now

$$\mathscr{F}_S\{u(x,0)\} = \mathscr{F}_S\left\{e^{-x}\right\} = \int_0^\infty e^{-x} \sin \alpha x \, dx = \frac{\alpha}{1 + \alpha^2} = U(\alpha, 0).$$

From this we obtain $c = \alpha/(1 + \alpha^2) - 1/\alpha$. Therefore

$$U(\alpha, t) = \left(\frac{\alpha}{1 + \alpha^2} - \frac{1}{\alpha}\right) e^{-k\alpha^2 t} + \frac{1}{\alpha} = \frac{1}{\alpha} - \frac{e^{-k\alpha^2 t}}{\alpha(1 + \alpha^2)}$$

and

$$u(x, t) = \frac{2}{\pi} \int_0^\infty \left(\frac{1}{\alpha} - \frac{e^{-k\alpha^2 t}}{\alpha(1 + \alpha^2)}\right) \sin \alpha x \, d\alpha.$$

9. (a) Using the Fourier transform we obtain

$$U(\alpha, t) = c_1 \cos \alpha a t + c_2 \sin \alpha a t.$$

If we write

$$\mathscr{F}\{u(x,0)\} = \mathscr{F}\{f(x)\} = F(\alpha)$$

and

$$\mathscr{F}\{u_t(x,0)\} = \mathscr{F}\{g(x)\} = G(\alpha)$$

we first obtain $c_1 = F(\alpha)$ from $U(\alpha, 0) = F(\alpha)$ and then $c_2 = G(\alpha)/\alpha a$ from $dU/dt \mid_{t=0} = G(\alpha)$. Thus

$$U(\alpha, t) = F(\alpha) \cos \alpha a t + \frac{G(\alpha)}{\alpha a} \sin \alpha a t$$

and

$$u(x, t) = \frac{1}{2\pi} \int_{-\infty}^\infty \left(F(\alpha) \cos \alpha a t + \frac{G(\alpha)}{\alpha a} \sin \alpha a t\right) e^{-i\alpha x} \, d\alpha.$$

Exercises 14.4 Fourier Transforms

(b) If $g(x) = 0$ then $c_2 = 0$ and

$$u(x,t) = \frac{1}{2\pi} \int_{-\infty}^{\infty} F(\alpha) \cos \alpha at\, e^{-i\alpha x}\, d\alpha$$

$$= \frac{1}{2\pi} \int_{-\infty}^{\infty} F(\alpha) \left(\frac{e^{\alpha at i} + e^{-\alpha at i}}{2} \right) e^{-i\alpha x}\, d\alpha$$

$$= \frac{1}{2} \left[\frac{1}{2\pi} \int_{-\infty}^{\infty} F(\alpha) e^{-i(x-at)\alpha}\, d\alpha + \frac{1}{2\pi} \int_{-\infty}^{\infty} F(\alpha) e^{-i(x+at)\alpha}\, d\alpha \right]$$

$$= \frac{1}{2} \left[f(x-at) + f(x+at) \right].$$

10. Using the Fourier sine transform we obtain

$$U(\alpha, t) = c_1 \cos \alpha at + c_2 \sin \alpha at.$$

Now

$$\mathscr{F}_S\{u(x,0)\} = \mathscr{F}_S\left\{ xe^{-x} \right\} = \int_0^{\infty} xe^{-x} \sin \alpha x\, dx = \frac{2\alpha}{(1+\alpha^2)^2} = U(\alpha, 0).$$

Also,

$$\mathscr{F}_S\{u_t(x,0)\} = \left. \frac{dU}{dt} \right|_{t=0} = 0.$$

This last condition gives $c_2 = 0$. Then $U(\alpha, 0) = 2\alpha/(1+\alpha^2)^2$ yields $c_1 = 2\alpha/(1+\alpha^2)^2$. Therefore

$$U(\alpha, t) = \frac{2\alpha}{(1+\alpha^2)^2} \cos \alpha at$$

and

$$u(x,t) = \frac{4}{\pi} \int_0^{\infty} \frac{\alpha \cos \alpha at}{(1+\alpha^2)^2} \sin \alpha x\, d\alpha.$$

11. Using the Fourier cosine transform we obtain

$$U(x, \alpha) = c_1 \cosh \alpha x + c_2 \sinh \alpha x.$$

Now the Fourier cosine transforms of $u(0,y) = e^{-y}$ and $u(\pi, y) = 0$ are, respectively, $U(0, \alpha) = 1/(1 + \alpha^2)$ and $U(\pi, \alpha) = 0$. The first of these conditions gives $c_1 = 1/(1 + \alpha^2)$. The second condition gives

$$c_2 = -\frac{\cosh \alpha \pi}{(1+\alpha^2) \sinh \alpha \pi}.$$

Hence

$$U(x, \alpha) = \frac{\cosh \alpha x}{1 + \alpha^2} - \frac{\cosh \alpha \pi \sinh \alpha x}{(1 + \alpha^2) \sinh \alpha \pi} = \frac{\sinh \alpha \pi \cosh \alpha \pi - \cosh \alpha \pi \sinh \alpha x}{(1 + \alpha^2) \sinh \alpha \pi}$$

$$= \frac{\sinh \alpha (\pi - x)}{(1 + \alpha^2) \sinh \alpha \pi}.$$

and

$$u(x, y) = \frac{2}{\pi} \int_0^\infty \frac{\sinh \alpha(\pi - x)}{(1 + \alpha^2) \sinh \alpha\pi} \cos \alpha y \, d\alpha.$$

12. Since the boundary condition at $y = 0$ now involves $u(x,0)$ rather than $u'(x,0)$, we use the Fourier sine transform. The transform of the partial differential equation is then

$$\frac{d^2U}{dx^2} - \alpha^2 U + \alpha u(x,0) = 0 \quad \text{or} \quad \frac{d^2U}{dx^2} - \alpha^2 U = -\alpha.$$

The solution of this differential equation is

$$U(x, \alpha) = c_1 \cosh \alpha x + c_2 \sinh \alpha x + \frac{1}{\alpha}.$$

The transforms of the boundary conditions at $x = 0$ and $x = \pi$ in turn imply that $c_1 = 1/\alpha$ and

$$c_2 = \frac{\cosh \alpha\pi}{\alpha \sinh \alpha\pi} - \frac{1}{\alpha \sinh \alpha\pi} + \frac{\alpha}{(1 + \alpha^2) \sinh \alpha\pi}.$$

Hence

$$U(x, \alpha) = \frac{1}{\alpha} - \frac{\cosh \alpha x}{\alpha} + \frac{\cosh \alpha\pi}{\alpha \sinh \alpha\pi} \sinh \alpha x - \frac{\sinh \alpha x}{\alpha \sinh \alpha\pi} + \frac{\alpha \sinh \alpha x}{(1 + \alpha^2) \sinh \alpha\pi}$$

$$= \frac{1}{\alpha} - \frac{\sinh \alpha(\pi - x)}{\alpha \sinh \alpha\pi} - \frac{\sinh \alpha x}{\alpha(1 + \alpha^2) \sinh \alpha\pi}.$$

Taking the inverse transform it follows that

$$u(x, y) = \frac{2}{\pi} \int_0^\infty \left(\frac{1}{\alpha} - \frac{\sinh \alpha(\pi - x)}{\alpha \sinh \alpha\pi} - \frac{\sinh \alpha x}{\alpha(1 + \alpha^2) \sinh \alpha\pi} \right) \sin \alpha y \, d\alpha.$$

13. Using the Fourier cosine transform with respect to x gives

$$U(\alpha, y) = c_1 e^{-\alpha y} + c_2 e^{\alpha y}.$$

Since we expect $u(x, y)$ to be bounded as $y \to \infty$ we define $c_2 = 0$. Thus

$$U(\alpha, y) = c_1 e^{-\alpha y}.$$

Now

$$\mathscr{F}_C\{u(x,0)\} = \int_0^1 50 \cos \alpha x \, dx = 50 \frac{\sin \alpha}{\alpha}$$

and so

$$U(\alpha, y) = 50 \frac{\sin \alpha}{\alpha} e^{-\alpha y}$$

and

$$u(x, y) = \frac{100}{\pi} \int_0^\infty \frac{\sin \alpha}{\alpha} e^{-\alpha y} \cos \alpha x \, d\alpha.$$

14. The boundary condition $u(0, y) = 0$ indicates that we now use the Fourier sine transform. We still have $U(\alpha, y) = c_1 e^{-\alpha y}$, but

$$\mathscr{F}_S\{u(x,0)\} = \int_0^1 50 \sin \alpha x \, dx = 50(1 - \cos \alpha)/\alpha = U(\alpha, 0).$$

This gives $c_1 = 50(1 - \cos \alpha)/\alpha$ and so

$$U(\alpha, y) = 50 \frac{1 - \cos \alpha}{\alpha} e^{-\alpha y}$$

and

$$u(x, y) = \frac{100}{\pi} \int_0^\infty \frac{1 - \cos \alpha}{\alpha} e^{-\alpha y} \sin \alpha x \, d\alpha.$$

15. We use the Fourier sine transform with respect to x to obtain

$$U(\alpha, y) = c_1 \cosh \alpha y + c_2 \sinh \alpha y.$$

The transforms of $u(x, 0) = f(x)$ and $u(x, 2) = 0$ give, in turn, $U(\alpha, 0) = F(\alpha)$ and $U(\alpha, 2) = 0$. The first condition gives $c_1 = F(\alpha)$ and the second condition then yields

$$c_2 = -\frac{F(\alpha) \cosh 2\alpha}{\sinh 2\alpha}.$$

Hence

$$U(\alpha, y) = F(\alpha) \cosh \alpha y - \frac{F(\alpha) \cosh 2\alpha \sinh \alpha y}{\sinh 2\alpha}$$

$$= F(\alpha) \frac{\sinh 2\alpha \cosh \alpha y - \cosh 2\alpha \sinh \alpha y}{\sinh 2\alpha}$$

$$= F(\alpha) \frac{\sinh \alpha(2 - y)}{\sinh 2\alpha}$$

and

$$u(x, y) = \frac{2}{\pi} \int_0^\infty F(\alpha) \frac{\sinh \alpha(2 - y)}{\sinh 2\alpha} \sin \alpha x \, d\alpha.$$

16. The domain of y and the boundary condition at $y = 0$ suggest that we use a Fourier cosine transform. The transformed equation is

$$\frac{d^2 U}{dx^2} - \alpha^2 U - u_y(x, 0) = 0 \qquad \text{or} \qquad \frac{d^2 U}{dx^2} - \alpha^2 U = 0.$$

Because the domain of the variable x is a finite interval we choose to write the general solution of the latter equation as

$$U(x, \alpha) = c_1 \cosh \alpha x + c_2 \sinh \alpha x.$$

Now $U(0, \alpha) = F(\alpha)$, where $F(\alpha)$ is the Fourier cosine transform of $f(y)$, and $U'(\pi, \alpha) = 0$ imply $c_1 = F(\alpha)$ and $c_2 = -F(\alpha) \sinh \alpha \pi / \cosh \alpha \pi$. Thus

$$U(x, \alpha) = F(\alpha) \cosh \alpha x - F(\alpha) \frac{\sinh \alpha \pi}{\cosh \alpha \pi} \sinh \alpha x = F(\alpha) \frac{\cosh \alpha(\pi - x)}{\cosh \alpha \pi}.$$

Using the inverse transform we find that a solution to the problem is

$$u(x, y) = \frac{2}{\pi} \int_0^\infty F(\alpha) \frac{\cosh \alpha(\pi - x)}{\cosh \alpha \pi} \cos \alpha y \, d\alpha.$$

17. We solve two boundary-value problems:

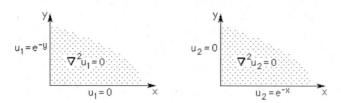

Using the Fourier sine transform with respect to y gives

$$u_1(x,y) = \frac{2}{\pi} \int_0^\infty \frac{\alpha e^{-\alpha x}}{1+\alpha^2} \sin \alpha y \, d\alpha.$$

The Fourier sine transform with respect to x yields the solution to the second problem:

$$u_2(x,y) = \frac{2}{\pi} \int_0^\infty \frac{\alpha e^{-\alpha y}}{1+\alpha^2} \sin \alpha x \, d\alpha.$$

We define the solution of the original problem to be

$$u(x,y) = u_1(x,y) + u_2(x,y) = \frac{2}{\pi} \int_0^\infty \frac{\alpha}{1+\alpha^2} \left[e^{-\alpha x} \sin \alpha y + e^{-\alpha y} \sin \alpha x \right] d\alpha.$$

18. We solve the three boundary-value problems:

Using separation of variables we find the solution of the first problem is

$$u_1(x,y) = \sum_{n=1}^\infty A_n e^{-ny} \sin nx \quad \text{where} \quad A_n = \frac{2}{\pi} \int_0^\pi f(x) \sin nx \, dx.$$

Using the Fourier sine transform with respect to y gives the solution of the second problem:

$$u_2(x,y) = \frac{200}{\pi} \int_0^\infty \frac{(1 - \cos \alpha) \sinh \alpha(\pi - x)}{\alpha \sinh \alpha \pi} \sin \alpha y \, d\alpha.$$

Also, the Fourier sine transform with respect to y gives the solution of the third problem:

$$u_3(x,y) = \frac{2}{\pi} \int_0^\infty \frac{\alpha \sinh \alpha x}{(1+\alpha^2) \sinh \alpha \pi} \sin \alpha y \, d\alpha.$$

The solution of the original problem is

$$u(x,y) = u_1(x,y) + u_2(x,y) + u_3(x,y).$$

19. Using the Fourier transform, the partial differential equation equation becomes

$$\frac{dU}{dt} + k\alpha^2 U = 0 \quad \text{and so} \quad U(\alpha,t) = ce^{-k\alpha^2 t}.$$

Now

$$\mathscr{F}\{u(x,0)\} = U(\alpha,0) = \sqrt{\pi}\,e^{-\alpha^2/4}$$

by the given result. This gives $c = \sqrt{\pi}\,e^{-\alpha^2/4}$ and so

$$U(\alpha,t) = \sqrt{\pi}\,e^{-(\frac{1}{4}+kt)\alpha^2}.$$

Using the given Fourier transform again we obtain

$$u(x,t) = \sqrt{\pi}\,\mathscr{F}^{-1}\{e^{-(1+4kt)\alpha^2/4}\} = \frac{1}{\sqrt{1+4kt}}\,e^{-x^2/(1+4kt)}.$$

20. We use $U(\alpha,t) = ce^{-k\alpha^2 t}$. The Fourier transform of the boundary condition is $U(\alpha,0) = F(\alpha)$. This gives $c = F(\alpha)$ and so $U(\alpha,t) = F(\alpha)e^{-k\alpha^2 t}$. By the convolution theorem and the given result, we obtain

$$u(x,t) = \mathscr{F}^{-1}\{F(\alpha)\cdot e^{-k\alpha^2 t}\} = \frac{1}{2\sqrt{k\pi t}}\int_{-\infty}^{\infty} f(\tau)e^{-(x-\tau)^2/4kt}\,d\tau.$$

21. Using the Fourier transform with respect to x gives

$$U(\alpha,y) = c_1 \cosh \alpha y + c_2 \sinh \alpha y.$$

The transform of the boundary condition $\partial u/\partial y\,|_{y=0} = 0$ is $dU/dy\,|_{y=0} = 0$. This condition gives $c_2 = 0$. Hence

$$U(\alpha,y) = c_1 \cosh \alpha y.$$

Now by the given information the transform of the boundary condition $u(x,1) = e^{-x^2}$ is $U(\alpha,1) = \sqrt{\pi}\,e^{-\alpha^2/4}$. This condition then gives $c_1 = \sqrt{\pi}\,e^{-\alpha^2/4}\cosh \alpha$. Therefore

$$U(\alpha,y) = \sqrt{\pi}\,\frac{e^{-\alpha^2/4}\cosh \alpha y}{\cosh \alpha}$$

and

$$U(x,y) = \frac{1}{2\sqrt{\pi}}\int_{-\infty}^{\infty} \frac{e^{-\alpha^2/4}\cosh \alpha y}{\cosh \alpha}\,e^{-i\alpha x}\,d\alpha$$

$$= \frac{1}{2\sqrt{\pi}}\int_{-\infty}^{\infty} \frac{e^{-\alpha^2/4}\cosh \alpha y}{\cosh \alpha}\,\cos \alpha x\,d\alpha$$

$$= \frac{1}{\sqrt{\pi}}\int_{0}^{\infty} \frac{e^{-\alpha^2/4}\cosh \alpha y}{\cosh \alpha}\,\cos \alpha x\,d\alpha.$$

22. Entries 42 and 43 in the Table of Laplace transforms imply

$$\int_{0}^{\infty} e^{-st}\frac{\sin at}{t}\,dt = \arctan \frac{a}{s}$$

and

$$\int_{0}^{\infty} e^{-st}\frac{\sin at \cos bt}{t}\,dt = \frac{1}{2}\arctan \frac{a+b}{s} + \frac{1}{2}\arctan \frac{a-b}{s}.$$

Identifying $\alpha = t, x = a$, and $y = s$, the solution of Problem 14 is

$$u(x,y) = \frac{100}{\pi} \int_0^\infty \frac{1 - \cos\alpha}{\alpha} e^{-\alpha y} \sin\alpha x\, d\alpha$$

$$= \frac{100}{\pi} \left[\int_0^\infty \frac{\sin\alpha x}{\alpha} e^{-\alpha y}\, d\alpha - \int_0^\infty \frac{\sin\alpha x \cos\alpha}{\alpha} e^{-\alpha y}\, d\alpha \right]$$

$$= \frac{100}{\pi} \left[\arctan\frac{x}{y} - \frac{1}{2}\arctan\frac{x+1}{y} - \frac{1}{2}\arctan\frac{x-1}{y} \right].$$

23. Using the definition of f and the solution in Problem 20 we obtain

$$u(x,t) = \frac{u_0}{2\sqrt{k\pi t}} \int_{-1}^1 e^{-(x-\tau)^2/4kt}\, d\tau.$$

If $v = (x-\tau)/2\sqrt{kt}$, then $d\tau = -2\sqrt{kt}\, du$ and the integral becomes

$$v(x,t) = \frac{u_0}{\sqrt{\pi}} \int_{(x-1)/2\sqrt{kt}}^{(x+1)/2\sqrt{kt}} e^{-v^2}\, dv.$$

Using the result in Problem 9 of Exercises 14.1 in the text, we have

$$u(x,t) = \frac{u_0}{2} \left[\operatorname{erf}\left(\frac{x+1}{2\sqrt{kt}}\right) - \operatorname{erf}\left(\frac{x-1}{2\sqrt{kt}}\right) \right].$$

24.

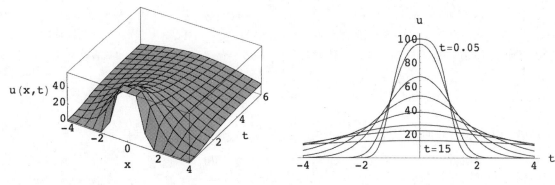

Since $\operatorname{erf}(0) = 0$ and $\lim_{x\to\infty}\operatorname{erf}(x) = 1$, we have

$$\lim_{t\to\infty} u(x,t) = 50[\operatorname{erf}(0) - \operatorname{erf}(0)] = 0$$

and

$$\lim_{x\to\infty} u(x,t) = 50[\operatorname{erf}(\infty) - \operatorname{erf}(\infty)] = 50[1-1] = 0.$$

1. The partial differential equation and the boundary conditions indicate that the Fourier cosine transform is appropriate for the problem. We find in this case

$$u(x,y) = \frac{2}{\pi} \int_0^\infty \frac{\sinh \alpha y}{\alpha(1+\alpha^2)\cosh \alpha \pi} \cos \alpha x \, d\alpha.$$

2. We use the Laplace transform and undetermined coefficients to obtain

$$U(x,s) = c_1 \cosh \sqrt{s}\, x + c_2 \sinh \sqrt{s}\, x + \frac{50}{s + 4\pi^2} \sin 2\pi x.$$

The transformed boundary conditions $U(0,s) = 0$ and $U(1,s) = 0$ give, in turn, $c_1 = 0$ and $c_2 = 0$. Hence

$$U(x,s) = \frac{50}{s+4\pi^2} \sin 2\pi x$$

and

$$u(x,t) = 50 \sin 2\pi x \, \mathcal{L}^{-1}\left\{\frac{1}{s+4\pi^2}\right\} = 50 e^{-4\pi^2 t} \sin 2\pi x.$$

3. The Laplace transform gives

$$U(x,s) = c_1 e^{-\sqrt{s+h}\, x} + c_2 e^{\sqrt{s+h}\, x} + \frac{u_0}{s+h}.$$

The condition $\lim_{x\to\infty} \partial u/\partial x = 0$ implies $\lim_{x\to\infty} dU/dx = 0$ and so we define $c_2 = 0$. Thus

$$U(x,s) = c_1 e^{-\sqrt{s+h}\, x} + \frac{u_0}{s+h}.$$

The condition $U(0,s) = 0$ then gives $c_1 = -u_0/(s+h)$ and so

$$U(x,s) = \frac{u_0}{s+h} - u_0 \frac{e^{-\sqrt{s+h}\, x}}{s+h}.$$

With the help of the first translation theorem we then obtain

$$u(x,t) = u_0 \mathcal{L}^{-1}\left\{\frac{1}{s+h}\right\} - u_0 \mathcal{L}^{-1}\left\{\frac{e^{-\sqrt{s+h}\, x}}{s+h}\right\} = u_0 e^{-ht} - u_0 e^{-ht} \operatorname{erfc}\left(\frac{x}{2\sqrt{t}}\right)$$

$$= u_0 e^{-ht}\left[1 - \operatorname{erfc}\left(\frac{x}{2\sqrt{t}}\right)\right] = u_0 e^{-ht} \operatorname{erf}\left(\frac{x}{2\sqrt{t}}\right).$$

4. Using the Fourier transform and the result $\mathscr{F}\left\{e^{-|x|}\right\} = 1/(1+\alpha^2)$ we find

$$u(x,t) = \frac{1}{2\pi}\int_{-\infty}^{\infty}\frac{1-e^{-\alpha^2 t}}{\alpha^2(1+\alpha^2)}e^{-i\alpha x}\,d\alpha$$

$$= \frac{1}{2\pi}\int_{-\infty}^{\infty}\frac{1-e^{-\alpha^2 t}}{\alpha^2(1+\alpha^2)}\cos\alpha x\,d\alpha$$

$$= \frac{1}{\pi}\int_{0}^{\infty}\frac{1-e^{-\alpha^2 t}}{\alpha^2(1+\alpha^2)}\cos\alpha x\,d\alpha.$$

5. The Laplace transform gives

$$U(x,s) = c_1 e^{-\sqrt{s}\,x} + c_2 e^{\sqrt{s}\,x}.$$

The condition $\lim_{x\to\infty}u(x,t) = 0$ implies $\lim_{x\to\infty}U(x,s) = 0$ and so we define $c_2 = 0$. Thus

$$U(x,s) = c_1 e^{-\sqrt{s}\,x}.$$

The transform of the remaining boundary condition is $U(0,s) = 1/s^2$. This gives $c_1 = 1/s^2$. Hence

$$U(x,s) = \frac{e^{-\sqrt{s}\,x}}{s^2} \quad \text{and} \quad u(x,t) = \mathscr{L}^{-1}\left\{\frac{1}{s}\frac{e^{-\sqrt{s}\,x}}{s}\right\}.$$

Using

$$\mathscr{L}^{-1}\left\{\frac{1}{s}\right\} = 1 \quad \text{and} \quad \mathscr{L}^{-1}\left\{\frac{e^{-\sqrt{s}\,x}}{s}\right\} = \mathrm{erfc}\left(\frac{x}{2\sqrt{t}}\right),$$

it follows from the convolution theorem that

$$u(x,t) = \int_0^t \mathrm{erfc}\left(\frac{x}{2\sqrt{\tau}}\right)d\tau.$$

6. The Laplace transform and undetermined coefficients give

$$U(x,s) = c_1\cosh sx + c_2\sinh sx + \frac{s-1}{s^2+\pi^2}\sin\pi x.$$

The conditions $U(0,s) = 0$ and $U(1,s) = 0$ give, in turn, $c_1 = 0$ and $c_2 = 0$. Thus

$$U(x,s) = \frac{s-1}{s^2+\pi^2}\sin\pi x$$

and

$$u(x,t) = \sin\pi x\,\mathscr{L}^{-1}\left\{\frac{s}{s^2+\pi^2}\right\} - \frac{1}{\pi}\sin\pi x\,\mathscr{L}^{-1}\left\{\frac{\pi}{s^2+\pi^2}\right\}$$

$$= (\sin\pi x)\cos\pi t - \frac{1}{\pi}(\sin\pi x)\sin\pi t.$$

7. The Fourier transform gives the solution

$$u(x,t) = \frac{u_0}{2\pi} \int_{-\infty}^{\infty} \left(\frac{e^{i\alpha\pi} - 1}{i\alpha}\right) e^{-i\alpha x} e^{-k\alpha^2 t} d\alpha$$

$$= \frac{u_0}{2\pi} \int_{-\infty}^{\infty} \frac{e^{i\alpha(\pi-x)} - e^{-i\alpha x}}{i\alpha} e^{-k\alpha^2 t} d\alpha$$

$$= \frac{u_0}{2\pi} \int_{-\infty}^{\infty} \frac{\cos\alpha(\pi - x) + i\sin\alpha(\pi - x) - \cos\alpha x + i\sin\alpha x}{i\alpha} e^{-k\alpha^2 t} d\alpha.$$

Since the imaginary part of the integrand of the last integral is an odd function of α, we obtain

$$u(x,t) = \frac{u_0}{2\pi} \int_{-\infty}^{\infty} \frac{\sin\alpha(\pi - x) + \sin\alpha x}{\alpha} e^{-k\alpha^2 t} d\alpha.$$

8. Using the Fourier cosine transform we obtain $U(x,\alpha) = c_1 \cosh\alpha x + c_2 \sinh\alpha x$. The condition $U(0,\alpha) = 0$ gives $c_1 = 0$. Thus $U(x,\alpha) = c_2 \sinh\alpha x$. Now

$$\mathscr{F}_C\{u(\pi,y)\} = \int_1^2 \cos\alpha y\, dy = \frac{\sin 2\alpha - \sin\alpha}{\alpha} = U(\pi,\alpha).$$

This last condition gives $c_2 = (\sin 2\alpha - \sin\alpha)/\alpha \sinh\alpha\pi$. Hence

$$U(x,\alpha) = \frac{\sin 2\alpha - \sin\alpha}{\alpha \sinh\alpha\pi} \sinh\alpha x$$

and

$$u(x,y) = \frac{2}{\pi} \int_0^{\infty} \frac{\sin 2\alpha - \sin\alpha}{\alpha \sinh\alpha\pi} \sinh\alpha x \cos\alpha y\, d\alpha.$$

9. We solve the two problems

$$\frac{\partial^2 u_1}{\partial x^2} + \frac{\partial^2 u_1}{\partial y^2} = 0, \quad x > 0, \quad y > 0,$$

$$u_1(0,y) = 0, \quad y > 0,$$

$$u_1(x,0) = \begin{cases} 100, & 0 < x < 1 \\ 0, & x > 1 \end{cases}$$

and

$$\frac{\partial^2 u_2}{\partial x^2} + \frac{\partial^2 u_2}{\partial y^2} = 0, \quad x > 0, \quad y > 0,$$

$$u_2(0,y) = \begin{cases} 50, & 0 < y < 1 \\ 0, & y > 1 \end{cases}$$

$$u_2(x,0) = 0.$$

Using the Fourier sine transform with respect to x we find

$$u_1(x,y) = \frac{200}{\pi} \int_0^{\infty} \left(\frac{1 - \cos\alpha}{\alpha}\right) e^{-\alpha y} \sin\alpha x\, d\alpha.$$

Using the Fourier sine transform with respect to y we find

$$u_2(x, y) = \frac{100}{\pi} \int_0^\infty \left(\frac{1 - \cos\alpha}{\alpha}\right) e^{-\alpha x} \sin\alpha y \, d\alpha.$$

The solution of the problem is then

$$u(x, y) = u_1(x, y) + u_2(x, y).$$

10. The Laplace transform gives

$$U(x, s) = c_1 \cosh\sqrt{s}\,x + c_2 \sinh\sqrt{s}\,x + \frac{r}{s^2}.$$

The condition $\partial u/\partial x \big|_{x=0} = 0$ transforms into $dU/dx \big|_{x=0} = 0$. This gives $c_2 = 0$. The remaining condition $u(1, t) = 0$ transforms into $U(1, s) = 0$. This condition then implies $c_1 = -r/s^2 \cosh\sqrt{s}$.

Hence

$$U(x, s) = \frac{r}{s^2} - r\frac{\cosh\sqrt{s}\,x}{s^2 \cosh\sqrt{s}}.$$

Using geometric series and the convolution theorem we obtain

$$u(x, t) = r\mathscr{L}^{-1}\left\{\frac{1}{s^2}\right\} - r\mathscr{L}^{-1}\left\{\frac{\cosh\sqrt{s}\,x}{s^2 \cosh\sqrt{s}}\right\}$$

$$= rt - r\sum_{n=0}^\infty (-1)^n \left[\int_0^t \text{erfc}\left(\frac{2n+1-x}{2\sqrt{\tau}}\right) d\tau + \int_0^t \text{erfc}\left(\frac{2n+1+x}{2\sqrt{\tau}}\right) d\tau\right].$$

11. The Fourier sine transform with respect to x and undetermined coefficients give

$$U(\alpha, y) = c_1 \cosh\alpha y + c_2 \sinh\alpha y + \frac{A}{\alpha}.$$

The transforms of the boundary conditions are

$$\frac{dU}{dy}\bigg|_{y=0} = 0 \quad \text{and} \quad \frac{dU}{dy}\bigg|_{y=\pi} = \frac{B\alpha}{1 + \alpha^2}.$$

The first of these conditions gives $c_2 = 0$ and so

$$U(\alpha, y) = c_1 \cosh\alpha y + \frac{A}{\alpha}.$$

The second transformed boundary condition yields $c_1 = B/(1 + \alpha^2)\sinh\alpha\pi$. Therefore

$$U(\alpha, y) = \frac{B\cosh\alpha y}{(1 + \alpha^2)\sinh\alpha\pi} + \frac{A}{\alpha}$$

and

$$u(x, y) = \frac{2}{\pi}\int_0^\infty \left(\frac{B\cosh\alpha y}{(1 + \alpha^2)\sinh\alpha\pi} + \frac{A}{\alpha}\right) \sin\alpha x \, d\alpha.$$

12. Using the Laplace transform gives

$$U(x, s) = c_1 \cosh\sqrt{s}\,x + c_2 \sinh\sqrt{s}\,x.$$

The condition $u(0, t) = u_0$ transforms into $U(0, s) = u_0/s$. This gives $c_1 = u_0/s$. The condition $u(1, t) = u_0$ transforms into $U(1, s) = u_0/s$. This implies that $c_2 = u_0(1 - \cosh \sqrt{s}\,)/s \sinh \sqrt{s}$. Hence

$$U(x, s) = \frac{u_0}{s} \cosh \sqrt{s}\, x + u_0 \left[\frac{1 - \cosh \sqrt{s}}{s \sinh \sqrt{s}} \right] \sinh \sqrt{s}\, x$$

$$= u_0 \left[\frac{\sinh \sqrt{s} \cosh \sqrt{s}\, x - \cosh \sqrt{s} \sinh \sqrt{s}\, x + \sinh \sqrt{s}\, x}{s \sinh \sqrt{s}} \right]$$

$$= u_0 \left[\frac{\sinh \sqrt{s}\,(1 - x) + \sinh \sqrt{s}\, x}{s \sinh \sqrt{s}} \right]$$

$$= u_0 \left[\frac{\sinh \sqrt{s}\,(1 - x)}{s \sinh \sqrt{s}} + \frac{\sinh \sqrt{s}\, x}{s \sinh \sqrt{s}} \right]$$

and

$$u(x, t) = u_0 \left[\mathscr{L}^{-1} \left\{ \frac{\sinh \sqrt{s}\,(1 - x)}{s \sinh \sqrt{s}} \right\} + \mathscr{L}^{-1} \left\{ \frac{\sinh \sqrt{s}\, x}{s \sinh \sqrt{s}} \right\} \right]$$

$$= u_0 \sum_{n=0}^{\infty} \left[\mathrm{erf} \left(\frac{2n + 2 - x}{2\sqrt{t}} \right) - \mathrm{erf} \left(\frac{2n + x}{2\sqrt{t}} \right) \right]$$

$$+ u_0 \sum_{n=0}^{\infty} \left[\mathrm{erf} \left(\frac{2n + 1 + x}{2\sqrt{t}} \right) - \mathrm{erf} \left(\frac{2n + 1 - x}{2\sqrt{t}} \right) \right].$$

13. Using the Fourier transform gives

$$U(\alpha, t) = c_1 e^{-k\alpha^2 t}.$$

Now

$$u(\alpha, 0) = \int_0^{\infty} e^{-x} e^{i\alpha x}\, dx = \frac{e^{(i\alpha - 1)x}}{i\alpha - 1} \Big|_0^{\infty} = 0 - \frac{1}{i\alpha - 1} = \frac{1}{1 - i\alpha} = c_1$$

so

$$U(\alpha, t) = \frac{1 + i\alpha}{1 + \alpha^2} e^{-k\alpha^2 t}$$

and

$$u(x, t) = \frac{1}{2\pi} \int_{-\infty}^{\infty} \frac{1 + i\alpha}{1 + \alpha^2} e^{-k\alpha^2 t} e^{-i\alpha x}\, d\alpha.$$

Since

$$\frac{1 + i\alpha}{1 + \alpha^2} (\cos \alpha x - i \sin \alpha x) = \frac{\cos \alpha x + \alpha \sin \alpha x}{1 + \alpha^2} + \frac{i(\alpha \cos \alpha x - \sin \alpha x)}{1 + \alpha^2}$$

and the integral of the product of the second term with $e^{-k\alpha^2 t}$ is 0 (it is an odd function), we have

$$u(x, t) = \frac{1}{2\pi} \int_{-\infty}^{\infty} \frac{\cos \alpha x + \alpha \sin \alpha x}{1 + \alpha^2} e^{-k\alpha^2 t}\, d\alpha.$$

14. Using the Laplace transform the partial differential equation becomes

$$\frac{d^2U}{dx^2} - sU = -100$$

so

$$U(x,s) = c_1 e^{-\sqrt{s}\,x} + c_2 e^{\sqrt{s}\,x} + \frac{100}{s}.$$

The condition $x \to \infty$ implies $\lim_{x \to \infty} U(x,s) = 100/s$ and the condition at $x = 0$ implies $U'(0,s) = -50/s$. thus $c_2 = 0$ and $c_1 = 50/s\sqrt{s}$, so

$$U(x,s) = \frac{100}{s} + 50\frac{e^{-x\sqrt{s}}}{s\sqrt{s}}$$

and by (4) of Table 14.1 in the text,

$$u(x,t) = 100 + 100\sqrt{\frac{t}{\pi}}\,e^{-x^2/4t} - 50x\,\mathrm{erfc}\left(\frac{x}{2\sqrt{t}}\right).$$

15. The Fourier cosine transform of the partial differential equation gives

$$\frac{dU}{dt} + k\alpha^2 U = 0 \qquad \text{so} \qquad U(\alpha,t) = c_1 e^{-k\alpha^2 t}.$$

The transform of the initial condition gives

$$U(\alpha,0) = \frac{\alpha}{\alpha^2 + 1} = c_1$$

$$U(\alpha,t) = \frac{\alpha}{\alpha^2 + 1}\,e^{-k\alpha^2 t}$$

and so

$$u(x,t) = \frac{2}{\pi}\int_0^\infty \frac{\alpha e^{-k\alpha^2 t}}{\alpha^2 + 1}\cos\alpha x\,d\alpha.$$

16. Using the Fourier transform with respect to x we obtain

$$\frac{d^2U}{dy^2} - \alpha^2 U = 0.$$

Since $0 < y < 1$ is a finite interval we use the general solution

$$U(\alpha,y) = c_1\cosh\alpha y + c_2\sinh\alpha y.$$

The boundary condition at $y = 0$ transforms into $U'(\alpha,0) = 0$, so $c_2 = 0$ and $U(\alpha,y) = c_1\cosh\alpha y$. Now denote the Fourier transform of f as $F(\alpha)$. Then $U(\alpha,1) = F(\alpha)$ so $F(\alpha) = c_1\cosh\alpha$ and

$$U(\alpha,y) = F(\alpha)\frac{\cosh\alpha y}{\cosh\alpha}.$$

Taking the inverse Fourier transform we obtain

$$u(x,y) = \frac{1}{2\pi}\int_{-\infty}^{\infty} F(\alpha)\frac{\cosh\alpha y}{\cosh\alpha}e^{-i\alpha x}\,d\alpha.$$

But

$$F(\alpha) = \int_{-\infty}^{\infty} f(t)e^{i\alpha t}\, dt,$$

and so

$$u(x, y) = \frac{1}{2\pi} \int_{-\infty}^{\infty} \left(\int_{-\infty}^{\infty} f(t)e^{i\alpha t}\, dt \right) \frac{\cosh \alpha y}{\cosh \alpha} e^{-i\alpha x}\, d\alpha$$

$$= \frac{1}{2\pi} \int_{-\infty}^{\infty} \int_{-\infty}^{\infty} f(t)e^{i\alpha(t-x)} \frac{\cosh \alpha y}{\cosh \alpha}\, dt\, d\alpha$$

$$= \frac{1}{2\pi} \int_{-\infty}^{\infty} \int_{-\infty}^{\infty} f(t)(\cos \alpha(t - x) + i \sin \alpha(t - x)) \frac{\cosh \alpha y}{\cosh \alpha}\, dt\, d\alpha$$

$$= \frac{1}{2\pi} \int_{-\infty}^{\infty} \int_{-\infty}^{\infty} f(t) \cos \alpha(t - x) \frac{\cosh \alpha y}{\cosh \alpha}\, dt\, d\alpha$$

$$= \frac{1}{\pi} \int_{0}^{\infty} \int_{-\infty}^{\infty} f(t) \cos \alpha(t - x) \frac{\cosh \alpha y}{\cosh \alpha}\, dt\, d\alpha,$$

since the imaginary part of the integrand is an odd function of α followed by the fact that the remaining integrand is an even function of α.

15 Numerical Solutions of Partial Differential Equations

Exercises 15.1   Laplace's Equation

1. The figure shows the values of $u(x, y)$ along the boundary. We need to determine u_{11} and u_{21}. The system is

$$u_{21} + 2 + 0 + 0 - 4u_{11} = 0$$

$$1 + 2 + u_{11} + 0 - 4u_{21} = 0$$

or

$$-4u_{11} + u_{21} = -2$$

$$u_{11} - 4u_{21} = -3.$$

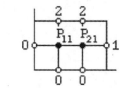

Solving we obtain $u_{11} = 11/15$ and $u_{21} = 14/15$.

2. The figure shows the values of $u(x, y)$ along the boundary. We need to determine u_{11}, u_{21}, and u_{31}. By symmetry $u_{11} = u_{31}$ and the system is

$$u_{21} + 0 + 0 + 100 - 4u_{11} = 0$$

$$u_{31} + 0 + u_{11} + 100 - 4u_{21} = 0$$

$$0 + 0 + u_{21} + 100 - 4u_{31} = 0$$

or

$$-4u_{11} + u_{21} = -100$$

$$2u_{11} - 4u_{21} = -100.$$

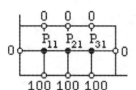

Solving we obtain $u_{11} = u_{31} = 250/7$ and $u_{21} = 300/7$.

3. The figure shows the values of $u(x, y)$ along the boundary. We need to determine u_{11}, u_{21}, u_{12}, and u_{22}. By symmetry $u_{11} = u_{21}$ and $u_{12} = u_{22}$. The system is

$$u_{21} + u_{12} + 0 + 0 - 4u_{11} = 0$$

$$0 + u_{22} + u_{11} + 0 - 4u_{21} = 0$$

$$u_{22} + \sqrt{3}/2 + 0 + u_{11} - 4u_{12} = 0$$

$$0 + \sqrt{3}/2 + u_{12} + u_{21} - 4u_{22} = 0$$

or

$$3u_{11} + u_{12} = 0$$

$$u_{11} - 3u_{12} = -\frac{\sqrt{3}}{2}.$$

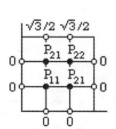

Solving we obtain $u_{11} = u_{21} = \sqrt{3}/16$ and $u_{12} = u_{22} = 3\sqrt{3}/16$.

4. The figure shows the values of $u(x, y)$ along the boundary. We need to determine u_{11}, u_{21}, u_{12}, and u_{22}. The system is

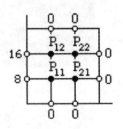

$$u_{21} + u_{12} + 8 + 0 - 4u_{11} = 0 \qquad\qquad -4u_{11} + u_{21} + u_{12} = -8$$

$$0 + u_{22} + u_{11} + 0 - 4u_{21} = 0 \qquad\qquad u_{11} - 4u_{21} + u_{22} = 0$$

or

$$u_{22} + 0 + 16 + u_{11} - 4u_{12} = 0 \qquad\qquad u_{11} - 4u_{12} + u_{22} = -16$$

$$0 + 0 + u_{12} + u_{21} - 4u_{22} = 0 \qquad\qquad u_{21} + u_{12} - 4u_{22} = 0.$$

Solving we obtain $u_{11} = 11/3$, $u_{21} = 4/3$, $u_{12} = 16/3$, and $u_{22} = 5/3$.

5. The figure shows the values of $u(x, y)$ along the boundary. For Gauss-Seidel the coefficients of the unknowns u_{11}, u_{21}, u_{31}, u_{12}, u_{22}, u_{32}, u_{13}, u_{23}, u_{33} are shown in the matrix

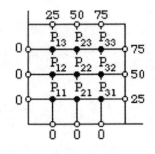

$$\begin{bmatrix} 0 & .25 & 0 & .25 & 0 & 0 & 0 & 0 & 0 \\ .25 & 0 & .25 & 0 & .25 & 0 & 0 & 0 & 0 \\ 0 & .25 & 0 & 0 & 0 & .25 & 0 & 0 & 0 \\ .25 & 0 & 0 & 0 & .25 & 0 & .25 & 0 & 0 \\ 0 & .25 & 0 & .25 & 0 & .25 & 0 & .25 & 0 \\ 0 & 0 & .25 & 0 & .25 & 0 & 0 & 0 & .25 \\ 0 & 0 & 0 & .25 & 0 & 0 & 0 & .25 & 0 \\ 0 & 0 & 0 & 0 & .25 & 0 & .25 & 0 & .25 \\ 0 & 0 & 0 & 0 & 0 & .25 & 0 & .25 & 0 \end{bmatrix}$$

The constant terms in the equations are 0, 0, 6.25, 0, 0, 12.5, 6.25, 12.5, 37.5. We use 25 as the initial guess for each variable. Then $u_{11} = 6.25$, $u_{21} = u_{12} = 12.5$, $u_{31} = u_{13} = 18.75$, $u_{22} = 25$, $u_{32} = u_{23} = 37.5$, and $u_{33} = 56.25$

6. The coefficients of the unknowns are the same as shown above in Problem 5. The constant terms are 7.5, 5, 20, 10, 0, 15, 17.5, 5, 27.5. We use 32.5 as the initial guess for each variable. Then $u_{11} = 21.92$, $u_{21} = 28.30$, $u_{31} = 38.17$, $u_{12} = 29.38$, $u_{22} = 33.13$, $u_{32} = 44.38$, $u_{13} = 22.46$, $u_{23} = 30.45$, and $u_{33} = 46.21$.

7. **(a)** Using the difference approximations for u_{xx} and u_{yy} we obtain

$$u_{xx} + u_{yy} = \frac{1}{h^2}(u_{i+1,j} + u_{i,j+1} + u_{i-1,j} + u_{i,j-1} - 4u_{ij}) = f(x, y)$$

so that

$$u_{i+1,j} + u_{i,j+1} + u_{i-1,j} + u_{i,j-1} - 4u_{ij} = h^2 f(x, y).$$

(b) By symmetry, as shown in the figure, we need only solve for u_1, u_2, u_3, u_4, and u_5. The difference equations are

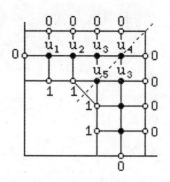

$$u_2 + 0 + 0 + 1 - 4u_1 = \frac{1}{4}(-2)$$

$$u_3 + 0 + u_1 + 1 - 4u_2 = \frac{1}{4}(-2)$$

$$u_4 + 0 + u_2 + u_5 - 4u_3 = \frac{1}{4}(-2)$$

$$0 + 0 + u_3 + u_3 - 4u_4 = \frac{1}{4}(-2)$$

$$u_3 + u_3 + 1 + 1 - 4u_5 = \frac{1}{4}(-2)$$

or

$$u_1 = 0.25u_2 + 0.375$$

$$u_2 = 0.25u_1 + 0.25u_3 + 0.375$$

$$u_3 = 0.25u_2 + 0.25u_4 + 0.25u_5 + 0.125$$

$$u_4 = 0.5u_3 + 0.125$$

$$u_5 = 0.5u_3 + 0.625.$$

Using Gauss-Seidel iteration we find $u_1 = 0.5427$, $u_2 = 0.6707$, $u_3 = 0.6402$, $u_4 = 0.4451$, and $u_5 = 0.9451$.

8. By symmetry, as shown in the figure, we need only solve for u_1, u_2, u_3, u_4, and u_5. The difference equations are

$$u_2 + 0 + 0 + u_3 - 4u_1 = -1 \qquad u_1 = 0.25u_2 + 0.25u_3 + 0.25$$

$$0 + 0 + u_1 + u_4 - 4u_2 = -1 \qquad u_2 = 0.25u_1 + 0.25u_4 + 0.25$$

$$u_4 + u_1 + 0 + u_5 - 4u_3 = -1 \quad \text{or} \quad u_3 = 0.25u_1 + 0.25u_4 + 0.25u_5 + 0.25$$

$$u_2 + u_2 + u_3 + u_3 - 4u_4 = -1 \qquad u_4 = 0.5u_2 + 0.5u_3 + 0.25$$

$$u_3 + u_3 + 0 + 0 - 4u_5 = -1 \qquad u_5 = 0.5u_3 + 0.25.$$

Using Gauss-Seidel iteration we find $u_1 = 0.6157$, $u_2 = 0.6493$, $u_3 = 0.8134$, $u_4 = 0.9813$, and $u_5 = 0.6567$.

Heat Equation

1. We identify $c = 1$, $a = 2$, $T = 1$, $n = 8$, and $m = 40$. Then $h = 2/8 = 0.25$, $k = 1/40 = 0.025$, and $\lambda = 2/5 = 0.4$.

TIME	X=0.25	X=0.50	X=0.75	X=1.00	X=1.25	X=1.50	X=1.75
0.000	1.0000	1.0000	1.0000	1.0000	0.0000	0.0000	0.0000
0.025	0.6000	1.0000	1.0000	0.6000	0.4000	0.0000	0.0000
0.050	0.5200	0.8400	0.8400	0.6800	0.3200	0.1600	0.0000
0.075	0.4400	0.7120	0.7760	0.6000	0.4000	0.1600	0.0640
0.100	0.3728	0.6288	0.6800	0.5904	0.3840	0.2176	0.0768
0.125	0.3261	0.5469	0.6237	0.5437	0.4000	0.2278	0.1024
0.150	0.2840	0.4893	0.5610	0.5182	0.3886	0.2465	0.1116
0.175	0.2525	0.4358	0.5152	0.4835	0.3836	0.2494	0.1209
0.200	0.2248	0.3942	0.4708	0.4562	0.3699	0.2517	0.1239
0.225	0.2027	0.3571	0.4343	0.4275	0.3571	0.2479	0.1255
0.250	0.1834	0.3262	0.4007	0.4021	0.3416	0.2426	0.1242
0.275	0.1672	0.2989	0.3715	0.3773	0.3262	0.2348	0.1219
0.300	0.1530	0.2752	0.3448	0.3545	0.3101	0.2262	0.1183
0.325	0.1407	0.2541	0.3209	0.3329	0.2943	0.2166	0.1141
0.350	0.1298	0.2354	0.2990	0.3126	0.2787	0.2067	0.1095
0.375	0.1201	0.2186	0.2790	0.2936	0.2635	0.1966	0.1046
0.400	0.1115	0.2034	0.2607	0.2757	0.2488	0.1865	0.0996
0.425	0.1036	0.1895	0.2438	0.2589	0.2347	0.1766	0.0945
0.450	0.0965	0.1769	0.2281	0.2432	0.2211	0.1670	0.0896
0.475	0.0901	0.1652	0.2136	0.2283	0.2083	0.1577	0.0847
0.500	0.0841	0.1545	0.2002	0.2144	0.1961	0.1487	0.0800
0.525	0.0786	0.1446	0.1876	0.2014	0.1845	0.1402	0.0755
0.550	0.0736	0.1354	0.1759	0.1891	0.1735	0.1320	0.0712
0.575	0.0689	0.1269	0.1650	0.1776	0.1632	0.1243	0.0670
0.600	0.0645	0.1189	0.1548	0.1668	0.1534	0.1169	0.0631
0.625	0.0605	0.1115	0.1452	0.1566	0.1442	0.1100	0.0594
0.650	0.0567	0.1046	0.1363	0.1471	0.1355	0.1034	0.0559
0.675	0.0532	0.0981	0.1279	0.1381	0.1273	0.0972	0.0525
0.700	0.0499	0.0921	0.1201	0.1297	0.1196	0.0914	0.0494
0.725	0.0468	0.0864	0.1127	0.1218	0.1124	0.0859	0.0464
0.750	0.0439	0.0811	0.1058	0.1144	0.1056	0.0807	0.0436
0.775	0.0412	0.0761	0.0994	0.1074	0.0992	0.0758	0.0410
0.800	0.0387	0.0715	0.0933	0.1009	0.0931	0.0712	0.0385
0.825	0.0363	0.0671	0.0876	0.0948	0.0875	0.0669	0.0362
0.850	0.0341	0.0630	0.0823	0.0890	0.0822	0.0628	0.0340
0.875	0.0320	0.0591	0.0772	0.0836	0.0772	0.0590	0.0319
0.900	0.0301	0.0555	0.0725	0.0785	0.0725	0.0554	0.0300
0.925	0.0282	0.0521	0.0681	0.0737	0.0681	0.0521	0.0282
0.950	0.0265	0.0490	0.0640	0.0692	0.0639	0.0489	0.0265
0.975	0.0249	0.0460	0.0601	0.0650	0.0600	0.0459	0.0249
1.000	0.0234	0.0432	0.0564	0.0610	0.0564	0.0431	0.0233

2.

(x,y)	exact	approx	abs error
(0.25,0.1)	0.3794	0.3728	0.0066
(1,0.5)	0.1854	0.2144	0.0290
(1.5,0.8)	0.0623	0.0712	0.0089

3. We identify $c = 1$, $a = 2$, $T = 1$, $n = 8$, and $m = 40$. Then $h = 2/8 = 0.25$, $k = 1/40 = 0.025$, and $\lambda = 2/5 = 0.4$.

TIME	X=0.25	X=0.50	X=0.75	X=1.00	X=1.25	X=1.50	X=1.75
0.000	1.0000	1.0000	1.0000	1.0000	0.0000	0.0000	0.0000
0.025	0.7074	0.9520	0.9566	0.7444	0.2545	0.0371	0.0053
0.050	0.5606	0.8499	0.8685	0.6633	0.3303	0.1034	0.0223
0.075	0.4684	0.7473	0.7836	0.6191	0.3614	0.1529	0.0462
0.100	0.4015	0.6577	0.7084	0.5837	0.3753	0.1871	0.0684
0.125	0.3492	0.5821	0.6428	0.5510	0.3797	0.2101	0.0861
0.150	0.3069	0.5187	0.5857	0.5199	0.3778	0.2247	0.0990
0.175	0.2721	0.4652	0.5359	0.4901	0.3716	0.2329	0.1078
0.200	0.2430	0.4198	0.4921	0.4617	0.3622	0.2362	0.1132
0.225	0.2186	0.3809	0.4533	0.4348	0.3507	0.2358	0.1160
0.250	0.1977	0.3473	0.4189	0.4093	0.3378	0.2327	0.1166
0.275	0.1798	0.3181	0.3881	0.3853	0.3240	0.2275	0.1157
0.300	0.1643	0.2924	0.3604	0.3626	0.3097	0.2208	0.1136
0.325	0.1507	0.2697	0.3353	0.3412	0.2953	0.2131	0.1107
0.350	0.1387	0.2495	0.3125	0.3211	0.2808	0.2047	0.1071
0.375	0.1281	0.2313	0.2916	0.3021	0.2666	0.1960	0.1032
0.400	0.1187	0.2150	0.2725	0.2843	0.2528	0.1871	0.0989
0.425	0.1102	0.2002	0.2549	0.2675	0.2393	0.1781	0.0946
0.450	0.1025	0.1867	0.2387	0.2517	0.2263	0.1692	0.0902
0.475	0.0955	0.1743	0.2236	0.2368	0.2139	0.1606	0.0858
0.500	0.0891	0.1630	0.2097	0.2228	0.2020	0.1521	0.0814
0.525	0.0833	0.1525	0.1967	0.2096	0.1906	0.1439	0.0772
0.550	0.0779	0.1429	0.1846	0.1973	0.1798	0.1361	0.0731
0.575	0.0729	0.1339	0.1734	0.1856	0.1696	0.1285	0.0691
0.600	0.0683	0.1256	0.1628	0.1746	0.1598	0.1214	0.0653
0.625	0.0641	0.1179	0.1530	0.1643	0.1506	0.1145	0.0617
0.650	0.0601	0.1106	0.1438	0.1546	0.1419	0.1080	0.0582
0.675	0.0564	0.1039	0.1351	0.1455	0.1336	0.1018	0.0549
0.700	0.0530	0.0976	0.1270	0.1369	0.1259	0.0959	0.0518
0.725	0.0497	0.0917	0.1194	0.1288	0.1185	0.0904	0.0488
0.750	0.0467	0.0862	0.1123	0.1212	0.1116	0.0852	0.0460
0.775	0.0439	0.0810	0.1056	0.1140	0.1050	0.0802	0.0433
0.800	0.0413	0.0762	0.0993	0.1073	0.0989	0.0755	0.0408
0.825	0.0388	0.0716	0.0934	0.1009	0.0931	0.0711	0.0384
0.850	0.0365	0.0674	0.0879	0.0950	0.0876	0.0669	0.0362
0.875	0.0343	0.0633	0.0827	0.0894	0.0824	0.0630	0.0341
0.900	0.0323	0.0596	0.0778	0.0841	0.0776	0.0593	0.0321
0.925	0.0303	0.0560	0.0732	0.0791	0.0730	0.0558	0.0302
0.950	0.0285	0.0527	0.0688	0.0744	0.0687	0.0526	0.0284
0.975	0.0268	0.0496	0.0647	0.0700	0.0647	0.0495	0.0268
1.000	0.0253	0.0466	0.0609	0.0659	0.0608	0.0465	0.0252

(x,y)	exact	approx	abs error
(0.25,0.1)	0.3794	0.4015	0.0221
(1,0.5)	0.1854	0.2228	0.0374
(1.5,0.8)	0.0623	0.0755	0.0132

4. We identify $c = 1$, $a = 2$, $T = 1$, $n = 8$, and $m = 20$. Then $h = 2/8 = 0.25$, $h = 1/20 = 0.05$, and $\lambda = 4/5 = 0.8$.

TIME	X=0.25	X=0.50	X=0.75	X=1.00	X=1.25	X=1.50	X=1.75
0.00	1.00	1.00	1.00	1.00	0.00	0.00	0.00
0.05	0.20	1.00	1.00	0.20	0.80	0.00	0.00
0.10	0.68	0.36	0.36	1.32	-0.32	0.64	0.00
0.15	-0.12	0.62	1.13	-0.76	1.76	-0.64	0.51
0.20	0.56	0.44	-0.79	2.77	-2.18	2.20	-0.82
0.25	0.01	-0.44	3.04	-4.03	5.28	-3.72	2.25
0.30	-0.36	2.70	-5.41	9.07	-9.37	8.26	-4.33
0.35	2.38	-6.24	12.67	-17.26	19.49	-15.91	9.20
0.40	-6.42	15.78	-26.40	36.08	-38.23	32.50	-18.25
0.45	16.47	-35.72	57.33	-73.35	77.80	-64.68	36.94
0.50	-38.46	80.48	-121.66	152.12	-157.11	130.60	-73.91
0.55	87.46	-176.38	259.07	-314.28	320.44	-263.18	148.83
0.60	-193.58	383.05	-547.97	652.17	-654.23	533.32	-299.84
0.65	422.59	-823.07	1156.96	-1353.07	1340.93	-1083.25	606.56
0.70	-912.01	1757.48	-2435.09	2810.16	-2753.61	2207.94	-1230.53
0.75	1953.19	-3732.17	5115.16	-5837.05	5666.65	-4512.08	2504.67
0.80	-4157.65	7893.99	-10724.47	12127.68	-11679.29	9244.30	-5112.47
0.85	8809.78	-16642.09	22452.02	-25199.62	24105.16	-18979.99	10462.92
0.90	-18599.54	34994.69	-46944.58	52365.51	-49806.79	39042.46	-21461.75
0.95	39155.48	-73432.11	98054.91	-108820.40	103010.45	-80440.31	44111.02
1.00	-82238.97	153827.58	-204634.95	226144.53	-213214.84	165961.36	-90818.86

(x,y)	exact	approx	abs error
(0.25,0.1)	0.3794	0.6800	0.3006
(1,0.5)	0.1854	152.1152	151.9298
(1.5,0.8)	0.0623	9244.3042	9244.2419

In this case $\lambda = 0.8$ is greater than 0.5 and the procedure is unstable.

5. We identify $c = 1$, $a = 2$, $T = 1$, $n = 8$, and $m = 20$. Then $h = 2/8 = 0.25$, $k = 1/20 = 0.05$, and $\lambda = 4/5 = 0.8$.

TIME	X=0.25	X=0.50	X=0.75	X=1.00	X=1.25	X=1.50	X=1.75
0.00	1.0000	1.0000	1.0000	1.0000	0.0000	0.0000	0.0000
0.05	0.5265	0.8693	0.8852	0.6141	0.3783	0.0884	0.0197
0.10	0.3972	0.6551	0.7043	0.5883	0.3723	0.1955	0.0653
0.15	0.3042	0.5150	0.5844	0.5192	0.3812	0.2261	0.1010
0.20	0.2409	0.4171	0.4901	0.4620	0.3636	0.2385	0.1145
0.25	0.1962	0.3452	0.4174	0.4092	0.3391	0.2343	0.1178
0.30	0.1631	0.2908	0.3592	0.3624	0.3105	0.2220	0.1145
0.35	0.1379	0.2482	0.3115	0.3208	0.2813	0.2056	0.1077
0.40	0.1181	0.2141	0.2718	0.2840	0.2530	0.1876	0.0993
0.45	0.1020	0.1860	0.2381	0.2514	0.2265	0.1696	0.0904
0.50	0.0888	0.1625	0.2092	0.2226	0.2020	0.1523	0.0816
0.55	0.0776	0.1425	0.1842	0.1970	0.1798	0.1361	0.0732
0.60	0.0681	0.1253	0.1625	0.1744	0.1597	0.1214	0.0654
0.65	0.0599	0.1104	0.1435	0.1544	0.1418	0.1079	0.0582
0.70	0.0528	0.0974	0.1268	0.1366	0.1257	0.0959	0.0518
0.75	0.0466	0.0860	0.1121	0.1210	0.1114	0.0851	0.0460
0.80	0.0412	0.0760	0.0991	0.1071	0.0987	0.0754	0.0408
0.85	0.0364	0.0672	0.0877	0.0948	0.0874	0.0668	0.0361
0.90	0.0322	0.0594	0.0776	0.0839	0.0774	0.0592	0.0320
0.95	0.0285	0.0526	0.0687	0.0743	0.0686	0.0524	0.0284
1.00	0.0252	0.0465	0.0608	0.0657	0.0607	0.0464	0.0251

(x,y)	exact	approx	abs error
(0.25,0.1)	0.3794	0.3972	0.0178
(1,0.5)	0.1854	0.2226	0.0372
(1.5,0.8)	0.0623	0.0754	0.0131

6. (a) We identify $c = 15/88 \approx 0.1705$, $a = 20$, $T = 10$, $n = 10$, and $m = 10$. Then $h = 2$, $k = 1$, and $\lambda = 15/352 \approx 0.0426$.

TIME	X=2	X=4	X=6	X=8	X=10	X=12	X=14	X=16	X=18
0	30.0000	30.0000	30.0000	30.0000	30.0000	30.0000	30.0000	30.0000	30.0000
1	28.7216	30.0000	30.0000	30.0000	30.0000	30.0000	30.0000	30.0000	28.7216
2	27.5521	29.9455	30.0000	30.0000	30.0000	30.0000	30.0000	29.9455	27.5521
3	26.4800	29.8459	29.9977	30.0000	30.0000	30.0000	29.9977	29.8459	26.4800
4	25.4951	29.7089	29.9913	29.9999	30.0000	29.9999	29.9913	29.7089	25.4951
5	24.5882	29.5414	29.9796	29.9995	30.0000	29.9995	29.9796	29.5414	24.5882
6	23.7515	29.3490	29.9618	29.9987	30.0000	29.9987	29.9618	29.3490	23.7515
7	22.9779	29.1365	29.9373	29.9972	29.9998	29.9972	29.9373	29.1365	22.9779
8	22.2611	28.9082	29.9057	29.9948	29.9996	29.9948	29.9057	28.9082	22.2611
9	21.5958	28.6675	29.8670	29.9912	29.9992	29.9912	29.8670	28.6675	21.5958
10	20.9768	28.4172	29.8212	29.9862	29.9985	29.9862	29.8212	28.4172	20.9768

(b) We identify $c = 15/88 \approx 0.1705$, $a = 50$, $T = 10$, $n = 10$, and $m = 10$. Then $h = 5$, $k = 1$, and $\lambda = 3/440 \approx 0.0068$.

TIME	X=5	X=10	X=15	X=20	X=25	X=30	X=35	X=40	X=45
0	30.0000	30.0000	30.0000	30.0000	30.0000	30.0000	30.0000	30.0000	30.0000
1	29.7955	30.0000	30.0000	30.0000	30.0000	30.0000	30.0000	30.0000	29.7955
2	29.5937	29.9986	30.0000	30.0000	30.0000	30.0000	30.0000	29.9986	29.5937
3	29.3947	29.9959	30.0000	30.0000	30.0000	30.0000	30.0000	29.9959	29.3947
4	29.1984	29.9918	30.0000	30.0000	30.0000	30.0000	30.0000	29.9918	29.1984
5	29.0047	29.9864	29.9999	30.0000	30.0000	30.0000	29.9999	29.9864	29.0047
6	28.8136	29.9798	29.9998	30.0000	30.0000	30.0000	29.9998	29.9798	28.8136
7	28.6251	29.9720	29.9997	30.0000	30.0000	30.0000	29.9997	29.9720	28.6251
8	28.4391	29.9630	29.9995	30.0000	30.0000	30.0000	29.9995	29.9630	28.4391
9	28.2556	29.9529	29.9992	30.0000	30.0000	30.0000	29.9992	29.9529	28.2556
10	28.0745	29.9416	29.9989	30.0000	30.0000	30.0000	29.9989	29.9416	28.0745

(c) We identify $c = 50/27 \approx 1.8519$, $a = 20$, $T = 10$, $n = 10$, and $m = 10$. Then $h = 2$, $k = 1$, and $\lambda = 25/54 \approx 0.4630$.

TIME	X=2	X=4	X=6	X=8	X=10	X=12	X=14	X=16	X=18
0	18.0000	32.0000	42.0000	48.0000	50.0000	48.0000	42.0000	32.0000	18.0000
1	16.1481	30.1481	40.1481	46.1481	48.1481	46.1481	40.1481	30.1481	16.1481
2	15.1536	28.2963	38.2963	44.2963	46.2963	44.2963	38.2963	28.2963	15.1536
3	14.2226	26.8414	36.4444	42.4444	44.4444	42.4444	36.4444	26.8414	14.2226
4	13.4801	25.4452	34.7764	40.5926	42.5926	40.5926	34.7764	25.4452	13.4801
5	12.7787	24.2258	33.1491	38.8258	40.7407	38.8258	33.1491	24.2258	12.7787
6	12.1622	23.0574	31.6460	37.0842	38.9677	37.0842	31.6460	23.0574	12.1622
7	11.5756	21.9895	30.1875	35.4385	37.2238	35.4385	30.1875	21.9895	11.5756
8	11.0378	20.9636	28.8232	33.8340	35.5707	33.8340	28.8232	20.9636	11.0378
9	10.5230	20.0070	27.5043	32.3182	33.9626	32.3182	27.5043	20.0070	10.5230
10	10.0420	19.0872	26.2620	30.8509	32.4400	30.8509	26.2620	19.0872	10.0420

(d) We identify $c = 260/159 \approx 1.6352$, $a = 100$, $T = 10$, $n = 10$, and $m = 10$. Then $h = 10$, $k = 1$, and $\lambda = 13/795 \approx 00164$.

TIME	X=10	X=20	X=30	X=40	X=50	X=60	X=70	X=80	X=90
0	8.0000	16.0000	24.0000	32.0000	40.0000	32.0000	24.0000	16.0000	8.0000
1	8.0000	16.0000	23.6075	31.3459	39.2151	31.6075	23.7384	15.8692	8.0000
2	8.0000	15.9936	23.2279	30.7068	38.4452	31.2151	23.4789	15.7384	7.9979
3	7.9999	15.9812	22.8606	30.0824	37.6900	30.8229	23.2214	15.6076	7.9937
4	7.9996	15.9631	22.5050	29.4724	36.9492	30.4312	22.9660	15.4769	7.9874
5	7.9990	15.9399	22.1606	28.8765	36.2228	30.0401	22.7125	15.3463	7.9793
6	7.9981	15.9118	21.8270	28.2945	35.5103	29.6500	22.4610	15.2158	7.9693
7	7.9967	15.8791	21.5037	27.7261	34.8117	29.2610	22.2112	15.0854	7.9575
8	7.9948	15.8422	21.1902	27.1709	34.1266	28.8733	21.9633	14.9553	7.9439
9	7.9924	15.8013	20.8861	26.6288	33.4548	28.4870	21.7172	14.8253	7.9287
10	7.9894	15.7568	20.5911	26.0995	32.7961	28.1024	21.4727	14.6956	7.9118

7. **(a)** We identify $c = 15/88 \approx 0.1705$, $a = 20$, $T = 10$, $n = 10$, and $m = 10$. Then $h = 2$, $k = 1$, and $\lambda = 15/352 \approx 0.0426$.

TIME	X=2.00	X=4.00	X=6.00	X=8.00	X=10.00	X=12.00	X=14.00	X=16.00	X=18.00
0.00	30.0000	30.0000	30.0000	30.0000	30.0000	30.0000	30.0000	30.0000	30.0000
1.00	28.7733	29.9749	29.9995	30.0000	30.0000	30.0000	29.9995	29.9749	28.7733
2.00	27.6450	29.9037	29.9970	29.9999	30.0000	29.9999	29.9970	29.9037	27.6450
3.00	26.6051	29.7938	29.9911	29.9997	30.0000	29.9997	29.9911	29.7938	26.6051
4.00	25.6452	29.6517	29.9805	29.9991	29.9999	29.9991	29.9805	29.6517	25.6452
5.00	24.7573	29.4829	29.9643	29.9981	29.9998	29.9981	29.9643	29.4829	24.7573
6.00	23.9347	29.2922	29.9421	29.9963	29.9996	29.9963	29.9421	29.2922	23.9347
7.00	23.1711	29.0836	29.9134	29.9936	29.9992	29.9936	29.9134	29.0836	23.1711
8.00	22.4612	28.8606	29.8782	29.9898	29.9986	29.9898	29.8782	28.8606	22.4612
9.00	21.7999	28.6263	29.8362	29.9848	29.9977	29.9848	29.8362	28.6263	21.7999
10.00	21.1829	28.3831	29.7878	29.9782	29.9964	29.9782	29.7878	28.3831	21.1829

(b) We identify $c = 15/88 \approx 0.1705$, $a = 50$, $T = 10$, $n = 10$, and $m = 10$. Then $h = 5$, $k = 1$, and $\lambda = 3/440 \approx 0.0068$.

TIME	X=5.00	X=10.00	X=15.00	X=20.00	X=25.00	X=30.00	X=35.00	X=40.00	X=45.00
0.00	30.0000	30.0000	30.0000	30.0000	30.0000	30.0000	30.0000	30.0000	30.0000
1.00	29.7968	29.9993	30.0000	30.0000	30.0000	30.0000	30.0000	29.9993	29.7968
2.00	29.5964	29.9973	30.0000	30.0000	30.0000	30.0000	30.0000	29.9973	29.5964
3.00	29.3987	29.9939	30.0000	30.0000	30.0000	30.0000	30.0000	29.9939	29.3987
4.00	29.2036	29.9893	29.9999	30.0000	30.0000	30.0000	29.9999	29.9893	29.2036
5.00	29.0112	29.9834	29.9998	30.0000	30.0000	30.0000	29.9998	29.9834	29.0112
6.00	28.8212	29.9762	29.9997	30.0000	30.0000	30.0000	29.9997	29.9762	28.8213
7.00	28.6339	29.9679	29.9995	30.0000	30.0000	30.0000	29.9995	29.9679	28.6339
8.00	28.4490	29.9585	29.9992	30.0000	30.0000	30.0000	29.9993	29.9585	28.4490
9.00	28.2665	29.9479	29.9989	30.0000	30.0000	30.0000	29.9989	29.9479	28.2665
10.00	28.0864	29.9363	29.9986	30.0000	30.0000	30.0000	29.9986	29.9363	28.0864

(c) We identify $c = 50/27 \approx 1.8519$, $a = 20$, $T = 10$, $n = 10$, and $m = 10$. Then $h = 2$, $k = 1$, and $\lambda = 25/54 \approx 0.4630$.

TIME	X=2.00	X=4.00	X=6.00	X=8.00	X=10.00	X=12.00	X=14.00	X=16.00	X=18.00
0.00	18.0000	32.0000	42.0000	48.0000	50.0000	48.0000	42.0000	32.0000	18.0000
1.00	16.4489	30.1970	40.1561	46.1495	48.1486	46.1495	40.1561	30.1970	16.4489
2.00	15.3312	28.5348	38.3465	44.3067	46.3001	44.3067	38.3465	28.5348	15.3312
3.00	14.4216	27.0416	36.6031	42.4847	44.4619	42.4847	36.6031	27.0416	14.4216
4.00	13.6371	25.6867	34.9416	40.6988	42.6453	40.6988	34.9416	25.6867	13.6371
5.00	12.9378	24.4419	33.3628	38.9611	40.8634	38.9611	33.3628	24.4419	12.9378
6.00	12.3012	23.2863	31.8624	37.2794	39.1273	37.2794	31.8624	23.2863	12.3012
7.00	11.7137	22.2051	30.4350	35.6578	37.4446	35.6578	30.4350	22.2051	11.7137
8.00	11.1659	21.1877	29.0757	34.0984	35.8202	34.0984	29.0757	21.1877	11.1659
9.00	10.6517	20.2261	27.7799	32.6014	34.2567	32.6014	27.7799	20.2261	10.6517
10.00	10.1665	19.3143	26.5439	31.1662	32.7549	31.1662	26.5439	19.3143	10.1665

(d) We identify $c = 260/159 \approx 1.6352$, $a = 100$, $T = 10$, $n = 10$, and $m = 10$. Then $h = 10$, $k = 1$, and $\lambda = 13/795 \approx 00164$.

TIME	X=10.00	X=20.00	X=30.00	X=40.00	X=50.00	X=60.00	X=70.00	X=80.00	X=90.00
0.00	8.0000	16.0000	24.0000	32.0000	40.0000	32.0000	24.0000	16.0000	8.0000
1.00	8.0000	16.0000	24.0000	31.9979	39.7425	31.9979	24.0000	16.0000	8.0000
2.00	8.0000	16.0000	23.9999	31.9918	39.4932	31.9918	23.9999	16.0000	8.0000
3.00	8.0000	16.0000	23.9997	31.9820	39.2517	31.9820	23.9997	16.0000	8.0000
4.00	8.0000	16.0000	23.9993	31.9687	39.0176	31.9687	23.9993	16.0000	8.0000
5.00	8.0000	16.0000	23.9987	31.9520	38.7905	31.9520	23.9987	16.0000	8.0000
6.00	8.0000	15.9999	23.9978	31.9323	38.5701	31.9323	23.9978	15.9999	8.0000
7.00	8.0000	15.9999	23.9966	31.9097	38.3561	31.9097	23.9966	15.9999	8.0000
8.00	8.0000	15.9998	23.9951	31.8844	38.1483	31.8844	23.9951	15.9998	8.0000
9.00	8.0000	15.9997	23.9931	31.8566	37.9463	31.8566	23.9931	15.9997	8.0000
10.00	8.0000	15.9996	23.9908	31.8265	37.7499	31.8265	23.9908	15.9996	8.0000

8. (a) We identify $c = 15/88 \approx 0.1705$, $a = 20$, $T = 10$, $n = 10$, and $m = 10$. Then $h = 2$, $k = 1$, and $\lambda = 15/352 \approx 0.0426$.

TIME	X=2	X=4	X=6	X=8	X=10	X=12	X=14	X=16	X=18
0	30.0000	30.0000	30.0000	30.0000	30.0000	30.0000	30.0000	30.0000	30.0000
1	28.7216	30.0000	30.0000	30.0000	30.0000	30.0000	30.0000	30.0000	29.5739
2	27.5521	29.9455	30.0000	30.0000	30.0000	30.0000	30.0000	29.9818	29.1840
3	26.4800	29.8459	29.9977	30.0000	30.0000	30.0000	29.9992	29.9486	28.8267
4	25.4951	29.7089	29.9913	29.9999	30.0000	30.0000	29.9971	29.9030	28.4984
5	24.5882	29.5414	29.9796	29.9995	30.0000	29.9998	29.9932	29.8471	28.1961
6	23.7515	29.3490	29.9618	29.9987	30.0000	29.9996	29.9873	29.7830	27.9172
7	22.9779	29.1365	29.9373	29.9972	29.9999	29.9991	29.9791	29.7122	27.6593
8	22.2611	28.9082	29.9057	29.9948	29.9997	29.9982	29.9686	29.6361	27.4204
9	21.5958	28.6675	29.8670	29.9912	29.9995	29.9970	29.9557	29.5558	27.1986
10	20.9768	28.4172	29.8212	29.9862	29.9990	29.9954	29.9404	29.4724	26.9923

(b) We identify $c = 15/88 \approx 0.1705$, $a = 50$, $T = 10$, $n = 10$, and $m = 10$. Then $h = 5$, $k = 1$, and $\lambda = 3/440 \approx 0.0068$.

TIME	X=5	X=10	X=15	X=20	X=25	X=30	X=35	X=40	X=45
0	30.0000	30.0000	30.0000	30.0000	30.0000	30.0000	30.0000	30.0000	30.0000
1	29.7955	30.0000	30.0000	30.0000	30.0000	30.0000	30.0000	30.0000	29.9318
2	29.5937	29.9986	30.0000	30.0000	30.0000	30.0000	30.0000	29.9995	29.8646
3	29.3947	29.9959	30.0000	30.0000	30.0000	30.0000	30.0000	29.9986	29.7982
4	29.1984	29.9918	30.0000	30.0000	30.0000	30.0000	30.0000	29.9973	29.7328
5	29.0047	29.9864	29.9999	30.0000	30.0000	30.0000	30.0000	29.9955	29.6682
6	28.8136	29.9798	29.9998	30.0000	30.0000	30.0000	29.9999	29.9933	29.6045
7	28.6251	29.9720	29.9997	30.0000	30.0000	30.0000	29.9999	29.9907	29.5417
8	28.4391	29.9630	29.9995	30.0000	30.0000	30.0000	29.9998	29.9877	29.4797
9	28.2556	29.9529	29.9992	30.0000	30.0000	30.0000	29.9997	29.9843	29.4185
10	28.0745	29.9416	29.9989	30.0000	30.0000	30.0000	29.9996	29.9805	29.3582

(c) We identify $c = 50/27 \approx 1.8519$, $a = 20$, $T = 10$, $n = 10$, and $m = 10$. Then $h = 2$, $k = 1$, and $\lambda = 25/54 \approx 0.4630$.

TIME	X=2	X=4	X=6	X=8	X=10	X=12	X=14	X=16	X=18
0	18.0000	32.0000	42.0000	48.0000	50.0000	48.0000	42.0000	32.0000	18.0000
1	16.1481	30.1481	40.1481	46.1481	48.1481	46.1481	40.1481	30.1481	25.4074
2	15.1536	28.2963	38.2963	44.2963	46.2963	44.2963	38.2963	32.5830	25.0988
3	14.2226	26.8414	36.4444	42.4444	44.4444	42.4444	38.4290	31.7631	26.2031
4	13.4801	25.4452	34.7764	40.5926	42.5926	41.5114	37.2019	32.2751	25.9054
5	12.7787	24.2258	33.1491	38.8258	41.1661	40.0168	36.9161	31.6071	26.1204
6	12.1622	23.0574	31.6460	37.2812	39.5506	39.1134	35.8938	31.5248	25.8270
7	11.5756	21.9895	30.2787	35.7230	38.2975	37.8252	35.3617	30.9096	25.7672
8	11.0378	21.0058	28.9616	34.3944	36.8869	36.9033	34.4411	30.5900	25.4779
9	10.5425	20.0742	27.7936	33.0332	35.7406	35.7558	33.7981	30.0062	25.3086
10	10.0746	19.2352	26.6455	31.8608	34.4942	34.8424	32.9489	29.5869	25.0257

(d) We identify $c = 260/159 \approx 1.6352$, $a = 100$, $T = 10$, $n = 10$, and $m = 10$. Then $h = 10$, $k = 1$, and $\lambda = 13/795 \approx 00164$.

TIME	X=10	X=20	X=30	X=40	X=50	X=60	X=70	X=80	X=90
0	8.0000	16.0000	24.0000	32.0000	40.0000	32.0000	24.0000	16.0000	8.0000
1	8.0000	16.0000	23.6075	31.6730	39.2151	31.6075	23.7384	15.8692	8.0000
2	8.0000	15.9936	23.2279	31.3502	38.4505	31.2151	23.4789	15.7384	7.9979
3	7.9999	15.9812	22.8606	31.0318	37.7057	30.8230	23.2214	15.6076	7.9937
4	7.9996	15.9631	22.5050	30.7178	36.9800	30.4315	22.9660	15.4769	7.9874
5	7.9990	15.9399	22.1606	30.4082	36.2728	30.0410	22.7126	15.3463	7.9793
6	7.9981	15.9118	21.8270	30.1031	35.5838	29.6516	22.4610	15.2158	7.9693
7	7.9967	15.8791	21.5037	29.8026	34.9123	29.2638	22.2113	15.0854	7.9575
8	7.9948	15.8422	21.1902	29.5066	34.2579	28.8776	21.9634	14.9553	7.9439
9	7.9924	15.8013	20.8861	29.2152	33.6200	28.4934	21.7173	14.8253	7.9287
10	7.9894	15.7568	20.5911	28.9283	32.9982	28.1113	21.4730	14.6956	7.9118

9. (a) We identify $c = 15/88 \approx 0.1705$, $a = 20$, $T = 10$, $n = 10$, and $m = 10$. Then $h = 2$, $k = 1$, and $\lambda = 15/352 \approx 0.0426$.

TIME	X=2.00	X=4.00	X=6.00	X=8.00	X=10.00	X=12.00	X=14.00	X=16.00	X=18.00
0.00	30.0000	30.0000	30.0000	30.0000	30.0000	30.0000	30.0000	30.0000	30.0000
1.00	28.7733	29.9749	29.9995	30.0000	30.0000	30.0000	29.9998	29.9916	29.5911
2.00	27.6450	29.9037	29.9970	29.9999	30.0000	30.0000	29.9990	29.9679	29.2150
3.00	26.6051	29.7938	29.9911	29.9997	30.0000	29.9999	29.9970	29.9313	28.8684
4.00	25.6452	29.6517	29.9805	29.9991	30.0000	29.9997	29.9935	29.8839	28.5484
5.00	24.7573	29.4829	29.9643	29.9981	29.9999	29.9994	29.9881	29.8276	28.2524
6.00	23.9347	29.2922	29.9421	29.9963	29.9997	29.9988	29.9807	29.7641	27.9782
7.00	23.1711	29.0836	29.9134	29.9936	29.9995	29.9979	29.9711	29.6945	27.7237
8.00	22.4612	28.8606	29.8782	29.9899	29.9991	29.9966	29.9594	29.6202	27.4870
9.00	21.7999	28.6263	29.8362	29.9848	29.9985	29.9949	29.9454	29.5421	27.2666
10.00	21.1829	28.3831	29.7878	29.9783	29.9976	29.9927	29.9293	29.4610	27.0610

(b) We identify $c = 15/88 \approx 0.1705$, $a = 50$, $T = 10$, $n = 10$, and $m = 10$. Then $h = 5$, $k = 1$, and $\lambda = 3/440 \approx 0.0068$.

TIME	X=5.00	X=10.00	X=15.00	X=20.00	X=25.00	X=30.00	X=35.00	X=40.00	X=45.00
0.00	30.0000	30.0000	30.0000	30.0000	30.0000	30.0000	30.0000	30.0000	30.0000
1.00	29.7968	29.9993	30.0000	30.0000	30.0000	30.0000	30.0000	29.9998	29.9323
2.00	29.5964	29.9973	30.0000	30.0000	30.0000	30.0000	30.0000	29.9991	29.8655
3.00	29.3987	29.9939	30.0000	30.0000	30.0000	30.0000	30.0000	29.9980	29.7996
4.00	29.2036	29.9893	29.9999	30.0000	30.0000	30.0000	30.0000	29.9964	29.7345
5.00	29.0112	29.9834	29.9998	30.0000	30.0000	30.0000	29.9999	29.9945	29.6704
6.00	28.8212	29.9762	29.9997	30.0000	30.0000	30.0000	29.9999	29.9921	29.6071
7.00	28.6339	29.9679	29.9995	30.0000	30.0000	30.0000	29.9998	29.9893	29.5446
8.00	28.4490	29.9585	29.9992	30.0000	30.0000	30.0000	29.9997	29.9862	29.4830
9.00	28.2665	29.9479	29.9989	30.0000	30.0000	30.0000	29.9996	29.9827	29.4222
10.00	28.0864	29.9363	29.9986	30.0000	30.0000	30.0000	29.9995	29.9788	29.3621

(c) We identify $c = 50/27 \approx 1.8519$, $a = 20$, $T = 10$, $n = 10$, and $m = 10$. Then $h = 2$, $k = 1$, and $\lambda = 25/54 \approx 0.4630$.

TIME	X=2.00	X=4.00	X=6.00	X=8.00	X=10.00	X=12.00	X=14.00	X=16.00	X=18.00
0.00	18.0000	32.0000	42.0000	48.0000	50.0000	48.0000	42.0000	32.0000	18.0000
1.00	16.4489	30.1970	40.1562	46.1502	48.1531	46.1773	40.3274	31.2520	22.9449
2.00	15.3312	28.5350	38.3477	44.3130	46.3327	44.4671	39.0872	31.5755	24.6930
3.00	14.4219	27.0429	36.6090	42.5113	44.5759	42.9362	38.1976	31.7478	25.4131
4.00	13.6381	25.6913	34.9606	40.7728	42.9127	41.5716	37.4340	31.7086	25.6986
5.00	12.9409	24.4545	33.4091	39.1182	41.3519	40.3240	36.7033	31.5136	25.7663
6.00	12.3088	23.3146	31.9546	37.5566	39.8880	39.1565	35.9745	31.2134	25.7128
7.00	11.7294	22.2589	30.5939	36.0884	38.5109	38.0470	35.2407	30.8434	25.5871
8.00	11.1946	21.2785	29.3217	34.7092	37.2109	36.9834	34.5032	30.4279	25.4167
9.00	10.6987	20.3660	28.1318	33.4130	35.9801	35.9591	33.7660	29.9836	25.2181
10.00	10.2377	19.5150	27.0178	32.1929	34.8117	34.9710	33.0338	29.5224	25.0019

(d) We identify $c = 260/159 \approx 1.6352$, $a = 100$, $T = 10$, $n = 10$, and $m = 10$. Then $h = 10$, $k = 1$, and $\lambda = 13/795 \approx 00164$.

TIME	X=10.00	X=20.00	X=30.00	X=40.00	X=50.00	X=60.00	X=70.00	X=80.00	X=90.00
0.00	8.0000	16.0000	24.0000	32.0000	40.0000	32.0000	24.0000	16.0000	8.0000
1.00	8.0000	16.0000	24.0000	31.9979	39.7425	31.9979	24.0000	16.0026	8.3218
2.00	8.0000	16.0000	23.9999	31.9918	39.4932	31.9918	24.0000	16.0102	8.6333
3.00	8.0000	16.0000	23.9997	31.9820	39.2517	31.9820	24.0001	16.0225	8.9350
4.00	8.0000	16.0000	23.9993	31.9687	39.0176	31.9687	24.0002	16.0392	9.2272
5.00	8.0000	16.0000	23.9987	31.9520	38.7905	31.9521	24.0003	16.0599	9.5103
6.00	8.0000	15.9999	23.9978	31.9323	38.5701	31.9324	24.0005	16.0845	9.7846
7.00	8.0000	15.9999	23.9966	31.9097	38.3561	31.9098	24.0008	16.1126	10.0506
8.00	8.0000	15.9998	23.9951	31.8844	38.1483	31.8846	24.0012	16.1441	10.3084
9.00	8.0000	15.9997	23.9931	31.8566	37.9463	31.8569	24.0017	16.1786	10.5585
10.00	8.0000	15.9996	23.9908	31.8265	37.7499	31.8270	24.0023	16.2160	10.8012

10. (a) With $n = 4$ we have $h = 1/2$ so that $\lambda = 1/100 = 0.01$.

(b) We observe that $\alpha = 2(1 + 1/\lambda) = 202$ and $\beta = 2(1 - 1/\lambda) = -198$. The system of equations is

$$-u_{01} + \alpha u_{11} - u_{21} = u_{20} - \beta u_{10} + u_{00}$$

$$-u_{11} + \alpha u_{21} - u_{31} = u_{30} - \beta u_{20} + u_{10}$$

$$-u_{21} + \alpha u_{31} - u_{41} = u_{40} - \beta u_{30} + u_{20}.$$

Now $u_{00} = u_{01} = u_{40} = u_{41} = 0$, so the system is

$$\alpha u_{11} - u_{21} = u_{20} - \beta u_{10}$$

$$-u_{11} + \alpha u_{21} - u_{31} = u_{30} - \beta u_{20} + u_{10}$$

$$-u_{21} + \alpha u_{31} = -\beta u_{30} + u_{20}$$

or

$$202 u_{11} - u_{21} = \sin \pi + 198 \sin \frac{\pi}{2} = 198$$

$$-u_{11} + 202 u_{21} - u_{31} = \sin \frac{3\pi}{2} + 198 \sin \pi + \sin \frac{\pi}{2} = 0$$

$$-u_{21} + 202 u_{31} = 198 \sin \frac{3\pi}{2} + \sin \pi = -198.$$

(c) The solution of this system is $u_{11} \approx 0.9802$, $u_{21} = 0$, $u_{31} \approx -0.9802$.

11. (a) The differential equation is $k \dfrac{\partial^2 u}{\partial x^2} = \dfrac{\partial u}{\partial t}$ where $k = K/\gamma\rho$. If we let $u(x,t) = v(x,t) + \psi(x)$, then

$$\frac{\partial^2 u}{\partial x^2} = \frac{\partial^2 v}{\partial x^2} + \psi'' \quad \text{and} \quad \frac{\partial u}{\partial t} = \frac{\partial v}{\partial t}.$$

Substituting into the differential equation gives

$$k \frac{\partial^2 v}{\partial x^2} + k\psi'' = \frac{\partial v}{\partial t}.$$

Requiring $k\psi'' = 0$ we have $\psi(x) = c_1 x + c_2$. The boundary conditions become

$$u(0,t) = v(0,t) + \psi(0) = 20 \quad \text{and} \quad u(20,t) = v(20,t) + \psi(20) = 30.$$

Letting $\psi(0) = 20$ and $\psi(20) = 30$ we obtain the homogeneous boundary conditions in v: $v(0,t) = v(20,t) = 0$. Now $\psi(0) = 20$ and $\psi(20) = 30$ imply that $c_1 = 1/2$ and $c_2 = 20$. The steady-state solution is $\psi(x) = \frac{1}{2}x + 20$.

(b) To use the Crank-Nicholson method we identify $c = 375/212 \approx 1.7689$, $a = 20$, $T = 400$, $n = 5$, and $m = 40$. Then $h = 4$, $k = 10$, and $\lambda = 1875/1696 \approx 1.1055$.

TIME	X=4.00	X=8.00	X=12.00	X=16.00
0.00	50.0000	50.0000	50.0000	50.0000
10.00	32.7433	44.2679	45.4228	38.2971
20.00	29.9946	36.2354	38.3148	35.8160
30.00	26.9487	32.1409	34.0874	32.9644
40.00	25.2691	29.2562	31.2704	31.2580
50.00	24.1178	27.4348	29.4296	30.1207
60.00	23.3821	26.2339	28.2356	29.3810
70.00	22.8995	25.4560	27.4554	28.8998
80.00	22.5861	24.9481	26.9482	28.5859
90.00	22.3817	24.6176	26.6175	28.3817
100.00	22.2486	24.4022	26.4023	28.2486
110.00	22.1619	24.2620	26.2620	28.1619
120.00	22.1055	24.1707	26.1707	28.1055
130.00	22.0687	24.1112	26.1112	28.0687
140.00	22.0447	24.0724	26.0724	28.0447
150.00	22.0291	24.0472	26.0472	28.0291
160.00	22.0190	24.0307	26.0307	28.0190
170.00	22.0124	24.0200	26.0200	28.0124
180.00	22.0081	24.0130	26.0130	28.0081
190.00	22.0052	24.0085	26.0085	28.0052
200.00	22.0034	24.0055	26.0055	28.0034
210.00	22.0022	24.0036	26.0036	28.0022
220.00	22.0015	24.0023	26.0023	28.0015
230.00	22.0009	24.0015	26.0015	28.0009
240.00	22.0006	24.0010	26.0010	28.0006
250.00	22.0004	24.0007	26.0007	28.0004
260.00	22.0003	24.0004	26.0004	28.0003
270.00	22.0002	24.0003	26.0003	28.0002
280.00	22.0001	24.0002	26.0002	28.0001
290.00	22.0001	24.0001	26.0001	28.0001
300.00	22.0000	24.0001	26.0001	28.0000
310.00	22.0000	24.0001	26.0001	28.0000
320.00	22.0000	24.0000	26.0000	28.0000
330.00	22.0000	24.0000	26.0000	28.0000
340.00	22.0000	24.0000	26.0000	28.0000
350.00	22.0000	24.0000	26.0000	28.0000

We observe that the approximate steady-state temperatures agree exactly with the corresponding values of $\psi(x)$.

12. We identify $c = 1$, $a = 1$, $T = 1$, $n = 5$, and $m = 20$. Then $h = 0.2$, $k = 0.04$, and $\lambda = 1$. The values below were obtained using *Excel*, which carries more than 12 significant digits. In order to see evidence of instability use $0 \leq t \leq 2$.

TIME	X=0.2	X=0.4	X=0.6	X=0.8
0.00	0.5878	0.9511	0.9511	0.5878
0.04	0.3633	0.5878	0.5878	0.3633
0.08	0.2245	0.3633	0.3633	0.2245
0.12	0.1388	0.2245	0.2245	0.1388
0.16	0.0858	0.1388	0.1388	0.0858
0.20	0.0530	0.0858	0.0858	0.0530
0.24	0.0328	0.0530	0.0530	0.0328
0.28	0.0202	0.0328	0.0328	0.0202
0.32	0.0125	0.0202	0.0202	0.0125
0.36	0.0077	0.0125	0.0125	0.0077
0.40	0.0048	0.0077	0.0077	0.0048
0.44	0.0030	0.0048	0.0048	0.0030
0.48	0.0018	0.0030	0.0030	0.0018
0.52	0.0011	0.0018	0.0018	0.0011
0.56	0.0007	0.0011	0.0011	0.0007
0.60	0.0004	0.0007	0.0007	0.0004
0.64	0.0003	0.0004	0.0004	0.0003
0.68	0.0002	0.0003	0.0003	0.0002
0.72	0.0001	0.0002	0.0002	0.0001
0.76	0.0001	0.0001	0.0001	0.0001
0.80	0.0000	0.0001	0.0001	0.0000
0.84	0.0000	0.0000	0.0000	0.0000
0.88	0.0000	0.0000	0.0000	0.0000
0.92	0.0000	0.0000	0.0000	0.0000
0.96	0.0000	0.0000	0.0000	0.0000
1.00	0.0000	0.0000	0.0000	0.0000

TIME	X=0.2	X=0.4	X=0.6	X=0.8
1.04	0.0000	0.0000	0.0000	0.0000
1.08	0.0000	0.0000	0.0000	0.0000
1.12	0.0000	0.0000	0.0000	0.0000
1.16	0.0000	0.0000	0.0000	0.0000
1.20	−0.0001	0.0001	−0.0001	0.0001
1.24	0.0001	−0.0002	0.0002	−0.0001
1.28	−0.0004	0.0006	−0.0006	0.0004
1.32	0.0010	−0.0015	0.0015	−0.0010
1.36	−0.0025	0.0040	−0.0040	0.0025
1.40	0.0065	−0.0106	0.0106	−0.0065
1.44	−0.0171	0.0277	−0.0277	0.0171
1.48	0.0448	−0.0724	0.0724	−0.0448
1.52	−0.1172	0.1897	−0.1897	0.1172
1.56	0.3069	−0.4965	0.4965	−0.3069
1.60	−0.8034	1.2999	−1.2999	0.8034
1.64	2.1033	−3.4032	3.4032	−2.1033
1.68	−5.5064	8.9096	−8.9096	5.5064
1.72	14.416	−23.326	23.326	−14.416
1.76	−37.742	61.067	−61.067	37.742
1.80	98.809	−159.88	159.88	−98.809
1.84	−258.68	418.56	−418.56	258.685
1.88	677.24	−1095.8	1095.8	−677.245
1.92	−1773.1	2868.9	−2868.9	1773.1
1.96	4641.9	−7510.8	7510.8	−4641.9
2.00	−12153	19663	−19663	12153

Exercises 15.3   Wave Equation

1. (a) Identifying $h = 1/4$ and $k = 1/10$ we see that $\lambda = 2/5$.

TIME	X=0.25	X=0.5	X=0.75
0.00	0.1875	0.2500	0.1875
0.10	0.1775	0.2400	0.1775
0.20	0.1491	0.2100	0.1491
0.30	0.1066	0.1605	0.1066
0.40	0.0556	0.0938	0.0556
0.50	0.0019	0.0148	0.0019
0.60	−0.0501	−0.0682	−0.0501
0.70	−0.0970	−0.1455	−0.0970
0.80	−0.1361	−0.2072	−0.1361
0.90	−0.1648	−0.2462	−0.1648
1.00	−0.1802	−0.2591	−0.1802

(b) Identifying $h = 2/5$ and $k = 1/10$ we see that $\lambda = 1/4$.

TIME	X=0.4	X=0.8	X=1.2	X=1.6
0.00	0.0032	0.5273	0.5273	0.0032
0.10	0.0194	0.5109	0.5109	0.0194
0.20	0.0652	0.4638	0.4638	0.0652
0.30	0.1318	0.3918	0.3918	0.1318
0.40	0.2065	0.3035	0.3035	0.2065
0.50	0.2743	0.2092	0.2092	0.2743
0.60	0.3208	0.1190	0.1190	0.3208
0.70	0.3348	0.0413	0.0413	0.3348
0.80	0.3094	-0.0180	-0.0180	0.3094
0.90	0.2443	-0.0568	-0.0568	0.2443
1.00	0.1450	-0.0768	-0.0768	0.1450

(c) Identifying $h = 1/10$ and $k = 1/25$ we see that $\lambda = 2\sqrt{2}/5$.

TIME	X=0.1	X=0.2	X=0.3	X=0.4	X=0.5	X=0.6	X=0.7	X=0.8	X=0.9
0.00	0.0000	0.0000	0.0000	0.0000	0.0000	0.5000	0.5000	0.5000	0.5000
0.04	0.0000	0.0000	0.0000	0.0000	0.0800	0.4200	0.5000	0.5000	0.4200
0.08	0.0000	0.0000	0.0000	0.0256	0.2432	0.2568	0.4744	0.4744	0.2312
0.12	0.0000	0.0000	0.0082	0.1126	0.3411	0.1589	0.3792	0.3710	0.0462
0.16	0.0000	0.0026	0.0472	0.2394	0.3076	0.1898	0.2108	0.1663	-0.0496
0.20	0.0008	0.0187	0.1334	0.3264	0.2146	0.2651	0.0215	-0.0933	-0.0605
0.24	0.0071	0.0657	0.2447	0.3159	0.1735	0.2463	-0.1266	-0.3056	-0.0625
0.28	0.0299	0.1513	0.3215	0.2371	0.2013	0.0849	-0.2127	-0.3829	-0.1223
0.32	0.0819	0.2525	0.3168	0.1737	0.2033	-0.1345	-0.2580	-0.3223	-0.2264
0.36	0.1623	0.3197	0.2458	0.1657	0.0877	-0.2853	-0.2843	-0.2104	-0.2887
0.40	0.2412	0.3129	0.1727	0.1583	-0.1223	-0.3164	-0.2874	-0.1473	-0.2336
0.44	0.2657	0.2383	0.1399	0.0658	-0.3046	-0.2761	-0.2549	-0.1565	-0.0761
0.48	0.1965	0.1410	0.1149	-0.1216	-0.3593	-0.2381	-0.1977	-0.1715	0.0800
0.52	0.0466	0.0531	0.0225	-0.3093	-0.2992	-0.2260	-0.1451	-0.1144	0.1300
0.56	-0.1161	-0.0466	-0.1662	-0.3876	-0.2188	-0.2114	-0.1085	0.0111	0.0602
0.60	-0.2194	-0.2069	-0.3875	-0.3411	-0.1901	-0.1662	-0.0666	0.1140	-0.0446
0.64	-0.2485	-0.4290	-0.5362	-0.2611	-0.2021	-0.0969	0.0012	0.1084	-0.0843
0.68	-0.2559	-0.6276	-0.5625	-0.2503	-0.1993	-0.0298	0.0720	0.0068	-0.0354
0.72	-0.3003	-0.6865	-0.5097	-0.3230	-0.1585	0.0156	0.0893	-0.0874	0.0384
0.76	-0.3722	-0.5652	-0.4538	-0.4029	-0.1147	0.0289	0.0265	-0.0849	0.0596
0.80	-0.3867	-0.3464	-0.4172	-0.4068	-0.1172	-0.0046	-0.0712	-0.0005	0.0155
0.84	-0.2647	-0.1633	-0.3546	-0.3214	-0.1763	-0.0954	-0.1249	0.0665	-0.0386
0.88	-0.0254	-0.0738	-0.2202	-0.2002	-0.2559	-0.2215	-0.1079	0.0385	-0.0468
0.92	0.2064	-0.0157	-0.0325	-0.1032	-0.3067	-0.3223	-0.0804	-0.0636	-0.0127
0.96	0.3012	0.1081	0.1380	-0.0487	-0.2974	-0.3407	-0.1250	-0.1548	0.0092
1.00	0.2378	0.3032	0.2392	-0.0141	-0.2223	-0.2762	-0.2481	-0.1840	-0.0244

2. (a) In Section 12.4 the solution of the wave equation is shown to be

$$u(x,t) = \sum_{n=1}^{\infty} (A_n \cos n\pi t + B_n \sin n\pi t) \sin n\pi x$$

776

where

$$A_n = 2 \int_0^1 \sin \pi x \sin n\pi x \, dx = \begin{cases} 1, & n = 1 \\ 0, & n = 2, 3, 4, \dots \end{cases}$$

and

$$B_n = \frac{2}{n\pi} \int_0^1 0 \, dx = 0.$$

Thus $u(x,t) = \cos \pi t \sin \pi x$.

(b) We have $h = 1/4$, $k = 0.5/5 = 0.1$ and $\lambda = 0.4$. Now $u_{0,j} = u_{4,j} = 0$ or $j = 0, 1, \dots, 5$, and the initial values of u are $u_{1,0} = u(1/4,0) = \sin \pi/4 \approx 0.7071$, $u_{2,0} = u(1/2,0) = \sin \pi/2 = 1$, $u_{3,0} = u(3/4,0) = \sin 3\pi/4 \approx 0.7071$. From equation (6) in the text we have

$$u_{i,1} = 0.8(u_{i+1,0} + u_{i-1,0}) + 0.84u_{i,0} + 0.1(0).$$

Then $u_{1,1} \approx 0.6740$, $u_{2,1} = 0.9531$, $u_{3,1} = 0.6740$. From equation (3) in the text we have for $j = 1, 2, 3, \dots$

$$u_{i,j+1} = 0.16u_{i+1,j} + 2(0.84)u_{i,j} + 0.16u_{i-1,j} - u_{i,j-1}.$$

The results of the calculations are given in the table.

TIME	x=0.25	x=0.50	x=0.75
0.0	0.7071	1.0000	0.7071
0.1	0.6740	0.9531	0.6740
0.2	0.5777	0.8169	0.5777
0.3	0.4272	0.6042	0.4272
0.4	0.2367	0.3348	0.2367
0.5	0.0241	0.0340	0.0241

(c)

i,j	approx	exact	error
1,1	0.6740	0.6725	0.0015
1,2	0.5777	0.5721	0.0056
1,3	0.4272	0.4156	0.0116
1,4	0.2367	0.2185	0.0182
1,5	0.0241	0.0000	0.0241
2,1	0.9531	0.9511	0.0021
2,2	0.8169	0.8090	0.0079
2,3	0.6042	0.5878	0.0164
2,4	0.3348	0.3090	0.0258
2,5	0.0340	0.0000	0.0340
3,1	0.6740	0.6725	0.0015
3,2	0.5777	0.5721	0.0056
3,3	0.4272	0.4156	0.0116
3,4	0.2367	0.2185	0.0182
3,5	0.0241	0.0000	0.0241

3. (a) Identifying $h = 1/5$ and $k = 0.5/10 = 0.05$ we see that $\lambda = 0.25$.

TIME	X=0.2	X=0.4	X=0.6	X=0.8
0.00	0.5878	0.9511	0.9511	0.5878
0.05	0.5808	0.9397	0.9397	0.5808
0.10	0.5599	0.9059	0.9059	0.5599
0.15	0.5256	0.8505	0.8505	0.5256
0.20	0.4788	0.7748	0.7748	0.4788
0.25	0.4206	0.6806	0.6806	0.4206
0.30	0.3524	0.5701	0.5701	0.3524
0.35	0.2757	0.4460	0.4460	0.2757
0.40	0.1924	0.3113	0.3113	0.1924
0.45	0.1046	0.1692	0.1692	0.1046
0.50	0.0142	0.0230	0.0230	0.0142

(b) Identifying $h = 1/5$ and $k = 0.5/20 = 0.025$ we see that $\lambda = 0.125$.

TIME	X=0.2	X=0.4	X=0.6	X=0.8
0.00	0.5878	0.9511	0.9511	0.5878
0.03	0.5860	0.9482	0.9482	0.5860
0.05	0.5808	0.9397	0.9397	0.5808
0.08	0.5721	0.9256	0.9256	0.5721
0.10	0.5599	0.9060	0.9060	0.5599
0.13	0.5445	0.8809	0.8809	0.5445
0.15	0.5257	0.8507	0.8507	0.5257
0.18	0.5039	0.8153	0.8153	0.5039
0.20	0.4790	0.7750	0.7750	0.4790
0.23	0.4513	0.7302	0.7302	0.4513
0.25	0.4209	0.6810	0.6810	0.4209
0.28	0.3879	0.6277	0.6277	0.3879
0.30	0.3527	0.5706	0.5706	0.3527
0.33	0.3153	0.5102	0.5102	0.3153
0.35	0.2761	0.4467	0.4467	0.2761
0.38	0.2352	0.3806	0.3806	0.2352
0.40	0.1929	0.3122	0.3122	0.1929
0.43	0.1495	0.2419	0.2419	0.1495
0.45	0.1052	0.1701	0.1701	0.1052
0.48	0.0602	0.0974	0.0974	0.0602
0.50	0.0149	0.0241	0.0241	0.0149

4. We have $\lambda = 1$. The initial values of n are $u_{1,0} = u(0.2, 0) = 0.16$, $u_{2,0} = u(0.4) = 0.24$, $u_{3,0} = 0.24$, and $u_{4,0} = 0.16$. From equation (6) in the text we have

$$u_{i,1} = \frac{1}{2}(u_{i+1,0} + u_{i-1,0}) + 0u_{i,0} + k \cdot 0 = \frac{1}{2}(u_{i+1,0} + u_{i-1,0}).$$

Then, using $u_{0,0} = u_{5,0} = 0$, we find $u_{1,1} = 0.12$, $u_{2,1} = 0.2$, $u_{3,1} = 0.2$, and $u_{4,1} = 0.12$.

5. We identify $c = 24944.4$, $k = 0.00020045$ seconds $= 0.20045$ milliseconds, and $\lambda = 0.5$. Time in the table is expressed in milliseconds.

TIME	X=10	X=20	X=30	X=40	X=50
0.00000	0.1000	0.2000	0.3000	0.2000	0.1000
0.20045	0.1000	0.2000	0.2750	0.2000	0.1000
0.40089	0.1000	0.1938	0.2125	0.1938	0.1000
0.60134	0.0984	0.1688	0.1406	0.1688	0.0984
0.80178	0.0898	0.1191	0.0828	0.1191	0.0898
1.00223	0.0661	0.0531	0.0432	0.0531	0.0661
1.20268	0.0226	-0.0121	0.0085	-0.0121	0.0226
1.40312	-0.0352	-0.0635	-0.0365	-0.0635	-0.0352
1.60357	-0.0913	-0.1011	-0.0950	-0.1011	-0.0913
1.80401	-0.1271	-0.1347	-0.1566	-0.1347	-0.1271
2.00446	-0.1329	-0.1719	-0.2072	-0.1719	-0.1329
2.20491	-0.1153	-0.2081	-0.2402	-0.2081	-0.1153
2.40535	-0.0920	-0.2292	-0.2571	-0.2292	-0.0920
2.60580	-0.0801	-0.2230	-0.2601	-0.2230	-0.0801
2.80624	-0.0838	-0.1903	-0.2445	-0.1903	-0.0838
3.00669	-0.0932	-0.1445	-0.2018	-0.1445	-0.0932
3.20713	-0.0921	-0.1003	-0.1305	-0.1003	-0.0921
3.40758	-0.0701	-0.0615	-0.0440	-0.0615	-0.0701
3.60803	-0.0284	-0.0205	0.0336	-0.0205	-0.0284
3.80847	0.0224	0.0321	0.0842	0.0321	0.0224
4.00892	0.0700	0.0953	0.1087	0.0953	0.0700
4.20936	0.1064	0.1555	0.1265	0.1555	0.1064
4.40981	0.1285	0.1962	0.1588	0.1962	0.1285
4.61026	0.1354	0.2106	0.2098	0.2106	0.1354
4.81070	0.1273	0.2060	0.2612	0.2060	0.1273
5.01115	0.1070	0.1955	0.2851	0.1955	0.1070
5.21159	0.0821	0.1853	0.2641	0.1853	0.0821
5.41204	0.0625	0.1689	0.2038	0.1689	0.0625
5.61249	0.0539	0.1347	0.1260	0.1347	0.0539
5.81293	0.0520	0.0781	0.0526	0.0781	0.0520
6.01338	0.0436	0.0086	-0.0080	0.0086	0.0436
6.21382	0.0156	-0.0564	-0.0604	-0.0564	0.0156
6.41427	-0.0343	-0.1043	-0.1107	-0.1043	-0.0343
6.61472	-0.0931	-0.1364	-0.1578	-0.1364	-0.0931
6.81516	-0.1395	-0.1630	-0.1942	-0.1630	-0.1395
7.01561	-0.1568	-0.1915	-0.2150	-0.1915	-0.1568
7.21605	-0.1436	-0.2173	-0.2240	-0.2173	-0.1436
7.41650	-0.1129	-0.2263	-0.2297	-0.2263	-0.1129
7.61695	-0.0824	-0.2078	-0.2336	-0.2078	-0.0824
7.81739	-0.0625	-0.1644	-0.2247	-0.1644	-0.0625
8.01784	-0.0526	-0.1106	-0.1856	-0.1106	-0.0526
8.21828	-0.0440	-0.0611	-0.1091	-0.0611	-0.0440
8.41873	-0.0287	-0.0192	-0.0085	-0.0192	-0.0287
8.61918	-0.0038	0.0229	0.0867	0.0229	-0.0038
8.81962	0.0287	0.0743	0.1500	0.0743	0.0287
9.02007	0.0654	0.1332	0.1755	0.1332	0.0654
9.22051	0.1027	0.1858	0.1799	0.1858	0.1027
9.42096	0.1352	0.2160	0.1872	0.2160	0.1352
9.62140	0.1540	0.2189	0.2089	0.2189	0.1540
9.82185	0.1506	0.2030	0.2356	0.2030	0.1506
10.02230	0.1226	0.1822	0.2461	0.1822	0.1226

6. We identify $c = 24944.4$, $k = 0.00010022$ seconds $= 0.10022$ milliseconds, and $\lambda = 0.25$. Time in the table is expressed in milliseconds.

TIME	X=10	X=20	X=30	X=40	X=50
0.00000	0.2000	0.2667	0.2000	0.1333	0.0667
0.10022	0.1958	0.2625	0.2000	0.1333	0.0667
0.20045	0.1836	0.2503	0.1997	0.1333	0.0667
0.30067	0.1640	0.2307	0.1985	0.1333	0.0667
0.40089	0.1384	0.2050	0.1952	0.1332	0.0667
0.50111	0.1083	0.1744	0.1886	0.1328	0.0667
0.60134	0.0755	0.1407	0.1777	0.1318	0.0666
0.70156	0.0421	0.1052	0.1615	0.1295	0.0665
0.80178	0.0100	0.0692	0.1399	0.1253	0.0661
0.90201	-0.0190	0.0340	0.1129	0.1184	0.0654
1.00223	-0.0435	0.0004	0.0813	0.1077	0.0638
1.10245	-0.0626	-0.0309	0.0464	0.0927	0.0610
1.20268	-0.0758	-0.0593	0.0095	0.0728	0.0564
1.30290	-0.0832	-0.0845	-0.0278	0.0479	0.0493
1.40312	-0.0855	-0.1060	-0.0639	0.0184	0.0390
1.50334	-0.0837	-0.1237	-0.0974	-0.0150	0.0250
1.60357	-0.0792	-0.1371	-0.1275	-0.0511	0.0069
1.70379	-0.0734	-0.1464	-0.1533	-0.0882	-0.0152
1.80401	-0.0675	-0.1515	-0.1747	-0.1249	-0.0410
1.90424	-0.0627	-0.1528	-0.1915	-0.1595	-0.0694
2.00446	-0.0596	-0.1509	-0.2039	-0.1904	-0.0991
2.10468	-0.0585	-0.1467	-0.2122	-0.2165	-0.1283
2.20491	-0.0592	-0.1410	-0.2166	-0.2368	-0.1551
2.30513	-0.0614	-0.1349	-0.2175	-0.2507	-0.1772
2.40535	-0.0643	-0.1294	-0.2154	-0.2579	-0.1929
2.50557	-0.0672	-0.1251	-0.2105	-0.2585	-0.2005
2.60580	-0.0696	-0.1227	-0.2033	-0.2524	-0.1993
2.70602	-0.0709	-0.1219	-0.1942	-0.2399	-0.1889
2.80624	-0.0710	-0.1225	-0.1833	-0.2214	-0.1699
2.90647	-0.0699	-0.1236	-0.1711	-0.1972	-0.1435
3.00669	-0.0678	-0.1244	-0.1575	-0.1681	-0.1115
3.10691	-0.0649	-0.1237	-0.1425	-0.1348	-0.0761
3.20713	-0.0617	-0.1205	-0.1258	-0.0983	-0.0395
3.30736	-0.0583	-0.1139	-0.1071	-0.0598	-0.0042
3.40758	-0.0547	-0.1035	-0.0859	-0.0209	0.0279
3.50780	-0.0508	-0.0889	-0.0617	0.0171	0.0552
3.60803	-0.0460	-0.0702	-0.0343	0.0525	0.0767
3.70825	-0.0399	-0.0478	-0.0037	0.0840	0.0919
3.80847	-0.0318	-0.0221	0.0297	0.1106	0.1008
3.90870	-0.0211	0.0062	0.0648	0.1314	0.1041
4.00892	-0.0074	0.0365	0.1005	0.1464	0.1025
4.10914	0.0095	0.0680	0.1350	0.1558	0.0973
4.20936	0.0295	0.1000	0.1666	0.1602	0.0897
4.30959	0.0521	0.1318	0.1937	0.1606	0.0808
4.40981	0.0764	0.1625	0.2148	0.1581	0.0719
4.51003	0.1013	0.1911	0.2291	0.1538	0.0639
4.61026	0.1254	0.2164	0.2364	0.1485	0.0575
4.71048	0.1475	0.2373	0.2369	0.1431	0.0532
4.81070	0.1659	0.2526	0.2315	0.1379	0.0512
4.91093	0.1794	0.2611	0.2217	0.1331	0.0514
5.01115	0.1867	0.2620	0.2087	0.1288	0.0535

Chapter 15 in Review

1. Using the figure we obtain the system

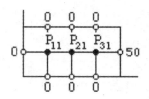

$$u_{21} + 0 + 0 + 0 - 4u_{11} = 0$$

$$u_{31} + 0 + u_{11} + 0 - 4u_{21} = 0$$

$$50 + 0 + u_{21} + 0 - 4u_{31} = 0.$$

By Gauss-Elimination then,

$$\begin{bmatrix} -4 & 1 & 0 & | & 0 \\ 1 & -4 & 1 & | & 0 \\ 0 & 1 & -4 & | & -50 \end{bmatrix} \xrightarrow[\text{operations}]{\text{row}} \begin{bmatrix} 1 & -4 & 1 & | & 0 \\ 0 & 1 & -4 & | & -50 \\ 0 & 0 & 1 & | & 13.3928 \end{bmatrix}.$$

The solution is $u_{11} = 0.8929$, $u_{21} = 3.5714$, $u_{31} = 13.3928$.

2. By symmetry we observe that $u_{i,1} = u_{i,3}$ for $i = 1, 2, \ldots, 7$. We then use Gauss-Seidel iteration with an initial guess of 7.5 for all variables to solve the system

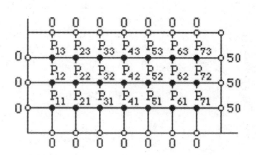

$$u_{11} = 0.25u_{21} + 0.25u_{12}$$

$$u_{21} = 0.25u_{31} + 0.25u_{22} + 0.25u_{11}$$

$$u_{31} = 0.25u_{41} + 0.25u_{32} + 0.25u_{21}$$

$$u_{41} = 0.25u_{51} + 0.25u_{42} + 0.25u_{31}$$

$$u_{51} = 0.25u_{61} + 0.25u_{52} + 0.25u_{41}$$

$$u_{61} = 0.25u_{71} + 0.25u_{62} + 0.25u_{51}$$

$$u_{71} = 12.5 + 0.25u_{72} + 0.25u_{61}$$

$$u_{12} = 0.25u_{22} + 0.5u_{11}$$

$$u_{22} = 0.25u_{32} + 0.5u_{21} + 0.25u_{12}$$

$$u_{32} = 0.25u_{42} + 0.5u_{31} + 0.25u_{22}$$

$$u_{42} = 0.25u_{52} + 0.5u_{41} + 0.25u_{32}$$

$$u_{52} = 0.25u_{62} + 0.5u_{51} + 0.25u_{42}$$

$$u_{62} = 0.25u_{72} + 0.5u_{61} + 0.25u_{52}$$

$$u_{72} = 12.5 + 0.5u_{71} + 0.25u_{62}.$$

After 30 iterations we obtain $u_{11} = u_{13} = 0.1765$, $u_{21} = u_{23} = 0.4566$, $u_{31} = u_{33} = 1.0051$, $u_{41} = u_{43} = 2.1479$, $u_{51} = u_{53} = 4.5766$, $u_{61} = u_{63} = 9.8316$, $u_{71} = u_{73} = 21.6051$, $u_{12} = 0.2494$, $u_{22} = 0.6447$, $u_{32} = 1.4162$, $u_{42} = 3.0097$, $u_{52} = 6.3269$, $u_{62} = 13.1447$, $u_{72} = 26.5887$.

3. (a)

TIME	X=0.0	X=0.2	X=0.4	X=0.6	X=0.8	X=1.0
0.00	0.0000	0.2000	0.4000	0.6000	0.8000	0.0000
0.01	0.0000	0.2000	0.4000	0.6000	0.5500	0.0000
0.02	0.0000	0.2000	0.4000	0.5375	0.4250	0.0000
0.03	0.0000	0.2000	0.3844	0.4750	0.3469	0.0000
0.04	0.0000	0.1961	0.3609	0.4203	0.2922	0.0000
0.05	0.0000	0.1883	0.3346	0.3734	0.2512	0.0000

(b)

TIME	X=0.0	X=0.2	X=0.4	X=0.6	X=0.8	X=1.0
0.00	0.0000	0.2000	0.4000	0.6000	0.8000	0.0000
0.01	0.0000	0.2000	0.4000	0.6000	0.8000	0.0000
0.02	0.0000	0.2000	0.4000	0.6000	0.5500	0.0000
0.03	0.0000	0.2000	0.4000	0.5375	0.4250	0.0000
0.04	0.0000	0.2000	0.3844	0.4750	0.3469	0.0000
0.05	0.0000	0.1961	0.3609	0.4203	0.2922	0.0000

(c) The table in part (b) is the same as the table in part (a) shifted downward one row.

Appendices

Gamma Function

1. **(a)** $\Gamma(5) = \Gamma(4+1) = 4! = 24$

   (b) $\Gamma(7) = \Gamma(6+1) = 6! = 720$

   (c) Using Example 1 in the text,

   $$-2\sqrt{\pi} = \Gamma\left(-\frac{1}{2}\right) = \Gamma\left(-\frac{3}{2}+1\right) = -\frac{3}{2}\Gamma\left(-\frac{3}{2}\right).$$

   Thus, $\Gamma(-3/2) = 4\sqrt{\pi}/3$.

   (d) Using (c)

   $$\frac{4\sqrt{\pi}}{3} = \Gamma\left(-\frac{3}{2}\right) = \Gamma\left(-\frac{5}{2}+1\right) = -\frac{5}{2}\Gamma\left(-\frac{5}{2}\right).$$

   Thus $\Gamma(-5/2) = -8\sqrt{\pi}/15$.

2. If $t = x^5$, then $dt = 5x^4\,dx$ and $x^5\,dx = \frac{1}{5}t^{1/5}\,dt$. Now

   $$\int_0^\infty x^5 e^{-x^5}\,dx = \int_0^\infty \frac{1}{5}t^{1/5}e^{-t}\,dt = \frac{1}{5}\int_0^\infty t^{1/5}e^{-t}\,dt$$

   $$= \frac{1}{5}\Gamma\left(\frac{6}{5}\right) = \frac{1}{5}(0.92) = 0.184.$$

3. If $t = x^3$, then $dt = 3x^2\,dx$ and $x^4\,dx = \frac{1}{3}t^{2/3}\,dt$. Now

   $$\int_0^\infty x^4 e^{-x^3}\,dx = \int_0^\infty \frac{1}{3}t^{2/3}e^{-t}\,dt = \frac{1}{3}\int_0^\infty t^{2/3}e^{-t}\,dt$$

   $$= \frac{1}{3}\Gamma\left(\frac{5}{3}\right) = \frac{1}{3}(0.89) \approx 0.297.$$

4. If $t = -\ln x = \ln \dfrac{1}{x}$ then $dt = -\dfrac{1}{x} dx$. Also $e^t = \dfrac{1}{x}$, so $x = e^{-t}$ and $dx = -x\,dt = -e^{-t}\,dt$. Thus

$$\int_0^1 x^3 \left(\ln \frac{1}{x}\right)^3 dx = \int_\infty^0 (e^{-t})^3 t^3 (-e^{-t})\,dt$$

$$= \int_0^\infty t^3 e^{-4t}\,dt$$

$$= \int_0^\infty \left(\frac{1}{4}u\right)^3 e^{-u} \left(\frac{1}{4}du\right) \qquad [u = 4t]$$

$$= \frac{1}{256} \int_0^\infty u^3 e^{-u} du = \frac{1}{256}\Gamma(4)$$

$$= \frac{1}{256}(3!) = \frac{3}{128}.$$

5. Since $e^{-t} \geq e^{-1}$ for $0 \leq t \leq 1$,

$$\Gamma(x) = \int_0^\infty t^{x-1} e^{-t} dt > \int_0^1 t^{x-1} e^{-t} dt \geq e^{-1} \int_0^1 t^{x-1} dt$$

$$= \frac{1}{e}\left(\frac{1}{x}t^x\right)\Big|_0^1 = \frac{1}{ex}$$

for $x > 0$. As $x \to 0^+$, we see that $\Gamma(x) \to \infty$.

6. For $x > 0$ we integrate by parts:

$$\Gamma(x+1) = \int_0^\infty t^x e^{-t} dt$$

$u = t^x$	$dv = e^{-t}\,dt$
$du = xt^{x-1}\,dt$	$v = -e^{-t}$

$$= -t^x e^{-t}\Big|_0^\infty - \int_0^\infty xt^{x-1}(-e^{-t})\,dt$$

$$= x \int_0^\infty t^{x-1} e^{-t} dt$$

$$= x\Gamma(x).$$

Appendix II   Matrices

1. (a) $\mathbf{A} + \mathbf{B} = \begin{pmatrix} 4-2 & 5+6 \\ -6+8 & 9-10 \end{pmatrix} = \begin{pmatrix} 2 & 11 \\ 2 & -1 \end{pmatrix}$

   (b) $\mathbf{B} - \mathbf{A} = \begin{pmatrix} -2-4 & 6-5 \\ 8+6 & -10-9 \end{pmatrix} = \begin{pmatrix} -6 & 1 \\ 14 & -19 \end{pmatrix}$

   (c) $2\mathbf{A} + 3\mathbf{B} = \begin{pmatrix} 8 & 10 \\ -12 & 18 \end{pmatrix} + \begin{pmatrix} -6 & 18 \\ 24 & -30 \end{pmatrix} = \begin{pmatrix} 2 & 28 \\ 12 & -12 \end{pmatrix}$

2. (a) $\mathbf{A} - \mathbf{B} = \begin{pmatrix} -2-3 & 0+1 \\ 4-0 & 1-2 \\ 7+4 & 3+2 \end{pmatrix} = \begin{pmatrix} -5 & 1 \\ 4 & -1 \\ 11 & 5 \end{pmatrix}$

   (b) $\mathbf{B} - \mathbf{A} = \begin{pmatrix} 3+2 & -1-0 \\ 0-4 & 2-1 \\ -4-7 & -2-3 \end{pmatrix} = \begin{pmatrix} 5 & -1 \\ -4 & 1 \\ -11 & -5 \end{pmatrix}$

   (c) $2(\mathbf{A} + \mathbf{B}) = 2\begin{pmatrix} 1 & -1 \\ 4 & 3 \\ 3 & 1 \end{pmatrix} = \begin{pmatrix} 2 & -2 \\ 8 & 6 \\ 6 & 2 \end{pmatrix}$

3. (a) $\mathbf{AB} = \begin{pmatrix} -2-9 & 12-6 \\ 5+12 & -30+8 \end{pmatrix} = \begin{pmatrix} -11 & 6 \\ 17 & -22 \end{pmatrix}$

   (b) $\mathbf{BA} = \begin{pmatrix} -2-30 & 3+24 \\ 6-10 & -9+8 \end{pmatrix} = \begin{pmatrix} -32 & 27 \\ -4 & -1 \end{pmatrix}$

   (c) $\mathbf{A}^2 = \begin{pmatrix} 4+15 & -6-12 \\ -10-20 & 15+16 \end{pmatrix} = \begin{pmatrix} 19 & -18 \\ -30 & 31 \end{pmatrix}$

   (d) $\mathbf{B}^2 = \begin{pmatrix} 1+18 & -6+12 \\ -3+6 & 18+4 \end{pmatrix} = \begin{pmatrix} 19 & 6 \\ 3 & 22 \end{pmatrix}$

4. (a) $\mathbf{AB} = \begin{pmatrix} -4+4 & 6-12 & -3+8 \\ -20+10 & 30-30 & -15+20 \\ -32+12 & 48-36 & -24+24 \end{pmatrix} = \begin{pmatrix} 0 & -6 & 5 \\ -10 & 0 & 5 \\ -20 & 12 & 0 \end{pmatrix}$

(b) $\mathbf{BA} = \begin{pmatrix} -4+30-24 & -16+60-36 \\ 1-15+16 & 4-30+24 \end{pmatrix} = \begin{pmatrix} 2 & 8 \\ 2 & -2 \end{pmatrix}$

5. (a) $\mathbf{BC} = \begin{pmatrix} 9 & 24 \\ 3 & 8 \end{pmatrix}$

(b) $\mathbf{A(BC)} = \begin{pmatrix} 1 & -2 \\ -2 & 4 \end{pmatrix}\begin{pmatrix} 9 & 24 \\ 3 & 8 \end{pmatrix} = \begin{pmatrix} 3 & 8 \\ -6 & -16 \end{pmatrix}$

(c) $\mathbf{C(BA)} = \begin{pmatrix} 0 & 2 \\ 3 & 4 \end{pmatrix}\begin{pmatrix} 0 & 0 \\ 0 & 0 \end{pmatrix} = \begin{pmatrix} 0 & 0 \\ 0 & 0 \end{pmatrix}$

(d) $\mathbf{A(B+C)} = \begin{pmatrix} 1 & -2 \\ -2 & 4 \end{pmatrix}\begin{pmatrix} 6 & 5 \\ 5 & 5 \end{pmatrix} = \begin{pmatrix} -4 & -5 \\ 8 & 10 \end{pmatrix}$

6. (a) $\mathbf{AB} = (5 \quad -6 \quad 7)\begin{pmatrix} 3 \\ 4 \\ -1 \end{pmatrix} = (-16)$

(b) $\mathbf{BA} = \begin{pmatrix} 3 \\ 4 \\ -1 \end{pmatrix}(5 \quad -6 \quad 7) = \begin{pmatrix} 15 & -18 & 21 \\ 20 & -24 & 28 \\ -5 & 6 & -7 \end{pmatrix}$

(c) $\mathbf{(BA)C} = \begin{pmatrix} 15 & -18 & 21 \\ 20 & -24 & 28 \\ -5 & 6 & -7 \end{pmatrix}\begin{pmatrix} 1 & 2 & 4 \\ 0 & 1 & -1 \\ 3 & 2 & 1 \end{pmatrix} = \begin{pmatrix} 78 & 54 & 99 \\ 104 & 72 & 132 \\ -26 & -18 & -33 \end{pmatrix}$

(d) Since $\mathbf{AB}$ is 1×1 and $\mathbf{C}$ is 3×3 the product $\mathbf{(AB)C}$ is not defined.

7. (a) $\mathbf{A}^T\mathbf{A} = (4 \quad 8 \quad -10)\begin{pmatrix} 4 \\ 8 \\ -10 \end{pmatrix} = (180)$

(b) $\mathbf{B}^T\mathbf{B} = \begin{pmatrix} 2 \\ 4 \\ 5 \end{pmatrix}(2 \quad 4 \quad 5) = \begin{pmatrix} 4 & 8 & 10 \\ 8 & 16 & 20 \\ 10 & 20 & 25 \end{pmatrix}$

(c) $\mathbf{A} + \mathbf{B}^T = \begin{pmatrix} 4 \\ 8 \\ -10 \end{pmatrix} + \begin{pmatrix} 2 \\ 4 \\ 5 \end{pmatrix} = \begin{pmatrix} 6 \\ 12 \\ -5 \end{pmatrix}$

8. (a) $\mathbf{A} + \mathbf{B}^T = \begin{pmatrix} 1 & 2 \\ 2 & 4 \end{pmatrix} + \begin{pmatrix} -2 & 5 \\ 3 & 7 \end{pmatrix} = \begin{pmatrix} -1 & 7 \\ 5 & 11 \end{pmatrix}$

(b) $2\mathbf{A}^T - \mathbf{B}^T = \begin{pmatrix} 2 & 4 \\ 4 & 8 \end{pmatrix} - \begin{pmatrix} -2 & 5 \\ 3 & 7 \end{pmatrix} = \begin{pmatrix} 4 & -1 \\ 1 & 1 \end{pmatrix}$

(c) $\mathbf{A}^T(\mathbf{A} - \mathbf{B}) = \begin{pmatrix} 1 & 2 \\ 2 & 4 \end{pmatrix}\begin{pmatrix} 3 & -1 \\ -3 & -3 \end{pmatrix} = \begin{pmatrix} -3 & -7 \\ -6 & -14 \end{pmatrix}$

9. (a) $(\mathbf{AB})^T = \begin{pmatrix} 7 & 10 \\ 38 & 75 \end{pmatrix}^T = \begin{pmatrix} 7 & 38 \\ 10 & 75 \end{pmatrix}$

(b) $\mathbf{B}^T\mathbf{A}^T = \begin{pmatrix} 5 & -2 \\ 10 & -5 \end{pmatrix}\begin{pmatrix} 3 & 8 \\ 4 & 1 \end{pmatrix} = \begin{pmatrix} 7 & 38 \\ 10 & 75 \end{pmatrix}$

10. (a) $\mathbf{A}^T + \mathbf{B}^T = \begin{pmatrix} 5 & -4 \\ 9 & 6 \end{pmatrix} + \begin{pmatrix} -3 & -7 \\ 11 & 2 \end{pmatrix} = \begin{pmatrix} 2 & -11 \\ 20 & 8 \end{pmatrix}$

(b) $(\mathbf{A} + \mathbf{B})^T = \begin{pmatrix} 2 & 20 \\ -11 & 8 \end{pmatrix}^T = \begin{pmatrix} 2 & -11 \\ 20 & 8 \end{pmatrix}$

11. $\begin{pmatrix} -4 \\ 8 \end{pmatrix} - \begin{pmatrix} 4 \\ 16 \end{pmatrix} + \begin{pmatrix} -6 \\ 9 \end{pmatrix} = \begin{pmatrix} -14 \\ 1 \end{pmatrix}$

12. $\begin{pmatrix} 6t \\ 3t^2 \\ -3t \end{pmatrix} + \begin{pmatrix} -t+1 \\ -t^2+t \\ 3t-3 \end{pmatrix} - \begin{pmatrix} 6t \\ 8 \\ -10t \end{pmatrix} = \begin{pmatrix} -t+1 \\ 2t^2+t-8 \\ 10t-3 \end{pmatrix}$

13. $\begin{pmatrix} -19 \\ 18 \end{pmatrix} - \begin{pmatrix} 19 \\ 20 \end{pmatrix} = \begin{pmatrix} -38 \\ -2 \end{pmatrix}$

14. $\begin{pmatrix} -9t+3 \\ 13t-5 \\ -6t+4 \end{pmatrix} + \begin{pmatrix} -t \\ 1 \\ 4 \end{pmatrix} - \begin{pmatrix} 2 \\ 8 \\ -6 \end{pmatrix} = \begin{pmatrix} -10t+1 \\ 13t-12 \\ -6t+14 \end{pmatrix}$

15. Since det $\mathbf{A} = 0$, $\mathbf{A}$ is singular.

16. Since det $\mathbf{A} = 3$, $\mathbf{A}$ is nonsingular.

$$\mathbf{A}^{-1} = \frac{1}{3}\begin{pmatrix} 4 & -5 \\ -1 & 2 \end{pmatrix}$$

17. Since det $\mathbf{A} = 4$, $\mathbf{A}$ is nonsingular.

$$\mathbf{A}^{-1} = \frac{1}{4}\begin{pmatrix} -5 & -8 \\ 3 & 4 \end{pmatrix}$$

18. Since det $\mathbf{A} = -6$, $\mathbf{A}$ is nonsingular.

$$\mathbf{A}^{-1} = -\frac{1}{6}\begin{pmatrix} 2 & -10 \\ -2 & 7 \end{pmatrix}$$

19. Since det $\mathbf{A} = 2$, $\mathbf{A}$ is nonsingular. The cofactors are

$$\begin{array}{lll} A_{11} = 0 & A_{12} = 2 & A_{13} = -4 \\ A_{21} = -1 & A_{22} = 2 & A_{23} = -3 \\ A_{31} = 1 & A_{32} = -2 & A_{33} = 5. \end{array}$$

Then

$$\mathbf{A}^{-1} = \frac{1}{2}\begin{pmatrix} 0 & 2 & -4 \\ -1 & 2 & -3 \\ 1 & -2 & 5 \end{pmatrix}^{T} = \frac{1}{2}\begin{pmatrix} 0 & -1 & 1 \\ 2 & 2 & -2 \\ -4 & -3 & 5 \end{pmatrix}.$$

20. Since det $\mathbf{A} = 27$, $\mathbf{A}$ is nonsingular. The cofactors are

$$\begin{array}{lll} A_{11} = -1 & A_{12} = 4 & A_{13} = 22 \\ A_{21} = 7 & A_{22} = -1 & A_{23} = -19 \\ A_{31} = -1 & A_{32} = 4 & A_{33} = -5. \end{array}$$

Then

$$\mathbf{A}^{-1} = \frac{1}{27}\begin{pmatrix} -1 & 4 & 22 \\ 7 & -1 & -19 \\ -1 & 4 & -5 \end{pmatrix}^{T} = \frac{1}{27}\begin{pmatrix} -1 & 7 & -1 \\ 4 & -1 & 4 \\ 22 & -19 & -5 \end{pmatrix}.$$

21. Since det $\mathbf{A} = -9$, $\mathbf{A}$ is nonsingular. The cofactors are

$$\begin{array}{lll} A_{11} = -2 & A_{12} = -13 & A_{13} = 8 \\ A_{21} = -2 & A_{22} = 5 & A_{23} = -1 \\ A_{31} = -1 & A_{32} = 7 & A_{33} = -5. \end{array}$$

Then

$$\mathbf{A}^{-1} = -\frac{1}{9}\begin{pmatrix} -2 & -13 & 8 \\ -2 & 5 & -1 \\ -1 & 7 & -5 \end{pmatrix}^{T} = -\frac{1}{9}\begin{pmatrix} -2 & -2 & -1 \\ -13 & 5 & 7 \\ 8 & -1 & -5 \end{pmatrix}.$$

22. Since det $\mathbf{A} = 0$, $\mathbf{A}$ is singular.

23. Since $\det \mathbf{A}(t) = 2e^{3t} \neq 0$, $\mathbf{A}$ is nonsingular.

$$\mathbf{A}^{-1} = \frac{1}{2}e^{-3t}\begin{pmatrix} 3e^{4t} & -e^{4t} \\ -4e^{-t} & 2e^{-t} \end{pmatrix}$$

24. Since $\det \mathbf{A}(t) = 2e^{2t} \neq 0$, $\mathbf{A}$ is nonsingular.

$$\mathbf{A}^{-1} = \frac{1}{2}e^{-2t}\begin{pmatrix} e^{t}\sin t & 2e^{t}\cos t \\ -e^{t}\cos t & 2e^{t}\sin t \end{pmatrix}$$

25. $\dfrac{d\mathbf{X}}{dt} = \begin{pmatrix} -5e^{-t} \\ -2e^{-t} \\ 7e^{-t} \end{pmatrix}$

26. $\dfrac{d\mathbf{X}}{dt} = \begin{pmatrix} \cos 2t + 8\sin 2t \\ -6\cos 2t - 10\sin 2t \end{pmatrix}$

27. $\mathbf{X} = \begin{pmatrix} 2e^{2t} + 8e^{-3t} \\ -2e^{2t} + 4e^{-3t} \end{pmatrix}$ so that $\dfrac{d\mathbf{X}}{dt} = \begin{pmatrix} 4e^{2t} - 24e^{-3t} \\ -4e^{2t} - 12e^{-3t} \end{pmatrix}$.

28. $\dfrac{d\mathbf{X}}{dt} = \begin{pmatrix} 10te^{2t} + 5e^{2t} \\ 3t\cos 3t + \sin 3t \end{pmatrix}$

29. (a) $\dfrac{d\mathbf{A}}{dt} = \begin{pmatrix} 4e^{4t} & -\pi\sin \pi t \\ 2 & 6t \end{pmatrix}$

(b) $\displaystyle\int_0^2 \mathbf{A}(t)\,dt = \begin{pmatrix} \frac{1}{4}e^{4t} & \frac{1}{\pi}\sin \pi t \\ t^2 & t^3 - t \end{pmatrix}\Bigg|_{t=0}^{t=2} = \begin{pmatrix} \frac{1}{4}e^8 - \frac{1}{4} & 0 \\ 4 & 6 \end{pmatrix}$

(c) $\displaystyle\int_0^t \mathbf{A}(s)\,ds = \begin{pmatrix} \frac{1}{4}e^{4s} & \frac{1}{\pi}\sin \pi s \\ s^2 & s^3 - s \end{pmatrix}\Bigg|_{s=0}^{s=t} = \begin{pmatrix} \frac{1}{4}e^{4t} - \frac{1}{4} & \frac{1}{\pi}\sin \pi t \\ t^2 & t^3 - t \end{pmatrix}$

30. (a) $\dfrac{d\mathbf{A}}{dt} = \begin{pmatrix} -2t/(t^2+1)^2 & 3 \\ 2t & 1 \end{pmatrix}$

(b) $\dfrac{d\mathbf{B}}{dt} = \begin{pmatrix} 6 & 0 \\ -1/t^2 & 4 \end{pmatrix}$

(c) $\displaystyle\int_0^1 \mathbf{A}(t)\,dt = \begin{pmatrix} \tan^{-1}t & \frac{3}{2}t^2 \\ \frac{1}{3}t^3 & \frac{1}{2}t^2 \end{pmatrix}\Bigg|_{t=0}^{t=1} = \begin{pmatrix} \frac{\pi}{4} & \frac{3}{2} \\ \frac{1}{3} & \frac{1}{2} \end{pmatrix}$

(d) $\displaystyle\int_1^2 \mathbf{B}(t)\,dt = \begin{pmatrix} 3t^2 & 2t \\ \ln t & 2t^2 \end{pmatrix}\Bigg|_{t=1}^{t=2} = \begin{pmatrix} 9 & 2 \\ \ln 2 & 6 \end{pmatrix}$

(e) $\mathbf{A}(t)\mathbf{B}(t) = \begin{pmatrix} 6t/(t^2+1)+3 & 2/(t^2+1)+12t^2 \\ 6t^3+1 & 2t^2+4t^2 \end{pmatrix}$

(f) $\dfrac{d}{dt}\mathbf{A}(t)\mathbf{B}(t) = \begin{pmatrix} (6-6t^2)/(t^2+1)^2 & -4t/(t^2+1)^2+24t \\ 18t^2 & 12t \end{pmatrix}$

(g) $\displaystyle\int_1^t \mathbf{A}(s)\mathbf{B}(s)\,ds = \int_1^t \begin{pmatrix} 6s/(s^2+1)+3 & 2/(s^2+1)+12s^2 \\ 6s^3+1 & 6s^2 \end{pmatrix} ds$

$\quad = \begin{pmatrix} 3\ln(t^2+1)+3t-3\ln 2-3 & 2\tan^{-1}t+4t^3-\pi/2-4 \\ 3t^4/2+t-5/2 & 2t^3-2 \end{pmatrix}$

31. $\begin{pmatrix} 1 & 1 & -2 & | & 14 \\ 2 & -1 & 1 & | & 0 \\ 6 & 3 & 4 & | & 1 \end{pmatrix} \Longrightarrow \begin{pmatrix} 1 & 1 & -2 & | & 14 \\ 0 & -3 & 5 & | & -28 \\ 0 & 6 & 1 & | & 1 \end{pmatrix} \Longrightarrow \begin{pmatrix} 1 & 0 & -1/3 & | & 14/3 \\ 0 & 1 & -5/3 & | & 28/3 \\ 0 & 0 & 11 & | & -55 \end{pmatrix}$

$\Longrightarrow \begin{pmatrix} 1 & 0 & 0 & | & 3 \\ 0 & 1 & 0 & | & 1 \\ 0 & 0 & 1 & | & -5 \end{pmatrix}$

Thus $x=3$, $y=1$, and $z=-5$.

32. $\begin{pmatrix} 5 & -2 & 4 & | & 10 \\ 1 & 1 & 1 & | & 9 \\ 4 & -3 & 3 & | & 1 \end{pmatrix} \Longrightarrow \begin{pmatrix} 1 & 1 & 1 & | & 9 \\ 0 & -7 & -1 & | & -35 \\ 0 & -7 & -1 & | & -35 \end{pmatrix} \Longrightarrow \begin{pmatrix} 1 & 0 & 6/7 & | & 4 \\ 0 & 1 & 1/7 & | & 5 \\ 0 & 0 & 0 & | & 0 \end{pmatrix}$

Letting $z=t$ we find $y=5-\frac{1}{7}t$, and $x=4-\frac{6}{7}t$.

33. $\begin{pmatrix} 1 & -1 & -5 & | & 7 \\ 5 & 4 & -16 & | & -10 \\ 0 & 1 & 1 & | & -5 \end{pmatrix} \Longrightarrow \begin{pmatrix} 1 & -1 & -5 & | & 7 \\ 0 & 1 & 1 & | & -5 \\ 0 & 9 & 9 & | & -45 \end{pmatrix} \Longrightarrow \begin{pmatrix} 1 & 0 & -4 & | & 2 \\ 0 & 1 & 1 & | & -5 \\ 0 & 0 & 0 & | & 0 \end{pmatrix}$

Letting $z=t$ we find $y=-5-t$, and $x=2+4t$.

34. $\begin{pmatrix} 1 & 1 & -3 & | & 6 \\ 4 & 2 & -1 & | & 7 \\ 3 & 1 & 1 & | & 4 \end{pmatrix} \Longrightarrow \begin{pmatrix} 1 & 1 & -3 & | & 6 \\ 0 & -2 & 11 & | & -17 \\ 0 & -2 & 10 & | & -14 \end{pmatrix} \Longrightarrow \begin{pmatrix} 1 & 0 & 5/2 & | & -5/2 \\ 0 & 1 & -11/2 & | & 17/2 \\ 0 & 0 & -1 & | & 3 \end{pmatrix}$

$$\implies \begin{pmatrix} 1 & 0 & 0 & | & 5 \\ 0 & 1 & 0 & | & -8 \\ 0 & 0 & 1 & | & -3 \end{pmatrix}$$

Thus $x = 5$, $y = -8$, and $z = -3$.

35. $\begin{pmatrix} 2 & 1 & 1 & | & 4 \\ 10 & -2 & 2 & | & -1 \\ 6 & -2 & 4 & | & 8 \end{pmatrix} \implies \begin{pmatrix} 1 & 1/2 & 1/2 & | & 2 \\ 0 & -7 & -3 & | & -21 \\ 0 & -5 & 1 & | & 4 \end{pmatrix} \implies \begin{pmatrix} 1 & 0 & 2/7 & | & 1/2 \\ 0 & 1 & 3/7 & | & 3 \\ 0 & 0 & 22/7 & | & 11 \end{pmatrix}$

$$\implies \begin{pmatrix} 1 & 0 & 0 & | & -1/2 \\ 0 & 1 & 0 & | & 3/2 \\ 0 & 0 & 1 & | & 7/2 \end{pmatrix}$$

Thus $x = -1/2$, $y = 3/2$, and $z = 7/2$.

36. $\begin{pmatrix} 1 & 0 & 2 & | & 8 \\ 1 & 2 & -2 & | & 4 \\ 2 & 5 & -6 & | & 6 \end{pmatrix} \implies \begin{pmatrix} 1 & 0 & 2 & | & 8 \\ 0 & 2 & -4 & | & -4 \\ 0 & 5 & -10 & | & -10 \end{pmatrix} \implies \begin{pmatrix} 1 & 0 & 2 & | & 8 \\ 0 & 1 & -2 & | & -2 \\ 0 & 0 & 0 & | & 0 \end{pmatrix}$

Letting $z = t$ we find $y = -2 + 2t$, and $x = 8 - 2t$.

37. $\begin{pmatrix} 1 & 1 & -1 & -1 & | & -1 \\ 1 & 1 & 1 & 1 & | & 3 \\ 1 & -1 & 1 & -1 & | & 3 \\ 4 & 1 & -2 & 1 & | & 0 \end{pmatrix} \implies \begin{pmatrix} 1 & 1 & -1 & -1 & | & -1 \\ 0 & 0 & 2 & 2 & | & 4 \\ 0 & -2 & 2 & 0 & | & 4 \\ 0 & -3 & 2 & 5 & | & 4 \end{pmatrix} \implies \begin{pmatrix} 1 & 0 & 0 & -1 & | & 1 \\ 0 & 1 & -1 & 0 & | & -2 \\ 0 & 0 & 2 & 2 & | & 4 \\ 0 & 0 & -1 & 5 & | & -2 \end{pmatrix}$

$$\implies \begin{pmatrix} 1 & 0 & 0 & -1 & | & 1 \\ 0 & 1 & 0 & 1 & | & 0 \\ 0 & 0 & 1 & 1 & | & 2 \\ 0 & 0 & 0 & 6 & | & 0 \end{pmatrix} \implies \begin{pmatrix} 1 & 0 & 0 & 0 & | & 1 \\ 0 & 1 & 0 & 0 & | & 0 \\ 0 & 0 & 1 & 0 & | & 2 \\ 0 & 0 & 0 & 1 & | & 0 \end{pmatrix}$$

Thus $x_1 = 1$, $x_2 = 0$, $x_3 = 2$, and $x_4 = 0$.

38. $\begin{pmatrix} 1 & 3 & 1 & | & 0 \\ 2 & 1 & 1 & | & 0 \\ 7 & 1 & 3 & | & 0 \end{pmatrix} \implies \begin{pmatrix} 1 & 3 & 1 & | & 0 \\ 0 & -5 & -1 & | & 0 \\ 0 & -20 & -4 & | & 0 \end{pmatrix} \implies \begin{pmatrix} 1 & 0 & 2/5 & | & 0 \\ 0 & 1 & 1/5 & | & 0 \\ 0 & 0 & 0 & | & 0 \end{pmatrix}$

Letting $x_3 = t$, we find $x_2 = -\frac{1}{5}t$ and $x_1 = -\frac{2}{5}t$.

39. $\begin{pmatrix} 1 & 2 & 4 & | & 2 \\ 2 & 4 & 3 & | & 1 \\ 1 & 2 & -1 & | & 7 \end{pmatrix} \Longrightarrow \begin{pmatrix} 1 & 2 & 4 & | & 2 \\ 0 & 0 & -5 & | & -3 \\ 0 & 0 & -5 & | & 5 \end{pmatrix} \Longrightarrow \begin{pmatrix} 1 & 2 & 0 & | & -2/5 \\ 0 & 0 & 1 & | & 3/5 \\ 0 & 0 & 0 & | & 8 \end{pmatrix}$

There is no solution.

40. $\begin{pmatrix} 1 & 1 & -1 & 3 & | & 1 \\ 0 & 1 & -1 & -4 & | & 0 \\ 1 & 2 & -2 & -1 & | & 6 \\ 4 & 7 & -7 & 0 & | & 9 \end{pmatrix} \Longrightarrow \begin{pmatrix} 1 & 1 & -1 & 3 & | & 1 \\ 0 & 1 & -1 & -4 & | & 0 \\ 0 & 1 & -1 & -4 & | & 5 \\ 0 & 3 & -3 & -12 & | & 5 \end{pmatrix} \Longrightarrow \begin{pmatrix} 1 & 0 & 0 & 7 & | & 1 \\ 0 & 1 & -1 & -4 & | & 0 \\ 0 & 0 & 0 & 0 & | & 5 \\ 0 & 0 & 0 & 0 & | & 5 \end{pmatrix}$

There is no solution.

41. $\begin{bmatrix} 4 & 2 & 3 & | & 1 & 0 & 0 \\ 2 & 1 & 0 & | & 0 & 1 & 0 \\ -1 & -2 & 0 & | & 0 & 0 & 1 \end{bmatrix} \xrightarrow{R_{13}} \begin{bmatrix} -1 & -2 & 0 & | & 0 & 0 & 1 \\ 2 & 1 & 0 & | & 0 & 1 & 0 \\ 4 & 2 & 3 & | & 1 & 0 & 0 \end{bmatrix}$

$\xrightarrow[\text{operations}]{\text{row}} \begin{bmatrix} 1 & 0 & 0 & | & 0 & \frac{2}{3} & \frac{1}{3} \\ 0 & 1 & 0 & | & 0 & -\frac{1}{3} & -\frac{2}{3} \\ 0 & 0 & 1 & | & \frac{1}{3} & -\frac{2}{3} & 0 \end{bmatrix}; \quad \mathbf{A}^{-1} = \begin{bmatrix} 0 & \frac{2}{3} & \frac{1}{3} \\ 0 & -\frac{1}{3} & -\frac{2}{3} \\ \frac{1}{3} & -\frac{2}{3} & 0 \end{bmatrix}$

42. $\begin{bmatrix} 2 & 4 & -2 & | & 1 & 0 & 0 \\ 4 & 2 & -2 & | & 0 & 1 & 0 \\ 8 & 10 & -6 & | & 0 & 0 & 1 \end{bmatrix} \xrightarrow[\text{operations}]{\text{row}} \begin{bmatrix} 1 & 2 & -1 & | & \frac{1}{2} & 0 & 0 \\ 0 & 1 & -\frac{1}{3} & | & \frac{1}{3} & -\frac{1}{6} & 0 \\ 0 & 0 & 0 & | & -2 & -1 & 1 \end{bmatrix}; \quad \mathbf{A}$ is singular.

43. $\begin{bmatrix} -1 & 3 & 0 & | & 1 & 0 & 0 \\ 3 & -2 & 1 & | & 0 & 1 & 0 \\ 0 & 1 & 2 & | & 0 & 0 & 1 \end{bmatrix} \xrightarrow[\text{operations}]{\text{row}} \begin{bmatrix} 1 & -3 & 0 & | & -1 & 0 & 0 \\ 0 & 1 & 1 & | & 1 & 1 & 0 \\ 0 & 0 & 1 & | & -1 & -1 & 1 \end{bmatrix}$

$\xrightarrow[\text{operations}]{\text{row}} \begin{bmatrix} 1 & 0 & 0 & | & 5 & 6 & -3 \\ 0 & 1 & 0 & | & 2 & 2 & -1 \\ 0 & 0 & 1 & | & -1 & -1 & 1 \end{bmatrix}; \quad \mathbf{A}^{-1} = \begin{bmatrix} 5 & 6 & -3 \\ 2 & 2 & -1 \\ -1 & -1 & 1 \end{bmatrix}$

44. $\begin{bmatrix} 1 & 2 & 3 & | & 1 & 0 & 0 \\ 0 & 1 & 4 & | & 0 & 1 & 0 \\ 0 & 0 & 8 & | & 0 & 0 & 1 \end{bmatrix} \xrightarrow[\text{operations}]{\text{row}} \begin{bmatrix} 1 & 0 & 0 & | & 1 & -2 & \frac{5}{8} \\ 0 & 1 & 0 & | & 0 & 1 & -\frac{1}{2} \\ 0 & 0 & 1 & | & 0 & 0 & \frac{1}{8} \end{bmatrix}; \quad \mathbf{A}^{-1} = \begin{bmatrix} 1 & -2 & \frac{5}{8} \\ 0 & 1 & -\frac{1}{2} \\ 0 & 0 & \frac{1}{8} \end{bmatrix}$

45. $\begin{bmatrix} 1 & 2 & 3 & 1 & | & 1 & 0 & 0 & 0 \\ -1 & 0 & 2 & 1 & | & 0 & 1 & 0 & 0 \\ 2 & 1 & -3 & 0 & | & 0 & 0 & 1 & 0 \\ 1 & 1 & 2 & 1 & | & 0 & 0 & 0 & 1 \end{bmatrix} \xrightarrow[\text{operations}]{\text{row}} \begin{bmatrix} 1 & 2 & 3 & 1 & | & 1 & 0 & 0 & 0 \\ 0 & 1 & \frac{5}{2} & 1 & | & \frac{1}{2} & \frac{1}{2} & 0 & 0 \\ 0 & 0 & 1 & -\frac{2}{3} & | & \frac{1}{3} & -1 & -\frac{2}{3} & 0 \\ 0 & 0 & 0 & 1 & | & -\frac{1}{2} & 1 & \frac{1}{2} & \frac{1}{2} \end{bmatrix}$

$$\xrightarrow[\text{operations}]{\text{row}} \begin{bmatrix} 1 & 0 & 0 & 0 & | & -\frac{1}{2} & -\frac{2}{3} & -\frac{1}{6} & \frac{7}{6} \\ 0 & 1 & 0 & 0 & | & 1 & \frac{1}{3} & \frac{1}{3} & -\frac{4}{3} \\ 0 & 0 & 1 & 0 & | & 0 & -\frac{1}{3} & -\frac{1}{3} & \frac{1}{3} \\ 0 & 0 & 0 & 1 & | & -\frac{1}{2} & 1 & \frac{1}{2} & \frac{1}{2} \end{bmatrix}; \quad \mathbf{A}^{-1} = \begin{bmatrix} -\frac{1}{2} & -\frac{2}{3} & -\frac{1}{6} & \frac{7}{6} \\ 1 & \frac{1}{3} & \frac{1}{3} & -\frac{4}{3} \\ 0 & -\frac{1}{3} & -\frac{1}{3} & \frac{1}{3} \\ -\frac{1}{2} & 1 & \frac{1}{2} & \frac{1}{2} \end{bmatrix}$$

46.
$$\begin{bmatrix} 1 & 0 & 0 & 0 & | & 1 & 0 & 0 & 0 \\ 0 & 0 & 1 & 0 & | & 0 & 1 & 0 & 0 \\ 0 & 0 & 0 & 1 & | & 0 & 0 & 1 & 0 \\ 0 & 1 & 0 & 0 & | & 0 & 0 & 0 & 1 \end{bmatrix} \xrightarrow[\text{interchange}]{\text{row}} \begin{bmatrix} 1 & 0 & 0 & 0 & | & 1 & 0 & 0 & 0 \\ 0 & 1 & 0 & 0 & | & 0 & 0 & 0 & 1 \\ 0 & 0 & 1 & 0 & | & 0 & 1 & 0 & 0 \\ 0 & 0 & 0 & 1 & | & 0 & 0 & 1 & 0 \end{bmatrix}; \quad \mathbf{A}^{-1} = \begin{bmatrix} 1 & 0 & 0 & 0 \\ 0 & 0 & 0 & 1 \\ 0 & 1 & 0 & 0 \\ 0 & 0 & 1 & 0 \end{bmatrix}$$

47. We solve

$$\det(\mathbf{A} - \lambda\mathbf{I}) = \begin{vmatrix} -1-\lambda & 2 \\ -7 & 8-\lambda \end{vmatrix} = (\lambda - 6)(\lambda - 1) = 0.$$

For $\lambda_1 = 6$ we have

$$\begin{pmatrix} -7 & 2 & | & 0 \\ -7 & 2 & | & 0 \end{pmatrix} \implies \begin{pmatrix} 1 & -2/7 & | & 0 \\ 0 & 0 & | & 0 \end{pmatrix}$$

so that $k_1 = \frac{2}{7}k_2$. If $k_2 = 7$ then

$$\mathbf{K}_1 = \begin{pmatrix} 2 \\ 7 \end{pmatrix}.$$

For $\lambda_2 = 1$ we have

$$\begin{pmatrix} -2 & 2 & | & 0 \\ -7 & 7 & | & 0 \end{pmatrix} \implies \begin{pmatrix} 1 & -1 & | & 0 \\ 0 & 0 & | & 0 \end{pmatrix}$$

so that $k_1 = k_2$. If $k_2 = 1$ then

$$\mathbf{K}_2 = \begin{pmatrix} 1 \\ 1 \end{pmatrix}.$$

48. We solve

$$\det(\mathbf{A} - \lambda\mathbf{I}) = \begin{vmatrix} 2-\lambda & 1 \\ 2 & 1-\lambda \end{vmatrix} = \lambda(\lambda - 3) = 0.$$

For $\lambda_1 = 0$ we have

$$\begin{pmatrix} 2 & 1 & | & 0 \\ 2 & 1 & | & 0 \end{pmatrix} \implies \begin{pmatrix} 1 & 1/2 & | & 0 \\ 0 & 0 & | & 0 \end{pmatrix}$$

so that $k_1 = -\frac{1}{2}k_2$. If $k_2 = 2$ then

$$\mathbf{K}_1 = \begin{pmatrix} -1 \\ 2 \end{pmatrix}.$$

For $\lambda_2 = 3$ we have

$$\begin{pmatrix} -1 & 1 & | & 0 \\ 2 & -2 & | & 0 \end{pmatrix} \implies \begin{pmatrix} 1 & -1 & | & 0 \\ 0 & 0 & | & 0 \end{pmatrix}$$

so that $k_1 = k_2$. If $k_2 = 1$ then

$$\mathbf{K}_2 = \begin{pmatrix} 1 \\ 1 \end{pmatrix}.$$

49. We solve

$$\det(\mathbf{A} - \lambda\mathbf{I}) = \begin{vmatrix} -8 - \lambda & -1 \\ 16 & -\lambda \end{vmatrix} = (\lambda + 4)^2 = 0.$$

For $\lambda_1 = \lambda_2 = -4$ we have

$$\begin{pmatrix} -4 & -1 & | & 0 \\ 16 & 4 & | & 0 \end{pmatrix} \implies \begin{pmatrix} 1 & 1/4 & | & 0 \\ 0 & 0 & | & 0 \end{pmatrix}$$

so that $k_1 = -\frac{1}{4}k_2$. If $k_2 = 4$ then

$$\mathbf{K}_1 = \begin{pmatrix} -1 \\ 4 \end{pmatrix}.$$

50. We solve

$$\det(\mathbf{A} - \lambda\mathbf{I}) = \begin{vmatrix} 1 - \lambda & 1 \\ 1/4 & 1 - \lambda \end{vmatrix} = (\lambda - 3/2)(\lambda - 1/2) = 0.$$

For $\lambda_1 = 3/2$ we have

$$\begin{pmatrix} -1/2 & 1 & | & 0 \\ 1/4 & -1/2 & | & 0 \end{pmatrix} \implies \begin{pmatrix} 1 & -2 & | & 0 \\ 0 & 0 & | & 0 \end{pmatrix}$$

so that $k_1 = 2k_2$. If $k_2 = 1$ then

$$\mathbf{K}_1 = \begin{pmatrix} 2 \\ 1 \end{pmatrix}.$$

If $\lambda_2 = 1/2$ then

$$\begin{pmatrix} 1/2 & 1 & | & 0 \\ 1/4 & 1/2 & | & 0 \end{pmatrix} \implies \begin{pmatrix} 1 & 2 & | & 0 \\ 0 & 0 & | & 0 \end{pmatrix}$$

so that $k_1 = -2k_2$. If $k_2 = 1$ then

$$\mathbf{K}_2 = \begin{pmatrix} -2 \\ 1 \end{pmatrix}.$$

51. We solve

$$\det(\mathbf{A} - \lambda\mathbf{I}) = \begin{vmatrix} 5 - \lambda & -1 & 0 \\ 0 & -5 - \lambda & 9 \\ 5 & -1 & -\lambda \end{vmatrix} = \lambda(4 - \lambda)(\lambda + 4) = 0.$$

If $\lambda_1 = 0$ then

$$\begin{pmatrix} 5 & -1 & 0 & | & 0 \\ 0 & -5 & 9 & | & 0 \\ 5 & -1 & 0 & | & 0 \end{pmatrix} \implies \begin{pmatrix} 1 & 0 & -9/25 & | & 0 \\ 0 & 1 & -9/5 & | & 0 \\ 0 & 0 & 0 & | & 0 \end{pmatrix}$$

so that $k_1 = \frac{9}{25}k_3$ and $k_2 = \frac{9}{5}k_3$. If $k_3 = 25$ then

$$\mathbf{K}_1 = \begin{pmatrix} 9 \\ 45 \\ 25 \end{pmatrix}.$$

If $\lambda_2 = 4$ then

$$\begin{pmatrix} 1 & -1 & 0 & | & 0 \\ 0 & -9 & 9 & | & 0 \\ 5 & -1 & -4 & | & 0 \end{pmatrix} \implies \begin{pmatrix} 1 & 0 & -1 & | & 0 \\ 0 & 1 & -1 & | & 0 \\ 0 & 0 & 0 & | & 0 \end{pmatrix}$$

so that $k_1 = k_3$ and $k_2 = k_3$. If $k_3 = 1$ then

$$\mathbf{K}_2 = \begin{pmatrix} 1 \\ 1 \\ 1 \end{pmatrix}.$$

If $\lambda_3 = -4$ then

$$\begin{pmatrix} 9 & -1 & 0 & | & 0 \\ 0 & -1 & 9 & | & 0 \\ 5 & -1 & 4 & | & 0 \end{pmatrix} \implies \begin{pmatrix} 1 & 0 & -1 & | & 0 \\ 0 & 1 & -9 & | & 0 \\ 0 & 0 & 0 & | & 0 \end{pmatrix}$$

so that $k_1 = k_3$ and $k_2 = 9k_3$. If $k_3 = 1$ then

$$\mathbf{K}_3 = \begin{pmatrix} 1 \\ 9 \\ 1 \end{pmatrix}.$$

52. We solve

$$\det(\mathbf{A} - \lambda\mathbf{I}) = \begin{vmatrix} 3-\lambda & 0 & 0 \\ 0 & 2-\lambda & 0 \\ 4 & 0 & 1-\lambda \end{vmatrix} = (3-\lambda)(2-\lambda)(1-\lambda) = 0.$$

If $\lambda_1 = 1$ then

$$\begin{pmatrix} 2 & 0 & 0 & | & 0 \\ 0 & 1 & 0 & | & 0 \\ 4 & 0 & 0 & | & 0 \end{pmatrix} \implies \begin{pmatrix} 1 & 0 & 0 & | & 0 \\ 0 & 1 & 0 & | & 0 \\ 0 & 0 & 0 & | & 0 \end{pmatrix}$$

so that $k_1 = 0$ and $k_2 = 0$. If $k_3 = 1$ then

$$\mathbf{K}_1 = \begin{pmatrix} 0 \\ 0 \\ 1 \end{pmatrix}.$$

If $\lambda_2 = 2$ then

$$\begin{pmatrix} 1 & 0 & 0 & | & 0 \\ 0 & 0 & 0 & | & 0 \\ 4 & 0 & -1 & | & 0 \end{pmatrix} \implies \begin{pmatrix} 1 & 0 & 0 & | & 0 \\ 0 & 0 & 1 & | & 0 \\ 0 & 0 & 0 & | & 0 \end{pmatrix}$$

so that $k_1 = 0$ and $k_3 = 0$. If $k_2 = 1$ then

$$\mathbf{K}_2 = \begin{pmatrix} 0 \\ 1 \\ 0 \end{pmatrix}.$$

If $\lambda_3 = 3$ then

$$\begin{pmatrix} 0 & 0 & 0 & | & 0 \\ 0 & -1 & 0 & | & 0 \\ 4 & 0 & -2 & | & 0 \end{pmatrix} \implies \begin{pmatrix} 1 & 0 & -1/2 & | & 0 \\ 0 & 1 & 0 & | & 0 \\ 0 & 0 & 0 & | & 0 \end{pmatrix}$$

so that $k_1 = \frac{1}{2}k_3$ and $k_2 = 0$. If $k_3 = 2$ then

$$\mathbf{K}_3 = \begin{pmatrix} 1 \\ 0 \\ 2 \end{pmatrix}.$$

53. We solve

$$\det(\mathbf{A} - \lambda\mathbf{I}) = \begin{vmatrix} -\lambda & 4 & 0 \\ -1 & -4-\lambda & 0 \\ 0 & 0 & -2-\lambda \end{vmatrix} = -(\lambda+2)^3 = 0.$$

For $\lambda_1 = \lambda_2 = \lambda_3 = -2$ we have

$$\begin{pmatrix} 2 & 4 & 0 & | & 0 \\ -1 & -2 & 0 & | & 0 \\ 0 & 0 & 0 & | & 0 \end{pmatrix} \implies \begin{pmatrix} 1 & 2 & 0 & | & 0 \\ 0 & 0 & 0 & | & 0 \\ 0 & 0 & 0 & | & 0 \end{pmatrix}$$

so that $k_1 = -2k_2$. If $k_2 = 1$ and $k_3 = 1$ then

$$\mathbf{K}_1 = \begin{pmatrix} -2 \\ 1 \\ 0 \end{pmatrix} \quad \text{and} \quad \mathbf{K}_2 = \begin{pmatrix} 0 \\ 0 \\ 1 \end{pmatrix}.$$

54. We solve

$$\det(\mathbf{A} - \lambda\mathbf{I}) = \begin{vmatrix} 1 - \lambda & 6 & 0 \\ 0 & 2 - \lambda & 1 \\ 0 & 1 & 2 - \lambda \end{vmatrix} = (3 - \lambda)(1 - \lambda)^2 = 0.$$

For $\lambda = 3$ we have

$$\begin{pmatrix} -2 & 6 & 0 & | & 0 \\ 0 & 0 & 0 & | & 0 \\ 0 & 1 & -1 & | & 0 \end{pmatrix} \implies \begin{pmatrix} 1 & 0 & -3 & | & 0 \\ 0 & 1 & -1 & | & 0 \\ 0 & 0 & 0 & | & 0 \end{pmatrix}$$

so that $k_1 = 3k_3$ and $k_2 = k_3$. If $k_3 = 1$ then

$$\mathbf{K}_1 = \begin{pmatrix} 3 \\ 1 \\ 1 \end{pmatrix}.$$

For $\lambda_2 = \lambda_3 = 1$ we have

$$\begin{pmatrix} 0 & 6 & 0 & | & 0 \\ 0 & 1 & 1 & | & 0 \\ 0 & 1 & 1 & | & 0 \end{pmatrix} \implies \begin{pmatrix} 0 & 1 & 0 & | & 0 \\ 0 & 0 & 1 & | & 0 \\ 0 & 0 & 0 & | & 0 \end{pmatrix}$$

so that $k_2 = 0$ and $k_3 = 0$. If $k_1 = 1$ then

$$\mathbf{K}_2 = \begin{pmatrix} 1 \\ 0 \\ 0 \end{pmatrix}.$$

55. We solve

$$\det(\mathbf{A} - \lambda\mathbf{I}) = \begin{vmatrix} -1 - \lambda & 2 \\ -5 & 1 - \lambda \end{vmatrix} = \lambda^2 + 9 = (\lambda - 3i)(\lambda + 3i) = 0.$$

For $\lambda_1 = 3i$ we have

$$\begin{pmatrix} -1 - 3i & 2 & | & 0 \\ -5 & 1 - 3i & | & 0 \end{pmatrix} \implies \begin{pmatrix} 1 & -(1/5) + (3/5)i & | & 0 \\ 0 & 0 & | & 0 \end{pmatrix}$$

so that $k_1 = \left(\frac{1}{5} - \frac{3}{5}i\right) k_2$. If $k_2 = 5$ then

$$\mathbf{K}_1 = \begin{pmatrix} 1 - 3i \\ 5 \end{pmatrix}.$$

For $\lambda_2 = -3i$ we have

$$\begin{pmatrix} -1 + 3i & 2 & | & 0 \\ -5 & 1 + 3i & | & 0 \end{pmatrix} \implies \begin{pmatrix} 1 & -\frac{1}{5} - \frac{3}{5}i & | & 0 \\ 0 & 0 & | & 0 \end{pmatrix}$$

so that $k_1 = \left(\frac{1}{5} + \frac{3}{5}i\right) k_2$. If $k_2 = 5$ then

$$\mathbf{K}_2 = \begin{pmatrix} 1 + 3i \\ 5 \end{pmatrix}.$$

56. We solve

$$\det(\mathbf{A} - \lambda\mathbf{I}) = \begin{vmatrix} 2 - \lambda & -1 & 0 \\ 5 & 2 - \lambda & 4 \\ 0 & 1 & 2 - \lambda \end{vmatrix} = -\lambda^3 + 6\lambda^2 - 13\lambda + 10 = (\lambda - 2)(-\lambda^2 + 4\lambda - 5)$$

$$= (\lambda - 2)(\lambda - (2 + i))(\lambda - (2 - i)) = 0.$$

For $\lambda_1 = 2$ we have

$$\begin{pmatrix} 0 & -1 & 0 & | & 0 \\ 5 & 0 & 4 & | & 0 \\ 0 & 1 & 0 & | & 0 \end{pmatrix} \implies \begin{pmatrix} 1 & 0 & 4/5 & | & 0 \\ 0 & 1 & 0 & | & 0 \\ 0 & 0 & 0 & | & 0 \end{pmatrix}$$

so that $k_1 = -\frac{4}{5}k_3$ and $k_2 = 0$. If $k_3 = 5$ then

$$\mathbf{K}_1 = \begin{pmatrix} -4 \\ 0 \\ 5 \end{pmatrix}.$$

For $\lambda_2 = 2 + i$ we have

$$\begin{pmatrix} -i & -1 & 0 & | & 0 \\ 5 & -i & 4 & | & 0 \\ 0 & 1 & -i & | & 0 \end{pmatrix} \implies \begin{pmatrix} 1 & -i & 0 & | & 0 \\ 0 & 1 & -i & | & 0 \\ 0 & 0 & 0 & | & 0 \end{pmatrix}$$

so that $k_1 = ik_2$ and $k_2 = ik_3$. If $k_3 = i$ then

$$\mathbf{K}_2 = \begin{pmatrix} -i \\ -1 \\ i \end{pmatrix}.$$

For $\lambda_3 = 2 - i$ we have

$$\begin{pmatrix} i & -1 & 0 & | & 0 \\ 5 & i & 4 & | & 0 \\ 0 & 1 & i & | & 0 \end{pmatrix} \implies \begin{pmatrix} 1 & i & 0 & | & 0 \\ 0 & 1 & i & | & 0 \\ 0 & 0 & 0 & | & 0 \end{pmatrix}$$

so that $k_1 = -ik_2$ and $k_2 = -ik_3$. If $k_3 = i$ then

$$\mathbf{K}_3 = \begin{pmatrix} -1 \\ 1 \\ i \end{pmatrix}.$$

57. Let
$$\mathbf{A} = \begin{pmatrix} a_{11} & a_{12} \\ a_{21} & a_{22} \end{pmatrix}.$$

Then
$$\frac{d}{dt}[\mathbf{A}(t)\mathbf{X}(t)] = \frac{d}{dt}\begin{pmatrix} a_1 & a_2 \\ a_3 & a_4 \end{pmatrix}\begin{pmatrix} x_1 \\ x_2 \end{pmatrix} = \frac{d}{dt}\begin{pmatrix} a_1x_1 + a_2x_2 \\ a_3x_1 + a_4x_2 \end{pmatrix} = \begin{pmatrix} a_1x_1' + a_1'x_1 + a_2x_2' + a_2'x_2 \\ a_3x_1' + a_3'x_1 + a_4x_2' + a_4'x_2 \end{pmatrix}$$

$$= \begin{pmatrix} a_1 & a_2 \\ a_3 & a_4 \end{pmatrix}\begin{pmatrix} x_1' \\ x_2' \end{pmatrix} + \begin{pmatrix} a_1' & a_2' \\ a_3' & a_4' \end{pmatrix}\begin{pmatrix} x_1 \\ x_2 \end{pmatrix} = \mathbf{A}(t)\mathbf{X}'(t) + \mathbf{A}'(t)\mathbf{X}(t).$$

58. Assume $\det \mathbf{A} \neq 0$ and $\mathbf{AB} = \mathbf{I}$, so that
$$\begin{pmatrix} a_{11} & a_{12} \\ a_{21} & a_{22} \end{pmatrix}\begin{pmatrix} b_{11} & b_{12} \\ b_{21} & b_{22} \end{pmatrix} = \begin{pmatrix} 1 & 0 \\ 0 & 1 \end{pmatrix}.$$

Then
$$a_{11}b_{11} + a_{12}b_{21} = 1 \qquad a_{11}b_{12} + a_{12}b_{22} = 0$$
$$\text{and}$$
$$a_{21}b_{11} + a_{21}b_{21} = 0 \qquad a_{21}b_{12} + a_{21}b_{22} = 1$$

and by Cramer's rule
$$b_{11} = \frac{a_{22}}{\det \mathbf{A}} \qquad b_{12} = \frac{-a_{12}}{\det \mathbf{A}}$$
$$b_{21} = \frac{-a_{21}}{\det \mathbf{A}} \qquad b_{22} = \frac{a_{11}}{\det \mathbf{A}}.$$

Thus
$$\mathbf{A}^{-1} = \mathbf{B} = \frac{1}{\det \mathbf{A}}\begin{pmatrix} a_{22} & -a_{12} \\ -a_{21} & a_{11} \end{pmatrix}.$$

59. Since $\mathbf{A}$ is nonsingular, $\mathbf{AB} = \mathbf{AC}$ implies $\mathbf{A}^{-1}\mathbf{AB} = \mathbf{A}^{-1}\mathbf{AC}$. Then $\mathbf{IB} = \mathbf{IC}$ and $\mathbf{B} = \mathbf{C}$.

60. Since
$$(\mathbf{AB})(\mathbf{B}^{-1}\mathbf{A}^{-1}) = \mathbf{A}(\mathbf{BB}^{-1})\mathbf{A}^{-1} = \mathbf{AIA}^{-1} = \mathbf{AA}^{-1} = \mathbf{I}$$
and
$$(\mathbf{B}^{-1}\mathbf{A}^{-1})(\mathbf{AB}) = \mathbf{B}^{-1}(\mathbf{A}^{-1}\mathbf{A})\mathbf{B} = \mathbf{B}^{-1}\mathbf{IB} = \mathbf{B}^{-1}\mathbf{B} = \mathbf{I}$$
we have
$$(\mathbf{AB})^{-1} = \mathbf{B}^{-1}\mathbf{A}^{-1}.$$

61. No; consider
$$\mathbf{A} = \begin{pmatrix} 1 & 0 \\ 0 & 0 \end{pmatrix} \quad \text{and} \quad \mathbf{B} = \begin{pmatrix} 0 & 0 \\ 1 & 0 \end{pmatrix}.$$

62. (a) $\mathbf{A}^{-1} = \begin{bmatrix} 1/a_{11} & 0 \\ 0 & 1/a_{22} \end{bmatrix}$

(b) $\mathbf{A}^{-1} = \begin{bmatrix} 1/a_{11} & 0 & 0 \\ 0 & 1/a_{22} & 0 \\ 0 & 0 & 1/a_{33} \end{bmatrix}$

(c) For any diagonal matrix, the inverse matrix is obtaining by taking the reciprocals of the diagonal entries and leaving all other entries 0.